INJECTION MOLDING HANDBOOK
The Complete Molding Operation
Technology, Performance, Economics

INJECTION MOLDING HANDBOOK
The Complete Molding Operation
Technology, Performance, Economics

Edited by

Dominick V. Rosato, P.E.
University of Lowell

and

Donald V. Rosato, Ph.D.
Borg-Warner Chemicals/International Center
and University of Lowell

VNR VAN NOSTRAND REINHOLD COMPANY
————————————————————— *New York*

Copyright © 1986 by Van Nostrand Reinhold Company Inc.

Library of Congress Catalog Card Number: 84-25769
ISBN: 0-442-27815-2

Manufactured in the United States of America

Published by Van Nostrand Reinhold Company Inc.
115 Fifth Avenue
New York, New York 10003

Van Nostrand Reinhold Company Limited
Molly Millars Lane
Wokingham, Berkshire RG11 2PY, England

Van Nostrand Reinhold
480 La Trobe Street
Melbourne, Victoria 3000, Australia

Macmillan of Canada
Division of Canada Publishing Corporation
164 Commander Boulevard
Agincourt, Ontario M1S 3C7, Canada

15 14 13 12 11 10 9 8 7 6 5

Library of Congress Cataloging in Publication Data

Main entry under title:

Injection molding handbook.

   Includes index.
   1. Injection molding of plastics—Handbooks, manuals,
etc. I. Rosato, Dominick V. II. Rosato, Donald V.
III. Society of Plastics Engineers. Injection Molding
Division.
TP1150.I55   1985   668.4'12   84-25769
ISBN 0-442-27815-2

Foreword

For some time, a strong need has been felt by the plastics industry for a book that provides an understanding of the complete injection molding operation and covers the latest advancements in technology. We have had some books in the past that provided fairly good coverage of the subject and fulfilled industry's need at the time.

The industry has come a long way since. Its changing needs reflect the dynamic environment of the last quarter of this century. Apart from its own normal growth, injection molding technology has been impacted by other exploding technologies like electronics, computers, and robotics together with the demands of exotic molding materials. This book is an attempt to satisfy the industry's needs for today.

The Society of Plastics Engineers is divided into seventeen technical divisions. Being the largest with over 6,000 members worldwide, the Injection Molding Division is its strongest technological arm. The first item on the Division's charter is to promote the knowledge of injection molding technology. With that in mind, the Division Board of Directors undertook the task of publishing a much needed piece of literature; the result is presented to you in this book.

The concept of this book evolved in a Division Board of Directors' meeting when a committee was created for the purpose and Dr. Robert E. Nunn appointed its chairman. This responsibility, along with editorial responsibilities, was later turned over to Prof. Dominick V. Rosato. As editor of the book, Prof. Rosato has done an outstanding job of putting it all together. On behalf of the Injection Molding Division of SPE, I would like to express my sincere gratitude to all those who volunteered their services for the book. I would also like to thank the contributing authors who shared their intimate knowledge of their respective fields of expertise with all of us.

Kishor S. Mehta
Chairman
Injection Molding Division
Society of Plastics Engineers

Manager, Design Engineering Services
Mobay Chemical Corporation
Pittsburgh, PA

Preface

To make injection molded plastic parts, many important processing steps are involved—steps that must come together properly to produce parts consistently meeting performance requirements at the lowest cost. Lack of knowledge or control of these important processing steps will result in less-than-desirable molded parts. Many plants are not taking full advantage of using their injection molding operation properly. Part of the problem is that information on how to use the complete molding operation effectively is scattered and difficult to locate, particularly for the new molder.

This one-source handbook clearly explains all aspects of the complete molding operation. It provides information based on today's technology as well as what is ahead. Readers will find out how to interrelate the important processing steps that start with designing the part to be molded then proceed to designing the mold, to properly operating the injection molding machine, to handling plastic material properly, to setting up clearly defined troubleshooting guides, to setting up testing and quality controls on plastic materials downstream as well as upstream and they will be given a thorough orientation to the basic injection molding machine with its mold. The complete molding operation is summarized in Fig. 1-1, Chapter 1.

We have provided a complete review on injection molding to ensure that all the different machine setting operations can be understood and put into practical use. It fulfills the important need to bring the reader up to date and to review what is ahead in injection molding. This handbook has been prepared with an awareness that its usefulness will depend largely upon its simplicity. The guiding premise has been to provide only information which is essential to injection molding. Chapters are organized to best present a methodology for molding.

The emphasis of this book is on understanding injection molding from both the practical and theoretical viewpoints. For those involved in practicalities, the theoretical aspect (which is kept separate) will be useful. In turn, the theorist will gain an insight into limitations owing to that equipment and plastic materials are not "perfect"— they have plus-to-minus limitations which the molder understand in practice (or will understand after reading this book).

Information is presented that is unique to injection molding and not contained in any other hardback book, for example, how to start up and operate an injection molding machine (Chapter 4), how to buy a mold (Chapter 10), relating in simplified terms molding variables to part performance (Chapters 19, 25, others), role of rheology in injection molding (Chapter 26), financial management of molding plant (Chapter 31).

This book integrates the following:
a. Designing a part to meet performance and manufacturing requirements at the lowest cost.
b. Specifying the proper plastic molding material that provides part performance after processing the plastics.
c. Specifying equipment requirements by:
   1. Designing the mold "around" the part.
   2. Putting the proper operating injection molding machine "around" the mold.
   3. Setting up auxiliary equipment to match all other hardware in the complete molding operation.
   4. Setting up complete controls (testing, quality control, troubleshooting, preventative maintenance, safety, cost, etc.).
   5. Purchasing and warehousing plastic materials.

Important to recognize, the major cost in the production of injection molded parts, taking into account all equipment in the complete operation plus plastic material and utilities, is that of the plastic material itself, ranging from 60 to 90 percent of the parts' cost. Thus it is important to understand how best to process plastics. This book reviews different molding plastics based on their specific machine settings and mold design parameters. The objective is meeting performance requirements with the least amount of plastic, resulting in significant cost savings.

This comprehensive handbook covers and explains every aspect of injection molding technology. It provides essential information on principles of operation, machine performance, part design, machines of special design (for injection blow mold, stretching techniques to increase performance, "continuous" molding, RIM, LIM, coinjection, etc.), material handling, chillers/coolers, granulators, parts handling equipment, robots, quick mold changes, testing/quality controls, process controls, CAD/CAM, plastic melt with flow characteristics; and on complete plant operations, including technical and financial management. Machine process controls are presented in their simplest form. Easy-to-follow guidelines cover all aspects of controls, enabling personnel with a minimal understanding of control theory and computer technology to develop an optional process control strategy.

With plastics, to a greater extent than with other materials and in particular injection-molded production parts, an opportunity exists to optimize design by focusing on material composition, structural orientation during molding, and molding conditions. Throughout this book, reviews are given of problems that can occur during molding: what causes them, how to eliminate them, and/or how to take corrective action. Chapter 27 provides a summary on troubleshooting the injection molding machine and auxiliary equipment. All important terms, definitions, and explanations are discussed throughout the book.

Injection molded plastic products continue to make tremendous strides (since 1872) with vigorous growth rates ahead; it now represents a multibillion dollar annual business. Against this background is the emerging new injection molding technology, reviewed in this book, that relates to new molding equipment, new plastic materials, product and mold design development parameters, and processing control techniques.

The plastics industry is characterized by its rapid expansion not only in output but also in the design and production of all shapes, sizes, and weights of plastic products to meet continuous new market requirements. To satisfy this demand, most of the processing equipment has been and will continue to be injection molding machines.

Within the book reviews may appear to be repetitious; this design is deliberate, to allow approaches to specific subjects or topics from different perspectives. An example is using a plastic according to (1) product performance requirements (Chapter 6), (2) mold design (Chapter 7), (3) molding cycle (Chapter 11), (4) granulating (Chapter 17), (5) choice of material (Chapter 18), (6) flow characteristics (Chapter 26), (7) quality control (Chapter 28), (8) cost control (Chapter 31), and so on. The reviews integrate the different parameters to make for a better understanding of injection molding. What may be a logical approach to designing a part or a mold may make it difficult (or impossible) to mold unless the *complete* molding operation is understood.

Information contained in this book may be covered by U.S. and foreign patents. No authorization to utilize these patents is given or implied; they are discussed for information only. Any such disclosures are not a license to operate under, or a recommendation to infringe, any patent. With a few exceptions, no attempt has been made in this book to refer to patents by number, title, or ownership.

In the preparation of this handbook, the editor has been assisted and encouraged by many friends and business associates. Special acknowledgment must be made to the contributors listed in the chapter headings, in sections within chapters, and in the references at the ends of the chapters. All have contributed to advance the state-of-the-art in injection molding.

D. V. ROSATO
Editor

Contents

SECTION VII. PROBLEM SOLVING

SECTION VIII. DIFFERENT MOLDING TECHNIQUES

SECTION IX. SUMMARY

INJECTION MOLDING HANDBOOK
The Complete Molding Operation
Technology, Performance, Economics

Section I
OVERVIEW

Chapter 1
The Complete Injection Molding Operation

D. V. Rosato

INTRODUCTION

Plastics is one of the world's fastest growing industries, ranked as one of the few billion-dollar industries. Its two major processing methods are injection molding and extrusion. Approximately 32 percent by weight of all plastics processed goes through injection molding machines and 36 percent through extruders.

With all this activity injection molding plants continue to expand, new plants are built, and modernization of plants continues to occur. The target of these plants is to produce injection molded parts that meet the customers' performance requirements at a cost that insures a profit for the plant. Although economic success essentially depends upon external influences such as the market situation, competition, and ecological considerations, attention must also be directed toward the possibilities existing within the plant.

This book will review different technical and organizational elements in the plant. Understanding how to obtain the maximum performance of each individual operation in the complete molding operation and properly integrating each step to meet product performance at the lowest cost are important to all plant personnel, not just to management. A block diagram of the complete molding operation according to the Fallo approach (Follow All Opportunities) is shown in Fig. 1-1.

Basically, the Fallo approach consists of: (1) designing a part to meet performance and manufacturing requirements at the lowest cost; (2) specifying the proper plastic molding material that provides part performance after processing the plastics; (3) specifying equipment requirements basically by (a) designing the mold "around" the part, (b) putting the "proper performing" injection molding machine "around" the mold, (c) setting up auxiliary equipment (material handling, chiller, granulator, decorator, etc.) to "match" all other hardware in the complete molding operation, and (d) setting up "complete" controls (testing, quality control, troubleshooting, maintenance, etc.) to produce "zero" defects; and (4) purchasing and warehousing materials (1).

Injection molding has the advantage that molded parts can be manufactured economically in unlimited quantities with little or practically no finishing operations. It is principally a mass-production method, and because of the high capital investment in machines, molds, and auxiliary equipment, it operates most economically only as such. The surfaces of injection moldings are as smooth and bright or as grained and engraved as the surfaces of the mold cavity in which they were prepared. Among the special developments have been insertion of printed film in the mold to avoid the need for subsequent decoration, foam

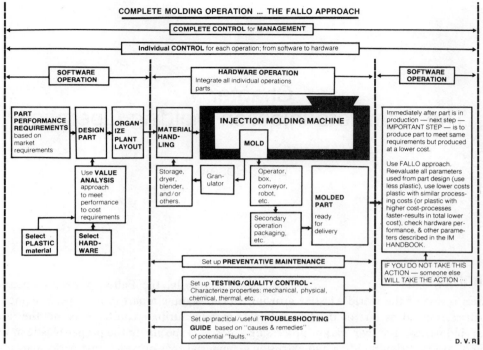

Fig. 1-1   Complete molding operation.

molding, blow molding, and others that will be reviewed in this book.

The first patent for injection molding was granted in the United States in 1872 to John Hyatt, for what was termed a stuffing machine plunger injection molding machine (details in Chapter 32). The next major development occurred in 1951 when William H. Willert, also in the United States, developed the reciprocating screw plasticizer for injection molding machines; Willert's U.S. patent 2,734,226 was issued February 14, 1956. The most recent major development concerns process controls that permit controlling the plastic melt. This subject will be reviewed in detail throughout the book.

Productivity

Injection molding of plastics is characterized by its rapid expansion not only in output, repeatedly molding to tight tolerances and meeting performance requirements, but also in processing of new types of plastics to continue the expanding use of plastics in all markets. The total plastics industry in the United States

has swelled into a $90 billion business that offers 1.4 million jobs, nearly half a million more than the burgeoning computer industry. Plastics are among the nation's most widely used materials, surpassing steel on a volume basis (see Figs. 1-2 and 1-3) and world consumption of materials by volume and by weight (2).

The U.S. plastics industry has grown at an average rate of 12 percent during the past 25 years, even taking into account 1975 and 1980, years in which no growth was recorded. It has become the nation's fourth largest industry in annual dollar volume. Plastic products and materials cover the entire spectrum of the nation's economy, so that "fortunes" are not tied to any particular business segment. Thus plastics are in a position to benefit by a turnaround in any one of a number of areas: packaging, transportation, housing, automotive, and many more industries.

In any particular technological sphere such as injection molding, an appreciable advance in knowledge is hardly possible in the long run without related progress in other fields. Indeed, the incentive for further development

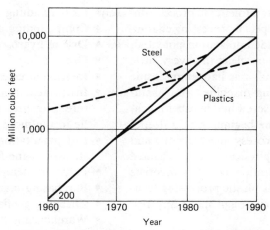

Fig. 1-2 World consumption of raw materials by volume.

is often provided by forces not subject to scientific laws. In some branches of plastics technology, particularly the field of processing techniques and machine construction, it was originally sufficient to adapt existing equipment, at that time devoted mainly to rubber and thermoset plastics processing. The new thermoplastics had a wider range of working temperatures and viscosities compared with rubber, but it was soon recognized that they

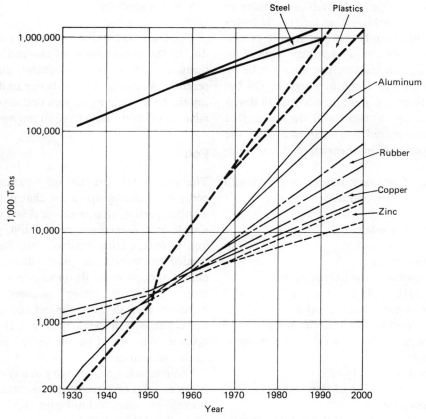

Fig. 1-3 World consumption of raw materials by weight.

required individual treatment in process design because of their special processing characteristics and their wide application potential.

It eventually became clear that very little was known about what was happening inside the familiar processing machines—for example, in the flights of a screw and in other mechanisms. Only with the beginning of a deeper understanding of process mechanisms and their underlying physical laws (gained through close cooperation between theorists and technologists) has plastic processing technology and machinery design made any real progress.

The 1940s and 1950s were the most productive periods in the latest phase of injection development, which at last became the province of the scientific engineer rather than the craftsman. The many publications of recent years describing investigations of the rheological and thermodynamic phenomena occurring during the injection, with their considerable use of mathematics, can be appreciated only by a limited circle of specialists. In spite of what has been achieved so far, the industry has surmounted only the first hurdle of systematic development. The present state of injection design and technology must not be regarded as the last word in progress. On the contrary, there are great possibilities in development, many of them still dormant, that must be recognized and examined with the close cooperation of theorists and technologists.

Subjects of primary interest are reviewed in this book. They include:

- Injection pressure
- Mold temperature
- Molding cycle
- Temperature of injected plastic
- Flow rate of plastic in mold
- Shear rate of plastic in mold
- Melt viscosity of plastic in mold
- Elastic shear stresses of plastic in mold
- Plasticizing
- Screw design
- Developing melt and flow control
- Process control
- Increased production through fine tuning

- Parts handling equipment
- Compounding and coloring in-plant
- Drying hygroscopic plastics
- Granulating
- Internal stresses in molded parts
- Instrumentation for measuring temperature, pressure, and flow viscosity of plastic in the mold during injection
- Instrumentation to measure internal stresses in the finished molded part
- Polymer chemistry
- Purchasing materials
- Checking goods received
- Warehousing
- In-plant transportation of materials
- In-plant transportation of parts
- Production control of quality
- Machine and plant safety
- Economic control of equipment
- Costing
- Working smarter—innovating
- Analyzing failures to achieve more success
- Value analysis

The increased use of injection molding is due to the development of the reciprocating screw as well as process control, and, more recently, a combination of better understanding the basic molding factors that involve cost advantages and market requirements.

People

This review will provide information on the complete molding operation that is useful to all plant personnel: nontechnical and technical people in management, production, purchasing, design, quality control, etc. People are needed to operate the plant efficiently. Machines, process controls, upstream and downstream equipment, design of parts, material handling, and all the technical and organizational elements in the plant (Fig. 1-1) can only operate efficiently if people set the plant into its correct pattern.

The recipe for productivity in any company includes a list of ingredients: research and development, new technologies, updated machinery, automated systems, and modern facilities, to name a few. But the one ingredient that ties the recipe together is people; none of the other factors has much impact without the right individuals. Without people who can do the research, who know the technologies, and who can use the machinery, you are not going to be productive no matter how large your capital expenditures are.

Management controls are only as good as the input they recieve. To operate efficiently one must understand how to obtain the maximum performance for each individual operation and—what is very important—to integrate the steps properly through planning. In that way molded products that meet performance requirements at the lowest costs are produced.

Conversion

Because English units are still widely used in the U.S. plastics industry, measurements in this book are given in English and, if it can be done conveniently, in SI (metric) units. In many instances, particularly in tables and graphs, only the system most used currently is followed, to achieve a simple, compact, uncluttered presentation. The following relationships enable the reader to make the necessary conversions:

Length
1 mm = 0.0394 in.
1 cm = 10 mm = 0.394 in.
1 m   = 100 cm = 39.4 in. = 3.28 ft
1 in. = 2.54 cm
1 ft = 0.305 m

Area
1 cm^2 = 0.155 in.^{2}
1 m^2 = 10.8 ft^2
1 in.^{2} = 6.45 cm^2
1 ft^2 = 0.0929 cm^2

Volume
1 cm^3 = 0.0610 in.^{3}
1 m^3 = 35.3 ft^3
1 in.^{3} = 16.4 cm^3
1 ft^3 = 0.0283 m^3

Force
1 N = 102.0 g = 0.225 lb
1 g = 9.81×10^{-3}N = 2.20×10^{-3} lb
1 lb = 4.45 N = 453.6 g

Stress
1 N/m^2 = 1 Pa = 1 dyne/cm^2
$\qquad$ = 1.02×10^{-5} kg/cm^2 = 1.45×10^{-4} lb/in.^{2} = 2.08×10^{-2} lb/ft^2
1 kg/cm^2 = 9.81×10^4 N/m^2 = 14.2 lb/in.^{2} = 2.05×10^3 lb/ft^2
1 lb/in.^{2} = 6.89×10^3 N/m^2 = 7.03×10^{-2} kg/cm^2 = 144 lb/ft^2
1 lb/ft^2 = 4.79×10 N/m^2 = 4.88×10^{-4} kg/cm^2 = 6.94×10^{-3} lb/in.^{2}

Unit Weight
1 N/m^3 = 1.02×10^{-4} g/cm^3 = 6.37×10^{-3} lb/ft^3
1 g/cm^3 = 9.81×10^3 N/m^3 = 62.4 lb/ft^3
1 lb/ft^3 = 1.57×10^2 N/m^3 = 1.60×10^{-2} g/cm^3

Temperature
1°C = 1°K = 1.8°F
1°F = 0.555°C = 0.555°K
0°K = −273°C = −460°F

$T_C = (5/9)(T_F - 32°) = T_K - 273°$
$T_K = T_C + 273° = (T_F + 460)/1.8$
$T_F = (9/5)T_C + 32° = 1.8 T_K - 460°$

Basic Process Description

The term injection molding is an oversimplified description of a quite complicated process that is controllable within specified limits. This section summarizes the basic process. Details on the machine are reviewed in Chapter 2. Melted or plasticized plastic material is injected or forced into a mold where it is held until removed in a solid state, duplicating the cavity of the mold (see Figs. 1-4 and 1-5). The mold may consist of a single cavity or a number of similar or dissimilar cavities, each connected to flow channels or "runners" that direct the flow of the melted plastic to the individual cavities. The process is one of

Fig. 1-4a   The basic injection molding cycle.

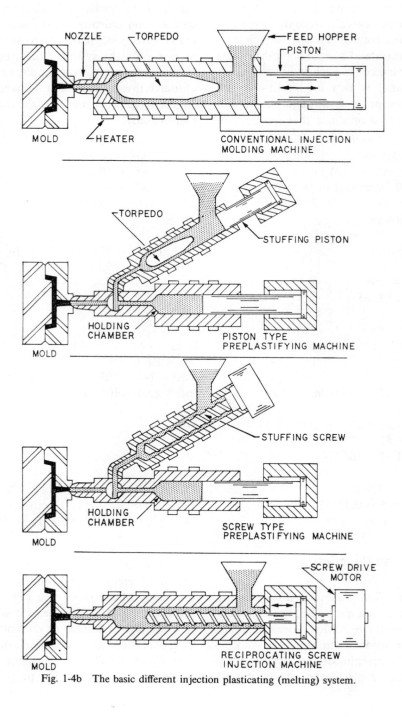

Fig. 1-4b   The basic different injection plasticating (melting) system.

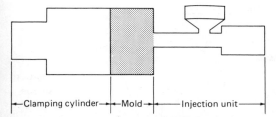

Fig. 1-5 Basic elements of injection molding process. Three basic mechanical units work together in the injection molding process. The injection molding machine provides (1) a mold clamping device and (2) an injection unit. Into the machine is placed a removable (3) mold.

the most economical methods for mass-producing a single item. Three basic operations exist:

1. Raising the temperature of the plastic to a point where it will flow under pressure. This is usually done by simultaneously heating and masticating the granular solid until it forms a melt at an elevated and uniform temperature and uniform viscosity. Today, this is accomplished in the cylinder of an injection molding machine equipped with a reciprocating screw. The latter provides the mechanical working required in conjunction with the heating of the material. This overall process is called plasticizing (or plastication) of the material.

2. Allowing the plastic to solidify in the mold, which the machine keeps closed. The liquid, molten plastic from the injection cylinder of the injection machine is transferred through various flow channels into the cavities of a mold, where it is finally shaped into the desired object by the confines of the mold cavity. What makes this apparently simple operation complex is the limitations of the hydraulic circuitry used in the actuation of the injection plunger and the complicated flow paths involved in the filling of the mold and cooling action in the mold.

3. Opening the mold to eject the plastic after keeping the material confined under pressure as the heat (which was added to the material to liquefy it) is removed to solidify the plastic and freeze it permanently into the shape desired, for thermoplastics.

These three steps are the only operations in which the mechanical and thermal inputs of the injection equipment must be coordinated with the fundamental properties of the plastic being processed. These three operations are also the prime determinants of the productivity of the process, since manufacturing speed will depend on how fast we can heat the plastic to the molding temperature, how fast we can inject it, and how long it takes to cool the product in the mold. A simplified description of the molding machine functions is shown in Fig. 1-6.

The other operations in the injection process, such as clamping the mold, ejecting the part, feeding the machine, etc., are also important in the injection process. Their speed and their effect on productivity will normally depend on factors other than the material being processed—for example, the hydraulics of the clamp system and design of an injection system. See Fig. 1-7 for a description of a toggle injection molding machine.

The basic components of an injection molding system are: (1) blending, (2) drying, (3) hoppering, (4) metering, (5) plastication, (6) injection, (7) cooling, and (8) ejection. Plastic is usually purchased in pellet form and heated in the injection heating chamber until it reaches a viscous state in which it can be forced to flow into the mold cavities. Each plastic differs in its ability to flow under heat and pressure. For the best result, the correct melting temperature, injection pressure, and mold filling speed must be determined by trial for the particular plastic and mold used. Some molding conditions require that both the speed of injection and injection pressure vary during the filling process. A heat-sensitive plastic may be degraded if too rapid a fill rate is used. Forcing the plastic through the orifices at too high a velocity may increase the shear and temperature enough to cause overheating and burning.

However, thin-walled parts require a fast fill rate to prevent chilling of the plastic before the cavity has properly filled. Some molded parts carry both thin and thick sections, plus such interrupted flow patterns as are required to move around cored holes. Demanding requirements such as these necessitate consider-

MOLDING MACHINE FUNCTION

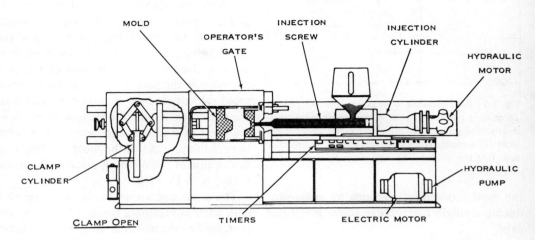

CLAMP OPEN

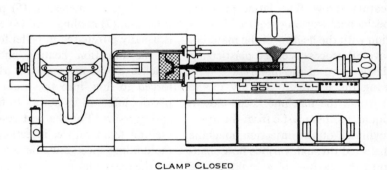

CLAMP CLOSED

ELECTRICAL	HYDRAULIC	MECHANICAL
1. MOTOR SWITCH "ON"	PUMP PRESSURIZES SYSTEM	
2. CLOSE OPERATOR'S GATE AND START CYCLE	OIL FLOWS TO CLAMP CYLINDER FROM HYDRAULIC MANIFOLD	CLAMP CLOSES
3. CLOSING CLAMP TRIPS LIMIT SWITCH DIRECTING OIL TO INJECT	OIL FLOWS TO INJECTION CYLINDER	INJECTION RAM FORWARD TO INJECT
4. INJECTION TIMES OUT	OIL FLOWS TO SCREW DRIVE MOTOR	SCREW PUMPS ITSELF BACK AS PARTS COOL IN MOLD
5. CLAMP COOLING TIMES OUT AND SCREW TRIPS SHOT SIZE LIMIT SWITCH	OIL FLOWS TO CLAMP CYLINDER ROD	SCREW STOPS ROTATING AND CLAMP OPENS
6. EJECTION LIMIT SWITCH IS TRIPPED	OIL FLOWS TO EJECTOR CYLINDER	PART IS EJECTED FROM MOLD
7. RECYCLE TIMER TIMES OUT	START CYCLE ETC.	

Fig. 1-6 Molding machine function.

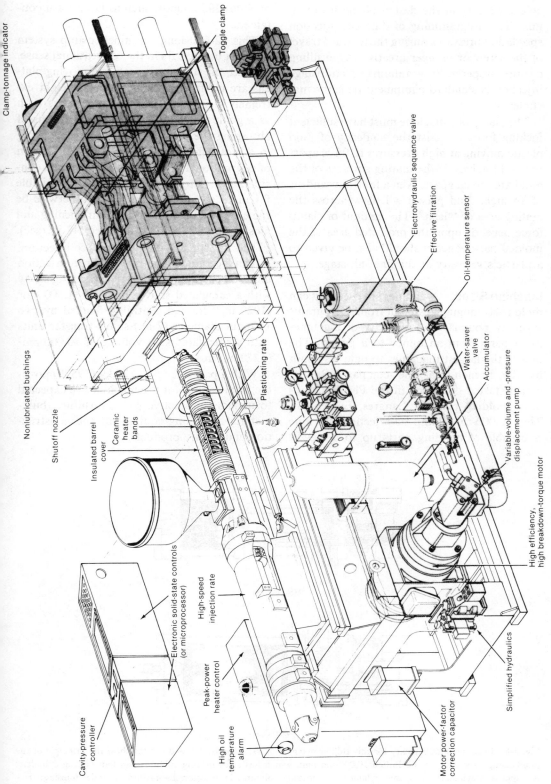

Clamp-tonnage indicator

Toggle clamp

Nonlubricated bushings

Shutoff nozzle

Insulated barrel cover

Ceramic heater bands

Plasticating rate

Electrohydraulic sequence valve

Effective filtration

Oil-temperature sensor

Water-saver valve

Accumulator

Variable-volume and -pressure displacement pump

High efficiency, high breakdown-torque motor

Electronic solid-state controls (or microprocessor)

High-speed injection rate

Cavity-pressure controller

Peak-power heater control

High oil temperature alarm

Motor power-factor correction capacitor

Simplified hydraulics

Fig. 1-7   Layout for the toggle injection molding machine.

able versatility in the design of the injection unit. The programming of different injection speeds and pressures during the forward travel of the screw or plunger greatly aids in filling cavities properly. Programming or multistage injection is standard equipment on some machines.

The clamp of a machine must have sufficient locking force to resist the tendency of fluid plastic moving at high pressures to force apart the mold halves. If the mating surfaces of the mold are forced apart, even a few thousandths of an inch, fluid plastic will flow across the mating area or "flash." The amount of clamp force depends upon the projected area of the molded part; concern should also be given to a plastic's viscosity at the final fill stage.

Machine Sizes. Machines are used that can mold parts ranging in weight from a few grams to many pounds. The majority of machines mold parts from an ounce to a pound. The first of the largest machines to date was built by the French machine company Billion, which started operating in the Plastic Omnium's molding plant at Langres, France (3). This 10,000-ton clamp co-injection machine is capable of molding parts up to a weight

of about 395 pounds, such as large trash containers.

This 92-foot-long machine's clamp system does not use tie bars in the conventional sense. Instead, the stationary, rear, and moving platens are mounted within a series of eight extremely rigid steel frames, which serve both as a guide for the platens and as a means of absorbing the reaction forces upon clamping in a most effective manner. This elimination of conventional tie bars also means that a considerably greater platen area is made available for mold mounting than would otherwise be the case (see Fig 1-8). Maximum opening and closing speeds are 1200 mm/sec (47 in./sec).

The machine is designed for co-injection. It can accommodate up to three injection units: one 200-mm ($7\frac{3}{4}$-in.) reciprocating screw with a calculated shot volume of 31,500 cm³ (1920 in.³) located centrally, flanked by two 180-mm diameter (7-in.) screw-transfer units with a calculated shot volume of 70,000 cm³ (4270 cu. in.).

With a part size up to 170 kg (395 lb) possible, applications for the machine are expected to include automotive parts, furniture, bulk-materials-handling equipment, and construction and leisure products.

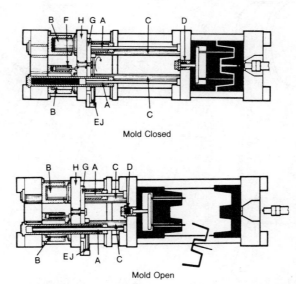

Fig. 1-8 Injection shot capacity for this Billion machine is 395 pounds. These diagrams show the principle of the mold-clamping system on the Billion 10,000-ton unit, which has eight locking columns and four closing cylinders. A—approach cylinders; B—clamp cylinders; C—locking columns; D—ejector; E—pivoting cylinder (closing); F—return cylinder; G—pivoting plate; H—support plate; J—pivoting cylinder (opening).

Injection Rates. Machines can operate to move the plastic melt into a mold at different injection rates. Generally the faster rate permits molding thinner parts and reducing cycle time. With the typical reciprocating screw injection molding machines in use today, the injection rate capability varies with the machine size, particularly the injection unit shot size.

Basically the larger shot size machines have a greater injection rate capacity. This is required because larger machines have larger platen areas for larger molds with increasing projected areas and melt flow lengths. If the injection rate did not increase with injection unit shot size, large parts would have to have even thicker walls to enable the cavity to be completely filled before melt freeze-off in the area near the gate. Typcial injection rate specifications for standard machines are 8 to 16 in.^{3}/sec for a 150-ton machine with 6- to 12-ounce shot capacity, 25 to 45 in.^{3}/sec for a 500-ton machine with 48 to 76-ounce shot capacity, on up to 70 to 90 in.^{3}/sec for a 1000-ton machine with a 160- to 180-ounce shot capacity. The injection rate capacity of the machine is a direct function of the hydraulic pumping capacity of the machine. Hydraulic flow rate in gallons per minute (gpm) determines the injection rate, and hydraulic pressure controls the injection pressure.

The first step normally taken to increase the injection rate capacity of the machine is to add additional hydraulic pumping capacity, either by replacing existing motors and pumps with larger ones, or by adding an additional motor and hydraulic pumps. Available from most machine builders, this option is commonly called a power pack. Typical power pack options may increase the machine's injection rate capacity by 20 to 30 percent, depending on the size of the added pumps. For a 500-ton machine with a 48- to 76-ounce shot size, this means the injection rate would go up from 25 to 45 in.^{3}/sec to about 35 to 60 in.^{3}/sec.

To be energy-efficient, the machine sequence can be arranged to utilize this additional pumping capacity only as needed during the injection step. For parts with longer cycles and lower injection rate requirements, the additional motor and pumps can be turned off to save energy and reduce the molded part cost.

In the past few years, the need for a still higher injection rate capacity in the molding process has become more widely recognized and is being satisfied. A review of alternate methods to accommodate this need, to find the best way significantly to increase the hydraulic flow capacity on the molding machine, indicated that simply adding more pumping capacity to the machine was not the best alternative. The hydraulic accumulator is a more attractive and energy-efficient method.

Specification Information. An injection molding machine is generally only "quickly" identified by clamp tonnage and shot size. There are many more parameters to include when specifying a machine, such as:

1. Injection Specifications
   Injection capacity (in.^{3})
   Injection capacity (oz)
   Injection rate (in.^{3}/sec)
   Screw recovery rate (oz/sec)
   Injection pressure (psi) Max.
   Screw diameter (in.)
   L/D ratio
   Screw speed (RPM) Max.
   Screw drive motor (hp)
2. Clamp Specifications
   Clamping force (tons)
   Clamp stroke (in.) Max.
   Daylight opening (in.) Max.
   Mold thickness (in.) Min.-Max.
   Distance between tie-rods (in.)
   Clamp closing speed (in./sec)
   Clamp opening speed (in./sec)
3. Hydraulics and Motor
   Hydraulic line pressure (psi)
   Pump delivery
   Motor(s) total connected HP
4. Features
   Screw speed adjustment
   Barrel lining
   Injection unit pivot
   Removal of screw
   Safety stop bar

Torque selection
Convertible to themoset
5. Cost Comparison
Price-standard machine
Low pressure mold close
Hydraulic knockout
Motorized mold height adjustment
Xaloy barrel
Nozzle temperature control
Automatic cycle (4th timer)
Screw speed tachometer
Screw decompression
Hopper magnet
Precision leveling mounts
Others

STANDARD NOMENCLATURE FOR INJECTION MOLDING MACHINES

Plasticizing

The function of the injection molding machine heating cylinder is to thoroughly and uniformly convert (plasticize) the plastic feed material into a homogeneous heated plastic melt of controlled viscosity, and then force it into the clamped mold where the end-product is formed (see Fig. 1-9). This section is a summary of the subject. Details are reviewed in Chapter 3.

The heating cylinder is a simple heat exchanger. Most cylinders have heavy steel walls with highly polished and hardened inner surfaces. For some purposes, the cylinder may be lined inside with a special corrosion-resistant material designed to resist the possible degradation products of thermally unstable resins.

It is important to note that only the cylinder temperature is directly controlled. The actual temperature of the plastic melt within the screw and as it is ejected from the nozzle can vary considerably, depending on the efficiency of the screw design (see Chapter 3) and the method by which it is operated. Factors that affect the melt temperature include: the time the material remains in the cylinder; the internal surface heating area of the cylinder and screw per volume of material being heated; the thermal conductivity of both the cylinder and screw wall and the plastic material; the differential in temperature between the cylin-

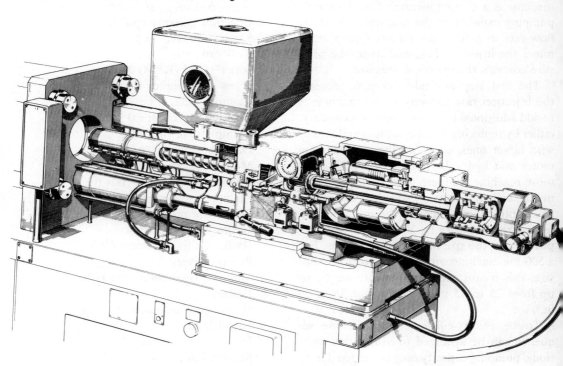

Fig. 1-9 Cutout of injection unit (plasticizing unit). Courtesy of Negri Bossi.

der and the plastic; the wall thickness of the cylinder and of the stationary film (on the inner cylinder wall) of the plastic being heated; and the amount of turbulence in the cylinder.

Because of their molecular structure, plastics have low thermal conductivities; thus it is difficult to transmit heat through them rapidly. In addition, plastic melts are very viscous, and it is difficult to create any turbulence or mixing action in them without the positive application of some form of mechanical agitation in the screw. The problem is further complicated by limitations of the length of time the plastic may be allowed to remain in the cylinder. In designing the screw, a balance must be maintained between the need to provide adequate time for proper heat exposure of material in the cylinder and the need to process maximum quantities of materials for the most economical operation.

In general, the heat transfer problems have led injection screw designers to concentrate on making more efficient heat transfer devices. As a result, the internal design and performance of these units vary considerably, based on the material to be processed.

Screw Design.   The primary purpose for using a screw is to take advantage of its mixing action. Theoretically, the motion of the screw should keep any difference in melt temperature to a mimimum. It should also permit materials and colors to be blended better, with the result that a more uniform melt is delivered to the mold.

The design of the screw is important for obtaining the desired mixing and melt properties as well as output rate and temperature tolerance in the melt. (See Chapter 3 for details on screw designs.) Generally most machines use a single, constant-pitch, metering-type screw for handling the majority of plastic materials. A straight compression-type screw or metering screws with special tips (heads) are used to process heat-sensitive thermoplastics, etc.

The helix angle affects the conveying and the amount of mixing in the channel. A helix that advances one turn per nominal screw diameter gives excellent results. This corresponds to an angle of 17.8 degrees, which has been universally adopted. The land width is 10 percent of the diameter. The radial flight clearance is the clearance between the screw flight and the barrel; it is specified considering the following effects:

1. Amount of leakage flow over the flights.
2. Temperature rise in the clearance. Heat is generated in shearing the plastic, with the amount of heat generated related to the screw speed, the design of the screw, and the material.
3. The scraping ability of the flights in cleaning the barrel.
4. The eccentricity of the screw and the barrel.
5. Manufacturing costs.

The length of the screw is the axial length of the flighted section. An important criterion of screw design is the ratio of the length over the diameter of the barrel (L/D). Long screws with a $20:1$ L/D are generally used. An advantage of using a long screw can be that more of the shear heat is uniformly generated in the plastic without degradation.

Basically, a screw has three sections: feed, melting (transition), and metering (see Fig. 1-10). The feed section, which is at the back end of the screw, can occupy from zero to 75 percent of the screw length. Its length essentially depends upon how much heat has to be added to the plastic in order to melt it. The pellet or powder is generally fed by gravity into this section and is conveyed some distance down the barrel, during which time it becomes soft. Heating is accomplished by both conduction and mechanical friction.

The melting (transition) section is the area where the softened plastic is transformed into a continuous melt. It can occupy anywhere from 5 to 50 percent of the screw length. This "compression" zone has to be sufficiently long to make sure that all of the plastic is melted. A straight compression-type screw is one having no feed or metering sections.

In the metering section, the plastic is smeared and sheared to give a melt having a uniform composition and temperature for

GENERAL PURPOSE SCREW

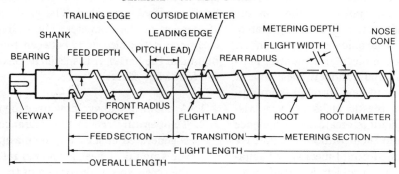

VENTED SCREW

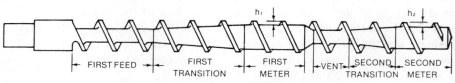

Fig. 1-10   Sections of a conventional screw.

delivery to the mold. As high shear action will tend to increase the melt's temperature, the length of the metering zone is dependent upon the resin's heat sensitivity and the amount of mixing required. For heat-sensitive materials, practically no metering zone can be tolerated. For other plastics, it averages about 20 to 25 percent of the total screw length. Both the feed and the metering sections have a constant cross section. However, the depth of the flight for the feed zone is greater than that in the metering zone. The screw's compression ratio can be determined by dividing the flight depth in the feed zone by that in the metering zone.

The plastic material in the screw channel is subjected to different experiences as the screw operation changes during the cycle (see Fig. 1-11). Each operation of the screw, whether moving forward during injection, rotating and retracting during shot preparation, or static during an idle period, subjects the plastic to different thermal and shear situations. Consequently, the injection molding plasticating process is rather complex.

In addition to melting during screw rotation, melting occurs during the static idle period as heat is conducted from the hot barrel. This causes the melt film between the barrel and the solids region to grow in thickness as some solids are consumed. Subsequently, during the injection stroke, still further melting occurs as a result of heat conduction from the hot barrel and shear heating due to the

Dimension	Rigid PVC	Impact Polystyrene	Low-density Polyethylene	High-density Polyethylene	Nylon	Cellulose Acet/Butyrate
diameter	4½	4½	4½	4½	4½	4½
total length	90	90	90	90	90	90
feed zone (F)	13½	27	22½	36	67½	0
compression zone	76½	18	45	18	4½	90
metering zone (M)	0	45	22½ (450)	36	18	0
depth in (M)	.200	.140	.125	.155	.125	.125
depth in (F)	.600	.600	.600	.650	.650	.600

Fig. 1-11   Typical screw designs.

screw forward movement in the barrel. This results in widening of the melt pool.

At a fixed screw speed, the pitch, diameter, and depth of the channels determine output. A deep-channel screw is much more sensitive to pressure changes than a shallow channel. In the lower pressure range, a deep channel will mean more output; however, the reverse is true at high pressures. Shallower channels tend to give better mixing and flow patterns.

The flow pattern in the screw flights changes with the back pressure. With the flow of a particle in the flights with open discharge and in blocked flow, there is a similar circulatory motion between the flights, but no forward motion because the open end is closed. The greatest mixing occurs when the flow is blocked. This is an important flow concept: the more blocked the flow, the better the mixing in the screw. The higher the pressure, the greater the pressure flow and the lower the output. In injection molding, this pressure corresponds to the back pressure setting of the machine. Because of the better mixing, color dispersion is improved and homogeneity increased by raising the back pressure. Often warpage and shrinkage problems can be overcome in this manner.

Basically the mechanism for melting starts after the plastics move from the hopper to the screw. Plastic touches the barrel to form a thin film of melted plastic on the barrel surface. The relative motion of the barrel and screw drags the melt, which is picked up by the leading edge of the advancing flight of the screw. This edge flushes the polymer down in front of it, forming a circulating pool. Heat is first conducted from the barrel through the film of plastic attached to it. Heat then enters the plastic by shearing action, the shear energy being derived from the turning of the screw. The width of the melted polymer increases as the width of the solid bed decreases. Melting is complete at the point where the width of the solid bed is zero.

The reciprocating screw machine uses the screw as a plunger. During forward motion of the plunger, the material could flow past the screw head and back into the flights. For more viscous materials, such as PVC, a ta-

pered tip on the front of the screw is sufficient to permit the screw to act as a plunger. The rapid forward motion of the plunger does not let too much material flow back. Moreover, the plain tip is also good for molding heat-sensitive material such as PVC, because this type of screw front provides the least opportunity for hangup and material degradation.

The less viscous materials require a valve to prevent back flow over the screw tip. Screw tips, in either case, are a varying source of frictional heat, material hangups, intermittent malfunctioning, and potential high maintenance costs.

The ring-type non-return valve is generally used. It is a three-part assembly. The check ring and seat are slipped on the main body, which contains the tip. The assembly is then screwed into the reciprocating screw. The sliding ring fits snugly in the barrel. When the screw rotates, the nozzle end permits the plasticized material to flow under it through flutes or grooves on the main assembly. The screw slides back until the amount of material necessary for the shot is plasticized. On the forward or injection stroke, the ring slides toward the seat and seals the rear of the screw from the front so that material cannot leak by as the plunger comes forward.

Productivity and People

Instructions for operating machines can be simply stated by issuing the usual guidelines, such as these "start-up procedures" (details of which are reviewed in Chapter 4):

- Preset the heat controllers on the barrel and nozzle.
- Start the machine motor and screw motor when heat controllers indicate proper heat has been reached.
- While the equipment is in manual operation, close the safety gate and close the press to lock.
- Check to see that the resin feed hopper gate is closed, and adjust the flow control valve down to zero.
- Turn the plunger switch to the out position. Adjust the flow control valve until

the screw rotates. (If it will not rotate, the heat has not been on long enough; so shut down the machine and try again in 10 or 15 minutes.)

- As the screw rotates, open the feed (off and on) to allow small amounts of resin to feed into the screw. Watch the screw load, and if it exceeds 100 percent, reduce the screw rpm.
- Continue opening and closing the feed hopper until the machine is pumping well and the load is holding fairly even.
- Open the feed to the screw and let it purge until the melt appears to be consistent (adjust the back pressure valve to hold the screw in the forward position).
- Etc.

However, there is more to productivity than listing guidelines and checklists (see Table 1-1). Trained operators are needed. This section is a summary of the entire subject. (See Chapter 4 for details.)

Today's emphasis on latest-generation machinery and space-age controls often makes the individual seem less important than he or she used to be. However, the men and women on the machine lines now have a more important role than ever. They add a critical capability to a line; they give it versatility.

The more one visits plants of all types, the more one finds that totally dedicated lines are not as common as might be expected. In fact, they are the exception rather than the rule—in the context of the full range of lines running today. The obvious reason for the growing emphasis on versatility is that market fragmentation, product proliferation, and all that they imply are bringing shorter runs and more variations to most lines in the typical types of plants making molded products.

Assuming that you do have a well-rounded, ongoing training program and genuine, continuing, two-way communication with plant personnel, ask yourself this question: Have you taken time to think of all possible ways to team the people with whatever machines you have, to add versatility? For instance, if you are not sure whether a commitment to a fully automatic operation will pay off, why

not train or retrain a group of people to team up with a semi-automatic loading sequence? For another, do you make the most of varying the numbers of people, to speed up or slow down a given line? Assume you have a powered belt and two tables where crew members assemble combinations, complete packages, or otherwise complement the machinery running ahead of them on the line. Assume you can get the speed, say, from very slow to quite fast, from 1 to 25 lineal feet a minute. When you get a rush order or special priority, do you add crew members and speed up the belt beyond what you think should be the norm for day-to-day running?

Such suggestions as these may seem too elemental to deserve your attention, but today's packaging lines reveal a growing preoccupation with variety. If you are trying to reconcile the often nearly irreconcilable goals of peak efficiency and peak versatility in the face of short runs and dramatically varying combinations of products, containers, and sizes, take a second look at what plant personnel can contribute.

By 1990, one estimate predicts a shortage of 100,000 technicians in all U.S. industries to fill such jobs as maintaining microprocessors, electronic controls, robots, and the like. The problem will grow as more plants automate, computerize, and robotize. However, a number of avenues are available to attack the problem. At the core of the problem is the fact that high technology advances too rapidly for support services to keep pace. Many of our technicians are older, are becoming lost through retirement, and usually are not trained in the new technologies. Training programs, especially in vocational schools and for in-plant people, have trouble keeping pace.

The situation cries out for improved training, particularly for in-plant workers to maintain and repair their plant's own high-tech equipment. Training in-plant people is crucial because most plants cannot tolerate equipment downtime, and the time delays associated with relying on independent service technicians.

Fortunately improved training is becoming available from more and more sources, including trade associations, continuing edu-

Table 1-1 Molding data record.

JOB
OPERATORS
ENGINEERS

MOLD DESCRIPTION
SCREW USED
PRESS NO.

MACHINE SETUP INSTRUCTIONS
NOZZLE #

SPECIAL INSTRUMENTATION
SAFETY CHECK

DATE	TIME	Resin		TEMPERATURES (°F) (°C)				MOLD			PRESSURES (PSI) (kp/cm²)				CYCLE TIMES (SEC)							Pad (inches)	RPM	WEIGHTS (GMS)		REMARKS
		Run Number	Lot Number	Rear	Center	Front	Nozzle	Fixed	Movable	Melt	Injection 1st Stage	Injection 2nd Stage	Clamp (Tons)	Back	Injection	Hold	Open	Overall	Booster	Ram in Motion	Screw Retraction			Full Shot	Part Only	

COMMENTS ON MOLDING OPERATION, START-UP, ETC.

cation in colleges, vocational schools, and on-site training by equipment manufacturers. New training media are available, including packaged video tape/programmed learning courses, and computer-assisted instruction, in which the computer terminal actually instructs the trainee. Improved training can help speed apprenticeship programs from, say, four to two years to keep pace with technology.

Another answer involves getting away from craft specialization, and making craftspeople proficient in more than one area. In this multi-craft concept, if trouble occurs in an electro-mechanical-pneumatic system, for example, one person troubleshoots the system instead of three. It is reported that such "job enrichment" sparks new enthusiasm in workers, but implementing it requires cooperation from labor.

Another solution is to locate the plant where the high-tech technicians are, that is, near military bases, and employ technicians who have been discharged from the service. Some companies are doing this. The military, foremost the Navy, trains technicians on state-of-the-art equipment, and provides years of hands-on experience. The crisis is surmounta-ble, and calls for solutions that feature ingenuity, flexibility, improvisation, and willingness to do things in new ways—on the part of management, craftspeople, and labor.

Process Control

As injection molding becomes more complex, molders require greater accuracy and increased variations in the types of cycles that could be adapted to their machines. These include variations in core pull sequences, different ejection sequences, and changes in the timing of high pressure application after clamp close in combination with injection, screw rpm, back pressure during melting, etc. This section is a summary of the subject. (See Chapters 5, 11, 12, and 13 for details.)

Different types of machine process controls can be used to meet the requirements, based on the molders' operating needs (see Fig. 1-12). Control systems available can monitor (alarm buzzes or lights flash on deviation), feedback (deviation sets up corrective action), and program the controller (minicomputers interrelate "all" machine functions and "all"

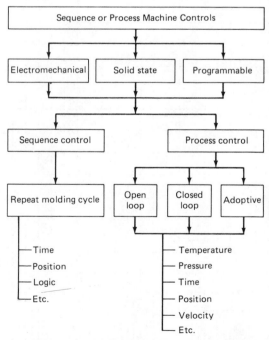

Figure 1-12 Simplified overview of process controls for injection molding machines.

melt process variables). Knowledge of your machine and its operating needs is a prerequisite before an intelligent process control program can be developed and used.

Most controls are of an open-loop type. They merely set a mechanical or electrical device to some operating temperature, pressure, time, or travel. They will continue to operate at their setpoints even though the settings are no longer suitable for making quality parts. During molding the problem is that the total process is subject to a variety of hard-to-observe disturbances that are not compensated for by open-loop controls. Process control closes the loop between some process parameter and an appropriate machine control device to eliminate the effect of process disturbances.

With controls properly installed and applied, the performance of the plastics in the machine can be controlled within limits to produce zero defect parts meeting performance requirements at the lowest molding cost. The limits have to be set on the basis of testing and evaluation of molded parts. See Figs. 1-13, 1-14, and 1-15 for the basic analysis of effects of specific injection molding machine and plastic material variables (4 & 5). The next important aspect to analyze is the effect

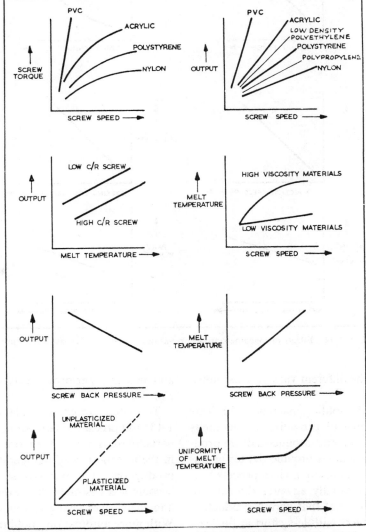

Fig. 1-13   Effects of injection molding machine and plastics material variables.

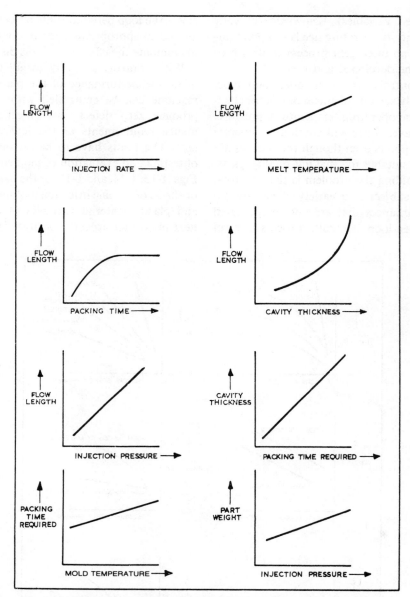

Fig. 1-14 Effects of machine settings on various properties such as flow length.

of interfacing the different variables, as shown in Fig. 1-16.

Practically all molding machines with little effort are capable of providing useful melts that go into molds and produce salable products. Certain machines provide tighter operational controls (thanks to modern process controls which continually advance the state of the art in processing) that permit production of quality products with less effort at the least cost. The interrelationship between a plastic and machine performance is summarized in Fig. 1-17.

To reduce molding cycle time (see Fig. 1-18) and produce quality-controlled or useful parts, there is a need for more precise control in the injection molding operation. At higher production rates, excessive scrap and rejects become less desirable than ever, and molders find themselves trying to reduce these levels. With more automation, molding optimizaiton is further complicated by automated opera-

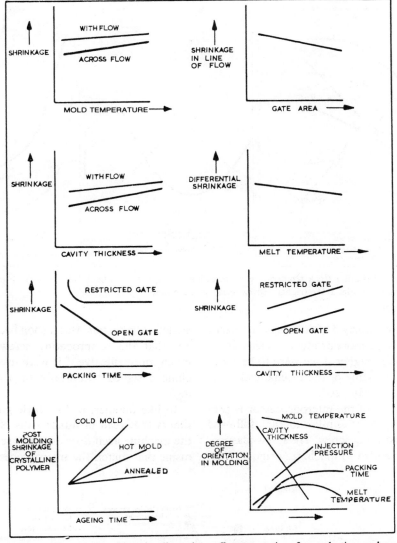

Fig. 1-15 Machine settings and mold cavity dimensions affect properties of any plastics, such as shrinkage.

tions that move the products directly from the molding machine to the assembly stations. Effective process control, therefore, is essential to maintain the benefits of modern process technology.

Purchasing a more sophisticated process control system is not a foolproof solution to molding-quality problems. Solving part-reject problems requires a full understanding of the real cause, which may not be as obvious as they first appear. The conventional place to start troubleshooting a problem is the melt temperature and pressure. But often, the problem is a lot more subtle; it may involve mold design, faulty control devices, and other machine components.

Problems in mold design can cause pressure differences and temperature differences between cavities. Sometimes factors not directly related to the process may be influencing quality, such as an operator making random adjustments of control devices and the rate that plastic moves into the screw. Process control systems usually cannot compensate for such extraneous conditions.

Studies have shown that compared to the most efficient plastics molding machine of the 1970s, a new microprocessor-controlled ma-

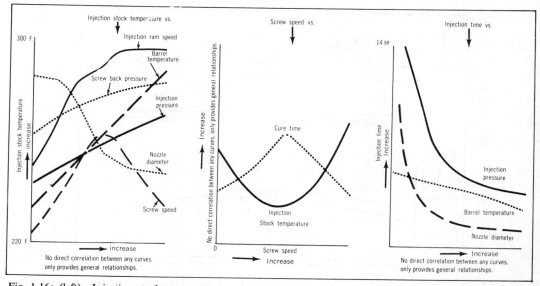

Fig. 1-16a (left) Injection stock temperature vs. injection ram speed. Fig. 1-16b (center) Injection time vs. injection pressure. Fig. 1-16c (right) Screw speed vs. cure time.

chine can save nearly $1,000 a year in energy alone while being more productive. That figure would be much higher if we were to base our comparison on some of the mechanical relics that are still widely used.

Development of the microprocessor is proceeding along lines similar to that followed by the reciprocating screw. The screw plasticator was first added to machines originally de-

signed for plungers. But it soon became apparent that the reciprocating screw would be much more effective if it were used on a machine designed specifically to accommodate it.

In like manner, it does little good to have timers that read in hundredths of seconds if the machine itself does not have servo-mechanisms that match the microprocessor's preci-

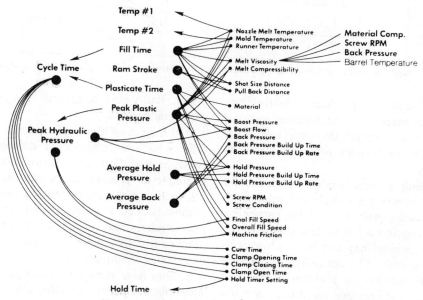

Fig. 1-17 Interrelationship of plastic and machine.

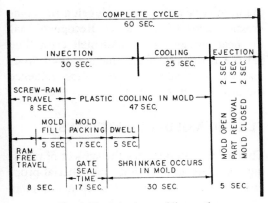

Fig. 1-18 Injection molding cycle.

sion. The machinery must be as good as the control device, or there is little sense in having a microprocessor.

Developing Melt and Flow Control

The mechanism for melting was described above in the section on "Plasticizing," under "Screw Design." The reciprocating screw machine uses the screw as a plunger, and the screw tip may or may not require a valve to prevent back flow. (See under "Screw Design" for details.)

Addition of back pressure is a means of creating an additional workload on the screw by restricting its ability to pump the plastic. In the process it increases melt temperature and uses more drive energy. Its benefit is to increase working of the plastic to improve color blending and improve melt quality. The proper heat profile does these things as well. Therefore, consider using no back pressure until the heat profile is obtained, and then add only that back pressure necessary. The rate of fill of the mold is determined by many factors such as viscosity of the melt, the gate size, the machine's capabilities, the mold temperature, etc. (For more details see Chapters 3, 11, and 13.)

Increase Production through Fine Tuning. The computerized molding machine increases productivity. It does this in several ways. First, it enables the molder to fine-tune all the relationships that exist in both the clamp and the injection end of the machine.

At a digital control panel, the setup technician can stroll through all the machine functions, cutting off a tenth of a second here, three tenths of a second there, throughout the entire job.

Furthermore, once the optimum settings have been determined for a particular mold, they can be repeated simply by entering them in the control each time the job is run. An exacting setup job need only be performed once during the life of a mold.

But cycle time is not the lone determinant of productivity. Fast cycle times may easily be negated by high rejection rates. The computer's ability to fine-tune the machine to a given mold also results in the highest-quality part and the lowest rejection rate. Part quality is also enhanced through the use of process control devices that are computer-driven.

The microcomputer is far more dedicated than the human worker. It can continually make adjustments within the machine and make it perform more efficiently. Energy consumption is an important concern of the molder. The microcomputer is by far the best energy-saving tool known to modern industry. (Details on computer simulation are given in Chapters 11, 12, and 13.)

During molding there is the need for a good-quality melt at the lowest possible temperature. The hotter the melt is on injection, the longer the mold must be held closed in order to cool the part. This action effects cycle time. It also effects part quality, and energy use. The proper heat profile will not only reduce cycle time, but will save energy as well. If a reduced amount of heat is put into the plastic through the heater bands, the energy required to take the heat out of the mold will be reduced as well.

Behavior of Plastics. To obtain the best melt, start with the plastics manufacturers' heat profiles, which are starting points for various types of materials. Or you can start with the profile used on another machine, but it is important to continue changing the barrel temperature profile to obtain the best heat profile that is repeatable during the plasticizing process. Details on polymers, properties of plastics, and effect of polymer structure on injection molding are given in Chapters 25 and 26.

An amorphous material, such as PS, PVC, SAN, ABS, cellulosic, acrylic, etc., usually needs a fairly low feed end heat, as the purpose is to preheat the material but not really melt it, before it reaches the compression zone of the screw. Crystalline materials, such as PE, PP, nylon, acetals, etc., require a higher heat load in the rear to ensure melting by the time the material reaches the compression zone.

Molded part properties and machine cycle times are very strongly influenced by the plasticating process. The success of injection molding as a production technique to date is largely due to the efficiency of reciprocating screw plasticating units, in both melting the plastic feedstock and simultaneously providing sufficient mixing to ensure good melt uniformity. The melt quality desired essentially is good-quality melt (based on visual observation) coming out of the nozzle, which gives good moldings without welds, sinks, etc. Molds and materials are uniquely related to each other, so one cannot generalize about what makes a good melt. Experience of the molder and knowledge of needs constitute the final determining factor. There are several ways that you can determine the efficiency of the heat profile that you are using. One is to observe the screw drive pressure. If the right heat profile is being used, the screw drive pressure should be at about 75 percent of maximum. If it is below that level, lower the rear zone heat until the drive pressure starts to rise.

With melt quality changing, raise the center zone to bring quality in. Changes to the temperature should be made in increments of $10°$ to $15°F$, with a 10- to 15-minute stabilizing time allowed before the next change. Once the rear zone is set, lower the front zones to whatever level that will still give you good molding conditions. The target is a good-quality melt at the lowest practical temperature. With crystalline materials such as nylon, propylene, etc., watch the screw return. If the screw is moving backward in a jerky manner, there is not enough heat in the rear zone, and

the unmelt is jamming or plugging the compression zone of the screw. Recognize that the heat energy required to melt crystalline plastics is different from that for amorphous plastics. Also additives and fillers influence the heat energy required.

PLASTIC MOLDING MATERIALS

This section summarizes the subject of molding materials used. (Details on material properties and performance are given in Chapters 18 through 26.)

Practically all molding machines with little effort are capable of providing useful melts that go into molds and produce salable products. The interrelationship between plastic and machine is summarized in Fig. 1-17 (see above).

On the average, at least half of the costs in plastics processing are incurred in raw materials and services; wages, utilities, and capital costs account for the rest. Thus it is important to purchase the raw materials at favorable prices, to have them delivered punctually, to use as little as possible (do not overpack material in cavity if not necessary, and mold to tight tolerances), and to ensure that their quality remains constant. Savings may be effected by judicious selection of the form in which materials are supplied.

The system for ordering materials depends on the production program. It may be based on requirements, stocks, or agreed-upon deadlines. Costs can be saved by finding out the qualities that can be supplied on the most favorable terms. Decreases in price effected by purchasing larger amounts must be balanced against the extra costs for storage and the larger amount of tied-up capital; a certain amount must represent an optimum. Purchasers must also allow for delivery times. Frequently materials in the natural color can be supplied direct from stocks.

Checking Materials Received

An important factor in the production of parts is that the quality of the raw materials must always conform to specification. Certain properties must be checked when the goods are received. In view of the wide variety of applications for plastic articles, a testing schedule of general validity cannot be submitted here. Each case must be treated individually.

Over the years, many hours have been devoted to designing methods for testing materials to develop values for their properties. These tests, conducted under procedures established by organizations such as the American Society for Testing and Materials (ASTM), are a means of extracting basic knowledge about materials. (See Chapter 28 for testing.)

Although raw materials of constant properties are essential for high-quality moldings, they do not suffice for this purpose by themselves. In particular, mistakes in processing could adversely affect the properties. If possible, allowance must be made for this potential problem in the testing schedule.

The first task in checking goods received is to make sure that they conform to type. In other words, they must be checked to ensure that they agree with samples of former deliveries. This check includes examination for contamination, and is followed by specific tests such as simply determining bulk density. Often samples are sent in advance of materials dispatched in tankcars or large containers. In this case, statistical rules must be observed in taking the random samples.

The preliminary check must proceed without loss of time; so rapid tests with specific aims are frequently used. Since injection molding has been caught up in the automation trend, it is feasible for checking the goods received to become part and parcel of the actual production process. However, this entails that any deviations from standard must remain within narrow limits. For technical and economic reasons, this adaptive process control, as it is called, is still a long way from being realized.

Compounding and Coloring In-Plant

Compounding or mixing is an important stage in the production of the raw materials. The way it is performed can affect injection molding, especially if the compound is in the form of a powder, and the ingredients (which have different weights) are not mixed together until shortly before molding.

Great significance has been attached to adding all kinds of masterbatches, for example. There are color masterbatches, reinforcing-fiber masterbatches, flame-retardant and antistatic masterbatches, and masterbatches containing foaming agents and other additives. Since the importance has been recognized of what are known as plastics alloys, which widen the field of application of thermoplastics, different pellets or powders are also mixed with one another.

Synergistic effects can be developed when certain plastics are combined. Some property improvements with alloying are shown in Fig. 1-19 and Tables 1-2 and 1-3.

A distinction based on the stirrer speed is drawn between gravity mixers and stirrers (slow and high-speed). The peripheral velocity in slow stirrers is 30 ft/sec, and in high-speed stirrers from 30 to 150 ft/sec (10 to 50 m/sec).

In-plant blending of the molding compounds offers some advantages. It dispenses with some of the fabrication costs and poten-

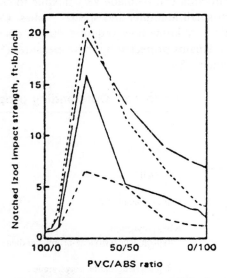

Curves reflect four different polyblends.

Fig. 1-19 Example of how alloying affects resin properties. Compounding and alloying technology makes it possible to combine two or more polymers into alloys with distinctive, often unique properties.

Table 1-2 Upgrading PVC by blending.

UPGRADED PROPERTY	BLENDING POLYMER
Impact resistance	ABS, methacrylate-butadiene-styrene, acrylics, polycaprolactone, polyimide, polyurethanes, PVC-ethyl acrylate
Tensile strength	ABS, methacrylate-butadiene-styrene, polyurethanes, ethylene-vinyl acetate
Low-temperature toughness	Styrene-acrylonitrile, polyurethanes, polyethylene, chlorinated polyethylene, copolyester
Dimensional stability	Styrene-acrylonitrile, methacrylate-butadiene-styrene
Heat-distortion temperature	ABS, methacrylate-butadiene-styrene, polyimide, polydimethyl siloxane
Processability	Styrene-acrylonitrile, methacrylate-butadiene-styrene, chlorinated polyethylene, PVC-ethyl acrylate, ethylene-vinyl acetate, chlorinated polyoxymethylenes (acetals)
Moldability	Acrylics, polycaprolactone
Plasticization	Polycaprolactone, polyurethanes, nitrile rubber, ethylene-vinyl acetate, copolyester, chlorinated polyoxymethylenes (acetals)
Transparency	Acrylics, polymide
Chemical/oil resistance	Acrylics
Toughness	Nitrile rubber, ethylene-vinyl acetate
Adhesion	Ethylene-vinyl acetate

tial problems due to heat history, and greatly reduces inventories. Purchasing one type in bulk reduces the costs of the raw materials. Production can be made very flexible to cope with small amounts and special wishes. Certainly any know-how acquired during operation remains protected by these measures (See Chapter 15.)

One of the most important mixing tasks in the injection molding factory is in-plant coloring. The advantages are obvious: saving costs incurred by the higher prices of colored grades, a wider selection of colors, adaptability, and reduced inventories. However, these advantages are balanced by the responsibility for selecting suitable colorants (for instance,

Table 1-3 Outstanding properties of some commercial alloys.

ALLOY	PROPERTIES
PVC/acrylic	Flame resistance, impact and chemical resistance
PVC/ABS	Flame resistance, impact resistance, processability
Polycarbonate/ABS	Notched impact resistance, hardness, heat-distortion temperature
ABS/polysulfone	Lower cost
Polypropylene/ethylene-propylene-diene	Low-temperature impact resistance and flexibility
Polyphenylene oxide/polystyrene	Processability, lower cost
Styrene acrylonitrile/olefin	Weatherability
Nylon/elastomer	Notched Izod impact resistance
Polybutylene terephthalate/polyethylene terephthalate	Lower cost
Polyphenylene sulfide/nylon	Lubricity
Acrylic/polybutylene rubber	Clarity, impact resistance

types that are resistant to heat or to ultraviolet radiation or are physiologically harmless). Moreover, the colorants must not impair the properties of the molding compound. At any rate, the demands imposed on the quality of the shades and their reproducibility from one machine to another and from one batch to another can never be so severe as those imposed on molding compounds supplied by the raw materials manufacturer. The cycle may become longer, and the shrinkage may change. Consequently, the workers entrusted with in-plant coloring are chiefly responsible for its quality.

Formerly, mixers were set aside in special rooms for in-plant coloring. They are now being supplemented by devices that allow coloring on the injection molding machine. They can proportion as many as three types of colorant, the molding compound in the natural color, and the regrind (i.e., a total of five ingredients), and are usually fitted with a mixer. The colorants are in the form of pellets, ground masterbatches, free-flowing and non-free-flowing pigment powders, pigment dispersions, and pumpable liquids. Great value must be attached to their dispersibility.

The quality of coloration obtained with in-plant techniques depends not only on the proportioning and mixing in the feeding device but also on plastification in the injection molding machine. Frequently, screws with mixing attachments in the metering section or with static mixers connected in series behind the metering zone are indispensable.

In-plant blending of virgin plastics (plastics that have not been processed in the plant) with granulated or recycle plastics is important to proper control. If not controlled, performance of the part can be below requirements. (Details on granulating are given in Chapter 17.)

PRODUCTION CONTROL OF QUALITY

Quality control is a complex task in injection molding. The quality and serviceability of a molding depend on many factors, starting from the raw materials and embracing the processing and application conditions.

Quality control begins with the design of the part, design of the mold, and capability of the injection molding machine. The number of cavities, the type of location of the sprue, the size of the machine, the allowances to be made for inserts, demolding, finish, and the tolerances laid down—are all factors that decide the quality and govern the price. In the early mold design stage, the tests to be adopted for quality control should already be decided upon and drawn up in the form of a checklist that will be accepted by the customer concerned.

The optimum injection conditions are determined in trial runs and noted in a report. The moldings thus produced are tested according to the checklist. The acceptance tests for the raw materials are a part of quality control.

The live production run is usually controlled by continuous visual inspections of the moldings and by checking their weight and a few dimensions. Measuring the dimensions at this stage is only of relative value because processing shrinkage is not always completed after the moldings have cooled. This applies particularly to partially crystalline molding compounds.

The outlay described clearly reveals that thorough quality control is expensive and can greatly influence the total costs. (Details on testing procedures used to set up quality control are given in Chapter 28.)

ECONOMIC CONTROL OF EQUIPMENT

In view of continuously rising costs, the main consideration in investing capital must be the ratio of earnings to costs. Production aids can make a considerable contribution to reducing costs. The most important are those required for feeding the raw material, deflashing, regrinding and recycling scrap, sorting the moldings from the sprues, demolding, stacking, packing, automatic machining, and bonding with adhesives (See Chapters 14 through 17.)

The only item that does not rise in cost is the machine performance. There are always new machines that will provide lower cost to melt the plastics.

Factors to be considered in the acquisition

of new injection molding machines are the criteria set up by the intended production program. For the injection molding of packaging, the main factors are the injection rate, the dry cycle time, the plastification rate, and the price of the machine. As opposed to this, the quality of the melt, process control aspects, and the clamping force are the factors that predominate in the production of machine precision parts.

Other requirements that are imposed on an injection molding machine for economic running are favorable starting-up characteristics, constant production characteristics, ease of operation, ease of retooling, and a long life.

Savings can be achieved in tooling by standardizing the platens, the radii of curvature, the fittings, and the electrical circuit. Machinery costs can be reduced by parts that do not require maintenance. This applies particularly to the hydraulic system.

Practically any step involved in processing the plastics contributes to cost and can easily be evaluated with respect to cost reduction. Consider, for example, when you should replace your machines as well as upstream and downstream equipment. Various methods can be used to replace old equipment, and perhaps should be. In the United States today, a lot of molders are losing money with old equipment, and they do not even know it. Not only are the new machines more productive; they also create less waste, use less energy, and are smaller, quieter, and safer.

Savings may also be possible in costs for fresh water and effluents, which have increased rapidly. There is generally a shortage of water in periods of dry weather, and water consumption in factories is growing as a result of increasing mechanization. Consequently, many injection molding factories have their own cooling water supplies. The main types are (1) open-circuit water cooling systems with an evaporation-type cooling tower, (2) closed-circuit water cooling systems with compression-type refrigeration machines, and (3) composite systems.

Open-circuit cooling systems operating exclusively with cooling towers were very popular in the past but have now lost their efficiency. As a result of evaporation and slime formation, 1.5 to 2.5 percent of the water circulated is lost and must be replenished. The temperature and humidity of the ambient air impose limits on the temperature that can be attained by the cooling water. At most, the temperature of the cooling water can be reduced to a value of 3°C above the wet bulb temperature. This is quite unsatisfactory, especially in the summer.

The compressor-type refrigerating machines in the closed-circuit systems operate with air-cooled or water-cooled condensers. Reciprocating machines and turbocompressors predominate. The main refrigerant is liquefied fluorohydrocarbon under pressure.

Combinations of open-circuit and closed-circuit cooling water systems also operate with evaporation-type cooling towers. Normally, the temperature of the cooled water in the closed refrigerating machine circuit is between 5°C and 20°C. This water is used for cooling the mold. A second system of pipes carries the water that is cooled by flowing over the cooling tower. This water is used for the condenser of the refrigerating machine and for the hydraulic system. The twin-circuit system saves great amounts of energy because it can function as a single-circuit system in winter with the evaporation-type cooling tower. In summer, it is refitted as a twin-circuit system.

Machines Save Energy

It is important to evaluate how much energy a machine requires for its operation. There are two types of machines, those that require a great deal of energy and those that require less. (See Chapters 3, 14, 15, 16, and 17 for details on how to evaluate energy efficiency in machines.)

Injection molding is one of the most energy-intensive methods employed for converting plastics resin to a finished product. It requires not only the energy used by the machine to drive the motor or motors for hydraulic power, but the energy to the heater bands to melt the resin, as well (6).

Then there is the problem of removing the heat that is generated in the hydraulic system by using water in the heat exchanger, and water is also needed to cool the mold to remove the heat from the plastic. This water can be from a city system, and depending on the machine size and mold and the water temperature available, as much as 20 to 30 gallons per minute could be required, thus creating a sizable water bill. Most plants have acquired their own wells, or closed systems using cooling towers, chillers, and the like. These require pumps and motors, plus, in the case of chillers, compressors as well.

Machine grinders are quite often used, plus materials handling equipment, conveyors, etc. In all, a lot of energy is used for the process. We will confine ourselves here to the machine alone, as it is the single largest user of energy.

It is estimated that the cost of energy will double in the next five years. This being the case, the molder is faced with two problems. First, of course, on new machinery purchases, he should buy the most energy-efficient machine available. This is a long-term investment, so that price alone or any other single reason is not justified when the long-term use of energy is considered. Also, a machine that is not energy-efficient may be difficult, if not impossible, to resell a few years later. No one can go out and replace all of his machines with energy-efficient ones, a situation that leads to the second problem: we must, if possible, reduce the energy used on present equipment. Is it possible to do so? In many ways it is.

Energy consumption in a molding machine is directly related to the hydraulic pressure used. The higher the pressures, the more power—and thus the more energy—needed. So the basic approach is to determine how to reduce the pressures required to do the job. Let's explore the possibilities.

First of all, consider the clamp. The more tonnage that is required to lock up the mold, the higher the hydraulic pressure must be to accomplish this. Whether we are talking about a hydraulic ram machine or a toggle machine, the problem is the same.

Basically we are trying to hold the mold closed against the force of injection to prevent flashing. The first consideration is the mold.

Is the mold base relieved to minimize the area of the mold that must be clamped to ensure a good shut-off? This is a relatively inexpensive adjustment which would allow using less clamp. Less clamp tonnage translates into less energy used, but also improves running conditions, as the vents are more effective, etc. So spend a few dollars on the mold to ensure good operating conditions. This is a periodic expense that reduces your monthly expense for power—a good trade-off.

The greatest use of energy occurs at the injection end. There energy is used to produce the melt and force it into the mold. The heater bands draw electrical energy to melt the plastic along with the screw drive, which provides some heat to the plastics through shear. Putting the plastics into the mold requires high pressures and a large pump capacity. What can we do about this high energy use?

There are quite a few ways to help reduce energy costs in this area. First, consider the screw recovery or plasticating. Probably the most efficient way to run the screw is with about 60 to 70 percent of the heat being provided by the heater bands and the remaining 30 to 40 percent by shear. To accomplish this one needs to know something about the screw and how it works in order to arrive at a heat profile suitable to the resin being processed and the rate at which it is processed.

A starting point would be to set the rear zone of heat at about 50° F above the softening point of the resin to be run, the center zone about 50° F above the front zone, and the front zone at the stock temperature that one desires to run at. Watch the screw drive pressure during recovery. It should be at about 50 to 65 percent of the maximum available. It it is below 50 percent, no shear heat is being used and the mix of the melt is not very good, particularly if coloring is being used. Above 65 percent, too much of the heat is being put in through shear, a condition that is not energy-efficient.

Heater bands have received a lot of publicity recently as a possible energy-saving source. The bands touted as energy savers are the ceramic element bands with one-half inch of insulation over them. Reed Prentice has done considerable testing in this area and obtained some sound data on the subject. The testing was done with a melt thermocouple in the nozzle to establish a target melt temperature. The only change that we made was in the heat profile, to maintain the target temperature when we changed heater band conditions.

Our standard arrangement used mica heater bands with a 360-degree cover of shiny aluminum. With this as a base we then tested without a cover, with insulation fastened to the cover, with ceramic bands and with an insulation blanket over the cover, and with the blanket over the ceramic bands themselves.

Removing the heater cover showed a 44 percent increase in energy required. Insulating the inside of the cover showed a 12 percent increase in energy required. The ceramic bands showed a 22 percent increase in energy required. We then ignored the melt temperature and put the same heat profile on the ceramic heater bands as on the first test. We noted a 10°F swing in melt temperature, overriding of the heat controllers in the center and front zones, and a 15 percent increase in energy.

We found that a heat sink problem occurs, in that the insulation directly on the heater band does not allow for a modulation of the heat at the surface, which is greater than at the thermocouple; so this greater heat has no place to go but down through the steel to the plastic. The full cover with uninsulated bands provides an oven effect which eliminates that condition.

The variation in melt temperature, on amorphous materials particularly, can affect molding conditions to the point of providing slight non-fills or sinks to slight flashing due to viscosity change with the temperature change.

Our recommendation is that uninsulated bands with a full cover be used as the most energy-efficient arrangement, which provides the best control over the melt. Testing of the blanket over the cover is not complete, but this idea would be good for air-conditioned operations.

The force required to put the plastic material into the mold consumes the most energy. Contributors to this problem are the viscosity of the melt, the size of the gate, the settings of the pressures, and the speed of fill, as well as the duration of the boost or delay unload.

The viscosity of the melt must be carefully controlled to get the best quality of melt possible. (See the section on "Plasticizing" earlier in this chapter.) This size of the gate is another matter. Gate sizes are usually smaller than is necessary because it is easier to open up a gate than it is to make it smaller. Once a mold is filling, the gate size cannot be changed. Although gate marks are an appearance problem on some products, a .040-in. gate is pretty common in this country and a one-millimeter (.039-in.) gate is common in Europe.

A very small change, in thousands of an inch, can have significant bearing on the cross-sectional area of the gate. For example, going from a .040-in. gate to a 0.50-in. gate, results in a 56 percent increase in area. That would have a decided effect on the pressure required to fill the mold. It could also mean a reduction in melt temperature, which translates into faster cycling due to there being less heat to remove from the mold.

So gating has a significant part to play in the energy used. We have not, as on industry, spent enough time considering this factor and the effect it has on part quality, cycle time, and energy consumption.

These are some of the considerations that will affect your operating costs (though not all of them, as space does not permit such an examination).

In conclusion, we can make the following points:

1. Do not try to run a tool that is not in good condition. A few hundred dollars spent on tool maintenance can save thousands of dollars spent on wasted energy.
2. Do not use more clamp than necessary.
3. Learn how to use the screw to best advantage. Talk to your supplier, and get his recommendations. Do not use the same heat profile on different sizes or models of machines.
4. Use as low an injection pressure as possible. Open up gate sizes where possible to reduce the pressure required. Do not

hold boost pressure on after ram stops moving. (Note: Boost or delay unload does not mean pressure, but volume of oil.) Use second stage in place of first stage injection pressure.

5. Reduce melt temperature if the gate size will let you. You save energy on melt preparation and on removing it from the mold, and improve cycle time as well.

Plastics Save Energy

There are always improvements to be made in machines and equipment in the plant whereby energy savings can be made with a net savings in total production costs. But sometimes equipment can be made more energy-efficient and a condition during molding will cause a total increase in cost (as, for example, if cycle time increases).

But if we study the relationship of plastics and energy savings vs. the use of practically any other materials (see Table 1-4), plastics conserve energy in significant ways. Energy is saved in the service life of the plastics product. Energy is also saved in shipping and maintenance, since plastics are lightweight and require less fuel for shipping and are inherently inert to chemicals, rot, mildew, corrosion, and

hostile environments. Another important aspect of their use is that as new markets for plastics are developed, new ways to save energy are found in all phases of the manufacturing process and in performance.

Of the many uses of petrochemicals, the production of plastics materials is the most ingenious. The versatility of these long-chain macromolecules of basic elements combined to make diversified products is testimony to the imagination and talents of the industry. As compared to more than occasional serendipity or accidental discovery of new products just a few decades ago, today's research and development emphasis is on materials engineering and processing innovations. The plastics industry's research frontiers continue to be in multipolymer alloys, conductive polymers, biomaterials, and high-strength/lightweight reinforced plastics/composites. These goals reflect the industry's commitment to conservation of energy and resources and will contribute significantly to the quality of life in tomorrow's society (7).

SAFETY STANDARDS/GUIDES

Since the injection molding process is a high-pressure, high-speed process, it is reasonable

Table 1-4 Energy requirements for different materials.

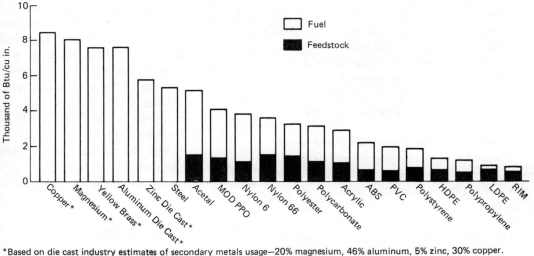

*Based on die cast industry estimates of secondary metals usage—20% magnesium, 46% aluminum, 5% zinc, 30% copper.

to assume that a great deal of force and heat are generated by the equipment. Thus machine safety is a must to ensure operator safety. A machine without adequate safety guards is significantly dangerous to the operator and the personnel working in the area.

There are procedures that outline how to operate your machines and all other plant operations, including plastic storage procedures, moving plastics around the plant, etc. Throughout this book different procedures are reviewed in operating the machines (in Chapters 4, 5, and others) as well as other equipment (Chapters 15, 16, 17, etc.). An important guide to machine and plant safety procedures can be obtained from the Society of the Plastics Industry–Committee on Occupational Safety and Health, New York City (8, 9, 10).

Operators of machines should consider steps to be taken that will ensure personal safety. An example is in the proper lockout of the machine's electrical circuit. Properly locking out a machine's electrical circuit before starting repairs protects the maintenance worker from accidental start-ups, which could cause severe injury. The National Safety Council offers the following steps for the proper lockout procedure:

- Shut down all possible switches at the point of operation, then open the main disconnect switch.
- Snap your own lock on the locking device. An ordinary padlock can be used for most electrical lockouts.
- Check the lockout device to make sure the switch can't be operated.
- Place a name tag on the shank of the lock to indicate that the machine has been locked out.
- Notify the supervisor when the repair work has been completed. Only he should give the go ahead to remove your lock.
- Take off the name tag and remove the lock.

The National Safety Council, a voluntary organization located in Chicago, provides all industries important information on safety and accident prevention. Their data on the

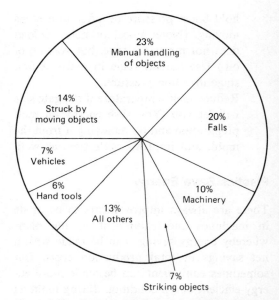

Figure 1-20 General statistics on where accidents occur in all types of manufacturing plants, including plastics plants (10).

source of work injuries in all U.S. industries (including the plastics industry) based on various state labor department accident facts are shown in Fig. 1-20. Note that handling of materials ranks the highest at 23 percent, next it falls at 20 percent, and so on (10).

American National Standard

The American National Standard B151.1–1976 was issued specifically for the construction, care, and use of horizontal injection molding machines (11). This standard, prepared by the American National Standards Institute of New York City, implies a consensus of those substantially concerned with its scope and provisions. ANSI B151.1–1976 is intended as a guide to aid the manufacturer, the consumer, and the general public. This standard will be periodically revised so users are cautioned to obtain the latest edition from ANSI (11).

REFERENCES

1. "Injection Molding," University of Lowell Plastic Seminars Workbooks, 1982, 1983, 1984.
2. Rosato, D. V., Fallon, W. K., and Rosato, D. V.,

Markets for Plastics, Van Nostrand Reinhold, New York, 1969.

3. "World's Largest Injection Molding Machine," *Plastics Machinery & Equipment,* Sept. 1980, p. 60.

4. Glanvill, A. B., *The Plastics Engineer's Data Book,* The Machinery Publ. Co., 1971.

5. "Structural Plastics Design Manual," U.S. Government Printing Office, FHWA-TS-79-203, 1979.

6. Olmsted, B. A., "Energy Savings in Injection Molding," SPE-IMD Conference, Oct. 20–22, 1981.

7. Driscoll, S. B., "Polymeric Material Systems," University of Lowell Plastic Seminars Workbook, 1984.

8. Rosato, D. V., and Lawrence, J. R. *Plastics Industry Safety Handbook,* Cahners Publ., 1973.

9. Safety and Health Reference Guide "PlasTIPS" for the Plastics Processor, Society of Plastics Industry, 355 Lexington Ave., New York, NY 10017, 1984.

10. "Accident Prevention Manual for Industrial Operations," National Safety Council, 444 N. Michigan Ave., Chicago, IL 60611, Annual.

11. ANSI B151.1–1976, American National Standard Institute, Inc., 1430 Broadway, New York, NY 10018, 1976.

Section II
MACHINE OPERATION

Chapter 2
The Injection Molding Machine

William T. Flickinger

HPM Corporation
Mt. Gilead, Ohio

An injection molding machine is a machine for the converting, processing, and forming of raw plastic material of powder, pellets, or regrind into a part of desired shape and configuration. The process of injection molding consists of heating plastic material until it melts, then forcing this melted plastic into a mold where it cools and solidifies.

Figures 2-1, 2-2, and 2-3 show a toggle, a straight hydraulic, and a hydro-mechanical injection molding machine, respectively.

MACHINE NOMENCLATURE

Before we go further, it would be well to review the nomenclature and terms used in the injection molding industry.

Clamping System Terminology

Clamping Unit: that portion of an injection molding machine in which the mold is mounted, and which provides the motion and force to open and close the mold and to hold the mold closed with force during injection. When the mold is closed in a horizontal direction, the clamp is referred to as a horizontal clamp. When closed in a vertical direction, the clamp is referred to as a vertical clamp. This unit can also provide other features necessary for the effective functioning of the molding operation.

Moving Platen or Plate (Figs. 2-4 and 2-5): that member of the clamping unit which is moved toward a stationary member. The moving section of the mold is bolted to this moving platen. This member usually includes the ejector (knockout) holes and mold mounting pattern of bolt holes or "T" slots. A standard pattern is recommended by the SPI Proposed Testing Method (Injection Machinery Division Methods and Procedures, September 11, 1958).

Stationary Platen or Plate (Figs. 2-4 and 2-5): the fixed member of the clamping unit on which the stationary section of the mold is bolted. This member usually includes a mold mounting pattern of bolt holes or "T" slots. A standard pattern is recommended by the SPI Proposed Testing Method (Injection Machinery Division Methods and Procedures, September 11, 1958). In addition, the stationary platen usually includes provision for locating the mold on the platen and aligning the sprue bushing of the mold with the nozzle of the injection unit.

Tie Rods, Bars, or Beams (Figs. 2-4 and 2-5): those members of the clamping force actuating mechanism that serve as the tension members of the clamp when it is holding the mold closed. They also serve as a guide member for the movable platen.

Ejector (Knockout): a provision in the clamping unit that actuates a mechanism within the mold to eject the molded part(s)

39

Fig. 2-1   Toggle injection molding machine.

from the mold. The ejection actuating force may be applied hydraulically or pneumatically by a cylinder(s) attached to the moving platen, or mechanically by the opening stroke of the moving platen.

Full Hydraulic Clamp (Fig. 2-4): a clamping unit actuated by a hydraulic cylinder that is directly connected to the moving platen. Direct fluid pressure is used to open and close the mold, and to provide the clamping force to hold the mold closed during injection.

Toggle Clamp (Hydraulic Actuated, Mechanical Actutated) (Fig. 2-5): a clamping unit with a toggle mechanism directly connected to the moving platen. A hydraulic cylinder, or some mechanical force device, is connected to the toggle system to exert the opening and closing force and hold the mold closed during injection. The clamping force to hold the mold closed during injection is provided by the mechanical advantage of the toggle.

Slow Mold Breakaway: a provision in the machine designed to provide slow platen movement for an adjustable distance during the initial opening of the mold.

Clamp Close Slow Down: a provision in the machine designed to slow down the moving platen for an adjustable distance before the mold faces come in contact.

Clamp Open Slow Down: a provision in the machine designed to slow down the moving platen for an adjustable distance before it reaches its maximum open position. This sequence is often employed to reduce the effect of knockout impact when mechanical knockouts are used. It is sometimes referred to as the ejector or clamp open cushion.

Clamp Close Stroke Interruption: a com-

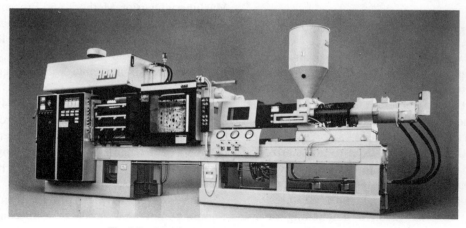

Fig. 2-2   Straight hydraulic injection molding machine.

Fig. 2-3   Hydro-mechanical injection molding machine.

plete stop of the clamp closing stroke to allow an auxiliary operation(s) before completion of the closing stroke.

Clamp Open Stroke Interruption: a complete stop of the clamp opening stroke to allow an auxiliary operation(s) before completion of the opening stroke.

Clamp Close Preposition: a provision in the machine circuit to allow the clamp to open fully and then close to a predetermined position. It is generally used to allow the ejector (knockout) mechanism to retract so that inserts can be placed in the mold.

Low Pressure Mold Closing: a provision in the machine to lower the clamp closing force during the clamp closing cycle. The lower clamp forces minimize the danger of mold damage caused by parts caught between the

mold faces. Provisions are also provided for parting mold faces at a timed interval in case of an obstruction.

Injection System Terminology

Injection Plasticizing (Plasticating) Unit: that portion of an injection molding machine which converts a plastic material from a solid phase to a homogeneous semi-liquid phase by raising its temperature. This unit maintains the material at a preset temperature and forces it through the injection unit nozzle into a mold.

Plunger Unit (Fig. 2-6): a combination in-

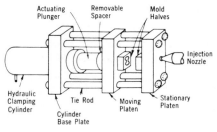

Fig. 2-4   Hydraulic clamp.

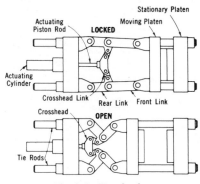

Fig. 2-5   Toggle clamp.

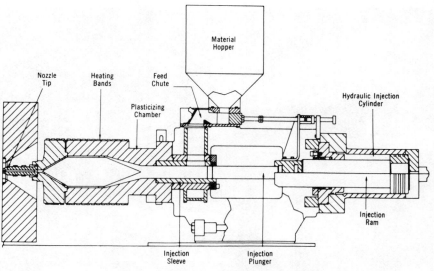

Fig. 2-6   Plunger unit.

jection and plasticizing device in which a heating chamber is mounted between the plunger and the mold. This chamber heats the plastic material by conduction. The plunger, on each stroke, pushes unmelted plastic material into the chamber, which in turn forces plastic melt at the front of the chamber out through the nozzle.

Two-Stage Plunger Unit (Fig. 2-7): an injec-

tion and plasticizing unit in which the plasticizing is performed in a separate unit. The latter consists of a chamber to heat the plastic material by conduction and a plunger to push unmelted plastic material into the chamber into a second stage injection unit. This injection unit serves as a combination holding, measuring, and injection chamber. During the injection cycle the shooting plunger forces the

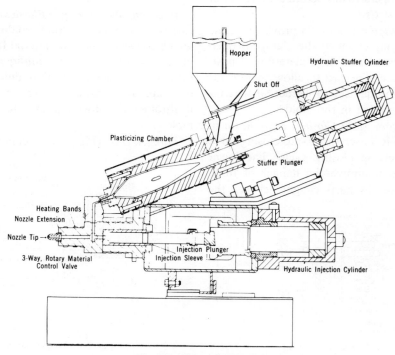

Fig. 2-7   Two-stage plunger unit.

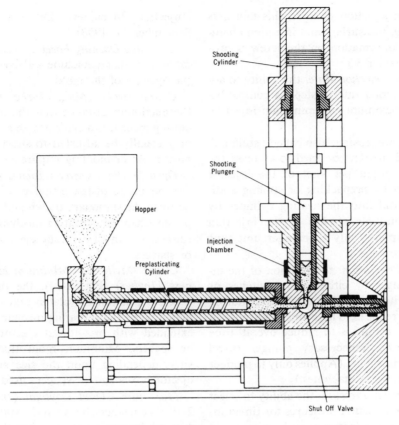

Fig. 2-8 Two-stage screw unit.

plastic melt from the injection chamber out through the nozzle.

Two-Stage Screw Unit (Fig. 2-8): an injection and plasticizing unit in which the plasticizing is performed in a separate unit that consists of a screw extrusion device to plasticize the material and force it into a second stage injection unit. This injection unit serves as a combination holding, measuring, and injection chamber. During the injection cycle a plunger forces the plastic melt from the injection chamber out through the nozzle.

Reciprocating Screw (Fig. 2-9): a combination injection and plasticizing unit in which an extrusion device with a reciprocating screw is used to plasticize the material. Injection of material into a mold can take place by direct extrusion into the mold, or by using the reciprocating screw as an injection plunger, or by a combination of the two. When the screw

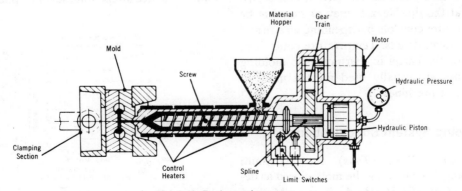

Fig. 2-9 Reciprocating screw.

serves as an injection plunger, this unit acts as a holding, measuring, and injection chamber. (More information on the screw is provided in Chapter 3.)

Adjustable Injection Rate: the ability to adjust the injection rate in stepless control between the maximum and minimum injection rate.

Prepack: prepacking, also called "stuffing," is a method that can be used to increase the volumetric output per shot of the injector plunger unit by prepacking or stuffing additional material into the heating cylinder by means of multiple strokes of the injection plunger. (Applies only to plunger unit type injection machines.)

Plunger Preposition: the position of the injection plunger by either limit switches or pressure switches, so that total travel during injection is reduced. The primary purpose of plunger preposition is to reduce overall time by eliminating unnecessary plunger travel time during injection. (Applies only to plunger unit type injection machines.)

Dual Injection Pressure: the ability to select two separate injection pressures for timed intervals during the injection cycle.

Nozzle Retraction Stroke: the maximum stroke of the hydraulic cylinder, used to separate the injection unit from the sprue bushing of the mold for cleaning and/or purging purposes.

Thermoset Addendum: the basic change between the two machines (thermoplastic and thermoset) is in the barrel, screw, and nozzle. Thermoset barrels generally use water jackets for temperature control. Screws in thermoset machines are shorter (in the range of 13:1 to 16:1 *L/D*) and do not have a non-return valve at the tip. Nozzles may or may not be temperature-controlled, depending on size and other design details. Temperature controllers used for the barrel in thermoplastic molding machines are usually used for the mold in thermoset machines.

Clamping System Specifications

Clamping Force (*Tons*): the maximum clamping force holding the mold closed as determined by SPI Standards, Testing Method (Injection Machinery Division Standards, September 11, 1958).

Clamping Opening Force (*Tons*): the maximum force that a machine will exert to initiate the opening of the mold.

Clamp Stroke (*Max.*) *Inches* (Fig. 2-10): the maximum distance that the opening and closing mechanism can traverse a platen. This may usually be adjusted to shorter travel to meet mold or molding requirements.

Open Daylight (*Max.*) *Inches* (Fig. 2-10): the maximum distance that can be obtained between the stationary platen and the moving platen when the actuating mechanism is fully retracted with or without ejector box and/or spacers.

Closed Daylight or Minimum Mold Thickness (*Inches*) (Fig. 2-10): the distance between the stationary platen and the moving platen when the actuating mechanism is fully extended with or without ejector box and/or spacers. Minimum mold thickness will vary, depending upon the size and kind of ejector boxes and/or spacers used.

Maximum Closed Daylight (*Inches*) (Fig. 2-10): the distance between the stationary platen and the moving platen when the actuating mechanism is fully extended without ejector box and/or spacers.

Minimum Closed Daylight (*Inches*) (Fig. 2-10): the distance between the stationary platen and the moving platen when the actuating mechanism is fully extended with standard ejector box and/or spacers.

Injection System Specifications

Injection Capacity (*Theoretical*): the maximum calculated swept volume (or trapped vol-

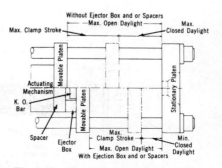

Fig. 2-10 Clamp die space nomenclature.

ume in a plunger unit) in cubic inches that can be displaced by a single stroke of the injection plunger or screw, assuming no leakage and excluding the use of a rotating screw to displace additional volume.

Thermoset Injection Capacity: Injection capacity can be measured in cubic inches of swept volume, but as there is no non-return valve on the thermoset screws, this figure cannot be used to convert to true shot weight because some material flows back over the screw during injection. The amount of back flow is dependent on variables in both the machine and molding material; hence actual injection capacity cannot be absolutely defined.

Plasticizing (Plasticating) Capacity (Continuous): the maximum quantity of a specified plastic material that can be raised to a uniform and moldable temperature in a unit of time. This is usually expressed in pounds per hour as available from a plunger unit. In the case of a screw unit, plasticizing (plasticating) capacity is generally expressed in pounds per hour as calculated from the recovery rate.

Recovery Rate: the volume or weight of a specified moldable material discharged from the screw per unit of time, when operating at 50 percent of injection capacity as determined by the SPI test procedure (effective January 1, 1968). It can be expressed as cubic inches per second or ounces per second.

Injection Pressure (Max. psi): the maximum theoretical pressure of the injection plunger or screw against the material expressed in psi (assuming no loss of pressure due to frictional drag of the plunger or screw) at maximum force acting on the injection piston.

Maximum Injection Rate (cu in./sec.): the maximum calculated rate of displacement of the injection plunger or screw, expressed in cubic inches per second, computed at maximum injection pressure as specified.

Minimum Injection Rate (cu in./sec): the minimum calculated rate of displacement of the injection plunger or screw, expressed in cubic inches per second computed at maximum injection pressure as specified.

Screw Terminology and Specifications

Screw (Fig. 2-11): a helically flighted shaft that when rotated within the barrel mechanically works and advances the material being processed.

Screw Flight: the helical metal thread of the screw.

Screw Root: the continuous central shaft, usually cylindrical or conical in shape.

Flight Land: the surface at the radial extremity of the flight constituting the periphery or outside diameter of the screw.

Screw Shank: the rear protruding portion of the screw to which the driving force is applied.

Feed Section of Screw: the portion of a screw that picks up the material at the feed opening (throat) plus an additional portion downstream. Many screws have an initial constant lead and depth section, all of which is considered the feed section.

Transition Section of Screw: the portion of a screw between the feed section and metering section in which the flight depth decreases in the direction of discharge.

Metering Section of Screw: a relatively shallow portion of the screw at the discharge end with a constant depth and lead.

Screw Diameter: the diameter developed by the rotating flight land about the screw axis.

Helix Angle: the angle of the flight at its

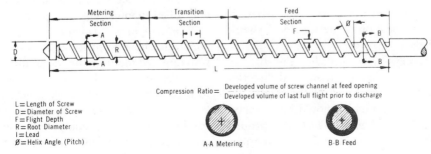

Compression Ratio = (Developed volume of screw channel at feed opening) / (Developed volume of last full flight prior to discharge)

L = Length of Screw
D = Diameter of Screw
F = Flight Depth
R = Root Diameter
I = Lead
Ø = Helix Angle (Pitch)

A-A Metering   B-B Feed

Fig. 2-11 Screw.

periphery relative to a plane perpendicular to the screw axis. The location of measurement should be specified.

Axial Flight Land Width: the distance in an axial direction across one flight land.

Normal Flight Land Width: the distance across one flight land in a direction perpendicular to the flight.

Flight Lead: the distance in an axial direction from the center of a flight at its outside diameter to the center of the same flight one turn away. The location of measurement should be specified.

Flight Pitch: the distance in an axial direction from the center of a flight at its periphery to the center of the next flight. In a single flighted screw, "pitch" and "lead" will be the same, but they will be different in a multiple flighted screw. The location of measurement should be specified.

Flight Depth: the distance in a radial direction from the periphery of the flight to the root. The location of measurement should be specified.

Full Flighted Length of Screw: overall axial length of flighted portion of the screw, excluding non-return valves, smear head, etc.

Constant Lead Screw, Uniform Pitch Screw: a screw with a flight of constant helix angle.

Constant Taper Screw: A screw of constant lead and uniformly increasing root diameter over the full flighted length.

Decreasing Lead Screw: a screw in which the lead decreases over the full flighted length (usually of constant depth).

Screw Channel: with the screw in the barrel, the space bounded by the surfaces of flights, the root of the screw, and the bore of the barrel. This is the space through which the stock is conveyed and pumped.

Screw Channel Depth: the distance in a radial direction from the bore of the barrel to the root. The location of measurement should be specified.

Axial Screw Channel Width: the distance across the screw channel in an axial direction measured at the periphery of the flight. The location of measurement should be specified.

Normal Screw Channel Width: the distance across the screw channel in a direction perpen-

dicular to the flight measured at the periphery of the flight. The location of measurement should be specified.

Axial Area of Screw Channel: the cross-sectional area of the channel measured in a plane through and containing the screw axis. The location of measurement should be specified.

Developed Volume of Screw Channel: the volume developed by the "axial area of screw channel" in one revolution about the screw axis. The location of measurement should be specified.

Enclosed Volume of Screw Channel: the volume of screw channel starting from the forward edge of the feed opening to the discharge end of the screw channel.

Compression Ratio: the factor obtained by dividing the developed volume of the screw channel at the feed opening by the developed volume of the last full flight prior to discharge. (Typical values range from 2 to 4, also expressed as a ratio $2:1$ or $4:1$.)

Channel Depth Ratio: the factor obtained by dividing the channel depth at the feed opening by the channel depth just prior to discharge. (In constant lead screws this value is close to, but greater than the compression ratio.)

Screw Efficiency (Volumetric): the volume of material discharged from the machine during one revolution of the screw expressed as a percentage of the developed volume of the last turn of the screw channel.

Screw Speed: number of revolutions of the screw per minute.

Machine Operation Terminology

Manual Operation: operation in which each function and the timing of each function are controlled manually by an operator.

Semi-Automatic Operation: operation in which machine performs a complete cycle of programmed molding functions automatically and then stops. It then requires an operator to manually start another complete cycle.

Automatic Operation: operation in which machine performs a complete cycle of programmed molding functions repetitively, and stops only if there is a machine or mold malfunctions or it is manually interrupted.

Barrel and Screw *L/D*

One of the most widely used terms applied to the injection unit screw and barrel is *L/D,* called "*L* over *D.*" The following definitions and examples explain this term.

Plasticizing Unit L/D Ratio. The common denominator used for the comparison of all plasticizing units, regardless of screw diameter, is the plasticizer length to diameter ratio, normally referred to as *L/D*. This ratio is applicable to extruders, recriprocating screws, and two-stage screw units alike. Barrel *L/D,* however, must not be confused with screw *L/D.* (See Fig. 2-12.)

Barrel Length to Diameter Ration—L/D: the distance from the forward edge of the feed opening to the forward end of the barrel bore divided by the bore diameter and expressed as a ratio wherein the diameter is reduced to one, such as 20/1.

Screw Length to Diameter Ratio—L/D: the distance from the forward edge of the feed opening to the forward end of the screw flight (*not* including tips, pressure cones, and non-return valves) divided by the diameter of the screw. It is not based on the full flighted length of the screw.

STANDARDS FOR USING INTERCHANGEABLE MOLDS

The Society of Plastics Industries have published standards in regard to mounting molds in the injection molding machine. The obvious advantage of these standards is that they easily allow molds to be designed to run on more than one brand of machine. These standards are listed here for descriptive purposes. Always check the latest revision of the SPI Standard for actual use.

Platen Bolting Pattern

These standards specify the location and size of tapped holes in the stationary and moving platen for the attachment of the mold. They cover three sizes of machines:

- Up to 750-ton clamping capacity
- Over 750- to 1600-ton clamping capacity
- Over 1600- to 4100-ton clamping capacity

Knockout Pin Locations

Knockout pins are used to push the molded part off the mold during the portion of the machine cycle called ejection. Since the location of these pins is critical to the mold design, the Society of Plastics Industries has written a standard for locating the knockout pins. The location and size of the knockout pin holes are classified according to machine sizes identical to those used for mold mounting holes standards.

Machine Nozzle and Die Locating Ring

The die locating ring is used to align the mold to the machine platens. It is commonly affixed to the stationary platen. This alignment is needed for the knockout pins and for the injection unit to line up with the mold. The machine nozzle provides the mating necessary between the injection unit and the mold.

OTHER INJECTION MOLDING MACHINE STANDARDS

Many other SPI, ANSI, NFPA, etc., standards apply directly or indirectly to injection

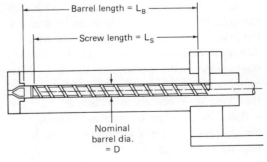

Barrel L/D = L_B/D
Screw L/D = L_S/D

Fig. 2-12 *L/D* ratio.

molding machines. One standard that should be followed by the molder is ANSI B151.1, "Safety Standards for Horizontal Injection Molding Machines." Many other standards are currently being developed. The injection molding machine manufacturers are excellent sources of information on machine standards. (See Chapter 5 on safety.)

BASIC MACHINE OPERATION

The injection molding machine must be installed with the proper electrical, water, and air connections. The machine hopper must be filled with plastic material either manually or with automatic material handling systems. The mold must be properly clamped to the movable and stationary platens with its electrical or water connections properly made. Assuming all systems have been checked out for stationary operation, the machine is ready to cycle.

In the starting position, the clamp is open, the electrical power system is providing power to the hydraulic system, the electrical sequence system is turned on, and the injection barrel heater band system is turned on and up to temperature.

Normally plastic material would be melted and ready for injection into the mold prior to closing of the clamp. Melting occurs by turning the plasticizing screw, which pumps and shears the plastic material while the heater bands add heat. This combination of shearing and heat melts the plastic. After a preset amount of plastic has been melted, the plasticating screw is stopped.

The clamp then comes forward and builds tonnage. When the proper tonnage is obtained, the injection cycle can take place. The hydraulic system directs oil to the injection cylinders, pushing the melted plastic into the mold. Normally the injection cycle continues for a preset time period. When the injection process has ended, the plasticating process begins, with melted plastic again being prepared for injection into the mold.

If a rotary shut-off valve is used, the clamp can open during the plasticating process. If not, the clamp opens after the plasticating cy-

cle. As the clamp opens, ejection takes place. The ejection of the molded part can occur by ejector pins in the mold being actuated mechanically or hydraulically by the machine, while some parts can be blown off with air. Once the machine is open, the next cycle can take place.

CLAMPING CONCEPTS

The injection molding machine clamp is used to close the mold, hold it closed during the injection and curing of the plastic material, and open the mold for the removal of the formed part.

There are three different types of clamp designs:

1. Straight hydraulic clamp
2. Linkage or toggle clamp
3. Hydromechanical clamp

Straight Hydraulic Clamp

This design uses hydraulic fluid and pressure to open and close the clamp and to develop the force required to hold the mold closed during the injection of plastic.

The basic concept is to direct hydraulic fluid to the booster tube to move the clamp ram forward. Oil fills the main area by flowing from the tank through the prefill to the main area. As the ram moves forward, a slight vacuum is developed in the main area, pulling fluid from the tank into this chamber. Once the clamp is closed, the prefill valve is closed, trapping the oil in the main cylinder area. High pressure fluid is put into this area, compressing this volume of oil and thus raising the pressure in this area. The maximum pressure is controlled by a pressure control valve, which closely controls the clamp tonnage (the maximum hydraulic pressure times the area it pushes against).

To open the clamp, hydraulic fluid is directed to the pull back side of the cylinder while the prefill valve is open, with fluid from the main cylinder being returned to the tank. One of the major advantages of the straight hydraulic clamp is its very precise control of the clamp tonnage.

Linkage or Toggle Clamp

This concept uses the mechanical advantage of a linkage to develop the force required to hold the mold closed during the plastic injection portion of the cycle. Normally the linkage design is done in such a way that slowdowns are built in. The advantage of a toggle clamp is that less hydraulic fluid is required to open and close the clamp. A disadvantage is that the clamp tonnage is not precisely known.

A small hydraulic cylinder is used to close the clamp. This cylinder travels at a constant speed with the slowdown for mold close built into the linkage. The mechanical advantage of the linkage is extremely high, so a relatively small closing cylinder can develop high tonnage.

Hydromechanical Clamp

This design uses a mechanical means for high-speed close and open. A short stroke cylinder is used to develop tonnage identical to the straight hydraulic design. This concept is said to offer the advantage of toggle clamps for high-speed close and open, and the advantage of a straight hydraulic for precise control of the clamp tonnage. The hydromechanical design normally has a high-speed clamp close and open device which is usually a hydraulic cylinder or actuator. The closing and opening mode occurs with relatively low force. Once the clamp is closed, a blocking action takes place allowing a large-diameter hydraulic cylinder to build tonnage similar to the straight hydraulic design. When the clamp is to be opened, the blocking member is removed, and the clamp opens rapidly. The blocking action is normally a mechanical device, and the tonnage action is done by hydraulics; hence the name hydromechanical.

Comparison of Clamp Designs

Over the years many arguments have been presented showing each clamp design concept to be superior to the others. In reality, each concept has merit.

The straight hydraulic design has proved over the years to have long-term reliability, excellent control of low pressure mold protection, and exact control of tonnage, and it will not allow the clamp to be overstressed due to high injection forces.

The toggle clamp has extremely fast closing and opening actions, and is typically lower in cost than the straight hydraulic. The energy required to hold the developed tonnage is less than for the straight hydraulic, but this energy is small compared to the total energy usage of the machine. With good lubrication, the toggle bushings and pins last well, but they still must be reworked after several years of service. The toggle design will also develop higher than lockup tonnage if the clamp is overpowered by the injection end, or due to temperature buildup in the mold.

The hydromechanical tends to have the advantages of the straight hydraulic, whereas the toggle is more complex because of the block action required.

The debate over the three clamp concepts will continue for many years.

PLASTICIZING UNITS

The evolution of plasticizing units occurred in the following order:

1. Plunger
2. Two-stage plunger
3. Plasticizing screw (nonreciprocating) and plunger
4. Reciprocating screw

Single-Stage Plunger

The first single plunger units (Fig. 2-6) were heated with hot oil and eventually with electrical heating bands. A torpedo section in the rear zone spread the plastic feed material out into a thin section as the plunger moved forward. This thin section of material then entered the center section where most of the plasticizing took place. On later designs the center section consisted of a series of small holes drilled parallel to the horizontal center-line of the plasticizing chamber. In the front section of the plasticizing chamber the torpedo was reduced in diameter, and the cross section of the melt was reduced to better plasticize any unmelt coming out of the center section. The front section had to be of sufficient length to ensure a constant temperature throughout the melt as it was forced out of the plasticizing chamber and into the mold.

Some of the problems experienced with these units were leakage between the center section and front end, color change, pressure drop through the chamber, scored plungers, poor shot control, and difficulty in running heat-sensitive materials. As the designs improved, the mechanical slide type of feeder was replaced with weigh feeders which substantially improved the shot size control. Pressure switches and counters were added to the electrical and hydraulic circuits to permit packing material between the plunger and torpedo and to allow prepositioning of the plunger to be past the feed opening before the next shot. These two features increased the maximum shot size on a unit and also reduced the cycle time.

One thing these units could do best was to run marbelized textured material such as plastic wall tile.

Two-Stage Plunger

With the development of the two-stage plunger units (Fig. 2-7), cycle times were reduced, injection rates (cu in./sec) were increased, and clamp tonnage requirements were reduced.

The first two-stage units consisted of a stuffer unit for plasticizing and an injection unit. The stuffer unit was the conventional single plunger unit set piggyback over the injection unit. Material was stuffed through the plasticizing chamber, and the melt from the plasticizing unit was channeled down in front of the injection plunger, causing the injection plunger to be pushed to the back of the injection shot chamber. Shot size was controlled by the use of an adjustable positive stop and limit switches, which controlled the position of the injection plunger when the required amount of melt was in front of the injection plunger. A rotary valve was placed in the melt channel between the stuffer unit and injection chamber and the injection unit and nozzle tip. In one position, the valve directed melt from the stuffer unit into the injection shot chamber. When the melt was to be forced into the mold, the valve was shifted, connecting the injection shot chamber to the nozzle tip and closing off the passage to the stuffer unit. These units were a great improvement over the single-stage units, but they still had problems with color changes and running heat-sensitive material.

Two-Stage with Fixed Screw

The two-stage plunger units were then replaced by a two-stage unit with a fixed screw for the plasticizing unit (Fig. 2-8). The injection shot unit was essentially the same as used with the plunger plasticizer. These units could run any type of material, and with design improvements on the injection plunger configuration and the washing action obtained as the first melt for each shot entered the shot chamber, color-change problems were greatly reduced.

Reciprocating Screw

One of the most significant breakthroughs in plastics occurred with development of the concept of reciprocating screw injection units (Fig. 2-9). These units allowed high melting rates of plastic material, close tolerance on shot size, and the ability to control temperature of the plastic melt and perform reliably. By reducing the L/D of the barrel and the screw, these units can run thermoset materials including rubber and BMC (Bulk Molding Compound). Longer-L/D units (24:1 to 30:1) are used for hard-to-melt materials and vented machine application. Chapter 3 discusses these units in more detail.

THE MACHINE HYDRAULICS

The hydraulic system on the injection molding machine provides the power to close the clamp, build and hold tonnage, turn the screw to plasticate material, inject the plastic material into the mold, and so on. A number of hydraulic components are required to provide this power, including pumps, valves, hydraulic motors, hydraulic fitting, hydraulic tubing, and hydraulic reservoirs.

Reservoirs

The reservoir or tank provides hydraulic oil to the system for use in powering the machine. The reservoir must be sized to ensure that an adequate supply of oil is available to the system and also allow sufficient capacity for the system to return oil to the reservoir.

Suction lines should be placed near the bottom of the reservoir to ensure an ample oil supply. Return lines from the system should discharge beneath the oil level to avoid spraying into the air and foaming, and some type of antisiphon device should be used to stop the backflow of oil through the return lines in case of line breakage or the removal of a component for service. A standard guideline for sizing a reservoir is that it be three times the pump output in one minute, but all system requirements should be carefully considered before the final reservoir size is determined.

Suction Strainers

A suction strainer should be placed before the inlet to the pump to be sure that particles of sufficient size to damage the pump will be removed. One should consult the pump manufacturer when determining what size strainer to use. It is important that both the flow capacity and the maximum removable particle size be correct.

Pump

The hydraulic pump receives oil from the tank at low pressure and increases the pressure to that required by the system. The pump provides hydraulic flow and pressure to the system.

Several different types of hydraulic pump are used. The most common is a fixed displace-ment type pump, which provides an almost constant output at various pressures. Various designs are available to provide this constant output, the most common being:

1. Vane
2. Piston
3. Gear

Variable volume and pressure compensated pumps are being used more frequently in an attempt to conserve energy. These pumps are capable of varying the output to meet a particular flow requirement, or will put out only enough flow to develop a particular pressure requirement.

Figure 2-13 shows a power unit on an injection molding machine utilizing a fixed vane pump system.

Directional Valves

Directional valves are used to direct the hydraulic oil coming from the pump to where it is needed. Spool, check, and cartridge valves are commonly used for this control.

The spool type directional valve is commonly used on injection molding machines. Spool valves can be either two-position or three-position. In a two-position valve, a solenoid is energized for one position, and normally a spring will return the spool to the second position when the solenoid is de-energized. The three-position valve is obtained by adding a second solenoid.

Small valves can be directly operated by the solenoid; on larger valves solenoid-operated pilot valves direct pilot flow to the main spool for shifting. Figure 2-14 shows a cutaway view of a directional spool valve.

A check valve is a single one-direction valve that allows flow in only one direction.

An extension of the check valve that is beginning to find greater use is the cartridge valve. It is essentially a check valve that is powered open normally by a small spool directional valve. Cartridge valves are grouped to provide the same directional flow capability as spool valves. Figure 2-15 shows a schematic of a cartridge valve; note that the sleeve and

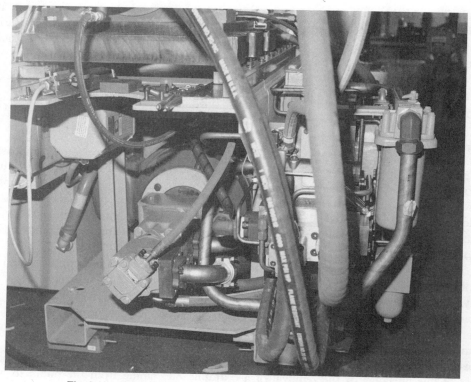

Fig. 2-13 Fixed vane pump power unit on an injection molding machine.

its internal parts are mounted within the manifold.

Servo and Proportional Valves

With the advent of more advanced microprocessor control systems with process control, greater use is being made of proportional valves and servo valves. The valves can be used to control flow and pressure. The main difference in performance between the proportional and servo valves is speed of response, with the servo being much faster than a proportional valve.

Injection Screw Drive Hydraulic Motor

The most common drive unit for the injection end plasticating screw is the hydraulic motor. It receives pressurized fluid from the hydraulic pumps and in turn converts this energy to the hydraulic motor drive shaft in the form of torque and rpm.

The most common type of injection system uses a direct drive hydraulic motor. This motor is directly coupled to the screw; thus the hydraulic motor torque and speed are also the plasticating screw's torque and speed. A second type of injection unit uses a gear box in conjunction with a hydraulic motor. In this way a smaller hydraulic motor can be used operating at a high speed, with the gear box reducing the speed and increasing the torque. With the gear box, it is also possible to use a variable-speed electric motor to provide power to the screw.

Fig. 2-14 Cutaway view of a directional spool valve.

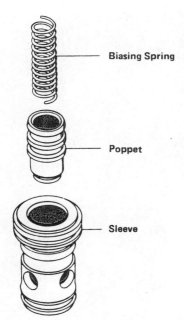

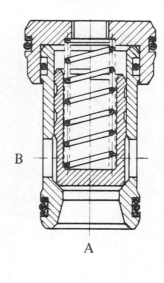

Fig. 2-15   Schematic of cartridge valve.

Hydraulic System Assembly

The assembly of the hydraulic system and the components used for the assembly are very important in reducing hydraulic leakage. Manifolds should be used wherever possible for the mounting of directional valves. Internal drilled holes in the manifold route the hydraulic fluid to the proper valves and units. Straight thread O-ring fittings should be used as the connector between the manifolds and the hydraulic actuators. These fittings have a positive O-ring seal at the threaded connections, thus eliminating the need to rely on the metal-to-metal seal found in the older pipe thread fittings.

HYDRAULIC CIRCUIT PRINCIPLES

With the significant usage of hydraulic components on injection molding machines, it is important that one learn some basic circuit symbols and how a circuit performs. Figure 2-16 illustrates how a circuit is put together to open and close a toggle machine.

The circuit is shown in the idle mode in Fig. 2-16. The pump flow is going to the tank

through the directional valve (P to T) noted as position 2. If solenoid A is energized, the spool in the directional valve will slide to position 1, thus allowing flow from P to A and into the main area of the cylinder. The pull back area will then be directed to the tank through the directional valve (B to T). With flow directed to the main area of the cylinder and the pull back side directed to the tank,

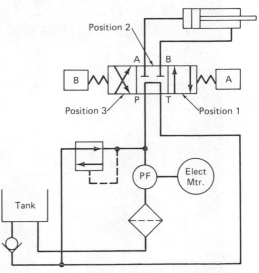

Fig. 2-16   Hydraulic circuit for a toggle machine.

Fig. 2-17 Typical injection barrel with heater bands.

the cylinder will go forward. If solenoid *B* is energized, the spool will slide to position 3, allowing the pump to flow to the pull back area, and the main area will be directed to the tank. This will allow the cylinder to open.

In a simplified form, this circuit could close a toggle clamp.

The basic areas of the injection molding machine that the hydraulic circuit normally operates are:

Fig. 2-18 Pyrometers that control heater bands.

1. Clamp
2. Ejector
3. Injection
4. Plasticating screw
5. Cores
6. Pull-in cylinder

BARREL TEMPERATURE CONTROL SYSTEMS

An important part of melting the plastic material is the heat that is added to the material from heater bands placed around the injection unit barrel. Grouped in zones in the screw, these heater bands are controlled by pyrometers that receive temperature information from thermocouples located in the injection unit barrel and turn the heater bands on or off,

depending on where the indicated temperature is in relation to the pyrometer set point. Figure 2-17 shows a typical injection barrel with heater bands, and Fig. 2-18 shows the pyrometers that control these heater bands. On the larger microprocessor systems, the temperature control capability is built into the software, eliminating the need for separate pyrometers.

MAINTENANCE AND SERVICE

Most injection molding machines will run reliably if they are properly serviced and maintained. It is important that a scheduled maintenance program be established. Common problems on machines are lack of lubrication, insufficient cooling water, failure to change filters and strainers, and general housekeeping.

Chapter 3
The Reciprocating Screw Process

Robert E. Nunn, Ph.D., P.E.

HPM Corporation
Mt. Gilead, Ohio

The widespread acceptance of the reciprocating screw injection unit since its introduction some two decades ago has caused a revolution in the versatility and productivity of the thermoplastic injection molding process.

In its infancy, many of the design concepts for the reciprocating screw were simply adapted from the similar continuous single-screw extrusion process. However, as the reciprocating screw process has matured, a much greater understanding of its own unique characteristics has evolved and been applied in process design.

This chapter reviews the basic functional elements of the reciprocating screw process, and examines their synthesis into an overall model. From this, certain process limitations are identified, and some screw design concepts that overcome these limitations are explained. Consideration is given to the effects of various parameters upon the process, from which the performance of reciprocating screws in a real-life environment can be optimized.

Throughout the chapter, the emphasis is upon developing a qualitative understanding of the process, as opposed to a rigorous analytical study. For more detailed study, the literature sources cited may be reviewed.

THE RECIPROCATING SCREW INJECTION UNIT

Physical Construction

The main elements of a typical reciprocating screw injection unit are shown in Fig. 3-1: a screw occupying the bore of a cylindrical barrel, a motor used to rotate the screw, and an injection ram and cylinder used to provide axial movement of the screw relative to the barrel.

For processing thermoplastic materials, the barrel is generally equipped with electrical resistance heater bands around its circumference, and thermocouples are used to monitor the barrel temperature for control purposes. In some cases, where extremely precise control of barrel temperature is required, air blowers may be provided for additional cooling capability, or combination heating/cooling bands may be used. These have provision for circulation of a cooling fluid, normally water or oil, to provide additional cooling capability.

Hydraulic or electric motors are normally used for screw rotation, driving the screw either directly or through a transmission system. Direct hydraulic drive motors are particularly

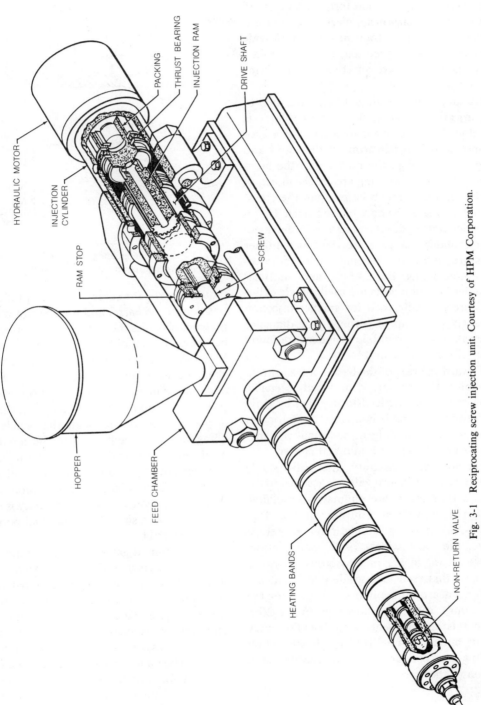

HYDRAULIC MOTOR

PACKING

THRUST BEARING

INJECTION RAM

DRIVE SHAFT

INJECTION
CYLINDER

RAM STOP

SCREW

HOPPER

FEED CHAMBER

HEATING BANDS

NON-RETURN VALVE

Fig. 3-1 Reciprocating screw injection unit. Courtesy of HPM Corporation.

popular, for in addition to a relatively low capital cost they have the capability of providing an essentially constant torque over a broad speed range. Conversely, electric motors utilizing a change gear transmission can provide a range of useful speed and torque combinations, and often have an advantage in terms of energy efficiency.

By supplying pressurized hydraulic fluid to the injection cylinder, the necessary force is developed on the screw to pressurize a shot of molten polymer in front of the screw and displace it through the nozzle into the mold cavity. The main working area of the injection ram is generally much larger than the cross-sectional area of the bore of the barrel. Therefore moderate hydraulic pressures in the injection cylinder can generate the much higher pressures in the polymer shot needed for injection mold filling. Both single- and multiple-cylinder injection units are used. Single cylinders are generally used for smaller injection units (typically up to perhaps 3-in.-diameter screws), whereas multiple cylinders are found primarily on larger units.

Toward the rear of the barrel, a feed throat or material entry port is provided to enable entry of the polymer feedstock, either by gravity feed from a simple hopper or by the action of an auxiliary feeding device such as a metered starve feeder or a material stuffer.

For low melt viscosity, thermally stable polymers, a non-return valve is attached to the front of the screw to prevent material backflow along the screw channel during injection. Figure 3-2(a) shows a sliding ring non-return valve, the most widely used configuration. However, for higher melt viscosity polymers or those that tend to degrade due to stagnation, a smearhead is often substituted for the non-return valve, as shown in Fig. 3-2(b). With this arrangement, some backflow may occur, but it is minimized by the reduction in flow area obtained from the elevated land region of the smearhead.

Sequence of Operations

The sequence of operations for a reciprocating screw injection unit is shown schematically in Fig. 3-3. At the commencement of the

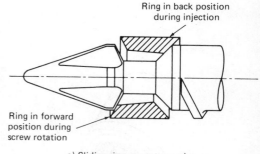

Ring in back position
during injection

Ring in forward
position during
screw rotation

a) Sliding ring non return valve.

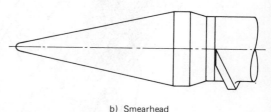

b) Smearhead

Fig. 3-2   Injection molding screw tip configurations. (a) Sliding ring non-return valve. (b) Smearhead.

molding cycle, the screw occupies a retracted position in the barrel and a charge of molten polymer occupies the region of the barrel bore between the front of the screw and the nozzle. When the mold halves have been closed and clamp pressure applied, hydraulic fluid is supplied to the injection cylinder causing the injection ram and the screw to advance, thereby displacing material through the nozzle and filling the mold cavity. When the mold has filled, pressure is maintained in the injection cylinder until the material in the mold gates solidifies; during this time, contraction of the solidifying polymer in the mold cavity is compensated by the supply of additional polymer from the barrel.

Once the mold gates have frozen, thus isolating the polymer in the mold cavity, a new charge of polymer melt can be prepared for injection into the mold in the subsequent molding cycle. Screw rotation commences, and material is conveyed along the screw. During its passage along the screw, the material is melted and mixed, and it is discharged from the forward end of the screw. Pressure generated in the discharged polymer is transmitted by the screw to the injection ram, which displaces hydraulic fluid from the injection cylinder and allows the screw to retract in the barrel. By throttling the discharge flow of fluid from the injection cylinder, the deliv-

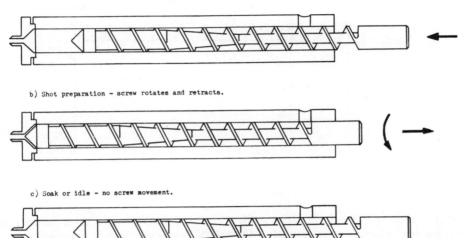

a) Injection - screw moves axially forward.

b) Shot preparation - screw rotates and retracts.

c) Soak or idle - no screw movement.

Fig. 3-3   Reciprocating screw sequence of operations. (a) Injection: screw moves axially forward. (b) Shot preparation: screw rotates and retracts. (c) Soak or idle: there is no screw movement.

ery pressure of the molten polymer can be varied; this procedure is termed application of "back pressure." When the desired volume of polymer melt has been discharged from the screw, this typically being established by monitoring the axial displacement of the screw, screw rotation ceases. The process of shot preparation is often termed "screwback."

Once the new charge of polymer melt has been prepared, and when the solidified part in the mold cavity has been ejected and the mold reclamped, the subsequent molding cycle can begin. In general, there can be a period of time between the end of screw rotation and the start of injection; any such delay is termed "soak" or "idle time."

The overall process of converting the polymer from a solid feedstock to a melt is termed "plastication." Since the overall reciprocating screw process involves a sequence of different events, the overall plasticating process becomes quite complex. In subsequent sections of this chapter, the interrelationship between the various events and their effects upon the plasticating process will be evaluated in detail.

Screw Terminology and Basic Concepts

A plasticating screw is normally classified by its nominal diameter and flighted length, the latter usually being expressed as the ratio of flighted length to diameter. Hence a 2-in.-diameter screw with a flighted length of 40 in. would be classified as a 20:1 L/D screw. Generally, with the screw in its fully forward position, the full flighted length is available for plasticating. However, as the screw retracts during operation, its effective flighted length decreases. The retraction stroke of the screw is similarly expressed as a length:diameter ratio; consequently, a 20:1 L/D screw operating with a 4:1 L/D stroke will have an effective length between 20:1 and 16:1 L/D, depending on the instant in the cycle being considered.

A typical single-stage screw is shown in Fig. 3-4. The flighted portion consists of three geometrically distinct sections: a relatively deep constant-depth feed section, an intermediate tapered transition section, and a relatively shallow constant-depth meter section. These sections approximately correspond to the principal processing functions of the screw: solids conveying or "feeding," polymer melting, and melt conveying or "metering" and mixing.

The ratio of feed section depth to meter section depth is usually called the "compression ratio" of the screw, and typically lies in the range from 2:1 to 4:1 for most commercial thermoplastics. A screw with a low compression ratio is adequate for easily melted

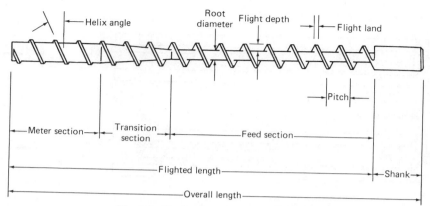

Fig. 3-4 Single-stage screw nomenclature.

polymers, particularly if they have a high melt viscosity, whereas a higher compression ratio is preferred for hard-to-melt polymers, particularly those exhibiting low melt viscosities, since a higher compression ratio enables a greater mechanical work input to the polymer to enhance its rate of melting.

In addition to compression ratio, the absolute depth of the screw channel is of importance in its effects upon the polymer. A relatively deep screw provides a high conveying capability but necessitates a high torque for rotation—hence, for a limited drive power, requiring relatively slow rotational speed. Conversely, a relatively shallow screw will require a higher screw speed, but lower torque, for a corresponding output rate. Shear rates imposed on the polymer by a deep-channel screw at a slow rotational speed are lower than those imposed by a shallow-channel screw at higher rotational speeds. Hence shear-sensitive polymers tend to be processed using deeper screws than those favored for extremely hard-to-melt materials.

The relative lengths of the feed, transition, and meter sections play important roles in screw performance. Hard-to-feed materials can often be accommodated by extending the length of the feed section, whereas if improved melt homogenization is required, a useful approach is to extend the length of the meter section. The length of the transition section is important in assuring stability of the plasticating process. An insufficiently long transition section can result in unstable melting and a lowered melt homogeneity.

Most conventional screws utilize a square pitch; that is, the pitch is equal to the diameter. However, many newer and more specialized screw designs make use of different pitches to enhance different facets of performance. Similarly, although most conventional screws are single-flighted, many new designs use double flights (and hence two separate screw channels in parallel). This can be of advantage in reducing screw and barrel wear because it imposes a degree of axial symmetry to reduce the tendency toward sideways deflection of the screw under load.

ELEMENTS OF THE PLASTICATING PROCESS

When the reciprocating screw injection unit was originally introduced, screw designs were based upon the similar, but conceptually simpler, continuous single-screw extrusion process. More recently, the significance of the cyclic nature of reciprocating screw plasticating has been recognized, and injection molding screw design concepts have evolved accordingly (as reviewed, for example, in reference 1).

It is convenient to separate the process of reciprocating screw plastication into three elements as follows:

1. Screw rotation
2. Soak
3. Injection stroke

These elements may then be subsequently

combined into an overall model of the reciprocating screw plasticating process.

Screw Rotation

During screw rotation, polymer is conveyed along the screw channel owing to the velocity difference between the screw and the barrel. Although in practice the screw rotates inside a fixed barrel, the process is more easily analyzed from the viewpoint of an imaginary observer rotating with the screw. This situation is equivalent to an apparently stationary screw inside an apparently rotating barrel (where the apparent rotation of the barrel is in the opposite sense to that of the actual direction of screw rotation). The apparent circumferential velocity of the barrel relative to the screw, as shown in Fig. 3-5, is given by:

$$v = \pi DN \qquad (3-1)$$

where D = screw diameter and N = Screw rotational speed.

This circumferential relative velocity can be resolved into two orthogonal components, one being directed in the helical down channel direction and the other in the cross channel direction as follows:

Down channel component:

$$V_z = V \cos \theta \qquad (3-2)$$

Cross channel component:

$$V_x = V \sin \theta \qquad (3-3)$$

where θ = screw helix angle measured at the flight tips.

These two components possess a certain physical significance. The conveying action of a screw results from drag forces imposed upon the material in the direction of the down channel component, V_z, whereas a recirculating flow is established by the cross channel component, V_x.

In fact, for a reciprocating screw, screw rotation is accompanied by axial motion as the screw retracts. Consequently, the apparent velocity of the barrel relative to the screw is not exactly circumferential, but is modified by an additional component, V_r, as shown in Fig. 3-6. Generally, however, the retraction velocity, V_r, is sufficiently small in comparison with the rotational velocity, V, that its effect can be neglected, thus approximating the situation to that of the single-screw extruder.

As the polymer is conveyed along the screw channel, it is subjected to a series of different effects. In the early part of the screw channel, the polymer becomes compacted and moves as a solid elastic plug. In the middle region of the screw, melting of the polymer takes place from the combined effects of heat transfer from the heated barrel and conversion of mechanical energy from the screw drive into thermal energy by the processes of frictional working and viscous dissipation. When melting is complete, melt conveying occurs in the final stage of the screw channel. These processes are summarized in Fig. 3-7, which depicts an idealized series of cross-sectional views through the screw channel at different stages along the screw.

Solids Conveying. The function of the feed section of the screw is to accept and convey the solid polymer feedstock to the melting region of the screw (Fig. 3-7).

Initial acceptance and conveying of the ma-

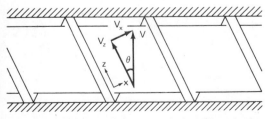

Fig. 3-5 Resolution of apparent velocity of barrel relative to screw.

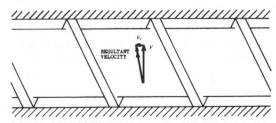

Fig. 3-6 Modification of apparent velocity of barrel relative to screw by retraction component.

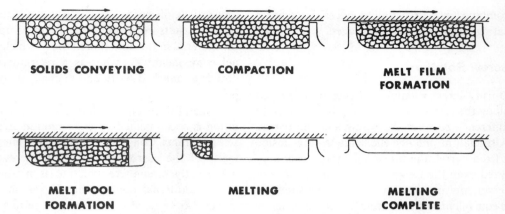

SOLIDS CONVEYING   COMPACTION   MELT FILM
                                                FORMATION

MELT POOL   MELTING   MELTING
FORMATION   COMPLETE

Fig. 3-7 The plasticating process during screw rotation.

terial appear to be governed primarily by gravity forces, in much the same way as in a simple Archimedian screw, although at high screw speeds and large screw diameters, centripetal forces may also play a role. In the initial part of the feed section, perhaps for a turn or so of the screw, the particles of material are generally not closely packed, and a great deal of rolling, tumbling, and general rearrangement appears to occur. Because of the initial loose packing of the particles, conveying rates can often be improved by provision of a tangentially tapered feed pocket in the feed throat to allow a volumetric reduction as the particles become more closely packed. Alternatively, closer packing can often be accomplished by the use of an auxiliary force feeding device, or stuffer; this approach, however, is currently more widely applied in extrusion than injection molding.

In any event, as the particles of material are conveyed along this initial portion of the feed section, they rapidly orient into closer-packed arrangements with less free volume until general recirculation ceases, and the material conveys as a solid elastic plug. As large stresses become generated in the solid plug, the conveying mechanism is predominantly governed by Coulomb friction at the interfaces between the polymer and the barrel and between the polymer and the screw.

Various analytical approaches have been proposed to the flow of an elastic solid plug through the use of certain simplifying assumptions. A major contribution was made by Darnell and Mol (2), in which they proposed an equation of the following form for estimating the solids conveying capacity of a screw:

$$Q/N = \pi^2\, Dh(D - h)\frac{\tan\theta \cdot \tan\phi}{\tan\theta + \tan\phi} \quad (3\text{--}4)$$

where Q = volumetric output rate, h = local height of the channel, ϕ = angle of movement of the outer surface of the solid plug, and other symbols are as previously defined.

Figure 3-8 indicates the angle of movement of the outer surface of the solid plug, ϕ, relative to the barrel.

As may be inferred from equation (3–4), provided that θ and ϕ do not vary, output increases with increasing screw speed, increasing screw diameter, and increasing channel depth.

Unfortunately, solids conveying theory, such as equation (3–4), is rarely applied directly in screw design to calculate potential output because a realistic estimate of ϕ, the

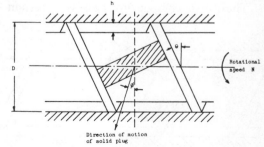

Fig. 3-8 Angle of movement of solid plug relative to the stationary barrel.

angle of movement of the solid plug, is particularly difficult to determine—primarily owing to problems in assigning adequate values to the coefficients of friction between the polymer plug and the adjacent metal surfaces. In a typical feed section, both temperature and pressure variations occur, directly affecting the frictional characteristics of the materials.

However, output does depend upon the angle of movement of the solid plug. For a zero value of ϕ, the polymer rotates with the screw, and there is no output. As ϕ is progressively increased, output increases until a practical maximum occurs at $\phi = \pi/2 - \theta$, where the direction of motion is normal to the screw flights. Consequently, the task of the screw designer is to ensure as large a value of ϕ as possible in a practical situation. Solids conveying theory provides qualitative insights that are widely applied in screw design.

Solids conveying capacity can be optimized by providing a high coefficient of friction between the polymer and the barrel, and a low coefficient of friction between the polymer and the screw. In practice, this can be achieved by providing a somewhat rough barrel surface and a smooth screw surface. An extreme case of barrel roughness may be the provision of axial grooves in the barrel wall, tending to provide very large effective coefficients of friction between the polymer and the barrel. This approach is quite common on single-screw extruders, but, at present, rare on injection molding machines.

From the viewpoint of the machine user, these considerations are important. Feed zone barrel temperatures should be set to provide high coefficients of friction between the polymer and the barrel, whereas the screw should be at an appropriate temperature to provide low polymer-to-screw friction; this can often be accomplished through the use of a circulating fluid to provide screw cooling. A rather common problem that this approach can overcome is known as "melt plugging," where an excessively hot screw in the feed zone results in very high friction between the polymer and the screw and a corresponding drastic reduction in output. Indeed, in extreme cases the polymer may melt and fuse to the

screw before any significant barrel-to-polymer frictional forces are generated, in which case output may cease entirely.

Another implication of solids conveying analysis is that output is sensitive to the level of pressure generated at the end of the feed section. A high delivery pressure results in a low angle of movement of the solid plug, and hence a low output. In fact (as shown in reference 3, for example), a pressure buildup of only perhaps 3 to 5 lbf/sq in. in the first two turns of an extruder feed section may result in a drastic reduction in output, whereas with no pressure buildup and low friction between the polymer and the screw, a very close agreement can be found with equation (3–4) for many feedstocks.

Pressure buildup in solids feeding is exponential; so only moderately long feed sections may be capable, in principle at least, of generating very high pressures. The magnitude of the final generated pressure depends upon its initial value at the formation of the solid plug. This is generally not readily calculable because pressure appears to be initiated by frictional and centripetal forces, which in turn may vary cyclically as the screw rotates, resulting in cyclic variation in delivery pressure. However, neglecting cyclic effects, the pressure-generating capability of a feed section at a known output rate is often estimated by using equation (3–4) to back-calculate the angle of motion of the solid plug, and then to assume "reasonable" values of coefficients of friction and initial plug pressure. This is often of use in estimating wear characteristics of screws and barrels when processing abrasive materials.

In addition to the magnitude of the pressure generated, its rate of generation with respect to channel distance, referred to a pressure gradient, affects output. Because a high pressure gradient results in a low output, most injection molding screws are made with fairly long feed sections. This is of particular importance when screw retraction effects are included, since for large shot sizes, a significant length of feed section may be inactive toward the end of screwback.

A further consideration in solids feeding involves the effects of helix angle upon the angle of movement of the solid plug and output rate. Depending on material properties and pressure conditions, an optimum helix angle exists at which output is maximized. Generally, the optimum value of the helix angle lies in the range of 10 to 30 degrees, with the lower end of the range being optimum for high coefficients of friction between the polymer and the screw and the upper end of the range being optimum for low coefficients. A reasonable compromise is made for most general-purpose screw designs, where a square pitch is selected for the feed zone. In this case, the screw pitch is equal to the screw diameter, with a helix angle of 17.6 degrees. This approach provides adequate feeding for the broadest range of materials.

In terms of practical application, if Coulomb friction can be maintained over a sufficient initial length of a screw, generally for perhaps six to eight turns, then sufficiently high pressures can be generated to dominate any resistance offered by the later melting and melt conveying zones of the screw. This has been accomplished by intensively cooling the feed section of the barrel and screw to avoid any melting of the polymer, but is rarely applied in practice. More generally, after perhaps four or five turns a melt film forms at the interface between the polymer and the barrel, often accompanied by the formation of a melt film between the polymer and the screw. When this occurs, the force driving the solid plug in the down channel direction ceases to be Coulomb friction and instead becomes the shear force generated in the melt film between the plug and the barrel, and the conveying rate becomes more sensitive to changes in feed section delivery pressure.

As a result, solids feeding tends not to dominate the performance characteristics of most injection molding screws, although inadequate or problematic solids feeding can have profound effects upon the subsequent melting and melt conveying processes. Rather, the melting process generally is of greatest significance, and is considered in the following section.

Melting. Beyond the zone of solids conveying, the main function of the screw becomes that of melting the polymer feedstock. Understanding and analysis of the melting process has been based upon experimentally observed melting mechanisms. The experimental method employed is to operate the machine under steady conditions, either continuously in the case of single-screw extrusion or under conditions of stabilized and repeatable cycles for injection molding, and then at an appropriate point in time stop the screw and rapidly cool the barrel to freeze the polymer in the screw channel. When the barrel has cooled, the screw is removed, and the frozen strip of polymer occupying the screw channel can be unwrapped and cut into sections from which the melting mechanism may be observed. Identification of the various regions within the sections can be aided by coating the feedstock particles with a contrasting dry colorant. Using this technique, the composition of the solid bed, as the solid plug is generally named during melting, is very evident from individual particles. In regions of sheared melt, particle identity is lost, but flow patterns can be observed from streak lines in the rather uniformly colored melt. Intermediate "softened" particles can be seen to deform, and one can gain an indication of strain fields from studying these deformations. A major benefit is that it becomes possible to distinguish between areas of high and low (or essentially zero) deformation and thereby identify areas where melting is purely by conductive heat transfer or involves shear-induced viscous dissipation.

As noted previously, pure solids conveying ends with the formation of a thin melt film at the interface between the solid bed (the plug of solid material) and the barrel. The formation of this film occurs when the surface temperature of the solid polymer is raised to the melting point by a combination of frictional work between the sliding polymer and the barrel, and conductive heat transfer from the normally heated barrel into the solid polymer. In the instances when barrel cooling is applied, its function is to remove the frictional heat and thereby delay the formation of the film. Once the film has formed, viscous dissipation of energy within the film becomes predominant due to the high shear rates and high melt viscosities normally encountered.

The initial process of melting, then, becomes that of the development of the melt film in the down channel direction. High fluid pressure within the film causes melt to infiltrate to some extent into the voids between the closely packed particles where heat transfer to the solid particles plays a role in generally increasing the bulk mean temperature of the solid bed. Experimental observations indicate that for small screws almost complete penetration of melt into the solid bed may be expected. However, for larger screws, owing to the physically longer flow paths involved, penetration may be only partial. Analysis of the initial development of melting, often termed the "delay zone," until the establishment of a subsequent different melting mechanism, can be performed by considering viscous drag upon the solid bed which is melting due to heat transfer from the intensively sheared thin melt film.

Beyond the delay zone, experimental observations show that a radically different melting mechanism exists. Once the melt film reaches a certain critical value, generally somewhat larger than the clearance between the barrel and the screw flight tip, the action of the advancing flight becomes that of a scraper. Melt is scraped from the continuous film adhering to the barrel and is deposited in an ever-widening pool of recirculating melt adjacent to the advancing flight. An idealized cross section is shown in Fig. 3-9(a). Melting proceeds by entrainment of freshly melted material from the upper surface of the solid bed into the melt film and its subsequent deposition into the melt pool. As a result, the solid bed progressively decreases in width, and the melt pool becomes progressively wider with respect to position along the channel until complete melting occurs with the consumption of the solid bed.

A simplified Newtonian analysis of such a melting mechanism was first provided by Tadmor and Klein (4). Based on certain simplifying assumptions, their model indicated that the rate of melting, ω, for a given solid bed width, X, can be expressed as:

$$\omega = \Phi \, X^{1/2} \qquad (3\text{-}5)$$

where ω = rate of melting per unit down channel distance, X = solid bed width, and Φ = melting parameter, where:

$$\Phi \equiv \left\{ \frac{V_x \, \rho_m \, [k_m \, (T_b - T_m) + (\mu/2) \, V_j^2]}{2 \, [Cp_s \, (T_m - T_s) + \lambda]} \right\}^{1/2}$$

$$(3\text{-}6)$$

where V_j = barrel velocity relative to moving solid bed, ρ_m = melt density, k_m = melt thermal conductivity, μ = melt (Newtonian) viscosity, Cp_s = solid specific heat, λ = latent heat of fusion, T_b = barrel temperature, T_m = polymer melting temperature, and T_s = feedstock temperature.

The form of these equations is particularly useful in understanding the melting process. For an initially wide solid bed, the rate of melting is high, but as the width of solid bed decreases, a reduction in the melting rate may be expected. In practice, the final stages of

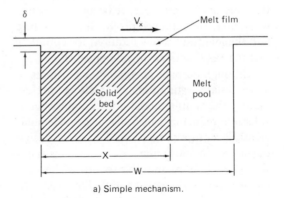

a) Simple mechanism.

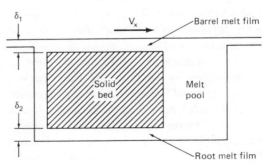

b) Modification by root melt film.

Fig. 3-9 Observed melting mechanisms. (a) Simple mechanism. (b) Modification by root melt film.

melting appear to be primarily governed by thermal conduction into the remaining small portion of solid from the melt pool and from the screw rather than by the above mechanism. In equation (3–6), the numerator is a measure of the energy available for melting and the denominator a measure of the enthalpy change required for the polymer to melt. A large value of the numerator or a small value of the denominator results in a high rate of melting. Hence a heated barrel, high melt viscosity, or low specific enthalpy results in rapid melting, whereas low barrel temperature, low melt viscosity, and high specific enthalpy produce slow melting rates. In general, amorphous polymers melt more rapidly than crystalline polymers. The two terms in the numerator of equation (3–6) represent the two sources of energy for melting, thermal conduction from the barrel and viscous dissipation in the melt film, respectively. In most injection molding situations, the latter is predominant.

From knowledge of the rate of melting and the polymer output rate, it becomes possible to estimate the solid bed profile during melting and the length of screw required for complete melting (as described in reference 4).

Although perhaps adequate for purposes of approximate estimation of melting performance, the simple model above requires considerable refinement in order to provide sufficiently accurate predictions for screw design purposes. In particular, as previously mentioned, melt films may be found at the root of the screw and against the flights, as shown in Fig. 3-9(b). Also, significant deformation and acceleration of the solid bed may occur with important effects upon the stability of the process. In addition, the actual non-Newtonian characteristics of the melt result in modifications to the flow and heat transfer considerations. Consequently, many improved melting models have been developed in recent years with increasing subtlety and accuracy. For example, the model proposed in reference 5 takes into account these various refinements.

Inevitably, as models develop in complexity, methods of solution become more tedious. Rather than striving for exact analytical solutions, the more normal approach is to use a high-speed computer to generate a solution using numerical approximation techniques. With sufficient expenditure of computational time, the required degree of accuracy can, in principle, be obtained.

Melting of the polymer is the primary purpose of an injection molding screw. However, as will be considered later, the melting process is extremely sensitive to the necessary cyclic effects that are imposed by the injection molding process, and the main task of the screw designer is to provide a design capable of adequate melting over a wide range of different conditions. It is in this context that the main variations occur between screws designed for injection molding and continuous single-screw extrusion.

Melt Conveying (Metering). In single-screw extrusion, long metering sections are widely used to stabilize the output of molten polymer, and also provide some degree of homogenization. This is necessary because a variation in output will produce a variation in extruder product at exit from the die, due to the coupling of screw and die behavior in the continuous process. In the case of injection molding, the shot preparation and product forming processes are independent, so that a moderate variation in screw output with time will not necessarily reduce molded product quality and may be tolerable. In addition, the high delivery pressures required in extrusion, and provided by a long metering section, are not normally encountered in the injection molding process.

Consequently, metering is relatively less important in the injection molding process, the main function of the metering section being to provide a throttle to ensure that adequate melting takes place, and to generate sufficient pressure to overcome the resistance imposed by the non-return valve and by the back pressure. Metering sections tend to be short, generally less than one-fourth of the overall flighted length of the screw and often much less, particularly for screws designed for easily melted high-viscosity polymers.

Melt conveying has been extensively studied

and is now well understood. Comprehensive treatments are available in the literature (for example, in references 6 and 7.)

A very simple analysis of melt conveying, as derived in most standard texts, shows that for Newtonian flow in a single shallow channel, where channel curvature and flight thickness are neglected:

$$Q = \frac{\pi^2 D^2 h N \sin \theta \cos \theta}{2}$$
$$- \frac{\pi D h^3 \sin \theta}{12\mu} \left(\frac{\partial p}{\partial z}\right) \quad (3\text{--}7)$$

where p = pressure, z = down channel coordinate, and other symbols are as previously defined. The first term on the right-hand side of equation (3–7) is often referred to as the "drag flow" and represents the flow induced by the down channel component of the velocity of the barrel relative to the screw as derived in equation (3–2). The second term on the right-hand side of equation (3–7) is often referred to as the "pressure flow"; and may be considered to represent the change in output due to any pressure gradient existing along the metering section. In the absence of any pressure gradient, the pressure flow term vanishes, and the output is equal to the drag flow. However, if a positive (i.e., increasing in the down channel direction) pressure gradient is imposed, the output will be reduced, and if a negative pressure gradient is imposed, the output will increase. Figure 3-10 shows the effect on the down channel drag flow velocity distribution of superimposing a pressure flow.

Application of back pressure corresponds to an increase in the magnitude of the pressure flow term in equation (3–7). As a result, some additional flow complexity may be introduced to enhance mixing, and the corresponding reduction in output provides a longer residence time of polymer in the screw channel and thus enhances melting. This versatility enables a general-purpose screw design to better accommodate a broad range of materials. However, screws designed for more dedicated application over a narrow range of materials can be designed for optimum output by suitable

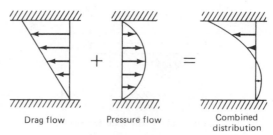

a) With positive pressure gradient.

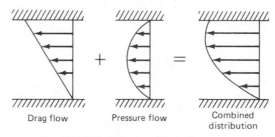

b) With negative pressure gradient.

Fig. 3-10 Down channel velocity distribution. (a) With positive pressure gradient. (b) With negative pressure gradient.

choice of the geometric parameters in equation (3–7), namely, channel depth, helix angle, and length. Generally, meter section design becomes a compromise between output and homogeneity considerations.

Soak

Owing to the cyclic nature of injection molding, the plasticizing process is influenced by phenomena that occur at times other than when the screw is rotating.

During any soak periods—although conveying does not occur, and thus no melting takes place due to shearing—the presence of the heated barrel in contact with the polymer does give rise to some additional conductive melting. In the early stages of the screw channel, this can result in the formation and growth of melt films where previously none existed. In later stages of the screw channel, it can produce additional growth in film thicknesses, even to the extent in the later stages of the screw of completely melting any thin regions of the solid bed.

In reference 1, a simple analysis of growth of a melt film due to heat transfer from the

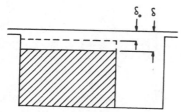

Fig. 3-11 Melting during soak.

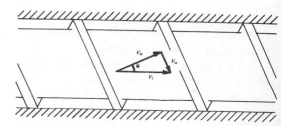

Fig. 3-12 Resolution of injection motion into orthogonal components.

barrel, as shown in Fig. 3-11, provided the following solution:

$$\delta^2 = \frac{2k_m(T_b - T_m)t_s}{\rho_s[\lambda + Cp_s(T_m - T_s)]} + \delta_0^2 \quad (3\text{-}8)$$

where δ = film thickness, δ_0 = initial film thickness, ρ_s = solid density, and t_s = soak time. Of interest here is the general form of equation (3-8) because it indicates that for an initially thin or zero thickness film the film growth is large, but for an initially thick film the film growth is small. Consequently, the rate of melting by conduction (in units of mass per unit time) progressively decreases as the melt film grows in thickness.

For long soak times, the degree of melting by conduction can be particularly significant. However, even for short soak times, where the degree of melting may be rather small, other profound effects may occur, in particular in the feed section where the formation of a melt film may significantly affect feeding performances during subsequent screw rotation, and in the melting region where thicker melt films will significantly reduce shear rates and hence viscous dissipation rates.

Injection Stroke

During the injection stroke, in addition to melting by conduction, the relative motion that exists between the screw and barrel provides a contribution to melting.

The axial motion of the barrel relative to the screw can be resolved into two components as follows, for a uniform injection velocity (see Fig. 3-12):

$$V_i = S_i/t_i \quad (3\text{-}9)$$

where V_i = injection velocity, S_i = injection stroke, t_i = injection time, and hence:

Cross channel component:

$$V_{ix} = V_i \cos \theta \quad (3\text{-}10)$$

Up channel component:

$$V_{iz} = V_i \sin \theta \quad (3\text{-}11)$$

The fact that an up channel component exists directed toward the feed end of the screw channel indicates that a potential exists for backflow along the screw channel. However, experimental studies indicate that backflow does not occur in practice. Apparently the presence of the solid plug in the feed section tends to lock into place and resist backflow, at least for normally encountered injection stroke lengths.

The transverse component, V_{ix}, however, does provide a cross-channel melting mechanism very similar to that encountered during screw rotation, but with the important difference that simultaneous conveying does not occur. In this case molten material does transfer from the melt film into the melt pool, during which time experimental observations indicate that a constant film thickness is maintained. Thus reduction in the width of the solid bed occurs.

Reference 1 presents an analysis from which the reduction in width of the solid bed may be determined during the injection stroke due to combined effects of viscous dissipation in the barrel melt film and conductive heat transfer, as shown in Fig. 3-13. The relationship may be expressed in the following nondimensional form:

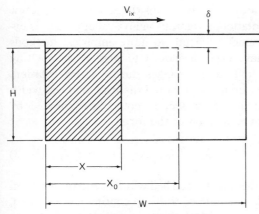

Fig. 3-13 Injection stroke melting.

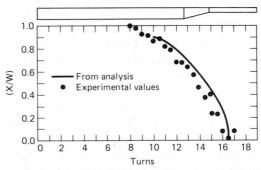

Fig. 3-14 Comparison of actual and predicted reduced solid bed width profie at end of injection. From reference 1.

the solid bed profile at the end of the injection stroke, as shown in Fig. 3-14.

OVERALL PLASTICATING PROCESS

Combined Model

$$(X/W) = (X_0/W) \cdot \text{Exp}\left\{ \frac{-1}{t_i \delta H \rho_m} \right.$$
$$\left. \cdot \frac{2 k_m t_i^2 (T_b - T_m) + \mu (s_i \cos \theta)^2}{2 C p_s (T_m - T_s) + 2\lambda} \right\} \quad (3\text{–}12)$$

where X_0 = initial solid bed width, W = channel width, H = solid bed height, and other symbols are as previously defined.

The form of equation (3–12) indicates the contributions of both conduction and viscous dissipation. Almost invariably in practice, the viscous dissipation term is small in comparison with the conductive term: hence the melting is dominated by conductive melting. Even so, the degree of melting is generally significantly greater than would occur in a static soak situation because the melt film is maintained at a constant thickness and thereby provides a high degree of conductive heat transfer.

A consequence of the exponential form of equation (3–12) is that a characteristic time constant may be evaluated for a given practical situation, during which a reduction in the solid bed width by a factor of 0.632 occurs. Typically, practical time constants lie in the range of 5 to 50 seconds for most injection stroke melting situations, which when compared to the injection time indicate the significance of the injection stroke melting.

The validity of this model for injection stroke melting has been demonstrated, and it can provide a useful method for estimating

In steady extrusion, ignoring any instabilities in the process, a melting profile is set up that is invariant with respect to time. Consequently, extruder screw designs can be optimized making the reasonable assumption of a unique melting profile. However, in the case of reciprocating screw injection molding, the melting profile can change dramatically during the cycle, the degree of variation depending on operating conditions, cycle proportions, and machine capacity, and the assumption of a constant melting profile is not realistic.

Figure 3-15 indicates the cyclic variation in melting profile at an imaginary cross section in the melting zone. As shown in Fig. 3-15(a),

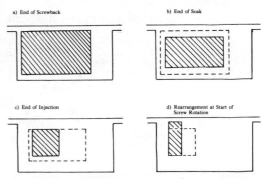

Fig. 3-15 Cyclic melting model. (1) End of screwback. (b) End of soak. (c) End of injection. (d) Rearrangement at start of screw rotation.

at the end of screw rotation the channel is occupied by a large proportion of solid bed. Generally, a period of time elapses between the end of screw rotation and the start of the injection stroke, and during this soak period, melting takes place due to conduction, as seen in Fig. 3-15(b). During injection, further melting takes place, and the solid bed reduces in width to the condition shown in Fig. 3-15(c). Then, when screw rotation commences, some slight rearrangement occurs, shown in Fig. 3-15(d), as the barrel melt film reverts to a thinner dimension as material transfers to the melt pool and the rotational melting mechanism is reestablished.

Thus, at the start of screw rotation, the contents of the channel have been subjected to soak and injection stroke melting, so that any given screw channel cross section now contains a relatively high proportion of melt.

Subsequently, however, as the screw rotates and material is conveyed along the screw channel, a progressive increase in the width of the solid bed occurs as hotter, soaked material is replaced with cooler material being delivered from the feed hopper, until the condition shown in Fig. 3-15(a) is reestablished at the end of screw rotation. Experimental measurements showing the change in melting profile during screw rotation are given in Fig. 3-16. In fact, if the shot size is sufficiently large that all material present in the melting region is replaced during screw rotation, the resulting melting profile closely resembles that of a comparable single-screw extruder, as also seen in Fig. 3-16.

In view of the complexity of the transient nature of behavior of the screw during screw rotation, a rigorous analytical approach is extremely difficult; so more empirical methods have been developed to predict the melting profile as it changes during screw rotation. Reference 1 proposed that the solid bed width at any cross section could be estimated by an expression of the form:

$$X = X_i + \{X_e - X_i\} \cdot \{1 - e^{-\beta N t}\} \quad (3\text{--}13)$$

where X_i = solid bed width at end of injection, X_e = solid bed width in continuous extrusion, β = empirical constant, t = time coordinate, and other symbols are as previously defined.

This expression implies an exponential change in the solid bed width during screw rotation from a width, X_i, at the start of screw rotation tending to the limiting value, X_e, which would prevail under continuous extrusion conditions with the screw in the retracted position with the same screw speed and delivery pressure. The empirical exponent, β, appears to have a roughly constant unique value for each polymer over a range of processing conditions, with typical values of 0.0067 for polypropylene and 0.0075 for ABS. (Acrylonitrile Butydiene Styrene).

This approach is useful for practical screw design because analytical models of extrusion melting are quite successful in predicting melting profiles, which can be used as initial approximations in iterative development of reciprocating screw melting profile variations.

A combined melting model that includes provision for melting during soak and melting during injection enables a firm understanding of the reciprocating screw plasticating process, and experimental studies have confirmed its validity. Figure 3-17 shows the variation in melting actually observed during an actual molding cycle, as described in reference 1, as verification of the model.

Practical Performance Considerations

Since a considerable variation in melting profile may occur during the screw rotation period, it is not surprising that significant variation in melt quality can exist through the shot.

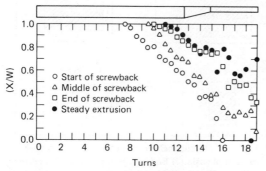

Fig. 3-16   Solid bed width during screwback. From reference 1.

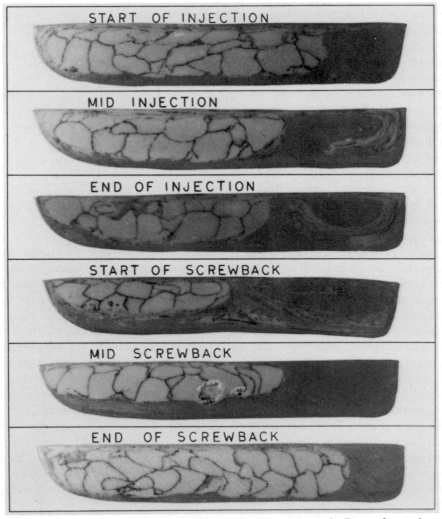

Fig. 3-17   The melting mechanism during the injection molding cycle. From reference 1.

The initial position of the shot is likely to be fully melted and well mixed. However, subsequent material delivered from the screw is likely to exhibit poorer homogeneity, both physically and thermally, and in an extreme case, for an inadequate screw, unmelted polymer may be discharged from the end of the screw toward the end of screw recovery. The presence of an inadequate melt quality can often be inferred from molded product, which may exhibit symptoms of warping, poor color dispersion, or the presence of unmelted particles, or may require excessive injection or packing pressure.

Alternatively, the uniformity of temperature throughout the shot can be monitored, either by the use of a fast response recording thermocouple situated in the melt stream in the nozzle area or by probing a collected air shot of polymer with a low thermal mass thermocouple or thermometer; a poor-quality melt generally has poor thermal homogeneity. Two approaches have been utilized to counteract the tendency for melt quality to decline through the shot: (1) the use of regions of elevated barrel temperatures, to induce higher conductive melting rates during the soak period of material that will form the latter stages of the shot; or (2) the use of variable back pressure.

Other important effects are also evident during screw rotation. As the screw retracts, the effective length of the feed zone decreases with the result that output rate may exhibit some reduction, as shown in Fig. 3-18. In addition, the screw torque may exhibit variation due to the combined effects of two countervailing processes. First, as feed rate decreases owing to the reduction of effective feed zone length, the power, and hence torque, required for conveying will decrease. Second, however, as relatively well-melted material in the channel is replaced with less well-melted material, the power and torque required for melting will increase. The resulting variation in torque will depend upon the relative magnitudes of these two effects, as shown in Fig. 3-19.

Changing screw speed has profound effects upon the plasticating process. As screw speed increases, output rate increases, and the rate of conversion of mechanical energy to thermal energy through the mechanism of viscous dissipation increases. However, most commercial polymers exhibit pseudoplasticity with decreasing melt viscosity at increasing shear rate. Consequently, the proportional change in viscous dissipation due to an increase in screw speed is generally less than that in output. In addition, the melting process is governed by the rate at which thermal energy can be conducted into the solid bed. Thus, an increase in screw speed generally results in an increase in the physical length of screw channel required to complete melting, since throughput is higher and energy conversion by viscous dissipation, on a unit mass basis, is lower. In

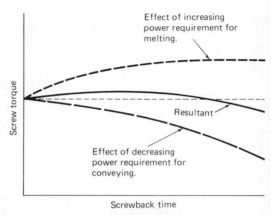

Fig. 3-19   Torque variation during screwback.

practice, this length may exceed the actual physical length of the screw, in which case the output from the screw will contain a proportion of unmelted material.

In order to accommodate higher screw speeds, and higher outputs, the trend in recent years has been toward the use of increasingly long injection molding screws to provide such additional length for melting. In terms of operating practice, it can also be concluded that optimum melting performance generally can be obtained by using the slowest screw speed possible consistent with other cycle constraints, rather then using a very high-speed, short-duration screwback followed by a lengthy soak period until the molding cycle is completed.

For relatively short overall cycle times, the melting mechanism is generally dominated by viscous dissipation. However, if overall cycle time is increased, the contribution of conductive heat transfer to melting becomes more significant, and can become predominant if cycle times are sufficiently extended. In practice, typical injection molding screws are capable of generating higher output rates if supplied with a molten feedstock than when they are processing the more usual solid feedstock of the same material. If conductive melting when the screw is not rotating becomes significant, the performance of the screw will tend toward that of the melt-fed situation, and a higher output rate will generally be observed, as shown in Fig. 3-20. Conversely, as cycle times are reduced, as usually dictated by prac-

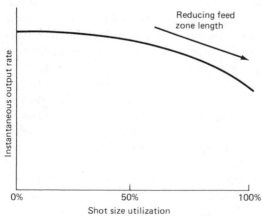

Fig. 3-18   Output rate change during screwback.

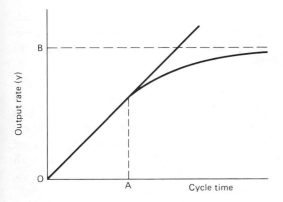

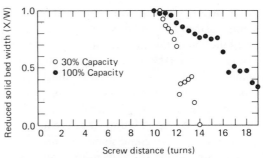

Fig. 3-21   Effect of shot size on melting profile (at end of screwback). From reference 1.

OA = minimum cycle time with screw rotating 100% of cycle
       (i.e., continuous extrusion)
OB = Maximum output at infinite cycle time (i.e., melt-fed
       condition)

Fig. 3-20   Effect of cycle time on output rate, for fixed shot size and screw speed. OA = minimum cycle time with screw rotating 100% of cycle (i.e., continuous extrusion). OB = maximum output at infinite cycle time (i.e., melt-fed condition).

tical economics, some reduction in output rate, and hence some increase in screw rotational time to convey the shot, should be anticipated.

Conductive melting also increases in relative importance if only a small proportion of the shot capacity of the machine is utilized. Typically, the total contents of the channel of a reciprocating screw are of the order of two to three times the nominal shot size of the machine. Consequently, if the machine is being operated at full shot size utilization, an element of polymer is resident in the screw channel for two to three cycles, on average. At lower shot size utilization, an element of

polymer will experience a proportionally greater residence time. Under these circumstances, the contribution of conductive heat transfer from the heated barrel gains in importance, as shown in Fig. 3-21.

Finally, a phenomenon often observed experimentally that can have significant effects in practice is an instability in the melting process often referred to as "solid bed breakup." When this occurs, the solid bed periodically fractures toward the end of the melting process, usually in the tapered transition region of the screw, into a series of discrete agglomerations of particles separated by regions of melt, as shown in Fig. 3-22, generally accompanied by pressure surges. These agglomerations appear subsequently to melt primarily by conduction rather than in the more controlled fashion of a continuous solid bed, giving a somewhat less homogenous product at exit from the screw. This process apparently is due to acceleration of the solid bed in the transition region, which causes the stresses within the

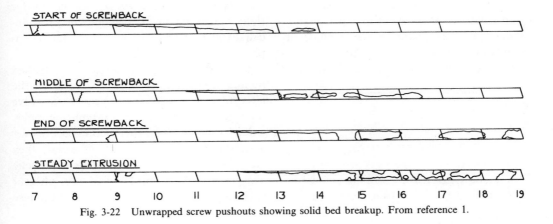

Fig. 3-22   Unwrapped screw pushouts showing solid bed breakup. From reference 1.

solid bed to deviate from an ideal uniform compression. The rate of acceleration, in turn, is a function of throughput rate and the severity of taper in the transition region. Consequently, high-compression-ratio, short-transition screws may be expected to be quite prone to this phenomenon. Further discussion of solid bed breakup is found in reference 8.

In order to overcome many of the performance limitations of conventional screws, a variety of specialized designs have evolved in recent years, generally with emphasis upon specific facets of the plasticating process, as described in the following section.

SPECIALIZED SCREW DESIGNS

Local Mixing Sections

Local mixing sections are typically short devices, often incorporated at the delivery end of the screw channel, that generate locally high rates of melting or mixing. The use of areas of high shear rates assists melting, and regions of complex flow assist mixing. Most mixing sections provide both functions, although one may be emphasized relative to the other.

A particularly widely applied mixing device is the Union Carbide mixing section shown in Fig. 3-23. This geometry provides a degree of flow recirculation within the inlet and outlet flutes, in combination with a high shear barrier zone to complete melting and enhance dispersive mixing. Its primary advantages include the yield of a particularly high degree of thermal homogeneity, as indicated in reference 9, together with a high level of dispersive mixing.

Other common mixing devices include radial mixing pins, slotted rings and discs, undercut rings, and slotted single or multiple

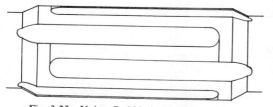

Fig. 3-23   Union Carbide local mixing section.

flights. The choice of device is normally dictated by the process requirements of the application, and whether melting or mixing is to be primarily emphasized.

Local mixing sections may incur some potential disadvantages, including some loss of positive conveying capability, close clearances that may exhibit accelerated wear from abrasive feedstocks or contaminants, or areas of flow stagnation that preclude their use with thermally unstable feedstocks. As a result, local mixing sections are generally used in dedicated applications requiring their specialized advantages, rather than in general-purpose screw designs.

Extended Mixing Screws

Extended mixing screws differ from those incorporating local mixing sections by virtue of the physical extent of the mixing section. One example is HPM Corporation's Double Wave screw, depicted in Figs. 3-24 and 3-25. In this approach, two long melt channels are provided in parallel, each with periodically varying depths. The shallower regions of each channel provide moderately high shear rates to increase melting, accompanied by a periodic variation in melt pressure. By arranging the channel geometries to be out of phase with each other, the resulting alternating pressure difference between the channels produces material interchange from channel to channel across the secondary flight, which is undercut for this purpose. The interchange of material between the channels, each with a different velocity distribution, gives rise to distributive (long range) mixing. This is in contrast to localized mixing sections, which primarily emphasize dispersive (short range) mixing.

In most molding applications, extended mixing screws are used both to accelerate the later stages of melting and to provide extensive mixing, as discussed, for example, in references 10 and 11. Since their action occurs over a substantial physical length of screw, they are well suited to accommodate the cyclic charges in melting profile that are inherent in the injection molding process. Also because

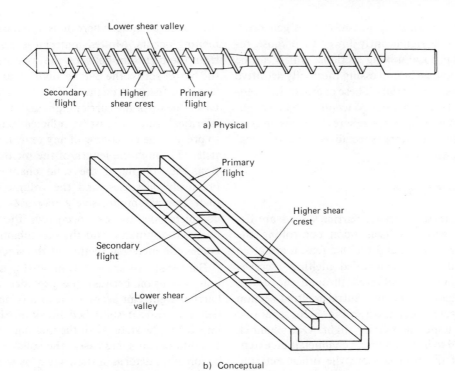

a) Physical

b) Conceptual

Fig. 3-24 HPM Double Wave extended mixing screw. (a) Physical. (b) Conceptual.

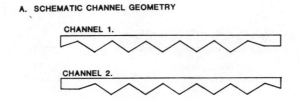

A. SCHEMATIC CHANNEL GEOMETRY

CHANNEL 1.

CHANNEL 2.

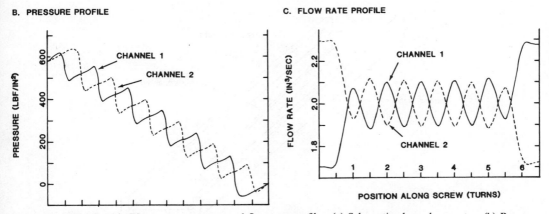

Fig. 3-25 HPM Double Wave screw pressure and flow rate profiles. (a) Schematic channel geometry. (b) Pressure profile. (c) Flow rate profile.

of their length the peak shear rates generated tend to be significantly lower than those required in local mixing sections, which can be an advantage in processing thermally sensitive or abrasive materials. These characteristics enable extended mixing concepts to be incorporated in general-purpose screw designs, in contrast to the more specialized use of local mixing sections.

Barrier Screws

Unlike mixing screws, barrier screws are designed with emphasis upon controlling the melting process, and for this reason are often referred to as "controlled melting" screws. This control is achieved through the use of twin channels, in the central melting section of the screw, as shown in Fig. 3-26, separated by an undercut barrier flight. One channel, referred to as the "solids" channel, is an extension of the feed section; the other, referred to as the "melt" channel, extends into the metering section. Solid material is conveyed into the solids channel where it is retained until melting occurs. Molten material is conveyed from the melt film between the solid bed and the barrel across the barrier flight, and is collected in the melt channel. The clearance between the tip of the barrier flight and the barrel is critical, since it must be sufficiently narrow to prevent the cross-flow of any unmelted particles. Through the length of the melting section, the volume of the solid channel progressively decreases, and the volume of the melt channel progressively increases. When complete melting has occurred, the solids channel terminates, and the melt channel becomes the metering section of the screw.

Barrier screws are widely applied in continuous extrusion because the presence of the barrier flight can serve to restrain the solid bed and prevent solid bed breakup, thereby improving the stability of the melting process. In addition, in some cases, the solid bed can be forced to deform in such a way as to generate locally high rates of melting as the solids channel reduces in volume. Hence, high out-

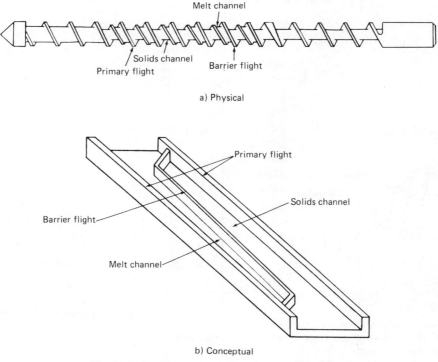

a) Physical

b) Conceptual

Fig. 3-26 Barrier screw. (a) Physical. (b) Conceptual.

put rates can be achieved. However, the use of barrier screws in injection molding applications is not widespread, perhaps because barrier screws are designed to accommodate a constant melting profile, as may reasonably be anticipated in a steady extrusion process but not in an injection molding process, where significant changes in melting profile are normal during the cycle.

Low Shear Screws

Some injection molding applications may require complete melting but with minimal strains or stresses applied to the melt. Minimizing induced strains is required when an otherwise high level of mixing would destroy some desired inhomogeneous feature of the material; a typical example occurs in the injection molding of mottled or marbelized products using a polymer feedstock consisting of dissimilarly colored components. In this example a high degree of melt mixing can result in uniformly colored product. Minimizing applied stresses may be required to avoid physical degradation of the feedstock, as, for example, in the injection molding of polymers reinforced with long glass fibers. In this case, breakage of the reinforcement during processing may result in an insufficiently strong product.

A specialized shear screw design, after reference 12, is shown in Fig. 3-27, consisting of a flighted section just long enough to provide adequate conveying and moderate compaction to initiate melting, followed by a deep flightless section to provide extensive material residence which enables conductive heat transfer from the barrel to provide a major contribution to melting. The absence of screw flights in the latter section of the screw in effect substitutes an essentially two-dimensional simple strain field for the more complex three-dimensional strain field encountered in a fully flighted screw channel, and significantly reduces mixing. The localized high shear stresses associated with recirculatory flow in a fully flighted section are similarly avoided.

VENTED BARREL INJECTION MOLDING

The hygroscopic nature of many widely used thermoplastics can result in severe molding problems unless entrained moisture is removed prior to molding. Excessive moisture levels can result in appearance defects, such as splay, or even losses in physical properties. One approach to removing entrained moisture is to predry the material, but in most cases a more viable approach is the use of a vented barrel molding machine without predrying. In this case, the polymer is devolatilized after it has been melted, and because the vapor pressure of water at typical melt temperatures is high, devolatilization can be accomplished rapidly. Moreover, at typical melt temperatures other undesirable nonaqueous volatiles may also be removed by using a vented barrel molding machine.

Devolatilization from the melt stream is made possible by the use of a two-stage screw and a barrel incorporating a vent port as shown in Fig. 3-28. The first stage of the two-stage screw accomplishes the basic plasticating functions of solids feeding and melting. During this process, significant material pressures are generated.

Molten polymer leaving the first stage of the screw enters a decompression section with a large cross-sectional area such that the channel does not completely fill with melt. As a result, the melt pressure drops to essentially atmospheric pressure, and volatiles are released from the exposed surface of the melt

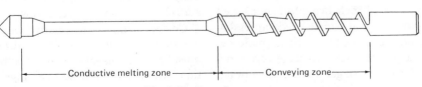

Conductive melting zone — Conveying zone

Fig. 3-27 Low shear screw.

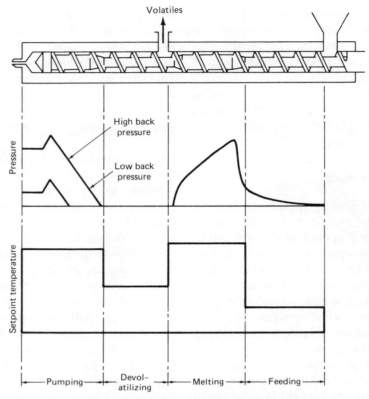

Fig. 3-28   Vented barrel pressure and temperature profiles.

by diffusion. At the end of the second stage, the melt is again compressed to generate the pressure necessary for material to flow through the non-return valve and provide the force necessary for screw retraction.

The molding operation for a vented barrel machine is the same as for a conventional machine. However, since the traditional function of melt devolatilization occurs, the shot preparation process involves factors not otherwise encountered, which must be addressed in order to gain the best possible performance, as described in reference 13.

Two-Stage Screw Designs

Since the vented barrel molding process involves the steps of melt decompression, melt devolatilization, and melt recompression which do not occur in conventional molding, it is not surprising that two-stage screw designs are typically longer than conventional single-stage screws so as to provide necessary

additional physical length. Although two-stage screws can be designed of the same overall length as a typical single-stage screw, some sacrifice in performance is inevitable. Unfortunately, excessively long two-stage screws can experience mechanical difficulties due to buckling instabilities when subjected to the high axial load of injection, which may lead to accelerated wear due to contact loading between the moving screw and the barrel. Consequently, the length of a two-stage screw for injection molding involves a compromise between the higher output of longer screws and their lower mechanical strength. Typically, a 26:1 L/D two-stage screw provides adequate strength, and with recent advances in screw design can provide output rates equivalent to a 20:1 L/D conventional screw.

Correct sizing of the relative lengths of the two stages is critical. The devolatilization zone must be at least as long as the maximum injection stroke, and the second-stage pumping zone must be sufficiently long to provide a filled channel capable of generating the pressure necessary to retract the screw without backfilling the devolatilization zone. Consequently, the available length for the first stage of the two-stage screw is generally somewhat less than that of a comparable single-stage screw, perhaps reduced by as much as 35 percent.

Early two-stage screw designs were often limited by the maximum melting capability of a short conventionally designed first stage. Recent designs, however, have overcome this limitation by incorporating special melting or mixing devices in the first stage, and thus are capable of significantly higher output rates than those obtained with the earlier designs.

The overall performance of the two-stage screw depends upon correctly balancing the conveying characteristics of the two stages. The second stage must be capable of conveying all material delivered by the first. In general, the second stage is designed to provide high melt conveying rates with a high degree of stability. If the output of the first stage exceeds that of the second, the screw channel in the devolatilization zone will become filled, and material will be forced out through the vent port as vent bleed. If this occurs, the filled channel is incapable of devolatilizing the melt. The use of an auxiliary metered starve feeding device enables the output of the first stage to be regulated so as to prevent filling of the decompression section, thus avoiding vent bleed.

Beyond the obvious requirement of maintaining an unfilled screw channel in the devolatilization section, the degree of fill can have profound effects. The degree of fill generally should be as low as possible in practice, for the following reasons:

1. The rate of surface renewal through flow recirculation increases as the degree of fill is reduced. Since the mechanism of melt devolatilization involves evaporation from the free surface of the melt, a high rate of surface renewal improves the devolatilization rate.
2. The degree of fill affects the stability of conveying during screw rotation. A highly

filled channel is sensitive to perturbations in the flow that may induce vent bleed.
3. Excessive melt expansion and foaming in the decompression zone can result in channel fill. The amount of tolerable expansion is inversely proportional to the initial degree of fill. A low degree of fill offers flexibility.
4. Backflow of material in the devolatilization zone during the injection stroke, due to barrel drag, can result in channel filling due to accumulation of this material ahead of the end of the first stage. A low initial degree of channel fill is an advantage in minimizing this effect.

Effects of Process Parameters

Although the vented barrel injection molding process is different from conventional injection molding, many processing considerations are common to both. Where differences do occur, they do not cause any particular difficulty in setup and operation, provided a systematic approach is maintained. The effects of certain key process parameters are considered in the following paragraphs.

Barrel Temperature Profiles. In general, barrel temperature setpoints should be chosen to reflect the process functions occurring in the screw channel at each barrel zone. As shown in Fig. 3-28, four basic temperature zones correspond to the four major functional zones of the screw: feeding, melting, devolatilizing, and pumping. If a particular machine has more than four barrel temperature zones, then intermediate temperatures will normally fall between those of the neighboring zones.

Feed zone temperature can be critical if an auxiliary feeder is not being used. Often, feed zone temperature can be used to change the specific conveying rate (i.e., mass flow rate per screw revolution) to alter the degree of fill in the devolatilization section. If an auxiliary feeder is used, however, feed zone temperature is less critical.

The melting zone temperature should be set to provide complete melting at exit from the first stage with a sufficiently high melt temperature to ensure adequate devolatilization. Often, this may be accomplished at a significantly lower melt temperature than that required for the molding operation. Excessively high melting zone temperatures should be avoided, since material degradation or excessive melt foaming may result.

Since the function of the devolatilization zone is to provide residence to enable volatiles to escape from the recirculating melt, a high heat input is not necessary. Consequently, the barrel temperature should be set at a level sufficient only to maintain the desired melt temperature, this being generally a lower setpoint than in the melting zone.

The final pumping zone temperature is selected to adjust the final melt temperature to that required for molding, and to provide a lower melt viscosity to reduce the pressure drop through the non-return valve to avoid any loss in pumping capacity.

Screw Speed. Since devolatilization is a rate-sensitive diffusion process, a long devolatilization time ensures a high reduction in volatiles. Consequently, unless an auxiliary starve feeder is used, the slowest screw speed that maintains an adequate throughput, consistent with cycle time requirements, should be used. This will provide the slowest transition of material through the devolatilization zone, and hence the greatest devolatilization time. However, when an auxiliary feeder is used, higher screw speeds may be advantageous:

1. Since higher shear rates, hence higher levels of viscous dissipation, occur in the melting zone, higher melting rates can be generated.
2. When extremely low melt viscosity prevents the second stage from generating sufficient pressure to retract the screw without material backup into the devolatilization zone, a higher screw speed can provide a higher drag flow component to counteract the reverse pressure flow.
3. Since the conveying rate in the devolatilization section is a product of screw speed and degree of fill, when the rate is con-

trolled externally by the auxiliary feeder a higher screw speed will reduce the degree of fill, and hence provide better devolatilization.

Back Pressure. In conventional molding, application of back pressure is used to improve the melting characteristics of an otherwise marginally performing screw. However, as shown in Fig. 3-28, the first stage of a two-stage screw is hydraulically isolated from the second stage by the unfilled devolatilization zone. Consequently, back pressure cannot be used to affect melting.

Applying back pressure affects the second zone only, and serves to increase the reverse pressure flow component. This will necessitate a longer filled length of the second stage to produce adequate conveying, and thus the length of unfilled channel will be reduced and devolatilization impaired. In an extreme case, back filling can progress to the vent port, and vent bleed will occur.

The only practical advantage of back pressure lies in the additional mixing it induces in the second stage. In rare instances, this additional mixing may be advantageous. However, the additional length of a two-stage screw is almost always sufficient to ensure adequate mixing without application of back pressure.

Residence Time. Certain polymers, notably polycarbonate and thermoplastic polyesters, are hydrolytically degradable and may suffer undesirable depolymerization effects due to chemical reaction of moisture with the polymer prior to devolatilization. Consequently, the residence time of material in the first stage of the screw should be minimized, and, in practice, this implies that a high throughput rate is required. Average residence time is long for extended cycle times and small shot utilization. Consequently, care is necessary in correctly sizing the injection unit for the application. In cases where the potential for significant hydrolytic degradation exists, process conditions may be altered to compensate, for example, by reducing melt temperature in the melting and devolatilization zones.

ENERGY CONSIDERATIONS

Injection molding is an energy-intensive process, since the central operation is the conversion of the feedstock from the solid phase to a physically and thermally homogeneous melt followed by a forming operation and a subsequent resolidification of the polymer. The phase change from solid to melt requires an input of thermal energy from two main sources, the heated barrel and the drive mechanism; a second energy input is then required for the forming process; and, finally, energy must be removed in the resolidification process.

A certain minimum energy input is inevitable and is determined by the thermodynamic properties of the polymer being processed. From reference 14, this minimum energy can be defined as:

$$E_{\min} = M \int_{T_s}^{T_M} Cp \cdot dT + \int_0^{t_i} p \cdot \dot{v} \, dt \quad (3\text{--}14)$$

where Cp = specific heat, M = mass of shot, T_M = required melt temperature, T = temperature, $\dot{v}$ = volumetric injection rate, and other symbols are as previously defined.

The first term on the right-hand side of equation (3–14) is the energy required for shot preparation, and the second is the energy required for forming. Typically, in practice, the energy required for shot preparation is perhaps an order of magnitude greater than that required for forming.

An overall process efficiency for injection molding can therefore be defined as:

$$\eta = E_{\min}/E_A \quad (3\text{--}15)$$

where E_A = actual total energy supplied per cycle.

On this basis, typical efficiencies of machines now in use range from 10 to 25 percent. Efficiencies of this order indicate significant energy losses in practice. However, since injection molding machines are required to provide useful motion, such as clamp opening and closing, product ejection, sprue break, etc.,

a certain expenditure of energy is necessary beyond the theoretical minimum for shot preparation and forming. A major design objective, therefore, is to produce required motions with minimum energy expenditure. In contrast to injection molding, process efficiency for commercial single-screw extruders typically ranges from 35 to 75 percent, as indicated in reference 15.

Since shot preparation involves the larger component of the theoretical minimum process energy, it is useful to define a plasticating efficiency as:

$$\eta_p = \left(\int_{T_s}^{T_M} Cp \cdot dT \right) \bigg/ E_p \quad (3\text{--}16)$$

where E_p = actual plasticating energy supplied per cycle.

Plasticating efficiency of standard injection units under typical molding conditions has been observed in the range of 25 to 50 percent. However, significantly higher efficiency is possible with injection units designed to minimize energy losses and equipped with high performance screws.

An energy balance approach is particularly useful in the study of injection unit performance. As shown in reference 14, this approach can be used to determine the magnitude of energy losses at different points in the system; for example, feed throat cooling losses, drive motor return line losses, inherent back pressure losses, mechanical friction losses, thermal losses from the barrel, and drive train losses. In particular, application of the energy balance approach to each barrel zone provides particularly useful information concerning screw performance. Barrel insulation has considerable potential for minimizing thermal losses, but the insulation should not be used unless screw design is adequate. If, because of incorrect screw design, areas of the barrel experience a net heat transfer from the polymers, the temperature of the barrel in those areas will increase until an equilibrium is achieved in which thermal conduction to neighboring barrel zones is equal to the heat transfer from the polymer. This heat can produce undesirable effects in plasticating, and in particular, may lead to thermal degradation of the polymer.

In more general terms, particular care should be taken in selecting mechanical components that will operate at high efficiency in the given application. This is particularly true in the case of the drive mechanism. An oversized drive train will tend to operate at lower efficiency than an optimally sized one, since higher fixed losses are inevitable. Experimental observations indicate that the efficiency of a typical hydraulic drive system can drop from 80 percent in a high loading condition to 50 percent in a low load condition.

High performance screw designs offer significant energy-saving potential in addition to improved product quality. Often, improved melt mixing affects product quality more than the average temperature of the final melt, making it possible to save energy by reducing melt temperature. This provides the additional advantage that mold cooling time can be reduced, thereby increasing productivity and also reducing the contribution of fixed losses, such as thermal convective losses, to overall energy losses.

NOTATION

Cp	Specific heat
Cp_s	Specific heat of solid polymer
D	Screw diameter
E_A	Actual total energy supplied per cycle
E_{min}	Minimum theoretical energy utilized per cycle
E_p	Actual plasticating energy supplied per cycle
H	Solid bed height
h	Local height of the channel
k_m	Melt thermal conductivity
L	Flighted length of screw
N	Screw rotational speed
p	Pressure
Q	Volumetric output rate
S_i	Injection stroke
T	Temperature
T_b	Barrel temperature
T_M	Required melt temperature
T_m	Polymer melting temperature
t	Time coordinate
t_i	Injection time
t_s	Soak time
V	Apparent circumferential rotational velocity of barrel relative to screw
V_i	Injection velocity
V_{ix}	Cross channel injection velocity component
V_{iz}	Up channel injection velocity component
V_r	Screw retraction velocity
V_x	Cross channel rotational velocity component
V_z	Down channel rotational velocity component
$\dot{v}$	Volumetric injection rate
W	Channel width
X	Solid bed width
X_e	Solid bed width in continuous extrusion
X_i	Solid bed width at end of injection
X_0	Initial solid bed width
x	Cross channel coordinate
z	Down channel coordinate
β	Empirical constant
δ	Film thickness
δ_0	Initial film thickness
η	Process efficiency
η_p	Plasticating efficiency
θ	Screw helix angle
λ	Latent heat of fusion
μ	Newtonian viscosity
ρ_m	Melt density
ρ_s	Solid density
Φ	Melting parameter
ϕ	Angle of movement of outer surface of solid plug
ω	Rate of melting

REFERENCES

1. Nunn, R. E., "Screw Plasticating in the Injection Moulding of Thermoplastics," Ph.D. thesis, University of London, 1975.
2. Darnell, W. H., and Mol, E. A. J., "Solids Conveying in Extruders," *S.P.E.J.*, *12*, 20 (April 1956).
3. Kruder, G. A., and Nunn, R. E., "Applying Basic Solids Conveying Measurements to Design and Operation of Single Screw Extruders," *Tech. Papers*, S.P.E. ANTEC, *26*, 136 (1980).
4. Tadmor, Z., and Klein, I., *Engineering Principles of*

Plasticating Extrusion, Van Nostrand Reinhold, New York, 1970.

5. Edmondson, I. R., and Fenner, R. T., "Melting of Thermoplastics in Single Screw Extruders," *Polymer, 16,* 49 (1975).

6. Bernhardt, E. C., Ed., *Processing of Thermoplastic Materials,* Van Nostrand Reinhold, New York, 1959.

7. Fenner, R. T., *Extruder Screw Design,* Iliffe, London, 1970.

8. Fenner, R. T., Cox, A. P. D., and Isherwood, D. P., "Surging in Screw Extruders," *Tech. Papers,* S.P.E. ANTEC, *24,* 494 (1978).

9. Gudermuth, C. S., and Bishop, T. G., "Consider Special Screws for Injection Machines," *Plast. Eng., 30,* 46 (May 1974).

10. Nunn, R. E., Takashima, S., and Kruder, G. A., "New Injection Screw Improves Melt Quality at High-Output Rates," *Mod. Plast., 58,* 74 (April 1981).

11. Nunn, R. E., and Bell, J., "Application of Double Wave Screws in Injection Molding," *Tech. Papers,* S.P.E. Injection Molding RETEC, 23 (1981).

12. U.S. Patent 4,299,792, "Injection Molding Process Utilizing Low Shear Screw" (1981).

13. Nunn, R. E., "Vented Barrel Injection Molding has Come of Age," *Plast. Eng., 36,* 35 (February 1980).

14. Nunn, R. E., and Ackerman, K., "Energy Efficiency in Injection Molding," *Tech. Papers,* S.P.E. ANTEC, *27,* 786 (1981).

15. Kruder, G. A., and Nunn, R. E., "Extruder Efficiency: How you Measure it—How you get it," *Plast. Eng. 37,* 20 (June 1981).

Chapter 4
The Injection Molding Machine Operation

Joseph B. Dym, P.E.

INTRODUCTION

The instructions presented in this chapter are intended to develop a training program in steps conducive to easy learning, which over time will result in full knowledge of the molding operation. The author has used this approach for many years, and it has been instrumental in developing qualified personnel for the plastic molding establishment. Many courses and seminars are conducted in injection molding, but most of them, perhaps because of time limitations, do not cover the detailed information required for the most Productive and economical operation.

The program outlined here provides instruction for performance that can be made to fit any time span, in order to suit individual abilities to absorb information while actively engaged in learning by doing.

Suggestions are included for the substitution of calculated values for those obtained by the cut-and-try method, in the interest of conserving time of personnel and minimizing the loss of material. The main object of the instructions is to give each worker in the injection molding operation a good understanding of every element that goes into the operation; the worker, in turn, having gained the needed knowledge, should take full advantage of it byputting it to constructive and productive use.

Under practical operating conditions, learning of the injection molding process takes place in stages:

- The first stage covers the running of an injection molding machine.
- Stage two involves setting molding conditions to a prescribed number of parameters for a specific plastic material and a specific mold that will produce acceptable parts.
- The final stage is devoted to problem solving and fine tuning of the operation, which will lead to favorable productivity and part quality.

FIRST STAGE: RUNNING AN INJECTION MACHINE

In the injection molding operation a granular plastic material is softened by heat so that it will flow under the influence of pressure and can be delivered to a tightly closed mold where it is held for a specified time. The mold is maintained at a temperature that will permit the injected material to become solid in a short time. After a prescribed time interval, the mold is opened and the injected material released as a finished product.

If we can clearly picture the basic process, the more involved actual operations will be easy to understand and remember. The general description just given applies to thermoplastic as well as thermoset materials (see Chapters 18 through 27 for details on materials).

Now let's describe the operation of a machine in greater detail, starting with an *injection screw machine* for thermoplastics arranged for semi-automatic operation. (See Chapters 2 and 3 for views and details on the injection molding machine.) Hard plastic granules are delivered to a hopper, from which they are fed through a throat onto a rotating screw. The screw moves and compresses the material through a heated chamber where the granules soften to such a degree that they can become fluid and can be delivered to a section of the heating chamber known as the measuring chamber. In addition to turning, the screw will on proper signal stop its rotation and move in a forward or reverse direction as a plunger. When enough material for the mold cavity is supplied to the measuring chamber, as determined by the controlled distance of the backward-moving screw, an electrical command is given to it to act as a plunger and inject the fluid material into the tightly closed mold. The mold is maintained at a relatively cool temperature that will cause the plastic to become rigid after a set length of "curing" time. At the end of cure time the mold opens, and at the same time the operator causes the gate to open, the parts are ejected from the mold (sometimes into the hands of the operator), the mold is checked to see that it is fully clear of plastic, the gate is closed again, and a new cycle is started.

The molded parts are briefly checked for quality and consistency in appearance and disposed of either for storage or for other operations, such as gauging, auxiliary operations (hot stamping, etc.), or packaging. Normally the work at the press is planned so that the attendant is kept occupied during the cycle; in this way a consistent results in part quality, cycle time, and safety of operation can be anticipated. The operation of cycles becomes repetitive, and the attendant should exert every effort to have the motions organized and coordinated so that variables will not be introduced that could influence consistency of quality and uniformity in the cycle. The best re-sults are obtained when all elements in a cycle are repeated uniformly from shot to shot.

In spite of all precautions taken by the operator and setup person with respect to all machine parameters, there is occasionally some need to interrupt the cycle. While an open gate will prevent the starting of a cycle, there may be other reasons for stopping the motor and pumps, disconnecting electrical units on the chamber from the power, and so on. Therefore, the operator should know how to activate some of the switches of the control panel. Only those switches that should be activated in an emergency by the machine operator will be described here.

The "emergency" button when activated and held in the proper position will cause the clamp to return to the starting position. Opening the gate to correct the problem that necessitated the use of the "emergency" button should reset the machine to the operating condition that existed before the interruption. Some machine may require the pushing of a "cycle reset" button before normal operation can be restarted. If the correction, for example, requires removing an obstruction between mold halves or correction of a minor mold malfunction that can be accomplished in about two minutes, it should be possible to continue running the machine in a normal manner by simply closing the gate and, if appropriate, pushing the "cycle reset" button.

A sudden oil leak in the hydraulic system would call for pushing the "motor stop" button in order to eliminate pressure in the hydraulic system and thus keep the loss of oil to a minimum and its spread to the shop floor, where it may cause safety hazards. The "motor stop" may also have to be activated if unusual noises develop on the injection end of the machine, indicating some problems with the running mechanisms.

A sizable leak of plastic material between machine nozzle and mold or anywhere on the front portion of the injection cylinder would indicate an undesirable condition that could lead to variation in the feed to the mold, thus causing defective parts. The reasons for leakages of plastic material must be determined and eliminated. The machine operator would turn the selector switch of the extruder to "extruder-off," stopping all action and leaving it to the supervisor to take corrective measures.

Any change of a button position on the control panel will bring about a modification in the sequence of operations in the electrical circuit, and consequently in the hydraulic circuit, that cause an orderly movement of components during a cycle. Definite information cannot be provided here for restoring machine operation after an interruption because considerable variation exists among makes and models of machines. This information is best obtained from supervising personnel, who have access to instructional manuals and wiring diagrams for each machine. When given an explanation about restarting a machine after a specified interruption, the operator should make notes and save them for future reference. One should not rely on memory alone for such vital instructions.

We can better appreciate the above observations by considering a condensed version of sequences in the molding operation. Early in the chapter the principle of molding was described with reference to the plastic material and its movement in and out of the mold. Here we will concern ourselves with switches and timers that accurately control the sequence of every action performed by the machine.

The Sequence in a Cycle

(a) Closing the gates actuates a limit switch that in turn brings about rapid forward movement of the clamp-ram.

(b) When the clamp reaches a position a couple of inches before closing, it activates another limit switch which causes clamp slowdown; and finally, at a distance of about $\frac{1}{16}$ before tight closing, a third limit switch is activated that signals the high pressure (2000 to 3000 psi) needed to squeeze the mold shut.

(c) When pressure is fully built behind the clamp, a pressure switch closes its contacts and initiates the following: the nozzle valve (if used) is opened, the "injection high pressure timer" is started, and injection high pressure

movement of the extruder-plunger action is instigated.

(d) When the "injection high pressure timer" times out, it initiates the "injection overall timer," which for several seconds maintains pressure on the material in the cavity.

(e) When the "injection overall timer" times out, the melt "decompression timer" starts. When melt decompression times out, the nozzle valve (if used) closes, the extruder starts turning, preparing the plastic for the following shot, and the clamp high pressure drops to "low hold."

(f) While turning and feeding the plastic into the shot chamber, the extruder moves backward (to provide space for the shot) until it contacts a limit switch that causes it to stop.

(g) The "overall timer" or "clamp timer" times out, bringing about slow opening of the clamp.

(h) The opening clamp activates a limit switch that causes its rapid reverse movement until another limit switch is contacted that causes slowing down of the clamp travel to the point at which the final limit switch contact provides the "stop" for the open position.

(i) A clamp open timer is provided that either sets a time for removal of parts from the mold, or in case of automatic (continuous) molding can be energized by the reverse "stop" limit switch to perform the same electrical function as is performed by manual gate closing and its activation of the limit switch by the gate.

All the limit switches and timers carry out their commands in an orderly manner, and any interference with this systematic arrangement by pushing a control button will throw the plan out of order. There are certain steps required to restore the orderly working of the machine, but unfortunately these steps vary almost from machine to machine. When we recognize that each timer alone can have three modes of operation for resetting to zero upon timing out for the following restart, we realize that extreme care must be exercised in restarting a machine after interruption. Close attention to the details of machine operation is very much in order here.

Repeating the cycle in a consistent manner is obviously the major responsibility of the machine operator. Also, certain observations must be made that will lead to a better understanding of the process, and will aid the worker's advancement in the field.

Certain details require attention:

(a) A machine in good working order should produce no unusual noises. It should close the mold by rapid movement of the ram, slow down as the mold faces comes within $\frac{1}{16}$ of each other, and finally shut the mold by squeezing action under high pressure (no banging). During mold opening, about the first half inch should be done slowly, followed by rapid movement up to the distance at which ejection begins and then slowing down to a stop at the open position.

(b) Tie rods of the machine and leader pins of the mold should be adequately lubricated to prevent excessive wear and associated problems.

(c) Temperature of the hydraulic oil should be within limits of the gauge mounted on the machine; overheated oil will bring about higher leakage in hydraulic pumps and valves, thereby making it difficult to maintain the required pressure for injection and clamping cylinders. Maintenance of constant pressure on the components is an important factor in producing acceptable parts.

(d) The travel of the extruder screw should repeat its distance for volume, and its forward position for filling of the mold. If screw travel is not in the normal forward position, not all the required volume of material is injected into the mold, with the result that parts are not dense, excessive shrinkage takes place, and the surface is not smooth. An increase in the back-travel of the screw may cause an excess of material to be delivered and may overpack and flash the parts, causing enlarged dimensions, waste material, and possibly a deflashing operation. Adequacy of the supply of material in the hopper should be checked when it is expected to reach a low level.

(e) The injection high pressure, which can be read by depressing a button (at the hydraulic panel for injection pressure) during the injection time, should be checked for deviation from the required setup reading. Uniform pressure on the plastic in the mold is a very important determinant of product quality.

(f) Temperature settings for the injection cylinder at each zone should be recorded and checked at intervals of about 4 hours to see that unexpected variations are not introduced. A plastic material is not a pure chemical of certain description, but encompasses all kinds of additives (colorants, plasticizers for flow, flame retardants, UV stabilizers, anti-oxidants, etc.); so the heating temperature must be confined to limits, not just because of the basic plastic but also in the interests of protecting the additives. Excessive temperature and/or prolonged exposure to that of the normal melt heat can cause gassing, degradation of the material, and change in flow properties, all of which can have a most undesirable effect on parts.

Automatic operation of the thermoplastic injection screw machine is in every respect the same as that outlined for the semi-auto method except that the "stop" limit switch for the clamp-ram will initiate the "clamp open" timer, which in turn will restart the cycle while the gate stays closed.

Molds that have been designed and tested for automatic operation require only intermittent observation of behavior to ensure that everything is working in an approved manner.

The details requiring attention in the semi-auto operation also apply to this mode, but the operator in this case will be concerned with checking product quality, ensuring an adequate supply of granular plastic material, and removal of the molded parts to a designated station, in addition to these details. The duties of an operator can be to perform auxiliary operations, if necessary, at a single press or to attend to a number of cavities and required checking for quality. An operator can attend 4 to 16 presses.

A slight modification in the way a mold functions can lead to automatic operation and thus improve productivity. Automatically operated molds usually result in better and more consistent quality and fewer rejects of parts. In most cases mold life is also enhanced.

The thermoset injection machine is, from the operator's point of view, very similar to a thermoplastic machine. There are, however, some additional points of concern:

(a) The material content in the hopper should not fall below the half-point so that there is always a sufficient weight of material to exert a pressure that will ensure good flow to the throat.
(b) The temperature in the cylinder is critical. It must be observed that no increase in the setting occurs that could cause hardening of the plastic in the chamber, since this could cause the operation to be interrupted.
(c) The nozzle of the cylinder must be maintained at a low temperature to prevent hardening of the material in it. This is usually accomplished by retracting the nozzle from the sprue bushing of the mold. The mold is usually at temperatures of 300°F and up, depending on the material.

The nozzle can also be maintained at a low temperature by incorporating a circulating coolant in it. Whatever the method employed, it must be seen that the material in the nozzle is maintained in soft condition to ensure free flow for each shot.

Handling Plastic Materials

A machine attendant may be involved in occasionally supplying plastic material to the hopper. However, in most cases he or she will deal with defective parts, runners, and sprues to be reground for future use. It must be recognized that plastic materials are chemicals that can be easily contaminated unless proper precautions are taken to assure chemical cleanliness. The following is an explanation how to keep plastic materials protected from contamination.

In addition to machine variables, there is one major source of problems in controlling quality plastic parts, namely, the cleaniness and conditioning of the material as it is placed in the hopper. If we keep the material free of contamination—that is, free of foreign matter as well as free of other plastic—our chances of making good products are enhanced. It takes only a few parts per million of contamination to affect the properties of some materials. The way contamination will influence properties is not known without extensive research. Even when materials are intentionally combined, the component ingredients lose some of their original characteristics while gaining some new ones. Take, for example, ABS, an alloy of acrylic, butadiene (synthetic rubber), and styrene. While ABS itself has desirable properties, the styrene part in it has lost its rigidity and clarity, the butadiene has lost chemical resistance, and the acrylic has lost resistance to ultraviolet rays and elements of nature. The combination, however, has toughness, impact resistance, and good moldability, entitling it to a vital place in the plastic family. (See Chapter 19 for deails on ABS.)

It must be remembered that the ABS combination is achieved under predetermined favorable conditions. Accidentally contaminated materials may not look objectionable, but properties may be adversely affected. Think for a minute that one cubic foot of material contains about two million small cubes of the material, and it only takes ten to twenty similar cubes of another material to cause contamination. To make matters still worse, these small cubes in many instances cannot be distinguished from each other, nor can they be uncovered in the molded part if it happens to be opaque.

A greater variety of materials will be used in the future, and the products that they will be applied to will be more intricate and functionally more important. Thus it behooves us to seek immediately a foolproof manner for handling the materials so that all dangers of contamination are eliminated, and the chances of weakened parts are avoided. Above all, care, and more care, will be needed. (See Chapter 15 on material handling and Chapter 17 on size reduction/granulating.)

STAGE TWO: PARAMETER SETTING AND STARTING A JOB

Principles of Machine Operation

During the process of converting a plastic raw material into a finished molded product, three basic elements in modling—time, temperature, and pressure—must be correlated in a way that will produce a part with anticipated properties. Most deviations in product quality can be traced to variations from established values in time, temperature, or pressure. Changes in any of these individually or in combination spell problems in product properties and performance characteristics.

Time involves these elements: time beginning with material entering the heating cylinder until injected into the mold (also called residence time in the cylinder); time of injection into the mold; time of maintaining pressure in the mold cavity; time of solidification or cure time; press open time; press opening and closing time; time of part ejection in relation to mold opening time.

Temperature is affected by: temperature of material entering the hopper; throat temperature; heat contributed by screw compression and speed of rotation; heat absorbed from the cylinder and the setting arrangement of pyrometers in the heat zones; averaging of heat by continuous mixing and homogenizing up to injection time; mold temperature readings; flow control of coolant in mold passages for

desired temperatures; temperature of the environment.

Pressures that require consideration are: injection high pressure, or the pressure needed to fill cavities to proper part density; hold pressure, or the pressure that is maintained on material during solidification and prevents backflow into the nozzle area; back pressure, which influences mixing and feeding of material into the measuring chamber; clamp pressure, which indicates effective mold closing.

Principle of the Molding Operation

The molding machine has the function of injection molten plastic material into a tightly closed mold where the shape of a product is formed. The mold is kept closed for a specified time, or cure time, during which the fluid material becomes solid and rigid. A coolant circulates through passages in the mold, so that heat from the fluid plastic is transferred to the mold and from there to the circulating fluid, a process that accelerates the curing or solidification of the part. At the end of curing time the mold is opened, and the parts are ejected ready for packaging or other operations if required. At this point, a new cycle begins. Now, let us see in detail how the machine carries out its job. (See Fig. 4-1.)

The cavity half of the mold is attached to the stationary platen (7), where it is centered by means of the locating ring. The core half of the mold is mounted on the moving platen (8). When the press gate in front of the mold is closed, a hydraulic circuit is activated that causes the main ram (9) to move forward at a fast rate. This movement is brought about by supplying a large volume of oil from pumps directly into the booster ram (10). This oil exerts a pressure on the body of the main ram (9), causing it to slide over the booster ram (10) and move forward until at a designated position the moving main ram actuates a limit switch that sends a signal to the hydraulic circuit ordering the high-volume pump to dump its oil at low pressure into the prefill tank (11), while at the same time the low-volume pump keeps supplying its oil to the

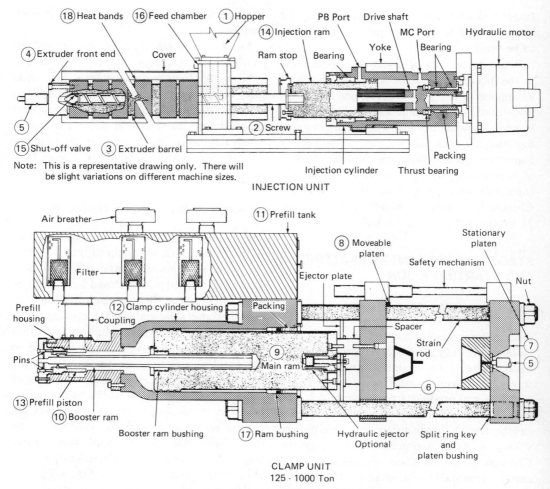

Note: This is a representative drawing only. There will be slight variations on different machine sizes.

INJECTION UNIT

CLAMP UNIT
125 - 1000 Ton

Fig. 4-1 Crossection of HPM-machines.

booster ram (10), thus causing slow main ram movement.

The pressure at which this slow movement takes place is controlled by a mold protection valve. The pressure of this valve is set at a low figure (around 200 psi) so that the pressure exerted on mold halves, if something is caught between them, will be low and not cause damage to the mold. The space vacated in the clamp cylinder housing (12) is filled with oil by gravity from the prefill tank (11) through the opening of the prefill piston (13) in its retracted position. The mold halves make contact at the low speed of the ram movement, and at this point another limit switch closes the prefill piston (13) and activates a high pressure pump (2000 to 3000 psi), which will apply its full pressure over the main ram (9), holding

the mold halves tight and resisting opening when plastic material is injected into the mold at pressures up to 20,000 psi. This second limit switch also initiates the movement of the injection ram (14), which injects the plastic into the mold.

Injection is carried out by the front of the screw (2), which contains a shut-off valve (15) that prevents any possible backflow of the fluid plastic. The screw is firmly attached to the injecting ram (14), whose movement takes place at a fast rate (usually in about 1 to 2 seconds for the full shot capacity).

The injection time is controlled by a timer (the injection high timer), and the ability to respond to the timer setting is determined by the pressure of injection and the fluidity of the material.

The speed of injection can be varied by means of a flow control valve that can bypass a desired amount of the pump oil and thereby reduce the speed of injection. This valve usually has ten bypassing positions, thus providing a considerable degree of injection speed variation.

Once the shot is completed, the high-volume oil injection pump is ordered by a signal from the timer to dump its oil into the prefill tank (11) at low pressure; at the same time, a low-volume pump (hold pump) maintains pressure on the material in the cavity until the gate through which the material was fed freezes and prevents backflow to the cylinder. (Backflow can be caused by the pressure within the cavity if the feed gate is open). The hold pump duration is set by the injection hold timer. At the expiration of this timer, the screw starts rotating, picks up material from the throat in the cooled chamber (16), and moves, compresses, and shears it in the extruder chamber (3), where it absorbs heat and liquefies before entering the measuring portion of the injection chamber.

The extruder barrel is heated by strip heater bands (18). A group of heaters is divided into zones, with each zone having a pyrometer for controlling the temperature. There are usually three or four zones on the extruder chamber. The extruder work—represented by feeding, compressing, and shearing of the material—shows up as partial heat induced in the plastic. The heat needed to fluidize the plastic is derived partially from the work of the screw, the balance coming from the strip heaters of the extruder chamber.

As the material comes off the extruder screw (2), it creates pressure on the front face of the screw, causing it to retract so that a space is created for the incoming material required for the shot. This backward movement of the screw makes it necessary to push oil out from behind the injecting ram (14).

The displaced oil passes through a controlled valve, which can be adjusted to provide varying degrees of resistance for the screw's backward travel. This resistance, known as the back pressure, is utilized to provide good mixing and homogenizing of the material in the injection chamber. When a slight temperature adjustment is needed for the material that is to be injected, a small increase in the back pressure will accomplish this requirement. The duration of screw rotation is determined by a limit switch, which is activated by the backward-moving screw at a position where the necessary volume of material required for the shot has been reached. The screw limit switch may also start a melt decompress timer, which will cause continued limited backward movement of the screw. This additional screw movement creates a space in front of the screw that permits the built-up pressure to decrease enough that, when the mold opens, no drooling of plastic takes place.

The final stop of the screw movement usually coincides with the expiration of the cure time as determined by the corresponding cure timer. On a signal from the cure timer, the press starts opening the mold. This is accomplished by feeding oil from a small-volume pump into the space behind the ram bushing (17). This starts the press to open slowly; then another limit switch is actuated by the ram movement, which orders a large volume of oil to be fed into the space so as to speed up the press opening time. Since the area between the clamp cylinder (12) and the main ram (9) is small, and this area multiplied by the pump psi gives the force for mold opening, this force is small in comparison with the clamping tonnage (usually around 5 percent of the clamping tonnage). Before stripping (ejecting) starts, the ram is slowed down by actuating still another limit switch for gentle action of the knockout pins, to prevent the pins from punching through the parts while pushing them off the cores. With hydraulic ejection the slowdown can be so delayed that no banging takes place when the ram returns to the starting position. After ejection, the parts are removed from the press, and the cycle starts all over again. All limit switches have numbers that tie them to specific actions.

Plastic Processing Data

The processing data for a material family—and specifically for a grade within a family as supplied by the producer of the raw material—are of utmost importance to the setup person. These data provide guidelines for setting parameters that will safeguard the properties described in the data sheet for the particular grade.

Plastics are man-made materials known by the general name polymers. Each of them has a different prescription for processing or converting it into a finished product. (See Fig. 4-2.)

Polymers are created from atoms that are joined to form a molecule. The atom is the smallest element in a chemical compound. The molecules resulting from joined atoms are caused to combine with each other to create a long chain (the molecular chain), in a process called polymerization. These chains fold, intertwine with each other, and are held together by forces that exist between them. The molecular chain mixture becomes a plastic raw material called a polymer.

This oversimplified definition is not intended to mislead anyone into believing that the science of polymers can be learned easily. It is one of the most complex branches of chemistry, and a great deal of skill is required to master it. (See Chapters 18, 26, and 27.)

Fundamentally these materials are chemicals, but the molder hardly ever works with the pure polymer because modifications are needed to make conversion of these materials into useful products practical. A variety of additives are compounded into the materials before they are marketed.

Individual additives are essential for: reduction in heat sensitivity during molding; stability during exposure to ultraviolet light; color; anti-oxidization during exposure to the air; lubrication during molding; reduction of flam-

Material _____

Mold Shrinkage _____ Specific Gravity _____

Coefficient of Thermal Expansion, in./in./°F _____ Heat Deflection Temp. @ 264 psi _____

Water Absorption % (24 hrs. @ 73° F) _____ Drying Temp. _____

Melt Temp. _____ Mold Temp. _____

Specific Heat, btu/lb/°F _____ Injection PSI on Material _____

Back Pressure PSI on Material _____ Screw RPM _____

Screw Torque _____ Runners _____

Vents (depth) _____ Gates _____

Land _____ Nozzle _____

Note:

Fig. 4-2 Material processing data.

mability; acting as reinforcement to improve properties; serving as extenders to reduce the cost of material; and many others uses, as requirements may demand. In working with materials of such complex composition, it is imperative to follow the exact specifications for processing outlined by the manufacturer for each one of the grades.

Variations in processing requirements not only exist from one type of plastic material to another, but also may prevail from one grade to another within a specific family type. A good example is the various grades of ABS, in which the prescription for processing changes with most of the grades. Thus it should not be taken for granted that setup conditions for similar materials apply to any grade unless careful investigation verifies the original assumption. This precaution should not be neglected if quality problems are to be avoided.

Certain terms that are frequently used in connection with plastics need explanation. The word *polymer* means consisting of many mers. Mers are the repeating units (in a molecular chain) that under time, temperature, and pressure are caused to combine with each other to form a molecular chain. The plastic material consists of many such mers; hence the name polymer. A *monomer* (mer) is a chemical compound in itself and also a constituent of a polymer (e.g., styrene in polystyrene, ethylene in polyethylene). There are, however, many plastic materials that are not involved in intermediate steps of forming mers, but are created from molecules bonded together to form a polymer. *Copolymers* are materials in which mers (molecules) of two different materials are caused to combine, with the formation of a new material. *Homopolymers* are polymers that contain a *single* repeating mer, such as polyethylene, in which ethylene is the mer.

Crystalline and *amorphous* structures of plastics exist. Thermoplastics at room temperature have either a crystalline or an amorphous structure. In the crystalline arrangement the molecules appear orderly and compact as if they were stacked behind each other, and this structure prevails throughout the length of the molecular chains. The amorphous structure, however, follows no particular pattern of organization and could be compared to felt or fiberglass filters in promiscuous arrangement of fibers, except that the molecular chains may be miles long. (Details are given in Chapter 26.)

It should be noted that at melt temperature all plastics are amorphous. After molding, some of the properties of both crystalline and amorphous structures should be observed.

Examples of crystalline materials are polyethylene, nylon, and polypropylene. Amorphous plastics are polystyrene and polycarbonate.

Crystalline materials have a high shrinkage, with the component in the direction of flow usually greater than that perpendicular to it. When a symmetrical part such as a round cup is fed in its center, the shrinkage is uniform and is usually the average of the two components. Amorphous materials have a low shrinkage, which is the same in all directions.

Crystalline plastics require more heat than amorphous ones to bring them to the desired flow, because of heat of fusion. After the material is brought to melt temperature, additional heat is needed so that it will flow properly. (Heat of fusion is the heat necessary to bring about a change of state; e.g., the heat necessary to melt ice at 32°F to water at the same temperature.

When plastics flow through gates and runners, their molecules tend to be oriented in the direction of flow. A smaller gate area will cause a higher orientation, except that there is a low limit for the gate area for amorphous, heat-sensitive, and long-molecular-chain materials such as polycarbonate. Oriented plastics gain in strength in the direction of flow.

The application of this feature is the living hinge in polypropylene, where the gate opening is .020 in. thick, and the direction of material flow is perpendicular to the hinge action. Under these production conditions the living hinge will not crack.

Shear-Rate-Sensitive and Insensitive Materials. In order to understand the term "shear rate," we have to use our imagination and visualize a plastic flowing through a pipe as consisting of minute layers parallel to each other in the direction of flow. The layer that is next to the wall sticks to it and does not move. The next layer moves and slides over the layer adhering to the wall. The remaining layers move with respect to each other at an increasing rate as the distance from wall to center increases. This imaginary layer movement is known as shearing. The speed of movement of layers in relation to each other is called *shear rate*. The pressure that is exerted on the fluid and brings about the shearing action is called *shear stress*. The relation between shear stress and shear rate is the *viscosity* of the flowing material, and viscosity is also defined as the internal resistance of a material to flow. Increased pressure will increase viscosity, therefore increasing ease of flow.

Shear-rate-sensitive materials respond to pressure by having their molecules readily shifted and aligned with the direction of flow. The molder's concern is with the shear-rate-insensitive plastics, which consist of long-chain molecules (polycarbonate, for example) so intertwined that an increase in shear stress will only cause greater entanglement. The net result is that the viscosity will not change, and the danger exists that the entanglements created in polymerization can be disturbed and the polymer properties damaged. In practical terms, gates cannot be too small, and back pressures should be low, passages of material to the cavity from the cylinder rather large, and speed of screw rotation rather low. (See Chapter 19.)

Melt index, a term used mainly for polyethylene, indicates how much material can be pushed through a set orifice with other conditions controlled. It shows the "flowability" of a material. The higher numbers indicate easier flow of a compound.

Some processing physical properties of plastics are worth noting:

Melt Temperature. Values given provide a range of temperatures within which necessary adjustments can be made in order to get favorable fluidity of material.

Mold temperature is another range of values within which adjustments may be made if pyrometer readings indicate that such a step will improve quality and productivity.

Injection pressure, or, more accurately, the injection pressure needed in the cavity that will produce consistent quality of parts, is a very important processing element. The reading on the gauge of "injection pressure" denotes a pressure that is composed of several increments of pressure drops, such as: pressure loss within the heating cylinder, through the nozzle, through the sprue bushing and runners, through the gate, and then through the cavity, giving finally the pressure required at the end of flow to produce a dense part with a smooth surface. The pressure at the end of the flow in the cavity need only be 2000 psi for many materials, and this value may only be $\frac{1}{5}$ to $\frac{1}{10}$ of the gauge pressure reading, depending on the size of pressure drops that were listed. The most important reading is the one that determines the quality of the part, which is made at the end of the material flow and is in many cases about 2000 psi.

Process control devices are made that limit cavity pressure to a specified predetermined value, and they have proved very successful in minimizing rejects. The consistency of injection pressure in the cavity is an essential element in producing uniform parts. The values shown in processing sheets refer to gauge readings, and are intended to indicate whether or not the material flows easily and is readily compressible.

Back Pressure on Material. This pressure indicates resistance to the backward movement of the screw during preparation for a subsequent shot. This pressure is exerted by the material on the screw while it is being fed into the shot chamber. During rotation of the screw and the material under pressure, thorough mixing of the polymer is achieved, and some temperature increase also results. In dealing with heat-sensitive and shear-rate-insensitive materials, care must be taken to keep this value within prescribed limits.

Screw Torque. There are two torque settings available on the machine. In practice it has been found that the high torque setting is hardly ever changed; the low torque would be applicable only if a highly liquid melt material were molded, requiring high speeds of screw rotation.

Screw RPM. This is related to work input

into a material; higher speeds are applied only when there is insufficient heat to be absorbed from the cylinder for a particular shot. Heat-sensitive and shear-rate-insensitive materials do not tolerate the higher speeds.

Vents. There is a maximum vent depth through which flow of material will not take place. However, this depth will be located away from the gate (at least 90 to 180 degrees).

Type of Nozzle. Two types of nozzles are available: the general purpose and the nylon types. With the advent of screw type injection machines and effective utilization of the melt decompress feature, the drooling present with a general purpose nozzle while molding nylons can be effectively controlled. This is based on the fact that the check-ring shut-off system fits the barrel properly to produce effective suction at the point of the nozzle outlet.

Drying Temperature. Materials that are moisture-sensitive and those that may pick up moisture for some other reason will have to be dried before molding. A drying temperature is used that will permit the removal of moisture without causing the granules to adhere to each other, behavior that could cause bridging over the throat where the screw picks up the material. It is also useful to set the water valve for cooling the throat so that its temperature will not be too low, causing condensation of the plastic, or too high, causing bridging. Attention to the correct setting of the water valve can yield saving in water and heat of plastiction in the chamber. The preferred method of drying is the dehumidifying process where the humidity is removed and dry air is supplied at the specified conditions for each material. Also available are so called vented injection machines that are capable of removing moisture during the processing of the material. A simple test for moisture content has been developed by General Electric and is known as the T.V.I. test.

Mold Shrinkage. These data can be used in checking dimensions of parts, thus giving indirect verification that the setting of all parameters has been properly executed.

Specific Gravity. This value will be used in the data section of this chapter for such purposes as evaluating machine capacity in relation to polystyrene, screw travel, rate of injection, etc.

Purging Information and Precautionary Notes. If purging procedure or shut-down steps or any other precautionary move is indicated for a specific material, that should be suitably indicated under this heading.

Clamping and Moving the Mold

Attaching of molds to platens should be done in a manner that will ensure retention of the mold in position without danger of shifting or loosening. Any change of position of a mold half will place an excessive burden on the leader pins and bushings to keep the halves aligned, thus causing wear on the pins and bushings and in time affecting the quality of the parts being molded.

The conventional method of holding mold halves in place is by employing mold clamps. The platens are tapped for bolts ranging from one-half to one inch in diameter, and the holes are laid out to an SPE standard pattern.

The forces holding a mold in the press have been analyzed and the result is the following table.

BOLT SIZE (IN.)	ENGAGEMENT IN PLATEN (IN.)	SLOT IN CLAMP (IN.)	HOLDING POWER IN CLAMP (LB)	TORQUE WRENCH (IN. LB)
$\frac{1}{2}$	.75 to 1.0	$2\frac{13}{16}$	32	210
$\frac{5}{8}$	$\frac{15}{16}$ to $1\frac{1}{8}$	$3\frac{3}{8}$	45	340
$\frac{5}{8}$	$\frac{15}{16}$ to $1\frac{1}{8}$	5	35	340
$\frac{3}{4}$	$1\frac{1}{8}$ to $1\frac{5}{16}$	$3\frac{3}{8}$	50	450
$\frac{3}{4}$	$1\frac{1}{8}$ to $1\frac{5}{16}$	5	40	450
1	1.5 to 1.75	$5\frac{5}{16}$	80	900

Only forged bolts with a yield strength of 120,000 psi should be used. In order for each clamp to hold with equal force, a torque wrench is indicated.

When the calculations are made for an actual clamping system, the number of clamps should be divisible by 4 since there are four clamping faces. For example, mounting a 300 lb mold with $\frac{1}{2}$-in. bolts would give $\frac{300}{32}$, or 9.37 bolts. To be divisible by 4, 12 clamps, or 3 on each side, are required. In all cases, the clamp surface should be paralled to the clamping slot and platen. The closer the holding bolt is placed to the flange of a mold, the higher is its holding power.

Handling of a mold to the press and removing it to storage is normally done by means of chains or wire rope slings. These auxiliary means for hoisting a weight are treated in technical handbooks under "Crane Chain and Hooks" and under "Strength and Properties of Wire Rope." Additionally, the Federal government's OSHA agency provides information which prescribes certain regulations for weight handling and make the user liable to stiff penalties if they are not followed.

Under average conditions of a molding shop, the responsibility of frequent inspection of the hoisting means as well as the approaching end of their useful life should be assigned to one responsible person. This person should obtain literature from the suppliers of these devices and become familiar with the above described literature and use the information to instruct others in the safe handling of molds. Improperly lifted molds can be a hazard to workers and damage presses in the event they fall. They can also be damaged and rendered unusable.

Note: At this point of instructions it is desirable to become familiar with the hydraulic system of the machine so that the following descriptions will be easier to comprehend.

Guidelines for Molding Parameters

The literature on processing of plastics usually suggest limits within which the controlling instruments should operate, but seldom do we find explanations for the prescriptions.

The setup sheet, which is expected to contain all the needed information for starting a job and getting ready for a production run, is in itself a very useful explanatory tool.

The implementation of this information will be good if there is an understanding all the items on the list and their variables, as well as the factors surrounding them. It will be the aim of this section not only to list every item considered vital to the successful operation but also to provide information that will aid in proper interpretation of such items. The systematic arrangement and listing of the items is the setup record. (Fig. 4-3).

Description of Setup Record. The setup record is made in order to establish the most favroable operating conditions for each mold in a particular press. These favorable conditions pertain to good product quality, minimal rejects, and shortest possible cycles. Once these favorable operating conditions are established and approved by management, they should be faithfully executed. Should any modifications become necessary during some future run, they must be implemented only with the approval of the authorized individual in charge of plant management. In such an event, the setup record should be suitably revised, or an additional one made indicating the reason for modification and the elements affected.

The setup record is a most important document in starting a job. If properly interpreted and precisely carried out, this record should result in the same quality, consistent cycles, and low quantity of rejects every time the job is in operation. For these reasons, it is desirable to describe each column of the record and point out what factors enter into the determination of a particular setting. Thus, those filling out the record and those applying it to the setup will have the same understanding of the information at hand. The goal is to acquaint those involved with the setup and the running of the operation with this descriptive information so that the job is carried out in a standardized manner leading to good performance.

If we consider that one second of machine

Item	Part No.		Mold No.		Mold Type			Pcs./Mold		Mold Drwg. No.	
Material	Pc. Wt.		Shot Wt.		Overall Cycle	Pcs/Hr.		Clamp & Shot Size		Make & No. of Press	

SET / **CLAMP CYCLE**	Clamp Fwd. slow LS-2 Mold Protection		**TIMERS IN SEC.**	Injection High or Total Injection		**MOLD MOUNTING DATA**	Eye Bolts Size & No.		**WATER DATA**	Water Hose Size & No.	
	Pressure Buildup LS-3 Prefill Closed			Injection Low			Mold Shot Height-In.				
	LS-5 Clamp Fast Reverse			Cooling—Cure			Horizontal-in (eye Bolt Side			Water Temp.—Cavity	
	LS-6 Clamp Slow down			Melt—Decompress			Vertical-In.			Water Temp.—Core	
	LS-7 Clamp Reverse stop			Clamp—Open			Spacer size		**TEMP. CAVITY POSITION**	1	
	LS-8 Clamp Overstroke			Air Ejection—on/off			Pull backs or K.O. Rods size & no.			2	
DATA	LS-20 Hyd. Eject		**NOZZLE VALVE**	Front, Zone #1			Mold Weight			3	
	Press Daylight			Middle, Zone #2			Time to place Mold			4	
	Clamp High—Clamp Low			Rear, Zone #3			Sling Type Size & No.		**TEMP. CORE POSITION**	1	
	Clamp Open—Slow/Fast			Rear, Zone #4			Lift Size			2	
	Injection High		**TEMP. CYLINDER, NOZZLE & NOZZLE VALVE**	Time to reach settings		**MOLD PLACEMENT**	Type of Hold Down Clamps			3	
	Injection High-Squeeze			Melt Temp.			Heels			4	
PRESSURES, FEED & SPEEDS	Injection Hold			Nozzle Temp.			No. of Hold Down Clamps		**MISC. COLUMNS AS NEEDED**		
	LS-25 Setting			Nozzle Valve Temp.			Hold Down Clamp Spacing—Cavity				
	Back Pressure			Nozzle V. actuated			Hold Down Clamp Spacing—Core				
	Injection Feed			Throat Temp.			"Screw Jacks" Bottom Supports				
	Cushion		**FIRST SHOT**	First shot feed for runnerless mold			Torque on Clamp Bolts				
	Injection Speed Setting			Mold warm-up time		**REMOVAL**	Core & Cavity Temp.				
							Cover with Moldsaver				
	Screw RPM			Hot runner warm-up time			Time to Remove Mold				

Fig. 4-3   Setup record for injection molding.

time alone is worth between three-fourths and one and one-half cents (depending on machine size), and each machine produces at least two shots per minute, or 720,000 shots a year, we can see that a single second wasted during one shot can amount to about $7,200 per year. With these kinds of values in mind, exact reproduction of the settings indicated on the setup record becomes imperative.

Discussion of Injection Molding Parameters. The most productive setup sheet will implement the basic principles of molding: time, temperature, and pressure (see earlier discussion). Only if these elements are fully explored in relation to machine specifications and material processing characteristics can we be assured that the molding operation has been optimized. For example, if the mold temperature is kept at the low end of the range, the part is .065 in. or less, the material temperature must be in the medium to upper range so it can be injected at the full speed of the machine, and the pressure will be just high enough to do the filling of the cavity without opening of the mold.

Let us look at some of the details connected with the setup sheet specifications.

An Improved Approach to Setup. In the past many parameters for mold setup were determined by the cut-and-try method, thus wasting considerable time and losing expensive material. At present, we have inexpensive and easy-to-manipulate electronic calculators that enable us to figure out many of the parameter settings to the correct values in a few seconds so that only minor adjustments are necessary. To go this route, we shall provide certain formulas that will enable the setup person to calculate in a few seconds what would take quite a few minutes and a considerable amount of material to accomplish by the cut-and-try method.

The machine data along with material data must be compiled on the available molding machines and materials in use at the plant, so that the needed factors will be at hand when formulas are applied to a specific problem. Some data for which formulas will be given

are available from a few of the machinery manufacturers; on the other hand, many machines on the market lack detailed data that their manufacturers think the customers will not use.

The following formulas will be useful in establishing the time of material injection, rate of injection, and related information.

Determination of Cubic Inch Machine Capacity (known as shot capacity in ounces). This is generally based on polystyrene. The formula for cubic inches is $1.734 \times 32/1.06$. A 32-ounce, 250-ton press will have a capacity in cubic inches of:

$$\frac{1.734 \times \text{ounces (32)}}{\text{specific gravity (1.06)}} = 52.35 \text{ cu in.}$$

This is the theoretical required capacity.

Screw Travel. In the 32-ounce machine, the plasticating screw has a diameter of 2.75 in. and an area of 5.94 sq in. Dividing 52.35 cu in. by the area of the cylinder, 5.94 sq. in., we obtain 8.81 in. of screw travel. The usual marking of screw travel is by mounting an inch scale in front of the screw travel pointer. In the interest of simplification, the scale is usually in whole inches, and machine cubic inches are correspondingly rounded out to the nearest suitable number. In this case, the travel distance was selected as 10 in., which made the cubic content 5.94×10 or, rounded off, 59 cu in. The 10-in. selection was also dictated by the need for melt decompress travel, which normally is one inch or more above the shot travel requirement.

The shot travel requirement for the 32 ounces is 8.81 in.; by providing a 10-in. travel we have a 1.19-in. allowance for melt decompress action. It should be noted that the ounces of machine indicated on the specification sheet are nominal, but the actual travel distance for a specific weight of shot can be figured as indicated above. These calculations show the theoretical cubic inches which correspond to shot capacity, as well as the practical values, derived by multiplying the area of the screw by its actual travel distance as shown on the scale ($5.94 \times 10 = 59$), thus giving a volume of 59 cu in.

If shot size is given in grams, the conversion is:

$$\text{cu in.} = \frac{.0611 \times \text{grams}}{\text{specific gravity}}$$

$$= \frac{.0611 \times 28.35 \times 32}{1.06} = 52.29 \text{ cu in.}$$

(Note: one ounce = 28.35 grams.) This value for all practical purposes is the same as the one obtained in the "ounce" calculations.

Let us take a practical example and apply the above information.

Example. A shot of polypropylene with a specific gravity of .905 weighs 14 ounces or 396.7 grams. How many cubic inches will that be, and what screw travel will it involve?

$$\text{cu in.} = \frac{1.734 \times \text{ounces}}{\text{specific gravity}} = \frac{1.734 \times 14}{.905}$$

$$= 26.78 \text{ cu in.}$$

To establish travel distance, we take the 10-in. travel for a 59-cu in. volume and set up a proportion as follows:

$$\frac{\text{Example "cubic inch"}}{\text{Actual "cubic inch"}} = \frac{x}{10}$$

or

$$x = 10 \times \frac{\text{Example "cubic inch"}}{\text{Actual "cubic inch"}}$$

or

$$10 \times \frac{26.78}{59} = 4.5 \text{ in. of screw travel}$$

will be needed to fill the shot of polypropylene for 14 ounces of material. If the job requires a melt decompress action of one inch, the total screw travel will be 4.5 + 1.0 = 5.5 in.

Injection Rate (i.e., cubic inches per second). Many machine suppliers show this information as part of their specification sheet. In case of 250 tons × 32 ounces, the rate is shown as 22.5 cu in./sec, or the time required to fill the complete shot of 59 cu in. is:

$$\frac{59}{22.5} = 2.62 \text{ sec}$$

The number of cubic inches of a plastic material injected per second if not given in the machine specification can be established by determining how many gallons per minute are fed into the injection cylinder by the pump, or pumps, and the diameter of the shooting piston. The gpm and piston diameter are as a rule shown on the hydraulic diagram of the machine. Since the injecting piston and the screw are connected to each other, for each inch of poston travel there will be an inch of screw travel and a corresponding displacement of cubic inches of plastic.

The rate of speed of piston travel therefore is the gallons of oil per minute (gpm) delivered by the pump divided by the area of the piston. Converting the gpm into cubic inches per second and the piston area into square inches, we obtain:

$$\text{speed} = \frac{231 \text{ cu in./gallon} \times \text{gpm}}{60 \text{ sec/min} \times \text{ area of piston}}$$

Using the 32-ounce, 250-ton machine as an example, in which the injection pump capacity is 60 gpm and the piston has a diameter of 8.75 in. or an area of 60.132 sq in.:

$$\text{speed} = \frac{231 \times 60}{60 \times 60.132} = 3.84 \text{ in./sec}$$

or the full stroke would take:

$$\frac{10 \text{ in.}}{3.84 \text{ in./sec}} = 2.6 \text{ sec}$$

and will displace in this machine:

$$\text{cu in./sec} = \frac{59}{2.6 \text{ sec}} = 22.7 \text{ cu in./sec}$$

(The proportion was set up from the figures obtained above, under "Screw Travel."

Since the screw speed is equal to piston speed, or 3,84 in./sec, the cubic inches in that second will be found using the ratio:

$$\frac{3.84}{10} \text{ of 59 total cu. in.} = 22.7 \text{ cu. in./sec}$$

In the shot of 14-ounce polypropylene, we have established that screw travel including melt decompress will be 5.5 in. The time involved for this travel will be:

$$\frac{\text{total distance}}{\text{in./sec}} = \frac{5.5}{3.84} = 1.43 \text{ sec}$$

This is the total injection time needed for the above shot and is the guide for the injection high timer setting.

In some cases the shots are not filled with the screw traveling at full speed because if the injection pressure is set high enough to do this, it may cause flashing at the parting line. In such cases the pressure is set to have the screw travel 90 to 95 percent of the distance at full speed, with the remainder slowed down to take more time for filling, causing the material to be less fluid and thus have less tendency to flash. This action will increase the time shown above of 1.43 sec by some amount that can be determined by the use of a stopwatch for the complete injection time.

Another way to accomplish the filling of the last 5 to 10 percent of the shot is to add a limit switch at the desired distance from the end of the screw travel, and electrically signal the hold timer to take over the job of completing the shot as well as maintaining a pressure on the material in the cavity until the gate is frozen shut.

Some machines have this type of limit switch for the injection stroke, so that a low-volume pressure hold pump can be signaled to replace the high volume pump and thereby slow down the screw travel at the end of the filling action of the cavities. In effect, the limit switch cuts on the injection low timer, instead of the injection high timer. It is a desirable feature and can easily be added if not provided on an existing machine.

This hold pump is of particular value when machine clamping capacity in relation to projected area of the molded part does not provide a reasonable margin for viscosity variation in the plastic, and thus can cause flashing

at the parting line. The pressure on the injection hold pump under its normal usage is considerably lower than the injection high pressure pump. However, for the application of completing cavity filling, the pressure setting may have to be equal to or even higher than the injection high pressure. Use of the limit switch system permits the calculaton of time required to fill the cavities with the low volume pump.

The 250-ton press selected as an example has a low volume pump of 17 gpm capacity. Therefore, the speed of screw travel with this pump will be slower in proportion of pump capacities:

$$\text{speed} = \frac{17}{60} \times 3.84 = 1.09 \text{ in.}$$

Seventeen gallons per minute is the capacity of the hold pump; 60 gpm is that of the high pressure pump; and 3.84 is the inches per second when the 60 gpm is active.

In this type of an application the hold pressure pump should be activated over a distance of 0.25 in. or less. If we use in our example 0.25 in. distance, the time involved will be $0.25/1.09 = 0.23$ sec for the hold pump during the screw travel distance, of 0.25 in. The travel at high pressure pump capacity will now be 0.25 in. less; therefore the time will be $5.25/5.5 \times 1.43 = 1.36$ sec, in proportion to distances, or the corrected time is 1.36 sec plus the 0.23 sec for the hold pump, giving $1.36 + 0.23 = 1.59$ sec for the total injection time.

In this case, the injection high timer is in effect bypassed, and the injection hold timer instigates the subsequent machine functions.

There are molds with part configurations in which the rate of injection must be reduced to a value that will permit trouble-free filling of cavities. For this purpose the injection machines are equipped with a flow control valve that is rated in gallons per minute and normally has ten settings; each one of them represents one-tenth of the valve capacity. When the valve is set at a number other than zero, it indicates the number of one-tenth units of pump oil that will be bypassed to the tank.

In the machine chosen as an example, the control valve has a capacity of 45 gpm, and each division represents 4.5 gpm. Let us assume a setting of "3"; then the bypassed oil will be $3 \times 4.5 = 13.5$ gpm. The high pressure pump will now deliver an effective volume of 60 gpm $- 13.5$ gmp $= 46.5$ gpm.

The rate of injection with the control valve setting at 3 will be:

$$\frac{46.5}{60} \times 3.84 = 2.98 \text{ in./sec}$$

and the time required for the screw to travel the 5.5 in. in the example will be:

$$\frac{5.5}{2.98} = 1.85 \text{ sec}$$

The cubic inches per second will be:

$$\frac{59}{10} = 2.98 = 17.58, \quad \text{or} \quad 17.6 \text{ cu in./sec}$$

Occasionally there are jobs that require a certain number of cubic inches per second to be injected into a mold to ensure a good-quality product. If the mold is run in the same press, the recorded settings can be repeated. Frequently, it becomes necessary to transfer a mold to a press with different specifications, in which the requirement of a specified number of cubic inches per second must be repeated. The following example points out how this can be accomplished.

Let us assume that we wish to maintain the 17.6 cu in./sec in a press that will have a capacity of 30 cu in./sec. Injection pump capacity is 75 gpm; control valve capacity is 75 gpm.

We can set up a proportion as follows:

$$\frac{17.6 \text{ cu in./sec}}{30.0 \text{ cu in./sec}} = \frac{x}{75}$$

or

$$x = \frac{17.6 \times 75}{30.0} = 44.0 \text{ gpm}$$

Thus 44.0 gpm are needed to deliver 17.6 cu in./sec in the new press. Subtracting 44.0 from 75 we have to dispose of 31 gpm. The control valves with settings of 7.5 gpm each will call for $31.00/7.5 = 4.1$ divisions on the control valve, which will result in the desired 17.6 cu in./sec rate of injected material.

The information developed above is not only useful for setup purposes, but it can also be instrumental in diagnosing potential problems in machine performance. For example, if the time of screw travel is well above the established value, that would indicate a decrease in volume of oil delivered to the injection cylinder and suggest possible pump wear.

In reviewing the discussions of the various calculations, this may appear to be a lengthy process. However, all the needed information can be organized in chart form for each machine and thus be readily available for application to a specific job.

In practice, the following will be needed:

1. Converting machine shot capacity into cubic inches.
2. Finding the screw diameter and its area, to give the theoretical travel distance of the feed screw (melt decompress *not* included).
3. Rate or speed of screw travel.
4. Rate of cubic inches per second.
5. Converting the weight of a shot for a job into cubic inches.

With the above information applied to the job at hand, we can determine the distance that screw travel is increased by the distance of melt decompress; the time needed to inject material; the timer setting for injection high pressure; adjustments in pressure or speed of injection if necessary.

For materials that are known to be shear-rate-insensitive and frequently heat-sensitive, the setting of the back pressure is important. It is also significant for other materials, but to a lesser degree. For a better understanding of this problem, let us first explain how injection pump pressure is reflected in the material pressure in front of the screw plunger and in the mold cavity.

The force that causes the piston in the injection cylinder to move forward is the same force that moves the plasticating screw, since they are connected to each other. The force that moves the cylinder piston is the area of the piston in square inches multiplied by the pounds per square inch of pump pressure. The above force is also equal to the area of the plasticating screw multiplied by the injection pounds per square inch on the material. Putting this information in equation form, we have:

Area of piston × pump psi

$\quad$ = Area of screw × psi on material

If we use the 250/32 press as an example, where the screw diameter is 2.75 inc., the piston diameter is $8\frac{3}{4}$ in., and the pump pressure is 2100 psi, we obtain, substituting the values in the above formula:

(area of piston) 60.132 × 2100 = (screw
$\quad\quad$ area) 5.9396 × psi on material

$$\text{psi on material} = \frac{60.132 \times 2100}{5.9396} = 21{,}260$$

Since the piston has to overcome its own friction and that of the screw, the actual psi on the material will be reduced from 21,260 to about 21,000. We can say that the multiplier of pump psi against the cavity and material psi is about 10 for a machine with the above specifications.

The setting of the back pressure as read on the gauge for injection pressure is on the order of 50 to 100 psi. Using the multiplier of 10, the pressure on the material in front of the screw plunger will be 500 to 1000 psi. With the material in a highly fluid condition, these pressures are adequate for mixing the material thoroughly, driving the gases out, and measuring a reasonably accurate volume for a shot. The pressures on the material can go as high as 5000 psi (500 psi gauge reading), but higher pressures than absolutely necessary can cause excessive drooling at the nozzle, overheating the material in the measuring chamber with their by-products and creating

molding problems. It is a setting that should be used with care, especially when we consider that the readings are made on the dial portion of the gauge, which may not be very accurate.

It was mentioned that the injection high timer setting should correspond to the maximum rate of injection of the machine. In the case of the 250-ton press, according to press specification the time of injection would be equal to volume of the injection chamber divided by injection rate.

$$\frac{59 \text{ cu in.}}{22.5 \text{ cu in./sec}} = 2.62 \text{ sec}$$

If the material is injected within this period, it will be quite fluid all over the cavity, and for practical purposes the solidification and cooling should occur in a uniform manner all over the part. Pressure would also be applied uniformly over the molding surfaces. Both of these conditions would result in good flow-welds, minimal stresses in the part, and favorable appearance. On the other hand, when filling of the cavity takes 3 sec or more, the portion around the gate starts solidifying before the forward-moving material has filled the cavity, and this causes a decrease in the opening for material flow as well as a differential rate of cooling of part surfaces. In practical terms higher injection pressures are needed, which cause stresses in the part and unfavorable conditions for self-welding of the flow, thereby creating poor and visible welds and a finished product whose appearance does not reflect the finish of the mold.

If the injection speed is such that the material is fluid all over the cavity, even for a very short time, that may tend to cause mold opening and flashing. This indicates that the practical values of clamping pressure for mold projected area do not hold—for example, the 2 tons/sq in. of cavity projected area for polyethylene. Since a fast rate of injection offers many advantages to product properties, let us analyze such undesirable side-effects as flashing, poor dimensional control, and waste of materials. All of these occur because the pressure generated in the cavity exceeds that of the clamp.

Mold Clamping Pressure. Let us take as an example a part molded in the 250-ton press; the material used is polyethylene. The clamping pressure that is available for keeping the mold closed, in actual terms, is not 250 tons, but on the average is 10 percent less, or about 225 tons. The reason for this is that molding conditions are never perfect; for example, the press platens are not perfectly parallel, the mold thicknesses from front to back are not exactly the same at all points, the fit between guide pins and bushings may not be perfectly aligned, or other small deviations are such as to require a certain part of the clamping force to get the mold tightly closed, so that both mold halves make intimate contact to prevent material leakage. Observations under actual operating condition indicate that 10 percent of clamp capacity may be considered a reasonable waste of force, used to straighten mold faces and bring them to the close condition. In the case of polyethylene, the usual requirement of clamp force is 2 tons/sq in. of projected mold area. In the selected example, the projected mold area should be $225/2 = 112.5$ or, in round figures, 110 sq in. The force that can develop in the cavity should be around 220 tons maximum in order to prevent leakage from the cavity (flashing). This means that 220 tons or 440,000 pounds $= P \times 110$, or:

$$P = \frac{440,000}{110} = 4000 \text{ psi}$$

P = Pressure in cavity

Gate Size. The parts we are molding will be .090 in. thick in the shape of a box and the material content will be 25 cu in. The recommended gate depth size is two-thirds the part thickness, and 2 gate width is twice this depth. The gate area will be .060 × .120 sq in. What should be injection pressure gauge setting? The pressure that is indicated on the injection gauge represents the reading in front of the screw when the material is being injected from the measuring chamber into the mold. This pressure on the average molded product is about *50 percent higher* than the

average pressure in the cavity, because of the pressure drop in the nozzle, sprue bushing, runner, and gate. This would make the injection pressure gauge reading 6000 psi. The injection time would be:

$$\frac{25 \text{ cu in. (size of our shot)}}{22.5 \text{ cu in./sec (from machine data)}} = 1.1 \text{ sec}$$

Let us now assume that the prescribed pressure and time of filling did not produce complete parts. That would indicate that the gates could not accommodate this much material in the 1.1 sec.

We shall apply the Newtonian flow formula, which reads as follows:

$$Q = \frac{\pi P R^4}{8\mu L} \text{ (for cylindrical shapes)}$$

$$Q = \frac{P h^4}{9\mu L} \text{ with } w = 2h$$

(for rectangular shapes)

in which: Q = material flow in cubic inches per second; R = radius of cylinder (gate) through which flow takes place; L = length of cylinder (gate) in inches; μ = viscosity, lb·sec/in. h = height of rectangular duct (gate) in inches; W = width of rectangular duct (gate) in inches, mostly $= 2h$; and P = pressure, psi.

The flow formula applies to viscoelastic materials such as thermoplastics when under one set of conditions, such as the same pressure and same viscosity. In the molding conditions that we have set up, the pressure and viscosity will be the same as on the first trial run, and we shall change gate dimensions to improve the gate's ability to accommodate twice the amount of material in the same time span. Since the volume per second increases with the fourth power of the gate depth, raising this dimension 19 percent will double the capacity of flow in the same time period. All other factors will remain the same. Thus the gate will now be .071 × .143 sq in. This small change in size should have no affect on degating or any other aspect of the molding parameters.

This modification should result in filled-out cavities; and if a small cushion is available and the hold pressure is set at about 1000 psi higher than the injection high pressure, our parts should be of the desired quality.

This example points out that an analysis of machine specifications and moldability features of the mold can lead to an arrangement that will produce quality products, saving on power as well as wear and tear on machines, by using lower injection pressure.

Applying pressure transducers in strategic mold locations can lead to a more accurate determination of prevailing molding conditions. (See Chapter 11, on process control technology.)

Force on Mold Faces. In the discussion of clamp size vs. counteracting pressure generated in the cavity during injection molding, it was remarked that the average force used to straighten out mold faces amounts to about 10 percent of clamp capacity.

The question arises, how do we determine the actual force for full contact of mold faces if the suspicion exists that the case under investigation wastes a higher percentage of clamp force than the 10 percent cited?

The following steps will provide a reasonably close answer:

In order to maintain the integrity of the land area outside the cavity, a pressure of $3\frac{1}{2}$ tons/sq in. is allowed for steels of 300 Bn and 5 tons/sq in. for H13 heat-treated steel (or similar tool steels). These values not only lead to long tool life but also provide enough concentrated pressure to give the mold effective closing force. To test the size of the force needed to obtain good contact between faces of the mold halves, we first see that the land area is so dimensioned as to give approximately the $3\frac{1}{2}$ or 5 tons/sq in. (depending on the steel).

Having verified this, we take a paper whose area is the same as the mold base, of .003 to .005 in. thickness, cut out the shape of the cavity, and place it between the mold halves. Applying a force of $\frac{1}{3}$ ton/sq in. by reducing the clamp pressure, we close the press, and upon opening it we check to see if the impres-

sion is uniform all over the contact area of the paper. If contact is lacking in any part of the land circumference, the test should be repeated at increased pressure. The increase should be made in increments of 5 tons of clamp size until complete contact is established. The tonnage read when the impression on the paper covers the full circumference of the cavity is the tonnage wasted and used to straighten the mold. The difference between it and rated capacity is the amount left to keep the mold from opening during injection of the fluid plastic.

Let us continue with the example in which we decided that the clamp would keep a mold closed with 110 sq in. of projected area. The part of 110 sq in. of projected are. The part of 110 sq in. will have dimensions of 10" × 11"—rectangular—and a circumference of 2" × 11" plus 2" × 10", or 42". We are working with a mold of 300 Bn hardness. The square inches are calculated as follows:

$$\text{Tons} = \text{Area} \times 3.5 \text{ (tons), or } 250 = A \times 3.5$$

and:

$$A = \frac{250}{3.5} = 71.4 \text{ sq in.}$$

A is expressed in circumference times width of land, from which width is calculated:

$$71.4 = 42 \times W$$

$$W = \frac{71.4}{42} = 1.7 \text{ in.}$$

When contact of the 1.7" × 42" land area is uniform after being compressed with $\frac{1}{3}$ of a ton on 71.4 sq in. or 23.8 tons clamping force, we have obtained the tonnage needed to straighten out the mold halves. Otherwise the clamp size must be increased in steps of 5 tons until good contact is observed at the reading that produced it. If, for example, this reading were 33.8 tons, only about 216 tons would be available to prevent the mold from opening and the 110 sq in. cavity from flashing—a pressure that under normal conditions would be expected to keep the mold closed shut.

Residence Time. A time element that deserves more consideration than it normally receives is residence time in the heating chamber to which a material is exposed during molding.

The average chamber with *L/D* ratio (length to diameter) of 20/1 has cubical content of two times its rated capacity. Thus a 32-ounce nominal machine with about 59 cu. actual chamber volume would have about 118 cu in. capacity with the screw in the full forward position. If the full shot (32 ounces) had a cycle time of 60 sec (1 minute) the material on the screw would be exposed to the full heat for 2 minutes. If the shot were only 16 ounces and the cycle half a minute, the exposure would still be only 2 minutes because of the reduced cycle time. With a shot of 8 ounces and the cycle again half a minute, the exposure would be double the 2 minutes, or a total of 4 minutes. This length of time may be excessive for some materials, and can cause degradation of properties. Whenever residence time is on the high side and danger of polymer damage exists, corrective measures must be taken.

The most important corrective step is to keep the heat derived from the work of plastication to a minimum. This means that screw rpm should be at the low end, back pressure should be as low as practical, and pressure drops (from such sources as small nozzle diameter, small sprue, small runners, small gates, rough finish in runner, sharp corner at bends, rough surfaces of cavity, and core) should be minimized, so that mechanical energy that is converted into heat will be at the lowest possible level. In addition, cylinder temperatures should be arranged to have the lowest possible setting in the lead section area, with a gradual increase to the metering portion, to the level required for adequate melt temperature.

If all these measures do not remedy the problem, then the only relief can come from a machine with a cylinder of lower shot capacity.

Mold Placement and Job Starting

Descriptions of placing the mold in the press and the sequence of other moves necessary to start a job are based on the general operating manual of the machine manufacturer. Any information contained herein is intended to act as a supplement to:

1. Machine instructional manuals
2. Local plant and shop safety rules and codes
3. Federal and other government safety laws and regulations

Wherever there may appear to be a contradiction between the three instructional sources, one should clarify and reconcile the points in question before proceeding with the setup instructions involving the case under consideration. If one knows the machine functions, safety features, and operating procedures, and observes them with concentration and attention to detail, successful and safe molding operation will result.

All warning signs on the machine are for the benefit of persons at or near the machine, and should be faithfully adhered to. The standard requirements for dress and appearance around running machinery should be strictly observed. These requirements have been established over a period of many years and found to be most effective in eliminating accidents. Plant safety regulations provide for the wearing of protective devices applicable to specific operations and the maintenance of a safe and orderly workplace.

Whatever the work performed, the guidelines should include safety and caution. Extenuating circumstances may dictate deviation in procedures for any operation in an individual shop. Individual plants may have particular preferences regarding setup. Generally speaking, one must be sure that nothing is done to jeopardize manufacturers' warranties while at the same time satisfying governmental regulations.

From the instant a mold is picked up from a storage shelf up to the time production is initiated, the setup personnel should have as their main concern the safety of people working around the press and protection of the mold and press against damage. One should not actuate electrical buttons or selector switches without assuring that the "deck" is clear for the contemplated action. When work is performed between platens and one's arms are extended into the area between mold halves, it is best to have the main power disconnect-switch open, to be sure that no accidental pressing of a push button can initiate any press movement.

All safety gates are to be in place before any machine movement is initiated.

1. Daylight.

When daylight adjustment requires removal or addition of spacer blocks between moving platen and ram piston, the clamp should be in the extreme open position (i.e., maximum daylight) before any bolt removal takes place on existing joints. If the clamp piston is being moved while disconnected, one should be on the lookout for a tendency to slight rotation of the piston. Such rotation if not controlled could cause damage to limit switches or the limit switch bar. A simple jig could be made to prevent the possible rotation, and such a jig could be applicable to a variety of clamp sizes available at the plant.

The minimum mold size in relation to platen size should be one-half the distance between strain rod centers. On the 250-ton press that distance is $24'' \times 24''$, and the smallest mold size would be $12'' \times 12''$. A smaller mold would cause excessive platen deflection; if full clamp pressure of 250 ton were applied, this could endanger the integrity of the platen. A reduction in the psi pressure on the clamp would permit the use of smaller molds, provided the tons per square inch of mold area was reduced accordingly. For example, a $10'' \times 10''$ mold should reduce the clamp to the same proportion as:

$$\frac{250}{12 \times 12} = \frac{x}{10 \times 10} \text{ or } x = \frac{250 \times 10 \times 10}{12 \times 12}$$

which equals 173.6 tons mold clamp setting.

2. Mold Protection.

The usual setting for pressure in connection with mold protection is 200 psi. This 200 psi will generate a pressure or force on the mold that can be calculated as follows:

First we have to determine the area of the booster opening in order to obtain the force that is active during mold protection.

From machine specifications we know that the clamp ram speed at "fast close" is 2000 in./min. According to the Hydraulic sequence, in this operation we have a 60 gpm pump plus 17 gpm plus 6 gpm active on the booster area, which brings about the high speed movement of the ram. Expressing this mathematically we have;

cubic inches of oil per minute
$$= \text{area} \times \text{inches per minute}$$
$$231 (60 + 17 + 6)$$
$$= \text{area} \times 2000 \text{ in./min}$$
$$\text{area} = \frac{231 \times 83}{2000} = 9.5865 \text{ sq in.}$$

The 231 factor is for cubic inches per gallon.

The pressure exerted on the platen is:

$$9.5865 \times 200 \text{ psi} = 1917 \text{ pounds}$$

Part of this force is used to move the platen, which is estimated to be about 350 pounds, thus giving a net force of $1917 - 350 = 1567$ pounds for mold protection.

The force needed to move the platen can be figured by obtaining the weight of platen and ram and multiplying it by the coefficient of friction. Calculation of the force for mold protection is used for a condition in which springs of considerable resistance to deflection are employed in a mold, as, for example, when the stripper plate is required to return to the original position when the press is closing. Let us assume that we have a mold in which four 100-pound springs are used in conjunction with the stripper system. These springs will reduce the mold protection force to 1167 pounds, which may not be adequate for proper mold closing. To correct this condition we should find out by what amount the pressure setting in psi has to be changed from 200 in order to have a condition comparable to the mold without springs. The force of the four springs is 400 pounds; dividing this quantity by the area of the booster opening, we obtain the additional psi for mold protection. Thus:

$$400/9.5865 = 41.73, \text{ or about 42 psi}$$

The new setting of the pressure valve will be 242 psi. Thus correct mold protection can be calculated when springs in a mold counteract the force of mold closing.

3. Mounting the Mold.

Having considered the principles pertaining to mold setup, we shall put the knowledge gained to practical use.

We must have: (1) the material processing sheet for the grade of material that will be used; (2) the machine specifications, which give pertinent performance data of the press; (3) the setup records marked with all the needed settings for operating the machine: (4) a copy of the push-button control panel and a description of each selector switch and each push-button function, which should be attached to press specifications so that a setup person can readily refer to them and not rely on memory (there is considerable variation in the function of switches and pushbuttons, not only between various machines from different manufacturers but also on machines from the same source made at different times; and (5) special instructions that are applicable to the overall performance of the job. A review of this information should disclose what preparation will be needed to have all the accessory items in place before any machine action is begun. Remember that in most instances the hourly machine cost exceeds the man-hour cost of the setup person by several hundred percent; so machine utilization is a most important cost consideration

One of the first operations to be performed is the placement of the mold. The mold should be inspected before it is mounted in the press. Some of the mold features that need close examination are:

- *Vents,* which are used to permit the displacement of air and gases from the cavity so that the incoming plastic material will form a solid part free from included gas pockets. If vents are not of proper size, number, and location, there is a tendency to form gas pockets, to fill parts improperly, to need higher injection pressures, to have weak weld lines, to produce a burnt part, and to encounter other deficiencies, depending on the shape of the part.
- *Land of gate and gate size.* Each part and its method of molding require analysis for gate and land. Thus, for example, a long land may cause a part to stick to the cavity instead of the core.
- *Cooling cavity and core.* This is a very important feature, not only for cycle control but also for maintaining quality. The connection of water lines and their division into several circuits (depending on the cooling system) can make the difference between a smooth-running job and one that requires constant nursing. It is to be remembered that a core absorbs about two-thirds of the heat from the plastic; so it is this half that requires more care in the water hookup. There is a tendency for water passages to become rusty around the wall, and thus reduce the heat conductivity from the molding surfaces. This condition must be corrected by circulating a rust-removing substance until the passages are clean (see Chapter 16).
- *Weight of mold.* Molds that weigh more than 500 pounds will tend to slip under the constant vibration of machine operation. Such slippage will cause excessive wear on the guides of mold halves and in the long run cause uneven walls in the parts and present problems in filling cavities. This condition can be prevented or minimized by a horizontal clamping method, using jack screws attached to the platen.
- *Molding surfaces.* Check these surfaces for cleanlines or corrosion areas that may impair appearance and/or removal of parts.

- *Parting line* edges of cavities and runners. These should be checked for peening, especially when the mold is constructed of semi-hard steel. Such peened edges are a result of repeated light hammering of mold halves when pressures on the parting line are higher than required to keep the mold closed during injection. The peened edges may cause a poor appearance and make parts stick to the wrong mold half.
- *Moving sections* within the mold. These should be examined for possible burrs or other impediments to smooth operation. The seat of the sprue bushing should be checked for smoothness and the opening for absence of burrs or other impediments to sprue removal. The opening size in the sprue bushing should be $\frac{1}{32}$ in. larger than the nozzle opening. If that is not the case a nozzle with correct opening must be installed.
- Each mold has its own design and performance characteristics which might require checks prior to use in addition to those enumerated. Another preparatory move may be necessary if the molding material calls for drying prior to molding. Still another case may require a hot stamping operation during the cycle. In addition to preparing those potential operations for the job, there are such items as clamps, bolts, water hoses, etc., that have to be checked to make sure that they are available for the efficient installation and starting of the mold.

In summary, reviewing the informational data sheets will alert the setup person to the implementation of all auxiliary moves before he or she approaches the press for mounting the mold and starting the operation. It should be reemphasized that press time is very expensive and that nonproductive time should be reduced to a minimum.

In a later section we shall describe the mounting of the mold in the press; but first, we must become familiar with the operational functions of the press by learning what push buttons to press and in what sequence, so that

the machine will function properly. For this we copy from the *Operators Manual* the following:

1. Push bottom stations and their description.
2. Limit switch arrangements and their function.
3. Starting procedure and condensed description of cycle.
4. Machine specifications.

These copies for each machine should be inserted at this point of instruction.

Placing the Mold and Machine Start-up

The preceding copies from "Operator Manual" deal with the function of selector switches, push buttons, limit switches, pressure adjustments, etc. This section will elaborate on the placement of the mold in the press and the start of machine operation.

There are two basic modes of machine operation. One is setup, which includes manual operation. In this mode every press action requires manual pushing of an appropriate button to bring about the desired action. The second mode is semi-auto or full-auto. In automatic operation timers, relays, and limit switches are electrically coordinated to produce the proper sequence of operations, so that each cycle is repeated from shot to shot, and the end result is a finished product with consistent characteristics.

No buttons should be actuated if the machine is in a shut-down condition. In that case, the first move should be to open the water lines to all connections, such as the heat exchanger, the hopper throat, and any other component requiring water coolant. Next one should make sure that all pump suction valves are open.

One should check the setup record, which indicates not only the settings of parameters for the job but also the accessories needed.

Having assembled all the items needed for mold mounting and setup, one can proceed to manipulate the machine.

Before any mode is selected, power has to be available in the control circuit so that the individual control settings can be operative.

This power is applied by turning the control OFF–ON selector switch to the "on" position. The "control" light indicates power availability. The next move is to energize all the electric motors, so that pumps driven by these motors can supply oil which will actuate the appropriate hydraulic circuit and bring about desired action. The motors are energized by pushing the motor start button, and a light indicates that motors are running. Should it be necessary to stop the motors, as could happen with a severe oil leak, pushing the motor stop button will accomplish this. For the Cincinnati Milacron 250-ton machine the following moves are necessary (the cycle reset push button must be depressed to activate any of the modes listed below).

With the electric motors running, the operator selects the setup mode by turning the mold-set selector switch to the "on" position. This switch position brings about a slow movement of the clamp, and the pressure that the pump will generate is determined by the mold protection pilot seat (about 200 psi). The manual push button has to be depressed in order to instigate any machine action.

The clamp open–close selector switch, if held in position, will bring about ram opening or closing, depending on the switch position. One should open and close the clamp three or four times to gain confidence in performing these actions.

Now the operator is ready to start placing the mold in position. However, because of the preceding run, certain actions are necessary to prevent possible interference with the mold location. (1) The positive stripping bars should be adjusted to zero ejection action by screwing the bars to a position in which the stripping plate cannot be actuated. (2) The plasticizing chamber should be in the retracted position. This is accomplished by operating a manual detent lever of a valve that admits oil to the cylinder that carries the injection assembly. The detent lever in the extreme left position causes the flow of oil in the cap end, which causes the injection assembly to be retracted. The opposite position of the detent lever will cause a forward movement of the chamber until the nozzle contacts the sprue bushing

seat of the mold. The speed of movement can be controlled with the aid of a needle valve, also hand-operated. This needle valve changes the volume of oil to the activating cylinder and thereby changes its speed. During operation the injection assembly must be in the forward position and make good contact with the sprue bushing seat. For this reason constant pressure must be maintained on the forward position of the cylinder; so the restricting needle valve must be open at least one turn in order to ensure that there is an opening for the pressurized oil during the entire operating period. (3) The limit switch that causes a build-up of high pressure to close the mold should be moved out of the way (in the direction of the stationary platen) so that there is no high pressure generated before the full operation is started.

The operator should move the injection assembly back and forth several times to get a proper feel for it, leaving it in the retracted position.

With the potential interferences out of the way, it may now be the right time to heat the injection cylinder so it will be ready for manipulation of the screw when the mold is clamped in position. The heat OFF–ON selector switch turned to "on" will supply power to all heater zone pyrometers. Each pyrometer should be set to suit the material and conditions of the contemplated job as outlined in setup record. The pyrometers are located in the main electrical enclosure.

The operator is now ready to handle the mold. Eyebolts screwed into appropriate tapped holes are used for lifting the mold out of storage and placing it in the press. Only forged steel eyebolts should be used for the purpose. Before usage they should be checked to see that their threads are in good condition and that the threaded portion is not bent, to be sure the bolt has not been unduly stressed.

The standard sizes and capacities of eyebolts are:

- $\frac{1}{2}''$ will support 2,600 lb; has a thread engagement of $\frac{3}{4}''$.
- $\frac{3}{4}''$ will support 6,000 lb; has a thread engagement of $1\frac{1}{4}''$.

- 1″ will support 11,000 lb; has a thread engagement of $1\frac{1}{2}″$.

The eyebolts are hooked by means of rope or chain slings onto a lifting device of certain capacity such as a hoist, lift, or crane. (See directions for slings, and be sure to follow practices of hoisting outlined therein.)

The safe handling procedure of the mold is now established, and the steps for placing the mold can be as follows (not necessarily in the same sequence):

1. Set the clamp opening to required daylight. The approximate daylight opening is: mold thickness plus two times core height. Setting limit switch for "clamp open stop" will establish the extreme backward movement of the platen.
2. Lower the mold between platens while lining up the locating ring of the mold with the corresponding opening in the stationary platen. The clamp is moved slowly (setup) forward to hold the mold firmly in position. The size and number of clamps have been determined by the mold weight and are being placed in position and tightened with a torque wrench. The clamp attachment is for the stationary half of the mold only. The moving half of the platen may have to be backed away from the mold in order to attach stripping rods to the stripper plate. If this operation is not needed, or when it is completed, the platen is moved forward to contact the mold; clamping for the moving half is completed in the same way as for the stationary half. *Caution: It is safest to have the main power supply disconnected while fastening the clamps to the mold.*
3. Ejection rods should be adjusted in a uniform manner for effective operation of the ejection system.
4. Limit switches and other settings should be made in accordance with the setup copies of "Operators Manual" to have opening and closing of press in the desired manner.
5. Change selector switches: "mold set" to

"off" and "auto-hand" to "hand." With the press in the "hand" mode, open and close the mold to see that everything is functioning properly. Closing of the mold must be such that no banging or hammering will take place; the parting line edges of the mold halves must be protected against peening if flash-free parts are to be molded. The clamp should start "fast forward," followed by "slow down" at low pressure as the mold halves approach closing, and, finally, "slow" at high pressure. The opening of the clamp should start slowly until mold halves are separated about 0.5 in., continue "fast," and change to "slow" when stripping starts so that the chance of marking or punching of the plastic is prevented. With these settings, the approximate limit-switch settings can be checked.

6. The "extruder reverse stop" should be set to a position that can be calculated as shown in setup.
7. The extruder "speed," "torque," and "back pressure" are indicated on the material processing sheet and should be set according to the setup record.
8. Check cylinder temperatures to determine whether settings have been reached. Also check the nozzle temperature. With the extruder unit in the retracted position and the extruder selector switch in "run OFF–ON" turned to the "on" position, depress the extruder "run" button until the "extruder reverse stop" limit switch is actuated, indicating that the shot zone is filled with material.
9. Depress the "injection forward" button to purge material into a suitable container, making sure it does not splatter. Repeat this operation until all new clean material is coming through.
10. The needle valve that controls the movement of the heating cylinder by means of the "pull-in" cylinder is opened, and the "seal valve" is moved so that it will cause the "pull-in" cylinder to seat the nozzle against the sprue bushing. Depressing the "clamp forward" button will apply the full pump pressure to the "pull-in"

cylinder and thus bring about a good seat between the nozzle and the bushing.

11. Set "injection high pressure," "speed of injection," and "low pressure injection," as indicated on the setup record.

12. Set the "full-semi-auto" switch to "semi," set the extruder switch to "on," and change "hand" to "auto." The press is now ready for normal operation.

13. After a final check of pyrometers to see that they are up to the setting, the press may be operated by opening and closing the gate.

Note: There may be slight variations in designations of switches or preferred sequences on presses of different manufacturers; however, the ultimate objectives are the same.

The press is now ready for the semi-auto mode of operation except that final adjustments for settings—where needed—must be made.

Operating the Machine

The job can now be run, and the result should be a smooth cycle along these lines:

1. The safety gate is open, and its limit switch is not activated.

2. Closing the safety gate activates its limit switch, and the clamp closes fast.

3. As the clamp approaches mold closing, the mold protection limit switch is activated and causes slow movement of the ram.

4. When mold halves make contact, the high pressure limit switch is activated and brings about high pressure buildup in the main ram area.

5. The clamp pressure switch is operated. The nozzle valve (if present) opens, and "injection forward" at high speed takes place.

6. The "injection high pressure" timer times out, and the "injection low pressure" timer takes over to control the duration of the injection hold pressure.

7. The "injection low pressure" timer times

out, and the extruder starts running. The clamp goes to low pressure hold.

8. The "Extruder reverse" limit switch is operated, and the extruder stops. It may run an additional distance for melt-decompress if desired.

9. The curing timer times out, and the clamp opens slowly.

10. The "clamp fast reverse" limit switch is operated, and the clamp opens fast.

11. The "clamp reverse slow-down" limit switch is operated, and the clamp slows down for stripping action.

12. The "clamp reverse stop" limit switch is operated, and the clamp stops in its open position.

13. If the clamp open timer is used and it times out, the press is ready for the next cycle.

FINAL STAGE: OPTIMIZING MOLDING PRODUCTION

Anyone involved in this part of training should have a detailed knowledge of machine operation, be familiar with all molding parameter settings and their tolerable variations, have an understanding of mold components, and, finally, have the knowledge of processing data for materials that may be under reivew on some specific analysis of a molding problem.

One of the major concerns to a person in this program is to insure that the products of molding are equal to or exceed the expectation of the designer not only for appearance but, mostly, for performance characteristics. It means that in those cases all parameter settings are accurately carried out, but in addition one must be on the lookout for some external causes that may contribute to variation in properties. For example, a change in ambient temperature can affect the heating chamber and, since reaction to heat is relatively slow, we will find a considerable number of parts being molded to a substandard quality. The worst aspect of this occurrence is that there are no external signs of the malfunction taking place. A similar case can also exist when there is a voltage fluctuation in which most electrical parts are affected but the results are not, in most cases, detected on the surfaces of the product. When a product is made by an operator attending a machine, a variation from cycle to cycle can cause property inconsistencies which are also not visible to the naked eye.

Another source on the injection machine of considerable property and appearance variation is the pumps used to actuate machine performance.

A few years ago it was demonstrated that when a process control keeps the cavity pressure of each cycle at a consistent value the properties of parts are not only of the same value but the reject rates are practically negligible. This in turn means that if the fluctuation of the pump pressure is kept to a very minimum a similar result can be obtained on a press of standard design. Uniformity of cavity pressure is one of the most important parameters that reproduces properties in a molded product.

Here are two major factors that can keep the pump from the usual fluctuations and thereby approach a reasonable consistency resulting in similar properties from each cycle.

(a) It was pointed out in previous discussions how to run a job at a low pressure. If this condition is attained, there will be less heat generated in the pump and also the leakage of various hydraulic components will be less; the consequence of all this will be lower variation in pump pressure.

(b) Another factor in this scheme is the variation of the oil temperature. The pump supplier requires a variation of the oil from

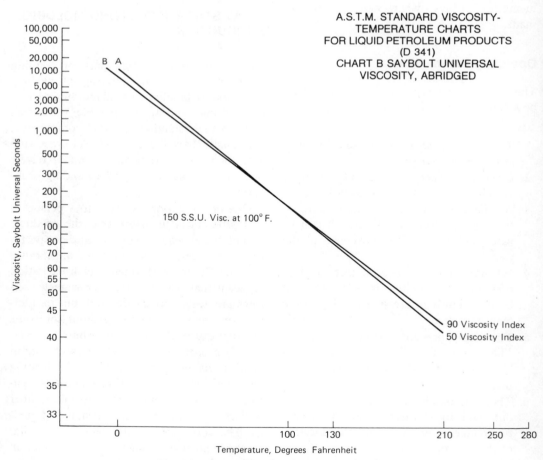

Fig. 4.4. Curve effect of temperature on viscosity.

120 to 150°F. In practice this is very often exceeded. If this recommendation is kept to a tolerance of ±5 or a total of 10°F instead of the present 30°F, there would be another helpful element promoting low fluctuation. The viscosity of the oil under these conditions would change very little; this consistency would provide another benefit (see Figure 4–4).

The major assignment for oil temperature control is given to the heat exchanger that is part of the machine. Its efficiency can be maintained only if the water side as well as the oil side is kept free of foreign substances that tend to insulate the copper tubes and thus reduce their intended function. A systematic cleaning schedule should be arranged based on local observations so that the highest cooling efficiency is always maintained.

Another aid to oil temperature control is to keep the level of oil in the tank to about $\frac{1}{4}$ of the prescribed value. The oil itself should be of the prescribed grade, and supposedly interchangeable property oil from other suppliers should not be mixed in because of proprietary formulations. Cleanliness of oil in all stages of use and handling is essential, so that it must be safeguarded against dust, dirt or other contamination.

In spite of all these positive actions to keep the pump pressure from fluctuating, there are still other malfunctions in the hydraulic system that can cause variation in pump performance. Some of these are:

1. Clogged intake openings such as filters, therefore not permitting a free flow of oil supply.

2. Air leakage in the intake side.
3. Defective pilot control valve-seats or malfunctioning springs, causing leakage in the system.
4. Air bubbles in the intake of oil.
5. Reservoir air-vent not open.
6. Tank surfaces covered with an insulating coating that prevents radiation of heat.
7. Oil leakage on the outside.

Corrective steps to remedy the foregoing malfunctions have been incorporated in a press, and not only have the results in terms of properties and appearance of the product been uniform but the rate of rejects has been a small fraction of that for a normal operation.

While discussing the pump performance, we must recognize that another important element in efficient machine functioning is a screw in good condition. Thus we must insure that the screw and check valve diameters have the proper clearance with respect to the injection chamber. The screw itself should have no gouges nor irregularities so that constant delivery of material is assured. And finally there should be an accurate limit switch for the back movement of the screw so that the same volume is delivered for each cycle during continuous operation.

All the information cited herein are merely some points that bear on the fine tuning of molding operations.

More areas exist in which molding operations can be optimized. They may pertain to the maintenance of equipment, to the condition of the mold or to the ability to reproduce test-bar properties of materials as product characteristics. This phase of training is intended to develop an analytical attitute that permits nothing to be taken for granted; the outcome due to such analysis will most likely be simple and satisfactory solutions.

Chapter 5
Machine Safety

John C. Rexford

HPM Corporation
Mt. Gilead, Ohio

Safety, as defined by Webster, is the condition of being safe from undergoing or causing hurt, injury, or loss. Relating this definition to an injection molding machine means that it must be designed, constructed, and used in a manner to prevent hurt, injury, or loss from occurring. This chapter considers how we obtain this condition, and who has this responsibility.

Quite often an accident is blamed on the carelessness of the injured party, when in fact the carelessness is a predictable human error. The effectiveness of predicting such human error and guarding against it will determine the safety of the machine.

RESPONSIBILITY THROUGH THE DEVELOPMENT STAGE

Looking first at the responsibility for machine safety, we find it cannot be delegated to any individual or group. Through the design and manufacturing stage, input is provided by many individuals, each one affecting machine safety.

- Marketing must determine needs of the industry, providing input to others without overstating the requirements.
- Research and development must convert these needs into workable ideas without creating unrealistic demands that inherently create hazards.
- Design engineering must convert these

ideas into workable concepts that guard against the predictable human error.
- Detail design must turn the concepts into reliable components and assemblies.
- Manufacturing and assembly must create and combine these components in a manner that ensures that the design concept has been maintained.
- Quality control must ensure that the design integrity is intact.
- Sales must match the needs of users with the features of the design without misrepresenting the product's abilities and features.
- Service must be aware of the machine's abilities and features to provide needed communication on the product's use.

RESPONSIBILITY OF THE MACHINE USER

When the machine leaves the manufacturer's possession, input for its safety is not complete. A new set of individuals must continue the process of maintaining machine safety and guarding against predictable human error.

- Installation may be critical in ensuring proper conditions for reliable performance over the machine's life.
- Training often prevents accidents due to unexpected or unknown occurrences.
- Maintenance will provide preventive ac-

tion that may eliminate hazards from developing, and corrective action that not only reduces the possibility of unexpected occurrences but also maintains the safe integrity of the machine.

- Supervision of the operator's actions and incentives might reduce the predictable human error that results in accidents.
- Employee vigilance, involving the individual who is closest to the machine and knows its characteristics, is essential. The presence or lack of a noise, changes in speed, or changes to the finished product might be signs of a developing hazard. These changes should be identified and corrective action taken if necessary.

Seeing that responsibility for safety covers the entire machine life, we need to analyze those areas that effect the condition of being safe.

IDENTIFYING HAZARDS

Hazards are those things that move, pinch, rotate, become hot, contain electricity, or merely exist that cause hurt, injury, or loss. Some hazards on injection molding machines are obvious (e.g., the clamp closing). Others (e.g., a component failure due to contamination) may not be obvious or even predictable. It therefore becomes the responsibility of each person associated with the machine to be alert to potential hazards.

Hazards that are obvious must be evaluated as to their probability of occurrence and the danger they pose. This evaluation begins in the development of the initial concepts and continues throughout the product's life.

As an example of this evaluation process, consider the clamp motion of the injection molding machine. The traditional injection molding machine consists of a mold that is opened and closed under great force. This motion creates a hazard that cannot be eliminated. Historically, parts have been removed by human operators. This action coupled with predictable human error creates a high probability that serious injury could occur.

Since we cannot eliminate this hazard and still have a usable tool, we must explore the second alternative in creating safety—that is, removing the human and his or her predictable errors from the hazard. This can be done by incorporating devices such as conveyors or robots for part removal. The use of automatic part loading devices for setup, or the remote placement of operating devices, away from the hazard area, may also help in preventing a serious injury.

Some applications may require the presence of the individual at the hazard site. We must then turn to the third alternative, which is to guard the hazard by placing a physical barrier between it and the operator. Safety gates with interlocking devices are used for this purpose.

As a final alternative, where physical barriers cannot be used, warning signs notify the operator that a hazard exists. The necessity of part removal requires a part exit area. Reaching into this area might only be prevented by warning the operator against this hazard. Only after all other alternatives have been exhausted should the machine design rely on warning signs. The operator's reacting to situations before thinking of the consequences is one human error that is predictable.

The design of the machine reflects the intended safety, but improper assembly, variations of critical part tolerances, loose belts, etc., can all destroy design integrity. Thorough testing and inspection of the injection molding machine must be performed and documented to maintain this integrity.

The analysis process of the manufacturer must also be used by the machine user. He has taken control of this machine and must also take the responsibility for maintaining its safety. Use of a checklist (as shown later in this chapter) might help in maintaining safety.

Auxiliary equipment, often added to improve productivity or safety, might in fact create additional hazards. Pinch points, obstacles that cause tripping, or carelessly wired devices are examples of such hazards. The actions of personnel in the area might also create new hazards. One way to guard against new hazards would be to establish and enforce safety rules for the molding department. Such rules are included at the end of this chapter.

SAFETY BUILT INTO THE MACHINE

This section discusses some of the injection molding machine's inherent hazards along with appropriate preventive measures.

Clamp Area

The closing and opening of most machines is accomplished through the use of either a hydraulic clamp or a toggle linkage. As the hydraulic clamp opens or the linkage operates, pinch areas can be created. Sheet metal or expanded metal shields are typically used to guard the area behind the movable plate. Similar guards may be necessary across the top of smaller machines. Care must be taken to ensure that the guards to not themselves create pinch hazards. These guards should be electrically interlocked to prevent machine operation if they are not in place.

Front Safety Gate

The front safety gate is used to deter entry into the mold parting line during the machine closing and injection portion of the machine cycle. Gates include a window for viewing the clamp motion. The window should conform to the American National Standard Safety Performance Specifications and Methods of Test for Safety Glazing Material Used in Buildings, Z97.1–1975. Gates are designed to be in a fully closed position before the clamp can be closed.

Power Safety Gates

On larger machines safety gates are often closed and opened with hydraulic or pneumatic power. The pressure and speed used in these systems should be kept metered down so that the gate itself does not create a pinch or strike hazard.

The leading edge of the powered gate should be constructed with some form of resilient padding. If the closing force or inertia of the gate creates a pinch or strike hazard greater than can be cushioned with padding, a leading edge safety strip such as the type used on elevator doors should be provided.

During opening of the power gate, the rear edge could strike anyone in its line of travel. The gate should be designed so there is no pinch point. The rear edge of the powered gate should be padded with a resilient material. Safety strips along the rear edge are not normally considered necessary.

Interlocking the Safety Gate

Because of predictable human error that normally causes the accident, the safety gate should be interlocked to prevent the operator from entering the hazard area created by the clamp.

The primary interlock used on the safety gate is an electrical device such as a normally open limit switch, held closed when the gate is fully closed. The device should be positioned so that it cannot be inadvertently operated. The limit switch is wired into the circuit in such a way that the clamp will stop its motion or reverse to an open position when the electrical device is released. The reaction of the clamp is determined by the portion of the cycle the machine is in. The clamp should not be allowed to open during the injection portion of the cycle because the molten plastic being forced into the mold could escape from the mold cavity, creating additional hazards to the operator or damage to the mold. The limit switch is also positioned so that it will be released before the gate is opened one inch. Allowing the gate to open a greater distance might allow an operator to reach into the hazard zone before it is safe. The machine operator will depend upon the position of the gate to tell him when it is safe to reach into the mold area.

As a backup to the electrical interlock, a hydraulic or pneumatic interlock is used. This device provides redundancy, should there be a failure of the electrical interlock. The hydraulic or pneumatic device has been incorporated into circuits in different ways, the most common being to interrupt the flow of pilot oil to the main clamp four-way valve, preventing the valve from shifting to a closing position. Some circuits block the pilot flow, while others divert it away from the valve. Another method is to provide a blocking piston on one end of the spool which physically prevents the valve spool from shifting to the clamp close position. A less desirable method is to dump the entire volume of oil through the hydraulic interlock valve to the tank. This method is normally not practical because of the large volume of oil present.

Mechanical Safety Devices

A mechanical safety device is a mechanical bar used to physically prevent the clamp from closing when the safety gate is open.

Initially mechanical safety devices were used on toggle machines to guard against inadvertent closure of the mold due to a mechanical failure of the traversing cylinder. Later, hydraulic presses, which did not have this mechanical failure problem, began appearing with mechanical safety devices. This device became a third interlock for the safety gate.

Three basic design types of mechanical safeties are commonly used in the industry: an interference type, a drop-bar type, and a rack pawl type. The type is usually determined by both the design and the size of the machine. The interference type consists of an adjustable safety bar attached to the moving platen. A safety pawl is attached to the stationary platen and is engaged by a camming device on the safety gate. As the gate is closed, the pawl is lifted, removing the mechanical interference and allowing the mold to close. The drop-bar type consists of an adjustable safety bar attached to the stationary platen. This bar pivots into and out of the die space area. A cam attached to the safety gate engages an actuating arm on the bar to lift it from between the platens when the safety door is closed. This type of bar is normally limited to the smaller type machines. The mass of the bar required on larger machines makes it difficult to lift.

One drawback of these two types of safety bars is that they must be properly adjusted as the mold height changes. Improper adjustment could make the safety device inoperative. Therefore, it is recommended that some type of interlock device be added to prevent operation, should the bar be out of adjustment. Mechanical or electrical interlocks are commonly used for this purpose.

The rack-pawl-type mechanical safety bar is a third alternative. This consists of a ratcheted or notched bar attached to the moving platen. A safety pawl attached to the stationary platen is lifted by an air cylinder when the safety gate is closed. If the gate is opened during the clamp opening stroke, the pawl ratchets back in the bar. This type of safety bar has the advantage that it will prevent clamp closure along the opening stroke and not merely in the full open position. This feature is particularly beneficial in toggle type machines where the breakage of a small traversing cylinder could cause a repeat stroke during the clamp opening cycle. The disadvantage of this type of device is that a safe condition exists only when the safety pawl is positioned into a notch. On small, short-stroke machines, the condition might exist in which the safety pawl is never positioned into a notch.

Each of these mechanical safety devices places a mechanical obstruction between the stationary and moving platens. This mechanical obstruction in itself can create a new pinch point that may need guarding.

Rear Guard

The clamp area opposite the machine operator must be guarded to prevent access to the closing hazard. This area is normally used only for maintenance or during mold setup. It is often visually blocked from the operator, who might close the clamp believing the rear of the machine to be clear. It is therefore recommended that the rear guard be electrically interlocked to shut off the motors when it is opened.

The rear guard is typically constructed with a metal frame supporting an expanded metal screen. It should be so placed on the machine as to leave an opening between the guard and the platens or machine frame. This allows clearance for water lines and other necessary items that are connected to the molds.

Top Guard

The top of the machine, or the area directly above the die space, can allow access to the clamp closing hazard. The need for a guard in this area depends on predictable human error. On machines where it would be possible for the operator, standing on the floor, to reach over the top of the front or rear guard down into the hazard zone, a guard should be provided. If this guard is portable or movable for purposes other than maintenance, then the top guard must be interlocked.

If, on the other hand, the top access area to the hazard zone is remotely located from the operator standing on the floor, a top guard may not be required. This might be the case on large machines or on machines where the front and rear guards are high enough to prevent the operator from reaching over the top. It must be assumed that if the operator or another person makes a conscious effort to climb onto the machine or another object, he is also conscious of the hazard he is encountering. This conscious effort will generally eliminate the predictable human error.

Bottom or Drop Through Guard

The bottom of the machine, or the area where completed parts drop out, can give access to the clamp closing hazard. A normal operating practice today is for the operator to sit on a stool and inspect or remove and package parts. These parts are ejected from the mold and drop onto a conveyor or chute that brings them to the operator. The predictable human error is that the operator will reach up into the hazard area, should a part become hung up. To guard against this, the machine should be so constructed that the distance the operator must reach is greater than the normal reaching distance. This meets the design objective of removing the operator from the hazard. If this is not possible, guards should be provided to prevent access. The guard design is critical because part removal is essential to

the molding operation. If the guards restrict part removal, they become targets for removal.

Maintenance of Guards

The guards for the clamp, when properly designed and maintained, will normally protect the operator. The users of the injection molding machine must keep these guards in good repair, reconstruct them when necessary, and keep them installed on the machine.

Feed Openings

Material for injection molding machines is loaded through hoppers into the plasticating barrel. The rotating and reciprocating screw, within the barrel, creates a hazard for anyone inserting a hand into the opening. This hazard must be guarded against. If guarding is not possible, then warning signs should be used. Bridging of the plastic in the feed opening or trapped foreign matter might necessitate work in this area. In that case, the power to the machine should be shut off and a soft metal rod used to remove unwanted parts. Hands should never be inserted into the opening.

Injection Cylinders

Rotating rams and reciprocating cylinders create hazards at the injection end of the machine. Access to this part of the machine is necessary only for maintenance; so fixed permanent guards should be used. Interlocking of these guards is not considered necessary.

Purging Protection

During a material change or during a shutdown, material should be purged from the barrel. This should be done with a purging compound compatible with the material being used. Improperly mixed materials can cause violent reactions.

During normal purging, a shield must be provided to protect the front, top, and rear of the purging area behind the stationary platen. The material being shot into the air may splatter onto the operator if the purge shield is not available or in use. This shield should be interlocked to prevent purging when it is not in place.

The machine circuitry should be designed so that purging cannot take place unless the safety gate is closed. This will protect against molten plastic passing through the sprue hole, into the mold area, and out on the operator.

Control Location

The location of controls on the machine should be such that the operator has visual access to the device he is controlling. They should also be located consistently to avoid confusing the operator as he moves between machines. A remote location for the controls might be used to remove the operator from a hazard area. Each operator's situation must be considered individually to determine the best location for controls.

Limit Switch Devices

Limit switches are used to control machine movements and to determine that safety devices are in place. They must never be deactivated with tape, wire, or other unauthorized means. Supervisors must be instructed to check machines regularly and enforce this rule strictly.

In some cases the machine control can be designed to check whether the limit switch is actuated, and released, during each cycle. It is recommended, if possible, that this control verify that switches are not tied down. This control can also be used to check for defective switches.

Machine Closing Control

In some industries and on some early injection molding machines, the clamp close function is accomplished by a dual hand control. This control was accepted because it occupies both hands of the operator, thus protecting him. This practice is neither recommended nor necessary in the injection molding industry. Under no circumstances should the clamp be allowed to close without the safety gate's being fully closed.

Toxic Fumes

Under certain situations, with some plastics, toxic fumes may be released during the molding process. Operating supervisors should be aware of this possibility and know the steps required to protect the operating personnel. The material supplier should provide adequate warnings on materials subject to this problem.

Warning Signs

The American National Standards Institute has supplied a list of signs in its B151.1 standard. This list suggests wording to point out the hazards to be covered instead of giving actual wording to be used. The design of the machine will usually dictate the actual wording required.

The nameplates reviewed later in this chapter are given as examples of signs used on one manufacturer's machines.

As discussed earlier, signs should be used only after all other types of safety devices have been considered. Never replace an acceptable device with only a sign. Signs should be used to complement existing safety features.

CURRENT AND FORMER INSTALLATIONS

The use of safety devices has been an evolutionary process. Early machines were built to make a product. Only after operators were injured did manufacturers realize that safety devices were required. Early attempts to add safety devices covered areas identified from a history of accidents. As time passed and different types of accidents occurred, new and better safety devices were added to machines.

The machines being produced today are generally considered safe. However, with changes in technology new sets of circumstances could result in accidents totally unpredictable by today's manufacturers.

This evolutionary process had meant that machines in use today are operated with varying degrees of safety. Machine owners, familiar with their machine condition, their operating procedures, and their personnel, must take the responsibility for updating those machines not fully equipped with current safety devices.

INJECTION MOLDING MACHINE SAFETY CHECKLIST

This checklist is not intended to be all-inclusive. It is meant to serve as a guide to help individual injection molding companies establish comprehensive lists that will satisfy their individual needs.

	O.K.	NEEDS REPAIR

CLAMP:
1. Hydraulic Cylinders:
   a. Are packing glands tight?
   b. Are bolts tight?
   c. Is the packing leaking?
   d. Are tie rods tight?
   e. Are tie bar nuts tight?
2. Toggle Machine Linkage:
   a. Are all bolts tight?
   b. Are retainer washers on properly?
   c. General condition (pins and links).
3. Plates:
   a. Are mold clamps tight?
   b. Are cylinder mounting bolts tight?
   c. Are there any loose parts lying on plates?
4. Safety Bar:
   a. Are anchor blocks anchored securely to plates?
   b. Is the bar properly guarded?
   c. Is the bar adequately guided?
   d. Does the safety pawl move freely?
   e. Does the safety pawl camming work?
   f. Is the proper air pressure used if necessary?

INJECTION:
1. Hydraulic Cylinders:
   a. Are packing glands tight?
   b. Are bolts tight?
   c. Is the packing leaking?
   d. Are the tie rods tight?
   e. Are the tie bar nuts tight?
2. Screw Drive:
   a. Are mounting studs/bolts tight?
   b. Is screw secure to drive device?
3. Barrel and Front End:
   a. Is barrel securely mounted to feed device?
   b. Are front end parts securely mounted to barrel?
   c. Does nozzle tip properly align with die?
   d. Are heating bands properly secured and functioning?
   e. Are thermocouples properly secured and functioning?

HYDRAULICS:
1. Hoses:
   a. Are hoses properly used and installed?
   b. Is the proper hose being used?
   c. Does hoses show any sign of wear?
   d. Are connectors tight?
2. Piping:
   a. Are pipes and tubing properly supported?
   b. Are weld repairs made properly?
   c. Are flange bolts tight?
   d. Are tubing connections tight?
3. Hydraulic Leaks:
   a. Welds.
   b. Hoses and/or fittings.
   c. Pipes and/or connections.
   d. Ball joints.

	O.K.	NEEDS REPAIR

    e. Packing.

    f. Are leaks cleaned up?

SAFETY GATES AND GUARDS:

1. Safety Gate:

    a. Are the rail support brackets tight?

    b. Are the rails secure to the brackets?

    c. Are the roller/trolley, etc., tight?

    d. Are the gates secure to the trolley?

    e. Are the gates/windows in good condition?

    f. Is the door edge safety working properly?

    g. Does the hydraulic/pneumatic interlock work properly?

    h. Does the electrical interlock work properly?

    i. Does the gate prevent access to pinch points?

2. Rear Guard:

    a. Are the mounting brackets secure and tight?

    b. Are guards secure to the brackets?

    c. Are the guards in good condition?

    d. Does the electrical interlock work properly?

    e. Does the rear guard prevent access to pinch points?

Fixed Guards:

    a. Are the guards securely mounted to the machine?

    b. Do the guards prevent access to pinch points?

    c. If guards are removed for reasons other than maintenance, are they interlocked to prevent machine operation?

4. Top Guards:

    a. Is the top of the machine adequately protected either by a guard or by height to prevent someone standing on the floor from reaching over the top of the safety gate?

    b. Is the top guard, if needed, properly interlocked?

5. Purge Guard:

    a. Is purging prevented by machine circuitry when the safety gate is open?

    b. Is the purge guard securely mounted to the machine?

    c. Is the purge guard in good condition?

    d. Does the purge guard contain a safety glazed window in good condition?

    e. Does the purge guard protect the front, rear, and top of the purging area?

6. Pump Coupling Guards:

    a. Are guards in place?

    b. Do guards adequately cover rotating shaft?

7. Feed Openings:

    a. Are feed openings guarded against accidental insertion of hands?

SAFETY TAGS:

1. Are tags properly located?

2. Are tags legible and understandable?

ELECTRICAL:

1. Controls and Operator's Panel:

    a. Is the inside clean and neat?

    b. Is the disconnect properly working?

    c. Is the panel door kept closed?

    d. Are there any uncovered openings?

    e. Are all tags legible?

    f. Are all buttons and switches working properly?

    g. Do all components work freely?

2. Wire Ways and Junction Boxes:

    a. Are all covers on boxes and connectors?

    b. Is any sealtite broken, or are connectors loose?

3. Switches:

    a. Are all covers in place?

	O.K.	NEEDS REPAIR

b. Are switches free of oil and water?
c. Are all switches working freely?
4. Electrical Circuit:
  a. Are circuit drawings legible?
  b. Are the circuit drawings up-to-date for the machine?
  c. Have any circuit changes been made, and have they been approved by the machine builder?
  d. Does the circuit conform to the latest state of the art?
5. Machine and Auxiliary Equipment:
  a. Is electrical interface wiring done safely?
  b. Is there duplication or confusion of terms on various pieces of equipment?
  c. Is the overall electrical circuit safe?
  d. Has the interface created any electrical, hydraulic, or mechanical safety hazards?
Operator Safety:
1. Has the operator been trained?
2. Can the operator read all tags?
3. Can the operator understand the tags?
4. Has the operator had time to become familiar with the machine?
5. Is the operator's manual easily accessible to the operator?

SAFETY RULES FOR MOLDING DEPARTMENT

1. Do not operate the machine unless you have been instructed in its operation and safety devices.
2. Be certain all safety devices are working properly before operating the machine.
3. If any safety equipment is missing, damaged, or inoperative, notify your supervisor immediately and do not operate the machine.
4. Report any hazard to your supervisor, no matter how minor it is.
5. Report any open receptacles, junction boxes, bare wires, oil leaks, or water leaks to your supervisor.
6. The operator must keep oil and water off the floor around the machine.
7. Keep the platform and work area clean.
8. Use safety devices provided, and do not bypass, change, or otherwise make inoperative any such safety device or equipment.
9. Shouting or horseplay is strictly forbidden.
10. Never block fire extinguishers, fire exits, or other emergency equipment.
11. Use only tools and equipment that are in good condition.
12. When lifting, keep your back straight and lift with your legs. If the load is too heavy, get help or notify your supervisor.
13. Report all injuries to your supervisor immediately.
14. Wear safety shoes and safety glasses at all times.
15. Follow directions for mold setup as posted on the setup sheet. No unauthorized deviations are to be made.
16. Be sure barrel and mold temperatures are maintained. Report deviations to your supervisor.
17. Maintain correct hydraulic oil temperature and level.
18. Check to see that the nozzle tip is properly seated in the mold before starting.
19. Check pressure gauges for proper settings.
20. When in doubt, ask your supervisor.
21. Never climb on the machine while it is running.
22. Whenever you leave your machine, be sure it is turned off.
23. At the start of each shift be sure the machine is operating properly and that molding parameters are set properly.
24. If the machine must be shut down, plastic materials should not be left in a plasticizing cylinder heated to operating temperatures.
25. Material should never be left in the mold.

Remove the molded parts and the sprue before shutting the machine down.

26. Before working on the machine or between plates, be sure proper lockout procedures have been followed.

27. When purging material from the plasticizing cylinder or changing materials, be sure of the compatibility of materials being used. Check with your supervisor for this information.

28. Follow all posted danger and caution signs.

WARNING SIGNS

Examples of warning signs are shown here. Actual wording used for a particular machine should be dictated by the design of that machine. The location of the sign should be as close to the actual hazard as practical.

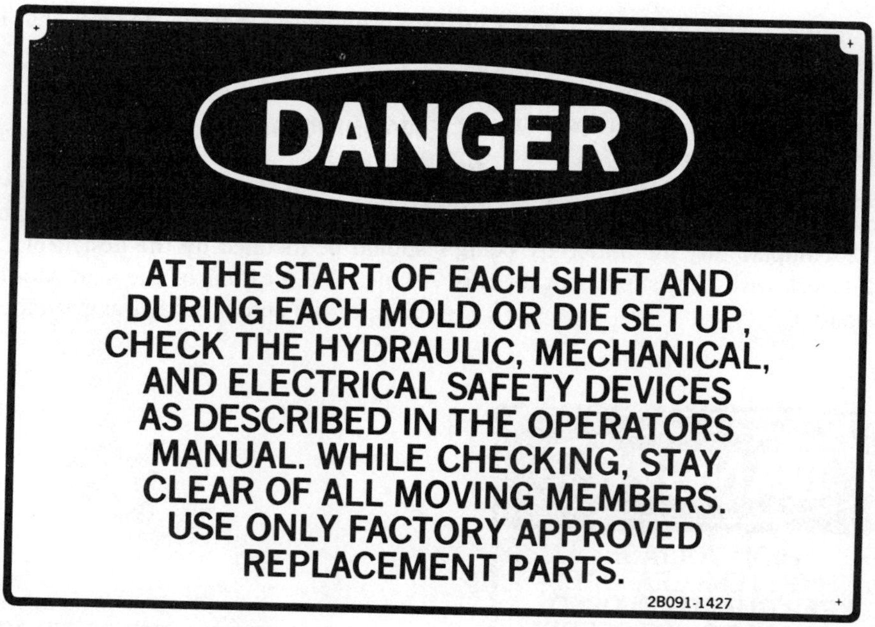

DANGER

AT THE START OF EACH SHIFT AND
DURING EACH MOLD OR DIE SET UP,
CHECK THE HYDRAULIC, MECHANICAL,
AND ELECTRICAL SAFETY DEVICES
AS DESCRIBED IN THE OPERATORS
MANUAL. WHILE CHECKING, STAY
CLEAR OF ALL MOVING MEMBERS.
USE ONLY FACTORY APPROVED
REPLACEMENT PARTS.

2B091-1427

DANGER

DO NOT OPERATE MACHINE
UNLESS THOROUGHLY
INSTRUCTED ON SAFETY
RULES AND OPERATION.
FAILURE TO FOLLOW
POSTED WARNINGS MAY RESULT
IN SEVERE PERSONAL INJURY.

2B091-1428

DANGER

TO AVOID SEVERE PERSONAL
INJURY NEVER REACH OVER,
UNDER, OR AROUND MACHINE
GUARDS. NEVER REACH INTO
OPENINGS WHILE MACHINE IS
CYCLING.

2B091-1498

DANGER

TO AVOID SEVERE PERSONAL
INJURY SHUT OFF MOTORS AND
TURN DISCONNECT TO OFF
POSITION WHEN PERFORMING
MAINTENANCE.

2B091-1432

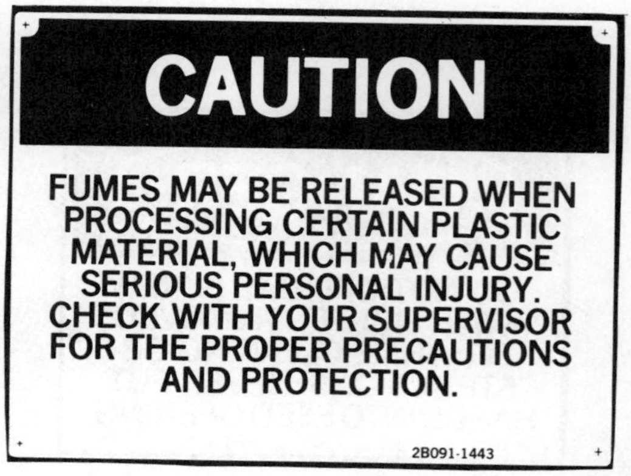

CAUTION

FUMES MAY BE RELEASED WHEN
PROCESSING CERTAIN PLASTIC
MATERIAL, WHICH MAY CAUSE
SERIOUS PERSONAL INJURY.
CHECK WITH YOUR SUPERVISOR
FOR THE PROPER PRECAUTIONS
AND PROTECTION.

2B091-1443

Section III
PRODUCT AND MOLD DESIGN

Chapter 6
Designing Products with Plastics

J. B. Dym and D. V. Rosato

BASICS IN DESIGNING

There is a practical and easy approach in designing with plastics; it is essentially no different from designing with other materials—steel, aluminum, wood, concrete, and so on. This chapter presents design information based on properties of plastics, structural responses, performance characteristics, part shape, process variables, and economics (1–4).

Plastics have been designed into many different products for over a century. They have been used successfully and have provided exceptional cost advantages compared to other materials. Unfortunately, some people think plastics are new because the industry has an endless capability of producing new plastics to meet new performance or processing requirements. This does not mean that they will replace other materials (metals, aluminum, wood, glass, concrete, titanium, etc.); each material will be used where it offers cost-to-performance advantages. As of 1980 (Fig. 1-2) more plastics were used worldwide on a volume basis than any other material except wood and concrete. Before the end of this century there will be more plastics used on a weight basis, again not including wood and concrete (Fig. 1-3). About one third of this plastic is consumed in the United States.

With plastics, more than with other materials, the opportunity exists to optimize design by focusing on material composition and orientation, as well as on structural member geometry. With plastics especially, there is an important interrelationship between shape design, material selection (including reinforcement orientation, etc.), and manufacturing selection. This interrelationship is different from that of most other materials, for which the designer basically is limited to obtaining specific prefabricated forms or profiles that are bent, welded, etc.

The job of designing is becoming more difficult as more materials become available—with plastics constituting the major portion of those materials. There are over 10,000 different plastics, only a few hundred of which are used in large quantities. Plastics are not a single type of material, but are a family of materials, each having its special advantages (Figs. 6-1 to 6-7). Details on plastic materials used in molding are given in Chapters 18 through 26.

Many different products can be designed using plastics—to take low to extremely high loads and to operate in widely differing environments, ranging from highly corrosive to electrical-insulation conditions. They provide the designer with a combination of often unfamiliar and unique advantages and limitations that can challenge his or her ability. By understanding their many different structures, properties, design freedoms, and fabrication techniques, the designer can meet this challenge.

Metals—Basic: aluminum, magnesium, beryllium, titanium, copper, nickel, gold, iron, etc.

Metals—Alloys: steel, brass, and all metals that are not basic metals.

Nonmetals—Ceramics: aluminas, beryllias, carbides, cordierites, nitrides, titanias, steatites, zirconias, etc.

Nonmetals—Glasses: silicas, soda limes, leads, borosilicates, aluminosilicates, etc.

Nonmetals—Others: fluids, greases, lubricants, oils, papers, etc.

Nonmetals—Plastics:[1]

 THERMOPLASTICS: ABS, acetates, acrylics, cellulosics, chlorinated polyethers, fluorocarbons, nylons (polyamides), polycarbonates, polyethylenes (low density, high density, etc.), polypropylenes, polyimides, polyphenylene oxides, polystyrenes, polysulfones, polyurethanes, polyvinyl chlorides, etc.

 THERMOSETS: alkyds, diallyl phthalates, epoxies, melamines, phenolics, polyesters, polyurethanes, silicones, etc.

---

1. In addition to the many different types of plastics (or polymers) that are made up of only a single plastic (or a homopolymer), there are many different combinations of plastics that provide many different cost-to-performance advantages. A major example is ABS (acrylonitrile-butadiene-styrene).

Fig. 6-1   Examples of materials for the designer.

There is no scarcity of materials for the designer, who has plastics, metals, aluminum, etc., available. In the area of steels alone, one can select from any number of plain steels, as well as many alloy steels, the superalloys, etc. Among plastics, there are reinforced or composite plastics incorporating fibers or filaments of glass, as well as those with carbon-graphite, aramid/organic, boron, single crystal/whisker, and flake type reinforcement (Figs. 6-8 to 6-10).

The major consideration for a designer is to analyze properly what is available and develop a logical selection process to meet performance requirements, which generally are related to cost factors. The range of plastic properties is indicated in most of the accompanying figures, particularly Figs. 6-2 and 6-8. Figure 6-2 follows the "pie" approach, identifying broad mechanical properties for all plastics; however, the "pie" can be used for individual plastics such as polyvinyl chlorides and polyurethanes. The "pie" approach for selecting plastics can also be used to devise separate "pies" for physical properties, chemical resistance, electrical resistance, static or dynamic loads, creep (to no creep) behavior, heat resistance, directional properties, etc.

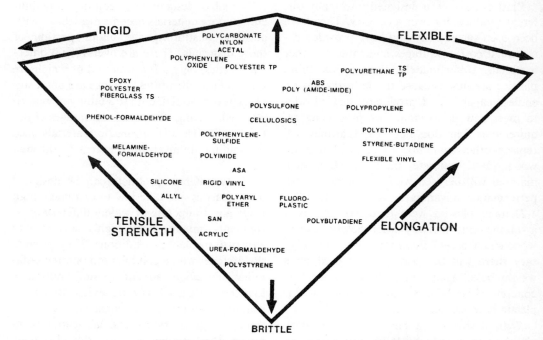

Fig. 6-2   Example of range of mechanical properties for plastics.

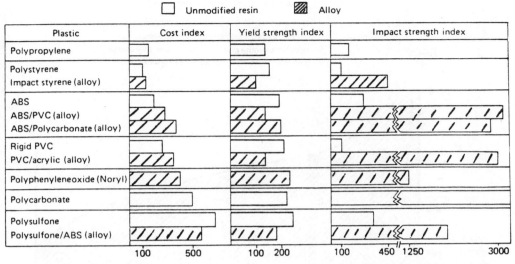

Fig. 6-3 Different plastics can be combined ("alloyed") to provide cost-to-performance improvements.

The range of properties literally encompasses all types of environmental conditions, each with its own individual broad range of properties for the different plastics. These properties can include wear resistance, integral color, impact resistance, transparency, energy absorption, ductility, thermal or sound insulation, weight, and others. Unfortunately no one plastic can meet all property requirements, but the designer has the option of combining different plastics (keeping them as separate materials by processes such as co-extrusion or co-injection). Plastics are also combined with other materials such as steel. Any combination requires that certain aspects of compatibility exist, such as thermal coefficient of expansion or contraction, etc.

Combining or mixing plastics to produce what are generally referred to as "alloys" creates plastic compounds with properties that differ from those of the components (Fig. 6-3). Compounding certain plastics can produce synergistic results, enhancing properties (Fig. 6-11). With the right knowledge, designers can create their own alloys. With metals you may have to vary a structure to do a job and get strength where you want it; with plastics you can produce a composite to meet performance requirements.

The designer can use conventional plastics that are available in sheet form, I-beams, etc., as is done with steel and most other materials, but this approach is rarely used with plastics, since their real advantage lies in their process-

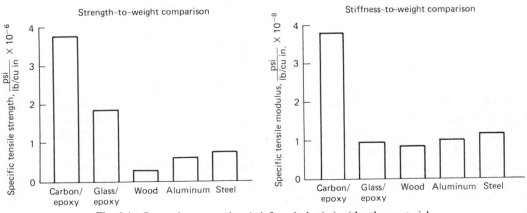

Fig. 6-4 Comparing composites (reinforced plastics) with other materials.

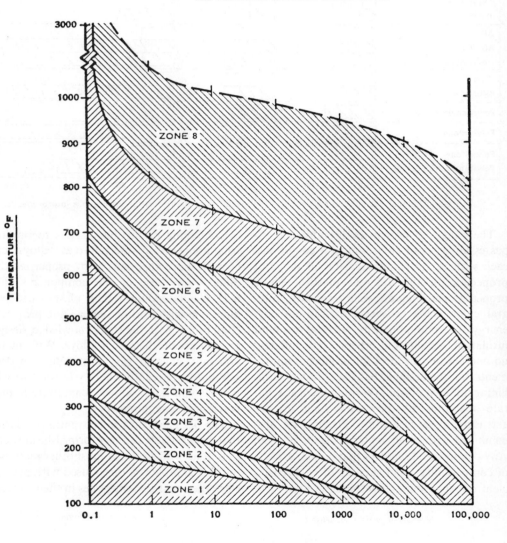

. . . . . BASED ON TEMPERATURE AND TIME

. . . . . RETAINING 50 PERCENT OF MECHANICAL
OR PHYSICAL PROPERTIES OBTAINABLE AT
ROOM TEMPERATURE, WITH RESIN EXPOSURE
AND TESTING AT ELEVATED TEMPERATURE.

Fig. 6-5 Heat-resistant properties of plastics.

ability. Thus the designer must be familiar with the techniques used in processing plastics. Another important advantage, then, is the ability to design complex shapes with plastic. Unlike the case of most other materials, it is possible to combine two or more individual plastic parts into one integrated processed part. Parts can include mechanical connections, living hinges, color, etc.

In regard to flammability, all plastics, like other materials, can be destroyed by hot enough fires. Some burn readily, others slowly, others with difficulty, while still others do not support combustion upon removal of the flame. Certain plastics are used to resist the reentry (2500°F) heat that occurs when outer space vehicles return into the earth's atmosphere. Different industry standard codes can

Examples

ZONE 1

Acrylic
Cellulose Acetate (CA)
Cellulose Acetate-Butyrate (CAB)
Cellulose Acetate Propionate (CAP)
Cellulose Nitrate (CN)
Cellulose Propionate
Polyallomer
Polyethylene, Low-Density (LDPE)
Polystyrene (PS)
Polyvinyl Acetate (PVAC)
Polyvinyl Alcohol (PVAL)
Polyvinyl Butyral (PVB)

Polyvinyl Chloride (PVC)
Styrene-Acrylonitrile (SAN)
Styrene-Butadiene (SBR)
Urea-Formaldehyde

ZONE 2

Acetal
Acrylonitrile-Butadiene-Styrene (ABS)
Chlorinated Polyether
Ethyl Cellulose (EC)
Ethylene Vinyl Acetate Copolymer (EVA)
Furan
Ionomer
Phenoxy
Polyamides
Polycarbonate (PC)
Polyethylene, High-Density (HDPE)
Polyethylene, Cross-Linked
Polyethylene Terephthalate (PETP)
Polypropylene (PP)
Polyvinylidene Chloride
Urethane

ZONE 3

Polymonochlorotrifluoroethylene (CTFE)
Vinylidene Fluoride

ZONE 4

Alkyd
Fluorinated Ethylene
   Propylene (FEP)
Melamine-Formaldehyde
Phenol-Furfural
Polyphenylene oxide (PPO)
Polysulfone

ZONE 5

Acrylic (Thermoset)
Diallyl Phthalate (DAP)
Epoxy
Phenol-Formaldehyde
Polyester
Polytetrafluoroethylene (TFE)

ZONE 6

Parylene
Polybenzimidazole (PBI)
Polyphenylene
Silicone

ZONE 7

Polyamide-imide
Polyimide

ZONE 8

Plastics now being developed
using intrinsically rigid linear
macro-molecules' principle
rather than usual crystallization
and cross-linking principles.

be used to rate plastics at various degrees of combustibility. Behavior in fire depends upon the nature and scale of the fire, as well as the surrounding conditions. Fire is a highly complex, variable phenomenon; and designing in this environment requires an understanding of all variables, so that the proper plastic is used.

Design Shape

One advantages of plastics is their formability into almost any conceivable shape (see above).

PLASTIC MATERIAL	Aromatic solvents		Aliphatic solvents		Chlorinated solvents		Weak bases and salts		Strong bases		Strong acids		Strong oxidants		Esters and ketones		24-hr. Water absorption
temperature F	77	200	77	200	77	200	77	200	77	200	77	200	77	200	77	200	% change by weight
Acetals*	1-4	2-4	1	2	1-2	4	1-3	2-5	1-5	2-5	5	5	5	5	1	2-3	0.22 - 0.25
Acrylics	5	5	2	3	5	5	1	3	2	5	4	4-5	5	5	5	5	0.2 - 0.4
Acrylonitrile-Butadiene-Styrenes (ABS)	4	5	2	3-5	3-5	5	1	2-4	1	2-4	1-4	5	1-5	5	3-5	5	0.1 - 0.4
Aramids (aromatic polyamide)	1	1	1	1	1	1	2	3	4	5	3	4	2	5	1	2	0.6
Cellulose Acetates (CA)	2	3	2	3	3	4	2	3	3	5	3	5	3	5	5	5	2 - 7
Cellulose Acetate Butyrates (CAB)	4	5	1	3	3	4	2	4	3	5	3	5	3	5	5	5	0.9 - 2.0
Cellulose Acetate Propionates (CAP)	4	5	1	3	3	4	1	2	3	5	3	5	3	5	5	5	1.3 - 2.8
Diallyl Phthalates (DAP, filled)	1-2	2-4	2	3	2	4	2	3	2	4	1-2	2-3	2	4	3-4	4-5	0.2 - 0.7
Epoxies	1	2	1	2	1-2	3-4	1	1-2	1	2	2-3	3-4	4	4-5	2	3-4	0.01 - 0.10
Ethylene Copolymers (EVA) (Ethylene-Vinyl Acetates)	5	5	5	5	5	5	1	2	1	5	1	5	1	5	2	5	0.05 - 0.13
Ethylene/Tetrafluoro-ethylene Copolymers (ETFE)	1	1	1	1	1	1	1	1	1	1	1	1	1	1	1	1	<0.03
Fluorinated Ethylene Propylenes (FEP)	1	1	1	1	1	1	1	1	1	1	1	1	1	1	1	1	<0.01
Perfluoroalkoxies (PFA)	1	1	1	1	1	1	1	1	1	1	1	1	1	1	1	1	<0.03
Polychlorotrifluoro-ethylenes (CTFE)	1	1	1	1	3	4	1	1	1	1	1	1	1	1	1	1	0.01 - 0.10
Polytetrafluoroethylenes (TFE)	1	1	1	1	1	1	1	1	1	1	1	1	1	1	1	1	0
Furans	1	1	1	1	1	1	2	2	2	2	1	1	5	5	1	1	0.01 - 0.20
Ionomers	2	4	1	4	4	4	1	4	1	4	2	4	1	5	1	4	0.1 - 1.4
Melamines (filled)	1	1	1	1	1	1	2	3	2	3	2	3	2	3	1	2	0.01 - 1.30
Nitriles (high barrier alloys of ABS or SAN)	1	4	1	2-4	1-4	2-5	1	2-4	1	2-4	2-5	5	3-5	5	1-5	5	0.2 - 0.5
Nylons	1	1	1	1	1	2	1	2	2	3	5	5	5	5	1	1	0.2 - 1.9
Phenolics (filled)	1	1	1	1	1	1	2	3	3	5	1	1	4	5	2	2	0.1 - 2.0
Polyallomers	2	4	2	4	4	5	1	1	1	1	1	3	1	4	1	3	<0.01

Fig. 6-6   Effects of elevated temperature and chemical agents on stability of plastics.

ENVIRONMENT

PLASTIC MATERIAL	Aromatic solvents		Aliphatic solvents		Chlorinated solvents		Weak bases and salts		Strong bases		Strong acids		Strong oxidants		Esters and ketones		24-hr. Water absorption
	77	200	77	200	77	200	77	200	77	200	77	200	77	200	77	200	% change by weight
Polyamide–imides	1	1	1	1	2	3	1	1	3	4	2	3	2	3	1	1	0.22 - 0.28
Polyarylsulfones (PAS)	4	5	2	3	4	5	1	2	2	2	1	1	2	4	3	4	1.2 - 1.8
Polybutylenes (PB)	3	5	1	5	4	5	1	2	1	3	1	3	1	4	1	3	<0.01 - 0.3
Polycarbonates (PC)	5	5	1	1	5	5	1	5	5	5	1	1	1	1	5	5	0.15 - 0.35
Polyesters (thermoplastic)	2	5	1	3-5	3	5	1	3-4	2	5	3	4-5	2	3-5	2	3-4	0.06 - 0.09
Polyesters (thermoset-glass fiber filled)	1-3	3-5	2	3	2	4	2	3	3	5	2	3	2	4	3-4	4-5	0.01 - 2.50
Polyethylenes (LDPE-HDPE — low-density to high-density)	4	5	4	5	4	5	1	1	1	1	1-2	1-2	1-3	3-5	2	3	0.00 - 0.01
Polyethylenes (UHMWPE — ultra high molecular weight)	3	4	3	4	3	4	1	1	1	1	1	1	1	1	3	4	<0.01
Polyimides	1	1	1	1	1	1	2	3	4	5	3	4	2	5	1	1	0.3 - 0.4
Polyphenylene Oxides (PPO) (modified)	4	5	2	3	4	5	1	1	1	1	1	2	1	2	2	3	0.06 - 0.07
Polyphenylene Sulfides (PPS)	1	2	1	1	1	2	1	1	1	1	1	1	1	2	1	1	<0.05
Polyphenylsulfones	4	4	1	1	5	5	1	1	1	1	1	1	1	1	3	4	0.5
Polypropylenes (PP)	2	4	2	4	2-3	4-5	1	1	1	1	1	2-3	2-3	4-5	2	4	0.01 - 0.03
Polystyrenes (PS)	4	5	4	5	5	5	1	5	1	5	4	5	4	5	4	5	0.03 - 0.60
Polysulfones	4	4	1	1	5	5	1	1	1	1	1	1	1	1	3	4	0.2 - 0.3
Polyurethanes (PUR)	3	4	2	3	4	5	2-3	3-4	2-3	3-4	2-3	3-4	4	4	4	5	0.02 - 1.50
Polyvinyl Chlorides (PVC)	4	5	1	5	5	5	1	5	1	5	1	5	2	5	4	5	0.04 - 1.00
Polyvinyl Chlorides-Chlorinated (CPVC)	4	4	1	2	5	5	1	2	1	2	1	2	2	3	4	5	0.04 - 0.45
Polyvinylidene Fluorides (PVDF)	1	1	1	1	1	1	1	1	1	2	1	2	1	2	3	5	0.04
Silicones	4	4	2	3	4	5	1	2	4	5	3	4	4	5	2	4	0.1 - 0.2
Styrene Acrylonitriles (SAN)	4	5	3	4	3	5	1	3	1	3	1	3	3	4	4	5	0.20 - 0.35
Ureas (filled)	1	3	1	3	1	3	2	3	2	3	4	5	2	3	1	2	0.4 - 0.8
Vinyl Esters (glass fiber filled)	1	3	1-2	2-4	1-2	4	1	3	1	3	1	2	2	3	3-4	4-5	0.01 - 2.50

temperature F

Note: A rating of "1" equals greatest stability.

133

RELATION OF STRESS-STRAIN-TIME; CREEP

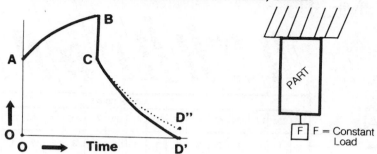

O-A: *Instantaneous loading* produces *immediate strain.*
A-B: *Viscoelastic deformation* (or creep) gradually occurs with sustained load.
B-C: Instantaneous *elastic recovery* occurs when load is removed.
C-D: Viscoelastic recovery gradually occurs; where no permanent deformation (D') o with a permanent deformation (D''-D'). Any permanent deformation is related to type plastic, amount & rate of loading and fabricating procedure.

RELATION OF STRAIN-STRESS-TIME; STRESS-RELAXATION

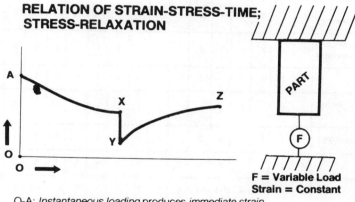

O-A: *Instantaneous loading* produces *immediate strain.*
A-X: With strain maintained gradual *elastic relaxation* occurs.
X-Y: *Instanteous deformation* occurs when load is removed.
Y-Z: *Viscoelastic deformation* gradually occurs as residual stresses are relieved. Any permanent deformation is related to type plastic, amount & rate of loading and fabricating procedure.

Fig. 6-7 Viscoelastic behavior. A property that has made plastics extremely useful for the past century is their viscoelastic nature. (See text, section on "Stress Relaxation.")

Plastic shapes can be almost infinitely varied in the early design stages, and for a given weight of material can provide a whole spectrum of strength properties, especially in the most desirable areas of stiffness and bending resistance (5).

In all materials, elementary strength-of-material theory demonstrates that some shapes resist deformation resulting from external loads or residual processing stresses better than others. Deformation in beam and sheet sections depends upon the product of the modulus and the second moment of inertia, commonly expressed as *EI.* Physical performance can be changed by varying the moment of inertia or the modulus or both.

Using thick plastic panels to meet stiffness requirements is an expensive design method because inefficiently large quantities of material are used, and the long heating and cooling

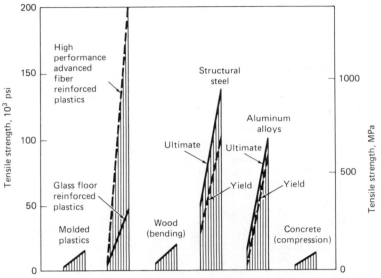

Fig. 6-8a. Comparison of strengths of plastics and other materials (5).

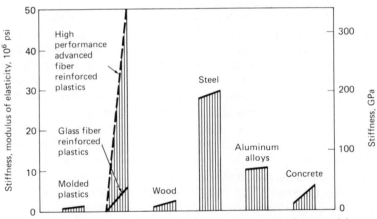

Fig. 6-8b. Comparison of elastic moduli of plastics and other materials.

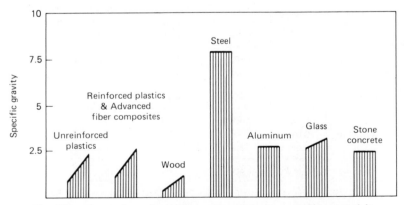

Fig. 6-8c. Comparison of specific gravities of plastics and other materials.

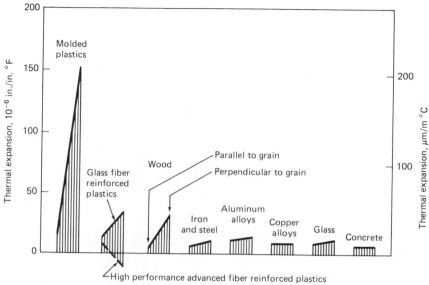

Fig. 6-8d. Comparison of thermal expansion of plastics and other materials.

times required affect the economics of production considerably by increasing cycle times.

Therefore, the desired stiffness is better obtained by the use of ribbing, double curvature, or shaping. Although these devices tend to increase tool costs, in long production runs these investments are more than offset by shortened molding cycles and the use of less raw material. Also such techniques as using integral skin foam cores further extend the stiffness of the thinner sheet.

Adding material to a structural component does not always make it stronger. For example, ribs are often added to stiffen a structure, particularly in large molded plastic parts

which normally use many ribs for weight savings. However, the addition of a rib to a plate loaded in bending may cause stress to be increased rather than decreased. Although a rib increases the moment of inertia of the structure, the distance from the structure's center of gravity to its outer edge may become relatively greater, and stress increases accordingly. The key to reducing stress is adding ribs in the correct proportions.

Lengthy equations for moments of inertia, deflection, and stress normally are required to determine the effect of ribs on stress. However, nondimensional curves have been developed to allow a quick determination of proper

FIBER	DENSITY LB/IN^3	TENSILE STRENGTH 10^3 PSI	STRENGTH SPECIFIC 10^6 IN	TENSILE MODULUS 10^6 PSI	SPECIFIC MODULUS 10^8 IN
Glass	0.092	500	5.43	10.5	1.14
	0.090	665	7.39	12.4	1.38
Boron	0.095	450	4.74	58	6.11
Graphite	0.065	400	6.15	38	5.85
	0.070	300	4.29	55	7.86
	0.063	360	5.71	27	4.29
Aramid	0.052	400	7.69	18	3.46

Fig. 6-9 Properties of High Performance Fibers Used in Composites or Reinforced Plastics.

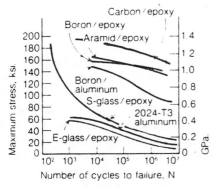

Fig. 6-10 Fatigue behavior of different unidirectional composites and aluminum.

rib proportions, while a corresponding program for a pocket calculator permits direct calculation if greater precision is required.

With potential limitations of rib panels, the use of foamed material with or without a solid outer skin in structural applications can be a way to offer a high stiffness-to-weight ratio with a wide range of depths for improvement of the moment of inertia and the torsional stiffness. Integral skin molding techniques offer the additional advantage that the surface of the panel of the material is effectively unfoamed.

With the latitude that exists in designing shapes with plastic materials, designs using large amounts of materials do not necessarily work the best or give the best physical performance per unit weight of material. Sometimes quite minute amounts of material judiciously placed in, say, injection-molded bottle crates

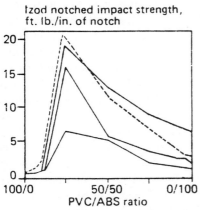

Fig. 6-11 Synergistic gains can occur when alloying certain plastics.

can make an important difference to the behavior of the crates when stacked. In one example, the removal of eight corner ribs weighing 0.7 pound from a crate originally weighing 14 pounds can reduce the crushing strength by 25 percent and increase compressive creep by about 40 percent.

SHAPE AND STIFFNESS

Plastic parts, in most cases, can take advantage of a basic beam structure in their design. Much of the conventional design with other materials is based on single rectangular shapes or box beams because in timber and in steel these are commonly produced as standard shapes. Their use in plastics components is often accompanied by a wasteful use of material, as in large steel sections.

Hollow-channel, I-, and T-shapes designed with generous radii (and other basic plastic flow considerations) rather than sharp corners are more efficient on a weight basis in plastics because they use less material, providing a high moment of inertia. The moments of inertia of such simple sections, and hence stresses and deflections, can be fairly easily calculated using simple theories.

Such nonrectangular sections are common in many thermoplastics articles. Channels, T-sections, and hollow corner pillars are found in many parts such as crates and stacking containers, and inverted U-sections and cantilevers are common in items such as street lamp housings and aircraft.

Processing any plastics (unreinforced or reinforced) into curved panels is relatively easy and inexpensive. Such panels conform to recognized structural theories that curved shapes can be stiffer in bending than flat shapes of the same weight. Putting it differently, a square-section component built to withstand external pressure will usually be heavier than one of circular section and the same volume. Both single and double curvature designs are widely used to make more effective use of plastics materials.

An example of single curvature in a structural element is a corrugated roofing panel, which is inherently much stiffer than the same volume of material would be in a flat sheet. Some calculation can be made of the stiffness of corrugated panels under certain loading conditions. To improve the stiffness further, the corrugated panels can sometimes be slightly curved along the length of the corrugations.

Double curve shells can take the form of spherical domes, be saddle-shaped, or use hyperbolic shapes. Some of these features are found in well-known architectural designs using plastics materials. These domes can be made in modular form molded of composite/ reinforced plastics, providing an efficient structural shape with a higher buckling resistance than spherical domes of comparable curvature and thickness.

The following brief and partial list tells what reinforcements can do in plastics design:

1. Increase tensile strength.
2. Increase flexural strength.
3. Increase torsional strength.
4. Increase impact strength.
5. Increase the modulus.
6. Increase creep resistance.
7. Decrease the coefficient of thermal expansion.
8. Increase thermal conductivity.
9. Extend the available supply of resin.
10. In many cases, lower the cost of the compound, as with glass fibers and other low-cost reinforcements.

STRESS RELAXATION

The properties of plastics are strongly dependent on temperature and time, compared to those of other materials such as metals. This dependence is due to the viscoelastic nature of plastics (Fig. 6-7). Consequently, the designer must know how the product is to be loaded with respect to time.

In structural design it is important to distinguish between various failure modes in the product. The behavior of any material in tension, for example, is different from its behavior in shear. So it is with plastics, metals, or concrete. For viscoelastic materials such as plastics the history of deformation also has an effect on the response of the material, since viscoelastic materials have time- and temperature-dependent material properties.

Whether the part is deformed in tension or shear, continued straining under a sustained load is called *creep*. Either type of deformation leads to a material property called the *creep compliance*. There is a creep compliance that describes the tensile behavior of the material, and there is a creep compliance for shear behavior.

On the other hand, if the material is subjected to a sustained constant strain rather than constant stress, the force or torque necessary to sustain this constant strain decays with time. This process is called *relaxation*. As in creep, there is a fundamental material property associated with relaxation, the *relaxation modulus*. There are practical problems in which relaxation is important, just as there are practical problems where creep matters. A designer must be able to recognize which of these fundamental modes is involved in a particular situation.

These stress relaxation modes can occur in all types of design and in different materials, including the low performance plastics. Note that with the very high strength or high modulus of elasticity composite plastics, this stress relaxation condition is easier to analyze than those of the high performance steels. Creep and stress relaxation behavior are illustrated in Fig. 6-7.

The creep behavior of a plastic can go through certain specific phases. Load can be applied instantaneously, resulting in strain point A (Fig. 6-7), a reaction that can be called the *elastic response*. With the stress maintained, the plastic deforms or creeps with time to point B. The *viscoelastic deformation* that occurs identifies the strain from A to B. When the load is removed, the plastic strain is reduced immediately to point C; this phase can be identified as the *elastic recovery*. The next phase has the plastic gradually relieving its strain (with time) to point D (*viscoelastic recovery*). Based on the type of plastic used and the amount of load applied, the final recovery could be a return to zero strain.

Stress relaxation behavior of a plastic can

be analyzed after load is applied instantaneously to stress point *A* (*elastic response*). As the strain is maintained, the plastic relaxes with time to point *X*. With load removed, the plastic recovers elastically (immediately) to point *Y*, less than point *X*. With time, residual stress induces *viscoelastic deformation* to point *Z*. The amount of strain due to this type of stress relaxation behavior will depend on the type of plastic used and amount of load applied.

PREDICTING PERFORMANCE

Avoiding structural failures can depend, in part, on the ability to predict performance for all types of materials (plastics, metals, glass, etc.). Design engineers have developed sophisticated computer methods for calculating stresses in complex structures using different materials. These computational methods have replaced the oversimplified models of materials behavior formerly relied upon. The result is early comprehensive analysis of the effects of temperature, loading rate, environment, and material defects on structural reliability. This information is supported by stress–strain behavior data collected in actual materials evaluations.

With computers the finite element method has greatly enhanced the capability of the structural analyst to calculate displacement, strain, and stress values in complicated plastic structures subjected to arbitrary loading conditions. In its most fundamental form, the finite element technique is limited to static, linear elastic analyses. However, there are advanced finite element computer programs that can treat highly nonlinear dynamic problems efficiently. Important features of these programs include their ability to handle sliding interfaces between contacting bodies and the ability to model elastic–plastic material properties. These program features have made possible the analysis of impact problems that only a few years ago had to be handled with very approximate techniques. Finite element techniques have made these analyses much more precise, resulting in better and more optimum designs.

Nondestructive testing (NDT) is used to assess a component or structure during its operational lifetime. Radiography, ultrasonics, eddy currents, acoustic emissions, and other methods are used to detect and monitor flaws that develop during operation (see Chapter 28).

The selection of the evaluation method(s) depends on the specific type of plastic, the type of flaw to be detected, the environment of the evaluation, the effectiveness of the evaluation method, the size of the structure, and economic consequences of structural failure. Conventional evaluation methods are often adequate for baseline and acceptance inspections. However, there are increasing demands for more accurate characterization of the size and shape of defects that may require advanced techniques and procedures, and may involve the use of several methods.

CHOOSING MATERIALS AND A DESIGN

The procedure for designing a product in a plastic follows the same logic that applies in any product design cycle; that is, in order to engineer a new design and select the proper materials for the component parts, a series of questions must be addressed:

Design Concept.   (1) What are the end-use requirements for the part or product (aesthetic, structural, mechanical)? (2) How many functional items can be designed into the part for cost effectiveness? (3) Can multiple parts be combined into one large part?

Engineering Considerations.   (1) What are the structural requirements? (2) Are the loads static, dynamic, cycling? What are the stress levels? (3) What deflection can be tolerated? (4) Is the part subject to impact loads? (5) What tolerances are required for proper functioning and assembly? (6) What kind of environment will the part see? (7) What operating temperature will it have? (8) What will its chemical exposure be? (9) What is the expected life of the product? (10) How will the product be assembled? (11) What kind of finish will be required on the parts? (12) Are agency requirements or codes involved, such as UL, DOT, MIL? (13) Can the proposed product be molded and finished economically?

Once the above questions have been considered, the next step is usually to consult data property sheets to compare material. Properties presented in these sheets are for comparative purposes and not generally for design. Seldom will a part's design conditions match the conditions used for generating the data on the property sheets, but the standardized tests are a valuable tool. Without standardized data properties, fair comparisons could not be made. The standardized information on mechanical strength, impact, chemical resistance, etc., must be adjusted for the end-use environments and life of the product.

After one selects the proper material for the part, calculations of wall thicknesses and part geometry are made, followed by the next design step, which is to improve the effectiveness of the design. In the case of injection molded parts, the design should be reviewed in terms of the following questions: (1) Can a tool be built and the part molded? (2) Are the wall thicknesses adequate for the flow of the material to fill the part? (3) Have all internal corners been "radiused" to reduce all high localized stress points? (4) Do all changes in wall thickness have smooth transitions? (5) Are heavy wall sections cored out to give a uniform wall where possible? (6) Is the ratio of rib or boss thickness to adjacent wall thickness proper? (7) Is it possible to gate into the thicker wall sections and flow to the thinner sections? (8) Are weld lines going to present strength or appearance problems? (9) Have adequate draft angles been included on all surfaces? (10) Have reasonable tolerances been selected for all parts?

If you have properly evaluated the needs of the product, chosen the proper material, optimized the design for that material, and, finally, carefully considered proper manufacturing practices, you will be on the way to having a part that works.

DESIGN CONSIDERATIONS

Design in the plastics context is a concept with many connotations. Essentially, however,

it is the process of devising a product so that it fulfills as completely as possible the total requirement of the end user. The needs of the producer from the standpoints of both sales and cost effectiveness should be automatically fulfilled if the design function is properly performed. The economic use of available materials and production processes, including the all important tooling aspects, must be effected in one design effort. Considerations include:

1. *General:* (a) performance requirements (structural, aesthetic, etc.); (b) possible combination of multiple parts or functions; (c) structural load requirements (static, dynamic, cycling, impact, etc.); (d) environment (temperature, time, chemical, etc.); (e) tolerance requirements; (f) life of product; (g) quantity of product vs. fabrication process; (h) secondary operations; (i) others.

2. *Environmental Conditions That Principally Affect Plastics:* (a) temperature; (b) time; (c) load; (d) other environments (chemical, water, etc.).

3. *Engineering Design Facts:* (a) type of load (viscoelastic concept); (b) frequency of load; (c) stress rate; (d) strain amplitude; (e) load deformation (tensile, compression, shear, etc.); (f) apparent modulus (includes strain due to creep); (g) direction of load; (h) sound dampening (foam); (i) others.

4. *Plastics Characteristics That Affect Engineering Design:* (a) polymer structure; (b) molecular weight; (c) molecular orientation; (d) plastic (with or without reinforcement) orientation; (e) types of additives, fillers, and/or reinforcements used; (f) heat history; (g) glass transition [mechanical properties change during the glass transition (T_g), as the plastic basically changes from a hard, stiff, glass-like polymer to a soft, elastic polymer]; (h) process to fabricate; (i) thermal stress (coefficient of expansion, frozen stresses); (j) economics; (k) others.

5. *Tests (ASTM, LP-406, etc.):* (a) tension (takes into account flows); (b) compression (tends to represent a pure polymer with no flow); (c) flexural, shear, etc.;

(d) creep; (e) dynamic/fatigue/torsion; (f) impact; (g) Poisson's ratio; (h) heat distortion; (i) others.

6. *Stress–Strain Data:* (a) as influenced by viscoelastic polymer behavior; (b) directional loading; (c) models in understanding shape of stress–strain using Maxwell's "spring and dashpot" concept [in which (1) short-time constant behavior (impact) is more influenced by the elastic component of polymer behavior (spring concept), and (2) time-dependent properties for long-time constant forces (cold flow, creep) are influenced by the viscous component (dashpot concept)]; (d) different curves for different tests [(1) compression modules usually higher than tensile modules; (2) compression strength of a brittle plastic higher than tensile strength by a factor of $1\frac{1}{2}$ to 4; (3) flexural strength tending to be greater than tensile strength; (4) tensile strength generally less than twice shear strength]; (e) correlating test data with end use (most tests are too limited, such as a single temperature condition); (f) others.

7. *Relating Properties to Materials and Processes:* (a) directional flow of plastic; (b) directional layout of reinforcements; (c) frozen stresses; (d) regrind; (e) prototype (machining, casting, molds, etc.); (f) others.

Design Parameters

In contrast to conventional materials such as metals, design in plastics cannot, except in rare elementary design cases, be based on one key property (as, for example, tensile or shear stress with a steel or metal). The designer rarely can call on standard sections (as in structural steel work) or on standard metal components for mechanical engineering applications. Also, plastics are constrained by a much larger set of production variables, all of which must be examined by the designer. Moreover, plastics materials vary greatly from each other in their property spectrum, so that their selection is one of the key early decisions in a design.

Figure 6-12 provides a simplified flow dia-

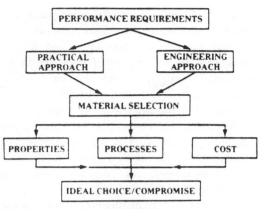

Fig. 6-12 Flow diagram for setting up selection design procedure.

gram for setting up a design program. It shows what can be referred to as the practical engineering approach (predominantly used) and the advanced engineering approach ("practical approach" and "engineering approach" in the figure).

Practical Engineering Approach. Most plastic products are required to withstand mechanical load without static or dynamic loads (appliance housings, containers, etc.). Property requirements could include mechanical strength and stiffness, chemical resistance, heat resistance, electrical insulation, etc. Short-term, conventional static tests generally suffice.

Advanced Engineering Approach. Many plastic products that have been in use since 1940 have been exposed to long-term static or dynamic loads based on varying environmental conditions such as time and temperature (large chemical tanks, boat hulls, gears, automobile parts, pressure pipe, aircraft principal structures, etc.). When a part is exposed to any load for a long period, the design engineer is provided long-time data, such as creep, fatigue, tensile/temperature/time, and other data. These data provide common stress analysis, using creep or other data for determination of permissible stresses or strains.

Types of Plastics

Plastics, like other materials, exhibit many different properties. Plastics have many different names; and they can also be called polymers, resins, reinforced plastics, and composites.

Their properties, advantages, and limitations must be understood if they are to be used intelligently. No quick and easy definition is possible; the materials called plastics—like metals, for example—cover a wide range of behaviors. However, plastics share some common properties. They are "plastic" (flow) at some stage—that is, they are soft and pliable and can be shaped, usually by the application of heat, pressure, or both, into desired forms. Some can be cast, requiring no pressure. Some plastics can be resoftened and rehardened by heating and cooling; others, once they have hardened, cannot be resoftened. (See Chapters 18 through 27 for details on plastic materials.)

LONG-TERM BEHAVIOR OF PLASTICS— CREEP

Meaningful ASTM and UL tests are conducted and used in the design of plastic products. Static and dynamic tests are used efficiently, and when necessary are properly related to time–temperature or any other time–environment condition. The ASTM tests include D149, D150, D256, D570, D621, D632, D638, D648, D671, D696, D746, D785, D790, D792, D955, D1003, D1044, D1435, D1525, D1693, D2863, and D2990; there are also UL94 and other UL tests, (see Chapter 29 for more information on testing and quality control).

Certain plastics provide long-term behavior (creep) data when exposed continuously to stresses, the environment, excessive heat, abrasion, and continuous contact with liquids. Tests outlined by ASTM D2990 are intended to produce consistency in observations and records by various manufacturers, so that they can be correlated to provide meaningful information to product designers. This long-time creep and stress-relaxation test procedure provides useful data.

The creep developed is also called cold flow. (2). In this test when a load is initially applied to a specimen, there is an instantaneous strain or elongation. Subsequent to this, there is the time-dependent part of the strain, called creep, which results from the continuation of the constant stress at a constant temperature. In terms of design, creep means changing dimensions and deterioration of product strength when a product is subjected to a steady load over a prolonged period of time.

All the mechanical properties described in tests for data sheet properties represent values of short-term application of forces; and, in most cases, the data obtained from such tests are used for comparative evaluation, or as controlling specifications for quality determination of materials along with short-duration and intermittent-use design requirements.

A stress–strain diagram is a significant source of data for a material. In metals, for example, most of the data needed for mechanical property considerations are obtained from such diagrams (see Fig. 6-13). In plastics, however, the viscoelaticity causes an initial deformation at a specific load and temperature, followed by a continuous increase in strain under identical test conditions until the product is either dimensionally out of tolerance or fails in rupture as a result of excessive deformation (see Fig. 6-7).

This viscoelastic behavior can be explained with the aid of a Maxwell model (see Fig. 6-14). With a load applied to the system, shown diagrammatically, the spring will deform to a certain degree. The dashpot will at first remain in a stationary position under the applied load; if the same load continues to be applied, the viscous fluid in the dashpot will slowly leak past the piston, causing the dashpot to move. The dashpot movement corresponds to the strain or deformation of the plastic material.

When the stress is removed, the dashpot does not return to its original position, as the spring does. Thus we can visualize a viscoelastic material as having dual actions: one of an elastic material, like the spring, and the other like the viscous liquid in the dashpot. The properties of the elastic phase are independent of time; however, the properties of the viscous phase are very much a function of time, temperature, and stress. The phenomenon is further explained by looking at the dashpot again, where we can visualize that a thinner fluid

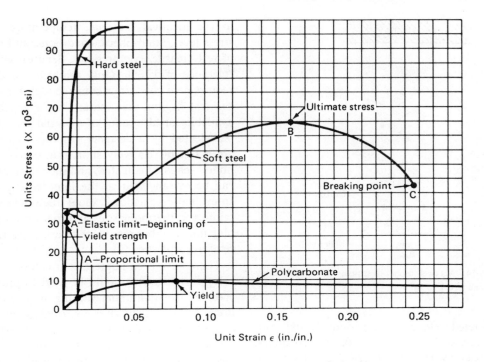

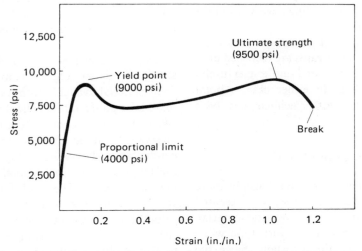

Fig. 6-13   Tensile stress–strain diagrams (top) refer to hard and soft steel, and also polycarbonate. With polycarbonate diagram (bottom) on extended scale, specific characteristics can be used in design.

resulting from increased temperature under a higher pressure (stress) will have a higher rate of leakage around the piston during the time that the above conditions prevail. Translated into plastic creep, this means that at higher use temperature and higher stress levels the strain will be higher, therefore resulting in greater creep.

The visualization of the reaction to a load (without time) by such a dual-component interpretation is valuable to our understanding of the creep process, but basically meaningless for design purposes. For this reason, the designer is interested in actual deformation or part failure over a specific time span. Observations of the amount of strain at certain time intervals must be made, which will make it possible to construct curves that can be ex-

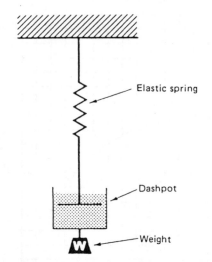

Fig. 6-14   Maxwell model to explain viscoelastic behavior. (See also Fig. 6-7.)

trapolated to longer time periods. Initial readings are at 1, 2, 3, 5, 7, 10, and 20 hours, followed by readings every 24 hours up to 500 hours, and then readings every 48 hours up to 1000 hours.

The time segment of the creep test is common to all materials. Strains are recorded until the specimen ruptures or is no longer useful because of yielding. In either case, a point of failure of the test specimen has been reached.

Designing with Creep Data

The strain readings of a creep test can be more convenient to a designer if they are presented as a creep modulus. In a viscoelastic material, strain continues to increase with time while the stress level remains constant. Since the modulus equals stress divided by strain, we have the appearance of a changing modulus.

The creep modulus, also known as apparent modulus or viscous modulus when graphed on log–log paper, is normally a straight line and lends itself to extrapolation for longer periods of time. The apparent modulus should be differentiated from the modulus given in the data sheets, because the latter is an instantaneous value derived from the testing machine (per ASTM D638).

Generally creep data application is limited to the identical material, temperature use, stress level, atmospheric conditions, and type of test (tensile, compression, flexure) with a tolerance of ±10%. Only rarely do product requirement conditions coincide with those of the test or, for that matter, are creep data available for all grades of material that may be selected by a designer. In those cases a creep test of relatively short duration—1000 hours—can be instigated, and the information can be extrapolated to the long-term needs. In evaluating plastics it should be noted that reinforced thermoplastics and thermosets display much higher resistance to creep than unreinforced plastics (2).

There have been numerous attempts to develop formulas that could be used to predict creep information under varying usage conditions. In practically all cases, suggestions are made that the calculated data be verified by actual test performance. Furthermore, numerous factors have been introduced to apply such data to reliable predictions of product behavior. (More information on creep is given in Chapter 28).

Creep data can be very useful to the designer. The data in Fig. 6-15 have been plotted from material available from or published by material manufacturers. The first point is the 100-hour time interval. The data for shorter intervals do not, as a rule, fit the straight-line configuration that exists on log–log charts for the long-term duration beyond the first 100-hour test period. The circled points are the 100-, 300-, and 1000-hour (and other observed values) test periods, and a straight line is fitted either through the circles or tangent to them to give the line a slope for long-term evaluation.

From this line, it can be estimated at what time the strain will be such that it will cause tolerance problems in product performance; or, by using the elongation at yield as the point at which the material has attained the limit of its useful life, we can estimate the time at which this limit is reached.

The formula "modulus (apparent) = stress/strain" enables us to locate the modulus that corresponds to the test stress and strain (strain is obtained by using the dimensional change

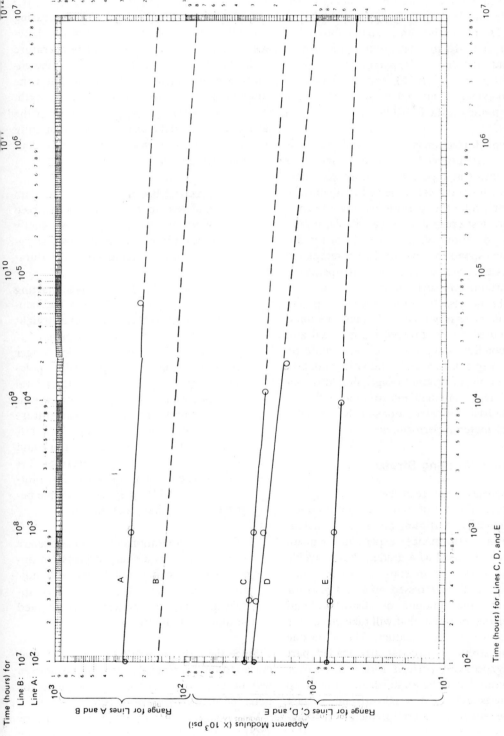

Fig. 6-15 Apparent modulus vs. time. (*A*) Merlon, polycarbonate, 2000 psi at 73°F (23°C). (*B*) Extrapolation of (*A*) beyond 10⁷ hours. (*C*) Noryl 731, modified PPO, 2000 psi at 73°F (23°C). (*D*) Delrin 500, acetal, 1000 psi at 73°F (23°C). (*E*) Zytel 109, nylon, 50% relative humidity and 1000 psi at 73°F (23°C). The broken lines represent extrapolated values. The circles are actual test reading points. Note: The log–log graph sheets are 9″ × 15″ containing 3 × 5 cycles. The end of the first time cycle represents 1000 (10³) hours.

145

or elongation limit) where it intersects the straight line leading to an appropriate time value. The polycarbonate creep line shows that a limit of 0.010 in elongation is reached at the end 10^5 hours (apparent modulus = 200,000 psi in Fig. 6-15), and an elongation of 0.06 (yield) is arrived at after 10^7 hours, or indefinitely if the 0.010 limitation does not exist.

Creep information is not as readily available as short-term property data. From a designer's viewpoint, it is important to have creep data available for products subjected to a constant load for prolonged periods of time. Even if standard test creep data are available, probably the conditions of the test will not reasonably correspond to those of the contemplated product use, such as stress level, temperature, or environmental surroundings.

In the interest of sound designing procedure, the necessary creep information should be procured on the prospective material and under conditions of product usage. In addition to the creep data, a stress–strain diagram, also at conditions of product usage, should be obtained. The combined information will provide the basis for calculations of the predictability of material performance.

Allowable Working Stress

The viscoelastic nature of the material requires not merely the use of data sheet information for calculation purposes, but also the actual long-term performance experience gained, which can be used as a guide. The allowable working stress is important for determining dimensions of the stressed area and also for predicting the amount of distortion and strength deterioration that will take place over the life-span of the product. The allowable working stress for a constantly loaded part that is expected to perform satisfactorily over many years has to be established, using creep characteristics for a material with enough data to make reliable long-term prediction of short-term test results.

The creep test data when plotted on log–log paper usually form a straight line and lend

themselves to extrapolation. The slope of the straight line, which indicates a decreasing modulus, depends on the nature of the material (principally its rigidity and temperature of heat deflection), the temperature of the environment in which the part is used, and the amount of stress in relation to tensile strength.

The author (Dym) has plotted most of the available creep data test results (amounting to over 600 charts), from which certain conclusions can be drawn:

1. For practical design purposes, the data accumulated up to 100 hours of creep are of no real benefit. There is usually too much variation during this test period, which is of relatively short duration.
2. The apparent modulus values, starting with a test period of 100 hours and continuing up to 1000 hours, form a straight line when plotted on log–log paper.
3. This line may be continued for longer periods on the same slope for interpolation purposes, provided the stress level is one-quarter to one-fifth that of the ultimate strength and the test temperature is no greater than two-thirds of the difference between room temperature and the heat deflection temperature at 264 psi. This conclusion was verified by plotting the available creep data for time periods greater than 1000 hours.

When the limitations outlined above are exceeded, there is a sharp decrease in apparent modulus after 1000 hours, with indications that failure due to creep is approaching (i.e., the material has attained the limit of its usefulness).

Since the designer will be expected to plot curves to suit his requirements, some examples will be cited that can serve as a guide for potential needs. (See Fig. 6-16.)

One example (an ABS) uses creep data for 1000 psi stress at 73°F (23°C). When the line is extended to 10^5 hours, the apparent modulus is 140,000 psi. If the product is designed

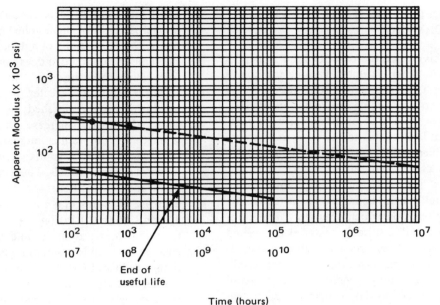

Fig. 6-16 Creep data for ABS.

for the duration of 10^5 hours and calculations are made for part dimensions, the modulus of 140,000 psi should be inserted into any formula in which the modulus appears as a factor.

At 10^5 hours total strain is:

$$E = \frac{\text{stress}}{\text{strain}}$$

$$140,000 = \frac{1000}{\text{strain}}$$

$$\text{strain} = \frac{1000}{140,000} = 0.007 \text{ or } 0.7\%$$

Based on this calculation, if the product can tolerate this type of strain without its affecting performance, then the dimensional requirements are met.

The elongation at yield for this particular ABS is 0.0275, which could be considered the end of the useful strength of the material. The apparent modulus corresponding to this strain at 1000 psi and 73°F (23°C) is:

$$E = \frac{1000}{0.0275} = 36,364 \text{ psi}$$

In the lower part of the graph in Fig. 6-16, we draw at the point of 56×10^3 on the left side a line parallel to the original creep line and find that it intersects the apparent modulus line at a time of $10^9 \times 0.5$ hours. The product would fail at that time owing to loss of strength even if dimensional changes permitted satisfactory functioning of the product.

Some charts show creep test data beyond the 1000-hour duration; under most conditions the straight line between the 100- and 1000-hour points is continued into the 10,000- and 20,000-hour range. Even in these charts a deviation from the straight line occurs occasionally, which should not be considered unreasonable because of all the variables that enter into the test data.

Selection of an allowable continuous working stress at the required temperature must be such that we can make an estimation of the elongation at the end of the product life. For example, if a product will be stressed to 1700 psi at a temperature of 150°F (66°C), and data are available for 2000 psi stress at 160°F (71°C), this information plotted on log–log paper should allow us to extrapolate the long-term behavior of the material.

DESIGN FEATURES THAT INFLUENCE PERFORMANCE

So far this chapter has described the properties of plastic materials under various use conditions, and the extent to which they may be applied to product design requirements. Before we proceed to put ideas on paper (or computer), it is advisable to review some features that influence the performance characteristics of most plastic materials.

One of the earliest steps in product design is to establish the configuration of the parts; it is the basis on which strength calculations are made and a suitable material is selected for anticipated requirements. During the development of shape and cross sections, certain things must be remembered in order to avoid degradation of test data properties. Some features may be called *property detractors*, since they are responsible for internal stresses that reduce the available stress level for load-bearing purposes. In this section we shall consider those features as well as *precautionary measures* that may influence the favorable performance of a part if properly incorporated.

Product Design Points

The product drawing does not specifically spell out the way many of its details will be carried out in the mold design. Some features adversely affect the strength and quality of the molded product. In most cases, these problem details can be modified by the designer in order to minimize their adverse effects on the properties of the part.

The following is a general summary of how to reduce problems to tolerable limits:

1. Inside sharp corners are normally shown as two intersecting straight lines without specific indications as to the functional requirement or the degree of sharpness. Inside square corners are stress concentration areas. They are quite similar to a notch in a test bar. The "Izod impact strength" of notched and unnotched test bars shows the relative impact strength of each material at the two conditions.

Thus, for example, polycarbonate has an impact strength of the notched $\frac{1}{8}$-in. test bar of 12 to 16 ft-lb/in. of notch, whereas the same bar unnotched does not fail the test. Polypropylene has an impact strength 30 times greater in the unnotched than the notched bar. Nylon shows a drastic increase in impact strength as the radius increases from sharpness to $\frac{3}{64} R$. A similar trend exists for most other materials. These examples point out that brittleness increases with decrease of radius in a corner. Visually, a radius of 0.020 in. on a plastic part may be considered sharp, and its influence on strength is much more favorable than a radius of 0.004 in. To the moldmaker, a sharp corner is usually easier to produce; to the plastic part, however, it is a source of brittleness and, in most cases, highly undesirable. Inside sharp corners on plastic part drawings are a frequent occurrence. It is the mold designer's responsibility to call attention to such strength degradation and invite appropriate corrective measures.

2. Varying wall thicknesses from thick and thin sections in a part lead to problems in molding. A uniform wall throughout a part gives it good strength and appearance. Thick and thin sections will have molded-in stresses, different rates of shrinkage (causing warpage), and possibly void formation in the thick portion. Since the parts in a mold solidify from the outer surfaces toward the center, sinks will tend to form on the surface of the thick portion. When thick ($\frac{3}{16}$ in. and over) and thin ($\frac{1}{8}$ in. or less) portions are unavoidable, the transition should be gradual, and, whenever possible, coring should be utilized.

3. Sinks are not only due to the causes listed above; they also occur whenever supporting or reinforcing ribs, flanges, or similar features are used in an attempt to provide functional service without changing the basic wall thickness of a product. If the appearance of a sink on the surface is objectionable, then the ribs

and transition radius should be proportioned so that their contribution to the sink is minimal. Figure 6-17 is a guide to the dimensioning of ribs.

4. Molded-in metal parts should be avoided whenever alternate methods will accomplish desired objectives. If it is essential to incorporate such inserts, they should be shaped so that they will present no sharp inside corners to the plastic. The effect of the sharp edges of a metal insert would be the same as explained in (1), namely, brittleness and stress concentration. The cross section that surrounds a metal insert should be heavy enough that it will not crack upon cooling. A method of minimizing cracking around the insert is to heat the metal insert prior to mold insertion to a temperature of 250°F to 300°F so that it will tend to thermoform the plastic into its finished shape. The thickness of the plastic enclosure varies from material to material. A reasonable guide is to have the thickness 1.75 to 2 times the size of the insert diameter.

5. Most plastic parts are used in conjunction with other materials. If the use temperature is other than room temperature, certain compensatory steps must be taken to avoid problems arising from the difference in the thermal coefficient of expansion of the different materials. Most of the plastic materials expand about ten times as much as steel. Thus, careful analysis of the conditions under which the metallic materials are co-employed for functional uses is called for. In the automotive industry, many long plastic parts are used in conjunction with

metal frames. If proper compensation is not made for the difference in thermal coefficients of the materials, buckling or looseness may result, causing noise or poor appearance.

6. Plastic threads have a very limited strength and may be further degraded if the thread form is not properly shaped. The V-shaped portion at the outside of a female thread will present a sharp inside corner that will act as a stress concentrator and thereby weaken the threaded cross section. The rounded form that can be readily incorporated in the molding insert will appreciably improve the strength over a V-shaped form. When self-tapping, thread-cutting, or thread-form screws are used, their holding power can be increased if, at joining time, either the screws or plastic parts are heated to a temperature of 180°F to 200°F. This will provide thermoforming action to some degree and keep the stress level caused by the joining action at a low point.

These possible sources of problems in the molded part should be marked on the part drawing and explained to the product designer, for corrective action or awareness of possible product defects due to design limitations. This is a necessary step in a chain of events in which the aim is to produce a tool that will provide parts for a good working product. Even if the mold design, mold workmanship, and molding operation are carried out to the highest degree of quality, they cannot overcome built-in weakness of product design.

Sharp Corners

When the part drawing does not show a radius, the tendency is for the toolmaker (while making a mold) to leave the intersecting machined or ground surfaces as they are generated by the machine tool. The result is a sharp corner on the molded part. Such sharp corners on the inside of parts are the most frequent property detractors.

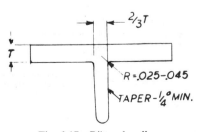

Fig. 6-17   Rib and wall.

The material data sheet shows the difference in impact strength between notched and unnotched test bars. In some materials this ratio is 1 to 30, while in others there is also a decided reduction in strength in the notched bars. Some, however, show no strength reduction. In a shaped product, an inside sharp corner is an indication that a certain specified tough material acts in a brittle manner. Sharp corners become stress concentrators.

The stress concentration factor increases as the ratio of the radius R to the part thickness T decreases. An R/T of 0.6 is favorable, and an increase in this value will be of limited benefit. If certain other details in this problem are properly counteracted, that will help in reducing stress concentration. In Fig. 6-18 we see that a concentric radius, in addition to eliminating the outside sharp corner, can play an important part in holding down the value of the stress concentration.

The ASTM Izod impact strength of nylon with various notch radii is shown in Fig. 6-19. Thus we see that with a radius of 0.005 in. the impact strength is about 1.3 ft-lb/in.; with R of 0.020 in., it is 4.5 ft-lb/in.; and with R of 0.040 in., it is 12 ft-lb/in. In most cases a radius of 0.020 in. can be considered a sharp corner as far as end use is concerned, and this size is a decided improvement over a 0 to 5 mil radius; therefore, it should be considered as a minimum requirement and should be so specified. If this radius of 0.020 in. caused interference, a corner such as shown in Fig. 6-20 should be considered.

The recommended radius not only reduces the brittleness effect, but also provides a streamlined flow path for the plastic in the mold. The radiused corner of the metal in the mold reduces the possibility of its breakdown and thus eliminates a potential repair

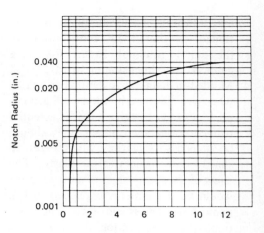

Fig. 6-19   Radius of notch versus notched Izod impact strength for nylon.

need. Too large a radius is also undesirable because it wastes material, may cause sink marks, and even contributes to stresses due to excessive variation in thickness.

Uniform Wall Thickness

Wall requirements are usually governed by load, support needs for other components, attachment bosses, and other protruding sections.

Designing a part to meet all these requirements while producing a reasonably uniform wall will greatly benefit its durability. A uniform wall thickness will minimize stresses, differences in shrinkage, possible void formation, and sinks on the surface; it also usually contributes to material saving and economy of production.

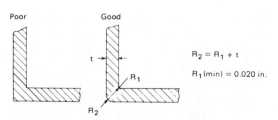

Fig. 6-18   An inside corner.

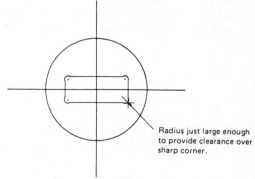

Fig. 6-20   Clearance radius for sharp corner.

Most of the features for which heavy sections are intended can be modified by means of ribbing, coring, and shaping of the cross section to provide equivalent strength, rigidity, and performance. Figure 6-21 shows a small gear manufactured from metal bar stock. The same gear converted to a molded plastic would be designed as shown in Fig. 6-22. This plastic gear design, compared to copying the metal gear, saves material; eliminates stresses due to having thick and thin sections; provides uniform shrinkage in teeth and the remainder of the gear; avoids the danger of warpage; with its thin web and tooth base, prevents bubble formation and potential weak spots; and having no sink in the middle of the thickness, provides full load carrying capacity of the teeth.

Figure 6-23 illustrates both well and poorly designed cross sections.

If a case exists where some thickness variation is unavoidable, the transition should be gradual to prevent sharp changes in temperature during solidification.

Ribs

Ribbing is one of the suggested means of creating uniform wall thicknesses. Ribs are also used to increase load bearing requirements when calculations indicate that wall thicknesses are above recommended values. They are provided for spacing purposes, for supporting components, etc.

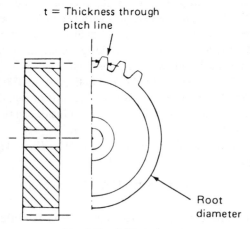

t = Thickness through pitch line

Root diameter

Fig. 6-21   Solid steel gear.

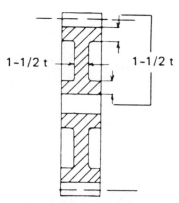

1-1/2 t   1-1/2 t

Fig. 6-22   Plastic design of steel gear shown in Fig. 6-21.

The first step in designing a rib is to determine dimensional limitations, followed by establishing what shape the rib will have to realize a part with good strength and satisfactory appearance that can be produced economically. Figure 6-24 shows proportional dimensions of rib vs. thickness. This arrangement will minimize voids (sinks), stresses, and shrinkage variations and lends itself to trouble-free molding.

If performance calculations indicate wall thicknesses well above those recommended for a particular material, one of the solutions to the problem is to achieve equivalent cross-sectional properties by ribbing. Heavy walls will cause a reduction in properties due to poor heat conductivity during molding, thus creating temperature gradients throughout the cross section with resultant stresses. Cycle times are usually exceptionally long, another cause of stress. Also, close tolerance dimensions are difficult to maintain, material is wasted, quality is degraded, and cost is increased. Solid plastic wall thicknesses for most materials should be below 0.2 in., and preferably around 0.125 in., in the interest of avoiding these pitfalls. In most cases ribbing will provide a satisfactory solution; in other cases reinforced material may have to be considered.

An example of how ribbing will provide the necessary equivalent moment of inertia and section modulus follows.

A flat plastic bar $1\frac{1}{2}$ in. wide, $\frac{3}{8}$ in. thick, and 10 in. long, supported at both ends and loaded at the center, was calculated to provide a specified deflection and stress level under

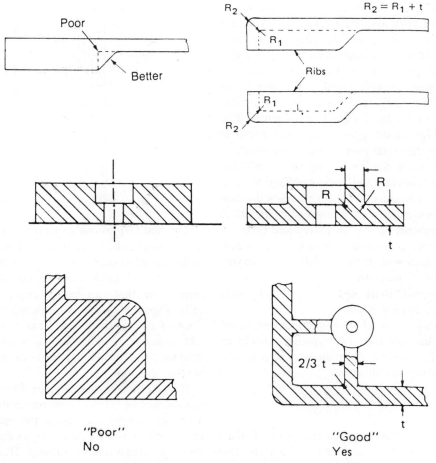

Fig. 6-23   "Poor," or problem, and "good," successful design features.

a given load. The favorable material thickness of this plastic is 0.150 in., and rib proportions would be as in Fig. 6-24.

Using judgement as a guide, it would appear that the $1\frac{1}{2}$ in. width would require about two ribs. So, as a starting point, we calculate the equivalent cross-sectional data as if we were dealing with two T-sections.

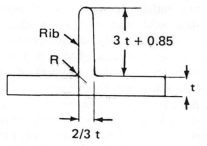

Fig. 6-24   Rib proportion vs. thickness.

According to engineering handbooks (under "Stress and Deflection in Beams," "Moments of Inertia," etc.), the resistance to stress is expressed by the moment of inertia and resistance to deflection by the section modulus. By finding a cross section with the two factors equivalent, we can assure equal or better performance in the ribbed design compared to the thick wall without ribs.

The stress = $Wl/4Z$ and deflection = $Wl^2/48EI$, where W = load, l = length of beam, Z = section modulus, E = modulus of elasticity, and I = moment of inertia.

For the flat bar $I = bd^3/12$, $b = 1\frac{1}{2}$, $d = \frac{3}{8}$; or:

$$I = (1.5 \times 0.375^3)/12 = 0.0066$$

for the rectangular bar. The section modulus is:

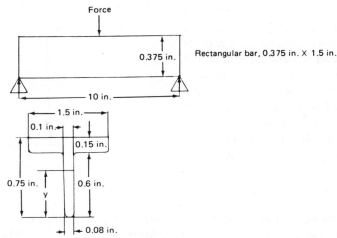

Fig. 6-25

$$Z = \frac{I}{Y} = \frac{0.0066}{0.1875} = 0.0352$$

The moment of inertia of a T-section is:

$$I = \tfrac{1}{12}[4bs^3 + h^3(3t + T)] - A(d - y - s)^2$$

where $b = 0.75$, $s = 0.15$, $h = 0.6$, $t = 0.08$, $T = 0.1$, $d = 0.75$, A = area = $bs + h(T + t)/2$.

$$y = d - [3bs^2 + 3ht(d + s) + l(T - t)(h + 3s)]/6A$$

where y is the distance from the neutral point to the extreme fiber. Substituting the values into the formulas we have:

$$A = (0.75 \times 0.15) + \frac{0.6(0.1 + 0.08)}{2} = 0.1665$$

$$y = 0.75 - \{(3 \times 0.75 \times 0.15^2) + [3 \times 0.6 \times 0.08(0.75 + 0.15)] + 0.6(0.10 - 0.08)(0.6 + 0.45)\}/ (6 \times 0.1665)$$

$$= 0.75 - \frac{0.0506 + 0.1296 + 0.0126}{0.999}$$

$$= 0.75 - 0.193 = 0.557$$

$$I = \tfrac{1}{12}[(4 \times 0.75 \times 0.15^3) + 0.6^3\{(3 \times 0.08) + 0.1\}] - 0.1665(0.75 - 0.557 - 0.15)^2$$

$$= \tfrac{1}{12}[0.0102 + 0.073] - 0.1665 \times 0.00185$$

$$= 0.00693 - 0.00031 = 0.00662$$

$$Z = \frac{0.00662}{0.557} = 0.0119$$

Two of the T-sections would provide a higher moment of inertia and decreased section modulus than a sandwich structure. When placed on the end, the two ribs would make a channel that would give a moment of inertia of 0.018 and a section modulus of 0.035, values that are more than adequate for the purpose. (See Fig. 6-25.) It should be noted that the two-rib construction forming a channel would require 70 percent of the material used in a solid bar. (See Fig. 6-26.)

Other means of stiffening surfaces can also be used—if appearance permits—where areas can be domed and corrugated. The basic goal in any action that leads to greater rigidity is

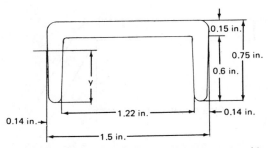

Fig. 6-25 (above) and Fig. 6-26 Shapes with good moldability that gives a section modulus and moment of inertia that are equivalent to a rectangular bar.

to specify practical wall thickness that will optimize strength and processing, and thus result in high-quality products. In addition to ribs, other features protruding from a wall, such as bosses or tubular shapes, should be treated similarly as far as transition radius, taper, and minimal material usage are concerned. The same principles are involved, and identical ill effects can be expected unless recommended practices are incorporated.

Tapers of Draft Angles

It is desirable for any vertical wall of a molded product to have an amount of draft that will permit easy removal from a mold. Figures 6-27 and 6-28 show two basic conditions in which draft is a consideration. Figure 6-27 is the most desirable application of the draft angle, and the amount of draft may vary from $\frac{1}{8}$ degree up to several degrees depending on what the circumstance will permit. A fair average may be from $\frac{1}{2}$ degree to 1 degree. When a small angle such as $\frac{1}{8}$ degree is used, the outside surface—the mold surface producing it—will require a very high directional finish in order to facilitate removal from the mold.

In another example, as shown in Fig. 6-28, there is a separating inside wall which generally should be perpendicular to the base. The draft in this case should be on the low side ($\frac{1}{8}$ degree) so that additional material usage is small, the possibility of voids close to the base is avoided, and increased cycle time in manufacture is minimized. Here again the vertical molding surfaces will demand a much higher surface finish, with polishing lines in the direction of part withdrawal. On shallow

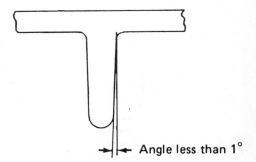
Fig. 6-28   Internal wall taper.

walls, the draft angle can be considerably larger, since the influence of the enumerated drawbacks will be minor. The designer should be cognizant of the need for drafts on vertical walls. If problems are encountered during removal of parts, stresses can result, the shape of the product can be distorted, and surface imperfections can be introduced.

Weld Lines

With molded parts that include openings (holes), problems develop. In the process of filling a cavity, the flowing plastic is obstructed by the core, splits its stream, and surrounds the core. The split stream then reunites and continues flowing until the cavity is filled. The rejoining of the split streams forms a weld line that lacks the strength properties that exist in an area without a weld line. The reason for the reduction in strength is that the flowing material tends to wipe air, moisture, and lubricant into the area where the joining of the stream takes place and introduces foreign substances into the welding surface. Furthermore, since the plastic material has lost some of the heat, the temperature for self-welding is not conducive to the most favorable results. A surface that is subjected to load bearing should not contain weld lines; if this is not possible, the allowable working stress should be reduced by at least 15 percent. (Weld lines are reviewed in Chapters 4, 13, and 18.)

Gate Size and Location

Because of the high melt pressure the area near a gate is highly stressed, by the frictional

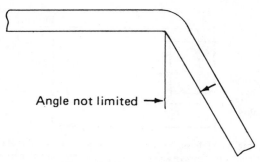
Fig. 6-27   External wall taper.

heat generated at the gate and the high velocities of the flowing material. A small gate is desirable for separating the part from the feed line, but is not desirable for a part with low stresses. Gates are usually two-thirds of the part's thickness; and if they are that large or larger will reduce frictional heat, permit lower velocities, and allow the application of higher pressures for densifying the material in the cavity. The product designer should caution the tool designer to keep the gate area away from load bearing surfaces and to make the gate size such that it will improve the quality of the product. (See Chapters 4 and 13 for more details.)

Wall Thickness Tolerance

When relatively deep parts are designed, a tolerance for the wall thickness on the order of ± 0.005 in. is usually given. What this tolerance should mean is that a product will be acceptable when made with this tolerance, but that the wall thickness must be uniform throughout the circumference.

If we analyze the molding condition of this part and assume that one side is made to minimum specifications and the opposite to maximum specifications, we find the following taking place: The resistance to plastic flow decreases with the third power of the thickness, which means that the thick side will be filled first, while the thin side is filled from all sides instead of the gate side alone. This type of filling creates a pocket on the thin side and compresses the air and gases to such a point that rising temperature due to compression causes the material to be charred while the pocket is being filled up.

The charred material will create a porosity, a weak area, and an electrically defective surface. Furthermore, filling of the thick side ahead of the thin side creates a pressure imbalance generated by the 5 to 10 tons/sq in. injection pressure that can cause the core to deflect toward the thin side and further aggravates the difference in wall thickness. This pressure imbalance will contribute to mold damage and make part production difficult if not impossible. We may conclude from this

discussion that the wall uniformity throughout the circumference must be within narrow limits such as ±0.002 in., whereas the thickness in general may vary from the specified value by ±0.005 in.

Molded-in Inserts

If metal inserts are to be molded into a product, their shape should present no sharp edges to the plastic, since the effect of the edges would be similar to a notch. A knurled insert should have the sharp point smoothed, again to avoid the notch effect. The practice of molding inserts in place is usually employed to provide good holding power for plastic products, but there are certain drawbacks to this method: it is dangerous to have an operator place an arm between the mold halves while the electrical power to the machine is turned on; it normally takes a pin to support the insert, and since this pin is small in relation to the cored hole for the insert, it is easily bent or sheared under the influence of injection pressure; should the insert fall out of position, there is danger of mold damage; the hand placement of inserts contributes to cycle variation and with it, potentially, product quality degradation. Some of these problems can be overcome by higher mold expenditures, for example, shuttling cavities.

On the other hand, desired results in fastening can be attained by other means; for example, (a) coring holes in the part that will permit ultrasonic welding of inserts in place, (b) coring a hole in the part that will be of a size when the part is removed from the mold that will permit slight press fit plus a gain in the holding power from postmolding shrinkage, or (c) coring a hole in the part that will permit dropping the insert and providing a retaining shoulder by spinning or ultrasonic forming.

All these assembly methods require the same time to perform as placing inserts in the mold, but they also lower machine time. There are probably several other means of accomplishing the desired result that depend on the circumstances at hand. In any event, molded-in inserts, in the long run, prove costlier and so should be avoided.

Internal Plastic Threads

The strength of plastic threads is limited, and when molded in a part involving either an unscrewing device or a rounded shape of thread—similar to bottle cap threads—they can be stripped from the core. Screw threads, when needed, should be of the coarse type and have the outside of the thread rounded so as not to prevent a sharp V to the plastic, which can produce a notch effect.

If a self-threading screw can be substituted, it will not only appreciably decrease the mold maintenance and mold cost, but most likely, with proper type selection, will give better holding power. A screw that has a thin thread with relatively deep flights can give high holding power. If the screw or plastic is preheated to about 250°F (121°C), a condition of thermoforming in combination with material displacement will exist, thereby improving the holding power. When male plastic threads are being considered, the coarser threads are again preferred, and the root of the thread should be rounded to prevent the notch effect.

Blind Holes

A core pin forming blind holes is subjected to the bending forces that exist in the cavity due to the high melt pressures. Calculations can be made for each case by establishing the core pin diameter, its length, and anticipated pressure conditions in the cavity. From technical handbooks, we know that a pin supported on one end only will deflect 48 times as much as one supported on both ends. This suggests that the depth of hole in relation to diameter should be small in order to maintain a straight hole. Sometimes a deep and small-diameter hole is needed, as in the example of pen and pencil bodies. In this case the plastic flow is arranged to hit the free end of the core from four to six evenly spaced gates that will cause a centering action, and the plastic will continue flowing over the diameter in an umbrella-like pattern to balance the pressure forces on the core. When this type of flow pattern is impractical, an alternative may be

a through hole (or tube formation) combined with a postmolding sealing or closing operation by spinning or ultrasonic welding.

At the other extreme, let us consider a $\frac{1}{4}$ in. diameter core, exposed to a pressure of 4000 psi with an allowance for deflection of 0.0001 in., and see how deep a blind hole can be molded under these conditions.

According to engineering handbooks the deflection, a, may be used to calculate this depth, l, as follows:

$$a = \frac{Wl^3}{8EI}$$

$$= \frac{1000l^4}{8 \times 30,000,000 \times 0.049 \times 0.0039}$$

$$I = \frac{\pi d^4}{64} = 0.049d^4$$

$$d^4 = 0.0625 \times 0.0625 = 0.0039$$

where W = total load

$\qquad$ = psi $\times d \times l$ (projected area of pin)

$\qquad$ = $4000 \times \frac{1}{4} \times l = 1000l$

$\qquad l$ = length of pin; d = diameter of pin

$\qquad a$ = deflection = 1/10,000

$$\frac{1}{10,000} = a$$

$$= \frac{1000l^4}{8 \times 30,000,000 \times 0.049 \times 0.0039}$$

$$= \text{deflection}$$

$$l^4 = 0.0045864$$

$$l = 0.26'' \text{ or slightly over diameter size}$$

If a hole deeper than 0.26 in. is needed, we can calculate the amount of deflection that will be present and whether the calculated deflection will produce an opening of the necessary tolerance, as well as the kind of stress that will be generated in the pin along with its corresponding life expectancy. Let us now assume that the desired depth of hole is $\frac{3}{8}$ in.

The deflection is calculated as follows:

$$a = \frac{Wl^3}{8EI}$$

$$= \frac{1000l^4}{8 \times 30,000,000 \times 0.049 \times 0.0039}$$

$$l^4 = 0.0198$$

$$a = \frac{0.0198}{45,864} = 0.00043'' \text{ deflection}$$

The maximum stress, S, is found by:

$$S = \frac{Wl}{2Z} = \frac{1000l^2}{2 \times 0.006125} = \frac{1000 \times 0.1406}{2 \times 0.006125}$$

$$Z = 0.098d^2 = 0.006125$$

$$S = 11,480 \text{ psi}$$

These results indicate that a hole with 0.0004 in. variation may be satisfactory, and if the pin is made of a springlike material, properly heat-treated, it should last a long time.

Undercuts

Undercuts, whether external or internal, should be avoided if possible. In cases where it is essential to incorporate them in part design, a great many can be realized by appropriate mold design in which either sliding components on tapered surfaces or split cavity cam actions will produce the needed undercut. This obviously means an increased tool cost, in the neighborhood of 15 to 30 percent.

Some conditions, however, will permit incorporating undercuts with conventional stripping of the part from the mold. Certain precautions are necessary in order to attain satisfactory results. First, the protruding depth of the undercut should be two-thirds of the wall thickness or less. Second, the edge of the mold against which the part is ejected should be radiused to prevent shearing action, and, finally, the part being removed should be hot enough to permit easy stretching and return to the original shape after removal from the mold.

How easily the task can be accomplished depends on material elasticity and spring back.

Many threaded plastic caps are stripped from the cores instead of being unscrewed. Coarse threads with the crest of the core thread rounded and a material with good elongation and ability to spring back make it feasible to apply conventional part stripping. The undercut problems can be solved by the cooperation of the designer, moldmaker, and processor, since each product configuration presents different possibilities.

Thermoplastic Hinge

Hinge dimensions of lids and boxes made of thermoplastics such as polypropylene have been well established, and are shown in Fig. 6-29. The successful operation of such a "living" hinge depends not only on design, but also on ensuring that the flow pattern of the plastic through the hinge stretches the molecules to give a strong and pliable hinging section. In larger-sized parts, there is a tendency to place a gate at the center of the box and another at the center of the lid, with the result that the flow pattern is not conducive to creating favorable hinge strength.

Functional Surfaces and Lettering

Surfaces of plastics may be provided with designs that can give a good grip or that can stimulate wood, leather, etc. These types of surfaces should be specified in a manner that will not create undercuts during withdrawal from a mold. The undercut effect can be responsible for stresses and marring. A similar condition applies to lettering, and the location of such lettering should conform to smooth withdrawal requirements.

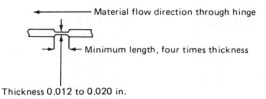

Thickness 0.012 to 0.020 in.

Fig. 6-29 Thermoplastic hinge.

Tolerances

There is a Society of the Plastics Industry standard that specifies limits for certain dimensions, and each material supplier converts the data to suit the specific material. Table 6–1 shows this information. (See Chapter 22 for such information on polycarbonate.) This type of information is intended to give the designer a guide for tolerances that are to be shown on the drawing; these tolerances include variation in part manufacture and some variation in tooling.

Tolerances on dimensions should be specified only where absolutely necessary. Too many drawings show limits of sizes when other means of attaining desired results would be more constructive. For example, if the outside dimensions of drill housing halves were to have a tolerance of ±0.003 in., this would be a tight limit. Yet if half of the housing were to be on the minimum side and the other on the maximum side, there would be a resulting step that would be uncomfortable to the feel of the hand while gripping the drill. A realistic specification would call for matching of halves that would provide a smooth joint between them, with the highest step not to exceed 0.002 in. The point is that limits should be specified in such a way that those responsible for the manufacture of a product will understand the goal that is to be attained. Thus we may indicate "dimensions for gear centers," "holes as bearing openings for shafts," "guides for cams," etc. This type of designation would alert a mold maker as well as molder to the significance of the tolerances in some areas and the need for matching parts

Table 6-1   Commercial tolerances for plastics based on standard set up by the Society of the Plastics Industry.

NOTE: The **commercial** values shown below represent common production tolerances at the most economical level. The **fine** values represent closer tolerances that can be held but at a greater cost.

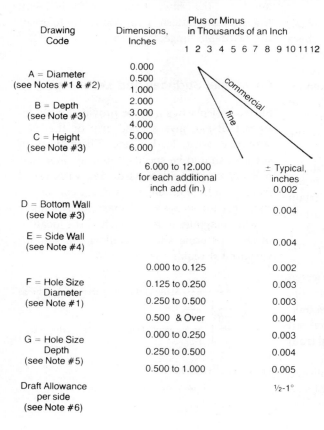

Drawing Code	Dimensions, Inches	Plus or Minus in Thousands of an Inch 1 2 3 4 5 6 7 8 9 10 11 12
A = Diameter (see Notes #1 & #2)	0.000 0.500 1.000	
B = Depth (see Note #3)	2.000 3.000 4.000	
C = Height (see Note #3)	5.000 6.000	
	6.000 to 12.000 for each additional inch add (in.)	± Typical, inches 0.002
D = Bottom Wall (see Note #3)		0.004
E = Side Wall (see Note #4)		0.004
F = Hole Size Diameter (see Note #1)	0.000 to 0.125	0.002
	0.125 to 0.250	0.003
	0.250 to 0.500	0.003
	0.500 & Over	0.004
G = Hole Size Depth (see Note #5)	0.000 to 0.250	0.003
	0.250 to 0.500	0.004
	0.500 to 1.000	0.005
Draft Allowance per side (see Note #6)		½-1°

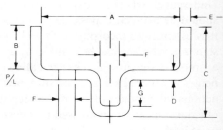

Reference Notes

1. These tolerances assume a mold temperature ≥ 200°F (93°C). Annealing at 300°F (150°C) will cause ≤ 0.1% overall dimensional change.

2. Tolerances based on ⅛" (3.2 mm) wall section.

3. Parting line must be taken into consideration.

4. Part design should maintain a wall thickness as nearly constant as possible. Complete uniformity in this dimension is impossible to achieve.

5. Care must be taken that the ratio of the depth of a cored hole to its diameter does not reach a point that will result in excessive pin damage.

6. These values should be increased whenever compatible with desired design and good molding technique.

SI conversion
Inches × 25.400 = millimeters

in other places and clearance for assembly in still other locations.

Most of the engineering plastics faithfully reproduce the mold configuration, and when processing parameters are appropriately controlled, they will repeat with excellent accuracy.

We see plastic gears and other precision parts made of acetal, nylon, polycarbonate, and Noryl whose tooth contour and other precision areas are made with a limit of 0.0002 in., and the spacing of the teeth is uniform to meet the most exacting requirements. (See Chapters 18 through 26 for information on different plastic materials.)

The problem with any precise part is to recognize what steps are needed to reach the objective and to follow through every phase of the process in a thorough manner to safeguard the end product. Throughout this book "shrinkage" is reviewed based on material characteristics. Different factors can cause variation in shrinkage; indeed, the way processing parameters can influence dimensional variation is very important. Some materials perform better than others in that respect.

Generally, if we approach tolerances according to their purposes ((a) functional requirements, such as running it, sliding fit, gear tooth contour, etc.; (b) assembly require-ments—that is, to accommodate parts with their own tolerances; and (c) matching parts for appearance or utility), we should come up with feasible tolerances that will be reasonable and useful. This will be more productive than trying to apply tolerances strictly on a dimensional basis.

Tolerances should be indicated only where they are needed, carefully analyzed for their magnitude, and of proven usefulness.

Adaptation of metal tolerances to plastics is not advisable. The reaction of plastics to moisture and heat, for example, is drastically different from that of metals, so that pilot testing under extreme use conditions is almost mandatory for establishing adequate tolerance requirements.

REFERENCES

1. "Advanced Engineering Design with Plastics," University of Lowell Plastic Seminars Workbooks, 1984/1985.
2. Dym, J. B., *Injection Molds and Molding,* Van Nostrand Reinhold, New York, 1979.
3. "Structural Design Manual," Superintendent of Documents, Government Printing Office, Washington, DC, 023–000–00495–0.
4. Beck, R. D., *Plastic Product Design,* Van Nostrand Reinhold, New York, 1980.
5. Lubin, G., *Handbook of Fiberglas and Advanced Composites,* Van Nostrand Reinhold, New York, 1982.

Chapter 7
Injection Mold Design

D. V. Rosato

INTRODUCTION

The function of a mold is twofold: imparting the desired shape to the plasticized polymer and cooling the injection molded part. It is basically made up of two sets of components: (1) the cavities and cores and (2) the base in which the cavities and cores are mounted. Figures 7-1 through 7-3 and Table 7-1 show typical layouts and descriptions of parts in the mold, including the cavities and cores. Different mold constructions are used. Other examples are shown in Fig. 9-2 (a–d). (See also references 1–22.)

The mold, which contains one or more cavities, consists of two basic parts: a stationary mold half on the side where the plastic is injected, and a moving half on the closing or ejector side of the machine. The separation between the two mold halves is called the parting line. In some cases the cavity is partly in the stationary and partly in the moving sections.

These factors are interrelated, but the size and weight of the molded parts limit the number of cavities in the mold and also determine the machinery capacity required. In the case of large molded parts, such as an auto radiator grille or a one-piece bucket chair, the large exterior dimensions of a single-cavity mold require a correspondingly large clearance between the machine tie-rods. Similarly, the machine tie-rod clearances basically limit the number of cavities that can be installed in a multicavity mold.

From the general concept of the molding operation it is important to design a mold that will safely absorb the forces of clamping, injection, and ejection. Furthermore, the flow conditions of the plastic path must be adequately proportioned in order to obtain uniformity of product quality in cycle after cycle. Finally, effective heat absorption from the plastic by the mold has to be incorporated for a controlled rate of solidification prior to removal from the molds.

The mold designer should become thoroughly familiar with the processing information of the plastic material for which the mold is being built, so that the molding factors that enter into the design are fully taken into account. (See Chapters 18 through 27 for information on material processing, particularly Chapters 19 and 26.)

The mold determines the size, shape, dimensions, finish, and often the physical properties of the final product. It is filled through a central feed channel, called the sprue. This sprue, which is located in the sprue bushing, is tapered to facilitate mold release. In single-cavity molds, the sprue usually feeds the polymer directly into the mold cavity, whereas in multicavity molds it feeds the polymer melt to a runner system, which leads into each mold cavity through a gate.

The mold is aligned with the injection cylin-

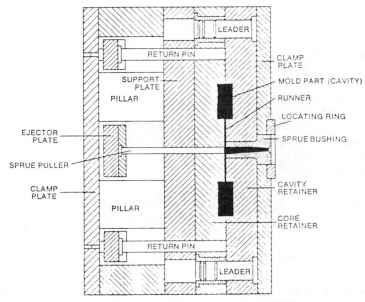

Fig. 7-1 General configuration of a mold (DME).

der by means of a ring in the stationary mold half, into which the cylinder nozzle seats. The locating ring surrounds the sprue bushing and is used for locating the mold in the press platen concentrically with the machine nozzle. The opening into which the ring fits is made to a tolerance of −0.000 and +0.002 in. The ring itself is made 0.010 in. smaller than the open-

ing, providing a clearance of 0.005 in. per side. A clearance above this amount may cause misalignment with the nozzle, which in turn would entrap part of the sprue, causing sprue sticking on the wrong side. The sprue bushing on the locating ring end has a spherical radius of $\frac{1}{2}$ or $\frac{3}{4}$ in. to fit the machine nozzle radius. The hole through the length of the sprue has

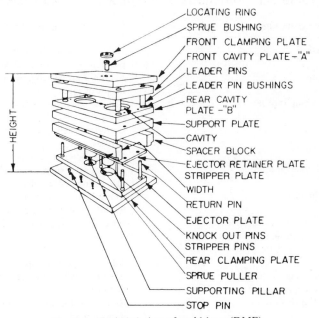

Fig. 7-2 Exploded view of mold base (DME).

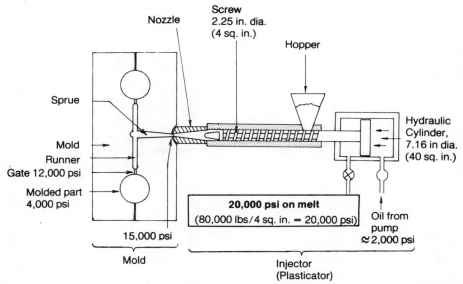

Fig. 7-3 Example of pressure loading on plastic melt; melt moves from the injection unit, through the sprue, runners, and gates, and into the cavities.

a $\frac{1}{2}$ in./ft taper of 1 degree, $11\frac{1}{2}$ minutes per side. This hole must have a good reamed and polished finish to prevent sprue sticking (2).

The parting line is formed by cavity plates A and B. Cavity plate A retains the cavity inserts and supports the leader pins, which maintain the alignment of cavity halves during operation. These guide pins are preferably mounted in the stationary mold half to ensure that the molded product(s) will fall out of the mold during ejection without being fouled. One of the four leader pins is offset by an amount of about $\frac{3}{16}$ in. to eliminate the chance of improper assembly of the two halves.

Leader pins are made 0.001 in. below nominal and to a tolerance of +0.0000 and −0.0005

Table 7-1 Functions of the injection mold (DME).

MOLD COMPONENT	FUNCTION PERFORMED
Mold Base	Hold cavity (cavities) in fixed, correct position relative to machine nozzle.
Guide Pins	Maintain proper alignment of two halves of mold.
Sprue Bushing (Sprue)	Provide means of entry into mold interior.
Runners	Convey molten plastic from sprue to cavities.
Gates	Control flow into cavities.
Cavity (female) and Force (male)	Control size, shape, and surface texture of molded article.
Water Channels	Control temperature of mold surfaces, to chill plastic to rigid state.
Side (actuated by cams, gears, or hydraulic cylinders)	Form side holes, slots, undercuts, threaded sections.
Vents	Allow escape of trapped air and gas.
Ejector mechanism (pins, blades, stripper plate)	Eject rigid molded article from cavity or force
Ejector Return Pins	Return ejector pins to retracted position as mold closes for next cycle.

in. The bushings are made 0.0005 in. above nominal and to a tolerance of +0.0000 and −0.0005 in. This provides a maximum clearance between leader pin and its bushing of 0.0025 in. and a minimum of 0.0015 in. In actual practice, this amount of play may never exist because of the four pins and bushings and the tolerance of their location. If by chance the space materializes and the job requires extreme accuracy of alignment, a set of tapered plugs and sockets can be incorporated into the base.

Mating with A plate is B plate, which holds the opposite half of the cavity or the core and contains the leader pin bushings for guiding the leader pins. The core establishes the inside configuration of a part. B plate has its own backup or support plate. The backup B plate is frequently supported by pillars against the U-shaped structure known as the ejector housing. The U-shaped structure, consisting of the rear clamping plate and spacer blocks, is bolted to the B plate, either as separate parts or as a welded unit. This U-shaped structure provides the space for the moving ejector plate or ejection stroke, also known as the stripper stroke. The ejector plate, ejector retainer, and pins are supported by the return pins. When in an unactivated position, the ejection plate rests on stop pins. When the ejection system becomes heavy because of required high injection forces, additional supporting means are provided by mounting added leader pins in the rear clamping plate and bushings in the ejector plate.

The overall height of the mold should correspond to the open space in between the machine platens. In the moving mold half, spacers are used to create space for the ejector system, which consists of two ejector plates with ejector pins. The open space should be such as to permit the ejector pins to complete their ejection stroke. Note that the mold height, or dieheight, in the usual horizontal operating machine is the horizontal dimension of the mold. When the mold is removed and placed "upright" on a workbench, its mold height is vertical.

All the mold plates (excluding the ejector parts) and spacer blocks are ground to a thickness tolerance of ±0.001 in. Conceivably a combination of tolerances could build up to cause an unevenness at the four corners. If great enough, such a condition would damage a platen when under full ram pressure. It is advisable to check the uniformity of all four corners prior to preparing the base to receive cavities.

Both mold halves are provided with cooling channels filled with coolant to carry away the heat delivered to the mold by the hot thermoplastic polymer melt. With thermosets, electric heaters are located in the mold.

When the mold opens, molding and sprue are carried on the moving mold half; subsequently, the central ejector is activated, causing the ejector plates to move forward, so that the ejector pins push the article out of the mold. Ejector pins have a tendency to produce a very slight flash line, which in some areas of a part may be objectionable; therefore, their location and the amount of recess formed by them in the part should be agreed upon with the product designer.

In the smallest injection molding machines, the mold may be completely demountable, and while being filled is held in a simple vise. This can be vertically or horizontally acting to suit the cylinder; some cylinders are down-stroking and some horizontally acting. With a horizontally acting cylinder and a vertical clamp, the runners and sprue bushing are in the same plane; and often because the pressures involved are not very great, the hardened sprue bushing is replaced by a simple runner cut into one or both halves of the mold.

With the larger horizontal clamping, thought should always be given to whether a horizontal or a vertical flash line is either possible or desirable. In Fig. 7-4 a vertical flash line is shown, whereas Figs. 7-1 and 7-3 show the more common horizontal flash line. With the vertical flash line, sometimes called a positive mold, it can be seen that material cannot escape from the parting line of the mold until considerable opening movement has taken place. If an over-full shot were to be made, the mold would not "flash" although the shot weight might be too great. This design also has the advantage that over-

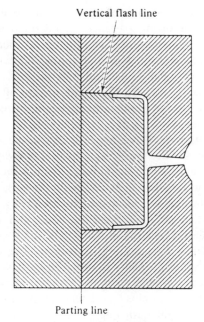

Fig. 7-4   Mold designed with vertical flash line or a positive mold system.

size moldings can be made with a relatively small mold locking force. If the correct amount of material is injected into the mold it may open slightly (a few thousandths of an inch) but no material escapes; and as the molding cools, the mold closes again, thus "compression-molding" the part. This "positive mold" is the principal type used in compression molding. It is also the design used with certain "structural foam" molding and in "combining" injection molding with compression.

TYPES OF MOLDS

There are basically six types of injection molds in use today, constituting about a two-billion-dollar market in the United States alone. These types are: (1) the cold-runner two-plate mold; (2) the cold-runner three-plate mold; (3) the hot-runner mold; (4) the insulated hot-runner mold; (5) the hot-manifold mold; and (6) the stacked mold. Figures 7-5 through 7-7 illustrate these six basic types of injection molds (10).

A two-plate mold consists of two plates with the cavity and cores mounted in either plate. The plates are fastened to the press platens,

and the moving half of the mold usually contains the ejector mechanism and the runner system. All basic designs for injection molds have this design concept. A two-plate mold is the most logical type of tool to use for parts that require large gates. This cold-runner system results in the sprue, runners, and gates solidifying with the cavity plastic material.

The three-plate mold is made up of three plates: (1) the stationary or runner plate, which is attached to the stationary platen and usually contains the sprue and half of the runner; (2) the middle or cavity plate, which contains half of the runner and gate and is allowed to float when the mold is open; and (3) the movable or force plate, which contains the molded part and the ejector system for the removal of the molded part. When the press starts to open, the middle plate and the movable plate move together, thus releasing the sprue and runner system and degating the molded part. This type of cold-runner mold design makes it possible to segregate the runner system and the part when the mold opens. The die design makes it possible to use center-pin-point gating.

With the hot runner, the runners are kept hot in order to keep the molten plastic in a fluid state at all times. In effect this is a "runnerless" molding process, and is sometimes called that. In runnerless molds, the runner is contained in a plate of its own. Hot-runner molds are similar to three-plate injection molds, except that the runner section of the mold is not opened during the molding cycle. The heated runner plate is insulated from the rest of the cooled mold. Other than the heated plate for the runner, the remainder of the mold is a standard two-plate die.

Runnerless molding has several advantages over conventional cold runner-type molding. There are no molded side products (gates, runners, or sprues) to be disposed of or reused, and there is no separating of the gate from the part. The cycle time is only as long as is required for the molded part to be cooled and ejected from the mold. In this system, a uniform melt temperature can be attained from the injection cylinder to the mold cavities. Shot size capacity and clamp tonnage required

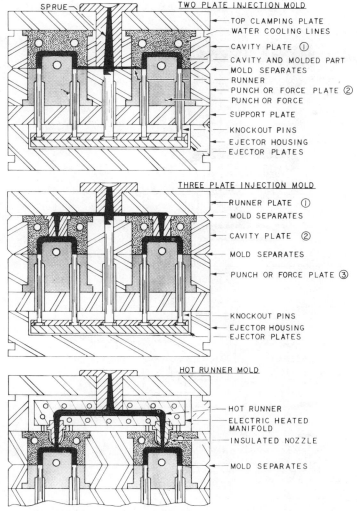

Fig. 7-5 Types of injection molds illustrated here are cold-runner two-plate, cold-runner three-plate, and hot-runner (10).

in the injection molding machine are decreased by the size of the sprue and runners.

The insulated hot-runner is a variation of the hot-runner mold. In this type of molding the outer surface of the material in the runner acts like an insulator for the molten material to pass through. In the insulated mold, the molding material remains molten by retaining its own heat. Sometimes a torpedo and a hot probe are added for more flexibility. This type of mold is ideal for multicavity center-gated parts. The size of the runner is almost twice the diameter when compared to the cold-runner system.

The hot-manifold is a variation of the hot-runner mold. In the hot-manifold die, the runner, and not the runner plate, is heated. This is done by using electric-cartridge-insert probes in sprue, runners, and gates.

Basically a stacked mold is a multiple two-plate mold, with the molds placed one on top of the other. This construction can also be used with three-plate, hot-runner, and insulated hot-runner molds. A stacked two-mold construction doubles the output from a single press and only requires the same clamping force on the mold if a duplicate set of cavities is used or the maximum clamping cross-sectional area is not exceeded. The machine will require additional shot capacity. Stacked

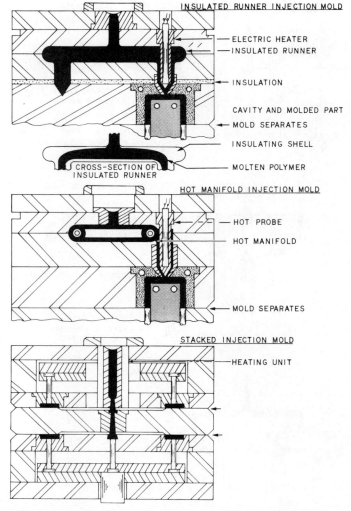

Fig. 7-6   Types of injection molds illustrated here are insulated runner, hot-manifold, and stacked (10).

molds are also being used with more than "two-plates" (17).

FLOW OF PLASTIC MELT

The design of molding is governed first by its intended function, and second by the specific limitation of the injection molding process. The properties of the plastic to be used and the engineering aspects of the mold design are added factors.

Consequently, the designing of injection moldings requires not only a thorough knowledge of plastics properties but a sound insight into the problems of injection molding and

mold design. For this reason, close cooperation among the experienced product designer, raw material supplier, processor, and mold designer is a prerequisite for a product that satisfies the particular requirements of its function and of the injection molding process, and can be produced economically. (Other sections of this book underscore the importance of this interdependence of materials, molder, etc.)

Although the injection molding process offers a wide degree of freedom of design, optimum results can be obtained only if the product designer takes the numerous processing factors into account and realizes that the de-

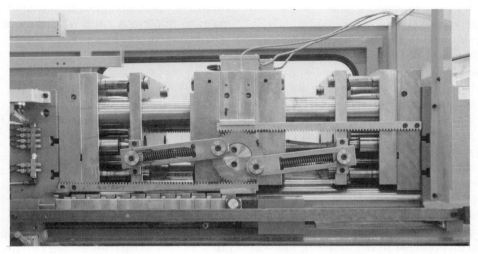

Fig. 7-7 Stacked injection mold with stripper plates in the open position. Courtesy of Husky.

sign will greatly influence the eventual mold construction.

Whether or not a certain part can be made by injection molding depends first of all on the flow properties of the plastic. Thus, as far as size and shape are concerned, the designer is often faced with certain limitations. Even under optimum molding conditions, very long flow paths, large surfaces, or excessively thin sections may result in short shots.

The extent to which mold cavity dimensions should be larger than the required product dimensions will depend on the total shrinkage of the plastic. For crystalline plastics, total shrinkage may be taken to be the sum total of mold shrinkage and after-shrinkage.

Mold shrinkage is the difference between the dimensions of the mold cavity and the molding immediately after injection molding and cooling in air. The degree of mold shrinkage depends on the plastic type, processing conditions as they relate to the flow of the melt, and product shape. Moreover, there is a difference between shrinkage in the flow direction of the plastic and shrinkage across this flow direction. This difference may be substantial, particularly in the case of glass fiber reinforced plastics.

The design of a molding must satisfy the functional requirements of the final product, but full allowance must be made for the spe-

cific nature of the injection molding process. Curved, grooved, or corrugated surfaces are preferred to flat ones, as the latter are always liable to warp. Warping of flat surfaces can be prevented by means of ribs, but these ribs have a tendency to show up on the other side of the wall as light sink marks. Corners must be rounded, to reduce the risk of notch sensitivity and stress concentrations. Also rounded corners offer less flow resistance.

In the design of injection moldings, wall thickness should be kept as thin and uniform as possible. This ensures: (1) minimum plastic consumption, (2) minimum cycle time resulting from shorter solidification time, (3) uniform shrinkage throughout the molding, (4) uniform mold filling, and (5) minimum risk of internal stresses.

When design requires differences in wall thickness, the transitions must be gradual. As a general rule, reinforcing ribs must be thinner than the wall they reinforce (about two-thirds of the wall thickness), and their height must not exceed about three times the wall thickness. Wall thickness is governed not only by the functional requirements in service but also by the size of the molding and, more important, the length of the flow path.

The flow of the plastic melt in the mold depends on various factors: plastic used, temperature, mold temperature, length and diam-

eter of sprue and runners, gate type, etc. Together, these factors determine a certain
minimum wall thickness. It is understandable
that for easy-flow, low viscosity injection
molding materials the minimum wall thickness that can be filled is smaller than for stiffer-
flowing materials having higher viscosity
(lower melt index).

Factors differ for practically each different
design and plastic, so that an exact specification of minimum wall thickness in relation
to flow path is not easily given. However, there
is a certain relationship between wall thickness
and length of flow path that can be used for
most plastics. The length of flow path attainable is proportional to the square of the wall
thickness ratio in the range of 0.020 to 0.080
in. thickness. Thus if a plastic melt has a flow
path of 4 in. with 0.040 in. wall thickness,
an increase in the wall thickness to 0.060 in.
will increase its flow path to:

$$\left(\frac{0.06}{0.04}\right)^2 \times 4 = 9 \text{ in.}$$

Typical flow-path-to-cavity-thickness ratios
of general-purpose grades of thermoplastics,
based on a cavity thickness of 0.1 in. (2.54
mm) and conventional molding techniques,
are given in Table 7-2.

As the material flows through the mold,
its condition in the cavity is determined to a
major degree by the injection pressure that
compresses it into the desired shape. The effective pressure that exerts the densifying force
on the molded product is the component that
can be recorded in the cavity by a transducer
placed, for example, under the head of an injection pin. This cavity pressure component
is part of the total injection pressure indicated
on the hydraulic machine pressure gauge minus all the pressure drops of the numerous
passages (Fig. 7-3).

Cold-Slug Well

When we consider the heat condition between
the nozzle and sprue bushing, we find a nozzle
heated to about the same temperature as the
front of the cylinder contacting a relatively

Table 7-2 Approximate maximum flow-path-to-thickness ratio of thermoplastics.

ABS	175:1
Acetal	140:1
Acrylic	130–150:1
Nylon	150:1
Polycarbonate	100:1
Polyethylene	
low density	275–300:1
high density	225–250:1
Polypropylene	250–275:1
Polystyrene	200–250:1
Polyvinyl chloride, rigid	100:1

cool sprue bushing. As a result, the temperature at the nozzle tip is lower than the required
melt temperature. There is a gradual rise in
heat for about 0.5 to 1 in. depth of the nozzle,
to the point at which the normal melt temperature exists. The material lying in the nozzle
zone that is not fully up to heat does not have
good flow properties; therefore, if it entered
a cavity, it would produce defective parts.

To overcome this situation, a well is provided as an extension of the sprue to receive
the cool material, thus preventing it from entering into the runner system. The well is equal
in diameter to the sprue at the parting line,
and is about 1 to 1.5 times the diameter in
depth. These sizes may vary considerably, but
the important thing is to have the inside of
the nozzle of such shape and so heated that
the volume of cool material is less than the
cold-slug well.

In some materials, it is desirable also to
have smaller cold-slug wells at the end of the
runners or even their branches to prevent
some of the runner-cooled material from getting into the cavity.

A cold slug also performs the function of
providing the means of extracting the sprue
from its bushing, thereby acting as a retainer
for the sprue with runners on the moving half
of the mold. During stripping, a pin, which
is attached to the stripper plate and also forms
the bottom of the well, moves to eject the
sprue with runners from the mold. If you are
not sure that these wells are unnecessary, plan
to leave appropriate space for their inclusion
at a later date.

NUMBER OF CAVITIES

Once the plastic has been selected and the design of the article decided upon, a decision must be made as to whether a single- or a multiple-cavity mold should be used. Points to be taken into consideration include:

1. Number of moldings and period of delivery.
2. Quality control requirements (dimensional tolerances, etc.).
3. Cost of the moldings.
4. Polymer used (influencing location and type of gate).
5. Shape and dimensions of molding (influencing position of mold parting line and mold release).
6. Injection molding machine (determining shot capacity, plasticizing capacity, and mold release).

It is logical that the decision will aim to ensure optimum economy of production; however, there should be a sufficient guarantee of the quality of the product. Advantages and disadvantages must be weighed carefully (15).

Advantages of single-cavity molds are:

1. Their simple and compact construction; thus, lower cost and quicker construction, compared to multicavity molds.
2. That the shape and dimensions of moldings are always identical. In multiple-cavity molds, it could be extremely difficult to make intricate cavities exactly alike. Consequently, if technical articles are to be produced within very close dimensional tolerances, a single-cavity mold often may be preferred.
3. Their better process control, since processing conditions need only be adjusted to suit one molding.
4. That single-cavity molds allow greater latitude in design, both for product and material. The technical requirements regarding gating system, ejector system, cooling system, and mold parting line can, in the majority of cases, be met without compromise.

The complexity of multiple-cavity molds not only makes such molds expensive but increases the risk of faults in fully automatic operation. Moreover, it is often very difficult to set up a cooling system that provides effective cavity cooling without impairing the mold's operating reliability. This generally causes longer molding cycles.

Nevertheless, for long production runs, multiple-cavity molds are often the more profitable type. (Most molds produced are multicavity.) If large numbers of products must be molded in a relatively short period of time, the use of multiple-cavity molds offers distinct advantages. If very small articles must be molded, and no suitable machine is available, a multiple-cavity mold is the only possibility.

As a starting point, the number of cavities must be established. This is usually determined by the customer, who balances the investment in the tooling against part cost. From the molder's point of view, the number of cavities can be determined as follows:

1. The maximum number of mold cavities follows from the ratio of shot weight (S) and weight of molding (W) including sprue and runners. S is generally taken to be 80 percent of the shot capacity of the machine. If S is 200 grams and W 50 grams, the maximum number of mold cavities is:

$$\frac{S}{W} = \frac{200}{50} = 4$$

2. Additionally, the number of cavities is governed by the plasticizing capacity (P) of the machine and the estimated number of shots per minute (X). If P is 18 kg/hr = 300 g/min and X is 2, the number of cavities is:

$$\frac{P}{X \times W} = \frac{300}{2 \times 50} = 3$$

In this example the mold should not contain more than three cavities.

Other factors also affect the number of mold cavities, and the location of the mold cavities is also subject to restrictions. The limitation is that the distance between the outer cavities and the primary sprue must not be so long that the plastic melt loses so much heat in the runners that it is no longer sufficiently fluid to fill the outer cavities properly. Speed of filling tends to minimize this, provided the melt viscosity and runner cross section are adequate, and gates are not too restrictive.

The layout of the mold, dictated by product design, may also restrict the number of cavities, so that the capacity of the injection molding machine cannot be fully used. For instance, if side actions are used, the cavities must, of necessity, be situated in one or more parallel rows.

One of the most important aspects of multiple-cavity mold design is the layout of the feed to the cavities. The cavities should be so arranged around the primary sprue that each receives its full and equal share of the total pressure available, through its own runner system (so-called balanced runner system). This requires the shortest possible distance between cavities and primary sprue, equal runner and gate dimensions, and uniform cooling. When practical, a correct arrangement of cavities will avoid differences in product dimensions, stress buildup, mold release problems, flash, etc.

Multiple-cavity molds preferably should contain cavities of identical shape. In principle, different parts of an article should not be produced by means of one multiple-cavity mold, although this is sometimes done for reasons of economy. In this case, the largest cavities should be nearest to the sprue, and the runner and gate dimensions should be checked by test molding. If necessary, corrections should be made, by balancing the feed to each cavity first by using appropriate runner sizes; this subject will be discussed later. Figure 7-8 shows balanced and unbalanced cavity layouts. The example of the lower left layout in Fig. 7-8 has three runners, which divide and balance the cavities. This arrangement provides a flow pattern with cold-slug wells at the ends of the runners that serve to trap partly cooled plastic before it enters the cavities.

It is not good practice (although it is very often done) to have cavities of greatly differing size in a single mold. Many so-called family molds require this, and almost without exception they cause difficulties. It may be that some of the moldings warp, while others show excessive frozen-in strain, and any attempt at balancing for melt flow and cooling the cavities leads to long and tedious trails.

CLAMPING FORCE

The clamping force required to keep the mold closed during injection must exceed the force given by the product of the live cavity pressure and the total projected area of all impressions and runners. The projected area can be defined as the area of the shadow cast by the molded shot when it is held under a light source, with the shadow falling on a plane surface parallel to the parting line.

With cold-runner systems for thermoplastics (or so-called hot-runner systems for thermosets), the projected areas of runners and sprue are included with the cavity(s). When hot-runner systems are used for thermoplastics (cold-runner for thermosets), the force to move the melt in the runner does not "push" apart the molds at the parting line; instead, it "floats" within the mold.

As an example, if the total projected area is 132 sq in. and a pressure of 5000 psi is required in the cavity(s), based on the plastics being processed, clamping force required is:

Minimum clamping force = Projected area × plastic pressure cavity
= 132 sq in. × 5000 psi = 660,000 lb.

or:

$$= \frac{660,000 \text{ lb}}{2,000 \text{ lb}} = 330 \text{ tons}$$

Consider including a safety factor of about 10 to 20 percent to ensure sufficient clamping pressure, particularly when one is not familiar with the operation. Thus the standard injection molding machine maximum clamping

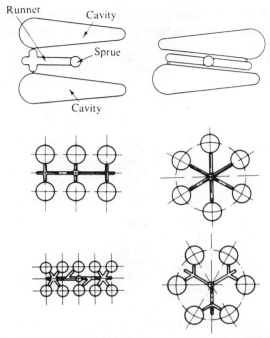

Fig. 7-8 Examples of cavity layout in multicavity molds. Right views show balanced systems, and left views are unbalanced systems except for top left that is balanced.

force could be 330, 375, or 400 tons. A guide to clamping pressure requirement of polyolefins and polystyrenes based on flow path length is given in Table 7-3.

For a true hydraulic fluid such as water, the clamping force required for each square inch of projected area would be equal to the unit pressure applied by the injection plunger. However, owing to the partial hardening of the plastic as it flows through the sprue and runners and into the cavity, the actual pressure exerted by the plastic within the cavity is much less than the applied plunger pressure. For this reason, an applied plunger pressure of 20,000 psi would seldom require a clamping force of more than 5 or 6 tons/sq in. of projected area of the plastic shot (2000 psi = 1 ton).

For a given plunger pressure, the actual pressure developed within the cavity varies directly with the thickness of the molded section, and inversely with the melt viscosity. Thick sections require greater clamping force than thin sections because the plastic melt in a thick section stays semifluid for a longer time during the cavity-filling injection stroke.

Similarly, a higher stock temperature, a hotter mold, larger gates, or a faster rate of injection will require a higher clamping pressure. As a general rule, good molding practice requires about three tons of clamp for each square inch of projected area of the molded shot (2).

By proper mold design and careful adjustment of molding conditions, it is sometimes possible to mold satisfactory parts with as little as one ton of clamp per square inch of projected area. However, it is unwise to attempt to operate a mold on this basis, as the range of permissible molding conditions would be seriously limited, and a long flow cavity fill could not be achieved.

It is also important to avoid applying too much clamping force to the mold. If a small mold is installed in a large machine and closed under full clamp, the mold can actually sink into the machine platens. Also, if the area of mold steel in contact at the parting line is insufficient, the mold may be crushed under the excessively high clamping force. Steel molds will begin to crush when the unit clamp pressure exceeds 10 tons per square inch of contact area. In less severe cases, the mold

Table 7-3   Clamping pressure required of polyethylene, polypropylene, and polystyrene based on flow path length and section thickness.

AVERAGE COMPONENT SECTION THICKNESS		CLAMPING PRESSURE REQUIRED (PSI OR KGF/CM² OF PROJECTED AREA)				
		RATIO OF FLOW PATH LENGTH TO SECTION THICKNESS				
IN.	MM	200:1	150:1	125:1	100:1	50:1
0.04		—	9,960	9,000	7,200	4,500
	1.02	—	706	633	506	316
0.06		12,000	8,500	6,000	4,500	3,000
	1.52	844	598	422	316	211
0.08		9,000	6,000	4,500	3,800	2,500
	2.03	633	422	316	267	176
0.01		7,000	4,500	3,500	3,000	2,500
	2.54	492	316	246	211	176
0.12		5,000	4,000	3,100	3,000	2,500
	3.05	352	281	218	211	176
0.14		4,500	3,500	3,100	3,000	2,500
	3.56	316	246	218	211	176

components may be distorted, or may fracture prematurely from fatigue.

While the projected area determines the clamping force required, the weight or volume of the molded shot determines the capacity of the injection machine in which the mold must be operated. Note that the shot weight or shot volume includes the weight of the sprue and runners, except in hot-runner molds. Capacities of injection machines are commonly rated in ounces of polystyrene that can be injected by one full stroke of the injection plunger.

Calculated against the mold clamping force, the total projected surface area of the moldings must not exceed the machine's maximum permissible molding area, when subjected to injection pressure. Machinery manufacturers usually provide this information.

Contact Area at Parting Line

Another item that requires attention is the contact area of the spacer blocks (2). The stress on these areas should be such as to prevent the embedding of these blocks into the plates, thus causing a change of the space for the ejection system. The safe tonnage that a mold base will take as far as spacer blocks are concerned can be calculated in this way. Let us take a $9\frac{7}{8}'' \times 11\frac{7}{8}''$ standard mold base

made of low carbon steel. The weakest section of the spacer bar is at the clamping slot area. For this size mold base, the width of the block is $1\frac{7}{16}''$ and the width of the clamping slot is $\frac{5}{8}''$. The area will be $1\frac{7}{16}''$ minus $\frac{5}{8}''$, or $\frac{13}{16}''$, $\times 11\frac{7}{8}'' \times 2$, since there are two blocks.

$$\text{Area} \times \text{allowable stress} = \text{Compressive force}$$

The allowable stress for low carbon steel is 25,000 psi. Thus:

$$\frac{13}{16}'' \times 11\frac{7}{8}'' \times 2 \times 25,000 \text{ psi} = 482,000 \text{ lb}$$
$$\text{or } 241 \text{ tons}$$

Higher-strength steel throughout the base can double or even triple the ability to absorb the compressive force. The addition of supporting pillars will also increase the compressive force in proportion to the area that they provide. Thus, two 2-in. diameter pillars would add the following force:

$$\text{Area of a 2-in.-diam. pillar} = 3.14 \text{ in.}^2 \times 2 = 6.28 \text{ in.}^2$$
$$6.28 \times 25,000 = 157,000 \text{ lb or } 78.5 \text{ tons}$$

The embedding problem does arise, especially when changes are made in supporting bar dimensions or supporting blocks.

SPRUE, RUNNER, GATE SYSTEMS

Complete runner systems, sprue, gates, etc., are shown in Figs. 7-1, 7-3, 7-8, and 7-9. The sprue, which forms the transition from the hot molten thermoplastic to the considerably cooler mold, is part of the flow length of the plastics and has to be of such dimension that the pressure drop is minimal and its ability to deliver material to the extreme "out" position is not impaired. The starting point for sprue size determination is the main runner, and the outlet of the sprue should not be smaller than the runner diameter at the meeting section. Thus, a $\frac{1}{4}$-in.-diameter runner would call for a $\frac{7}{32}$-in.-diameter "O" opening, for an average sprue length of 2 to 3 in. It has been established experimentally that for shots of 6 in.³ up to 20 in.³, the $\frac{7}{32}$-in. "O"

dimension will satisfy the low pressure-drop needs. For larger shots, a $\frac{9}{32}$-in. "O" opening would be indicated.

The material processing data give a range of runner sizes for each material. The smaller sizes can be applied for cases when the length of runners does not exceed 2 in. and the volume of material is less than 15 in.³ For economical reasons, it is preferable to keep the runners on the smaller end, since that not only reduces the amount of regrind but also accelerates the freezing of the gate, thus affecting cycle time. The pressure drop must be kept in mind. It becomes a matter of proportioning runners in relation to spacing of cavities, wall thickness of parts, length of cavities, and corresponding gate sizes.

Basically the distance from the injector (melt plasticator) of the injection machine to

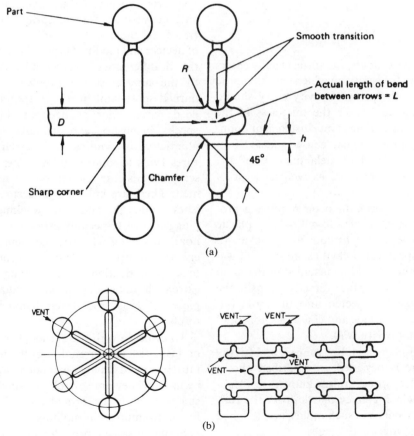

(a)

(b)

Fig. 7-9 (a) Effective length of runner bends; example for $R = \frac{1}{3}$ to $\frac{1}{2}$ of D, effective length; for sharp corners, effective length is 25 L; for chamfer $h = \frac{1}{3}$ of D, effective length is 2.5 L. (b) Balanced spoke runner layout (left) and balanced "H" runner layout (right).

the mold cavity(s) should be as short as possible. However, different factors must be considered that could require longer distances. One factor involves what was earlier discussed as number of cavities. Another factor relates to mold side actions that require longer runners. It is very important to allow sufficient space for cooling channels.

Perhaps the least understood and least well applied factor is the inclusion of cooling channels to meet proper heat transfer from the plastic melt to the cooling liquid (for thermoplastics). Usually insufficient space is allowed between cavities, particularly in molding the crystalline polymers (polyethylene, polypropylene, nylon, etc.) General information on cooling is reviewed later in this chapter. (See Chapters 12 and 13 for details on how to properly analyze mold cooling as well as mold flow.)

Sprue

In single cavity molds, the sprue usually enters directly into the cavity, in which case the sprue diameter at the point of cavity entry should be approximately twice the thickness of the molded article at that point. Insufficient diameter of the sprue gate can cause excessive frictional heating and/or delamination of the plastic at the gate area, as well as wear of the metal.

Too large a sprue diameter requires a prolonged molding cycle, to allow the plastic sprue sufficient time to cool for removal. In all direct-sprue-gated cavities, an internal water "fountain" should be installed in the mold to cool the mold surface directly opposite the gate. All plastic injected into the mold impinges on this surface and causes a "hot spot" on the metal cavity wall.

In three-plate and hot-runner molds, the main sprue is designed as described above. The smaller sprues (also known as "subsprues"), which convey plastic from the runners to the cavities in such molds, are designed to converge toward the gates.

The sprue area has been the location of more than its share of problems in the injection-molding process. The cause of most of these problems is the great temperature difference (about 300°F) between the nozzle and the sprue. The nozzle is a transfer system and must maintain a temperature to keep the plastic in the liquid state, whereas the sprue is part of the mold-fill system and maintains a temperature conducive to solidifying the plastic.

The devices applied in the area of the sprue do not address a graduated temperature change between nozzle and sprue. Among the more frequent problems are nozzle freeze-off, materials degradation, and nonuniform melt. These problems are aggravated when the materials are highly crystalline or temperature-sensitive. The usual approach to solving sprue problems is to design tools that minimize the length and size of the sprue, use a heated sprue, or eliminate the sprue altogether.

Efforts to overcome the temperature difference between nozzle and sprue have concentrated on the nozzle, resulting in a variety of devices and modified types of nozzles. When the fill difference is overcome by adding heat to the nozzle, severe problems can exist: burned spots, knit lines, gas trapping, weakened parts, color change, streaking, black specks, blemishes, and increased scrap. The alternative of running with cooler temperatures leads to almost an equal generation of scrap, in this case related to cold spots in the melt. There are knit lines and surface blemishes, and, in addition, sticking sprues, plugged gates (especially using pin gates), and nozzle freeze-off. This situation tempts the operator to resort to crude on-the-spot remedies to keep production going. Among the more extreme have been cardboard insulators, long pieces of brass rod—even hammers and a torch.

To dispense with the sprue when using hot or insulated runner molds or to feed direct into the mold cavity, extended nozzles can be useful. They are suitable for single-impression work or, in the form of a manifold nozzle, for multi-impression molding.

Sprue bushings provide an interface between the injection-machine nozzle and the runner system in the mold, and their design will vary greatly with the type of mold and the type of injection machine required for a particular molding job. Sprue bushings are generally pre-engineered catalog items, and it is usually a good idea to examine a large number of designs from various manufacturers before deciding on a bushing for a particular mold.

Runner Systems

Cavities should be placed so that (1) the runner is short and, if possible, free of bends, and (2) the supply of material to each cavity is balanced. This means that the shape and size of the runner—its length as well as the gate size—are for all practical purposes identical. This becomes especially important for precision parts.

A balanced supply ensures that any change made in any one of the molding parameters will affect all cavities in the same direction. It is good practice to use a runner plate of the same grade of steel as the cavities, that has a surface machined to 50 rms. In some applications, especially in cases of mild usage of a mold, there is a tendency to use the cavity plate for machining the runner in it. If a cavity protrudes on one side above the plate, a runner plate on that side is a must. Runner systems will vary in size and shape.

The surface finish of the runner system should be as good as that in the cavity, for example, machined to 50 rms. A good surface finish not only keeps the pressure drop low but also prevents a tendency of the runner to stick to either half of the mold. Such sticking would aggravate the highly stressed area of the gate portion to an even higher stress level.

The runners in multicavity molds must be large enough to convey the plastic melt rapidly to the gates without excessive chilling by the relatively cool mold. Runner cross sections that are too small require higher injection pressure and more time to fill the cavities. Large runners produce a better finish on the molded parts, and minimize weld lines, flow lines, sink marks, and internal stresses. However, excessively large runners should be avoided, for the following reasons:

1. Large runners require longer to chill, thus prolonging the operating cycle.
2. The increased weight of a large runner system subtracts from the available machine capacity, not only in terms of ounces per stroke that can be injected into the cavities but also in plasticizing capacity of the heating cylinder in pounds per hour.
3. Large runners produce more "scrap," which must be ground and reprocessed, resulting in higher operating cost and an increased possibility of contamination.
4. In two-plate molds containing more than eight cavities, the projected area of the runner system adds significantly to the projected area of the cavities, thus reducing the effective clamping force available.

Note that these objections do not apply to hot-runner or runnerless molds.

Various shapes of runners are used (Fig. 7-10). A full round (i.e., circular cross-section) runner is always preferred over any other cross-sectional shape, as it provides the minimum contact surface of the hot plastic with the cool mold. The layer of plastic in contact with the metal mold chills rapidly, so that only the material in the central core continues to flow rapidly. A full round runner requires machining both halves of the mold, so the two semicircular portions are aligned when the mold is closed.

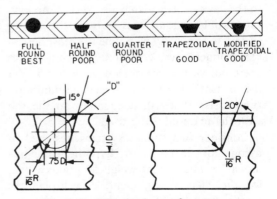

Fig. 7-10 Different shapes of runners.

There are, however, many mold designs that make it desirable to incorporate the runner in one plate only. In that case, a trapezoidal cross section is used, of a size that will surround a corresponding round diameter. In fact, if the trapezoid can be cut so that it would exactly accommodate a fully round runner of the desired diameter with the sides of the runner tapered at 5 to 15 degrees from vertical above the halfway line, this is almost as good as the round runner.

Design of Runner System.   Designing the smallest adequate runner system will maximize efficiency in both raw materials use and energy consumption in molding. At the same time, runner size is constrained by the amount of pressure drop and the injection capacity of the machine. Molders often seem unaware of how to balance these two equally important considerations (16).

Since molding a runner system does cost money, it makes sense to minimize the amount of nonsalable material molded into the runner. Even though the runner system will probably be reground and recycled, it is still important to keep its weight and size to an absolute minimum because some plastics tend to degrade during repetitive processing. A properly designed runner will not only help reduce costs but will also help preserve part quality.

Traditionally, there have been a number of misconceptions about proper runner design, many of which are still prevalent in molding shops. In the past, many injection molders and tool builders felt that the larger the runner, the faster the melt would be conveyed to the cavity. They also believed that the lowest possible pressure loss through the runner system to the cavity would be the most desirable. Runners were commonly machined into the mold with these objectives in mind. Yet it is in fact important to select the minimum runner size that will adequately do the job with the material being used.

Consider two runner systems designed in nylon, for example. A hypothetical "tradional" runner might weigh 50 grams while, for the sake of argument, a well-planned, smaller yet adequate runner weighs 20 grams.

Assume the mold produces 750,000 shots per year. At an electrical cost of 5¢/kWh and an energy requirement of 350 Btu/lb to plasticate nylon, the cost of molding the extra material in the overweight runner system is about $300/yr. The latter figure assumes close to 100 percent mechanical and electrical efficiency. Given the actual efficiency factors typical of many molding machines, however, an added cost of $1000/mold/yr with a poorly designed runner is not unreasonable. Multiply this amount by the number of machines in your shop, and you have an idea how much energy and money can be wasted by not carefully considering runner size.

Although properly sizing a runner to a given part and mold layout is a relatively simply task, it is often overlooked because the basic principles are not widely understood. For one thing, few processors are comfortable with using the straightforward arithmetical calculations involved. Also, the rules of runner design can be easily neglected in the rush to commit a part design to the tool maker. Lack of familiarity with the rules of optimum runner design undoubtedly leads processors to think there is some mystery involved, which there is not.

There are techniques for computing the minimum runner size required to convey melt at the proper rate and pressure loss to achieve optimum molded part quality. As a result, runner design has evolved from pure guesswork into an engineering discipline based on fundamental plastic flow principles. The molder who neglects the opportunity to "engineer" his runner systems is likely to miss a major opportunity to lower costs and improve productivity.

The computations are based on a key rheological property of the material to be molded. This property is the material's shear rate vs. its melt viscosity at several commonly encountered melt temperatures for the material. (Explanations of these terms are given in Chapters 26, 27, and 28.) Usually, this information is available from your resin supplier, and it is frequently displayed in molding manuals for individual materials. Figure 7-11 provides an example of such data.

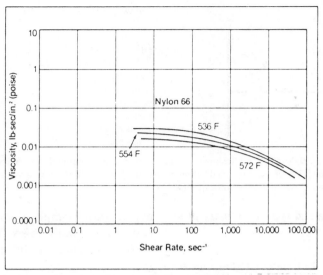

Fig. 7-11 Typical viscosity curve is available from most resin suppliers (or can be determined with proper equipment; see Chapter 29) and is essential to calculating optimum runner diameter.

Since no single calculation will do the job, it is necessary to start with a reasonable runner size, estimated on the basis of prior experience, which can then be refined with the aid of calculations. Initial considerations include the part weight and configuration and its performance or appearance requirements. For example, it is desirable when molding nylon to fill the part within 2 to 3 seconds. In fact, the same is true of the majority of injection molded parts made from crystalline thermoplastics, though not necessarily for amorphous resins.

Engineering a runner system requires an understanding of the pressure drop of the plastic melt as it passes through a channel. This pressure drop is controlled primarily by the volumetric flow rate or injection speed, the melt viscosity, and the channel dimensions. While it is possible to reduce the melt viscosity by increasing the melt temperature—hence reducing the pressure drop—most injection molding materials have an "ideal" melt temperature that provides fast cycles and optimum part quality. Thus, runner engineering should start by assuming an "ideal" melt temperature. This temperature can be found in the resin supplier's molding manual.

The other assumption that must be made intially is the amount of pressure drop that can be tolerated. The injection molding machine is usually capable of delivering 20,000 psi pressure. Since common sense rules against designing a mold to demand the absolute pressure limit of the machine, the mold should be designed so that the pressure required is somewhat less than the machine's capacity. A good value to assume is 10,000 to 15,000 psi. For the example shown here, a 15,000 psi injection pressure is assumed.

Unless the part design is unusual—such as in long, thin parts—or experience dictates otherwise, a pressure of 5000 psi is usually adequate to fill and pack out most parts. This means, in our example, that the runner system can be designed for 10,000 psi pressure drop. How is this done? The starting point is our hypothetical eight-cavity, balanced runner layout, shown in Fig. 7-12. We assume that all runners are the full-round type, that material specific gravity is 1.0, and that part weight is 15 grams. For eight cavities together, the total amounts to 120 grams or 7.31 in.³ Lengths of the primary, secondary, and tertiary runners are shown in the figure. We also assume a typical fill or injection time of 3 seconds. The foregoing are all fixed parameters; what remains to be determined is the optimum runner diameter; but to start with, we estimate the diameters as shown, going

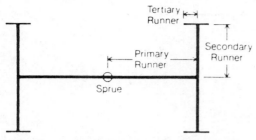

Part Weight: 15 gm   Specific Gravity: 1.0

	Length in.	Assumed Diam, in.	Revised Diam, in.
Primary Runner:	5	0.250	0.160
Secondary Runner:	3	0.200	0.150
Tertiary Runner:	1	0.150	0.125
Total Runner Volume, cu in.		1.01	0.510
Total Pressure Drop Into Cavity, psi		2400	9875

Fig. 7-12 Example of an eight-cavity mold shows how calculation proved that optimum runner system (and weight) was in this case only half that of the "assumed" runner system. Pressure drop in the assumed runner utilized only a fraction of the machine's available injection capacity.

by prior experience and typical industry practice.

Runner volume, V, is calculated as follows:

$$V = \pi r^2 L$$

where r is the runner radius and L is the length. Thus:
Primary runner: $V_p = \pi(0.125)^2(10)$
$$= 0.49 \text{ cu in.}$$
Secondary runner: $V_s = \pi(0.100)^2(12)$
$$= 0.38 \text{ cu in.}$$
Tertiary runner: $V_t = \pi(0.075)^2(8)$
$$= 0.14 \text{ cu in.}$$
Total shot volume (runner + parts)
$$= 7.31 + 0.49 + 0.38 + 0.14 = 8.32 \text{ cu in.}$$

Since the flow splits at the intersection of the sprue and primary runner into two identical halves of the runner system, we need only calculate the pressure loss through one-half of the mold. The volume of melt, therefore, that must be conducted through the primary runner in this half of the system is 4.16 cu in. Given our specified 3-second fill time, the desired flow rate is 1.39 cu in./sec. This is the volumetric flow rate, Q.

Now the shear rate, S_r, can be calculated:

$$S_r = \frac{4Q}{\pi r^3} = \frac{4(1.39)}{\pi(0.125)^3} = 906 \text{ sec}^{-1}$$

The melt viscosity at this shear rate and at the specified melt temperature must be read from a chart similar to Fig. 7-11. For this hypothetical example, the apparent melt viscosity is $\mu = 0.016$ lb-sec/in.^{2} (poise).

Next, we calculate the shear stress, S_s:

$$S_s = \mu S_r = (0.016)(906) = 14.5 \text{ psi}$$

Finally, the pressure drop, P, through that runner segment is calculated:

$$P = \frac{S_s(2L)}{r} = \frac{14.5(2)(5)}{0.125} = 1160 \text{ psi}$$

Now the next runner segment must be considered. The total volumetric flow through each secondary runner is 4.16 cu in. minus the volume in the primary runner:

$$\frac{4.16 - 0.25}{2} = 1.95 \text{ cu in.}$$

(Remember that the flow splits in half again at the secondary runner.)

The volumetric flow rate in each secondary runner segment is 1.95/3 or 0.65 cu in./sec. Thus:

$$S_r = \frac{4(0.65)}{\pi(0.100)^3} = 827 \text{ sec}^{-1}$$

The melt viscosity at the shear rate is 0.017 poise. Therefore:

$$S_s = (0.017)(827) = 14.0$$
$$P = \frac{(14)(2)(3)}{0.100} = 840 \text{ psi}$$

Volumetric flow through each tertiary runner can be calculated by subtracting the volumes of primary and secondary runners, or simply by adding together the total tertiary runner volume and the total part volume and dividing by eight cavities:

$$\frac{0.14 + 7.31}{8} = 0.93 \text{ cu in.}$$

The volumetric flow rate is thus 0.93/3 or 0.31 cu in./sec, and:

$$S_r = \frac{4(0.31)}{\pi(0.075)^3} = 936 \text{ sec}^{-1}$$

The viscosity corresponding to this shear rate is 0.016 poise, and:

$$S_s = (0.016)(936) = 15.0$$
$$P = \frac{(15)(2)(1)}{0.075} = 400 \text{ psi}$$

The total pressure loss from the sprue to each gate is the sum of the pressure losses through each segment:

$$\text{Pressure loss (total)} = 1160 + 840 + 400 = 2400 \text{ psi}$$

This preliminary calculation shows that much smaller channels can be designed to accommodate a 10,000 psi pressure loss. By repeating the calculations for progressively smaller runner diameters until we reach the targeted pressure loss, we eventually obtain the assumed runner diameters shown in Fig. 7-12.

In calculating and recalculating optimum runner diameters, the question may arise as to what is the appropriate relationship between the diameters of primary, secondary, and tertiary runners. In fact, there is no hard-and-fast rule for this, and the relationship can be relatively arbitrary. It is logical, however, that since each successive stage of the runner system carries less melt than the previous stage, the successive runner diameters normally run smaller. At times it is necessary to build molds where the number of cavities is not two, or where it is not possible to balance the cavity layout for equal flow distances to all cavities. While this type of design presents no particular problem in molding parts with loose tolerances, the effect on dimensions and part quality must be considered carefully when designing runner systems for critical parts. The primary objective in the latter case is to design a runner system so that all cavities fill at the same rate. This is necessary to ensure that they cool at the same rate and provide uniform shrinkage; surface gloss can also be affected. Molders will frequently try to balance the fill rates of individual cavities by changing the gate size. While this has some utility, it is a relatively ineffective way of making up for unbalanced runner layouts. The land length of the gate is too short to make any significant difference in pressure drop from one cavity to another. It is much better to vary the runner diameters and control fill rate.

Figure 7-13 shows an actual six-cavity mold that was used to make a large automotive part, in which the sprue was offset from the center of the runner system. Since we want all the cavities to fill at the same rate, what is required

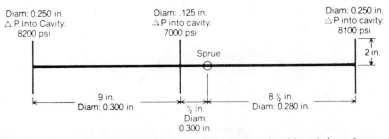

Fig. 7-13 By sizing the runner diameters (and not gate sizes) differently, this unbalanced runner system for an automotive job was adjusted for uniform cavity fill rates, indicated by the similar pressure-drop values. Pressure drop need not necessarily be identical in all cases for best results.

is a computation of the runner diameters that will provide the same pressure drop from the sprue bushing to the gate of each cavity. Clearly, since the runner lengths are different for each pair of cavities, different runner diameters will be required as well. As shown by a previous equation, pressure drop is proportional to runner length; so it is evident that the longer runner segments will need to be slightly wider. Figure 7-13 shows the actual lengths and diameters for each segment of the runner system. Note that the total pressure drop into the various cavities is similar though not identical for each; it is often impractical (and unnecessary) to exactly balance the pressure drop into each cavity. In this case, it was considered impractical to go smaller than $\frac{1}{8}$ in. for the diameter of the secondary runners closest to the sprue in order to raise the pressure drop there to a level closer to the level of the other secondary runners. In actuality,

the parts all filled uniformly, despite some degree of disparity in pressure drop leading into the cavities.

Figure 7-14 shows an extreme case of how runner diameter, not gate size, can be used to balance flow and pressure drop in an unbalanced cavity layout. Here again, we have an actual ten-cavity family mold, which produced dissimilar parts ranging in size from 2 in. in diameter by 1 in. long to $\frac{1}{4}$ in. in diameter by $\frac{1}{2}$ in. long. Nonetheless, as the numbers in the drawing show, it was possible to balance the pressure drops into the cavities quite closely.

The principles used in calculating optimum diameter of the final runner segments of a three-plate mold with multiple drops into the cavity are the same as those discussed above. Yet for most three-plate molds with multiple drops, it is frequently difficult to design them so that an equal volume of melt passes through

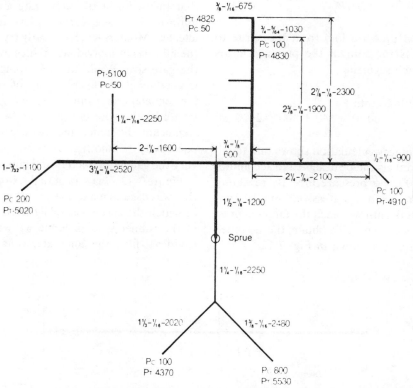

Fig. 7-14   An automotive family-mold runner system with ten cavities producing parts of widely varying dimensions and sizes. Sequence of three numbers shows runner-segment length and diameter, respectively, in inches, followed by pressure drop in psi. P_c = pressure drop in cavity and P_t = total pressure drop, the sum of all the pressure drops in the runner segments and cavity.

each drop. For circular parts with tight tolerances, it is nonetheless highly desirable that the part fill equally from each gate in order to minimize out-of-roundness. The answer is to use the procedures already described to calculate the pressure loss through each drop, and size the runner drop accordingly. Since the drops are usually tapered, the diameter is not constant. This difficulty can be circumvented by using the diameter at half the length as a basis for this calculation.

Sucker pins in the drop area will obviously influence the pressure loss and can provide additional restrictions to help equalize flow into each drop. Both the length and diameter of the sucker pin can be used to regulate the flow. However, it is seldom necessary to calculate the pressure loss across a sucker pin exactly; a resonable assumption will usually prove adequate.

For those who do not go through the calculations, industry-recommended runner diameters for different plastics are shown in Table 7-4.

The Hot Runner. There is nothing new about the runnerless molding process. Tools for this type of molding have been in use since the 1940s, with most of the activity starting during the early 1960s. Yet because of problems these molds have encountered (drooling,

Table 7-4 Recommended runner diameters, for use if runner size not calculated.

RECOMMENDED RUNNER SIZES INJECTION MOLDING

MATERIAL	DIA. IN IN.	METRIC M.M.
ABS, SAN	0.187–0.375	4.7–9.5
Acetal	0.125–0.375	3.1–9.5
Acrylic	0.312–0.375	7.5–9.5
Cellulosics	0.187–0.375	4.7–9.5
Ionomer	0.093–0.375	2.3–9.5
Nylon	0.062–0.375	1.5–9.5
Polycarbonate	0.187–0.375	4.7–9.5
Polyester	0.187–0.375	4.7–9.5
Polyethylene	0.062–0.375	1.5–9.5
Polypropylene	0.187–0.375	4.7–9.5
PPO	0.250–0.375	6.3–9.5
Polysulfone	0.250–0.375	6.3–9.5
Polystyrene	0.125–0.375	3.1–9.5
PVC	0.125–0.375	3.1–9.5

freeze-off, leakage, high maintenance, and others), runnerless molding has been used with some irregularity. However, new design concepts and tool-building methods have overcome these objections, and today's tools for runnerless molding are highly efficient and relatively fault-free.

The term "runnerless" refers to the fact that the runner system in the mold maintains the plastic resin in a molten state. This material does not cool and solidify, as in a conventional two- or three-plate mold, and is not ejected with the molded part. It is a logical choice for any high-speed operation where scrap cannot be reused.

There are two design approaches for tools used in runnerless molding: the insulated runner and the hot runner. Insulated-runner molds have oversize passages formed in the mold plate. The passages are of sufficient size that, under conditions of operation, the insulating effect of the plastic combined with the heat applied with each shot maintains an open flow path. Runner insulation is provided by a layer of chilled plastic that forms on the runner wall.

Hot-runner molds, which are the more popular of the two types, are generally built in two styles. The first is characterized by internally heated flow passages, the heat furnished by a probe or torpedo located in the passages. This system takes advantage of the insulating qualities of the plastics to avoid heat transfer to the rest of the mold.

The second, more popular system consists of a cartridge-heated manifold with interior flow passages. The manifold is designed with various insulating features to separate it from the rest of the mold, thus preventing heat transfer.

Of the two basic systems, the insulated runner has seen less attention in recent years. While the insulated-runner molds are generally less complicated in design and less costly to build than hot runners, they also have a number of limitations, including freeze-up at the gates, fast cycles required to maintain the melt state, long start-up periods to stabilize melt temperature/flow, and problems in uniform mold filling.

A great deal of interest has centered on the hot-runner molds since the industry has improved the distribution of heat and the level of temperature control. Furthermore, the industry has developed numerous components that enhance the design and construction of hot-runner molds. These "standard" components include a variety of cartridge-, and band-, or coil-heated machine nozzles, sprue bushings, manifolds, and probes; heat pipes; gate shut-off devices; and electronic controllers for the various heating elements. Because of this interest, the remainder of this section will focus on hot-runner molds.

The design of hot-runner molds should take into account the thermal expansion of the various mold components; this applies mainly to the center distances between the nozzles, supports, set bolts, and centering points. The bends in the hot runners to the nozzles should be generously radiused to prevent dead corners. In the design each nozzle contains a capillary to act as a valve to prevent plastic leakage. Heating elements positioned around the nozzles provide proper temperature control. When thick-walled articles are molded, the long after-pressure time may necessitate the use of nozzles with needle valves, as capillaries tend to freeze up rather quickly.

Heater loading in hot-runner manifolds is:

1. For general-purpose materials (polystyrene, polyolefins, etc.):

   15–20 W/in³ of manifold (0.09–0.12 W/cm³)

2. For high-temperature thermoplastics (nylon, etc.):

   20–30 W/in³ of manifold (0.12–0.18 W/cm³)

Heater loading in the gate torpedo for insulated runner molds is 35 W.

Not all materials or all parts are equally adaptable to runnerless molding; so each case must be judged individually.

Here is a checklist of considerations:

1. *Material:* Has it been processed by runnerless molding before? What does the materials supplier recommend? Not all of the thermoplastics have been molded via runnerless techniques, and the major problems are encountered with heat-sensitive materials, where the time/temperature relationship can be a problem. However, with today's technology, even the acrylonitriles and polyethylene terephthalate are being run successfully on hot-runner molds.

2. *Part:* Is the part weight sufficient? With current technology, a very small part may not require sufficient material to be purged through the nozzle tip, and degradation may occur from excessive residence time in the heated channel. Does the part require a runner? For instance, in the case of a family mold, it might be desirable to leave the parts together on a runner system until they reach the assembly station.

3. *Process:* Is the viscosity of the material (nylon, for example) such that a positive, drool-free shut-off is required?

4. *Volume:* Does the run justify the additional expense of a hot-runner system? Although there is no firm figure on how much more runnerless molding will cost than cold-runner molds, the tooling cost could run 5 to 7 percent more for standard tooling and applications and substantially more for nonstandard tooling. The additional mold cost must be compared with the anticipated savings in machine hours, scrap, etc.

To clarify a point, the term "runnerless mold" is a misnomer. With the exception of a mold with a single cavity which is fed directly from the machine nozzle, all injection molds have a runner system. This term originated in the use of insulated or heated runner channels in which the resin does not cool and solidify. No plastic is ejected from the runner channel when the mold is opened and the mold part ejected. Thus, the term "runnerless" is indicative of the absence of scrap from the runner system; and, more accurately, "runner-less molding" should be used.

Gates

The gate is given a smaller cross section than the runner so that the molding can be easily degated (separated from the runners). The positioning and dimensioning of gates is critical, and sometimes the gates must be modified after initial trials with the mold. Feeding into the center of one side of a long narrow molding almost always results in distortion, the molding being distorted concave to the feed. In a multicavity mold, sometimes the cavities closest to the sprue fill first and the farther cavities later in the cycle. This condition can result in sink marks or shorts in the outer cavities. It is corrected by increasing the size of some gates so that simultaneous filling of all cavities will result (10).

The location of the gate must be given careful consideration, if the required properties and appearance of the molding are to be achieved. In addition, the location of the gate affects mold construction. The gate must be located in such a way that rapid and uniform mold filling is ensured. In principle, the gate will be located at the thickest part of the molding, perferably at a spot where the function and appearance of the molding are not impaired. In this respect, it should be noted that large-diameter gates require mechanical degating after ejection, and always leave a mark on the product. It is for this reason that in small or shallow moldings the gate is sometimes located on the inside. However, this necessitates mold release from the direction of the stationary mold half, which interferes with effective cooling and generally increases mold cost.

Furthermore, the location of the gate must be such that weld lines are avoided. Weld lines reduce the strength and spoil the appearance of the molding, particularly in the case of glass fiber reinforced plastics.

Also, the gate must be so located that the air present in the mold cavity can escape during injection. If this requirement is not fulfilled, either short shots or burnt spots on the molding will be the result.

During the mold filling, thermoplastics show a certain degree of molecular orientation in the flow direction of the melt (as previously reviewed), which affects the properties of the molding. Important factors in this respect are the location and type of the gate.

The flow is largely governed by the shape and dimensions of the article and the location and size of the gate(s). A good flow will ensure uniform mold filling and prevent the formation of layers. Jetting of the plastic into the mold cavity may give rise to surface defects, flow lines, variations in structure, and air entrapment. This flow effect may occur if a fairly large cavity is filled through a narrow gate, especially if a plastic of low melt viscosity is used.

Jetting can be prevented by enlarging the gate or by locating the gate in such a way that the flow is directed against a cavity wall.

The hot plastic melt entering the cavity solidifies immediately upon contact with the relatively cold cavity wall. The solid outer layer thus formed will remain in situ and forms a tube through which the melt flows on to fill the rest of the cavity (Fig. 7-15). This accounts for the fact that a rough cavity wall adds only marginally to flow resistance during mold filling. Practice has shown that only very rough cavity walls (i.e., sand-blasted surfaces) add considerably to flow resistance.

For gate type and location, the points where two plastic flow faces meet must also be taken into consideration. If in these places flow comes to a standstill, which may be the case for flow around a core, premature cooling of the interfaces may cause weak weld lines. Although in practice sufficient strength may be obtained in such cases by good molding venting, high injection speed, and proper polymer and mold temperatures, the weld line can only be eliminated entirely by ring gating. Partial improvement is provided by a design in which the weld line has been shifted to a tab on the molding. This tab must be removed later, a step that involves additional cost, unless it is included in the design.

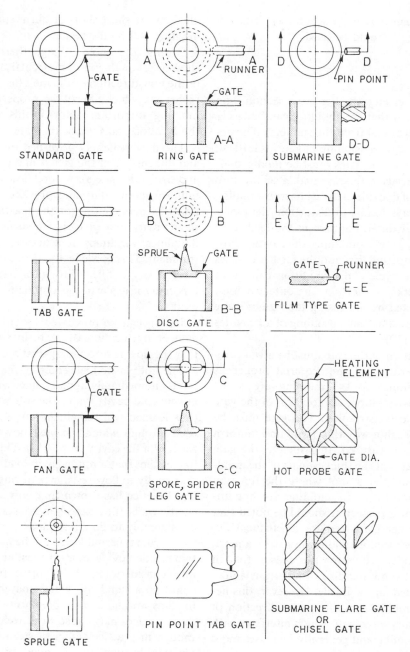

Fig. 7-15 Different gate types.

Weld lines may also be formed at places where the plastic flow slows down, for example, at a place where wall thickness increases suddenly. In grid-shaped articles, weld lines are mostly inevitable. By correct gate location, the plastic flows may be arranged so as to meet on an intersection, in which case the plastic continues to flow, so that better strength is obtained than if the weld line were situated on a bar between two intersections.

The following gate types are usually employed, and each has its own advantage for application (Fig. 7-15):

1. *Direct gate:* For single-cavity molds where the sprue feeds material directly

into the cavity, a direct gate is applied. A standard bushing, a bushing for an extended nozzle or a heated bushing, may be used. Good rapid mold filling occurs.

2. *Pinpoint gate:* Generally used in three-plate and hot-runner mold construction, it provides rapid freeze-off and easy separation of the runner from the part. The size of such gates may be as great as $\frac{1}{8}$ in., provided that the part will not be distorted during gate breaking and separation. A further advantage of pinpoint gating is that it can easily provide multiple gating to a cavity (for thin-walled parts), should such a move be desired for part symmetry or balancing the flow. It also lends itself to automatic press operation if the runner system and parts are arranged for easy drop-off. For a smooth and close break-off, it is best to have the press opening at its highest speed at the moment when the plates causing the gate to snap are separating.

3. *Submarine gate/Tunnel gate:* Often used in multicavity molds, it degates automatically, so is particularly suitable for automatic operation. For multiple cavities, an angular gate entrance requires special care in machining during moldmaking, in order to ensure uniformity of the gate opening and consistency in the angular approach for a balanced runner system. The angle of approach is determined by the rigidity of material during ejection and the strength of the cavity at the parting line affected by the gate. A flexible material will tolerate a greater angle of entrance than a rigid one. The rigid material may tend to shear off and leave the gate in place, thus defeating its intended purpose. On the other hand, the larger angle will give greater strength to the cavity, whereas a smaller angle may give a cleaner shearing surface than a larger angle.

4. *Tab gate:* This gate is used in cases where it is desirable to transfer the stress generated in the gate to an auxiliary tab, which is removed in a post-molding operation. Flat and thin parts require this type of gate.

5. *Edge gating:* Edge gating is carried out at the side or by overlapping the part. It is commonly employed for parts that are machine-attended by an operator. Normally it is possible to remove the complete shot with one hand and in a rapid manner. The parts are separated from the runner system by hand with the aid of side cutters or, if an appearance requirement demands it, by such auxiliary means as sanders, millers, grinders, etc. When degating is performed with the aid of auxiliary equipment, it becomes necessary to construct holding devices.

6. *Fin or flash gate:* This gate is used where the danger of part warpage and dimensional change exists. It is especially suitable for flat parts of considerable area (over 3×3 in.).

7. *Diaphragm-and-ring gate:* This gate is used mainly for cylindrical and round parts in which concentricity is an important dimensional requirement and a weld-line presence is objectionable.

8. *Internal ring gate:* This gate is suitable for tube-shaped articles in single-cavity molds.

9. *Four-point gate/Cross gate:* This is also used for tube-shaped articles, and offers easy degating. Disadvantages are possible weld lines and the fact that perfect roundness is unlikely.

10. *Hot-probe gate:* This may also be called an insulated runner gate, and is used in runnerless molding. In this type of molding, the molten plastic material is delivered to the mold through heated runners, thus minimizing finishing and scrap costs.

Gates should always be made small at the start; they can easily be made larger but cannot so easily be reduced in size. Gate dimensions are important. Since the pressure drop in a system is proportional to the length of the channel, the land length of the gate should be as short as possible, but the strength of the metal may be a limiting factor, as may its machining method (with EDM, a "razor" edge can be used). On the average, 0.040 to 0.060 in. is a suitable length. The cross-sectional area for thin wall parts generally has a width and height of 50 to 100 percent of the runner cross section. (An example of a gate for thicker walls is shown in Fig. 7-16.) Equations are available for determining gate sizes of different shapes based on the plastic shear rate and volumetric flow rate.

Where cavities are of different shot weights, the gate size of one cavity may be established arbitrarily as follows:

1. For round gates:

$$d_2 = d_1 \left(\frac{W_2}{W_1}\right)^{1/4}$$

2. For rectangular gates (assuming gate width constant):

$$t_2 = t_1 \left(\frac{W_2}{W_1}\right)^{1/3}$$

where d_1 = gate diameter of the first cavity (in. or cm)
d_2 = gate diameter of the second cavity (in. or cm)
t_1 = depth of gate in first cavity (in. or cm)
t_2 = depth of gate in second cavity (in. or cm)
W_1 = weight of first cavity component (oz or g)
W_2 = weight of second cavity component (oz or g).

MOLD VENTING

Every mold contains air that must be removed or displaced as the mold is being filled with a plastic material. The air present in the mold cavity must be allowed to escape freely during injection. At high injection speeds, insufficient mold venting may produce a considerable compression of the air, with consequent slow mold filling, premature plastic pressure build-up, and, in extreme cases, burning of the plastic (brown streaks on the molding).

Venting is done by small gaps or vents (dimensions are shown in Fig. 7-17) provided in the mold parting lines, or by other small channels in the mold (i.e., around ejectors pins, cores, etc.). Vents must be provided at the end of the flow path(s). A center-gated mold cavity, for instance, must be vented all around, whereas in an edge-gated cavity the vents must be provided at the cavity end generally or at the point where the flow path is expected to end. In gate design, and even in article design, allowance should be made for mold venting.

Vacuum venting of molds has not yet found widespread acceptance in injection molding of thermoplastics. However, in view of the

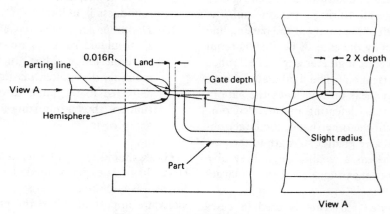

Fig. 7-16   Gate detail requirements.

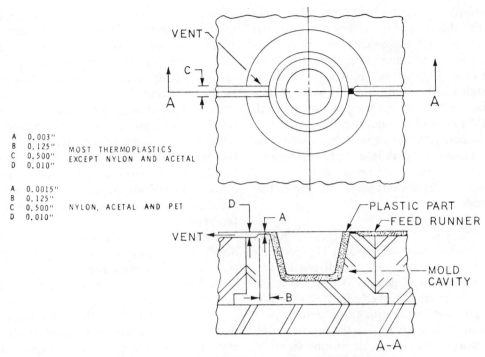

A   0.003"
B   0.125"   MOST THERMOPLASTICS
C   0.500"   EXCEPT NYLON AND ACETAL
D   0.010"

A   0.0015"
B   0.125"   NYLON, ACETAL AND PET
C   0.500"
D   0.010"

Fig. 7-17   A method of venting thermoplastic injection molds.

present trend toward higher injection speeds, it is most probable that in the future vacuum molds will be generally used to prevent venting problems.

In venting parts, the most minute flash may be objectionable—such as may occur with gears. While the depth of venting specified for each material is obtained after extensive testing by suppliers of raw materials, one must remember, in addition to the measured depth, to consider the peaks and valleys from the surface roughness of machining. This roughness measurement plus the "micrometer depth" should be considered as the value indicated in the tabulation. In the case of gears and similar parts, it may be advisable to adopt the following procedure for venting: (1) vent the runner system thoroughly, (2) vent all ejector pins as indicated on "material processing," (3) water-blast mating surfaces at the parting line with 200-grit silicon carbide abrasive, and (4) polish the vent in the direction of flow (it is also important to polish the cavity in the direction of melt flow—to eliminate problems on parts such as rough surface, sticking in the mold, etc.).

Vents are more important in thermosets than in thermoplastics. First of all, runners should be vented prior to approaching the gate. The vents should be the full width of the runner and 0.005 in. deep. The circumference of the cavity should be vented, and the vents should be spaced about 1 in. apart and be 0.25 in. wide and 0.003 to 0.007 in. deep, depending on the flow characteristics of the material.

A softer material would call for a lower value. Knockout pins should be as large as possible, and in most cases they should have 0.002-in.-deep flats—three or four of them ground on the circumference of the diameter, with the grinding lines parallel to the length of the pins. The grind should be with a fine-grit wheel. The end of the pin should have the corner broken by 0.005 in. so that if any flash is formed, it will adhere to the part.

Occasionally, it is necessary to place knock-out pins at the vent slots to ensure that the flash from the vents is physically removed, thereby assuring open vents for the following shot.

Waterline venting is a relatively new technique based on negative-pressure coolant technology. (Mold coolant is being pulled, not pushed, through the coolant system, producing a negative pressure in the coolant system.) One easy way to vent into a waterline is through ejector pins. The only twist to this method is that the ejector pin runs through a waterline, and the molding gases vent into the coolant rather than into the atmosphere. Coolant does not leak into the cavity because it is under atmospheric pressure.

Two distinct advantages are unique to this method:

1. The pin is kept cool because it passes right through the coolant. This prevents it from overheating, which can lead to a build-up of gummy deposits that exudes from the overheated plastic. The gummy deposits, in turn, tend to plug up the vent, and make the problem even worse: the increased compression of gases further heats the pin, resulting in even more gummy deposits. If the pin is kept cool, the problem does not even begin.
2. Placement of venting pins (and just plain pins, for that matter) has one less constraint: the location of water channels no longer dictates that a pin cannot be placed there. (The reverse, of course, is also true: the location of a critical pin no longer means that a water channel cannot be put there.)

Before we leave the subject of vented ejector pins, here is another way to easily provide more venting: vent the entire circumference of the pin. The tip of the pin is ground down to the proper vent depth and land, and then a pick-up groove is ground around the base of the vent. This vents back to the atmosphere or coolant via a few large flats ground in the major O.D. of the pins. This vent provides a great deal more venting area than a number of small flats and is easier to machine.

Porous metal provides another method of venting into the coolant. The chief advantage of this method is that a tremendous venting area is gained.

Figure 7-18 illustrates the primary concepts of the technique. It is particularly advantageous to install the vent at the top of a core such as the one shown, but it may be used in virtually any place where it can be attached in the mold and there is a waterline nearby.

Pressed metal, of course, will leave a texture on the plastic, but usually it can be located either where it is not seen or in a place where the texture does not matter; or it can be blended into a texture on the rest of the mold. Also, very finely woven porous metal is available that leaves only a very faint texture on the plastic.

Nor is improved venting per se the only benefit. Many new mold constructions are possible, such as gating a part as shown in Fig. 7-18. Normally this mold design would not be feasible because a bad burn would result at the top of the part. In this particular case, the tool would typically be designed as a three-plate mold, with its concomitant expense and complications. Many other unique designs are now possible with these techniques that reduce tool cost and improve molding efficiency and/or part appearance.

The water transfer process was designed primarily to cool long, thin cores, such as those for pen barrels, that have a hole in either end. The process is illustrated in Fig. 7-19. Coolant passes from one half of the mold to the other half, right through the part, when the mold is closed. When the mold is open, the supply to the mold is shut off, and both ends are

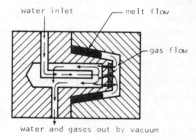

Fig. 7-18 Porous metal can be used to provide a very large venting area. The gases exhaust into the coolant. The key to the process is that the porous metal is directly cooled by the water behind it, which keeps it from overheating (porous metal has very poor heat conductivity) and plugging up with hot plastic. Coolant does not leak out because it is held at subatmospheric pressure. Pat. No. 4,091,069 of Logic Devices, Inc.

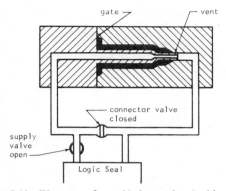

Fig. 7-19 Water transfer mold shown closed with coolant passing through part. Before mold opens, both valves shift and coolant is evacuated from mold. Venting occurs at tip of core. Pat. No. 4,151,243 of Logic Devices, Inc.

subject to subatmospheric pressure that evacuates the remaining coolant so that none leaks into the tool. Note that the design in Fig. 7-19 results in a parting-line vent at the end of the part opposite the gate, just where it is needed. In this case, the gases vent into the coolant rather than into the atmosphere.

With this type of construction, which is typical of water transfer applications, it is in fact almost difficult not to vent as well as to cool the part, since this would require a virtually perfect mate between the tip of the core and the cavity. Obviously, if venting does need to be added, it would require very little effort to grind an adequate venting clearance in the tip of the core.

MOLD COOLING

One of the most important aspects of mold design is the provision of suitable and adequate cooling arrangements. In all injection molding, even though it may involve having a heated mold, the purpose of the mold is to cool the molten plastic. If a mold had no means of cooling and was insulated to prevent any escape of heat by conduction, convection, or radiation, it would quickly reach the temperature of the material being molded and would no longer fulfill its function.

The cooling system is an essential mold feature, requiring special attention in mold design. It should ensure rapid and uniform cooling of the molding. In the design of mold components and the layout of guides and ejectors, allowances should be made for proper size and positioning of the cooling system.

Rapid cooling improves process economics, while uniform cooling improves product quality by preventing differential shrinkage, internal stresses, and mold release problems. In addition, uniform cooling ensures a shorter molding cycle. Rapid and uniform cooling is achieved by a sufficient number of properly located cooling channels. The location of these channels should be consistent with the shape of the molding, and should be as close to the cavity wall as allowed by the strength and rigidity of the mold.

Increasing the depth of the cooling lines from the molding surface reduces the heat transfer efficiency, and too wide a pitch gives a nonuniform mold surface temperature. (See Fig. 7-20.) Straight-drilled lines are preferred to bubblers. When it is necessary to use bubblers, they should be designed so that the cross-sectional area remains constant for the entire circuit. For tube bubblers, areas on both sides of the tube should be equal. Materials with higher thermal conductivity should be used if all of the heat cannot be removed with a steel mold. Examples of layouts usually employed for cooling systems are shown in Fig. 7-21.

The desired location of the heating–cooling passages is in the mold inserts themselves; they should be located close to where most of the heat has to be dissipated—that is, where most of the material is located.

The inclusion of fluid passages in the A and B plates as well as in their supporting plates adds to the ability to control cavity temperature, but not to the extent that one might expect. Although steel surfaces always have some heat-insulating film, the contact between them is never such as to induce the best conductivity. This was verified in practice by interposing a sheet of soft copper or brass between the B plate and its supporting plate on the core side, and checking for temperature while all other conditions remained the same. The average drop in temperature was found to be 25°F, and the core came close to the temperature of the cavity (2). Prior to this

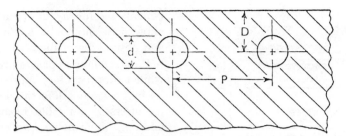

Recommended depth and pitch
d = Diameter of Water Line = 7/16 to 9/16 inch
D = Depth of Water Line = d to 2d
P = Pitch = 3d to 5d

Fig. 7-20   Recommended depth and pitch of mold cooling channels (consider depth of 1 diameter for steel, 1½ diameters for beryllium copper, and 2 for aluminum).

change, the core was running considerably hotter than the cavity; this made it possible to reduce the cycle by 30 percent. When a core consists of numerous thin sections that are difficult to arrange for individual control, the addition of a good heat conductor between plates may accomplish the desired result, provided that there are enough passages in the plates to make a good heat exchange possible.

Fluid passages for effective mold and part cooling should be placed to cover most of the molding surface and to be close to the mold face. The distance between mold face and fluid passage opening has to be large enough to resist distortion or flexing of the metal under injection pressures. The inlets and outlets for each cavity should be connected in parallel to their source of supply, thereby ensuring uniform heat transfer. The dimensions of the fluid passages should be such as to create a turbulent flow, since turbulent flow will dissipate about three times as many Btu's as laminar flow.

Assuming that a satisfactory cavity cooling–heating system has been provided, we now have to concern ourselves with thermally isolating the mold insert from the mold base. One practical way to do so is to provide circulating passages in the supporting plates and to maintain temperatures in them that will enable the cavity inserts to perform their function properly (i.e., consistently dissipate the heat introduced by the molded part).

Whenever there is a specification for straight, smooth, and dimensionally correct

openings, the cores making them will require special attention to temperature control. The nature of cores is such that material shrinking over them produces an intimate contact, and the bulk of the heat from the plastic is conducted into them. This condition necessitates an efficient way for dissipating the heat from the cores (2).

Cavity and core temperature control are also important to the proper functioning of the mold base. If cores are permitted to exceed the temperature of the cavities, the high heat of the cores will ultimately transfer onto the plate containing them. The B plates also hold the bushings for the leader pins. It is conceivable, and frequently occurs, that there is a difference of 30°F between mold halves. What would this mean to a 24-in. mold base? the expansion of the hotter side will be:

Expansion = Linear expansion × Length of mold × Temperature difference, °F

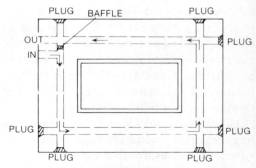

Fig. 7-21   Examples of cavity cooling systems.

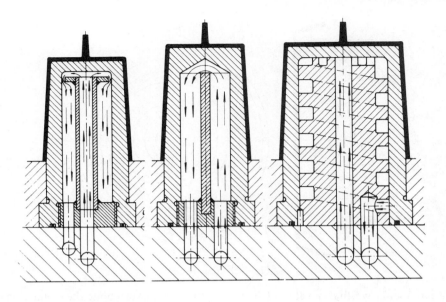

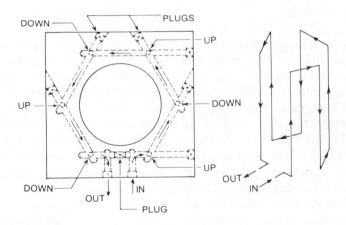

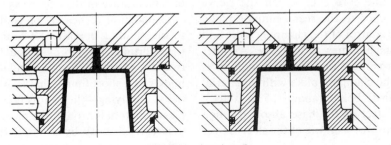

Fig. 7-21   (continued)

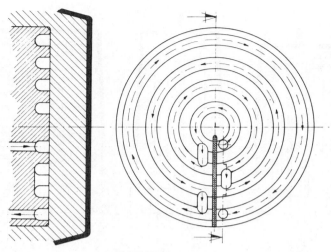

Fig. 7-21   (continued)

Using a handbook, we find a value of linear expansion for steel. Substituting, we have:

$$\text{Expansion} = 6.33 \times 10^{-6} \times \text{in.} \times {}^\circ\text{F}$$
$$= 6.33 \times 10^{-6} \times 24 \times 30$$
$$= 0.0046 \text{ in.}$$

This difference in expansion will cause binding, misalignment, difficulty in mold opening and closing, and, in the long run, excessive wear on the components that work together.

Another mold component that is affected by the temperature of the mold halves at the parting line is the stripper plate. In most cases, the stripper plate is near—or at—room temperature. For the majority of cases, the temperature difference in the stripper plate is compensated for by having an adequate clearance between plate, pin diameter, and pinhead to allow them to move freely to whatever position conditions dictate. There are cases, however, where this clearance provision does not apply. For example, when sleeve ejection is needed and the core over which the sleeve slides is attached to the rear clamping plate, temperature control of the clamping plate and stripper plate becomes a necessity. Another way to approximate the desired condition is to insulate the clamping plate with a material such as transite board (about 0.5 in. thick), and let the steel of the base absorb enough heat to permit free working of the sleeves over the core pins.

It is best to calculate the elongation of plates under the particular condition and to decide on the basis of the calculation what action should be taken.

Basic Principles in Heat Flow

Heat flows from a body of higher temperature to one of a lower temperature. It is the temperature difference—and not the amount of heat contained in separate bodies—that determines the flow of heat. The greater the difference in temperatures between two bodies, the greater the rate of heat flow between them. Heat can be transferred from one medium to another in three ways: radiation, conduction, and convection.

In the molding process of thermoplastics, we are not concerned with radiation heat, except for some postmolding operations involving stress relieving, straightening, or similar operations. In those operations, some source of infrared heat is utilized for rapid change of temperature of defined areas.

Conduction of heat is of vital importance to the molding operation. The mold material is performing the job of conducting the heat from the plastic to the mold and its circulating lines. For a short time—that is, while the plastic is solidifying—the heat is flowing from the plastic material through the mold; from there, it is carried away by a circulating medium. After temperature equilibrium is attained, heat flows in the reverse direction in order to maintain the temperature of the molded part at the desired level.

Extensive tests of various substances have led to verification of a formula for calculating the heat of conduction to be transferred in order to maintain certain temperature conditions (2). This formula is:

$$H = \frac{KAT(t_2 - t_1)}{L}$$

where A = area of the cavity in contact with the molding material.

T = time in hours, from the instant the plastic enters the cavity until the time ejection starts, i.e., cycle time per shot.

t_2 = temperature of the injection plastic.

t_1 = temperture of the circulating medium.

L = distance from the face of the mold to the start of the hole in which the circulation of the medium takes place.

Quantity H in the above formula is the heat in Btu conducted through a substance (mold) with a surface area A in square feet, during a time T in hours, when the difference in temperature is t_2 minus t_1 °F, the length or thickness through the substance (mold) is L in feet, and K is the thermal conductivity factor of the substance, expressed in Btu per hour, per square foot, per °F, per foot of length. Temperatures of the plastic material and mold surface, respectively, are expressed by t_2 and t_1.

The conductivity factor, which is related to molding conditions, may be characterized as follows: Conductivity factor K changes with the temperature at which conduction takes place. Most of the materials of interest to the plastics processor have been tested at 100°C (212°F). Since this may be considered a reasonable temperature for the molding operation, values for K are listed (as follows) at 100°C (212°F), in units of Btu/ft-°F-hr:

METALS		OTHER MATERIALS	
Stainless steel	10	Polystyrene	0.07
Tool steel (H 13)	12	Polypropylene	0.07
Tool steel (P-20)	21	Air	0.14
Beryllium copper	62	Nylon	0.14
Kirksite	62	Polyethylene	0.18
Brass (60–40)	70	Water	0.39
Aluminum	100		
Copper (pure)	222		

In the conduction formula, two factors will influence the heat conductivity. The K factor may be viewed as a consideration for cavity material selection, but in most cases the significant characteristics of the material are performance over a large number of pieces, integrity of shape, and controllability of fabrication; therefore, it is evident that during selection of a cavity material, its K factor is not a major consideration. However, one must keep the K factor of the selected material in mind while designing the heat-transfer system. The cooling arrangement will be more elaborate for a material with a low K value in comparison to a material of a higher K value.

The K value can be used most advantageously for small-size deep cores where a straight and uniform opening is needed. For this application, it can be calculated whether a beryllium copper pin or a steel pin with a copper core can be most effective in conducting the heat away. The area exposed to the plastic material times the K value will provide a comparative figure for the preferred selection, since all other factors are common.

For example a $\frac{3}{8}$-in.-diameter pin × 1.5-in. deep cored hole will have for beryllium copper a value of:

$$\text{Area} \times K = \tfrac{3}{8} \times 3.14 \times 1.5 \times 62 = 109$$

For a steel pin with $\frac{5}{16}$-in.-diameter copper insert, the value is:

$$\frac{5}{16} \times 3.14 \times 1.5 \times 222 = 326$$

The relative heat conductivity using beryllium copper, 109, against that using the steel pin inserted with a copper rod, 326, points to the decided advantage of the latter. It is assumed that during insertion of the copper, an intimate contact with the steel is established.

The L distance of the circulating fluid opening is determined by strength considerations, namely, by limiting the deflection to very low values and by thermal fatigue.

L can be calculated by viewing the condition shown in Fig. 7-22 as a beam fixed at both ends with a load in the middle. Using a handbook (under "Beams, Stresses in"), we find the stress in the middle is:

$$S = \frac{Wl}{8Z} = \frac{WD}{8Z}$$

where W = load on 1 in.^{2} of hole opening = 20,000 psi

l = length of beam = D = 0.4375

Z = section modulus

$$= \frac{bd^2}{6} = \frac{bL^2}{6} = \frac{2.29L^2}{6}$$

$b = \dfrac{1}{0.4375} = 2.29$

$d = L$

S = safe load stress = 10,000 psi

$Z = \dfrac{WD}{8S} = \dfrac{20,000 \times 0.4375}{8 \times 10,000} = 0.1094$

Substituting 0.1094 for Z we have:

$$0.1094 = \frac{2.29L^2}{6}$$

$$L^2 = 0.2862$$

$$L = 0.535$$

Most circulating holes are many inches long, and so the chances of a drill "running out" exist; therefore, it would be more practical to consider the L distance as $\frac{9}{16}$ to $\frac{5}{8}$ in. for H-type steels and $\frac{3}{4}$ in. for steels that are not formulated for thermal fatigue (2).

The next step is to account for the heat transfer by convection. Heat transfer by convection takes place when a medium, be it liquid or gas, moves from one place to another and carries heat with it; thus, the water or synthetic oil moved by a mold circulator through the circulating passages in the mold carries away or introduces heat for maintaining a desired temperature. Heat is taken on or given off by the circulating medium according to the basic heat equation:

$$H = MS(T_2 - T_1)T$$

Quantity H is the heat content in Btu, and M is the weight of material that circulates in time T hours, and produces a temperature difference of $T_2 - T_1$ in °F at the inlet and outlet of the mold, respectively. S is the specific heat value for the circulating medium. T hours is meant for the "curing" cycle duration. This basic heat formula can be expressed in terms related to mold design requirements.

Substituting the appropriate values for M, we have:

$$M = \text{Volume} \times \text{Specific gravity}$$

Volume = Q = Area of passage × Velocity × Duration of flow (curing cycle)

$$= A \times V \times T_f$$

where $A = 0.7854 \times \left(\dfrac{11}{32}\right)^2$ or 0.7854

$$\times \left(\frac{7}{16}\right)^2 \text{ or } 0.7854 \times \left(\frac{19}{32}\right)^2$$

= 0.073 in.^{2}; 0.118 in.^{2}; 0.217 in.^{2}

V = velocity of fluid in ft/sec

T_f = time in hours during curing

Thus:

$$M = A \times V \times T_f \times \text{Specific gravity}$$

For a given passage opening in the mold and a given cycle, the only element that is

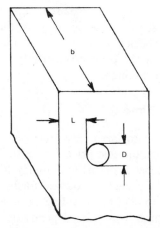

Fig. 7-22   Coolant hole distance from molding surface.

Table 7-5   Kinematic viscosity; centistokes.

Water Temperature (°F)	Viscosity$_n$
32	1.79
50	1.30
68.4	1.00
100	0.68
150	0.43
212	0.28

controllable is the velocity of the circulating fluid. It has been established that within a certain range of velocities the heat transfer is decidedly improved. This is the range in which turbulent flow takes place as contrasted with predominantly laminar flow. It is this turbulent flow that is most effective in heat transfer because of the transverse movement of the liquid particles. In laminar flow, the liquid particles arrange themselves in parallel layers with velocities at their highest in the center and decreasing in parabolic shape as they approach the wall of the passage; thus the layer next to the wall is moving very slowly, with little capacity to pick up heat.

The flow conditions in passages, whether laminar or turbulent, are characterized by a ratio known as the Reynolds number. The formula for the Reynolds number is:

$$R = \frac{7740\ VD}{n} \ \text{or} \ \frac{3160\ Q}{Dn}$$

V = fluid velocity, ft/sec

D = diameter of passage, in.

n = kinematic viscosity, centistokes, for 50°F water = 1.3 (see Table 7-5)

Q = flow rates, gpm

A Reynolds number of 2000 or less covers laminar flow, and 3500 and 5500 or even higher covers turbulent flow. In between 2000 and 3500, there is a transition stage. Viscosity appears in the formula for the Reynolds number, and because viscosity changes with temperature, we have to relate the flow to a specific water temperature and establish the gpm for the selected water temperature and range of Reynold numbers. For 50°F entrance water and $\frac{7}{16}$-in.-diameter opening, the gallons required for turbulent flow will be:

$$R = 3500 = \frac{3160\ Q}{0.4375 \times 1.3}$$

$$Q(\text{min}) = \frac{3500 \times 0.4375 \times 1.3}{3160} = 0.63 \text{ gpm}$$

For a Reynolds number of 5500, we will have an average flow rate of:

$$Q(\text{avg}) = \frac{5500 \times 0.4375 \times 1.3}{3160} = 1.00 \text{ gpm}$$

By substituting in the formula the $\frac{1}{8}$- or $\frac{3}{8}$-in. pipe-size holes, we can figure the corresponding flows. For $\frac{1}{8}$-in. pipe-size holes ($\frac{11}{32}$-in. diameter), the values of flow rate are:

$$Q(\text{min}) = 0.5 \text{ gpm}$$
$$Q(\text{avg}) = 0.75 \text{ gpm}$$

For $\frac{3}{8}$-in. pipe-size holes ($\frac{19}{32}$-in. diameter), the values are:

$$Q(\text{min}) = 0.855 \text{ gpm}$$
$$Q(\text{avg}) = 1.34 \text{ gpm}$$

At a Reynolds number of 3500, the heat conductivity is 1.5 times that of the laminar flow; at 5500, the conductivity is practically 3 times better.

For water temperatures above 50°F, the

corresponding viscosity will be substituted in the formula, and a new gpm will be obtained.

When the flow in each line approaches the gpm values shown, a decided improvement in heat conductivity will result. (See Chapters 12, 13, and 16.)

Heat Transfer by Heat Pipes

A heat pipe is a means of heat transfer that is capable of transmitting thermal energy at near isothermal conditions and at near sonic velocity. The heat pipe consists of a tubular structure closed at both ends and containing a working fluid. For heat to be transferred from one end of the structure to the other, the working liquid is vaporized; the vapors travel to and condense at the opposite end, and the condensate returns to the working liquid at the other end of the pipe. The heat transfer ability of saturated vapor is many times greater than that of solid metallic material.

Heat pipes can be used either to remove or to add heat. The smaller heat pipes, which can be used to operate against gravity, are equipped with thick homogeneous wicks and have higher thermal resistance, so that the heat transfer will not be quite as fast as in the case of gravity-positioned pipes. Even with the higher thermal resistance, these heat pipes still have a very high heat-transfer rate in comparison with solid metals.

Heat Balance of Halves

Some products are so shaped that the heat from the plastics is equally absorbed by each mold half. The vast majority of parts are such that they have a core of some depth and a cavity that surrounds the core. In this type of mold, the heat absorption of each half is different from the other.

Mold Connection for Fluid

Mold temperature connections should be placed away from the operator side and recessed wherever feasible so that danger of damaging them is eliminated. Whenever quick disconnect couplings are used, care should be taken to see that the openings in the fittings will not restrict the flow to the mold and to ensure that the proper velocity for turbulent flow is maintained.

Cooling Time

In addition to mold, raw material, and machine costs, the cost price of injection molded articles depends on the molding cycle. A large part of the cycle is accounted for by the time required to cool the molding to mold release temperature. This time depends on the heat of the molding.

In principle, the molding may be released from the mold as soon as its outer layer is sufficiently rigid, at a temperature called the mold release temperature. The inside of the molding will often still be considerably hotter than the outer part. Minimum cooling time required to reach mold release temperature is governed by:

1. Wall thickness of the molding.
2. Difference between polymer and mold temperature.
3. Difference between mold release temperature of the article and mold temperature.

The minimum cooling time may be estimated from the following equation (4):

$$S = \frac{-t^2}{2\pi\alpha} \log_e \left[\frac{\pi}{4} \frac{(T_x - T_m)}{(T_c - T_m)} \right]$$

where S = minimum cooling time (sec)
t = thickness of molding (in. or cm)
α = thermal diffusivity of material (in²/sec or cm²/sec)
T_x = ejection temperature of molding (often heat distortion temperature is used) (°F or °C)
T_m = mold temperature (°F or °C)
T_c = cylinder temperature (°F or °C)

Based on this formula, Fig. 7-23 illustrates the effect of mold temperature on cooling time.

REMOVAL OF PARTS FROM THE MOLD

Adherence of parts on the ejection half requires the placing of cores and other retaining means on the moving half of the mold so that there is no chance of parts hanging up in the cavity. Even a slight tendency to stick in any portion of the cavity will cause warpage, stresses, and dimensional distortion of parts. Such a tendency may indicate a need for additional taper, polish lines in the direction of withdrawal, or manipulation of the mold temperature.

In dimensioning the cavity and core, close attention is to be given to ensuring the unstressed retention of the part on the ejection side. This is normally accomplished by the plastic shrinking tightly over the cores and adhering to them. In such cases, it is desirable to have a relatively rougher surface on the cores than is incorporated in the cavities. In some configurations, it becomes desirable to provide narrow undercuts of 0.002 to 0.005 in. deep in the area of the ejection pins.

All surfaces in line with the mold opening direction, such as sidewalls, etc., must have a certain draft to facilitate ejection of the molding from the mold. Insufficient draft can cause deformation or damage. The draft required for mold release is primarily dependent on the depth of the cavity: the deeper the cavity, the more draft necessary. In the determination of the draft required, shrinkage (which differs for each plastic) must also be taken into account. If metal inserts are used, shrinkage can have an adverse affect on mold release,

which can be prevented by using more draft.

Another factor affecting mold release is the rigidity of the molding; rigid moldings require less draft than more flexible ones. In general it is recommended that a minimum draft of one degree be used. For small moldings, a draft of one-half to one degree may be sufficient in some cases, whereas for large moldings drafts up to three degrees may be required.

The ejection of a molding is generally by ejector pins, which are commercially available in various designs and qualities, or by strips, bushings, plates, or rings. The choice of ejector system is largely governed by article shape, and by the rigidity or flexibility of the plastic used. Whatever ejector system is chosen, ejection must never cause damage to, or permanent deformation of, the molding. The mold preferably should be fitted with ejectors at those spots around which the molding is expected to shrink (i.e., around cores or male plugs).

The force required to strip a molding off a male core may be determined approximately from:

$$P = \frac{S_t \times E \times A \times \mu}{d\left(\dfrac{d}{2t} - \dfrac{d}{4t} \times \gamma\right)}$$

where P = ejection force required (lbf or kgf)
E = elastic modulus (lbf/in.^{2} or kgf/cm^2)
A = total area of contact between molding and mold faces in line of draw (in.^{2} or cm^2)
μ = coefficient of friction between plastic and steel
d = diameter of circle of circumference equal to length of perimeter of molding surrounding male core (in. or cm)
t = thickness of molding (in. or cm)
γ = Poisson's ratio of the plastic
S_t = thermal contraction of plastic across diameter d
= coefficient of thermal expansion × temperature difference between softening point and ejection temperature × d (in. or cm)

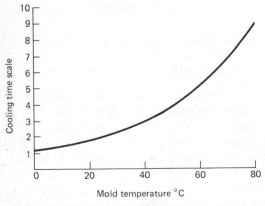

Fig. 7-23   Effect of mold temperature on cooling time.

At high mold temperatures allowance must be made for thermal expansion of the mold platens. These platens will expand more than the plates of the ejector mechanism. It is recommended that the ejectors be provided with a cylindrical head, and they should be mounted with some clearance to allow correction of possible variations in center distances during machine operation.

Ejection of articles with large cylindrical or flat surfaces may sometimes be hampered, as such surfaces tend to create a vacuum between the article and the cavity wall. In such cases, mold release may be improved and the vacuum broken by an air ejection system operated by air valves or channels operating independently or in conjunction with ejector pins.

Air ejection is effected by compressed air through channels provided in the mold and valve stem.

Related to an ejector is a sprue puller, which is a device used to pull or draw a molded sprue out of the sprue bushing. It is generally a straight round pin with the end machined in the form of an undercut. There are different methods in general use for providing an anchor to pull the sprue. The sprue lock pin is located where the small depression at the mold entrance meets the runners. It is fastened to the knockout or ejector mechanism and runs through the movable part of the mold in direct line with the mold entrance (10).

UNDERCUTS

Basically when the mold separates into two or more sections, the part can be removed. In order for the part to be removed from the mold, certain geometrical considerations must be met. First, the part should have no undercut sections that will lock if it is pulled from the mold.

If the shape is essential to function, a much more complicated mold is required where a portion of the mold is retracted to permit the undercut to be removed. This complicates the molding procedure and the mold, and may result in higher costs as well as a poorer quality part. The part will usually have some surfaces that are nearly parallel and perpendicular to the opening surface of the mold (referred to as the parting line), and pulling the part against these parallel surfaces could result in sticking and drag that would make removal difficult and damage the product.

The product designer should restrict the number of undercuts to a minimum and should consider carefully whether any undercuts in the design will present major problems in mold design. Moldings made from flexible plastic with small undercuts often allow forced mold release; that is, during mold opening the molding distorts sufficiently, because of its flexibility, to jump free of the undercut. This method is not recommended without experience. In such cases a certain degree of deformation may have to be accepted. Generously rounded corners are a must if this method of mold release is used.

For rigid plastics and large undercuts, use must be made of movable or rotating side cores, which obviously influence mold constructions. Screw threads are an example of an undercut frequently met. To eliminate undercuts, consider tapering a wall so that a sliding shut-off can be used (Fig. 7-24).

Molded parts with undercuts (i.e., articles that cannot be released in the direction of the mold opening) require molds with more than one parting line. For such articles, various methods have been developed that may be operated manually, mechanically, hydraulically, pneumatically, or electromechanically:

1. Molds with side cores (Fig. 7-25 to 7-28)
2. Molds with wedges (Figs. 7-27 and 7-28)
3. Molds with rotating cores (Fig. 7-29 to 7-31)
4. Molds with loose cores or inserts

The choice of method, or of a combination of these methods, is not only governed by the shape of the article and the properties of the polymer (flexibility, rigidity, shrinkage, etc.) but also by the standards of quality to be met by the article (15). For articles with external screw thread, for instance, either method 1 or 3 can be used. However, if method 1 is

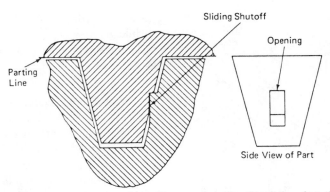

Fig. 7-24 Eliminate undercut, requiring side cores, by using this sliding shut-off system.

used, the mold parting line shows, which may be undesirable for aesthetic or design reasons. The method used should depend entirely on the circumstances of the article.

MOLD SHRINKAGE/TOLERANCE

By shrinkage or tolerance is meant the dimensions to which a cavity and core should be fabricated in order to produce a part of desired shape and size. The usual way to decide on the amount of shrinkage is to consult data supplied by the material manufacturer. The supplier's information is obtained from a test bar molded according to an ASTM standard. (See Chapters 6 and 28 on ASTM tests.) The test bar is molded at a specific pressure, mold temperature, melt temperature, and cure time. The thickness of the test bar is normally $\frac{1}{8}$ in. However, molded parts are very rarely produced under conditions and sizes that are the same as or even similar to those used for test bars (2).

For precision parts with close tolerance dimensions, shrinkage information from test bars as furnished by material suppliers can be inadequate but useful as a guide. We must become familiar with the factors that influence shrinkage so that we may arrive at more exact dimensions for a specific part. (See Chapter 19 on process effects.) According to compiled data, shrinkage is a function of mold temperature, part thickness, injection pressure, and melt temperature.

Shrinkage is influenced by cavity pressure to a very large degree. Depending on the pressure in the cavity alone, the shrinkage may vary as much as 100 percent.

Part thickness will cause a change in shrinkage. A thicker piece ($\frac{1}{8}$ in. or more) will have a shrinkage value on the high side of the data; whereas a thin one ($\frac{1}{20}$ in. or less) will have a lower shrinkage value.

The mold and melt temperature also influence shrinkage. A cooler mold will result in less shrinkage, whereas a hotter melt will cause more shrinkage, compared to the supplier's information.

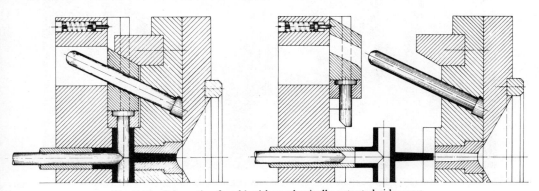

Fig. 7-25 Schematic of mold with mechanically actuated side cores.

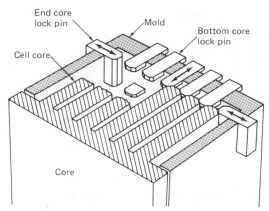

Fig. 7-26   Schematic of mold with side core actions (battery case molded of PP).

The longer the time in the cavity, the closer the part comes to mold dimensions, which means a lessening of shrinkage.

Openings in the part will cause variation in shrinkage from section to section because the cores making these openings act as temporary cooling blocks, which prevent change in dimension while the part is solidifying. A relatively large gate size will permit higher cavity-pressure buildup, which brings about a lower shrinkage.

The shrinkage problem can be categorized as follows:

1. Amorphous materials with a shrinkage rate of 0.008 in./in. or less have readily predictable shrinkage, which is not difficult to adjust with molding parameters such as cavity pressure and mold or melt temperature or, as a last resort, with the cycle.

2. Parts made of crystalline materials with high shrinkage (above 0.010 in./in.), but which are symmetrical and suitable for center gating, will also have a readily predictable shrinkage, adjustable with molding parameters.

3. Parts made of materials with a high shrinkage rate that are symmetrical but cannot be center-gated, may approximate a center-gate condition if multiple gating close to the center (three, four, or six gates) is possible. In this case, the

Fig. 7-27   Cavity blocks for PVC pipe elbow uses CSM414 prehardened stainless steel. Courtesy of Crucible Speciality Metals Division of Colt Industries, Syracuse, NY.

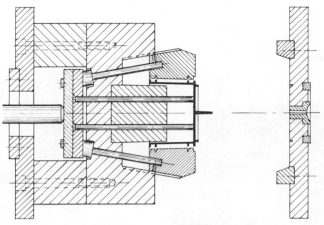

Fig. 7-28 Mold with wedge side core action.

prediction of shrinkage is somewhat more difficult but still presents a chance of success.

4. The major problem exists with materials that have a high shrinkage rate of about 0.015 to 0.035 in./in. In most of these cases, the material suppliers either show nomographs (see Fig. 7-32 and Chapter 20) in which all contributing factors are drawn and coordinated to supply reasonably close shrinkage information, or they point to examples with actual shrinkage information and molding parameters so they can be used for comparative interpolation. With most high-shrinkage crystalline materials (i.e., nylon, polyethylene, acetal), when the material is edge- or side-gated, a larger shrinkage occurs in the direction of flow and a smaller one perpendicular to it. (See Chapter 25 regarding amorphous and crystalline plastics.)

If, upon review of the shrinkage information, there is still doubt about whether the precision dimensions will be attained, then there is one way left for establishing accurate

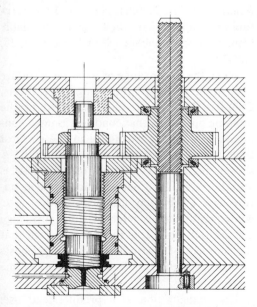

Fig. 7-29 Mold with rotating core that operates during mold opening and closing. The drive gear rotates via the worm shaft, which in turn transmits rotation to the geared core. The core unscrews threaded molded part.

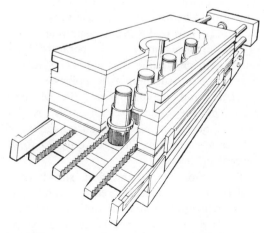

Fig. 7-30 Cores are positioned in rows, as this cutaway view of a closed mold frame indicates. Each core resides within a gear that, when engaged by one of the parallel racks, causes the core to rotate and unscrew the molded caps. Courtesy of Newark Die.

Fig. 7-31   Cap unscrewing mold that has a rotating action for removal of molded caps. When mold opens, the pin extending to the right side of the mold follows a guide plate and provides the rotating action for ejection.

shrinkage data: prototyping. In this method, a single cavity is built, and the critical dimensions are so calculated that they will allow for correction after testing, by providing for metal removal (machining). The test sample should be run for at least half an hour and under the same conditions as a production run. Only the last half-dozen pieces from the run should be used for dimensioning.

It is best to make the measurements after a 24-hour period at room temperature. However, with crystalline thermoplastics such as acetal, nylon, thermoplastic polyester, polyethylene, and polypropylene, the ultimate shrinkage may continue for days, weeks, months, or even a year. The shrinkage noted 1 hour after molding may be only 75 to 95 percent of the total.

The reason for postmolding shrinkage is that there is a molecular rearrangement and stress relaxation going on until equilibrium is attained, at which point shrinkage stops; both the moelecular rearrangement and stresses are brought about by molding conditions. The conditions that are most favorable for reaching the ultimate shrinkage in the shortest time are relatively high mold temperature and a lower rate of freezing of the material. Each material has its own rate of postmolding shrinkage as a function of time. Curves showing the rate of shrinkage as a function of time for changing mold temperature and varying part thickness are available from material suppliers. (See Chapters 19, 25, and 26.)

When the upper range of mold temperature (shown in material processing data sheets, Chapters 18 through 28) is applied to the molding operation, that condition is most conducive to stopping shrinkage in the shortest time after part removal from the mold. A slower rate of heat removal from the part is also desirable. Thin parts, which by their nature have a relatively fast heat-removal rate, consequently have a relatively long shrinkage stabilization time. It is to be emphasized that the problem of postmolding shrinkage exists principally with crystalline or semicrystalline thermoplastics.

The configuration of a product, end use, and assembly will determine to what degree the postmolding shrinkage will be a factor and what steps have to be taken to overcome a potential problem.

NOMOGRAPH IN SI UNITS

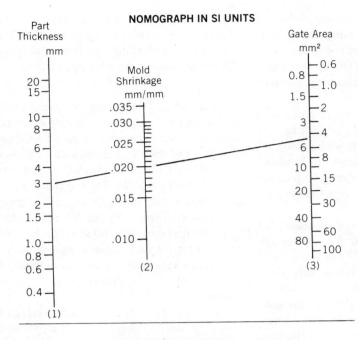

Example: A straight line connecting a part thickness of 3 mm (.12 in.) on Line 1 to a gate area of 4.5 mm² (.0072 sq. in.) (Gate: 1.5 mm (.06 in.) thick x 3 mm wide (.12 in.) on Line 3 intersects Line 2 at mm/mm (.020 in./in.), the estimated mold shrinkage at a mold temperature of 93°C (200°F).

Important: See the detailed discussion of molding conditions used, limitations, and correction factors for variations and compositions in the text. Nomographs are to be used for "DELRIN" 500, 500 CL, 900 and 8010.

NOMOGRAPH IN BRITISH UNITS

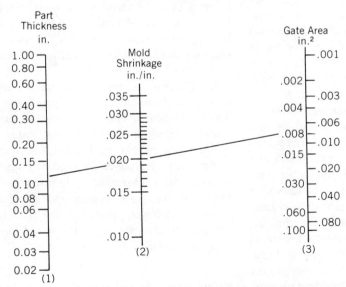

Fig. 7-32   Estimating mold shrinkage for Du Pont Delrin acetal plastics.

On some critical parts, an annealing–stress-relieving operation may be necessary to offset possible dimensional changes. Remember that each crystalline material has a different post-molding shrinkage stabilization time.

Shrinkage vs. Cycle Time

Reducing cycles on injection machines means not only greater part output but in some cases less total capital investment in tooling and manpower for a given job. One important way to achieve such savings is to optimize cycles by properly sizing a tool based on realistic shrinkage rates for the material being run (19).

While this sounds like a simple concept, implementation often requires considerable experience. It also demands a willingness to take more time before production start-up to employ the traditional engineering method of "cut-and-try." It is not for molders who are working within tight capital constraints. Generally, its payoff is only worthwhile for those high-volume jobs where the fast cycling characteristics of crystalline engineering thermoplastics are encountered.

The basic idea is simple: before going into actual production, you try to cycle a part as fast as possible commensurate with quality requirements, measure the material shrinkage at the optimum cycle, and size your tool accordingly. There are several ways to approach this situation.

Almost all plastic material data sheets list shrinkage values, calculated for an average conservative molding cycle. They are only of marginal practical use, since it is usually possible to speed up cycles. And because shrinkage goes up as cycle time goes down—owing to faster cooling—every cycle time below a certain "conservative" value quoted on the data sheet has a specific shrinkage value. There are other factors: mold design, method of cooling, wall thickness, and more. However, it is wrong just to depend on published data for sizing your tool unless you plan to run the part at the exact same cycle suggested by the resin supplier for a specific wall thickness.

For all trials a starting point must be established. The imperfect rule of thumb is that for crystalline materials such as nylon and acetal, shrinkage will run at about 0.015 in./in. and 0.020 in./in., respectively, for an average $\frac{1}{8}$ in. of wall thickness, at a conservative cycle of 45 seconds. Except for wall thickness, overall part size is not much of a limiting factor. With this in mind, Du Pont's Technical Services Laboratory, in Wilmington, Delaware, developed a nomograph, accounting for a number of variables in the molding process, that will provide some refined information on shrinkage for a select number of materials (Fig. 7-32). Such a nomograph—along with experience with certain parts and resins—can be helpful in arriving at a starting point in sizing the tool.

From then on it is mostly trial and error, but careful analysis of part design and optimum cycles also is very important. There are no tables you can use to determine meaningful

"typical" shrinkage values at very fast cycles. You must determine them yourself at different cycle times. Following are two actual cases, illustrating several possible approaches.

Acetal Part. In this case, we were dealing with a flat part with precise hole-to-hole openings and a slide area that must be held open and flat. The resin was acetal with a shrinkage range from 0.02 to 0.035 in./in. The objective for the molder was to size the cavity so that a minimum cycle could be attained at proper dimensions. Typical cycles for this material might range from 30 to 60 seconds. However, in order for the part to run at optimum cycle, mold shrinkage must be chosen carefully for the part to stay in dimensions at very fast cycles.

When the cycle is longer than average for this material, shrinkage on such a part as described above will normally be about 0.02 in./in. at a cycle of about 60 seconds, allowing for a long time of cooling. If the part is to be molded as quickly as possible, shrinkage increases to about 0.035 in./in. for a 30-second cycle. (See Table 7-6.)

In this case the molder was in a position to use the fastest cycles. His prototype tool was sophisticated enough to run at various cycles, including the 30-second rate. The molder then cut his production tool to 0.035 in./in.—after having established this figure as accurate for a 30-second cycle in trials—and the parts were all on-size and acceptable. However, that also meant that the cycle could not be slowed down, or parts would tend to become oversize.

Nylon Part. In this case, the part is circular with 2-in. O.D. and I.D. of 1.6 in., leaving a wall thickness of 0.2 in. Let us assume that the tolerance requirements on the I.D. are such that the part must be held within 0.006 in. Therefore, the I.D. dimension is 1.6 in. ± 0.003 in. Tolerance on the 2-in. O.D. dimension is somewhat wider: ±0.005 in.

The resin used is a toughened nylon. The range of mold shrinkages applicable to this resin and part could be as low as 0.018 in./

Table 7-6 Acetal flat part with precise hole-to-hole openings.

	LONG CYCLE: 60 SEC LOW SHRINKAGE	SHORT CYCLE: 30 SEC[b] LOW SHRINKAGE	SHORT CYCLE: 30 SEC HIGH SHRINKAGE
Core Pin Spacing, in.	7.142	7.142	7.253
Required Hole Spacing, in.	7.0 ± 0.015	7.0 ± 0.015	7.0 ± 0.015
Corresponding Hole Spacing, in.	7.0	6.892	7.0
Designed Mold Shrinkage, in./in.	0.020	0.020	0.035
Actual Mold Shrinkage, in./in.	0.020	0.035	0.035

[a] Resin is a grade of Delrin acetal with a 0.15-in. nominal wall thickness; [b] Part is undersized;

in. or as high as 0.032 in./in., depending on the molding cycle.

In order to achieve the lower shrinkage— 0.018 in./in.—the cycle would have to be fairly long, about 45 seconds. On the other hand, at a 25-second cycle—well within the capability of this particular resin—shrinkage could be as high as 0.032 in./in. (See Table 7-7.) Therefore the molder must choose the cycle time first, then determine the shrinkage at that cycle.

For instance, should the molder choose to run the fastest cycle—25 seconds—he would have to cut his mold for a 0.032-in./in. shrinkage. This, of course, assumes that there are no artificial limitations to achieving the fast cycle—lack of screw plasticating capacity, mechanical function of the mold, slow machine function, or improper cooling. The other consideration is the inability to predict the shrinkage value accurately at the given cycle time. Ideally, the molder would have experience

with similar type parts; thus he would be able to predict shrinkage at a specific cycle without cut-and-try methods.

Unfortunately, in most cases this is impossible. The part may be new, and the molder may have to go through a series of tests to establish the optimum cycle and the corresponding shrinkage in order to size his tool properly. There are several approaches he can use.

With this particular part, the molder solved the shrinkage problem at the prototype tool stage. A prototype cavity, properly cored, that runs automatically at various cycles, is the most reliable means of arriving at the minimum cycle.

Such a prototype cavity provides highly reliable data, provided the cooling for the prototype equals the cooling on the production tool.

Cooling of the prototype tool is important. The more information you need from your prototype tool, the more sophisticated it must

Table 7-7 Nylon "doughnut"-shaped part.

	LONG CYCLE: 45 SEC LOW SHRINKAGE	SHORT CYCLE: 25 SEC[b] LOW SHRINKAGE	SHORT CYCLE: 25 SEC HIGH SHRINKAGE
Core Diam, in.	1.629	1.629	1.652
Cavity Diam, in.	2.036	2.036	2.065
Corresponding Part Size			
I.D., in.	1.6	1.577	1.6
O.D., in.	2.0	1.971	2.0
Designed Mold Shrinkage, in./in.	0.018	0.018	0.032
Actual Mold Shrinkage, in./in.	0.018	0.032	0.032

[a] Resin is a toughened nylon, Zytel ST 801 with a 0.2-in. nominal wall thickness; [b] Part is undersized;

be—a simple Kirksite tool will not do for any extensive evaluation. A prototype tool probably should be made of P-20 steel, the same steel used for many production tools. The only difference is that the prototype tool is un-hardened—vs. C 55 Rockwell for production tools—permitting machining ease. Although all this evaluation work is costly, a good prototype can save money in the long run; apart from optimizing cycles, with a tool that closely approximates the production tool, problems such as molded-in stresses due to improper gating or unequal rate of cooling can be detected and solved before full production start-up. As far as shrinkage is concerned: if you are able to cut 20 seconds off a 60-second cycle, this means a 33 percent productivity improvement.

A second, similar approach involves constructing a so-called lead cavity for the production mold. (For instance, on a four-cavity tool you would cut only one cavity for trial runs.) The part is then molded at various cycles to determine the shrinkage. This information can be translated into sizing the remaining cavities. The advantage is that this lead cavity will be designed to operate as a production tool, and all information learned is fully applicable to the other cavities.

Often, insufficient time is allowed for constructing the production tool, so use of a lead cavity is impossible. In spite of the frequent problem of insufficient lead time, serious thought should be given to the advisability of slowing down the initial development work. The time and money invested to develop cycle/shrinkage information can be amply rewarded, in higher productivity for the life of the part.

A third possible approach is to size the core piece for maximum shrinkage and size the cavity section of the production mold for minimum shrinkage. Referring to the circular part described above, we assume that the I.D. will shrink 0.032 in./in.; applying this figure to the 1.6-in. I.D. indicates that the core size should be cut to 1.652 in. If the cavity is sized to minimum shrinkage—0.018 in./in.—its size should be 2.036 in. With this technique it is possible to remove steel from the core and

cavity after the mold has been fully tested, and the actual rate of shrinkage at a given cycle has been established. While this is a somewhat time-consuming approach, it is by far less costly and time-consuming than other methods described previously.

A fourth approach would be to choose a reasonable shrinkage value—based on available data similar to Fig. 7-32—and cut both core and cavity to that value and cycle the part accordingly. For example, we might choose a shrinkage of 0.026 in./in. and pick a cycle time also in the mid range. If there are surprises—as can happen with any new part—the cycle and shrinkage can be adjusted in molding to achieve the required part size. In this particular case, good cycle information is not available from the prototype tool. It is possible to produce parts to the required size at the given cycle, but there is no way of knowing whether this cycle represents the full potential of the resin and tool. It would make more sense economically to cut both cores and cavity to achieve the final part dimension at a fast cycle rather than at a middle value. The cavity work is rather easy to accomplish, since it entails increasing its size by cutting steel. The cores would have to be remachined from scratch because they must be larger.

Shrinkage can vary with cycle time, so that final part dimensions depend on the precise cycle time for a mold. Therefore, before you size your tool, determine how fast you can effectively run the part. Find out the shrinkage at that cycle. Based on these findings, your final tool dimensions will then be keyed to production at optimum cycle time.

In some cases, of course, a too-rapid cycle can interfere with some secondary operation such as an operator assembly procedure required at every ejection. This situation would require that the molder go the other way, re-sizing the cavity to slow down the cycle.

As a molder you have to be aware of potential quality defects caused by fast cycles. For example, very fast cycles may prevent thick sections from fully packing out, resulting in voids. Before you decide to push a cycle to its limit, you must decide if part quality will suffer. Good candidates for pushing up cycles are usually thin-walled parts and those with uniform wall thickness.

Standard Tolerances

The tables on standard tolerances (Tables 7-8 and 7-9) were prepared by the Custom Molders of The Society of the Plastics Industry. These tables are to be used only as a guide. The dimensions are based on a hypothetical molded article, whose cross section is explained by the tables. These tables pertain only to specific polycarbonate and phenolic materials.

MOLD-STEEL SELECTION

Overview

The steel used in the manufacture of a mold base varies, depending on the requirements of the application. (As an example, polyvinyl chloride requires stainless steel to eliminate corrosion.) The structural sections of the mold base are usually made from both medium carbon (SAE 1030) and AISI 4130 type steels. Among the steels selected for cavity and core plates are P-20 and H-13 type steels, as well as stainless steel (T-420).

The available spectrum of modern tool

Table 7-8   Standard tolerance chart for a polycarbonate.

STANDARDS AND PRACTICES OF PLASTICS CUSTOM MOLDERS	Engineering and Technical Standards POLYCARBONATE

NOTE: The Commercial values shown below represent common production tolerances at the most economical level. The Fine values represent closer tolerances that can be held but at a greater cost.

Drawing Code	Dimensions (Inches)	Plus or Minus in Thousands of an Inch
A = Diameter (see Note #1)	0.000 0.500 1.000 2.000	
B = Depth (see Note #3)	3.000 4.000	
C = Height (see Note #3)	5.000 6.000	

		Comm. ±	Fine ±
	6.000 to 12.000 for each additional inch add (inches)	.003	.0015
D=Bottom Wall (see Note #3)		.003	.002
E = Side Wall (see Note #4)		.003	.002
F = Hole Size Diameter (see Note #1)	0.000 to 0.125	.002	.001
	0.125 to 0.250	.002	.0015
	0.250 to 0.500	.003	.002
	0.500 & Over	.003	.002
G = Hole Size Depth (see Note#5)	0.000 to 0.250	.002	.002
	0.250 to 0.500	.003	.002
	0.500 to 1.000	.004	.003
Draft Allowance per side (see Note #5)		1°	½°
Flatness (see Note #4)	0.000 to 3.000	.005	.003
	3.000 to 6.000	.007	.004
Thread Size (class)	Internal	1B	2B
	External	1A	2A
Concentricity (see Note #4)	(T.I.R.)	.005	.003
Fillets, Ribs, Corners (see Note #6)		.015	.015
Surface Finish	(see Note #7)		
Color Stability	(see Note #7)		

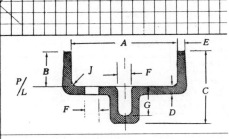

REFERENCE NOTES

1 – These tolerances do not include allowance for aging characteristics of material.

2 – Tolerances based on ⅛" wall section.

3 – Parting line must be taken into consideration.

4 – Part design should maintain a wall thickness as nearly constant as possible. Complete uniformity in this dimension is impossible to achieve.

5 – Care must be taken that the ratio of the depth of a cored hole to its diameter does not reach a point that will result in excessive pin damage.

6 – These values should be increased whenever compatible with desired design and good molding technique.

7 – Customer-Molder understanding necessary prior to tooling.

Table 7-9   Standard tolerance chart for a general-purpose phenolic.

STANDARDS AND PRACTICES OF PLASTICS CUSTOM MOLDERS	Engineering and Technical Standards
	GENERAL PURPOSE PHENOLIC

NOTE: The Commercial values shown below represent common production tolerances at the most economical level. The Fine values represent closer tolerances that can be held but at a greater cost.

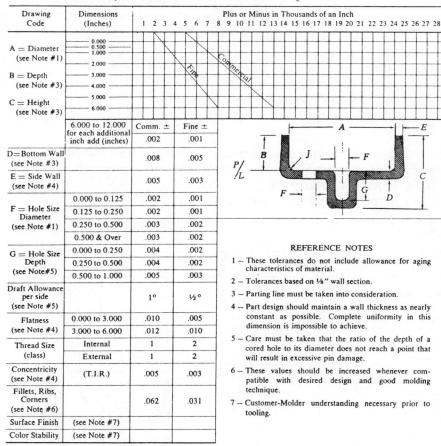

Drawing Code	Dimensions (Inches)	Comm. ±	Fine ±
A = Diameter (see Note #1)	0.000 / 0.500 / 1.000 / 2.000		
B = Depth (see Note #3)	3.000 / 4.000		
C = Height (see Note #3)	5.000 / 6.000		
	6.000 to 12.000 for each additional inch add (inches)	.002	.001
D = Bottom Wall (see Note #3)		.008	.005
E = Side Wall (see Note #4)		.005	.003
F = Hole Size Diameter (see Note #1)	0.000 to 0.125	.002	.001
	0.125 to 0.250	.002	.001
	0.250 to 0.500	.003	.002
	0.500 & Over	.003	.002
G = Hole Size Depth (see Note #5)	0.000 to 0.250	.004	.002
	0.250 to 0.500	.004	.002
	0.500 to 1.000	.005	.003
Draft Allowance per side (see Note #5)		1°	½°
Flatness (see Note #4)	0.000 to 3.000	.010	.005
	3.000 to 6.000	.012	.010
Thread Size (class)	Internal	1	2
	External	1	2
Concentricity (see Note #4)	(T.I.R.)	.005	.003
Fillets, Ribs, Corners (see Note #6)		.062	.031
Surface Finish	(see Note #7)		
Color Stability	(see Note #7)		

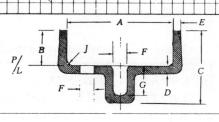

REFERENCE NOTES

1 – These tolerances do not include allowance for aging characteristics of material.

2 – Tolerances based on ⅛″ wall section.

3 – Parting line must be taken into consideration.

4 – Part design should maintain a wall thickness as nearly constant as possible. Complete uniformity in this dimension is impossible to achieve.

5 – Care must be taken that the ratio of the depth of a cored hole to its diameter does not reach a point that will result in excessive pin damage.

6 – These values should be increased whenever compatible with desired design and good molding technique.

7 – Customer-Molder understanding necessary prior to tooling.

steels offers properties in numerous combinations and to widely differing degrees. Fortunately, the needs of the vast majority of tool-steel users can be satisfied with a relatively small number of these steels, the most widely used of which have been given identifying numbers of the American Iron and Steel Institute (AISI).

With pre-engineered molds and components the manufacturers can provide performance and capabilities based on the mold design requirements. As reviewed in Chapters 7 through 10, it is important in the process of mold purchasing to develop professional forms that detail special mold design features as well as steel types, heat treatments, and surface finish requirements. Sample forms have been developed (see Table 7-10). Information presented in Tables 7-11 through 7-16 characterizes materials that are useful in work with pre-engineered molds and components, as well as in your own mold (20).

Tool steels (or mold steels) may be defined as highly alloyed steels. The chemistry and method of manufacture determine the use of the final product. As high-performance alloys such as cobalt, vanadium, and chromium become more difficult and more expensive to

Table 7-10 Guide for mold quotation. Courtesy of The Moldmakers Division of The Society of The Plastics Industry.

THE MOLDMAKERS DIVISION
THE SOCIETY OF THE PLASTICS INDUSTRY, INC.
3150 Des Plaines Avenue (River Road), Des Plaines, Ill. 60018, Telephone: 312/297-6150

TO _____ FROM _____ QUOTE NO. _____
_____ _____ DATE _____
_____ DELIVERY REQ _____

Gentlemen:
Please submit your quotation for a mold as per following specifications and drawings:

COMPANY NAME _____
Name 1. _____ B/P No. _____ Rev. No. _____ No. Cav. _____
of   2. _____ B/P No. _____ Rev. No. _____ No. Cav. _____
Part/s 3. _____ B/P No. _____ Rev. No. _____ No. Cav. _____

No. of Cavities: **Design Charges:**   **Price:**   **Delivery:**

Type of Mold: ☐ Injection ☐ Compression ☐ Transfer ☐ Other (specify) _____

Mold Construction
☐ Standard
☐ 3 Plate
☐ Stripper
☐ Hot Runner
☐ Insulated Runner
☐ Other (Specify) _____

Mold Base Steel
☐ #1
☐ #2
☐ #3

Special Features
☐ Leader Pins & Bushings in K.O. Bar
☐ Spring Loaded K.O. Bar
☐ Inserts Molded in Place
☐ Spring Loaded Plate
☐ Knockout Bar on Stationary Side
☐ Accelerated K.O.
☐ Positive K.O. Return
☐ Hyd. Operated K.O. Bar
☐ Parting Line Locks
☐ Double Ejection
☐ Other (Specify) _____

Material
Cavities   **Cores**
☐ Tool Steel   ☐
☐ Beryl. Copper   ☐
☐ Steel Sinkings   ☐
☐ Other (Specify) _____

Press
Clamp Tons _____
Make/Model _____

Hardness
Cavities   **Cores**
☐ Hardened   ☐
☐ Pre-Hard   ☐
☐ Other (Specify) _____

Finish
Cavities   **Cores**
☐ SPE/SPI   ☐
☐ Mach. Finish   ☐
☐ Chrome Plate   ☐
☐ Texture   ☐
☐ Other (Specify) _____

Cooling
Cavities   **Core**
☐ Inserts   ☐
☐ Retainer Plates   ☐
☐ Other Plates   ☐
☐ Bubblers   ☐
☐ Other (Specify) _____

Ejection
Cavities   **Cores**
☐ K.O. Pins   ☐
☐ Blade K.O.   ☐
☐ Sleeve   ☐
☐ Stripper   ☐
☐ Air   ☐
☐ Special Lifts   ☐
☐ Unscrewing (Auto)   ☐
☐ Removable Inserts (Hand) ☐
☐ Other (Specify) _____

Side Action
Cavities   **Cores**
☐ Angle Pin   ☐
☐ Hydraulic Cyl.   ☐
☐ Air Cyl.   ☐
☐ Positive Lock   ☐
☐ Cam   ☐
☐ K.O. Activated Spring Ld. ☐
☐ Other (Specify) _____ ☐

Type of Gate
☐ Edge
☐ Center Sprue
☐ Sub-Gate
☐ Pin Point
☐ Other (Specify) _____

Design by: ☐ Moldmaker ☐ Customer
Type of Design: ☐ Detailed Design ☐ Layout Only
Limit Switches: ☐ Supplied by _____ ☐ Mounted by Moldmaker
Engraving: ☐ Yes ☐ No
Approximate Mold Size: _____
Heaters Supplied By: ☐ Moldmaker ☐ Customer
Duplicating Casts By: ☐ Moldmaker ☐ Customer
Mold Function Try-Out By: ☐ Moldmaker ☐ Customer
Tooling Model/s or Master/s By: ☐ Moldmaker ☐ Customer
Try-Out Material Supplied By: ☐ Moldmaker ☐ Customer

Terms subject to Purchase Agreement. This quotation holds for 30 days.

Special Instructions: _____

The prices quoted are on the basis of piece part print, models or designs submitted or supplied. Should there be any change in the final design, prices are subject to change.

By _____ Title _____

Distribution: Use of this 3 part form is recommended as follows: 1) White and yellow - sent with request to quote.
Pink - maintained in active file. 2) White original - returned with quotation. Yellow - retained in Moldmaker's active file.

Table 7-11   Materials used in molds; arranged in order of surface hardness. Courtesy of Stokes-Trenton, Inc.

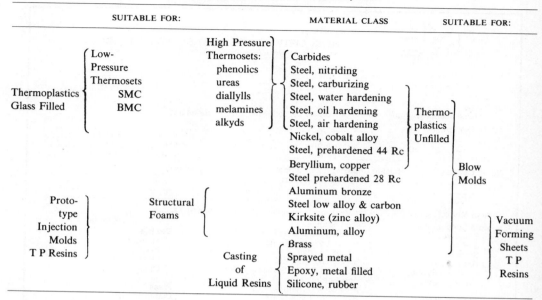

obtain, it is safe to say that quality of the product and better usage will be the key to the future of mold steels.

Proper materials selection and proper combination of alloys in varying percentages are required for finished tools. Characteristics of machinability, wear, shock, and anti-galling properties, resistance to corrosion, and, of

course, hardenability are directly attributable to alloy type and content.

Combining the chemistry of alloys with the best melting, rolling, and annealing techniques allows steel mills to consistently produce fine-quality tool steels. Tool steels are electric-furnace-melted and, where beneficial, are vacuum-degassed on pouring. Detailed annealing

Table 7-12   Applications for the typical mold steels.

TYPE OF STEEL	TYPICAL USES IN INJECTION MOLDS	TYPE OF STEEL	TYPICAL USES IN INJECTION MOLDS
4130/4140	General mold base plates	A6	Cavities, cores, inserts for high-wear areas
P-20	High-grade mold base plates, hot-runner manifolds, large cavities and cores, gibs, slides, interlocks	A10	Excellent for high-wear areas, gibs, interlocks, wedges
4414 SS, 420 SS (prehardened)	Best grade mold base plates (no plating required), large cores, cavities and inserts	D2	Cavities, cores, runner and gate inserts for abrasive plastics
P5, P6	Hobbed cavities	420 SS	Best all-around cavity, core and insert steel; best polishability
01	Gibs, slides, wear plates	440C SS	Small to medium-size cavities, cores, inserts, stripper rings
06	Gibs, slides, wear plates, stripper rings	250, 350	Highest toughness for cavities, cores, small unsupported inserts
H-13	Cavities, cores, inserts, ejector pins and sleeves (nitrided)	455M SS	High toughness for cavities, cores, inserts
S7	Cavities, cores, inserts, stripper rings	M2	Small core pins, ejector pins, ejector blades (up to $\frac{5}{8}$ in. diam)
A2	Small inserts in high-wear areas	ASP 30	Best high-strength steel for tall, unsupported cores and core pins

Table 7-13 General characteristics of typical mold steels.

AISI type[1]	Trade designation	General characteristics	Properties rankings[2]				Typical applications
			Toughness	Dimensional stability in heat-treatment	Machinability (annealed)	Polishability (heat-treated)	
P-20	CSM-2	Medium carbon (0.30%) and chrome (1.65%). Available prehardened (300 Bn) or annealed (200 Bn).	10	7	9 (pre-hardened)	8 (pre-hardened)	Excellent balance of properties for injection and compression molds of any size.
H-13	NuDie V	Hot-work die steel; 5% chrome. Hardenable to about 50 Rc.	9	8	9	9	Higher hardness than P-20; good toughness and polishability. Used for abrasion resistance in RP molds and high-finish injection molds.
A-2	Air Kool	Cold-work die steel, high carbon (1.0%) 5% chrome. Hardenable to about 60 Rc.	8	9	8	7	High hardness for abrasion-resistant, long-wearing compression and injection molds. Limited to small sizes.
D-2	Airdi 150	Cold-work die steel; high carbon (1.55%) 11.5% chrome. Hardenable to about 60 Rc.	7	9	5	6	Highest abrasion resistance. Difficult to machine. Susceptible to stress cracking. Small molds only.
414	CSM 414	Stainless steel; 12% chrome, 2% Ni, 1% Mn, low carbon (0.03%). Available prehardened (300 Bn).	10	10	9	9	"Stainless version" of P-20; similar properties and uses.
420	CSM 420	Stainless steel; 13% chrome, 0.80% Mn, medium carbon (0.30%). Hardenable to about 50 Rc).	9	10	8	10	"Stainless version" of H-13; similar properties and uses. Very stable in heat treatment, takes high polish
4145	Holder Block	Medium carbon (0.50%) and chrome (0.65%). Available prehardened.	10	10 (pre-hardened)	10 (pre-hardened)	6 (pre-hardened) 7 (fully hardened)	Low cost steel, for mold bases and large molds. Not suited to high quality finish.

[1] Crucible Steel designations
[2] On scale of 1 to 10 (10 = best)

Table 7-14 Important properties of the major mold steels and beryllium copper.

POINT RATINGS, 1 TO 10 (10 IS HIGHEST)

TYPE	AISI DESIGNATION	RECOMMENDED HARDNESS, ROCKWELL C[b]	WEAR RESISTANCE	TOUGHNESS	COMPRESSIVE STR.	HOT HARDNESS	CORROSION RESISTANCE	THERMAL CONDUCTIVITY	HOBBABILITY	MACHINEABILITY	POLISHABILITY	HEAT TREATABILITY	WELDABILITY	NITRIDING ABILITY
Prehardened	4130/41040	30–36	2	8	4	3	1	5	1	5	5	10	4	4
	P-20	30–36	2	9	4	3	2	5	1	5	8	10	4	5
Prehardened Stainless	414 SS	30–35	3	9	4	3	7	2	1	4	9	10	4	6
	420 SS	30–35	3	9	4	3	6	2	1	4	9	10	4	7
Carburizing	P-5	59–61	8	6	6	5	2	3	9	10	7	6	9	8
	P-6	58–60	8	7	6	5	3	3	8	10	7	6	8	8
Oil Hardening	01	58–62	8	3	9	5	1	5	5	8	8	6	2	3
	06	58–60	8	4	8	5	1	5	7	10	8	7	2	3
Air Hardening	H-13	50–52	6	7	7	8	3	4	6	9	5	6	2	10
	S7	54–56	7	5	8	8	3	4	6	9	8	8	5	8
	A2	56–60	9	3	9	7	3	4	4	9	8	8	3	8
	A5	56–60	8	4	8	7	2	5	5	8	7	9	2	8
	A10	58–60	9	5	9	7	2	5	5	10	7	7	4	7
	D2	56–58	10	3	8	8	4	2	4	8	6	7	2	NA[d]
Stainless	420 SS	50–52	6	6	6	8	7	2	4	4	6	9	1	10
	440C SS	56–58	8	3	8	7	8	2	3	7	10	8	6	8
Maraging	250	50–52	5	10	10	7	4	3	4	4	7	9	5	9
	350	52–54	6	10	10	7	4	3	4	4	7	9	5	9
Maraging Stainless	455M	46–48	5	10	5	7	10	2	3	4	8	9	5	NA[d]
High-Speed	M2	60–62	10	2	10	10	3	3	2	4	6	8	2	10
	ASP 30[c]	64–66	10	4	10	10	4	3	1	4	7	8	2	NA[d]
Beryllium Copper	Be Cu	28–32	1–2	1	2	4	6	10	10	10	8–9	7	7	NA[d]

Table 7-15 Anticipated range of size changes in heat treatment of various mold steels. Courtesy of Stokes-Trenton, Inc.

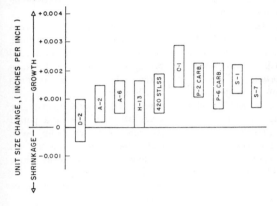

procedures produce structures that also enhance fine quality. This quality is of vital importance to the moldmaker, who must produce molds of the highest caliber repetitively.

The following is an indication of where most common mold steels and beryllium copper find application in injection molds, along with comments on each material's particular strengths or weaknesses.

Type 4130, 4140: Most commonly used in a prehardened state at a hardness of 30–36 RC for load- or pressure-bearing mold base plates, such as cavity and core retainer plates, or extra-large cavities and cores, which have no special surface-quality requirements.

Type P-20: Same as 4130/4140; however, its cleaner manufacturing requirement results in a more homogeneous microstructure and thus good polishability. It is used for large cavities and cores requiring good polish, and for hot-runner manifolds.

Type 414 SS/420 SS prehardened: Most commonly used at 30–35 RC hardness; excellent for large cavities and cores that require a good polished finish and corrosion resistance. They are also very good for cavity and core retainer mold base plates, providing toughness and corrosion resistance without a need for plating.

P-5 and P-6: Carburizing steel available in annealed condition. They are easy to hob and/or machine for making cavities, and can be carburized to a depth of 60 mils and a

case hardness of 58–61 RD. The relatively soft core hardness (15–30 RC) lowers the overall compressive strength, which is a key quality in modern mold-steel requirements. In the long run, it is often more economical to pay for the higher mold manufacturing cost of EDMed tool steel cavities of the through-hardened type, rather than using the hobbing process, because of the much longer life expectancy.

01 oil-hardening: Available in annealed condition; is capable of attaining a maximum of 62 RC hardness. It is excellent for gibs, slides, wear plates, and the like, but not recommended for cavity or core components or mold base plates.

06 oil-hardening: Same applications as 01, but provides better machinability and especially good wear characteristics in applications with metal-to-metal contact, because of the presence of free graphite in its microstructure.

H-13 air-hardening: One of the most useful steels for moldmaking, providing good all-around steel qualities for cavities and cores as well as inserts.

S-7 air-hardening: Same as H-13 but providing the often required higher hardness of 54–56 RC. Extreme care is required in the heat-treating process; and, to prevent cracking in the quench, a double draw is highly recommended. It is very important also that a hardness of 55 RC be achieved accurately, because of this steel's very sharp break-off point in impact strength or toughness (highest at 55 RC, lowest at 57–58 RC).

Types A2, A6, A10 air-hardening: Medium-alloy tool steels available in annealed condition. A2 is the most abrasion-resistant steel of this group under molding conditions, because of its higher chrome content. A10 has remarkable wear and nonseizing qualities in metal-to-metal contact applications, because of its free graphite content. All three are easy to machine and very high in compressive strength. Welding, however, can create cracking problems.

D2 air-hardening: In a class by itself with respect to excellent abrasion resistance, and recommended for severe molding conditions, such as when glass or mineral fillers are used.

Table 7-16   Effects of alloying materials.

Aluminum (Al)—Combines with nickel and titanium to form an intermetallic compound, which precipitates on aging and provides strength and hardness. Also used as a deoxidizer and to produce fine grain size.

Carbon (C)—Very influential in controlling hardness, depth of hardness, and strength.

Chromium (Cr)—A carbide-forming element that contributes strongly to hardenability and abrasion and wear resistance. Additional amounts of chromium, greater than are needed for carbide formation, remain in solution and enhance corrosion resistance.

Cobalt (Co)—An element added to the maraging steels to improve strength.

Manganese (Mn)—Combines with free sulfur to form discrete sulfide inclusions and improve hot workability. It is also a deoxidizing agent. In larger quantities, it increases hardenability by decreasing the required quenching rate. It is the principal element used to obtain quenching by air cooling, which minimizes distortion.

Molybdenum (Mo)—Promotes hardenability in mold steels. The elevated tempering requirement increases the steel's strength at higher operating temperatures and provides more complete relief of residual stresses for greater dimensional stability.

Nickel (Ni)—Usually added to improve hardenability of low-alloy steels. In maraging steels, nickel combines with aluminum and titanium to form an intermetallic compound that increases hardness and strength on aging. Larger amounts of nickel also assist in corrosion resistance.

Silicon (Si)—Principal function is as a deoxidizing agent during melting. In higher quantities, it retards tempering, thus allowing higher tempering and operating temperatures (hot hardness).

Titanium (Ti)—Found in maraging steels, where it acts as a potent strengthener by combining with nickel and/or aluminum to form an intermetallic compound, which precipitates on aging.

Tungsten (W)—Increases hardness, strength and toughness.

Vanadium (V)—A strong carbide-forming element, which is usually added to control grain size and to increase wear resistance.

It is not recommended for welding and is somewhat sensitive to cracking, owing to its low toughness.

Types 420 and 440C stainless: Good choices for corrosion resistance where corrosive plastics are used or where moisture or humidity could affect cavity surface finish or cooling-channel corrosion. Type 440C is somewhat better in wear resistance and compressive strength, owing to its higher hardness, while 420 SS represents the true mold cavity steel with good to very good all-around qualities and exceptionally high and consistent polishability, provided that it is manufactured by vacuum degassing and/or electroslag remelting. Relatively low thermal conductivity compared with other mold steels is only a minor factor in the first few days or weeks of processing a new mold. As soon as corrosion inside cooling channels takes hold, thermal conductivity of other mold steels, with respect to cooling-channel effectiveness, will be worse than that of stainless steels!

Maraging types 250, 350, 440M: Excellent mold cavity and insert steels. They are by far the best performers when toughness is the number one priority. This could be a very important factor in cases of very thin cross sections or small, fragile, and unsupported cavity or core inserts. Their resistance to cracking could, in the long life of the mold, be a crucial factor in mold-repair expenses, offsetting the much higher initial price of these steels (five to ten times that of other tool steels). Dimensional stability and simplicity of heat treating these steels are valuable considerations for the moldmaker.

M2 and ASP 30 high-speed steels: Probably the most useful of all the many high-speed steels for moldmaking, M2 is by far the most useful steel for good-quality, long-lasting round core pins or blade ejectors, and is also readily available. ASP 30 is an advanced-generation steel manufactured by Uddeholm, using a new powder-metallurgy process. Its extremely high density gives it remarkable rigidity, which can be very important in resisting deflection of tall, unsupported cores.

Beryllium copper: When heat-treated, the strongest of all copper-based alloys. It is not usually recommended for high-production molds, because of its relatively low wear resistance, toughness, and compressive strength, compared with tool steel. It does, however, have a special place in moldmaking when economy in cavity manufacturing and injection molding cycle time (the latter minimized by Be/Cu's high thermal conductivity) are of the utmost importance. But one must take into consideration that, over the lifetime of a mold, periodic cavity replacement costs can become a great disadvantage.

Kirksite: In processes where pressures are relatively low and short runs are anticipated (usually in the thousands), Kirksite molds can be used. Because the material pours so well, it is generally cast, and Type A Kirksite is usually used. Since the pouring temperatures are low (800°F for Kirksite as compared to 3000°F for steel and 2000°F for beryllium copper castings), it is possible to cast copper tubing cooling lines directly into place in Kirksite. More important, the low casting temperatures (and the retention of fluidity for a relatively long period of time) enable Kirksite to pick up fine detail from the pattern over a very large casting area. This means that Kirksite molds will reproduce pattern detail in the molded parts (it falls somewhere in between aluminum and beryllium copper in this regard) and thus has found application where fine patterns such as wood grains are required (e.g., furniture parts). Shrinkage is about 0.008/in.

Kirksite is lower in cost than most other metals and machines well. It is nonmagnetic and therefore may need clamping for grinding. It has a tendency to load grinding wheels badly. Kirksite is also heavier than aluminum and only slightly lower in weight than steel or beryllium copper. It is not as strong as either of these other metals and therefore will require heavier wall sections, making it more difficult to handle. Cycles with Kirksite molds are usually shorter than steel, but longer than aluminum molds.

Heat Treating

As progress has been made in the quality of tool steels and in mold construction, so have advances in heat treating. Knowledge has expanded, and development of new equipment such as vacuum furnaces, fluid bed furnaces, and finer tempering facilities has made the heat-treating operation much more of a science than ever before.

Many times this procedure appears to be taken for granted; yet it is one of the most important operations. In investigative analysis, 70 percent of all tool failures are related to heat treating, and it is not always the fault of the hardener. Of greater concern is the fact that one-half of these failures are due to poor surface conditions, the bane of all molders.

Requirements To Be Met by Mold Steel

Machinability: Molds are usually formed by machining of steel blocks. In view of the high cost, the steel must possess good machinability, which depends on the composition and structure of the steel as supplied.

Ability to harden: In general, hardening of small molds or mold components does not present great problems. However, hardening of large and complicated molds may cause deformation, dimensional variances, or even cracks, if in the selection of the tool steel insufficient allowance was made for the hardening treatment, the machining techniques, and the dimensions of the mold components (size and shape). The material must be capable of being hardened without any risk. Depending on the hardening process, the following steel grades may be used: oil-hardening steel, air-hardening steel, pre-heat-treated steel, case-hardened steel, and nitrided steel.

Ability to take a polish: The surface finish of the molding is first and foremost governed by the mold cavity finish. A cavity polished to a mirror finish produces a glossy molding surface and assists polymer flow in the mold. Polishing ability depends on the hardness, purity, and structure of the tool steel used. High-carbide steel grades are hard to polish to a mirror finish and thus require additional labor.

Corrosion resistance: If corrosive plastics are processed, proper corrosion resistance of the mold steel is a must. Even the slightest corrosion of the mold cavity will interfere with mold release.

Pre-engineered Molds

Standardized mold components, such as ejector pins, guide pins, guide pin bushings, sprue bushings, bolts, etc., and complete standardized mold assemblies have been commercially available since 1943. In these assemblies only the cavities and cooling channels need to be machined and the ejectors fitted.

Such standardized mold assemblies are available in various types. Advantages include: low cost, quick delivery, interchangeability, and promotion of industrial standardization (see Chapter 8).

For the construction of injection molds (pre-engineered or otherwise), various grades of tool steel are available, each of which has its own specific set of properties. This often makes selection of the proper grade for a particular purpose difficult, especially since a thorough knowledge is required of the properties and possibilities of the various commercially available steel grades and the machining techniques (1, 6, 15, 21–25).

POLISHING

Plastic molds usually require a high polish. Though the operation seemingly is gentle, polishing can still damage the steel unless it is properly done. A common defect is "orange-peel," a wavy effect that results when the metal is stretched beyond its yield point by overpolishing and takes a permanent set. Attempts to improve the situation by further vigorous polishing only make matters worse; eventually the small particles will break away from the surface. The more complicated the mold, the greater the problem.

Hard carburized or nitrided surfaces are much less prone to the problem. Orange-peel results from exceeding the yield point of the steel. The harder the steel, the higher the yield point and therefore the less chance of orange-peel.

The surest way to avoid orange-peel is to polish the mold by hand. With powered polishing equipment, it is easier to exceed the yield point of the metal. If power polishing is done, use light passes to avoid overstressing.

Orange-peel surfaces usually can be salvaged by the following procedure: Remove the defective surface with a very fine-grit stone; stress-relieve the mold; re-stone; and diamond-polish. If orange-peel recurs after this treatment, increase the surface hardness by nitriding with a case depth of no more than 0.005 in., and repolish the surface.

A large part of mold cost is polishing cost, which can represent from 5 to 30 percent of the mold cost. Recognize that an experienced polisher could polish from 2 to 5 sq in./hr. Certain shops can at least double this rate if they have the equipment.

Polishing is rarely done for appearance alone. It is done either to get a desired surface effect on the part, to facilitate the ejection of the product from the mold, or to prepare the mold for another operation such as etching or plating. If a part is to be plated, it is important to remember that plating doesn't hide any flaws—it accentuates them. Therefore, on critical plated-part jobs the mold polish must be better than for nonplated parts (6).

Another purpose of polishing is to remove the weak top layer of metal. It may be weak from the stresses induced by machining or from the annealing effect of the heat generated in cutting. When it is not removed, this layer very often breaks down, showing a pitted surface that looks corroded.

The techniques used to get a good and fast polish are basically simple, but they must be followed carefully to avoid problems. The first rule is to have the part as smooth as possible before polishing. If electrical-discharge machining is used, the final pass should be made with a new electrode at the lowest amperage. If the part is cast or hobbed, the master should have a finish with half the roughness of the desired mold finish.

When the mold is machined, the last cut should be made at twice the normal speed, at the slowest automatic feed, and at a depth of 0.001 in. No lubricant should be used in this last machining, but the cutting tool should be freshly sharpened, and the edges should be honed after sharpening. The clearance angle of the tool should be from 6 to 9 degrees, and if a milling cutter or reamer is used, it should have a minimum of four flutes. A steady stream of dry air must be aimed at the cutting tool to move the chips away from the cutting edge.

Polishing a mold begins when the designer puts the finishing information on the drawing. Such terms as "mirror finish" and "high polish" are ambiguous. The only meaningful way is to use an accepted standard to describe what has to be polished and to what level. It is also important that the designer specify a level of polish no higher than is actually needed for the job because going from just one level of average roughness to another greatly increases the cost of the mold. Figure 7-33 illustrates the cost of polishing as a function of the roughness of the finish.

Roughness is given as the arithmetical average in microinches. A microinch is one-millionth of an inch (10^{-6} in.) and is the standard term used in the United States. Sometimes written as MU inch or μ inch, it is equal to 0.0254 micron. A micron, one-millionth of a meter (10^{-6} meter), is the standard term in countries that use the metric system. It is often written as micrometer or μm and is equal to 39.37 microinches.

There are several ways to specify a certain level of roughness. One common and important standard is the "SPI-SPE Mold Finish Comparison Kit." It consists of six steel disks finished to various polish levels and covered with protective plastics caps processed in molds with those finishes. Unfortunately, because of the levels selected, the kit is of little practical use. One disk has a roughness of 0 to 3 microinches, another has a roughness of 15 microinches, and all the others are coarser. Because 15-microinch roughness is acceptable in fewer than 10 percent of all jobs, the result is that the disk with zero to 3-microinch roughness, or the highest level of polish, must be selected in almost all other cases.

Another standard that has come into general use is the specifying of a finish as produced by a polishing compound containing diamond particles within a certain microinch range (Table 7-17). It is not a perfect system but works in most cases.

A near-perfect system is the American Standard Association's standard ASA B 46.1. This corresponds to the Canadian standard CSA B 95 and the British standard BS 1134. For a definition of the terms used in ASA B 46.1 and how to apply them to mold drawings, see Figs. 7-34, 7-35 and 7-36. The use of this standard in specifying finishes leaves no room for disputes about what is alled for, and since all quotes apply to the same standard, the polishing costs tend to be more uniform. The biggest drawback to its use is that many tool shops do not have the necessary test equipment.

The term "roughness cutoff width" refers to the distance the instrument checks to get the roughness values (which do not include the wave values). This distance should be long enough to measure all the irregularities except the waves. The standard values specified are 0.003, 0.010, 0.030, 0.100, 0.300, and 1 in. If no value is specified, then 0.030 in. is assumed. The metric equivalents are 0.075, 0.250, 0.750, and 7.5 mm.

To determine the wave width and height, both of which are caused by the cyclic instability of the machine doing the cutting, it is necessary to measure at least one wave width, which sometimes is as much as 1.5 in., or 40 mm. The wave height is measured as the maximum peak-to-valley distance. When only one value is specified for a particular characteristic, it should always be taken as the maximum.

Both the moldmaker and the designer should be aware of the types of finishes that can be obtained using various manufacturing methods. Table 7-18 contains a list of these methods.

Conditions Required for Polishing

Cleanliness is a crucial factor in improving the productivity of the polishing department, and standards should apply both to the workpiece and the surrounding area. Between each polishing step, the workpiece must be cleaned thoroughly with soft tissue, soft rags, or a soft brush, and kerosene. An effective tool for keeping the workpiece free of metal particles and dust is a vacuum cleaner—as long as its

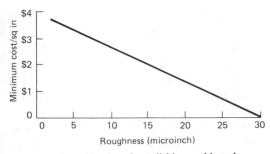

Fig. 7-33 Cost for polishing mold steels.

Table 7-17 Diamond compound specifications.

DIAMOND COMPOUND	FINISH	PARTICLE SIZE, MICROINCH	APPROXIMATE MESH	COLOR
1–8M	Super	0 to 2	14,000	Ivory
3–7M	Very high	2 to 4	8,000	Yellow
6–48M	Mirror	4 to 8	3,000	Orange
9–6M	High	8 to 12	1,800	Green
15–5M	Fine	12 to 22	1,200	Blue
30–4H	Lapped	22 to 36	600	Red

brush is not used for general cleaning and is kept free of dust. Kerosene may also be used to thin the diamond compound when it dries.

In each step of polishing, the "lay" (direction of polishing) should be changed. All traces of the previous lay must be completely removed before the next step is begun. Changing the lay makes it very easy to determine visually when the marks from the previous step have been completely removed.

The best polishing units are the flexible-shaft units. These incorporate $\frac{1}{10}$-hp universal motors with speeds up to 14,000 rpm. The speed control can be of the rheostat, carbon-pile, or electronic type. A standard shaft is 39 in. long, and a standard hand piece has a chuck that can accommodate shanks up to $\frac{5}{32}$ in. in diameter.

The first polishing step is a lapping operation, and a No. 15–5M blue diamond compound (12–22-micron range and a mesh of approximately 1200) can be used. To hold the lap, which is round, $\frac{1}{2}$- and $\frac{3}{4}$-in. nylon bob holders are employed. Spring-loaded mandrels allow the hand piece to be slightly tilted and slight contours to be polished without fear of damage to the workpiece.

Short pieces of $\frac{1}{4}$- and $\frac{1}{2}$-in. brass nipples, with the outside cut down to fit the bobs, or round pieces of cast iron, with holes drilled through the centers, can be used as laps. The laps must fit snugly into the bob holders so that the hand piece can be lifted off the work without dropping the lap.

In areas inaccessible for lapping, a 600-grit silicone-carbide stone may be used, but the work area has to be kept wet with kerosene. On areas that have been EDMed, a hard scale is left on the surface; and resin-bonded stones do a better job than silicone carbide stones. If the EDMing was properly done, however, nothing coarser than a 600-grit stone should be used.

The next steps are all done with hard-bristle—not brass or steel-wire—brushes on $\frac{1}{8}$-in.-diameter mandrels. Successively finer-grit diamond, starting with No. 15–5M, then No. 6–48M, 3–7M, and finally 1–8M, can be employed until the desired finish is achieved. A different brush should be used for each grade of compound.

The last step involves using a soft brush and the last grade of diamond compound used in the preceding step and covering the workpiece with a protective spray. Mold spray should also be used when the piece is left overnight, because it is not unusual for a piece of steel to start rusting within a matter of hours.

Sometimes, when a moldmaker has sent a mold in for polishing before it has been properly machine-finished, the polisher starts with stones as coarse as 100 grit in order to avoid sending the mold out for another setup. That is nearly always a serious mistake and will cost a lot of extra time.

When it is decided that the polisher must remove this excess roughness, he should use

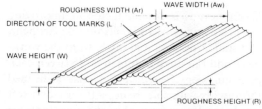

Fig. 7-34 Terms used in standard ASA B 46.1 for measuring mold surface roughness.

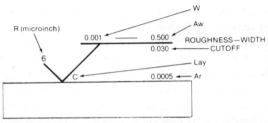

Fig. 7-35 Method of illustrating roughness at a given point on a mold surface per ASA B 46.1.

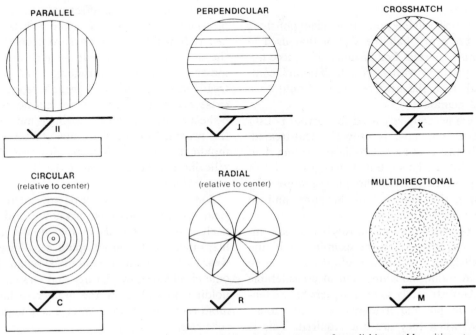

Fig. 7-36 Symbols used on mold drawings to indicate the lay pattern for polishing mold cavities, etc.

Table 7-18 Surface roughness produced by various manufacturing methods.

PROCESS	ROUGHNESS HEIGHT IN MICROINCHES	PROCESS	ROUGHNESS HEIGHT IN MICROINCHES
Flame cutting	250–2000	Lapping	1–32
Snagging	125–2000	Superfinishing	1–32
Sawing	63–2000	Polishing	1–32
Drilling	63–500	Sand casting	250–2000
EDM & CM	32–1000	Ceramic casting	32–250
Milling	16–1000	Investment casting	32–250
Turning	16–1000	Pressure casting	32–125
Boring	16–1000	Forging	63–500
Reaming	16–250	Diecasting	16–250
Tumbling	2–63	Injection molding	2–63
Grinding	2–125	Stamping	16–250
Honing	2–63		

a No. 2 riffler or file instead of a stone. The only time a stone coarser than 600 grit should be used is when the mold has already been hardened. This is a very bad practice, since it is much easier to polish a soft mold.

Sometimes an inexperienced polisher makes orange-peel and pit marks on the surface. These defects are caused by the metal flowing and tearing under heat and stresses beyond its elastic limit. The cure is to repeat the previous step.

The most reliable guide to efficient polishing is to use a grit half the size of that used in the previous step. For instance, if a piece has been polished with a No. 6 diamond compound, then the next one used should be a No. 3 diamond.

There is rarely any need for further polishing after the mold is properly heat-treated. The brown coating left by heat treating does not affect the finish, but it does give a lot of rust protection. Even though the shine may be gone, the finish is still the same, and no further work is needed.

As reviewed throughout this book, the complete operation has to be examined initially. Definitely tight tolerances of all types can be met, including polishing, but at considerable cost. Unfortunately some designers have a tendency to put requirements to "tight tolerances" when they are not required. This is particularily true for polishing. Complex molds can be designed to permit a practical approach in polishing so that performance requirements of the molded part can be met at the lowest cost.

PLATING, COATING, AND HEAT TREATING CAVITIES/MOLDS

Molds are generally plated, coated, and/or heat-treated to resist wear, corrosion, and release problems. Treatments such as those reviewed in Table 7-19 that reduce wear are especially helpful in gates, runners, ejector pins, core pins, inserts, and cavity areas opposite the gate. Other treatments resist the corrosion damage inflicted by chemicals such as hydrochloric acid when processing PVC, ammonia with acetals, and oxidation caused by

interaction between molds and moisture in the plant atmosphere. Release problems require treatments that decrease friction and increase lubricity in mold cavities.

No single mold treatment is ideal for solving all of these problems. The molder must determine which mold problems are causing the greatest loss of productivity (or could cause loss) and then select the mold treatments that will be most effective in solving the problems.

Plating and coating affect only the surface of a mold or component, while heat treating generally will affect the physical properties of the entire mold. Treatments such as carburizing and nitriding are considered to be surface treatments, and although heat is applied in these processes, they are not considered to be heat treatments. Heat treating is more often the province of the steel manufacturer and moldmaker than the molder. However, stress relieving is a heat treatment that the molder can perform.

Some mold wear cannot be prevented. This wear should be observed, acknowledged, and dealt with at intervals in the mold's useful life; otherwise the mold could be allowed to wear past the point of economical repair. Periodic checks of how platings and coatings are holding up will allow the molder to have a mold resurfaced before damage is done to the substrate.

While a poorly finished mold is being used for the first time, its surface is actually being reworked by heat, pressure, and plastic. Fragmented metal is pulled out of the metal fissures, and plastic is forced into them. While the fissures are plugged with plastic, the molder may actually be molding plastic against plastic. "Breaking in" a poorly finished mold can be haphazard without proper presurfacing. If the underlying mold surface is unsound (and no prior treatment was used although required), a thin layer of metal plating, particularily chrome plating, will not make it correct. A poorly prepared surface makes for poor adhesion between treatment and the base metal.

Most molders or moldmakers will not find it practical to do very much surface treatment for themselves. Few of the plating, coating,

Table 7-19   Mold surface plating and coating treatments (23).

PROCESS	MATERIAL	APPLIED TO	PURPOSE
Coating by impingement—molecularly bound	Tungsten disulfide	Any metal	Reduce friction and metal-to-metal wear with dry film—nonmigrating
Coating by impingement—organically bound	Graphite	Any metal	Reduce sticking of plastics to mold surface—can migrate
Electrolyte plating	Hard chrome	Steel, nickel, copper alloys	Protect polish, reduce wear and corrosion (except for chlorine or fluorine plastics)
	Gold	Nickel, brass	Corrosion only
	Nickel	Steel and copper alloys	Resist corrosion except sulfur-bearing compounds, improve bond under chrome, build up and repair worn or undersized molds
Electroless plating	Nickel	Steel	Protect nonmolding surfaces from rusting
	Phosphor nickel	Steel and copper	Resist wear and corrosion
Nitriding	Nitrogen gas or ammonia	Certain steel alloys	Improve corrosion resistance, reduce wear and galling; alternative to chrome and nickel plating
Liquid nitriding	Patented bath	All ferrous alloys	Improve lubricity and minimize galling
Anodizing	Electrolytic oxidizing	Aluminum	Harden surface, improve wear and corrosion resistance

and lubricating treatments lend themselves to being done in a molding shop. With few exceptions, treatments involve processes and chemicals that should not be used anywhere near an injection machine (because of corrosiveness), and they are best handled by custom plating and treating shops that specialize in their use.

The distinction between platings and coatings is not entirely clear. Generally thin layers of metals applied to the surface of mold components are considered platings. Application of alloys, fluorocarbons, or fluoropolymers (such as TFE) or dry lubricants are considered coatings.

The effectiveness of a surface treatment depends not only on the material being applied but also on the process by which it is applied. For any plating or coating to stick to the surface of a mold component, it has to bond to

the surface some way. The bonding may be relatively superficial, or it may be accomplished by a chemical/molecular bond. The nature and strength of the bond directly affects the endurance and wear characteristics of the plating or coating. The experience of the plater is an important factor in applications where cut-and-dried or standard procedures have not been developed.

Nickel

Electroless nickel plating deposits up to 0.001 in. of nickel uniformly on all properly prepared surfaces. The term electroless is used to describe a chemical composition/deposition process that does not use electrodes to accomplish the plating. Plating by use of electrodes is called electrodeposition. Electroless nickel provides a good protective treatment for mold components, including the holes for the heating media. It prevents corrosion; also any steel surface that is exposed to water, PVC, or other corrosive materials or fumes benefits from this minimal plating protection. Electroless nickel has the characteristic of depositing to the same depth on all surfaces, which eliminates many of the problems associated with other metallic platings. Grooves, slots, and blind holes will receive the same thickness of plating as the rest of the part. This allows close tolerances to be maintained.

The surface hardness of electroless nickel is 48 Rockwell C, and the hardness can be increased by baking to 68 Rockwell C. Plated components will withstand temperatures of 700°F.

Chrome

Chromium is a hard, brittle, tensile-stressed metal that has good corrosion resistance to most materials. As it becomes thicker, it develops a pattern of tiny cracks because the stresses become greater than the strength of the coating. These cracks form a pattern that interlaces and sometimes extends to the base metal. A corrosive liquid or gas could penetrate to the base metal. This action can be prevented in three ways: a nickel undercoat can be applied to provide a corrosion-resistant barrier; the chrome plating can be applied to a maximum thickness; or a thin dense chrome can be substituted.

Hard chrome can be deposited in a rather broad range of hardnesses depending upon plating-bath parameters. Average hardness is in the range of 66 to 70 Rockwell C. A deposit of over 0.001 in. thickness is essential before chromium will assume its true hardness characteristics when used over unhardened base metals. Over a hardened base, however, this thickness is not necessary because of the substantial strength/backing provided. Precise control of thickness tolerances can be achieved in a particular type of chrome plating generally referred to as thin, dense chrome. This kind of plating provides excellent resistance to abrasion, erosion, galling, cavitation, and corrosion wear.

Adhesion between a chrome layer and the base metal is achieved by a molecular bond. The bond strength is less upon highly alloyed steels and on some nonferrous metals; however, bonds in excess of 35,000 psi tensile strengths are common. Although it is often said that the chromium layer will reproduce faithfully every defect in the base material, actually as the chrome deposit thickens, it will level imperfections. For smooth chrome plating, the base material should be at least as smooth as the expected chrome. Imperfections can be ground or polished after plating to erase them.

In some instances, cracks in the base material that are not visible through normal inspection techniques may become apparent only after plating. This phenomenon is attributed to the fact that grinding of steel often causes a surface flow of material, which spreads over cracks and flaws. However, this cover is dissolved during the preplating treatment, and the cracks become apparent as the coating thickness increases. The deposited layer of chromium, although extremely thick, will not bridge a large crack.

With electrodeposition the current distribution over different areas of a component greatly varies, depending on its geometrical shape. Elevations and peaks, as well as areas directly facing the anodes, receive a higher current density than depressions, recesses, and areas away from the anodes that do not directly face the anodes. The variation in current density over different areas produces a corresponding variation in the thickness of the deposited metal.

Hard chrome plating is recommended to protect polished surfaces against scuffing and to provide a smooth release surface that will minimize sticking of the parts in the mold. Some precautions are necessary. Hydrochloric acid created in the injection molding of PVC will attack chrome. Chrome that is stressed and cracked under adverse conditions will permit erosion from water and gas penetration into the steel. To deal with hydrogen embrittlement created when hydrogen is absorbed by steel during the plating process, chrome-plated components should be stress-relieved within a half hour of completion of the plating. To protect against galling, chrome should not be permitted to rub against chrome or nickel.

While it is not uncommon for a mold to be sampled in the machine under pressure before plating, this is an undesirable practice. The effects of moisture on the steel can cause chrome plating to strip later. Carelessness can also result in scratching an unplated mold. Dimensional checks can readily be made outside the press using wax or other sampling materials. Chrome can be stripped from a mold after sampling in order to make essential changes. Periodic checks after a chrome-plated mold is in full production are desirable to find evidence of wear, which will show up first in the mold corners and high flow areas. A simple check can be made by swabbing a copper sulfate solution in the mold areas. If the copper starts to form a plating on the surface, the chrome is gone and must be replaced.

Other Plating Treatments

Numerous metals other than chromium and nickel have been used at one time or another to coat the components of plastics molds. Gold plating can be used to create a protective surface for PVC and some fluorocarbon materials. It prevents tarnishing, discoloration, and oxidation of the mold surface. Gold will protect the original finish and provide a hydrogen barrier. Fifty-millionths of an inch of gold plating (0.00005 in.) is adequate for the purpose. Gold can also be used as a primer under polished chrome. Platinum and silver have also been used to plate molds, and they share with gold the notable drawback of high cost.

Coating Treatments

Composites of ultrahard titanium carbides distributed throughout a steel or alloy matrix are used very effectively for coating. The carbides are very fine and smooth, and the coatings are applied by sintering, a process in which the component to be coated is preheated to sintering temperature, then immersed in the coating powder, withdrawn, and heated to a higher temperature to fuse the sintered coating to the component.

Flexibility in selecting and controlling the composition of the matrix alloy makes it possible to tailor the qualities of the alloy according to the requirements of the application. When heat-treatable-matrix alloys are used, the composite can be annealed and heat-treated, permitting conventional machining before hardening to 55 to 70 Rockwell C.

Standard alloy matrices have been developed with quench-hardenable tool steel, high-chromium stainless steels, high-nickel alloys, and age-hardenable alloys. Special alloys can be formulated for even the most corrosive conditions.

These types of coatings are effective in combating the severe wear that occurs with abrasive plastic compounds. Ceramic/metal composites provide the hardness and abrasion resistance to withstand wear by the most damaging glass-fiber-, mineral-, or ferrite-filled compounds. With the right metal-matrix selection, resistance to corrosion and heat is also obtained.

Treatments are available via a chemical process that utilizes thermal expansion and contraction to lock PTFE (fluoropolymer) particles into a hard electrodeposited surface such as chromium. Surfaces treated by the process are reported to have the sliding, low-friction, nonstick properties of PTFE, along with the hardness, thermal conductivity, and damage resistance of chrome. Core rods and other internal mold components benefit from this surface treatment. The process builds a thickness of 0.002 in. of electrodeposited chromium on component surfaces and is available for ferrous, copper, and aluminum-alloy parts.

Impregnation processes are available that provide continuous lubrication to metal parts by impregnating fluoropolymers into the surface pores of the metal. Reduced friction, wear, and corrosion are reported as benefits, along with improved plastics flow and release characteristics.

Impregnation is applied to mold cavities, runners, core pins, and ejector pins, enabling easier ejection of parts and minimizing the need for release agents for most hard-setting thermoplastics. Pin galling and metal-to-metal sliding friction on core and ejector pins are also eliminated.

Titanium carbonitride is the material most commonly applied to provide wear resistance to tools and wear parts made from tool steels. The coating can often be applied in a layer thick enough to allow a stock for a surface-finishing operation after coating. The coating inhibits galling and results in a favorable coefficient of friction. The process protects against wear from abrasive fillers and corrosion from unreacted polymers that release acids. The coating is said to be 99 percent dense, that is, to have virtually no porosity, and is inert to acids.

Note that the processes used for hardening screws and barrels are not necessarily the same as those applicable to mold components. A spin-casting process is used to coat the inside of barrels under heat, after which the barrels are rebored. The outsides of screws are hardened with gas or ion nitriding. Screws are also hardened by plasma-welding a bead of stellite on the screw surface, which is then ground to the desired shape and finish. The processes used on mold components are different in part because brittle tool steel cannot be twisted like a screw.

Heat Treatments

Most mold steels are subjected to heat treatment in some form to obtain the hardness necessary for their intended use. The routine heat treating that is part of the moldmaking process is not normally performed by molders, unless they also run their own moldmaking shop.

Exceptional wear resistance of mold surfaces can be accomplished by carburizing or nitriding; both processes are also known as case hardening, which refers to the fact that the surface layer of the material being treated is made considerably harder than the interior. The depth of penetration of the treatments can vary from a few thousandths up to one-sixteenth of an inch.

Carburizing is accomplished by heating steel to between 1600°F and 1850°F in the presence of a solid carbonaceous material, a carbon-rich atmosphere, or liquid salts. Nitriding consists of subjecting parts to the action of ammonia gas at temperatures of 950°F to 1000°F, or to contact with nitrogenous materials in order to impregnate the surface with nitrogen.

These treatments can produce skin hardnesses considerably above the maximum hardness obtainable in heat-treated tool steels and provide excellent resistance to abrasion. General guidelines for heat treating are readily available, but specific heat-treating data must always be obtained from the supplier of a particular grade of steel.

Anyone dealing with mold steels should be aware of potential sources of stress that can be detrimental to the life and usefulness of the finished tool. These sources of stresses may be a result of tool design, heat treating, machining, grinding, EDMing, welding or brazing, or anything else that contributes to heating of the steel in a nonuniform manner. Heat treating a mold always introduces risks of distortion and cracking; permanent linear movements in steel during heat treatment are to be expected. It is impossible to predict accurately the extent or direction of movement, since chemical composition, mass, geometry, design, and heat-treating techniques all affect the final dimensions of a mold.

Carburizing temperatures and hardening temperatures, which vary with the type of steel being used, are provided in technical data furnished by steel producers. Care must be taken during heat treating to protect the mold surface against oxidation. This is done by packing the mold into spent cast-iron chips or pitch coke, by heating in a controlled-atmosphere furnace, or by heating in a vacuum furnace.

After the mold is heated to the hardening temperature, it is either quenched in a liquid such as oil or allowed to cool in air, depending upon the analysis of the steel. High-alloy steels harden sufficiently when cooled in air from the hardening temperature.

A mold can benefit greatly from periodic stress relieving, a form of heat treatment highly recommended by metallurgists. This process can extend the life of questionable mold sections, even though they may not yet have exhibited cracks. The stress-relieving process consists of heating parts of the mold in question to the same temperature or just below the temperature at which the mold sections were tempered originally. The plating must be stripped before this annealing operation, since the stress-relief temperature is usually above the plating stability point.

Experience is the only available teacher to suggest the desired interval for this operation. Expensive mold sections clearly merit this form of preventive care, on the premise that an ounce of prevention is worth a pound of cure. Mold components with sharp or nearly sharp fillets and cores with a high ratio of length to cross section are vulnerable to cracking and therefore will require the most frequent stress relief.

There are practical limits to the use of mold treatments to solve molding problems. It makes more sense, for example, to replace three-dollar nitrided ejector pins than to spend great sums of money increasing their wear resistance further. For example, pins may break—no matter how much their surfaces are hardened—more often than they wear out. Furthermore, a pin nitrided to 70 Rockwell C will cause the hole around it to wear before the pin does. The solution to that problem is to ream the hole out and install an oversize pin. Judicious use of the various forms of plating, coating, and heat-treating processes available to today's molder will go a long way toward maximizing mold life and productivity. Well-treated molds, closely observed for wear and diligently maintained, are likely to provide the molder with fast, long, and profitable operation.

STRENGTH REQUIREMENTS FOR MOLDS

The forces involved with the molding operation are compressive; they are exerted by the clamping ram and the internal melt pressure. Forces exist inside a cavity as a result of injecting the plastic material under pressure. (For details on mold strength requirements see reference 2.)

Clamp Tonnage and Mold Size

If we inspect the stationary platens of injection machines in an operating plant, we will find that a number of them have indentations and impressions. These are a result of some projections from the mold base, and, in some cases, of the mold base being too small for the clamping force, thus causing a concentrated stress in the platen that brings about the flow of the platen metal. The platen impressions are dangerous because they reduce the contact area for the mold, thereby increasing the potential for further indentation. These indentations, if permitted to increase in number, may ultimately cause cracking of a platen, which would not only take the press out of operation but also cause a large and expensive replacement.

Practically all presses have provisions for reducing the clamp tonnage, but the problem is to recognize the danger and the limits within which it is safe to concentrate a load on the platen. Most platens are made of cast steel with a yield strength of about 25 ton/in.^{2} Allowing a safety factor of 7, we have a permissible load of 3.5 ton/in.^{2} With this information, we are able to calculate the minimum number of square inches a certain press size will safely accommodate or to determine to what tonnage to reduce the clamp in order to protect the platen against damage.

A mold $10'' \times 12''$ is to be placed in a 500-ton press. First, we will establish the minimum square inches of mold base needed to safely absorb the clamp force:

$$500 \text{ tons} = \text{in.}^2 \times 3.5 \text{ (permissible stress)}$$
$$\text{in.}^2 = \frac{500}{3.5} = 143 \text{ in.}^2$$

The mold in question is $10'' \times 12'' = 120$ in.^{2} This shows that the mold should be operated with a reduced tonnage, calculated as follows:

$$\begin{aligned} \text{Tonnage} &= (\text{in.}^2 - \text{Locating hole area}) \times 3.5 \\ &\quad \text{(permissible stress)} \\ &= (120 - 12.5) \times 3.5 \\ &= 107.5 \times 3.5 = 376 \text{ tons} \end{aligned}$$

To protect press platens against damage, it is advisable to check the contact area of the mold and the platen to see that the safe permissible load is not exceeded.

Stress Level in Steel

When we examine the great variety and complexity of plastic parts, we realize that the molds in which they are produced are even more complex. A great many factors should be remembered when a design layout for a mold is made, but none is more important than maintaining a low stress level in the steel of all the components of the cavity and core. Highly stressed parts mean short tool life.

Heavy and high-speed cuts during metal removal, severe grinding action, and (to a much lesser degree) electric discharge machining all produce some amount of stress in various tool steels. Stress relieving will minimize the danger of failure.

Molds that are built for long life and high activity are heat-treated either initially or whenever the intermediate hardness of the cavities begins to affect the quality of the product unfavorably. From a heat-treatment standpoint, the tool designer should be on the lookout for the following:

1. The parts should be so shaped that they will heat and cool as uniformly as possible. A part that may heat so that a temperature difference exists between two points, will produce a harmful strain when quenched.
2. A balanced section will heat and cool more uniformly than an unbalanced one, thus guaranteeing a much lower stress level.
3. Addition of holes may be used to reduce the mass of metal in one area to offset the lower mass in an adjacent point.
4. Sharp angles and corners are a most common error that, with a little effort, could be minimized. Sharp corners and angles are points of high stress concentration. When a rectangular insert is being made for the cavity, the sharp corner in the plastic will most likely tolerate a

radius of 0.020 in. If this is not permissible, the insert portion that is even with the cavity can be made larger and can have a generous radius; the portion that is molding can be of whatever shape is required.
5. A thin section will cool faster than a thick one during quenching and will set up stresses. A larger radius or an even taper in the transition area will minimize stresses.
6. For whatever purpose they are intended, blind holes should be eliminated. The through-hole makes for greater uniformity in cooling and eliminates the stress concentration from the sharp corner at the bottom of the hole. Junctions of holes, such as might be planned for fluid circulation, should be avoided in favor of drilled-through-holes, since the intersection of holes will act to raise stress.

The designer should be aware that the best choice in material coupled with the best effort of the heat treater cannot overcome faulty design. When layouts for the cavity and core are made, an outline of the components should be presented to a heat treater for a recommendation of design modification that will lead to parts with low-level stresses. This matter deserves serious consideration. An order to the heat treater should specify: "To be stress-relieved if heat-treating steps will not reduce stress."

Pillar Supports

The general construction of a mold base usually incorporates the U-shaped ejection housing. If the span between the arms of the U is long enough, the forces of injection can cause a sizable deflection in the plates that are supported by the ejector housing. Such a deflection will cause flashing of parts. To overcome this problem, the span between supports is reduced by placing pillar supports at certain spacings so that the deflection is negligible.

For the determination of pillars and their spacing, the beam formula can be applied. For this purpose, we consider a 1⅞-in.-thick plate (Fig. 7-37) as beam-supported at 8.5-in. centers with a uniform load. For this loading system, we consult a handbook (under "Stresses of Beams at the Center") to find the stress (2):

$$\text{Stress at center} = \frac{WL}{8Z} = S$$

where W is the load that the plate can support, L is the length between supports = 8.5 (Fig. 7-37), and Z is the section modulus or a property of the cross section that resists flexure.

In the handbook (under "Section Modulus"), we find the following formula:

$$Z = \frac{bd^2}{6} = \frac{bB^2}{6} = \frac{15 \times 1.875^2}{6}$$

in which $d = B = 1.875$
$$b = 15$$
$$Z = \frac{15 \times 3.52}{6} = 8.80$$

S is the allowable safe stress, and the value suggested by the mold base manufacturer is 12,000 psi. Referring to Fig. 7-37, and using the formula for S:

$$12,000 = \frac{W = 8.5}{8 \times 8.8}$$

$$W = \frac{12,000 \times 8 \times 8.8}{8.5} = 99,275 \text{ lb}$$

where W is the permissible load on the support plate.

When the mold is closed, the cavities will exert a concentrated pressure on the support plate. For this condition, a safe concentrated stress in compression of 7000 psi is allowed. The compression formula from the handbook is:

$$S = \frac{P}{A}$$

where S = 7000 psi allowable stress
$P = W = 99,275$

Thus:

$$A = \frac{P}{S} = \frac{99,275}{7000} = 14.13 \text{ in.}^2$$

Thus, the total area of the back of the cavities can be only 14.13 in.² If one row of support pillars is added, the L dimension is 4.25 in the load formula, thus doubling the load capacity of the plate and also doubling the permissible cavity area to 28.26 in.², and at the same time maintaining the allowable stress of 7000 psi. Increasing the number of rows of pillar supports decreases the distance between the resting points of the beam, thereby increasing the area for the concentrated pressure of the cavities.

Steel and Size of Mold Base

The size and type of mold base are determined by placement of the cavities, by the method of feeding the cavities, by the ejection employed, by the type of pockets desired, by temperature control, by type of cam action, or by any unusual factor that becomes necessary for a specific part. A layout of these and the other elements so far established will indicate the type and overall size of the mold base.

Taking a four-cavity mold as an example, we obtain the outside width and length of the cavities. To the width of the cavities, we add 1.75 in. per side, so that they are placed close to support blocks of the ejector housing. On the end, we add an additional inch for the return pins, or 2.75 in. per end. These overall dimensions are checked against standard

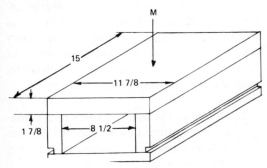

Fig. 7-37 Schematic for pillar requirements.

available mold bases, and the selection is made to satisfy the outline of the layout.

There are normally three grades of steel employed in mold bases, as follows:

1. The lowest-priced steel grade in a mold base is a medium carbon type with tensile strength of 55 to 75×10^3 psi. This grade is suitable for application where the cavities are in themselves strong enough to withstand the conditions of application. The main function of a mold base in the preceding case is to keep the two halves aligned and the ejection side rigid enough to permit ease of ejection on a cycle of two or more times a minute. Where the cavities are mounted in a cut-through plate, care must be exerted that the surrounding frame is thick and wide enough to safeguard the guiding features of the halves. Where blind pockets are employed, this steel is suitable for a majority of applications.

2. The next higher grade of steel employed for bases is an AISI type 4130, heat-treated to a hardness of 300 Bhn with a tensile strength of 126 to 155×10^3 psi. This grade is usually considered for cases where the cavities are constructed in sections, and it is the function of the plates to retain these sections without allowing them to separate under the forces of injection pressures. It is also applied in cases where cooling lines and other machining requirements weaken the cavity plate to a point where a material with higher physical properties is prescribed.

3. There are occasions when it is desirable to machine cavities into the cavity plates instead of fabricating cavities and inserting them into mold-base plates. This may be the case for a product with a yearly activity of less than 10,000 pieces and a configuration that is relatively easy to machine from a mold-base plate. For such application, the mold-base cavity plates may be specified to be an AISI 4135 steel heat-treated to 300 Bhu with a physical strength of 129 to 155×10^3

psi. It is a suitable steel for polishing and higher hardness heat-treating if necessary.

DEFORMATION OF MOLD

The function of a mold is to receive molten plastic material from the plasticator (injection unit) ranging in temperature from 350°F to 900°F at pressures between 4000 and 20,000 psi. In the injection process, the plastic comes from a heated nozzle, and passes through a sprue bushing into feed lines (runners) to a gate into a cavity. The cavities are maintained at temperatures generally ranging from 30°F to 350°F for thermoplastics and 250°F to 600°F for thermosets, at which solidification takes place. They are provided with a means for controlling the temperature.

At the end of the injection stroke, and during after-filling, pressure is built up in the mold cavity. This pressure, which depends on the type of molding and plastic, is generally one-third to one-half of the injection pressure set on the machine. In normal cases pressure in the mold cavity will be up to 4000 to 8000 psi. However, in exceptional instances pressure may rise to 15,000 psi in certain mold components, usually when close dimensional tolerances need to be held. The consequences of such pressures must be appreciated. They cause elastic deformations, such as bending of cavity retainer plates and cores, that are virtually unavoidable.

The use of a sturdy construction (sufficiently thick cavity retainer plates, and support pillars in the open gap for the ejector system) may reduce elastic deformation to a minimum. Such possibilities are, however, often restricted, since light construction is also required, for efficient cooling, for necessary spaces for guide pins and ejector system, etc. Elastic deformation of weak or insufficiently solid mold components may result in:

1. Differences in wall thickness with consequent excessive dimensional variations as well as insufficient dimensional stability and rigidity of the molding.
2. Nonuniform melt flow in the mold. In

the case of thin-walled moldings, this may give rise to flow lines, weld lines, internal stresses, or even trapped air.
3. In weak molds, the bearing surfaces or other components being forced apart by the plastic pressure, causing flash formation that may interfere with proper mold release. Moreover, the subsequent deflashing operation is a considerable cost-raising factor.
4. Faulty operation of the ejector system and guide pins. It is even possible that the mold will jam.

The general principles of molding are similar regardless of the type of press employed. All presses must meet the basic elements of molding: time, temperature, and pressure, the range of temperatures (Table 28-5) and pressures depending on the type of plastic material. The plastic is held in the cavity for a prescribed time until full solidification takes place; at this point, the mold opens, exposing the part to the ejection or removal action.

Mold Filling

The effect of mold dimensions and resin viscosity on pressure requirements is expressed as follows:

$$Q = \frac{P}{K\eta}$$

or:

$$p = K\eta Q$$

where Q = volumetric rate of mold fill in.^{3}/sec or cm^3/sec)
p = pressure at mold entrance (lbf/in.^{2} or kgf/cm^2)
η = resin viscosity (lbf sec/in.^{2} or kgf sec/cm^2)
K = mold flow resistance factor

For end-gated rectangular cavity section fill out:

$$K = \frac{12L}{W_t^3}$$

For end-gated annular cavity section fill out:

$$K = \frac{12L}{\pi D_m t^3}$$

For end-gated cylindrical cavity section fill out:

$$K = \frac{128(L + 4D_c)}{\pi D_c^4}$$

In these equations,

L = mold cavity length (in. or cm)
W = mold cavity width (in. or cm)
t = mold cavity thickness (in. or cm)
D_m = mean diameter of annulus mold cavity (in. or cm)
D_c = mean diameter of cylindrical mold cavity (in. or cm)

From the above it may be seen that at constant flow rate and resin viscosity:

1. The pressure required to fill is directly proportional to the mold length.
2. The pressure required to fill is inversely proportional to cavity width or diameter.
3. The pressure required to fill is inversely proportional to the cube of the mold thickness.
4. The pressure required to fill radial fill patterns (i.e., center gated) is exponential.

The pressure required is proportional to the resin viscosity, and is reduced by an increase in temperature and/or shear rate as these equations indicate:

Shear rate for rectangular section

$$= \frac{6Q}{t^3(W + t)}\text{sec}^{-1}$$

Shear rate for annular section

$$= \frac{6Q}{t^2(\pi D_m + t)}\text{sec}^{-1}$$

Shear rate for cylindrical section

$$= \frac{32Q}{\pi D_c^3}\text{sec}^{-1}$$

Thus, shear rate is increased and resin viscosity decreased by a decrease in mold cavity dimensions (i.e., an effect opposite to that which such cavity dimensions have on the mold flow resistance factor K) (4).

Deflection of Mold Side Walls

Rectangular Cavities. The maximum deflection commonly allowed in such molds is 0.005–0.01 in. (0.13–0.25 mm), depending upon the size of the tool. Of this, 0.004–0.008 in. (0.1–0.2 mm) may be due to clearances between the blocks of the built-up mold and elongation of the bolster or register faces. For stress design purposes, therefore, a maximum deflection of 0.001–0.002 in. (0.025–0.05 mm) is usually taken. The approximate thickness of the side wall required may be calculated from the following formulas:

$$y = \frac{Cpd^4}{Et^3}$$

or:

$$t = \sqrt[3]{\frac{Cpd^4}{Ey}}$$

where y = deflection of side walls (in. or cm)
C = constant (see Table 7-20)
p = maximum cavity pressure (lbf/in.² or kgf/cm²)
d = total depth of cavity wall (in. or cm)
E = modulus of elasticity for steel (30×10^6 lbf/in.² or 2.1×10^6 kgf/cm²)
t = thickness of cavity wall (in. or cm)

Cylindrical Cavities. The increase in radius due to the internal pressure of the injected material can be determined approximately as follows:

$$r_1 = \frac{rp}{E}[\{(R^2 + r^2)(R^2 - r^2)\} + m]$$

where r_1 = increase of inside radius (in. or cm)

Table 7-20 Constant used in calculating deflection of mold side walls.

RATIO OF THE LENGTH OF CAVITY WALL TO THE DEPTH OF CAVITY WALL	VALUE OF C
1:1	0.044
2:1	0.111
3:1	0.134
4:1	0.140
5:1	0.142

r = original inside radius (in. or cm)
R = original outside radius (in. or cm)
m = Poisson's ratio (0.25 for steel)

The strength requirements for the two configurations are satisfactorily met. In all the calculations, it was taken for granted that the ram pressure was applied to the cavities only. This was accomplished by having the cavity insert protrude above the A or B plate about 0.005 in.

Let us now assume that for some valid reason a two-cavity mold is ordered, and the press in which it is to be run is still the 200-ton size. In this case, the width of the cavity face would be unchanged except that cavity inserts would be mounted flush with A and B plates so that the plates would absorb part of the force.

The problem of mounting a cavity will be favorably met in either a machined-through picture-frame pocket or a blind pocket, whichever is more suitable from the standpoint of mold temperature control as well as other considerations. Based purely on strength considerations, the calculated dimensions will incorporate in the cavity itself the ability to safely absorb all the forces to which it may be subjected during molding (2).

During injection of the fluid plastic into the cavity, we find pressures existing there between 4000 and 10,000 psi close to the point of exit from the gate. As the flow approaches the outside extreme point, these pressures may be 2000 psi. The difference in readings between those of the pressure gauge and 2000 psi at the end of flow is found in the pressure drops coming from the screw acting as a plunger, the nozzle, the sprue bushing, the runners, the gate, and resistance to flow within the cavity. The average pressure in the cavity may be 4000 to 10,000 psi. Even the lower pressures in a cavity will cause a sizable deflection in a cavity wall unless it is made heavy enough to keep such deflection within acceptable values. As J. Dym explains, the following takes place in the cavity: The projected area of the side wall times the psi in the cavity creates a force that will bring about a movement of the side wall of, say, 0.003 in. After the material is cooled and the inside pressure drops to a negligible value, we have a force from the deflected steel tending to return to zero deflection. This force is comparable to that which caused the original deflection.

If the part is made of a thickness that will shrink 0.003 in. in the cavity, then the steel will merely go back to its original position without any ill effects upon the operation. If, however, the plastic will only shrink 0.001 in., the steel pushes into the plastic by 0.002 in., causing difficulty in mold opening, possibly marring the surface, and adversely influencing dimensions and properties of the plastic. Last but not least, the large forces involved will gradually cause movement of cores with respect to cavities, with additional complications and ultimately mold damage.

The problem of mold deflection deserves attention and must be resolved in a way that will eliminate these difficulties. A cavity must be looked upon as a very high pressure vessel in these considerations. Some of the formulas related to the subject are found in: *Formulas for Stress and Strain* by Raymond J. Rark and Warren G. Young (McGraw Hill, 1975).

These formulas have been modified and rearranged here to suit conditions that exist in molds. The formula for deflection in a cavity *without* the restraining effect of top and/or bottom is:

$$d_1 = \frac{p\,a\,c_1}{E} = \text{deflection}$$

A cavity *with* restraining effect in which the bottom is either an integral part of the cavity or is so interlocked that it will act as an integral part of the unit shows this deflection:

$$d_2 = \frac{p\,a\,c_2}{E} = \text{deflection}$$

where p = average pressure in the cavity in psi, a = half the width of a part as viewed from the top, c_1 and c_2 are factors from Fig. 7-38 (equal to $n \times a$), and E = modulus of elasticity for steel (30×10^6 psi).

Factor c_2 is applicable only in cases when the depth of the cavity is equal to a. For greater depths there is a gradual transition to the deflection that exists in the case of an unrestrained bottom or top. The distance at which the c condition will be reached is:

$$L = a\,\sqrt{\frac{\text{unstricted deflection}}{\text{restrained deflection}}}$$

In practice the requirements for creating equivalent conditions that will correspond to an integral solid bottom or top are:

1. The cavity insert must have metal-to-metal contact with its retainer; that is, the O.D. of the insert and I.D. of the retainer must have the exact same dimensions and when assembled result in a light press fit.
2. The body of the cavity and the bottom or top must be so interlocked that there is no chance that the insert will move with respect to its retainer.
3. The clamping pressure should be 25 percent higher than would be the case if there were no deflection problem.

Let us take an example: The average cavity pressure is 6000 psi, and the construction will be of the restraining type.

Part depth = 8.5 in.
Average diameter = 8.0 in.
Wall thickness = 0.070 in.
Shrinkage = 0.020 in./in.

The total shrinkage on the part will be about 0.0014 in., but when the pressure in the cavity

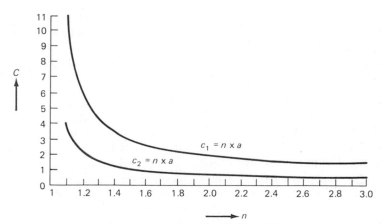

Fig. 7-38   Factors to use in determining deflection in a mold cavity.

is decaying, it may be only half that amount, or 0.0007 in. If we allow for a maximum deflection in the center of the cavity of 0.0006 in., there should be no ill effects on part performance. If the cavity is constructed of sections, no plastic material will flow between sections at selected deflection. Thus the mold should work properly.

Therefore:

$$d_2 = \frac{p\,a\,c_2}{E} = \frac{6000 \times 4 \times c_2}{30 \times 10^6} = 0.0006$$

$$c_2 = \frac{30 \times 10^6 \times 0.0006}{6000 \times 4} = \frac{18}{24} = 0.75$$

From the Fig. 7-38 curve we find $n = 2.4$. The outside cavity dimension will be $n \times a = 2.4 \times 4 = 9.6$, giving a wall of 5.6 in. Since water lines will be incorporated and will cause a decrease in strength, we will add $1\frac{1}{2}$ in. for 1-in. baffled lines and $\frac{3}{4}$ in. for $\frac{7}{16}$-in. lines. In this case the addition will be $1\frac{1}{2}$ in., or the new thickness will be 7.1 in. This will be satisfactory for the 4-in. depth of cavity, and since the cavity is 8 in. in diameter, we will arrange the core end of the mold for interlocking and restricting action. Thus the 4 in. distance from each end will satisfy the depth restricting requirement. (See Fig. 7-39)

EYEBOLT HOLES

Eyebolt holes are normally on the side of the clamping slots and should be provided on both halves opposite each other; they should be placed in areas where balanced lifting of the mold base is possible. Holes should also be tapped on surfaces perpendicular to the slots.

The forged steel eyebolts have a safe load-carrying capacity as listed:

$\frac{1}{2}$ in.	2600 lb
$\frac{3}{4}$ in.	6000 lb
1 in.	11,000 lb

Only forged steel eyebolts should be specified for safety reasons, preferably those with a shoulder for better stability.

MOLD PROTECTION

A tool that has received all the necessary attention and care from the designer and moldmaker should be handled with extreme care so that the expanded effort is fully protected. Any protruding parts should be protected against damage in transfer. The mold surfaces, especially cavities and cores, should be covered with a protective coating against surface corrosion. The coating should be easily removable before the molding operation starts.

The protection of mold surface applies equally to the time after a run, when the mold is ready to be removed from the press and to be stored for the next run. In some areas

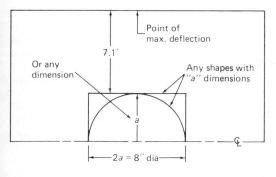

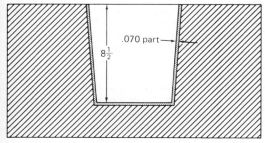

Fig. 7-39   Cavity of part where average pressure is 6000 psi.

where the atmosphere is highly corrosive, the mold must be protected while in the press for anticipated operation. This is especially important over a long holiday weekend of 72 hours or more. Commercial coatings are available for this purpose; before they are used, however, they should be carefully evaluated for their ability to protect the area involved. Also vacuum containers are used after molds are properly dried.

CONCLUSION: DESIGNING FOR EFFICIENCY

The design of molds should be given much greater attention than it receives. It is particularly important to recognize that the forces involved, resulting from high injection pressures, are very large, and their affects on cavity strength must be carefully analyzed. The temperature of molding and its influence on cavity configuration must also be taken into account.

The questions of strength are not only related to compressive, tensile, shear, and bonding strengths, but also to heat expansion–contraction, wall deflections, and similar oc-

currences. It appears logical that after the general design concept has been established, a detailed analysis of strength requirements should be undertaken. Only after this phase has been satisfactorily resolved, should one proceed with dimensions and proportions of vital components.

Mold design should not be confined to drafting the layout of the complete mold with the aim of supplying the necessary information to the shop for making the mold, but should be preceded with calculations for runner size, position of gate(s), strength that will assure safe and satisfactory performance of the very costly tool, and so on.

Quick mold changes are now made to minimize downtime. Captive and custom molders can make frequent mold changes to accommodate short runs, and thus take advantage of at least one principal benefit—lean inventories. One measure of productivity in a molding operation involves making maximum use of available molding-machine capacity, expressed as the sum of productive time and downtime. Downtime can be further viewed as that time attributable to normal maintenance and unanticipated machine failure, and to the time used to change molds. Until recently, most efforts to improve machine productivity (and profitability) have concentrated on the productive period. To improve maximum yields, the industry has sought to reduce cycle times, use multicavity tooling, and improve quality control.

Another approach to improved profitability lies in the concept of operating with reduced inventories of both work-in-process and finished goods. But the consequences of reduced inventories are shorter production runs, often creating a situation requiring more frequent mold changing. The apparent contradiction between improved machine utilization (longer runs and less-frequent mold changes) and leaner inventories (shorter runs and more-frequent mold changes) could be resolved if mold changing could be streamlined.

Close scrutiny of the mold-changing operation reveals at least seven distinct steps that must be performed:

1. Disconnecting and removing the existing mold.
2. Locating and clamping the replacement mold.
3. Determining the length and proper placement of ejector bars.
4. Connecting and providing water circuits for core and cavity.
5. Connecting ancillary hydraulic systems.
6. Connecting the electrical systems for manifold heating or mold operation interlocks.
7. Purging the machine and replacing the material with another type or color.

The time required to complete these steps can range from 4 to 16 hours with larger machines. Obviously, not all seven steps are needed for every mold change, since they are dependent upon mold complexity. Beyond efficiency, integrated mold changing means improved safety for mold setters. The use of safe procedures becomes even more critical when a decrease in changeover time is the main objective.

Systems are available, however, that speed up mold changing by integrating the steps into a single operation. This requires planning in targeting to standardize size and shapes of molds to be used. Extended plates are attached to molds for the quick mold changes. The result is mold changes made in minutes.

Computer-Aided Mold and Part Design

Mold designers and builders can benefit from the use of Computer-Aided Design (CAD), Computer-Aided Manufacturing (CAM), and Computer-Aided Engineering (CAE) techniques. Computer programs permit analyzing the flow of plastics into cavity(s), mold cooling system design, and mechanical stress in molded parts. These programs can simplify the design of molds with lower stress levels, less warpage, shorter cycle time, etc. (See Chapters 12 and 13 on mold designing with CAD/CAM/CAE.)

REFERENCES

1. Rosato, D. V., "Fundamentals and CAD/CAM in Mold & Part Design," University of Lowell Plastic Seminar Workbooks, 1984, 1985.
2. Dym, J. B., *Injection Molds and Molding*, Van Nostrand Reinhold, New York, 1979.
3. Sors, L., Bardocz, L., and Radnoti, I., *Plastic Molds and Dies*, Van Nostrand Reinhold, New York, 1981.
4. Glanvill, A. B., *The Plastics Engineer's Data Book*, Machinery Publ. Co., 1971.
5. Dym, J. B., *Product Design with Plastics*, Industrial Press Inc., 1983.
6. "Injection Molding Operations, A Manufacturing Plan," Husky Injection Molding Systems Ltd., 1980.
7. Schwartz, S. S., and Goodman, S. H., *Plastics Materials and Processes*, Van Nostrand Reinhold, New York, 1982.
8. Frados, J., *Plastics Engineering Handbook of SPI*, Van Nostrand Reinhold, New York, 1976.
9. Patton, W. J., *Plastics Technology: Theory, Design and Manufacture*, Reston Publ. Co., 1976.
10. Beck, R. D., *Plastic Product Design*, Van Nostrand Reinhold, New York, 1980.
11. Brown, J., *Injection Molding of Plastic Components*, McGraw-Hill Book Co., New York, 1979.
12. Lubin, G., *Handbook of Composites*, Van Nostrand Reinhold, New York, 1982.
13. Rosato, D. V., Fallon, W. K., and Rosato, D. V., *Markets for Plastics*, Van Nostrand Reinhold, New York, 1982.
14. Benjamin, B. S., *Structural Design with Plastics*, Van Nostrand Reinhold, New York, 1982.
15. "Injection Moulds for Thermoplastics," Akzo Plastics bv, 1980.
16. Glenn, W. B., Jr., "Calculating Mold Shrinkage for Optimum Injection Cycles," *Plastics Technology* (Sept. 1980).
17. "Second Generation Stack Mold (4-Plate) Increases Speed, Flexibility," *Plastics Technology*, p. 15 (May 1983).
18. "Injection Molding Operations—A Manufacturing Plan," Husky Injection Molding Systems Ltd., 1980.
19. "Computer Aided Mold Design," University of Lowell Seminar Workbook, 1984.
20. Hoffman, M., "What You Should Know About Mold Steels," *Plastics Technology* (Apr. 1982).
21. "Polishing," *PM&E* (1979).
22. "The Master Unit Concept," General Catalog, Master Unit Die Products, Inc., 1979.
23. DuBois, J. H., and Pribble, W. I., *Plastics Mold Engineering Handbook*, Van Nostrand Reinhold, New York, 1978.
24. "Customs and Practices of the Moldmaking Industry," Society of Plastics Industry, Inc., 1983.
25. "Classification of Injection Molds for Thermoplastic Materials," Society of Plastics Industry, Inc., 1983.

Chapter 8
Pre-engineered Mold Bases and Components

D. V. Rosato

INTRODUCTION

The mold can be considered to be the most critical component of the injection molding system. Mold design has a tremendous impact on productivity and product quality, and therefore the overall economics of the injection molding operation. Continual advances are being made designing the part, designing the mold, and producing the mold. (Details on these aspects of injection molding are given in Chapters 6, 7, 9, 10, 13, and 27.)

Within the industry, some manufacturers have developed mold standardization programs. In choosing the number of cavities per mold, consideration should be given to the standard molds available. There are benefits to mold standardization; for example: (1) high-quality manufacturing techniques result in consistent quality and reduced mold cost; (2) there is improved delivery time, with only the core and cavity having to be machined, as other components can be inventoried by the manufacturer of the pre-engineered parts; and (3) mold performance can be closely predicted, based on the past experience of the manufacturer or the molder. Thus you can obtain the required dimensional accuracy, close tolerance, high-quality steels, and interchangeability. Table 8-1 provides information on some of the manufacturers that produce pre-engineered mold components (Fig. 8-1). Some of these companies specialize in specific components: Mold-Masters Ltd., hot-runner systems; Incoe, special gate controls to automatically control resin flow into the cavity(s) (Fig. 8-2); Erico, special mold clamps and retainers; Master Unit Die Products (MUD), quick change cavity mold (see Figs. 9-2a and 9-2b); Logic Devices, mold venting devices (Fig. 8-3); 3M, custom-molded cavities (Fig. 8-4), etc.

Certain companies such as Husky have extensive pre-engineered mold capability but can also "package" the mold around the complete injection molding machine with all types of parts handling equipment, such as robots (see Chapter 14 on parts handling systems). Husky is a major world manufacturer of injection molding machines, specializing in producing the complete injection molding line (mold, machine, parts handling, etc.), with components that operate at very fast rates producing quality products. D-M-E Company provides the industry with almost all types of pre-engineered mold bases and components. Some of the D-M-E products will be reviewed later in this chapter to give examples of what is available. The manufacturers each have their own booklets or manuals describing their products and how they best can simplify and operate the mold with tight control.

STANDARDIZED MOLD BASE ASSEMBLIES

Of the thousands of standardized mold base assemblies offered by D-M-E, the most popu-

Table 8-1   Manufacturers of pre-engineered mold bases and components.

ABA Tool & Die Co., Manchester, CT
Alliance Mold Co., Inc., Rochester, NY
Bermer Tool & Die Co., Southbridge, MA
Chromalloy Div., Sintercast, West Nyack, NY
Columbia Engineering, Red Lion, PA
D-M-E, Sub. of VSI, Madison Heights, MI
Erico Products, Solon, OH
Ethyl/VCA Marland, Inc., Pittsfield, MA
Fast Heat Element Mfg., Elmhurst, IL
G-W/Carborundum, Bethel, VT
Husky Injection Molding Systems Ltd., Bolton, Ontario
IMS Co., Cleveland, OH
Incoe Corp., Troy, MI
Industrial Heater, New York, NY
ITT-Vulcan Electric, Kezar Falls, ME
Kona Corp., Gloucester, MA
Logic Devices, Bethel, CT
3M Custom Molded Products, St. Paul, MN
Master Unit Die Products, Inc., Greenville, MI
Mold-Base Industries, Inc., Harrisburg, PA
Mold Masters Ltd., Toronto, Canada
National Tool & Mfg., Kenilworth, NJ
Newark Die Co., Springfield, NJ
Parker-Hannifin, Quick-Coupling Div., Elyria, OH
Sno-Trik, Solon, OH
Stilson Div., Stocker & Yale, Roseville, MI
Value Molding Corp., Loveland, CO

lar is the "A" Series (Fig. 8-5). This most frequently used assembly is available in 42 sizes from $7\frac{7}{8}''$ x $7\frac{7}{8}''$ to $23\frac{3}{4}''$ x $35\frac{1}{2}''$ (Fig. 8-6). It has been pre-engineered in cooperation with experienced mold designers to accommodate the widest variety of injection molding applications. The "A" Series mold base suits most plastics part requirements, simplifies mold design, increases moldmaking productivity, and gives the molder the most economical, high-performance mold construction. The "A" Series mold base permits through-pocket machining for cavity and core inserts in the cavity retainer plates, reducing mold machining time and costs.

The "B" Series mold bases are a modification of the "A" Series design made with the same steels, interchangeable component parts, and precision manufacturing (Fig. 8-5). The fundamental difference is its two-plate design vs. the four-plate assembly provided in the "A" Series. The "B" Series mold base uses the cavity retainer plates for a dual purpose, eliminating the need for a separate top clamping plate and support plate. Multiple-cavity molds designed into the "B" Series require that the cavities and cores be inserts into blind pockets machined into the cavity retainer plates. The "B" Series is sometimes specified

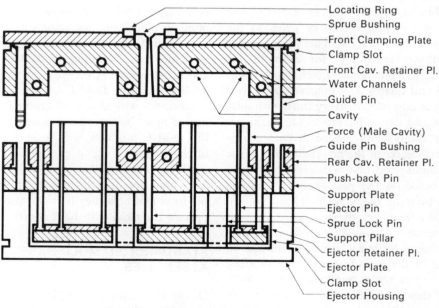

Fig. 8-1   Pre-engineered standard mold (D-M-E).

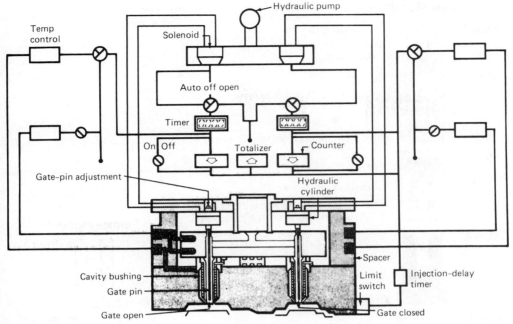

Fig. 8-2 Gate and resin flow control (Incoe).

for single-cavity plastics molds where the cavity and core are machined directly in the cavity retainer plate, or where overall mold height is critical.

For applications requiring stripper plates for part ejection, D-M-E "X" Series mold bases can be used (Fig. 8-5). Two versions of this mold base are available—the six-plate series, with a support plate, and the five-plate series, without a support plate.

Another variation of the standard "A" Series mold base is called the "AX" Series (Fig.

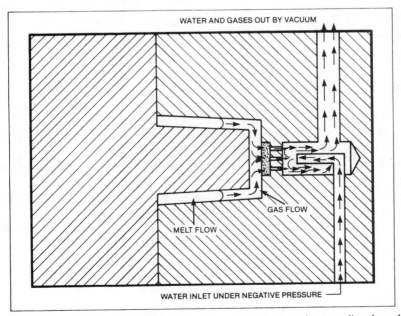

Fig. 8-3 Negative pressure can be used to pull gas from the mold cavity into the water line through a hole in the mold surface covered by a porous metal filter (Logic Devices).

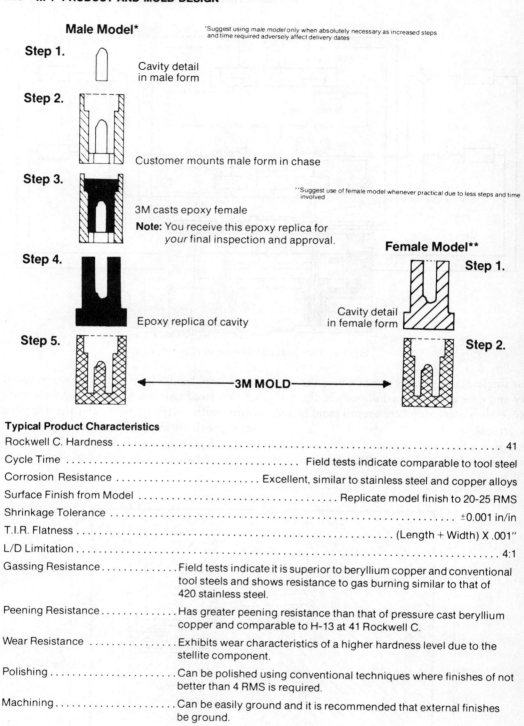

Male Model*

*Suggest using *male model* only when absolutely necessary as increased steps and time required adversely affect delivery dates

Step 1. Cavity detail in male form

Step 2. Customer mounts male form in chase

Step 3. 3M casts epoxy female

Note: You receive this epoxy replica for *your* final inspection and approval.

Female Model**

**Suggest use of female model whenever practical due to less steps and time involved

Step 1. Cavity detail in female form

Step 4. Epoxy replica of cavity

Step 5. **3M MOLD** **Step 2.**

Typical Product Characteristics

Rockwell C. Hardness 41

Cycle Time . . . . Field tests indicate comparable to tool steel

Corrosion Resistance . . . . Excellent, similar to stainless steel and copper alloys

Surface Finish from Model . . . . Replicate model finish to 20-25 RMS

Shrinkage Tolerance . . . . ±0.001 in/in

T.I.R. Flatness . . . . (Length + Width) X .001″

L/D Limitation 4:1

Gassing Resistance . . . . Field tests indicate it is superior to beryllium copper and conventional tool steels and shows resistance to gas burning similar to that of 420 stainless steel.

Peening Resistance . . . . Has greater peening resistance than that of pressure cast beryllium copper and comparable to H-13 at 41 Rockwell C.

Wear Resistance . . . . Exhibits wear characteristics of a higher hardness level due to the stellite component.

Polishing . . . . Can be polished using conventional techniques where finishes of not better than 4 RMS is required.

Machining . . . . Can be easily ground and it is recommended that external finishes be ground.

Fig. 8-4   Custom-molded cavities (3M).

"A" AND "AR" SERIES ASSEMBLIES

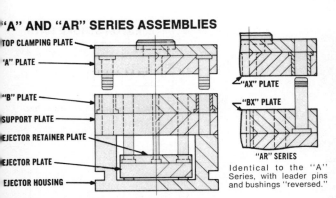

TOP CLAMPING PLATE
'A" PLATE
"B" PLATE
SUPPORT PLATE
EJECTOR RETAINER PLATE
EJECTOR PLATE
EJECTOR HOUSING

"AX" PLATE
"BX" PLATE

"AR" SERIES
Identical to the "A"
Series, with leader pins
and bushings "reversed."

The most frequently used Standard Assembly, the "A" Series Mold Base is available in 42 sizes from 7⅞ x 7⅞ to 23¾ x 35½.

"B" SERIES ASSEMBLY

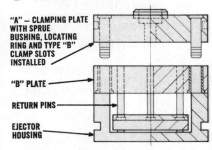

"A" — CLAMPING PLATE WITH SPRUE BUSHING, LOCATING RING AND TYPE "B" CLAMP SLOTS INSTALLED
"B" PLATE
RETURN PINS
EJECTOR HOUSING

When cavities and cores are to be inserted into blind pockets, or machined directly into the "A" and "B" plates, the "B" Series Assembly is sometimes used. The Top Clamping Plate and Support Plate are omitted from the assembly.

"X" SERIES (STRIPPER PLATE) ASSEMBLY

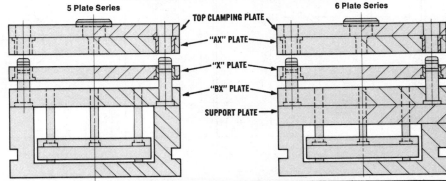

5 Plate Series   6 Plate Series

TOP CLAMPING PLATE
"AX" PLATE
"X" PLATE
"BX" PLATE
SUPPORT PLATE

Most frequently used for molds requiring stripper plate ejection, the "X" Series Assembly is available with a Support Plate (6-plate series) or without a Support Plate (5-plate series).

"AX" SERIES ASSEMBLY

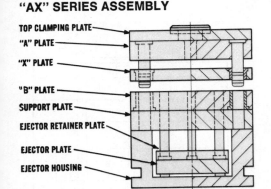

TOP CLAMPING PLATE
"A" PLATE
"X" PLATE
"B" PLATE
SUPPORT PLATE
EJECTOR RETAINER PLATE
EJECTOR PLATE
EJECTOR HOUSING

The "AX" Series Assembly is used when the mold requires a floating plate to remain with the upper or stationary half of the assembly. It is basically an "A" Series Assembly with a floating plate ("X") added.

"T" SERIES ASSEMBLY

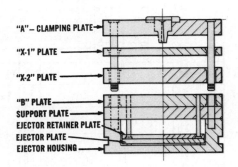

"A" — CLAMPING PLATE
"X-1" PLATE
"X-2" PLATE
"B" PLATE
SUPPORT PLATE
EJECTOR RETAINER PLATE
EJECTOR PLATE
EJECTOR HOUSING

The "T" Series Assembly is used for top runner molds that require two floating plates ("X-1" — runner stripper plate, "X-2" — cavity plate) to remain with the upper or stationary half of the assembly.

Fig. 8-5   Pre-engineered standard mold base terminology (D-M-E).

GENERAL DIMENSIONS

D = DIAMETER OF LOCATING RING
   Cat. No. 6501 (D = 3.990) Standard
   Cat. No. 6504 (D = 3.990) Clamp Type
   (For other rings, see pages K 19-21)
E = LENGTH OF EJECTOR BAR
   $7\frac{7}{8}$, $11\frac{7}{8}$, 16" or 20"
O = SMALL DIA. OF SPRUE BUSHING ORIFICE
   $\frac{5}{32}$, $\frac{7}{32}$ or $\frac{9}{32}$
R = SPHERICAL RADIUS OF SPRUE BUSHING
   $\frac{1}{2}$ or $\frac{3}{4}$

EJECTOR STROKE DATA					
C	$2\frac{1}{2}$	3"	$3\frac{1}{2}$	4"	$4\frac{1}{2}$
S	$1\frac{3}{16}$	$1\frac{11}{16}$	$2\frac{3}{16}$	$2\frac{11}{16}$	$3\frac{3}{16}$

C = Height of Riser

S = Maximum Stroke of Ejector Bar

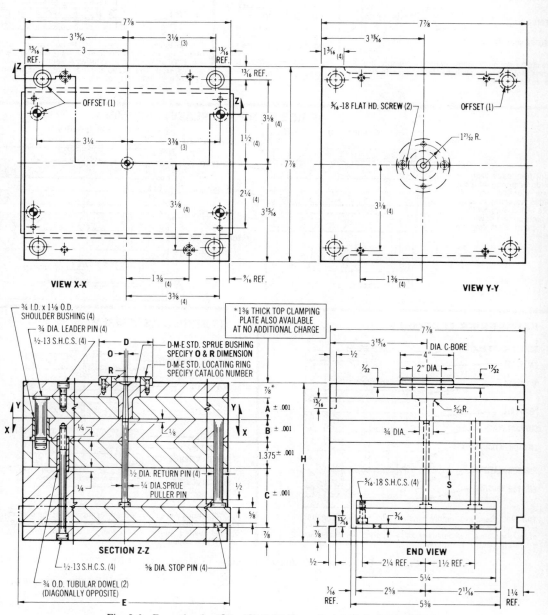

Fig. 8-6   Example of a $7\frac{7}{8}$" x $7\frac{7}{8}$" D-M-E standard "A" Series mold base.

8-5). The "AX" is basically an "A" Series type base with a floating plate ("X" plate) added between the cavity plates. This type assembly is used when it is desirable to have the floating plate remain with the upper half of the assembly, for example, when runners are top-mounted. Another group of mold bases are designated the "T" Series (Fig. 8-5). These bases are used for top runner molds that require two floating plates ("X-1"—runner stripper plate; "X-2"—cavity plate) to remain with the upper or stationary half of the assembly.

In addition to the standard mold bases described above, there are designed and engineered custom mold bases available for specific molding machines. These custom-designed bases include a variety of configurations, including the shuttle type and the universal type with adapter plates.

Mold base component parts such as cavity retainer sets, mold plates, die blocks, spacer blocks, and ejector housings are available for those cases where standard assemblies cannot be used (Fig. 8-7). The broad line of standard mold base component parts permits the design and construction of custom mold assemblies, while retaining the important advantages of interchangeability. Plates and components for large mold assemblies up to $45\frac{3}{4}''$ x $66''$ are also available, providing the benefits of standardization for large tooling applications as well.

MOLD COMPONENTS

Pre-engineered mold components provide the same important economic and technical advantages as standard mold bases—dimensional accuracy, interchangeability, availability, etc. These components can be divided into various categories, including basic mold components, alignment and registry components, heating and cooling items, and specialty components (Fig. 8-1).

Basic mold components include items such as ejector pins and sleeves, used to eject plastics parts from the mold; leader pins and bushings, used to maintain mold alignment when the mold is removed from the press; sprue bushings, installed in the mold to accept the plastic melt from the molding machine nozzle; locating rings, installed in the stationary half of the mold to "locate" the machine nozzle with the sprue bushing; and support pillars, used to increase the capacity of the mold to support the projected area of the cavities, runner, and sprue.

Alignment and registry components include tubular dowels, used in mold base assemblies to accurately align the "B" plate, support plate, and ejector housing; and round and rectangular tapered interlocks, used when very accurate registration of mold halves, mold plates, or individual cavities and cores is required.

Pre-engineered heating items include components such as helical tubular heaters, used for heating injection molding machine nozzles; band heaters, designed for heating mold plates, probes, injection cylinders, and nozzles; and thermocouples, used to monitor temperatures.

Cooling components include a variety of brass items for controlling water temperature and flow within the mold. These components take various forms such as bubbler tubes, cascade water junctions, plug baffles, pressure plugs, diverting plugs and rods, and "Jiffy-Matic" connectors—plug and socket components used to provide quick connect/disconnect of water lines.

One of the newest pre-engineered components for heating/cooling of mold cavities, cores, etc. is the thermal pin, a heat transfer device that can be used in place of bubblers, baffles, fountains, or blades. Designed specifically to accept cyclic heat loads found in molds, the thermal pin has a wick and nontoxic working fluid sealed inside a thin copper shell. These heat conductors transfer heat rapidly to the coolant, rather than requiring the coolant to flow into the heated area. They are also used to transfer heat to a cooler portion of the mold (which serves as a heat sink) or to open air, thereby permitting cooling of otherwise inaccessible areas and eliminating potential coolant leakage.

When installed and operational, the opposite ends of the thermal pin heat conductor—

Cavity Insert Blocks and Rounds

Cavity Insert Blocks are stocked in over 90 standard sizes, from 3″ × 3″ to 6″ × 8″; ⅞ to 4⅞ thick. They are available in your choice of D-M-E No. 3 (P-20 type) or No. 5 (H-13 type) steel. The more popular sizes are also available in D-M-E No. 6 (T-420 type) stainless steel.

Cavity Insert Rounds are available in 41 standard sizes, from one to four inches in diameter, ⅞ to 3⅞ long. They are stocked in both D-M-E No. 3 and No. 5 cavity steels.

Die Blocks and Plates — No. 5 Steel

Available in over 300 standard sizes from 7⅞ × 7⅞ to 23¾ × 35½; 1⅜ to 11⅞ thick (depending on length and width). They are supplied in milled condition, with approximately .060″ stock allowance.

Extra Thick Mold and Die Blocks

Available in D-M-E No. 1, No. 2 or No. 3 Steel, in over 100 standard sizes, from 14⅞ × 17⅞ to 23¾ × 35½; 6⅞ to 11⅞ thick. These blocks are supplied in milled condition with approximately .060″ stock allowance.

Mold Plates and Plate Items

Mold Plates are available in over 400 standard sizes, from 6″ × 7″ to 23¾ × 35½ in D-M-E No. 1, No. 2 or No. 3 Steel. They are finish ground top and bottom to a thickness tolerance of plus or minus .001″ with all edges finished square and parallel.

A wide range of other mold plate items available as standard include:

Spacer Blocks

Plain, Slotted and Angle Spacers are all made from D-M-E No. 1 Steel. Riser height (C dimension) is finish ground to plus or minus .001″.

Ejector Housings

Rigid one-piece construction is made from D-M-E No. 1 Steel. Available in over 150 standard sizes, corresponding to D-M-E Standard "A" Series Mold Bases. The riser height (C dimension) is finish ground to plus or minus .001″.

Fig. 8-7   Mold base component parts (D-M-E).

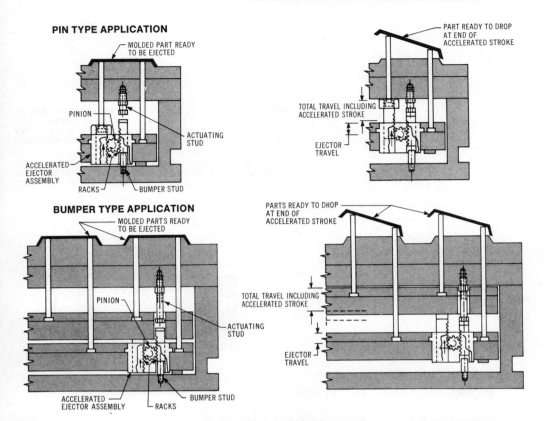

PIN TYPE APPLICATION

MOLDED PART READY
TO BE EJECTED

PINION

ACTUATING
STUD

ACCELERATED
EJECTOR
ASSEMBLY

RACKS

BUMPER STUD

PART READY TO DROP
AT END OF
ACCELERATED STROKE

TOTAL TRAVEL INCLUDING
ACCELERATED STROKE

EJECTOR
TRAVEL

BUMPER TYPE APPLICATION

MOLDED PARTS READY
TO BE EJECTED

PINION

ACTUATING
STUD

ACCELERATED
EJECTOR ASSEMBLY

RACKS

BUMPER STUD

PARTS READY TO DROP
AT END OF
ACCELERATED STROKE

TOTAL TRAVEL INCLUDING
ACCELERATED STROKE

EJECTOR
TRAVEL

D-M-E Accelerated Ejectors use a rack and pinion mechanism to provide up to ⅝'' additional ejector stroke. Their simple, linear movement can be used to increase the speed and stroke of ejector pins, ejector sleeves or entire ejector assemblies. The flanges and rounded corners on these units facilitate installation within the ejector assembly. The rectangular cross section of the racks prevents them from rotating. Included with each unit is a bumper stud which assures positive return of the racks when the ejector assembly is fully returned.

D-M-E Accelerated Ejectors are available in two sizes (small or regular) and two types (pin or bumper). The pin type units are used for individual ejector pin acceleration (one unit per pin). Bumper type units are used for accelerating the entire upper ejector assembly in a dual ejector assembly mold (a minimum of four units are normally used in this application).

Fig. 8-8   Standard accelerated ejectors (D-M-E).

evaporator and condenser sections—have separate yet dependent functions. The evaporator section captures the heat of the molded material, which in turn vaporizes the working fluid. The vapor then travels the length of the heat into the cooling source (water, air, or mold component). Finally, the condensed fluid travels by capillary action along the wick back to the evaporator section. The thermal pin heat conductors are available in a variety of standard lengths and diameters to suit many heating and cooling applications.

Specialty Components

Specialty components are those that have been pre-engineered to improve the performance of particular mold functions. These functions can be as straightforward as returning the ejector assembly early to time or as sophisticated as a runnerless molding system. The important point is that all these devices have been standardized for installation in a variety of molding applications, and as a result do not have to be designed and built from scratch by individual mold designers and moldmakers.

Accelerated ejectors use a rack and pinion mechanism to provide up to $\frac{5}{8}$ in. additional ejector stroke (Fig. 8-8). Their simple, linear movement can be used to increase the speed and stroke of ejector pins, ejector sleeves, or entire ejector assemblies. The flanges and rounded corners on these units facilitate installation within the ejector assembly. The

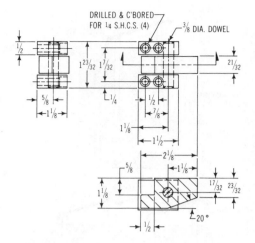

The D-M-E accelerated knock-outs are simple in design, using a pivot-type motion for accelerated ejection. Mechanical advantage is 1:1. They will accommodate ejector pins up to ⅜″ in diameter. (Pins with head diameters over ⅝″ can be ground down to fit.)

Simplicity of design permits D-M-E Accelerated Knock-outs to be either inserted into the ejector plate (as shown below) or top mounted, depending on space available for the ejection movement.

TYPICAL APPLICATIONS

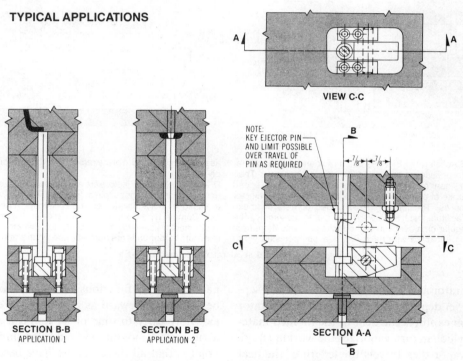

Fig. 8-9   Standard accelerated knock-outs (D-M-E).

rectangular cross section of the racks prevents them from rotating. Included with each unit is a bumper stud that assures positive return of the racks when the ejector assembly is fully returned. Accelerated ejectors are available in two sizes (small or regular) and two types (pin or bumper). The pin type units are used for individual ejector pin acceleration (one unit per pin). Bumper type units are used for

accelerating the entire upper ejector assembly in a dual ejector assembly mold (a minimum of four units are normally used in this application).

Accelerated knock-outs are simple in design, using a pivot-type motion for accelerated ejection; the mechanical advantage is 1:1. Simplicity of design permits accelerated knock-outs to be either inserted into the ejec-

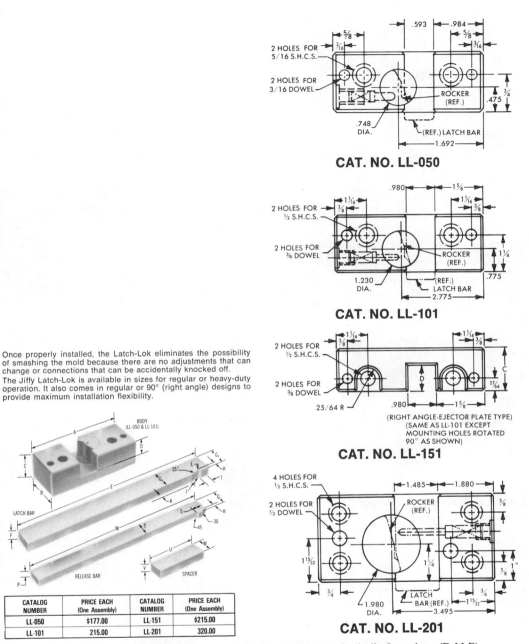

CAT. NO. LL-050

CAT. NO. LL-101

CAT. NO. LL-151

(RIGHT ANGLE-EJECTOR PLATE TYPE)
(SAME AS LL-101 EXCEPT
MOUNTING HOLES ROTATED
90° AS SHOWN)

CAT. NO. LL-201

Once properly installed, the Latch-Lok eliminates the possibility of smashing the mold because there are no adjustments that can change or connections that can be accidentally knocked off.
The Jiffy Latch-Lok is available in sizes for regular or heavy-duty operation. It also comes in regular or 90° (right angle) designs to provide maximum installation flexibility.

CATALOG NUMBER	PRICE EACH (One Assembly)	CATALOG NUMBER	PRICE EACH (One Assembly)
LL-050	$177.00	LL-151	$215.00
LL-101	215.00	LL-201	320.00

Fig. 8-10   Standard assemblies to easy-to-install a device to mechanically float plates (D-M-E).

tor plate or top-mounted, depending on space available for the ejector movement (Fig. 8-9).

The Jiffy Latch-Lok (D-M-E) provides new freedom in design to mechanically float plates (Fig. 8-10). There is not need for electric switches, pneumatic controls, or timing devices with delicate adjustments. The action of the Latch-Lok is positive. Once properly installed, it eliminates the possibility of smashing the mold because there are no adjustments that can change or connections that can be accidentally knocked off. The Jiffy Latch-Lok is available in sizes for regular or heavy duty

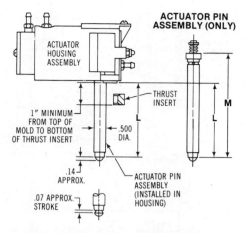

The pneumatically controlled and operated D-M-E Jiffy-Jector is a compact and powerful device for positively ejecting runners from three-plate molds. It moves the runner system away from the X-1 Plate and then by means of a short positive stroke and a blast of compressed air, ejects the runner system down, and out of the mold—putting an end to costly hang-up problems.

The basic requirement for proper operation of the Jiffy-Jector is a clear, unobstructed path out of the mold for the runner system. It is adaptable to most three-plate molds and can be designed into new molds or retrofitted to existing molds.

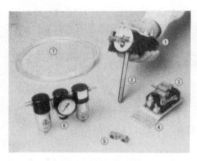

Each Jiffy-Jector package includes:

① Actuator Housing Assembly
② Actuator Pin Assembly
③ Air Logic Manifold Assembly
④ Manifold Bracket
⑤ Thrust Insert
⑥ Dual Air Filtration System
⑦ Polyurethane Tubing (⅛ I.D. x 9 feet long)
⑧ Installation Instructions

D-M-E JIFFY-JECTOR OPERATING SEQUENCE

STEP 1
☐ Parting lines ① and ② open.

STEP 2
☐ Parting line ③ opens and actuates the pneumatic limit switch.
☐ Air cylinder moves actuator pin assembly and runner system away from the X-1 Plate.

STEP 3
☐ Near the end of the 1″ air cylinder stroke, the depressed ball gives the valve a short positive stroke, loosening the runner system from the actuating pin. Simultaneously, a blast of compressed air is released, ejecting the runner system down and out of the mold.
☐ After the adjustable time delay valve times out, the air cylinder returns the actuator pin assembly to the X-1 Plate.
☐ The mold begins the normal mold close sequence.

NOTE:
Actuator Pin Assemblies are interchangeable for replacement purposes or for subsequent applications where a longer or shorter pin is required.

For best results, it is recommended that the end of the actuator pin be located as close to the sprue as possible, especially when molding flexible resins. (Select actuator pin length accordingly.)

Fig. 8-11 Positive runner ejection from three-plate mold (D-M-E).

operation. It also comes in regular or 90-degree (right angle) designs to provide maximum installation flexibility.

The slide retainer provides a compact and economical means of slide retention that makes obsolete the cumbersome external spring or hydraulic methods. Its simple and positive operation makes it equally suitable for new tooling design or retrofitting existing molds. Available in three sizes with increasing

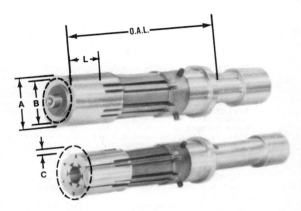

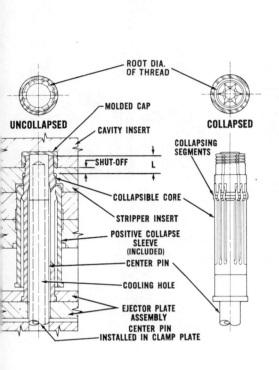

The Collapsible Core is a major breakthrough for molding plastics parts requiring internal threads, undercuts, protrusions or cut-outs. The patented design incorporates only three moving parts . . . which utilize conventional mold movements.

In addition to being automatic in operation . . . the Collapsible Core makes it possible for you to produce parts that, until now, have been considered impossible to mold. Parts with internal protrusions, dimples, interrupted threads and cutouts can now be economically produced on a high or low volume basis.

For conventional threaded parts, The Collapsible Core could cut your cycle time up to 30% when compared with unscrewing or other complex actuating mechanisms.

CATALOG NUMBER	A Max. O.D. of Thread or Configuration		B Min. I.D. of Thread or Configuration		Center Pin Dia. (At Top of Collapsible Core)		L Max Molded Length (Incl. Mold Shut-Off				C Collapse per Side at Tip of Core††				O.A.L. Overall Length of Collapsible Core (Only)	
	inch	mm	inch	mm	inch	mm	inch	mm	*inch	*mm	inch	mm	*inch	*mm	inch	mm
CC-200-PC	1.270	32.25	.910	23.11	.785	19.93	.975	24.76	1.150	29.21	.043	1.09	.048	1.21	7.315	185.80
†CC-250-PC	1.270	32.25	.910	23.11	.785	19.93	.975	24.76	1.150	29.21	.043	1.09	.048	1.21	5.440	138.17
CC-202-PC	1.390	35.30	1.010	25.65	.885	22.47	.975	24.76	1.150	29.21	.055	1.39	.064	1.62	7.315	185.80
†CC-252-PC	1.390	35.30	1.010	25.65	.885	22.47	.975	24.76	1.150	29.21	.055	1.39	.064	1.62	5.440	138.17
CC-302-PC	1.740	44.19	1.270	32.25	1.105	28.06	1.225	31.11	1.400	35.56	.068	1.72	.083	2.10	7.315	185.80
†CC-352-PC	1.740	44.19	1.270	32.25	1.105	28.06	1.225	31.11	1.400	35.56	.068	1.72	.083	2.10	6.065	154.05
CC-402-PC	2.182	55.42	1.593	40.46	1.388	35.25	1.535	38.98	1.700	43.18	.090	2.28	.103	2.61	7.815	198.50
CC-502-PC	2.800	71.12	2.060	52.32	1.750	44.45	1.750	44.45	1.900	48.26	.115	2.92	.125	3.17	9.625	244.47
CC-602-PC	3.535	89.78	2.610	66.29	2.175	55.24	2.125	53.97	2.400	60.96	.140	3.55	.148	3.75	11.250	285.75

Fig. 8-12 Standard collapsible core (D-M-E).

weight-holding capacities, the slide retainers can be used individually or in multiples for larger or heavier slides. Generally mounted behind and below the slide, the slide retainer is a compact unit that is entirely contained within the mold. Interference with machine tie bars or safety gates is not a problem. It can even be installed completely underneath the slide if space is limited.

As the mold opens, the dowel pin installed in the slide positively locks into the retainer until disengaged by the mold's closing action. The small spring placed crosswise in the retainer maintains the gripping force required to keep the dowel pin in the socket when the mold is open. The slide retainer is designed with a generous lead-in at the socket opening so the dowel pin will enter the socket even if there is a slight misalignment between the retainer and the pin. The investment cast unit

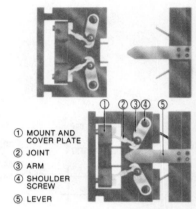

The D-M-E Toggle-Lok is an externally mounted device used to positively return the ejector assembly before the mold completely closes. Employed when ejector pins or other moving components interfere with normal mold closure, the Toggle-Lok returns and locks the ejector assembly firmly in place—preventing expensive mold damage. It uses a positive cam linkage and activator lever rather than springs, pneumatics, hydraulics or other more cumbersome methods. Since it is mounted externally, the Toggle-Lok allows more space for cavities, slides or other required mold components. It is available in three styles to suit a broad variety of applications. Each style is available in a Toggle-Lok Package which corresponds to a specific mold base requirement.

OPERATING SEQUENCE
Upper photo shows mold in open position, with ejector assembly fully forward.
Before the mold has fully closed, the Toggle-Lok levers engage and move the arms to return the ejector assembly (as shown in lower photo).

① MOUNT AND COVER PLATE
② JOINT
③ ARM
④ SHOULDER SCREW
⑤ LEVER

U.S. Patent No. 3,226,771

STYLE X

STYLE Y

Each Standard Toggle-Lok Package includes components and hardware necessary to equip both ends of the mold.

STYLE X PACKAGE INCLUDES:
1. (4) Arms with:
   (4) Shoulder Screws (TLSS-12)
   (4) Retainers (TLR-10)
   (4) Button-Head Screws
2. (4) Joints
3. (4) Side Mounts (standard) or Top Mounts (if specified) with:
   (4) Cover Plates
   (4) Socket Head Cap Screws
   (8) Dowels
4. (2) Levers with:
   (2) Lever Spacers (TLS-20)
   (4) Socket Head Cap Screws
   (4) Dowels
5. (1) Hole Location Template
STYLE Y PACKAGE:
Same as Style X, with half the quantities listed in items 1, 2 and 3.
STYLE Z PACKAGE:
Same as Style X, with twice the quantities listed in item 4.

STYLE Z

Fig. 8-13 Standard positive early ejector return (D-M-E).

includes an integral protective cover over the spring, preventing foreign objects from interfering with the spring's action.

The pneumatically controlled and operated Jiffy-Jector (D-M-E) is a compact, powerful device for positively ejecting runners from three-plate molds (Fig. 8-11). It moves the runner system away from the X-1 plate and then, with a short positive stroke and a blast of compressed air, ejects the runner system down and out of the mold—ending hang-up problems. The basic requirement for proper operation of the Jiffy-Jector is that the runner system have a clear, unobstructed path out of the mold. It is adaptable to most three-plate molds, and can be designed into new molds or retrofitted to existing molds.

The collapsible core is a major improvement for molding plastic parts requiring internal threads, undercuts, protrusions, or cutouts (Fig. 8-12). The D-M-E patented design incorporates only three moving parts, which utilize conventional mold movements. Besides operating automatically, the collapsible core makes it possible to produce parts that, until now, were considered impossible to mold. Parts with internal protrusions, dimples, interrupted threads, and cutouts can be economically produced on a high- or low-volume basis. For conventional threaded parts, the collapsible core can cut cycle time up to 30 percent when compared with unscrewing or other complex actuating mechanisms.

The early ejector return Toggle-Lok is an externally mounted device used to positively return the ejector assembly before the mold completely closes (Fig. 8-13). Employed when ejector pins or other moving components interfere with normal mold closure, the Toggle-Lok returns and locks the ejector assembly firmly in place, preventing mold damage.

Chapter 9
Prototyping

Steve Galayda

Consultant
Auburn, Washington

WHY PROTOTYPE?

The reasons for prototyping are as numerous as the companies that use prototyping as an integral step in their product development programs. The five most common reasons for prototyping are: to save money (the primary reason), to improve product quality, to enhance product reliability, to make production tools more reliable, and to shorten the response time in the market place. These five major reasons encompass most of the miscellaneous reasons for prototyping, several of which we will look at more closely later.

Prototyping is one of the steps taken by a company to convert engineering and feasibility studies into a product that the sales department can sell. The traditional prototype has been one or two engineering models of a product fabricated, usually at high cost, using standard machining techniques and equipment. In order to machine an engineering prototype, it has often been necessary to strike a compromise between the desired production material and that material's machining characteristics. The result is an exact model of the part which can be used to evaluate cosmetic appeal and potential fit problems. The result is also a part that, owing to its high cost and possible material compromise, is not usually suited to any type of destructive property testing. Although this last statement is not always true, it is invariably true that, except in the simplest

cases, parts machined from bar or block stock do not give the same test results as parts produced in a prototype or production mold. Also parts produced in prototype tooling can reveal many potential molding problems before the fabrication of a production mold—problems that would never show up in a machined model.

Prototype molding provides a powerful and cost-effective tool for a designer to use when questions about a new product or potential new material arise. Questions about a part or material early in the design process can be answered most definitively and cost-effectively in prototype mold. Although the cost of a prototype mold insert set is a function of each part's individual design and requirements, the cost usually will run between 10 and 30 percent of the cost of the production mold. Another way to look at the cost of prototype tooling is to compare the information and data that it will provide with the information and data that a top-notch design staff can provide for the same cost.

Questions that require good hard answers do not always lend themselves to traditional analytical engineering solutions: questions concerning cosmetic qualities, such as finish, sink marks, witness lines from parting planes or slides, ejector pin marks, knit or weld lines, and different styles of texturing; questions concerning the moldability of a part such as flow-through thin sections, the location of gates and vents, flow into bosses or around pins, the location of the parting plane, and potential ejection problems; questions concerning product quality and reliability such as shrinkage, mechanical strength of bosses and knit or weld lines, pullout resistance of molded-in inserts, electrical properties, and component fit or mating subassemblies.

The majority of the data provided from parts molded in a prototype mold can be obtained in no other more reliable or cost-effective manner. Although it is possible by using specialized computer programs to predict a materials flow path, or the location of gates, vents, knit or weld lines, or the effects of parting plane location, with a good degree of accuracy, these programs have limitations. The highly complex parts that would benefit most by the use of these programs now overwhelm most of them; and on simpler parts the programs are not cost-effective for the data provided vs. a prototype mold. Also, any part data provided from prototype tooling can be obtained from parts produced in a production tool; but, again, this is not a cost-effective route to follow.

Prototyping is used because it is both cost-effective and time-effective. The majority of Fortune 100 companies that have divisions engaged in injection molding are now using prototyping programs to reduce design and development costs. Nor is prototyping cost-effective only for megacompanies. The small custom molder or captive shop can achieve the same cost-effectiveness ratio as any large company. Also, as will be seen, a program of prototyping can enhance any custom molder's competitive position.

ADVANTAGES OF PROTOTYPE TOOLS VS. PRODUCTION TOOLS

The prototype tool has several distinct advantages over a production tool for the purpose of producing parts for engineering and market evaluation. Prototype tooling will always cost less and take less time to build than a production tool. In the hands of a good-quality prototype mold shop, the time required to build, run, inspect, and deliver to print parts will generally be between 4 and 8 weeks, after receipt of a purchase order and approved part prints. On the other hand, a good-quality tool shop building a multiple-cavity production tool will require from 10 to 20 or more weeks to design and build the tool alone, much less run parts. Generally the time required just to design and get design approval on a production mold will run from 3 to 6 weeks. As can be seen, a prototype mold insert will be well on its way to producing parts while the production tool is still in the design phase. Remember that a prototype tool saves money by keeping unforeseen problems out of costly production tools.

A simple example of this is a tool built recently. An order was placed with a prototype shop to build a prototype mold insert set for a flanged wave guide. The tool was kept simple by using hand-pulled cores in place of mechanical core pulls (see Fig. 9-1). The mold insert set and its cores were run in a standardized insert base. The insert set cost $3,845 and was producing parts to print six weeks after the order was placed. A 100 percent inspection was run on a sample of 20 percent of the order. All of the inspected parts met the design tolerances.

While this tool was being fabricated, a two-cavity production tool was designed. Engineering took this step to ensure that each shop quoted the same thing. The quotes estimated an average delivery time of 10 weeks and an average cost of $18,000.

The first batch of parts produced by this tool did not meet expectations. Several other new materials were then tested to find the ideal choice. Although the insert set had been fabricated to run one specified material, the cost of modifying it to run one of the other tested materials would have been about $1,000. No prices were requested for a potential tool rework on the quoted production tool, but considering that the cores would have to be replaced and the hardened cavities enlarged by grinding, it would not have been cheap. A point of interest is that even though the mold insert set is brass, several hundred parts have been run with the tool showing very little wear.

Fig. 9-1.   Electronic wave guide mold insert set fabricated by A1-Ko Industries, Dallas, TX for Texas Instruments.

Prototype Tools for Short Runs

Another good cost-effective use of the prototype tool is in test market evaluation. Although engineering evaluation and testing require a relatively small number of sample parts, a good market study may require several thousand parts. Product-intensive markets, such as consumer electronics, small household tools or appliances, general home-use items, medical products, and automotive components, require thousands and often tens of thousands of parts for successful market research and evaluation.

Once the final part modifications have been incorporated into the prototype mold insert set, the final batch of parts for engineering evaluation and testing is produced. This final set for changes generally is a result of market and engineering studies that have revealed some unforeseen flaw or problem with the part. At this point the mold insert set is taken apart and used to generate a detailed set of core and cavity designs. These will then become the basis for a multiple-cavity production tool design. There is still one last use for the prototype mold insert set before its final relegation to a storage shelf.

As pointed out earlier, a multiple-cavity production mold can require upwards of 20 or more weeks to build. It can take another 2 to 4 weeks to get that tool into a production press to start producing parts. In the ultra-competitive consumer electronics and medical markets, for example, a three- to five-month lag between good prototype parts and market entry can be critical in terms of market share and profit. Because of the effort to eliminate this lag, many companies build and modify costly production tools early in the design process. Although this procedure gives a production tool that is ready to go when good parts are finally made, it also gives a mold that, owing to multiple reworks, is not of the same quality as it would have been if it had not been modified after fabrication. A prototype tool, on the other hand, is ideally suited to both minor and major modifications, with respect to both cost and effects on the quality of the mold insert set.

Therefore the last service a prototype mold insert set will provide is to fill the lag-time gap while a production mold is built. Obviously this delay will limit the initial market entry effort, but the improved product and production tool will help to offset this disadvantage. Also, a second insert set could be fabricated if needed to increase output. Enhanced products result from using prototype tools compared to modified production tools because part modifications in a production mold with hardened cores and cavities require compromises between the cost and ability to change the mold and the minimum changes that will give a functional part. Although the part is functional, it may not be the best it could be; and too many less-than-optimum parts can result in a finished product that is not all that it could be—a very costly outcome in terms of decreased market share and profits.

THE PROTOTYPE TOOL

The Standardized Insert Base

Most suppliers of standard mold bases and components sell some variation of a standard insert base. Also most prototype mold shops have developed insert bases that are suited to the size and type of part they tend to specialize in. The larger prototype shops generally do not limit the type of work they do and therefore have a good variety of bases to cover both general and specialized work or anything in between (See Fig. 9-2, a–d.) No single insert base will be able to handle every job.

Once a company has made the decision to start a prototype program, the next step is crucial. A high-quality set of mold insert bases will not be cheap. Ample time should be spent indentifing part size ranges and materials. These data, when given to the mold designers, will enable them to either design a set of custom bases or recommend an existing set of base designs. If you are planning to work closely with one or two prototype shops, then they should be involved with your tooling decision. Compatibility between their tools and yours will be crucial when the tools are used to fill the gap between the final prototype modifications and the production mold.

The prototype mold insert base must be considered to be a long-run, long-life, precision tool. The temptation to save money on the base must be resisted. Unlike production tools, which are custom-built with the cores

and cavities finely tuned to ensure a perfect fit, sometimes at the cost of absolute interchangeability, a prototype insert base must have absolute interchangeability. Also a prototype mold insert base must be able to stand up to the rigors of countless assemblies and disassemblies without becoming sloppy in its function.

Design of the Prototype Insert Base

To begin with, a pair of round tapered interlocks must be installed at the parting plane. Grooved bushings with grease fittings and extra-long leader pins are a minimum requirement; ball bearing bushing would be an even better investment. The sprue bushing must be pinned in place to prevent rotation, as constant assembly and disassembly require a few tenths of an inch additional clearance. Hollow dowel pins must be used between the top clamp plate and the top cavity plate (A plate). The thickness of the top cavity plate should never be less than 1.875 in. The bottom cavity plate (B plate) should never be less than 1.875 in. thick. The support plate should be a minimum of 1.875 in. thick. The ejector housing should have a least 3.500 in. of height. Although the bottom cavity plate, support plate, and ejector housing are aligned with hollow dowel pins, these should be reworked by the addition of hardened bushings to provide a good fit and wear-resistant surface. The ejector plate should be .875 in. thick and the ejector pin retainer plate should be .500 in. thick. The entire ejection system must be guided using guide pins and bushings, and must have at the very minimum a spring-loaded return system. (See Fig. 9-3.)

Extreme care must be taken to ensure adequate heat removal from the insert base. In larger inserts and insert bases it is possible to provide water in the inserts themselves. On the other hand, the small inserts and insert bases depend on heat transfer for temperature control. The biggest problem in any water system is that of corrosion in the channels; this can be taken care of by plating them with nickel. Also coolant should be manifolded so that, whenever possible, coolant moves from

Fig. 9-2a   Standardized insert mold base by Master Unit Die Product Co. with single base plate.

Fig. 9-2b   Similar to Fig. 9-2a except it has a double base with runner located between the sets of plates.

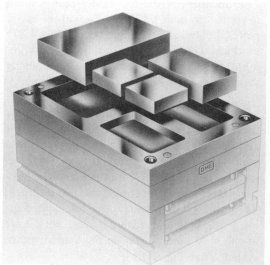

Fig. 9-2c   Standardized inserts with mold base by D-M-E using rectangular inserts.

Fig. 9-2d—Similar to Fig. 9-2c except circular inserts are used.

the center of the mold to the edges to prevent temperature fluctuations in the tool. All direct connections to the mold should be by brass or stainless steel fittings. A nice extra is to have a pair of thermocouples on the mold with digital readouts. If used, they are best placed on the mold centerline between the water lines and 1.0 in. from the edge. These devices can prove very useful in setting up new tools and repeating previous setups.

The information from an in-depth mold cooling analysis can be used to adjust coolant temperature and give some compensation for the lack of cooling channels in the smaller inserts. In the larger insert sizes, the cooling information can be used to determine whether

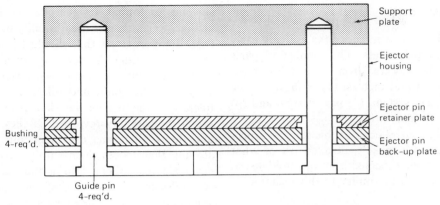

Fig. 9-3 Guide ejector system using standard components.

an insert needs internal cooling channels or not. As will be seen later, this will have an effect on the cost of both basic and advanced insert sets. Just as mold flow analysis is used for balancing run systems and gates in a production, mold cooling analysis should be used in design of the production mold.

Cavity insert pockets, support plate pockets, and ejector plate pockets must be fabricated to an LC1 fit or better. This will facilitate and ensure the interchangability and ease of assembly between the inserts and insert base. Another area that needs attention is pocket finish and pocket squareness. The pockets should be finished to RMS 16 or less, or an SPE/SPI finish between number 2 and 3. It is also extremely important that the pockets be finished in the direction of assembly and disassembly. Another area that is critical to the insert base is the relationship of the pocket sidewalls to the parting plane. The pocket sidewalls should be perpendicular to the parting plane within 15 seconds of a degree. Finally, each cavity insert pocket must also have two set screws for locking the inserts in place.

Insert Base Abilities

If the base is constructed as described, the only limiting factors to achieving the maximum in mold performance are the insert design and the ability to remove heat and control the insert temperature. As will be shown, there are three primary types of prototype insert design. Each of the three types has its own

effect on mold performance. Since the inserts can be designed to perform as desired, the ability to remove heat becomes the controlling factor in an insert base's performance.

When used with the basic insert style, the insert base will rarely approach its design limitations. In over 80 percent of these applications the mold insert by design prohibits automatic operations. Also in over 70 percent of basic mold insert design, no consideration is given to the installation of independent cooling channels in the insert itself. Thus the insert base is used for its interchangeability and structural rigidity.

On the other hand, when the insert base is used with an advanced mold insert, the only tool limiting factor to achieving maximum base and insert performance is money. As has already been shown, these inserts are generally used for test marketing or to fill the gap between a prototype tool and the production tool. The advanced insert is also an inexpensive test of the production tool. By building an advanced insert that duplicates a cavity of the production mold, all functions of the production tool can be examined except for runner reaction. It is at this stage that computer programs for predicting optimum mold cooling and material flow will be most cost-effective, as they are being used at a stage when the part probably will no longer be changed. This is not to say that computer modeling is a must for good prototyping; it is just another new tool of the trade and should be used to the best possible advantage.

The Basic Insert

The basic insert is intended for use very early in the design process. The number of parts that can be expected from a basic insert depends on several factors: (1) money, (2) the desired resin system, (3) part design, and (4) cores or molded-in inserts.

To help control costs at this stage of a program, you can make a detailed list of what is required and expected from the mold insert. The list should be similar to this example:

ITEM	VALUE TO PROGRAM	COST TO PROGRAM
1) Quantity (100)	High	$xxxx
2) Finish (SPE #3)	Medium	xxx
3) Part tolerance	High	xxx
4) Witness lines		
a) In sight	Medium	xxx
b) Out of sight	Low	xxx
5) Total cost all options above:		$xxxx

When given to a prototype shop along with a part drawing and the type of material to be molded, this list will enable that shop to supply a quote with few surprises. The prototype shop fills in the "cost to program" column. Whenever necessary, you should sit down with your prototyper and go into detail about the tool he is to build for you. Remember, a good working relationship with your prototyper is as important as a low price, and a detailed list is useful in achieving and maintaining both.

The basic mold insert is usually a nondesign tool. Several items must be supplied to the prototype shop before work can start on the insert. These include a good part print or detailed sketch, the desired part material, the material shrinkage, the number of parts to be produced, the finish that is required, the location of gates, the location of ejector pins or parting planes (if important cosmetically), and any other details you or your prototyper think are necessary.

Depending on the stage of a program when a basic insert is ordered and built, the quality of part drawings will range from hand sketches to formal designs at some level of change. Regardless of the formality of the design, there are areas that should be attended to before turning the drawings over to the prototyper. One area that is overlooked at this stage is corner radii and fillets. Another is tolerance—the use of restrictive tolerances should be carefully reviewed. If they are required, by all means use them; but if they are not, then loosen them within reason, but never eliminate them. Still another area that is overlooked is that of draft angles. Draft angles are critical to the easy removal of parts from the mold and therefore should be incorporated into the design instead of being squeezed into the straight side tolerance. If an area of a part has cosmetic requirements, that area should be pointed out on the drawing.

Once a purchase order and part drawing are received by the mold shop that will build the tool, the drawing is given to a moldmaker, who will sketch in the details of the core, cavity, and runner system, on his copy of the part drawing. Having done this, he will then select an appropriate size standard insert into which he will cut the core and cavity. The moldmaker then calculates the dimensions of the part to include the material shrinkage. The choice of which moldmaker will build a specific tool is based upon an evaluation of the difficulty of the mold and the skill and knowledge of the moldmaker. It should be noted that the moldmaker is not alone when questions about the building of a mold insert set arise. He can call upon the expertise of the shop's mold designer and, of course, the other moldmakers in the shop.

The basic insert style of tooling is employed either to save money or to produce a very limited number of parts. As a result, these inserts are fabricated with a minimum of attention to incidental details. If no finish is specified, then the tool will be built to the shop standard, which, depending on the shop, can vary greatly. Also, if the part requires removable cores, these will, in most instances, be hand-loaded and hand-pulled. Parts will also be edge-gated unless otherwise specified. Also, only the very largest of basic mold inserts will have self-contained water channels. Another source of surprise is the location of ejector pins and parting planes. As mentioned earlier, if these areas are not spelled out on the print or purchase order, the moldmaker will place them where he believes they should go. Because of this small oversight alone, countless molds have been built that molded parts with ejector pin marks or witness lines from parting planes on the wrong side or in the wrong place. Remember your prototype tool builder will give you only what is on the print or in the purchase order. It is easy to remove too much information from a print, but it can be a real burden to remove it from a tool, even a prototype tool.

A basic insert, depending on the part design and the material, will produce as few as 50 parts to as many as 4000 parts or more. A part that has one, two, or more hand-loaded and -pulled cores can at best produce only 200 to 300 parts before wear on mating surfaces takes its toll. On the other hand, a moderately complex part with no cores may produce 2000 or 3000 parts without showing signs of excessive wear, depending on the resin system used. The longest runs in basic tools therefore come from simple parts made from unfilled resin systems. Unfortunately, few parts are truly simple. They are usually moderately to extremely complex and almost always made from an engineering grade resin system. The variety of fillers that modern resins contain also affect the life of a mold insert set. Glass fillers, for example, are more abrasive than mineral fillers but less abrasive than metal fillers. Thus for the basic insert no exact quantity of parts can be predicted. An experienced prototype tool builder can predict the life of a tool fairly accurately based on previous experience.

The material of construction of the prototype mold insert also has an effect on the number of parts that will be produced. Prototype mold inserts are usually built from brass or aluminum, or, on rare occasions, steel. Brass and aluminum are the materials of general choice for several reasons, especially the ease with which both of these materials can be machined and their ability to be modified. Both brass and aluminum inserts lend themselves to having entire sections cut away and replaced by welded, brazed or pressed-in inserts without adverse effects. Also, because they are not hardened, it is relatively easy to straighten any warpage that occurs during welding or brazing of new sections. Another reason for preferring these materials is their ability to transfer heat. As noted earlier, over 70 percent of the basic inserts do not have water channels in them; therefore, the better the rate of heat transfer, the more effectively the mold will operate. Also, in very large insert sets aluminum is the only real material of choice, because of its lighter weight. It is obvious that due to the softness of both brass and aluminum, these inserts will show more wear than even one produced from cold roll steel.

Since the purpose of prototyping is both to solve potential tool problems and to find part flaws or deficiencies early in a program, tool life is not a prime concern. The prime concern is whether the tool will produce enough parts for a proper evaluation. The data and information that a prototype mold can provide depend on the part to be molded. In one example, parts were made with very thin walls: a cone was molded from 40 percent mineral-filled nylon, less than .016 in. thick, and a dome was molded from 20 percent glass-filled nylon, less than .020 in. thick. The material information available offered little encouragement that these parts could be molded satisfactorily. Basic mold inserts were quoted for each part; the cone cost less than $900, and the dome less than $1,300. In addition to proving that the parts could be molded, we now had parts that could be destructively tested to confirm our calculations dealing with the performance of these parts. As stated earlier in this chapter, many questions are better answered in a mold than on paper. The moldability of both parts was not the only question of importance, nor were the mechanical properties; we were also trying to improve the appearance of the part by molding it. As can be seen, only one of our three questions could be answered by traditional analytical engineering solutions, and even that was in doubt because of the thinness of the part.

The Advanced Insert

Whereas the basic insert is meant to provide parts primarily for engineering evaluation and testing, the advanced mold insert is meant to provide enough parts for test market research and production tool cavity evaluation, and to fill the gap between a basic mold insert and a production tool. The advanced insert style is often used early in a program because of special part requirements, such as restrictive tolerances, special finishes, multiple core pulls of an exacting nature, or an extremely complex part configuration.

The advanced insert, depending on the part, may vary very little or greatly from the basic insert. For example, the small dome in Fig. 9-4 was molded in a basic nondesign insert that proved we could make the part. An advanced insert would be built only to prove the production tool design. The changes would be minor, dealing mainly with gate style and quantity to facilitate ultrasonic degating methods. Unfortunately, a good example of ex-

Fig. 9-4   Thin-walled cone and dome insert sets fabricated by Al-Ko Industries, Dallas, for Texas Instruments.

treme variation between a basic insert and an advanced insert was not available, partly because many of the components used in the basic insert can be salvaged and used in the advanced insert. Such would be the case with the dome in Fig. 9-4; only moderate rework would be needed to change the tool as described.

Fabrication of the Advanced Insert

The advanced insert should, whenever possible, be fabricated from a minimal design. This will ensure that the advanced insert set does what is expected of it. The design should, if possible, be done from a fully detailed part print. Remember that a good clear part print can serve as the background for a minimal design, not only saving time but also lessening the chance of error.

As stated earlier, any print should include but not be limited to corner radii, fillets, and draft angles. Also the habit of restrictive tolerancing must be watched carefully. Moreover, the first material of choice should be noted on the drawing, along with the anticipated material shrinkage value. The desired location of the parting plane should be shown, as well as the desired location of the ejector pins. If slides are required, the location of their parting of movement plane should be shown if possible. Also the amount of mark-off from ejector pins or mismatch at parting plane locations should be noted on the drawing whenever possible. All finishes should be noted and clearly marked on the drawing. Also if texturing is required other than that which the prototype shop can provide, then the drawing should name the vendor who can supply the finish and his code for identifying the finish. Also, if a particular style of runner system or gate is desired, this should be noted. There are extra details on the drawing; in reality a plastic part drawing should cover details in much the same way that a casting drawing does.

The insert design as far as possible should be done using section views alone. These views should show whatever detail the moldmaker will require to fabricate the mold insert. The only dimensions needed on these views are those of the part, and those dimensions should include the material shrinkage. The fit of cores and slides and heels should be shown but not detailed. (See Fig. 9-5.) The moldmaker will set the fits required to ensure ease of assembly while maintaining tolerance and preventing flash between components. The LC class of fits in the machinery handbook is ideal for this purpose.

As with the basic insert, the number of parts that can be produced varies. The same factors that affect the number of parts produced by a basic insert will also affect an advanced insert. But, owing to better finishes and the judicious use of steel for slides and wear plates, the number of parts that an advanced mold insert can produce will run as high as 8000 to 10,000. Also as with the basic insert, the less complex the part, the more pieces the insert will produce. But even complex parts with multiple core pulls have run over 5000 parts.

The advanced inserts, like the basic inserts, are generally fabricated from brass or aluminum and on occasion from a prehardened tool steel such as P-20. It should be noted that if P-20 is used as the material of construction of an advanced mold insert, there will be two effects. First, because of the difficulty of machining P-20 vs. brass or aluminum, the insert will cost 25 to 35 percent more, depending on the shop. For example, a $4000 brass insert will cost $5200 fabricated of P-20. The second effect is one that justifies and offsets the first. By fabricating the insert from P-20, it will

be possible to produce 25,000 to 30,000 parts, even if the part is very complex or the material is moderately filled (up to 25 percent).

As with the basic insert, the prime use of the advanced insert is to obtain data. But whereas the basic insert was used primarily for engineering, sales, and corporate evaluation, the advanced insert is used primarily for tool and market evaluation, or to produce parts for a short-run program or to fill the gap while a production tool is being fabricated. These four needs are best filled by the advanced insert because it can be built to do exactly what is required. For example, in evaluating a production tool it is better to err in a single cavity then in 4, 8, 12, etc. The advanced insert can be used to test computer cooling calculations or predictions of resin flow in a mold cavity; but for production of parts for market evaluation, for short-run programs, or to serve until a production tool can be built, the advanced insert is ideal. It can be tailored to do the job at hand (e.g., market evaluation), and in a short-run program may have to offer only volume, not speed; but filling the gap before production may require both volume and the best rate an advanced insert can be designed to deliver.

CHOOSING A PROTOTYPE TOOL SHOP

As stated earlier in the chapter, the main reasons for prototyping are to save money and answer questions that cannot be answered by standard analytical engineering methods. This requires that a cost-effective, quality tool shop be found to build the prototype tooling. But remember when looking for the prototype shop that cost-effective and cheap do not mean the same thing. A cost-effective tool is one that has been quoted in a manner that ensures no monetary shocks in midprogram. The quote is a price that includes but is not limited to: design if needed, materials (both tool and part), inspection (of both tool and part), and the cost of setting-up and running parts plus a part piece price. This is how a cost-effective tool is quoted; it is not cheap, but it is not necessarily expensive.

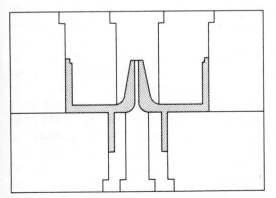

Fig. 9-5 Dimensions required for shaded part area only. Example of minimal insert design.

Reaction Time

In prototyping the single most important item that both vendor and customer deal in is time. In my experience, shops that are able to react quickly to a customer's wishes are generally staffed by top-notch personnel.

There are four reaction time requirements that the prototype shop must meet in order to provide cost-effective service. The first is the amount of time it needs to reply to a request for quotation. A quality shop will be able to give a verbal reply to an R.F.Q. in seven days or less (this will of course be supported by a written quote). The second is the time required to fabricate the tool after a purchase order has been received. In general, unless the part is very complex, a quality prototype shop should be able to build the tool in eight weeks or less. The third is the time required to produce deliverable parts from the tool. Generally this should be five working days or less after the tool is finished. These are not the parts that are shot to test the mold; they are the parts that are to be delivered to the customer. The fourth reaction time is that required to rework the insert as required. A good shop should be able to get the tool back to work within seven working days. This time will depend upon the changes made, but rework should not take longer than the original tool took to build.

The customer also has reaction times he must meet in order to support the prototype tool shop. The first is the decision to accept the quote and issue a purchase order. This should take seven working days or less. The second reaction time deals with answers to questions the tool shop may have. These should be answered within 24 hours or less. The third and last is the time required to decide if changes are needed in the tool. I can place no set value on this other than to say that time is money: the longer it takes to decide, the longer it will be before changes are made and parts received.

The Moldmakers

If time is the single most important item in a shop, then the moldmakers are the most important people there. A tour of the shop will give some insight into the quality of the tool makers. Things to look for are: adequacy of work space; neatness of each person's area; whether the tool makers are well equipped, both with personal and shop tools and with supplies; and how they feel about their work. This last concern is vital to a quality shop. If the tool makers take pride in their work, they produce better tools for the money.

Although this approach will tell much about the shop and its tool makers, the only true test of skill and ability comes from having a tool built. This is the final criterion upon which I base my decision. One other item to look for is the average experience of the tool makers, although this can be misleading in a prototype shop, where the demands of the job require the tool makers to work under more pressure than those working in a production mold shop face. It is a job for younger people; it is not unusual for average experience to run from 11 to 15 years.

The Design Staff

A good design staff is required for two reasons. The first is to provide good efficient designs, including designs for production tools requiring totally detailed complete tool drawings, and designs for advanced inserts requiring effective minimal designs. The second is to keep the shop and thus the moldmakers abreast of the latest developments in tooling and molding. The best way to assess the design staff is to order a production tool and see how well it is designed. This can be very costly if more than one shop is being considered, but there is another, more cost-effective method. A review of the proposed tool with the shop designers will generally be sufficient to disclose any weaknesses the staff may have. If it is possible, ask to see a couple of advanced insert designs. Look for detail that pertains to the part, as well as a lack of detail regarding the fits of components inside the insert itself. Remember that too much detail on a minimal design could indicate a lack of skill in the shop or an overbearing design staff, both of which could cost you money later. The design staff and the moldmakers must mesh like a precision gear set, or a prototype shop cannot be cost-effective.

The Molding Department

In order for the prototype shop to provide a complete service, an area is required for testing the tools and producing parts. Of the three shop elements discussed so far, this is the easiest to evaluate. There are examples of the molding department's work everywhere—parts lying about in display cases and hanging from the walls. Examine these parts closely. If ribs are thin and long, they are difficult to fill; assuming the tool is ready for it, the part should be completely filled. Look for spray, orange peel, burn marks, jetting, coarse weld and knit lines, a loss of sharpness on corners at the end of the flow path, et cetera. Remember a tool can be the best tool money can buy and still not produce parts if the people running it are not of the proper skill level.

This discussion has assumed that the person reviewing the prototype shop is well versed in mold design, moldmaking, and injection molding. If you do not have such a person on your staff hire a consultant to perform this review. Picking the wrong shop can be disastrous, and will make the selection process even more difficult the next time.

The Equipment

A quality prototype shop is not made of men alone. The machines in mold-building and test areas must be of a quality to complement the people in the shop. The shop will have general equipment such as mills, surface and tool grinders, general-purpose drilling equipment, and finishing equipment. A quality prototype shop will also have much of the following equipment: jig bores, jig grinder, inside- and outside-diameter grinders, gun and radial drills, tracer mills, electrical discharge machines, and pantographic equipment. All equipment with movement in the X-, Y-, or Z-axis should be equipped with digital readouts. Many prototypes shops also have numerically controlled machine tools.

Equipment age is also a factor. Late-model modern equipment is an indication of a commitment to quality by the prototype shop. There is less risk that a machine problem will cause an error during tool fabrication with new equipment (seven years old or less) than with older equipment.

The molding area, like the tooling area, requires a minimum of standard equipment to be efficient. It requires a molding machine capable of holding the largest mold the prototype shop can build. It also requires a chiller, water or oil heater, material dryer, and granulator. In addition, quality prototype shops usually contain a variety of molding machines. Generally these start at about 50 to 75 tons and 4- to 6-ounce shot size, increasing in about 100-ton increments to about 550 tons. Beyond 550 tons most prototype shops contract test time for molds from a proven vendor or the customer himself. Also, as with machine tools, varying levels of control are now available that allow quicker setup, ease repeatability problems, and generally provide better control of all molding machine functions. This can be very useful for testing an advanced insert that is proving-out a cavity design, for example.

Quality Control

Finally, there is the matter of quality control. A quality prototype shop provides quality by controlling it; and for a prototype shop to provide the level of quality control that modern manufacturers require, the quality department must be answerable only to upper management.

A quality control department must under all circumstances maintain its credibility. This is best done by having a written quality control manual that sets up minimum standards for tool inspection, molded part inspection, incoming material inspection, and documentation of data.

Quality starts in the machine shop. If you build a bad or marginal tool, problems will appear on every run. Quality in the tool room starts with the machine tools themselves. The machine tools should be on a 75- to 90-day formal quality inspection routine, which should include, but not be limited to, checking actual travel of moving components against digital readout units or micrometer cranks.

For example, if a mill table readout is 6.555 in., this reading should then be proved or disproved by using certified gauge blocks. Other checks should include rotation of rotary tables, movement ratios on the pantographs, movement of cross slides and lead screws on lathes, the squareness of moving components with respect to each other, and so on. All readings taken by quality control during this routine should be recorded and saved. At the same time that the routine check of machine tools takes place, the personal measuring equipment of the moldmakers should be checked. If any problems are uncovered in either inspection, they should be noted and corrected as soon as possible. This routine, if followed, will ensure the ability of the tool maker to build an accurate mold.

Inspection of the Prototype Insert. Having ensured that the machine tools will not be a source of trouble, quality control merely has to inspect the insert to confirm that the tool will make the part that is ordered. The insert inspection should be limited to the part dimensions only, for which quality control requires at least one part print. For this, I prefer to assign a number to each dimension on the print and then generate a chart from that information. This chart has seven columns labeled as follows: dimension number, print dimension (actual), dimension with shrinkage added, tolerance, dimension as measured, go, and no go. As can be seen, the inspector merely has to measure each dimension and check go or no go. If all dimensions are go, then the tool can be sent to the molding department. On the other hand, a no go dimension must be fixed, or use of the tool with a bad dimension must be approved by the customer. Quality control is also responsible for inspecting and approving any call-out finish or texturing the customer may have ordered.

Inspection of the Molded Part. The molded part is inspected in the same manner as the prototype insert. The same chart is used, but the "dimension with shrinkage" column is crossed out. The inspector then goes through the checks either go or no go. If all the dimensions are go, then the parts are shipped. On the other hand, a no go dimension first requires a check to see if it was no go on the insert inspection also. If this is the case, it will be all right to ship the parts. If, however, the dimension was go on the insert inspection, there are two options. The first is to return the insert to the moldmaker so that he can make any possible adjustments while remaining within the part tolerance. If a correction cannot be made within the limits of the part tolerance, the customer must be notified and any decisions to modify print dimension left to him. The second option is to contact the customer and ask if the parts can be used as molded. It is the customer's responsibility to tell the prototyper how many parts to inspect, as it costs a great deal to 100 percent inspect a run of parts, even a short run of 20 or 30 parts. The prototyper needs only to inspect two or three parts to confirm that they are to print.

Inspection of Materials. Inspection of incoming materials covers both the material the tool is built from and the material the part is molded from. In its most basic state, this involves checking the shipper invoices against the shop purchase orders. At its most complex, it can involve mechanical and chemical testing of both tool material and part material. Almost all prototypers keep to the first method. The main thing that quality control is concerned with is molding the part of the proper material as designated by the customer on his print and in the purchase order.

Documentation of Inspection Results. This will vary from shop to shop. At the very least it should include placing a copy of the tool and part inspection reports in the job file and the shop file with the originals kept by quality control. The inspection reports on the machine tools should be maintained by quality control and control to standards of the U.S. Bureau of Standards. Minimum requirements also call for providing tracability of all quality control standards back to U.S. standards; this can only be done by having the quality control standards certified at least once a year by a government-approved inspection facility.

Chapter 10
How to Buy a Mold

Stuart A. Scace
Professional Consultant
Pittsfield, Massachusetts

INTRODUCTION

It is not necessary to know the exact details of house construction to purchase a home wisely, or all the manufacturing techniques used to produce an automobile to get a "good deal"; nor must you be a class A moldmaker to buy a mold intelligently. In all three cases, however, a good general knowledge of the items that you purchase will help to ensure that you get what you are paying for.

While you should not pay more for a mold than it is worth, you also should not pay less. Buying a mold that was quoted much lower than others may be false economy. Like you, the moldmaker is in business to make a reasonable profit. When the moldmaker sees that the cost of your low-price job is going over budget, he may very well look for shortcuts and, invariably, will be late on mold delivery. Because the major cost of building a mold is labor, the moldmaker probably will have underestimated the time required to manufacture the mold. Working overtime may help meet the delivery deadline shortfall, but it will also increase the cost of producing the mold. For a moldmaker to keep his shop equipped with the best facilities and attract and keep the highly skilled people necessary, he must make a profit. If you go for a very low price, you may well buy an unreliable mold that will require longer molding cycles, extra secondary operations, a press operator rather

than automatic operation, higher mold maintenance, and probably a shorter overall mold life. If you return the mold for repairs, the moldmaker may no longer be in business.

Just as you would ask a moldmaker to take a second look at a quotation that is over your budget, it is good business to ask him to take a second look at a quote that is well below competitive prices.

This chapter will attempt to assist those who buy molds, either directly or indirectly, to understand the normal procedures used to purchase molds. Some factors that may increase cost and lengthen the delivery date will be pointed out.

Mold Nomenclature

Figure 10-1 is an illustration of a mold, indicating the common names of the various components. Unfortunately, not everyone will use the same name for the same part. Guide pins may be leader pins; ejector pins may be K.O. or knockout pins; the cavity retainer may be a top retainer or chase; but the functions will still be the same.

REQUESTS FOR MOLD QUOTATIONS

In sending out requests for mold quotations, the more complete the information is that accompanies the request, the more accurate the price and delivery schedule are, and the less

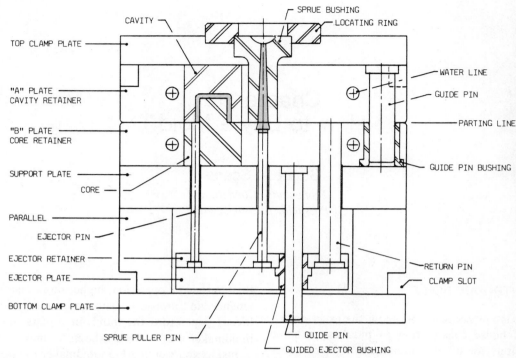

Fig. 10-1   Mold nomenclature.

Table 10-1   Injection mold data sheet.

Request # _____ Date _____ Reply _____

Customer _____

Part
Name
1. _____ Dwg.# _____ Rev.# ____ #Cav. ____

2. _____ Dwg.# _____ Rev.# ____ #Cav. ____

3. _____ Dwg.# _____ Rev.# ____ #Cav. ____

4. _____ Dwg.# _____ Rev.# ____ #Cav. ____

5. _____ Dwg.# _____ Rev.# ____ #Cav. ____

6. _____ Dwg.# _____ Rev.# ____ #Cav. ____

#Cavities	Price	Delivery
_____	_____	_____
_____	_____	_____
_____	_____	_____

-Type of Mold-	-Special Features-	-Ejection-

		Cavities	Core
___ Std. 2 plate	___ Guided K.O. Bar	___ K.O. Pin	
___ 3 plate	___ Spring loaded K.O. Bar	___ Sleeve	___
___ insulated runner	___ Hyd. Cyl. K.O. Bar	___ Stripper	___
___ Hot Manifold	___ 2 Stage eject	___ Blade	___
___ Other	___ Positive K.O. Return	___ Air	___
	___ Parting Line Interlocks	___ Unscrewing	___
Mold Base Steel	___ Accelerated K.O.	(Auto)	
___ #1	___ Spring box core pins	___ Other	___
___ #2	___ Hand Load inserts		
___ #3			

-MATERIAL-	-HARDNESS-	-SIDE ACTION-
Cavities Cores	Cavities Cores	Cavities Cores
		___ Angle Pin ___
___ P-20 ___	___ FINISH	___ Cam
___ BeCu ___	Cavities Cores	___ Hyd.Cyl. ___
___ Tool Steel ___	___ SPE#	___ Air Cyl. ___
(Type)	___ Texture ___	Spring Loaded
___ Other ___	___ Other	___ Cam
		___ Other

Table 10-1   (*Cont.*)

Molding Material _____

Shrinkage Factor _____

TYPE OF GATE
(Size)
____Edge
____Direct Sprue
____Sub-Gate
____Pin Point
____Other

PRESS

Make _____   K.O. Location
                                     Horizontal _____
Clamp_____   Vertical _____
(Tons)
                                     Max. K.O. Stroke _____
Shot Size_____
                                     Nozzle Radius _____
Between Tie
Bars   Sprue Dia. _____
Horizontal _____
Vertical_____   Loc. Ring Dia. _____

Min. Mold Height_____

Max. Mold Height _____

Max. Mold Open _____

COMMENTS: _____
_____
_____
_____

PREPARED BY: _____

chance there is for inaccurate assumptions to be made by the mold builder. At a later date, misinterpretations would be costly in both time and money required to correct them. Be reasonable with your request. The mold builder should not have to second guess you if the shot size, number of cavities, or projected area would exceed the molding press limitations.

It is generally accepted practice to "balance" the runner system by designing the mold so that the flow length from the sprue to each part in a multicavity mold is equal. Therefore; 2, 4, 8, 16, 32, 64, 128, etc., cavity molds are the easiest and most economical numbers to balance.

Mold requirements can be in the form of a letter; or, in the interest of saving time and assuring that information is not omitted, a standard mold specification form can be used. One may be devised for your company and used both for quotation requests and as a guide for mold design when the order is placed. An example is shown in Table 10-1.

EVALUATION OF MOLD QUOTATIONS

While price is always an important factor, it should not be the only one for choosing the mold builder for a particular project. The following factors may also affect the decision.

Delivery.   Does the mold builder normally meet the promised date of completion, or is he consistently late? In your consideration, were there revisions to the part drawings but not to the original delivery schedule? If the builder has never built a mold for you, ask for some of his other customers' names so you may inquire about the mold builder's past performance.

Cooperation.   Is the mold builder willing to work with all parties involved to realize a successful completion of a project, or does he have a "take it or leave it" attitude?

Workmanship.   In the past, have molds been returned to the moldmaker for corrections because of their errors, such as: cavity or cores cut incorrectly that mold a part out of tolerance; ejector pins that mold a flash; inadequate mold polish and lack of sufficient ejection causing the part to distort or stay in the mold?

Mold Builder Confidence.   Has the mold builder manufactured similar molds for you or others? Does he have the equipment and expertise needed for your project? There are several mold-building companies that have become specialists in different types of molds. Some of the specialties are: molds for close tolerance gears and screw threads, automatic unscrewing closure molds, and large three-dimensional shape molds such as automobile fenders and bumpers. You should not expect a company that usually builds molds to produce dashboards to be able to make a mold that produces a fine-pitch spur gear within the tolerances of AGMA quality No. 12, or vice versa.

Site Visitation.   If at all possible, before a purchase order is issued to a new vendor, visit the plant. This visit will give you the opportunity to examine the quality and type of work in process as well as the general appearance and condition of the mold builder's facilities. An experienced mold buyer observes several key items, such as:

1. *General Building Appearance* (*inside and out*): The shop conditions tend to reflect the general work attitude of management and moldmakers. An operation that is well maintained shows that they care.
2. *Condition and Age of Equipment:* Condition and accuracy are more meaningful than age. It is not uncommon for machine tools to be rescraped, square, flat, and parallel with lead screws replaced. This restores the machine tool so that it is capable of maintaining its original tolerances.
3. *Capacity of Equipment:* Will the mold be able to be machined efficiently on their equipment? What is the maximum mold weight the shop can handle?
4. *Mold Design Facilities:* Does the mold builder use in-house designers, outside designers, or both? If both, that usually indicates there will be an early start on the mold design.
5. *Size of Work Force:* Are there enough people available for the necessary manhours to produce the mold project on time? What is the normal work week, and do they have occasion to work reasonable overtime (not over 55 hours per week)? Do they have more than one shift?
6. *Quality Control:* What procedures and equipment are available for checking and measuring mold components?
7. *Mold Sampling:* Does the mold builder have facilities available either in house or locally to sample the mold? Customarily, mold sampling is an additional charge and is well worth the cost. Even precision-built molds may require several debuggings and samplings before the mold is capable of production runs.

PLACING THE PURCHASE ORDER

Once the decision has been reached to place an order with a mold builder, it is common practice to place the purchase order and its number by phone with a follow-up in writing. It is not good practice to use only verbal orders, which often lead to misunderstandings and unneeded delays. The order should include:

The mold specification sheet (Table 10-1)
Total price
Firm date of shipping (not ASAP)
Terms of payment
An acknowledgment copy

ACCEPTANCE OF THE ORDER

The acknowledgment copy of the purchase order should be received within seven days of the date when the order was placed. This will confirm to the buyer that the mold builder agrees to the specifications and conditions of the order or his exceptions to them.

MOLD PROGRESS REPORTS

To help ensure that the mold project stays on schedule, some mold builders have charts that show important milestones in the mold construction. These milestones would be the start and completion of:

Design
Models and hobs
Cavities
Cores and inserts
Mold base
Mold polish
Mold assembly

An example of a detailed mold progress report with these milestones is shown in Table 10-2.

Table 10-3 shows a general mold progress report. The amount of detailed information required and frequency of the report are usually agreed upon when the order is placed.

MOLD DESIGN

Preliminary Design

It is good practice to request a preliminary layout of the mold while it is being designed. To avoid delays, the layout should be reviewed and returned promptly with any questions or

Table 10-2   Detailed Mold Progress Report

COMPANY NAME

Address

+MOLD PROGRESS REPORT+

Part Name _____   P.O. Number _____   Job Number _____

Customer _____Scheduled Delivery: Original __/__/__   Date of Report __/__/__

Attention of: _____   Current __/__/__   Report by: _____

Estimated Completion	DESIGN	MODEL/HOBS	CAVITIES	CORES/INSERTS	MOLD BASE	POLISH	ASSEMBLY
	__/__ % complete	__/__ % complete	__/__ % complete	__/__ % complete	__/__ % complete	__/__ % complete	__/__ % complete
Week							
1. __/__							
2. __/__							
3. __/__							
4. __/__							
5. __/__							
6. __/__							
7. __/__							
8. __/__							
9. __/__							
10. __/__							
11. __/__							
12. __/__							
13. __/__							
14. __/__							
15. __/__							
16. __/__							

Table 10-3   General mold progress report.

TO: Date: __/__/__

JOB NUMBER _____   YOUR P.O. NUMBER _____

DESCRIPTION: _____

_____

_____

SCHEDULED DELIVERY: Original Date: __/__/__

                     Current Date: __/__/__

COMMENTS: As of this date, the mold is _____% complete.

_____

_____

_____

_____

_____

                         Signed:_____

                         Title: _____

comments. Questions you should consider with a preliminary layout are:

1. Will the mold fit the intended press, not only between the tie bars but also the minimum and maximum mold height?
2. Are press ejector holes shown?
3. Is sufficient ejection stroke provided?
4. Are there enough ejector pins, and are they placed properly?
5. Will the part stay on the ejection side of the mold?
6. Is there sufficient temperature control?
7. On side draw molds, is there enough stroke to permit part ejection?
8. Is the cavity outline the reverse of the part?
9. Is there sufficient mold base steel supporting the cavities and cores on all sides?

Final Design

As soon as the mold design is complete, two copies should be sent to the mold buyer—one for the buyer's file, the other to be returned to the mold builder with an approval signature. A design checklist is shown in Table 10-4.

PRODUCT DESIGN FACTORS THAT AFFECT MOLD COST

Without intending to discourage innovative product designs, we must point out that product design has a direct relationship to mold cost. Some important factors in product design are:

Unclear Product Drawing. One of the most common causes of excess mold cost is failure to provide the mold builder with a clear product drawing. Missing lines and conflicting views not to a scale may cause inaccurate assumptions. Usually, the mold builder will consider a worse-case situation and quote a higher price accordingly.

Undercuts and Several Parting Line Levels. Molding an undercut in a part usually

Table 10-4   Mold design check list.

1. ___ Was latest issue part drawing used?
2. ___ Will mold fit press for which intended? Are press ejectors specified?
3. ___ Are daylight and stroke of press sufficient for travel and ejection?
4. ___ Are reverse views correct?
5. ___ Are one guide pin and one return pin offset?
6. ___ Do guide pins enter before any part of mold?
7. ___ Can mold be assembled and disassembled easily?
8. ___ Has allowable draft been indicated?
9. ___ Is plastic material and shrinkage factor specified?
10. ___ Are mold plates heavy enough?
11. ___ Are mold parts to be hardened clearly specified?
12. ___ Are sufficient support pillars located and specified?
13. ___ Are waterlines, steam lines, thermocouple holes, or cartridge holes shown and specified?
14. ___ Does water in/out location clear press tie bars and clamp locations?
15. ___ Is ejector travel sufficient?
16. ___ Are stop buttons under ejector bar specified?
17. ___ Are ejector pins sufficient? Specified?
18. ___ Is the steel type for mold parts specified?
19. ___ Have eyebolt holes been provided?
20. ___ If stripper type, does stripper plate ride on guide pins for full stroke?
21. ___ Do loose mold parts fit one way only? (Make foolproof.)
22. ___ Will molded part stay on ejector side of mold?
23. ___ Can molded part be ejected properly?
24. ___ Have trademarks and cavity numbers been specified?
25. ___ Has engraving been specified?
26. ___ Has mold identification been specified?
27. ___ Has plating or special finish been specified?
28. ___ Is there provision for clamping mold in press?
29. ___ Are runners, gates, and vents shown and specified?

increases the mold cost because additional movements in the mold other than a straight opening and closing are required. These movements may be mechanical, hydraulic, or air-operated components that have to be made and fitted to the mold. If the product design allows, undercuts may be molded by changing the parting line levels. Although this procedure is generally less expensive than slides, cams, and core pulls, it still increases mold cost.

Threads, Gears, Splines. Generally, molds that produce threads, gears, or splines are relatively expensive because special machines and highly skilled people are required to produce their components.

Surface Finish of the Plastic Part. The surface finish of the plastic part is a reflection of the surface finish of the mold cavity and core. A high-gloss finish on the plastic part requires the same high gloss in the mold. A clear part that requires transparency will require a high gloss on both cavity and core. Although several mechnical devices developed for mold polishing are commercially available, there is still a considerable amount of hand work and time required to remove machine tool marks and work the molding area to the desired finish with the use of polishing stones, aluminum oxide cloth, and diamond polishing compound. To avoid misunderstandings, the surface finish should be specified either on the part print or on the mold data sheet. Almost every mold requires some degree of polishing so that the plastic part will release from the mold.

The Society of the Plastics Industry and The Society of Plastics Engineers has approved a system that assigns numbers to various levels of a mold polishing technique. Although this system has helped reduce many misunderstandings between the mold buyer and mold builder, common practice has shown that the gap between level numbers is too great. Table 10-5 shows a chart with the SPI-SPE numbers and suggested numbers between them.

Texture Finish. Another common method of finishing mold cavities is use of a special chemical photo etch commonly called texturing. Generally, most plastic resins require $\frac{1}{2}$ degree to 2 degrees draft in the mold for release. A mold with texture on the side walls will require an additional degree of draft for every .001 in. of texture depth. Many standard texture patterns are available. Sample plastic plaques of standard designs are available on request from companies that perform this service. It is best to sample molds before any texturing is applied to ensure proper fit of mating parts and smooth operation of the mold. If the part will not release from the cavity

Table 10-5   Mold finish numbers.

Finish Number	
SPI-SPE #1	Nat. Bur. Std. Grade Diamond Compound No. $\frac{1}{4}$ to 3
1.3	3
1.6	6
1.9	9
SPI-SPE #2	15
2.3	30
2.45	45
2.5	500 Grit Aluminum Oxide Cloth Plus Buffing Wax
2.6	1000 Grit Abrasive Stone
2.7	500 Grit Aluminum Oxide Cloth
2.8	500 Grit Abrasive Stone
SPI-SPE #3	320 Grit Abrasive Cloth
SPI-SPE #4	320 Grit Abrasive Stone
SPI-SPE #5	240 Grit Blast
5.5	180 to 80 Grit Blast
SPI-SPE #6	24 Grit Blast

before texturing, it will be necessary to provide additional draft before texturing. If the cavity was textured before the first sampling and insufficient draft caused the part to stick or scuff in the cavity, the texture will be removed when increased draft is added. Retexturing means an unnecessary cost and a time delay.

Close Product Tolerances. Each dimension that has a tolerance limit should be carefully analyzed by the mold builder and the molder to determine if the dimension can be held within tolerance by molding. A more costly secondary operation may be necessary. Several variables in the injection molding process can affect the dimensions of the plastic part; therefore, the mold must be manufactured within a small percent of the tolerance allowed on the plastic part. Table 10-6 shows the tolerance allowed to the moldmaker according to the tolerance and size of the plastic part.

Part Geometry That Requires Two Ejection Systems. Occasionally, plastic parts are designed with deep sections molded in the cavity or the stationary side of the mold. These sections are molded by standing members in the cavity. After the plastic has entered the cavity and solidified, it will shrink around

Table 10-6   Dimensional tolerances allowed to moldmakers.

PRINT DIMENSION	PLASTIC TOLERANCE	MOLD TOLERANCE
	± .002 in.	± .0004 in.
.0 to 1.0 in.	.005	.00125
	.010	.0025
	.020	.006
	.005	.0005
1.0 to 2.0	.010	.003
	.020	.006
	.005	.001
2.0 to 3.0	.010	.003
	.020	.006
	.005	.001
	.010	.003
3.0 to 5.0	.020	.006
	.030	.009
	.010	.003
5.0 to 8.0	.020	.006
	.030	.009
	.010	.003
8.0 to 12.0	.020	.006
	.030	.009
	.040	.012
	.020	.006
12.0 to 16.0	.030	.009
	.040	.012
	.020	.006
16.0 to 20.0	.030	.009
	.040	.012

these standing members. When the mold opens, some of the plastic part probably will stay in the cavity, frustrating the molder and causing possible damage to the mold when the plastic is removed from the cavity. An experienced mold builder will foresee this potential problem and provide means of extracting the standing members before the parting line opening, by a method commonly referred to as a "spring box" (see Fig. 10-2).

Another way to ensure that the part ejects from the moving half is to provide an ejection system on both halves of the mold (see Fig. 10-3). This design is often used when the part is symmetrical to the parting line (e.g., a cylinder lying on its side) and it is uncertain in which side of the mold the part will remain. Both methods will make the mold more ex-

pensive initially, and almost always have to be designed in the original mold. Because much of the mold would have to be modified, it is more expensive to add these designs at a later date.

Gate and Ejector Location Restrictions. Function and general appearance of the plastic part will sometimes prevent the most economical placement of gate and ejector locations. The additional cost will depend on the product design and its restrictions.

Three-Dimensional Contours. With the possible exception of the very simplest parts, all plastic parts are three-dimensional outside and inside. Large radii and compound angles are much more difficult and expensive to produce in a mold because additional operations for making models or special electrodes for EDM (Electrical Discharge Machine) or generating tool paths for N/C or CNC equipment are necessary. The blending of generous radii and compound angles that are possible by molding have given plastics designers opportunities to create items that are streamlined and pleasing to the eye, and thus more marketable. However, it should be understood that three-dimensional contours are very expensive to produce.

Sharp Outside Corners, Inside Corners with a Radius. Generally, when a plastic part is designed, there is little, if any, input or consideration about how the mold will be made to produce the part. Many times a model may be produced to simulate the molded part design and to establish whether there are possible interferences that would not be readily seen on a drawing. Drawings for models normally would place sharp corners and radii opposite where they could be economically produced in a mold. Figure 10-4 shows examples of each.

High Mold Maintenance Areas. Molded parts that have small, deep holes are an example of a mold that would be prone to damage and require high mold maintenance charges. Even if all possible precautions are taken, the

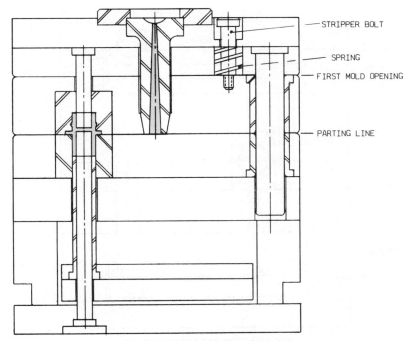

STRIPPER BOLT

SPRING

FIRST MOLD OPENING

PARTING LINE

Fig. 10-2 "Spring box" mold design.

mold should be built for ease and interchange-ability of the cores. This requires that the interchangeable mold components be produced to much closer tolerances, usually ±.0001 in. nonaccumulative, than normally would be necessary. More zeros after the decimal point mean more dollars for the mold.

OTHER FACTORS THAT AFFECT MOLD COST

Molds are described generally by the number of cavities, the type of plastic runner system, and the method of part ejection from the mold. Although most molds are assembled using several plates, the most common is called a "two-plate" mold. The stationary half, the "A" plate, and the movable half, the "B" plate, usually contain the cavity and core respectively.

Unless otherwise stated, a two-plate mold is assumed and not usually needed for the general mold description. It should be obvious that as the number of cavities is increased, the mold cost will increase. Machine tool set-ups and special fixturing for manufacturing components are significant contributing fac-tors toward the cost of producing molds. Because of this, as the number of like cavities is increased, the cost per cavity should de-crease.

The most common and economical method of injecting the plastic into the cavity is to use a sprue, runner, and gate system. Unless stated otherwise, this is also assumed. Other runner systems such as the three-plate, insu-lated runner, and hot runner manifold are more expensive to manufacture than this sys-tem, but usually parts are produced more effi-ciently with those systems.

The third part of the general description is the method of ejecting the part from the mold. The most common, and usually the most economical, way is to use standard ejec-tor pins. The use of ejector pins is assumed and not ordinarily included in the general mold description. Other means of ejection such as sleeve or stripper plate may be used in the general mold description even if ejector pins are also used. Assuming the same plastic part will be produced from each mold, the least expensive mold would be a single-cavity mold (sprue, runner, edge gate, and pin ejec-tors). The most expensive would be a multi-

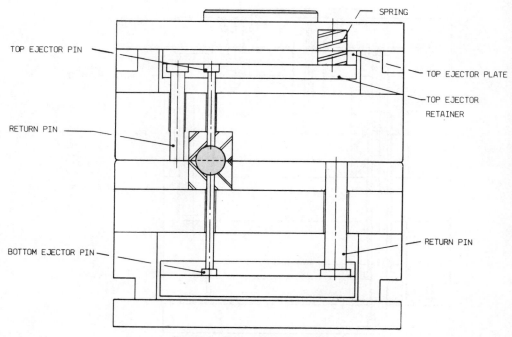

Fig. 10-3   Top and bottom ejection.

cavity mold, with hot runner manifold and stripper plate ejection.

Cavity Inserts vs. a Complete Mold. A method of keeping mold cost down is to use a master frame that will accept different interchangeable mold cavities. This method is good to use when a product is undergoing design developments or when the production quantities are low, but it is not without some problems. Usually, there is a limit to the size of the plastic part that will fit into the cavity insert, and there is a compromise with waterlines because of the placement of the various ejector pin locations. If high injection pressures are required to fill the cavity, the mold

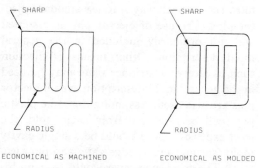

Fig. 10-4   Sharp and radius corners.

may flash because support pillars and plates are usually inadequate.

Another problem is that the master frame may be scheduled to use several other cavities or plastic materials ahead of yours, which will increase the lead time for the delivery of parts. Usually, there is a setup charge to change the mold cavities.

Anyone who is purchasing plastic parts from cavity insert tooling should be aware that the cavity inserts most likely will not be universal to the master frames of other molders.

Nonstandard, Company "Standards." While there may be very good reasons for your particular mold specifications and requirements, it would be advisable periodically to review and analyze them and the effect they have on the mold price.

Revised Delivery Schedules. A mold buyer who is continually revising the scheduled delivery of the mold by requiring a shorter delivery than quoted will most likely be charged a premium for overtime. The buyer who places a "hold" on mold construction should realize that he has increased the moldmaker's cost. Depending on the stage of construction, machine setups may have to be taken down and work rescheduled. A "hold" for one week could very well delay the mold delivery by four weeks or more, depending on the circumstances.

Revising the Product. One sure way of increasing the mold cost is to make product revisions after the construction of the mold has started. Even if the product is revised so it will be easier to produce the mold components, some parts may have to be scrapped or reworked extensively. Also, there will be the cost of revising the mold drawings.

Financial Terms. Most moldmakers now require a deposit with the purchase order. The deposit normally is one-third to one-half of the total order. There are several reasons for the deposit. The majority of moldmaking companies have fewer than 20 employees, so they do not have capital available to them to finance the cost of building a mold. Their suppliers of services and materials require payment long before the mold is completed, and the employees must be paid weekly. Even with a sizable deposit, the moldmaker is still financing part of the mold construction. When a mold buyer makes a financial commitment, he is indicating trust in the moldmaker, and vice versa.

Usually a second payment is due upon completion of the mold.

The final payment should be made when the mold has been approved for production, but no longer than 30 days after the mold has been shipped.

Two common payment terms are: one-third down, one-third when complete, one-third 30 days after completion; or one-half down, one-quarter when complete, one-quarter when approved for production.

Agreement on the financial terms should be made when the purchase order is placed.

Freight Charges. It is common practice for the mold buyer to be responsible for the shipping charges when a new mold is first sent to the mold buyer's designation point. This should be indicated on the original quotation. A difficult situation may arise if product changes are required on the new mold and the moldmaker has to make corrections. The question is, who pays the freight charges? A possible solution would have both parties ship the mold "freight collect." If the mold is returned strictly for mold revisions, the mold buyer should be expected to pay freight charges both ways. If the mold is returned for corrections in the mold, normally the moldmaker will pay both freight charges. To avoid possible conflicts, these areas should be discussed and agreed upon when the original order is placed.

MOLD MATERIAL COMPARISONS

A general cost analysis of common steels used in mold components, namely, H-13, 420 stainless steel, and P-20 prehardened vs. 6060T6 aluminum indicates that the steels cost about three to four times more per cubic inch than aluminum. On a typical mold size, this would mean a difference of only $150 to $250. Also, comparing the cost of a typical mold base, a #2 steel (4130/4140) base is about 20 percent more, a #3 steel (P-20) about 33 percent more. This would mean an increase over a #1 steel mold base of approximately $300 for a #3 mold base. While these figures appear to be significant at first glance, the cost of material in a mold is usually only 5 to 20 percent of the total mold cost. The majority of the cost in building a mold is the skilled labor required to set up and machine the various components. Consideration must be given to the life expectancy of the mold. Almost all of the engineering types of thermoplastics require high injection pressures and melt temperatures to fill the mold properly. Aluminum, even when used in a prototype mold, would not be recommended for these plastics. It has much lower tensile strength than tool steel, it tends to cool the plastics too quickly, and it requires even higher injection pressures to fill the cavity.

The cost for steel used in the construction of the mold components will vary according to the number of pounds ordered and whether saw cutting is required. An example is shown in the table.

STEEL	100 LB.	400 LB.
H-13	$.99/in.³	$.95/in.³
420 SS	.83/in.³	.69/in.³
P-20	.76/in.³	.63/in.³

There are several considerations in choosing the mold steel: the life expectancy of the mold, the type of plastic material, whether the components will be subjected to parting line shutoff or internal slide movement, and heat treatment stabilities, just to name a few. The table below compares the common steels used in mold construction, showing the normal hardness recommendations and their general application.

The mold buyer should be aware that there are other methods and materials available to the moldmaker. There are usually size, geometric, and hardness limitations to each pro-

	COMMON MOLD STEELS	
STEEL	HARDNESS, R_c	APPLICATION
A2	56–60	Cavities, cores, gibs,
A6		wear plates
A10		
O1	58–60	Gibs, wear plates
O6		
H-13	50–52	Cavities, cores, nitrided ejector pins and sleeves
P-20	32–35	Large cavities, cores, mold plates
S7	54–56	Cavities, cores
420 SS	50–52	Cavities, cores

cess; therefore, before specifying a particular process, a thorough investigation of the various alternatives should be made. Cast beryllium-copper, 3M cast cavities, hobbing, Kirksite, Cavaform, and metal spray buildup all require a model, master, or mandrel to be produced. The masters must include not only the shrinkage factor of the plastic to be molded, but also any shrinkage associated with the specific process.

Section IV
CONTROLS

Chapter 11
Process Control Technology

D. V. Rosato

Process controls for injection molding machines can range from unsophisticated to extremely sophisticated devices. As this chapter will review, they can: (1) have closed loop control of temperature and/or pressure; (2) maintain preset parameters for the screw ram speed, ram position, and/or hydraulic position; (3) monitor and/or correct the machine operation; (4) constantly fine-tune the machine; and (5) provide consistency and repeatability in the machine operation. Recognize that process control is not a toy or a panacea (see references 1–25). The need for the basic controls is illustrated in Fig. 11-1.

Process control demands a high level of expertise from the molder. The price that must be paid for the use of process control is not always just the capital cost of the equipment. There is also the price of responsibility for using the control correctly—and that takes time, patience, and a willingness to learn new ways of molding good parts.

Since their introduction process controls have been maligned as expensive contraptions that don't make the injection-molding machine run any better. The industry cliche is the story of the molder who spent thousands of dollars on process control and then disconnected it six months after installation. While this story may be true, the problem with the controls was more than likely created by their misuse and misapplication. And there are now many success stories—stories of parts that could not be molded without process control and of six-month paybacks permitted by material savings alone.

The problem may be that process control was oversold when it was first introduced. Potential customers did not understand how it worked, but they reacted when told that process control could help them produce a better, more consistent part, tighten tolerances while saving resin, reduce mold wear by eliminating overpacking, and save energy by cutting the time the hydraulic booster pump is operated. What many molders failed to realize is that you must be able to mold the part well in the first place before you can gain the advantages of process control.

ON-MACHINE MONITORING

There are different available means of monitoring. First, for clarity, let us separate "monitoring" from "controlling." Monitoring means watching/observing—in our case, the performance of a molding machine. Traditionally, on an injection molding machine, this is done in a variety of ways: by time and temperature indicators, screw speed tachometers, hour-meters, mechanical cycle counters, and the like. Controlling means just that: command/directional capability (4).

Often a control function is combined with monitoring in one instrument. These devices may be called "indicating controllers."

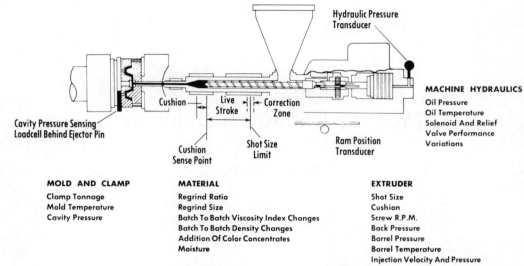

Fig. 11-1   Why injection process controls are needed. Courtesy of Hunkar Lab.

This review will focus on monitoring as opposed to controlling; specifically, monitoring parameters such as cycle time, down time, rate, and totals as opposed to temperature, pressure, and other process parameters.

There are several levels of sophistication available in monitoring devices for molding machines. The crudest, and the one in most frequent use, is the mechanical stopwatch. There are two serious deficiencies with monitoring by mechanical stopwatch:

1. The worst indictment is that it is impossible to monitor the machine with sufficient frequency to be certain that the cycle time has not changed, because stopwatch monitoring is very time-consuming for whoever is taking the readings. There is also often a conflict between responsibility for operation and responsibility for monitoring: the person running the machine may be supposed to monitor it, but may not do so very often.
2. The accuracy of mechanical stopwatch measurements is notoriously poor. The major variable is the human one. Monitoring of machine cycle times is too important

to allow human error in measurement to contribute to poor productivity.

More sophisticated "electronic stopwatches," or monitors, are available that take advantage of the fact that molding machines have numerous signals that are specifically indicative of the cycle. These signals can be utilized to trigger the electronic watch by direct electrical connection to molding machine contacts. With these direct connections, accurate cycle times are assured. For example, measuring from the injection forward relay (a frequent choice) can provide an accurate, continuous display of overall machine cycle times (Fig. 11-2).

There are two proven benefits from monitoring the cycle time on a continuous basis:

1. Production can be maintained at the established optimum cycle time. Display resolution to .01 second quickly shows changes. For example, if a mechanical or hydraulic problem is developing, it can be detected before it progresses to a breakdown. If unauthorized people are meddling with machine settings, they can be observed.

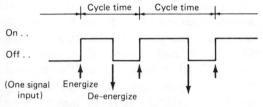

Fig. 11-2   Overall machine cycle.

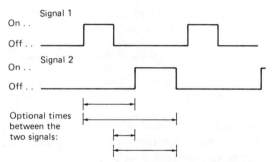

Fig. 11-4   Two-signal input cycle time display.

When changes are easily seen, unauthorized people are deterred from making them.

2. Product quality can be kept high because cycle variations due to the previously suggested potential causes are minimized. Futher, material changes that contribute a small cycle effect but have a significant product effect can be picked up with continuous, accurate monitoring.

Implicit in maximizing these benefits is having the cycle time displayed on the machine. Many users post the standard cycle time in large numerals next to the digital display. This enables engineers, operators, mechanics, supervisors, foremen—anyone walking by the machine—to see and compare the current cycle with the desired one and to respond appropriately to deviation.

In addition to monitoring overall cycle time, elapsed time displays can yield precise information about the individual elements that comprise the overall cycle. For instance, with a single-signal input cycle time display, the time a specific relay, switch, valve, etc., is energized can be measured and displayed (Fig. 11-3). Other digital electronic stopwatches are available that accept input signals from two independent sources and can measure a variety of times between them (Fig. 11-4).

An electronic stopwatch that accepts two input signals adds analytical capability beyond

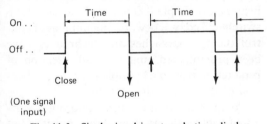

Fig. 11-3   Single-signal input cycle time display.

that available with one-signal input. For example, an engineer wants to set the optimum time for every element of a cycle. First, he must accurately determine where they are now. Then by "tweaking" the times down—while monitoring for verification—and checking product quality, he can "set" each segment as fast as possible while maintaining good quality. If all active segments are optimized, and there is no "dead time" between segments, the cycle will, by definition, be as fast as it can be and still produce the desired product.

Note that "dead time" between active portions of the cycle must be eliminated. A dual input digital display enables this to be done by switching between various signal sources in the machine. Once eliminated, dead time must also be kept out of the cycle in order to keep production up. By continuous monitoring of the most likely areas for dead time, it can be minimized. For instance, improper material additives have affected screw retraction adversely, to the extent of extending cycles because of screw slippage. With continuous monitoring of the retraction time, as measured between two limit switches or their equivalent, this problem could be detected quickly so the material could be changed as soon as it occurred.

Before discussing more sophisticated monitors, let's examine why it is critical to optimize. It is simply a case of economics. As inflation has driven up the cost of everything, time has become more valuable—smaller time increments are worth more. A cycle that is, on the average, 1 percent slow, in one year will cost more than one man-week's production (Table 11-1). Note that 1 percent on the

Table 11-1 Lost time: 1 percent slow.

Std. cycle	10.00 sec	30.00 sec	59.0 sec
Actual	10.1 sec	30.3 sec	59.6 sec
Lost time (5000-hr year)	50 hr	50 hr	50 hr

standard cycles listed is only a fraction of a second; many injection molding machines are slower vs. standard than this.

Machine-hour rates, determined in a variety of ways, currently range from $15 to $50 per hour. With the 1 percent slow losses shown in Table 11-1, the dollar losses in production would be $750 to $2,500 per year in machine time. In general, these figures are conservative. Our experience in the field with elapsed time displays has shown payback on investment periods as short as one day, more typically one week to one month.

The most sophisticated level of monitoring takes advantage of the evolution that has occurred in electronics. With the microprocessor, it is possible to add economical memory and multifunction capability to a machine display.

"Multifunction" means that in addition to the important "cycle-time measuring" component, additional data can be acquired, stored, and displayed. Unless it is separately available on the machine, all monitors of this type for injection molders should include a cycle measurement function. This may be either cycle time directly (in seconds or minutes) or production rate (in shots per hour, cycles per minute, etc.). The availability of a production rate display is important because in many companies the "shop floor language" is shots per hour, and a digital display of these numbers directly is more meaningful than a time readout; for example, the change in rate from 120 to 119 shots per hour compared to the cycle time changing from 30.0 seconds to 30.25 seconds. The successful use of monitors hinges on operating personnel understanding them as an aid to production. Therefore, the display should be scaled and read out in the user's particular terminology.

Additional data that may be compiled with these powerful monitors include totals, run time, down time, etc. "Down time" may be defined in several ways. It can be as simple as a machine set on the manual switch setting (for setup) instead of automatic or semi-automatic; or, as complicated as the monitor "learning" a good cycle and comparing every subsequent cycle to it, then accumulating down time for any cycle that is not at least 90 percent of the "good" cycle. The "learning" of the good cycle may be via a user-set switch identifying a desired cycle, or by the monitor calculating an average cycle. The ability to specifically accumulate and record down time on the machine changes a notoriously inaccurate data source—down time is usually guessed—to a precise record that is used to improve performance of machines and people.

Monitors may also be obtained with outputs to drive typical machine audio/visual alarms. These outputs can be energized when down time occurs, when a slow cycle occurs, when a rate is below a standard the user inputs. (The latter is only available with the most sophisticated type of monitor—one that communicates bidirectionally to a keyboard/computer.)

These more sophisticated, powerful monitors can provide multiple functions displayed on the machine; in addition, they can communicate directly with a centrally located computer. The central computer eliminates the manual collection of production data; it summarizes data, prints reports, calculates efficiencies and utilization, etc., automatically and immediately, not hours or days later.

TEMPERATURE CONTROL OF BARREL AND MELT

Twenty years ago, controlling the temperature on injection machinery was limited to a choice of either manual selection of the power to the heaters or simple "ON–OFF" closed loop switches. Today, a bewildering number of control theory approaches and techniques have been promulgated, and a broad selection of products is available to implement the application of the chosen theory (16).

Comprehensive literature exploring a variety of control theory is available, but it is not the intention of this presentation to explore or summarize this body of literature. This discussion will acquaint you with the latest developments in the quality of temperature control, present both the component and systems approaches, and provide an insight into what the future holds.

The viscosity of the melt and the speed and pressure of injection determine whether an acceptable molded part is produced. Viscosity is a function of the temperature of plastics, and temperature is a result of the forces of screw rpm, back pressure, and externally applied heat. Injection machine control specialists are generally agreed that one-third of the melt temperature is derived from external heat. Closed loop temperature control thus deserves in-depth attention.

Many excellent instruments are available today as a result of reliable and cost-effective solid state and digital technologies. The temperature control result is, of course, no better than the quality of other components and installation practices employed on the machine. Too many times we find the advantages of a sophisticated temperature control (TC) instrument completely negated by poor installation techniques. Before deciding prematurely that the instrument is at fault, you should make the following checks:

1. Is the thermowell too big for the TC protection tube? Air is an excellent insulator.
2. Is there contamination inside the thermowell? Rust, scale, and residue prevent proper contact of the protection tube with the thermowell.
3. Is the TC junction partially open?
4. Are there oxidation and corrosion inside the protection tube?
5. Is the proper extension wire being used: Copper wire allows another thermocouple junction.
6. Is extension wire polarity observed? A single reversal will give a downscale reading; a double reversal will result in an erratic input to the controller.
7. Are wire terminations properly isolated? False cold junctions are a common problem.

8. Is the cold junction compensation at the extension wire termination on the controller working properly? A poorly positioned or poorly connected compensation component will allow the input to vary.
9. In the panel, are the thermocouple leads isolated from the ac wiring as required? Are the TC wiring and the ac wiring run in separate conduits from the control cabinet to the machine as required?
10. Is the control cabinet thermal environment within the specification of the controller? Excessive cabinet temperatures can cause a controller to drift.
11. Examine the power contactor. If it is a mechanical contactor, deterioration of the contacts can result in reduced power delivered to the heaters.
12. Are the heaters sized correctly? Modern temperature controllers can compensate for limited missizing, but cannot substitute for proper design.
13. Heater bands must be secured tightly to the barrel; again, air is an excellent insulator.
14. Check the voltage being supplied to the heaters. High voltage leads to premature heater failure.
15. Inspect wiring terminations at the heater band; connections must be secure.

If the integrity of the heating system has been verified, your attention can now be turned to the advantages of modern temperature control instrumentation. To demonstrate the improvements available during the past 15 years, a comparison of three basic instrument designs is helpful. Millivoltmeter designs can hold setpoint within 20 to 30 degrees; solid state designs can hold within 10 to 20 degrees; microprocessor-based designs typically hold setpoint within 2 to 5 degrees.

Microprocessor-based designs provide several distinct advantages. Already mentioned is the inherent ability to control the temperature at setpoint. Sometimes they do too well. There are reports, in fact, where the customer claimed the controller was not working, because the process reading was the same as setpoint for an entire shift. Microprocessors do not drift; they either work perfectly, or they experience a catastrophic failure. They are absolutely repeatable, allowing the operator to duplicate a log of setpoint temperatures perfectly the next time that particular job is run. Microprocessors allow a natural avenue to provide digital displays of process information. Values are not subject to inaccurate interpolations and misreadings. In new installations, the precision of the digital readout has sometimes proved to be a two-edged sword. We have provided start-up assistance in plants where the operator reports the process reading to be several degrees different from the setpoint. An investigation usually will discover a problem in one of the control loop segments previously outlined. One has to conclude that the problem had existed for some time; the customer just never knew it because the process drum meter on his old instrument could not be read to any finer resolution than perhaps ten degrees.

Microprocessors allow the implementation of PID (proportional, integral, derivative) control at little or no cost. PID has been shown to reduce process variations by as much as three or four degrees. Discussions of PID advantages are available from all major temperature control suppliers.

Microprocessor technology is relatively trouble-free—about six times more reliable than analog solid state designs, and about twelve times more reliable than millivoltmeter designs. Based on customer data, the maintenance costs on an analog instrument average $100 annually; on a microprocessor design, the costs are reduced to $12.

Another significant cost reduction effort being implemented recently with excellent results focuses on the controller output and power handling. Although an analog controller output accepted by a phase angle or zero angle SCR power controller is ideal in terms of power factor and heater life, it is a relatively costly arrangement. A more acceptable method, in line with cost restraints and providing very nearly the same advantages, is to use a controller with a solid state time-proportioned pilot duty output along with inexpensive mercury contactors or solid state relays.

The controller output cycle time can then be reduced to ten seconds or less, thus approaching the same constant temperature and heater life advantages available with the more costly design.

Many more advantages are available when the microprocessor is used as the core component for temperature control. Automatic tuning, introduced recently, has already established an enviable track record. Its benefits fall into three major areas:

1. The unit will identify varying thermal behavior and adjust its PID values accordingly. Variables affecting viscosity include screw rpm, back pressure, variations in heater supply voltage, resin melt index, resin contamination, room ambient temperature, percent colorant, screw wear, barrel lining wear, heater and thermocouple degradation, percent regrind, hygroscopic characteristics, and feed zone instabilities. (The effects of these variables on melt temperature are thoroughly developed in several of this chapter's references.)
2. Savings in management and maintenance activity will result from auto-tuned temperature control. Documentation of PID values for various jobs and machines can be eliminated. Individual operator preference for PID values that vary from the norm is precluded. Maintenance personnel are not required to dedicate a particular unit to a specific zone; instruments can be interchanged at will, and spares can be installed with no attention other than selection of the appropriate setpoint. A payback through reduction of overhead costs alone can generally be expected in six to eight months.
3. Energy savings is another major benefit. One customer study showed a 50 percent reduction in power consumed by the heaters, solely because the automatic tuning feature eliminates the cycling around setpoint normally associated with ineffectively tuned instruments.

Microprocessors also provide a means to communicate digital data to information collection stations. Although the economic feasibility of including the function with an individual temperature control instrument has not been demonstrated in the plastics industry, the feature is beginning to enjoy significant exposure on multiple zone injection machine controllers because of the low cost of adding another digital card to an existing rack. More commonly found on discreet controllers is an analogy communications output that provides a signal to remote recorders.

The ultimate implementation of the microprocessor has been its design in systems installations. Available systems include multizone temperature control and multipoint, multiloop control of sequence. Systems that depend on a single central processing unit (CPU) are available from many suppliers to control temperature, sequence, position, velocity, or pressure. Even more cost-effective are the total machine controllers, such as the Barber-Colman MACO IV, which control all machine parameters from a single keyboard. As compared to individual instruments, these systems typically reduce the per-zone cost of control, and provide unlimited future control flexibility as needs change. As production professionals discover the need to manage the process at the least possible cost, machine control systems that can communicate with a central management computer are of increasing importance. Central control systems are available that can simultaneously receive information from the injection machine and transmit required parameter changes or complete job setups at the same time. Many injection machines can thus be interfaced with a single control location. If central on-line control is not justified, but one-way machine reporting is required, a choice of several management information systems is available.

PID INJECTION PRESSURE CONTROL

A trend to faster-acting, more precise, and more energy-efficient hydraulic systems and components is one response by injection molders and machine builders to a business climate that demands higher productivity and more consistent product quality (17). Examples of this trend are found in the growing popularity of accumulators, which can deliver a large amount of oil at high pressure, making possible very high injection speeds without the need for an extremely large, energy-consuming pump; of variable-volume pumps, either single or multiple, which provide just the amount of flow that is needed at any point in the cycle, for energy-efficient molding; of servovalves, whose fast response is necessary to control the high injection speeds that the more efficient hydraulic systems can provide; and of multistep injection speed and pressure profiling, providing more sensitive control of the process so as to improve part quality.

One thing that all the above have in common is the tendency for changes in hydraulic pressure during a machine cycle to occur faster than ever before, and this in turn necessitates application of pressure controls that are responsive enough to keep pace. Fortunately, meeting this need does not require inventing new control technology, but rather, more thorough application of what we already have.

Hydraulic pressure-control logic is, in fact, the same as that used for temperature control; its most sophisticated form uses three modes of control, known as PID, for proportional, integral and derivative (also called gain, reset and rate, respectively). Each of these mutually interrelated modes of control has an adjustable "tuning constant" that permits the operator to adjust the sensitivity of the pressure controls to the dynamics of the particular machine's hydraulic system.

Some molders may not realize that these tuning adjustments are variables that are just as important to good process control as the setpoints for the actual pressure values that the controller is asked to achieve.

Most commerical process-control systems for injection molding to date have not provided full PID pressure control—usually only proportional, or perhaps proportional-plus-reset (integral), control. Furthermore, these systems have commonly offered at most a gain adjustment, or else no tuning adjustment at all. Consequently, the concept of PID pressure control is probably unfamiliar to most molders, as is the role of tuning in obtaining the maximum benefit from three-mode controls.

Yet it is our feeling that, in order to get the kind of cycle-to-cycle repeatability that today's market demands and that today's microprocessor-based control systems are designed to provide, molders should understand the value of PID control logic and must know how to keep such controls properly tuned. Fortunately, current microprocessor know-how can make full PID control available at little or no extra cost, and makes tuning an easy task for the average setup man or technician.

PID Tuning: What It Means. The following is a brief explanation of the three control modes and their tuning constants. It is important to remember that the three terms are not independent, but mutually interactive, and that both the order and magnitude of adjustments made to the tuning constants can affect the settings of the others.

- *Proportional control (gain)*: With this type of control, the magnitude of the control output is proportional to the difference between the actual pressure and the desired pressure—in other words, the magnitude of the error signal. The "proportional band" is the range of error above and below setpoint, within which the control output is proportioned between zero and 100 percent.

Usually the proportional band is expressed in terms of its inverse, the gain. If the proportional band is set too wide (low gain), the controller will probably not be able to achieve the setpoint within the time frame of that segment of the cycle. On the other hand, if the proportional band is too narrow (high gain), it will cause violent oscillation of pressure around the setpoint, leading to intense machine vibration, shaking of hoses, and rapid movement of valve spools back and forth, all of which are hard on your machine's hydraulic system and can shorten the life of its components. In either case, inconsistent cycles will result.

The proportional band, or gain, setting is the most fundamental part of the tuning process, which strongly influences everything else. For that reason, it is usually performed first, although subsequent adjustment of the other tuning constants may require some readjustment of the gain.

- *Integral (or reset) control:* Unfortunately, it is a characteristic of purely proportional control that, in response to changing load conditions, it tends not to stabilize the process at setpoint, but rather, some distance away from it. Integral or reset control responds to this steady-state error, or "proportional droop," by shifting the proportional band up or down the pressure scale (without changing the band's width) so as to stabilize the process at setpoint. The amount of reset action to use, expressed in repeats per minute, is the second tuning constant.
- *Derivative (rate) control:* This type of control action responds to changes in error, or the rate at which the actual pressure approaches the setpoint. The faster the change in the magnitude of the error, the greater the rate control signal, and vice versa. It serves to intensify the effect of the proportional corrective action, causing the process to stabilize faster. Rate control's main effect is to prevent the undershoot/overshoot oscillation that may never be completely eliminated with proportional-plus-reset control alone. The amount of rate action, expressed in percent, is the third tuning constant, usually the last to be set.

Rate Control Necessary on High Speed Machines. Until recently, it was not always necessary for an injection process controller to have rate or derivative control in addition to proportional and reset. Rate control has, however, become essential on newer, faster cycling machines with updated hydraulics.

For example, the high injection speeds of accumulator-assisted machines can create extremely fast changes in the conditions governing hydraulic pressure. In order to smooth out the resulting pressure fluctuations, rate control responds only to fast changes in hydraulic pressure, such as when the ram begins to feel resistance of the melt pushing through the runners and gates of the mold. Changing from one pressure setpoint to another, as in multistep injection profiling, can require the same fast stabilizing action, so the derivative control will help to bring about a faster setpoint change, with minimal overshoot.

A multiple-pump machine will experience a momentary drop in hydraulic pressure when the high-volume pump "drops out" and the smaller holding pump continues injection. This drop in pressure is sometimes so large that the injection ram will actually back up. Derivative control will help to lessen this short dip in pressure and smooth out the injection pressure curve.

RELATING PROCESS CONTROL TO PRODUCT PERFORMANCE

Monitoring the molding system can show the effects of mechanical and thermal strains. Strains are imposed upon the material as it is conveyed through the machine and mold. Instrumentation to sense, measure, and display changes in molding parameters helps to determine process consistency (18).

Monitoring helps relate the process to the product. The sensed molding parameters can show the relationship between pressure, temperatures, and position (movement) during the process.

Monitoring can also establish whether additional machine control is needed. The forgiving nature of the molding process and liberal product dimensions allow most parts to be produced with conventional "open loop" machine control systems. As product demands become more stringent, both dimensionally and physically, "closed loop" machine control may become advantageous.

Sensor Requirements

Any sensor used requires a power supply and an amplifier. A sensor is driven by an input voltage, usually called an "excitation" voltage.

A resultant output signal is generated as the sensor responds to the monitored parameter. An amplifier is used to boost the output signal's strength. Increased signal strength or amplitude is needed for recording capabilities.

Sensors and electrical systems should be tested and calibrated before actual use. Variances do occur between sensors of the same type. Sensors should be maintained at a "zero" reference if precise monitoring or measuring is to be done. Electrical "drifting" destroys the accuracy of the information being obtained.

Molding Parameters

Pressure

• *Machine hydraulic pressure transducer:* A hydraulic pressure transducer is used to generate a signal. Monitoring the hydraulic pressure profile can help diagnose many machine problems. The hydraulic pressure transducer should be placed as close to the injection ram as possible; this location gives the most accurate pressure profile. Hydraulic pressure profiles can determine the following:

1. Hydraulic pressure relief valve setpoint consistency
2. Timer accuracy for switching cutoff pressures
3. Hydraulic back pressure setting during screw return
4. Screw return time consistency
5. Hydraulic pressure changes during injection, reflecting material viscosity changes

• *Machine material pressure transducer:* Monitoring the material pressure can be done with a transducer in the machine nozzle. The material pressure profile will be similar to the machine hydraulic pressure profile. The pressure of the material and the hydraulics in the machine barrel become similar as the mold is filled. Sensing of material pressure at the machine nozzle can be done, but its usefulness is questionable.

• *Mold material pressure:* Material pressure transducers can be installed in the mold's runner system and in the cavity. Indirect and direct material sensors are available. The type of transducer selected depends upon the product configuration in the mold, mold construction, and type of runner system.

Pin-loaded-type material pressure transducers must be designed and installed with care. The use of pins to transmit material pressure can cause errors; the pins can stick, bend, and induce thermal effects during cure time. Location and pin diameter must be considered for monitoring. Because of the "select point" pressure sensing, the transducer output may be poor.

Direct material pressure transducers are now available. The accuracy of pressure sensing is much better, but there is a problem in selecting the location to sense and monitor the material pressure. Monitoring at a point located halfway into the cavity is a good general rule. Maintenance of built-in transducers should be considered when designing a mold.

The mold material pressure profile can determine the following:

1. Material filling time
2. Material peak pressure consistency
3. Machine nozzle contamination or freeze-off

Temperature

• *Machine barrel temperature:* Barrel temperatures are sensed and controlled with thermocouples (T.C.). One T.C. is needed for each zone that is being controlled. Usually three zones (front, middle, and rear) are sensed and controlled. The nozzle usually has its own control. For accurate temperature control and temperature setpoint, current-proportioning controllers should be used, not the time on–off type of temperature setpoint controllers.

Monitoring barrel temperatures can determine:

1. Temperature controller performance
2. Barrel heater failure

• *Mold temperature:* The control of mold temperature is usually done with an independent heater/chiller unit(s). The controller has temperature setpoints, and the mold usually balances out at some temperature around the setpoint. If the controller supply lines, mold water lines, and pressure losses are minimized, the control is acceptable.

Monitoring of the mold temperature is usually done with T.C.'s. Their accuracy depends on the T.C. placement. The T.C. location must be tried to determine the optimum location. This area of monitoring temperature in the mold is difficult because of the high thermal inertia in the heater/chiller/mold system.

• *Material temperature:* Material temperature can be measured in the machine nozzle. Commercial T.C. sensors are available to measure the material melt temperature. The T.C. devices are the simplest and most stable to install; infrared and ultrasonic systems are also available, but are much more complex.

Material temperature variances can exist in the melt because of screw mixing, barrel heating, and a varying shot-to-barrel ratio. Sensing the nozzle melt can show:

1. Material melt consistency
2. A change in machine plasticating
3. Heater failure on the barrel

Position

• *Machine ram position:* The ram position is monitored from a potentiometer mounted on the machine, either linear or rotary. The sensor indicates the ram during the molding process and can show the following:

1. Injection rate of material into the mold
2. Consistency of ram profile during "open loop" or "closed loop" machine control.
3. Screw position during return to back position
4. Screw return time consistency

• *Machine tie bars:* Machine tie bars "stretch" when the mold is clamped. This mechanical strain or elongation can be measured with strain gauges, dial indicators, and linear variable displacement transducers (LVDT's). The LVDT's eliminate the need to drill holes in the tie bars or clamping on small indicating devices. Monitoring tie bar strain can show:

1. Balance of tie bar strain during clamp
2. Mold clamp tonnage
3. Machine/mold clamp tonnage changes occurring because of thermal effects of machine cycling and mold heating or cooling

• *Mold part line:* Mold part line separation can be measured with indicator gauges and LVDT's. As material is packed into the mold, the part line can open. There is a direct relationship between machine clamp on the mold, material viscosity, and material injection rate. Monitoring for a mold's part line separation can show the following:

1. Dimensional changes in the product
2. Mold flashing

Display of Monitored Molding Parameters

Analog Display.   Analog devices include:

1. Chart recorder
2. Voltmeter with a sweep needle
3. Oscilloscope

Analog signals are useful for seeing a continuous profile of the parameter being sensed. This profile is useful because it is time-related. Chart recordings show a continuous profile but are limited in the type of information that may be interpreted. Total span and peak changes are shown, but comparisons of one cycle to another are difficult.

Digital Display.   Digital devices include:

1. Controllers with numerical setpoints
2. Sensing devices with numerical readout display

Digital monitoring devices give a numerical readout. The sensor's output signal is conditioned to give a discrete numerical readout(s). Data loggers are used to monitor multiple parameters digitally. Digitizing (displaying discrete numerical values at a certain rate) of analog signals can be a useful technique, but the rate at which information can be digitized must be considered. If any rapidly occurring events are being considered, this system can give erroneous or insufficient information.

CRT Display.   Cathode Ray Tube (CRT) displays include:

1. Oscilloscopes (scope)
2. Storage scope
3. Analog/digital scope
4. Television

Storage scopes can be utilized to monitor repeating cycles. A selected starting point is used to "trigger" the scopes. The storage scope display shows the excursion of a parameter over a period of time. A multichannel storage scope is very useful to relate more than one molding parameter on a single display.

Machine Control

"Open Loop" Machine Sequence Control.   In a conventional "open loop" machine sequence control system, input commands are set, and there is an unknown machine output response.

Monitoring of machine hydraulic pressure and ram position relates:

1. Screw return profile consistency
2. Hydraulic pressure profile consistency
3. Ram injection rate consistency

An open loop machine control system *cannot* compensate for changes in material viscosity. Material viscosity changes result in:

1. Increased viscosity (increased stiffness)
   (a) Higher initial hydraulic pressure profile

(b) Slower ram injection rate

(c) Lower final in-mold material pressures

2. Lower viscosity (more fluid)

   (a) Lower initial hydraulic pressure profile

   (b) Faster ram injection rate

   (c) Higher final in-mold material pressure

The ram injection rate is controlled by the metering of oil into the hydraulic injection ram cylinder. Material viscosity establishes the hydraulic pressure profile during mold filling and packing. The hydraulic pressure profile is a valuable parameter to monitor for establishing mold/machine consistency.

"Closed Loop" Machine Sequence Control.

In a "closed loop" machine sequence control system, input commands are set, and corrections are made to the machine output response. The correction can be either of the following:

1. *Real time:* a sensed deviation is corrected in cycle, as quickly as the machine electrohydraulic valve and fluid system can respond.
2. *Adaptive:* a sensed deviation is adjusted for on the next cycle. The system's ability to adjust depends upon how sensitive the molding process is and controller capability to correct the deviation.

A closed loop machine control system *can* compensate for changes in material viscosity. This capability improves the consistency of initial mold filling but does not fully address final packing pressure in the mold.

The ram position is programmed to establish a material filling rate into the mold. The hydraulic pressure compensates for material viscosity changes during the controlled filling of the mold's sprue, runner, and cavity.

The final packing pressure is controlled by switching from the ram position (velocity) profile to a hydraulic packing pressure.

Control of the molding process is better, but actual improvement in the product is not always realized. Monitoring the molding system can help us to:

1. Improve mold setup consistency
2. Resolve molding problems
3. Determine the effectiveness of the equipment
4. See the process working

Microprocessor Advantages

Microprocessor-based process controllers have been achieving more widespread acceptance as their cost has come down. Whereas a few years ago these controls were used only for applications that required their precise control, now we find advantages in their application on almost any job.

- Setup time reduction: Time for setup can be greatly reduced by the ability to record and store timer settings, limit switch positions, and pressure levels. The data can then be fed to the controller in seconds to preadjust the machine to the new setup.

- Easier operator "tuning": Since the microprocessor inputs can be located at the operator station, adjustments can be made without crawling around the machine.

- Smoother operation: This is achieved through ramping of the control signals. We can now eliminate many of the readjustments necessary as the machine temperature changes, simply by setting these ramps such that the time is longer than the response under conditions of start-up. Since the signal is now slower than valve response, the signal is always in control, yielding a more uniform cycle.

- Less downtime: The constant monitoring of machine performance made possible with these systems can allow lower pressures and eliminate shock peaks, thereby extending component life. A properly applied system will also have fewer components to troubleshoot when a problem does occur, and diagnostic programs can be included.

- Input energy reduction: By programming the hydraulic system to respond to the varying demands of the circuit, we have the potential to reduce input power requirements.

This is what the "brains" of process control can do for us, but how we interface the controller can affect these advantages. Ideally, we want a system consisting of as few components as possible, and these components should be maintainable, tolerant of the industrial environment, and repairable by general maintenance personnel (19).

ADAPTIVE RAM PROGRAMMER

The injection molding process has a number of variables in material and machine conditions that tend to change during production. All these variables affect the critical properties of the molded part. When material properties change or the machine drifts outside the ideally preset operating parameters, the operator must reestablish the conditions best suited for making the part (21). He is faced with a complex situation as the interdependency of machine functions and material conditions requires a thorough understanding of the process, and a series of complex adjustments on the machine must be made to maintain part quality. Often the variables are not controllable to the necessary degree, and he has to contend with imperfect production and a high rejection rate.

The Spencer and Gilmore equation developed nearly 20 years ago is now widely utilized to predict the relationships that must be maintained to keep the critical functions that affect part quality constant. This equation indicates that plastic pressure and volume are inversely related if temperature (or material viscosity) is constant. During molding, filling, and packing, the plastic temperature drops only slightly because of the short time interval involved. The material viscosity tends to change, however, as a function of composition or long-term temperature conditions of the machine.

The shrinkage of the plastic during mold cooling is primarily determined by the number of molecules in a given cavity under a given pressure. For this reason, cavity pressure controls have been utilized in an effort to control the shrinkage parameters of the part. As viscosity changes, however, it is important to adjust the plastic volume so that the number of molecules packed in a mold cavity will remain constant. In order to accomplish this, the precompressed shot size must be adjusted so that when the desired pressure in the cavity is reached, the total volume under pressure that exists between the tip of the ram and the cavity will be held constant. As the two parameters, pressure and volume, are highly interdependent, continual adjustments must be made (on each shot) following the trends in material parameters.

Another critical condition to be maintained is plastic flow rate. The Poiseuille equation for fluid flow shows the significance of pressure on flow rates. Plastic viscosity deviates considerably from constant during flow. The effect is to make flow behavior dependent upon pressure. As the operator desires to maintain the flow surface velocity for the plastic constant, or adjust the flow in accordance with the requirements of the mold, the injection velocity together with material volume and pressure form the most important parameters that have to be controlled to maintain part quality.

Heretofore, individual parameters such as cavity pressure, ram oil pressure, and ram velocity have been measured and even controlled. The interdependence of these three functions, however, demands that a control system be utilized that can control all three parameters simultaneously, and is capable of automatic adjustments and decision making to maintain the equations in balance during the molding process.

The Hunkar Model 315 adaptive ram programmer system is designed to perform these functions totally automatically, having the capability of continually adjusting critical parameters and maintaining the process constant. Figures 11-5 through 11-9 explain some of the control functions.

SIMPLIFIED APPROACH TO UNDERSTANDING PROCESS CONTROL (MOOG INC./22)

The process of making an injection molded plastic part has many dynamic fragments that must come together properly for a successful

Cavity Pressure Program

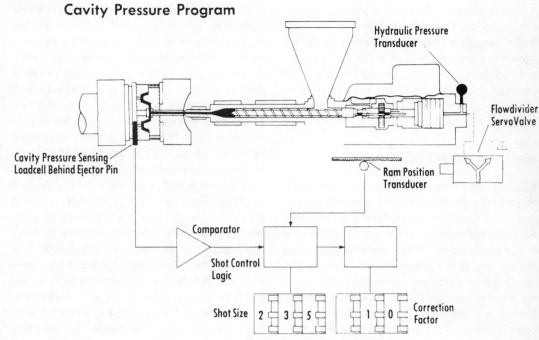

Fig. 11-5 Shot control function.

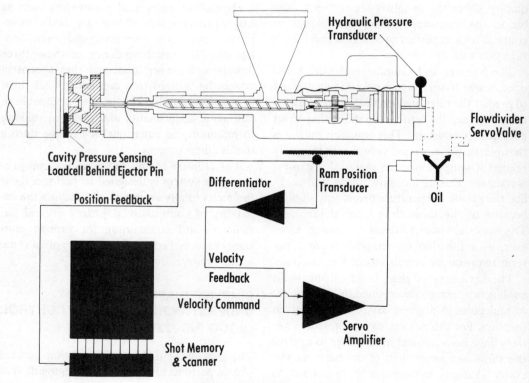

Fig. 11-6 Injection velocity control.

With Cavity Pressure Sensor

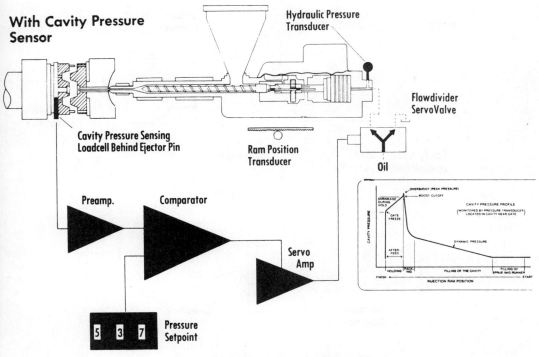

Fig. 11-7 Injection pressure cutoff control.

The system also includes an automatic velocity override to prevent part flashing and a programmable pressure transition slope eliminating ram bounce, thus ensuring a uniform stress profile throughout the part.

With Hydraulic Pressure Sensor

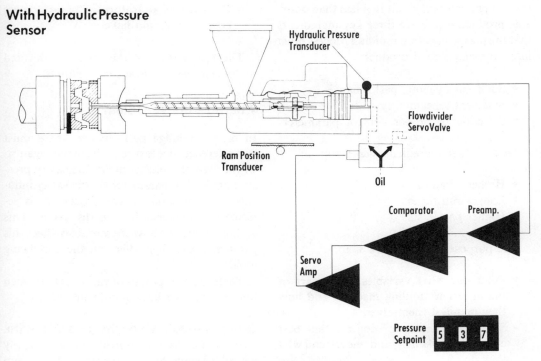

Fig. 11-8 Injection pressure cutoff control.

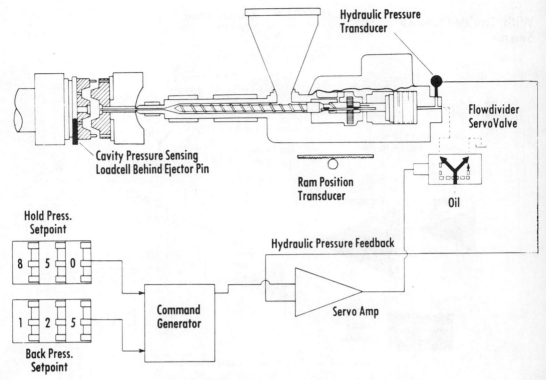

Fig. 11-9 Holding and plasticize pressure control.

result. Lack of sufficient control over each of these fragments will result in a less than desirable product. There are three key ingredients that the plastic injection molding process must have to make a good product:

- Sufficient dynamic performance
- Sufficient repeatability
- Selection of proper control parameters

A lack of these ingredients can result in:

- Higher scrap rates
- Longer run times
- Higher part costs

The purpose of this review is to:

- Point out what variables are a part of the injection molding machine and how they manifest themselves.
- Select parameters for control that best eliminate variability, and understand why they do.

- Discover what enables controllability.
- Discover what features a basic process controller should have.

The last portion of the review is devoted to applications of those basic features.

What Are the Variables?

In order to judge performance, there must be a reference to measure performance against. In the case of a plastic mold, the cavity pressure profile is a parameter that is easily influenced by variations in the process. It is selected as a reference for this discussion. This section points out how the variables affect this parameter and their effect on the part being molded.

There are four groups of variables that when lumped together have similar influences:

Group 1—Melt Viscosity and Fill Rate. Typical non-process-control machines apply a fixed injection hydraulic pressure to the ram piston. The resultant force in turn is counteracted by the speed of the ram in the viscous plastic melt. The result is a fill rate inversely proportional to the viscosity of the melt and proportional to the hydraulic pressure. The lower the viscosity and/or the higher the hydraulic injection pressure, the faster the fill rate. Fill rate variations with a constant boost time are shown in Fig. 11-10. If the fill rate is too fast (curve a), the cavity pressure increases long before boost time out. The result is overpacking of the part. Some of the effects are flashed and/or out-of-tolerance parts on the (+) side. If the fill rate is too slow (curve c), just the opposite happens; cavity pressures indicate underpacked parts, resulting in poor surface finish, voids, and/or dimensional problems. Group 1 variables (injection pressures, melt temperatures, melt viscosity, and fill rate) are clearly interrelated and have dramatic effects on part characteristics, as evidenced by the cavity pressure variations of Fig. 11-10.

Group 2—Boost Time.

Typical non-process-control machines have a boost timer to terminate the fill and pack cycle. Even with good fill rate repeatability, variations in peak cavity pressures can result from variations in the time the ram is in the boost mode (see Fig. 11-11). These variations typically result from valve and solenoid response times from one cycle to the next, as well as long-term drifts of these components. Cavity pressure variations that occur when coming out of boost have the same effect on parts as the Group 1 variables. The problem is addressed separately here because its solution is different from the solution for Group 1 variables.

Group 3—Pack and Hold Pressures.

Typical non-process-control machines use the same ram pressure setting during the packing of the mold as was used during the filling of the mold. The level of the pressure setting is that which gives good mold fill-out without flashing the mold. Variations in this pack pressure result in cavity pressure profile variation (see Fig. 11-12). These cavity pressure variations indicate an inconsistancy that can be causing dimensional and surface finish problems. These pressure variations are a result of relief valve repeatability problems caused by valve wear and temperature conditions as well as shot-to-shot variations. In addition, the final pack pressure setting may be limiting you to a less than time-optimal part fill ability.

After the part has been packed, the boost

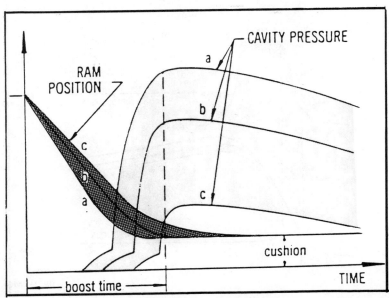

Fig. 11-10 Cavity pressure variations resulting from different melt viscosities and different fill rates.

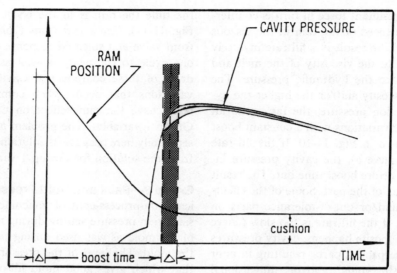

Fig. 11-11   Cavity pressure variations resulting from boost time variations.

timer reduces the applied hydraulic pressure to a hold pressure while the part cools. At this point the cavity pressure sensor starts to lose accurate plastic pressure readings because the part surface is beginning to harden. Further deductions made from this signal would be inaccurate. Experience has shown, however, that in switching from pack to hold, a

pressure dip (see Fig. 11-12) can cause sinks in the part surface.

Group 4—Recovery or Plastication. The variables that are involved during recovery do not appear on the cavity pressure profile until the next fill cycle. Recovery has much to do with the viscosity of the melt (see under

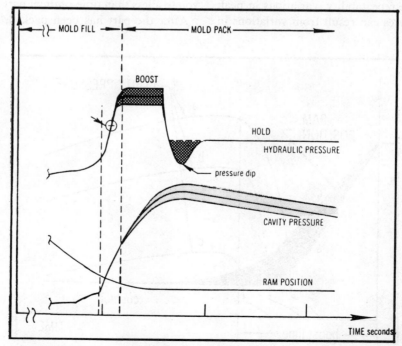

Fig. 11-12   Cavity pressure variations resulting from boost pressure variations.

Group 1 variables). Recovery variables can be identified, however. These variables have to do with how much energy is added to the plastic material; this energy and the resulting viscosity will vary.

The three main variables in descending order of importance are:

- Screw torque times speed product
- Back pressure times rate of ram withdrawal product
- Barrel temperature

Efforts to control these variables typically have to do with flow and/or relief valves, which have their own short- and long-term problems.

Why Have Process Control?

Simply stated, there are three reasons for a process controller:

- To select a group of controlled parameters that will gain control over the process variables.
- To improve the parameter repeatability.
- To improve the parameter setability.

Control of Which Parameters Can Best Eliminate Variability?

Fill Cycle. To eliminate variations of mold fill resulting from all the Group 1 variables, a control scheme that modulates the hydraulic pressure as viscosity variations and mold reaction forces occur would be desirable. This is known as fill velocity control. Velocity is the independent parameter, and hydraulic pressure is the dependent variable. Figure 11-13 shows how the hydraulic pressure may have to vary to keep the independent velocity (rate of ram position change) at the commanded level. The result is a more consistent filling of the mold and a more consistent cavity pressure profile. One added benefit is that injection pressures may exceed pack pressures if necessary to achieve a desired fill rate.

Fill-to-Pack. Elimination of boost time variation is possible by simply removing the boost timer. However, something must replace it that is sensitive to the occurrence of the operation following fill. At this portion of the cycle the mold will be essentially filled, and any further filling will result in extensive compression of the plastic melt. Plastic compression is necessary for good part qualities, and the

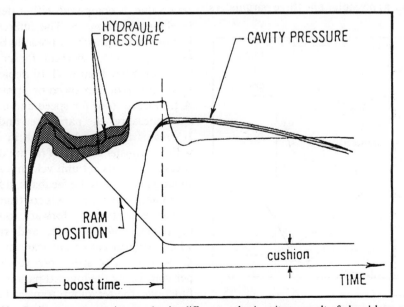

Fig. 11-13   Cavity pressure consistency despite different melt viscosity, a result of closed loop control.

extent of compression must be properly controlled. When this event is close at hand, a
dramatic rise in hydraulic and cavity pressure
is experienced, as seen in Fig. 11-14. Sensing
the dramatic rise in hydraulic pressure will
place the end of fill at its proper time without
the use of a timer. Connecting the detection
of this event to a specific region (see Fig. 11-
14) allows higher injection pressures during
fill if they are needed.

Pack and Hold. In the case of pack and
hold, the proper parameter has already been
selected—pressure. However, the methods of
pressure control can be improved. The level
of pressure in pack or hold and the dynamic
performance are important, and ways to improve them are discussed later.

Plastication. Proper melt viscosity is the
desired end for the plastication phase of a machine cycle. There is not yet a good way to
tell if the plastication phase has done its job
properly until the next part is made. This
might be fertile ground for the development
of a transducer to measure viscosity at the
tip of the screw as feedback to an algorithm
for control of screw speed and back pressure.
In the absence of such a device, attempts are
made to keep the energy added to plastication
as repeatable as possible. The three parameters

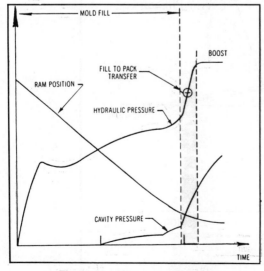

Fig. 11-14   Fill-to-pack transfer.

that are controllable in the addition of energy
to the melt are screw speed, back pressure,
and barrel temperature. Speed and pressure
control on standard machines imply flow and
pressure valves, and each of these devices
bring with them short- and long-term variations. Ways to improve flow and pressure control are discussed later. Temperature, the third
controllable parameter, appears to have sufficient control with the state-of-the-art devices.

In summary, most variables can be eliminated through the use of two parameters, velocity and pressure. The more repeatable, the
more dynamically controllable these parameters are made to be, the better the ability an
injection molding machine will have to mold
a part. Figure 11-15 seems to demonstrate the
repeatability brought to making a part with
improved parameter control. This figure
shows the difference in cavity pressure repeatability with open loop and closed loop machine control.

What Enables Parameter Controllability?

Closed loop servo control is the best known
way to control a parameter. Closed loop theory says that a parameter is measured with
a sufficiently accurate transducer. The signal
from the transducer, representing the parameter's value, is compared with a desired signal
level for the parameter. The difference or error
is amplified as much as possible before being
sent to a control element for correction of
the parameter. Figure 11-16 depicts a closed
loop control of ram speed or pressure (force).
A transducer (one for speed and one for pressure) measures the parameter under control.
It creates a feedback voltage in accordance
with its transfer function (H) in volts per unit
pressure or volts per unit velocity. A summing
junction compares the feedback voltage to one
commanded by the process controller. The difference is sent to the forward loop elements
(amplifier, control valve, and ram piston)
whose lumped parameter transfer function is
G, with units of speed per volt or pressure
per volt. Using the lumped parameter transfer
functions, the servoloop transfer function can
be expressed mathematically as equation (11-

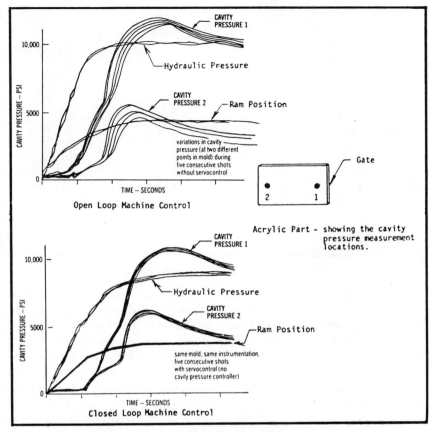

Fig. 11-15   Repeatability data (five shots).

1). Using differential calculus, it can be shown that as G gets very large, the servoloop transfer function becomes that of the transducer:

$$\frac{\text{OUTPUT}}{\text{COMMAND}} = \frac{G}{1 + GH}\bigg|_{\lim G \to \infty} = \frac{1}{H} \quad (11\text{-}1)$$

This is an important concept because it eliminates all the anomalies of the forward loop elements such as drift and nonlinearity. The controlled parameter's value is now a function of the transducer used, not the controlling elements of the forward loop. This, then, is what

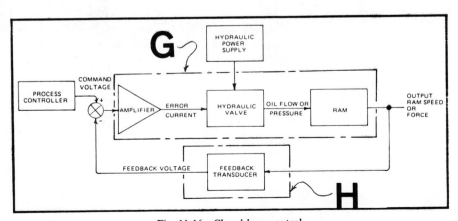

Fig. 11-16   Closed loop control.

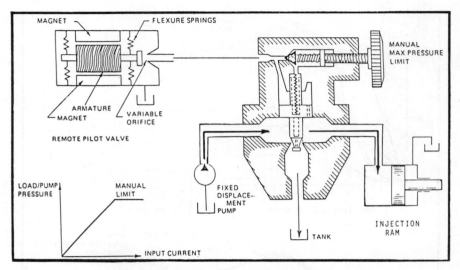

Fig. 11-17   Remote piloted pressure relief valve.

gives closed loop control the ability to provide better control over the parameters of velocity and pressure as well as others not used here.

The ability to make *G* large depends upon three factors:

- Load natural resonant frequency
- Valve response and load flow characteristics
- The type of frequency compensation used in the amplifier.

Load Resonance.   Load natural resonant frequency is a physical phenomenon resulting from the ram piston and screw mass interacting with a hydraulic oil spring. The desirable condition is one of high load resonance. Little can be done with the mass involved to improve the resonant frequency, but the oil spring can be influenced. The oil spring is made up from all oil volume directly influencing the ram piston. Figures 11-17 and 11-18 are hydraulic schematic representations that use piloted relief and flow divider valves. When ram piston pressure is to be raised or lowered for control purposes, all the oil volume from pump to piston must be pressurized or depressurized. This larger volume of oil ends up causing a

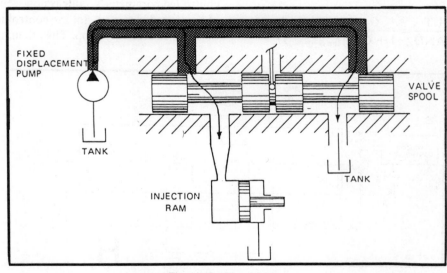

Fig. 11-18   Flow divider.

less than maximum possible load resonance. The closed center servovalve, depicted schematically in Fig. 11-19, provides the higher load resonance because the only oil to see load pressure variation is that between the valve and the piston. The best results are obtained when the valve is mounted closest to the ram piston. The higher the load natural resonant frequency, the better the dynamic performance exhibited by the servo. The key to which method is used is the type of performance that is needed from the machine.

Control Valve Response and Load Flow. The ability of a control valve to react to an input from the forward loop amplifier directly affects the amount of forward loop gain (G) that can be added to a servoloop. Since there are various manufacturers of valves that exhibit different performances, the valve selected should have sufficient performance for what is expected of the servoloop. Figure 11-20 shows the response capability for three different valve manufacturers.

The type of load flow a control valve exhibits can also limit the amount of forward loop gain (G) that can be achieved in a given servoloop. As the valve is required to deliver flow to the ram, its flow gain (change of load flow for a change of input command) will vary as the load pressure varies. This characteristic is shown in Fig. 11-21 for a closed center servovalve. As the load pressure increased from 0 to 1500 psi, the valve drop went from 2000 psi to 500 psi. This 4:1 drop in available valve pressure was accompanied by a 2:1 drop in valve flow (see dotted line of Fig. 11-21). It is a square root relationship. As the load pressure increases and the valve drop deceases, the valve flow gain drops as the square root of the change in valve drop. This reduces the valve's contribution toward forward loop gain (G) but is *not* a destabilizing effect. In the case of the flow divider whose characteristics are depicted in Fig. 11-22, the opposite is true. As the ram load pressures increase, the slope of the flow vs. input curves increases. This apparent flow gain increase has a destabilizing effect on the servoloop and limits the forward loop gain (G) at lower loads, since the maximum loop gain must be set for stable operation at higher loads.

The added performance a closed center servovalve can provide for a servoloop suggests that it should be a favored device. With today's technology, there is little cost difference between the closed center servovalve and the flow divider.

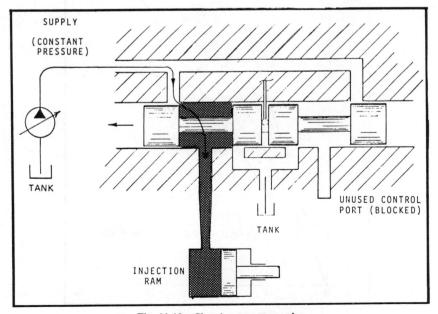

Fig. 11-19   Closed center servovalve.

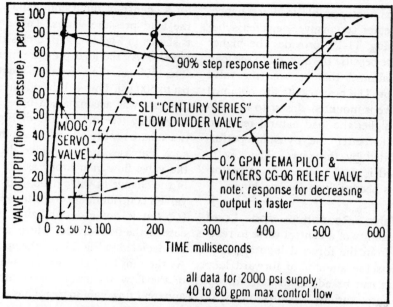

Fig. 11-20   Response capability of three valves.

Amplifier. The servoloop amplifier is where the remaining amount of forward loop gain (*G*) is added, for it is easiest to adjust. It is also good practice to put only as much gain as necessary into other elements of the forward loop and add the remainder to the amplifier, for it has the fewest problems with drift, resolution, repeatability, response, and adjustabil-

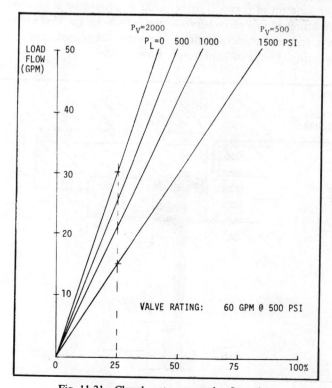

Fig. 11-21   Closed center servovalve flow plot.

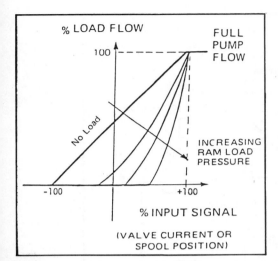

Fig. 11-22 Flow divider valve load flow vs. input signal.

ity. It is also the best spot to add frequency compensation such as integration. Integration adds high static gain to the velocity loop. Adding this feature to the amplifier will not add the problem of integration deadband, as does happen when this feature is added to the control valve. Other frequency compensation is added at the amplifier to improve the performance of the pressure servoloop also.

Figure 11-23 shows the two servoloop configurations for velocity and pressure control. Each of the elements of the forward loop is shown, and the transducer associated with each loop as well. Changing each servoloop configuration to the other is done by electrical switching of setpoint and feedback sources.

Where Does the Process Controller Go?

Now that control over the variables has been accomplished through the proper selection of control parameters and the use of closed loop servocontrol, can a process controller be of any value? By itself it is just a toy, but combined with the technology just mentioned it can be of great value. Figure 11-24 shows where the process controller fits in the overall machine schematic. The process controller has the servoloop electronics in it to combine with the transducers and servovalve of the machine. The process controller is now set to take input

setpoints and do a meaningful job of controlling the variables.

What Are the Basic Features a Process Controller Should Have?

There are four portions of an injection molding machine cycle that have to do with the injection operations:

- Fill
- Fill-to-pack transfer
- Pack and hold
- Plastication

There are many levels of sophistication each of these areas can have in numbers of setpoints, operator presentation interface, and additional controllability such as cavity pressure pack cutoff or adaptive shot size control. The following are essential.

Fill Control. A fill control as depicted in Fig. 11-25 should break the shot up into several segments. The speed of injection for each segment should be easily and repeatably setable. With this feature, the best speed for each area of the mold will be setable regardless of how fast or slow the injection levels are elsewhere. This feature is valuable, since it allows the mold to be filled as quickly as possible and still eliminates problems of burning, splay, flow lines, and voids. Filling quickly makes it possible to have hotter material in the mold, which helps surface finish, weld lines, and dimension control later in the pack segment. Also helpful would be a group of indicators to show where in the inject phase the programmer is, as well as a group of indicators to indicate that a given segment is not meeting programmed speed.

Transition Control. When the mold is essentially filled, it is desirable to switch to a packing control because any further attempt to fill the mold with the fill control could result in excessive cavity pressures. The transition control depicted in Fig. 11-26 should provide a setable position in the shot stroke where no high injection pressures are expected. This

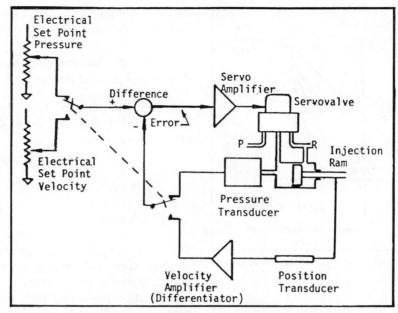

Fig. 11-23 Loop control diagram.

point serves to arm the pressure sensing system so that it may detect a rapid rise in ram pressure, indicating that the mold is essentially filled. This feature will allow higher injection pressures earlier in the fill cycle where needed and provide a sufficient indication that it is time to switch the servoloop configuration to a packing operation. Should the transition not occur as a result of setup conditions or a stuck cavity in the mold, an alternate means of transition should prevail such as time, to complete the injection cycle. This condition should be indicated as an incomplete stroke, and this logic condition should be output to the machine sequence controller to initiate appropriate actions.

Pack and Hold Control. Once transition has occurred from fill to the pack portion of the cycle, it is desirable to be able to set the packing pressure to fill out the part completely for surface features and density. The amount of time spent in packing should also be adjustable. Figure 11-27 depicts these features. After

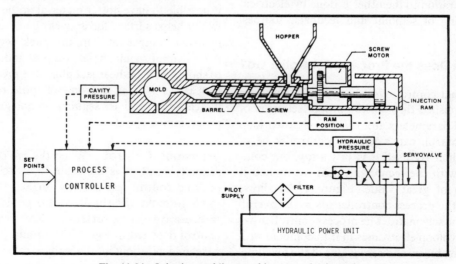

Fig. 11-24 Injection molding machine control schematic.

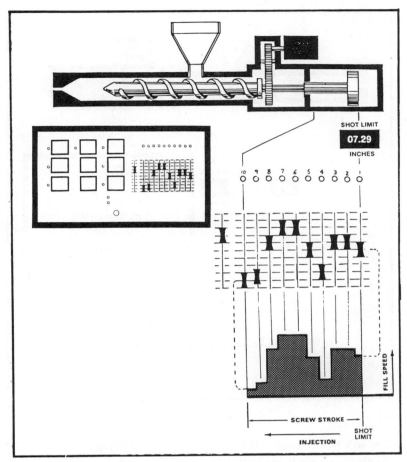

Fig. 11-25 Fill control.

pack time (T1) it will be desirable to switch to a hold pressure independent of the pack pressure. The rate of transition from pack to hold pressure should also be adjustable. This feature will help control warpage that can be caused by abrupt changes in part densities resulting from abrupt pack to hold pressure changes during cooling. The time the hold pressure is applied should be setable independent of pack time.

Plastication Control. As mentioned earlier, there are three sources of energy input to the plastication process. Temperature is already well controlled. Of the two remaining, it would be very convenient to control back pressure because there already exists a control loop for that purpose. The pack and hold pressure control can be used to control back pressure as well. Simple electronic switching can

implement this feature. Figure 11-28 depicts what would be desirable for this portion of the injection cycle. The amount of back pressure should be setable and should apply for the complete plastication phase. The phase is terminated when the proper shot limit is reached. The shot limit will control the shot volume plus the desired amount of cushion. A decompression feature is also desirable and should have a fine range of setability. In most cases, settings will be between 0.05 in. and 0.20 in. Excessive decompression can cause streaks and splay as air is introduced into the barrel and injected into the mold.

Applications

To better demonstrate the four basic features of a process controller, examples citing where each was instrumental in solving a specific

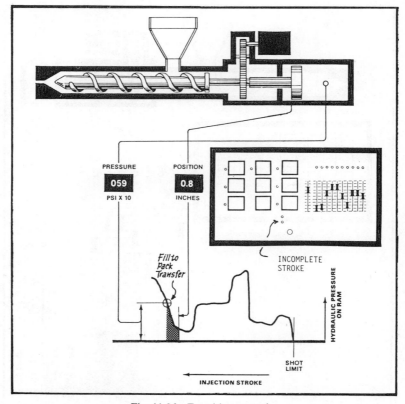

Fig. 11-26   Transition control.

molding problem will be presented, although any given application can most likely utilize more than one of these features to optimize the process. In all cases the control had the ability to affect a particular portion of the cycle, fine-tune it, and hold the setpoints accurately and repeatably as indicated on the operator's panel.

Fill Control/Speed Profiling. An optical part, in this case an edge-gated acrylic collimator-lens having dissimilar convex faces, was required to maintain exacting surface curvature and focal point tolerance. Considerable effort had been expended to ensure that the molds were accurate, and indeed acceptable parts were being made, but inconsistently. With a control, it was found that the best fill was a very slow, steady injection of 0.013 in./sec. As little as 0.0054 in./sec variance could be seen to throw the focal length off specification. The success of the application

was demonstrated by molding 100 pieces with a single reject. It would have been totally impossible to achieve such consistency without closed loop control of injection velocity.

A second example of fill speed control is an all-plastic valve for use in a chemical plant. The one-piece body was molded around the ABS ball and stem with Teflon seals being supported by spin-out cores. A degradation in the surface of the ball, opposite the gate, due to impingement of the melt, made a low rate of fill desirable. However, a weld line resulted on the other side of the cores, from the gate, if injection was not kept fast enough to prevent a cold melt front. The solution was to begin injection quite slowly and then accelerate quickly to complete the part. This allowed a skin to form, insulating the ball from the heat of the following material. The weld line could now be prevented by maintaining a high fill rate throughout the remainder of the shot. Leakage from an imperfect ball sur-

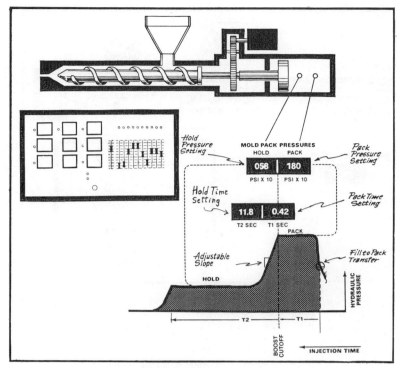

Fig. 11-27   Pack and hard control.

face was all but eliminated, and burst pressure was increased more than threefold by improving the weld of the material around the cores.

Fill-to-Pack Transfer.   An example of a part requiring precise fill-to-pack transfer is a carburetor body for a two-cycle engine. Molded in a mineral-filled nylon, it required 100 percent inspection due to flashover of two small holes in the throat area, among other problems. A characteristic of the material was that it required a great deal of pressure to fill the cavity, but once it was filled, the nylon easily flashed over the pins, kissing off against the throat core. It was essential here that the control switch out of the speed loop, to stop the fill without overpacking at the instant the transfer pressure was reached. In this case, accurate transfer allowed a reduction in Q.C. inspections to once per shift, and a reduction in clamp tonnage was required because it was no longer needed to hold against flashing.

A dramatic demonstration of transfer control is found on machines running materials

such as a polyamide–imide requiring a very high speed of fill, possibly boosted by a hydraulic accumulator. While capable of impressive fill rates, they are subject to a wide variance in shot size because of their inability to stop the fill at a repeatable point. Our experience has been on machines injecting at over 40 in./sec with hydraulic flows of more than 500 gpm; and while this is a rather extreme example of the fill-to-pack transfer feature, the consistency attained here can be beneficial to any molder trying to hold a tighter control over shots or flash.

Pack and Hold Control.   Process controls permit added flexibility over the standard molding machine in that the molder can sense the actual point at which the cavity is filled. This now becomes the most important event in the cycle, since it signals the separation between filling of the cavity and densification of the part. On the standard machine control, the boost pressure is used to fill the cavity, but only after the cavity has completely filled

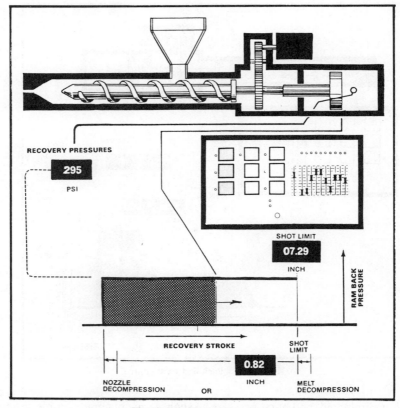

Fig. 11-28 Plastication control.

is this pressure actually felt by the material. As the fill time varies, so does the length of time in which boost pressure is applied to densify the melt in the cavity. This is a chief source of part weight inconsistency. Process controls are capable of holding pack and hold pressures to within a few psi for a very specific length of time.

Returning to the example of the carburetor body, part weight was held consistent despite changes in hydraulic fluid temperature, environment, material lot, etc. Further, the parts were shipped by weight, and it was found that 12 percent more parts were being shipped in a given container weight because dimensions could be held without overpacking.

Another example of a common problem that is easily resolved by process control is ram bounce. Whereas it may be desirable to go from a rather high boost pressure down to a much lower hold pressure, the resulting decompression on the ram causes material to be withdrawn throughout the gate, putting stress into the part and possibly affecting adequate pack-out. Any thin flat part susceptible to warpage can benefit from a control's ability to change pressure gradually.

Plastication. As noted earlier, a process control should be able to maintain hydraulic pressure to within a few psi of setpoint. However, it can allow far more control of the energy being put into the melt. Since many molds provide a separate path back to the tank that bypasses the machine's manifolds, a process control can allow lower back pressures than found in the standard machine. This can be of use with highly filled materials.

Although once they may have been oversold as a panacea for all the molder's problems, process controls have come of age through maturity of technology and application, and the understanding by a growing segment of the industry that many of the so-called process controls are more placebo than performance.

The latest generation of controls offers the molder performance and repeatability at a price he can afford, and are simple enough to be understood and utilized.

PROCESS CONTROL IS NOT A FOOLPROOF SOLUTION

Shorter cycle times and thinner or more complex parts continually increase the need for precise control in injection molding. At these higher production rates, excessive scrap and rejects become less desirable than ever, and the molder finds himself trying to reduce these levels (1, 23, 24, 25).

Molding optimization is further complicated by highly automated operations that move the products directly from the molding machine to the assembly stations. Effective process control, therefore, is essential to maintain the benefits of modern process technology.

Purchasing a more sophisticated process control system is not a foolproof solution to molding-quality problems, though. To solve part-reject problems requires a full understanding of the real causes, which may not be as obvious as first appear. Failure to identify the contributing factors may send the molder on a time-consuming quest for "the perfect part."

The conventional place to start troubleshooting a problem is the melt temperature and pressure. But often, the problem is a lot more subtle; it may involve mold design, faulty control devices, and other machine components. Sometimes, factors not directly related to the process may be influencing quality, such as an operator making random adjustments of control devices. Process control systems usually cannot compensate for such extreneous conditions.

Control

All processes are under some degree of control. All molding machines are equipped with a variety of controls. However, most of these controls are of an open-loop type. They merely set a mechanical or electrical device to some operating temperature, pressure, time, or travel. They will continue to operate at their setpoints even though the settings are no longer suitable for making quality parts. The problem is that the total process is subject to a variety of hard-to-observe disturbances which are not compensated for by open-loop controls.

"Process control" closes the loop between some process parameter and an appropriate machine control device to eliminate the effect of process disturbances.

There are several levels of process control sophistication; each uses different control parameters. One level employs cavity-pressure measurement, which is the single most useful control parameter for injection molding.

The most efficient application of a process control strategy requires an understanding of the various aspects of process variation and their relationship to product quality.

There are two basic approaches to solving a molding quality problem: (1) correct the basic problem, or (2) overpower it with an appropriate process control strategy.

The approach that is selected depends upon the nature of the processing problem and whether time and money are available to correct the problem. Process control may, in some cases, provide the most economical solution. To make the decision one must systematically measure the magnitude of these normal process disturbances, relating them to product quality and identifying the cause whenever possible.

Before investing in a more expensive system, the molder must methodically determine the exact nature of the problem in order to decide whether or not a "better" control system will solve it.

Tie-Bar Growth

An example of a problem that most controls do not consider involves the effect of heat on tie bars, which can directly influence mold for performance. The following information provides the calculations for tie-bar elongation and mold thermal growth.

Tie-Bar Elongation. The change in tie-bar length, e, can be calculated as follows:

$$e = \frac{F \times L}{E \times A}$$

where F = force per tie bar, L = bar length, E = modulus of elasticity, and A = cross-sectional area of bar.

At maximum die height (178 in.) on a 500-ton injection-molding machine with a tie-bar diameter of 6 in. (or a cross-sectional area of 28.27 sq. in.), tie-bar elongation equals:

$$e_{max} = \frac{250,000 \text{ pounds} \times 178 \text{ in.}}{30 \times 10^6 \text{ psi} \times 28.27 \text{ in.}^2} = 0.0524 \text{ in.}$$

At minimum die height (146 in.), the elongation is:

$$e_{min} = \frac{250,000 \text{ pounds} \times 146 \text{ in.}}{30 \times 10^6 \text{ psi} \times 28.27 \text{ in.}^2} = 0.0430 \text{ in.}$$

To calculate the effect of a small change in elongation on the force on a tie bar, we solve for F:

$$F = \frac{eEA}{L}$$

At maximum die height, the change in force per 0.001-in. elongation equals:

$$F_{max} = \frac{0.001 \text{ in.} \times 30 \times 10^6 \text{ psi} \times 28.27 \text{ in.}}{178 \text{ in.}}$$
$$= 4764 \text{ pounds}$$

At minimum die height, the change in force for the same elongation is:

$$F_{max} = \frac{0.001 \text{ in.} \times 30 \times 10^6 \text{ psi} \times 28.27 \text{ in.}}{146 \text{ in.}}$$
$$= 5808 \text{ pounds}$$

Thermal Mold Growth. Uneven mold growth can occur with a temperature differential across the mold. Mold growth, G, is calculated by the following formula:

G = Mold length
   $\times$ Coefficient of linear expansion
   $\times$ Mold temperature

In a 20-in.-long mold, where the temperatures are 100 and 120°F, mold growth equals:

G_{100} = 20 in. $\times$ 6 $\times$ 10^{-6} in. per in. per degree
   $\times$ 100°F = 0.0120 in.

G_{120} = 20 in. $\times$ 6 $\times$ 10^{-6} in. per in. per degree
   $\times$ 120°F = 0.0144 in.

The difference in growth on different sides of the mold, then, is:

$$G_{120} - G_{100} = 0.0144 - 0.0120 = 0.0024 \text{ in.}$$

PROCESS CONTROL TECHNIQUES

This section will review a control system (the Barber-Colman system) in which selected critical molding variables are measured and controlled to maintain part consistency. This approach to injection molding process control involves the measurement and control of two critical molding parameters; ram position and mold (or cavity) pressure, during the mold filling and packing phases of the injection cycle (Fig. 11-29). It is during the filling and packing phases that most variations in molding conditions make themselves evident and therefore can be easily detected. For example, a change in material viscosity is reflected as a change in ram speed and can be detected by measuring ram position with respect to time. A change in material viscosity also reflects itself as a change in plastic pressure and can be detected by measuring mold or cavity pressure with respect to time. Other variations in molding conditions, such as hydraulic pressure, oil temperature, melt temperature, etc., display themselves similarly and can be detected by monitoring ram position and plastic pressure with respect to time. Since it is possible to detect variations in molding conditions, it is also feasible to compensate or correct for these variations.

Primary injection pressure can be divided

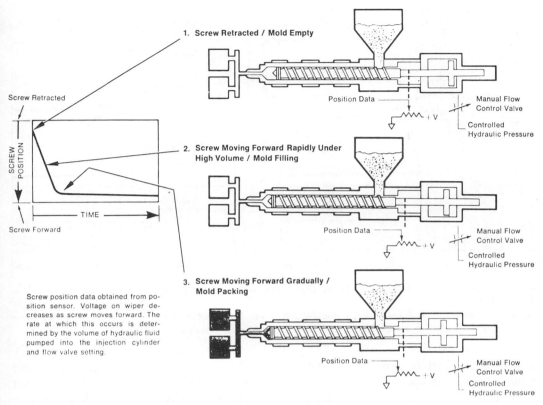

1. **Screw Retracted / Mold Empty**

Screw Retracted

SCREW POSITION

TIME

Screw Forward

Position Data

Manual Flow
Control Valve

Controlled
Hydraulic Pressure

2. **Screw Moving Forward Rapidly Under
   High Volume / Mold Filling**

Position Data

Manual Flow
Control Valve

Controlled
Hydraulic Pressure

3. **Screw Moving Forward Gradually /
   Mold Packing**

Screw position data obtained from position sensor. Voltage on wiper decreases as screw moves forward. The rate at which this occurs is determined by the volume of hydraulic fluid pumped into the injection cylinder and flow valve setting.

Position Data

Manual Flow
Control Valve

Controlled
Hydraulic Pressure

Fig. 11-29   Screw travel during rotation.

into two major phases: mold filling and mold packing. As shown in Fig. 11-30, screw movement occurs primarily during the mold filling phase while mild pressure buildup takes place in the mold packing phase. The association of screw movement with mold filling and mold pressure with mold packing is important and should be remembered.

Attempting to control the molding process using ram position only or mold pressure only as the measured variable and adjusting primary injection pressure as the control function is not satisfactory because both mold filling and mold packing take place with the same injection pressure value. Changing the primary injection pressure affects both phases of the injection cycle. For example, assume that a new batch of material with a lower melt index has been put into the feed hopper. Since the melt index is lower (apparent viscosity is higher), ram screw speed under a given primary injection pressure will decrease. This means that the mold will fill more slowly, and

mold pressure buildup will occur later. Depending on the change in the material, the mold pressure will also decrease. The effect of the material change on ram screw speed and mold pressure is shown in Fig. 11-30.

To compensate for the material change and get the average injection rate back to normal, the primary injection pressure could be increased. Increasing primary injection pressure will increase the average injection rate but at the same time lowers the apparent viscosity by the effect of shear stress/shear rate applied to the material. Dynamically, this apparent lower viscosity carries over to the packing phase to a point that would result in increased mold pressure, overpacking, and a heavier part with stresses. The effects of the increased injection pressure are shown in Fig. 11-31.

The ram curve and thus the average injection rate have returned to their original value, but mold pressure due to the change in apparent material viscosity peaks out at a higher value, resulting in an overpacked, more

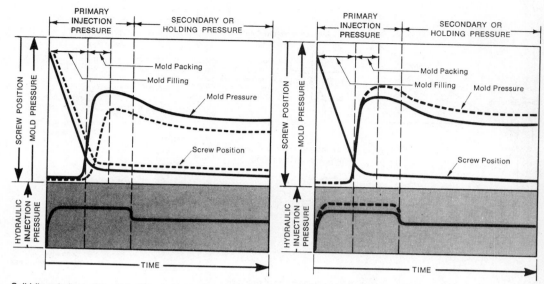

Solid lines indicate signals before material change. Dashed lines indicate signals after material change (Fig. 11-30) or after correction (Fig. 11-31).

Fig. 11-30 Effect of melt index on screw position and mold pressure. Courtesy of Barber-Colman.

Fig. 11-31 Effect of injection pressure on screw position and mold pressure.

densely molded part. In all probability the part would also stick in the mold.

Thus it can be seen that control of the average injection rate cannot properly compensate for a change in material viscosity and produce an acceptable molded part. Without going into detail, it can also be shown that a control system based solely on the measurement and control of peak mold pressure will not successfully do the job either. In the case of a material with a lower melt index, peak mold pressure control will produce a part that is lighter and less densely packed—in other words, the opposite of that achieved with only ram position control.

The control philosophy of the Barber-Colman Molding Process Controller is based on independent measurement of screw ram travel and mold pressure during injection. Deviation of either of these variables, when compared to preset adjustment limits, initiates control action to bring the out-of-tolerance variable back within limits. This is accomplished in such a way that correction of one variable has minimal influence on the other variable.

To provide independent ram position and mold pressure control and eliminate interaction between the two, the mold filling and

mold packing phases of the injection cycle must be isolated. In other words, you must be able to control the average injection rate during mold filling without creating a change in mold packing pressure. Conversely, you must be able to control mold packing pressure without affecting the average injection rate.

The first step in implementing control is to identify that point-in-time when the transition from mold filling to mold packing takes place. The second step is to identify that point-in-time when proper mold packing pressure is reached. These points are readily identified, as shown on the ram position and mold pressure curves in Fig. 11-32. The identification of these points establishes were the control limits should be implemented. Each point of control has a high and low limit, thus providing an operational bandwidth that allows for minor variations and also provides directional control for increasing or decreasing control pressures as required.

To isolate the mold filling and mold packing phases and prevent interaction, you must be able to control the primary injection pressure value for each phase. To accomplish this, a built-in timer which can be set to transfer primary injection pressure to a new setpoint

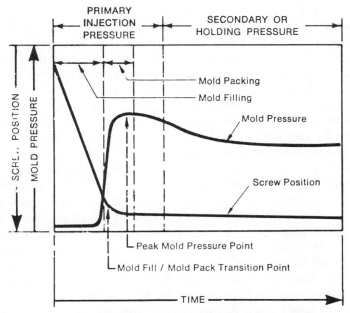

Fig. 11-32   Fill/pack transition and peak mold pressure points.

value is provided. In this way control action can be applied to the mold filling setpoint value independently of the mold packing setpoint value, or vice versa.

Fig. 11-33 again shows the screw position and mold pressure curves. This time, however, the figure shows that primary injection pressure time is split into filling pressure time and packing pressure time segments.

Control of each segment is achieved via

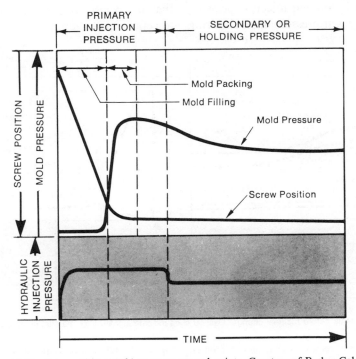

Fig. 11-33   Ram position/mold pressure control points. Courtesy of Barber-Colman.

point-in-time sampling of both screw position and mold pressure and comparing the value of the sampled signals to adjustable high and low limits. Violation, by either variable, of the respective limits will cause the controller to generate a corrective signal that will add to or subtract from the corresponding setpoint, thereby incrementing a change in injection pressure. The pressure change, however, is applied only to that time segment where the violation occurred. In some cases, violations will take place on both variables, and a correction will be generated for each segment.

For example, assume that the same conditions exist as used previously where a new batch of material with a lower melt index is put into the hopper. Since the screw moves more slowly, it violates its upper limit. Mold pressure, because it doesn't reach the same peak value, will violate its lower limit. Both conditions call for an increase in injection pressure. The effect on the ram position and mold pressure curves is shown in Fig. 11-34.

The amount of correction applied to each pressure setpoint is independently adjustable, thus allowing mold filling and mold packing to take place with different injection pressure values. The increased pressure applied during

fill pressure time will cause the ram screw speed to approach the original value. Just how fast it approaches the original screw speed depends on two factors: the amount of deviation and the amount of correction applied by the controller.

The mold pressure low limit violation called for an increase in injection pressure during packing pressure time. Here again the rate at which it returns to the original value is dependent on the amount of deviation and the amount of correction applied by the controller. The effects of applying independent corrective signals are shown in Fig. 11-35.

Even though we have segmented the mold filling and mold packing phases by using separate pressure setpoints and a timer, we cannot totally remove the interactive effects caused by the machine dynamics. The response of the hydraulic system and the machine mechanical parts must be considered, particularly in setting the filling pressure/packing pressure transfer point. This is normally set to transfer just prior to the ram position sample point to accommodate the dynamics of the hydraulic system.

Pressure correction should be applied such that the respective process variable (average injection rate or mold pressure) returns to its

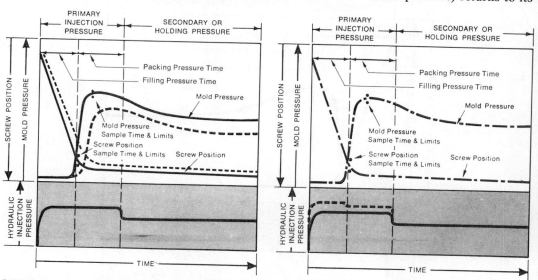

Solid lines indicate signals before material change. Dashed lines indicate signals after material change (Fig. 11-34) or after correction (Fig. 11-35).

Fig. 11-34 Effect of increased viscosity on ram position and mold pressure. Courtesy of Barber-Colman.

Fig. 11-35 Effect of pressure correction applied to ram position and mold pressure.

original value in small incremental steps rather than in one abrupt rather large step. Approaching control in adjustable incremental steps allows time for the machine dynamics to respond and eliminates the possibility of cycling above and below the control points. Independent control of mold filling and mold packing allows each segment to be brought back in control independent of the other in the least amount of time without interaction.

MOLDING DIAGRAM TECHNIQUES

By plotting injection pressure (ram pressure) vs. mold temperature, a molding area diagram (MAD) is developed that shows the best combinations of pressure and temperature that will produce quality parts. The size of the diagram (Fig. 11-36) shows the molder's latitude in producing good parts. To mold parts at the lowest cycle time, the molding machine would be set at the lowest temperature and highest pressure location on this two-dimensional diagram (MAD). If due to machine and plastic variables rejects develop, then move the machine controls so that higher heat and/or lower pressures are set to produce only quality

parts. This is a simplified approach to producing quality parts, since only two variables are being controlled. (This MAD approach example uses a thermoplastic; with a thermoset, to reduce cycle time the highest temperature and pressure would be used, etc.)

The next step in the molding area technique is to use a three-dimensional diagram (Fig. 11-37). By plotting melt temperature vs. injection pressures vs. mold temperature, one obtains a molding volume diagram (MVD), providing more precision control in setting the machine.

Developing the actual data involves slowly increasing the ram (injection) pressure until a value is obtained where the mold is just filled out. This is referred to as the minimum fill pressure for that combination of material, mold temperature, and melt temperature. Ram pressure is then increased until the mold flashes. This is logged as the maximum flash pressure. These two pressure values then represent a set of data points for one combination of melt and mold temperatures.

Next, melt temperature is changed (leaving mold temperature constant), and a new set of minimum and maximum pressures is deter-

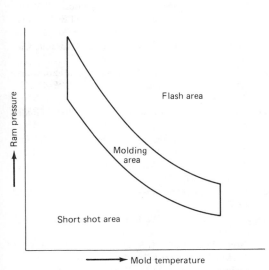

Fig. 11-36 Two-dimensional molding area diagram (MAD) that plots injection pressure (ram pressure) vs. mold temperature. Within this area all parts meet performance requirements. However, at the edge of the diagram rejects can develop because of machine and plastic variations.

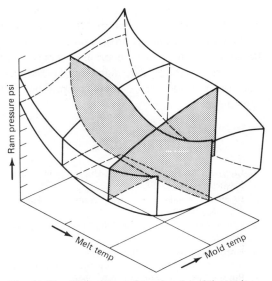

Fig. 11-37 After a three-dimensional molding volume diagram (MVD) is constructed, it is possible to analyze it to find the optimum combination of melt temperature, mold temperature, and injection pressure (ram pressure).

mined. This is continued until the maximum and minimum melt temperatures are found.

Then, mold temperature is changed, and all the above repeated until the maximum and minimum mold temperatures are found. Once the data are obtained, three-dimensional molding volume diagrams are constructed.

Molding volume diagrams show that melt temperature for injection molding plastic is an important variable that was not evident in two-dimensional molding area diagrams. MVD's are used with any thermosets or thermoplastics.

The significance of the MVD approach lies in the fact that one ends up with a dramatic and easily comprehended visual aid to analyzing three of the most important variables for injection molding, namely, injection pressure, mold (or barrel for thermoplastics) temperature, and melt temperature.

Using this two- and three-dimensional approach for making molding diagrams, you can analyze injection rate, cavity pressure, etc. Also consider whether to use manual or automatic process controls. As discussed in this chapter and Chapters 12 and 13, use of automatic controls makes it easier to set controls and ensure quality. Of course some molds produce quality parts just with manual controls; most of the 70,000 injection molding machines in the United States use only manual controls. But major changes are occurring because the automatic controls can significantly reduce cost and provide zero defects (or practically zero defects).

REFERENCES

1. Injection Molding Workbooks, University of Lowell Plastics Seminars, 1983.
2. Injection Molding Technical Conference Papers, SPE, October 20–23, 1983.
3. SPE-ANTEC Injection Molding Division Papers, SPE, May 1983.
4. Huebner, W. A., "On-Machine Monitoring of Injection Molding Machine Performance," Production Process Div., Londonderry, NH, 1983.
5. Schott, N. R., and Orroth, S. A., "Process Transducer Measurements in Injection Molding," University of Lowell Plastics Engineering Dept., 1983.
6. Cinco, F. R., "Comparing Today's Temperature Measurement and Controls with Tomorrow's Requirements," Package Machinery Co., East Long Meadow, MA, 1973.
7. Omega Temperature Measurement Handbook, Omega Engineering Company, Stanford, CT, 1978, p. B-16.
8. EID Staff, Electronic Instrument Design, p. 34 (May 1969).
9. Omega Temperature Measurement Handbook, p. L-2, Omega Engineering Company, Stanford, CT.
10. Schott, N. R., Berg, J., and Flynn, R. H. 32nd SPE ANTEC, 20, 133, San Francisco (May 1974).
11. Orroth, S. A. Jr., Schott, N. R., and Flynn, R. H., 32nd SPE ANTEC, 20, 56, San Francisco (May 1974).
12. Holman, J. P., Experimental Methods for Engineers, 2nd ed., McGraw-Hill, New York, 1971, p. 246.
13. Trans-Tek Instruction Bulletin LI32, 74 Eastern Blvd., Glastonbury, CT 06033.
14. OPTACH Bulletin No. 2–0, H H Controls Co., Inc., 16 Frost St., Arlington, MA 02174.
15. "How to Understand Plastics Temperature Control Instruments," West Instrument, Gulton Ind., RI, 1981.
16. Chostner, R., Barber-Colman Co., "Temperature Control of Barrel and Melt," SPE Technical Conference-IMD, October 20–22, 1983.
17. Oliversen, G. L., Barber-Colman Co., "PID Injection Pressure Control: Why Today's Machines Need Them," Plastics Technology (January 1983).
18. Buja, F. J., "Instrumentation to Monitor and Verify Control of the Molding Process," SPE Injection Molding Technical Conference, October 20–22, 1983.
19. Andres, P., "Process Control and the Hydraulic Circuit," Parker Hannifan, MI, 1983.
20. Johnson, M., "Practical Applications of Microprocessors to Injection Molding Machines," SPE-ANTEC (May 1983).
21. Adaptive Ram Programmer 315–7 Users Guide, Hunkar Lab's., 1981.
22. Schmidt, H., and Gleason, G., Moog Inc., NY, "The Whats and Whys of Injection Molding Process Control," SPE Injection Molding Technical Conference, October 20–22, 1983.
23. Injection Molding Division Newsletter, SPE-IMD 1982.
24. Ansorge, A., "Can We Learn Something from Die-Casting?," Plastics Engineering (June 1981).
25. Sercer, M. J., Catic, I. J., and Zoric, J. B., Univ. of Zagreb, Yugoslavia, "Trend Regulation of Injection Molding Process," SPE-ANTEC (May 1983).

Chapter 12
CAD/CAM for Plastic Part and Mold Design

John Gosztyla

Computervision Corporation
Bedford, Massachusetts

INTRODUCTION

The technology of plastic part and mold design has evolved over a period of years to become a multifaceted problem. Many of today's mold designs are the results of many years of trial and error techniques practiced by a relatively small number of artisans, craftsmen, and specialists involved in an ever growing plastics industry. This period of sustained growth of the plastics industry has been accompanied by an equally sustained growth period in the level of understanding of polymer materials and how they react to various changes in processing parameters. The net result has been the development of a technology base that can explain many of the phenomena that hitherto were considered the "art" of successful plastic design. The "rules of thumb" that governed designs of the past are now giving way to sophisticated analysis tools used on high-speed digital computers which are enabling major advances in the speed, productivity, accuracy, and quality of plastics designs.

In a parallel evolution to that of the plastics industry, CAD/CAM* has emerged to the point that it now shows the promise of being one of the most significant technology advances of the century. The same high-speed digital computers used for analysis and data

processing activities of the past are now generating graphical engineering information, and are helping to automate a domain once thought of as totally a creative discipline, the design process itself. The use of computers in manufacturing operations dates back to early work in the 1950s in which the dream was to control metal-cutting machine tools by computer. This was hoped to eliminate the requirement for many tooling aids such as tracer templates that helped the accuracy and repeatability of machining operations on the shop floor. During this period of development of machine tool control, the only types of computers available were extremely expensive "mainframe" computers whose cost required time sharing of this valuable resource. Programming was accomplished via a punched card medium and was tedious and time-consuming to develop and debug. Additionally, the only means to check cutter paths developed by the computer was to do a "prove out" run on the shop floor, and thus occupy another expensive piece of equipment with a tedious and time-consuming task.

The concept of using a graphic display device to visually display the programmed cutter path was proposed and developed during the 1960s, as one of the solutions to these problems. This important development was actually the predecessor to today's CAD/CAM systems. During this same period of display device and software developments for numeri-

---

*CAD/CAM = computer aided design/computer aided manufacturing.

cal control of machine tools, an important hardware development was also occurring, the development of the minicomputer. This newcomer to the computer field brought in a totally new price and performance spectrum which created a drastic increase in the acceptance of computers in general, and in the use of computers for scientific, engineering, and manufacturing functions in particular. While the 1970s brought a continued development in the hardware and software products available, they also brought about a change in the business climate toward companies that could supply a total systems approach to computer graphics problems. This spawned the "turnkey" CAD/CAM suppliers of today—companies that could supply both the computer hardware and user-friendly software, ready to run, in the customer's place of business.

The first predominant applications of CAD/CAM technology were in the area of two-dimensional printed circuit board (PCB) and integrated circuit (IC) design. Both of these applications were relatively easy to capitalize upon, as they can be described by geometries on planar surfaces, and generally use many repetitions of standard electrical components such as resistors, capacitors, etc. These two key elements made computer assistance to the design process a logical and natural step. CAD/CAM developments continued, eventually resulting in systems that could produce representations of three-dimensional objects. This single development implied a complete change of scope in the capabilities of CAD/CAM systems, moving them from two-dimensional drafting tools into the realm of producing a true spatial mathematical modeling tool.

The technology of CAD now implies a completely different methodology of engineering design. As such, there has been some resistance to the acceptance of CAD. A completely different method of representing information is hard to grasp and accept for many of the "old-timers" of the drafting-board era who are used to having information presented to them in a specific manner. Nevertheless, CAD/CAM is revolutionizing the speed and efficiency of the plastic design functions. The more the entire design function is studied, the more repetitive tasks are uncovered in that function. The computer's ability to perform these tasks untiringly and with blazing speed is the basis for these productivity gains.

At this point, many readers may disagree with the concept that something as complicated as a mold design is repetitive in nature. To illustrate the point, let us review some of the major elements of a plastic design.

1. Product Model: the first step of the process is creating a model of the product to be molded. Today this is provided via the product drawing. Plastic product design is similar to design in other materials. The product designer seeks to provide a solution to a product need at minimum cost and with the greatest quality and speed. A design is prepared and detailed, a model or prototype is built, and the design is revised until acceptable product function and cost are achieved. This process involves many repetitions of creation of drawings and models. Many times the products being designed incorporate features that exist in other products such as bosses, ribs, snap fits, etc., which incorporate standard design practices.
2. Shrink Corrected Product Model: a type of repetition of the original product model, this model is the "pattern" used to develop the cavity and core and allow for material shrinkage in the processing operation.
3. Incorporating Model into a Standard Mold Base: the concept of a "standard" mold base implies a repetitive element in standardization (and thus repetitive design), to promote economies of scale in manufacturing.
4. Installing Standard Mold Components: this is a repetitive operation where components (i.e., ejector pins) are used over and over again within the design and between designs.
5. Generation of Mold Drawings: this is repetition of all the information that may be contained in the mold layout.
6. Machining of the Mold Geometry: this is the repetition by a machine tool of all ge-

ometry previously described by the shrink corrected product model and the mold layout.

The basic point of the preceding illustration is that there is a great amount of repetitive information flow in the plastic design process. As a result, what is perceived to be an extremely creative process is actually very repetitive in nature.

The types of analytical problems that are encountered in the mold design process generally fall into the sciences of fluid mechanics and heat transfer theory. These fields encompass many complicated mathematical functions and relationships that were too time-consuming to evaluate in manual or conventional mold designs. The ability of the computer to remember and execute these computations quickly now adds a new dimension to the mold design process, allowing prospective design alternatives to be evaluated and simulated by computer rather than in the molding press. CAD/CAM is enabling the creative energies of plastic part and mold designers to be spent in producing better designs in shorter time periods rather than in doing repetitive mold design tasks.

The Benefits of CAD/CAM for Mold Design

Without going further into the details of how CAD/CAM software is being applied to mold design applications, it is appropriate to discuss what benefits can be derived from successful application of the technology. Those firms and individuals who have invested time and money in learning and implementing CAD/CAM systems have identified the following primary benefits of their use:

Productivity improvement
Quality enhancement
Turnaround time improvements
More effective utilization of scarce resources

Each of these primary benefits is further described in detail in the paragraphs that follow.

Productivity

Many types of productivity benefits have been documented and verified that result from CAD/CAM being effectively applied to plastic design tasks. The first benefit is an actual increase in the productivity of the mold design itself. CAD/CAM software provides the basic tools to provide productivity increases from a low ratio of 2 to 1, up to firms with specific applications demonstrating 10-to-1 increases or more. Much of the achievable benefit depends on the degree of commonality or standardization present in the type of molds or parts a particular firm designs. The CAD/CAM technique known as group technology can be of great benefit in the capture of similar part designs. This technique shows great promise as a tool to enhance the overall productivity of the design process. Other tools such as finite element modeling can help to reduce the amount of prototyping and testing required to successfully design plastic products. Other software capabilities such as creation of shaded images allow the aesthetic appeal of a product to be evaluated without requiring the construction of physical models or prototype molded parts.

A second productivity benefit is obtained in mold manufacture. CAD/CAM technology facilitates the use of numerically controlled (NC) machining technology in the fabrication of the molds themselves. As such, much more of the mold can be cut in single setups on a single machining center. In turn, this reduces the number and complexity of manual setup operations, and thus more time is spent "making chips." Additionally, the ability to build three-dimensional product models in the data base and to automatically generate tool paths from these models reduces the amount of effort spent in defining section views, calculating pickup points, and the like. The availability of graphic tool path generation and verification eliminates the manual data entry steps and proofing cuts previously required on the shop floor. The availability of communications software for direct numerical control (DNC) of machine tools eliminates the time-consuming and error-prone process of punching paper tape to drive the machine tools in the shop.

An additional productivity benefit is a reduction in the amount of time required for mold start-up, a substantial cost element of running a molding plant. Quite often, management either overlooks or chooses to accept this cost element as a fact of life. In fact the debug and "fine tuning" time for a mold can be greatly reduced by effective utilization of CAD/CAM. It has been demonstrated that analysis programs can be of great assistance in enhancing the quality of first-time mold designs. Additionally, they provide a great means to define the molding "process window," or extremes of acceptable process conditions, before the mold is put on the production floor. These factors result in a better use of molding press time and less disruption of production activities in the molding plant. Molders have reported differences of as much as 10 to 1 in start-up time, attributed to the better accuracy and quality of molds designed and constructed using CAD/CAM.

The last great area of productivity improvement is the one with the greatest financial payback, the part production environment. Integration of analysis tools into the design process provides much of this production benefit. One major plastics product manufacturer reported a 20 percent improvement in the cooling time requirements for 80 percent of the molds that were analyzed. High production molds have yielded cost savings in excess of $100,000 per year by application of cooling analysis techniques. Flow analysis techniques provide the benefits of being able to safely design runner systems of smaller diameters. They also allow design of unbalanced runners while avoiding problems of overpacking, and at the same time provide molds with higher production yields. All of these results yield substantial reductions in material usage and manufacturing costs. Economic analysis tools allow true optimization of the molding operation within a "real world" operating environment. Trade-offs between real-world variables such as: number of cavities to build vs. press size and capability, product tolerance requirements, and product quality requirements, can be evaluated. The result is a minimization of the total cost per piece at a given production volume level.

Quality

The quality benefits of CAD/CAM are perhaps the most underrated of all benefits. Drawings produced by CAD/CAM systems from three-dimensional models have been shown to be of a consistently higher quality than those produced manually. Dimensions are totally defined by the geometry in the data base, and as such are never incorrect. Tolerance stack-ups and other tolerance-related issues can be calculated by the CAD/CAM system, resulting in far fewer tolerancing errors. Complex geometries such as sculptured surfaces and blending radii are totally described in the data base, and thus not subject to ambiguities in drawing interpretation.

Another of the means of enhancing quality is in mold designs and the molds themselves, by a reduction in the number of errors caused by redefinition of the product geometry. This geometry is transferred first to the shrink corrected geometry, then to the cavity and core details, and finally to the machined components. Each of these steps is driven directly from the original product model with little possibility for error other than operator-induced error. Even the probability of operator error is reduced as many of the tedious tasks of redefinition have been eliminated, with the result that the operator does a more consistent job.

The accuracy of the mold cavity and core with respect to the product model is also a great aid in yielding quality enhancements. By utilizing N/C in moldmaking, and by eliminating the dependence on second- and third-generation tool-making aids such as die models, the conformance of the mold core and cavity to the product requirements is virtually guaranteed. The limits of accuracy obtainable are often those of the N/C machine tool plus the error created by the hand finishing operations.

Accuracy of the part is another quality aspect to be considered. Many of the analysis packages promote a better understanding of molding process parameters and the interrelationships between process variables. This contributes to a better ability to control previously mysterious phenomena (such as warpage) by better process control and cooling system design. The final quality benefit is the benefit of reproducibility. This includes reproducibility cavity to cavity, and mold to mold. Additionally, if cavities are severely damaged and require newly built cavities from new shop setups, the use of N/C will result in cavities closer to the originals than those constructed manually. The ultimate result of the previously mentioned benefits is a more consistent product quality.

Turnaround Time

Companies that are designing and using plastic products in today's fiercely competitive business environment are sorely aware of the time it takes to design, manufacture, and debug tooling for injection molded products. The ability of CAD/CAM to speed up the plastic part and mold design process, the mold manufacturing process, and the start-up and debugging process makes it a great aid in solving these age-old lead-time problems. Teamed with other prototype moldmaking techniques, CAD/CAM is helping to revolutionize the plastics product development cycle. Plastic products are now designed, analyzed, and evaluated for both technical and economic feasibility entirely without paper or physical models. Prototype molds, when required, are now turned out in one-fourth to one-half the time of their production counterparts, resulting in greater degrees of product design quality from testing of molded rather than machined prototypes. Additionally, lead times for production molds are being pared down by many of the same techniques, resulting in more effective utilization of assets such as cash and inventory.

Resource Utilization

The final benefit of CAD/CAM for moldmaking is that it allows us to effectively utilize scarce resources, especially skilled labor. It is well known that while the usage of plastics materials is increasing at a steady rate, the population of skilled moldmakers is on the decline. This creates a problem for those firms intending to continue manufacturing injection molds, either as a primary business or as a part of a larger business. Since sociological changes are reducing the skilled moldmaker labor pool available for employment, a new means to continue mold production without dependence on that skilled labor base must be developed. CAD/CAM provides the opportunity to reduce the elements of moldmaking that require high skill levels, thereby enabling skilled moldmakers to be used on tasks that cannot be accomplished by any other means. It is possible, and indeed probable, that use of CAD/CAM for the repetitive and routine tasks of moldmaking will enhance the quality of work life for those skilled moldmakers by providing more challenge and job satisfaction.

INTRODUCTION TO CAD/CAM MODELING

Now that we have discussed the major benefits of CAD/CAM for plastic design applications, it is appropriate to discuss the technology by which these benefits may be gained. In the normal engineering environment, products and mold designs are usually presented as series of orthographic projections in the form of engineering drawings. These projections allow us to represent a three-dimensional world as if we had photographed it and reduced it to a planar image. If we define on a drawing that we are working with orthographic views of the product, the mind is able to synthesize a three-dimensional image of the product. In many cases, however, orthographic images fall short of the mark in their ability to describe a complex geometry. This necessitates the creation of auxiliary and section views of the product in order to communicate the geometric description accurately.

The methodology behind producing product designs and mold designs via CAD/CAM technology is analogous to that in the normal method, except that the description of the product (or mold) is contained in a product model data base. The product model differs from a drawing in that the model is generally a three-dimensional representation of the real-world object. This representation can completely describe that object without auxiliary views or supplementary information. In this sense, the product model fulfills the requirements of a classic definition of a model. A model has been defined as: "a representation to show the structure of something; an image to be reproduced in more durable material; a pattern or mode of structure or formation." By these definitions one can anticipate that while moldmaking has always relied on the production of wood patterns, die models, or other tool-making aids to be able to accurately reproduce geometry defined via paper media, CAD/CAM technology is eliminating many of those requirements. It is therefore appropriate to include in this discussion on CAD/CAM how the concept of a product model applies to the area of mold design. We should first begin with a basic discussion on the types of modeling tools available today, and how these tools are used to construct the data base used in our product and mold designs.

In order to construct an effective data base, it is important to note that the data base must integrate and consolidate all of the information requirements used by engineering and manufacturing operations subsequent to the initial product or mold design. This is indeed a sizable task, and it is the reason why data base integrity and completeness should be of major concern to CAD/CAM system purchasers and users. To consider a few of the many application areas using the data base, the following list is provided:

Engineering functions
 Design
 Drafting
 Analysis
 Technical publications
Manufacturing functions
 Machine control applications
 Robotic applications
 Tool and mold design applications
 Quality management applications
 Communication with other business functions
 Materials management (bills of material)
  Cost estimating

A graphic representation of these relationships of sharing a common data base is depicted in Fig. 12-1 and 12-2. As can be imagined, in order to drive the number of applications areas previously discussed, an extremely robust data base must be constructed and maintained. The comprehensiveness of the data base is one of the primary differences between CAD/CAM systems today. Each application that reuses this data base recognizes the benefits of improved productivity and quality. This is attributable to the fact that already existing information in the product model does not need to be redefined, which would present the chance of introducing errors.

Many readers are probably asking themselves at this point, "Aren't there a number of 2-D CAD/CAM systems on the market today, and if so why aren't they being discussed?" The answer to this question is that certainly systems are being sold which are called CAD/CAM systems that basically satisfy the need for automating the drafting function. However, they also fall short of being able to satisfy the information requirements of the previously listed functions. This applies particularly to the area of mold design, where the problems of clearly representing complex surface geometries and other spatial relationships lead one to the conclusion that 2-D systems are not adequate to fully reap all of the benefits that CAD/CAM has to offer.

The three-dimensional product model provides the means by which many of the benefits can be obtained. The 3-D model clearly depicts all of the spatial relationships between items of interest in the product model. Terms such as "blend to suit," which were prevalent on drawings of the past, are now replaced by exact and reproducible mathematical descriptions of the surface curvatures desired. This has led to a new era in quality and dependability of product representation, now possible by these 3-D modeling approaches.

Modeling Methods Applied to Part and Mold Design

There are currently three predominant methods of building 3-D models of products and storing them in a data base. Each of these methods has associated costs and benefits. The

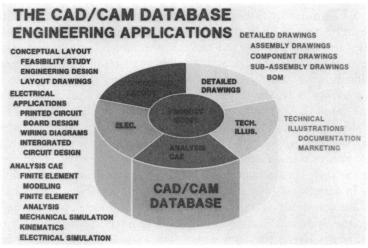

Fig. 12-1 Engineering functions that use the product model data base.

following paragraphs provide a brief overview of each of these modeling techniques, in addition to summarizing benefits and costs associated with each technique. It is interesting to note that because major changes in the power of computing hardware and software are occurring rapidly, these costs and benefits are appropriate only to today's "snapshot" of the state of the art in CAD/CAM. As this evolutionary process continues, it will bring other benefits not yet imaginable, along with a hardware base whose cost-to-performance ratio makes realization of these benefits easier to achieve.

The three major methods of modeling in common use today are:

Wire frame modeling
Surface modeling
Solids modeling

Each of these modeling methods and their potential application to mold design will be discussed in detail.

Wire Frame Modeling

Wire frame modeling is the simplest of the CAD/CAM modeling methods. The product model is constructed as a collection of geometric entities. Typical entities used in wire frame construction are points, lines, arcs, b-splines, strings, and the like. The wire frame method

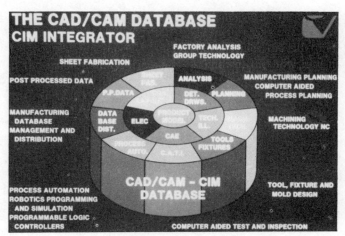

Fig. 12-2 Manufacturing functions using the same product model data base.

of modeling is very similar to orthographic projection drawing in that each of the lines and other entities represent the edges of the physical surfaces of the product. Unlike orthographic projection, however, the three-dimensional data base allows the graphic display devices (terminals) of the CAD/CAM system, such as the one depicted in Fig. 12-3 to automatically display isometric views of the product from any perspective the user desires, thus communicating the three-dimensional nature of the product. In these views of the object, lines and such are connected at their ends, thus portraying the appearance of a frame built of wire elements. Hence the term wire frame modeling was coined to describe the technique.

The simplicity of this modeling method also implies simplicity in the data base, thus resulting in superior system performance. The computing power required and storage requirements are minimal. Manipulation of the geometry to display different views is relatively fast and easy, owing to the small number of data elements that require mathematical transposition. The major disadvantage of this

approach is that all of the entities to describe the geometry are simultaneously displayed. For products with complicated geometries, this may result in many lines being visible on the screen at the same time, often resulting in a complex and confusing image on the screen. Hidden lines may not be removed unless a surface or plane bounded by the wire frame geometry is described to the computer. This then allows the computer to determine which entities are behind that surface and therefore invisible. Additional CAD/CAM features and techniques are described later in this chapter to assist you in organizing the wire frame entity information more effectively.

The wire frame modeling approach is best applied to rectilinear objects without complex surface geometries. This applies to a reasonable portion of plastic products, especially those that are more functional than aesthetic in nature. Such products might include gears, cams, brackets, and other non-appearance-type items. In mold design this applies to objects such as mold bases, cavity blocks, leader pins, etc. The surfaces of these objects are flat or circular, and generally need no subsequent

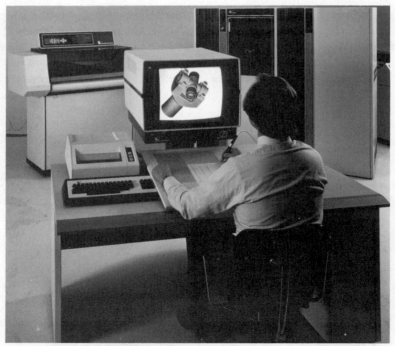

Fig. 12-3 Typical CAD/CAM system components including workstation (foreground), central processing unit (background right), and plotter (background left).

manufacturing operations. Hence, addition of surface information to the data base description of these items only adds to storage space requirements without any tangible benefit. Wire frame modeling is also useful for describing mold features that require relatively simple manufacturing operations, such as drilling, and pocket and profile machining. Machining of this type is referred to as two and one-half axis machining. Machine cuts are made at positions (or contours) relative to the X and Y axes, and at a fixed position relative to the Z axis. Since a contour relative to the Z axis is not being machined, it is only counted as a half axis. Wire frame geometry suffices to clearly describe the cutter path boundary as a series of two-dimensional curves. The cutter depth can be established from the distance between two curves which are known to be parallel (e.g., the top and bottom of a pocket). An example of wire frame geometry is shown in Fig. 12-4, where a standard mold base is described via a collection of wire frame entities. Since these entities are not ambiguous, positions such as ends of the lines or the center of the circles may be later used in the construction of additional geometry or for machining operations.

Surface Modeling

The next modeling technique discussed is surface modeling, a technique that can describe the total outer boundary of a part. Surface modeling differs from wire frame modeling in that not only does it allow you to describe the edges of the geometry to be represented, but also all of the faces. The surface model is used to represent all of the outside faces or the boundaries of the product. By using a surface model, ambiguities that are present in the wire frame approach are immediately eliminated. With a surface model, every point on the product surface is definable, either by explicit coordinates of a key point or by interpolation using an explicit set of parametric equations to specify those points between key points. Most of today's surface modeling methods use a number of key points at prescribed intervals to bound or define the sur-

face. Through these points a curve is fitted, in the same way that a french curve is used on the drawing board to fit a curve through a number of random points. As the CAD/CAM system must use a precise set of mathematical equations to define such a curve, this method makes every point along the length of the curve definable. If you can imagine these curves as a type of spatial grid, by understanding the mathematical base upon which the set of curves was built, it is possible to uniquely identify any point on the surface by interpolation using the same set of mathematics used in the construction of the curves themselves. Surface modeling, therefore, is particularly appropriate when complex 3-D geometries need to be described without any ambiguities.

Many plastic products fall into a category of this type. Enclosures, covers, and consumer items often require not only basic function but considerable aesthetic appeal to enhance market attractiveness. CAD/CAM and surface modeling provides the means not only to define the surface, but also to evaluate the aesthetic appeal of the product, and to directly control the finished product by an unambiguous description of all surface curvatures and radii.

The practical implications of surface modeling in mold manufacturing are manifold. First,

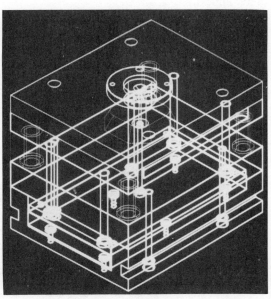

Fig. 12-4   A mold base modeled in wire frame geometry.

since every point on the surface may be explicitly defined, it is possible to create a three- to five-axis tool path for a milling machine, to follow the surface contour. This is of great practical significance in the area of mold design. Because it can generate three-axis contour tool paths, surface modeling is particularly appropriate for describing the geometries of cavities and cores within the mold. Other benefits include the ability to calculate angles of incidence and angles of refraction for light rays, which allow you to compute shaded images. These in turn greatly enhance the ability of humans to comprehend the surface geometry. An example of shaded imagery is illustrated in Fig. 12-5 and 12-6, which show a surface modeled plastic molding (a cover for an automobile ignition computer) as a surface model. Figure 12-5 shows the primary construction of the surfaces which are displayed for speed and efficiency reasons as sort of a wire frame representation. Stretching between the wires, however, like canvas on an airplane wing, is a mathematically defined surface geometry. It is the existence of this geometry that allows us to display the same product in a shaded image, as in Fig. 12-6. The shaded image representation is preferable to the wire frame in terms of ease of comprehension of the geometry of the product, as well as aesthetic appeal.

If we think for a moment about the concept of a surface, we should realize that it has no inside or outside. It is merely a spatial curve of zero thickness. This is another primary reason why the surface is very important in the mold design area. Assuming that the product designer does an effective job of defining a product model using the surface modeling technique, the CAD/CAM system can mathematically adjust this surface to account for the effects of plastic material shrinkage. The resultant surface model after accounting for shrinkage is both the surface of the part and the surface of the mold. Since the surface is totally described at every point, we can select a distance between successive cuts from a ball type and mill, and then calculate a tool path that always keeps the cutter tangent to the surface. The final benefit of the surface model is that it allows mass-related properties to be computed for the product model (e.g. volume, surface area, and moments of inertia), and also allows section views to be automatically generated. This is due to the fact that the intersection of a cutting plane with the surface can be calculated with certainty, resulting in another curve that becomes the section view.

Because of these benefits, surface modeling is quickly becoming the workhorse of the CAD/CAM field for describing geometries typical of those of molded products. There is a cost that accompanies the benefits of surface modeling, however. The system now requires far more computing resources than those required by the wire frame modeling techniques, in the form of both computing power and data base storage space. This is due to the necessity for more complex computations for such operations as automatically intersecting, trimming, and filleting entire surfaces. The state of the art allows users to do this type of construction interactively, using the wire frame type of displays shown in Fig. 12-5. As computing power and speed increase, however, it will become possible to do surface

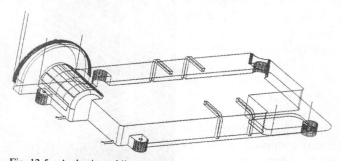

Fig. 12-5   A plastic molding, an automobile ignition computer housing, modeled in surfaces.

Fig. 12-6 The same ignition housing described in Fig. 12-5, but shown in shaded image representation.

manipulation with the shaded image type displays shown in Fig. 12-6, which will greatly enhance the user-friendliness of CAD/CAM systems.

Solids Modeling

The concept of solids modeling has become the ultimate concept in modeling of real objects today. The solids model takes the concept of the surface model one step further in that it assures that the product being modeled is valid and realizable. A solids model of a product may be created in a variety of methods, such as Boolean addition and subtraction of primitive shapes. These geometric primitives would include objects such as cubes, spheres, cylinders, etc. Other solids modeling

techniques include sweeping of 2-D profiles through space, and even the possibility of "sewing" together the edges of surface models. Via a solids model, the mass and boundaries of the product are represented in totally defined terms. An example of solids modeling is shown in Fig. 12-7, in which not only the geometry and surface boundaries are known, but mass properties of the object are instantly available upon creation of the model. In surface modeling, the system cannot distinguish between the inside and the outside of the product being modeled. In solids modeling, the system recognizes the difference, and during construction can avoid inconsistencies possible using surface mathematics. One famous example of such inconsistencies is the famed Mobius strip, in which a surface turns over on itself. An additional benefit of solids modeling is that mass properties for the product are generally immediately available from the data base.

The key concept of the solids model states that every point in space can be determined to be either inside or outside the object being modeled. While surface models define every point on a surface itself, solids models can determine every point within the solid object.

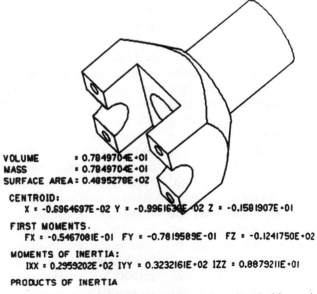

VOLUME   = 0.7849704E+01
MASS   = 0.7849704E+01
SURFACE AREA = 0.4895278E+02

CENTROID:
    X = -0.6964697E-02 Y = -0.9961636E-02 Z = -0.1581907E+01

FIRST MOMENTS.
    FX = -0.5467061E-01 FY = -0.7819589E-01 FZ = -0.1241750E+02

MOMENTS OF INERTIA:
    IXX = 0.2959202E+02 IYY = 0.3232161E+02 IZZ = 0.8879211E+01

PRODUCTS OF INERTIA

Fig. 12-7 A solid model, with hidden lines removed, and with associated mass properties.

It is this characteristic that brings about a great interest in solids modeling. Solids modeling enables us to determine solutions to some very complex problems of practical interest, such as automatic interference checking. With such a set of important benefits, solids modeling may appear very attractive. However, these benefits are not without a price. Interactive solids modeling demands a great deal of computing power to accommodate the practical designs of today. In many instances it is perceived to be too slow for widespread application. In many cases, solids models are based on the addition and subtraction of simple primitive geometric shapes that are easily defined. While this speeds up the design process, it is inadequate to describe solid objects of a class defined by sculpted surfaces. Given these circumstances, the additional benefits gained by using solids modeling in plastic part and mold design applications are often outweighed by the costs of the additional time and computing power required to accomplish the task. Remember that hardware and software technologies are changing so rapidly today that solids modeling may not require such a drastic trade-off in the very near future.

ADDITIONAL CAD/CAM FEATURES USED IN PLASTIC PART AND MOLD DESIGN

Group Technology

Group technology is a technique by which parts may be characterized and classified into parts of similar geometric features. Once the part is classified, a code number is assigned to it. The numbers within the code numbering system are significant; that is, they each have a significance related to a description of the product.

Once a data base of designs is encoded, when the requirement for a new design is introduced, the existing data base may be searched for exact or close matches of part description, which is provided by the significant code number. This tool prevents accidental duplication of existing parts, and encourages using existing parts that are close to the design needs. Not only are design costs reduced by this method, since existing designs

need only be "edited" to product new designs, but substantial manufacturing costs may be saved as well. Parts can sometimes be "made from" existing parts by the addition of a machining or assembly operation, at a lower cost than manufacturing the new part from scratch. Additionally, it is often feasible to install interchangeable inserts in the existing part mold and thus allow a single mold to produce two or more similar components. Manufacturing process information may be similarly retrieved by a group technology code number, and thus group technology may sometimes be thought of as an artificial experience tool.

Finite Element Modeling

Finite element modeling is a technique whereby a material continuum is divided into a number of patches, or "finite elements," and the appropriate engineering theory is applied to solve a variety of problems. The initial (and probably dominant) use of finite element modeling was for solution of structural engineering problems. The technique is currently being applied by a number of companies and research institutions in the design of plastic products, CAD/CAM systems provide the means to create a "mesh" of finite elements directly from a product model data base, via automatic and semi-automatic means.

The model described by the mesh is then analyzed, and results can be displayed via graphic means. A mesh and the corresponding analysis results are shown in Fig. 12-8a and 8b. For structural design with plastic materials, several unique requirements exist. Since plastic materials may have nonlinear and anisotropic material properties, finite element programs for the analysis of polymer structures should possess the ability to characterize material nonlinearities and anisotropy. Additionally, materials that are reinforced by glass or other types of fibers will be influenced by the degree and direction of fiber orientation in the molded product. Several research institutions are working on computer programs to predict this orientation in molded parts.

Finite element modeling is receiving increasing interest in plastics not only for struc-

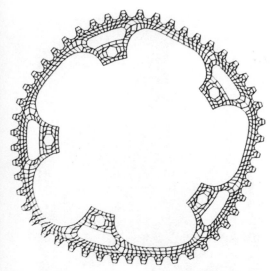

Fig. 12-8a   A typical finite element mesh, shown display-ing distortion caused by mechanical loads and predicted by analysis.

Fig. 12-8b   Maximum principal stress values in each element predicted by analysis and dis-played graphically.

tural design purposes, but in the area of pre-dicting plastic flow and heat transfer as well. Figs. 12-9a and 12-9b shows the finite element mesh and resulting pressure distribution for a bottom view of the computer housing previ-ously described in the discussion on surface modeling. In this case the isobar pattern helps to identify the total pressure required to fill the part, as well as the uniformity of pressure distribution, and the location of weld lines formed by flow around the holes in the part. The combination of ease of modeling provided by CAD/CAM systems and easier to use finite element systems will help to cause this com-puter aided engineering tool to proliferate in years to come.

Digitizing

Thus far we have discussed the methods used to define a product (or mold) geometry from scratch, or from a designer's concept. In many cases other work may have been done that can expedite the creation of a data base for the product. Traditional methods for product design and development have often involved the use of an appearance model or pattern where consumer testing or industrial design personnel may approve the appearance or functionality of the product prior to engineer-ing detail design work. The ability to reach

out and touch a physical model may be indis-pensable for certain design environments. If such a model exists, there is a strong possibil-ity that it may be used as a medium to expedi-ently create a product model data base and thus capture many of the benefits of CAD/CAM technology.

Digitizing describes the technique by which a physical model is re-created in digital form by in some way scanning the physical model and building a data base of points through which lines and curves are later constructed. This technique may be used in either a 2-D digitizing mode or a 3-D digitizing mode. In the 2-D mode, a set of 2-D models, such as drawing mylars, are placed on the table of a digitizing board or tablet. A number of points are created by using a tool positioned over the model at various locations; a pulse is sent to the computer, which records the instantane-ous X and Y coordinates of the device. This 2-D digitizing is useful when the input model or master exists in the form of a 2-D pattern.

If a substantial number of 2-D curves are available at varying Z depths, a 3-D model can be created by curve-fitting the points along the Z direction (axis) simultaneously with the X and Y directions. The most common way to create a digitized model for 3-D objects is from a 3-d model. In this method, a mea-surement device (such as a coordinate measur-

Fig. 12-9a   Finite element mesh applied to ignition computer housing shown in Figs. 12-5 and 12-6.

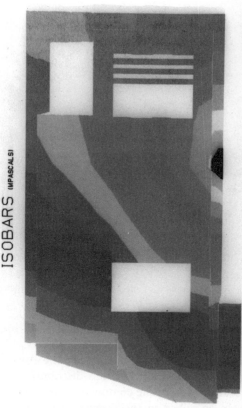

ISOBARS (IMPASCALS)

Fig. 12-9b   Results of flow analysis predicting isobars, or lines of constant pressure, at the instant of complete mold filling.

ing machine) is programmed to traverse to known X and Y locations, and the Z height of the model is measured and recorded at those locations. This series of points is then fitted with smooth curves such as b-splines, which are then used to develop surfaces such as b-surfaces.

While this method of geometry creation seems relatively straightforward, it is not without problems or pitfalls. The first consideration is whether the accuracy of the model and the digitizing method being used suffices for the application. If it is not sufficient, a more precise method of creating the geometry must be pursued. Additionally, digitizing works best for surfaces or geometries of relatively slowly changing curvatures. This means that geometries such as sharp corners or crease lines may cause problems. If a geometry such as a sharp crease in the product model exists, care must be taken to place a series of points

on that crease, and to stop the curve fitting occurring on either side of the crease at the crease line. This is due to the mathematics of fitting curves through a series of points, and the fact that the mathematical methods used do not handle drastic curvature changes between points very well. The only alternative is to increase the point density of the surface being modeled, which enlarges the size of the data base, causing a slowing of CAD/CAM system response. The digitizing method may be successfully applied, however, if the user is careful in his method of creating geometry and has a reasonable understanding of how the CAD/CAM system being used fits curves through points.

Layering

Another technique used in plastic part and mold design via CAD/CAM technology is the use of layering. Layering is a method by which the display of selected geometric entities may be turned on or off at the discretion of the user. The drawing-board analogy to layering is the overlaying of multiple tracings or mylars during the design process to build up assembly layouts from a series of detail drawings. Information can be selectively added or deleted from this layout by adding or removing tracings. Layering allows specific data base information to be turned on or off in the displayed image. Because the amount of information contained in view of a CAD/CAM product model is usually much greater than that displayed in a drawing view, layering is used to separate information into functional groups. This information may be turned on or off in combination with other groups to enhance the clarity of the model. Most CAD/CAM systems support some amount of layered geometry, many supporting 256 separate layers or more. An example of layering might be in the construction of a mold base. In order to avoid confusion in the model when designing a mold, the A half of the mold might be built on layer 1, while the B half might be built on layer 2. Each of the layers can be turned on or off independently, thus showing either mold half or the entire mold stack at once.

The real implications of layering go far beyond the graphic display benefits, however. Layering is a method within the product model that allows data to be efficiently organized for later use. For example, one mold design firm has built a set of engineering standards that are applied to each mold they design. One standard sets up specific conventions for constructing different mold components on specific layers. This allows us to preorganize information for downstream operations such as N/C machining. An example of how this standard applies is illustrated by the method for constructing ejector pins. The mold design firm using these standards assigns each size (diameter) of ejector pin to a separate work layer (e.g., $\frac{1}{8}$-in. pins on layer 11, $\frac{3}{16}$-in. pins on layer 12, $\frac{1}{4}$-in. pins on layer 13, etc.). In this manner the N/C programmer responsible for generating the tool paths required to drill holes in all of the ejector pin locations does not need to examine the model closely to generate tool paths. He would only need to call up layer 11 and request a point-to-point tool path with a $\frac{1}{8}$-in. drill, then call up layer 12 and repeat the process with a $\frac{3}{16}$-in. drill, and so on. The combined use of the system's layering capabilities as well as the procedural standard provides major productivity benefits.

Groups

Most CAD/CAM systems support a feature of creation of groups of geometric entities that may be selected or manipulated together. The concept behind a group is that several discrete entities (such as lines, circles, arcs, or points) may be associated one with the other as a unit. For example, the representation of an ejector pin might consist of four circles and four lines. These individual entities may be placed in a group, so that if we wanted to move the ejector pin, the entire group could be moved at once, rather than having to move each individual entity separately. Individual members of the group can generally be selected, to allow you to change or edit the geometry at a later time. For example, the lines and circles that make up the ejector pin may be altered, to "cut down" the length of the pin.

Patterns

The concept of a pattern is similar to the concept of a group with one important exception. The entities used to construct a pattern may not be selected or changed once the pattern is created. While this may be a disadvantage for using a pattern in place of a group, patterns require much less time to insert, translate, and otherwise manipulate the geometry they describe. Graphics systems are much faster in handling the graphic information of a pattern, and thus system performance will be enhanced in those instances where individual entity selection is not required. A practical example of how a pattern might be used in mold design would be in the creation of the graphics for a socket head cap screw. The complete component could be inserted far more quickly in the model if the screw was defined as a pattern in the data base, rather than a group of entities. Since we usually do not require additional changes to the geometry for machining or other operations, the use of a pattern is the best alternative in this circumstance.

Large-Scale Geometry Manipulation

This category describes the CAD/CAM system software that allows large blocks of geometric entities to be copied and moved to new positions, or to be rotated or mirrored about an axis. The major benefit of this type of system functionality is that common geometry only needs to be defined once, and can be reproduced in rotated positions, etc., in multiple locations. Most CAD/CAM systems allow multiple copies of a set of geometric entities to be created and later repositioned with a single command. The repositioning may be simply translating the model to a new location, or rotating the model about an axis, or a combination of both simultaneously. Also included in this category of system feature is the capability to create a "mirror image" of a model about a mirror plane, also with a single command. These operations may be performed error-free, and with great speed, thus allowing the mold designer to concentrate on creation of his mold design, rather than on the duplication of mold details. Tool paths for mirror image locations or multiple locations may also be generated without the need for manual calculation of the offset distances. The most practical use of this feature is in constructing multicavity molds or left- and right-hand geometries, etc. All of the common geometry used to describe each cavity case can be duplicated by the CAD/CAM system with great ease and speed.

Local Coordinates or Construction Planes

The concept of a local coordinate system or construction plane is one that makes modeling in 3-D very convenient. This feature allows the user to set an absolute coordinate system and build his model relative to that coordinate system. However, during the construction process, defining every point or entity relative to that coordinate system may become a cumbersome task. Since the computer can solve mathematical equations very quickly, a new coordinate system may be established at some location and orientation within the model. When the user defines new geometry relative to that local coordinate system, the computer can quickly perform the calculations required to properly position this new geometry within the product model. The practical implication of this feature may be illustrated by the following example. Suppose we define a mold base whose 0,0,0 (origin) location is at the top of the top clamping plate, and the center of the sprue bushing. If we wanted to locate certain points at the parting line of the mold, then without local coordinates we would have to keep track of the thickness of the top clamping plate and the A plate. These thicknesses can be summed and used as a Z coordinate in inserting points on the parting line. With the concept of local coordinates or construction planes, we can define a new coordinate system relative to the origin. This new coordinate system can be defined as being at the same X and Y location as the origin, but at a Z location of the top clamping plate thickness plus the A plate thickness (or at the height of the parting line). The new points may now be inserted relative to these construction coordinates, and the CAD/CAM system will correctly determine the position in space relative to the origin. This feature is particularly useful in constructing geometry on angled surfaces and the like, which would require trigonometry or other mathematics without local coordinates.

Model and Drawing Modes/Associativity

A number of CAD/CAM systems today support the concept of model and drawing modes of geometry creation. This allows those companies that require outputs of paper drawings to create them, while still maintaining the same high integrity of the product model data base. To explain this concept further, the product model generally contains all of the information required to program an N/C machine tool, for example. However, standard drafting practices dictate the use of additional graphic entities such as hole centerlines to describe the geometry in the paper drawing world. The concept of a drawing mode, illustrated in Fig. 12-10, was developed so the user could create numerous representations of the same product model, and add drafting entities such as centerlines and cross-hatching for sections.

The concept of associativity extends the power of the model/drawing mode further by providing the means by which all changes in the product model are automatically reflected in the drawings associated with the model, as changes in both pictorial appearance and attached dimensions. Further degrees of associativity can be provided such that changes in detail parts are automatically updated in the assembly. This is of great help in coping with design changes and in minimizing their costs in a product development environment. The drawing mode is as useful in the CAD/CAM applications of mold design as it would be in any other mechanical design application.

Verification of Geometric Relationships

Since the product model data base is fully described in 3-D space, then any relationship between two geometric entities in the data base may be quickly established by the computer. This means that rather than scaling a drawing or inserting a dimension in the data base, a function exists that allows us to understand the distances and angles between geometry existing in the data base. The importance of this type of feature is illustrated by the following example. If we are creating a large mold on a CAD/CAM system, the graphic display device will automatically scale the object so that it may be viewed on the screen. If we are designing in ejector pins, the location of the pin in relation to other geometry, such as a waterline, may not be very clear. By using the on-line verification functions, we can check the distance between the edge of a circle that defines the ejector pin, and the line that

defines the waterline. In this manner the existence of a safe steel condition can be verified without the need of a full-scale plot of the mold or some other measurement method.

Automatic Dimensioning/Automatic Tolerance Analysis

As previously mentioned, the product data base provides a complete description of each entity that makes up the product. Since this information is stored in the system, dimensions may be automatically generated directly from the data base model. If you simply indicate the entities to be dimensioned and where you want the dimensions to appear on the design, the CAD/CAM system will automatically calculate the dimension, draw in the leader lines, and put in the correct numbers. If the CAD/CAM system has "associative data base," then the dimension displayed will not be associated with the numbers it displayed, but with the geometry of the product model. As such, if changes are made to the product model geometry, these changes will be automatically updated and displayed on all drawings of that product model. Additionally, you have the option of also displaying manufacturing tolerances with dimensions. Automatic tolerance analysis allows the user to select a number of tolerances on the drawing, and the system will calculate a statistically derived sum of those tolerances and then minimum and maximum dimensions as a result of the tolerances applied. This is particularly useful in mold design applications for molds with multiple cavities or molds with interchangeable components, where tolerance accuracy must be verified in the design process to assure that mold components will fit together at final assembly.

On-Line Calculation Capabilities/Storage Areas

Any designer who has designed on the drafting board knows how useful a pocket calculator can be. It allows the user to accurately perform quick calculations such as dividing lines

Relationship Between a Model and a Drawing

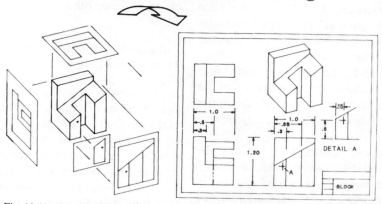

Fig. 12-10   The Model/Drawing Concept. The model is a three dimensional representation. The drawing documents the model through a series of two dimensional views.

of odd length into a number of equal segments, or calculating factors to correct for material shrinkage. In many CAD/CAM systems, a feature is provided to perform similar calculations on-line. In most cases the results of those calculations may be filed in a storage area so they may be accessed at a later time. This type of feature is useful to mold designers for such simple calculations as shrinkage allowances, clamp tonnage estimations, or storing the results of other system capabilities (e.g., projected area calculations from 2-D geometries).

FLOW OF THE MOLD DESIGN PROCESS, CONVENTIONAL VS. CAD/CAM

Because of the complexity of examining all of the possible variations of technique, style, and approach in plastic product design, we will focus on an illustration of how to apply CAD/CAM to the mold design process. In this illustration we will assume that the product design has been established and optimized for structural integrity, wall thickness, and cosmetic appeal. The task at hand is then to produce an optimum mold for a given level of production volume and to produce parts at an optimum quality level. CAD/CAM can be applied by the mold design team and the production team in order to successfully de-

liver the molded product at the lowest possible cost and in the shortest possible lead time.

In order to provide a better understanding of how to fully utilize the unique features of CAD/CAM systems in the mold design process, the following section discusses the differences between a manual method of designing molds and the corresponding method using CAD/CAM technology. The flow charts showing this comparison are provided in Figs. 12-11 and 12-12. Close examination of these flow charts indicates that a considerable difference exists in both the design methodology and the engineering emphasis for molds designed with and without the benefits of CAD/CAM technology. In general, molds that were manually designed depend heavily on the experience of the mold designer, and have a strong emphasis on paper as the communications and driver medium for subsequent manufacturing operations. These molds also rely on a number of iterations after the mold is built to achieve product function, and to fine-tune the performance of the mold. In the CAD/CAM approach, the product model data base serves as the communications and manufacturing driver medium, analysis programs augment mold designer experience, iteration is conducted more in the design phase prior to cutting cavity steel, and there are generally fewer molding trials. Each of these ap-

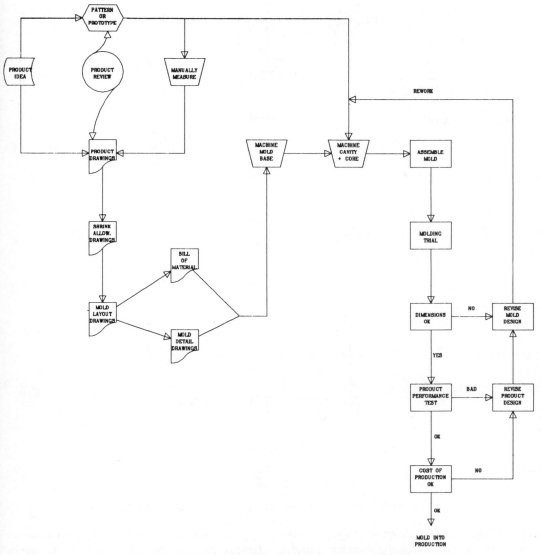

Fig. 12-11   Flowchart of conventional mold making activities.

proaches to mold design is described in further detail in the following paragraphs.

The Manual (Paper) Method

To analyze either of the mold design methods, we must begin with the information input to the mold designer, which is generally some type of product model. In the manual method the product model consists of a set of drawings that describe the product geometry. In many instances, however, three standard drawing views may not clearly describe the product, and quite often an additional description is included, appearing in the form of auxiliary and section views. Quite often even this "clarifying" information is still inadequate to describe the geometry of the product. In these instances, either approved patterns or prototype models must be used to uniquely define the product. This basic information is passed to the mold designer to begin the actual mold design layout.

The mold designer is next faced with the task of reconstructing the product geometry to account for material shrinkage during the

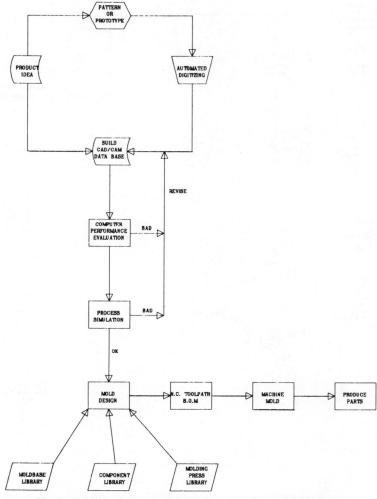

Fig. 12-12   Flowchart of mold making activities using CAD/CAM.

processing operation. Even with simple drawings, every dimension in the drawing is multiplied by a scaling factor (to account for shrinkage), and at least the resulting part periphery is redrawn to incorporate it into the mold layout. On more complex molds, with many dimensions, the scaling of every dimension is a tedious, time-consuming, and error-prone process. On products that are described by model, the drawing used to build the model must be corrected, and the model rebuilt with shrinkage accounted for, thus leading to the possibility of the incorporation of errors in the construction of the new model. Once the product drawings have been changed to reflect the shrinkage allowance, the mold layout process may begin.

In developing a mold layout, extraneous factors such as budget considerations begin to taint decisions such as the number of cavities to be included in the mold. Limited tooling budgets often cause molds to be designed with too few cavities to optimize the total costs of part production because a clear relationship between tooling investment and production cost was not evident at the mold design layout stage. Other technical decisions such as the number and location of gates for the part may be based solely on the experience of the mold designer. The quality of these decisions is directly proportional to the amount and quality of experience of the mold designer. Many of the practicing mold designers are competent mechanical designers, but they usually know very little of the intricacies and interrelationships in molding plastics. Nonetheless, after other features (such as ejection) are accounted for in the mold design, waterlines are squeezed in wherever they will fit. At this point the layout is generally approved by the customer, and planning begins for construction of die models and other tool-making aids. At the same time a bill of materials is generated from the design layout, and purchased components and cavity and core steel are ordered.

The designer then begins to break down the mold assembly into detail drawings of its component parts, and adds information such as the reference locations and pickup points that are required to fabricate the mold details. Again since all of these operations require reuse of the geometry, and manual use of trigonometry and other mathematical manipulations on that geometry, the probability of errors being introduced to the design is quite high. During the detail design period, the die models and tooling aids are being fabricated, often by another shop. Communications between the model maker and the end customer are generally poor, when they do exist, and again a high opportunity for errors appears.

At the end of mold detailing and model fabrication, the drawings and construction aids are released to the shop for machining of components. The mold is machined, hand-finished, assembled, fitted, and spotted. The next step is to hang the mold in the molding press for initial molding trials. If the trial progresses to a successful fill of the mold cavities, the parts created are inspected for dimensional compliance with the product drawing. If the parts do not comply, the mold is sent back to the tool room for reworking. This process continues until acceptable molded samples are created for evaluation. In many cases the product does not perform exactly as the designer anticipated, and the design is revised. This results in revisions to the mold design, rework to the mold, and an iteration of molding trials again. It is important to note that reworking the mold again and again may eventually result in the deterioration of the integrity of the mold, due to induced stresses from rework operations (such as welding). The final

test is to determine whether the functional product finally molded meets the cost targets required for consumer acceptance of the product. If the answer is no, the entire mold is scrapped, and the overall design cycle may begin again.

The preceding treatment of the manual method of mold design has highlighted many of the method's negative aspects. In fairness, it is quite unlikely that all of these problems would occur in a single product design. It is not uncommon, however, to have two or more of the above pitfalls occur in what appears to be even the simplest of plastic product designs. The high number of used mold bases in the industry attests to the many unsuccessful product and mold development activities. The following section will explain how the CAD/CAM system may be used not only to increase the probability of a successful design on the first molding trial, but also the probability of a mold design that is optimum in terms of quality, performance, and economy.

The CAD/CAM Method for Mold Design

Thus far, the effective uses of CAD/CAM technology for mold design applications have been shown to explain many benefits to the end user. In this discussion, we will follow the flow chart given in Fig. 12-12 to discuss one of the many possible scenarios for successful use of CAD/CAM technology for mold design. The first and foremost difference in a computer-aided approach to mold design is in the product model. The computer-based model is usually a complete 3-D representation, as compared to a 2-D model, which appears as a series of views on a paper drawing. The information contained in the 3-D model is organized quite differently from the information contained on a product drawing. Since the part description data base is always to scale, information may be directly extracted from the data base without fear of scaling errors or other mistakes. While conventional drafting views can be developed for the model (e.g. section views), they are often not required, as shaded images and other types of displays may be generated to adequately convey the information otherwise provided by those drafting views.

Since we are using the capabilities of some type of computer for the graphics creation and manipulation, we should also consider the advantage of using the same computer for analysis programs. We might begin our analysis by looking at some of the theoretical aspects of how the product is going to be produced. A rudimentary analysis of the product could evaluate whether the wall thickness and flow length are feasible, whether the part will run on available molding equipment, and what the anticipated production cost of the product is. Specific software programs exist to allow these types of analyses to be run prior to the actual mold design and development process. If, in fact, any potential problems appear, redesigning of the product should begin at this stage. If this preliminary check point is passed, then the next step is to determine the location and type of gating for the product. Many times gating is one of the last elements to be considered in a mold design. This can lead to major mistakes, which are avoidable when the gating is designed with the product performance as a prime consideration. There are such programs as flow analysis programs that can be used to determine gate locations to: (1) minimize molded stress in local areas, (2) help in determining gate location to provide the desired fill pattern, or (3) aid in positioning weld lines and meld lines in areas where they will have the least impact on product function. At the same time, these analysis programs can be used to identify the range of process parameters that must be used to increase probability of success of part fill and performance. Two key parameters whose impact can be evaluated at this stage are material melt temperature entering the cavity and the mold temperature of the cavity. The preceding paragraph describes the process simulation phase on the flow chart.

Once the gate location(s) that optimizes the product function has been established, the geometric relationship between the molded part and the nozzle of the molding machine is determined. The next step is to begin the mold layout work and select the appropriate mold base for the mold design. The first question that we need to ask ourselves is how many cavities we want to include in the mold, and the size of molding press that we want to mold the product on. Economic analysis programs are available to help us make these decisions with confidence, as all of the appropriate factors supplied to the analysis are being considered in the decision. Factors such as the required degree of process control to mold the part to tolerance, the increase in scrap loss due to a larger number of cavities, and the additional cost of making cavities and the larger mold base to contain them are all factored into the analysis. The end result is that cost per piece vs. tooling investment can be studied, so that parts may be produced at the lowest cost per part, including amortization of the tooling investment.

With the number of cavities and gate location(s) now optimized, computer software can be used to aid in establishing the minimum mold base size required to accomplish the proposed layout. Data such as the expected cavity insert sizes and other criteria such as the proposed type of runner layout (balanced, unbalanced, etc.) may be input, so that the computer software can make a recommendation on the size of mold base to be used. The criterion generally used in establishing the overall size of the mold base is that the ejector plate must completely cover, or contain within its bounds, all of the part cavity area in the design. Once the size of the mold base is selected, the user can call up the product model of the mold base from a standard library of mold bases. Now, rather than having to be traced from a pattern or catalog, the mold base will be automatically generated or retrieved by the CAD/CAM system. This standard mold base is now available for editing by using the CAD/CAM system, so that it can become the custom mold for the product.

The next step in the development process is to merge the part product model with the mold product model. The part geometry has usually been previously defined by a product designer, and is stored in the data base. All of this geometry may then be copied from the data base and automatically installed in the mold model at a location selected by the user, often with a single command. At this time a material shrinkage allowance may be incorporated in the product model. If the plastic material exhibits a uniform and consistent shrinkage, a simple scaling command can be used to rectify the model for the shrinkage factor. This is done by multiplying all of the dimensions of the part by the shrink factor, and then redisplaying the corrected graphics. In certain instances this calculation may be accomplished by the same command that allows the merge to be performed. If the material is more complex in nature, such as those materials with different shrinkage values in the flow and transverse directions, then a slightly more complex approach is required. In general, a more complex shrinkage factor is applied by defining a local coordinate system within the part, and the axes that correspond to the flow and transverse shrinkage directions. A command may then be issued to scale up the part, with independent scaling constants for the X, Y, and Z directions. While this type of manipulation does not precisely define real-world shrinkage, many R&D activities are currently under way to provide calculations that are far more accurate.

Once the molded product geometry has been successfully merged into the mold and corrected for anticipated material shrinkage, the cavity blocks can be defined. These blocks must include secondary split lines for slides, lifters, and other mechanical actions. If careful attention is given to layering when the split lines are inserted, subsequent extraction of the entities that describe these mold components is an easy task, and this facilitates the creation of detail drawings for those components. The mold designer is then free to focus his attention on finalizing the runner system design. Again the flow analysis packages come into play. These packages may be used to successfully create designs such as artificially balanced runner systems, family mold runner systems, and/or constant pressure drop runner systems. The analysis routines will provide the designer with the knowledge that his runner system design will now provide to the molder the widest possible "process window." Software is now available to analyze state-of-the-art designs such as artificially balanced hot runner designs for family molds. These user-friendly but powerful tools are bringing a new light to a previously mysterious design area.

With the runner design finished, the next step in the mold design process should be the design of the cooling system. Cooling system designs have always resulted in a classic real estate battle between waterline placements and ejector pin requirements. In the past, without the computer technology that we have today, the relationships between waterline geometry, water temperature, and waterline location were rarely studied or optimized. Because ejection requirements are better understood, the designer usually considered the ejection problem first, and then put waterlines as space permitted, to miss the ejection system. In the end, molds were created where the parts ejected properly, but where cooling cycle times were much longer than those of parts of similar materials and wall thicknesses. This low level of productivity increases operating and part costs unnecessarily. Tools are now available, in the form of cooling analysis programs, to recommend a cooling system design. The designer begins the process by obtaining a design recommendation of waterline size and depth from the mold surface. The designer then attempts to apply these design recommendations within the space constraints of the mold. The actual resulting cooling system design is then described to the computer. This design is then analyzed, and as a result its performance characteristics and economic impacts are understood before molding trials begin. From the description of the waterline geometries and locations, the computer will now be able to organize these cooling lines into flow circuits for the best combination of pressure drops and water temperature rises. It will then recommend the appropriate water temperatures and flow rates for proper mold cooling, as well as the projected cooling time. This projected cooling time can be evaluated against the theoretical minimum cooling time to determine whether changes in cooling system design will enhance productivity. Additional benefits of this type of analysis are determining whether the cooling circuits can prevent hot spots in the mold cavity, and whether the same water temperature can be used in both mold halves, so that parts may be molded at minimum cycle time without thermally induced warpage.

Once the cooling system has been designed, the remaining components such as ejector pins, support pillars, parting line locks, etc., may be incorporated into the mold design. Libraries of mold component parts are now available to either (1) create components, depending on their specific dimensions, or (2) retrieve components on file in the data base. Both of these library methods help to promote standardization of mold designs. This results in a higher quality of the design as well as creating cost reduction opportunities, as standard design practices can be combined with standard construction practices. Many CAD/CAM systems now support a method to add additional intelligence into the data base, in the form of nongraphic information, which is attached to a particular set of geometry. This capability allows text descriptions of individual component parts, such as part numbers and prices, to be stored in the data base with the geometric entities. The data base may be scanned at a later date, and this nongraphic information may be extracted into reports. This powerful technique may be used to automatically generate bills of material for the mold being designed. The library of parts concept facilitates this process because the program that creates the component can also create the text information, and that information can be extracted in different forms at a later date.

Once the basic design layout is complete, any moving mold components can be checked for functionality. The CAD/CAM system facilitates the translation and rotation of graphic entities on the screen, and these features can be used to verify the range of motion for components in the mold that must move relative to the rest of the mold. This would include classes of components such as slide mechanisms, lifter mechanisms, reverse ejection actuators, and the like. Some of the newest software available on CAD/CAM systems allow links and rotating joints to be specified to the system, and a true kinematic analysis can be performed. This allows the designer to check the mechanical action of the design to see that it follows the desired motion path, while simultaneously checking for interferences with other mold components.

The last steps of the process required include creating the detail drawings needed and generating N/C toolpaths to machine the actual mold. As mentioned in the section entitled "Introduction to CAD/CAM Modeling," drawing requirements may be significantly reduced by CAD/CAM technology. As a result, several mold shops are now producing molds from only a few views of the model and the N/C toolpaths derived from the product model data base. This methodology is contrasted with creating many detail drawings, which only serve to repeat information currently stored in the data base. Generating N/C toolpaths from the existing geometry is a relatively easy step, especially since the shrink-corrected product model contains a completely detailed description of the product. The types of N/C routines available are lathe routines and mill routines. Within the family of milling routines is the ability to generate tool paths for point-to-point machining operations, profiling operations, pocketing operations, and surface machining operations. Many systems now include other special features such as surface machining with containment boundaries and even automatic generation of cavity roughing. A brief description of the application of each of the N/C routines is given below.

- Lathe Routines: for cutting round cavities and cores, such as those used in container molds, etc. The geometry required is a 2-D profile of the object to be cut.
- Point to Point: for repetitive machining operations conducted at discrete points in space. It is especially useful for operations such as drilling waterlines and ejector pin holes that are performed with the table of the machine tool stationary at a particular point. The geometry required is simply a point in space.

- Profiling and Pocketing: used extensively in mold design applications for generating cavity insert pockets, slide retainer pockets, and constant-depth cuts on cavities and cores. The basic geometry required is a 2-D profile of the pocket or the profile to be cut.
- Surface Machining: one of the most powerful of the machining tools, used for cutting sculptured or other types of freeform surface geometry for products that involve styling. The cutting operation is generally conducted on a three- to five-axis machine tool that uses ball end mill type cutters. The geometery required to derive a tool path is either surface geometry or a large number of 2-D cuts through the surface at regular intervals.

The final steps of the mold design process are the actual fabrication operations and the molding trials. With the use of CAD/CAM technology, these steps have a higher chance of progressing smoothly than if manual design methods had been used. Greater attention has been paid to the details of the design, and scientific methods can now be applied in places where only experience was available in the past. When these methods are properly applied, they result in an optimally functioning mold and an acceptable molded product.

USE OF DESIGN DATA BASES

The Data Base Concept

One of the key factors required to implement a CAD/CAM program that will allow increased productivity is the effective use of a data base, especially design data bases. A design data base typically contains parts or components that are frequently used in mold designs, as well as standard design methods to be applied to new designs. There are at least two methods to create a design data base, both of which are described below. Before these methods are described, let us clarify the exact meaning of the data base. A data base can be defined as one of the largest elements in a hierarchy of information structure. In order

to illustrate this concept, let us consider the following types of information elements, ranked in order of increasing information content:

- Data Item: a single unit or specific piece of information. If we were to consider using a line as an illustration, a data item about a line might be the X coordinate of one of the ends of the line itself.
- Data Record: a collection of data items in a logical sequence. Again using the illustration of the line, the data record of the line might contain the X, Y, and Z coordinates of both ends of the line.
- Data File: a collection of data records, generally organized around some common attribute. Once again, a number of lines can be organized into a data file that describes the geometry of a specific standard part.
- Data Base: a collection of files, again usually organized around some common attribute or organized for a specific purpose. Continuing with our example, a number of data files that describe standard components can be further organized into a data base that describes all standard mold bases available from a particular manufacturer.

To further clarify the concept of a data hierarchy, whose highest tier is the data base, an analogy of a paper file system is presented in Fig. 12-13. In this analogy the data item is a specific line contained on a page of paper. The data record is the page of paper containing many data items. The next level of the hierarchy is the data file, which corresponds to the many sheets of paper contained in a file folder. The last tier of this structure is the data base, which corresponds to the entire filing cabinet.

Graphics Data Bases

The concept of a graphics data base is essential to understanding the nature of a CAD/CAM system. Such a system takes all of the design information that would normally reside on a

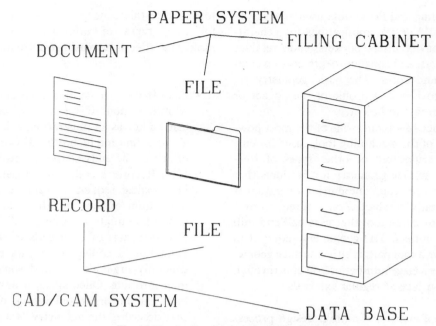

Fig. 12-13   Analogy of the data base concept.

piece of paper, and builds that information into files, to create the data base. Each graphic entity such as lines, arcs, circles, points, and surfaces is represented by a number of data items. These data items are further collected in a data record, and a number of these records are collected into a file, which generally describes a molded part or the mold itself. A number of these files may then be grouped together to form the data base. In the same concept of a hierarchy of information we may consider one of the key tools for boosting productivity of design, the library of parts. This tool is a data base of geometry, which is recalled and used as quickly as the computer can associate it with the particular design in question. A library of parts is particularly useful when items such as injection molds are created from standard components.

Defining the Library Data Base

The simplest method of creating a data base is by combining a collection of part files. A designer will generate all of the parts required on a CAD/CAM system, and then store them in a logical catalog structure so that they may be easily retrieved at a later date. Most CAD/

CAM software packages support a command that allows a part currently existing in the data base to be inserted into the part being designed. Additionally, many software packages support a method of preparing parts in a figure of associated graphics, rather than a collection of discrete entities, which speeds up the operation of inserting previously defined geometry into a new design. The major benefit of this overall technique is that once the part is created, it never needs to be redefined, and it can be used over and over again.

Unfortunately, there are also several disadvantages to this technique. The first constraint is that the parts must not change, or the geometry will require editing after insertion. Depending on the particular type of system, this may present problems, as some systems do not support editing of the graphics figures described in the preceding paragraph. Second, if there are a large number of unique combinations of geometry (i.e., mold bases), an extremely large data base will be required. This implies that a large amount of work is required to create each component, in addition to large amounts of storage required to keep the data base on-line. Since on-line storage is relatively expensive, another computer program, called a data base management system, is usually employed. The data base management system will provide the convenience and economics of storage off-line, but will allow rapid retrieval of components desired. One means to eliminate the problems caused by data bases of parts files is to use "graphics language programming."

The concept of a graphics language program is that a computer program builds the required geometry only when it is needed; and through interactive programming it allows the operator to intervene and change the geometry as required. The language in which this program is written in is called a graphics language, since the language facilitates creation of graphic entities automatically. An example of the use of such a program might be given by the need for graphics for a standard ejector pin. This ejector pin is available in a number of different lengths, but must be cut to a specific length to be properly installed in the mold. A graphics language program can then be written to prompt the user for the required diameter, length, and location of an ejector pin to be installed. The system will then read a data file of the corresponding head diameter and thickness, and will create the geometry and display the graphics to completely describe the part.

The advantage of the graphics programming technique is that it drastically reduces the amount of on-line storage space required to support large libraries of common geometry. Additionally, by allowing operator intervention, we may now build standard parts with unique characteristics, such as ejector pins "cut to length." Variable parameters such as A and B plate thicknesses and desired locating rings or sprue bushings can be interactively entered at the program run time, creating any of many unique mold bases. The addition of design standards can also be incorporated into the graphics programs, thus decreasing the probability of designer errors, and promoting standardization in both the design and manufacturing operations. An example of design standards might be the uniform recessing of screw heads in a design, which would allow cutting tools to be set once, and which would

help identify potential interferences caused by the shop inadvertently counterboring too deeply. The major disadvantage of the graphics programming technique is that the speed of inserting new geometry in the design is generally slower than an insert part type command. Depending on the complexity of the geometry and the software in question, this may or may not be a disadvantage. In general, the user needs to assess this performance issue for his particular application with the particular software package available on his system. It suffices to say, however, that graphics programming for family of parts applications such as components libraries has shown itself to be an extremely beneficial feature of CAD/CAM systems. Users considering the potential purchase or effective usage of said systems should give graphics programming languages and their ease of use careful consideration in their implementation activities.

COMPUTER INTEGRATED MANUFACTURING

Until now we have been discussing CAD/CAM and the various state-of-the-art techniques for its application in the field of mold design. Before closing, it might be appropriate to briefly discuss how CAD/CAM is changing in relation to computer integrated manufacturing (CIM). When addressing the field of computer aided manufacturing today, most authors speak solely of numerical control as the means of utilization of the product model data base. Obviously the field of manufacturing is composed of many processes that do not rely on numerical control as the control means. Computer integrated manufacturing is the natural evolution of CAM into serving an ever widening scope of manufacturing processes. The unique element of CIM, however, is that it tends to integrate or tie together all of the manufacturing processes into one coherent unit, all of which share a common data base. CIM involves the appropriate combination of hardware and software that allows the manufacturing processes and functions to draw information from, and contribute information to, the data base. This information is shared by all the respective functions, including business functions, as a primary business information system. As illustrated in Fig. 12-14, the CAD/CAM system can be envisioned as a central hub of primary information such as the product model and its attributes, and secondarily derived information such as tool-paths, bills of materials, etc. Just as in the manual method of doing business, the engineering information stored in a data base begins the product delivery cycle. This basic information about the geometry and attributes of the product is transferred and reused by many subsequent business functions, such as the materials planning function and the manufacturing engineering function. We are moving toward the realization of CIM when we are able to pass information efficiently to and from each of the manufacturing functions to the data base. We have discussed in this chapter how product model information for both the mold and the molded product is passed back and forth. We have also discussed how this model information is traditionally used to

drive N/C machine tools. There are a number of other manufacturing processes that can serve and be served by the data base.

Many of the process parameters we have established via analysis programs need to be expediently transferred to the shop floor. This can be accomplished in a number of ways. For those organizations still using paper as the primary communications medium, process planning software is available to build process plan documents automatically from the data base, and then distribute them to the shop floor. For those firms ready to take a further step in automation, it is now possible to send both graphic and nongraphic information to the shop floor at low cost. The same hardware and software systems that allow this type of distribution may also allow direct control of manufacturing equipment via high-speed, high-reliability communications links. Additionally, the capability to transfer other types of information, such as bills of materials, to other business functions, will make organizations less and less dependent on the expedient

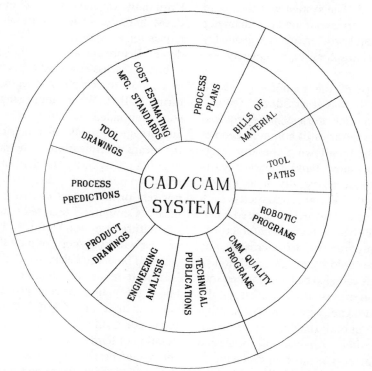

Fig. 12-14 Information that can be generated, stored, and maintained on a CAD/CAM system.

handling of paper. By communicating information such as actual molding cycle times and computer assisted inspection results back to the data base, it is now possible to close the loop on the entire manufacturing process. Utilization of data base management techniques such as group technology allows the quick retrieval of this information such that results of analysis programs may be tempered with a dose of real-world operating environment, with a sound statistical base. The net result of the entire effort will be a more productive manufacturing and product delivery environment, which allows products to be designed and delivered more effectively with less resources.

SUMMARY

In this chapter, we have given one treatment of the way CAD/CAM technology can be applied to the field of mold design. In that discussion, we have touched upon a number of technical details, involving software capabilities, integration of analysis results, and even user technique. The differences between models of the CAD/CAM world and drawings of the paper world have been discussed, as well as some of the common modeling techniques used today, and their implications for the design and manufacturing process. Some common features of CAD/CAM systems have also been listed, and their usefulness in mold design has been illustrated. An outline of the process by which CAD/CAM can be applied to a mold design, and how this process differs from the typical process of design in the paper world, has been presented. The major conceptual difference due to a reliance on computer analysis tools, as opposed to designer experience, in achieving designs of higher quality, dependability, and productivity has been discussed.

The results of successful application of these techniques were presented in the section on the benefits of CAD/CAM to the mold design process, those benefits being:

Productivity in design and production
Quality of design, mold, and molded product
Shorter turnaround time to product delivery
More effective utilization of skilled labor resources

It is hoped that this presentation has given the reader a taste of why many successful firms are pursuing the use of CAD/CAM as a means of increasing their overall business success. In closing, we would caution our readers that this is an ever changing field. During the useful life of this publication, changes in the technology are sure to occur that will affect the usefulness of the information presented. Certainly, what was perceived as mere dreams five years ago is in many cases a reality today. Potential users and purchasers of CAD/CAM systems should have high expectations of the capabilities of systems they use or purchase, and in many cases they will be pleasantly surprised by the breadth of automation possibilities these systems will afford. It is our hope that this work can in some way help to set that high level of expectation, which will eventually become reality.

Chapter 13
Process Analysis Tools for Plastics Processing

John Gosztyla

Computervision Corporation

Recent years have seen a tremendous growth in the number of software products available to serve the needs of manufacturing functions. The products that have evolved over this period are tools that serve to replace the "rules of thumb" of the past with analytical analyses based upon sound theoretical principles. These products combine the benefits of relative ease of use with the speed of the computer, resulting in tools that can be cost-effectively applied to a large number of problems. Over the last five to ten years specific software products have arisen to serve the needs of the injection molding industry. These tools are most effectively applied prior to construction of the mold, but they can be applied after the fact to solve process-related problems.

Three types of analysis tools have emerged recently, and are providing major benefits to molders. The types of analyses commercially available fall into three categories:

1. Flow analyses
2. Cooling analyses
3. Economic and plant operating analyses

In general, these analysis tools fall under the domain of CAE or computer aided engineering. The key word in CAE is *aided*. Analysis tools in no way replace skill or education in the basics of plastic material properties, mold design, or processing. What analyses do is supplement the knowledge of a trained individual, making him more productive and more accurate in his predictions.

The basic methodology behind the CAE technique is that a design or process is proposed as the first step. The engineer then constructs a model, or representation, of the specific design using a prescribed method. The computer is then used to rapidly evaluate the results of both the input conditions and the model that the engineer has described. The output conditions are listed by the computer, and the engineer evaluates the consistency of results with his experience, and then determines how the design must be modified to achieve acceptable results. The process is repeated until a successful design is achieved. In this manner the computer *aids* the engineer by calculating results much more rapidly and with greater precision than is humanly possible. The skill and experience of the designer are still reflected in the final results.

The CAE technique can be effectively applied in the injection molding field to:

1. Maximize the probability of first-time plastic part or mold functionality.
2. Solve process problems such as warping, dimensional inconstency, and long cycle times.
3. Reduce molding costs such as: mold start-up costs, part molding costs, material costs, mold rework costs, and scrap and regrind costs.

In this chapter two major types of analyses are reviewed, flow analysis and cooling analyses. The reader is encouraged to review the literature as more and more products are being introduced to serve the needs of the molding industry each year. Use of these CAE tools will assure that mold design and part processing can be accomplished with the greatest possible speed and effectiveness.

INTRODUCTION TO FLOW ANALYSIS

Over the last five years the field of flow analysis has gained increasing importance in injection molding. Flow analysis has provided rational solutions to many of the hard-to-understand effects that cause problems in the injection molding process. These effects have included warping, molded-in stress, excessive fill pressures, part flashing, and others. The interrelationships between part design and molding process parameters that cause problems of this nature were not well understood in the industry. Practical experience often was insufficient to identify potential problems, and was too limited to have encountered the full range of molding problems that can be addressed by techniques such as flow analysis. Hence much prototyping and mold "fine tuning" was necessary before successful molded product results can be achieved.

Computerized flow analysis has emerged as a powerful tool to aid in the implementation of applying injection molding as the production process of choice to a widening spectrum of products. The ability of modern digital computers to perform complex calculations in short periods of time has been the breakthrough that makes flow analysis a tool applicable to increasing numbers of new parts. (Examples Figs. 13-1a, 13-1b and 13-1c) Additionally, technology advances in computer hardware have allowed these flow models to increase in their sophistication and accuracy, while at the same time bringing the cost of the analyses into a range at which they can be applied to a large number of new designs. The flow analysis tools can be successfully applied and utilized by three different groups in the product development process: the product designer, the mold designer, and

finally the injection molder himself. Applications for each of these three groups are detailed below.

Product Designers

Product designers can apply flow analysis to the following questions:

Will the part fill at all? This age-old question concerns many designers, especially those who design large injection molded components such as covers, enclosures, furniture, and the like. The relationships among material structural properties, cosmetic properties, and processing properties are generally hazy in the designer's mind, and flow analysis provides a way to evaluate different materials in the design stage and to evaluate the processing-related characteristics in a scientific manner.

What is the minimum practical wall thickness for the part? This question is actually a primary consideration for the cost of the molded product. The ability to use thin walls on the product results in obvious savings in material (which many times comprises more than 40 percent of the finished product cost). A less obvious advantage, however, is the overall benefit in cycle time for the product using thinner walls. The cooling time of an injection molded part is known to be a function of the square of the wall thickness of the part; so reductions in wall thickness have substantial impact on the cycle time of the molded product. This is of great benefit in increasing the productivity of the molding plant, and thus is ultimately reflected in product costs.

Can gates be located acceptably? The ability of plastic materials to be formed into attractively styled shapes has long been recognized. This has led to an increasing use of plastic materials for applications requiring high degrees of aesthetical appeal. Proper use of flow analysis tools can help assure product designers that sufficient latitude exists in the design to allow gates to be positioned to protect the aesthetic properties of the design while at the same time allowing production of the item at reasonable cost.

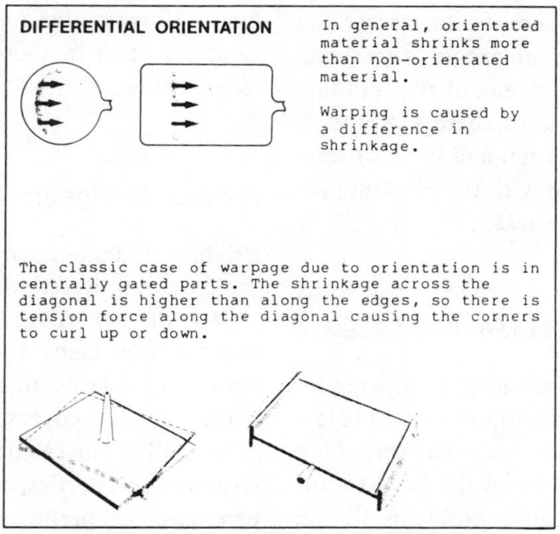

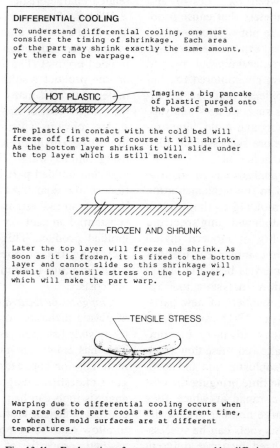

DIFFERENTIAL ORIENTATION

In general, orientated material shrinks more than non-orientated material.

Warping is caused by a difference in shrinkage.

The classic case of warpage due to orientation is in centrally gated parts. The shrinkage across the diagonal is higher than along the edges, so there is tension force along the diagonal causing the corners to curl up or down.

Fig. 13-1a   Explanation of part warpage caused by differing molecular orientation (Illustration courtesy Moldflow Australia Pty. Ltd.)

DIFFERENTIAL COOLING

To understand differential cooling, one must consider the timing of shrinkage. Each area of the part may shrink exactly the same amount, yet there can be warpage.

HOT PLASTIC
COLD BED

Imagine a big pancake of plastic purged onto the bed of a mold.

The plastic in contact with the cold bed will freeze off first and of course it will shrink. As the bottom layer shrinks it will slide under the top layer which is still molten.

FROZEN AND SHRUNK

Later the top layer will freeze and shrink. As soon as it is frozen, it is fixed to the bottom layer and cannot slide so this shrinkage will result in a tensile stress on the top layer, which will make the part warp.

TENSILE STRESS

Warping due to differential cooling occurs when one area of the part cools at a different time, or when the mold surfaces are at different temperatures.

Fig. 13-1b   Explanation of part warpage caused by differing cooling rates in the mold (Illustration courtesy Moldflow Australia Pty. Ltd.)

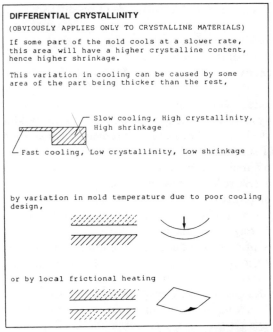

DIFFERENTIAL CRYSTALLINITY

(OBVIOUSLY APPLIES ONLY TO CRYSTALLINE MATERIALS)

If some part of the mold cools at a slower rate, this area will have a higher crystalline content, hence higher shrinkage.

This variation in cooling can be caused by some area of the part being thicker than the rest,

Slow cooling, High crystallinity, High shrinkage

Fast cooling, Low crystallinity, Low shrinkage

by variation in mold temperature due to poor cooling design,

or by local frictional heating

Fig. 13-1c Explanation of part warpage caused by differing crystalline content due to differing cooling rates within the mold (Illustration courtesy Moldflow Australia Pty. Ltd.)

Mold Designers/Moldmakers

Flow analysis can aid mold designers and moldmakers in obtaining the following objectives of a good mold design:

Good Fill Pattern. Of paramount importance in any injection molded component is control of the fill pattern of the molding so that parts may be produced reliably and economically. A good fill pattern for a molding is one that is unidirectional in nature, thus giving rise to unidirectional and consistent molecular orientation in the molded product. This approach helps to avoid warpage problems caused by differential orientation, an effect that is best demonstrated by the warpage that occurs in thin center-gated disks. In this case all the radials are oriented parallel to the flow direction, and the circumferences are oriented transverse to the flow direction. The difference in amounts of shrinkage exhibits itself in terms of warpage of the disk.

Optimal Gate Locations and Number of Gates. In order to achieve a controlled fill pattern, the mold designer must select the number and location of gates that will result in the desired pattern. Flow analysis can help by allowing the designer to try multiple options of gate locations and evaluate the impact on the molding process. This analysis often can be conducted with the product designer to achieve the best balance of gate location for cosmetic impact and molding considerations.

Successful Design of Variety of Runner Systems. In the practical world of mold design, there are many instances where design trade-offs must be made in order to achieve a successful overall design. While naturally balanced runner systems are certainly desirable, they may lead to problems in mold cooling or increased cost due to excessive runner-to-part weights. Additionally, there are many cases such as parts requiring multiple gates or family molds in which balanced runners cannot be used. Flow analysis tools allow successful designs of runners to balance for pressure, temperature, or a combination of both. They also allow an evaluation of the shear rates and degree of frictional heating that will be produced in the runner system, which can aid in avoiding problems of material degradation or excessive melt temperature variation delivered to the mold cavity.

Reduced Rework Costs. One of the major benefits of flow analysis to those who design and build molds is the increased probability that a mold will run successfully the first time in the press. Many mold builders and molders know how expensive repeated rework of the mold becomes in terms of both time and cost. By providing a means to establish the impact various mold design decisions have on the molding process prior to conducting molding trials, much of the time and cost to develop a successful molded product can be eliminated. Alternative designs that may cost weeks of time in rework can be evaluated in a matter of hours or days on the computer. Additionally, a mold that has not been repeatedly reworked will generally have a longer and more productive life.

Injection Molders

Injection molders can anticipate the following benefits of flow analysis of the quality, cost, and processability of the products they produce:

A Wider, More Stable Process "Window." Flow analysis can provide an objective view of the impact of changes of primary injection molding process parameters, namely, melt temperature, mold temperature, and injection speed. By conducting flow analyses on their molded products, molders can evaluate the correct values for each of the process variables and also determine the degree of latitude of the process for the part in question. In combination with the mold designer, they can establish the optimal mold design to allow production of the part on the most cost-effective equipment.

Reduced Stress and Part Warpage. Optimization of the process parameters allows the molder to produce parts with minimal levels of residual stress, which can result in post-molding warpage or even mechanical failure of the product.

Material Savings, Less Overpacking. Balanced flow applied to runner and cavity designs can help reduce the amount of material used in the molding process and eliminate problems such as warping that may be caused by local overpacking inthe cavity. Some molders have reported as much as a 5 percent material savings by using flow analysis techniques, which can lead to substantial benefits on high-production-volume components.

Minimization of Runner Size/Regrind Costs. Flow analyis can aid in optimizing the size of the runner system used in the mold. For the molder, this results in the benefit of minimizing the cycle time by possibly reducing the cooling time of the runner system, and in reducing the volume of runners to be reground or scrapped.

From the previous discussion, it is obvious why flow analysis has been receiving widespread attention in the literature, in research and development, and in practical application. What, then, is the underlying basis for the flow analysis programs in use today?

The basics of flow analysis involve simultaneously solving the equations describing non-Newtonian fluid flow, and the equations describing all of the heat transfer phenomena in the mold cavity. Programs that are currently available use a data base of resin properties developed for the application. This data base is constructed by testing the rheological properties of the material in a controlled manner, and then fitting curves to the data such that the viscosity of the resin can be predicted as a function of any combination of pressure, temperature, and shear rate. With the data in hand, a model of the geometry of the molded product is created, and a mathematical simulation of the filling process for a specific set of process conditions can then be performed. Two major methods of geometry modeling have been used to date, a simplified geometry method and a finite element method similar to that used in structural anlaysis.

In the simplified geometry method, the molding must be redefined into a combination of simple geometric shapes for which the equations of flow and heat transfer can be analytically obtained. These shapes are generally some combination of plates and tubes. Part of the model is also a guess at the initial flow pattern for the mold. Once the model is complete and the data have been entered, the analysis program is run for a specific set of process conditions. The accuracy of the modeler's flow predictions is then confirmed or rejected by the numerical results. The model is then revised and the analysis repeated until the predicted flow model and the numerical results are in agreement. If these results are unacceptable from a molding standpoint, then the geometry (such as gate locations, wall thicknesses, etc.) or the molding conditions are changed until an acceptable solution is found.

The second method in use is to describe the molded product in terms of a set of elements, generally triangular plates, that closely approximate the original geometry. Once the model is built, one or multiple gates can be positioned at "nodes" (vertices on the triangles) of the resulting "mesh." The analysis is then run, again for a specific set of process conditions. The computer now solves the equations, time and time again, until it reaches a consistent set of conditions within the entire mesh. Thus the computer, and not the analyst, has predicted the resulting filling pattern. The results of this type of analysis are still subject to a fair degree of interpretation as to their importance in the molding process.

Each of these two techniques has strengths and weaknesses as applied in practice. The simplified geometry method requires more judgment and experience than the finite element method in building correct flow models in a small number of tries. In applications such as runner system design, however, the flow is easily determined, and the simplicity and efficiency of the approach is readily apparent. The finite element method, on the other hand, allows a single model to be used for all analysis attempts. The drawbacks are that model creation may take slightly longer, and the computation time (and possibly cost) will be substantially greater than for the simplified geometry method. Technologies such as CAD/CAM are aiding in the time required to create finite element models (see Chapter 12), and continuing decreases in computing costs will certainly counteract the second drawback, leading to the conclusion that finite element flow analysis will continue to be an important technique.

In conclusion, there seems to be little doubt that flow analysis will continue to be an important element in injection molding technology. The following discussion on the mechanics of flow analysis will serve to illustrate the methods employed by flow analysis suppliers in solving flow-related problems. These methods involve creating software that is both simple and comprehensive enough to satisfy the modeling requirements for a variety of problems, while providing expedient solutions to the particular problem at hand.

FLOW OF PLASTICS INTO A MOLD

Colin Austin
Moldflow Ltd., Australia

Computer simulation of the injection molding process is not new. In fact, virtually since the introduction of the computer various attempts have been made to develop simulations. Almost all modern computer simulations are based on the work of the early pioneers. The method of analysis is based on a few simple fundamental laws of physics. Unfortunately, the inherent simplicity of the approach is easily lost in the mathematics. The basics are

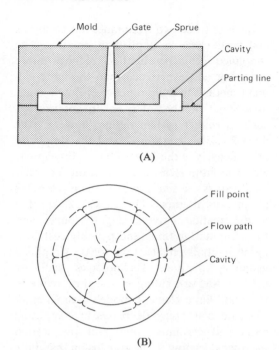

Fig. 13-2a Plastic does not flow uniformly through thin diaphragm of plate mold (A) in the compensating phase, but spreads in a branching pattern, (B).

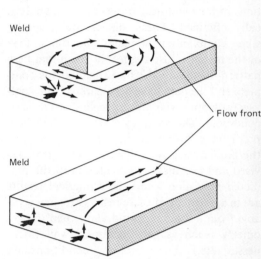

Fig. 13-2b Flow paths are determined by part shape and gate locations. Flow fronts that meet head on weld together, forming a weld line. Parallel fronts tend to blend, or mold, producing a less distinct weld line and a stronger bond.

described here in a simplified way, with due apologies to the original workers for perhaps making their contributions seem less significant than they actually are.

Basic Flow Analysis

There are two physical considerations in the flow of hot plastic into an injection mold: (1) the flow equations, and (2) the heat transfer equations. The method consists of the solution of the simultaneous equations of heat transfer and fluid flow. Classically, of course, the general equations are written as though they were to be solved by a Newtonian integration technique, and then an approximate numerical solution developed. (Figs. 13-2a, 13-2b, 13-3 and 13-4 show plastic flow.)

However, if we skip the stage of formulating actual equations, which is an approach we tend to adopt for historical reasons, and instead go straight into a numerical approach, the inherent simplicity becomes obvious.

The injection period is divided into three stages, filling, compression, and compensating flow. The approach is similar in each stage. In the filling stage either the pressure is set, and the flow rate calculated, or the flow rate is set, and the pressure calculated. In the compression and compensating stages, the holding pressure is set, and the resultant flow calculated.

Consider first the flow equations, and take the simple case, for demonstration only, of a thin rectangular section. Assuming that the section is symmetric about the centerline, the forces on a small block within the element can be balanced, and the pressure pushing the block along gives a force of $P*$width$*$thickness, which is resisted by the shear stresses acting on both faces; that is, shear stress$*$width$*$length. This gives the formula:

$$\text{stress} = \text{pressure drop} * \text{thickness} / (2 * \text{length})$$

This gives us a relation between pressure and shear stress across the section. It is a fundamental relation based solely on resolving forces, and is quite independent of material or flow characteristics.

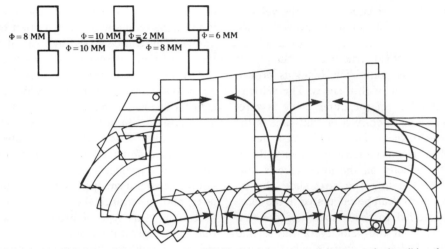

Φ = 8 MM   Φ = 10 MM   Φ = 2 MM   Φ = 6 MM
      Φ = 10 MM   Φ = 8 MM

Fig. 13-3 Describing the part to the computer. The Moldflow user's major task is accurately describing the geometry of the part to be molded so that the programs can analyze melt flow through the mold cavity. Simple parts present no problem, and sometimes flow length and part weight are all that are needed to balance the runner and gate system. (Moldflow conducts seminars to instruct users on all aspects of the system.) For complex parts, such as this headlight surround, a layflat graphic approach is used. The part is "flattened" to create a two-dimensional graphic representation. A series of circular flow fronts are drawn, sectioning the mold. These reflect the radial flow of the melt from the gate and could be thought of as the fronts of successively larger short shots.

Although in practice the viscosity is a complex relationship, of temperature, shear rate, pressure, etc., in a computer program it is usually read from some subroutine, which can be changed at will, depending on the required degree of precision. It could be a simple formula tying viscosity to shear rate and temperature, or it could be a complete matrix based on complex experimental data. All that matters for this demonstration is that if we know temperature and shear stress, the viscosity (or the shear rate) can be arrived at. If we know the viscosity and shear stress, the shear rate can then be calculated, since the definition of viscosity is shear stress/shear rate. Any errors from predicting shear rate come from the viscosity subroutine and not from any mathematical simplification.

(Note that the temperature of the element is taken as known at this point. This is dealt with later.)

Calculations using the marching approach

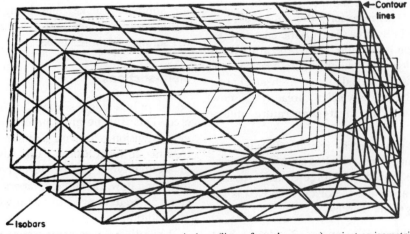

Fig. 13-4 Computer graphics display flow pattern as isobars (lines of equal pressure) against an isometric representation of the part being formed within the mold.

start from the outer edge, where the velocity of the plastic in contact with the wall is assumed to be zero. Ignoring abnormal effects such as jetting, this is true because the plastic is frozen at this point. The block is broken up into a number of thin slices. The velocity of the outer face of the outer block is zero. The increase in velocity over that slice is the shear rate*thickness of slice. (The basic definition of shear rate is increase in velocity/thickness.) The velocity on the inner face is now known. Moving into the next slice, the increase in velocity can be calculated in the same way, and added to the velocity of the first slice to arrive at the velocity of the second slice. If this is done for every block, the complete velocity distribution is known across the section. Multiplying the velocity of every slice by its cross-sectional area gives the volumetric flow in every slice, which can then be added together to give the total flow rate.

To summarize, if we know the temperature across the section and some relation between shear rate and shear stress, the flow rate can be calculated for a set pressure.

Now consider the heat flow equations by considering the heat flow into and out of each slice. There are four basic heat flows:

1. Heat in by conduction, which can be calculated as heat change (per time increment) equals thermal conductivity*area*temperature difference/slice thickness.

2. Heat out by conduction, which can be calculated in exactly the same way.

3. Heat in by flow. The plastic entering the section is at a different temperature from that leaving. The change in heat content again can be calculated as heat change equals velocity*slice area (i.e., flow rate)*specific heat*temperature difference.

4. Heat generated by friction. This is equal to work done, which is simply force*distance moved. The force is shear stress*slice area, and the distance moved is velocity*time increment.

One problem still remains. To do the heat transfer analysis, the flow calculations must have been completed, yet to do the flow calculations, the temperature distribution must be known. This can be solved by first calculating the velocity of the plastic as it starts to enter the section, when the temperature will be equal to the melt temperature, before any heat transfer has had a chance to occur. This gives a velocity distribution from which a new temperature distribution can be calculated, for a small time difference later. Based on this new temperature profile, a revised velocity distribution can be developed for that small time later, and so on until the mold is filled.

Running this type of program has shown that the temperature profile across the section reaches a semi-equilibrium state quite early in typical molding cycles. An alternative approach has been to calculate the equilibrium condition. This is mathematically simple and uses less computer time.

If the solution for the equilibrium condition is reached by an iterative procedure, it is possible to stop the iteration before full equilibrium has been reached. This is mathematically equivalent to the real-world situation, in which the stable temperature profile is not quite reached. Using this approach, quite good results can be obtained. The main errors are in the last section to fill, where the frozen layer has not had a chance to form, so pressures are overpredicted. This seems to be an intrinsic, if minor, weakness of the equilibrium approach which is yet to be solved.

There are definite practical advantages to this approach, the key ones being that significantly less computer power is required, and the solution is more stable.

Multi-sections

So far only a single section has been considered. This is of no great practical use, so the next extension is to string a number of sections together to form a flow path. Also a variety of section types can be developed. Sections for round flow, radial flow, tapers, trapezoids, etc., can be developed to suit a whole variety of conditions.

While this allows some practical problems to be analyzed, the majority of molds have a complex geometry with many different flow paths. The requirement therefore was for a system that could analyze the complex geometries found in practice.

The first system to be developed was the divided flow path method. The flow is considered to be rather like a tree, with plastic entering from one point, and dividing up into an increasing number of different flow paths (i.e., trunk, limbs, branches twigs, etc.). Because the flow can divide, the flow rate in each section of the tree is not immediately known. However, according to the basic flow laws, the flow rate into each section or element must equal the flow rate out of each section. This gives a boundary condition that allows the flow pattern to be calculated.

The divided flow method has proved very successful in practice, enabling a wide range of complex parts to be analyzed. The mold is first divided into a number of flow paths, then each flow path broken into sections. This modeling has to be done manually, usually by the mold designer. Some skill is required to recognize the various flow paths and to be able to align the sections in the direction of flow.

Finite Element Techniques

A further development of this concept is to use a full finite element method. The finite element technique is well known in stress analysis. The basic procedure is to break the structure up into a number of small elements, which are connected at points called nodes. Internal stresses and strains can be connected by a simple relation (i.e., a set of equations) to the forces and displacement at the nodes (element stiffness). Grouping all these equations together for every element gives a large family of equations, which can then be solved to give the nodal forces and displacements, which, when inserted into the stress and strain equations for each element, give the stresses and strains throughout the structure.

Similarly with flow—a relation between flow and pressure at each node can be written for each element, grouped together for the whole cavity, and the whole family of equa-

tions solved. To do this, boundary conditions must be established (i.e. either the pressure or the flow at every node must be known).

The crux of the finite element process is setting the boundary conditions. It is important to realize that one condition must be established at every node. Either a pressure is specified (the pressure is zero at the flow front), in which case the flow rate will be calculated; or a flow rate is specified (at all other nodes, except the flow front, the net flow is zero because flow in equals flow out), and the pressure is then calculated.

One approach is to do a step-by-step analysis in which a small mesh, an estimate of an initial flow pattern, is made. Total flow at each node, which is not part of the flow front, is put to zero (flow in equals flow out), and the pressure at each node, which is part of the flow front, is put to zero. This then allows the equations to be solved so that the flow rate at every frontal point is now known, as well as pressure at every other point.

Knowing the flow rate at the front permits a new flow front to be established, a new mesh drawn, and the procedure repeated until the mold is filled. A new mesh must be generated for each step in the flow. If the mesh is generated by hand, as is the general state of the art, significant manpower is required. With the development of automatic mesh generation, this will be one of the most sophisticated approaches.

A somewhat simpler method is to draw a mesh for the whole structure using small elements, and allow elements to be either full or empty (i.e. the flow front is developed such that it "jumps" from node to node).

A modification of this approach, devised by the author, is to use an iterative scheme, whereby the conditions at the instant of filling only are analyzed, but an iterative procedure is used to find a stable, final fill pattern. The mechanics of the iterative technique are as follows. The mold designer breaks the complete part up into a small number of triangular elements, to form a full finite element mesh. He then selects one node as an injection point and specifies the fill time. Since the volume is known, the flow rate into the cavity from that point is now set. Transparent to the designer, the program will run a finite element analysis assuming a nodal flow rate at each node, which will give a pressure distribution throughout the cavity. In other words the computer has guessed a filling pattern and is evaluating the feasibility of it. This process can be repeated until the flow pattern has stabilized to a true picture.

Simultaneously the temperature distribution is being developed along with the flow pattern. The final output is the pressure and temperature at each nodal point. In practice they can be plotted as isobars or lines of equal pressure.

Benefit Appraisal of the Flow Techniques

The reliability of analysis is evaluated as follows:

- All methods of flow analysis depend on solving the simultaneous equations of heat transfer and fluid flow. The solutions are based on fundamental laws of physics about which there is no dispute.
- The marching approach is the most accurate, using the minimum of assumptions; however, the equilibrium (or partial equilibrium) approach can still give very good results.
- The single flow path model is of value in improving the techniques, but is too limited for the typical complex geometry of practical plastic parts.
- The divided flow path, using a unidirectional element, is very powerful for analyzing runner systems, and used with skill in anticipating flow patterns it can be applied to a large number of complex parts. Since the element must be aligned in the direction of flow, it requires some skill in modeling. The results will quickly indicate if the anticipated flow model is incorrect, allowing a better model to be developed. Again this requires some skill in interpreting results.
- Using more complex elements such as tri-

angular elements, in which the flow can enter at one node, and leave the other nodes in any proportion, means the elements can be placed in any direction, without any concept of the flow pattern, which is then predicted by the program.

Moldflow Basic Technology

The Moldflow programs were developed in Australia, by Colin Austin over ten years ago. Moldflow is a registered trademark of Moldflow Australia. Using a number of different methods, we can describe the part's geometry to the computer. Then, by selecting from a data bank of tested material, we will have the information to run with the part description, through several subroutines within the program.

Within the main program for Moldflow, there is a simple procedure that gives a selection of 13 choices:

1) TO PRINT MOLD FILE
The computer works in meters, which is not the most readable system; so option 1 lets us print the dimensions of the mold file in meters, millimeters, or the inches that we are more familiar with.

2) TO GO TO REDIMENSION SUBROUTINE
As an analysis develops, some dimensions will need to be changed within the mold file. This option allows us to manually change any section.

3) TO CHANGE MATERIAL FILE
Often a molder will be looking at two or more materials. At this time there are more than 500 materials on file in the data bank. Because there are over ten times that number of materials, many suppliers are beginning to test their material to the Moldflow standards. This will help sell and support material choices.

4) TO CHANGE MOLD FILE
Because each part must be looked at by itself, a provision has been made to bring into the program any one of a number of different mold files. Also different runner systems can be viewed under the same conditions.

5) TO ANALYZE SINGLE FLOW

Within a single part, many different directions are taken by the plastic as it fills the cavity. This option will allow the study of these flows, one at a time.

6) TO ANALYZE ALL FLOWS

There is also the provision to look at the complete system, to see what happens to every flow within an analysis.

7) TO SCAN INJECTION TIME

As the designer is first developing his (or her) analysis, an average mold and melt temperature is first established. Using this as a trial setting, the correct fill time may be expeditiously brought about.

8) TO FLOW BALANCE

Once everything is known about the flow of the material through the system, speedy balancing of each flow is done with this option. Certain sections are set as being changeable by the computer to arrive at a total pressure for all of the flows.

9) TO MAKE EQUIVALENT RECTANGLE

In order to make it easier to balance a runner system, the part can be turned into one equivalent section that has the same pressure. This way, very long mold files are eliminated.

10) TO STORE RESULTS

As an analysis is progressing, some printouts are discarded as being too high or too low, too hot or too cold. Only the results that are of value to the analysis are saved and put into a store file, waiting to be printed out for the final report.

11) TO SPECIFY FLOW RATE

Once the optimum conditions have been established for the part, the flow rate is known. This then can be used to set the fill time for the runner file.

12) TO COPY CURRENT MOLD FILE

As the computer or the operator changes sections, by wall thickness or diameters or flow lengths, these descriptions can be saved under their own file name for later use.

13) TO END

Using a few subroutines, the designer has been able to do what has only been, up to this time, possible by trial and error. It was often at great expense, with many delays and sometimes with disastrous results.

INTRODUCTION TO MOLD COOLING ANALYSIS

Prior to the introduction of computer modeling tools, the cooling of injection mold cavities was a complex and often misunderstood phenomenon. The placement of cooling lines in the mold was often the last portion of the design to be considered, resulting in many molds that ran at low levels of productivity and producing molded parts of questionable quality. Modern techniques that allow the evaluation of a proposed cooling system design prior to construction of the mold are allowing mold designers to thoroughly evaluate the quality of their designs, and to revise those designs to achieve more efficient and balanced cooling of the mold. These techniques are generating large savings in cycle time, parts cost, and tryout and "tuning" time for new molds.

The importance of proper cooling system design can be summarized in two major areas of interest to the plastics industry: quality and productivity. Quality has become a major area of emphasis in the industry over the last few years. The quality of the product that can be produced from an injection mold can be directly attributed to three factors: the accuracy of fabrication of the mold cavity and core, the repeatability of the molding machine used, and the correct design of the mold to produce the part. Part of the correct design of the mold is a cooling system that will extract heat from the melt in the mold cavity at the maximum rate possible, and uniformly throughout the mold.

Several molding quality problems can be avoided if uniform cooling is designed into the mold. Warpage of the molded part is one problem whose roots may lie in the proper design of the cooling system for the mold. Another cooling problem that can be related to cooling system design is the homogeneity of properties in parts molded from crystalline resins. The morphology of the crystal structure developed in the molded part is strongly dependent on temperature history and cooling times. Minor changes in these variables can lead to major changes in crystal formation and thus in the mechanical properties of the molded product. Thus molds with uneven cooling profiles may produce parts with unacceptable mechanical properties, possibly resulting in premature mechanical failure of the molded product.

The second major benefit of effective cooling system design is the impact on productivity. Molds that can be designed with optimum cooling will produce parts in the shortest possible cycle time. This results in direct cost savings of several types. It has been shown many times that large energy savings on injection molding equipment can be gained by decreasing cycle time. Additionally, effective cooling system design can result in the use of higher coolant temperatures which imply less chiller capacity requirements or less plant water usage. There are also capital equipment savings accompanying these benefits. Capital outlays for new equipment can be minimized, since shorter cycle times reflect themselves in higher plant capacity without an increase in the number of molding machines. Likewise, maximizing the cooling water temperature needed to cool the mold results in a reduction in chiller requirements and corresponding savings in equipment costs.

The preceding paragraphs provide strong arguments for careful consideration of the cooling system design for injection molds. Successful application of cooling analysis tools can result in higher quality of the molded product as well as lower operating and production costs for the molder.

Fundamentals of Cooling Analysis and Design

In examining the problem of cooling of plastic parts being formed by injection molding, it is possible to separate the problem into three distinct elements:

1. Cooling of the melt
2. Conduction from the melt to the waterline
3. Convection cooling by the waterline

These elements must be considered in combination in order to understand the cooling performance of the mold. Improper design of the mold may result in a high thermal resistance that will lead to the quality and productivity problems previously mentioned. The three cooling elements are discussed below.

Cooling the Melt. If the first stage of the cooling process is considered, the heat contained in the melt must be transferred to the mold material via conduction. The basic theory for this type of problem is developed in solving the transient one-dimensional heat conduction equation for an impulse change in temperature corresponding to the entry of the melt into the mold cavity. The solution of the equation takes the form:

$$T_c = f\ (S^2,\ \text{Alpha}_{eff},\ \text{Delta}_{temp})$$

where T_c = theoretical minimum cooling time, S = plastic part wall thickness, Alpha_{eff} = effective thermal diffusivity of the plastic, and Delta_{temp} = ratio of differences between melt and mold temperature and part ejection and mold temperature.

There are several interesting points to be made about this relationship. First, the cooling time is a function of the square of the wall thickness of the molding. For the product designer who believes that bigger is better, this indicates that a severe penalty is being paid for unnecessary increases in wall thickness. Not only is the material usage of the part being affected, but also the productivity of the manufacturing process. Second, in considering the cooling time, the effective thermal diffusivity of the plastic must be taken into account. This value may differ significantly from a measured value, especially for crystalline resins, because of the latent heat of fusion. In highly crystalline materials such as polyethylene, as much as 40 percent of the total change in enthalpy between melt temperature and ejection temperature may be attributed to the latent heat of fusion. The accuracy of the diffusivity value used in the equation is therefore extremely significant in obtaining accurate predictions of cooling time. Last is the point that the melt, mold, and parts ejection temperatures all play an important part in determining cooling time, as might be expected.

Conduction in the Mold Wall.

The basic theory behind conduction in the mold wall is Fourier's law of heat conduction, which states:

$$Q = -KA \, dT/dX$$

where Q = the heat transfer rate (Btu/hr), K = the thermal conductivity of the mold material (Btu/hr-ft-degree F), and dT/dX = the temperature gradient in the wall (degree F/ft).

When this equation is solved for heat flow through a plate whose temperature on one side is represented by T_1 and on the other side by T_2, and whose area is A and thickness L, the governing equation becomes:

$$Q = -K(A/L)(T_1 - T_2)$$

The quantity A/L is solely a function of the geometry of the wall, and is sometimes called the conduction shape factor in handbooks of heat transfer.

In a similar manner to the above, many simple variations in geometry can be analytically solved. An example typical to the molding world would be the heat flow in a solid with a row of holes. In this case the geometric shape factor takes the form

$$S = \frac{2(Pi)}{\ln\left((2P/(Pi)D)\sinh\left(2(Pi)X/P\right)\right)}$$

where S = the shape factor, P = distance between holes (sometimes called "pitch"), D = diameter of the hole, and X = distance from the hole to the surface being cooled (or depth).

From the previous example it can be seen that the larger the value of the shape factor, the higher the rate of heat conduction through the mold wall. By solving equations such as the one describing the row of holes in a solid wall, it can be demonstrated that in order to increase the rate of conduction heat transfer,

one should decrease the distance between waterlines, decrease the depth of the waterlines to the molding surface, or increase the diameter of the waterlines. The last possibility for increasing conduction cooling effectiveness is by changing the mold construction material to one of a higher thermal conductivity value.

There are practical limitations to the implementation of the above suggestions that should be noted, however. Decreasing the depth from the waterlines to the molding surface to an extreme could result in very uneven cooling at the surface, thus causing other problems in the molding process. Also, increases in the diameter of the lines will result in higher coolant flow rates, which may require larger pumping systems in order to sustain adequate flow. Spacing waterlines too close together can result in decreasing the structural integrity of the mold, and could result in a mechanical failure of the mold due to withstanding the stress of injection pressures. Last, the selection of mold construction materials should be optimized by considering multiple factors, such as strength, wear resistance, polishability, corrosion resistance, and also conductivity.

Convection Cooling in the Waterline.

The last element in the mold cooling process is the convection cooling that takes place in the waterline. In this process, the water flowing through the line removes heat from the mold wall and carries it out of the mold to a point in the surrounding environment where it can be disposed of. In solving the convective cooling problem, the major variables that occur are related to the specific properties of the coolant being used, the nature of flow within the system, and finally the temperatures involved. The heat transfer coefficient for convection is established by the relationship:

$$h = (K/D) \, f \, (N_{Re}, Pr)$$

where h = the heat transfer coefficient, K = the thermal conductivity of the coolant, D = the diameter of the waterline, N_{Re} = Reynolds number, and Pr = the Prandtl number. The Reynolds number is a dimensionless quantity used to characterize the flow of coolant within the coolant channel. The flow is often characterized in three major flow regimes: laminar flow, transition turbulent flow, and fully developed turbulent flow. Laminar flow occurs for values of the Reynolds number less than 2100. The flow in this regime may be characterized by lamina, or layers of fluid moving at different velocities. Fully developed turbulent flow occurs for values of the Reynolds number greater than 10,000. In this regime the fluid is constantly mixing, and individual flow patterns are not distinguishable. Last is the intermediate, or transition, turbulent flow regime which exhibits characteristics different from the other two regimes. The Prandtl number is a measure of how rapidly momentum is dissipated compared to the rate of diffusion of heat through a fluid, or can be defined as the kinematic viscosity of the fluid divided by the thermal diffusivity of the fluid. By examining the many equations for the convective heat transfer coefficient proposed in the literature, we can observe that the impact of changes in the Prandtl number is small relative to changes in Reynolds number. Most references say that the heat transfer coefficient for laminar flow is a function of velocity to the $\frac{1}{3}$ power, while for turbulent flow it is a function of velocity to the 0.8 power. Thus there is great benefit in minimizing internal heat flow resistance in the mold by maintaining turbulent flow conditions in the mold. Despite this, many molders limit their productivity by "trickling" cold water through the mold at low flow rates rather than using a mold temperature controller to pump water at turbulent flow rates.

MOLD COOL ANALYSIS

Keith R. Schauer
AEC Corporation
Wood Dale, Illinois

Combining the Three Effects. Actual performance of the cooling system in the mold is a combination of all three heat transfer problems in combination. The overall heat transfer coefficient per unit length of cooling line in the mold is given by the relationship:

$$1/U = 1/KS + 1/(Pi)\,Dh$$

where U = the overall heat transfer coefficient, K = the thermal conductivity of the mold material, S = the conduction shape factor, D = the diameter of the cooling channel, and h = the convective heat transfer coefficient. The major objective in the mold design process is to design the mold cooling system such that this coefficient may be sufficiently large to allow cooling of the melt at its maximum rate, while avoiding any of the undesirable cooling effects previously mentioned. Commercially available software packages are designed to make this a relatively easy task in relation to the governing equations of fluid mechanics and heat transfer that describe the process. A well-written software package in the mold cooling area should make the modeling of the process sufficiently easy to be applied, as well as providing a means to optimize cooling designs to a reasonable level of precision. In this manner they may be successfully applied in the classical definition of computer aided engineering, proposing a design, analyzing its effectiveness, and revising the proposal. This process is continued until an acceptable compromise of all conditions is reached with the design.

Productivity Improvement

Effective heat transfer is the prime concern of mold analysis. This means that heat added to a mold by plastic and by any other sources must be removed as quickly as possible. Cooling is efficient if the heat removal is effected with the smallest possible expenditure of capital and energy. (See Chapter 16 for details on cooling equipment.) There are other concerns that are important and should be mentioned here. Maximum mold performance and proper chilling system design are two important results of proper mold analysis. The optimum operating conditions that are determined by this process include type of coolant, flow rate, supply pressure, and temperature. In fact, cycle time can be accurately predicted even before the mold is built.

Specifically, computerized mold analysis can:

- Improve part quality—even and balance cooling in the mold. This eliminates the thermal stresses that cause warping. Product impact strength is increased. Proper surface appearance is assured.
- Increase production. Optimum mold heat rejection reduces cycle time. Increases of at least 15 to 20 percent usually follow implementation of mold analysis recommendations. In some cases, machine productivity has increased by more than 80 percent.
- Reduce chilling costs. Mold heat transfer analysis provides the data needed to select the optimum flow, pressure, and temperature levels for the coolant used in molding machine chilling systems. This takes the guesswork out of sizing chilling systems and eliminates unnecessary expenditures on oversized chillers and pumps.
- Reduce capital investment. A properly designed mold, cooled by an effective chilling system, produces more parts; fewer molding machines are needed to meet production requirements. This major capital saving is augmented by reductions in chilling tonnage and mold purchases. The ability to get along with fewer machines means a big savings in plant floor space.
- Reduce operating and maintenance costs. Optimum operating conditions make the most efficient use of energy. Naturally the operation and maintenance costs are lower.

As previously mentioned, heat must be removed from the mold as quickly as possible.

To facilitate this the analysis has been formulated, and a discussion of the procedure follows. In the mold itself, resistance depends upon the heat conduction properties of the material of which the mold is made, the size and design of cooling passages, and the placement of the cooling passages with respect to the part being molded. The production rate of injection molded thermoplastic parts, and the percentage of those parts that conform to acceptable quality requirements, can be increased substantially by improving the heat transfer from the mold. This improvement can consist of (1) accelerating the rate of heat transfer and (2) balancing heat transfer evenly throughout the mold.

The opportunity to increase the rate of heat transfer stems from the basic character of the injection molding cycle, which is made up of three segments: (1) melt injection, (2) cooling, and (3) parts ejection. In a typical molding cycle, injection time and ejection time combined account for only 20 percent of total cycle time. The remaining time segment of the cycle, 80 percent, is consumed with cooling of the part. Therefore, reducing the duration of the cooling segment of the molding cycle can effect a significant reduction of total cycle time.

The potential for improving the consistency of quality of finished parts also is a function of heat transfer from the mold. When heat flow from the mold is not balanced, the result can be differential shrinkage, residual stress, and/or warping of the molded part.

At present, despite the importance of rapid, balanced cooling to cost-effective production, in many cases mold-cooling decisions are made on the basis of assumptions or habitual practices, without any assurance that the cooling system is the best that can be designed for a particular material, part, or mold.

The inadequacy of traditional methods becomes most evident when an existing mold with cooling passages designed by such a method is reevaluated with the aid of a CAD program. Such was the case with the mold illustrated in Fig. 13-5. Often, as in the case of this mold, the performance of an existing mold can be improved by a change in heat transfer operating conditions. It was discovered that total cycle time for this mold could

Slide Cube
Original Mold Design

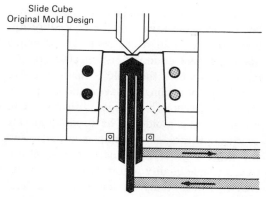

Fig. 13-5   Original design for slide cube mold, with inefficient bubbler. (Four-cavity mold using general-purpose
polystyrene.)

be reduced from 18 seconds to 16 seconds
simply by utilizing optimum coolant temperature and coolant flow conditions.

In addition, computer-aided evaluation of
the layout and circuiting revealed a few revisions that would make the mold more efficient.
The tube bubbler in the existing mold was
not efficient because the area inside the tube
was quite a bit smaller than the outside. A
larger-diameter tube would have been better.
The ground rule in selecting a tube should
be to divide the area of the drilled bubbler
equally so that the inside and outside areas
of the tube are equal. Replacing the tube with
a baffle did away with unequal areas, and heat
removal was considerably improved. Another
line was also milled to cool the gate area. The
revised mold, shown in Fig. 13-6, operated

Slide cube
revised mold design

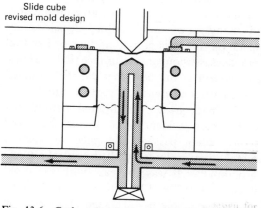

Fig. 13-6   Cycle reduced 12 seconds using design for
slide cube mold. Bubbler has been replaced by a baffle,
and an additional line has been milled to cool gate area.

at optimum conditions determined by the
computer, resulted in 12 seconds cycle time.
Table 13-1 shows an economic analysis. It is
evident that proper mold design and operating
conditions can result in savings of thousands
of dollars.

Computer Aided Design for Mold Cooling

A computer program has been developed. The
geometric data required are obtained from the
drawings of the mold being analyzed. These
data, and surface temperature data compiled,
are fed to the computer. The computer is programmed to provide a printout showing the
various flow rates per geometric segment of
the mold together with the corresponding temperature rise and heat removal information
for each segment. The printout results take
into account the previously mentioned variables together with Reynolds number and
Prandtl number calculations and their effect,
plus the overall coefficient of heat transfer.

The program is established to handle these
variables through a temperature range of from
$-4°F$ to $+150°F$ or higher for many different
coolants (pure water, a 25% ethylene glycol/
75% water solution, and a 50% ethylene gly-
col/50% water solution, etc.). The mold analysis is then completed by graphically interpreting the printout results showing the heat
removal level vs. the flow rate and corresponding pressure drop for each geometric segment.
A recommendation regarding proper circuiting is made, to be used along with the optimum operating conditions. Proper circuiting
is important because too many circuits in parallel can cause unnecessary high flow requirements for efficient heat removal, and too many
waterlines in series can cause high pressure
drop and uneven mold surface temperature
due to high temperature rise of the coolant.
(See Fig. 13-7.) Recommendations for proper
hose and manifold sizes are also included. As
a rule of thumb, the diameters of hoses and
connectors should be at least equal to the size
of the mold passage to avoid a pressure drop
and loss of cooling efficiency.

Each waterline is analyzed individually to

Table 13-1 Economic analysis of improved slide cube mold performance.

	①	②	③
Cycle Time (seconds)	18	16	12
Production Rate Parts/hr	800	900	1200
Improvement in Production Rate	—	12.5%	50%
Machine & Labor Cost $/hr	16	16	16
Hourly Savings	—	$2.00	$8.00
Yearly Savings ④	—	$12,000	$48,000

① Existing Cooling System.
② Cooling System recommended by computer.
③ Cooling System and mold revision as recommendation by computer.
④ Based on 6,000 operating hours per year.

maintain balanced cooling. The core and cavity halves are balanced against each other to achieve approximately equal heat removal from both halves. A combined graph is then obtained to represent the total mold. The optimum operating conditions for the mold are determined by considering slopes and magnitudes of various heat removal and pressure drop curves. A chilling system is then designed that will match the mold and produce optimum operating conditions. To make all the necessary calculations by hand, it is estimated, would require 52 hours for a typical six-circuit mold. With the computer the entire procedure is not only manageable but promptly available.

In the past, it was not practicable for mold designers or plastics processors to familiarize themselves with the complex mathematics involved in the science of heat transfer. Therefore, the kinds of modifications of coolant operating conditions and mold cooling passages described above were not made.

Computer aided design of mold cooling passages is based upon a mathematical analysis of unsteady heat transfer from the plastic in the mold to the surface of mold cooling passages. The term *unsteady state* refers to the fact that temperature at any point of the injected plastic part is continuously changing.

The variations involved in this kind of analysis are:

- The geometry and properties of the plastic.
- The initial and final temperature of the plastic and the mean temperature of the cooling medium.
- The overall heat transfer coefficient combining the basic elements of the materials involved (i.e., properties of the mold material and design and the coolant itself).

The properties of the plastic itself are its thermal conductivity, specific heat, viscosity, and density. The low conductivity of most plastics can be improved somewhat by the use of fillers, reinforcements, and other additives.

The cooling of the plastic is a function of

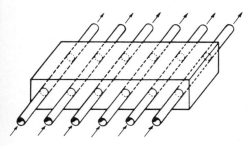

Water lines in parallel

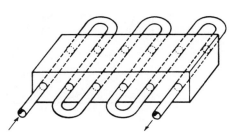

Water lines in series

Fig. 13-7 Waterlines in series and waterlines in parallel.

two dimensionless numbers. The unsteady state analysis gives the equation:

$$Y = \phi\left[\frac{hl}{2k} \times \frac{4\alpha\theta}{l^2}\right]$$

$$Y = \frac{T_0 - t}{T_i - t}$$

where Y is the ratio of final temperature difference and the initial temperature difference, h is the heat transfer coefficient, l is thickness of the material, k is thermal conductivity, α is thermal diffusivity, θ is cooling time, T_i is initial temperature of the material, T_0 is final temperature of the material, and t is temperature of the cooling medium.

As shown in Fig. 13-8, Y has been plotted as a function of $hl/2k$ and $4\alpha\theta/l^2$. The chart shows that Y decreases with increased cooling time. The initial rate of change in Y is slow; then it follows a steep slope, and finally again the slope becomes gradual. It should be noted that reduced thickness has the greatest effect on Y. Y decreases almost with the square of thickness reduction. Y also decreases with increased heat transfer coefficient. Note that Y tends to become constant for larger values of h; therefore, for each thickness there is a limit, and the cooling time cannot be reduced beyond that limit. The above equations and the graph, along with the boundary conditions for a particular problem, can be utilized to determine the minimum time required to cool different thicknesses of various plastics.

The foregoing equation is one among many formidable computations involved in understanding heat transfer from the mold and developing optimum cooling conditions for any given part. Such calculations are built into CAD programs for optimum cooling.

This section of the chapter is designed to familiarize mold designers and plastics processors with such programs, and also with the information they need to make best use of them for the design and circuiting of mold cooling passages.

System Operation—Overview

MOLDCOOL is a new system of computer-aided mold design that provides users with a variety of access methods. A user can access the program through a local phone call to a time-sharing network that allows the user to only pay for the computer time needed.

The user can choose from a menu of programs that (1) match the heat load with the cooling capacity of the mold so as to expedite the design of mold cooling channels; (2) discern areas of the mold in which cooling is deficient so that depths, pitch, and diameter

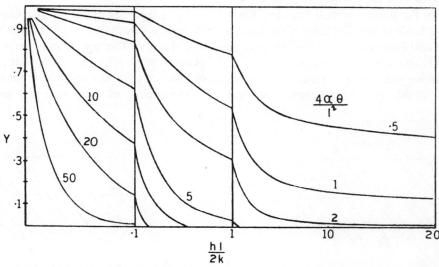

Fig. 13-8 The ratio of final temperature difference and initial temperature difference. Y plotted as a function of $hl/2k$ and $4\alpha\theta/l^2$.

of the cooling channels can be modified during the design process; and (3) select optimum coolant flow, temperature, and pressure conditions to balance heat removal from the mold.

If the program is loaded onto a CAD/CAM system, the user utilizes a digitizer to vary the design. the CAD programs automatically provide the user with optimum conditions and cycle time information for any given cooling design. The MOLDCOOL program also includes a design program that aids the mold designer in the optimization of cooling design.

The sequence of programs in the new computerized mold cooling design system is depicted in Fig. 13-9. All of the following discussion refers to that diagram.

MOLDCOOL uses interactive or "menu programming" whereby the computer actively assists the user in organizing the information needed to proceed with complex calculations. Interactive programming eliminates the need for a technical background or knowledge of special programming language. The computer asks a series of questions relating to the problem to be solved and helps the inexperienced user to organize the information needed to run the program.

Following computer access, the first step in MOLDCOOL analysis is to develop a mold file. The mold file identifies the material of which the mold is constructed. The computer lists the common materials used in mold con-

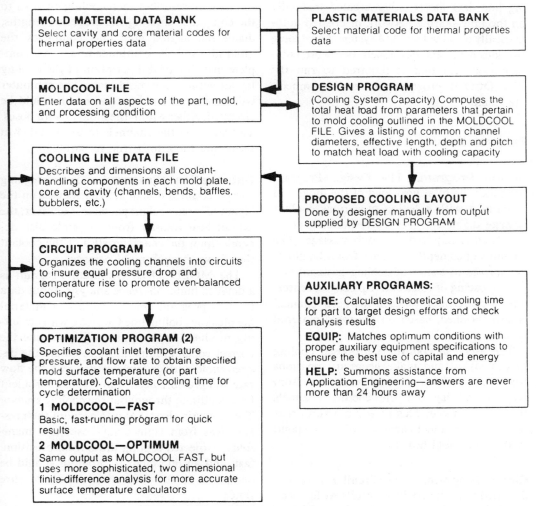

Fig. 13-9 MOLDCOOL computer-aided design system.

struction with corresponding codes. The user enters a code number for core and cavity that automatically describes the thermal properties of mold materials to the computer. This is also done for the plastic material type. AEC maintains its own plastic material testing service, which is available free of charge to authorized MOLDCOOL users.

The mold file is completed by entry of all other heat input variables such as hot-runner system capacity (if applicable), part geometry, part weight, total shot weight, and general processing conditions.

The user must also develop a cooling line data file. This file describes to the computer the interrelationships between the cooling line arrangement and part configuration. The cooling line data file is either created manually on the time-sharing network or created automatically on a CAD/CAM system. Once these two files have been developed, two-word commands are all that is required to run the MOLDCOOL programs and receive immediate results.

Program Menu

Design Program. The Design Program computes the total heat load of the mold and matches it with a listing of various coolant passage diameters with corresponding effective lengths required for each passage. The recommended depth (distance from the molding surface to center of cooling passage) and pitch (spacing of consecutive passages, centerline to centerline) for each diameter channel is also specified, based on the mold material used.

The user can confirm a proposed cooling system layout and make necessary revisions based on design's output; or, start from scratch and lay out the cooling channels through the mold. This program assures that the cooling channel capacity will correspond to the total mold heat load.

Circuit Program. The Circuit Program is designed to create cooling circuits within each mold plate so that each cooling channel's pressure drop is approximately equal. This pro-

motes even, balanced flow throughout the mold to avoid problems such as residual stress, differential shrinkage, and warping due to uneven cooling. Altogether, this program will handle up to six mold plates with up to 60 cooling channels.

Proceeding with the analysis, the user selects from a menu of channel types to be analyzed as follows:

1. Straight sections (round flow)
2. Straight sections (rectangular flow)
3. Circular sections (round flow)
4. Circular sections (rectangular flow)
5. Bubblers
6. Bafflers

The user describes each cooling channel to the computer, entering the branches, lengths, diameters, angles, etc. This completes the mold file, and the computer now has a complete mathematical description of the cooling layout within each mold plate. The computer will then print out how the mold should be circuited, either within the mold or with hoses. This includes the channels to be circuited in series and those to be run in parallel.

Optimization Program. The foregoing computer-aided design activity focuses on the layout of mold coolant passages. However, the rate of heat transfer from the mold also depends upon the characteristics of the coolant that flows through mold passages.

The MOLDCOOL Optimum Program is a detailed finite difference analysis. Input data for this program are the mold and material file plus the cooling line file along with circuiting results. The computer will prescribe the correct coolant temperature to obtain the desired mold surface temperature, and the flow rate and pressure needed to obtain a turbulent flow condition throughout the cooling system. This analysis also takes into account the pressure loss from connecting hoses and manifolds. Projected cooling times from precision, fast, and an intermediate program should be compared to the theoretical minimum cure time.

The MOLDCOOL Fast Program is a fast 1-D analysis for quick, economic results.

Auxiliary Programs. For any combination of plastic material, processing temperature, part configuration, and mold material, there is a theoretical minimum cooling time. The optional cure program enables the designer to compare the cooling time projected on the basis of the coolant passage layout and coolant characteristics with the theoretical minimum cure time. If there is a large difference, the designer would be well advised to close the gap, if possible, by modifying the material of which the mold is made, the mold cooling design, or the application of coolant.

The optional equipment program provides specifications for the auxiliary equipment that will be needed to support optimum cooling conditions and make the best use of capital and energy. This information will indicate whether available equipment will suffice to run the job or new equipment will be needed.

Practical Approach to Mold Cooling Design

Mold cooling passages do not function in isolation, but operate in what should be considered as a total heat transfer environment, comprised of a number of interrelated variables that affect cooling time. Some of these variables involve decisions that, in the normal sequence of mold design, are made prior to the layout and circuiting of mold cooling passages. In some cases, these decisions involve balancing the relative importance of functional considerations of plastic flow against improving a mold's function as a heat sink.

As an example, a mold designed with a short sprue and runner system might be considered desirable from the point of view of plastics flow; but it might be difficult to cool such a mold uniformly, because of an excess concentration of heat on the center of the mold. As a result, no conceivable cooling passage layout would be adequate for removal of heat from the mold at a rate suitable for economic production. The following factors are not all of equal importance in terms of practical choices available to the mold designer, mold builder, and processor. However, they do merit consideration whenever a part is to be molded.

Part Material. As a rule, the selection of the material of which a part is to be made is based upon the design and performance characteristics of that material and its cost. However, in cases where performance specifications permit the substitution of one material for another, it may prove cost-effective to make the substitution. In some cases, the price premium paid for the alternative material is far more than offset by a reduction in cycle time which translates into greater productivity and therefore greater revenue.

Part Thickness. It is generally recognized that resistance to heat transfer increases with part thickness. In some cases, cost of material permitting, it may be possible to substitute an alternative material that will provide the requisite physical and chemical properties and yet permit production of a thinner-walled part. Table 13-2 shows the relationships between cooling time and part wall thickness for several commonly used plastic resins. Various materials have considerably different cooling times for the same part thickness.

Nonuniform part thickness is sometimes an unavoidable aspect of the original part design. It can present problems with providing uniform rates of cooling and therefore result in shrinkage of the part with consequent molded-in stresses and warpage. There are compensating measures that can be taken in the design and operation of the mold to solve problems of differential thickness.

Mold Material. Heat transfer from the mold also is affected by the materials of which mold cores and cavities are made. Table 13-3 lists commonly used mold construction materials and the thermal conductivity of each material. Aluminum 2017 has the highest thermal conductivity. Steel has great structural strength. Most molds are made from tool steel, which offers the best compromise between strength and conductivity.

As a rule, the choice of material reflects the judgment of the moldmaker and the processor. In most cases, it is not practicable to alter the choice of structural material for the sake of heat transfer characteristics alone. It is important to avoid relying on heat transfer

Table 13-2 Cooling times for various wall thicknesses of commonly used plastic materials. (Equals pressure hold time plus $\frac{1}{2}$ injection time seconds.)

MAX WALL THICKNESS, MIL	IN.	ABS	NYLON	HDPE	LDPE	PP	PS	PVC
20				1.8		1.8	1.0	
30	$\frac{1}{32}$	1.8	2.5	3.0	2.3	3.0	1.8	2.1
40		2.9	3.8	4.5	3.5	4.5	2.9	3.3
50	$\frac{3}{64}$	4.1	5.3	6.2	4.9	6.2	4.1	4.6
60		5.7	7.0	8.0	6.6	8.0	5.7	6.3
70		7.4	8.9	10.0	8.4	10.0	7.4	8.1
80	$\frac{5}{64}$	9.3	11.2	12.5	10.6	12.5	9.3	10.1
90		11.5	13.4	14.7	12.8	14.7	11.5	12.3
100		13.7	15.9	17.5	15.2	17.5	13.7	14.7
125	$\frac{1}{8}$	20.5	23.4	25.5	22.5	25.5	20.5	21.7
150		28.5	32.0	34.5	30.9	34.5	28.5	30.0
175	$\frac{13}{64}$	38.0	42.0	45.0	40.8	45.0	38.0	39.8
200		49.0	53.9	57.5	52.4	57.5	49.0	51.1
225		61.0	66.8	71.0	65.0	71.0	61.0	63.5
250	$\frac{1}{4}$	75.0	80.8	85.0	79.0	85.0	75.0	77.5

between the mold walls and mold inserts. Metal-to-metal interfaces are poor conductors.

Computer-Aided Cooling Passage Design

About ten years ago, Application Engineering Corp., USA, developed computer software to optimize mold cooling. Originally, processors and mold designers would submit their own drawings to the AEC headquarters, where that company's engineers, working with the computer, would develop a set of recommendations that could then be implemented by the mold designer. This arrangement is still available.

It has been augmented by AEC's MOLD-

Table 13-3 Thermal conductivity of materials used to construct molds.

MATERIAL	THERMAL CONDUCTIVITY BTU/°F FT HR	RELATIONSHIP TO TOOL STEEL
Tool Steel P-20	21.0	1.0
Stainless Steel 316	9.4	.4
Kirksite	60.4	2.8
Beryllium Copper 25	64.0	3.0
Aluminum 2017	95.0	4.0

COOL, a program available on a time-sharing basis worldwide. The MOLDCOOL program also is licensed to companies with in-house computer systems and CAD/CAM systems. The program is interactive—that is, it involves active collaboration between the mold designer and the computer in the solution of cooling passage design problems.

Specifically, as shown in Fig. 13-9, the program handles or assists in the following design steps:

1. Calculation of the heat load for the application.
2. Calculation of various channel sizes, thermally equivalent lengths, and pitch (spacing) required to handle the calculated heat load.
3. Circuiting the mold designer's cooling channel layout to provide equal pressure drops and, therefore, uniform heat transfer characteristics throughout the mold.
4. Optimization of flow rate, pressure, and temperature of coolant.
5. Calculation of cooling times—both the theoretical minimum time and the projected time for the mold design.
6. Prediction of mold surface or part surface temperatures.

7. Provision of specifications for the auxiliary equipment needed to support optimum cooling conditions.
8. Provision of an economic analysis on the benefits available by using the program.

This section is concerned primarily with the layout and circuiting of mold cooling passages (items 1 through 3 above). However, the remaining aspects of the program are reviewed in other sections of this book.

Input Data. The following input data are required from the user in order to initiate a dialogue with the computer.

1. Identification of the plastic material.
2. Identification of the material of which the mold cavity and core are constructed.
3. Hot runner design and capacity (if any). If the full rated kW of the system is not available, it can be obtained by consulting the manufacturer of the system.
4. For molds with a running history—assuming that the existing mold is being redesigned for improved performance—the existing cooling time, which is defined as the total clamped closed time minus approximately one-half of the injection time.
5. Total parts surface area, which is the sum of both faces (top and bottom, and inside and outside) for one cavity. In the case of multicavity molds, parts surface area should be multiplied by the number of cavities of the mold.
6. Material injection temperature. Normally, injection temperature is measured at the nozzle. This measurement is adequate, even though, in conventional sprue and runners systems, the melt can increase in temperature owing to high friction shear as the material is pushed in; or, conversely, the material can lose heat if the sprue and runners are oversized. In the case of hot runner systems with probes, control setting temperatures should be used.
7. Desired parts surface temperatures (core and cavity). If this information is not available, it is best to consult a technical representative from the supplier of the material being used to produce the part. Surface temperatures are particularly important with respect to filling the mold, obtaining desired part surface appearance, and achieving complete cure of the mold in order to avoid postmolding warpage. (The core side of the mold can be considered to be the same as the ejection side; the cavity side should be considered to be the same as the nonejection side of the mold).

Important: Another means of obtaining surface temperature information is to make actual measurements of the temperatures of the parts surfaces of successfully running existing job material in which the part is made of the same material and has the same or similar configuration to the new part under consideration.

These measurements can be made using a digital pyrometer with surface contact thermocouple probes or an infrared heat scanning gun. The measurements must be performed very quickly. Three measurements should be taken for each mold half during the course of each of six cycles. The part's surface temperatures should be close to the part ejection temperature—i.e., approximately 50°F less than the desired average part ejection temperature.

8. Nominal and maximum thickness of parts. For purposes of entering data into the computer, the type of measurement made will be dependent upon the part configuration. If the wall sections are fairly uniform, measure the most common wall section and use that measurement as input to the computer. If the wall sections vary greatly, measurements should be made of varying thicknesses, and a mean average should be computed for use as input to the computer.
9. Total parts weight and total shot weight can be measured rather than calculated, whenever possible. If physical measurements cannot be done, the following for-

mulas can be used to calculate these variables:

$$\text{Volume (in.}^3) \times \text{Density (\#/in.}^3)$$
$$= \text{Weight}$$

or

$$\text{Volume (cm}^3) \times \text{Density (kg/cm}^3)$$
$$= \text{Weight}$$

As a result of entering these data, information of the type illustrated in Table 13-4 appears on the user's terminal screen. The code numbers opposite the first three numbers in Table 13-4 correspond to descriptions of plastic materials, mold cavities, and mold cores stored in the computer.

Using the input information illustrated in Table 13-4 and stored information on the thermal properties of mold materials and plastics materials, the computer calculates the specific heat load of the mold. The computer then lists on the screen possible combinations of coolant channel diameter, effective length, depth, and pitch to handle the heat load.

At this point, the mold designer can manually prepare a proposed cooling layout, selecting cooling channel characteristics from among those listed on the terminal screen. In drawing the cooling passage layout, the designer will find the following suggestions helpful:

1. All the horsepower of pumping power required to recirculate coolant through the mold passage adds heat to the process, which, in turn, must be removed by the cooling system. Thus, cooling passage design should take into consideration the desirability of minimizing pumping power requirements. Excessive bends in cooling circuit paths create the need for additional pumping power. Some of these bends are necessary, owing to physical restrictions. It is not always possible to drill straight-line passages through the mold. These are usually minor bends, each of which is responsible for a pressure drop. As a rule of thumb, it is advisable not to have more than 15 bends in one circuit path.

Excessively high velocities in the cooling manifold (greater than 2 to 3 ft/sec) will create undesirable pressure drops and may cause cooling lines at opposite ends of the manifold to have different flow rates.

2. The coolant flow path should begin in the cavity and gate areas of the mold in order to get more heat out faster. Short paths should be used in these hot areas, and longer paths in cooler regions of the mold.

3. Cooling line length should be no more than 4 to 5 feet at the maximum, depending on the efficiency of mold cooling. Where cooling is less efficient, the length should be shortened to avoid introducing thermal gradients across the mold surface.

4. The cross-sectional areas of the cooling lines in each circuit should be uniform. Changes in diameter cause imbalance and losses in the coolant stream.

If the cooling performance of an existing mold is being optimized with the aid of the MOLDCOOL program, the following suggestions can supplement information provided by the program:

- Coolant lines must be marked clearly to avoid confusion.
- Coolant channels should be cleaned before plumbing to the cooling system.
- The inside diameter of the hoses should be at least as large as, or preferably larger than, the coolant line diameter in the mold. The length of these hoses should be kept to an absolute minimum.

Table 13-4 Mold Cool Data File: Information needed by computer to calculate specific heat load of a mold in the form that it is displayed on a CRT screen.

(1)	Plastic Code	3
(2)	Mold Cavity Code	10
(3)	Mold Core Code	10
(4)	Hot Runner Capacity	6.000
(5)	Cooling Time	25.800
(6)	Injection Temperature	480.000
(7)	Parts Surface Temp (Core)	100.000
(8)	Parts Surface Temp (Cavity)	100.000
(9)	Nominal Thickness	0.150
(10)	Maximum Thickness	0.150
(11)	Total Parts Weight	0.2990
(12)	Total Shot Weight	0.2990
(13)	Total Parts Surface Area	215.580

- Counter-bore coolant lines for pipe nipples and fittings with inside diameter equal to or greater than the diameter of the coolant line.
- Quick disconnects must not restrict flow; oversize disconnects are required on all coolant lines.
- The supply and return manifolds should have low pressure drop for proper distribution of coolant in all coolant channels. The manifolds should be sized for velocities between 2 and 3 ft/sec.
- All piping should be insulated to reduce the ambient losses.

The efficiency of a large number of existing molds can be increased by checking them against the recommendations listed above and by optimizing coolant operating conditions.

Basic Analysis of Heat Flow

Internal resistance to heat flow depends upon the distance of waterlines from the molding surface and also on the pitch of the water passages. The shape factor determined by the depth and pitch of the water passages is a measure of the resistance offered by the mold itself to the heat flow. The depth of the centerline of the waterline from the molding surface should be one to two times the diameter of the waterline. The pitch, that is, the distance between the centerlines of two consecutive waterlines, should be three to five times the diameter of the waterline. A wider placement of waterlines would result in uneven temperature across the mold surface, and greater depth would increase the thermal resistance. Figure 13-10 graphically presents this concept.

Other things being equal, the greater the velocity of coolant flow, the greater the heat transfer from the mold to the coolant. This principle runs into the law of diminishing returns. After a certain point, increasing coolant velocity does not appreciably improve heat transfer from the mold. This fact can be observed in Fig. 13-11, where the heat flow curves gradually level off at higher coolant flow rates. The reason for this can be better understood by referring to Fig. 13-12.

At low coolant velocities, coolant flow will be laminar. In laminar flow, the coolant moves in layers parallel to the walls of the coolant passage; each thermal layer acts as an insulator, impeding heat transfer from the mold to the stream of coolant. In contrast, turbulent flow, the desired condition, creates random movement of coolant and substantially increases heat transfer. However, once turbulent flow conditions are reached, higher coolant velocities bring diminishing returns in improved heat transfer.

Turbulence of coolant flow is quantified by means of the dimensionless Reynolds number. It is directly proportional to cooling velocity and passage diameter, and is inversely proportional to coolant viscosity. The formulas is as follows:

$$N_{Re} = \frac{Vd\rho}{\mu}$$

where N_{Re} is the Reynolds number, V is the coolant velocity, d is the coolant passage diameter, ρ is the density of the fluid, and μ is the viscosity. Reynolds numbers of up to 2300 indicate laminar flow. There is a transition zone between laminar and turbulent flow that is defined by Reynolds numbers ranging from 2300 to 3500. Reynolds numbers above 3500 indicate turbulence.

The importance of coolant viscosity is considerable. The viscosity of water at 50°F is 1.5 cp; 25 percent ethylene glycol solution at 30°F has a viscosity of 3.6 cp; 50 percent ethylene glycol at 10°F is 15.6 cp. The unit of measure is cp—centipoise. Because the viscosity increases by a factor of 10 over this temperature range, the equation shows that it would be necessary to increase the velocity by approximately the same factor to maintain

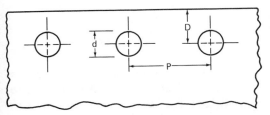

Fig. 13-10   Cooling passage design.

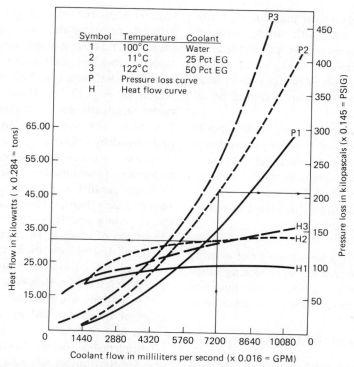

Fig. 13-11 Computer-generated cooling curves showing optimum combination of high heat transfer and low pressure drop for 25% ethylene glycol solution at 30°F.

the same Reynolds number and thus the same degree of turbulence. It would seem highly impractical to imagine buying a pump sufficient to increase the flow rate by a factor of 10 to make up for increased viscosity. This demonstrates mathematically why lower coolant temperatures and the resulting increase in viscosity should be approached with cau-

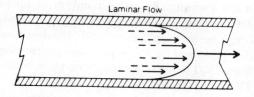

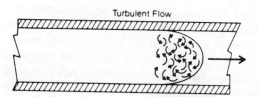

Fig. 13-12 Laminar and turbulent flow.

tion. Increased viscosity makes it more difficult to achieve turbulent flow.

The tremendous importance of flow conditions to heat removal can be appreciated by examining the effect on heat transfer. The heat transfer coefficient is a quantity that varies in relation to the Reynolds number, from which it is derived by a series of equations that need not be shown here. Suffice it to say that, under laminar flow conditions, the heat transfer coefficient in mold cooling passages might be in the range of 10 to 15 Btu/hr/ft²/°F; whereas with turbulent flow, the heat-transfer coefficient might be 300 to 1000 Btu/hr/ft²/°F. The importance of the heat transfer coefficient appears in the following equation, where it is symbolized by the letter U:

$$A = \frac{Q}{(U)(LMTD)}$$

A is the total area of the cooling passages; Q is the heat load (amount of heat to be removed); and $LMTD$ represents the temperature difference between the mold surface and the plastic involved in the process. Of course, the temperature of the mold surface is a function of the chilled water temperature. Since the heat load, Q, is a constant for any given application, the amount of channel area required is largely a function of the heat transfer coefficient. Note that there is a difference of a factor of 10 or 20 between the values of the coefficient at laminar and turbulent flow conditions. If use of a lower cooling temperature meant a loss in coolant velocity and a consequent loss of turbulent flow, the total channel area would have to increase 10 or 20 times to compensate. Given the limited amount of space in most molds for placement of cooling channels, it seems unlikely that the amount of channel area could be increased by such an amount. *Note:* Lower temperature has a higher *LMTD*.

All of this points to the importance of this fact. The most common mistake made by processors is to run their coolant at too low a temperature, which may actually reduce cooling effectiveness instead of increasing it. In a large number of cases, they would be better off running at a higher temperature and higher flow rate (gpm). This also illustrates why a careful cooling analysis is needed and not guesswork. If a molder simply goes out and buys a bigger pump to increase his flow rate, he may not achieve the desired results, because: (1) his mold may already be operating at turbulent flow conditions, or (2) he has no assurance that the larger pump will be sufficient to boost his coolant velocity into the

turbulent flow region—in which case, the improvement will also be marginal.

Describing Coolant Components to the Computer

In order to describe coolant passages to the computer, it is necessary to understand the various configurations and the standard terminology used to describe them. Basically, there are six types of flow paths in mold coolant channels that can be used in any combination to cover all types of design, as follows:

1. *Straight sections/round flow.* These are the most common cooling channels found in molds. They are normally a series of straight drilled passages that are round in diameter. (See Fig. 13-13.)

2. *Straight sections/rectangular flow.* Commonly found in backup plates or used in modular tooling, these are usually a series of straight passages cut in a rectangular shape that comprises the cross-sectional area. (See Fig. 13-14.)

3. *Circular sections, round flow.* Circular sections are used to cool mold plates, cylindrical/cone-shaped cores and cavities (spiral), etc. This arrangement, when properly interfaced, makes it possible to follow closely the radius of the round core or cavity so that the distance of the channel is kept at a uniform depth. This type of round or spiral pattern is differentiated by its round cross-sectional area. (See Fig. 13-15.)

4. *Circular section/rectangular flow.* These

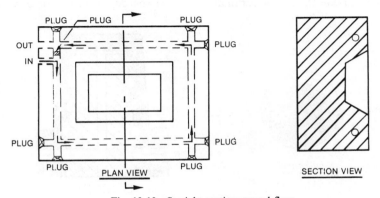

Fig. 13-13   Straight section, round flow.

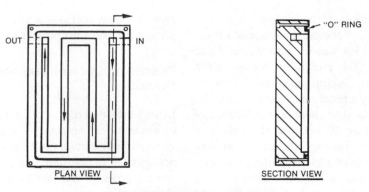

Fig. 13-14 Straight section, rectangular flow.

have rectangular cross-sectional areas for ease of machining. (See Fig. 13-16.)

5. *Bubblers* (*fountains*). These channel components are normally connected with straight sections/round flow. They are used in the cooling of pins, cores, and deep draw areas. Typically, two channels (straight section/round flow) are drilled parallel to the surface of the back face at different depths. In the bottom channel, tubes are screwed that go to the top of the area to be cooled (pin, core, etc.). The inlet water goes through the lower channel, fills the tube, and then overflows into the outlet. Each core tube receives the same cooling with maximum velocity. (See Fig. 13-17.)

6. *Baffles.* These constitute an alternative method for the cooling of pins, cores, and deep draw areas. Unlike bubblers, they are tied together in series, typically by straight sections/round flows. The coolant enters a straight section that intersects with all of the baffles in the channel. Each baffle is a round drilled section with a blade to divide the cross-sectional

area in half. The coolant flowing in the straight section runs into the baffle blade, and makes a 90-degree bend into the baffle. Inasmuch as the blade does not extend all the way up to the end of the baffle, the coolant can pass over the top of the blade at the end of the baffle. It then runs down the back side of the blade, makes another 90-degree turn back into the straight section, and goes on to the next baffle. This is an acceptable method for molds with a small number of cavities from side to side or very large-diameter channels and baffles. (See Fig. 13-18.)

In order to conduct a dialogue with the computer in a CAD program for cooling passages, the mold designer also must be familiar with the terms used to describe cooling channels, as follows:

1. *Depth* is calculated for each *cooling channel section*. It is the distance from the center of the cooling channel to the mold surface area cooled by the section. If the depth varies for a particular sec-

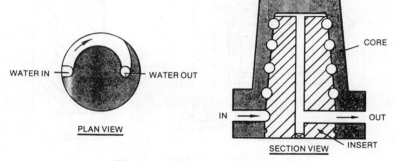

Fig. 13-15 Circular section, round flow.

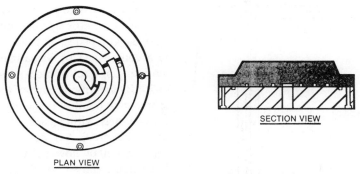

PLAN VIEW

SECTION VIEW

Fig. 13-16   Circular sections, rectangular flow.

tion, a weighted average must be used. (See Figs. 13-10 and 13-19.)

2. *Pitch* is calculated for each individual *cooling channel*. First, the total part surface area must be divided into separate sections for each cooling line. After the total part surface area has been divided, the cooling channel effective length must be determined. The part surface area cooled by the channel is then divided by the channel's effective length. (See Fig. 13-20.)

3. *Effective percent* is calculated for each cooling line section. It is determined by dividing the total channel section length by the channel section length that cools the part.

4. A channel *section* is defined as a straight coolant channel run up to the point where it changes direction by making a bend or turn, and/or a change in channel diameter. This is only applicable to straight sections, round and rectangular flow. (See Fig. 13-21.)

5. The number of *branches* is determined by section. If, for example, a section had only one flow path up to the next bend, this section would only have one branch. If a section were split up into three flow paths up to the next bend, this section would have three branches. (See Fig. 13-22.)

6. *Angle in degrees* denotes the geometrical angle of the bend at the end of a section. This can be derived using a protractor when one is extracting data from prints.

Also, in order to describe the manually prepared coolant passage layout to the computer, it is necessary to have some means of relating

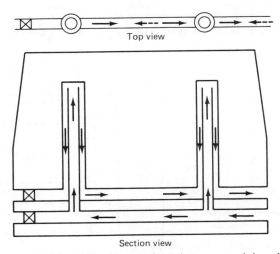

Top view

Section view

Fig. 13-17   Bubblers for cooling parts such as pins, cores, and deep draw areas.

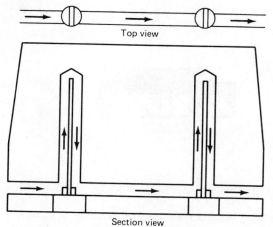

Top view

Section view

Fig. 13-18   Baffles for cooling.

cooling channels to their location in the core and cavity (ejector side as well as non-ejector side), backup plates, and slide plates. This can be accomplished by visualizing the mold as consisting of a set of blocks, as shown in Fig. 13-23. By identifying each of these blocks with a three-digit number as shown, and numbering the channels within each block in sequence (101, 102, etc.), it is possible to convey the organization of flow paths, cooling channel dimensions, and so on, to the computer.

Circuiting.   Circuiting is the process of developing individual cooling channels into circuits. It is done in order to achieve even cool-

Depth: Distance from mold surface to center of coolant channel

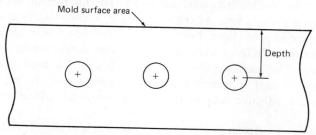

Mold surface area

Depth

Average Depth: Same as above only averaged for contoured shapes

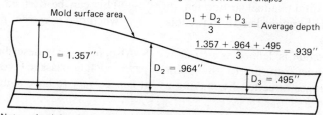

Mold surface area

$$\frac{D_1 + D_2 + D_3}{3} = \text{Average depth}$$

$$\frac{1.357 + .964 + .495}{3} = .939''$$

$D_1 = 1.357''$

$D_2 = .964''$

$D_3 = .495''$

Note — depth is calculated for individual sections

Fig. 13-19   Depth and average depth for cooler channels.

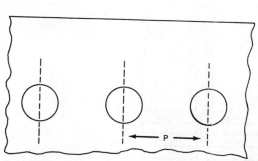

Fig. 13-20   Pitch—distance between individual cooling channels.

P

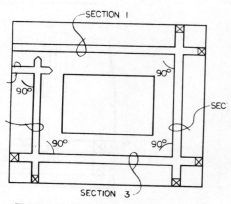

SECTION 1

90°

90°

SEC

90°

90°

SECTION 3

Fig. 13-21   Sections in the cooling lines.

The number of <u>branches</u> in a section is the number of flow paths
which a single section takes before reaching another section

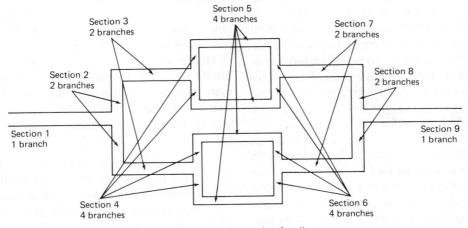

This is required input data for all
straight section flow paths.

Fig. 13-22 Branches in the cooling lines.

ant flow and therefore balanced rates of heat
flow throughout the mold. Unbalanced cooling can result in differential shrinkage, residual stress, and warping of the finished part.

As is the case with the design of cooling
channels, traditional rule-of-thumb procedures do not produce optimum results. Over
ten years of experience has shown that a mathematical approach should be utilized to determine heat flow conditions in each channel.

Such an approach is expedited by use of a
computer.

In order to determine the proper circulating
for a mold, the pressure drop must be calculated for each cooling channel. This calculation is performed by the computer. It is then
possible to determine which channels are to
be looped in series, and which channels need
to be run in parallel so that all circuits have
approximately the same pressure drop, result-

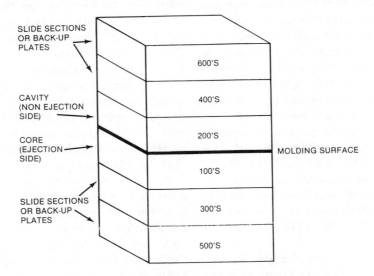

NOTE: CHANNELS MUST BE NUMBERED IN SEQUENCE.
EXAMPLE:
　　　RIGHT: 101, 102, 103, 501, 502, 503
　　　WRONG: 101, 103, 501, 503

Fig. 13-23 Mold visualized as consisting of a set of blocks.

ing in even-balanced cooling throughout the mold. This balancing is needed because water (coolant) flow will flow the path of least resistance. Flow rates must be kept within reasonable limits to optimize coolant circulating system (pump flow and pressure) requirements.

Injection molders have utilized two common techniques for selecting the proper circuitry arrangement for mold cooling passages. One technique, utilized in order to simplify mold change operations, consists of connecting or looping all of the cooling channels on the core or ejection half of the mold in series and all of the cavity half's channels in series.

The other common arrangement is to run all of the cooling channels in parallel by providing each channel with a separate inlet and outlet. This is done in an attempt to get more coolant through the mold.

In a majority of cases, these relatively simple approaches to circuiting do not result in optimum cooling time and uniform cooling.

The major problem with series circuit arrangements developed without the aid of a computer stems from an excessively high value of ΔT through cooling channels. This results in differential part surface temperatures, thereby creating differential shrinkage rates and consequent potential quality problems with finished parts. Moreover, when too many channels are connected in series, the resulting high pressure drops through the mold create the need for larger pumps to overcome the pressure loss. Larger pumps add heat to a system that must be removed. Figure 13-24 shows how a high ΔT was reduced in a problem circuit. The circuit's ΔT was reduced from 8.6°F to 2.6°F by a conversion to two separate circuits, each consisting of three channels connected in series. The major problem with parallel circuiting arrangements is that a high flow rate is required to achieve turbulence in the coolant. (Turbulent flow increases the rate of heat transfer from the mold into the coolant.)

Figure 13-25 shows how changing the circuiting arrangement of cooling passages in another mold from all channels running in parallel to all channels running in series made it possible to significantly reduce the flow rate

and therefore pumping horsepower requirements. Each circuit within a mold cools a specific mass of plastic, and each circuit should remove the required quantity of heat in order to balance the overall heat flow.

The determination of specific heat flow conditions in each channel involves complex mathematical calculations based upon three heat transfer dimensionless numbers (the first two of which were discussed earlier in this chapter):

1. Reynolds number—index of the flow characteristics (laminar, transitional, or turbulent) of fluid flowing in a passage.
2. Prandtl number—the ratio of fluid viscosity to thermal conductivity.
3. Nusselt number, which relates heat loss by conduction to the temperature difference and takes into account the configuration of each channel passage along with the thermal conductivity of the fluid.

It is almost always necessary to have turbulent flow in the coolant channels in the mold in order to meet the cooling requirement for minimum cycle time. In the majority of cases, the cycle time is limited by the cooling ability of the mold, and this is particularly true of parts with thickness of less than 150 mils. With the small-size coolant passages in the mold, turbulent flow is achieved at relatively low flow rates; however, the pressure drops through the small channels are high. As the channel size is increased, the pressure drop is reduced; however, high flow rates are required in order to get turbulent flow conditions. With large-size passages, the flow requirement for turbulent flow is often higher than the plant chilling system can provide.

At a Reynolds numbers less than 2100 to 2300, the flow is laminar, and the heat transfer is primarily by molecular motion in the fluid. As the Reynolds number is increased over 2300, turbulent cells start to form and cause the laminar fluid to be intermixed. At a Reynolds number of 3500 and above, the flow can be considered turbulent; complete turbulence occurs at a Reynolds number of 10,000 (see Fig. 13-12).

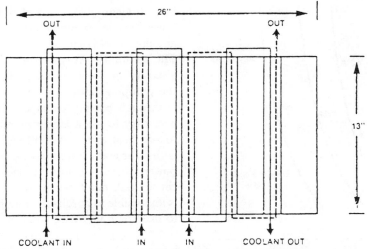

Fig. 13-24   Reduction of ΔT by changing circuit from six channels connected in series to two circuits, each of which consists of three channels connected in series.

In practice, a Reynolds number of at least 4000 should be achieved in the injection molds for efficient heat removal. Figure 13-26 shows gpm required to achieve a Reynolds number of 4000 with three different coolants flowing through different size passages.

To achieve fully turbulent flow, the flow required will be 2.5 times the flow obtained from the graph. Also, it is evident that as the diameters of the passages increase, the flow required to achieve the same turbulence increases proportionately. With the addition of ethylene glycol, the coolant solution becomes more viscous, and consequently a higher flow rate is required to achieve the same turbulence.

Optimization of Operating Characteristics

No matter how well mold passages are designed and circuited, their contribution to heat transfer from the mold is contingent upon the operating characteristics of a coolant—specifically, coolant temperature, pressure, and flow characteristics.

There is a widely held misconception that, other factors being equal, lowering of the coolant temperature will accelerate heat removal from the mold. Actually, the contrary is often the case, for the reason illustrated in Fig. 13-27. Commercially available water chillers by convention are rated for operation at 50°F

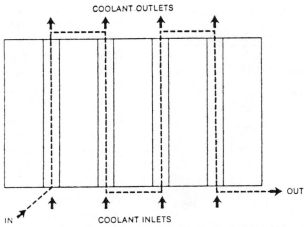

Fig. 13-25   Reducing coolant flow rate requirements by changing circuiting of channels from all in parallel to all in series.

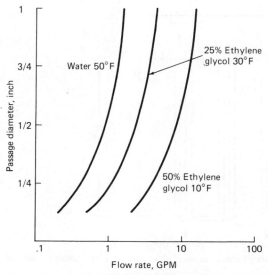

Fig. 13-26 Recommended minimum flow for various passage diameters (Re 4000).

leaving water temperature. For each degree by which water temperature is reduced below 50°F, the chiller loses about 2 percent of its capacity. At a coolant temperature of 10°F, the capacity of a 100-ton chiller is reduced to about 28 tons. Therefore, from the standpoint of cost-effectiveness, it is counterproductive to operate chillers at lower than their rated temperature. (There are situations, of course, in which this sacrifice of capacity is unavoidable.)

When coolant temperatures are reduced below the freezing point of water, the result can be counterproductive in terms of adding heat to a system that is designed to remove heat from the mold. At temperatures below freez-

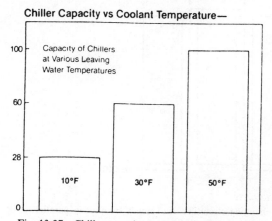

Fig. 13-27 Chiller capacity vs. coolant temperature.

ing, ethylene glycol is added to water. The lower the coolant temperature, the greater the amount of ethylene glycol that is required. The addition of ethylene glycol increases the viscosity of the solution and thus increases the pumping power that is needed to pump it through cooling passages. Therefore, it is necessary to use a larger pump, which tends to add heat to the system. (A rule of thumb is that for every 5-hp increase in pump horsepower, there is a loss of one ton of chiller capacity. A point can be reached at which it is impossible to purchase a pump large enough to overcome increased viscosity.)

With respect to coolant flow, other things being equal, the greater the velocity of coolant flow, the greater is the heat transfer from the mold to the coolant. But after a certain value of turbulence is reached, increasing cooling velocity does not appreciably improve heat transfer from the mold. The point at which this occurs varies for each mold. One example of this phenomenon is illustrated in Fig. 13-11. Here again, adding pumping power to increase flow rate after the optimum level of turbulence had been reached would be counterproductive from a cost standpoint, and also because increased pumping power would add heat to the system.

Determining optimum operating characteristics for the coolant involves mathematical computations similar to those involved in circuiting of the mold, which are quite literally impracticable without the aid of a computer.

There is more than one level of sophistication possible in the determination of these characteristics. The MOLDCOOL program, for example, offers processors a choice of two options—a simple, rapid, economical computer simulation used when it is adequate to maintain a single, average temperature over the entire surface of the mold, and a more precise option that takes into account temperature variations on the surface of the mold that might cause hot or cold spots. The output of this analysis consists of recommended coolant operating characteristics, including: ratio of water to ethylene glycol; coolant temperature; flow rate per circuit; pressure drop per circuit; total pressure drop; supply pressure;

average mold surface temperature; average part ejection temperature; cooling capacity of the mold (in tons); and total cooling time of the part.

In keeping with its character as a scientific (as opposed to empirical) approach to mold cooling, computer aided design of mold cooling passages has the capability of setting up a theoretical minimum cooling time based on resin type, processing temperature, part configuration, mold material, operating conditions, etc. If the duration of the projected cooling time for a new mold (or the actual cooling time for an existing mold that is being reevaluated with the aid of the computer) is longer than the minimum theoretical cooling time, the mold designer may wish to explore factors other than design and circuiting of cooling passages—e.g., materials of which the mold is constructed, coolant characteristics, cooling equipment—to determine whether cooling time can be reduced in this way.

Cooling Equipment Selection

How do you know whether or not your output is good? Only a thorough mathematical analysis can answer that with any confidence. However, Table 13-2 can give some idea of where you stand. For a part of a given material and maximum thickness, compare your actual cooling time with the minimum time shown in the table. If your cooling time is greater, then a cooling analysis may be able to improve your performance.

An analysis of cooling generally focuses on two parameters discussed above—mold passage design and coolant flow. Because of the mathematical complexity of such an analysis, we rely on a computer to perform the necessary computations. Specifically, these computations involve the following:

1. Based on a mold blueprint, mathematical analysis is made of the coolant circuits to ascertain whether there are equal pressure drops through each of them. Equal pressure drops mean equal flow distribution through all circuits. After making sure that individual circuits in each mold half are balanced, the computer then balances one mold half with the other.

2. The effects of alternative cooling parameters are established through computer simulation of injection molding conditions. This simulation utilizes a temperature profile of a mold established by providing input to the computer on the material being molded, such as its heat capacity, diffusivity, and conductivity, as well as the part thickness, initial injection temperature, and final ejection temperature. Typically, the temperature profile is plotted for three different coolants and temperatures—water at 50°F, 25 percent ethylene glycol solution at 30°F, and 50 percent ethylene glycol at 10°F. Coolant temperatures will vary with regard to material types and processing conditions. For each condition, the computer calculates and prints out data on the effect of coolant flow upon heat flow and pressure loss.

3. Curves for the three coolants and temperatures are superimposed on a single computer-generated graph, such as the one in Fig. 13-11. These curves make it possible to select the coolant temperature and flow rate that will provide the optimum balance of heat-flow and pressure-loss characteristics. In Fig. 13-11, a 30°F coolant temperature with a flow of 7200 ml/sec produces the desired combination of maximum heat flow with minimum pressure loss. *Note:* 7200 ml/sec = 115 gpm.

This type of cooling analysis also provides specific information on clamp closed time, total cycle time, and estimated processing rate utilizing the cooling parameters recommended by the computer. The recommendations and production data developed along with the curve shown in Fig. 13-11 are shown in Table 13-5.

4. If pricing information provided by the processor is supplied as input to the computer, the computer can translate cooling recommendations into an economic analysis, as shown in Table 13-6.

In the cooling analysis we just completed, computer simulation revealed that a change in existing coolant flow could achieve significant increases in productivity and profitability. No changes were required in mold design.

Table 13-5 Cooling analysis recommendations.

Material: High-Impact Polypropylene

Operating Conditions:
  Use 25% Ethylene Glycol at 30 F
  Flow Rate   115 gpm
  Pressure Drop Through Mold   29 psi
  Supply Pressure   60 psi

Benefits:
  Average Mold Surface
    Temperature   100 F
  Heat Rejection from Mold   9.1 tons
  Estimated Clamp Closed Time   42 sec
  Estimated Total Cycle Time   50 sec
  Estimated Processing Rate   346 lb/hr

Chiller:
  AEC Model WC-20 with 10-hp Pump

Remarks:
  Hoses $\frac{3}{4}$-in. I.D.
  Supply & Return Manifolds   1 Set,
                                         4-in. Diam

However, in some cases, it has been discovered, while analyzing mold drawings, that heat transfer might be improved by changing the design of mold passages. The effects of these changes in passage design are then tested by means of computer simulation, and appropriate recommendations are made.

The best way to be sure of optimum coolant passage design placement in the mold is to have the computer simulation made while the mold is still on the drawing board. In effect, the cooling analysis becomes a step in mold design. The moldmaker will then have additional technical input that will simplify and expedite the design of coolant passages and, in some cases, may help the moldmaker design a sprue and runner system that is better adapted to efficient heat removal from the mold.

Until recently, some processors interested in cooling analysis were concerned about time consumed by the cooling analysis procedure. There is also, sometimes, a reluctance to release mold drawings. To alleviate these and other problems, we have developed the capability to perform the analysis in the processor's plant. The consultant comes to the plant equipped with a computer that is preprogrammed with the necessary formulas for cooling analysis. This system offers a new dimension in cooling analysis.

Although the example used to explain the analysis procedure was the injection molding process, there are other applications. It has been used successfully in blow molding applications, and has also been used in compression molding applications. It is a tool that has proved its worth in the past, and its future uses look promising.

Table 13-6 Economic analysis.

PART 18136 HIGH-IMPACT POLYPROPYLENE

	BEFORE COOLING ANALYSIS	AFTER COOLING ANALYSIS	% IMPROVEMENT
Cycle Time, sec	60	50	16.7
Pieces Per Hour	60	72	20
Dollars Return Per Hour	$195	$234	20
Direct Cost Per Hour	$166	$186	12
Gross Profit Per Hour	$ 29	$ 48	66
Increased Gross Profit Per Hour		$ 19	
Based on:			
Selling Price			$3.25
Machine & Labor			$65/hr
Material			$0.35/lb
Operating Hours/Year			6000
Cooling Analysis Simple Payback			105 hours
Annual Increased Gross Profit After Cooling Analysis			$114,000

Conclusion

The major thrust within the molding industry today is toward a higher level of technological implementation. This would include CAD/CAM, plant automation, robotics, and computer technology in general. The reasoning behind this change is to increase productivity in order to stay competitive within an increasingly competitive industry. The MOLDCOOL computer-aided design program offers a scientific approach to the problem of mold cooling optimization and offers higher productivity gains than any other single mold analysis or mold design program. However, I believe that all of the productivity programs and new technology will be essential within the industry due to the ever increasing competition.

Traditionally, plastics processors have depended upon rule-of-thumb procedures to match cooling equipment capacity to the requirements of a given application. All too often, this approach has resulted in purchase of a cooling system that is either undersized, and therefore incapable of reducing cooling time to the required extent, or oversized, and therefore wasteful of investment capital and energy.

As indicated in the introduction to this section, the function of the computer in the design and circuiting of mold cooling passages is to perform complex calculations that are unfamiliar to and, in most cases, beyond the scope of mold designers. Moreover, the time involved in performing these calculations and the possibility of error are such that the CAD method is not only preferable but by far the practicable alternative.

Throughout the foregoing paragraphs, information presented provided an overview of the calculations involved and their relationship to the optimization of mold cooling. Whether the calculations are performed manually or with the aid of the computer, their justification is economic.

Economic analysis is a program that provides the user with the Return On Investments (ROI) associated with the cycle time reductions. The major function of this program is to determine whether the costs of a tool cooling design change will provide an adequate ROI.

Although the programs are all designed for easy use, should any problems develop, the user can access a HELP program. Problems are typed out and entered into the computer, and they will be answered every night via the computer by AEC's consultants.

In the case of a new job, where there is no standard of comparison, the economic analysis of benefits of CAD consists of a projection of mold productivity and profitability. However, analogical data that illustrate the benefits of a scientific approach to mold cooling can be obtained from economic analyses of existing molds. Such analyses pair mold performance prior to changes recommended by the computer and with that after those changes.

Section V
AUXILIARY EQUIPMENT

Chapter 14
Auxiliary Equipment—Overview

D. V. Rosato

It is important to recognize that there are different types of auxiliary equipment used to maximize overall molding plant productivity and efficiency. (See Chapter 7 on molds, Chapter 11 on process controls, Chapter 12 on computer-aided plastic flow, Chapter 15 for details on material conveying, Chapter 16 on chilling and cooling, Chapter 17 on size reduction/granulating, Chapter 28 on troubleshooting, and Chapter 29 on testing/quality control.)

The proper use and maintenance of auxiliary equipment that has been purchased to meet specific needs is as important as the selection and proper use of the injection molding machine and mold, and the plastic material to be processed. Equipment is available that provides many different functions to permit molding parts that meet performance requirements at the lowest cost. Unfortunately many molding plants just install "any type" of equipment without determining whether the best was purchased to meet specific requirements. The highest- or lowest-cost equipment does not necessarily equal the best performance. As an example, chillers for mold temperature control in many applications are not properly engineered for the molding cycle. The result generally is a higher energy cost that eliminates any cost reduction occurring due to reduced cycle time. Check your energy consumption; at least consider using an amp meter. Usually half of the users of chillers

do not obtain the correct unit, with the result that increased energy cost obliterates other gains.

Most molders use precompounded pellets, but the trend is for them to process powders and do their own in-plant compounding, which can reduce the cost of molding material from one to six cents per pound. Also improvement in molded part quality can occur. For successful in-plant compounding, equipment designed specifically for conveying non-pelletized materials is required. The dedicated line would include correct filtration systems and handling systems that are different from the lower cost system required for pellets.

All kinds of equipment exist to improve materials that are being fed, giving both accuracy and high throughput rates. Check both the accuracy and the life expectancy of your existing equipment (e.g., volumetric and gravimetric feeding/metering/proportioning equipment).

Throughout this book development of electrical energy efficiency in the molding plant is reviewed. With auxiliary equipment, major changes are occurring to reduce energy costs, particularly in drying and preheating units. More use of sophisticated controls is providing processors with improved diagnostics and temperature control to significantly help them to operate equipment at the lowest cost. As an exmaple, more integrated electronic controls on resin dryers now correlate air flow with the dewpoint and the amount of time heaters operate.

GRANULATING VS. PERFORMANCE

The process of granulating or recycling plastic is important in keeping the cost of operation down. Your first target is to eliminate the source of plastic that has to be granulated or at least reduce the amount of "scrap or rejects." The technology involved in the process of granulation is based on various factors: laws of physics concerning mass, velocity, kinetic energy, and gravity. Other factors involve type of plastic and size of part to be granulated (see Chapter 17).

With these factors it is important that you use the correct granulator. This process (Fig. 14-1) is not new; the correct granulator can be obtained and used in the plant. Recognize that you can literally destroy the granulated plastic if improper machines are used. It is very easy to overheat the plastic and destroy or at least reduce the plastic performance, particularly with heat-sensitive plastic.

Controls should be set for evaluating the performance of granulators or the effect they have when reused. Certain plastics, such as melt heat-sensitive types, can be completely degraded, unless special precautions are taken such as granulating when "frozen." The usual amount of regrind mixed with virgin plastic that would normally have no detrimental ef-

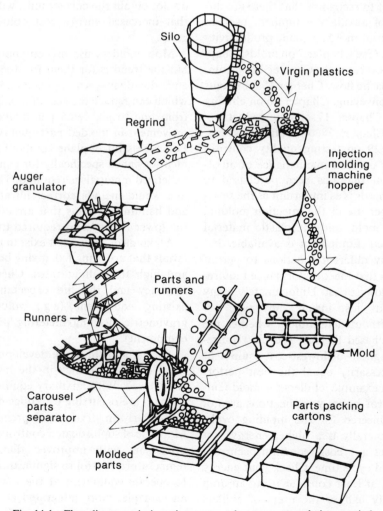

Fig. 14-1 Flow diagram: virgin resins, parts and runners, granulating, regrind.

fects ranges from 20 to 30 percent, by weight. With certain applications, 100 percent (all regrind) could be satisfactory. The size of regrind (usually not too uniform, including fines that will absorb moisture quickly) also influences performance. Regardless of what approach is used, the only way you know if there are any deteriorating effects is to run tests.

An example of what happens with regrind is shown in Fig. 14-2(a–c). Other tests that can be conducted, such as melt flow tests, are reviewed in Chapter 29.

The processing and economic advantages and possibilities of the automatic reuse of the sprue in the injection molding process and the direct recirculation at the machine are obvious only if an objective comparison with conventional sprue reprocessing is made. For many plants this represents an unused possibility for affecting and decisively improving profitability. The sprue reprocessing usually involves the following:

1. Separation of molding and sprue. This is generally a manual operation. However, should a mechanical separation step be used, care must be taken to ensure that a reliable separation occurs (part separation to be reviewed in this chapter).

2. Transport of the sprues to the storage point.

3. Storage of sprues, separated according to type of material and color.

4. Transport to a central granulator area.

5. Cleaning of the granulator after each material and color change.

6. Storage of regrind according to color and material type.

7. Transport of regrind to the mixing area.

8. Mixing of regrind with new material.

9. Storage of mixtures according to the color and material type.

10. Transport of the various mixtures to the individual processing machines.

11. Where hygroscopic material is used, redrying of the regrind before processing (to be reviewed in detail later in this chapter).

12. Feeding of the mixture to the processing machine or dispensing of the regrind where regrind is brought separately to the machine.

This usual method of recirculating is time-consuming and involves considerable personnel costs and significant working areas. Of even greater disadvantage are the problems arising from misplacing, contaminating, or mixing the various material types and colors. These problems often are evident only at the final control or after customer complaints.

The immediate reuse of the sprue at the processing machine may reduce reprocessing to the processing stages of sprue separation, sprue granulation at the machine, and direct recirculation into the manufacturing process. In addition to significant reductions in the recirculation processes, advantages of this method are that hygroscopic regrind does not

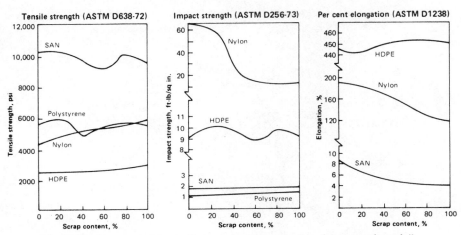

Fig. 14-2a   How regrind level affects mechanical properties after "once-through."

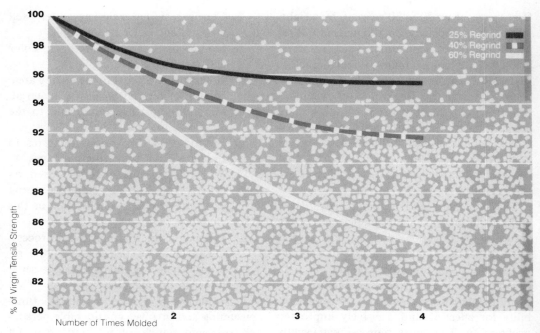

Fig. 14-2b   Tensile strength of regrind of thermoplastic polyester blends.

require redrying (because no moisture is taken up when regrind is immediately reused), and the problems resulting from contamination can be "reliably" forgotten. In contrast to the usual sprue reworking methods, the new method makes it possible to automate all the production stages by ensuring better use of all production means and improvement of product quality and consistency.

As an example, about two million pounds per year of ABS is recycled by Western Electric, and they are not recycling everything.

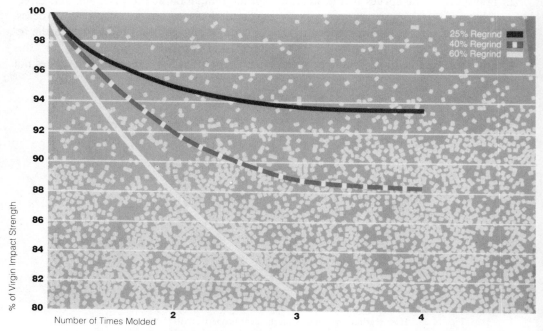

Fig. 14-2c   Impact strength of regrind of thermoplastic polyester blends.

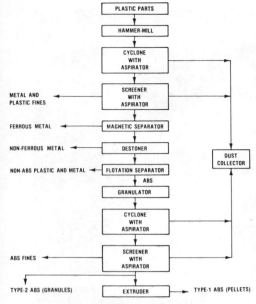

Fig. 14-3a ABS reclamation system (1).

who make parts for Western Electric. The higher-purity material, produced at Omaha, has undergone molding trials both at Western Electric and by outside suppliers under contract to Western Electric. The potential at Omaha is 1.5 million pounds per year for a two-shift, five-day-per-week production schedule (1).

This recycling at Western Electric involves the usual out-of-specification molded parts, as well as a large source of molded parts that are collected after having been in service and contain nonplastic materials (wire, metal inserts, etc.). The Omaha reclamation system (Fig. 14-3), except for the flotation-purification unit, consists of commercially available equipment, adapted from other industries. Obvious "junk" is removed from the ABS scrap at the sort conveyor, where such items as relays, terminals strips, plastic packaging materials, and even gum and candy wrappers have been found. The scrap is then sorted by color, with each reclaimed separately.

Next, hammermilling reduces the product to about a 2-in. size for easy handling by subsequent pieces of equipment. Here, metal brackets, inserts, and screws are broken loose from

Potentially they have about ten million pounds that could be recycled each year. Although most of the reclaimed ABS is molded by Western Electric, some of the recycled Atlanta product is sold by Nassau Recycle Corp. in Gaston, South Carolina, to outside molders

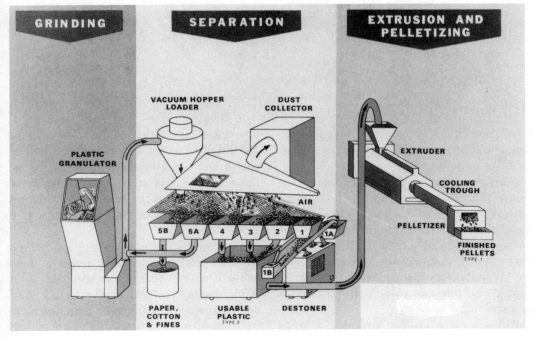

Fig. 14-3b Plastic reclamation system.

Fig. 14-3c   ABS telephone plastic at Western Electric service center.

the plastic scrap. Cotton is freed from the handles, and paper labels are torn away from plastic parts.

From the hammermill, material enters a cyclone, which slows down the high-velocity air from the blower that has just transported the material from the hammermill. Very light, airborne dust goes out the top of the unit, while the product falls (under the influence of gravity) out the bottom. The cyclone has a tapered, conical structure that reduces the air velocity in a controlled manner.

At every stage possible, aspiration rids the material stream of dirt and dust. The aspirator has sort of a vacuum-cleaning effect, removing cotton and paper labels as soon as possible. Later on in the process, these contaminants could get caught on the more ragged particles. At that point, it would be more difficult to extract them.

After the particles leave the aspirator, a two-deck vibrating screener removes metal and plastic fines that are smaller than $\frac{1}{16}$ in. Aspiration both before and after further eliminates airborne contaminants. Screened material then drops onto a moving magnetic belt, which removes ferrous metal and any steel-containing chopped plastic that is still present.

From the magnetic belt, the plastic enters the flotation unit. There, a sodium chloride solution, adjusted to a specific gravity of 1.15, is used to float the ABS particles. Heavier, non-ABS plastics sink. After a four-stage spray rinsing, then drying, the ABS is reduced to its final particle size with a knife-blade granulator.

Finally, a two-deck screener plus two aspirators remove any remaining airborne materials. The resulting product is a high-purity, granulated ABS, which is designated at Type 2. Without the flotation step, additional processing through a screening-extruder/pelletizer is necessary to remove the nonmelting impurities.

DRYING P.E.T. IS THE KEY TO MOLDING PRODUCTIVITY

A few years ago, P.E.T. resin (polyethelyne terephthalate) was introduced in the plastics marketplace. At that time, very few people knew anything about its drying requirements. We knew that it was hygroscopic, that it needed higher temperatures than most other resins, and that it contained something called acetaldehyde (2).

The two-liter P.E.T. beverage bottle (injection stretched–blow molded) has claimed over 80 percent of the market since its introduction in the late 1970s. The one-liter and half-liter P.E.T. beverage bottles have made substantial gains in their respective markets, and now manufacturers of products such as mouthwash and other personal care items are discovering P.E.T. containers to be an attractive alternative.

As the market for P.E.T. products continues to grow, so does the need for a better technical understanding of its processing requirements. The most critical of these requirements is proper drying.

The Drying Process

P.E.T. resin is hygroscopic, and when received, may contain about .05 percent moisture (five hundredths) by weight. To prevent loss of clarity and some physical properties the resin must be dried so that it contains less than .005 percent moisture (five thousandths) before entering the molding machine.

To achieve this very low moisture content, the resin must be heated to 350°F and exposed to air having a moisture content of −40°F dewpoint at a flow rate of about 1 ft/sec. This low dewpoint and high temperature create what we refer to as "pressure–dewpoint differential." More simply, in the drying hopper, we create conditions within and around the pellet that virtually boil the moisture out.

As illustrated (see Fig. 14-4), at ambient temperature and 50 percent relative humidity,

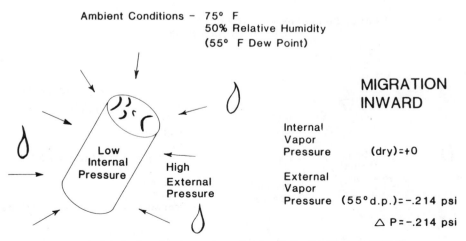

Ambient Conditions – 75° F
50% Relative Humidity
(55° F Dew Point)

MIGRATION INWARD

Internal Vapor Pressure (dry) = +0

External Vapor Pressure (55° d.p.) = −.214 psi

$\triangle$ P = −.214 psi

Fig. 14-4 Mechanics of moisture absorption in plastics (ambient conditions).

the vapor pressure of water outside the pellet is greater than that within. Moisture migrates into the pellet, thus increasing its moisture content until a state of equilibrium exists inside and outside the pellet.

However, inside the drying hopper, our controlled environment, conditions are very different (see Fig. 14-5). At a temperature of 350°F and −40°F dewpoint, the vapor pressure of water inside the pellet is much greater than that of the surrounding air, so that moisture migrates out of the pellet and into the surrounding airstream where it is carried away to the desiccant bed of the dryer.

As a result of prolonged exposure to high temperatures, the resin releases some small amount of acetaldehyde into the airstream. A special molecular sieve (Linde 13X) is used which has pore diameters of sufficient size to allow this substance to be trapped and later expelled from our closed loop system.

Acetaldehyde content must be controlled to meet established standards as well as to prevent flavor problems in some beverages. Most resins, when received from the supplier, contain less than 2.5 ppm acetaldehyde.

The Drying System

The strict moisture control required to successfully process P.E.T. resins is achieved by the use of a closed loop, hopper/dryer system (see Fig. 14-6). This system generally includes a self-regenerating desiccant dryer, high temperature process air heaters, a return air filter, an aftercooler (heat exchanger), and a well-insulated drying hopper.

A good rule of thumb used to determine the cfm (cubic foot per minute air flow) requirements for a P.E.T. system is: 1 cfm per pound per hour processed. As an example: If your machine processes 250 pounds of material per hour, you would select a dryer that produces an air flow of not less than 250 cfm at −40°F dewpoint.

To determine the capacity of the drying hopper (and thus the exposure, or residence time), we simply multiply the residence time by the use rate (expressed in pounds per hour). Using the 250 lb/hr use rate from the previous example, and a 4-hr residence time we have:

$$250 \text{ lb/hr} \times 4 \text{ hr} = 1000 \text{ lb usable capacity}$$

You would select a 1000-pound capacity hopper to ensure proper exposure time.

Now, let's quickly review. We have properly sized both the air flow and hopper capacity to enable us to successfully process P.E.T. resins (see Fig. 14-6). In the drying process, low-dewpoint air flows from the desiccant bed through the four-way valve to the process air heaters. There, its temperature is elevated to 350°F. The hot, dry air flows evenly up through the material, and removes moisture from the resin. The moisture-laden air flows through the return hose to the filter element where dust is removed from the airstream.

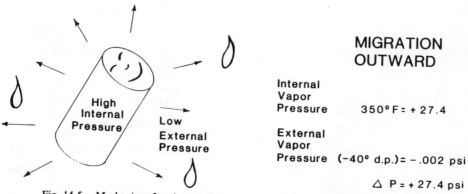

Hopper Conditions –   350° F
(−40 Dew Point)

High Internal Pressure

Low External Pressure

MIGRATION OUTWARD

Internal Vapor Pressure   350°F = + 27.4

External Vapor Pressure   (−40° d.p.) = − .002 psi

$\triangle$ P = + 27.4 psi

Fig. 14-5.   Mechanics of moisture absorption in plastics (controlled environment).

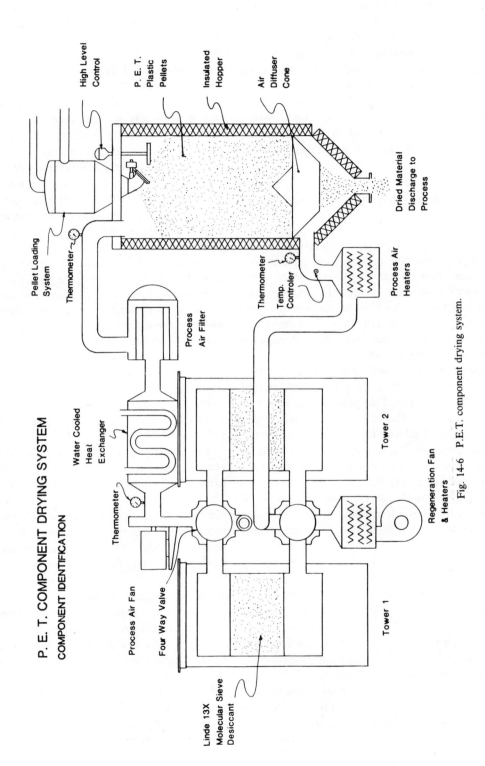

P. E. T. COMPONENT DRYING SYSTEM

COMPONENT IDENTIFICATION

High Level Control

P. E. T. Plastic Pellets

Insulated Hopper

Air Diffuser Cone

Pellet Loading System

Thermometer

Dried Material Discharge to Process

Process Air Filter

Thermometer

Temp. Controler

Process Air Heaters

Water Cooled Heat Exchanger

Thermometer

Tower 2

Process Air Fan

Four Way Valve

Regeneration Fan & Heaters

Tower 1

Linde 13X Molecular Sieve Desiccant

Fig. 14-6 P.E.T. component drying system.

393

Then the aftercooler (or heat exchanger), reduces the air temperature to about 90°F (for maximum desiccant efficiency) before it reenters the desiccant bed. There, the moisture is removed, later to be exhausted to the atmosphere during regeneration.

Because of the extremely low moisture levels necessary to process P.E.T. resins, and to maintain the efficiency of your drying system, any ambient air leaks into the system must be eliminated.

Things to Consider When Selecting a Vendor

The process air is heated by electrical heating elements, and the resulting air temperature is controlled by an electronic device. Two types of heaters are commonly used by dryer manufacturers: the Calrod, or encapsulated type; and the open coil type. The open coil type is more efficient and thus offers some definite energy savings.

Some dryer manufacturers locate the process air heaters directly on the air inlet of the drying hopper. This prevents the heat loss that normally occurs in systems where the process air heaters are located on or in the drying unit. Furthermore, direct mounting of the heaters on the drying hopper eliminates the need for a long insulated hose to connect the hopper to the dryer. The elimination of wide variations in process air temperatures at the drying hopper inlet is an added advantage.

The height-to-diameter ratio of the drying hopper is an important factor to consider when selecting your system. It should be designed so that the maximum air flow does not exceed 1 ft/sec. This is necessary to ensure that the air has sufficient time to transfer heat to the pellet and thus maximize the effective drying zone within the hopper.

The air flow must be evenly distributed across the diameter of the hopper. An air diffuser cone of proper design ensures the even distribution of drying air and causes the material in the hopper to mass-flow, that is; the first material to enter the hopper is the first material to exit.

The drying hopper must be well insulated to prevent both costly heat loss and injury to persons working near the equipment. The insulation is fitted around the circumference of the hopper and is retained by a riveted sheet metal covering. This insulation is generally applied to the straight wall section, the discharge cone, the air inlet tube, and the door, for maximum energy conservation and safety.

A very important part of your system—one that will require a *regular schedule* of maintenance—is the process air filter. Because it requires frequent inspection and cleaning, you should be certain that it is easily removed and that it is large enough to operate for a complete shift without cleaning.

The dust collected in P.E.T. drying systems is very fine and thus requires a high-quality filter with a large surface area. *The air filter must be maintained to prevent this dust from entering the desiccant beds.*

Should the beds become contaminated with dust, the heat of regeneration would cause it to fuse with the desiccant, which would thus lose its adsorption efficiency.

Proper maintenance makes good sense when compared to the cost of replacing desiccant—at $3.00/lb.

The aftercooler is a water-cooled heat exchanger and should be sized to lower the returning air temperature (usually 210°F to 250°F) to about 90°F to 110°F before entering the desiccant beds. The lower air temperature increases the adsorption efficiency of our desiccant, Linde 13X. Some manufacturers' models will vary, but 70°F water at a flow rate of about 15 gpm will enable the dryer to operate at peak efficiency.

Linde 13X, the most widely accepted desiccant in use, adsorbs less moisture at high temperatures than it does at low temperatures. So, the higher the air temperature entering the desiccant bed, the lower the efficiency of the drying system. Also, as shown on our graph (see Fig. 14-7), above 100°F Linde 13X molecular sieve adsorbs moisture far better than silica or alumina type absorbents.

The desiccant beds are regenerated alternately; when one bed is drying the process airstream, the other is being heated to around

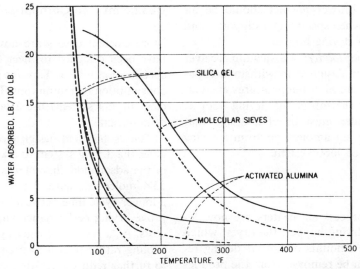

Fig. 14-7   Water vapor adsorption isobars, at 10 mm partial pressure (equilibrium data).

600°F to drive out the moisture previously adsorbed from the process. This cycle is controlled by a cammed timing unit that is preset at the factory. This cycle time enables the system to use the desiccant at optimum efficiency. It prevents desiccant bed saturation or breakthrough and ensures maximum regeneration.

Some Recommendations for Your System

1. *Dewpoint alarm/monitor.* Most manufacturers offer an alarm type that sets off a light or bell when the process air dewpoint rises above a preset level.
2. *Thermometers on the inlet and outlet of the aftercooler:* serve to monitor both the return air temperature from the drying hopper and the temperature of the air entering the desiccant bed; monitor the performance of your aftercooler.
3. *Thermometer on air inlet to drying hopper:* provides a direct readout of air temperature entering the drying hopper, and can be used to check the electrical temperature meter on the dryer control panel.
4. *Air flow indicator:* monitors air flow in system.
5. *Plugged filter indicator:* sounds an alarm or sets off a light when the process air filter needs cleaning.
6. *Proportioning heat control:* The process air heaters are electrically split into two banks and are individually controlled. During high temperature operation, one bank stays on continuously, while the other cycles on and off to maintain close control of the process air temperature. This offers an energy savings and provides closer control of the process air temperature.
7. *Amp draw indicators:* provide a visual indication of amperage draw in each leg (of a three-phase system) of the process and regeneration heaters. Should a heater problem arise, a change in amperage will be shown on at least one of the indicators.
8. *Support stand:* is used to provide additional support to the drying hopper when mounted on the throat of your machine. A flexible connection to the feed throat is strongly recommended. This enables the machine to move freely and allows the drying hopper to remain stationary.
9. *Drawer magnet:* prevents tramp metal from entering your machine. Clearance

for the drawer magnet should be included when specifying a support stand for your drying hopper.

10. *High heat loader:* is a vacuum receiver or loader designed to withstand exposure to the high temperatures encountered at the top of the drying hopper. The loader must be of a design that will prevent ambient air from entering the closed loop system.

New Ideas

Because the process air heaters are now located on the drying hopper, the dryer, which exhausts large amounts of hot air during regeneration, can be removed from the production area (see Fig. 14-8). The dryer can be located adjacent to the production area in an equipment room that houses the grinders, chillers, air conditioners, etc. This idea has two immediate advantages: (1) valuable floor space is saved in the production area, and (2) the heat of regeneration is no longer exhausted into an environmentally controlled area.

Desiccant Bed Comparison

More efficient dryers are now available that have been designed to meet the rigid drying specifications of P.E.T. resins and keep energy consumption to a minimum. These highly efficient drying systems use a large "solid" bed of desiccant.

The "solid" bed design (see Fig. 14-9) enables the drying system to take full advantage of the adsorption characteristics of the Linde 13× molecular sieve. Lower dewpoints and a longer cycle time between regenerations improve drying performance and reduce energy costs. As a result of longer cycle times, static cooling rather than forced air cooling is used to further reduce operating costs.

Smaller "shell type" desiccant beds are forced to operate on shorter cycle times than the "solid" bed design. The efficiency of these systems is reduced because ambient air must be used to cool the desiccant beds after regeneration. Regeneration occurs more frequently during a given period of time, and thus operating costs are affected.

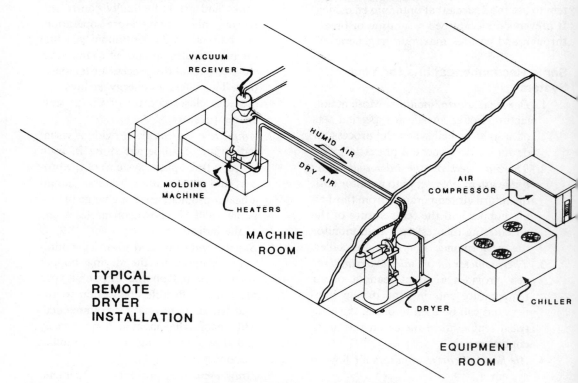

Fig. 14-8 Typical remote dryer installation.

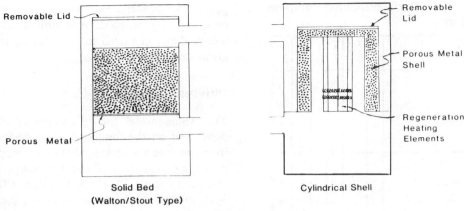

Fig. 14-9   Desiccant bed comparison.

Review—P.E.T. Requirements

P.E.T. resins, when received from the supplier, may contain approximately .05 percent moisture by weight. Successful processing requires that this moisture content be reduced to at least .005 percent by weight during the drying process.

Proper drying of P.E.T. resins is achieved through the use of a self-regenerating closed loop drying system. This system must provide a continuous supply of −40°F dewpoint air at a temperature of 350°F to the drying hopper. This hot, dry air is circulated through an insulated drying hopper that uses an internal air diffuser cone to provide an even flow of air through the resin. The air flow should be adequate for the material use rate. The residence time in the hopper should be no less than 4 hours.

A *well-maintained* drying system is the key to proper drying of P.E.T. resins. A good preventive maintenance program should include:

1. Regular filter cleaning.
2. Checking air hoses and gaskets for damage and leaks.
3. Monitoring dewpoint and temperature gauges on a regular schedule.

PARTS HANDLING EQUIPMENT (PHE)

The use of handling equipment has made major contributions to *moving plastic* materials in all types and sizes of processing plants. The next big change, which is now occurring, involves the use of handling equipment to *move fabricated parts* in these plants (3, 4).

Automation is one of the central events of our time. There has been much speculation in the past 20 years about what it will mean for society and the individual. Some of this speculation is motivated by fear; some by curiosity. And some individuals simply expect technical developments to pay off. At one time all seemed to agree that automation was an all-pervasive social phenomenon, and that its impact on the post-1975 world would be considerable. Look at what happened to moving material—it is now necessary for profit.

The payoffs *have* been significant in many industries, including the plastics industry. Processing plants are estimated to be spending at least $32 million a year in parts handling equipment (PHE). Most of the PHE goes into injection molding operations.

People vs. PHE

Among the many issues that have been raised by automation, two seem to constitute the focus of general concern. The first issue arises from a general sense of fear and so is badly stated. What it attempts to express is the extremely complex phenomenon of man's displacement in relationship to his perception of reality. In essence, man feels threatened by automation. The point to remember is that the full impact of automation might lead to a displacement comparable to the Darwinian revolution.

Table 14-1   Parts handling equipment functions.

	Type	Collect	Remove or pick	Place	Orient	Count/weight	Accumulate
Not integrated with IMM function	Manual	X	X	X	X	X	X
	Box	X	X				X
	Conveyor	X	X				X
	Unscramble/orient	X	X	X	X	X	X
Integrated with IMM	Sweep		X				
	Extractor	X	X	X			
	Cavity separator	X	X	X	X	X	X
	Robot/bang-bang	X	X	X	X	X	X
	Robot/sophisticated	X	X	X	X	X	X

The other issue concerns unemployment. Nothing yet has happened to cause massive unemployment. Its impact on the nature and structure of human work will tend to become cumulative as time passes. But like material handling, automation will become the way to operate within a plant.

What is happening in PHE, and in particular, robots, is analogous to what has happened in injection molding machines and their process controls. With PHE, productivity and quality go up, costs go down, and boring, simple, or hazardous tasks are lifted from the back of operators.

Parts automation, *like process control,* is im-

portant to molders. But no industrial enterprise can rely on modern technical achievements alone. The need for the manual inputs always remains.

Different Types

The "value-in-use" for the equipment used is dependent on the functions to be performed from the time a part is molded to its shipping/packaging operation. Functions range from manual to robot PHE manipulation (Tables 14-1 and 14-2). Types of operation include:

- *Manual.* Operator is used and in many cases necessary in automated systems.
- *Drop in the box.* Parts drop directly or via a slide into a box.
- *Conveyor.* Parts drop on a belt or through a tube and go directly to an accumulator (box, etc.). Systems can include part and runner separators.
- *Unscrambler/orientation:* directs parts with or without a runner through a collector, conveyor, and unscrambler to position and finally orient parts. Subsequent operations can include counting, weighing, stacking, and/or accumulation.
- *Sweep.* Mechanical arm clears molding area.
- *Extractor.* Mechanical arm removes parts and/or runners from the mold. The

Table 14-2   Parts handling equipment growth rate.

Type	Percentage used with IMM		No. of mold cavities	Part size	Cost/unit dollars
	Current	Future			
Manual	20	12	any	any	Does not include PHE
Box/collector	30	15	any	Sm, Med	50 to 500
Conveyor	30	30	any	any	500 to 3,000
Unscramble/orient	10	18	2 to 24	Med	2,000 to 40,000
Sweep	3	5	1 to 16	Sm, Med	200 to 1,500
Extractor	4	7	1 to 24	Sm, Med	500 to 5,000
Cavity separator	½	2	12 to 96	Sm.	5,000 to 50,000
Robot/bang-bang	2	8	4 to 10	Med	5,000 to 25,000
Robot/sophisticated	½	3	1 to 12	Lge	25,000 to 150,000

usual type derives power from the movable platen through a cam action.

- *Cavity separation.* A plate with pockets moves parts from a mold and places them in separate collectors. If one part from a 96-cavity mold is off specification, all those parts are quickly collected.
- *Robot/bang-bang:* is mechanically operated against preset stops with specific performance capabilities.
- *Robot/sophisticated:* has a high degree of flexibility, versatility, and capacity. Operations are rather unlimited with programmable point-to-point operation within mils (Fig. 14-10).

Use

In today's industry, PHE's are not the humanoids of science fiction. They are blind, deaf, dumb, and limited to a few preprogrammed motions. But in many production jobs, that is all that is needed. They are "solutions" looking for "problems."

Most plants can use some degree of parts handling automation which can substantially increase productivity. With short or long runs, PHE's pay off, particularly when they become more than product handlers or part removers. Think in terms of adding operations such as secondary operations, simplifying mold changes, improving quality, and dramatically reducing rejects, as well as OSHA compliance, inspection, orientation, and packaging. Also look into second and third shift operations.

The PHE is the *lowest-paid "employee"* in the industry. It can increase productivity by operating faster, is more reliable than operators, and can easily handle heavy or large parts with a long reach. It has the dexterity to maneuver large parts between the tie bars, the accuracy to seat parts in a drill press, and the ability to stock parts. Operators can be freed for reassignment to other operations.

Other accomplishments include constant repeatable positioning, handling small, delicate, or brittle parts, and increasing plant safety. PHE's can carry out certain functions too rapid or too complex for operators. Also, quality can be improved significantly. Cosmetic or noncontaminated medical parts can be easily handled.

Value-in-use

The concept of automatic molding has much appeal. Nevertheless, the ultimate justification for PHE (like material handling and process control) must be made on the basis of economics. Whether your goal is faster cycles, reduced scrap, or improved quality, the better process will always be one that gives the most acceptable product per total dollars invested and labor expended when all factors are considered.

This is an exciting and profit-motivating field that requires creativity with technical expertise in goodly amounts. Certain operations are simple in concept and relatively low in cost. This is why the conveyor belt system will continue to be used as the prime mover. Creativity with belts is unlimited.

Degree of justification requirements deter-

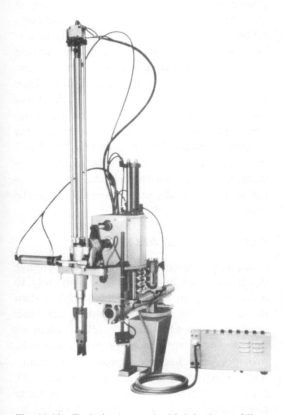

Fig. 14-10 Typical robot used with injection molding machines (AEC).

mine the equipment to be used. The prime differences are capability and flexibility. To be successful with PHE, proper planning is necessary.

PHE is possibly best understood in terms of what it does rather than what it is. A potential user must decide exactly what he wants, study the time cycle if it is important, see how apparently wasted time might be utilized in performing additional operations, determine availability of space for PHE, and set up the interface with IMM (Injection Molding Machine) as well as the plant facilities. Another important parameter is assigning reliable house personnel to coordinate installation and operation of equipment that comes from a reliable company.

Cavity separator systems as well as robots are relatively new to the plastics industry. Where used they receive unqualified approval. Typical of PHE, each has a unique blend of characteristics such as number of motion axes, arm configuration, load capacity, and type of program.

Detriments

In order to automate with PHE, a high degree of mold performance is necessary. Probably 30 percent (five years ago it was 45 percent) of the molds being used are not of sufficient quality to put out acceptable parts using PHE. They have blocked-off cavities, get die burns in a cavity, etc.—and are not capable of producing quality parts consistently.

Most applications will require a more skilled level of maintenance, the type that should be available to handle IMM and process controls. Potential problems also include mismatching with IMM or lack of understanding PHE.

A most important consideration is that in some operations superior product quality can only be achieved with an operator and PHE.

Top management can be expansion-oriented, not efficiency-oriented; industry is especially attuned to increased production through expanded facilities and faster machines. On the other hand, some top management requires total automation. But at the other end,

in the molding shop, people are not in tune with (or are afraid of) PHE.

Robots have proved themselves both technically and economically in a number of industrial applications, particularly in automotive-metal operations. But the number in use remains small compared to the potential applications where they will be affective.

Robots

No two are alike. Each is a unique blend of characteristics such as number of motion axes, arm configuration, load capacity, type of memory, ease of programming, teaching method, and (for less sophisticated types) ease of interfacing with a computer. Axes of motion vary from two, for the simple robots, to six, for the more complex machines. Three to five are used most commonly.

Most robot programs are open-loop in that they repeat the same set of functions continuously without modification. Robots are programmed in basically *two ways,* depending on the type of memory. In the simpler bang-bang robots, programs are set by physically inserting the program by fixing stops, setting switches, putting in a punched card or tape, arranging wires, or, in the case of an air logic system, plugging in air tubes. In the more complex, sophisticated robots that use magnetic tape, disc, or minicomputer memory, the program is established or taught by "walking" or leading the robot through the required motions. With the walk-through method, the robot arm is manually maneuvered through the cycle.

QUICK MOLD CHANGE

Completely automated quick mold change (QMC) devices have been used, but only to a limited degree. Cost is usually higher than the cost of the injection molding machine; however, it pays for itself "quickly" when required. QMC, with microprocessor controls, provides cost-effective approaches to plant-wide automation. Different designs are used, such as overhead or side loading/unloading platforms (Figs. 14-11, 14-12, and 14-13).

The concept is best suited to processors with

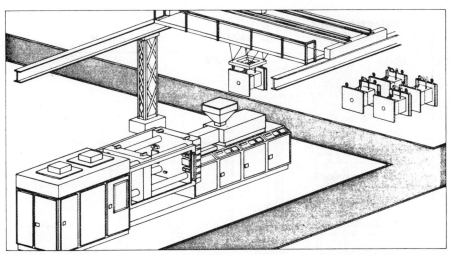

Fig. 14-11   Quick mold change system uses a single robotized overhead crane to service multiple injection molding machines from an inventory of 200 molds.

relatively short production runs and frequent mold changes. In such operations, the benefits of QMC are many: increased productivity, reduced inventory, increased scheduling flexibility, and more efficient processing.

Mold-changing time is wasted time. In the all-out effort to trim waste, increase productiv-ity, and reduce inventories, quick mold chang-ing will play an important role. Today, sys-tematic mold changing is a novelty; in a few years, mold-changing systems, include fully automatic changers, will be much more com-mon.

Systematic procedures to expedite mold

Fig. 14-12   Beside the machine, quick mold change.

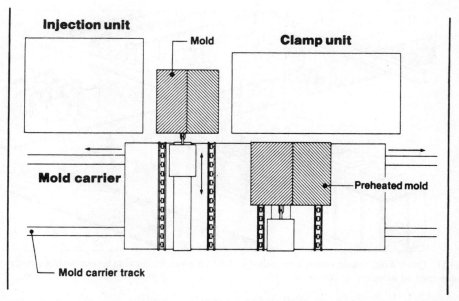

Fig. 14-13 Schematic of a fully automated quick mold change device; production interrupted from 1 to 2 minutes.

change can take many forms—from fully automated mold conveying devices to the addition of an extra overhead crane. Regardless of how it is accomplished, QMC means reducing mold changing time to roughly 1 to 10 minutes, and facilitating nearly instantaneous start-up on the new mold.

QMC goals are achieved by standardizing the construction of molds and machine mounting, raising operator training levels and awareness, and increasing reliance on microprocessor-based controls.

The software stores information regarding molding cycle times and temperatures for individual molds, platen spacing for each mold change, and orchestration of the mold changing devices. Increasingly important in software and controls will be ease of programming and the ability to change programs quickly, more microprocessor storage capabilities, and reliable self-diagnostics.

Completely automated systems contain (a) mold conveyors that propel the molds in and out of position on motorized rollers, and (b) mold carriers that index on a track parallel to the IMM to align the mold conveyors and the clamp unit. They can also convey molds to and from the machines from a central mold storage area.

Basically when the microprocessor-based control signals the end of a mold production run, the movable platen indexes to a preset position for mold removal. Automatic mold clamps are deactivated, releasing the old mold. Mold clamps or straps connect to lock mold halves together. The mold conveyor removes the old mold.

Computer control resets platens for the new mold, and the carrier table aligns the preheated, new mold with the clamp unit. As the new mold is inserted by mold conveyors, automatic mold positioners center the mold within seconds. Automatic clamps are activated by the control system, and new molding-cycle information stored in the microprocessor initiates the next production cycle. Estimated downtime between old- and new-mold shot is 2 minutes with a highly sophisticated QMC.

REFERENCES

1. "Recycling," *ME*, pp. 44–50 (February 1982).
2. Stout, P. E., "Drying PET," SPE-IMD Technical Conference, October 20–22, 1981.
3. "Robots Become More Desterious," *Compressed Air*, pp. 18–22 (July 1983).
4. "Robotics Committee Moves Into Action," *ASTM Standardization News* (May 1983).

Chapter 15
Material Handling

G. A. Butzen and G. A. Adams

AEC Corporation
Whitlock Inc.

INDUSTRY NEEDS

Material handling involves more than pneumatic conveying. It is railcar unloading and bulk storage, integrated blending, compounding, and dehumidifying drying—everything necessary to automatically supply resin to the processing machine. Most material manufacturers ship bulk materials, either pellet or powder, by railcar or truck trailer. Both of these bulk transporters are designed for easy unloading of the materials they carry. This chapter will present the systems available to transport bulk and reclaimed materials to the production machines that transform these materials into products (1).

Today, processors must be aware of every possible cost savings. Once the possibility of a cost savings is discovered, there is usually cost involved with the potential savings. To justify this cost, a return on investment must be calculated. Some considerations for the economic justification of bulk storage systems are:

1. Lower resin prices (2–10¢ per lb).
2. Reduced warehouse personnel.
3. Reduced lift truck operating hours.
4. Personnel not required to unload truck except for initial hose hook-up and final hose disconnecting.
5. Buying in bulk eliminates disposal problem of bags and cartons.

6. Elimination of material losses that could result from broken gaylords and bags.
7. Reduced plant duties of clean-up.
8. No inside warehouse space (warehouse space in an existing plant costs between $3 and $4 per ft³, new construction: $28 to $40 per ft²; a conservative figure for storage taking the space of an additional machine for production is $300 to $400 per ft²).
9. Consistency of lots in material.

A bulk storage system is the first phase of totally automated material handling systems. This review will explain all the components of a complete material handling system. It presents each component's function and operation. The final section, on "Material Handling Systems," ties all of the equipment together into the systems concept.

This chapter was written for plastics and chemical processors. It is not intended to be a substitute for professional material handling systems analysis, but rather a general overview to acquaint the reader with the many facets of this intricate art. Specific analysis and system design should be made in conjunction with the equipment manufacturer.

The manufacturer's system designers carefully plan conveying systems with high capacity filtering and conveying velocities that will permit consistent dust-free conveying. Engineers begin with careful evaluation of the conveyability of customer materials. The majority of difficult-to-handle materials have been successfully conveyed. Design and operating information is therefore available. Any doubt about the conveyability of a particular material is overcome by testing in the manufacturer's laboratory, where the material sample is actually test-conveyed over varying distances to establish a performance curve for that material. Filtration equipment is trial-tested to determine the amount and type needed for reliable dust-free conveying.

In many instances, sound enclosures and specialized electrical systems are incorporated into a system to conform to OSHA guidelines and regional or national electrical codes.

The manufacturer can provide complete "turnkey" service from system design through installation. Single-source responsibility is a must in today's material handling business because of the many economic factors and engineering variables that have to be considered in developing custom engineering conveying systems. Single-source responsibility and guaranteed performance ensure the successful operation and profitability of your material handling investment.

BASIC PRINCIPLES OF PNEUMATIC CONVEYING

Introduction

Bulk material handling and conveying is a combination of theory and experience. Almost any substance can be conveyed pneumatically.

Automation of the material handling system increases the advantage of bulk materials purchase and storage. A pneumatic conveying system can move materials from the storage area to the processing machine automatically, with little risk of contamination to the materials which commonly occurs when materials are moved to the processing machine manually.

The major problem with pneumatic conveying of materials is the absence of a standard set of formulas for calculating the equipment sizing and the flow characteristics of the material to be conveyed. Standard formulas do not yet exist because of the vast variety of plastic materials, additives, and their combinations. Most available information on the sizing of material handling systems is based on experiments conducted by equipment manufacturers and observation of successful conveying systems currently in operation.

Pellets, granules, and powders are pneumatically conveyed through what is classified as a pipeline conveying system. This type of system transports particles of solid materials through vertical and horizontal pipelines according to the physics principles of kinetic energy, pressure, and aerodynamic lift.

In their basic form, these formulas would state that if the particles were placed in a pipeline conveying system in a stream of flowing air at a higher velocity than the terminal velocity of the particles, these particles would move with the velocity equal to the difference between these velocities. The terminal velocity, in this case, would be the least amount of air flow needed to make the particles move.

Velocity is the key to transporting materials pneumatically. Velocity is defined as the rate of motion or speed. There are specific terms used when describing velocity in material conveying systems. *Critical velocity* is defined as the minimum superficial air velocity that will convey the material as specified. *Drop out velocity* is a term commonly used for the minimum amount of velocity needed to move particles through a vertical tube. *Settling velocity* is the minimum velocity needed to prevent material from falling out of the air stream.

Pipeline conveyors are commonly referred to as *dilute-phase systems*. *Dilute-phase conveying* is described as conveying of a small volume of material by a large volume of air. This is put into a ratio commonly called the material-to-air ratio and classified in pounds. The counterpart to dilute-phase is *dense-phase*. *Dense-phase conveying* is described as a pound of air moving its weight or more of material through the conveying tubes. In other words, material is conveyed in a dense-phase state that will have a low material-to-air ratio. Compactable powders are common materials that are conveyed in a dense state. This concept will be discussed in detail later in this chapter.

To understand pneumatic conveying, several terms and formulas must be discussed. To begin with, pneumatics is a branch of hydraulics power and not a separate form of power transmission, as many people think. The explanation of the term pneumatics is based on the fact that every liquid has a temperature and pressure point where it becomes a gas. A common example of this is water. When water is brought to a boiling point at 100°C–212°F, with atmospheric pressure at 14.696 psia, the water turns to steam, which is water in a gaseous state. The boiling point of liquids varies with the amount of pressure and temperature applied. For example, water can boil in a controlled situation at 0°C–32°F, which is the normal freezing point of water. However, with a pressure reading of .0885 psia, the reduction of atmospheric pressure will actually force the water to boil at that low temperature.

Air has the same characteristics as water. Air has a temperature and pressure point where it becomes a liquid substance. Liquids and gases have distinct characteristics that separate the two substances. Gases generally can be compressed easily, which make gases an excellent medium when a substance is needed to perform a task in a limited space. Air and other gases can be compressed and stored in containers to eventually be used to perform work. A gas can be compressed to an extreme pressure, where the temperature can be raised to force the substance to condense back into a liquid and be stored in a tank. Once the tank is opened and the substance in liquid form is released into the atmosphere, the liquid will return to a gaseous state because the pressure and temperature will immediately be lowered. Once a substance reaches a liquid state, it generally cannot be compressed into a smaller mass. The characteristics of a substance can be predicted and measured more easily when the substance is in a liquid state. It is this predictability of substances in a liquid that provides the basis for formulas used in pneumatic conveying.

The physics principles that play an important role in pneumatic conveying are gravity, pressure differntial, inertia, shear, and elasticity.

Gravity is an external force caused by the earth's rotation which attracts particles down toward the center of the earth.

Pressure differential is a force caused by a difference in pressure to initiate the movement of air and material.

Inertia is a force needed to overcome the natural resistance of movement for material.

Shear is the relative flow between adjacent particles of a viscous fluid.

Elasticity is the intrinsic tendency of a compressable gas to expand and flow from a high pressure to a low pressure.

The basic laws of physics affecting fluid flow are the laws of conservation of matter, the laws of the conservation of energy, the perfect gas properties, and the gas laws based on Boyle's Law, Gay Lussac's or Charles Law, and the Combined Gas Law.

The Law of Conservation of Matter

This law states that the mass flow through any cross section of a pipe is constant. This is shown by the following formula:

$$QM = pAv = \text{a constant}$$

where: QM = volume of discharge in cubic feet based on the molecular weight in slugs

p = absolute pressure in pounds per square inch

A = area in square feet

v = velocity in feet per second

An offshoot of this law is the continuity of flow through a pipeline equation. This equation states that the volume of discharge (Q) will be the same at any point within the pipeline. This equation is represented as follows:

$$Q = p_1 A_1 v_1 = p_2 A_2 v_2 \ldots = p_n A_n v_n$$

This law can be modified if there is no variation in the fluid density at any two points within a pipeline. This is commonly termed as the discharge equation:

$$Q = A_1 v_1 = A_2 v_2 \ldots = A_n v_n$$

Because of this law, the density of air is assumed to be a constant when the pressure differential between the beginning of a pipeline and the end is less than 1 percent based on measurements utilizing absolute pressure.

Absolute pressure is defined as gauge pressure plus atmospheric pressure.

The Law of Conservation of Energy

Energy is neither created or destroyed. Energy is simply transferred from one form to another. An example of this would be the burning of wood. Energy was originally transferred to the wood through the growth process, through photosynthesis. When the wood is cut down, it possesses what is termed potential energy. *Potential energy* is defined as a state where a substance has energy, and it needs another reactant to transfer that potential energy to kinetic energy which is the actual work being performed by energy. In the case of burning wood, fire is the reactant. The fire transfers the potential energy to kinetic energy. The kinetic energy in this case would be heat. If this heat energy is captured, it can be used to heat other materials and convert the kinetic energy to pressure energy, which will actually physically perform a working function.

To reiterate, energy in any form has the capacity to do work.

For pneumatic conveying, *kinetic energy* is defined as the energy retained by a mass due to its velocity. The formula for this is:

$$KE = \frac{W}{2} \times \frac{v^2}{g}$$

where: KE = Kinetic energy
W = Work in foot pounds
v = Velocity in feet per second
g = Gravitational acceleration

Potential energy is the possible work a mass can perform. This is described as work in pounds (w) falling from a specified height in feet (z) above a horizontal surface can perform (wz) foot pounds of work. The same amount of foot pounds would be required to lift the material or mass in an airstream.

Pressure energy is the energy of a fluid that actually performs a working function at a pressure above atmospheric pressure. The basic formula for a pressure balance is:

$$\frac{P}{w} = h$$

where: h = hydraulic pressure head
P = absolute pressure in pounds per square feet
w = weight per unit volume in pounds per cubic foot

By combining (W) work in this formula, we will have a formula for pressure energy:

$$\text{Pressure energy } W\frac{P}{w} = Wh$$

The total energy in a fluid can be calculated by combining these three formulas:

$$\frac{Wv^2}{2g} + W\frac{P}{w} + Wz = \text{a constant}$$

The energy per pound of fluid can be obtained by removing the work (W) aspect from this formula:

$$\frac{v^2}{2g} + \frac{P}{w} + Z = H = \text{enthalpy (total heat) in Btu per pound mass; total hydraulic head}$$

As mentioned before, these formulas are based on the flow of fluids. Some other formulas will have to be incorporated into these formulas based on the perfect gas laws properties. Unlike fluids, gases are compressable, which makes them the most likely candidates for conveying materials. The energy a gas contains is determined by the effective pressure, volume, and temperature. Should any change take place in one of these three values, a direct change should take place in the other two variables. Should the pressure increase by 5 lb/in.^{2}, the temperature should change by approximately 2 degrees. If another 5 lb/in.^{2} is added to that gas, the temperature should increase another 2 degrees. This is expressed by the formula:

$$\frac{T_1}{T_2} = \frac{P_1}{P_2} v \text{ constant} = \frac{V_1}{V_2} p \text{ constant}$$

The major problem with this formula is that the majority of gases used in pneumatic conveying are not perfect gases. Air is the most common medium used. Air contains many contaminants that prevent it from being classified as pure gas. A contaminant is defined as any substance that mixes with another substance that will cause any form of change in the molecular structure.

Pressure is a force exerted on a particular area. Pressure is measured on two scales: pounds per square inch (psi), which is commonly used by engineers: pounds per square foot (psf), which is used in pneumatic conveying. There are three different scales of pressure: Gauge pressure is pressure measured from a pressure gauge that starts at "0" psi. The "0" psi reading is equivalent to atmospheric pressure, which is 14.7 psi at sea level. Absolute pressure is equal to gauge pressure added to the atmospheric pressure.

Absolute = Gauge + atmosphere (14.7 psi)

Volume is expressed in cubic feet per pound. However, volume is rather tricky to measure because gas expands and contracts with changes in temperature and pressure. Volume is measured more easily by using the inverse of specific volume, which is density (D).

Temperature is a measure of the intensity of the molecular energy in a substance. The higher the temperature, the more molecular movement. With lower temperatures, the molecular movement will decrease. The temperature at which molecular movement ceases completely is "absolute zero." Absolute zero has not yet been reached in actuality, but in theory it appears possible.

Boyle's Law

Robert Boyle conducted various experiments with gas that demonstrated that if the temperature of a specified quantity of gas is held at a constant temperature, the volume of the gas will vary if different pressures are applied to the gas. This is shown by the following formula:

$$\frac{P_1}{P} = \frac{V}{V_1} \text{ or } P_1 V_1 = PV = \text{Constant}$$

Boyle's Law also states that the density of gas varies inversely with the volume proportionate to the pressure:

$$\frac{P}{w} = \frac{P_1}{w_1}$$

Working with a different combination of variables, Louis-Joseph Gay-Lussac conducted experiments that verified Jacques Charles's theory, which works along with Boyle's Law. Where Boyle used a constant temperature, Gay-Lussac and Charles used pressure as a constant. They verified that if the temperature is increased, the volume or density will increase and, inversely, if a lower temperature is used, the volume or density of the gas will decrease. This is shown with the following two formulas:

$$V_t = V_0(1 + \alpha_p t)$$

where: V = specific volume
t = ordinary temperature
p = absolute pressure in square inches
α = expansion coefficient for gases

$$P_t = P_0(1 + \alpha_v t)$$

where: P = pressure
t = ordinary temperature
v = specific volume
α = expansion coefficient for gases

The combined gas law combines gas laws from Boyle, Gay-Lussac and Charles. This law is based on the initial conditions that a gas is at $P_0 V_0 t_0 = 0$. Once heat is applied at the constant volume of V, the gas will arrive at the state of $P_1 V_0 t$ which interprets into the following formula:

$$P = P_0(1 + \alpha t)$$

Gas Processes

There are two gas processes utilized in pneumatic conveying, isothermal and adiabatic. Air compressors force air to do work isothermally by compressing air while retaining a constant temperature. This can take place by having the heads on the air compressor cooled by either air or water, which will remove the heat that is building up due to the increase in pressure. This is demonstrated by the formula:

$$PV = C \text{ or } P = \frac{C}{V}$$

where: P = pressure
V = volume
C = discharge coefficient
Ratio in compressors

The amount of work that is produced by this process can be measured by the following formula:

$$P_1 V_1 \left(\frac{P_2 - P_1}{P_1} \right) = (P_2 - P_1) \times V_1$$

Kinetic energy is used in the isothermal process to accelerate the air by using fans and low pressure blowers; thus the following formula must be added to the previous formula to achieve the amount of total work done:

Work = velocity head + static head

or $$= \frac{U_2^2 - U_1^2}{2_g} + (P_2 - P_1) \times V_1$$

The adiabatic gas process does not remove heat as in the isothermal process; however, the specific heat of the gas during the transfer of energy is assumed to remain constant. The adiabatic work formula is shown by:

$$\text{Work} = \frac{P_1 V_1}{K - 1} \left[1 - \left(\frac{V_1}{V_2} \right) \frac{K - 1}{K} \right]$$

when P_2 is unknown

$$\text{Work} = \frac{P_1 V_1}{K - 1} \left[1 - \left(\frac{P_2}{P_1} \right) \frac{K - 1}{K} \right]$$

when V_2 is unknown

The final basic formula used in pneumatic conveying is the horsepower formula. *Horsepower* can be defined as the rate of doing work. One horsepower is equal to the rate of 33,000 foot pounds of work per minute. For an isothermal process, the horsepower formula is:

$$\text{hp} = \frac{M P_1 V_1}{33,000} \ln \frac{P_2}{P_1}$$

where: M = pounds of gas delivered per minute
P_1 & P_2 = initial and final absolute pressures in pounds per square inch
V = specific volume (in cubic feet per pound)
ln = natural logarithm

33,000 is a constant that states one horsepower is equal to a rate of 33,000 pounds of work done per minute.

This formula can be changed because $M V_1$ is the volume of gas in cubic feet per minute at the compressor suction. The initial volume can be expressed as (Q) cubic feet per minute with P_1 and P_2 representing the initial and final absolute pressures in pounds per square inch. This formula is shown below as:

$$\text{hp} = \frac{144}{33,000} P_1 Q \ln \frac{P_2}{P_1}$$

144 is a conversion factor that converts feet-squared to inch-squared figures.

For fans, the air horsepower formula is stated as:

$$\text{Air horsepower} = \frac{CP}{33,000}$$

where: C = cubic feet of air discharged per minute
P = total pressure (static and velocity) in pounds per square feet

Fan pressure differential is usually given in inches of water; a conversion factor of 5.19 has to be used to convert one inch of water to pounds per square feet. This changes the formula to:

Air horsepower = Total inlet air, scfm ×

$$\frac{\text{pressure differential in } H_2O}{6356}$$

To compress air adiabatically for a quantity of gas for a volume (V_1), the formula is:

$$\frac{\frac{K}{K-1} MP_1 V_1 \left[1 - \left(\frac{P_2}{P_1}\right) \frac{K-1}{K} \right]}{33,000}$$

As with the isothermal formula, MV_1 can be expressed as cubic feet per minute (Q), and P_1 and P_2 are classified as the initial and final absolute pressures in pounds per square inch; thus the formula becomes:

$$hp = \frac{144}{33,000} \frac{K}{K-1} P_1 Q \left[1 - \left(\frac{P_Q}{P_1}\right) \frac{K-1}{K} \right]$$

k = the ratio of specific heats. The specific heat of air is 1.4, thus simplifying this formula even more:

$$hp = 0.0153 \, P_1 Q \left[1 - \left(\frac{P_2}{P_1}\right) 0.286 \right]$$

This formula can be modified for a multistage air compressor by adding (n) for the number of stages if they have the same inlet temperature at each stage.

$$hp = 0.0153 \, P_1 Q \left[1 - \left(\frac{P_2}{P_1}\right) \frac{0.286}{n} \right]$$

These formulas explain the flow and work capabilities of air in a pneumatic system. However, these characteristics of air are only one factor in the art of pneumatic conveying. The other factor concerns the actual material that is to be conveyed pneumatically.

Various properties and characteristics of materials used in the plastics industry that can be conveyed pneumatically affect the sizing and design of the conveying system. The following explains the characteristics that affect material flow in a system, along with a few tests that can be conducted to determine factors affecting the material flow.

Specific gravity is one of the more important characteristics that pertain to conveying a material in an airstream. Specific gravity is defined as the ratio of the material's density compared to the density of water. The test for determining specific gravity is a basic test to see how much water the material displaces when placed in a container containing a specific amount of water.

Powders are commonly used in the plastics industry. The specific gravity test for powders is done by vibrating them in a container until the powders form a densely packed mass. The volume and density are compared to the known density of an equal volume of water.

Specific gravity is an aid in determining how much air flow is needed to lift a particle in an airstream.

Particle size is also a consideration in pneumatic conveying systems. The material has to be tested to determine the amount of fines and dust that may be contained in the material. This will help determine the type of air flow in a system, whether it be a vacuum or pressure system, along with the type of filters that will be utilized in the system. Particle size is measured by using sieves that are made to standards set by the American Society of Testing Materials or the United States Standards Institute.

A common method of test for the particle size and range would be to place the material in a sieve with a large mesh opening and shake the material until all of the particles that are small enough pass through the screen. This procedure is repeated, with screens of a smaller mesh size, until all of the particulate is separated.

Tackiness is another characteristic that must be examined. If the material is extremely tacky, it may not be suitable to be conveyed pneumatically. Tacky material may smear against conveying pipes and cause a buildup of material which will eventually clog a line. A simple test for material tackiness would be to take some material into your hand and squeeze it into a compact ball. If the material sticks together, it is classified as a tacky material.

The only way to determine if a tacky material can successfully be conveyed through a pneumatic system is to run the material in an actual pneumatic system to determine if the material will flow properly through the conveying lines. This test can be performed by the manufacturer.

The melting point of a material should be determined. There are plastic materials that melt at low temperatures. If these materials are conveyed at a faster rate than necessary, they may slide against the walls of the conveying tubes and heat up by friction which in turn will cause them to begin to melt, producing what is called "angel hair." This commonly takes place in a bend of a conveying tube due to the centrifugal force that is placed on the pellets forcing them to slide along the outer periphery of the tube. Angel hair is caused by the melting plastic pellet running along the wall, leaving a thin trail of plastic along the tube wall. This thin plastic will partially peel away from the wall as the pellet moves back toward the center of the airstream, leaving what appears to be a fine hair. If enough of this occurs with other pellets in a particular area in a system, the angel hair will clog the system, thus preventing material from flowing through the system.

Abrasiveness of a material is another concern in pneumatic conveying. Materials that are abrasive may cause the conveying tubes to wear through quickly. Abrasive materials may have to be conveyed at a lower rate than other materials if at all possible. There are other modifications that can be made to a system to combat premature failure of the conveying tubes such as using wear-resistant materials, which will be discussed later. The only real test which can be performed on a prospective abrasive material is a run in a test system to determine how quickly the material can wear through a conveying tube.

Corrosiveness is a characteristic of powders or other materials that contain acids. There are very few of these types of powders used in the plastics industry. A material can be tested for acid content by testing the material for pH factor. A pH of 7 is neutral. Any reading below 7 is an indication of acid. A pH reading above 7 would indicate that the material is alkaline. Powdered materials with strong acid indications will have to be conveyed through special pneumatic systems in order to prevent any corrosion from taking place within the system.

Aeration and de-aeration are additional factors to be considered. If a material can be continuously saturated by air in a free-flowing state, the material is aerated material. Should a material clump together and block the air flow, the material is a de-aerated material. De-aerated material can still be conveyed pneumatically. The manufacturer can make recommendations for handling this type of material.

A test can be performed on materials to determine aeration characteristics. The material can be placed in a container with a lid on it. The container is shaken for a few moments. If the volume of the material appears to have increased and is taking a long time to settle to the bottom of the container, the material is said to be an aerated material. If the material settles to the bottom of the container quickly, it is said to be a de-aerated material.

Another test for the aeration of material is the angle of repose. The material is put on a horizontal plane, and that plane is lifted at an angle until the material starts to flow (at the angle of repose). If the material starts to flow at a low angle, the material is said to be free-flowing or aerated material. If the material flows at a high angle, the material is said to be de-aerated or a hard flowing material. The angle of repose not only helps with sizing of pneumatic systems, but it also helps with choosing the proper storage system equipment.

Odors and the toxicity of materials should be considered when developing a conveying system. These two related factors are not very common in the plastics industry. However, these characteristics could be a common element when dealing with other chemicals that may be related to the plastics industry. These elements have an effect on the type of conveying system and filtration system that should be incorporated in a plant.

System Sizing

Despite the vast amounts of formulas and testing methods, empirical formulas for pneumatic conveying have not been established because of the almost unlimited supply of materials that can be conveyed by air. However, manufacturers of conveying and storage equipment have established simple sizing charts based on the most commonly used materials in the plastics industry. Most plastic materials have an average bulk density of 35 lb/cu ft. Polystyrene has a bulk density of 35 lb/cu ft, in both pellet and powder form. Various tests have been conducted on polystyrene pellets and powders to determine which conveying rates, distances, line sizes, and air pressures are appropriate to convey these two substances. It was discovered that most materials can be conveyed between 6 psi gauge pressure and 10 psi gauge pressure. From this information, graphs were designed for easier sizing of a pneumatic system. The following graphs (Figs. 15-1 through 15-5) are for vacuum and pressure systems that contain a total lift and run conveying loop with four elbows included.

There are some rules of thumb that can be included in this section on graphs and sizing. To make the sizing procedure less complicated, when other than straight tubing is encountered in the system, we convert the pressure drop of the bend, flex-hose, etc., into an equivalent length of straight tubing.

When a graph is used, a 10 percent conversion factor is added to the conveying rate for each elbow under the amount of four that is not used. The 10 percent should be subtracted for every elbow up to six total elbows. If more than six total elbows are used, the manufacturer should be contacted for specific conversion factors relating to a particular material. Other rules of thumb include:

1 foot horizontal tubing = 1 equivalent
conveying foot

1 foot vertical tubing	= 2 equivalent conveying feet
1 long radius bend	= 20 equivalent conveying feet
1 foot of any type flex hose	= 4 equivalent conveying feet

A derating factor of 3 percent should be used for every "y" tube used in a conveying system. If a proportioning hopper is used, the system should also be derated by 20 percent.

Another chart will show research that has been completed on various pipe sizes and related factors such as volume pressure drops for different cubic feet per minute ratings, along with velocities that can be obtained with different horsepower outputs.

AIR MOVERS

Vacuum Units

Vacuum conveyors are used for the automatic pneumatic conveying of most free-flowing dry granular materials. With an addition of a low head separator or a filter chamber combination, fine powders can also be conveyed with ease.

Vacuum conveyors consist of six basic components:

1. Vacuum power unit
2. Vacuum hopper
3. Material pick-up device
4. Tubing (between power unit and vacuum hopper)
5. Material tubing
6. Filter chamber

The vacuum power unit consists of a control enclosure, a motor, and a positive displacement blower as the key elements in the system. The vacuum power unit will normally fall in a size range from $1\frac{1}{2}$ hp to 30 hp. The positive displacement blowers are constant-speed machines that deliver a relatively constant volume of air over a varying range of discharge pressures (see Fig. 15-6). Positive displacement blowers consist of two rotary lobes that

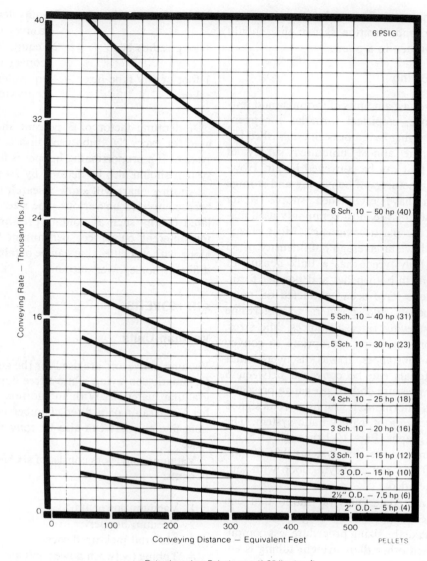

Rates based on Polystyrene @ 35 lbs./cu. ft.

Fig. 15-1 Continuous pressure—pellets.

rotate and intermesh with each other to force air through the blower. Positive displacement blowers normally have a pressure range from 5 psi gauge to 18 psi gauge. There are blowers manufactured with higher pressure ratings.

The control box in vacuum conveying units utilizes various designs for different conveying systems. Vacuum conveyors are generally classified as a single material line or a multiple material line system. Control boxes can be a simple single material line system that uses electro-mechanical control devices such as a

main disconnect switch, a motor starter, and either a fused overload protection system or a heater element protection device.

Multiple material line control systems are slightly more complex due to the addition of relays, control switches, and timers that permit the conveyor to switch the material flow through up to 12 different lines.

Control boxes are available in either electro-mechanical control or solid state circuitry which utilizes solid state control modules having various control circuits such as automatic

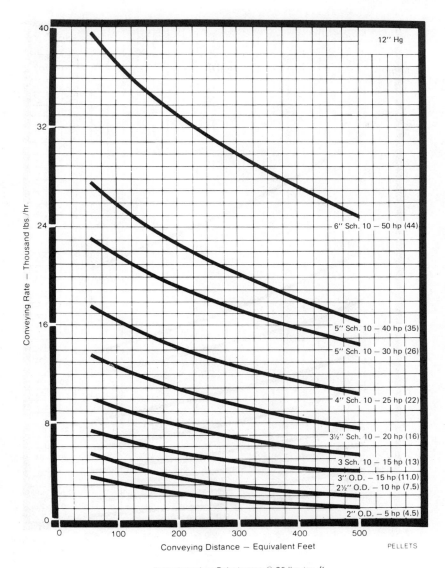

Fig. 15-2 Continuous vacuum—pellets.

shutdown, binary counting control for multiple systems, binary reset control, individual blowback timers, and individual vacuum timer circuits. Up to 20 different solid state control circuits are available for solid state control boxes.

A secondary section of the vacuum conveyor is the piping and valving system. A common air piping system on a vacuum conveyor is a nonreversing valving setup. The air flow direction in this type of system is controlled by two positive seating air directional control valves that eliminate the conventional method of reversing the motor and blowers to reverse the flow direction. There are three different cycles that valves can be switched into: the vacuum cycle, the blowback cycle, and the idle position. In the vacuum cycle (Fig. 15-7), the upper valve is electrically energized and closed by the plant air supply. The lower valve is de-energized and allows air to pass through the positive displacement blower out through the muffler and back out to the atmosphere. At the end of the loading cycle, when

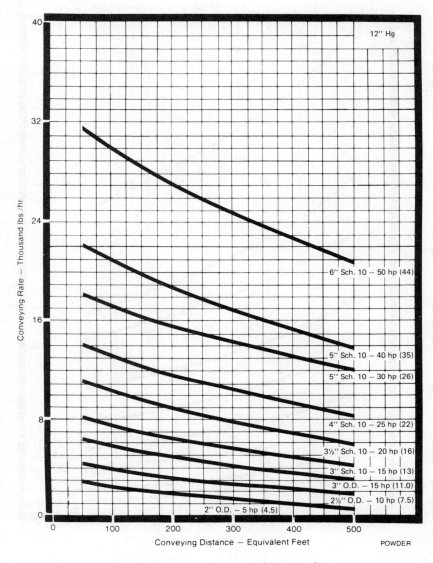

Rates based on Polystyrene @ 35 lbs./cu. ft.

Fig. 15-3 Continuous vacuum—PVC powder.

the vacuum hopper is filled, the vacuum power unit is switched into the blowback cycle by energizing either a timer switch or a high vacuum switch. The blowback cycle (Fig. 15-8), permits the hopper to unload the material into the machine and helps with filter cleaning. The blowback cycle takes place by simply switching the upper valve to the de-energized position and energizing the lower valve to reverse the flow in the system.

Should all of the vacuum hoppers have a full load, the vacuum will switch to the idle position (Fig. 15-9), in which both valves will be de-energized so that the air flow will cycle from the ambient air, through the filter, into the blower, through the muffler, and back into the ambient air.

This valve configuration also aids in the prevention of material clogging in the conveying system. The valves used in this system also act as pressure relief valves. Should excessive pressure build up within the system, the valves will work in combination with a time delay switch and a high vacuum switch.

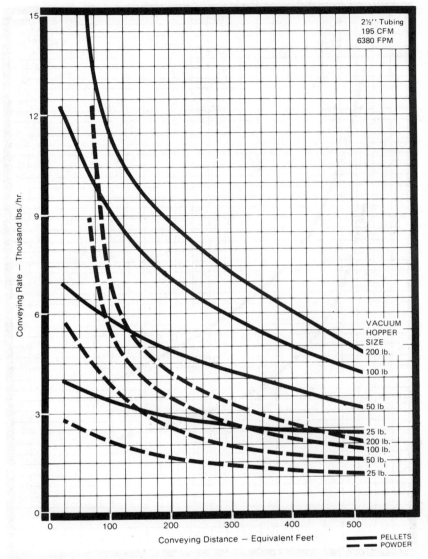

Rates based on Polystyrene @ 35 lbs./cu. ft.

Fig. 15-4   10-hp vacuum conveyor.

Should an obstruction occur in a conveying line causing a buildup of pressure, the valve will move off its seat to relieve a portion of the pressure buildup. At this point, a time delay switch will energize for 5 seconds. If the obstruction is not removed after the 5-second time limit, the time delay switch will de-energize, activating the high vacuum switch, which will switch the valving into the blowback cycle in an attempt to remove the obstruction by reverse force.

Pressure Units

Pressure power units are similar to vacuum power units, consisting of basically the same components as a vacuum power unit. A pressure power unit uses a positive displacement blower powered by a three-phase electrical motor. The pressure power unit works by taking ambient air from the plant and pulling it through a filter, through the positive displacement blower, and finally into the material

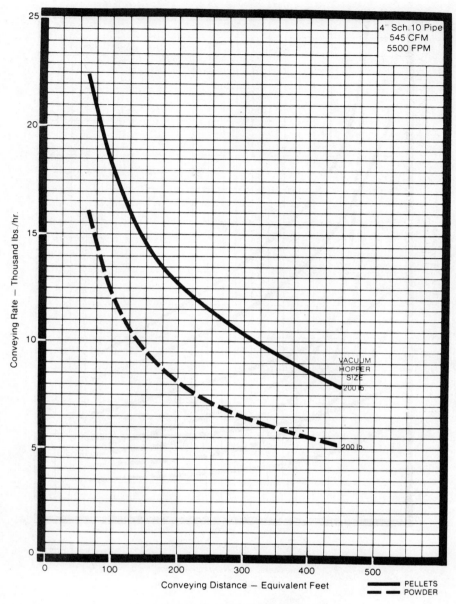

Rates based on Polystyrene @ 35 lbs./cu. ft.

Fig. 15-5   25-hp vacuum conveyor.

Incoming air (right) is trapped by impellers. Simultaneously, pressurized air (left) is being discharged.

As lower impeller passes wrap-around flange, Whispair jet equalizes pressure between trapped air and discharge area, aiding impeller movement and reducing power.

Impellers move air into discharge area (left). Backflow is controlled, resulting in reduction of noise relative to conventional blowers.

Fig. 15-6   Positive displacement blower operation (typical).

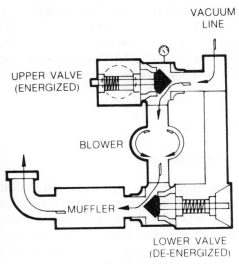

Fig. 15-7 Vacuum cycle.

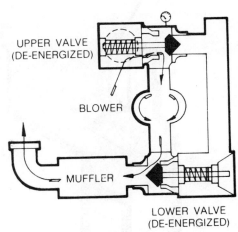

Fig. 15-9 Idle position.

line. Pressure power units, like their counterpart the vacuum power units, have a pressure range of 6 psi gauge to 10 psi gauge. Pressure power units have a horsepower range from 15 hp up to 60 hp. All pressure power units are available only as continuous pressure units, unlike their counterparts that have an idle cycle and blowback cycle.

Pressure units are commonly used in conjunction with vacuum units in order to convey material over long distances. Vacuum units are preferred over pressure units because if a leak does occur within the system, air will be pulled into the system, keeping dust in the tubes. If a leak develops in a pressure power

system, the dust will flow out into the plant air, and can be hazardous to the plant personnel.

Vacuum pressure systems have some distinct advantage when conveying materials over long distances. Using two separate power units in a system will reduce the pressure ratio across the blowers, which in turn reduces the heat that is normally generated by a single blower conveying over the same distance. The reduced heat buildup prevents premature failure due to excessive wear on the power units. The lower operating temperatures also prevent damage to the material. The damage that occurs to the material when high operating temperatures are present is in the form of premature melting or angel hair.

Vacuum and pressure conveying units can be purchased from the manufacturer with a sound enclosure to keep the power unit operating within OSHA sound limitations of 90 decibels or less within the workplace.

Rotary Valves

A rotary valve can best be described as a metering device for materials. The rotary valve consists of a vaned feeder rotor in a cast housing. The vaned feeder rotor is operated by a gear motor. Various types of rotary feeders are available to handle different types of materials. Rotary valves may be obtained with side entry openings (Fig. 15-10) that prevent shat-

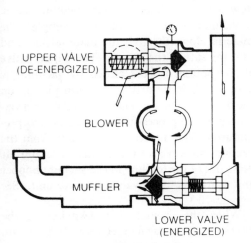

Fig. 15-8 Blowback cycle.

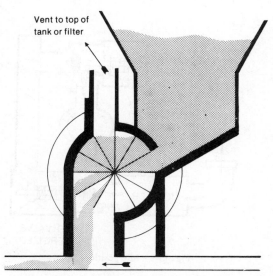

Fig. 15-10   Rotary valve with side entry opening.

tering of friable materials, which could happen if the material were to be dropped into the vaned rotor from the top of the unit (Fig. 15-11).

Rotary feeders serve a second function in a conveying system by separating the two different air flows (pressure and vacuum) (Fig. 15-12). Without the rotary feeder's action as an air lock, the two pressures would work against each other, thus counteracting the forces and reducing the efficiency of the system. Rotary feeders are basically used in two mediums. They are often used in conjunction with storage tanks; they are installed on the bottom of storage tanks and bins to meter the material. The other use would be to sepa-

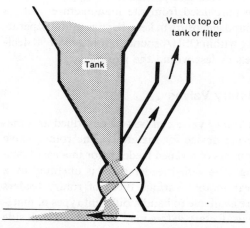

Fig. 15-11   Rotary valve with top entry opening.

rate the pressure system from a vacuum system, as discussed. The material would flow into a hopper in a vacuum system and then into a rotary valve, which would drop the material into the pressure system.

HOPPERS

Vacuum Hoppers

Vacuum hoppers are used in conjunction with vacuum conveyors as an easy, convenient means of loading plastic materials in molding machines or storage tanks. Various modifications have been made on vacuum hoppers to increase their versatility.

Vacuum hoppers are also used for dust removal in a conveying system by combining the vacuum hopper with a filter chamber and a low head separator.

Vacuum hoppers are available in sizes from 10- to 1000-pound capacity. They are manufactured from either steel, spun aluminum, or stainless steel. The vacuum hopper consists of a hopper body with a removable lid that seals the unit and prevents leaks in the vacuum hopper. Hoppers come complete with a paper or cloth filter that fits in the hopper lid, preventing dust and fines from flowing through the power unit and out into the plant air. The hopper has a free-hanging flapper valve that seals the unit when it is in the loading cycle; the flapper valve hangs freely to let material flow out of the hopper when vacuum ceases.

Vacuum hopper operation is simple. The material flows through the material line and into the side of the hopper. The flapper valve at the bottom is held closed by vacuum. A seal, which is formed by vacuum pulling on the flapper valve, prevents material from dropping into the processing machine or silo prematurely. The material is separated from the vacuum air stream by the filter. Vacuum air flows out of the top line in the vacuum hopper and continues to the vacuum power unit. (See Fig. 15-13.)

When the vacuum unit switches to the blowback cycle, the flapper valve drops down, and material flows out of the vacuum hopper

Load Cycle

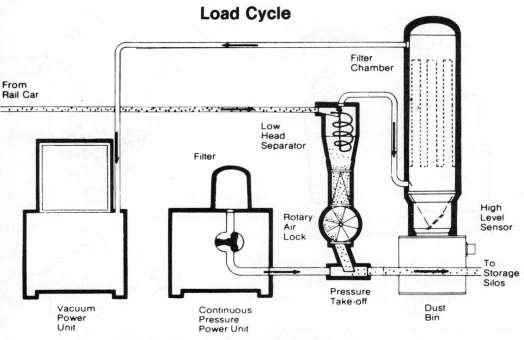

Pulse Jet Blowback Cycle

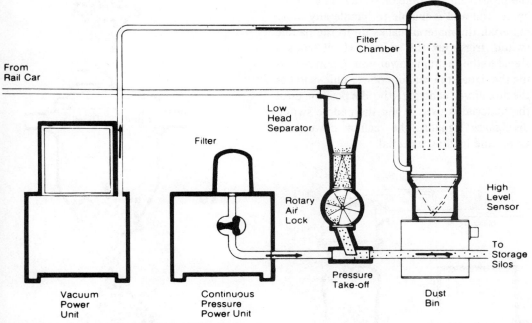

Rotary feeders are sized in accordance with the production rate of the system. The common sizing would in cubic feet of material per hour. See Figure 2-22 for specifications.

Fig. 15-12 System utilizing rotary air lock feeder to separate pressure and vacuum air flow.

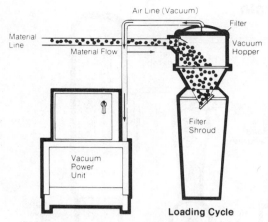

Fig. 15-13   Vacuum hopper loading cycle.

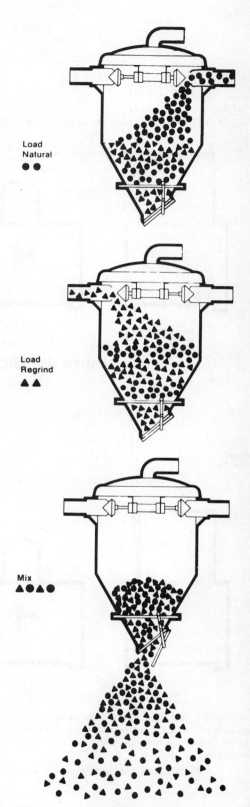

and into the receiving unit (silo or molding machine). (See Figs. 15-14 and 15-15.)

Vacuum hoppers may be modified, or accessories may be added to increase the efficiency of the units. A common accessory installed on vacuum hoppers is the paddle switch which is connected to the vacuum power unit. If the processing machine or storage facility becomes full and is unable to handle any more material, the material backs up to the paddle switch, triggering the switch, which sends a signal to the vacuum power unit. Upon receiving the signal, the vacuum power unit switches the directional control to the idle position until the material is used. Then the paddle switch disengages, signaling the vacuum power unit to resume loading material.

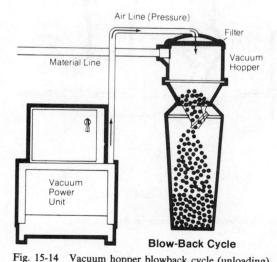

Fig. 15-14   Vacuum hopper blowback cycle (unloading).   Fig. 15-15   Vacuum proportioning hopper blowback cycle.

Modification has been made to the vacuum hopper so that two materials may be mixed together in the hopper. This type of vacuum hopper, called a proportioning hopper, is used to mix virgin plastic material with regrind material. The proportioning hopper utilizes a solid state panel to aid in maintaining the proper ratio between virgin and regrind plastic. This solid state panel has two adjustable time controls that permit the operator to choose the proper proportioning ratio. The positive seating valves are controlled by solenoids that trigger each valve for a preset time limit. Once the machine has made two complete loading cycles, the vacuum power unit will switch into the blowback mode, and material will flow into the processing machine or silo in the same manner as a standard vacuum hopper.

Hopper Loaders

Hopper loaders are similar to vacuum hoppers. The exception is that hopper loaders contain their own power unit on top of the hopper: Hopper loaders are also available in proportioning units for mixing virgin materials with regrind plastic. Proportioning hopper loaders work the same way as proportioning hoppers. Hopper loaders have the same options as vacuum hoppers.

FILTERS

Filters are commonly used in conveying systems when powders or materials that generate fines are conveyed. The reason for using filters in conveying systems is to convey these powders without clogging the air movers. There are different filters and methods utilized for handling powders and fines.

A basic type filter system has filter bags on a rack that is installed in a canister. The filter chamber has an inlet and an outlet for air. In a basic filter system, the air mixed with dusty material flows through the conveying lines on its way to the vacuum power unit. The air/dust mixture enters the filter chamber at the bottom of the unit. Air passes through the filter bags, which trap the dust. Air then moves out of the top opening in the filter chamber, and proceeds on to the vacuum power unit. The dust in the filter chamber must be cleaned manually by removing a hatch at the bottom of the chamber.

The filtration process is the same with all filter units. The only difference is that some units are modified for automatic cleaning. One modification is a flapper dump cone mounted on the bottom of the filter chamber. This permits the dust that is collected in the filter chamber to drop down into a collection bin or drum when the vacuum power unit switches to the blowback mode.

In situations where extremely dusty materials or fine powders are conveyed, automatic self-cleaning filter chambers should be used. Dusty materials and fine powders have a tendency to clog filter bags quickly. This condition causes the vacuum blower to switch to the blowback cycle more frequently, which cuts down on conveying time. An automatic, self-cleaning filter has a compressed air jet installed at the top of each individual filter bag. Jets are connected to a microprocessor, which controls the length of time the compressor will blow through the filter bag and the time between each cleaning of the individual bags. Microprocessors can be adjusted so the time limit can be varied for each bag. The length of time between blowback in each bag may be adjusted for the most efficient cleaning sequence. Bags can be timed to clean all at the same time, or individually, at a predetermined interval.

This type of filter chamber permits the vacuum power unit to run continuously while the blowback jets clean the powder from the filter bags.

The low head separator is another device used to separate fine powders and dust from the conveying air. This is done at the material collection point. A low head separator operates by bringing the air/dust into the separator; since the dust weighs more than air, the dust will drop down in the separator, and the air will exit through the top and travel on to the vacuum power unit.

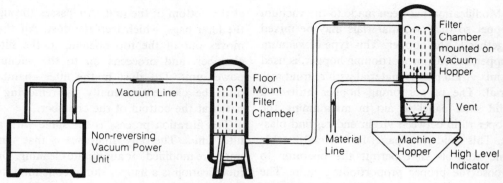

Fig. 15-16 Coarse powder filter system.

Low head separators are constructed of stainless steel, with the internal surface of the separator shell manfactured of tungsten carbide and epoxy-coated to resist wear.

There are three separate systems commonly used for conveying coarse powders, medium dusty materials, or pellets with excessive fines and fine powders.

The coarse powder system (Fig. 15-16) has a vacuum power unit that pulls the powder through two filters. One filter is mounted on a hopper which has a flapper valve and a dump cone. The air then travels to a basic filter to remove any fines that may have passed through the first filter. Then the conveying air passes through the vacuum power unit to the atmosphere.

When conveying medium dusty materials or pellets with excessive fines from a storage facility to a vacuum hopper, air containing the fines is moved through a filter chamber with a dump cone and flapper on the bottom. Fines are separated from the air, which then continues on to the vacuum power unit. Once the vacuum power unit shifts into the blow-back cycle, fines drop out of the filter into a dust drum (see Fig. 15-17).

When fine powders or extremely dusty materials are conveyed, material is brought from the storage facility through the material conveying line to a low head separator which is connected to a vacuum hopper. A filter cone keeps the material flowing directly into the machine hopper. Air with some fines mixed with it flows out of the separator to a self-cleaning filter with dump cone and flapper valve, which empties fines and dust in a dust drum. The air, as with the other systems, will continue through the vacuum power unit and out to the atmosphere (see Fig. 15-18).

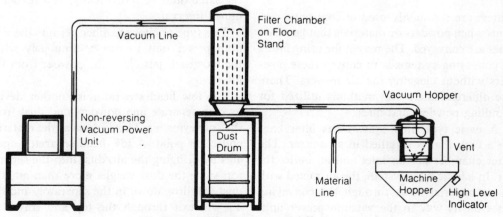

Fig. 15-17 Medium dusty material filter system.

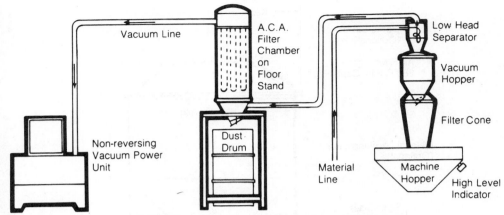

Fig. 15-18   Fine powder and extremely dusty material filter system.

BULK STORAGE

Delivery Container Size

Plastic material becomes less expensive when it is purchased in bulk quantities. In fact, the greater the quantity, the less expensive plastic will be on a cost-per-pound basis.

Plastic material comes in many different types of containers and quantities. The smallest quantities are bags or 55-gallon drums. The next larger size would be a gaylord, which may be up to 48 in. wide by 48 in. deep by 42 in. high. Plastic may also be purchased by the truckload. A truck tanker can hold approximately 24,000 pounds of plastic material. Plastic can be purchased in a railcar, which has a usable storage capacity of approximately 140,000 pounds of material. These figures will vary with the bulk density of the material.

Various types of storage systems are available for plastics plants (see, for example, Fig. 15-19). Most smaller systems are in-plant systems, but larger out-of-plant systems are also available.

There are three different types of tanks manufactured for outside storage use:

Welded
Bolted
Aluminum spun

Welded steel tanks are brought to the site complete with a primer coat of paint and lifted onto the pre-constructed concrete pad with a crane. Welded tanks come in sizes up to 5460 cubic feet, with a maximum diameter of 12 feet. Once the tanks are erected on the pad, various options such as filters, ladders, take-off boxes, etc., can be added, along with the final painting of the silo. Some manufacturers paint silos to meet user paint specifications. Welded silos can be purchased with a 45-degree or a 60-degree cone at the bottom of the tank.

Welded silos can come equipped with legs on the bottom of the tank or a skirt to support the silo. Skirts are recommended because they protect the valves, power units, and take-off compartments from weather elements.

A bolted tank may be preferred over its welded counterpart because it is available in larger sizes. Bolted tanks are brought to the site in pieces called staves. The staves are assembled to form a ring. Six staves make up a 9-foot-diameter ring, eight staves make up a 12-foot-diameter ring, and ten staves make up a 15-foot-diameter ring. These rings make up the tank. A bolted tank can be built from the bottom up by bolting each ring on top of the other, or from the top down by building a ring and raising it with jacks. The next ring will be built under the existing ring and, once complete, will be bolted to the first ring. Then both rings will be lifted in the air and the process repeated until the tank is complete. Tanks which are built from the bottom up require a lift, jack, or small crane to lift the

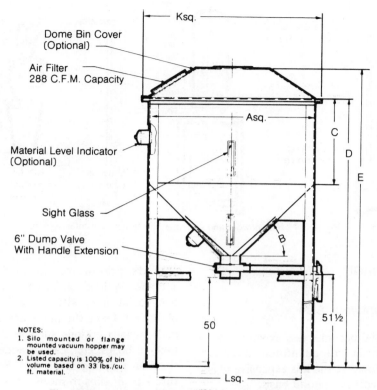

Ksq.

Dome Bin Cover
(Optional)

Air Filter
288 C.F.M. Capacity

Asq.

C

D

E

Material Level Indicator
(Optional)

B

Sight Glass

6" Dump Valve
With Handle Extension

51½

NOTES:
1. Silo mounted or flange
   mounted vacuum hopper may
   be used.
2. Listed capacity is 100% of bin
   volume based on 33 lbs./cu.
   ft. material.

50

Lsq.

Fig. 15-19 In-plant material storage bin with dump valve.

staves to the next level. The assemblers will be required to work above the ground. If the tank is built from the top down by the use of jacks, the work crew can stay on the ground unless extra equipment is added to the tank. Once the tank is completed, accessories can be added.

The last type of tank is a spun aluminum tank. Spun aluminum tanks can be prebuilt and brought directly to the plastics plant or spun right at the plant site. Spun aluminum tanks come in 12-foot and 15-foot diameters. They do not have the same storage capacity as a steel tank. The aluminum tank does not have to be painted, but it does have a tendency to oxidize.

On most systems, some accessories are necessary. OSHA-approved guardrails with toeboard are available if personnel are to have access to the top of the storage tank. For access to the top of the tank or tanks, ladders with safety cages and rest platforms are available.

Several important points must be taken into consideration with regard to storage tanks. First is local building codes. These codes vary greatly, and installations must conform to local specifications.

The location of power lines must also be taken into consideration. There are set standards for proximity of tanks to power lines. Local utility companies should be contacted before construction begins. It is also highly recommended to check for underground power, telephone lines, or other services in the vicinity of tank erection before the project is too far along. Again, local service companies can provide this information.

Sensors

Remote measuring devices are frequently used in storage tanks: high- and low-level sensors. The rotary mechanical level sensor, or "paddle-wheel," is a motor enclosed in a cast aluminum housing with a three-vane paddle-wheel attached to the motor shaft. A common arrangement for these rotary level sensors is to have one placed at the top and one at the bottom of a silo or storage bin. Alarm systems of either a visual lighted alarm or an audio alarm can be connected to operate with these two sensors. The rotary level sensor at the top of the silo or storage tank will rotate until the material transported into the tank reaches its level and stops the paddle-wheel from turning. This causes an alarm to go off, which signals the loader that the storage tank or silo is reaching its capacity. The rotary level sensor in the bottom of a tank or silo is set to sound an alarm when the material drops to a low level in the bin. The paddle-wheel starts to turn and energizes an alarm system to signal that the bin is low and it is time to order another shipment of material.

Another type of measuring device is the solid state capacitance level sensor. This device looks similar to a rotary mechanical level sensor, only without a paddle-wheel. The solid state capacitance level sensor has pads that send a signal to a control board when the material level is either too low or too high in the storage tank. The solid state capacitance level sensor is in a cast aluminum, hermetically sealed, epoxy-coated enclosure to prevent dust from contaminating the sensor.

Filtration

Dusty powders or granular materials with excessive fines can often foul filters and operating components, creating environmental dust and build-up to the point where they "plug" conveying systems. A.C.A. Series Filter Chambers are automatic self-cleaning filtration units that allow high conveying rates and increased filtration capacities for these materials with dust-free operation.

The A.C.A. Series Filters are adaptable to a variety of conveying system requirements and are designed with a high cloth-to-air ratio to provide complete filtering of the conveying air. They can be attached to silos or bins to operate as vents which automatically discharge collected fines. Numerous other applications are feasible, including powder-conveying system filtration, bag dumping station dust collectors, continuous vacuum system filters which provide the filter cleaning action when no blowback cycle is used, and material-receiving filters in vacuum systems.

During material conveying, dust and fines will accumulate on the outside surface of the filter bags. At preset time intervals a solenoid valve is activated that interrupts the compressed air flow to the filter chamber. This interruption causes an exhaust valve to open, allowing a burst of compressed air from an accumulator to be released down inside a filter bag. The burst of air momentarily stops the filtering process of the bag, flexes it, and causes the accumulated dust to drop off. Instantly the clean bag is filtering again. These air pulses are each directed to the individual filter bags in a sequence at a specified rate. The unique "sequence pulse" rate is adjustable for different conveying rates and material types. The higher the conveying velocities, and the more dusting that occurs during conveying, the more frequent the "sequence pulses" required.

Container Tilters

If the user decides to purchase material in gaylords, container tilters offer plastics processors an economical way to solve material unloading problems for large, hard-to-handle containers (see Fig. 15-20).

Without a tilter, material must be manually fed to conveying equipment pick-up tubes. A

Fig. 15-20   Container tilters.

container tilter can eliminate troublesome monitoring of material flow and assure an uninterrupted supply of material to machine hoppers or blending equipment.

Most tilter models can be designed to support the pick-up tube or wand supplied with conveying equipment, but are adaptable to most other types. Pick-up support stands are also available as an option on some models. A jogging system is also available as an option for hard-to-flow materials such as regrind. Each container tilter is ruggedly constructed of heavy-gauge welded angle iron and provided with convenient mounting holes. They are easily loaded with a lift truck and fully guarded with pinch guards on the model 30 and 40 according to OSHA guidelines. The units are built with safety in mind and constructed of heavy duty components to provide years of reliable service.

BLENDERS

Many applications in the plastic molding industry require the addition of compounds to the virgin material to obtain particular characteristics, such as color in the finished product. The additon of regrind material to virgin material as extenders can produce a high-quality part at a lower cost because of the lower cost of regrind material. Additive feeders or blenders are an efficient means of combining and mixing substances in order to produce a flawless product in the molding process.

There are different types of additive feeders and proportioning hoppers. Some are simple devices that mix two different solid substances; other can mix up to four or five different substances, including pellets, granular materials, powders, and liquids. Most units can be mounted directly on a molding machine. Some of the larger blenders are floor-mounted units with pneumatic take-offs that transfer blended material directly to the molding machine.

The most basic additive feeder is a machine-mounted, single-compartment unit. It is mounted directly on the molding machine in addition to a standard hopper. This additive feeder consists of an aluminum spun hopper

attached to a frame, with a rotating feed-screw powered by a variable-speed, direct current drive motor. This single-compartment additive feeder is used to add pelletized color additives into a tube that leads down into the processing machine hopper.

A dual-compartment additive feeder has a large compartment for virgin plastic material and a small compartment that leads into a rotating screw for adding either regrind plastic or color pellets. Material is mixed before it reaches the molding machine screw, in an operation similar to that of a single-compartment feeder.

A common characteristic of color pellets is that they melt at extremely low temperatures. This can be a problem if the virgin pellets and the grandular material have to be dried at high temperatures by a hot air dryer or a desiccant dryer. A special additive feeder is manufactured that is mounted on a plenum drying hopper to add the color additives after the plastic material is dried in the plenum hopper. This unit is a hopper made of mild steel with a tube that runs down to the rotating feed screw which mixes the colorant with the plastic material as it enters the molding machine.

Additive feeders may have a vacuum loader placed on top of the unit to produce a fully automated system that does not require manual refill of the hoppers or the additive feeders.

There are a variety of blenders available on the market today. When selecting a blender, the types of plastic material and colorants to be blended should be considered. Blenders can mix from 3000 to 9000 pounds of materials in an hour. This rate is based on 35 pounds per cubic foot bulk density.

Blenders can easily mix three or more different materials. A typical blender has three compartments which feed both pellets and regrind material. Materials are fed into the machine by a direct feed auger feeder or a vibratory feeder. The major components of a typical blender are:

1. Material supply hopper
2. Metering and mixing section
3. Machine supply hopper

Material supply hoppers are small storage compartments that hold virgin materials, regrind plastics, and/or color pellets. The material supply hopper can be filled manually or automatically with vacuum hopper loaders. Material supply hoppers will have material drainage or cleanout tubes, with a cap that can be taken off to remove material from the hopper. Each compartment should have a slide-gate to stop the flow of material.

After leaving the material supply hopper, the materials enter the metering and mixing area. Metering of pelletizied or granulate material can be done by three different methods:

1. Direct feed
2. Auger feed
3. Vibratory feed

Direct feed is a tube that runs directly to a rotating disc or conveying belt. The tube can be raised or lowered a specific height above the disc or belt to meter the amount of plastic to be mixed with the additives.

The auger feed is a rotating feed screw powered by a variable speed gear motor which meters material through a feed tube and onto a mixing disc. The material is commonly color concentrate or regrind.

A vibratory feeder is used for feeding color concentrate or regrind material through a tube and onto the mixing disc. Vibrator feeders control the amount of material that enters the machine by the frequency of vibrations per second. Frequency of vibrations is controlled by making a few minor adjustments to the vibratory feeder. Some vibratory feeders are not adjustable. Some feed devices must be calibrated in accordance with the amount of material the vibratory feeder conveys into the machine.

The rotating disc acts as both a metering device and a preliminary mixing apparatus. Some adjustable feed tubes are set a certain height above the disc which limits the amount of material fed into the machine. Since the disc rotates past each of the feed or metering tubes, some mixing takes place, but the actual mixing process takes place in the cascade mixing chute. Material is carried on the rotating disc to an opening where it drops through baffles in the mixing chute and into the machine supply hopper. The blended material is then fed into the processing machine.

Calibration

The proportion of virgin pellets to regrind material and color concentrate is called the let down ratio. Recommended proportions for colorant to plastic material can be obtained from the color manufacturer.

To determine the let down ratio, the slide gate to all of the material feed hoppers should be closed and the feed hoppers filled with the materials to be blended. The amount of material delivered through the blender is dependent upon the height of the adjustable drop tube above the rotating metering disc.

Virgin material is used as the constant when calibrating the remaining materials. To calibrate the virgin section of the supply hopper, open the access door of the machine hopper and baffle-box and place the calibration chute into position. Adjust the material drop tube to the midpoint of travel, start the blender, and let it run until the material moves down the calibration chute. Then turn the blender off. Empty the calibration chute and place it back into position. Start the blender and let it run. After a preset period of time, the material is weighed to determine whether the blender is producing the desired flow of material per hour. Should the material flow have to be changed, the drop tube can be moved up or down to adjust flow until the desired rate is achieved. The proportioning of the regrind material is adjusted in the same manner.

To adjust a vibratory feeder, slide gates to all of the machine supply hoppers should be closed and the blender run until all material left in the metered mixing section has run out of the machine. The short adjustable drop tube located between the slide gate and the feed pan should be set as close as possible to the bottom of the pan, allowing only a small amount of material flow into the pan. The tube controls the level of the material in the pan. The vibratory feed rate adjustment knob should be set to any fixed position for a point of reference. After starting the blender and allowing some material flow through the machine, turn the machine off. The calibration chute must be put into place and the blender turned on and run for the same length of time as the virgin and regrind materials. If the weight of the material is not the desired proportion, the vibratory feeder can be adjusted by the control knob and the test repeated until the proper weight is achieved.

The same procedure is followed for an auger feeder. The adjustment knob is set at a reference point. The blender is run to clear the other materials out of the mixing and metering chamber. The calibration chute is installed, and the machine is run for the same specified time as with the virgin materials. The material is weighed to ensure that the proper proportions will be achieved in the blender.

Once all of the materials are at their proper let down ratios, all the slide gates can be opened to let all of the material mix together.

Not all materials used in the plastics industry today are in pellets or granular form. In some cases, dry powder colorants are used. Dry colorants are generally less expensive than pelletized colorants. However, dry colorants can be extremely messy if not handled properly. Special blenders are available for blending dry colorants with pellets and granules.

Material feed hoppers for pellets and granulated materials are similar to those used in basic pellet, regrind, and pelletized colorant blenders. Color concentrate is fed into the mixing chamber by using a vibratory feeder connected to a material feed hopper. The dry colorant feeder is a cylindrical container with a disposable flexible container inside that promotes the flow of sticky materials that commonly "bridge" or "rathhole." The metering unit is an auger feed screw designed to keep the additive materials at a consistent density to avoid packing or erratic feed rates.

The mixing chamber has three mix zone separator pans. Each pan has an agitator to mix the materials. The materials drop into the first pan, where the agitator sweeps the material, causing the granulate and pellets to roll in the dry colorant or the color concen-

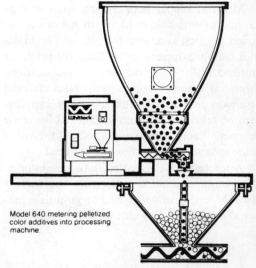

Model 640 metering pelletized color additives into processing machine.

Fig. 15-21   Additive feeder metering pelletized color additives into processing machine.

trate so that the materials will be mixed properly (Fig. 15-21). The material then drops to the next two levels, where the same mixing action continues. This ensures that the pellets and granulate are completely covered with the colorant before leaving the blender and entering the processing machine.

Vigin material is used for the constant with this blender. Virgin plastic flows at a continuous rate from the material feed hopper to the mixing assembly. The amount of material flowing through the machine can be determined by catching the material in the calibration chute for one minute and weighing the material to determine the flow rate.

Color Concentrate Let Down Ratio

The color concentrate or dry colorant feed rates are determined by using the following formulas:

$$\text{Pounds of concentrate required per minute} = \frac{\text{Pounds of natural material per minute}}{\text{Let down ratio (supplied by supplier)}}$$

or:

Concentrate #/minute

$$= \frac{\text{Natural \#/minute}}{\text{Let down ratio}}$$

The let down ratio of the color concentrate may be obtained from the manufacturer. To adjust the color concentrate, close all of the slide gates except the color concentrate material feed hopper. The color concentrate dial on the control panel should be set at a medium setting. The calibration chute should be put into place and the blender operated for one minute. The material collected in the calibration chute after one minute of operation should be weighed to determine if it is the amount of material needed for proper blending. The color concentrate dial can be adjusted to increase or decrease the amount of material flow.

Dry colorant calibration is done the same way as the other materials. The formula for calculating the amount of color concentrate is as follows:

Grams of colorant required per minute =
# of virgin plastic/min × grams/100#

The grams/100# is a figure that is supplied by the manufacturer of the dry colorant. The dry colorant adjustment dial should be set at a median point and the blender run for one minute. The amount of dry colorant is then weighed to determine the proper proportion of material. The dry colorant adjustment dial can be reset and the test repeated until the proper weight of material is obtained (Figs. 15-22 and 15-23).

Regrind material is fed through a vibration feeder. The regrind is adjusted by closing off all slide gates and running the blender for one minute, collecting the regrind in the calibration chute, and weighing the amount of material collected.

For increased blending capacities, a belt blender should be utilized. A belt blender can blend up to 7000 pounds of material per hour. This type of blender can be machine-mounted. It can also be used as a central blender for several molding machines.

The metering of material is done by raising or lowering the metering tubes certain distances above the belt. The unit is calibrated by reversing the belt movement so that material drops into calibration canisters at the back of the blending machine. Each material can be weighed to determine the proper let down ratios. When the proper adjustments have been made, the belt can be put back into its forward motion, thus causing the plastic materials to flow to a cascade mixing chute and into the supply hopper.

Two options used on blending equipment are low level indicators and sight glasses. Sight glasses are rectangular glass pieces in a rubber frame installed on the side of a material feed hopper; the material lever can then be checked visually without anyone's having to climb a ladder in order to look down inside the hopper. Low level indicators are commonly used to automate the system; a capacitance or paddle-wheel low level indicator can easily be combined with hopper loaders or vacuum hoppers to continually feed a molding machine.

Belt Flow Blenders

Belt flow blenders offer accurate and increased capacities for blending up to four different materials automatically. Free-flowing granular materials including regrind, color concentrates, and a wide variety of natural materials can be metered and blended at any desired ratios up to 7000 lb/hr @ 35 lb/cu ft.

Materials are gravity-fed from divided supply hoppers and are proportioned by volume on a moving conveyor belt in ratios determined by the elevation of material metering tubes. The material is then carried to the discharge end of the blender in its own trough. As the materials flow from the end of the belt, they "cascade" through a mixing chute that blends the material together before it reaches storage or processing supply. (See Fig. 15-24.)

A wide variety of material ratios can be selected by simply raising or lowering material metering tubes under the blender's supply hoppers. Proportions as small as one-eight of 1 percent can be produced (1 pound of one

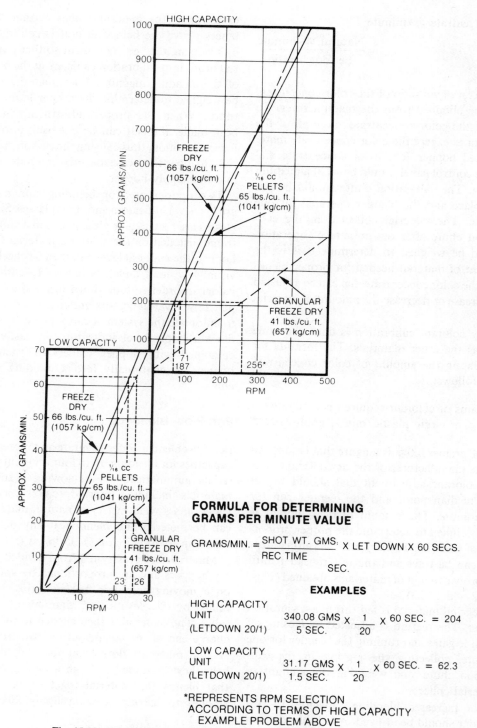

**FORMULA FOR DETERMINING
GRAMS PER MINUTE VALUE**

$$\text{GRAMS/MIN.} = \frac{\text{SHOT WT. GMS.}}{\text{REC TIME SEC.}} \times \text{LET DOWN} \times 60 \text{ SECS.}$$

EXAMPLES

HIGH CAPACITY
UNIT
(LETDOWN 20/1)

$$\frac{340.08 \text{ GMS}}{5 \text{ SEC.}} \times \frac{1}{20} \times 60 \text{ SEC.} = 204$$

LOW CAPACITY
UNIT
(LETDOWN 20/1)

$$\frac{31.17 \text{ GMS}}{1.5 \text{ SEC.}} \times \frac{1}{20} \times 60 \text{ SEC.} = 62.3$$

*REPRESENTS RPM SELECTION
ACCORDING TO TERMS OF HIGH CAPACITY
EXAMPLE PROBLEM ABOVE

Fig. 15-22 Selection chart and formula for determining grams per minute value.

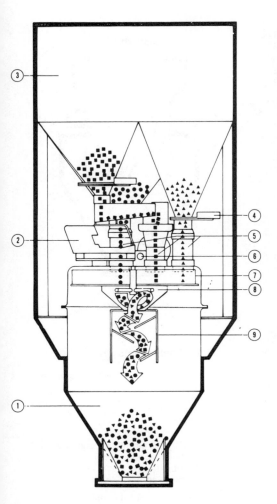

Fig. 15-24   Metering tubes and cascade mixing chute in a belt flow blender.

Major Components

1. Machine supply hopper
2. Vibratory additive feeder
3. Material supply hopper
4. Slide gate
5. Material metering tubes
6. Drive motor
7. Metering section
8. Rotating disc
9. Cascade mixing chute

● Virgin
■ Color Concentrate
▲ Regrind

Fig. 15-23   Blender components.

material to 800 pounds of another material). One of the material metering tubes is smaller than the others to facilitate accurate metering of color concentrates. The supply hoppers have removable dividers that allow the same

material to flow through two tubes simultaneously when required.

Material ratios are calibrated by reversing the direction of the belt and catching samples of each material in sampling containers. The containers are then weighed on a gram scale and the precise ratios of each material determined. The ratios are then set on easily read material tube calibration scales by dialing in the proper tube-over-belt height. Dial locks are provided to prevent changes in calibration settings.

The belt flow blender is a small, compact unit that may be mounted directly over a processing machine supply hopper, or may be floor-mounted and used as a central blending station. The unit's high capacity potential makes it possible to mix high volumes of a single blend continuously or to blend small amounts of various blends used on several processing machines.

Each material supply hopper has a level sensor that monitors the material level. If a material level falls too low, the blender automatically shuts off. No unmixed material is allowed to leave the blender. Each level sensor has a control switch and indicating light so that low hoppers can be immediately detected and unused hoppers cut out of the system.

DRYERS

Plastic materials used in the molding process, whether virgin pellets or regrind material, are subject to contamination by moisture. When moisture is present in a molding process, it tends to cause defects in the molded part such as irregular molded products, splay marks, and possibly brittleness. In most cases where close specifications must be followed, these types of defects cannot be tolerated in the finished product.

Plastics are classified in two categories in relation to moisture:

1. *Nonhygroscopic:* these are plastics in which moisture adheres to the surface of the pellets or granules. Polypropylene, polystyrene, and polyolefins are common nonhygroscopic plastics. These plastics are dried by blowing hot air over the material to evaporate the moisture and carry it out of the drying unit.

2. *Hygroscopic:* these plastics absorb moisture within the pellets or the granules and form a molecular bond within the material. Common hygroscopic materials are nylon, ABS, PET, and other rynite plastics. These plastics can only be dried by removing the moisture from the material using dehumidified hot air (see "Drying Hygroscopic Plastics" in Chapter 29).

Some common equipment used with both hot air dryers and dehumidifying dryers are the air diffuser cone and drying hoppers. An air diffuser assembly is designed to be used in existing loading hoppers. The air diffuser assembly consists of a football-shaped satellite with four legs that permit it to be placed in the hopper. The drying air is brought into the hopper by a flexible hose to a hood on top of the hopper that connects the flexible tube to the satellite diffuser. This satellite, in Fig. 15-25, is constructed of a solid sheetmetal cone at the top half of the diffuser and perforated sheetmetal cone at the bottom to disperse the hot air into the plastic material. An air trap cone is also incorporated into the system to prevent contaminated ambient air from entering the hopper. A vacuum hopper or proportioning hopper can be added to the top of the air trap cone to automatically load the hopper.

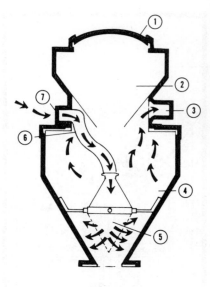

Air Diffuser Assembly

Major Components

1. Manual load plate with removable cover
2. Air trap cone
3. Bleed to atmosphere or return to dryer
4. Machine hopper
5. Perforated air diffuser cone and material diverter cone
6. Flexible hose
7. Delivery air from dryer

Fig. 15-25   Satellite diffuser.

Dryer hoppers can be purchased to replace standard hoppers when a greater capacity is needed. Drying hoppers range in size from 50 pounds, up to 6000 pounds capacity. This is based on an average weight of plastics which is 35 lb/cu ft bulk density. Plenum drying hoppers can be purchased as a machine-mounted unit that can have a capacity range from 50 pounds to 4000 pounds, or a floor-mounted unit that has a 50 pound to 6000 pound capacity range. Plenum hoppers have a diverter cone as an added design feature. This diverter cone serves two purposes. First, the diverter cone forces an even air distribution through the plenum drying hopper. Without this diverter cone, the heated air would take the easiest flow path, which would be straight up the middle of the hopper. Second,

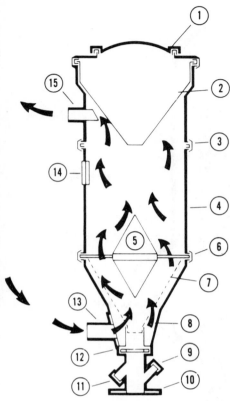

Plenum Drying Hopper

Major Components

1. Manual load plate with removable cover
2. Air trap cone
3. Upper swing clamp with gasket
4. Tank section
5. Diverter cone
6. Lower swing clamp
7. Perforated air diffuser cone
8. Lower outer cone
9. Purge inlet
10. Square mounting plate
11. Drain outlet
12. Slide gate
13. Delivery air from dryer
14. Sight glass
15. Return air to dryer

Fig. 15-26 Plenum drying hopper.

this diverter cone prevents an uneven flow of plastics through the hopper (see Fig. 15-26). Without the diverter cone, the plastic material would flow through the hopper in an hourglass effect with the plastic material at the center

of the hopper flowing through the hopper at a faster rate than the material at the sides. The disadvantage of this is that the material at the center of the hopper would not spend enough time (residence time) in the hopper to dry properly. The material at the sides of the hopper also has a tendency to plasticize due to excessive heat exposure. The diverter cone forces the plastic material at the center to mix with the plastic at the sides which ensures that the plastic material will remain in the hopper for the recommended drying time.

Recent advances have been made in plenum drying hopper technology with the introduction of a high efficiency plenum drying hopper. These new high efficiency plenum hoppers are insulated to prevent heat loss. The high efficiency plenum hopper retains 20 percent more heat than the average plenum hopper. This heat retention provides a significant savings in energy consumption. These high efficiency plenum hoppers have low watt density heaters installed in the base to heat the incoming air from the hot air blower and circulate this hot air through the hopper. The plastic material in the hopper is protected from excessive heat by an innovative system of baffles that channel the hot air from the heaters to the diffuser cones (see Fig. 15-27).

Hot Air Dryers

Nonhygroscopic plastics are dried by using a hot air dryer and plenum hopper or an air diffuser assembly. A hot air dryer is a relatively simple machine that consists of heaters and an air blower. Hot air dryers can deliver hot air thermostatically controlled up to 300°F at a capacity range from 60 to 1000 cubic feet per minute.

A hot air dryer works by pulling ambient air into the air drying filter, through the blower, then across the heating elements. The hot air is then blown to the hopper through a flexible tube. Once the hot air gets to the hopper, it is dispersed through the plastics. The hot air performs two functions. The hot air that passes through the plastic material evaporates the moisture, turning it into steam, and moves the steam out of the hopper back

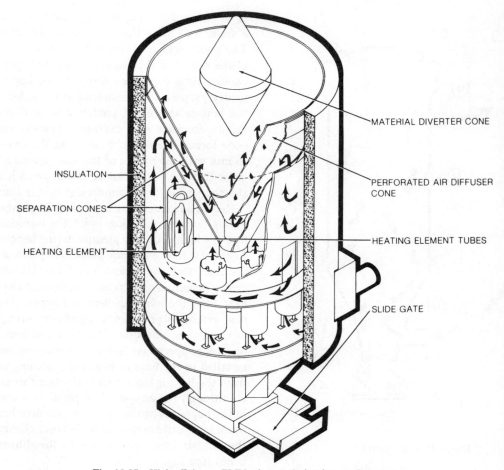

INSULATION

SEPARATION CONES

HEATING ELEMENT

MATERIAL DIVERTER CONE

PERFORATED AIR DIFFUSER CONE

HEATING ELEMENT TUBES

SLIDE GATE

Fig. 15-27 High efficiency (H.E.) plenum drying hopper base cutaway.

to the ambient air (see Fig. 15-28). The hot air also serves the function of preheating the plastic material. Preheat brings the material closer to the molding temperature. When this available heat is used, less heat is required for the molding process, and there is a reduction in energy consumption.

Many factors have to be considered when sizing a hot air drying system. The first is the plastic material. The material should have a specified residence time (length of time the material should be in a hopper dryer). The temperature of the drying air, as well as a certain temperature at which the material should be dried, is also critical to prevent melting or plasticizing of the material in the hopper. Another consideration when drying non-hygroscopic plastics is the production rate or, in simpler terms, the quantity of plastic in

pounds used in a one-hour time limit. Taking these two factors—residence time and production rate—into consideration, a proper plenum hopper can be selected. The proper selection permits the plastic to enter the hopper and slowly work its way down to the bottom of the hopper for $1\frac{1}{2}$ hours residence time (most hygroscopic plastics have a residence time of $1\frac{1}{2}$ hours) and keep up with a steady production rate. A hot air dryer can now be chosen based on the cubic feet per minute rating needed to dry the plastic.

For example, a system on an injection molding machine has a mold that uses 3 pounds of polyethylene plastic with a total cycle time of 1 minute. In 1 hour, this molding machine will use 180 pounds of material. The residence time for polyethylene, taken from Table 15-1 is $1\frac{1}{2}$ hours.

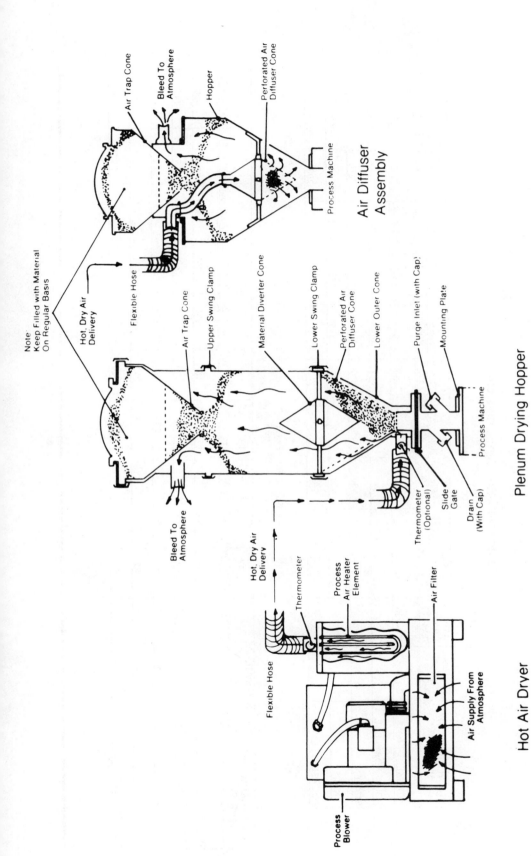

Air Trap Cone

Bleed To Atmosphere

Hopper

Perforated Air Diffuser Cone

Process Machine

Air Diffuser Assembly

Note:
Keep Filled with Material
On Regular Basis

Hot, Dry Air Delivery

Flexible Hose

Air Trap Cone

Upper Swing Clamp

Material Diverter Cone

Lower Swing Clamp

Perforated Air Diffuser Cone

Lower Outer Cone

Purge Inlet (with Cap)

Mounting Plate

Process Machine

Thermometer (Optional)

Slide Gate

Drain (With Cap)

Bleed To Atmosphere

Hot, Dry Air Delivery

Plenum Drying Hopper

Fig. 15-28   Hot air dryer flow diagram.

Flexible Hose

Thermometer

Process Air Heater Element

Air Filter

Air Supply From Atmosphere

Process Blower

Hot Air Dryer

435

Table 15-1 Residence time chart.

DRYING RATE CHART
HOT AIR DRYERS

NON-HYGROSCOPIC MATERIAL	RESIDENCE TIME IN HOURS	MATERIAL DRYING TEMP. °F.	HA60		HA150		330		350		HA1000	
			Plenum 200	Hopper 400	Plenum 600	Hopper 800	Plenum 1500	Hopper 2000	Plenum 2000	Hopper 3000	Plenum 4000	Hopper 6000
ACETAL	1.5	200–220	135	180	400	450	1000	1050	1335	1650	2670	3000
POLYPROPYLENE	1.5	220	135	140	345	345	805	805	1265	1265	2300	2300
STYRENE	1.5	185	135	180	400	450	1000	1050	1335	1650	2670	3000
VINYLS	1.5	200	135	240	400	535	1000	1335	1335	2000	2670	4000
POLYETHYLENE	1.5	185	120	120	300	300	700	700	1100	1100	2000	2000

HIGH EFFICIENCY
HOT AIR DRYERS

NON-HYGROSCOPIC MATERIAL	RESIDENCE TIME IN HOURS	MATERIAL DRYING TEMP. °F.	HA30/50	HA30/100	HA30/200	HA60/100	HA60/200	HA60/400	HA100/200	HA100/400	HA100/600	HA150/400	HA150/600
ACETAL	1.5	200–220	35	65	90	65	135	180	135	265	300	265	400
POLYPROPYLENE	1.5	200	35	65	70	65	135	140	135	230	230	265	345
STYRENE	1.5	185	35	65	90	65	135	180	135	265	300	265	400
VINYLS	1.5	200	35	65	120	65	135	240	135	265	400	265	400
POLYETHYLENE	1.5	185	35	60	60	65	120	120	135	200	200	265	300

In a continuous flow automated hopper system, an additional amount of 90 pounds of polyethylene material will be needed in the plenum drying hopper so the material that enters the top of the hopper will spend $1\frac{1}{2}$ hours in the hopper before entering the molding machine. The hopper with the capacity to handle 270 pounds of material would be a 400-pound capacity unit. A hopper must be filled to its capacity for proper operation of the air trap cone, which prevents contaminated air from entering the hopper. This hopper would have to be filled with 400 pounds of polyethylene material. The hot air dryer that would be able to handle this 400-pound load is a HA-150 with a rating of 150 cubic feet per minute. A thermometer would have to be installed in the hopper to get a true temperature reading, since a certain amount of heat is lost through the hose leading from the hot air dryer to the plenum drying hopper. There are two alternatives to this drying system. The first would be to substitute a high efficiency plenum hopper in place of the stan- efficiency plenum hopper in place of the standard plenum hopper. Heaters in the base of the high efficiency plenum hopper make up for the heat lost in the flexible tubing. The other alternative is a high efficiency dryer that combines a high efficiency plenum hopper which is an insulated unit with the heaters in the base of the unit. It also has an air blower attached to the plenum hopper. This high efficiency hot air dryer eliminates the hoses that air normally uses in conventional drying systems.

Dehumidifying Dryers

Hygroscopic plastics must be dried by using a dehumidifying dryer. Dehumidifying dryers absorb the moisture within the plastic material by using dry heated air brought down to a dewpoint of $-40°F$. This is done by the use of desiccant beads. These desiccant beads are molecular sieves which are synthetically produced crystalline metal aluminosilicates. All moisture is removed from the crystals during their manufacture. The main advantage to these crystals is that there is very little structural change when water is added or removed.

Molecular sieves can dry materials to moisture contents as low as 35 parts per billion.

Molecular sieves belong to a class of compounds called zeolites. Zeolites characteristically release water when heated and absorb water when cooled. Molecular sieves are purchased in beads that are $\frac{1}{16}$ or $\frac{1}{8}$ in. in size. These beads are a combination of 20 percent clay binding and 80 percent crystals. The crystals range in size from .1 to 10 microns. The clay and crystals are mixed together and formed into pellets by an extruder. The pellets are then dried in a kiln to force the remaining water out. The beads are packed in a desiccant bed canister, which is a round cylinder with a mesh screen on the bottom. A layer of $\frac{1}{8}$-in.-diameter beads is spread over the top of the screen. A layer of $\frac{1}{16}$-in. beads is spread over the top of the first layer. A layer of $\frac{1}{8}$-in. beads make up the third and final layer. A screen is placed on top of the beads. The $\frac{1}{8}$-in. beads next to the screen facilitate proper air flow. Smaller beads next to the screen would restrict air flow. Dessicant beds have more even air flow in the vertical position.

There are two classifications for drying systems: single-bed absorption systems, which use one dessicant bed, and multibed absorption systems, which use two or more desiccant beds. Dehumidifying dryers operate in a closed loop system (see Fig. 15-29). Air is brought in through a filter on the initial start-up and sent to the desiccant bed to absorb the water out of the air when the water molecules are absorbed by the desiccant beads. Approximately 1800 Btu per pound of moisture is released, causing the air temperature to rise approximately $19°F$. The air then travels to the heating unit where the air temperature is brought up to the drying temperature specifications. The dehydrated air is then circulated through the plastic in the drying hopper. Then the air is brought out of the hopper and recycled back through the unit, and the process is repeated.

Eventually the beads become saturated with moisture and have to be regenerated. This is done by blowing air heated to a temperature of $550°F$ through the desiccant beds. The elevated temperature drives the moisture out of

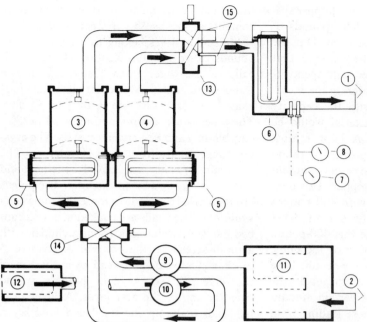

Major Components

1. process air to plenum hopper
2. moist air return from plenum hopper
3. "on stream" desiccant bed
4. regenerating desiccant bed
5. regeneration heaters
6. process air heater
7. process air thermostat
8. process of safety thermostat
9. process air blower
10. regeneration air blower
11. process air filter socks
12. regeneration air filter sock
13. upper air directional control valve
14. lower air directional control valve
15. regeneration air vents to atmosphere

Fig. 15-29—Dessicant dryer flow diagram.

the beds and into the ambient air. This process varies with the different types of dehumidifying dryers. Some dryer manufacturers have one desiccant bed in their dryers. These beds are regenerated by removing the desiccant bed and placing it in a special regeneration machine. The other type of single bed used is a rotating bed. The bed slowly rotates while one part of the bed is in the working cycle removing moisture; the other section of the bed is in the regeneration cycle. This type of dehumidifying dryer has a major disadvantage; leakage of moisture from the working part of the bed to the regeneration side of the bed. This prevents the desiccant from operating at full capacity.

A multiple dessicant bed absorption system is the most efficient method for drying. A common absorption bed setup is the double bed system. In a double bed system, one bed is on line drying material, while the other bed is in the regeneration cycle. There are two types of air flow direction to regenerate desiccant beds: counter-current and co-current. When the desiccant bed is in the working mode, the beads act like a sponge with water poured on one side of it. The water does not

get dispersed evenly through the bed. Beads that make contact with the wet air will become moist first. Once these beads reach the saturation point, other beads in close proximity become saturated. This process continues until all the beads are saturated. In counter-current regeneration, the air flows through the desiccant bed in the opposite direction of the working air flow. This forces the moisture out of the desiccant bed opposite the direction in which it entered. The advantage of this is that the bed can be regenerated faster. The possible disadvantage of this method is that some of the beads may not get used because the regeneration cycle and the working cycle are timed to switch over after a few hours of operation. The dry desiccant beads that do not get used in the working process will be hit with a blast of hot air that may break them down faster than beads that constantly get saturated.

Co-current regeneration flow has regenerating air flow in the same direction as the flow of the working air. This has the advantage of using all the beads in the bed. The disadvantage is that the regeneration time in increased. There are other factors that must be taken into consideration with counter-current and co-current flow, such as residual loading at the ends of the beds. Residual loading is the waste (dust, fines) that can get trapped in the ends of the bed and restrict air flow through the beds themselves.

The drying process is started by having the air drawn through process air filter socks through the process air blower into the base of the desiccant beds so the air will be brought down to a −40°F dewpoint. The air is then sent to the heaters where it is heated to a predetermined temperature depending on the type of plastic that is being used. The air is then sent through hoses to a plenum drying hopper. The dry, hot air is circulated through the plastic material and exits through the air trap system. The dry, hot air returns to the process air filter socks, and the process repeats itself (see Fig. 15-30).

Dehumidifying dryers are sized similar to a hot air drying system. The hopper is sized by the production rate multiplied by the residence time. The dryer is then sized by the corresponding figures from the dryer sizing chart. The dryers are sized on a flow rate of 50 ft/min. If the flow rate is more than 50 ft/min, the material will be blown around in the hopper. Any flow rates considerably less than 50 ft/min may not have enough velocity to dry the plastic material. For example, on sizing a dehumidifying dryer system, assume a production rate of 60 lb/hr of ABS material. ABS has a residence time of 4 hours. Sixty pounds per hour multiplied by 4 hours is 240 pounds. The amount of material that must remain in the hopper in order to achieve the correct residence time is 240 pounds. The correct hopper choice is a 400-pound capacity hopper. The dryer that would dry the plastics adequately is a 50 cfm dryer, which would be a DB-100 dehumidifying dryer.

Other factors have to be taken into consideration when setting up a total drying system. One factor is the total number of machines that are processing the same type of material. If more than one machine is running the same kind of material, it may be more advantageous to set up a central drying system with one large dryer and a central plenum drying hopper (see Fig. 15-31). The central system processes material for several processing machines. A central dehumidifying dryer can also be used with individual high efficiency plenum hoppers (see Fig. 15-32). When different types of plastics are used in several individual machines, an individual dryer and plenum hopper

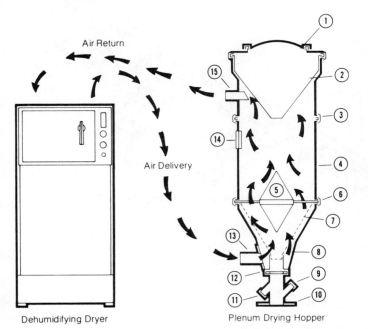

Major Components

1. Manual load plate with removable cover
2. Air trap cone
3. Upper swing clamp with gasket
4. Tank section
5. Diverter cone
6. Lower swing clamp
7. Perforated air diffuser cone
8. Lower outer cone
9. Purge inlet
10. Square mounting plate
11. Drain outlet
12. Slide gate
13. Delivery air from dryer
14. Sight glass
15. Return air to dryer

Air Return

Air Delivery

Dehumidifying Dryer

Plenum Drying Hopper

Fig. 15-30 Dehumidifying drying system flow diagram.

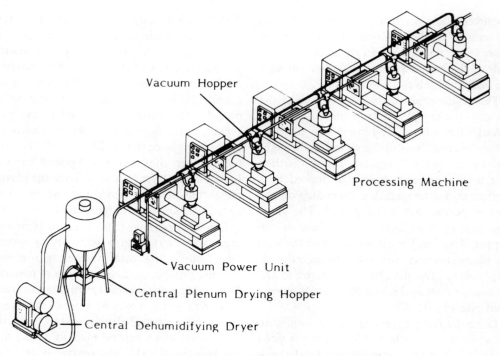

Fig. 15-31   Central dehumidifying dryer with central plenum hopper system.

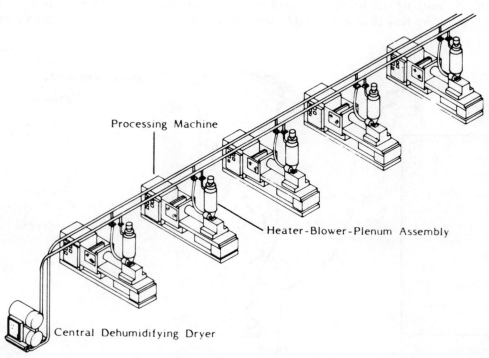

Fig. 15-32   Central dehumidifying dryer with individual high efficiency plenum hoppers.

for each separate machine is preferred. This configuration offers the maximum flexibility for drying a variety of hygroscopic materials (see Fig. 15-33).

PET Drying System

At this point we will take a look at a specific material and the spceial considerations that apply to its drying technique.

PET is hygroscopic material—that is, it absorbs a great deal of moisture from the air around it, and this moisture becomes tightly bound with the molecules of which the material is made. Unless most of this moisture is removed, the PET will not be of the proper viscosity to be injection-molded into the parisons from which beverage bottles are made.

It is not difficult to remove moisture from a hygroscopic plastic material such as PET. The material is simply passed through a drying system, where it is exposed to a stream of hot dry air for as long a time as is necessary to reduce its moisture content to the correct level. It is generally agreed that crystallized PET chips should be dried to a moisture level of 0.002 percent so that they will have the intrinsic viscosity of about 0.7 needed for injection molding.

Other things being equal, the higher the temperature of the hot air that passes through the material into the drying system, the more rapidly will drying be accomplished. For example, when a 300°F air temperature is used, a total of $3\frac{1}{2}$ hours of drying time is needed to reduce moisture level of PET material from 0.3 percent to 0.001 percent. If the temperature in the air is raised to 350°F, the same job can be completed in less than 2 hours.

On the basis of this kind of information, it is all too easy to assume that raising drying temperatures as high as possible is desirable, since higher temperatures mean faster drying. But this is not the case.

Excessively high drying temperatures can create serious problems. At temperatures of 400°F, PET polyethylene materials undergo a physical breakdown and become discolored. At unit temperatures higher than 350°F, acetaldehyde will be developed in excessively high quantities in the mold parisons. It will be carried over into the blow-molded bottles, where it can ultimately affect the flavor of the beverage.

Controlling Acetaldehyde Levels. The formation of excessive acetaldehyde can be prevented by avoiding excessively high temperatures and prolonged heating of PET resins during drying. PET material can be dried for injection molding purposes at a temperature of 350°F with the material held in the drying hopper for 3 to 4 hours.

But under these circumstances, acetaldehyde levels may still be too high. The acetaldehyde can be reduced by drying the material for a longer period. A residence time of 8 to 10 hours at 350°F may be required. (Air flow through the drying vessel should be 0.75 to 1 cfm for each pound of material per hour that is put through the drying vessel.)

After drying, the PET material is ready for injection molding into parisons. During the molding process, the level of acetaldehyde in the resin rises again because the resin is heated to the melting point in the molding machine barrel. However, the amount of acetaldehyde that results from this source is not excessive and will not create problems with the flavor of the beverage provided that acetaldehyde levels have not previously risen too high during the drying process.

In a typical drying system used in PET beverage bottle-making plants, hot dry air circulates from the dryer and enters the drying hopper at the bottom. It rises through the PET material, picking up moisture, and then returns to the dryer. This air circulation continues throughout the period that the PET material is in the drying hopper.

Inside the dryer itself, cool wet air that has picked up moisture from the PET material in the drying hopper enters the aftercooler, then passes through the cooling coil, where the air temperature is reduced prior to its entering the dehumidifying dryer. This assures maximum capacity of moisture removal from the wet air by the dehumidifying dryer. The air then goes through a filter, which removes unwanted particles; then passes through a de-

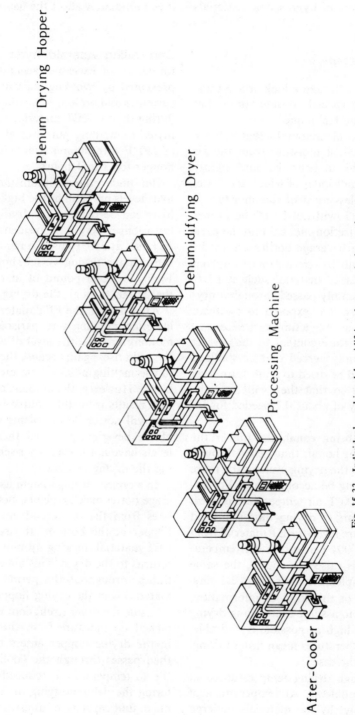

Fig. 15-33   Individual machine dehumidifying dryers and planum hoppers.

siccant bed, which removes its moisture; and then passes through a heater, which raises its temperature to the required level. The hot dry air then returns to the plenum drying hopper, and the cycle is repeated.

Eventually, the desiccant bed that removes moisture from the drying air becomes saturated. It can no longer hold moisture.

The saturated desiccant bed is disconnected from the system, and a second dry desiccant bed takes its place. While the second bed is in operation, the first one is regenerated—that is, it is dried out by heating elements. In this way, there is always a desiccant bed in operation, and drying can continue without interruption.

PET polyester is a relatively heavy material compared to other plastics. Therefore, a given weight of PET will occupy less space in the drying hopper than a comparable weight of another material. That means that drying hoppers can be smaller.

Since drying temperatures are relatively high, the drying hoppers must be well insulated. Fiberglass insulation is recommended for placement around the body of the hopper. This insulation layer in turn should be protected by an aluminum skin.

The molecular-sieve desiccants used to dry the air in some PET drying systems absorb acetaldehyde and they dissipate it, just as they dissipate moisture, when the desiccant is regenerated. The desiccant material is made of very small, round beads, which range in size from $\frac{1}{32}$-in. to $\frac{1}{8}$-in. Generally, the smaller the bead, the greater the service of the desiccant material and the higher the degree of moisture absorption.

A PET drying system should include an aftercooler to cope with the high temperatures (300°F or more) of air returning to the dryer at the initial start-up and whenever the processing machine is slowed down. The return air temperature should be no higher than 150°F in order to optimize the moisture absorption capacity of the desiccant material.

To ensure against excessively hot air reaching the desiccant, a thermostatic control for the aftercooler should be included in the system. If production slows down and the temperature of the air returning from the dryer rises, the thermostatic control valve will actuate the aftercooler to provide proper cooling in the air being returned to the dehumidifying dryer.

The aftercooler utilizes recirculating water, thereby minimizing water consumption and coil corrosion.

Molding Nylon—Special Moisture Problems

When processors refer to moisture problems in relation to nylon, usually the material is considered to be too moist. It may come as a surprise, therefore, to learn that serious problems also can arise if the nylon is molded too dry.

Overdrying has been found to be a factor in product failures. The widespread notion of "the drier, the better" often results in unsatisfactory flow characteristics, poor fill, and nonuniform crystallization. It is not uncommon in the winter months, when the humidity is low, for the nylon to be dried to 0.04 to 0.08 percent moisture content (Fig. 15-34).

For most injection molding applications, nylon 6/6 should be dried to a range of 0.15

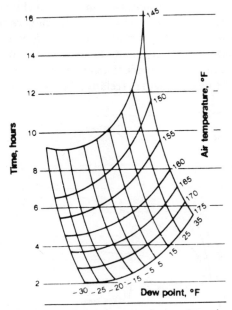

Fig. 15-34 Drying time to reduce moisture in nylon 6/6 from 0.2 to 0.08 percent.

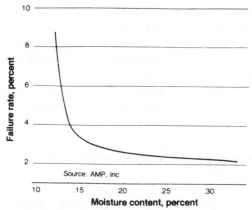

Fig. 15-35 Flexing-failure rate in connector legs vs. dryness of nylon before molding.

to 0.3 percent. A producer of nylon connectors found that, by controlling moisture content in that range, he could reduce brittle failure in gripper-fingers by 70 percent (Fig. 15-35).

Nylon 6/6 usually contains about 0.1 to 0.3 percent moisture before the package is opened and the material is exposed to the outside air. The resin often is used just as it comes out of the container.

However, after a container has been opened and the nylon is exposed to the air, it starts picking up moisture, and some type of drying may be required.

Figure 15-36 shows the moisture pickup on initially dry nylon 6/6 at room temperature while exposed for up to 4 hours at 75 percent relative humidity, and up to 8 hours at 25 percent and 50 percent relative humidity.

In damp weather (relative humidity of 75

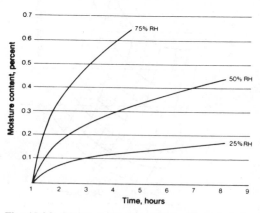

Fig. 15-36 Moisture-pickup rate for nylon at various relative humidity levels.

percent) the moisture level of the nylon 6/6 will rise by 0.35 percent in 1 hour. This additional moisture by itself is enough to cause part brittleness, irrespective of how little moisture was present in the nylon originally.

On the other hand, when it is very cold and dry outside and the relative humidity inside the plant is 25 percent or less, the time required to pick up 0.35 percent moisture would be about 40 hours.

Air Flow Is Critical. The air flow in the drying hopper is an important consideration in determining optimum drying time. The most practical flow rate appears to be 1 cu ft/min/hr for each pound of resin being dried. This is indicated by the fact that, at this rate of flow, the air temperature at various heights in the hopper is close to the temperature of the drying air (180°F) as it enters the hopper.

Unfortunately, the effectiveness of present-day dehumidifying dryer design increases the possibility of overdrying. In a typical design, air returning from the plenum drying hopper passes through a dry desiccant bed, which removes essentially all the water from the air before the air is reheated and returned to the plenum hopper. Even if the return air were not heated, its −40°F dewpoint could ultimately dry the nylon below 0.15 percent.

Protection against Overdrying. Neither conventional dehumidifying dryers nor hot air dryers can automatically limit the degree of drying and prevent overdrying. However, the problem can be solved. Dryers can be modified to achieve positive control over the dewpoint of the air being recycled to the drying hopper.

The dewpoint can be controlled by adjusting the proportion of dried and moist return air that is fed to the drying hopper. As shown in Fig. 15-37, the mixture is controlled by varying the amount of moisture-laden return air that is allowed to bypass the dryer beds, as directed by a moisture sensor at the outlet of the process-air heaters. This system prevents overdrying as well as maintaining an upper limit on moisture content. By adding moist air, as needed, it conditions the nylon during the drying process. Even when regrind

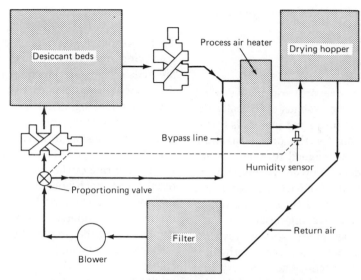

Fig. 15-37   Bypass system prevents overdrying of process air.

and/or other additives are combined with the virgin nylon, the extra drying time will not result in overdrying.

PNEUMATIC VENTURI CONVEYING

Pneumatic venturi conveying systems are an efficient means of transporting materials from drums and gaylords, and from granulator bases, up to hoppers on top of processing machines.

A pneumatic venturi conveyor is a pickup tube that leads to the venturi. The conveyor tube leads from the venturi to the hopper on the processing machine. The venturi operates on a compressed air system, which consists

of an air filter, air regulator, and air shut-off valve.

The venturi is a simple design, basically an hourglass with air blowing through it. Compressed air comes in through the air filter, the shut-off valve, and regulator. The regulator is set between 70 and 80 psig. The air then goes through the air hose and into the venturi. Air enters the venturi at the smallest section of the tube. The compressed air flows through the small section, which causes the air to remain compressed as it flows at a fast rate through the small middle section of the venturi. When the air reaches the end of the venturi, it expands and continues to flow at a fast rate through the conveying tube, thus causing a suction in the pickup tube that trans-

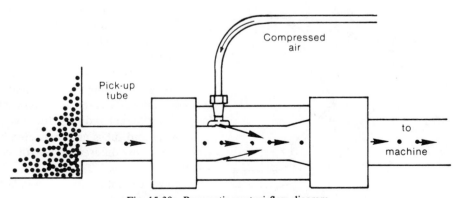

Fig. 15-38   Pneumatic venturi flow diagram.

ports the material to the processing hopper (see Fig. 15-38).

POWDER PUMPS

Powder pump conveyors are low-velocity, pneumatic conveying systems specifically designed to transport powdered materials by using compressed air for the power medium. Powder pumps are designed to handle hard-to-convey powders. Powdered material is compacted into "slugs" of packed powder, and the slugs are then moved through the tubing in a "dense-phase conveying mode."

This dense-phase mode allows dust-free conveying. A standard sock-type cloth filter is used with this system. There is very low degradation of the material because of the low conveying velocity. The velocity ranges from 1200 to 2000 ft/min, so abrasion wear in the conveying tubing is also low.

The powder pump operates on low volumes of compressed air at supply pressures of 60 to 80 psig. Conveying pressures vary from 25 to 80 psig and convey materials from 50 cu ft/hr up to 200 cu ft/hr, depending upon the conveying characteristics of the powders. Powder pumps have seven major components, as in Fig. 15-39.

The conveying sequence begins when the material valve opens, letting the powder flow by gravity into a pressurizing chamber. The valve closes, and injected compressed air pumps the powder, pushing it through the conveying line. The sequence is automatically repeated to pump the powder to its end use.

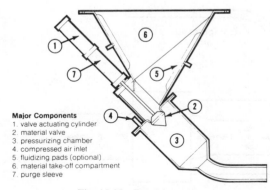

Major Components
1. valve actuating cylinder
2. material valve
3. pressurizing chamber
4. compressed air inlet
5. fluidizing pads (optional)
6. material take-off compartment
7. purge sleeve

Fig. 15-39 Powder pump.

PIPING

There are different types of piping used for conveying materials, along with different methods of hanging pipelines that should be discussed. Various elements should be considered when making a decision about the placement of conveying pipes in a plant.

The first consideration is the type of material to be conveyed. The second factor is the production rate or the specific amount of material needed each hour to satisfy the processing machines. The next factor is the distance the material will have to travel from the storage facility to the processing machines.

Once this information is determined, specific plant layout drawings will have to be made to determine exactly where the conveying pipelines will be placed to efficiently convey the material. The major objective when designing a system is for the piping to take the shortest route from the storage area to the processing machines without interfering with other plant fixtures, such as overhead cranes.

There are two types of piping used in pneumatic conveying:

1. Aluminum
2. Stainless steel

Aluminum is the most commonly used tubing for the majority of plastic materials and powders. Stainless steel piping, being more expensive than its counterpart, is used for abrasive and corrosive materials. Aluminum piping is available in a size range from $1\frac{1}{8}$-in. outside diameter with a wall thickness of .058 in. up to 6-in. schedule 10 tubing, available in 5- or 12-foot lengths. Stainless steel has a range from 1-in. schedule 40 up to 6-in. schedule 10, in 6-, 12-, and 20-foot lengths.

Both types of piping have premade bends that range from 30-degree to 90-degree angles, along with different radii for the bends.

The pipes and tubes are connected by couplers. There are different types of couplers used in the systems.

The most common type of coupler is the bolted coupler with a grounding strip. This type of coupler is a sleeve that slides over the tubes and is held in place by tightening the bolts on the sleeve. The grounding strip is a safety precaution that will transmit any electrical current to the next pipe length and to ground.

There are two types of socket couplers used to connect piping. One type uses epoxy as the sealing agent to keep the pipes together. The other socket coupler has O-rings inside the coupler to seal the piping.

Some accessories can be added to conveying tubing. A conveying tubing sight glass can be placed in a pipe line to observe the material flow in the line. This sight glass is an acrylic plastic tube that is placed between two sections of piping. This acrylic tube has a grounding strip running through the tube to transmit any electrical current on to the next tube and finally to ground.

In cases where several processing machines are being filled from the same storage facility, a single line system can be used utilizing "Y" tubes.

Single Material Line Sequence Conveying Systems

The material line for this type of system must be set up as shown in Fig. 15-40. Note that the "branch" arm of the special "Y" tubes carries the material to the next station and the straight portion must be connected to a short radius bend, as close to the "Y" as possible. The "Y" can be positioned in any one of three directions; horizontal left, horizontal right, or up—but never in the down position. If the main line is above the hopper entry elevation, a minimum of 2 feet straight on the horizontal must be connected to the short radius elbow before dropping into the vacuum hopper. The last station without a "Y" may have either a short or long radius bend. An internal flapper check valve on the fill tube inside the vacuum hopper opens when the vacuum power is applied to that station through the sequence control valve in the station's branch vacuum line. The internal check valve also prevents air leakage through this station when it is not being loaded.

Flexible metal and reinforced flexible plastic hoses are commonly used in pneumatic conveying, mainly in situations where lines will have to be switched from one silo to another or from a material line to a vacuum hopper. Flexible hoses are used for unloading railcars and tank trucks with ease, and are available in different sizes to fit the specific system needs.

Many bulk systems include provisions for switching the material or vacuum power through a quick-change station. The material and vacuum lines are fitted with flexible metal hoses and quick-change couplers for this purpose. The female coupler is usually fitted to the flexible metal hose, and the male is mounted on the rigid tubing.

Flexible metal material hoses are normally stainless steel. The flexible vacuum hose is a heavy-wall, reinforced hose that includes a nipple at each end to fit a standard tube coupler or a quick-change coupler, or may be supplied without nipples.

RAILCAR AND TANK TRUCKS UNLOADING

As discussed, materials used in the plastic industry are often purchased and delivered in bulk quantities, by railcars and tank trucks, and unloaded into silos or large storage bins within the plant. There are various requirements and considerations management and employees in a plastic plant should be aware of when unloading bulk shipments and planning unloading areas.

The first point that should be considered is access and distance of railroad tracks from the plant if the only alternative would be to have the material delivered by tank truck. In cases where a rail line is available, plant officials should contact the rail line to determine the cost of using the line.

A second point would be whether a spur line is required to prevent blocking the track during unloading. The owner of the railcar will also have to be consulted on such things as demurrage charges. A demurrage charge is a fine charged to the user for holding a

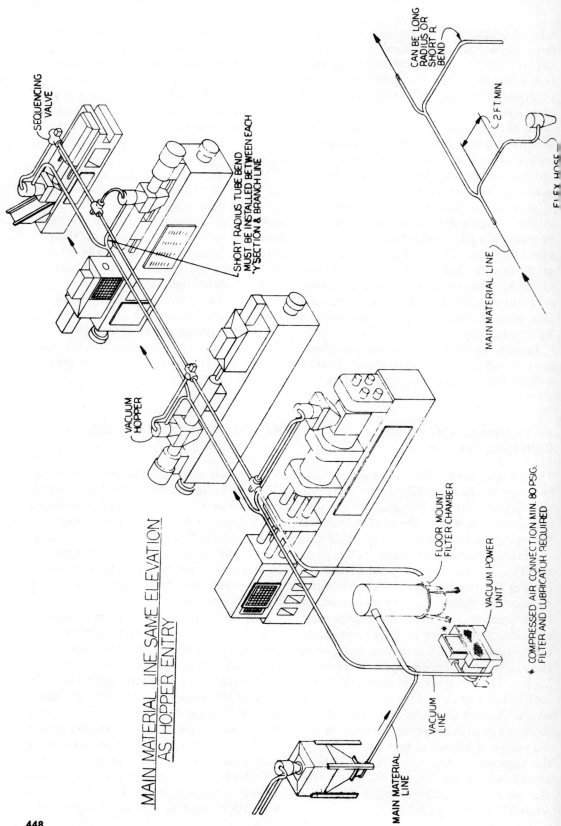

SEQUENCING VALVE

SHORT RADIUS TUBE BEND MUST BE INSTALLED BETWEEN EACH "Y"SECTION & BRANCH LINE

CAN BE LONG RADIUS OR SHORT R. BEND

2 FT. MIN.

FLEX HOSE

MAIN MATERIAL LINE

VACUUM HOPPER

MAIN MATERIAL LINE SAME ELEVATION AS HOPPER ENTRY

FLOOR MOUNT FILTER CHAMBER

VACUUM POWER UNIT

* COMPRESSED AIR CONNECTION MIN. 80 PSIG. FILTER AND LUBRICATOR REQUIRED

VACUUM LINE

MAIN MATERIAL LINE

448

railcar for longer than the agreed-upon time. Time for unloading a railcar should be determined in order to avoid these charges. Railcars can normally be unloaded in 10 to 30 hours, depending on the amount and type of material purchased. The type of railcars should also be taken into consideration, since various types of railcars need different adaptors for unloading the car.

Figure 15-41 shows the additional equipment needed to remove the material from the railcar. This equipment consists of a "Y" tube, manifold system located next to the railroad tracks along with a modular vacuum/pressure system to evacuate the plastic material from the car and push the material into the silo. Depending on the location of the equipment, it may be advisable to enclose the equipment with a fence or building to protect it from the elements.

When unloading a railcar, the filters should be placed on the railcar and the instructions followed for the particular railcar being used.

Bulk Railcar Unloading

An important adjustment for successful operation of the bulk system is setting the correct material-to-air ratio at the material pickup point in the car.

It is important that you familiarize yourself with the material pickup instructions for the various types of railcars your material may be shipped in. Each type of car will require a different unloading technique for successful operation of this part of your system.

A probe kit is used, consisting of: two aluminum nozzle caps; one filter; one adaptor assembly with a male quick-change coupler; five material pickup probes.

Also included for car unloading is a length of flexible material tubing for ease of switching from one car compartment to the next.

Because of the differences in railcar discharge nozzle designs, two styles of nozzle caps are provided, style A and style B. One of these will fit any of the commonly used

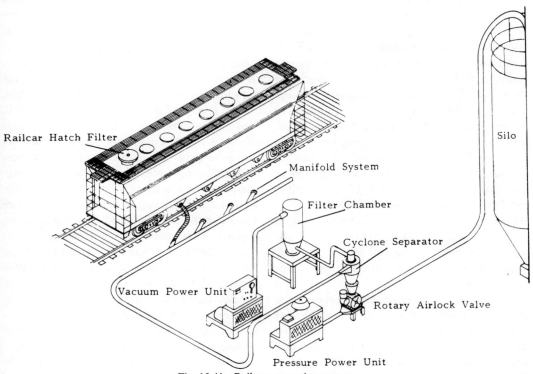

Fig. 15-41 Railcar evacuation system.

bulk railcars. They differ only in the location of the pilot hole for the adaptor assembly.

Style A's pilot hole is $\frac{1}{2}''$ from the bottom of the casting. Style B's pilot hole is flush with the bottom of the casting. The identifying letter is stamped on the front face of the casting.

The adaptor assembly includes a male quick-change connector. A female connector, on the end of system's flexible metal tube, clamps onto this male connector. This completes the system's connection from the car to the rigid conveying tubing.

MATERIAL HANDLING SYSTEMS

The degree to which a processing plant can utilize an automated material handling system is determined by a system design engineer, who must be familiar with the various systems, their components, and the physical properties of raw materials. The overall design will have a major impact on the ability of the plant to increase efficiency and production along with maintaining a high standard of housekeeping. Besides the various methods that material can be conveyed with, the designer should be aware of the overall system layout. Strategic processing machinery locations are essential for efficient automated systems. Their locations will determine the conveying piping layout and the number of bends required in the system. Pipe bends in a pneumatic conveying system should be kept to a minimum to prevent frictional resistance to flow. To ascertain the equivalent feet of piping bends, this guideline has been established: one foot of horizontal piping is equal to one equivalent foot, one foot of vertical piping is two equivalent feet, one long radius bend is twenty equivalent feet, and one foot of flexible tubing is four equivalent feet. These values must be utilized in designing a material handling layout to provide the most efficient and energy-conserving operation.

By comparing the relationship between the length of conveyor lines and the amount of material in pounds per hour that is conveyed through these lines, the proper power unit size and related equipment can be determined. Referring to Fig. 15-42, which compares con-

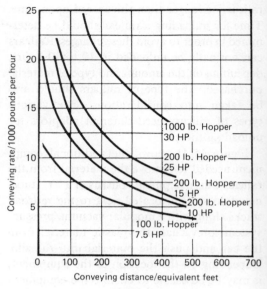

Fig. 15-42

veying footage to material pounds per hour in $7\frac{1}{2}$ through 30 hp vacuum power units, this relationship is clarified. This graph shows polyethylene with an average particle diameter of $\frac{1}{8}''$ and approximately 35 pounds per cubic foot bulk density and a conveying vacuum of $12''$ of mercury. Looking at the 25 hp curve on this graph we can see that this power unit will convey 18,000 pounds of polyethylene per hour if the material line is only 100 feet long. If the conveying line is 450 feet long, the same power unit will convey considerably less. There is a similar drop-off with the 15, 10, and $7\frac{1}{2}$ hp units, and you get the same kind of slope in the curves for the 5, 3, and $1\frac{1}{2}$ hp units (Fig. 15-43). Each of these curves is based on equivalent footage, which has previously been discussed. Thus, the longer the material lines, the more bends, and the more extensive the use of vertical lines, the less material you can convey per horsepower.

It is certainly worthwhile to consider the energy savings to reduce horsepower requirements, which can be translated into savings on electrical requirements. By rule of thumb, one horsepower is equal to .745 kW. For example, due to very long material lines, a processing plant may be using a 25 hp unit where, if the lines were shortened, a 10 hp unit could be utilized to convey the same amounts of

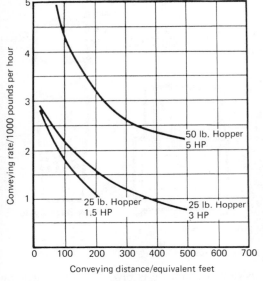

Fig. 15-43

mated system, basing their system designs on the following variables:

- The number of materials used in a plant
- Annual volume of each raw material
- Total number of processing machines
- Average length of the production runs

Although these variables correspond to the type of pneumatic conveying system to be utilized, the introduction of an automated system also depends largely on the plant site and its expansion requirements. Smaller processing plants can utilize the automated systems and design them for future growth. As processing machines are added, the material handling system can be expanded or simply repeated from the existing system.

There are some important benefits to consider when installing a bulk-handling pneumatic conveying system. Bulk material purchases are less expensive due to the higher packaging costs and freight rates that are associated with materials purchased in small containers or bags. Labor expenditures are reduced with the elimination of handling and storing of small material containers. Spillage is minimized and material contamination is prevented by handling in a closed system. The processing plant can obtain a neat appearance with improved housekeeping methods and increased floor space providing additional production area. Storage of bulk materials in large-capacity bins or silos in plant areas that might otherwise be unavailable for storage of packaged materials can be utilized, and the materials can be delivered to remote plant areas that would be economically inaccessible via mechanical type conveyors. The plant safety operations are also enhanced by eliminating manual handling of materials.

There are several alternative conveying systems that can be adapted to a particular processing plant's requirements. Negative pressure or vacuum systems are usually best when conveying from several locations to one discharge point with small flow rates (Fig. 15-44). In this system a high velocity air stream is established in the conveying line by air entering the suction side of a positive-displace-

material per hour. In that instance there is a difference of 11.2 kW per hour. Individual units or by the press vacuum loaders rated at $\frac{7}{8}$ hp equal .65 kW each. Six units equal 3.9 kW. A single 3 hp unit equals 2.2 kW, which is a savings of 1.7 kW over the six smaller units. Multiply that by monthly and yearly operation, and the difference adds up. By the press vacuum loaders certainly have advantages, since the material lines can be shorter, and they can move from a few hundred to even a thousand pounds per hour over distances up to 50 feet. But, from a cost and an energy standpoint, when your material handling requirements rise above this level, operating efficiencies dictate some type of central conveying system.

Many processing plants begin operations without serious planning of material management systems. As these manufacturing plants grow, they usually run into major cost increases, reduced efficiency, and deteriorated housekeeping methods. To avoid these problems and obtain the highest manufacturing performance, a raw material handling system must be designed into a processing plant with automation, flexibility, and expansion requirements being the key objectives. Large as well as small process plants can utilize an auto-

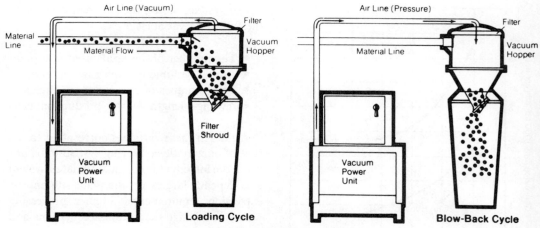

Fig. 15-44   Negative pressure system.

ment or centrifugal blower. Free-flowing material is drawn directly into the conveying line by air flow into a pickup nozzle, or hopper. A rotary air lock feeder can also supply material into this system by gravity or mechanical conveyors. The air stream then conveys material to a receiver-separator where the material is discharged. The conveying air passes through a dust filter and into the vacuum power unit and out into ambient air. Since this system utilizes negative pressure, all leaks are inward so air enters the system at leakage points rather than having dust leak out of the system. Blowback air is utilized in these systems to provide the filter cleaning action, and self-cleaning filters are required when the vacuum systems do not utilize the blowback cycle. The blowback air is directed back through the conveying lines and filters, causing dust accumulations to drop off. The cleaner the filter, the less the obstruction to flow of material and the more efficient the system.

A vacuum system is cost- and energy-efficient in conveying free-flowing material at rates up to 15,000 pounds per hour to one of several use points over distances of up to 500 feet. The efficiency of this system stems from the fact that it only uses one power unit to service several processing stations. In a multiple material line vacuum system (Fig. 15-45), there is a separate material line for each station so that it is possible to feed a variety of materials or colors in whatever sequence

is desired. Control is provided by sequencing valves placed in the common vacuum line running to the vacuum power unit. Electrical signals open these valves in turn to permit the vacuum to be diverted to one station at a time. Any station not requiring material is cut out of the sequence. If several processing stations are using the same material, a multiple vacuum system can have a common material line. The single line "Y" system, as it is referred to, delivers a single material to processing machines through specially formed "Y" tubes in a common material line. The size of the vacuum power unit is determined by the total capacity requirements of the system.

Positive-pressure systems work well when conveying from one pickup point to several discharge points with large flow rates. In such a system a high velocity air stream is established in the conveying line by the blower. Raw materials are gravity-fed from storage or transport vessels into this rapid air stream via a rotary air lock valve. The material is transported through the conveying line to one or more vented receivers. Due to the higher pressures and smaller pipe lines utilized with this system the product-to-air ratio is higher, which increases conveying distance ability and flow rates. Rotary air lock valves are required at material inlets in these systems along with venting systems at material discharge points. Receivers on pressure systems are generally bins with vent connections to atmosphere or

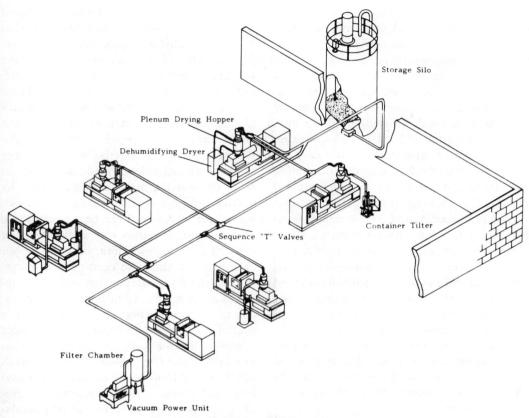

Fig. 15-45 Multiple material line vacuum system.

to ordinary single or multicompartment dust filters, depending on material characteristics. In some installations, the dust filters mounted on receiving bins are self-cleaning filtration units that automatically discharge dust and fines directly back into the receiving bin. These venting filters prevent pressure buildup and possible discharge of dust into surrounding process areas or into the environment. Venting systems are also required at or near the rotary air lock feeder that discharges the material into the conveying line. This relieves pressure buildup at this point by venting the air leaking back into the material feeding system from the air conveying line.

If conveying lines are long or have extensive bends and materials have to be delivered from a silo or storage container to several widely separated points of use, pressure drops will go up, and more horsepower and energy are required. When the pressure drop is greater than 12″ mercury, it is necessary to utilize a pressure system that operates at pressures up to 10 psi.

Positive and negative pressure systems may be combined whenever it is desirable to obtain the advantages from both systems, pickup from several sources of material supply and discharge to several delivery locations (see above, Fig. 15-12). The vacuum/pressure system is also useful when vacuum unloading of transported material is required with pressure delivery to several locations. In a combination system the vacuum unit conveys materials from storage or transport areas to a small intermediate receiver (cyclone separator or filter receiver) which feeds the material into a rotary air lock valve that acts as a barrier between the negative and positive sides of the system. The material is then delivered into the rapidly moving air stream of the pressure system where it is conveyed to receiving points. The combination system is very suitable for conveying light and dusty materials and unloading transport carriers.

If a processing plant is running two or more different materials that are transported in bulk quantities and must be unloaded and conveyed to storage areas, a single vacuum/pressure unit can be utilized. By using single vacuum and pressure units with dual intermediate receivers (cyclone separators) with rotary air lock valves and dual filter chamber assemblies, two different materials can be unloaded and conveyed to separate storage areas without mixing or contamination. This system is feasible when controlling vacuum or pressure conveying air from single power units with butterfly valves in vacuum and pressure air lines. Either material can be unloaded and conveyed by switching the butterfly valve and directing the vacuum or pressure power unit air stream. The cyclone separators act as a separation device for vacuum conveying air and material, the rotary air lock valve transfers the material into the pressure side of the system, and the filter chamber assembly filters out contaminants from returning conveying air, preventing damage to the vacuum power unit.

Pneumatic venturi conveying systems are usually good for machine side loading operations or conveying short distances at low flow rates. Pneumatic venturi conveyors utilize compressed air, which is directed through a venturi that induces a vacuum and draws material into the conveying line. These are economical, efficient systems for machine side loading operations, but floor space is reduced due to material containers and equipment that has to be located near processing machinery. Another type of machine side loader is the vacuum hopper loader, which supplies material to processing machines on demand. These units are self-contained conveyors with multiple stage centrifugal fans that induce a vacuum on the inside of a hopper chamber. This negative pressure draws material into the hopper until the desired amount is obtained, the fan shuts off, and the material is discharged by gravity through a vacuum seal device. Individual hopper loader units can be converted into central vacuum systems with a simple higher capacity vacuum power unit, since the hopper chambers are similar to central vacuum system vessels. When scrap regind and virgin materials are proportioned at the processing machine, a proportioning vacuum loader or a central vacuum system proportioning hopper is used to automatically proportion materials in varying ratios (Fig. 15-46). Virgin materials can be loaded from beside the press from gaylords and drums or centrally conveyed. Regrind can be conveyed from regrind granulation collection bins or centrally stored and conveyed to the proportioning hoppers. These materials are automatically supplied to the processing machine on a demand basis and can be adjusted to blend infinitely variable proportions.

With the knowledge of several pneumatic conveying methods, one must make an analysis of the plant structure and the material that

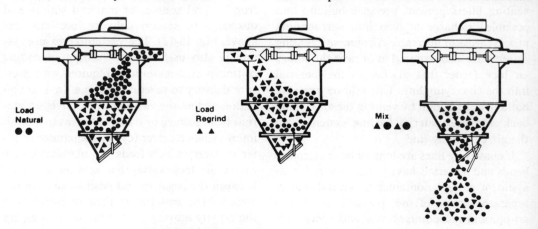

Load Natural

Load Regrind

Mix

Fig. 15-46   Automatic vacuum proportioning hopper.

is to be conveyed when selecting the conveying equipment. The actual requirements of a processing plant and the material properties and characteristics will determine the equipment to be used. Some important aspects that influence the equipment selection and overall design include the following:

- The electrical voltage available in the plant.
- How many pounds per hour are required at each processing machine.
- Type of material to be conveyed and its form.
- The bulk density of the material.
- Whether regrind is to be proportioned and from where.
- Total horizontal conveying distances.
- Total vertical conveying distances and number of bends.
- What we are conveying from.
- Where we are conveying to, and whether there are overhead obstructions.

Energy conservation is also an important factor that should be considered in the design of an automated system. Fortunately, any steps that are taken to save energy will usually save money and reduce the cost of electricity, gas, or oil, as well as contributing to overall operation efficiency. One of the best opportunities to save energy in a material handling system occurs when you are designing a new plant, expanding a plant, or putting in a new material handling system. Locate processing machinery as close as possible to material storage areas, and use the shortest possible route for conveying line and air line connections. Using long conveying lines with extensive bends and vertical lines reduces the quantity of material that can be conveyed per horsepower of a conveying unit.

Processors can also make substantial savings in material costs by in-plant blending operations. Automatic blending contributes not only to direct savings in material costs but to lowered material inventories, lowered labor, and more precise proportional control. Precolored material can be purchased from material suppliers already color formulated for spe-

cific requirements, and this method provides very accurately and consistently colored materials; but the cost of material per pound is significantly higher than for uncolored material. The other method is to buy uncolored material and blend the colorant in-plant with color concentrates or use dry powdered colors with drum tumblers or continuous automatic blending equipment. The blending equipment is usually mounted directly on the processing machine or floor-mounted and utilized as a central blending station to service an entire processing area. In a central blending station, materials are loaded and blended, then pneumatically conveyed to appropriate storage or processing areas. Dry color-blended materials must be blended at the machine, since they cannot be pneumatically conveyed. Central blending stations are usually adapted if there are many processing machines with relatively long production runs, and if only a few material and color combinations are run at the same time. Central blending systems require the least amount of floor space, are the least costly to operate, are best for plant housekeeping and appearance, and have the lowest initial capital cost when compared to individual station blending systems. Individual station blending must be utilized if processing machines are using different materials and colors.

Selection of automatic continuous blending equipment for each particular processing plant is made on the basis of:

- The electrical voltage available.
- Pounds per hour to be blended.
- What types of materials are to be blended and the ratios of each.
- The form of the material to be blended (powder, pellet, etc.).
- The mounting application of the equipment, machine mount or floor-mounted.
- Amount of regrind to be blended and its source.

REFERENCE

1. Butzen, G. A., and Adams, G. A., Engineering and Application Manual for AEC Companies—Material Handling Division, Whitlock/AEC Corp., Elk Grove, IL 60007, 1982.

Chapter 16
Chilling and Cooling

G. A. Butzen

AEC Corporation
Elk Grove, Illinois

CHILLING AND COOLING NEEDS

Today's water-chilling and water-recovery techniques for the plastics industry have reached a high level of sophistication—so high, in fact, that we refer to them as a means of increasing profits through water management (1).

Processors no longer wonder whether such systems are a desirable option but, rather, how best to relate available methods and equipment to their specific needs. What's more, processors no longer strive to minimize their investment in water-chilling and recovery equipment. Instead, they now accept the fact that some type of mechanically refrigerated chilling system is needed to eliminate variables in cooling water temperature, and that evaporative cooling devices, or their equivalent, are vital in eliminating or minimizing unnecessary water waste.

Some Basics for You. Water temperature levels are essential to an analysis of refrigeration and cooling equipment—types, where each fits in, water-treatment systems, and how to evaluate chilling and cooling equipment. What's necessary first is to give you a basic course in cooling so that you won't be buying or using this equipment in a vacuum.

First, let's consider water temperature levels for various plastics processing applications.

They fall into three general categories: (1) 80–85°F, mostly for equipment that can be cooled with water obtained from a cooling tower or other evaporative cooling devices; (2) 45–55°F, the most common temperature range used and one that requires refrigerated or chilled water; and (3) from about 45°F to below 0°F, a range that makes it necessary to add antifreeze to the chilled water to avoid damaging the refrigeration equipment.

The chilled water ranges are usually governed by the efficiency of the process, and each should be evaluated carefully.

A typical plastics process will require cooling water at more than one temperature. Thus, while it may be necessary to refrigerate the water used to chill injection and blow molding processes, it's more economical to cool hydraulic oil coolers, vacuum pumps, and air compressors with higher temperature water available from an evaporative cooling device such as a cooling tower.

The determination of heat-transfer requirements prior to the selection of the equipment or system is contingent upon the application (i.e., the type of equipment and/or process to be cooled and/or type of material that's being processed).

Heat-Transfer Calculations. Another important factor in helping you to understand chilling and cooling needs is knowing the basic heat-transfer calculation formulas. Two basic ones must be kept in mind in any heat-transfer problem.

First and most common: $Q = W \times Cp \times dT$, where W = the weight per unit time, Cp = the specific heat of the material in question, and dT = the temperature difference between entering and leaving conditions.

Variations of this formula will be used for calculating heat-transfer requirements for the coolant being used and the material being processed, that is, the requirements of a body or substance going through a change in temperature of heat being absorbed or released from any medium.

Second and equally important: $Q = U \times A \times LMTD$, where A = the area, $LMTD$ = the log mean temperature difference between mediums, and U = the overall heat-transfer coefficient of the process.

Bear in mind that the expression Q is the same for both formulas and is expressed in Btu/hr.

It is important to recognize that while the first formula tells us how much heat removal is required, the second in effect tells us if it is possible to satisfy those requirements with the tools or devices at hand. In other words, it tells us how effective the devices being used are in terms of the efficiency of area available for heat transfer in molds, heat exchangers, etc. This is obviously an oversimplification and is intended only to alert you to the existence of two formulas involved in the heat-transfer problem as found in plastics-processing plants.

Requirements Vary with Materials. Material chilling requirements are a function of the thermal characteristics of the processed material, expressed as specific heat (British thermal units or Btu's per pound per degree F) plus latent heat (Btu/lb) or enthalpy (Btu/lb for a specific condition), the quantity of material being processed (lb/hr), and the entering and leaving temperatures of the materials involved as required by the process.

Probably the simplest way to regard the heat content of a plastic material in process is to think in terms of balance. On the one hand we have the heat supplied and/or generated, and on the other we have the heat removed. A proper heat balance is achieved, of course, when both sides are equal.

The heat content (or enthalpy) of virtually all materials passing through the injection molding process consists of latent heat and sensible heat. Latent heat (relevant for all materials except polystyrene) is the heat gained by the material as a result of a change in state in the mold without any change in its temperature. Sensible heat develops from a change in the temperature of the materials and is relevant for all materials.

Table 16-1 shows the values of a number of commonly processed materials and processes in terms of heat-removal requirements.

The cooling of thermoplastic materials essentially is the removal of all the heat that was previously put in. The amount of heat put in or removed is expressed in Btu's. As a rule, a nearly continuous process is involved, so we speak of the removal of Btu/hr.

The Btu/hr that must be removed from process constitutes the load (Q), which the chilling equipment must have the capacity to remove. Refrigeration capacity is measured in tons. One ton of refrigeration capacity equals a heat load of 12,000 Btu/hr and is commonly used for refrigeration equipment rating.

Table 16-1 is of value in determining refrigeration capacity requirements. For example, assume that polystyrene is being processed at a rate of 50 lb/hr. (It's pretty well established that processing 50 lb/hr of polystyrene requires one ton of chilling.) The heat removal required for the process is identified as shown—175 Btu/lb. Multiplying 175 Btu/lb by 50 lb/hr results in a total of 8750 Btu/hr. In sum, this suggests that we fall short of a full one-ton refrigeration requirement. However, practically speaking, the additional 3250 Btu/hr is lost through the surfaces of the mold, cycling of the process, heat loss to atmosphere, interconnecting piping, and other parts of the system.

In selecting the properly sized chilling unit to handle the load as calculated, its capacity in Btu's or tons must be corrected for the

Table 16-1   Heat-transfer values for plastics processing.

PROCESS	TEMP OF MATERIAL (T_1, INTO MOLD), °F	GUIDELINE/TON OF CHILLING	ENTHALPY BTU/LB (FINAL T_1, 125 °F)
Injection Molding			
Polyethylene	425	30 lb/hr	300
Polypropylene	400	35 lb/hr	275
PVC	375	45 lb/hr	200
ABS	425	50 lb/hr	175
Polystyrene	425	50 lb/hr	175
Sheet Extrusion, Calendering			
ABS		60 lb/hr	
Polystyrene		60 lb/hr	
Pipe, Profile Extrusion			
Polyethylene		50–65 lb/hr	
PVC		60–75 lb/hr	
ABS		60–75 lb/hr	
Compounding			
PVC (Cool from 230°F to 100°F.)		170 lb/hr	

operating temperature of the chilling system to determine proper equipment size. Generally a 20 percent loss in rated capacity per 10°F drop below the standard 50°F rating point is allowed for safety purposes. (Think of a refrigeration compressor as you would an air compressor. A given horsepower will provide a balance of volume and pressure; that is, when the pressure increases, volume decreases. At lower operating temperatures in a refrigeration system, the compression ratio increases and reduces the volume of refrigerant passing through the system and thus total cooling capacity.)

Until recently it was assumed that the plastics process to be chilled constituted a given condition to which chilling equipment had to be adapted. Recent developments in the state of the art, however, have made it apparent that the process tools themselves can be modified advantageously in some cases to promote heat transfer. This is particularly the case in injection molding, where it is sometimes possible to reduce the internal heat-transfer resistance of the mold. Once this is accomplished, the rest of the system can be adjusted so as to obtain the best possible heat-transfer coefficient.

Because of the many mathematical calculations involved, it has proved advantageous to simulate alternate heat-transfer situations on a computer and thereby determine the opti-

mum combination of mold-cooling channel configuration, diameter, and coolant characteristics (i.e., flow-rate temperature and pressure). The target of such an analysis is reduction of cycle time and therefore greater productivity of injection molding machines. This is, of course, the critical thing to bear in mind.

Don't Neglect Water Recovery. While chillers make money, water-recovery systems save it for you—so they had better not be overlooked. This is a very important point.

The formulas applying to water recovery are identical to those described above. Generally, we are dealing with two basic heat inputs: the chiller condensing load, assuming that the chiller sized earlier is to be water-cooled, and the hydraulics or machine cooling loads. Also included on this type of system are air compressors, general extruder cooling, and, occasionally, temperature-control units, it should be noted.

In all cases, of course, we are concerned with the heat balance, and the input can be obtained from the amount of energy introduced to the system expressed in watts or horsepower. Remember that just because a motor is applied to a process (such as a hydraulic system), the total motor horsepower is not converted to heat. Work is performed by that motor, and in reality not all of the motor energy goes into heat to be removed by the water-recovery system. The load for the chiller condenser should be figured on a one-for-one basis, that is, one tower ton for one refrigeration ton.

We will not attempt here to show the mathematics involved in this process because of the many variables involved, which in actuality change for each machine considered. Age, duty cycle, design, and components all have a bearing. The point to remember is the need for a safe and practical allowance for all plastics processes normally encountered.

The importance of getting all the capacity you pay for is particularly evident when the nature of the plastics process being cooled limits heat-transfer efficiency. Cooling is more efficient in processes such as injection molding—which utilizes high pressures and cooling on two sides under controlled conditions—than processes that involve cooling of only one side of a thick object, such as a pipe that is immersed (under no pressure) in an uncontrolled cooling bath.

Table 16-2 shows the relative characteristics of various processes as they apply to cooling. Note that shell and tube heat exchangers are relatively high-efficiency devices when compared to molds; however, you should also keep in mind the fact that they are limited in their application.

Remember that in Table 16-1 the guidelines for processes which show the higher lb/hr/ton for a given material are really indicating efficiency losses; in other words, a pound of material can only contain so much heat and a higher lb/hr rule indicates less heat removal/lb.

GENERAL CONSIDERATIONS

Chillers fall into two basic physical categories: portable and stationary. Portable, self-contained chillers, which can be moved from one point in a plastics-processing plant to another, are generally used only where capacity requirements are 15 tons or less.

Portable chillers offer a number of advantages directly stemming from three small capacity increments: they minimize the effects of downtime, they can be purchased as needed, and they provide very flexible temperature control. In addition, they're advantageous for low-temperature cooling applications, which involve a higher cost per delivered ton whenever a chiller is used to circulate coolant at lower than design temperature. (Bear in mind that the condensing temperature of the refrigeration must be high enough to allow condensation by air or water, and the coolant temperature is a function of the vaporization of that liquid at a specific temperature. The lower that temperature is to be, the higher compression ratio the compressor must produce. A

Table 16-2 Cooling characteristics.

PROCESS OR SYSTEM	HEAT-TRANSFER COEFF. (BTU/°F/HR/SQ FT)	LIMITING FACTORS
Injection molding	30	Thickness and area available
Blow molding	25	One-side cooling and pinchoff cooling
Foam injection molding	20	Thickness and cellular structure
Jacketed vessels	50	Agitation and area available
Bath or trough	100	One-side cooling, thickness, and area available
Shell and tube exchanger		Area and temperature
Oil to water	100	difference
Water to water	250	Area and temperature difference

given horsepower can only provide so much work, and this work can be used to move a volume of refrigerant or provide a higher compression ratio.)

Portable chillers are available with both air-cooled and water-cooled condensers.

Go Central for Bigger Needs. For larger capacity requirements, some type of centrally located chilling system is used. One of the more practical central system designs is the energy-conserving air cooled type, which supplements plant heating during winter months by reusing the heat recovered from process. From an economical and ecological standpoint, this type of chilling system provides dual benefits: conservation of water and recycling of energy to conserve increasingly scarce heat-producing fuels. Through an adjustment in the duct work of this air-cooled chiller, it can be used to ventilate processing plants during the summer. This type of equipment is obviously not compatible with air-conditioned spaces.

Other types of systems separate the air-cooled condenser from the chiller and remote-mount the condenser outside the plant—perhaps on the roof. Such an arrangement is thought by some to be advantageous because of the cooling properties of outside air. On the other hand, fluctuations in ambient temperature may interfere with condensing efficiency, with a resultant adverse effect on the system's capacity and performance. Also, the installation of components requires other than normal plant personnel.

In general, very large process chilling requirements are most economically handled by a central water-cooled chilling system. Central chilling systems are generally limited to one temperature and must be designed and sized with production requirements in mind. Limited-duty equipment normally associated with air-conditioning use is generally less expensive than equipment designed for around-the-clock production. In a central system selection, the processor usually has his choice of open or semihermetic-type compressors, and he should make sure that the advantages and features of each are fully evaluated for his application.

Multiple-circuit chilling units can be selected to provide some of the downtime protection mentioned above as an advantage of portable chillers. Many configurations of both air- and water-cooled central systems are available. The correct choice requires careful evaluation.

In comparing air-cooled and water-cooled chillers, it is important to remember that water-cooled units are more efficient on a horsepower-to-ton basis and less expensive on a dollar/ton basis. In some cases, the factor may be counterbalanced, particularly in small- and medium-sized installations, by the cost of the evaporative cooler or other water source needed to cool the chiller.

Why Have a Cooling Tower. A cooling tower uses the evaporative-cooling principle to cool water that has picked up heat in plastics processing equipment. After passing through the tower, the water is recirculated through the system. Theoretically, the water can be cooled to a temperature equal to the prevailing wet-bulb temperature if enough surface area is available. (But, actually, an infinite area would be required.) The wet-bulb temperature is normally established at 78°F for most areas, and, practically speaking, the water will be cooled to a temperature approximately 7°F higher than the wet bulb (85°F). (Dry-bulb temperature is not a valid rating basis for cooling-tower efficiency or its sizing.)

Remember that because of evaporative cooling, the system will require make-up water. Also, evaporation of the water concentrates its mineral content.

Although cooling-tower design varies considerably, one of the preferred configurations is the combination spray and deck-filled counterflow unit (Fig. 16-1). The term counterflow refers to the flow of air in a direction opposite to that of the water flow. Spray filling means that the water is atomized through spray nozzles, and deck filling means that water is distributed over a surface area. This arrangement will provide for maximum heat transfer and also allow operation that is more independent of ambient air temperatures.

Cooling-tower design has been progressing rapidly during the past few years. One of the

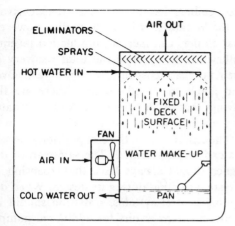

Fig. 16-1 Combination spray and deck-filled counterflow cooling tower of the blow-through type.

more significant developments is the industrial cooler, which is essentially a cooling tower with an inside coil through which processed water passes as it is being cooled by evaporation. In this arrangement, the processed water in never exposed to outside air and is therefore less vulnerable to contamination. (*A word of caution:* Control of this device is critical to prevent freeze-up during winter operation.)

Realistically, it is important not to overlook alternative sources of evaporative cooling, such as wells, ponds, and city water. But, in general, city water is prohibitively expensive for this purpose. Wells, particularly deep wells where there is only a 5°F spread in water temperature from winter to summer, constitute a good source of cooling provided they are available. Open ponds are another possible source, but only if the ponds are big enough to provide water in sufficient quantities and to sustain a relatively uniform temperature. Finally, it should be pointed out that maintenance requirements are usually higher with these alternative sources.

To sum up, then, it seems likely that, as the emphasis on water conservation increases, use of cooling towers and industrial coolers will increasingly displace these other sources.

Should You Treat the Water? Water treatment must be considered a valuable adjunct to water-cooling systems. Justification for water treatment is oriented to both performance and economics.

Untreated water, which results in scale deposits and corrosion, can be quite costly. In terms of production, a mere 0.006 in. of scale will reduce refrigeration compressor capacity by 30 percent. The thin deposit from scale, corrosion, slime, or algae in the cooling-tower system makes it necessary for a pump to work from 15 to 50 percent harder.

Maintenance to counteract fouling resulting from untreated water can be quite expensive and dangerous, since the acid used in cleaning heat-exchange equipment removes galvanizing and metal. It is also a hazardous process for personnel. What's more, the additional load imposed on pumps, motors, and compressors by deposits from untreated water tends to make them work overtime to compensate for the insulating effects of the deposits. The results are breakdowns, need for costly replacement parts, and labor costs, as well as lost revenues on production equipment because of downtime for service and repair.

As a rule, plastics-processing requirements can be met through use of a water softener or ion-exchange system. In the latter process, calcium and magnesium ions in the water are exchanged in a resin bed for more soluble sodium ions, resulting in water with minimum hardness and thereby preventing carbonate scaling.

Silicate scaling in such a system can be controlled by water system bleedoff. Essentially, this is a matter of diluting the mineral concentration to a safe level of less than 125 ppm silica. The same results will be achieved through hot lime softening and demineralizing.

It should be pointed out that even soft water is corrosive. Corrosion can be controlled by adding a corrosion inhibitor, such as certain commercially available polyphosphate chemicals fed into the system.

Slime and other organic growths can be controlled by addition of microbiocides and algaecides to the water.

Suspended solids collected in the evaporative-cooling device from the cooling air can be removed from the cooling-system water by filtration.

Water filtering is of particular importance where in-plant contamination of cooling systems is likely, as in the case of plants using

PVC powders, doing pipe extrusion, or in production of cast film. Plastics material contamination of cooling-system piping is worse than scale; it is far more difficult to remove because of its affinity for the piping. No matter how minor this type of contamination, it will build up to a point where it seriously interferes with heat transfer in the chilling system.

Secondary circuits should be considered to protect the refrigeration equipment against contamination, which can reduce its efficiency by as much as 50 percent. These secondary circuits can be cleaned in a way similar to the way hydraulic heat exchangers are cleaned. Many other process-control considerations strongly favor the secondary-circuit approach, and processors should definitely investigate it.

Evaluating Equipment before You Buy. The heat-transfer calculations are usable only if it can be assumed that the capacity of the chilling and cooling equipment is accurately described by the supplier. For various reaosns, this is not always the case. There has been a tendency to assume that the capacity of a chiller in tons is equal to the unit's compressor horsepower by a conversion factor of 1.0. But because of the differences in heat-transfer efficiencies from one unit to another—particularly in the heat-exchange devices—this is not necessarily the case. For example, a water-cooled shell-and-tube condenser is more efficient than an air-cooled condenser and, to go one step further, an air-cooled condenser is not as efficient at 100°F air temperature as it is at 80°F air temperature. (Standard condenser rating conditions are 85°F water and 105°F refrigerant for water cooling or 95°F air and 120°F refrigerant for air cooling.)

No less than four heat-transfer exchange processes are involved in any refrigeration system used in plastics processing. Efficiency is still the name of the game in heat transfer, and any exchange process or temperature deviation from standard conditions affects the performance of the equipment.

Payback in under a Year. Over the years the economic justification for use of refriger-ated chillers has been discussed extensively in the literature. It is worth noting, however, that chillers can increase production revenue sufficiently to produce the total recovery of their purchase cost in less than a year. Over and above increased revenue, there are the savings that result from conservation of water and maintenance of molds.

The customer is in the best position to assess his expansion potential. The salesman or designer should always have the expansion of a system in mind during its design. When designing central systems, provision should be made in the reservoirs for additional pumps; in addition, conservative pipe-sizing guides should be used, and the layout should be planned with expansion in mind.

Still another consideration is the possible addition of accessories at a later date. This should not be ignored even if the initial installation does not include them for budgetary reasons. Items such as water treatment and filtration are often invaluable from a maintenance-reduction standpoint. Their value is generally realized after a few months of operation. If provisions are made in the original design of the system for these accessories, then their installation cost at a later date is not excessive.

Finally, try to incorporate in your system monitoring capability for process water conditions even though this costs a bit more. The use, for instance, of mold analysis without assuring yourself that the operating conditions called for are met in actuality (i.e., temperture rise, flow and pressure drop) can be costly.

Improper system design can ruin the performance characteristics of the best-designed chilling equipment. Pipe sized for price or because it's already in place can be a false economy. All of the systems discussed are payback type, and the slight additional expense of properly sized pipe or drops to machines will not cause an appreciable lengthening of the payout period. Selection of chilling equipment, cooling equipment, treatment and filtration equipment should be governed by the value-engineering concept already adopted by many advanced-thinking companies in the plastics industry.

CALCULATION OF THE COOLING LOAD

Close approximation of the cooling load for the modern plastics processing plant is a necessity before any attempt is made to select the mechanical equipment which is to handle it. Most plastics plants require several water temperatures for their various processes, and the loads must therefore be categorized by individual temperatures.

The proper approach in making a cooling survey is to first determine the various water temperatures needed for each group of equipment. The cooling load can then be calculated for each temperature level and a determination made as to the type and size of the mechanical cooling equipment needed.

Cooling Water Temperature Requirements.

We can generalize the cooling water temperatures required sufficiently closely to enable us to select the equipment type required to do the job. Once this has been done, the individual needs of the process can be evaluated and the specific equipment selected. Table 16-3 gives temperature brackets for various equipment types.

A study of Table 16-3 will indicate that there is a temperature range that in most cases can be met with evaporative type equipment such as a cooling tower. Cooling towers can work efficiently to an approach of 7°F to the design wet bulb temperature. For example, in an area with a design wet bulb of 78°F, they will produce 85°F cooling water.

Table 16-3 Cooling temperature requirements.

Eddy current drive	80–85°F
Extruder barrels	80–85°F
Hydraulic oil coolers	80–85°F
Air compressors	80–85°F
Vacuum pumps	80–85°F
Refrigeration condensers	80–85°F
Temperature control units	80–85°F
Molds on injection molding machines	20–55°F
Molds on blow molding machines	20–55°F
Molds on bottle molding machines	50–55°F
Extruder troughs and cooling baths	50–55°F

Note: High-temperature cooling requirements for special compounds may require temperatures in excess of 250°F.

Mechanical refrigeration equipment is required to produce chilled water in the lower temperature ranges. Chillers can furnish water in the leaving water temperature range from 60°F to 20°F. Later chapters dealing with equipment selection and sizing will deal with these subjects.

As a general rule, supply water temperatures down to about 45°F can be furnished without the use of antifreeze solutions in the chilled water circuit. Lower temperatures make the use of antifreeze solutions mandatory to avoid the danger of freeze-up.

The Thermal Properties of Plastics.

A great deal of heat is required in plastics forming operations. The plastic material must be heated to its melting point and sufficiently beyond to ensure that the molded object is properly formed before it assumes its finished shape. Cooling should take place as quickly as possible within this limit for efficient production. About 80 percent of the injection molding cycle consists of mold cooling. The use of mechanical chillers speeds up this process, and the machine produces more pieces per hour.

Both sensible and latent heat are absorbed by plastic materials during the molding process. The plastic is first heated to its melting point, absorbing sensible heat. It then softens and melts, absorbing latent heat. Further heating past the melting point adds more sensible heat. Figure 16-2 shows the heat content of some of the more common plastic materials.

A study of this figure will show the abrupt increase in heat content as latent heat of fusion is added once the melting point is passed. A

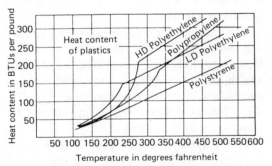

Fig. 16-2 Heat content of plastics.

notable exception is polystyrene, which simply softens and goes through a fluid phase without the addition of latent heat. The temperature–heat content relationship is a linear one as shown by the straight line.

A study of Fig. 16-2 will show that the heat added to various plastics during processing will vary. For example, HD polyethylene requires more heat than polystyrene to reach a given temperature. Therefore, more cooling will be required after the plastic has been processed to bring it back to its proper temperature.

It has become common practice to calculate plastics cooling needs in terms of the pounds of a given plastic that can be cooled from processing to handling temperature by one ton of refrigeration. One ton of refrigeration is equivalent to a heat-transfer rate of 12,000 Btu/hr, and the plastics cooling rate is also stated in terms of pounds per hour.

As an example of this type of calculation, assume that polystyrene is to be processed in an injection molding machine at 500°F initial temperature and then cooled to about 125°F final temperature. Figure 16-2 indicates that at 500°F polystyrene has a heat content of about 185 Btu/lb. At the final temperature of 125°F, the heat content is about 25 Btu/lb. This indicates a heat removal requirement of 160 Btu per pound of the plastic being cooled.

It is common practice to assume that one ton of refrigeration will handle the processing of 50 lb/hr of polystyrene. Simple arithmetic indicates that 50 lb/hr at 160 Btu/lb accounts for only 8000 Btu/hr. The difference between this figure and the actual 12,000 Btu/hr/ton is accounted for by factors other than the heat gained by the plastic.

For example, the centrifugal pump circulating the water through the system adds frictional heat. In addition, heat is gained through the chilled water piping and the mold surfaces themselves from the surrounding air. The 4000 Btu/hr difference is accounted for by these external heat gains.

HD polyethylene has considerably more heat added during its processing because of its latent heat requirements. At 440°F its heat

content is about 310 Btu/lb, while at 125°F its heat content is about 35 Btu/lb, a heat gain of 275 Btu/lb. A rule-of-thumb figure often used is one ton of cooling for each 30 lb/hr. This accounts for 8250 Btu/hr, with the balance of 3750 Btu/hr being accounted for by external heat gains.

The heat content or enthalpy of the plastic will vary, with the processing temperature dictated by a particular molding application. The higher the processing temperature, the greater the quantity of heat that must be removed. Following the guidelines established by experience ensures that adequate chiller capacity will be available to meet these varying requirements.

Experience and calculations similar to those just accomplished have established guidelines for chiller selection. These are listed in Table 16-4 for various plastics forming operations and the more commonly used plastics.

Manufacturers of plastics-processing equipment provide data on the number of pounds per hour of a given plastic their machines are capable of handling. For example, an injection molding machine may be rated at 175 lb/hr of polyethylene. Assuming there are three of these machines in a line-up, a total of 525 lb/hr of plastic requires cooling. (Actual rate is 50–60% of 175 lb/hr; 175 lb/hr requires

Table 16-4 Heat transfer values for plastics processing.

PROCESS	GUIDELINE/TON OF CHILLING
Injection Molding	
Polyethylene	30 lb/hr
Polypropylene	35 lb/hr
PVC	45 lb/hr
ABS	50 lb/hr
Polystyrene	50 lb/hr
Sheet Extrusion, Calendering	
ABS	60 lb/hr
Polystyrene	60 lb/hr
Pipe and Profile Extrusion	
Polyethylene	50–65 lb/hr
PVC	60–75 lb/hr
ABS	60–75 lb/hr
Compounding	
PVC (Cool from 230°F. to 100°F.)	170 lb/hr

continuous turning of the screw, which does not happen.)

From the table, one ton of chiller capacity can handle 30 lb/hr of polyethylene. Therefore, 17.5 tons of cooling are required:

$$525 \div 30 = 17.5$$

Once the load condition has been determined in terms of nominal tons, the supply water temperature required must be evaluated. Chillers have their nominal ratings based on 50°F supply water temperature. In the case of the job being discussed, an AC 25 chiller with a nominal capacity of 19 tons would be a good selection if 50°F water were required.

Chiller capacity decreases with supply water temperature. For the known load of our example, divide the load by .8. This establishes our load in terms of 50°F LWT, and standard rated chiller selections can be made—an AC 30S in this case. To determine standard chiller capacities at less than 50°F LWT, multiply standard rated capacity times .02 for each desired degree LWT. Below 50°F an AC 30S would have 22.08-ton capacity at 48°F LWT (100 − .04 = .96, 100 being 100% capacity, and at .02 × 2 degrees below 50, .96 being the correction factor; 23 × .96 = 22.08, 23 being capacity of AC 30 at 50°F or 100% rated capacity, .96 the determined correction factor, and 22.08 the capacity of AC 30 at 48°F LWT).

Both chillers selected in the above examples are of the air-cooled type. In some cases, water-cooled condensers may be used. A complete discussion on this subject will follow in the section of this manual applying specifically to water chillers.

Evaporative Cooler Load Calculation

Cooling Towers. Wherever required water temperatures permit, consideration should be given to the use of cooling towers. Their initial installed cost and operating economy make their use a logical choice. Reference to Table 16-3 indicates that there is much equipment in the 80–85°F range that can utilize this cooling method.

In the event that the plant has water-cooled chillers, the chiller condensers will be a part of the cooling tower load. Condensers require a cooling capacity of approximately 15,000 Btu/hr/ton due to the additional heat of compression put into the refrigerant by the compressor. Nominal cooling tower ratings are therefore based on this figure.

Hydraulic Drives. The cooling needs for hydraulic drives can usually be met by assuming the need for 0.10 ton capacity per drive horsepower. Using this yardstick, a 75-hp hydraulic drive would require 7.5 tons of cooling tower capacity. This holds true for injection molding machines.

Air Compressors and Vacuum Pumps. Air compressors will generally require 0.10 ton per brake horsepower for the air compressor and an additional 0.10 ton per brake horsepower if an aftercooler is used. On screw-type compressors, the oil cooler load is now 0.15 ton/hp. The aftercooler load is approximately .05 ton/hp.

The same rule holds true for vacuum pumps. These require no aftercoolers, and 0.15 ton per brake horsepower is used in all cases.

Magnetic Drives. Some extruders are equipped with dynamic or eddy current drives. In most cases, the consideration of a cooling tower load of 0.15 ton per brake horsepower of the drive motor will be sufficient. It is best to run a weighed water test where existing equipment is involved. This test will be described later in this chapter. The following example cooling tower capacity calculation illustrates the principles just discussed.

Example:
A new injection molding facility consisting of eight new 350-ton injection molding machines of 50 hydraulic horsepower each. The mold cooling requirement is rated at 50 lb/hr per machine of polyethylene material. A 40°F leaving water temperature is required to cool the molds.

Additional cooling is required for an air-cooled 15-hp air compressor having a water-cooled aftercooler.

Required: Select the water-cooled chiller and cooling tower required.

Solution:
Hydraulics:

$$8 \text{ machines} \times 50 \text{ hp/mach.} =$$
$$400 \text{ hp @ } .10 \text{ ton/hp} = 40.0 \text{ tons}$$

Air compressor aftercooler:

$$15 \text{ hp @ } .10 \text{ ton/hp} = 1.5 \text{ tons}$$

Mold cooling tons needed:

$$8 \text{ machines @ } 50 \#/\text{hr/ton} = 400 \#/\text{hr of PE}$$
$$400 \#/\text{hr} \div 30 \#/\text{hr/ton} = 13.3 \text{ tons @ } 40°\text{F water}$$

Choose a WC-20 chiller rated at 20 tons nominal less 20% or 16 tons.
Cooling tower capacity for chiller: 20.0 tons
Total cooling tower capacity
    required: 61.5 tons

Load Evaluation on Existing Installations

In some instances an existing plant requires updating of the cooling system for the installed equipment. The use of gauge readings on existing pumping equipment or weighed water testing will provide accurate flow information. In connection with flow determination, accurate temperature readings in and out of the equipment being cooled are required.

Where pump gauge readings are to be a part of the evaluation, it will be necessary that the pump performance curves for the installed pumping equipment be at hand at the time of the tests. If the customer does not have this information, check the pump nameplate. The pump model number, serial number, and manufacturer should appear here. Request copies of the pump curves from the manufacturer, providing the pump nameplate information for identification.

Determining Water Loads

The flow through a process and the temperatures entering and leaving must be accurately observed in order to calculate the cooling requirement.

Temperature. Temperatures and flow must be observed simultaneously. Carry three thermometers at all times, and submerge them in a glass of water to be certain that they agree before making any readings. Be sure to allow enough time for the thermometer to reach temperature when taking a reading. This may take several minutes if the temperature of the water is far above or below the initial reading on the thermometer. If there are no points in the system where the water flow is open to view and accessible for immersing a thermometer bulb, then some means must be found for bleeding water either onto the thermometer bulb or into a container in which the thermometer is immersed.

Flow. Flow may be determined in two general ways—from pumping curves or actual timed weight testing.

(a) *Flow Determined from Pumping Curves.* Pumping curves on all centrifugal pumps have the characteristic shape shown on 2518–3624 series curves (Fig. 16-3). If the pressure difference across the pump is accurately known, it becomes a simple matter to project that point over to the proper curve and read flow in gpm (gallons per minute) directly below.

Most pressure gauges read psi (pounds per square inch) positive and inches of mercury negative. Obtain a good serviceman's test gauge of the compound type. Do not rely on the gauge installed on the job. Readings must be taken at the suction and discharge connections of the pump with the same gauge. The preferred points are the actual test connections furnished on most pumps. Take several readings to be sure of having accurate observations. If the suction pressure and discharge pressure are both positive, the pumping head is the difference.

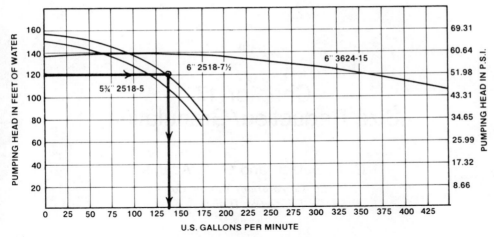

Fig. 16-3   Flow curves for centrifugal pumps.

Example #1—2518–7½:

Discharge pressure	57 psi
Suction pressure	5 psi
Pumping head	52 psi
Pumping head in feet of water	$52 \times 2.3 = 120$ ft

If the suction pressure is negative, the following procedure is followed. The pumping head is the sum.

Example #2—2518–7½:

Discharge pressure	47 psi
Suction pressure 10 in. convert inches Hg to psi 10/2	5 psi
Pumping head	52 psi
Pumping head in feet of water	$52 \times 2.3 = 120$ ft

Example #3—2518–7½:

Discharge pressure	52 psi
Suction pressure	0 psi
Pumping head	52 psi
Pumping head in feet of water	$52 \times 2.3 = 120$ ft

In each of the examples the pumping head was converted to feet of water. This was done because most pump curves are published with gpm vs. feet of water. In each case the 2518–7½ pump will handle 137 gpm and will require 7½ bhp. A given pump may have impellers with different diameters installed in the same casing. In order to check the performance of a pump, the diameter of the impeller must be known. If it is not noted on the nameplate, you will have to obtain the information from the manufacturer. (Nominal design flow = 3 gpm per ton for cooling tower, 2.4 gpm per ton for chilled water.)

(b) *Flow Determined from Measured Water.* In determining flow from measured water, the container (a 5-gallon bucket, for example) should be weighed empty and then filled. The weight of water will be the difference between the empty weight and the full weight. The accuracy of the quantity held in a 5-gallon container, for example, may be checked. Five gallons of water weighs $8.3 \times 5 = 41.5$ pounds. The difference between empty weight and full weight as observed on a scale should be 41.5 pounds if the container holds 5 gallons. Record the time required to fill the container with water leaving the process.

$$gpm = \frac{pounds}{8.3 \times minutes} \quad or \quad \frac{gallons}{minute}$$

Several tests should be run to be certain of the accuracy of observation. The flow must be known accurately.

REFERENCE

1. Butzen, G. A., Engineering and Application Manual for AEC Companies—Water Division, AEC Corp., Elk Grove, IL 60007, 1982.

Chapter 17
Size Reduction/Granulating

G. A. Butzen and G. A. Adams
AEC Corporation/Nelmor Co.
Elk Grove, Illinois

SIZE REDUCTION NEEDS

The economics of the plastics industry, where raw material costs form a significant part of the price of a finished article, has always provided an incentive for some recycling. As with the metal, glass, and paper industries, efforts have been concentrated on the reworking of clean and homogenous waste collected at the site of manufacture, but there is a basic difference between the nature of recycling of plastics and the recycling of metals, paper, glass, and some other materials (1). Metal production begins with an impure ore that is progressively concentrated, smelted, and refined, and has impurities removed from it. The production of plastics, however, begins with high-purity virgin polymer to which various additives, colorants, and reinforcing materials are added. In the metals industry, basic technology concerns itself with purifying and upgrading ores and concentrates. The plastics industry, by contrast, progressively modifies its raw material during its evolution into a finished product.

Technology to purify waste plastic has not yet been established. One major obstacle is that different polymers may not be compatible with each other and must be separated as part of the recycling process. The variety of different plastic materials is extensive and unlimited. Through polymer chemistry, new materials are appearing frequently. This increases

the markets for the plastics industry, as well as the amount of recycling possible.

The current estimate of post-consumer reclaim of plastics is less than one percent, which indicates that there is a vast amount of a valuable resource being wasted. We mentioned that a process of separation must be done because of the incompatibility of various materials. This approach was used to lay the foundation for this chapter. Plastics are quite complex compounds, and size reduction is a very complex art if we are to reclaim a useful product. This chapter presents equipment for size reduction of plastics, both for molders who wish to reclaim sprues, runners, or defective parts and for those who are in the scrap reclaim business.

Machines used for size reduction in the plastics and rubber industries may be broadly classified as granulators, dicers, pelletizers, die face cutters, hammer mills, attrition mills, pin mills, and pulverizers. These machines are universally used to produce a granulate, pellet, or powder as required, making it possible for the scrap to be reused or reclaimed into a processible material. Because granulators are very frequently referred to, we will give a brief explanation of what they are and how they work.

Granulators are machines that take scrap material that is too large in size to be reprocessed or reclaimed and mechanically reduce the size of the material to particles that can be easily recycled by molding machines. The scrap plastic is fed into a hopper on the granulator whereby it falls by gravity into rotating knives. Impact and shearing reduce the plastic until it is of a size capable of falling through a screen fitted beneath the cutting zone. The hole size of the screen enables the operator to vary the size of the resultant particle. Once granulated, the regrind, as the particulate is called, can be collected in either a bin or barrel or conveyed by air or mechanical means to a central collection station where it is reclaimed and reused again. Our aim is to familiarize the reader with size reduction equipment offering a granulate by-product.

Safety

The size reduction of plastics is probably the most hazardous operation in the plastics industry. Unfortunately, lack of knowledge of the machinery can make working with it a dangerous occupation. Rotary shears, densifiers, shredders, and granulators are designed with two main goals: safety and efficiency.

Granulators are equipped with a safety switch connected to the cutting chamber to terminate its operation if it is opened while the motor is running, to prevent a person from getting injured. At no time should anyone have to open a machine while it is running. Any adjustment to the knife or drive system should be made with the machine off and power disconnected.

When a serviceman or a machine operator has to perform any kind of work on the machine, that person should shut the machine off and disconnect the power at the main power source for that machine. In most cases this would be the lockout box or disconnect. A padlock should be put on the box, to ensure that the machine does not get energized accidentally. At no time should it be necessary for a person to put any part of his body into a granulator, rotary shear, densifier, or any piece of machinery. That includes climbing into a large granulator to free a jam-up or for cleaning purposes. The instructions for the machinery should be read very carefully. The warning label should be followed in its entirety.

The best safety device is a careful person.

GRANULATORS

The process of granulating and recycling plastics is not a new concept as many people think. The granulation and recycling of plastics has been done for the past 30 years. At first, plastics were ground up to reduce the size of scrap material to minimize the space needed during removal and disposal. With the rising cost of raw materials brought on by the past energy crisis, the plastics industry has found it more beneficial economically and environmentally to recycle most plastic material.

The technology involved in the process of granulation is based on laws of physics involving mass, velocity, kinetic energy, and gravity. Even though physics plays an important part in the granulation technology, it is extremely difficult to classify or simplify the process into a series of mathematical formulas, mainly due to the almost endless variety of plastics available on the market today. Instead, the design and sizing of granulators is largely based on experience and extensive laboratory testing of various types of plastics and machines.

Plastics have certain common characteristics in granulation that permit them to be placed into categories:

1. *Energy-absorbing plastics:* Most thermoplastic materials such as low-density polyethylene and thermoplastic eldstomers (TPR) are in this classification. These materials are relatively flexible and are able to withstand impact with very little damage. This material can be easily cut.
2. *High-impact plastics:* Materials such as ABS are common in this category. This family of plastic is extremely hard and does not fracture easily. When this type of material is exposed to great force it will have a tendency to shatter.
3. *Friable materials:* Phenolic resins and styrenes are in this class. This type or family of plastic will break apart easily with very

little impact. This type has a tendency to create a lot of dust when it is granulated.

Another factor that plays a very important part in the granulation of plastics is the size and cross section of the plastic components that are being granulated. The cross section is the profile of the various densities of a particular piece of plastic sample.

It is these factors that make it virtually impossible to establish rules or formulas for granulating plastics.

This section will discuss the four basic sections of the granulator, which is the primary size reduction device used for plastics reclaim (Fig. 17-1):

1. The hopper
2. The cutting chamber
3. The screen chamber
4. The base

In their basic form, granulators can be classified according to:

1. Type
   (a) Beside-the-press granulators
   (b) Under-the-press granulators
   (c) Central granulators
2. Cutting chamber dimensions: the width in inches by the depth in inches. Beside-the-press granulators will range in size from 4" × 4" up to 18" × 30". Under-the-press granulators have a size of 8" × 8" or 10" × 10". Central granulators have sizes from 14" × 24" to 30" × 72".

Hoppers

The hopper is the first part of the granulator to come in contact with the scrap plastic. Generally, the hopper is constructed of heavy gauge sheet metal stock that is welded together. Hoppers can be built in virtually any shape or size to fit the scrap that has to be fed into a machine. There are two main factors to be considered when a hopper is designed. The hopper must be of adequate size to accept the scrap being fed into the machine, but first the hopper must be designed with safety in

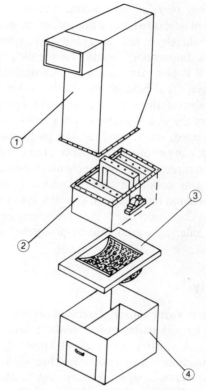

Fig. 17-1   Exploded view of a granulator.

mind to prevent someone from being injured while operating the granulator.

The most common type of hopper design is the manual feed hopper, which is a standard item on most granulators. Manual feed hoppers normally have a tray at the throat of the granulator so the scrap can be placed on the tray and pushed into the machine without the operator having to place a hand into the inner throat to load the granulator scrap material. The hopper is also designed to deflect any scrap material which may be thrown back up the hopper by the rotary knives. This will prevent the scrap material from flying out of the machine at full force and will help deflect the material back into the cutting chamber. There normally are doors located at the top and at the mouth of the hopper. The top door is a metal hinged type door that is used when cleaning a jammed hopper. The front door is a metal flap or several polymeric flexible flaps hinged from the top so they swing closed after material is pushed into the granulator.

Another style of hopper is the conveyor feed hopper. This type of hopper, as its name implies, can be combined with a conveyor belt system to automate the unit. This type of setup is desirable, since it prevents the machine operator from manually feeding in the granulator. This hopper has a door at the top of the hopper for cleanout purposes. The mouth of the hopper has a flexible polymeric material to reduce noise and prevent flyback.

A variety of other hopper designs can be specially built for the specific needs of a plastic-processing plant. A pipe hopper is an excellent example of a hopper built for a special need. A pipe hopper is designed with a long side load neck to handle long pieces of plastic pipe or thick profile materials that would normally have to be precut into smaller pieces in order to fit into a standard hopper.

The same design modification can be done for sheet plastic. The hopper is similar to the pipe hopper with the exception of having a smaller throat orifice designed to handle sheet plastic. In some cases where a long run of sheet plastic must be granulated, feed rolls can be installed on the hopper to feed the continuous thin plastic sheets into the granulator.

Combination hoppers can be built for virtually any type of plastic scrap part on the market today that requires size reduction.

Cutting Chamber

The cutting chamber is the heart of a granulator where the process of granulation actually takes place. The cutting chamber consists of six basic components:

1. The cutting chamber
2. Rotor
3. Rotor knives
4. Bed knives
5. Bed knife clamps
6. Bearings

The cutting chamber is the housing for the rotor, rotor knives, bed knives, bed knife clamps, and bearings. Cutting chambers are manufactured from a thick metal to withstand the force the granulator process will exert on the cutting chamber itself. The thick fabrication will prevent the other cutting chamber components from being damaged by warpage which would occur if a thin wall chamber were used. There are various forms of construction used to build a cutting chamber.

Some manufacturers use a cast iron cutting chamber. Cast iron chambers are relatively inexpensive to manufacture. One disadvantage of this style chamber would be a flaw in the design of the casting or in the actual casting itself. A flaw in the casting would cause the chamber to crack if any form of excessive force were exerted on the cutting chamber walls.

In the past, some of the machining processes were rather difficult to perform but, with the recent advances in technology, these machining operations can be done efficiently in a matter of minutes. There are two types of construction used in steel cutting chambers: bolted and welded. Welded steel chambers, while more expensive because of welding and polishing time, are preferred for granulator longevity. A chamber may be bolted in construction. However, the possibility of a bolt stripping or working itself loose by vibration and causing damage to the knives and rotor does exist.

Rotors. The rotor is the only moving part in the cutting chamber assembly. The primary function of the rotor is to hold the rotor knives in place during the granulating process. There are so many different types of rotors available today that it is difficult to classify the rotors in specific categories. However, the adopted form of classifying rotors is by the number of blades a rotor will hold and whether the rotor is a solid or an open type.

Rotors commonly fall into categories of two-blade (Fig. 17-2), three-blade (Fig. 17-3), five-blade, or staggered-blade rotors.

Two- and three-blade rotors are commonly used in the smaller beside-the-press and under-the-press granulators. Three-blade, five-blade, and staggered-blade rotors are usually found in central granulators.

Rotors can also be classified, as mentioned

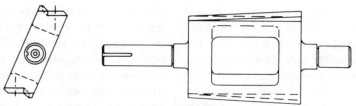

Fig. 17-2   Two-blade open rotor.

before, as open or solid. Two-blade rotors are commonly found as open rotors such as the rotor shown in Fig. 17-2. Three-blade and five-blade rotors can be purchased as solid or open rotors. Open rotors are also known as fin rotors in the plastics industry.

An open fin rotor would be used for granulating light loads of plastic material, whereas, in turn, a solid rotor would work better on heavy loads of scrap material. Open rotors use less power on start-up when compared to a solid rotor, but solid rotors will utilize the momentum gained from the extra weight to cut through heavy loads of scrap material, saving a considerable amount of energy when compared to an open rotor.

A solid rotor has an advantage over open rotors when cutting through heavyweight plastic material based on the physics principle of momentum. A solid rotor will cut through heavyweight plastic more easily than an open rotor due to the ability of the solid rotor to retain the majority of the initial force of the rotor coming in contact with a piece of plastic material with a heavy cross section.

Rotors have a few different designs for the mounting of knives. A tangentially mounted rotor knife is bolted to the top of the rotor (Fig. 17-4) and takes large cuts from the plastic material. Radially mounted knives are mounted from the bottom of the rotor fin (Fig. 17-5). A common example of a radially mounted rotor is a chipper rotor which is de-

signed to cut through scrap material with heavy cross sections.

In recent years there have been new innovations in rotor and knife design with the angled cut knife concept. The theory is based on the fact that a slicing or scissor cutting action is more efficient than the conventional chopping action. The rotors and the cutting chamber are machined to specified angles to achieve the proper slant for the cutting knives. The rotor itself is made of round and square metal stock that is welded together, then machined to strict specifications or forged and machined from raw stock. The rotor is both dynamically and statically balanced to produce a cylindrical spin. This will ensure that the rotor will spin free of vibration, thus saving the bearings from premature failure and permitting the knives to be set at close tolerances for a more efficient cutting action. Should there be an unacceptable difference, the rotor will have to be corrected to bring the rotor within specifications by welding weights to the rotor fin to make it heavier or by drilling holes in the fin to make that part lighter.

Knives.   The cutting knives are the most important part of the granulator. Cutting knives are made from alloyed tool steels which are a matrix of various types of steel used to form a higher-grade metal product. A typical alloyed steel cutting knife would be made from chrome vanadium steel. Higher grades of

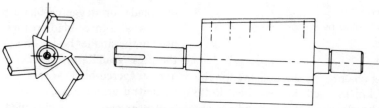

Fig. 17-3   Three-blade solid rotor.

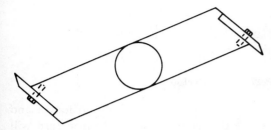

Fig. 17-4   Tangentially mounted rotor knives.

metal alloys are used to create a chrome vanadium steel. Higher grades of metal alloys are used to create a higher wear/toughness index in the knife.

There are different types of knives used with different types of plastic material. A high shear knife is used to granulate energy-absorbing plastics and friable materials. High shear knives have a flat surface on the bottom of the knife with the top cut at a compound angle (Fig. 17-6). High shear knives will cut the plastic material with a slicing action which will reduce the amount of fines produced in the granulating process.

Fifty-degree angle knives are used to granulate high impact plastic. This type of knife

has a reverse angle in addition to the obverse angle. The total of the two angles is 50 degrees, thus giving the knife its name (Fig. 17-7). This knife will cut the plastic with an ax effect, which will cause the material to shatter, rather than the cutting action of its counterpart the high shear knife.

Bed Knives.   The bed knife (Fig. 17-8) is bolted to the cutting chamber with bed knife clamps. The bed knife has elongated holes that permit the bed knife to be moved and adjusted to the proper clearance (Fig. 17-9).

Bed Knife Clamps.   As the name states, bed knife clamps hold the bed knives in place. The bed knife clamps have an angled bottom (Fig. 17-10) so that, once the bed knives and the bed knife clamps have been placed in the cutting chamber, the top of the bed knife clamps will be flat (Fig. 17-11). Figure 17-11 shows how the cutting chamber base is machined to accommodate the slant knife de-

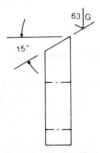

Fig. 17-7   Fifty-degree angle knife.

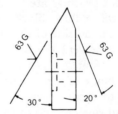

Fig. 17-8   Bed knife (profile).

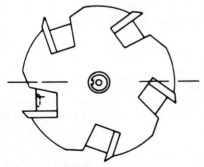

Fig. 17-5   Radially mounted rotor knives.

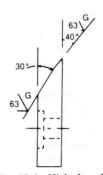

Fig. 17-6   High shear knife.

Fig. 17-9   Bed knife (top view).

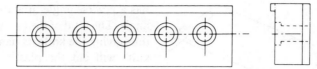

Fig. 17-10   Side and top view of bed knife clamps.

sign, and the manner in which the bed knife clamps are machined to hold the bed knives in place and be flat in the chamber. The bed knife adjusting screws can be turned in and out to set the gap between the bed knives and rotor knives.

Bearings.   The main function of the bearings is to support the rotor and keep it spinning freely during its course of operation. There are two types of bearings on the market today. One type are flanged mount bearings, which are bolted directly to the cutting chamber wall. This style of bearing must be checked and lubricated frequently to keep dust created in the granulating process from entering the bearing and causing premature wear.

Pillow block bearings are the other type of bearings used. Pillow block bearings are mounted on the outside of the cutting chamber. This will prevent the bearings from becoming contaminated. Even with this advantage, pillow block bearings must be checked periodically to ensure that the bearings are lubricated adequately. Granulators can be purchased with a central lubrication system that will make it easier to add grease to the bearings.

Cutting Chamber Assembly

The cutting chamber is assembled in the factory to close tolerances. The bearings are placed on the rotor and bolted to the cutting chamber. The clearance is checked on the rotor to be sure that it will turn freely. The surfaces on the rotor and the cutting chamber base are cleaned of any burrs so that the knives can be mounted on the chamber. If the burrs are not cleaned, the knives will not butt flat to the rotor or in the chamber. Once the cutting action starts and force is exerted on the

knives, it can force the bolts to loosen and cause the knives to slap, which in turn will cause the knives to chip or self-destruct. The rotor knives are then mounted into place and torqued to an ASME specification.

The torque specifications demonstrate the close tolerances used in the assembly and preventive maintenance of granulators. The operation and instruction manual for a particular granulator must be consulted for specific specifications to ensure that the machine will be properly serviced using the proper tools and procedures.

The bed knives are placed in the cutting chamber base, followed by the bed knife clamps and bolts. The bolts are snugged down to prevent the bed knives from moving easily. A gap of approximately .006 in. is now set between the bed knives and the rotor knives by moving the bed into position with the bed knife adjusting screws. The bed knives are torqued to an ASME specification.

Hard Face Welding

Granulating reinforced plastics can cause some serious wear problems in granulators. It has been discovered that reinforced plastics have a tendency to wear away cutting chamber walls and rotors quite rapidly. In order to combat this problem, cutting chambers and rotors can be purchased with a hard face welding option.

In hard face welding, a thick welded stellite bead is run horizontally and vertically on the cutting chamber walls and on the rotor to absorb most of the abuse from the reinforced plastics. When reinforced plastics are granulated, they will wear away the welded beads and not the cutting chamber walls or rotors. When the welded beads wear away, the cutting chamber can be sent back to the manufacturer where the hard face welding can be reapplied.

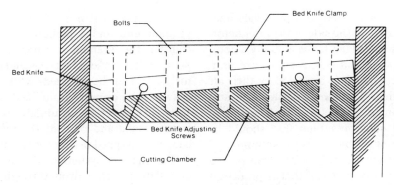

Fig. 17-11   Cross section of a cutting chamber.

Screen Chambers

Screen chambers consist of two parts: the screen cradle and the screen itself. The screen cradle is made from heavy-gauge metal stock that is welded together to fit on the top of the base. The center section of the cradle that supports the screen is machined to tight tolerances, permitting a snug fit between the screen and the cradle, thus preventing oversized material from leaking into the base.

The sole purpose of the screen is to classify the plastic granulate. This is simply done by not permitting any granulate that is larger than the screen hole size to fall through to the granulator base.

Screens are constructed of a metal stock that has holes drilled through the stock, then is rolled to its radial shape. Screen holes will range in size from $\frac{1}{8}$ in. up to 3 in. in diameter.

Most small granulators will have one screen that can be easily removed by one person. Larger granulators are constructed with two screens or three screens that are placed in tandem, permitting ease in removal and installation into the screen cradle of the granulator.

Screens are also available in two different types: cold rolled steel and stainless steel. The cold rolled steel is the standard screen offered in granulators and is an excellent screen for most plastics. Stainless steel screens are recommended for granulating thick walled components that require a greater toughness index.

Stainless steel screens are occasionally used in the medical and food industry because stainless steel resists corrosion and prevents the plastic granulate from becoming contaminated.

The Base

The base of a granulator could be used as a collection point utilizing a drawer or with an opening permitting a barrel or a gaylord to be placed under the machine to collect the plastic regrind. The base can also serve as a transfer point for a pneumatic dilute phase system employing either a material handling fan or a vacuum system.

Smaller granulators will have casters on the bottom of the machine, offering greater mobility of the unit. The large central station granulators are stationary machines.

The standard beside-the-press granulator base comes with a drawer that permits the granulator to be emptied manually. The drawer, like the rest of the base, is constructed of heavy sheet stock which is welded together.

As previously mentioned, granulator bases can be purchased with options enabling the granulator to be placed in an automated system so that the plastic material can be transported to another point in the plastics plant or directly back into a molding machine. The two methods of transporting the material are with a material handling fan or with a vacuum takeoff system.

The material handling fan system consists of a welded transition within the base. This is called a pneu-vey transition. The pneu-vey transition base is bolted to the bottom of the screen cradle.

The base is connected to a materials handling fan located on the side of the granulator. The material conveying fan is actually a combination negative and positive pressure unit. Material conveying fans are sized by their air handling rating. The fans will range in size from 100 cfm to 2000 cfm plus. The size of the material conveying fans that should be used is determined by the type, volume, and density of the material that will be transported and the amount of material that is processed hourly, along with the distance that the material will have to be conveyed.

The material handling system works simply by having the scrap plastic material drop through the screen and into the transition discharge. The material conveying fan forms a negative pressure in the transition base. The regrind drops into the air stream and is conveyed through tubing to a cyclone separator, which, in turn, separates the material from the air.

A vacuum system makes use of a separate vacuum unit and a pickup transfer tube to transfer the granulate material to another point in the plastics plant. A standard granulator base can be utilized with a modified drawer. The drawer is modified with sloping sides that direct the granulate to the pickup transfer tube and prevent the material from becoming trapped in the corners of the drawer.

The vacuum system works by having the granulate material drop through the screen to a modified storage drawer. The material is sucked through the pickup transfer tube and transferred to a holding point or to a molding machine by the vacuum unit.

Drive Systems

The drive system for granulators consists of a motor and two sheaves set up for a V-belt system. Motors operate on three phase current for 208, 230, 460, or 575 volts. The motors

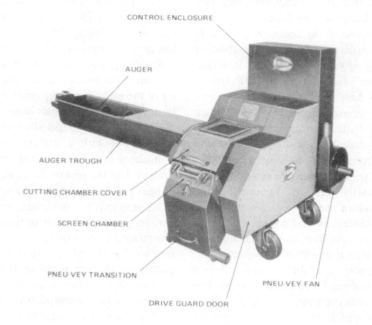

CONTROL ENCLOSURE

AUGER

AUGER TROUGH

CUTTING CHAMBER COVER

SCREEN CHAMBER

PNEU VEY TRANSITION

PNEU-VEY FAN

DRIVE GUARD DOOR

The auger granulator has five basic parts.
1. The auger
2. The cutting chamber
3. The screen
4. The screen cradle
5. The transition base

Fig. 17-12　Auger granulator.

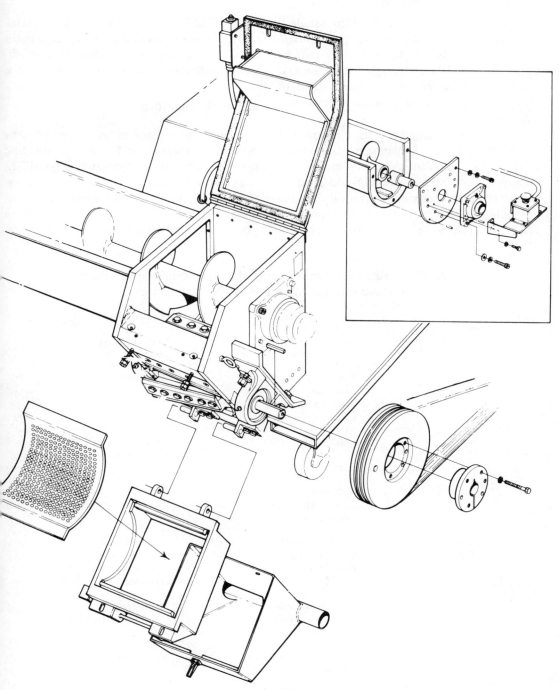

Fig. 17-13 Assembly view of an auger granulator.

have a drive sheave attached to the motor shaft. The driven sheave is attached to the rotor shaft.

The larger sheave can be replaced with a flywheel, which could give the rotor more iner-

tia for cutting heavily reinforced components. Some of the larger central granulators utilize a dual drive system.

Granulators are designed to handle plastics from different molding operations. Conventional granulators are designed to handle a variety of plastic materials. The plastic material falls directly on the rotor knives to be granulated and falls through the screen and into the base, which can be either a standard bin barrel, gaylord, or pneumatic conveying system.

The involute mounted cutting chamber is primarily used for granulating plastic material that is blow-molded, rotational-molded, or slush-molded. This chamber is designed so that the scrap material is dropped into the granulator and directed to the downstroke of the rotor so that the scrap has the effect of

being pulled into the machine. These granulators are commonly found with only pneumatic conveying systems due to their greater efficiency compared to conventional granulators.

Auger Granulators

The auger granulator is an under-the-press granulator designed to fit under a molding machine. The auger granulator (Figs. 17-12 and 17-13) combines a rotating auger screw that conveys the sprues and runners to a granulator at the end of the screw. The plastic is granulated in the cutting chamber and evacu-

Table 17-1   Specifications listing standard equipment and the optional equipment available for granulators.

ITEM	STANDARD	OPTIONAL
Hopper	Front feed with top clean-out door	Sheet chute, pipe chute, feed rolls
Cutting Chamber	Welded construction, slant knife seat-safety switch for electrical lockout	Wear resistant lining
Base	Caster mounted, bin type-optional front or rear bin removal	Pneu-Vey with 37″ base for barrel discharge
Rotor	High shear two blade slant design. One piece construction	Three blade slant design—open type rotor
Rotor Knives	(2) Chrome vandium alloy steel—(4) for Model G-12295M1	(2 + 3) Micro-Temp, V-7 Alloy, V-10 alloy—(4 + 6 for Model G12295M1)
Bed Knives	(2) Chrome Vandium alloy steel—(4) for Model G12295M1	Micro-Temp V-7 Alloy, V-10 Alloy
Screen	Choice of $\frac{1}{4}$, $\frac{5}{16}$ or $\frac{3}{8}$ inch dia. holes	$\frac{3}{16}$ or $\frac{1}{2}$ inch dia. holes
Motor	3 to 40 hp ODP Lincoln See model specifications	To customer specifications
Starter	Magnetic across-the-line in compliance with national electrical code	To customer specifications—115 volt controls
Drive	V-Belt	Flywheel

ated from the machine by vacuum or fan and transported to a collection bin or hopper to be blended with virgin resin.

The reverse flight reverses the rotation of sprue/runner movement, coupled with the involute rotor mounting to wipe clean the sprue/runner from the auger.

The cutting chamber is manufactured the same way as the standard granulator with a cutting chamber base, bed knives, bed knife clamps, the rotor, and rotor knives. All of the rotors that are used in auger granulators are solid three-blade slant rotors. The screen and screen chamber also function in the same capacity as the standard granulator, which permits the regrind to pass through the screen. Screen hole sizes for auger granulators will vary from $\frac{1}{4}$ in. to $\frac{3}{8}$ in.

The drive system for the auger granulator consists of two separate drive systems, one for the auger and the other for the granulator. The auger drive system utilizes a $\frac{1}{2}$ or $1\frac{1}{2}$ hp gear motor combined with two sprockets and heavy duty chains to drive the auger. The granulator is driven by a 2 or 5 hp motor via V-belts and sheaves. The energy source for the material handling removal fan can be directly coupled to the granulator drive in order to conserve energy and space.

The auger granulator can be altered to serve as a parts separator. The auger trough is modified to have an opening at the base to permit parts to drop through the trough. The runners and the sprues will be conveyed by the auger and carried to the granulator to be granulated in the same fashion as a standard auger granulator. The auger itself, when used in an auger separator, is available in different lengths along with different spacing between the screw fins to fit the needs of a particular plastics plant.

Auger granulators and auger separators are commonly used in molding operations that employ three-plate molds. Auger separators and auger granulators have also been com-

bined with robots for a new dimension in efficiency for separating parts from the runners and sprues. Auger granulators and auger sorters can also be purchased with hard face welding for granulating reinforced plastics.

Granulator Sizing

The selection for a granulator is, for the most part, centered around the plastic material that has to be granulated (Table 17-1). The first consideration would be to examine the molding process to determine what type of granulator should be utilized. For an injection molding machine, an under-the-press granulator in the form of a auger granulator may be used. A central granulator or a beside-the-press granulator can also be used for this type of process.

Another factor to be considered would be the production rate over one hour, commonly called "thruput." The hourly production rate is determined by the size of the cutting chamber and the rotor design.

The actual physical properties of the plastic material must be examined. The size of the part will have to be checked to ensure that the hopper and the cutting chamber will be able to accept that particular configuration. The basic properties of the plastic materials must be examined to determine if the material is a high-impact, energy-absorbing, or friable material so that the proper knives can be installed in the granulator. The size of the granulate must be determined so that the proper screen can be chosen for the machine, and a decision must be made on what type of material removal will be utilized with the granulator.

REFERENCE

1. Butzen, G. A., and Adams, G. A., Engineering and Application Manual for AEC Companies—Size Reduction Division, AEC Corp., Elk Grove, IL, 1982.

Section VI
PLASTIC MOLDING MATERIALS

Chapter 18
Plastic Molding Materials—Overview

D. V. Rosato

INTRODUCTION

Plastics may be made hard, elastic, rubbery, tough, crystal-clear, opaque, strong (see Fig. 18-1), stiff, outdoor-weather-resistant, electrically conductive, or practically anything that is desired, depending on the choice of starting materials and the method of molding. This chapter will review some of the properties and processing techniques of plastics as they relate to meeting the performance requirements after injection molding. Not all plastics will be reviewed. Information presented shows the typical different properties that can be obtained, based on the plastic used and how it is molded. As an example, a specific plastic can be molded using different injection molding machine settings so that dimensional tolerances on a part can vary after each molding, or the machine can be set so that extremely close tolerances can be met repeatedly. Chapter 19 provides important information on "molding variables vs. property responses" based on using ABS. However, the information presented is applicable to all plastics. (See references 1–23.)

Basically certain (very few) injection molded parts can be held to extremely close tolerances of less than a thousand of an inch or down to 0.0 percent. Tolerances that can be met can go from 5 percent for 0.020 in. thick, to 1 percent for 0.500 in., to $\frac{1}{2}$ percent for 1.000 in., to $\frac{1}{4}$ percent for 5.000 in., etc.

Economical production requires that tolerances not be specified tighter than necessary. However after production target is to mold to as tight as possible to be more profitable. Recognize that many plastics after molded change dimensions due to temperature, humidity and/or load. Heat treatment can significantly reduce or even eliminate these changes for certain plastics.

Dimensional accuracy that can be met depends on different factors, such as accuracy of mold and machine performance, properties of materials, operation of the complete molding cycle, wear or damage of machine and/or mold, shape/size/thicknesses of part, post shrinkage that can reach 3 percent for certain materials, degree of repeatability in performance of machine/mold/material, etc. (see Table 18-1).

Thermosets generally are more suitable to meet the tightest tolerances. With thermoplastics it can be more complicated. As is well known crystalline plastics (PE, PP, etc.) generally have different rates of shrinkage in the longitudinal and transverse direction of melt flow. In turn (but not recognized by most molders) these directional shrinkages can significantly vary due to changes in injection pressure, melt heat, mold heat and part thickness or shape. The changes can occur at different rates in the different directions. To minimize and control tolerances consider using the highest melt heat, keep gate surrounding area

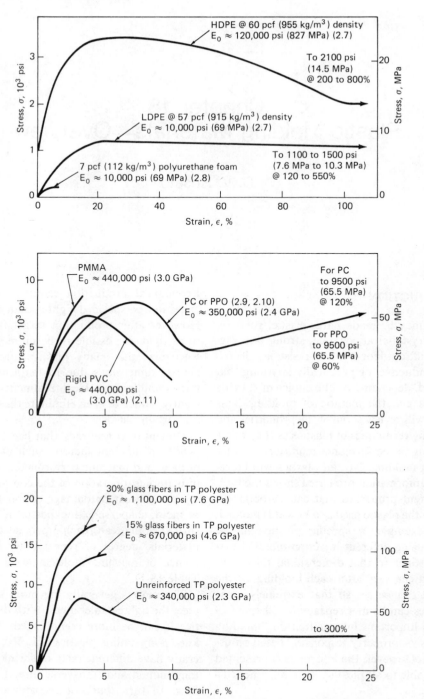

Fig. 18-1 Basic stress-strain relationships for several structural thermoplastics.

where tight tolerances are required, use machine that requires at least 70 percent of shot capacity, minimize time that melt is in the barrel and understand complete operation of machine/mold/material to ensure part tolerance repeatability. Not every material is suitable for molded parts requiring tight tolerances.

To understand plastics, one must first appreciate and accept the polymer chemist's

Table 18-1.   Parameters that influence part tolerance

PART DESIGN:	Part configuration (size/shape), relate shape to flow of melt in mold to meet performance requirements that should at least include tolerances.
MATERIAL:	Chemical structure, molecular weight, amount & type of fillers/additives, heat history, storage handling.
MOLD DESIGN:	Number of cavities, layout & size of cavities/runners/gates/cooling lines/ side actions/knockout pins/etc., relate layout to maximize proper performance of melt & cooling flow patterns to meet part performance requirements, preengineer design to minimize wear & deformation of mold (use proper steels), layout cooling lines to meet temperature to time cooling rate of plastics (particularly crystalline types).
MACHINE CAPABILITY:	Accuracy & repeatability of temperature/time/velocity/pressure controls of injection unit, accuracy & repeatability of clamping force, flatness & parallelism of platens, even distribution of clamping on all tie rods, repeatability of controlling pressure & temperature of oil, minimize oil temperature variation, no oil contamination (by the time you see oil contamination damage to the hydraulic system could have already occurred), machine properly leveled.
MOLDING CYCLE:	Setup the complete molding cycle to repeatedly meet performance requirements at the lowest cost by interrelating material/machine/mold controls.

ability to literally rearrange the molecular structure of the plastic or polymer to provide an almost infinite variety of compositions that differ in form, appearance, properties, and characteristics.

One must also approach the subject with a completely open mind that will accept all the contradictions that make it so difficult to pin common labels on the different families of plastics—or even on the various types within a single family. In the family of polyethylenes, for example, consumers are most aware of the so-called low-density polyethylenes, which are flexible materials most familiar in housewares, toys, trash bags, film overwraps, and the like. But there is another type of polyethylene, called high-density polyethylene, that is rigid and tough enough to be used in the manufacture of materials-handling pallets that can support thousands of pounds of stacked-up boxes. To compound the confusion even more, high-density polyethylene can also be produced in a flexible film form with properties quite different from low-density polyethylene film. Many of today's supermarkets use shopping bags based on high-density polyethylene film.

Finally, to fully understand plastics, one must be aware of the many different routes that the starting materials for plastics can take on the way to consumer or industry. Here

we are concerned primarily with those resins (or polymers) that are supplied to the processor in the form of granules, powder, pellets, flakes, or liquids and are transformed by him into solid or cellular plastics products, shapes, film or sheeting, or coatings and surfaces for various substrates.

However, the same starting materials used to make these resins can take another route and end up in the textile industry (nylon fibers share common roots with a molded nylon gear, acrylic fibers share common roots with acrylic sheet for glazing, etc.), the paint industry (the alkyd paints you use in your home and the alkyd resins that are processed into solid products for the electrical industry also have much in common), and the adhesives industry (the epoxy adhesives that you buy in your local hardware store are first cousin to the epoxy resin binders used for industrial reinforced plastic products).

Many plastics derive from fractions of petroleum or gas that are recovered during the refining process. For example, ethylene monomer (one of the more important feedstocks or starting materials for plastics) is derived, in a gaseous form, from petroleum refinery gas or liquefied petroleum gases or liquid hydrocarbons. Although petroleum or gas derivatives are not the only basic source used in making feedstocks for plastics, they are among the most popular and economical in use today. Coal is another excellent source in the manufacturing of feedstocks for plastics, and there are other materials—including such unique possibilities as agricultural oils such as castor oil or tung oil derived from plant life—that are also adaptable.

From these basic sources come the feedstocks we call monomers (that is, a small molecule or, from the prefix "mono," a single "mer"). The monomer is then subjected to a chemical reaction known as polymerization that causes the small molecules to link together into ever-increasing longer molecules. Chemically, the polymerization reaction has turned the monomer into a polymer (or, from the prefix "poly," many "mers"). The transformation from monomer to polymer may also help you understand why the names of so many plastics materials begin with the prefix "poly." Thus, a polymer may be defined as a high-molecular-weight compound that contains comparatively simple recurring units.

Outside of the plastics field, monomers can take different routes to produce a variety of other important products—from antifreeze to fertilizers. Even within the plastics field, a single monomer can contribute to the manufacture of a variety of different polymers, each with its own distinctive characteristics. When styrene monomer is polymerized, it becomes a styrene polymer or polystyrene, as it is more familiarly known in its plastic form. By a more direct route, the ethylene monomer can be polymerized to produce ethylene polymer or polyethylene, another popular plastic.

Basically, there is a great deal of flexibility in the plastic manufacturing process for creating a wide range of materials. The way in which the small molecules link together into larger molecules and the structural arrangement they take (e.g., packed closely together or separated by side protrusions or branches) is one determinant of the properties of the plastic. The length of molecules in the polymer chain is a second. The type of molecule is a third (e.g., substituting methyl groups or benzene rings or chloride atoms for hydrogen atoms, etc.). Polymerizing two or more different monomers together (a process known as co-

polymerization) is a fourth. And incorporating various chemicals or additives during or after polymerization is a fifth. Other modifying techniques are in use, and polymer chemists continue to come up with new ones.

The polymer or plastic resin must next be prepared for use by the processor, who will turn it into a finished product. In some instances, it is possible to use the plastic resin as it comes out of the polymerization reaction. More often, however, it goes through other steps that put it into a form that can be more easily handled by the processor and more easily run through processing equipment. The most popular solid forms for the plastic resin are as pellets, granules, flakes, or powder. In the hands of the processor, these solids are generally subjected to heat and pressure, melted, forced into the desired shape, then allowed to cure and set into a finished product. Plastics resins are also available as semi-solids (e.g., pastes) or as liquids, for casting.

Liquids can also be used to impregnate fibrous materials, which can then be allowed to harden into so-called reinforced plastics.

Another option available to processors is to use resin incorporating a blowing agent. Subjected to the heat of processing or the heat from the chemical reaction, these agents decompose and release gases that can turn a solid product into a foamed product (i.e., the gases create bubbles or cells within the plastic that impart a cellular construction).

No matter what direction is taken, however, it is important in understanding plastics to understand the general flow from basic feedstock to end-product—while also accepting the fact that there are probably exceptions to the rule all along the way (e.g., some monomers do not have to be polymerized before being processed into an end-product; rather, they can be cast to shape and polymerized in situ, that is, while the end-product is being shaped). In most cases, however, the flow will proceed along these lines: feedstocks, known as monomers, are polymerized by chemical reaction into polymers (or plastics resins); the resins are then made into forms useful for processing, sometimes called molding or extrusion compounds (this step can also involve the incorporation, or compounding in, of additives, plasticizers, modifiers, colorants, reinforcements, etc.,); the compound now goes to the processor, who has several techniques available to him for turning the resin into a finished product or part, into a secondary product (such as sheet, rod, tube shapes, etc.) that goes through subsequent fabricating operations, or into a coating or surfacing that can be applied to various substrates. At the processing level, plastics can also be turned into monofilaments for use in rope or household screening; or as binders, they are used with materials such as fibers (in "reinforced plastics") or sheets of paper (in "laminates") or sheets of wood (in "plywood") to turn out products such as boat hulls, tabletops, or airplane wingtips.

Definition for Plastics.   There is a generally accepted definition for plastics that goes like this: any one of a large and varied group of materials consisting wholly or in part of combinations of carbon with oxygen, hydrogen, nitrogen, and other organic and inorganic elements that, while solid in the finished state, at some stage in its manufacture is made liquid, and thus capable of being formed into various shapes, most usually (although not necessarily) through the application, either singly or together, of heat and pressure.

Plastics are a family of materials, not a single material, each member of which has its own distinct and special advantages. (See Table 18-2 for typical names of materials in the plastic families.) Whatever their properties or form, however, most plastics fall into one of two groups—the *thermoplastics* and *thermosets.*

Thermoplastic resins consist of long molecules, either linear or branched, having side chains or groups that are not attached to other polymer molecules. Thus, they can be repeatedly softened and hardened by heating and cooling. Usually, thermoplastic resins are purchased as pellets or granules that are softened by heat under pressure so they can be formed, then cooled, so that they harden into the final desired shape. No chemical changes generally

take place during forming. The analogy would be to a block of ice that can be softened (i.e., turned back into a liquid), poured into any shape of cavity, then cooled to become a solid again.

In thermosetting resins, reactive portions of the molecules form cross-links between the long molecules during polymerization. The linear polymer chains are thus bonded together to form a three-dimensional network. Therefore, once polymerized or hardened, the material cannot be softened by heating without degrading some linkages. Thermosets are usually purchased as liquid monomer–polymer mixtures or as a partially polymerized molding compound. In this uncured condition, they can be formed to the finished shape with or without pressure and polymerized with chemicals or heat. The analogy in this case would be to a hard-boiled egg, which has turned from a liquid to a solid and cannot be converted back to a liquid.

WELD LINE STRENGTHS AND MATERIALS

Weld lines can develop during molding, particularly if improper design of the part occurred. Different plastics can cause the problem. This review concerns tests conducted on specific materials by Monsanto Polymer Products Co. (9). Complex polymer (plastics) systems were used and evaluated.

This work involves the investigation of weld line strength of several rubber modified flame retardant polymers which include two styrene maleic anhydride (SMA) materials (one natural and one pigmented), three modified polyphenylene oxide (PPO) polymers, and a flame-retardant (FR) ABS (Table 18-3). Many other complex polymer systems that contain relatively high levels of plasticizers or inert solids such as pigments and fillers may display weld line performance like that described here.

In all but the simplest injection molding configurations, two or more melt streams will combine to form what is known as weld or knit lines. A weak line theory suggests that strength in the weld line region is more important than bulk material properties. A good

Table 18-2   Plastic families.

Acetals
Acrylics
  Polymethylmethacrylate
    (PMMA)
  Polyacrylonitrile
Alkyds
Alloys
Allyls
  Diallyl phthalate
  Diallyl isophthalate
Amino
  Urea formaldehyde
  Melamine formaldehyde
Cellulosic
  Cellulose acetate
  Cellulose triacetate
  Cellulose nitrate (celluloid)
  Ethyl cellulose
  Cellulose acetate propionate
  Cellulose acetate butyrate
  Hydroxypropyl cellulose
  Cellophane
  Rayon
Chlorinated polyether
Coumarone-indene
Epoxy
Fluorocarbon
  Polytetrafluoroethylene (PTFE)
  Fluorinated ethylene propylene
    (FEP)
  Copolymer perfluoroalkoxy
    (PFA)
  Polychlorotrifluoroethylene
    (PCTFE)
  Ethylene-chlorotrifluoroethylene
    (ECTFE)

Ethylene-tetrafluoroethylene (ETFE)
Polyvinylidene fluoride (PVDF)

Polyvinyl fluoride (PVF)
Furan
Hydrocarbon resins
Nitrile resins
Polyaryl ether
Polyaryl sulfone
Phenol-aralkyl
Phenolic
Polyamide (nylon)
Poly (amide-imide)
Polyaryl ether
Polycarbonate (PC)
Polyester
  Aromatic polyester
  Thermoplastic polyester
    Polybutylene terephthalate (PBT)
    Polytetramethylene terephthalate
      (PTMT)
    Polyethylene terephthalate (PET)
  Unsaturated polyester (SMC, BMC)
Polyimide
  Thermoplastic polyimide
  Thermoset polyimide
Polymethyl pentene
Polyolefins (PO)
  Low-density polyethylene (LDPE)
  Linear low-density polyethylene
    (LLDPE)
  High-density polyethylene (HDPE)
  Ultra high molecular weight
    polyethylene (UHMWPE)
Polypropylene (PP)
Ionomer

Polybutylene (PB)
Polyallomers

Polyphenylene oxide
Polyphenylene sulfide (PPS)
Polyurethanes
Poly p-xylylene
Silicones
  Silicone fluids
  Silicone elastomers
  Rigid silicones
Styrenics
  Polystyrene (PS)
  Acrylonitrile butadiene styrene
    (ABS)
  Styrene acrylonitrile (SAN)
  Styrene butadiene latricies
  Styrene base polymers
Sulfones
  Polysulfone
  Polyether sulfone
  Polyphenyl sulfone
Thermoplastic elastomers
Vinyl
  Polyvinyl chloride (PVC)
  Polyvinyl acetate
  Polyvinylidene chloride
  Polyvinyl alcohol
  Polyvinyl butyrate
  Polyvinyl formal
  Propylene–vinyl chloride
    copolymer
  Ethylvinyl acetate
  Polyvinyl carbazole

Table 18-3

MATERIAL	MEASURED HEAT DISTORTION AT 264 PSI, °C	SIDEWINDER FLOW AT MAX. RECOMMENDED STOCK TEMP.		FLAMMABILITY UL 94 AT 1.5 MM
		CM	°C	
FRABS				
Natural	86	33	238	V-0
Modified SMA				
Natural	101	34	249	V-0
Grey	102	33	249	V-0
Modified PPO				
A	89	43	288	V-0
B	99	36	288	V-1
C	103	26	288	V-0

deal has been written on how to adjust molding conditions to optimize weld line strength (10–13). It has also been shown that materials selection is very important if good weld line strength is desired (14–16).

Previous work emphasized the need to mold at high stock temperature to obtain good molecular entanglements across the weld line. Most of the previous work was done on relatively pure polymer systems. The presence of large concentrations of solid and liquid additives can greatly influence the formation of entanglements across the weld line interface and therefore affect weld line strength.

The objective of this study was to define variation in weld line strength with stock temperature for several complex polymer systems and identify a possible cause of the variation.

Tensile and tensile impact strength were used to study weld lines. The two tests measure different characteristics of weld lines. The tensile impact test was used to identify possible product weakness that cannot be seen from a routine tensile test run on samples containing weld lines. Dimensions of test specimens are shown in Fig. 18-2.

Electron micrographs were taken of a number of weld line profiles using a procedure developed by Warren Farnham (18). These micrographs show that weld line depth can vary considerably with small stock temperature changes.

Information on strength vs rate of testing is given in Table 18-4.

Figure 18-3 illustrates how the percentage of weld line breaks changes as molding temperature is changed. Dashed lines show where samples were molded above their recommended temperature limit. In all cases where impact weld line breaks occurred with a 5 cm³/sec injection rate, better results were seen with a 2.5 cm³/sec injection rate if the sample filled the mold at the lower rate.

Figure 18-4 shows absolute weld line strength as a function of molding temperature. Except for the PPO-C sample, all materials have a similar average weld line strength. However, when parts fail, it is usually the "weak link" that fails. Figure 18-5 shows weld line impact strength as a percentage of non–weld line impact strength vs. molding temperature. The FRABS, PPO-A, and PPO-B samples show weld line impact strength only 10 to 30 percent of non–weld line impact strength. This means that for these three materials, some parts may be produced with very weak weld lines.

The SMA samples show very good weld line strength at molding temperatures at or below the recommended maximum, while the PPO-C samples show good weld line strength at the two temperatures studied. It should be kept in mind that the PPO-C sample would not fill the mold until stock temperature was

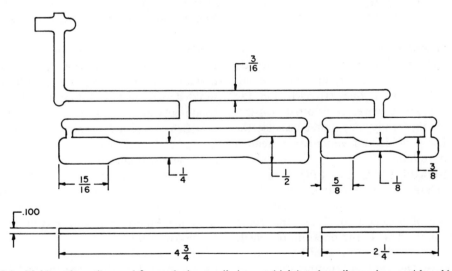

Fig. 18-2 Mold configuration used for producing tensile impact (right) and tensile specimens with weld lines.

Table 18-4 Weld line strength of FR polymers.

MATERIAL	SAMP. NO.	STOCK TEMP., °C	INJECTION RATE, CM³/SEC	TENSILE STRENGTH, MPA
FRABS	1	221	5	33.9
	1A	221	2.5	33.1
	2	241	5	32.4
	2A	241	2.5	33.9
SMA	6	236	5	26.8
natural	6A	236	2.5	26.8
	5	246	5	26.8
	5A	246	2.5	26.8
	9	260	5	28.9
SMA	7	237	5	26.5
grey	7A	237	2.5	26.5
	8	243	5	27.8
	8A	243	2.5	27.8
	10	260	5	29.6
	10A	260	2.5	30.3
PPO-A	11A	252	2.5	
	13	266	5	35.0
	13A	266	2.5	35.2
	15	274	5	35.0
	15A	274	2.5	35.2
	18	287	5	
PPO-B	16	277	5	32.0
	16A	277	2.5	
	19	289	5	32.1
	19A	289	2.5	32.1
PPO-C	20	290	5	—
	21	304	5	46.6
	21A	304	2.5	44.0
	22	312	5	46.8
	22A	312	2.5	46.8

increased well above the manufacturer's recommended stock temperature. Apparently the high molding temperature did not adversely affect weld line strength for the PPO-C sample. However, some other property outside of this study may suffer because the material was molded above its recommended temperature limit.

The FRABS, SMA, and PPO-A samples show a relatively low percentage of weld line breaks at low molding temperature and a high percentage of breaks at or above the upper molding limit. This is the opposite of tensile strength data, which show weld line strength to improve the molding temperature. For SMA and PPO-A samples, it appears that weld line breaks decrease at low molding temperature (Fig. 18-3). This seems probable based on previous studies. To identify the cause of this phenomenon, transmission electron micrographs were taken of a cross section of SMA samples 7, 8, and 10. Some very large differences were seen (Table 18-5 and Figs. 18-6 and 18-7). The notch depth value is approximate because some weld lines are irregular (Fig. 18-8). The deep 240-micron crack seen at the lowest molding temperature (sample 7) seems to explain why that sample had poor weld line strength.

Several electron micrographs (Fig. 18-6 SMA and Fig. 18-9 PPO samples) show that rubber particles are severely elongated in a direction parallel to the weld line. This effect will reduce impact resistance to fracture propagation parallel to and near the weld line.

The PPO-B sample showed 100 percent weld line breaks at both molding temperatures, while the higher melt viscosity sample PPO-C showed only 20 to 30 percent breaks (Fig. 18-3). Since notch depth correlates with high melt viscosity, poor weld line strength seen with the PPO-B sample at all molding temperatures and the FRABS, SMA, and PPO-A samples molded at or above the upper temperature limit must be due to something other than notch depth. Even chain entanglement and morphological differences should be changing in a direction that gives better weld line strength at high molding temperature (12, 20).

Some complex polymer systems are known to change composition when processed under certain conditions. Deposits of additives on molded part surface (bloom) and in molds (plate-out) are indications of this phenomenon. Samples displaying strong and weak weld line strength were analyzed under a scanning electron microscope to determine whether differences could be observed. Figure 18-10 is a scanning electron micrograph (SEM) taken of sample 7A (Table 18-4), which had a strong weld line. Figure 18-11 shows sample 9, which had a low weld line strength.

The reason for poor weld line strength can be seen from Fig. 18-11. At the high molding

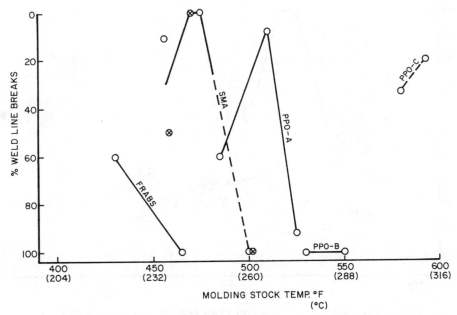

Fig. 18-3   Tensile impact, % breaks at weld line vs. stock temperature, 5 cm³/sec injection rate.

temperature, additives concentrate at the part surface and the weld line interface because of a phenomenon called fountain flow. This fluid flow behavior was described in a paper by R. L. Hoffman (21).

The presence of particles at the weld line will block movement of polymer molecules across the interface by what has been called repetition (22). The lack of chain entanglements across the interface will lead to reduced strength as shown by Jud, Hausch, and Williams (23).

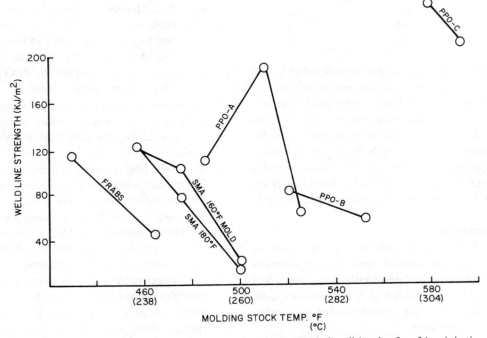

Fig. 18-4   Effect of stock temperature on average tensile impact strength for all breaks, 5 cm³/sec injection rate.

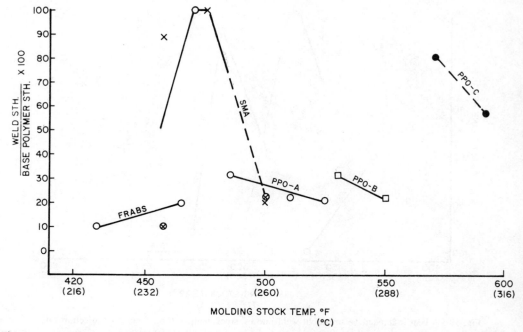

Fig. 18-5 Effect of stock temperature on tensile impact weld line strength as % base polymer strength, 5 cm³/sec injection rate.

Table 18-5 Depth of weld line of various SMA samples.

SAMPLE	MOLDING TEMP., °C	DEPTH OF NOTCH, MICRONS
7	237	240
8	243	5
10	260	0

Thermoplastics need to be molded at a high enough temperature so that molecules can replicate across weld lines giving good strength. For polymers such as flame retardant materials which contain significant amounts of solid additives, weld strength increases with melt temperature only to a point. Then particles concentrate at the interface, blocking chain movement and resulting in poor weld strength.

Conclusions

1. For complex polymer systems such as the ones studied here, impact strength at the weld line usually decreases when they are molded at a temperature near or above the recommended maximum temperature.

2. Material characteristics other than melt viscosity can have a great influence on weld line strength. Poor weld line impact strength seen during molding, especially at high temperature, is likely due to compositional changes at the weld line.

3. The rubber-modified SMA polymer is more apt to give good weld line strength when molded within its recommended temperature range than are the PPO and FRABS polymer samples studied.

4. An increased injection rate above some limit is apt to reduce impact strength at the weld line for high heat FR polymers.

5. The weld line strength of rubber-modified polymers is better characterized with impact testing than tensile testing.

6. The weld line notch depth varies greatly with mold temperature for a given polymer. A very deep notch will significantly reduce strength at the weld line.

7. In impact-modified polymers, rubber particles are apt to be compressed in the immediate region of the weld line so that they are elongated in the direction *paral-*

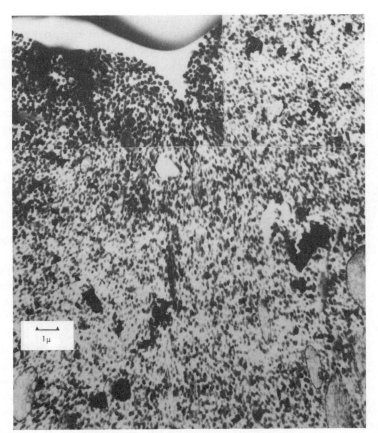

Fig. 18-6   Five-micron-deep SMA weld line from sample molded at 470°F (sample 8). Rubber particles are elongated parallel to weld line.

lel to the weld line. This indicates that the weld line region is highly stressed.

GUIDE FOR IDENTIFICATION OF PLASTIC MATERIALS: D 4000 STANDARD OF ASTM

A "classifying" plastic materials standard that can serve many of the industry needs has been issued by ASTM. This standard is designated as D 4000 and entitled "Standard Guide for Identification of Plastic Materials." It provides an easy means of identifying plastic materials used in the fabrication of parts.

Ever since classification systems were adopted many years ago for materials such as 1030 steel and elastomers, there had been an effort to issue this guide. The approach used follows the steel and elastomer unified classification systems of ASTM.

The guide provides tabulated properties for unfilled, filled, and reinforced plastic materials suitable for processing into parts. This standard is required to reduce the growing number of material specifications, paperwork, and man-hours used to ensure that parts of known quality are being produced from commercially available materials. The D 4000 standard will eliminate the many certifications required for the same material that a processor may have to obtain from several vendors for a customer or different customers.

Table 18-6 provides the basic outline that identifies the D 4000 line call-out.

The classification system and subsequent line call-out (specification) is intended to be a means of identifying plastic materials used

Fig. 18-7 Portion of 240-micron-deep SMA weld line molded at 237°C.

in the fabrication of end items or parts. It is not intended for the selection of materials. Material selection should be made by those having expertise in the plastics field after careful consideration of the design and performance required of the part, the environment to which it will be exposed, the fabrication process to be employed, the inherent properties of the material not covered in this document, and the economic factors.

This classification system is based on the premise that plastic materials can be arranged into broad generic families using basic properties to arrange the materials into groups, classes, and grades. A system is thus established that, together with values describing additional requirements, permits as complete a description as desired of the selected material. Note that Tables 18-7–18-10 provide only sections of the complete information contained in D 4000.

The format for this system (Table 18-6) was prepared to permit the addition of property values for future plastics. Plastic materials will be classified on the basis of their broad generic family. The generic family is identified by letter designations to be found in Table 18-7. These letters represent the standard abbreviations for plastics in accordance with Abbreviations D 1600. For example, PA = polyamide (nylon).

The generic family is based on the broad chemical makeup of the base polymer. By its designation, certain inherent properties are specified. The generic family is classified into groups according, in general, to the chemical composition. These groups are further subdivided into classes and grades as shown in the basic property table that applies. The letter

Fig. 18-8 Irregular surface of weld notch region can be seen in the center of this SEM taken of a flame retardant SMA weld line.

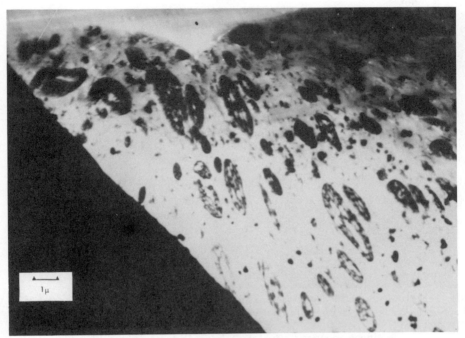

Fig. 18-9   PPO weld line shows oriented rubber particles at weld surface.

designation applicable is followed by a three-digit number indicating group, class, and grade.

The basic property tables have been devel-oped to identify the commercially available unreinforced plastics into groups, classes, and grades. These tables are found in the standards listed in Table 18-7. Where a standard does

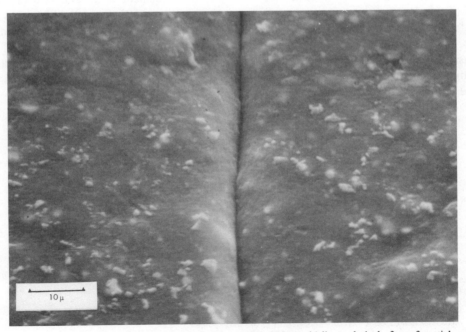

Fig. 18-10   Scanning electron micrograph of strong FR SMA weld line, relatively free of particles.

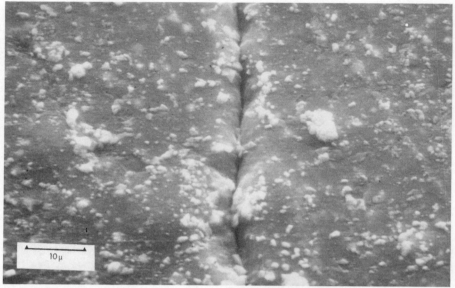

Fig. 18-11 SEM showing weak FR SMA weld line molded above the recommended stock temperature. A large amount of particles can be seen on part surface and in the weld notch.

not exist for this classification system, the letter designation for the generic family will be followed by three O's and the use of the cell table that applies. For example, PI000 would indicate a polyimide plastic (PI) from Table 18-7, with 000 indicating no basic property table and G12360 requirements from Cell Table G (Table 18-10).

To facilitate the incorporation of future materials, or where the present families require expansion of a basic property table, a number preceding the symbol for the generic family is used to indicate that additional groups have

been added to the table. This digit coupled with the first digit after the generic family will indicate the group to be found in the basic property table.

Reinforced versions of the basic material are identified by a single letter that indicates the reinforcement used, and two digits that indicate the quantity in percent by mass. Thus, the letter designation G for glass-reinforced and 33 for percent of reinforcement, G33, specifies a 33 percent glass-filled material. The reinforcement letter designations, with tolerance levels, are shown in Table 18-8.

Table 18-6   ASTM D 4000 line call-out.

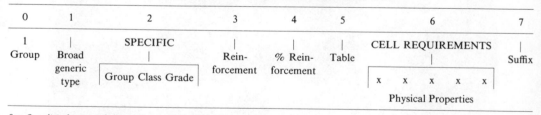

0	1	2	3	4	5	6	7
1 Group	| Broad generic type	SPECIFIC | Group Class Grade	| Rein- forcement	| % Rein- forcement	| Table	CELL REQUIREMENTS | x x x x x Physical Properties	| Suffix

0 = One digit for expanded group, as needed (see 5.2.2).
1 = Two or more letters identify the generic family based on Abbreviations D 1600.
2 = Three digits identify the specific chemical group, the modification or use class, and the grade by viscosity or level of modification. A basic property table will provide property values.
3 = One letter indicates reinforcement type.
4 = Two digits indicate percent of reinforcement.
5 = One letter refers to a cell table listing of physical specifications and test methods.
6 = Five digits refer to the specific physical parameters listed in the cell table.
7 = Suffix codes indicate special requirements based on the application, and identify special tests.

Table 18-7   Standard symbols for generic families with referenced standards and cell tables.

STANDARD SYMBOL	PLASTIC FAMILY NAME	ASTM STANDARD	SUGGESTED REFERENCE CELL TABLES FOR MATERIALS WITHOUT AN ASTM STANDARD	
			UNFILLED	FILLED
ABS	Acrylonitrile/butadiene/styrene	D___		
AMMA	Acrylonitrile/methylmethacrylate		E	
ASA	Acrylonitrile/styrene/acrylate		E	
CA	Cellulose acetate	D 706		
CAB	Cellulose acetate butyrate	D 707		
CAP	Cellulose acetate propionate		E	D
CE	Cellulose plastics, general		E	D
CF	Cresol formaldehyde		H	H
CMC	Carboxymethyl cellulose		E	
CN	Cellulose nitrate		E	D
CP	Cellulose propionate	D 1562		
CPE	Chlorinated polyethylene		F	
CS	Casein		H	H
CTA	Cellulose triacetate		E	D
CTFE	Polymonochlorotrifluoroethylene			
DAP	Poly(diallyl phthalate)		H	H
EC	Ethyl cellulose		E	D
EEA	Ethylene/ethyl acrylate		F	
EMA	Ethylene/methacrylic acid		F	
EP	Epoxy, epoxide		H	H
EPD	Ethylene/propylene/diene			
EPM	Ethylene/propylene polymer		F	D
ETFE	Ethylene-tetrafluoroethylene copolymer			
EVA	Ethylene/vinyl acetate		F	
FEP	Perfluoro (ethylene-propylene) copolymer			
FF	Furan formaldehyde		H	H
IPS	Impact styrene	(See PS)		
MF	Melamine-formaldehyde		H	H
PA	Polyamide (nylon)	D 4066		
PAI	Polyamide-imide		G	G
PARA	Polyaryl amide			
. . . etc., etc. . . .				

To facilitate the identification of new, special, and reinforced materials where basic property tables are not provided in a material specification, cell tables have been incorporated in this document. These tables should be used in the same manner as the cell tables that appear on the material specifications. Although the values listed in the cell tables include the range of properties available on existing materials, users should not infer that every possible combination of properties exists or can be obtained.

The requirements for special or reinforced materials will use the classification system as described by the addition of a single letter that indicates the proper cell table in which the properties are listed. A specific value is designated by the cell number for each prop-

Table 18-8 Reinforcement-filler symbols and tolerances.

SYMBOL	MATERIAL	TOLERANCE
C	CARBON AND GRAPHITE FIBER-REINFORCED	±2 PERCENTAGE POINTS
G	GLASS REINFORCED	±2 PERCENTAGE POINTS
L	LUBRICANTS (i.e., TFE, GRAPHITE, SILICONE, AND MOLYBDENUM DISULFIDE.)	BY AGREEMENT BETWEEN THE SUPPLIER AND THE USER.
M	MINERAL REINFORCED	±2 PERCENTAGE POINTS
R	REINFORCED-COMBINATIONS/ MIXTURES OF REINFORCEMENTS OR OTHER FILLERS/REINFORCE-MENTS.	±3 PERCENTAGE POINTS (BASED ON THE TOTAL REIN-FORCEMENT.)

Table 18-9 Suffix symbols and requirements.

SYMBOL	CHARACTERISTIC
A	Color (unless otherwise shown by suffix, color is understood to be natural)
	Second letter A = does not have to match a standard
	B = must match standard
	Three-digit number 001 = color and standard number on drawing
	002 = color on drawing
B	Not assigned
C	Melting point—softening point
	Second letter A = ASTM D 789 (Fisher-Johns)
	B = ASTM D 1525 Rate A (Vicat)
	C = ASTM D 1525 Rate B (Vicat)
	D = ASTM D 3418 (Transition temperature DSC/DTA)
	E = ASTM D 2116 (Fisher-Johns–high temperature)
	Three-digit number = minimum value °C
D	Deformation under load
	Second letter A = ASTM D 621, Method A
	B = ASTM D 621, Method B
	First digit 1 = total deformation
	2 = recovery
	Second and third digit × factor of 0.1 (deformation) = % min
	1 (recovery)
E	Electrical
	Second letter A = dielectric strength (short-time), ASTM D 149
	Three-digit number × factor of 0.1 = kV/mm, min
	B = dielectric strength (step by step), ASTM D 149
	Three-digit number × factor of 0.1 = kV/mm, min
	D = dielectric constant at 1 MHz, ASTM D 150, max
	Three-digit number × factor of 0.1 = value
	E = dissipation factor at 1 MHz, ASTM D 150, max
	Three-digit number × factor of 0.0001 = value
	F = arc resistance, ASTM D 495, min
	Three-digit number = value
	[Other methods under review, ASTM D 257 and D 1531]
F	Flammability (NOTE 1)
	Second letter A = ASTM D 635 (burning rate)
	000 = to be specified by user
	B = ASTM D 2863 (oxygen index)
	Three-digit number = value %, max

. . . etc., etc. . . .

Table 18-10 Cell table G detail requirements.*

DESIGNATION ORDER NUMBER	PROPERTY	CELL LIMITS									
		0	1	2	3	4	5	6	7	8	9
1	Tensile strength, ASTM D 638, MPa, min[A]	Unspecified	15	40	65	85	110	135	160	185	Specify value
2	Flexural modulus, ASTM D 790, MPa, min[A]	Unspecified	600	3 500	6 500	10 000	13 000	16 000	19 000	22 000	Specify value
3	Izod impact, ASTM D 256, J/m, min[B]	Unspecified	15	30	50	135	270	425	670	950	Specify value
4	Deflection (temperature, ASTM D 648, (1820 kPa), °C, min	Unspecified	130	160	200	230	260	300	330	360	Specify value
5	To be determined	Unspecified	...	...	...	...	...	...	...	...	...

[A] $MPa \times 145 = psi$
[B] $J/m \times 18.73 \times 10^{-3} = ft \cdot lbf/in.$
* Other cell tables are in D 4000.

erty in the order in which they are listed in the table. When a property is not to be specified, a zero is entered as the cell number.

REFERENCES

1. Frados, J., "The Story of the Plastics Industry," Society of The Plastics Industry, Inc., May 1977.
2. Rosato, D. V., and Schwartz, R. T., *Environmental Effects on Polymeric Materials,* Vols. 1 and 2, J. Wiley & Sons, New York, 1968.
3. Rosato, D. V., Plastics Industry/Injection Molding Workbooks, Univ. of Lowell Plastics Seminars, 1983, 1984.
4. Frados, J., *Plastics Engineering Handbook,* Van Nostrand Reinhold Co., New York, 1976.
5. Bikales, N. M., *Encyclopedia of Polymer Technology,* Vol. 1–16, Wiley, N.Y., 1972.
6. Lubin, G., *Handbook of Composites,* Van Nostrand Reinhold Co., New York, 1982.
7. Beck, R. D., *Plastic Product Design,* Van Nostrand Reinhold Co., New York, 1980.
8. Shue, R. S., Walker, J. H., and Brady, D. G., "HMW Injection Molding Polyphenylene Sulfide," *Plast. Engng.* (April 1983).
9. Remaly, L. S., and Sierodzinski, J., "Importance of Weld Line Strength," SPE–Injection Molding Div. Conference, October 20–22, 1983.
10. Hubbauer, P., "Controlling Injection Molding Parameters For Optimum Part Properties," *Technical Papers, 31st ANTEC,* SPE, 523 (1973).
11. Malguarnera, S. C., and Riggs, D. C., "Weld Line Morphology in Injection-Molded General Purpose and High Impact Polystyrene," *Poly. Plast. Technol. Eng., 17,* 193 (1981).
12. Malguarnera, S. C., and Manisali, A., "The Effects of Processing Parameters on the Tensile Properties of Weld Lines in Injection Molded Thermoplastics," *Polym. Engng. Sci., 10,* 586 (1981).
13. Malguarnera, S. C., and Manisali, A., "The Structure and Properties of Weld Lines in Injection Molded Thermoplastics," *Technical Papers, 38th ANTEC,* SPE, 124 (1980).
14. Malguarnera, S. C., "How to Strengthen Weld Lines in Injection Molded Parts," *Plast. Engng., 5,* 35 (1981).
15. Hagerman, E. M., "Weld-Line Fracture in Molded Parts," *Plast. Engng., 10,* 67 (1973).
16. Malguarnera, S. C., and Manisali, A. I., "The Tensile Properties of Weld Lines in Injection Molded Polypropylene," *Technical Papers, 39th ANTEC,* SPE, 775 (1981).
17. Ryan, J. T., and Matsuoka, S., "Development of Rigid PVC Materials for Telephone Applications," *Technical Papers, 32nd ANTEC,* SPE, 700 (1974).
18. Farnham, W. H., "Sample Preparation for Taking Transmission Electron Micrographs of a Polymer Surface Edge" (in preparation).
19. Isherwood, P., Williams, J. G., and Yap, Y. T., "Flow in Injection Molds," Imp. Coll. of Sci. & Technol., London, Engl. Rheology, *Proc.* of the Int. Congr. on Rheol., 8th, publ. by Plenum Press, New York, 37 (1980).
20. Hobbs, S. Y., "Some Observations on the Morphology and Fracture Characteristics of Knit Lines," *Polym. Engng. Sci., 9,* 621 (1974).
21. Hoffman, R. L., paper presented at the 55th annual meeting of the Society of Rheology, Knoxville, Tennessee, October 16–20, 1983.
22. deGennes, P. G., "Repetition of a Polymer Chain in the Presence of Fixed Obstacles," *J. Chem. Phys., 55,* 572 (1971).
23. Jud, K., Kausch, H. H., and Williams, J. G., "Fracture Mechanics Studies of Crack Healing and Welding of Polymers," *J. Materials Sci., 16,* 204 (1981).

Chapter 19
ABS Molding Variable–Property Responses

L. W. Fritch

Borg-Warner Chemicals
Washington, West Virginia

INTRODUCTION

There are many grades of ABS available. Each grade is tailored to provide a given property balance. This allows the finished product designer considerable freedom in selecting a particular grade to meet all the requirements of processing, end use demands, and cost effectiveness. However, overall success depends not only on selecting the correct ABS grade but also on being aware of how molding conditions can affect the mechanical properties and appearance of the finished article. In some situations the effects can be considerable. For example, on a general purpose ABS grade, by varying four molding parameters over typical commercial practice ranges the cross flow izod impact ranged from a low of 2 to a high of 8 foot pounds per inch. Direction of flow is also important; the izod broken *with* flow can be as low as 0.5 when the wrong molding conditions are used. Finally, position on the part can also influence a property such as impact; values can be significantly different near the gate compared to 12 in. downstream.

In short, it is just as important to pick the correct molding conditions as it is to pick the right polymer; both are critical in defining the part properties. The information that follows is to serve as a guide to help the processor select those molding conditions that represent the best set of compromises toward a given balance of end results. Trade-offs are inevitable when dealing with a complex operation such as injection molding. A lot of variables influence the end results, and some variables interact. The total picture is too complex for us always to predict results exactly. However, we will attempt to lay down some useful generalizations about the way things behave *most* of the time, recognizing that each particular part design and processing operation can present peculiar exceptions to these generalizations. To help the processor handle those cases that flaunt the generalizations, we will include information on *why* a given molding variable affects a property. Processors who know not only what can happen but also why can use that knowledge to develop sound approaches to process control on a day-to-day basis, and to work out their own problem-solving routes.

To this end some cause–effect behavior is reviewed before the generalizations for each molding variable and property are presented. Reference is also made to a number of published articles that provide additional background as well as specific experimental data. The reader is encouraged to review and keep up with the literature in this area to increase his or her skills in molding process control. It is good to know how these cause–effect mechanisms work; this knowledge lets you use them to your advantage or to counter unwanted effects that can occur.

MOLDING VARIABLES AND CAUSE-EFFECT LINKS

In order to understand cause–effect links it is helpful to consider the following relationships: *Machine settings* affect *molding variables,* which influence *cause-effect variables,* which determine *part properties* to determine *cause–effect links.* Let's look at each of these four elements in more detail. Notice that a distinction is made between machine settings and molding variables.

Machine settings are such things as:

- Cylinder temperature settings
- Screw rpm and back pressure
- Plunger injection speed
- Absence or presence of a cushion
- Hydraulic pressure during injection and hold
- Boost and hold time
- Mold temperature controller settings

Molding variables are more specific parameters than machine settings. They are related to machine settings, but sometimes in a nonobvious way.

- Melt temperature in the mold
- Melt front velocity in the mold
- Cavity melt pressure
- Mold surface temperature

The distinction between machine settings and molding variables is a necessary one if we are to avoid mistakes in using cause–effect relationships to our advantage (1). It is molding variables, properly defined and measured, not necessarily machine settings, that can be correlated with part properties (2). For example, if one increases cylinder temperatures, melt temperatures do not necessarily also increase. Melt temperature is also influenced by screw design, rpm, back pressure, and dwell times. It is much more accurate to measure *melt* temperature and correlate it with properties than to correlate cylinder settings with properties.

Another example concerns injection rate.

A ram speed of 1 in./sec in a 3-in.-diameter cylinder will produce a much faster mold fill rate than in a 2-in. unit. It is the local melt front velocity in the mold that directly influences properties. Cavity geometry also affects fill rates; thick sections fill more slowly than thin ones. Single-gated molds have local melt front velocities that are faster than with two gates. Thus, while it is not always easy to measure molding variables directly without special instrumentation, one does need to be aware of what is really being influenced (or not) when a machine condition is changed.

What are the cause–effect links that tie molding variables to part properties?

- Orientation
- Polymer degradation
- Free volume/molecular packing and relaxation
- Cooling stresses

The most influential of these four is polymer orientation, often erroneously called molded-in stress (or strain). Orientation warrants some elaboration; it will be covered in the next section, after a few comments on the other three cause–effect links.

Polymer degradation can occur from excessive melt temperatures, or abnormally long time at temperature (i.e., heat history). Very high shear rates can also be a cause. Conditions that create degradation are cylinder, nozzle, and hot runner heaters set too high, and high screw rpm—especially in combination with high screw back pressure. Also look for high melt residence times; shot size too small for the machine capacity, and hangup areas in the barrel end cap, nozzle, tip, and hot runner system. Excessive shear can result from poor screw design (3), too much screw flight to barrel clearance, cracked flights, restrictive check rings and nozzles, and undersized runners and gates.

Free volume relates to the spaces surrounding polymer chains that affect the mobility of the chain segments and their ability to relax. Packing pressure and the rate of cooling in the mold can affect relaxation to influence the unannealed heat deflection temperature and elevated temperature dimensional stability (4). (See reference 5 for a more comprehensive description of free volume.)

Stress, commonly called molded-in stress or strain, is a catch-all term frequently misused as being the cause of many molding variable–property problems. Stress is a totally different condition from orientation, yet many wrongly use the terms interchangeably (6). Stress is caused by either improper mold packing or from the inherently uneven cooling of the part in the mold after fill is completed. Generally, cooling stresses result in the surface of the part being under compressive stress while the core is in tension. It is entirely possible to have a section of a molded part oriented in the flow direction but in a compressive stress. This illustrates that residual stress and orientation are fundamentally different concepts. On a molecular level, stress is the result of short-range deformation of molecules at bonds between atoms.

ORIENTATION

Orientation in polymers simply refers to alignment of polymer chains, whether they are stretched ("stressed") or not. High residual stress is not a prerequisite for orientation. Polymer chains have a preferred relaxed state. If they are not frozen so stiff that they cannot move, it is their nature to randomly coil up into a "fuzz-ball" configuration (see Fig. 19-1a). When polymer melt is pushed through runners gates and mold cavities, these fuzz-balls distort from the stretching and shearing forces. This distortion creates alignment of chains parallel to each other, as shown in Fig. 19-1b. This parallel alignment creates strong and weak directions (anisotropy) in a molded part. The situation is somewhat similar to the way the grain in a piece of wood influences how easy it is to break the wood in the grain direction vs. cross grain (see Fig. 19-1c). Polymers are strong in the orientated direction because the atom-to-atom bonds (carbon-to-carbon in ABS) are much stronger than the weak forces attracting neighboring chains (7).

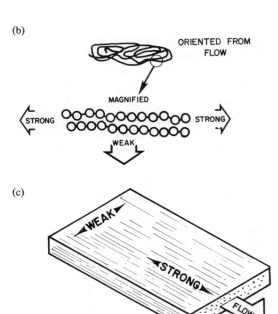

Fig. 19-1   Polymer orientations.

For example, an oriented specimen, broken across the flow direction, can have twice the impact of a nonoriented one. Similarly, it is possible for the broken-with-flow impact to be only 10 percent of the broken-cross-flow value on a strongly oriented specimen. Injection molded parts are not uniformly oriented. The degree of orientation varies considerably through the cross section from the surface of the part to the core. It also varies from the gate to the dead end (8). How pronounced these variations are also depends on the molding conditions—more precisely, the point-to-point flow, temperature, and pressure conditions at every location in the mold. It follows from this that certain mechanical properties sensitive to orientation will vary point-to-point in the molded part. The exact distribution of orientation will determine which properties are affected. (Some simple tests for the degree of surface and core orientation are described in reference 6.)

CAVITY MELT FLOW

A brief review of how polymer actually flows into the mold will help us see where the surface and core orientation comes from. It also helps us figure out some useful generalizations on how molding variables affect orientation patterns within the part.

Figure 19-2 shows the cross section of a mold cavity where the flow proceeds from left to right. We are looking at a cross section of the part thickness—typically 0.100 in. thick. The boundary between the advancing melt and the still-empty portion of the cavity is called the melt front. This melt front is a stretching membrance of polymer, like a balloon or bubble (9). Note that the direction of stretching at the front is at right angles to the main flow direction. This stretching creates considerable orientation of the polymer molecules. The melt front rolls out like a bulldozer tread onto the surface of the relatively cold mold, creating a zone of surface orientation on the part. There is no evidence, under normal molding behavior, that the melt slides along the old surface (10).

Behind the melt front more polymer is flowing—in a sense to keep the advancing melt front "inflated." In this zone orientation is caused by shearing of one polymer layer over another, which is a consequence of the unavoidable velocity difference resulting from the centerline flowing faster than the edges. This shearing flow creates another band of high orientation just under the surface layer

that came from the stretching front. One edge of this band is hung up on the frozen surface layer, while the other edge is trying to go along with the main flow. Finally, the core of the part is also oriented to some degree due to shearing and velocity gradations; the orientation gradually diminishes to nothing at the centerline. Thus cavity flow defines three layers of orientation: surface, subsurface, and core.

Molding variables affect the intensity and relative distribution of the layers because they can influence the two phases of cavity filling. In Phase I the melt actually flows into the cavity; phase II involves packing and cooling. Orientation is generated from stretching and shearing during phase I. However, when flow ceases, the stretching and shearing forces essentially disappear, and the polymer orientation can relax out to various degrees. How much relaxation takes place depends on melt temperature, mold temperature, and packing pressure. The net orientation retained in the part is the difference between what was generated during flow minus what relaxed out before the melt cooled down to the freeze temperature.

Let's take a look at how tne molding variables affect net orientation:

Fill Rate. Fast fill tends to put more orientation on the part surface and less in the core. This is so because ABS "shear thins"—mostly near the mold wall where the shear is maximum. As a result the core "plug flows" or

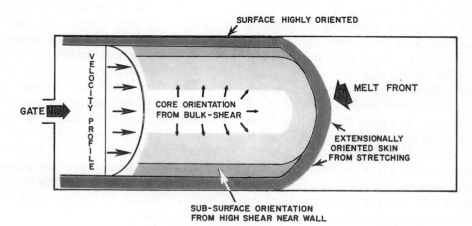

Fig. 19-2 Cavity melt flow model (looking at part thickness).

slips along under the shear-thinned subsurface layer. This mechanism reduces shear in the core to minimize orientation in the bulk of the part.

Conversely slow fill minimizes surface orientation and for several reasons allows the core to be more highly oriented than with fast fill. (Refer to Fig. 19-3.) With slow fill there is less shear thinning in the subsurface layer, and the mold has more time to cool the melt while it is flowing into the cavity. These circumstances cause a less locally intense but more evenly distributed orientation through the whole cross section of the part. So fill speed plays a large part in determining *where* in the part cross section the orientation is located—heavily concentrated in a thin layer at the surface, or spread out over the whole core. Fill rate can have additional effects because this variable interacts with melt temperature and packing pressure. Fast fill will cause the melt temperature to rise because of shear heating; slow fill can result in the mold actually cooling the melt. Fast fill also allows better transfer of packing pressure to the melt in the mold, provided there is a cushion present.

Melt Temperature.   Hotter melt yields less orientation in ABS than cold melt for a number of reasons. Hotter melt is less viscous,

so the stretching and shearing forces that generate orientation are reduced. Hotter melt also freezes more slowly and allows more time for melt relaxation (orientation decay) after flow ceases and before the part sets up. Figure 19-4 shows the combined response of fill rate and melt temperature on surface and core orientation.

Mold Temperature.   Generally mold temperature has a weaker influence on orientation than fill rate or melt temperature. There is little evidence that mold temperature has much effect on surface orientation. A hotter mold does tend to reduce core orientation because the melt freezes more slowly, allowing more time for orientation relaxation (11). (See Fig. 19-5.)

Packing Pressure.   If a cushion is present and injection hold time is sufficient, increased packing pressure generally increases orientation for two reasons. Creeping flow can occur during packing to compensate for the cooling and shrinking melt in the cavity (12). This slow creeping flow creates core orientation, particularly near the gate. Higher pressures also can reduce melt relaxation, so that more of the fill-induced orientation is retained. (See Fig. 19-6.)

Mold Geometry.   The geometry of the mold cavity can also influence the cause–effect links.

- For a given injection rate (ram travel or volumetric flow through the sprue) the local melt front velocity (MFV) will be higher for a thin part than a thick one. Thus local downstream variations in part wall thickness cause MFVs to change just as if the ram speed were being varied throughout the shot.
- A part with two gates can have one-half the local MFV of that with one gate.
- Undersize runners and gates create more shear heating, and thus higher melt temperatures.
- Thick parts cool more slowly than thin ones, providing more opportunity for relaxation of core orientation.

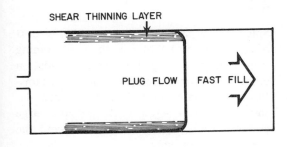

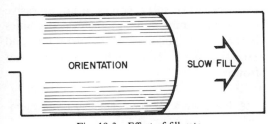

Fig. 19-3   Effect of fill rate.

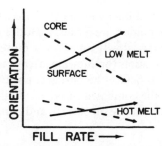

Fig. 19-4 Effect of melt temperature and fill rate on orientation.

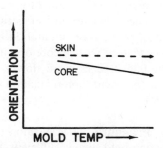

Fig. 19-5 Effect of mold temperature on orientation.

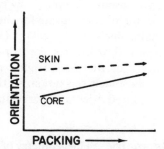

Fig. 19-6 Effect of packing pressure on orientation.

- The greater the distance from the gate, the lower the local packing pressure.
- Small perturbances in cavity surface geometry can also have curious but important effects on part surface properties such as electroplate adhesion and "paint soak." These will be discussed further in a later section.

MOLDING VARIABLE/PROPERTY RESPONSES

With this background on what affects what, and why, the following graphs and discussion summarize some generalizations about molding variable/property responses. As men-

tioned before, these generalizations have been gleaned from commercial experience and laboratory experiments. They hold true frequently enough to be useful as guidelines for part design, machine operation, and troubleshooting. However, contrary behavior can be observed in a particular isolated case from time to time. Processing technology has not yet advanced to the point where every aspect of what goes on in a machine or mold can be predicted with 100 percent accuracy.

The following graphs do not have values on the axes; they are *qualitative* responses only. The absolute numbers vary with each grade of ABS. The slopes of the curves and accompanying text do attempt to indicate whether the response is strong or weak. Also, references lead the reader to articles that contain actual data illustrative of how much a property can vary in a given case. Some of the curves are "envelopes" because responses can vary qualitatively, depending on the particular brand or grade of ABS in question. Also, on some graphs the "normal" operating range for a molding variable is indicated on the horizontal axis by a shaded bar to highlight responses that occur below and above this range.

APPEARANCE PROPERTIES

Splay

Splay (splash marks or silver streaks) is most often caused by bubbles in the melt coming from moisture, trapped air, and degradation gases (13). Proper predrying can avoid moisture-related splay (14). Improper screw design, insufficient screw back pressure, large polymer granules, high screw rpm, and the use of screw decompress can cause splay due to trapped air.

Degradation splay comes from excessive melt temperatures and/or long residence times in the cylinder, high nozzle temperatures, and excessive shear. Too much shear can result from:

- Poor screw designs that cause melt temperature override (3).

- Cracked screw flight or non-return valve.
- High screw rpm.
- Excessive screw back pressure.
- Restrictive runners and gates.
- Very fast injection rates.

Splay can also be packed out to various degrees, depending on the root causes, tool design, and machine conditions. Some machine conditions that enhance packout unfortunately aggravate gas bubble generation during cavity flow. The net results are not always easy to predict. Assuming that the nozzle and mold flow channels are properly designed, faster fill usually yields less splay (see Fig. 19-7a). This is so because the time is shortened for bubble growth, and fast fill enhances packout. Elevating melt temperature, while it often can also help packout, almost always results in more splay (Fig. 19-7a). Higher melt temperatures cause more bubble formation; the melt is less viscous, and the pressure inside the bubble is greater. So, depending on circumstances, at high melt temperatures faster filling can either decrease or increase splay. It will increase if the fast fill pushes the melt temperature too high from shear heating.

Figure 19-7b shows that increased packing

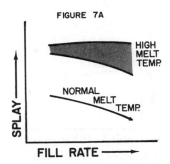

FIGURE 7A

Fig. 19-7a   Effect of fill rate on splay.

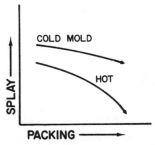

Fig. 19-7b   Effect of packing on splay.

can reduce net splay; this is most effective at high mold temperatures. Packing out splay is a "limp along gesture"—it is better to get rid of the root cause of the bubbles so that packout is not required. Packout should not be resorted to on parts to be electroplated or painted, or those that will be exposed to hot water or solvents.

Gloss

The melt front that rolls out to define the part surface is inherently lumpy on a micro scale because ABS contains two phases, one more deformable than the other. Optimizing gloss depends on pressing this lumpy "virgin" surface against the highly polished mold surface. Although one might expect high melt temperatures always to favor higher gloss, in most cases the opposite is true—especially with a cold mold (see Fig. 19-8a). Some ABS's are less sensitive so the response is flatter. It is also possible to have the melt so cold that low gloss will result because packing is hampered. The total response is then a humpback curve.

Mold temperature has a strong effect on gloss. Cold molds (under 140°F) reduce achievable gloss and also make gloss more sensitive to the other molding variables. Higher mold temperatures (150 to 180°F) promote gloss and flatten out the melt temperature effect (11).

Filling faster usually helps gloss, provided that the melt is not oversheared in restrictive runners and gates. The response to fill rate is greatest at low mold temperatures (see Fig. 19-8b).

Surprisingly, increasing packing pressure

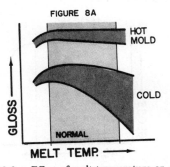

FIGURE 8A

Fig. 19-8a   Effect of melt temperature on gloss.

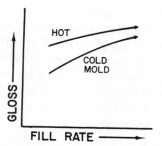

Fig. 19-8b   Response of gloss to fill rate vs. temperature.

beyond that needed to make a good full part does not always have the strong effect one might expect. The packing pressure effect can be weak and interactive with the mold temperature (see Fig. 19-8c). Cases have been noted where overpacking decreased gloss.

The best gloss is obtained with moderate melt temperature, "upper limit" mold temperature, fast fill, and sufficient but not excessive packing. Since hot molds trade off quick cooling and fast cycles, it is wise not to use any higher mold temperature than is needed. By manipulating these four molding conditions it has been shown that gloss values from 98 percent to 20 percent can be achieved on the same grade of ABS—the effects are that pronounced (11).

Warping

Molded parts can warp under no load conditions at elevated use temperatures for a number of reasons. One should also be aware the warpage is more likely to occur at high humidity (i.e., warping tendencies are greater in hot humid than hot dry conditions). Both molded-in core orientation and cooling stresses can cause parts to warp.

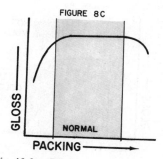

Fig. 19-8c   Effect of packing on gloss.

Elevating the melt temperature reduces molded-in orientation, thus reducing the tendency to warp. Filling faster also reduces core orientation, and consequently also generally reduces warping. The combined responses are shown in Fig. 19-9a.

Colder molds create more warping tendencies in several ways. There is less opportunity for relaxation of orientation, and the more rapid cooling sets up unwanted cooling stresses. Increased packing also creates more stresses, inhibits relaxation, and lowers unannealed heat deflection temperature (see section on HDT). Packing and mold temperature effects are shown in Fig. 19-9b.

To minimize warping, mold at the upper limit of melt temperature, mold temperature, and fill rate. Use only enough packing pressure to get a good full part.

MECHANICAL PROPERTIES AND MOLDING VARIABLES

Tensile Strength and Modulus

Tensile modulus is not significantly affected by any of the four molding variables. Tensile *strength* is primarily influenced by orientation;

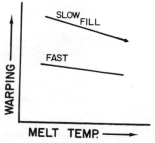

Fig. 19-9a   Effects of fill rate and melt temperature on warping.

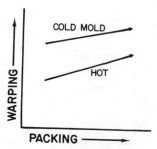

Fig. 19-9b   Effects of mold temperature and packing on warping.

parts are stronger in the oriented direction. There is also some indication that heat history has a deleterious effect. In contrast to impact properties, tensile property effects are of a lower order. Figure 19-10 shows the important qualitative responses.

Quantitatively, the tensile yield strength at room temperature might drop 5 to 10 percent of the nominal value over a typical molding condition range as one goes from cold melt–slow fill to hot melt–fast fill (15). Mold temperature and packing pressure have no significant effect.

Flexural Strength and Modulus

These properties respond to molding variables similarly to tensile properties, so the above-mentioned comments apply.

Flexural Creep

The limited data available indicate that there is no significant effect of molding variables on flexural creep.

Heat Deflection Temperature (HDT)

Studies have shown that unannealed and annealed HDT respond somewhat differently to molding variables. The unannealed HDT (UA-HDT) is affected by packing pressure and mold temperature. No effect due to melt temperature or fill rate has been noted. Attempts to correlate UA-HDT with orientation or cooling stresses have been unsuccessful (11). This in interesting in view of the well known fact that annealing normally increases

the HDT value by as much as 40°F. An explanation for molding variable effects based on a molecular relaxation concept is given in reference 4.

Figure 19-11a shows that overpacking the mold can result in a loss of 10 to 15°F in UA-HDT (15). Also cold molds (80°F) can reduce UA-HDT by 10°F (11). High packing and fast cooling inhibit molecular motion and hinder the preferred ordering of polymer molecules.

The annealed HDT, on the other hand, is *not* influenced by packing pressure, mold temperature, or fill rate. Some data have been accumulated showing that as melt temperature is increased the annealed HDT is reduced, perhaps by 10°F (see Fig. 19-11b). This effect is not always consistent and might depend on the exact ABS grade in question.

Izod Impact

Notched izod impact is strongly influenced by orientation, and the molding variable responses reflect this. Since orientation is directional, it is necessary to be specific about the direction of break relative to the flow direc-

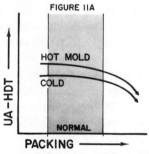

Fig. 19-11a Packing and mold temperature effects on UA-HDT.

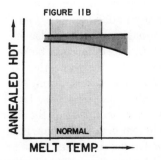

Fig. 19-11b Effect of melt temperature on annealed HDT.

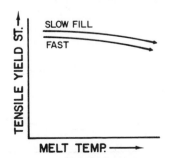

Fig. 19-10 Effects of fill rate and melt temperature on tensile strength.

tion. Orientation can be beneficial when izod is being broken *across* the flow (bAf). On the other hand, this same orientation weakens the part when broken *with* flow (bWf). It is possible for the bAf value to be two to five times greater than the bWf value (11). Some applications benefit by having as much impact in one direction as possible; the other direction is unimportant. Other applications require uniform impact (i.e., no directional preference). Molding variables can be manipulated to achieve either result to some degree.

Melt temperature affects izod impact through two possible mechanisms. Elevating melt temperature within the recommended range decreases the core orientation which strongly influences this property. This will cause bAf izod to decrease while the bWf value increases (see Fig. 19-12a).

Excessive melt temperatures not only yield even less orientation but can also degrade the polymer. This reduces both bAf and bWf izod impact. This is why the curves in Fig. 19-12a have a downward break above the recommended melt temperature range. Depending on their structure and stability, ABS's differ in their strength of response to melt temperature. Abusing the melt can cause several-fold reduction in izod bAf. The bWf generally responds less strongly; it can be improved about 50 percent by increasing the melt temperature within the recommended range (11, 15).

Filling faster decreases core orientation; consequently, the response is to reduce bAf izod impact and increase bWf. At low melt temperatures the fill rate effect can be 15 to 50 percent or more. At high melt temperatures the fill rate effect diminishes considerably because melt relaxation tends to erase any fill-induced orientation (see Fig. 19-12b). Increasing the mold temperature also has the effect of reducing the difference between the bAf and bWf values by promoting slow cooling and improving melt relaxation. The mold temperature effect (i.e., 80°F vs. 180°F mold) is not quite so pronounced as the fill rate effect. The mold temperature effect is strongest at low *melt* temperatures and slow fill rates (see Fig. 19-12c). Packing pressure does not have a strong or consistent effect on izod impact.

Weld Line Strength

Weld lines, formed by the rejoining or colliding of two melt streams, are typically weaker than nonweld areas for several reasons. There is a sharp notch at the weld that acts as a stress concentrator. Trapped air between the fronts can interfere with proper knitting. Orientation in the weld area is at right angles to the principal flow direction and comes from the elongational stretching of the melt front. This orientation is also thought to contribute to the weakness of the weld (16–19). It is important to avoid trapped air at the weld; so proper mold venting is imperative. Information on the effects of molding variables is not

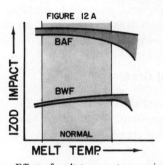
Fig. 19-12a   Effect of melt temperature on izod impact.

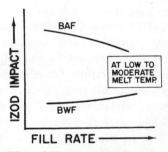
Fig. 19-12b   Effect of fill rate on izod impact at low to moderate melt temperatures.

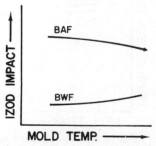
Fig. 19-12c   Effect of mold temperature on izod impact.

abundant, but what does exist gives us the following general guidelines.

Increasing both melt and mold temperature will frequently improve weld line strength (see Fig. 19-13a). Higher melt temperatures promote molecular knitting and entanglement at the weld and also yield less net orientation. Consequently, one can try elevating melt temperature within the recommended limits. Excessive melt temperatures will degrade the polymer causing a general weakness, including weakness at the weld. For this reason the melt temperature curves in Fig. 19-13a turn over above the recommended limit. Mold temperature elevation also helps because it promotes slow cooling, there is more time for packing out the weld notch and allowing the molecules to entangle, and more of the orientation relaxes away. However, in most cases the mold temperature effect, though positive, is not so pronounced as the melt temperature effect. Fill rate and packing pressure effects can be complex because of competing behavior and interactions. Because of these trade-offs, changing the variables can sometimes have no net effect, or they might go through a maximum. Exactly what happens can also depend on the particular grade of ABS, the part design, and the melt and mold temperature levels.

Increasing the fill rate, on the one hand, can promote knitting via the same mechanism as elevating the melt temperature. Fast fill will generate some heat as well as minimizing mold cooling during flow. On the other hand, fast fill can create more undesirable frontal orientation and aggravate venting problems, thus causing weld line weakness. (See Fig. 19-13b.)

Insufficient packing obviously can create more prominent and weaker welds. However, overpacking can also contribute to weaker welds by two mechanisms. Overpacking creates a sharper notch, which simply increases the stress concentration under service conditions. Also overpacking hampers melt relaxation and molecular entanglement (knitting). Figure 19-13c summarizes the probable situation.

There is an optimum packing pressure and fill rate that will depend on particulars rele-

vant to each part design. When troubleshooting, one can try going in both directions on these two variables and carefully noting the property response. Neither response is expected to be as strong as with melt or mold temperature changes. Also it should be realized that while manipulating these variables can improve welds to some degree, it is not likely that they can produce weld lines as strong as nonweld areas.

Missile Impact

The response of falling dart impact to molding variables can be significantly different in some respects when compared to bAf izod impact. In the case of bAf impact, molded-in core

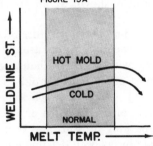

Fig. 19-13a Effects of melt and mold temperature on weld line strength.

Fig. 19-13b Effect of fill rate on weld line strength.

Fig. 19-13c Effect of packing on weld line strength.

orientation *increases* the impact. By contrast, orientation is almost always harmful to falling dart impact (FDI) because there will be weakness in the cross-flow direction. The FDI test causes biaxial deformation, and the part will be no stronger than the weakest direction— the high strength in the flow direction is of no help.

Generally the rule is to manipulate molding variables to minimize orientation without causing degradation. This means elevating melt temperature within the acceptable range. The heat stability of individual ABS grades will vary; some can tolerate increased melt temperature more than others before degradation takes away the gains made from decreased orientation. As a result of all these factors, the response is shown as an envelope (Fig. 19-14a).

An elevated mold temperature promotes relaxation of orientation and usually interacts significantly with fill rate. Changing fill rates has the most effect at low mold temperature (Fig. 19-14b). Quite often packing pressure has no effect on FDI. When an effect due to elevating packing pressure has been noted, it has almost always been detrimental. This has

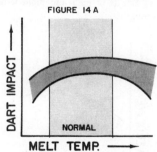

Fig. 19-14a   Effect of melt temperature on FDI.

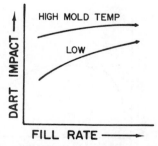

Fig. 19-14b   Effect of fill rate on FDI at low and high mold temperature.

been especially true in combination with low melt and mold temperature, and slow fill. Experiments and commercial practice have uncovered situations where overpacking has reduced FDI to half of the optimum value. Overpacking increases net orientation and could possibly upset the cooling stress balance to put the surface in tension, rather than the usual compression (6). Both of these conditions would be expected to reduce FDI.

MOLDING FOR ELECTROPLATING

Appearance, plate adhesion, and dimensional stability are all key quality factors when molding for electroplating. From a molding variable optimization standpoint, plating represents one of the most challenging cases because of tradeoffs and competing factors. For example, some molding variable settings that optimize plate adhesion are not the best choice for suppressing splay or part warping tendencies. Usually parameters are selected that give the best plate adhesion and thermal cycle performance. It is from this aspect that the following discussion is structured.

Assuming that the preplating and plating steps are properly carried out, the adhesion of the plate to the ABS is mainly determined by the strength of a thin layer of ABS just underneath the plate. Low adhesion and plate blistering seldom involve clean separation of the plate from the ABS. Rather there is a delamination of ABS from itself in the boundary layer. The boundary layer is conditioned by the orientation coming from the melt front. To optimize the strength of this critical layer, it is desirable to minimize the orientation there (20). As shown in Fig. 19-15a, the two key variables are melt temperature and fill rate. Slow fill rates should be used to minimize surface orientation and promote a strong ABS boundary layer for the plate to lock into. However, here is a good example of one of the aforementioned compromises. One might also want to minimize part warpage, since twisting or bending of the plated part could build up stresses that would pop or crack the plate. As mentioned earlier, warpage is minimized by filling fast, since it puts orientation on the surface rather than in the core. Fortunately, there is a reasonable way out of the situation. High melt temperatures favor relaxation of orientation, especially core orientation coming from slow fill. Thus one should use the high end of the melt temperature range without going so far as to degrade the polymer, causing splay or poor part appearance. Proper selection of fill rate and melt temperature can increase plate adhesion by 50 percent or more.

Mold temperature and packing pressure have a lesser effect. High mold temperatures will help to reduce orientation, especially core orientation coming from the required slow fill. Packing pressure should be sufficient only to get a full, good-looking part. Overpacking retains unwanted orientation and can build up unfavorable stresses. The mold temperature and packing responses are shown in Fig. 19-15b.

PROPERTY VARIATION WITH POSITION/MOLD GEOMETRY

The key molding variables of melt temperature, melt pressure, and fill rate are seldom

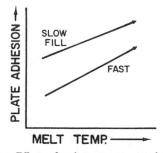

Fig. 19-15a Effects of melt temperature and fill rate on plate adhesion.

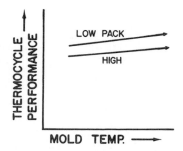

Fig. 19-15b Effects of packing and mold temperature on thermocycle performance.

the same point to point in the cavity. The first two particularly vary in the flow direction. The local velocity can vary in the flow direction even in a simple part, and in the cross-flow direction as well in complex ones. Local velocity is also affected by local thickness. Even mold surface temperature is seldom the same at each point in the cavity. Since these parameters vary point to point in the mold, their influence on properties does likewise. Indeed properties such as impact and electroplate adhesion do vary across and down the part flow path.

A full discussion of all the property vs. position effects and possibilities is beyond the scope of this chapter. A few key examples will make the point that the processor needs to be aware of this phenomenon.

Orientation is usually greatest at the gate end of the part and gradually lessens toward the end of the flow path. As a result, izod impact bAf will be higher at the gate end but lower at the dead end. Falling dart impact, because it suffers from uneven orientation, is less at the gate, higher at the end of flow. Reference 15 contains a good example of how much falling dart and izod impact can vary with position. On a 4-in.-wide slab, over a 15-in. flow length from the gate, the izod impact bAf dropped by half while the FDI *increased* fourfold! This example illustrates another point: molding variables, whether machine- or position-induced, can cause one property to increase while another falls off. Injection molding process control is full of such trade-offs. Even seemingly minor perturbances on the mold surface, such as knockout pins, part coding numbers, or embossing can produce surface orientation anomalies. These can affect properties that are sensitive to surface orientation; a good example is electroplate adhesion. It has been shown that scribing lines .005 in. deep in one mold surface crosswise to flow can reduce surface orientation on the noncorresponding part surface. With this technique one can locally improve plate adhesion. How these mold surface effects disrupt the stretching melt front and transfer their results to the opposite part surface is discussed in references 20 and 21.

REFERENCES

1. Allen, E. O., "Know All Your Molding Variables," *Plastics Engineering* (July 1974).
2. Paulson, D. C., "Measurements for Control in Injection Molding," *SPE ANTEC Papers* (1970).
3. "Injection Molding Zero-Meter Screws," *Plastics World* (March 1982).
4. Gaggar, S., and Wilson, J., "Factors Affecting the Heat Distortion of Rubber-Modified Amorphous Polymers," *SPE ANTEC Papers* (1982).
5. Hayward, R., *The Physics of Glassy Polymers,* Halstead Press, 1973.
6. Wintergerst, S., "Orientation and Stresses in Injection Moldings," *Kunstoffe,* 63 (October 1973).
7. Ruben, I. S., *Injection Molding Theory and Practice,* SPE monograph, Wiley-Interscience, New York, 1972.
8. Menges, G., and Wubken, G., "Influence of Processing Conditions, on Molecular Orientation in Injection Moldings," *SPE ANTEC* (1974).
9. Tadmor, Z., "Molecular Orientation in Injection Molding," *Journal of Applied Polymer Science, 18* (1974).
10. Fritch, L., "How Mold Temperature and Other Molding Variables Affect ABS Falling Dart and Izod Impact," *SPE ANTEC Papers* (1982).
11. Harland, W., et al., "Orientation in Injection Moldings," *Plastics and Rubber Processing* (June 1980).
12. Fritch, L., "Splay on ABS Injection Molded Parts—Causes and Mechanism of Formation," *SPE ANTEC Papers* (1975).
13. Fritch, L., "ABS Pellet Drying—A Practical Study of Key Variables," *SPE ANTEC Papers* (1974).
14. Fritch, L., "Injection Molding ABS for Properties," *SPE PACTEC V* (1980).
15. Malguanera, S. C., "Structure and Properties of Weld Lines in Injection Molded Thermoplastics," *SPE ANTEC Papers* 1981).
16. Malguanera, S., "The Effects of Processing Parameters on Tensile Properties of Weld Lines in Injection Molded Thermoplastics," *Polymer Engineering and Science* (July 1981).
17. Hagerman, E. M., "Weld Line Fracture in Molded Parts," *Plastics Engineering* (October 1973).
18. Brewer, G. W., "Treatments Influencing Weld Line Strength," *SPE-DIVTEC,* Columbus, Ohio (October 1981).
19. Fritch, L, W., "ABS Cavity Flow-Surface Orientation and Appearance Phenomena Related to the Melt Front," *SPE ANTEC* (1979).
20. Fritch, L. W., "Grooved Mold Improves Plate Adhesion," *Products Finishing* (February 1984).
21. Reilly, F., and Price, W., "Plastic Flow in Injection Molds," *SPE Journal* (October 1961).

Chapter 20
Injection Mold Nylon 66 for Optimum Performance

W. C. Filbert, Jr.

Du Pont Co.
Wilmington, Delaware

It has long been established that part and mold design requirements are strongly influenced by the processing characteristics of any plastic. Nylon 66 is no exception. As we will show, careful consideration of many processing-design factors is necessary to ensure that nylon 66 is molded under conditions that optimize its quality–function–cost potentials.

Pick the Right Nylon. Three important resin–mold–part interrelations must be considered at the outset by those specifying nylon: First, nylon 66 is a family of related resins, not just a single composition. As shown in Table 20-1, various additives or modifiers can be incorporated into nylon 66 that alter its processing-property characteristics.

Second, resin selection must be based on both processing and end-use requirements. Thus, it is important to establish carefully the process economics (cycle, number of cavities, heat removal, and mechanical operation of the tool) so that the mold can be built to handle the production goals.

Third, overall part design should be scrutinized for redundancy and simplified to require the least complicated mold design. This is a frequently overlooked way to reduce initial mold costs and to improve subsequent mold performance.

What's So Special about Molding Nylon 66? While nylon 66 must be considered as a family of resins, all compositions have certain common molding advantages:

- High flow and toughness in thin sections.
- Good weld strength and easy fill of complicated shapes.
- Predictable mold and annealing shrinkages; little tendency for warpage.
- Fast overall cycles. Resins can be molded in cold mold. Ejectability of parts from molds is good; undercuts are readily stripped from cores or cavities.
- Good rework stability. Property losses are minimal on remolding of dry, rework-virgin blends. Processing conditions are unaffected by recycling high levels of regrind.
- Moldability to close tolerances; multicavity tooling presents no unusual difficulties in achieving commercial tolerances.

Processing Characteristics Influence Mold and Part Design. We will review in detail each of the above processing keys and see how they influence part and mold design. Remember, the essence of profitable molding is seldom distilled from a consideration of a single processing variable but depends, rather, on optimizing at the same time as many factors as possible.

Design for Flow and Toughness in Thin Sections. The flow characteristics of nylon 66 molding-grade resins are outstanding. Con-

Table 20-1 Your guide to selecting nylon 66 injection molding compounds.

Your Guide to Selecting Nylon 66 Injection Molding Compounds		
If you need . . .	Specify . . .	What it is . . .
Good stiffness, strength, flow . . .	General purpose, unmodified.	Nylon 66 (melt point: 509°F).
and . . . Easier fill in plunger machines, or faster screw recovery in reciprocating screw machines . . .	General purpose, modified.	Lubricated.
and . . . Easy ejectability from the mold.	General purpose but with added mold release agent	Lubricated.
Or Greater heat stability in use (to 250°F)	Heat-stabilized grade.	Stabilizer retards embrittlement at high use temperatures; best thermal stability but poor electricals.
and . . . Greater heat stability plus improved ejectability.	Heat-stabilized grade with mold release.	Same as grade above but with mold release.
Or Outstanding weather resistance.	Weather-stabilized grade.	UV-stabilized grade with carbon black.
Or Fast molding, improved color retention on rework, lower mold shrinkage . . .	Color-stabilized, nucleated grade.	Rapid crystallization for fast cycles. Slightly stiffer than GP nylons but sacrifices some toughness.
and . . . Improved ejectability.	Color-stabilized, nucleated grade with mold release.	Same as grade above but with mold release.
Or Lower melting, lower mold shrinkage, good flow and color stability.	General-purpose nylon copolymer.	Copolymer with melt point of 445°F; processes easily and good for heavy sections, but sacrifices stiffness and high-temperature properties.
Or Outstanding impact toughness, especially at low temperatures.	General-purpose grade	Modified nylon 66 with good moldability (not UL SE Class II).

sidering that these resins exhibit a very sharp crystalline metling point at 509°F, the melt has good fluidity at temperatures as low as 520°F, as shown in Fig. 20-1. This figure also shows that the melt deviates from Newtonian behavior at all processing temperatures, meaning that melt viscosity decreases significantly as shear stress (injection pressure) or shear rate (injection speed) is increased. In other words, the melt gets more fluid as these molding variables increase.

Similarly, the temperature dependence of melt flow (Fig. 20-2), while not greatly different from other engineering resins, also serves to lower viscosity (i.e., improve flow) as the melt temperature is increased. For example, changing the melt temperature by 50°F will alter viscosity (flow) by a factor of about 2. (Incidentally, melt viscosities in Fig. 20-1 are typical values at injection pressures used in molding nylon 66.)

How Best to Fill Thin Sections. It can be argued that any cavity, regardless of dimensions, can be filled when the proper molding conditions are used. Assuming, for example, that melt solidification in the mold does not occur (if mold temperatures are held at or above the resin's freezing point), any cavity can be filled, provided sufficient injection time is available. From a practical standpoint, however, solidification does occur, and flow will terminate whether the cavity is filled or not. In many cases, this condition results from machine limitations. Thus, in estimating whether any part can be molded, one must first establish its minimum fill requirements (based on the resin's flow and freezing characteristics) and then determine whether this is within the capability of the molding equipment. That is, one must examine the pressure requirements for filling the cavity and the necessary injection rate imposed by solidification of the melt

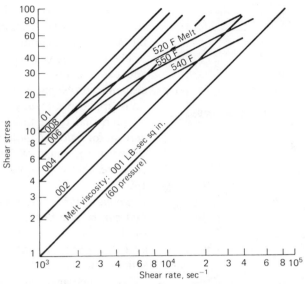

Fig. 20-1   This shows relationship of shear stress, shear rate and melt viscosity at indicated melt temperatures. (Zytel 101 type 66 nylon.)

in light of the molding machine's maximum hydraulic oil pressure and pump delivery rate.

Filling a cool cavity with molten nylon, or any thermoplastic for that matter, involves

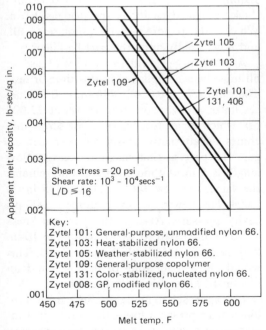

Fig. 20-2   Melt viscosity vs. temperature for different nylon 66 molding resins.

quite complicated fluid flow relationships. As Figs. 20-1 and 20-2 show, melt viscosity of plastics varies with shear stress, shear rate, and temperature. Accordingly, certain reasonable assumptions must be made to provide a practical guide for estimating fill.

Using a few basic melt-flow equations, we derive Fig. 20-3, which provides useful information, well within engineering reliability, for predicting mold flow and required cavity fill time for nylon 66 (general-purpose nylon, in particular). The data in Fig. 20-3 can be used to predict maximum flow length at specified thicknesses (or minimum thickness for a specified flow length) and to indicate the time available for filling the cavity of indicated thickness before melt freeze-off prevents additional mold penetration.

The simple, general assumptions we draw are these:

- The cavity can be considered a slab whose volume can be described as length × width × thickness.
- Cavity fill time is very fast, so that shear stress and rate are unaffected by melt cooling during the period of filling; that is, flow in the feed channels is isothermal,

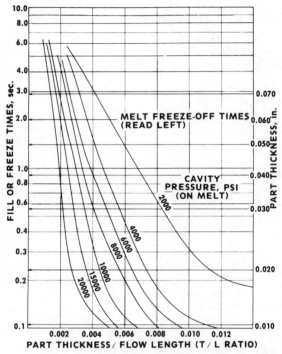

Fig. 20-3 Relationship of fill time, cavity dimensions, and pressure in estimating cavity fill at melt temperature of 550 ± 10°F and mold temperature at 120 ± 20°F. (Nylon 66 molding resins.)

and melt reaches the cavity at the same temperature at which it leaves the nozzle.
- Channel diameters remain constant until the cavity starts to fill.

Figure 20-3 is a generalized curve that we have derived specifically for nylon 66 resins. Based on rheological and thermal data, it is particularly useful for thin sections ($\frac{3}{32}$ in. or less). Knowing the solidification or freezing characteristics of nylon 66, you can compare the actual cavity fill time at various inlet pressures with the allowable fill time before the cavity freezes. (For estimating purposes, pressure drop through the cavity is assumed to be complete and is equal to the actual delivered melt pressure inside the gate.) The penetration or fill time for any cavity, expressed in terms of its thickness/length (t/L) ratio, depends on the available pressure, while the maximum allowable time is dependent on freezing of the part of specified thickness.

(Part width is not involved except, as discussed in the next section, when volumetric fill rate affects toughness.)

The following examples will illustrate how you can use Fig. 20-3. These are not mere academic problems, but real everyday situations applicable to any molding shop. Bear in mind that the data are valid within ± 10 percent for nylon 66 molding grade resins processed in the range of 520–580°F melt temperatures and with mold temperatures of 60–180°F.

Example 1. What is the maximum flow length that can be attained in a 0.030-in.-thick part, single-gated, with 8000 psi cavity pressure available, and what is allowable fill time before solidification occurs?

Look once again at Fig. 20-3 and observe that a 0.030-in.-thick part freezes in 0.55 sec; thus, fill time must be equal to or less than 0.55 sec. Intercept of 0.030 in. part thickness (horizontal line) freezing time (0.55 sec) with 8000 psi cavity pressure curve yields a t/L ratio of 0.0041 (at the bottom).

Since $t = 0.030$ in. and $t/L = 0.0041$, $L = 0.030$ in./0.0041 = 7.3 in. flow (total).

Example 2. The flow length of a 0.020-in.-thick end-gated cavity is specified as 10 in. What is the required cavity pressure to fill, and how rapidly must the part be filled?

A 0.020-in. part will freeze in 0.23 sec, so fill must be accomplished in this time or less. The ratio $t/L = 0.020$ in./10 in. = 0.002 necessitates a cavity pressure of about 22,000 psi, which is not feasible on most screw machines. An alternative exists, however. Center-gate the part so that maximum cavity flow length is 5 in. Although freeze time remains the same, the new $t/L = 0.020$ in./5 in. = 0.004. This new ratio now requires 12,000 psi cavity pressure. Also, if the part thickness could be increased to 0.030 in. for the 10-in. flow case, the t/L would become 0.003. This would allow a fill time of 0.55 sec and an identical 12,000 psi cavity pressure.

This approach to filling cavities can be combined with pressure loss calculations in runners, sprues, and gates to size the necessary "plumbing" of any mold. In Fig. 20-4, pres-

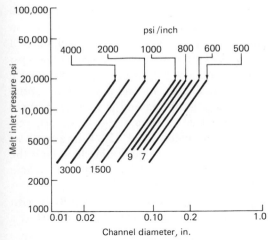

Fig. 20-4 Pressure drop (psi/in. of length) in sprues, round runners, and gates when molding nylon 66.

sure losses for various channel (bore or gate) diameters are plotted vs. upstream (entrance) injection pressures.

Example 3. You are given an edge-gated part 0.020 in. thick that has its most distant flow point 4 in. from the gate, and that is to be molded on a reciprocating screw machine capable of a maximum injection pressure of 20,000 psi; assume that the total feed runner length is 7 in. and that the total pressure drop across the nozzle, sprue, and gate is 3000 psi. You now must determine what the minimum feed runner diameter is that is necessary to fill the cavity in nylon 66.

Using Fig. 20-3 and working back from the cavity:

Thickness/
length, t/L = 0.020 in./4 in. = 0.005
P_c = cavity fill pressure (for 0.020 in. thick part) is 10,000 psi
$P_{(max)}$ = 20,000 (max available at entrance to nozzle)
$P_{(max)} - P_c$ = 20,000 − 10,000 = 10,000 psi (available to fill mold)
ΔP (nozzle/ sprue/gate) = 3,000 psi (typical pressure losses given)
ΔP_R = 10,000 − 3,000 = 7,000 psi (available to flow melt thru runner)

Runner length = 7 in.

$$\frac{\Delta P_R}{in.} = \frac{7000}{7} = 1000 \text{ psi/in. (max allowable pressure drop in runner)}$$

Inlet runner
pressure = 20,000 − 3,000 = 17,000 psi.

Using Fig. 20-4, the 1000 psi/in. pressure loss curve at 17,000 psi inlet pressure intercepts the minimum channel diameter coordinate at 0.135 in. Thus, to fill the cavity the 7 in. feed runner must be at least 0.135 in. in diameter (or equivalent area).

Cavity Fill Rate Important. In the preceding section we have pointedly not concerned ourselves with part width (actually, part volume or weight). As we saw in Fig. 20-3, there are specific minimum fill times for different thicknesses before cavity freeze-off occurs. Thus, the thickness establishes the allowable cavity fill time and establishes the maximum length of flow at various cavity pressures.

Equally significant, we must be concerned with the weight (volume) of the part, and the consequence that this weight must go through the gate in the time necessary to fill the cavity. This, in turn, establishes a volumetric fill rate through the gate. Based on a conservative premise that the fill rate through the gate should not exceed a critical shear rate for the resin (itself a function of melt temperature) in order to yield maximum part toughness, we can define the size of the gate necessary to fill a part without undue concern for flow-induced brittleness (assuming adequate cavity venting).

In Fig. 20-5, maximum fill rate (oz/sec/gate) for nylon 66 is shown as a function of gate diameter at several melt temperatures. Note that this is a maximum fill rate and is limiting in that faster fill rates through the opening may melt-fracture the nylon. This phenomenon often leads to brittleness, and it is good practice to design molds and to operate machines to avoid it.

A practical part design–melt flow consider-

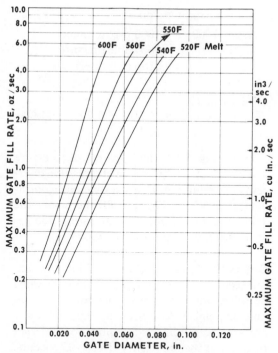

Fig. 20-5 Maximum fill rate through round gates. (Nylon 66 molding resins.)

ation is to combine the concepts of Figs. 20-3, 20-4, and 20-5. In Fig. 20-3 we have specified the allowable fill time (in sec) for a part having a given thickness. Using Fig. 20-4 and calculating pressure drops in the mold layout, we can arrive at the pressure available to fill the cavity and thus establish the maximum flow length. With Fig. 20-5, knowing the gate size (or assuming that it is equal to part thickness or some arbitrary fraction), we can determine the maximum allowable fill rate (oz/sec) that can be fed through the gate.

Multiplying by the fill time (sec) from Fig. 20-3, we fix the maximum weight (oz) of the part. Since the maximum weight is now known, we can calculate the volume of a solid part (density of nylon 66 at room temperature is 0.66 oz/in.^{3}). Since volume equals length × width × thickness, we can readily obtain the part width and thus have defined the limiting geometry for a tough part in nylon 66.

Let's take an example and solve it step-by-step:

Assume you have a rectangular part that is 0.040 in. thick, molded in a three-plate mold

with three gates, 0.030 in. in land and 0.040 in. in diameter (sub sprues are $\frac{1}{8}$ in. in dia and 1 in. long). The molding machine is a 20-oz, 300-ton unit and can deliver 18,000 psi (max) during injection at 5 gal/min; nozzle has a bore dia of $\frac{1}{4}$ in. and is 1 in. long; sprue averages $\frac{5}{16}$ in. in dia and is 2 in. long; three 5-in.-long feed runners are $\frac{3}{16}$ in. in dia; you are molding nylon 66 at 550°F in a mold at 120°F. You want to determine how long the part can be and what its maximum possible width and/or weight is.

1) Calculate pressure drops (Fig. 20-4) with machine at max pressure, 18,000 psi. Here's what we do step by step:

A) Nozzle: at 18,000 inj. pressure, nozzle pressure loss is 600 psi/in. × 1 = 600 psi.

B) $\frac{5}{16}$-in.-dia sprue: Inlet pressure = 17,400 psi.
ΔP_{sprue} = 500 psi/in. × 2 in. = 1000 psi

C) 5-in.-long feed runners, $\frac{3}{16}$ in. ida.: Inlet pressure = 17,400 − 1,000 = 16,400 psi.
ΔP_{runner} = 700 psi/in. × 5 in. = 3,500 psi

D) $\Delta P_{\text{sub sprues}}$ ($\frac{1}{8}$ in. dia × 1 in. long).
Inlet pressure = 16,400 − 3,500 = 12,900 psi.
$\Delta P_{\text{sub sprues}}$ = 900 psi/in. × 1 in. = 900 psi.

E) $\Delta P_{\text{gate, land}}$ = .030 in.
Inlet pressure = 12,900 − 900 = 12,000 psi
ΔP_{gate} = 1500 psi/in. × 0.03 ≅ 50 psi.

F) Effective cavity pressure 12,000 − 50 = 11,950 psi or 6 ton/in.^{2}.

2) From Fig. 20-5, with 0.040-in.-dia gate, max injection rate/gate is 1.0 oz/sec or 1.5 in.^{3}/sec. With three gates, 4.5 in.^{3}/sec (max) can be injected into the cavity without exceeding critical fill rate for a tough part.

3) Since machine can displace 5 gal/min or about 20 in.^{3}/sec, at 18,000 psi, and the 0.040-in. part must be filled in 0.9 sec, hydraulic oil flow must be throttled by a factor of 4 (0.9 × 20/4.5) from max so as not to melt-fracture resin during cavity fill.

4) Fig. 20-3 at 12,000 psi and fill time of 0.9

sec (freeze time for a 0.040 in. thick part) gives a max $t/L = 0.0027$.

5) Since $t/L = 0.0027$ and $t = .040$ in., $L = 14.8$ in.

6) We have established that the part is now 14.8 in. long $\times$ 0.040 in. thick and must be filled in 0.9 sec. Since the max combined fill rate three gates) is 3 oz/sec, it follows that 3×0.9 (or 2.7 oz) is max part weight. The corresponding volume is $2.7/0.66 \cong 4$ in.^{3}.

7) The part now can be defined as: 14.8 in. $\times$ 0.040 in. $\times w = 4$ in.^{3} ($w = 6.7$).

8) This hypothetical part could be molded in nylon 66 about $15 \times 0.040 \times 7$ and satisfy the machine and mold requirements for flow and toughness. However, projected cavity area is approximately $15 \times 7 = 105$ in.^{2}. As actual cavity pressure required to fill was 12,000 psi or 6 tons/in.^{2}, one would need about a 650-ton clamp to support this force. Obviously, a 300-ton clamp will not do, so in reality we are allowed then a part about 15 in. long $\times$ 3 in. wide in order not to flash the mold. Now the max part weight is $15 \times 3 \times 0.040$ in. $= 1.8$ in.$^3 \cong 1.2$ oz, not 2.7 oz, as originally determined in Step 7.

Avoid Common Pitfalls. Thin moldings of nylon 66 can exhibit strength differences in their flow and transverse directions. This phenomenon, called property anisotropy, is induced primarily by flow (pressure) orientation of the melt during flow.

Two complicating factors exist:

The first, part geometry, is extremely important. Parts of dissimilar shape, yet of the same thickness, will fill in different patterns, and thus differences in flow orientation can arise that may affect one part more than the other.

The second factor, the effect of flow length on orientation, is easier to define. At identical molding conditions, the longer the flow path (per given thickness), the more chance of induced orientation, since higher cavity pressures and injection rates are required. All injection moldings are produced under high-shear conditions, and, because cooling times are rapid, only partial recovery or relaxation of the oriented melt molecules can take place

before solidification occurs, leaving the part in a strained condition.

This situation acts to reduce the as-molded ductility in the direction of flow, since, in effect, some of the normally available elongation of the material has already been "used up." Usual ways of reducing flow orientation cause reduction in initial molecular stretching or give more time for stress relaxation.

Typical corrective actions are to increase part thickness to reduce injection pressures required for fill, use hotter molds (which lead to longer cycles and greater stress relaxation), or specify postannealing. But these remedies can lead to increased costs.

Minimize Flow Orientation During Design to Reduce Risk of Nonuniform Thickness. Parts requiring cavity fill pressures greater than 8000 psi are likely candidates for anisotropic property behavior, especially in thin sections. Accordingly, a critical flow length (at any thickness) can be calculated for 66 nylons, which is useful in planning gating and part design so as not to exceed this fill pressure requirement. Table 20-2 lists this maximum flow length at three mold temperatures for several section thicknesses.

Table 20-2   Critical flow distance for uniform physical properties of nylon 66 molding resins.

NOMINAL CAVITY THICKNESS, IN.	MOLD TEMP.		
	60°F	120°F	180°F
0.020	3.2	3.4	3.8
0.030	6.8	7.3	8.2
0.040	11.2	12.0	13.4
0.050	16.8	18.0	20.0

Welds Need Not Be Weak Links. In design of complicated shapes, weld lines are often unavoidable. Weld or knit lines are formed when more than one gate is used or wherever a divided stream of plastic joins after flowing around a pin or core. Thin sections are particularly prone to weak welds because of rapid melt solidification.

When welds are formed, they should be sweeping; when unavoidable, butt welds must be vented for maximum strength. In these cases, it is essential that air at the weld escape before the melt streams unite.

Evidence of poor venting at weld points in nylon 66 is usually burning or discoloration (e.g., yellowing). At such spots, strength of the welds will be inferior to the rest of the part. Obviously, good part and mold design calls for the least number of welds when extreme strength is necessary. The number of gates and internal shutoff cores should be considered an important aspect of the initial part and mold design problem.

The general formula for determining the number of welds is useful to know:

$$N = G - 1 + P$$

where N = No. of weld points, G = No. of gates, and P = No. of shutoff cores.

A double edge-gated bushing, for example, will have two welds because $2 - 1 + 1 = 2$:

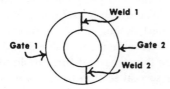

A double edge-gated disc has only one weld front ($2 - 1 + 0 = 1$);

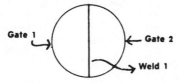

Weld Strength Can Be Maximized. Assuming adequate venting, conventional techniques to improve weld strength are to:

- Increase melt temperature.
- Increase mold temperature.
- Increase injection pressure.
- Avoid use of external release mold lubricant.

Often, mold release sprays are pushed into the weld area by the advancing front of molten polymer and prevent good fusion.

With the exception of the last point, the effect of the change is to increase the flow of melt to the junction, that is, to push the melt into the mold faster so that the two (or more) fronts can fuse or unite properly before resin solidification occurs. Typical weld strengths, in tension, range from 50 to 95 percent of base material strength, although poor weld strength usually shows up as failure in repeated flexure or shear, which is difficult to measure accurately.

To ensure maximum part strength when welds are involved in the design, look at Table 20-3. It assumes that cavity venting is not limiting, and that melt at 530°F is flowing into a cold mold at 60°F. (*Note:* Higher melt or mold temperatures improve welding.)

Table 20-3 Processing design conditions for maximum weld strength in thin sections (nylon 66).

PART THICK- NESS, IN.	MAX. FILL TIME, SEC	MAX. PART WT./GATE, OZ.
0.020	0.14	0.05
0.030	0.37	0.05
0.040	0.64	0.50
0.050	0.95	1.00
0.060	1.40	2.25
0.070	2.90	13.00

The basis for optimum welding is $\frac{2}{3}$ of the part-freeze time, as determined from Fig. 20-3. The maximum part weight (per gate) is based on Fig. 20-5 and on the assumption that the gate diameter is equal to part thickness. (Smaller gate diameters would further restrict weight.)

Dimensional Considerations—a Necessary Chore. The need for the end-user, part designer, and molder to establish and agree on the importance and number of critical dimensions is paramount to profitable molding. Many molds have been built with a certain plastic in mind only to have a poor mold shrinkage estimation or unexpected changes in dimensions after molding force the end-user or molder to try another resin with, usually, lower mold shrinkage to yield parts to print.

Frequently, property compromises are

made because (1) it's cheaper than reworking the mold to size, or (2) it's more advantageous to have a part that fits now (for a variety of reasons) than a part that gives better service over the long haul. Also, it is soon realized that employing unusual molding conditions or gate dimensions to alter mold shrinkage after the mold is built generally leads to poor-quality parts.

In a nutshell, then, use of longer cycles, shrinkage fixtures, or postannealing operations to compensate for bad mold shrinkage estimates can ruin the economics—and that means profits—of molding.

The simple question confronting the mold designer with respect to dimensions can be stated as follows: What size must the cavity be in order to produce a part to size when operating under end-use conditions? In order to answer this question, it is necessary to consider the dimensional changes in nylon 66 that are brought about by several factors:

Moisture and Temperature—Effect on Part Size.

Starting from the dimensions of the part under use conditions and working back to mold-cavity size, the first point to consider is effect of temperature and relative humidity on part size. Nylon, like other plastics and metals, gets larger as it is heated. It also absorbs moisture from the atmosphere, which results in an increase in part size. These factors are combined in Fig. 20-6, which shows changes in length (mils/in.) of a stress-free test specimen molded in nylon 66. These very predictable changes in the as-molded length represent steady-state values (equilibrium) with a given temperature and relative humidity.

In the typical exposure of a part to an environment of slowly varying humidity, no true moisture equilibrium is reached, but, rather, a balance is established with the average humidity. After initial moisture development has occurred, subsequent variations in relative humidity have little effect on total moisture content and dimensional changes in all but very thin sections. The time to equilibrium is highly dependent on temperature and part thickness (e.g., thin sections absorb water very rapidly

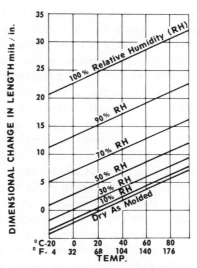

Fig. 20-6   Dimensional changes of GP nylon 66 (Zytel 101) vs. temperature at various humidities (annealed samples).

at higher temperatures). The combined effect of moisture content and thermal expansion causing dimensional changes in nylon 66 is easily shown. For example, assume that a part of unspecified length will be required to function at 104°F and 50 percent relative humidity. Using Fig. 20-6, it is easily determined that this part will grow to be 6.8 mils/in. longer in use than as molded.

Anneal for Maximum Stability.

Thus far we have determined the change in the size of the part under use temperature and humidity *after molding*. Another factor affects the size after molding: time.

Depending on part thickness and mold temperature employed during molding, dimensions can decrease with time, especially when parts are exposed to temperatures above 150°F. This is called postmolding shrinkage.

For greatest dimensional stability at elevated end-use temperatures, annealing is sometimes employed after molding to relieve molded-in stresses and to establish a uniform level of crystallinity in the part. (*Note:* The level of molded-in stresses in most 66 nylons is generally low because of their high melt fluidity right up to the onset of solidification. This permits relaxation of flow stresses and orientation effects. Nucleated nylons are sometimes prone to have higher residual stress levels.) Parts made in cold molds tend to be most affected because of rapid melt solidification. Flow-induced stresses in thin sections can be "frozen-in," and quasi-amorphous areas, often induced by cold molds, do not fully develop maximum crystallinity.

As-molded crystallinity depends largely on part thickness and mold temperature. The crystallinity of sections $> \frac{3}{16}$ in. molded in molds $> 175°F$ does not change greatly with time. However, thin sections molded in cold molds, $< 100°F$, can undergo appreciable postmolding crystallization (especially at elevated temperatures), which results in additional shrinkage. Parts molded under restraint (not free to shrink) may on exposure to temperatures $> 150°F$ shrink in the direction of restraint and expand perpendicular (transverse) to the restraint.

In general, articles molded of nylon 66 used at temperatures less than 130 to 150°F do not require annealing. Conversely, for parts exposed to higher temperatures, especially at low relative humidities in an application requiring stable dimensions, annealing is suggested. (Immersion in oil at 325°F to 350°F for 30 minutes is typical.)

Like moisture and temperature changes, the effect of annealing is very predictable. Figure 20-7 shows shrinkage during annealing of test specimens of varying thickness molded over a range of mold temperatures. These annealing changes result in contraction of the part. Often, they tend to negate the effect of moisture uptake at elevated temperatures which, as we said, leads to expansion, and, in many cases, total dimensional change after molding is negligible, since the opposing expansion/contraction effects often counterbalance each other.

To illustrate, if a nylon part were about 0.100 in. thick, molded with a 125°F mold and subsequently annealed, the shrinkage that would occur during annealing would be about 6.5 to 7 mils/in. As we have shown above, total expansion at 104°F and 50 percent relative humidity ultimately causes a 6.8 mil/in. increase in length. The net effect, then, would be little or no total change from the as-molded dimension.

Estimating Mold Shrinkage: A Critical Factor. The most critical factor in planning any injection molded part is the mold-shrinkage estimate. Molds are sized for a particular resin, usually after part design is finalized. It is common practice to leave metal for subsequent machining to final dimensions after trial moldings are made. This is costly, time-consuming, and not always good metallurgical practice, since many tool steels should be properly heat-treated before use. (Fortunately, EDM techniques allow machining of prehardened steel, which permits certain mold or cavity adjustments after trial shots are made.) Nonetheless, it is desirable to size the mold as closely as possible before use.

Parts injection molded from thermoplastics are smaller than the cavity in which they were molded. The reason for the difference in size between the cavity and the part is that the cavity is filled with a melt at high temperature

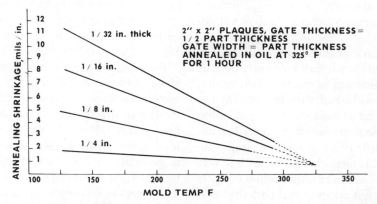

Fig. 20-7   Shrinkage during annealing vs. mold temperature for GP nylon 66 (Zytel 101 NC10).

that is less dense than the cooler solid. Actually, the difference between volume of the mold and volume of the part is the mold shrinkage. Traditionally, however, the difference between any linear dimension of the cavity and the corresponding linear dimension of the part is called the mold shrinkage of the plastic. Conventionally, this difference is expressed as a ratio of the original cavity dimension and is defined:

$$\text{Mold shrinkage (M.S.)} = (C - P)/C, \%,$$
$$\text{or mils/in.}$$

where $\left. \begin{array}{l} C = \text{cavity dimension} \\ P = \text{part dimension} \end{array} \right\}$ consistent units

The changes in density of a plastic during molding (actually, specific volume) depend largely on temperature of, and pressure on, the melt. As melt temperature increases, specific volume increases, and as pressure increases, specific volume decreases. At the freezing point, an abrupt decrease in specific volume occurs as the nylon changes from an amorphous liquid to a semicrystalline solid. As the temperature of the solid nylon is further decreased during cooling, the specific volume continues to decrease. Theoretically, the total volumetric change from melt to solid should approximate three times linear mold shrinkage. In the actual molding situation, nonuniform cooling spoils this simplified approach.

In practice, then, final mold shrinkage is determined by the temperature and pressure of the nylon melt in the cavity at the time of gate seal-off and the thickness and crystallinity of the frozen skin. Since the specific volume of a solid material is considerably less than that of any melt, the greater the thickness of the solid layer, the smaller will be the size change as the part comes to room temperature. Minimum shrinkage is obtained when the part is completely solidified when the gate freezes.

Nucleation of nylon raises the temperature at which solidification occurs and thereby hastens freezing of both the part and the gate. The usual effect of nucleation is to reduce mold shrinkage, but it also increases the amount of frozen-in flow orientation, which can lead to nonuniform shrinkage in flow and transverse directions and, at times, part warpage. (The transverse shrinkage will be greater.)

Pigmentation can also decrease the mold shrinkage of nylon. The greatest effect is seen with high loadings of TiO_2 and other inorganic pigments and salts which act to nucleate nylon. Organic pigments and dyes do not significantly affect shrinkage.

Part and mold geometry are also very important in determining mold shrinkage of a given dimension. If the cavity contains undercuts or cores that restrain the free shrinkage of the part, the as-molded shrinkage will be less than for an unrestrained part. The post-molding or aging shrinkage, however, will be greater for a part that is restrained from free shrinkage in the mold because of greater stresses retained within the part.

One may estimate mold shrinkage of unrestrained parts by using Fig. 20-8. Note that this nomograph is in two sections: mold variables and process variables. The net effect is additive, except notice that the signs are + in A and − in B. The nomograph is based on data obtained from parts of simple geometry. The injection speed and hold time (dwell under pressure) were adjusted to give maximum part weight with the injection pressure, melt temperature, and mold temperature as variables. For optimum predictability when using the nomograph, the importance of obtaining maximum part weight during molding cannot be overstated. If the cavity is not filled to the limit imposed by the gate seal time, then the measured shrinkage will be greater than predicted by the nomograph.

Use the Nomograph.   For illustrative purposes, we've imprinted a typical example on Fig. 20-8 for a part molded in general-purpose nylon (specifically, Zytel 101). Let's look at it:

Mold Variables:
Assume gate width   = 0.125 in.
Assume gate thickness = 0.090 in.
Assume part thickness = 0.125 in.

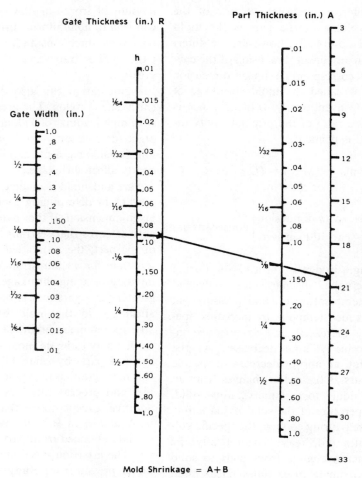

Section A, Mold Variables

Part Thickness (in.) A

Gate Thickness (in.) R

Mold Shrinkage = A+B

Connecting scales as shown, one obtains a value of about 20 for A (sign is +).

Process Variables

Assume melt temp. is 550°F.

Assume mold temp. is 150°F.

Connect points as shown to reference line R.

Using a screw injection machine, the required melt pressure to fill (for the example) is 15,000 psi. (*Note:* Injection gauge pressure on the machine must be converted to equivalent melt pressure; this factor varies with different machine manufacturers.)

Connect reference point (on R) with melt pressure, and read −6 on Scale B_1 (for Zytel 101). Mold shrinkage is A + B = 20 + (−6) = 14 mils/in.

Had a nylon resin other than Zytel 101 been molded, we would connect the point on B_1 horizontally to the specific B scale for the resin used. Note for Zytel 131 (Scale B_4) and Zytel 109 (Scale B_5) that both resins are nucleated and show a different B value, depending on whether measurement is made in the direction of flow or transverse to it.

To illustrate, had we selected Zytel 131 in the preceding example, the A value (+20) would be identical. However, the B value for flow direction shrinkage would be −14, and the mold shrinkage estimate would be:

$$A + B = 20 + (-14) = 6 \text{ mils/in.}$$

If transverse shrinkage were required:

$$A + B = 20 + (-9) = 11 \text{ mils/in.}$$

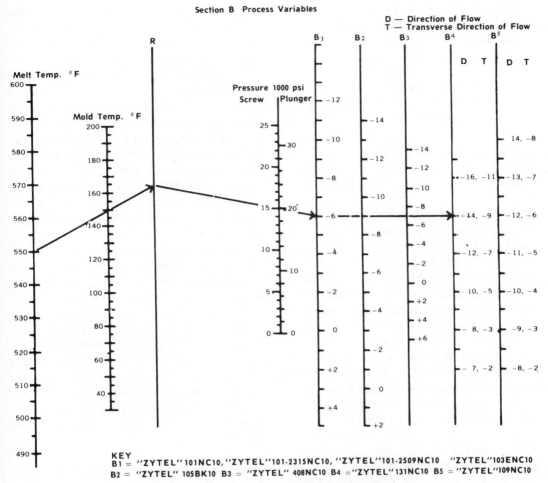

Fig. 20-8   Mold Shrinkage nomograph for nylon plastics.

Only the nucleated 66 nylons show different flow and transverse shrinkage. The other 66 nylons exhibit equal shrinkage in either direction. Surface lubrication and mold release agents do not affect mold shrinkage.

To summarize key points in this discussion, let's combine Figs. 20-6, 20-7, and 20-8 and estimate the mold shrinkage that would be necessary to produce a hypothetical part exposed to certain environmental conditions.

Assume that a part is molded in unmodified nylon 66. Part thickness is $\frac{3}{16}$ in., gate thickness 0.100 in., gate width 0.150 in., mold temp. 100°F, melt temp. 540°F, and injection pressure 12,000 psi. The part must be used at 30% RH @ 175°F. The part is to be annealed after molding for maximum dimensional stability.

The problem: What must be the size of the cavity to produce a part dimension (flow direction) of 1.000 in. in use?

From Fig. 20-6: Annealed nylon 66 will grow 8 mils/in. in use (30% RH @ 175°F).

From Fig. 20-7 (interpolating): For a $\frac{3}{16}$ in.-thick part molded at 100°F, the mold will shrink in length during annealing about 4 mils/in.

From Fig. 20-8: Using given mold and processing conditions, mold shrinkage is:

$$A + B = 22 + (-6) = 16 \text{ mils/in.}$$

Total shrinkage is $16 + 4 = 20$ mils/in. contraction, due to mold shrinkage and annealing.

Total growth is 8 mils/in. expansion, due to relative humidity and temperature.

Net shrinkage is $20 - 8 = 12$ mils/in. contraction or 0.012 in./in.

$$M.S. = (C - P)/C$$
$$0.012 \text{ in./in.} = (C - 1.000 \text{ in.})/C$$
$$C = 1.012 \text{ in.}$$

If cavity dimension is sized to 1.012 in., the corresponding annealed part dimension will be 1.000 in. at 30% RH and 175°F.

Mold to Close Tolerance. Nylon 66 possesses a number of processing characteristics that favor fast overall cycles and high production rates. Chief among them is rapid melt solidification or setup in the mold. In addition, rapid crystallization produces rigidity at elevated temperatures necessary for shape retention during part ejection from the mold.

These two factors can be further enhanced by nucleation. Mold release characteristics can be markedly improved by the addition of small amounts of surface-coated release agents.

The excellent flow characteristics of nylon 66 allow for easy mold penetration in thin sections, even in cold molds, without the need for unusually high melt temperatures or injection pressures. Cold molds in turn speed up melt solidification and minimize the force required to eject parts from molds.

Cure Time: An Important Factor. In this section, the actual injection of melt (ram-in-motion time) typically takes only a fraction of a second. As Fig. 20-9 demonstrates, most of the cycle involves curing or cooling the polymer, which, until gate freeze-off, can be considered to be packing of the cavity.

After gate seal, while the part is cooling until it is stiff enough to be ejected, the screw

normally rotates and retracts to produce the next shot. On very fast cycles, the screw sometimes must continue to rotate while the mold is open in order to produce the required melt.

Figures 20-10 and 20-11 give minimum cure times for nylon 66 at various melt and mold temperatures. Data are plotted for two thicknesses, 0.1 in. and 0.2 in. These times include both time to freeze and time to cool to a temperature where the modulus (stiffness) of the part is suitable for ejection. (Cooling time varies with part thickness approximately as the square of the thickness difference; e.g., if part thickness is reduced by half, cooling is four times faster.)

Cooling times for a nucleated nylon 66 (Zytel 131) are shown in both figures as dotted lines. About 12 to 15 percent cycle reductions are possible at any thickness using this resin.

Figures 20-10 and 20-11 can also be used to predict the changes in cooling time as melt or mold temperatures are varied. Obviously, the coldest melt and mold temperatures that can be used successfully result in the fastest cooling times and shortest cycles. These figures also indicate the necessity of careful planning for maximum mold cooling when fast cycles are involved.

What You Should Know about Mold Release. It can also be seen from Fig. 20-9 that part ejection affects overall cycle. The production cycle can be seriously lengthened if parts hang up and do not fall free from the mold, or if the operator must frequently remove stuck parts.

Mold release agents often help minimize these problems, but be aware that excessive mold release can cause mold vents to plug (leading to part burning) or can contribute to surface defects on the molded part. Also, the best mold releases cannot obviate serious mold-design limitations, such as excessive undercuts, too little draft or taper, improperly cooled cores, hot spots, overpacking of sprues, ejector pin penetration, insufficient knockout, etc.

Mold release is particularly affected by injection pressure and mold temperature. Higher injection pressures and mold tempera-

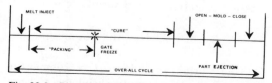

Fig. 20-9   Representation of a typical molding cycle for nylon 66.

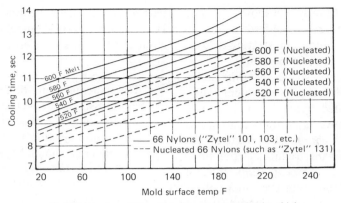

Fig. 20-10   In-mold cooling time for parts 0.1 in. thick.

tures usually necessitate higher ejection forces.

Nucleation Promotes Release. Nucleation of nylon 66 also improves part release. For example, you'll note in Table 20-4 that the nucleated nylon requires 40 percent less ejection pressure than the unnucleated nylon. As a result, it experiences less deformation on ejection than the unnucleated nylon when they are molded under identical conditions. (The lower ejection pressure stems from the fact that the nucleated nylon—Zytel 131—sets up faster and offers less drag resistance during part ejection.)

Being stiffer, nucleated compositions also resist pin penetrations. In the practical mold design case, nucleated nylons can be ejected using small-diameter pins, where frequently unnucleated nylons require sleeve, blade, or other types of more costly stripper knockouts,

Table 20-4   Mold release characteristics.

TYPICAL EJECTION PRESSURES[a], PSI	RESIN
1900–2000	Standard nylon 66 molding resins
1300–1400	Surface-coated with mold release
1100–1200	Nucleated
900–1000	Nucleated and surface-coated with mold release

[a] Based on mold release of single-cavity test specimen, using pressure transducer on hydraulic ejector mechanism.

especially on fast cycles, On balance, in molds that are prone to cause part sticking, nucleated nylon 66 will run on faster cycles.

Surface-coating nylon 66 resins with about 0.1 percent aluminum stearate reduces ejec-

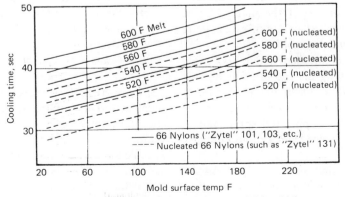

Fig. 20-11   In-mold cooling time for parts 0.2 in. thick.

tion pressure significantly. As shown in Table 20-4, surface-coating a standard nylon 66 molding resin with aluminum stearate reduces ejection pressure about 35 percent.

Although effective, surface-coating unnucleated nylons does not give a product equivalent to a nucleated nylon in mold release characteristics. However, surface-coating nucleated nylon with about 0.1 percent aluminum stearate can further reduce ejection pressures by 10 to 15 percent. Experimental compositions incorporating even more effective internal release agents are under evaluation.

Readily Stripped Undercuts Are a Key to Productivity. While not directly related to fast cycles in the sense of more rapid solidification or easier release, the fact that nylon 66 can be stripped readily from molded-in undercuts should increase production cycles.

A few precautions must be mentioned for you to take full advantage of this characteristic. With few exceptions, hotter mold temperatures permit a greater percentage of undercuts to be stripped. Unfortunately, higher mold temperatures also lengthen cycles, and so a compromise situation always exists. Moreover, nucleated nylon 66 resins and resins with high pigment or particular additive loadings tend to have less ductility (ability to be deformed) than unnucleated resins, and often cannot be specified in molds with deep undercuts.

An undercut is a projection or recess usually perpendicular to the angle of draw, or mold opening. A stripped undercut may be defined as any portion of the molded piece that is either stretched or compressed while being ejected from the mold.

The principle of molded undercuts, while most often involving an end-use function, is also used in mold design; undercuts for holding the molded part on the proper palte during mold opening, sucker pin, and sprue pullers are just a few examples from a mold builder's standpoint.

The design of a thermoplastic part (and the mold) where stripping of undercuts is involved must be approached with caution to prevent part breakage during ejection from the mold. Here are some general guidelines for stripping circular undercuts in thermopolastic materials:

- The undercut must be free to stretch or compress; i.e., the wall of the part opposite the undercut must clear the mold or core before ejection is attempted.
- The undercut should be rounded and well filleted to permit easy slippage of the plastic part over the metal and to minimize stress concentration during part ejection.
- Adequate contact area should be provided between the knock-out and plastic part to prevent pin penetration of the molded part or collapsing of thin wall sections during the stripping action.
- Figures 20-10 and 20-11 should be referred to for minimum mold cooling times (at indicated thickness) before ejection of undercuts is attempted.
- Mold release agents do not increase maximum allowable undercuts.

The method of calculating percent of undercuts, in tension or compression, is shown in Fig. 20-12. The calculation of a maximum allowable undercut is possible if we consider the stripping of an undercut as equivalent to an interference fit, or:

$$\left(\frac{I}{D}\right) = \left(\frac{S}{E}\right) \times (k)$$

where:

I = Interference $\Big\}$ express ratio as % of
D = Diameter $\quad$ original diameter
S = Yield Stress $\Big\}$
E = Modulus $\qquad$ consistent units
k = Constant $\Big\}$

Figure 20-13 gives maximum allowable circular undercuts in percentages for several nylon 66 molding resins as a function of mold temperature. Undercuts should always be specified as a percentage of the diameter being deformed during ejection rather than as a linear value. For instance, a 0.05-in. undercut on a 2-in. diameter can be easily stripped in

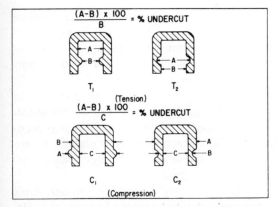

Fig. 20-12 Types of circular undercuts and calculations for maximum allowable undercuts when molding nylon 66.

general-purpose nylon 66 (e.g., Zytel 101) in an 80°F mold, but a 0.05 in. undercut on a 1-in. diameter would require a mold temperature of 180–190°F.

Close Tolerance Molding Can Be Achieved with Fast Cycles.

Like moldings of any thermoplastic resin, parts injection-molded from nylon 66 resins are subject to some variation in dimensions from shot to shot. The allowable variations in the dimensions of an injection-molded part are called the tolerances for the part.

Molding tolerances include the total variations in a part dimension which are caused by deviations in the overall molding operations. These deviations may be found in the mold or in the molding conditions and may be short or long term. Good quality-control

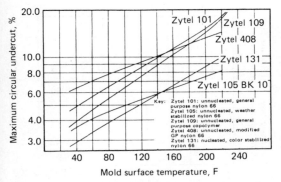

Fig. 20-13 Maximum allowable circular undercuts for different nylon 66 molding resins.

records are invaluable in determining the source of trouble.

The part or mold designer should be aware of a few general pointers in this regard.

First, tolerances set on any dimension by the designer usually represent a compromise between part function and its cost of manufacture.

Second, an important factor (often overlooked) is that plastic parts can usually operate satisfactorily with wider tolerances than can metal parts. It does not pay to specify closer tolerances than necessary.

Third, a part that has many critical dimensions will be more difficult to mold to tolerance than will a part with fewer such dimensions. Tight tolerances should not be put on every dimension, particularly those across a parting line, or on sections formed by movable cores or sliding cams.

Fourth, minimum tolerances are easiest to achieve in a single-cavity mold. Several sources of variability are introduced when multicavity molds are used (e.g., cavity-to-cavity differences and nonuniformity of runners and gates leading to the individual cavities). Fine tolerances usually cannot be achieved in molds that have more than one type of cavity.

In general, the greatest variation in part dimensions is introduced by the molding operation itself. Molding variables must be controlled closely if fine tolerances are specified because slight variations in molding conditions can affect part shrinkage. In order to attain high dimensional reproducibility, it is essential to mold on a fully repetitive cycle.

The ability to maintain close tolerances is dependent on part design, mold design, the injection molding equipment used, and, understandably, the ability of the molder. All areas must be optimized to maintain tight tolerances. (Processing problems that affect molding tolerances are outlined at the end of this chapter.) Without doubt, finer tolerances can be achieved in many cases by resorting to improved control of these problem areas.

Sprues, Runners, Reject Parts: What to Do with Them.

It is common practice for

injection molders to recycle reject parts, along with sprues and runners, through their molding machines. To the molder this reuse of material is frequently the difference between profit and loss on a job, and to the designer it is often the economic incentive to injection-mold a part.

In a typical mold design, it is an unusual occurrence when the sprues and runners amount to less than 25 percent of the shot weight. This percentage can occasionally run as high as 75 percent. It is possible to reuse previously molded nylon 66 without undue sacrifice in physical properties or quality, provided that proper precautions are taken in initial and subsequent moldings and, most important, during interim handling of the reground plastic.

Profitable use of rework demands adherence to three simple precautions:

- Protect regrind from moisture, since all nylons absorb moisture rapidly from the atmosphere. Regrind that is kept covered and reworked promptly (within one-half hour) will usually not require additional drying.
- Ensure that the regrind contains no degraded nylon. Burned or degraded nylon can form points of weakness when mixed with virgin and subsequently molded into new parts. Because a large quantity of virgin resin can be ruined by the inclusion of a small amount of degraded regrind, material held for long periods of time should be discarded and not reground.
- Prevent contamination of rework from other sources. Good housekeeping procedures and limited exposure of rework to dirty surroundings are keys to prevention of contamination.

In the latter regard, here are a number of easy-to-follow suggestions:

- The area and equipment in which the regrind is produced and handled should be kept as clean as possible.
- Grinders should be kept in close proximity to the molding machine. Sprues, run-

ners, and rejects should be reground as soon as they are removed from the machine; continuous reuse of material (blended with virgin to a fixed proportion) is the best policy.

- Runners, sprues, and parts that contain visible contamination must be discarded.
- Regrind should not be allowed to accumulate in an uncovered container. Whenever possible, eliminate intermediate storage of regrind.
- In any regrind-handling system, have some means to remove fine particles. Because of a large surface-to-volume ratio, fines absorb moisture very rapidly and present a large surface for static attraction of dust. Vibrating units equipped with 16–20-mesh screens have been found useful for separating fines. Keeping grinder blades sharp, with proper clearance and screen sizes, will also minimize fines.

How Much Regrind Should You Use? The ratio of reground nylon 66 that may be blended with virgin will depend on both the quality of the regrind and the specifications of the part. If careful regrind handling procedures are followed, high percentages of regrind can be used without difficulty.

The amount used should be established by the ratio of the weight of the sprue and runners to the weight of the parts. It is important that the regrind and virgin be mixed before molding, and that a constant proportion of regrind be maintained.

Figure 20-14 shows the number of succes-

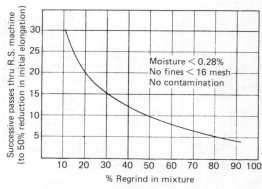

Fig. 20-14 Rework stability of regrind/virgin mixtures.

sive passes (in a screw-injection machine operating at normal conditions) that various blends of regrind to virgin can withstand before the as-molded elongation of nylon 66 is reduced in half (this corresponds to a drop from 60 percent to 30 percent). It is easily seen that as the percentage of regrind in the blend increases, the number of passes decreases.

Another practical way of looking at rework stability occurs where regrind is immediately fed back to the machine and proportioned to an exact sprue–runner/shot weight ratio. In this system, an equilibrium feed composition will be quickly attained.

For example, if a shot which is 50 percent sprue and runner is recycled with virgin resin, a condition rapidly exists where 25 percent of any shot will have been through the machine one time; 12.5 percent, two times; 6.25 percent, three times; 3.12 percent, four times; etc. Based on experimental data, this composition will have an as-molded percent elongation of 50 percent (vs. 60 percent if only virgin resin were used). An 80 percent regrind/20 percent virgin blend will yield on immediate recycle an elongation of 35 percent (vs. 60 percent), while a 20 percent regrind/80 percent virgin yields 58 percent (vs. 60 percent). Thus, using less than 50 percent regrind and recycling immediately results in minimal reduction in elongation. Runner layout should be designed with the weight ratio (to total shot weight) in mind whenever possible.

Chapter 21
Polyphenylene Oxide (PPO) and Injection Molding

George Feth

General Electric Co.
Selkirk, New York

INTRODUCTION

PPO® resin is a registered trademark of the General Electric Company and is the chemical abbreviation for polyphenylene oxide resin. The full chemical name of PPO resin is poly (1, 4-(2,6-dimethylphenyl) oxide.

PPO resin is a thermoplastic resin in powder form, and is primarily used to modify or alloy with other thermoplastics such as polystyrene. These thermoplastic alloys are manufactured by GE and sold under G.E.'s NORYL® Products Division trademark. They are the base of a large family of resins with typical heat distortion temperatures (DTUL) ranging from 180°F to over 300°F. While there are several dozen grades possible, there are currently 30 standard grades of PPO modified resins available for injection molding applications.

Because of the large number of resin formulations available in filled, unfilled, or flame-retarded versions, PPO modified resin has found a wide range of applications in appliances, electrical or telecommunications products, and automotive products, and as a replacement for flame-retardant and high-heat ABS, flame-retardant PS and ABS/PC and ABS/PVC alloys. PPO modified resin is a lower-cost alternative to polycarbonates, nylons, polyesters, and die-cast metals such as zinc, aluminum, brass, or bronze in the auto-

motive, electrical, construction, and liquid-handling markets.

PPO modified resins feature low specific gravity and low moisture absorption rates. The resins offer the maufacturer a combination of attractive molded-in finishes with a wide range of properties including heat resistance, good impact strength, and strong tensile and flexural properties at elevated temperatures.

The resins are noted for their resistance to hydrolysis in acids, bases, salts, and hot water. However, they may be attacked by certain hydrocarbons, esters, ketones, amines, and halogenated compounds.

Although some PPO modified resins contain polystyrene, normal solvents for styrene materials such as methylethyl ketone (MEK), tetrahydrofuran (THF), or cyclohexanone are not considered solvents of PPO resin. The resins are soluble in some halogenated solvents such as trichloroethylene (TCE), and chloroform. Xylene and toluene are fairly good solvents, but are slow evaporating. Most often, mixtures of trichloroethylene and dichloromethane are used as solvents, depending on the speed of evaporation needed.

Bonding agents such as cyanoacrylates, acrylics, epoxies, silicones, polyurethanes, and polymeric hot melt adhesives are available for PPO modified resins, but some adhesive formulation variations will cause adhesives to act as aggressive chemical agents rather than bonding agents, causing parts to crack and fail.

Physical, thermal, and environmental properties vary from resin to resin, and temperature, time, load, and resin concentration also play a role, both in the selection of the resin grade and in retaining physical properties. Other environmental considerations such as heat, ultraviolet radiation, or high energy radiation can also have an effect on the resin.

Thermal aging and outdoor weathering can cause some loss of impact properties, but the physical properties such as, tensile and flexural properties, change relatively little. In ultraviolet exposure, either simulated sunlight or indoor fluorescent light, the resins will yellow or darken somewhat, depending on the initial color and the length of exposure. However, UV-stabilized grades are available, which are more stable by a factor of three. High energy radiation, such as gamma or X-rays, causes little loss of properties, even at high dosages. Molded articles can be sterilized by autoclave, gas, or irradiation without harm to the material.

In processing, PPO modified resins provide excellent flow. Their shrinkage is low and predictable for consistent tolerance reproducibility. Their high stability overcomes warpage problems common to many other plastics, particularly in long, thin sections. PPO modified resins are also inherently stable and will resist degradation even under processing extremes.

While all types of injection molding machines have been used successfully with PPO modified resins, screw machines are preferred over ram or combination presses because of their rapid and more uniform heating of the resin, and reduced pressure loss through the cylinder. In addition, screw presses offer shorter cycle times, generally superior part appearance, and lower molded-in stress.

The length-to-diameter (L/D) ratios preferred on general-purpose screws are in the range of 20/1 to 24/1, and compression ratios of 2/1 to 3/1 are said to be most satisfactory. Vinyl screws and tips are not recommended for injection molding applications because they may cause shear degradation and give poor physical properties.

Short, free-flow nozzles having a minimum orifice of $\frac{3}{16}$-in. diameter or larger are recommended for molding PPO modified resin products. An orifice of $\frac{5}{16}$-in. is recommended for glass-reinforced resins. Shut-off nozzles or torpedo type heaters or mixing heads are not recommended, since they may produce severe localized heating or produce areas where materials could hang up and become degraded at the high molding temperatures. Extended nozzles must have uniform, adequate temperature control capabilities.

Clamping pressures depend on wall thickness and the length of flow. Generally, 3 to 5 tons per square inch of projected area is adequate to maintain tolerances and avoid flashing. Usually 40 to 80 percent of the barrel capacity of the machine is also recommended. This ensures part dimensional tolerances and avoids long residence times at high temperatures, which may cause degradation of the polymer material. Degraded material can cause gasses to become entrapped in the melt and result in surface splay or brittle parts.

As a rough guideline, to estimate the target melt temperature of a particular grade of PPO modified resin for injection molding applications, simply add 260°F to the heat distortion temperature. For example, the NORYL N190 Grade of PPO modified resin has a DTUL of 190°F. This plus 260°F would equal 450°F, a good starting temperature for this material. For N300 with a DTUL of 300°F, the starting temperature would be 560°F. In addition to temperature, the length of flow, part design, and mold temperature will also affect both the molded part and material flow.

Mold temperatures should be kept high when working with NORYL resins. High temperatures prevent poor knit-line integrity, reduce molded-in stress, and improve surface gloss. Most NORYL resins require a minimum mold temperature of 150°F, and for glass-filled and high heat resins, 200 to 250°F is not uncommon.

If the shot size and cycle times are not mismatched, barrel temperatures of up to 600°F can be used on some grades of NORYL resin without loss of properties. In general, an increase in melt temperature of up to 50°F is allowable, without loss of properties, using the guideline of 260°F plus the HDT of the resin.

Medium to fast ram speeds should be used with NORYL grades, speeds in the 2- to 15-second range. High booster speeds may cause frictional burning in marginally designed molds and show up as splay. NORYL PN235 resin requires very slow ram speeds to minimize the molded-in stress in the part. Screw speeds in the 40- to 80-rpm range provide excellent results. Cycle times are primarily dependent on part thickness. Small parts of 0.060 in. to 0.080 in. wall have overall cycles of 15 to 30 seconds. Sections up to .250 in. can be molded in 60 seconds or less. Specific cycle time is dictated by part and mold design, and by processing conditions.

One of the main advantages of working with PPO modified resin is its extremely low rate of moisture absorption. Generally speaking, NORYL grades can be molded without the need for drying. Without drying, while some surface splay can result from trapped moisture, physical properties are not affected. When optimum surface appearance is required, or when some kind of secondary finishing is to be done, drying of the resin is recommended.

Drying times and temperatures vary significantly with each resin grade. In most cases, especially during humid summer weather, the only moisture present is on the surface of the resin. This moisture is not absorbed and does not require prolonged drying for removal.

When air circulating ovens are used, NORYL pellets should be dried in shallow trays. Recommended drying temperatures range from 200 to 215°F for N190, up to 240 to 265°F on N300 or glass-filled resins. Recommended drying times are generally 2 to 4 hours. If the dryer is operating efficiently, up to 90 percent of the moisture is removed in the first hour of drying. Extended drying times will not provide improvements in surface appearance. Extended drying, in fact, may increase surface splay due to degradation of the polymers.

NORYL resins can be reground and used with virgin pellets without any loss of physical properties. The recommended use of regrind in injection molding systems is in the range of 25 percent regrind to 75 percent virgin pellets. Under proper conditions, NORYL resin has produced samples with very little loss of physical properties even after seven regrinds. Care should be taken, however, to keep degraded material away from regrind material and virgin material, as the usable materials could become contaminated and rendered useless.

The ability to be reground, in addition to its low specific gravity, makes NORYL Grades of resin more economically attractive when compared to some "low-cost" resins. For example, a NORYL N190 resin at $1.42/lb with a specific gravity of 1.08 compares favorably with a FR-ABS at $1.28/lb with a specific gravity of 1.21. By converting the cost of the resins per pound to cost per cubic inch, the apparent advantage of the FR-ABS disappears. To convert to cost per cubic inch, the following equation is used: (cost/pound in cents) × (specific gravity) × (.0361) = cost per cubic inch. Using the values cited, the equation is: (142 cents) × (1.08) × (.0361) = 5.54 cents/cu in. for N190, and: (128 cents) × (1.21) × (.0361) = 5.59 cents/cu in. for FR-ABS.

Similarly, if a glass-filled resin such as NORYL GFN3 resin with a specific gravity of 1.27 at $2.10/lb is compared to die cast zinc with a specific gravity of 7.10 at $0.15/lb, the cost per cubic inch of GFN3 is 9.63 cents vs. 3.84 cents for the zinc. However, the zinc part in all likelihood will require machining, deburring, drilling, or tapping operations, which add to the overall cost, while NORYL resin is ready as molded. In addition, the NORYL saves in shipping costs, since one zinc part typically weighs as much as five plastic parts.

Nearly all finishing and bonding techniques can be utilized with PPO modified resins. Some finishing, such as sputtering or plating, is done with special grades to meet specific requirements. PPO modified resins can be decorated with most thermal, chemical, or bonding techniques available to the general plastics industry.

MOLDING CONDITIONS

With the different PPO modified grades, different molding conditions occur. As an example, the flow of the resins is improved with increasing wall sections and melt temperatures (see Figs. 21-1 through 21-4). In these figures the effect of the wall section on flow length is clearly demonstrated. Not only is flow increased by larger sections, but also the slopes

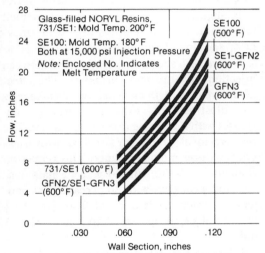

Fig. 21-1   Melt flow vs. wall section thickness.

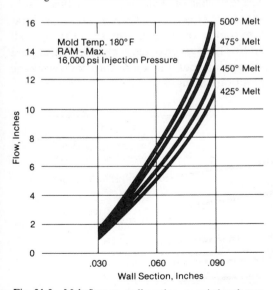

Fig. 21-2   Melt flow vs. wall section at varied melt temperatures—NORYL N225 resin.

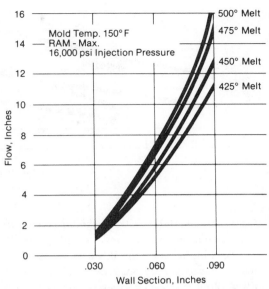

Fig. 21-3   Melt flow vs. wall section at varied melt temperatures—NORYL N190 resin.

of the curves are increased. As thickness is increased from .060 in. to .090 in., under the conditions shown, a 100 percent improvement in flow is possible. Although more viscous, glass-reinforced NORYL resins demonstrate proportionally similar improvements in flow with increased injection pressure and wall sections.

Small parts with flow lengths up to 2 in. have been filled in .025-in to .050-in. thick-

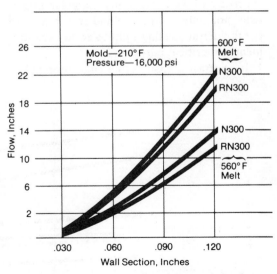

Fig. 21-4   Melt flow vs. wall section at varied melt temperatures—NORYL N300/RN300 resins.

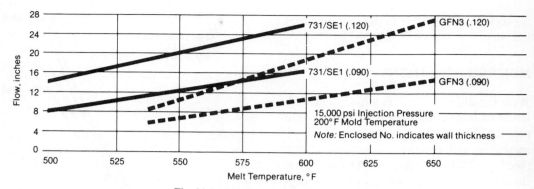

Fig. 21-5   Melt flow vs. melt temperature.

nesses. In general, 0.050 in. *should be considered a minimum thickness for all NORYL resins.* Thinner sections in parts whose flow length exceeds 2 in. may be troublesome to fill without the use of excessive temperatures and/or pressures.

Increased melt temperatures reduce resin viscosity. Figure 21-5 illustrates the effect of lower viscosity on flow. In this case, injection pressure and mold temperature are held constant. Depending upon thickness, increases in flow of from 10 to 50 percent are possible with 100°F increases in melt temperature.

EFFECTS OF MOLDING CONDITIONS ON SHRINKAGE

The low, predictable, and constant mold shrinkage of NORYL resins is a great advantage, not only to the toolmaker but also to the molder in meeting tight tolerance requirements on critical parts.

In contrast to many other engineering plastics, the shrinkage of NORYL resins varies only slightly with changes in pressure, melt temperature, mold temperature, wall thickness, and flow direction.

Figure 21-6 shows the relationship between injection pressure and shrinkage. The effect of different wall thicknesses on shrinkage is also illustrated. As expected, shrinkage decreases with increasing injection pressure. The reinforcement of NORYL resins with glass fibers reduces mold shrinkage. This reduction is about 3 mils/in. with 20 percent glass content and 4 mils/in. with 30 percent glass content. As thickness is doubled from $\frac{1}{8}$ in. to $\frac{1}{4}$ in., shrinkage is increased by only about 1 mil/in.

Like injection pressure, increased melt temperature has a tendency to reduce mold shrinkage (Fig. 21-7). Over a broad temperature range of 100°F, the shrinkage of all NORYL resins changes very little. This figure also points out the uniformity of shrinkage in all directions.

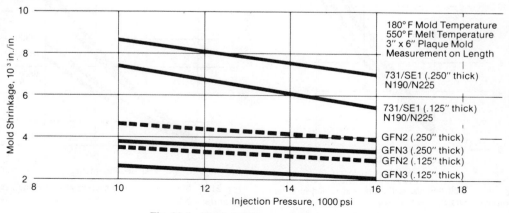

Fig. 21-6   Mold shrinkage vs. injection pressure.

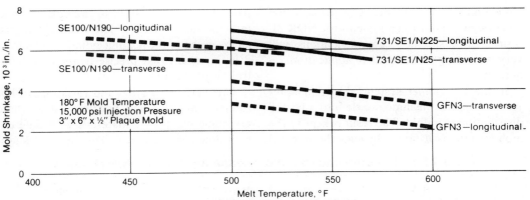

Fig. 21-7   Mold shrinkage vs. melt temperature.

Mold temperature over 125°F has essentially no effect on the shrinkage of NORYL resins. Typical shrinkage data for NORYL resins in parts of different geometries are illustrated in Fig. 21-8.

MOLD RELEASE

Sticking rarely occurs when parts are produced from well-designed molds. A light dusting or gently spraying of any of the mold release agents listed in Table 21-1 works well when required.

HOT RUNNER MOLDING

NORYL thermoplastic resins, because of their thermal stability and wide processing temperature ranges, are being successfully molded by this method for small parts in multicavity molds and for large multigated parts such as automotive grilles, instrument panels, and large business machine housings.

As in any operation, however, certain procedures should be followed to ensure that optimum performance is obtained from the molded part. The following information is in-

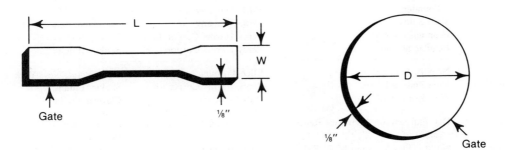

		731/SE1/N225	SE100/N190	All NORYL Reinforced Resins	
Molding Conditions					
Cylinder Temperature:		550°F	450°F	600°F	
Injection Pressure:		15,000 psi	15,000 psi	18,000 psi	
Mold Temperature:		200°F	180°F	200°F	
		Shrinkage, in./in.			
Part	Dimensions In.	731/SE1/N225	SE100/N190	GFN2/SE1GFN2	GFN3/SE1GFN3
Tensile Bar	L, 8.5	.006	.005	.002	.001
	W, 0.75	.005	.005	.0025	.002
Disc	D, 4.0	.005	.006	.002	.001

Fig. 21-8   Typical mold shrinkage data.

tended only to offer guidance in the selection and design of hot runner systems for NORYL resins, since specific recommendations will vary according to individual application requirements.

In the hot runner system, the resin is injected from the machine nozzle into a heated distribution manifold. The resin is kept in the molten state as it flows through the internal passages of the manifold to the bushings and into the mold cavities. The manifold is just an extension of the heating cylinder; therefore, it is important that precise temperature control be provided.

There are many types and variations of hot runner manifolds and nozzles just as there are numerous plastic materials with different melt-flow characteristics. The fact that a sys-

Table 21-1 Recommended mold release agents for NORYL resins.

TYPE	MOLD RELEASE AGENT	MANUFACTURER	ADDRESS
Silicone	Silicone Mist S220	IMS Company	Cleveland, Ohio
	Silicone Spray S512	IMS Company	Cleveland, Ohio
	Blue Label M416	IMS Company	Cleveland, Ohio
	Mixed Viscosity B516	IMS Company	Cleveland, Ohio
	Super 33	IMS Company	Cleveland, Ohio
	Sil-A Spray	Ellen Products	Stony Point, New York
	Silicone Resin 510	Ellen Products	Stony Point, New York
	Silicone Spray 500	Ellen Products	Stony Point, New York
	10/0 Silicone 3001	Crown Industrial Products	Hebron, Illinois
	20/0 Silicone 3012	Crown Industrial Products	Hebron, Illinois
	30/0 Silicone 3023	Crown Industrial Products	Hebron, Illinois
	40/0 Silicone 3034	Crown Industrial Products	Hebron, Illinois
	50/0 Silicone 3045	Crown Industrial Products	Hebron, Illinois
	Pure Silicone S100	Zip Aerosol Products	North Hollywood, California
	Bomb Lube-Blue Label	Price-Driscoll	Farmingdale, New York
	Dura Slick	American Durafilm Company	New Lower Falls, New York
	Midget Silicone S306	IMS Company	Cleveland, Ohio
	15% Silicone S712	IMS Company	Cleveland, Ohio
	2 Pounder	IMS Company	Cleveland, Ohio
	Verti Spray	IMS Company	Cleveland, Ohio
	Economist 41620	Percy Harms Corporation	Skokie, Illinois
	Regular Silicone 40120	Percy Harms Corporation	Skokie, Illinois
Stearates	No Stik Zinc #506	Ellen Products	Stony Point, New York
	Slide Zinc #41016	Percy Harms Corporation	Skokie, Illinois
	Zinc Stearate Z212	IMS Company	Cleveland, Ohio
Fluorocarbons	Dry Flurocarbon Lubricant Release #501	Crown Industrial Products	Hebron, Illinois
	No Stik Fluorocarbon Mold Release #507	Ellen Products	Stony Point, New York
	Slide Fluorocarbon	Percy Harms Corporation	Skokie, Illinois
	Fluorocarbon Dry Lubricant D5440	Zip Aerrosol Products	North Hollywood, California
Paintable	No Stick Silicone Mold #6075	Ellen Products	Stony Point, New York
	Pure Silicone S120	Zip Aerosol Products	North Hollywood, California
	Vydax Spray #V416	IMS Company	Cleveland, Ohio
	Slide Silicone #40020	Percy Harms Corporation	Skokie, Illinois
	Slide All Purpose	Percy Harms Corporation	Skokie, Illinois
Food and Drug	Anti-Stick for Food and Drug	Price-Driscoll	Farmingdale, New York
	Slide Silicone #40120	Percy Harms Corporation	Skokie, Illinois
	Pure Silicone Anti-Stick for Food Processing FS175	Zip Aerosol Products	North Hollywood, California

Table 21-2   Hot runner systems.

	NORYL RESINS	NORYL REINFORCED RESINS
Manifold		
Cartridge heaters outside of melt stream	Good	Good
Internal heaters in melt stream	Not recommended	
Nozzles		
External heaters;		
Cycle less than 20 seconds	Good	Good
Cycle more than 20 seconds	Good	Good
Nozzle shut-offs	Not required	Not required
Internal heaters;		
Cycle less than 20 seconds	Satisfactory	Not recommended
Cycle more than 20 seconds	Not recommended	Not recommended
Insulated		
Runner mold	Not recommended	Not recommended

tem performs satisfactorily with one material does not mean that it will perform with all materials.

Listed in Table 21-2 are the results of molding NORYL resins in several types of hot runner systems. Those classified as "good" have demonstrated good performance in all respects.

Figure 21-9 is a cross-section assembly diagram of a basic hot manifold system for delivering hot resin from the injection nozzle to the feeder bushings. The primary sprue bushing feeding the manifold should be as large as possible, up to $\frac{3}{8}$ in., for optimum fill rate and minimum pressure loss.

Manifolds are designed to suit each individual mold pattern, and therefore differ with each application. However, there are certain recommendations that must be followed to obtain optimum performance in the system:

- Manifold flow passages should be a minimum of $\frac{3}{8}$ in. and a maximum of $\frac{9}{16}$ in. in diameter. They should also be smooth and highly polished.
- Manifold ends must be rounded to eliminate possible material hang-up and degradation.
- For best heat distribution use four cartridge heaters for the manifold. (The manifold should also be controlled separately from the nozzle by variacs or solid state controls.)
- Use 75 watts per cubic inch of manifold

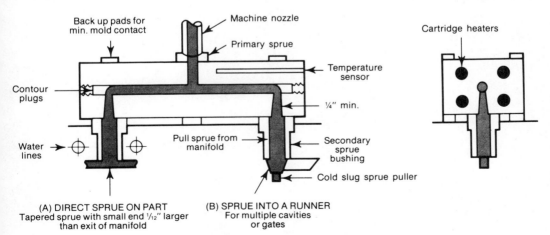

(A) DIRECT SPRUE ON PART
Tapered sprue with small end 1/12" larger
than exit of manifold

(B) SPRUE INTO A RUNNER
For multiple cavities
or gates

Fig. 21-9   Distribution manifold.

to estimate cartridge wattage. Normally this will allow heaters to operate at approximately 60 percent of full wattage for extended heater life.

Multiple secondary sprues can be dropped at convenient locations on a large part, or the sprue can feed a runner which in turn feeds a cluster of small parts as illustrated in the diagram.

A large part can be edge-gated while still being clamped centrally in the molding machine by utilizing only half of the system shown.

Chapter 22
Polycarbonate—Processing Influences on Properties

Jonathan M. Newcome

Mobay Chemical Corporation
Pittsburgh, Pennsylvania

INTRODUCTION

Chemistry and Physics

Commercial grade polycarbonate is a linear polyester of carbonic acid in which the carbonate groups recur in the polymer chain. This engineering thermoplastic is based on bisphenol A and has an aromatic structure as shown in Fig. 22-1. A product of the reaction of the sodium salt of bisphenol A with phosgene, its rigid aromatic rings coupled to the methylated carbon atom provide the polymer with its engineering properties (1). Although the linear polyesters may contain aliphatic, aliphatic–aromatic, or aromatic constituents, it is the aromatic type that is best recognized as polycarbonate. It is this structure that is responsible for the high softening temperature, broad temperature usage, rigidity complemented by toughness, resistance to creep, and other important properties.

Polycarbonate is basically an amorphous polymer and is therefore transparent. Although there are some grades of lower-molecular-weight polycarbonate that can be specially processed into a semicrystalline state, it is the amorphous polymer that is of the greatest commercial interest.

Although most basic grades of polycarbonate are linear polymers, polycarbonate can be produced with a limited degree of short chain branching which dramatically effects the low shear viscosity of the melt. The branched grades are generally used for extrusion although they have shown some application to injection molding where mold design and its effect on melt rheology have been taken into consideration.

Polycarbonates that are suitable for injection molding generally fall into the weight average molecular weight range of 26,000 to 35,000. Molecular weights higher than the upper limit of this range tend to be difficult to process because of high melt viscosity. Basic grade polycarbonate is typically available in three molecular weight grades: low, medium, and high. Since the viscosity increases as molecular weight increases, the molding application may dictate grade selection.

Properties

One of the important engineering attractions of polycarbonate is the limited creep it experiences when under design loads. Figure 22-2 describes the stress–strain relationship as a function of time (2). The temperature refer-

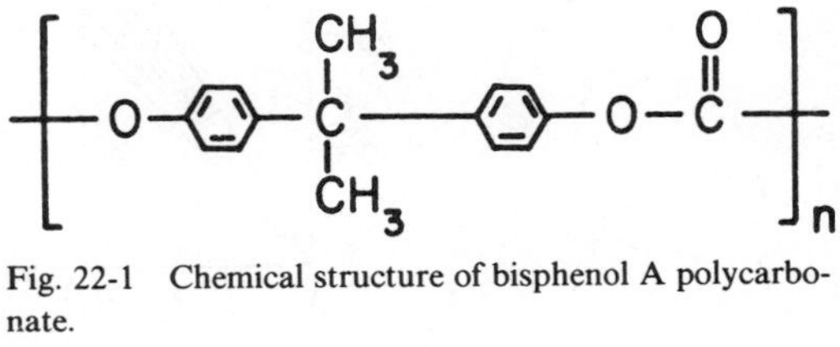

Fig. 22-1 Chemical structure of bisphenol A polycarbonate.

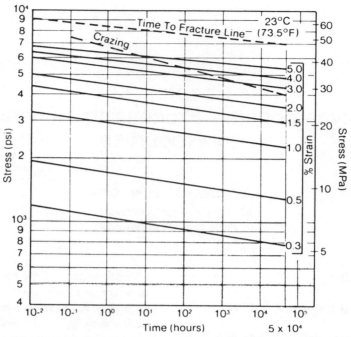

Fig. 22-2   Stress–strain–time correlation resulting from creep at 23°C.

ence for these data is 23°C. A given strain will produce a stress that decays with time as stress relaxation occurs. The time-to-fracture limit is also shown in the upper portion of the curve. This details the failure strength of the material as a function of time. However, before failure occurs, crazing is initiated at the surface. These small cracks may be likened to many small notches, and their effect would be to reduce impact strength and elongation.

Another way to represent these engineering data is to plot isochronous stress–strain curves as depicted in Fig. 22-3. With time as the parameter, creep may be determined as a func-

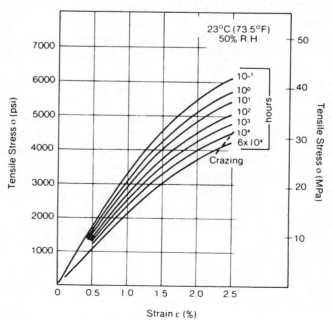

Fig. 22-3   Isochronous stress–strain curves at 23°C.

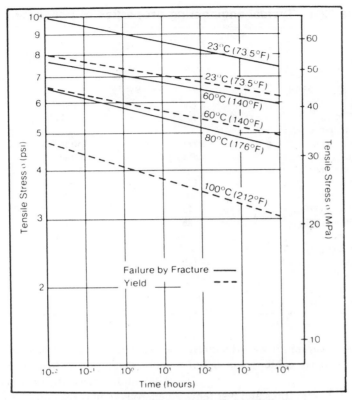

Fig. 22-4   Time-to-fracture lines of polycarbonate.

tion of a given stress level. The family of curves suggests that as stress level increases, resultant strain increases.

Creep is also influenced by temperatures. As temperature increases, the tensile stress at given strain decreases. A series of plots similar to Fig. 22-3 may be generated for test temperatures above 23°C. However, one concise way of representing the extreme in behavior at elevated temperature is to plot time-to-failure envelopes. This is done in Fig. 22-4, where both yield and failure are represented. At higher temperatures, lower stresses are required to promote yield and failure. Even at the elevated temperatures, polycarbonate offers design flexibility for engineering demands.

As with other polymers, some mechanical properties of polycarbonate are sensitive to molecular weight. Most noticeable are notched Izod impact strength, critical thickness, dart drop impact, and to a lesser extent heat deflection temperature. Table 22-1 lists these and other properties typical of various molecular weight polycarbonate as described

by ASTM D-2473 melt flow cell numbers. As molecular weight decreases, flow becomes easier and melt flow rate increases. In the same respect, some of the mechanical properties listed in Table 22-1 decrease with decrease in molecular weight. Even for low-molecular-weight polymer, these attributes of polycarbonate are the basis for its selection as an engineering resin.

Decisions on part wall thickness should take into account the influence of molecular weight on the notch sensitivity of polycarbonate. This phenomenon is represented in Fig. 22-5, where notched Izod impact strength is shown as a function of specimen thickness. The parameter for the curves is melt flow rate, a secondary indicator of molecular weight by an inverse relationship. The dramatic reduction in notched Izod impact strength at a specific thickness is dictated by molecular weight once specimen geometry, notch depth, and notch radius have been fixed. Melt flow rate increases, molecular weight decreases, and the thickness at which this transition occurs de-

Table 22-1 Typical engineering properties of polycarbonate.

	HIGH MOLECULAR WEIGHT	WEIGHT MOLECULAR WEIGHT	LOW MOLECULAR WEIGHT
ASTM D 2473			
Melt flow cell no.	6 and 7	4 and 5	2 and 3
Melt flow rate (g/10 min)	3 to 5.9	6 to 11.9	12 to 23.9
Notched Izod impact strength (J/m)			
3.2 mm thickness	800 to 900	750 to 850	640 to 750
6.4 mm thickness	120 to 150	100 to 130	95 to 120
Heat deflection temperature (°C)			
@ 1.82 MPa load	133	132	131
Brittleness temperature (°C)	−101	−101	−101
Tensile strength @ break (MPa)	72	70	69
Transmittance @ 550 mm (%)	86	87	88
Critical thickness (mm)	5.6	4.6	3.8
Dart drop impact strength (J)			
@ 3.2 mm thickness	190	150	120
25 mm diameter tip			
12 mm radius tup			
10 kg hammer			

creases. As a result, minimum wall thickness is recommended in parts exhibiting a notch potential.

The influence of notch radius and test temperature on notched Izod impact strength is shown in Fig. 22-6. Critical thickness is shown as a function of notch radius, the parameter being test temperature. Critical thickness is defined as that thickness at which notched Izod impact strength undergoes a significant reduction. From a design point of view, it represents the increase in notch sensitivity as thicker wall sections are considered. As test temperature increases, so does critical thick-

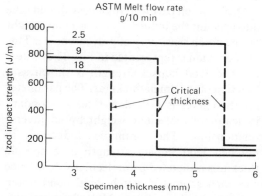

Fig. 22-5 The effect of specimen thickness on Izod impact strength of various molecular weight polycarbonates.

ness. The influence of notch radius is even more striking at radii in excess of 0.25 mm. The implications are obvious as the use of generous radii is recommended whenever possible.

Although polycarbonate has many outstanding engineering properties, it does have some limitations regarding use in some environment. It is generally not recommended for use in the presence of organic solvents such as chlorinated hydrocarbons, ketones, esters, aromatic hydrocarbons, etc. (3). Polycarbonate does have good room temperature resistance to water, dilute organic and inorganic acids, oxidizing and reducing agents, and neutral and acid salts, as well as mineral, animal, and vegetable oils and fats, and aliphatic and cyclic hydrocarbons.

Polycarbonate may be fabricated with a very smooth surface, but it has limited scratch and abrasion resistance. For outdoor use this problem is compounded by moisture and ultraviolet light. Recently, however, hard coating systems have become available to offer better protection against abrasion for both indoor and outdoor use. Some of the coating systems provided improved resistance to environments that would normally be aggressive to polycarbonate. Because of the design of these coat-

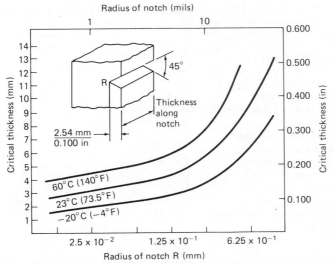

Fig. 22-6 Impact strength of polycarbonate at various temperatures vs. notch radius.

ings, each application has to be viewed individually to determine the best course of action for improvements in abrasion and environmental resistance. This can provide a means of overcoming an inherent limitation.

PREMOLDING PREPARATION

Drying

Similar to other polyesters, polycarbonate is hygroscopic and will absorb moisture from its surroundings. This characteristic often proves detrimental to the physical properties of the material when processing at high moisture levels. The result is a chemical reaction between the polymer and water that reduces the molecular weight of polycarbonate. As discussed later, this can have profound effects.

To ensure retention of engineering properties, moisture should not exceed 0.01 percent in the pellets (4). Since the equilibrium moisture content at 23°C and 50 percent relative humidity is 0.18 percent, drying in desiccant hopper dryers or forced convection ovens will be necessary. Although equilibrium moisture levels may be low, the polymer retains the water tenaciously as drying is diffusion-controlled. Hence, delivery air must be supplied from the desiccant unit to the hopper at −18°C dewpoint and 120°C. Drying may be accomplished in shallow trays using forced convection ovens operated at 120°C and with a fresh air makeup of 10 percent. In this case, pellet depth should be limited to $1\frac{1}{2}$ in. In either case, oven or desiccant hopper dryer, hot air contact time with the pellets should be 4 hours.

Regrind

Regrind usage is a concern for any injection molder with an eye on profit. Maximum utilization must be made of every pound of resin, particularly when engineering thermoplastics are involved. Scrap generated in the injection molding of polycarbonate may be reground and blended with virgin material and used successfully if certain precautions are observed.

The proportion of regrind blended with virgin resin will be influenced primarily by the shear and thermal histories of the melt. Long residence times in the barrel in combination with high melt temperatures may result in an increase in melt flow rate and subsequent change in physical properties. The sensitivity of these properties to processing also depends upon the grade of polymer. Table 22-2 compares physical properties according to regrind history for three different polycarbonate grades: two natural grades of different molecular weight and a flame-retardant grade. In comparing the change in melt flow rate, natural grades appear less sensitive to change than

Table 22-2 Properties of 100 percent polycarbonate regrind.

MATERIAL—PROPERTY	VIRGIN	REGRIND HISTORY 1ST	2ND	3RD
High Molecular Weight—Natural				
Melt flow rate (g/10 min)	4.6	4.9	5.0	4.9
3.2 mm Izod notched impact (J/m)	956	945	935	950
Yellowness index	2.8	4.8	7.3	10.1
Low Molecular Weight—Natural				
Melt flow rate (g/10 min)	15.2	15.2	16.2	16.0
3.2 mm Izod notched impact (J/m)	820	880	810	820
Yellowness index	1.8	3.5	5.1	6.6
Flame-Retardant—Natural				
Melt flow rate (g/10 min)	11.7	12.2	13.0	14.9
3.2 mm Izod notched impact (J/m)	110	105	95	100
Yellowness index	3.8	5.0	6.8	8.5

the flame-retardant or specialty type. In natural grades, higher melt flow rate material is slightly more sensitive to melt flow rate change than material with a lower melt flow rate. The most notable change in property occurs in the yellowness index, which is observed in all three grades. As pigmentation and other additives are introduced, physical properties would be expected to be influenced even more by regrind history.

In regard to retention of mechanical properties, the highest probability for successful regrind use will be found with high-molecular-weight natural polymer. In this case, color may be the discriminating factor. In certain applications, use of 100 percent one-time regrind may be acceptable, whereas a recycle stream of 25 to 30 percent regrind could prove detrimental to physical properties because of the regrind history distribution of the stream. The general recommendation of 20 percent regrind loading has been found to be suitable in most cases.

In certain applications, regrind usage is discouraged. High-quality optics demand stringent color and transmission standards that can be met by special grades of polycarbonate. Regrind blending of these grades, however, risks loss of their excellent optical properties.

PROCESSING

Processing Hardware Requirements

Injection molding of polycarbonate does require some special capabilities of the process-

ing equipment. Because of its high viscosity, polycarbonate is usually processed at a high temperature in order to obtain a less viscous melt. This requires barrel temperature capability of up to 350°C. Even at such reduced viscosity levels, a high-molecular-weight polycarbonate will require higher injection forces than lower-molecular-weight resin to fill certain part geometries. Therefore, injection molding equipment suited to processing polycarbonate should be capable of at least 138 MPa injection pressures.

All surfaces coming into contact with polycarbonate melt should be smooth, pore-free, and—for practical purposes—hard. High alloy steels are recommended for barrels; for barrel liners special alloys such as Xaloy* or Bernex** may be used. Screws should employ a hard steel to minimize wear but, most important, should be chrome-plated in order to reduce the tendency for polycarbonate to adhere and degrade on their surface.

Suggested screw designs for polycarbonate are illustrated in Fig. 22-7. Here a metering type screw is depicted. A generous feed length should be allotted to solids transport and melting. A rapid transition in the compression zone is not recommended, owing to the viscous nature of polycarbonate. Such a sudden compression could result in overloading of the screw or drive motor if melting were incomplete when the plastic entered this zone. In

---

* Registered trademark of Xaloy, Inc.
** Registered trademark of Bernex, Corp.

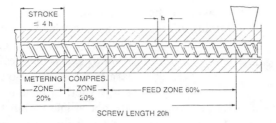

STROKE
≤ 4 h

h

METERING | COMPRES.
ZONE | ZONE | FEED ZONE 60%
20% | 20%

SCREW LENGTH 20h

SCREW DIAMETER (mm)	DEPTH OF ZONE		COMPRESSION RATIO
	FEED (mm)	METERING (mm)	
30	3.6	1.8	2.0:1
60	6.6	3.0	2.2:1
90	9.5	4.0	2.4:1
120	12.0	4.8	2.5:1
>120	Max. 14.0	Max. 5.6	Max. 3.0:1

SCREW PITCH

H = 1.0 D for Screw Diameter < 80mm
H = 0.9 D for Screw Diameter > 80mm

Fig. 22-7 Typical screw design for injection molding polycarbonate.

such a case, the high modulus of the pellets would create a sufficient resistance to deformation to cause degradation of the polymer and seizure of the screw. Screw pitch recommended for screws of diameter less than 80 mm is 1.0 D, and 0.9 D for diameter greater than 80 mm.

Minimum screw L/D is 15:1, and as indicated in Fig. 22-7, the compression ratio should be 2:1 for small screws, increasing to 2.5:1 to 3:1 for the larger diameters. The increase in compression ratio for larger diameters enhances back mixing of the melt which tends to offset the reduced efficiency of a deeper feed section.

Normally required, a shut-off valve should provide good flow characteristics, as illustrated in Fig. 22-8. The slip ring design employs a full channeled tip which aids in minimizing flow restriction. As with many other thermoplastics, polycarbonate will degrade when confined to dead spaces for long residence times. In addition, changing from a color to a natural or from color to color may

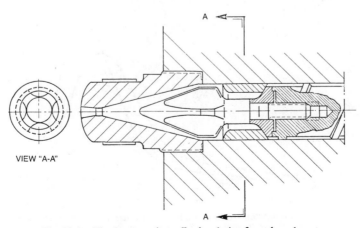

A

VIEW "A-A"

A

Fig. 22-8 Slip ring type shut-off value design for polycarbonate.

present the problem of color streaking if the shut-off design impedes self-cleaning of the tip.

Hydrolysis

When problems with performance or cosmetic features of polycarbonate are recognized after molding, the chances are very good that the cause was improper drying prior to molding. For this reason, drying should be given the highest degree of consideration when approaching the molding of polycarbonate. As will be discussed later, other process factors are of concern, but few are as important as drying.

Often the effects of inadequate drying emerge as visual defects in the molded part. The most common evidence in natural resins is the presence of silver streaks on the surface. If the moisture level is high enough, small bubbles may be seen in the body of the part. This is a result of vaporization of retained water and/or the generation of a gaseous degradation by-product, carbon dioxide. Visual identification of a "wet" polymer is, of course, subject to the limit of solubility of the gases in the resin, which is controlled by injection pressure and part geometry. In light of this, visual detection of moisture levels in excess of 0.06 percent moisture has been possible in some moldings.

Moisture may be detected by observing an open air purge shot. Even low levels of water in the melt can create small bubbles in the purge shot which are visible in clear material. Generally this is distinguishable from entrapped air, as bubbles of the latter are much larger. Moisture also creates a very frothy purge shot. In opaque materials, the bubbles are not visible within the melt; however, the surface of the purge shot appears pitted as if sprinkled with sand. A dry, opaque purge shot will have a very smooth, glossy surface. If degradation has occurred during processing of a wet colorless grade, a more pronounced yellowness may be observed in the part when compared to the preprocessed pellets. However, absence of these visible characteristics does not necessarily indicate that the polymer has been adequately dried prior to processing.

At processing temperatures, moisture promotes a hydrolytic attack on polycarbonate, resulting in a degradation of the polymer. Not only may there be a change in appearance, but there is also a chemical alteration involved. Hydrolysis breaks down the polycarbonate, forming a greater weight fraction of lower-molecular-weight chains. This degradation, due to inadequate pellet drying, may not be visually apparent, but will manifest itself in a reduction of impact strength, decrease in relative viscosity, broadening of the molecular weight distribution, and increase in melt flow rate.

Figure 22-9 indicates that change in molecular weight distribution is possible when adequate predrying measures are not observed

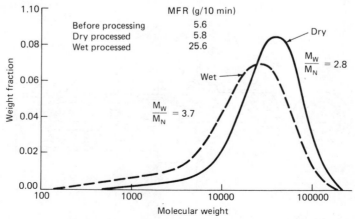

Fig. 22-9 Molecular weight distribution of a high-molecular-weight polycarbonate (gray).

(5). The same polymer, a pigmented polycarbonate, was subjected to two methods of conditioning. The dry material was tray-dried in a forced convection oven at 120°C for 3 to 4 hours. Wet pellets were left overnight in an open tray at room temperature and allowed to absorb moisture from the atmosphere. After processing, specimens of both conditions were subjected to gel permeation chromatography (GPC) and melt flow analysis. The polydispersity index, a measure of the breadth of the molecular weight distribution, had increased from 2.76 to 3.60. The number average molecular weight fell from 12,700 to 7,350, while the weight average molecular weight dropped from 35,000 to 26,450. Hydrolytic attack produced an increase in the lower-molecular-weight fraction of the polymer and a decrease in the higher-molecular-weight fraction.

Since the ease of flow of polycarbonate usually increases as degradation becomes more severe, melt flow rates were determined for the virgin pellets before processing and for parts after processing in the dry and wet conditions. These values are stated in Fig. 22-9. As expected, there is very little change between flow rates of the preprocessed pellets, 5.6 g/10 min, and the dry processed material, 5.8 g/10 min. However, there is a significant increase to 25.6 g/10 min in the wet processed polycarbonate. This comparison represents an extreme example of inadequate drying and should not be construed to represent the outcome of all drying irregularities. It does, however, suggest the degree of degradation that is possible if conditions permit it.

Figure 22-10 illustrates the effect of processing moisture on the relative solution viscosity. The dry and wet relative viscosities of three different lots of medium-moleular-weight polycarbonate are depicted. The equilibrium moisture content of the wet pellets was about 0.12% wt. The relative viscosity decreases, reflecting the increase in lower-molecular-weight fractions.

Degradation

The sensitivity of notch impact strength and elongation to the degree of degradation of

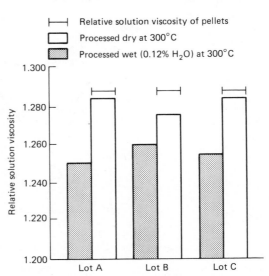

Fig. 22-10   Degradation of a medium-molecular-weight polycarbonate through inadequate drying.

medium-molecular-weight polycarbonate is shown in Fig 22-11. In both instances, the mechanical properties decrease as degradation becomes more pronounced. Here the decrease in relative solution viscosity is used to indicate increase in degradation. Note that although the higher-molecular-weight polycarbonate may be degraded to a relative viscosity corresponding to that of the lower-molecular-weight polymer, the mechanical properties of the degraded polymer are less than those of the lower-molecular-weight polymer. The same would be expected of high and medium molecular weight comparison.

Proper drying procedures are, therefore, always recommended in order to ensure maximum mechanical properties and acceptable part appearance. But merely attaining a good surface finish does not ensure maximum physical properties. The visual evidence of hydrolytic degradation may not be present or may be masked by pigment or other additives; a part may look acceptable and yet fail in use. Adequate drying will eliminate such problems related to moisture.

Molded part quality is also dependent upon melt temperature and residence time in the barrel. Molecular degradation can occur if either one or both of these are in excess of good molding practice. Excessive residence time can occur if shot size is lower than 20 percent

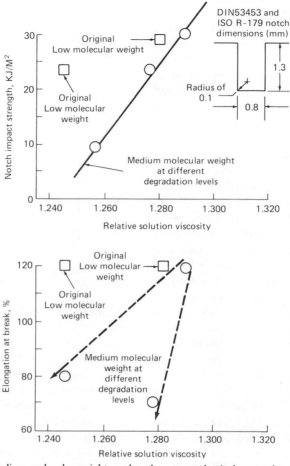

Fig. 22-11 Medium-molecular-weight polycarbonate mechanical properties after degradation.

of barrel capacity. If gating is too small, wall thickness too thin, or clamp size insufficient, very high melt temperatures may be required to fill the part. Figure 22-12 describes temperature and time effects on the melt flow rate of a flame-retardant polycarbonate. In this example, melt flow rate increase is a measure of increased molecular degradation. Although the tool used to generate the data was a hot manifold design, a similar trend would have been expected from conventional tooling. Comparing the curves, it is obvious that increase in barrel temperature is more significant than manifold temperature. At both conditions, melt flow rate increases with residence time. The slight difference in slope of the curves is due to the stronger influence of barrel temperature on melt flow rate compared to residence time. In this example, a pigmented,

flame-retardant polycarbonate would represent the extreme in melt flow rate change. A similar trend would have occurred for a natural, general-purpose polycarbonate, but the change in melt flow rate would not have been so severe.

Rheology

Selecting the proper molecular weight of polycarbonate will depend not only on performance requirement but also on the degree of difficulty in filling the cavity. Figure 22-13 describes the cavity-filling capablity in terms of the flow length vs. part thickness relationship for polycarbonate. At a given part thickness, a lower-molecular-weight polymer will have a longer flow length than a higher-molecular-weight polymer. The difference becomes even

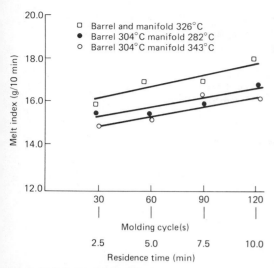

Resin Melt Index: 14.4 g/10 min
Color: 3301 White
Mold: 2-Cavity Tumbler 200 cm³ shot size hot runner mold
Machine: 273 metric ton–592 cm³ barrel

Fig. 22-12 Influence of molding cycle and barrel and manifold temperature on the melt index of flame-retardant polycarbonate.

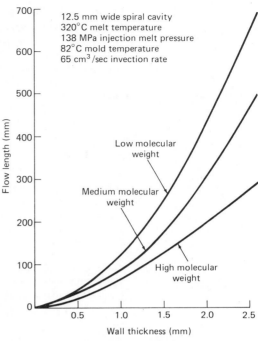

Fig. 22-13 Spiral flow comparison of polycarbonates typical of the three molecular weight ranges.

more acute as part thickness increases. These data were obtained at constant molding conditions. In order to assess the flowability of the polymer, one must consider the effect of the main processing variables.

A flowability study on a high-molecular-weight polycarbonate was performed in order to evaluate the effects of principal injection molding variables. Ranges of the variables were chosen to include those conditions commonly found in the injection molding of polycarbonate. A linear model with coded variables was applied to an experimental design of flow length in a spiral flow mold. It was felt that a comparison of the magnitude of the coefficients would yield information on the sensitivity of flow length to processing variables. Table 22-3 describes the four main variables considered and the range of investigation associated with each. Also depicted is the resultant model relating the four variables to the response of flow length at 2.5 mm wall thickness.

It is clear from the model that melt temperature and injection pressure are the two most important variables in determining flow length. Comparing the magnitude of the coefficients (variables are coded from −1 to +1), injection pressure is about three times as effective as injection speed, and melt temperature is four times as effective. Mold temperature has an almost insignificant effect on flow length.

These results can be explained by considering the effect of these variables on melt viscosity, the principal physical property affecting flow length. Temperature decreases viscosity, and because polycarbonate is non-Newtonian, an increase in shear rate or injection speed also decreases viscosity. Pressure has a slight hydraulic effect on flow and decreases viscosity through an increase in shear stress. Mold temperature is only slightly effective because of a low heat-transfer film coefficient beteen the mold wall and the polymer. However, the slight increase in melt temperature decreases viscosity and increases flow length. Not only the magnitude but also the direction or sign of the coefficients may be interpreted in regard to the variables' effect on viscosity.

Table 22-3 Flowability of high-molecular-weight polycarbonate at 2.5 mm wall thickness.

Process variables and associated levels of interest:

Melt temperature	300 and 340°C
Injection speed	16.9 and 50 cm³/sec
Injection pressure	69 and 124 MPa
Mold temperature	82 and 104°C

Experimental design: 4^2 (4 variables of 2 levels each)

Linear model where: $Y = A + BX_1 + CX_2 + DX_3 + EX_4$

Y = Flow length
A = Average flow length
B = Melt temperature coefficient
X_1 = Coded variable for melt temperature
C = Injection speed coefficient
X_2 = Coded variable for injection speed
D = Injection pressure coefficient
X_3 = Coded variable for injection pressure
E = Mold temperature coefficient
X_4 = Coded variable for mold temperature

Solution to model: $Y = 5.5 + 1.2X_1 + 0.3X_2 + 1.0X_3 + 0.1X_4$

Heat Transfer

Although melt temperatures may be high (280–340°C) for injection molding of polycarbonate in order to reduce viscosity, the high glass transition temperature of 150°C promotes short cooling times. In addition, the thermal diffusivity of polycarbonate is high in comparison to other polymers found in injection molding applications. Table 22-4 illustrates the thermal diffusivity of a number of polymers including polycarbonate. The defining relation for one-dimensional unsteady-state heat transfer is as follows (6):

$$\frac{\partial T}{\partial t} = \alpha \frac{\partial^2 T}{\partial x^2}; \alpha = \frac{k}{\rho C_p}$$

where T represents temperature, t is time, x is thickness, and α the thermal diffusivity. For a fixed part geometry, the cooling rate of the polymer is completely defined by α. The larger the value of α, the higher the cooling rate. High glass transition temperature and large thermal diffusivity allow for fast cycling of parts injection-molded of polycarbonate.

Figure 22-14 illustrates typical cooling times as a function of wall thickness for specific melt, mold, and part temperatures. Cool-

Table 22-4 Comparison of thermal diffusivities of various polymers.

POLYMER	THERMAL DIFFUSIVITY × 10^4 CM²/SEC
Polycarbonate	10.03
PBT	9.21
Nylon 6 glass reinforced	8.86
CAB foam	8.48
Polystyrene	7.73
ABS	7.43
SAN	7.30
Nylon 6 unreinforced	6.94
Polypropylene	5.88

ing time increases with the square of the thickness for any given temperature condition. At a given thickness, low mold temperature and high part temperature at ejection reduce cooling time. In addition, at a given thickness, cooling time is proportional to the logarithm of the temperature conditions so a linear change in mold or part temperature does not promote a corresponding linear change in cooling time.

The cooling curves were generated by the following relation (7):

$$t = \frac{x^2}{\alpha \pi^2} \ln \left[\frac{8}{\pi^2} \left(\frac{T_m - \bar{T}_w}{\bar{T}_p - \bar{T}_w} \right) \right]$$

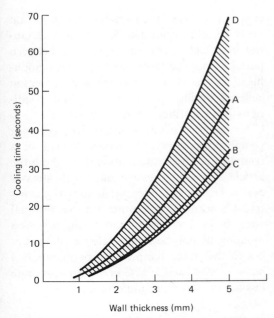

	Melt temp.	Mold temp.	Part Temp. (at ejection)
A	300°C	100°C	124°C
B	300°C	80°C	124°C
C	300°C	80°C	130°C
D	300°C	120°C	130°C

Fig. 22-14   Cooling time vs. wall thickness for polycarbonate.

where:

$$\overline{T}_w = 0.5 \left[\frac{b_w T_w + b_m T_m}{b_w + b_m} + T_w \right]$$

In the above relations, the following definitions apply:

t = cooling time (sec)
x = part thickness (cm)
α = thermal diffusivity of polymer (cm²/sec)
T_m = melt temperature (°C)
$\overline{T}_w$ = average mold wall temperature during injection cycle (°C)
$\overline{T}_p$ = average part temperature (°C)
b_w = thermal penetration number of mold $\left(\frac{\text{Joule}}{\text{cm}^2 - \text{sec}^{1/2} - °C} \right)$
b_m = thermal penetration number of polymer $\left(\frac{\text{Joule}}{\text{cm}^2 - \text{sec}^{1/2} - °C} \right)$

The cooling equations apply to other polymers as well as polycarbonate and may be useful in estimating cooling cycle times.

Residual Stress

Performance of parts injection-molded of polycarbonate will depend not only upon the grade (melt flow rate and presence of additive) of material but also part design, environment, and processing conditions. Failure of a polycarbonate article can often be traced to high residual or "frozen-in" stresses in the part. These stresses are a result of nonuniform cooling of the part while in the mold. Residual stress may also be promoted through overpacking of the mold cavity during injection hold. Thermally induced stresses occur when a given region cools more rapidly than its surroundings. Since shrinkage is temperature-dependent, cooler regions shrink in advance of hotter areas, giving rise to a nonuniform stress distribution.

A simplistic example of a thermally induced nonuniform stress distribution was investigated by So and Broutman (8). The effect of quenching polycarbonate sheet from the glass transition temperature, Tg, was analyzed. One-dimensional heat transfer was induced by restricting the thickness to a much smaller value then either lateral dimension. Polycarbonate sheet, 150 cm × 150 cm × 3 mm, was quenched from a 150°C oven to a 0°C ice bath. Residual stresses were determined by observing the curvature resulting from the repeated removal of thin layers from the surface. Through a stress analysis, illustrated in Fig. 22-15, compressive stresses were found in the surface, and tensile stresses were observed in the center of the sheet. This phenomenon was attributed to rapid cooling of the surface and slow cooling of the center regions.

The cooling condition described above is far more severe than any experienced in injection molding but does serve to illustrate the effect of steep thermal gradients on residual stress. However, most injection molded parts are not of such an ideal geometry and may not be subject to the same conclusion as drawn for polycarbonate sheet quenching. For exam-

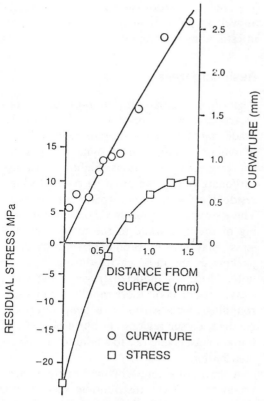

Fig. 22-15   Residual stress distribution of polycarbonate quenched from the glass transition temperature.

ple, a sharp-edged corner exposed to a mold surface will cool faster than its surroundings. The shrinkage of adjacent surfaces results in a tensile force being applied across the corner. Residual stress then becomes dependent not only upon cooling conditions but also on part geometry.

Failure associated with residual stress is thought to be influenced by the direction and magnitude of the local molded-in stress. For instance, if a region subject to tensile impact has a residual stress component that is tensile, the threshold for failure is reduced, and the part may fail under a lower impact force than if no residual stress were present. Similarly, a higher probability of failure may exist when residual tensile forces in a surface are exposed to a stress cracking environment.

Difficulty in part ejection is also related to a complicated residual stress distribution. In such a case, the residual stress may be manifested as nonuniform shrinkage which is influenced by mold and melt temperature, injection pressure, and part design. As a result, substantial frictional forces can be effected between part and mold surfaces due to the high modulus of polycarbonate. In addition, deformation suffered during ejection under these conditions can contribute to residual stress.

Nonuniform cooling applied to a complex geometry encourages the development of a complicated stress distribution throughout the article. Since compressive and tensile stresses cannot be induced independent of part geometry, it is normally recommended that residual stress be held to a minimum during injection molding of polycarbonate. Hence, it is suggested that mold temperature be maintained between 80°C and 105°C in order to promote less severe and more uniform cooling.

Annealing

Annealing can be employed to reduce the effects of residual stress, but this procedure is not recommended as an alternative to changes in effective processing variables such as mold temperature and injection hold pressure.

Annealing relieves molded-in stress but may result in a decrease in notched Izod impact strength and an increase in the brittle impact transition temperature. Figures 22-16 and 22-17 illustrate such changes when annealing is applied to the residually stressed, simple plate geometry described above. When the notched

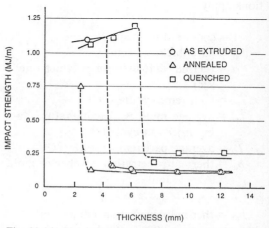

Fig. 22-16   Notched impact strength of polycarbonate as influenced by thermal treatment after processing.

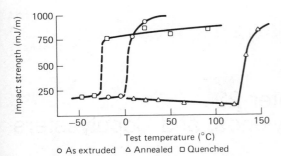

Fig. 22-17 Ductile–brittle transition of polycarbonate as influenced by thermal treatment after processing.

impact test adequately describes in-use behavior, the effects of annealing specimens can be anticipated; annealing increases the elastic modulus, initiates crazing after shorter periods of time at a given stress level, and can decrease ultimate elongation to as low as 10 percent. Thus, annealing a polycarbonate part generally can not be recommended if the article is exposed to continuous loading of significant magnitude in practical use.

CONCLUSION

When processed properly, polycarbonate can fulfill the engineering needs of demanding applications. This requires attention to detail in drying, molding, and recycling of regrind. In concert with proper part design, these considerations will yield the superior properties of one of the foremost of engineering resins.

REFERENCES

1. Schnell, H., *Chemistry and Physics of Polycarbonate,* Interscience Publishers, New York, 1964.
2. "Merlon Polycarbonate Design Manual," Mobay Chemical Corporation, Pittsburgh, PA, 1980.
3. "Merlon Polycarbonate—A General Reference Manual," Mobay Chemical Corporation, Pittsburgh, PA, 1979.
4. "Merlon Polycarbonate—A Processing Guide for Injection Molding," Mobay Chemical Corporation, Pittsburgh, PA, 1980.
5. Newcome, J. M., *Proceedings* SPE Regional Technical Conference, Boxborough, MA, 1977, p. 45.
6. Jenson, V. G., and Jeffreys, G. V., *Mathematical Methods in Chemical Engineering,* Academic Press, New York, 1969.
7. Menges, G., Wuebken, G., and Catic, I., *Plastverarbeitu, 25,* 1 (1974).
8. Broutman, L. V., and So, P., *Polymer Engineering Science, V6,* 12 (1976).

Chapter 23
Polyethylenes, Polypropylenes, and Copolyesters

B. V. Harris

Eastman Chemical Products, Inc.
Kingsport, Tennessee

LOW-DENSITY POLYETHYLENES

Low-density polyethylene (LDPE) is defined as polymerized ethylene having a nominal density of 0.910 to 0.925 g/cm³. However, medium-density polyethylene (MDPE), with a density of 0.926 to 0.940 g/cm³, is usually included with LDPE because their processing conditions and properties are quite similar. Therefore, the following information will encompass both low- and medium-density polyethylene unless otherwise stated. Specific polyethylene formulations used for illustration will be selected from the middle of the density ranges covered by LDPE and MDPE.

The injection molding of large, thin-walled items is one of the most difficult challenges in injection molding of LDPE. At the same time, one of the most commonly encountered items of injection molded LDPE is a lid. Therefore, injection molding of LDPE lids will be described as a fairly typical representation of injection molding of LDPE.

Injection Molding Polyethylene Lids

Injection molded polyethylene lids are used in a wide variety of closure applications. Many products, such as margarine, cream cheese, whipped topping, ice cream, and sandwich spreads, are packaged in plastic containers (which may be polyethylene) that have polyethylene lids for primary closure. Many other products, such as coffee, peanuts, and shortening, are packed in metal cans that are used after they are opened to store the unused portion of the contents. Most of these cans are sold with polyethylene overcaps that snap into place and furnish good closure for the cans after removal of the metal tops.

The characteristics demanded in polyethylene lids vary widely. Economy is always important; and in nearly every application, it is desirable that the lids be flat and that they snap tightly upon the container that they cover. Some applications demand some degree of clarity so that printed matter on a metal lid can be read through the overcap before the can is sold. Some require resistance to environmental stress cracking, so that the materials that may be in contact with them will not cause them to split. Some require still other characteristics. In addition, the polyethylene lid business has undergone significant technological advances in past years with most of the emphasis on processability or production rate. Extremely fast-cycling machines, stack molds, and larger tonnage presses all have contributed to an increase in the molder's productivity. Also, more sophisticated machine controls make present injection molding machines very sensitive to process and material change and/or variation.

Consequently, molding polyethylene lids for their many uses is an exacting process that requires good selection of molding machine, mold design, part design, plastic formulation, molding conditions, and other factors. This section presents detailed information about the important factors in the injection molding of lids and describes some polyethylene formulations widely used for this purpose.

Molding Machines

Screw-type molding machines are preferred to straight-ram machines for molding polyethylene overcaps because they produce more homogeneous melts and permit the use of shorter cycles. They also permit better control of such variables as injection pressure, injection speed, and melt temperature. It is very difficult to mold flat, acceptable lids on a straight-ram machine unless it is equipped with a screw preplasticator. In the preplasticator unit, melting of the plastic is performed in a simple extruder that pumps material into a secondary cylinder containing a ram that acts as the injection unit. Such an arrangement offers excellent control of shot size because the volume of the shot is measured while the material is in the molten state.

The size of the injection molding machine to be used to mold overcaps is intimately related to the diameter of the overcap and the number of cavities in the mold. Generally, molds with two to four cavities can be used on molding machines with capacities of two or three ounces and clamping forces of 75 to 150 tons, whereas molds with six or eight cavities frequently require machines of 5- to 16-ounce rating with 200 to 400 tons of clamping force available. However, these figures could vary considerably, depending on the size of the overcap.

There are advantages in both large and small machines. If several small machines are used rather than fewer larger ones, a machine breakdown or shutdown for routine maintenance will have less effect on production rates. Also, because there are normally fewer cavities in molds for small machines than in molds for large machines, small machines permit closer control of the molding variables in the individual cavities. On the other hand, large machines molding many lids per shot can have

lower direct molding costs per part produced, even though they require longer cycles. The cycle time increases as the number of cavities in a mold increases, but not proportionately. Thus, a four-cavity mold might run with a 5-second cycle, and an eight-cavity mold might require an 8-second cycle; but the larger mold and the longer cycle will produce more lids per minute of operation.

The clamping force usually necessary in a molding machine for producing an overcap 25 to 30 mils thick is $1\frac{1}{2}$ to 2 tons per square inch of projected area. A single-cavity mold for a 5-in. lid, therefore, would require 28 to 38 tons; a four-cavity mold for the same size lid, 110 to 150 tons. The projected area of the runner in an insulated or hot-runner mold need not be considered unless its total projected area is greater than that of the lids. If this should be the case, the projected area of the runner should be considered and that of the lids ignored.

Molding Conditions

When setting conditions for molding polyethylene, the objective should be to inject fairly hot material into a cold mold while subjecting the molded part to as little strain as possible. This is usually accomplished by using high injection pressures to ensure quick filling of the mold and by using very short plunger-forward times to avoid packing the mold.

Melt Temperature

High melt temperatures are used to permit the plastic to be injected quickly into the mold with minimum strain. High melt temperatures normally give maximum clarity, minimum sunburst, and minimum warpage in the molded parts.

If melt temperatures are too low, molding will be difficult, requiring excessive injection pressures and longer plunger-forward times. This combination of conditions can produce lids with poor clarity and excessive sunburst and warpage.

Melt temperature generally varies from 325°F to 550°F, depending on the machine used, the mold size and construction, and the plastic formulation.

Large machines with large material hold-up in the cylinder usually operate between 325°F and 475°F, whereas small machines with little hold-up generally operate between 425°F and 550°F. If the plastic moves through the cylinder rapidly, the cylinder temperatures may have to be set considerably higher than the above temperatures in order to maintain the desired melt temperature. In a machine operating near its plasticating limits, an indicated temperature of 480°F may be required to maintain a melt temperature of 450°F.

Figure 23-1 shows the relationship between injection molding melt temperatures and melt index of low-density polyethylene formulations with melt index ranging from 0.7 to 40.

Mold Temperature

The optimum mold temperature for lid production seems to be about 40 to 50°F. Temperatures in this range permit short cycles and produce lids with good clarity. Mold temperatures lower than 40 to 50°F can make mold filling difficult, but very clear lids are generally produced while higher mold temperatures result in slow lid cooling—usually causing excessively long cycles.

Cycle Time

The most important segments of the lid molding cycle are plunger-forward time and clamp time. Both factors significantly affect shrinkage and toe-in.

The plunger-forward time should be about 0.1 to 0.3 second longer than the actual mold-filling time. If it is significantly longer than this, the areas around the gates will be packed, and thus will shrink less than the areas around the outer edges of the lids, so that warpage could result. The plunger-forward time is generally determined by setting all temperatures for molding, decreasing the plunger-forward time in small increments until a short shot results, and then increasing the time about 0.1 to 0.3 second.

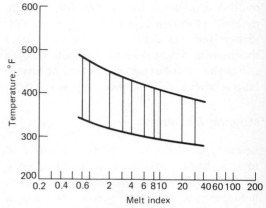

Fig. 23-1 Approximate injection molding temperature ranges for low-density polyethylene plastics.

The clamp time should be the absolute minimum setting at which a lid with acceptable flatness, toe-in, and shrinkage can be produced. The clamp time, which must be set after the plunger-forward time is fixed, must sufficiently exceed the plunger-forward time to allow the molten plastic to solidify in the cavities. Since toe-in is desirable but warpage is not, the clamp time must be set for each mold to give a satisfactory combination of these factors. If the cooling time necessary to produce acceptable lid flatness and shrinkage does not produce acceptable toe-in, the thickness of the lip can be increased so that this portion of the lid will be hotter and will shrink more after it has been removed from the cavity. This will increase toe-in.

Injection Pressure and Injection Speed

Since the main objective of lid moving is to fill the mold cavities as rapidly as possible, injection pressure and injection speed should be set as high as possible while still maintaining proper shot-size control. These settings should be at their maximum, if possible.

Shot Size

Proper adjustment of shot size ("starve feeding") is the preferred method for controlling packaging when molding lids; if possible, the shot size should be the exact amount of plastic needed to fill the mold cavities. On some small molding machines, and even on some larger ones equipped with screw preplasticators, shot-size control is sufficiently precise to make this possible. With precise shot-size control, the adjustments described for cycle time, injection pressure, and injection speed should be satisfactory.

On many large machines, it is not possible to control shot size with adequate precision—one shot might be short and the next one packed without changing machine settings. If necessary, keep a cushion of molten plastic in the injection machine to better control the rim action, but the cushion should be as small as possible. If satisfactory lids result, the adjustments described for cycle time, injection pressure, and injection speed should be satisfactory also.

Sometimes a cushion of molten plastic in the machine may cause excessive packing of the cavities, resulting in warped lids. If this occurs, it will be necessary to depart from the previously described adjustments of cycle time, injection pressure, and injection speed. The most common procedure is to reduce injection pressure first, then, if necessary, to reduce injection speed—but only to the extent necessary to prevent excessive packing. Either change will increase cycle length because the cavities will fill more slowly. Every effort should be made to keep the rate of flow of molten plastic into the mold as high as possible so that no appreciable solidification of material will occur until after the mold is filled.

Screw Speed

Maximum screw speed is usually used so that the time needed to pump material to the front of the screw will not delay the machine cycle. Fast screw speeds generate frictional heat in the plastic and help to produce a homogenous melt. If temperatures become too high and material degradation results, the screw speed should be reduced. The heat generated in the plastic by the screw rotation is a function of the square of the screw speed; thus, a small reduction in screw speed can result in an appreciable reduction in the heat generated.

Materials

Tenite polyethylene 18BOA (20 melt index, 0.923 g/cm³) is the Eastman formula most widely used for the production of thin, clear lids. This material is characterized by excellent processability, warpage resistance, and clarity, while exhibiting good toe-in characteristics and stress-crack resistance. A higher-melt-index version of 18BOA is 18DOA (40 melt index, 0.923 g/cm³). This material exhibits greater shrinkage and slightly better flow characteristics but does not exhibit the toe-in, processability, and stress-crack resistance of 18BOA. Tenite polyethylene 187OA (7 melt index, 0.923 g/cm³) exhibits exceptional stress-crack resistance. It provides a material with fast cycling characteristics for the lid molder interested in applications requiring high stress-crack resistance.

All three materials have been used extensively in the lid molding industry in a variety of closure applications and other related items. Because of their consistency, these materials perform especially well in stack molds.

To fully characterize this "family" of high-quality lid molding materials, an extensive study was performed to evaluate formulas 18BOA, 18DOA, and 187OA over a wide range of melt temperatures and injection pressures to determine lid characteristics at each of these conditions. The materials were evaluated on a 250-ton, two-stage Husky injection molding machine fitted with a three-cavity, 603-lid mold. A four-cavity, 401-lid mold was used in some cases to show the difference in characteristics between large and small lids. The results of this molding evaluation are given below.

Shrinkage

Figures 23-2, 23-3, and 23-4 show shrinkage as a function of melt temperature and injection pressure for the three materials involved. Three basic relationships exist. First, lid shrinkage increases with increasing melt temperature. Second, lid shrinkage increases with increasing injection pressure. The effect is most evident at pressures between 1000 and

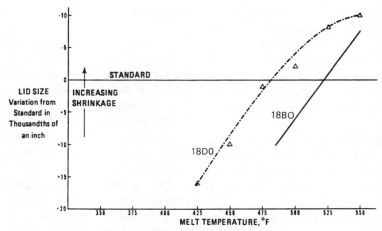

Fig. 23-2   Shrinkage vs. melt temperature (at 1000 psi).

1400 psi. This trend could be attributed to an increase in polymer temperature in the nozzle due to frictional heat generated by the high injection pressures. Third, lid shrinkage increases with increasing melt index. This trend is more evident at low injection pressure than at high pressures. Some deviation from this relationship appeared with formula 1870 at the higher melt temperature (greater than 500°F). This last trend was reversed with the small 401-lid size, as Fig. 23-5 indicates, with the lowest-melt-index material giving the greatest shrinkage, although only minor differences were observed among the samples. This would indicate that the cavity pressure and frictional heat generated in the material can be adjusted by careful control of the injection pressure, injection speed, and melt temperature to obtain the desired shrinkage. As opposed to the 603-lid results, no difference in shrinkage was observed with changing injection pressure with the smaller 401 lids.

Figures 23-2, 23-3, and 23-4 also show the relative flow properties of each material as functions of melt temperature and injection pressure. Summarizing, the minimum melt temperature at which each material could be molded without causing short shots is given in Table 23-1 for three different injection pressures.

As would be expected, these data indicate that proper mold filling is difficult at low melt

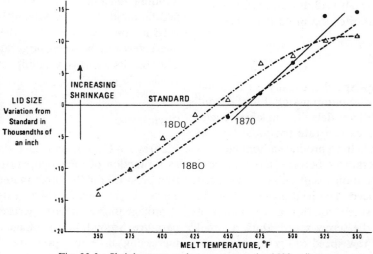

Fig. 23-3   Shrinkage vs. melt temperature (at 1400 psi).

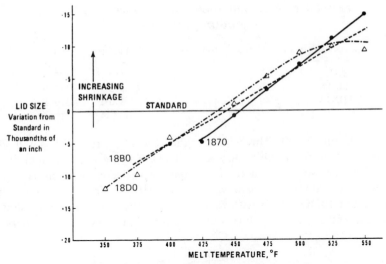

Fig. 23-4   Shrinkage vs. melt temperature (at 1800 psi).

temperatures, at low injection pressures, and with low-melt-index materials. This effect apparently lessens with increasing melt index and injection pressure. A greater difference exists in mold-filling capabilities between 187OA and 18BOA than between 18BOA and 18DOA at a given injection pressure and less difference among the three materials as the injection pressure increases.

Clarity

Figure 23-6 shows the effect of melt temperature on lid clarity. No difference in the clarity of lids molded at different melt temperatures or different injection pressures was observed.

All three materials formed lids with good clarity at all molding conditions.

Sunburst

Figure 23-7 shows lid sunburst effect vs. melt temperature for the three materials. Two basic relationships were found in this evaluation. First, sunburst decreased with increasing melt temperature; and second, sunburst decreased with increasing melt index. Injection pressure had no effect on sunburst.

Shot Weight

Shot weight vs. injection pressure and melt temperature for the three materials is shown

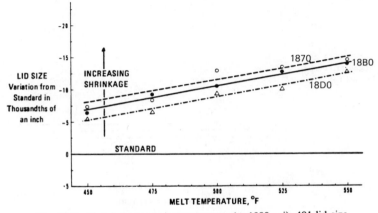

Fig. 23-5   Shrinkage vs. melt temperature (at 1800 psi), 401-lid size.

Table 23-1 Minimum melt temperature at indicated injection pressure.

MATERIAL	MELT INDEX	1000 PSI	1400 PSI	1800 PSI
187OA	7	Would not flow	450°F	400°F
18BOA	20	475°F	375°F	350°F
18DOA	40	425°F	350°F	350°F

in Figs. 23-8, 23-9, and 23-10. Three basic relationships are evident from these curves. First, shot weight increases with increasing melt index. This effect is more prominent with formulas 18BOA and 18DOA than with formulas 187OA and 18BOA in large lids, but the reverse is true with small lids (see Fig. 23-11). Second, shot weight increases to a point with decreasing melt temperatures; then, beyond this point, the material becomes too cold to flow properly, and shot weight decreases. Third, shot weight increases with in-

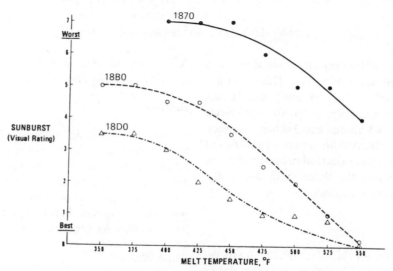

Fig. 23-6 Clarity vs. melt temperature.

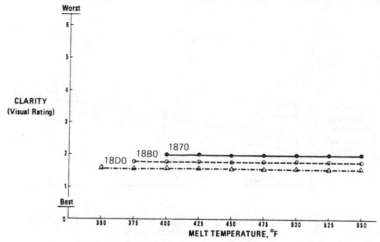

Fig. 23-7 Sunburst vs. melt temperature.

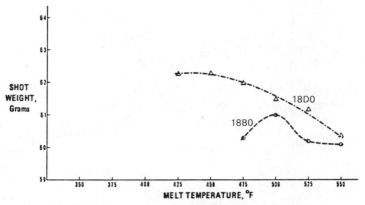

Fig. 23-8   Shot weight vs. melt temperature (at 1000 psi).

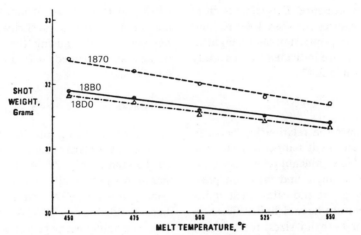

Fig. 23-9   Shot weight vs. melt temperature (at 1000 psi), 401-lid size.

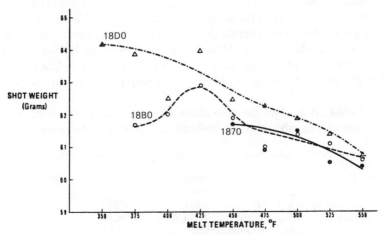

Fig. 23-10   Shot weight vs. melt temperature (at 1400 psi).

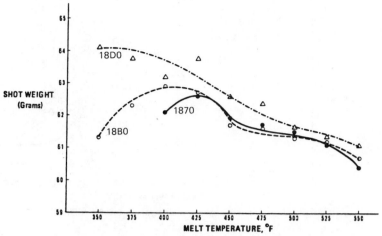

Fig. 23-11   Shot weight vs. melt temperature (at 1800 psi).

creasing injection pressure. This effect is more prominent at pressures between 1000 psi and 1400 psi. In general, optimum shot weight of the three materials are indicated by this study is as shown in Table 23-2.

Toe-in Angle

Figure 23-12 shows the relationship between lid toe-in angle and melt temperature for the three materials. No significant relationship exists between toe-in angle and injection pressure, but as this graph indicates, melt index and melt temperature have definite effects on toe-in angle. For both lid sizes, toe-in angle decreases as melt index increases. However, the relationship between melt temperature and toe-in depends on lid size—decreasing with increasing melt temperature for the large lids and exhibiting just the opposite effect for small lids. This difference is attributed to a shrinkage phenomenon. Large lids have more total shrinkage across their diameter than small lids, and this has the effect of pulling the top

of the lid rim inward around the circumference. This overrides the shrinkage in the skirt part of the lid, causing the toe-in angle to be less, as illustrated in Figure 23-13 for the 401-lid size.

Warpage

The warpage rating of the three materials noted on a standard cycle vs. melt temperature can be seen in Fig. 23-14. This graph shows that no warpage at all occurred in lids molded from formula 18BOA and 18DOA, whereas those molded from 1870A exhibited slight (but acceptable) warpage at the melt temperature studied. As only formula 1870A exhibited warpage, it is difficult to see any relationship between warpage and injection pressure. However, there is some evidence of decreasing warpage with increasing injection pressure for formula 1870A. The excellent warpage resistance exhibited by 18BOA and 18DOA over the wide range of injection pressures and melt temperatures used in the study indicates that

Table 23-2   Melt temperature at which optimum shot weight is obtained at indicated injection pressure.

MATERIAL	MELT INDEX	1000 PSI	1400 PSI	1800 PSI
1870A	7	Would not flow	450°F	425°F
18BOA	20	500°F	425°F	400°F
18DOA	50	425–450°F	350°F	350°F

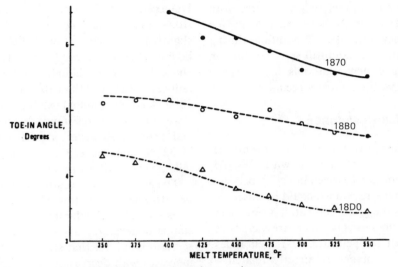

Fig. 23-12   Toe-in angle vs. melt temperature.

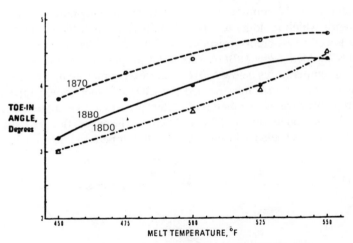

Fig. 23-13   Toe-in angle vs. melt temperature (401-lid size).

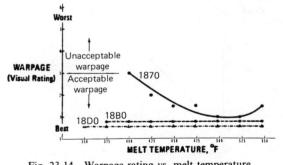

Fig. 23-14   Warpage rating vs. melt temperature.

warpage problems previously associated with lid molding materials have been overcome with these new formulas. In addition, during extensive field trials and full-scale production usage of these new materials, no warpage problems have been encountered.

Stress-Crack Resistance

The stress-crack resistance of the three materials vs. melt temperature is shown in Fig. 23-15. No relationship between injection pressure and stress-crack resistance could be observed. However, as the graph indicates, a significant relationship does exist between stress-crack resistance, melt temperature, and melt index. Stress-crack resistance is directly related to melt temperature and inversely related to melt index. Formula 187OA (7 melt index) exhibits a stress-crack resistance (time to 50 percent failures) of greater than 10 minutes for all melt temperatures at which molding was possible. Relatively lower stress-crack resistance was found in lids molded from 18BOA (20 melt index) at melt temperatures below 450°F than those obtained on 187OA lids molded over the same temperature range. At melt temperatures of 450°F and above, time to 50 percent failures of 18BOA exceeded 10 minutes.

However, analysis of the data indicates that stress-crack resistance of 18BOA is still slightly lower than that of 187OA, since some lid failures were recorded at melt temperatures above 450°F even though time to 50 percent failures was greater than 10 minutes. Formula 18DOA (40 melt index) exhibited significantly lower stress-crack resistance than both 187OA and 18BOA. However, at melt temperatures of 525°F and above, the time to 50 percent failures exceeded 10 minutes.

The apparent data scatter in the curves can be attributed to the test method; any type of stress-crack test exhibits a relatively large standard deviation. However, the curves do serve to point out that stress-crack resistance increases with decreasing melt index.

Minimum Cycle Time

Curves showing minimum cycle time (processability) vs. melt temperature are given in Fig. 23-16. Three basic relationships are shown in the graphs. First, cycle time increases with increasing melt temperature. Second, formula 18BOA (20 melt index) appears to exhibit better processability than the other materials. Third, the curves further indicate that the best melt temperature for lid molding from the

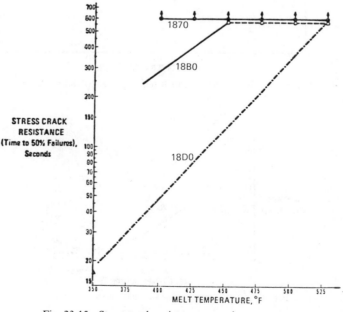

Fig. 23-15 Stress-crack resistance vs. melt temperature.

standpoint of processability would be the point just above where short shots occur. It is interesting to note that this point is similar to the region of maximum melt flow previously mentioned under the shot weight discussions. As would be expected, high injection pressures permit the material to fill the mold at low melt temperatures.

General Comments

Again, it should be emphasized that the data presented herein are pertinent to the type of mold and machine used in the study, and the specific numbers generated would not necessarily be valid for different machines and different molds. However, the basic trends illustrated here give an indication of the amount of variation that is possible in a molded lid due primarily to the effects of lid size, melt index, melt temperature, and injection pressure. No attempt has been made to establish optimum molding conditions for these materials, since each molder should determine them for his own equipment. The basic intent is to show the wide range of molding conditions at which the three lid materials could be molded and to determine lid characteristics at each of these conditions.

Summary

This molding study of Tenite polyethylene formulas 187O, 18BO, and 18DO indicates that the lid characteristics of shrinkage, shot weight, and low temperature mold-filling capabilities are affected by injection pressure, while all the lid characteristics except clarity are affected by changes in melt temperature. The study also indicates the significant effect melt index has on most of the lid characteristics. In general, it can be concluded that formula 18BOA (20 melt index) should be considered as the all-purpose, most versatile material of the three. It is characterized by excellent processability, warpage resistance, and clarity while exhibiting good toe-in and stress-crack resistance. The higher-melt-index material, 18DOA (40 melt index), exhibits higher shrinkage and slightly better low-tem-

perature flow characteristics than 18BOA. However, it does not exhibit the toe-in and stress-crack resistance properties of 18BOA and should be considered only when high shrinkage and/or exceptionally high flow properties are required. Although formula 187O (7 melt index) does not exhibit the flow characteristics and warpage resistance of 18BO, it does exhibit exceptional stress-crack resistance and should be considered in applications where this is the key property.

Specific properties obtained from injection-molded test specimens of these three materials are listed in Tables 23-3, 23-4, and 23-5, and a suggested start-up technique for injection-molding polyethylene lids from these three materials is presented below.

Start-Up Technique for Molding Polyethylene Lids

1. Inspect the mold and compare it with the engineering drawings. Particularly check the vents to be sure they are correct.
2. Mount the mold in the molding machine, set the mold temperature at 40 to 50°F, and operate on a dry cycle for a few minutes to see if all the mold parts are operating properly.
3. Adjust the machine to the clamp force required and continue the dry cycle.
4. Set the temperature controllers to obtain the desired melt temperature. A graduated temperature profile is suggested along the barrel of the extruder, increasing by 25°F increments from the throat to the injection portion of the machine. This will permit a good steady feed rate and uniform melting of the polymer. Melt temperature conditions depend considerably on the type of mold and machine being used.
5. Adjust the injection pressure to 12,000 to 15,000 psi on the plastic. This should be the maximum pressure that can be used without causing flashing and overpacking of the mold.
6. Set the injection speed fairly high. This is usually 0.4 to 0.6 oz/sec for each lid cavity; i.e., a four-cavity mold would re-

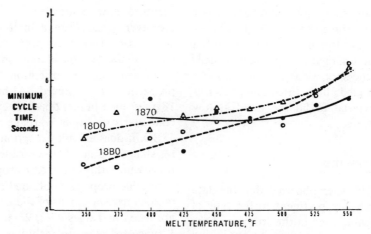

Fig. 23-16   Minimum cycle time vs. melt temperature.

quire an injection speed of 1.6 to 2.4 oz/ sec.

7. With the machine still cycling, start plastic feeding into the screw with the nozzle away from the mold. After about ten cycles, the injection unit can be brought up against the mold.

8. Interrupt the automatic cycle and operate

the machine manually until the runner system and all the cavities can be filled with plastic in a single shot, if possible. If the machine cannot fill the runner system and cavities with one shot, adjust the shot size so that all the cavities will be filled with the same material that completes filling the runner system. A short

Table 23-3   TENITE® Polyethylene 187O physical properties.

Injection Molding—Molding Temperature: 450°F (232°C)

PHYSICAL PROPERTIES:

PROPERTY†	UNIT	ASTM TEST METHOD	TYPICAL VALUE
Melt Index	g/10 min	D 1238	7.0
Density	g/cm³	D 1505	0.923
Softening Point, Vicat	°C	D 1525	94
Brittleness Temperature	°C	D 746	<-78
	°F		<-108
Tensile Strength at Yield	psi	D 638	1700
	MPa	Type IV	11.7
Tensile Strength at Fracture	psi	Specimen	1300
	MPa		9.0
Elongation at Fracture	%		200
Stiffness in Flexure	10⁵ psi	D 747	0.35
	MPa		241

TENITE Polyethylene 187O is a low-density formulation with excellent flow characteristics for injection molding. It is used for such applications as container closures that require stress-crack resistance.
A density of 0.923 g/cm³ is equivalent to 0.534 oz/in.³ or a yield of 29.99 in.³/lb.
Since some variation occurs in all plastics testing, there is no representation that the plastic in every individual order will conform exactly to all the properties shown above.
† *Unless noted otherwise, all tests are run at 23°C and 50% relative humidity (ASTM D618) conditions) on compression-molded specimens.*
TENITE is a registered trademark of Eastman Kodak Company.

Table 23-4 TENITE Polyethylene 18BO physical properties

Injection Molding—Melt Temperature: 425°F (219°C)

PHYSICAL PROPERTIES:

PROPERTY†	UNIT	ASTM TEST METHOD	TYPICAL VALUE
Melt Index	g/10 min	D 1238	20
Density	g/cm³	D 1505	0.923
Softening Point, Vicat	°C	D 1525	92
Brittleness Temperature	°C	D 746	−40
	°F		−40
Tensile Strength at Yield	psi	D 638	1650
	MPa	Type IV	11.4
Tensile Strength at Fracture	psi	Specimen	1300
	MPa		9.0
Elongation at Fracture	%		100
Stiffness in Flexure	10⁵ psi	D 747	0.35
	MPa		241

TENITE Polyethylene 18BO is a low-density formulation used for injection molding. It is an outstanding material for molding lids and other container closures because it processes easily on automatic molding equipment to produce molded items with excellent gloss, clarity, and warpage resistance.
A density of 0.923 g/cm³ is equivalent to 0.53 oz/in.³ or a yield of 30.0 in.³/lb.
Since some variation occurs in all plastics testing, there is no representation that the plastic in every individual order will conform exactly to all the properties shown above.
† Unless noted otherwise, all tests are run at 23°C and 50% relative humidity (ASTM D618 conditions) on compression-molded specimens.
TENITE is a registered trademark of Eastman Kodak Company.

Table 23-5 TENITE Polyethylene 18DO physical properties.

Injection Molding—Melt Temperature: 375°F (191°C)

PHYSICAL PROPERTIES:

PROPERTY†	UNIT	ASTM TEST METHOD	TYPICAL VALUE
Melt Index	g/10 min	D 1238	040
Density	g/cm³	D 1505	0.923
Softening Point, Vicat	°C	D 1525	91
Brittleness Temperature	°C	D 746	−35
	°F		−31
Tensile Strength at Yield	psi	D 638	1650
	MPa	Type IV	11.4
Tensile Strength at Fracture	psi	Specimen	1300
	MPa		9.0
Elongation at Fracture	%		90
Stiffness in Flexure	10⁵ psi	D 747	0.35
	MPa		241

TENITE Polyethylene 18DO is a low-density formulation suggested for the production of closures where a high-melt-index material is needed to meet shrinkage requirements and fast molding cycles. It molds with unusual ease and is suggested for large, thin items where stress-crack resistance is not a primary requirement. Formula 18DO is used as the base for color concentrates.
A density of 0.923 g/cm³ is equivalent to 0.53 oz/in.³ or a yield of 30.0 in.³/lb.
Since some variation occurs in all plastics testing, there is no representation that the plastic in every individual order will conform exactly to all the properties shown above.
† Unless noted otherwise, all tests are run at 23°C and 50% relative humidity (ASTM D618 conditions) on compression-molded specimens.
TENITE is a registered trademark of Eastman Kodak Company.

shot into the cavities on the first cycle could cause the gates to freeze while the shot is being removed manually. Return the machine to automatic cycling.

9. Adjust the plunger-forward timer so that the dead time is approximately 0.1 to 0.3 second.

10. Reduce the shot size until short shots appear; then slowly increase it until the mold cavities are just filled, without packing.

11. Reduce the clamp time until the snap rings begin to tear when the lid is ejected; then increase the clamp time slightly (0.1 to 0.3 second) to give each lid time to solidify. After molding has proceeded long enough for all molding conditions to become stable, reduce the gate timer as much as possible while still permitting the lids to clear the mold during ejection before the mold closes again.

12. Increase the injection pressure and injection speed while decreasing the plunger-forward time. This may necessitate increasing the cooling time sightly, but it should make possible a shorter total cycle.

13. If warpage, flash, or short shots are occurring erratically, shot-size control is probably not adequate, and a small cushion will have to be maintained to produce uniform lids.

14. If a cushion is used, it will probably be necessary to reduce the injection pressure and the injection speed and to control the amount of packing with the plunger-forward time.

15. If molding problems still occur, contact a technical service representative of the material supplier.

POLYPROPYLENES

Polypropylene and propylene copolymers are thermoplastic materials having the following characteristics:

Light weight	Ability to form an
Heat resistance	integral hinge
Hardness	Processability
Surface gloss	Chemical resistance
Stain resistance	Stress-crack resistance
Stiffness	Dimensional stability

These properties make polypropylene and propylene copolymers excellent choices for molding items such as housewares, appliance parts, automobile parts and accessories, closures, laboratory ware, hospital ware, toys, sporting goods, and miscellaneous items for home and industry.

Polypropylene is typically supplied in either cube cut or cylindrical $\frac{1}{8}$-in. pellets, the pellet shape depending upon the in-plant processing required for producing a particular formulation.

The plastic is offered in natural color and in a wide range of compounded colors custom-matched to the user's requirements and accurately controlled for uniformity between lots. It can also be colored in the user's plant with either dry colors or color concentrates.

With a nominal as-molded density in basic formulations of 0.902 g/cm^3, polypropylene is lighter than polyethylene and nonpolyolefin plastics and, therefore, produces more parts per pound than these other materials in any given mold. In addition, the high stiffness and excellent processability of polypropylene permit the molding of parts with thin sections that would often be too flexible or unmoldable with other thermoplastics.

Basic formulations of polypropylene are produced in flow rates ranging from less than 1 to 450 to meet a variety of processing and product performance requirements. Variations of basic formulas are available with additives to provide heat stability, weatherability, and ability to withstand some of the effects of radiation. Some formulas offer improved impact strength, while others contain fillers such as talc and calcium carbonate for applications requiring greater stiffness, tensile strength, and heat deflection temperature than are provided by general-purpose polypropylene. Processing and performance modifiers can be added such as antistatic, nucleating, mold release, and slip agents. Concentrates containing foaming agents designed to be used alone or mixed with other formulas can be supplied. Formulations lawful for use in contact with food under regulations of the U.S. Food and Drug Administration are also manufactured.

Physical Characteristics

In addition to low density and high stiffness, polypropylene has a high softening point and excellent chemical resistance, stress-crack resistance, electrical properties, and resistance to abrasion. Availability and a wide range of flow rates have promoted its use in a great variety of injection-molding applications. Tables 23-6 and 23-7 list the physical properties of two typical polypropylene formulations: a general-purpose material and an impact-modified polypropylene copolymer.

Injection Molding Machines

Polypropylene and copolymers are well adapted to molding in any of the commercially available molding machines. The fine points of the various types of molding machines are well described in the literature. For convenience, molding machines will be discussed only as screw-ram and plunger-type machines. These machines differ in the manner in which the plastic pellets are delivered from the feed hopper to the nozzle of the machine. The effect that screw-ram machines have on the plastic are different from those of the plunger machines.

Cylinder temperatures, injection pressures, and clamp pressures required for successful molding are normally lower for a screw-ram machine than for a plunger machine because the action of the screw results in better homogenization of the material and the development of frictional heat. The frictional heat added

Table 23-6 Physical properties of general-purpose TENITE® Polypropylene 4240.

Other formulations with the same physical properties:
   TENITE Polypropylene 4242—stabilized for extra life at high temperatures.
Grades offered:
   All formulas: A (molding).
   Formula 4240: G (modified pellet form).
Approximate processing temperatures: injection molding—melt temperature: 340–420°F (171–216°C).

PHYSICAL PROPERTIES:

PROPERTY		UNIT	ASTM TEST METHOD	TYPICAL VALUE
Flow Rate		g/10 min	D 1238L	9
Density		g/cm³	D 1505	0.902
Softening Point, Vicat		°C	D 1525	144
Deflection Temperature at 264-psi (1.82 MPa) load		°C	D 648	53
		°F		127
Tensile Strength at Yield		psi	D 638	4800
		MPa		33.1
Stiffness in Flexure		10⁵ psi	D 747	1.5
		MPa		1034
Flexural Modulus of Elasticity		10⁵ psi	D 790	1.7
		MPa		1172
Rockwell Hardness		R scale	D 785	90
Izod Impact Strength	Notched, at 23°C (73°F)	ft-lb/inch of notch	D 256	0.6
		J/m		32.1
	at 23°C (73°F)	ft-lb/inch of width		<25
	Unnotched,	J/m		<1335
	at −18°C (0°F)	ft-lb/inch of width		4
		J/m		214

TENITE Polypropylene 4240 and its related formula have the best combination of moldability and toughness of all unmodified TENITE Polypropylene formulations. Because of relatively low mold shrinkage and fast setup, they have found wide usage for closures; but large moldings with a fair degree of impact resistance are also possible. Typical applications: closures, battery caps, hair curlers, automotive air ducts.

Table 23-7 Physical properties of TENITE® Polypropylene 4E31 (see note).

Other formulations with the same physical properties:
  TENITE Polypropylene 4E3V—stabilized for moderate exposure to ultraviolet.
Grades offered: A (molding) and E (extrusion).
Approximate processing temperatures:
  Injection molding—melt temperature: 350–400°F (177–204°C).
  General-purpose extrusion—melt temperature: 400–450°F (204–232°C).

PHYSICAL PROPERTIES:

PROPERTY†	UNIT	ASTM TEST METHOD	TYPICAL VALUE
Flow Rate	g/10 min	D 1238L	4.5
Density	g/cm³	D 1505	0.90
Softening Point, Vicat	°C	D 1525	126
Deflection Temperature at 264-psi (1.82 MPa) load	°C	D 648	50
	°F		122
Tensile Strength at Yield	psi	D 638	3500
	MPa		24.1
Stiffness in Flexure	10⁵ psi	D 747	1.2
	MPa		827
Flexural Modulus of Elasticity	10⁵ psi	D 790	1.4
	MPa		965
Rockwell Hardness	R scale	D 785	67
Notched, at 23°C (73°F)	ft-lb/inch of notch		5.0
	J/m		267
Izod Impact Strength at 23°C (73°F)	ft-lb/inch of width	D 256	No break
Unnotched,	J/m		No break
at −18°C (0°F)	ft-lb/inch of width		>25
			>1335

TENITE Polypropylene 4E31 is a material with very unusual combination of stiffness and toughness. It is stiffer than conventional high-impact formulations of polypropylene and processes at relatively low temperatures into articles with high resistance to damage from impact. It is suggested for automotive components, both molded and extruded, that would benefit from its exceptional combination of properties.
A density of 0.90 g/cm³ is equivalent to 0.52 oz/in.³ or a yield of 30.8 in.³/lb.
Note: *Some properties of colored formulations may vary significantly from the values shown.*
Since some variation occurs in all plastics testing, there is no representation that the plastic in every individual order will conform exactly to all the properties shown above.
† *Unless noted otherwise, all tests are run as 23°C and 50% relative humidity (ASTM D 618 conditions) on specimens injection molded in accordance with ASTM Specification D 2146.*
TENITE is a registered trademark of Eastman Kodak Company.

by the work of the screw is proportional to the square of the screw speed—if the screw speed is doubled, the heat added is increased by a factor of four.

Faster molding cycles are generally possible with the screw-ram machines. Polypropylene and copolymers harden relatively fast when injection-molded, and with the lower melt temperature possible with the screw-ram machines, the cycle can be shortened.

The physical properties of items molded from polypropylene and copolymers on a screw-ram machine are generally better than those of identical items molded on a plunger-type machine. Usually, flexural strength, notched impact strength, and low-temperature toughness are increased while shrinkage is reduced. Articles molded in the screw-ram machine contain fewer stresses because the mold cavity can be filled at a lower injection pressure. Reduced molding stresses result in parts with better dimensional stability.

When polypropylene and copolymers are molded in colors, less time is required to change from one color to another when a screw-ram machine is used.

Polypropylene and copolymers behave in much the same way in processing operations, and conclusions drawn concerning one material generally apply to the other as well, except that copolymers appear to be better suited than polypropylene for insulated runner molding on a screw-ram machine.

The use of a preplasticating unit is not essential, but it is an advantage when polypropylene and copolymers are molded in a plunger-type machine. With such a unit, the high heat requirements of polypropylene and copolymers are partially provided before the material enters the cylinder. Therefore, the cylinder can be maintained at a lower temperature than when it is supplying the entire heat input, and the possibility of hot spots is greatly reduced.

Molding Thin Sections

Both polypropylene and propylene copolymers have excellent moldability, permitting small parts with a wall thickness of 0.010 in. to be molded satisfactorily. In general, these plastics show sharp decreases in viscosity at their melting point. This allows them to flow in the mold cavities more readily than do most other thermoplastics. Their superiority, not immediately apparent in molds that are easily filled, becomes quite obvious in difficult-to-fill molds as a result of this ease of flow. Thin sections can be molded more satisfactorily from polypropylene or copolymers than from almost any other thermoplastic.

Molding Thick Sections

The molding of thick sections from general-purpose polypropylene should be avoided if possible because of the formation of a coarse crystalline structure in the article caused by slow cooling of the plastic. Articles with such a structure usually have low impact strength. The toughness of articles molded of impact-modified polypropylene or propylene copolymers is much less dependent upon rapid cooling, and these formulations may be better than general-purpose polypropylene for molding thick sections.

The wall thickness of a part is usually dictated by the stiffness required in the molded piece and the material selected for the job. Because of the high stiffness of polypropylene, a part to be molded of this plastic may be designed with thinner walls than ordinarily would be required with other polyolefins.

When it is necessary to mold an item with a thick section, it is important to have the thick section near the gate with any reduction in thickness being made in the direction of flow. This makes it possible to maintain effective pressure on the thick sections for a longer time without having excess pressure on the thin sections which are farthest from the gate. Gating into thick sections minimizes sink marks and results in less tendency to warping than does gating into thin sections.

Another possibility to consider when molding thick sections is the use of a foam concentrate such as Eastman's Tenite polypropylene P2635–08AA. This cocnentrate has effectively demonstrated, through years of use, its capability for eliminating sinks and voids in heavy-sectioned parts. The specific use of this concentrate is described in an Eastman publication, MB-55, which is available upon request.

Molding Conditions

Table 23-8 gives ranges of conditions for injection molding articles of various thicknesses from polypropylene and copolymers. These are suggested start-up conditions. Final operating conditions may be different from those shown, as they vary with the application, the mold design, the formula selected, and the injection machine used. For molding the same part, plunger machines generally operate at temperatures 30°F higher than those for screw-ram machines.

Molding Techniques and Conditions Affecting Part

After the mold is constructed, the operating factors that affect the quality, quantity, and cost of the molded product must be determined. Although quality, quantity, and cost

Table 23-8 Ranges of injection-molding conditions for Tenite polypropylene and Tenite polyallomer formulations on screw-ram machines.

MOLDING CONDITIONS	THICKNESS OF SECTION		
	0.063 IN. (1.6 MM)	0.125 IN. (3.2 MM)	0.25 IN. (6.4 MM)
Temperatures			
Rear cylinder, °F	380–420	380–400	380–400
°C	193–216	193–204	193–204
Middle cylinder, °F	400–450	380–420	380–420
°C	204–232	193–216	193–216
Forward cylinder, °F	420–480	400–450	400–420
°C	216–249	204–232	204–216
Nozzle, °F	380–420	380–420	380–420
°C	193–216	193–216	193–216
Melt, °F	400–480	400–450	380–420
°C	204–249	204–232	193–216
Mold coolant, °F	50–80	50–80	50–80
°C	10–27	10–27	10–27
Hydraulic Injection Pressure,			
psi	600–1500	600–1500	600–1500
MPa	4–10	4–10	4–10
Typical Cycle Time, sec			
Plunger forward	5–10	10–15	15–20
Total cycle	15–25	25–35	35–60
Shrinkage, %	1–2	1–2	1–2

are primarily dependent on the quality and type of tooling and machine employed, proper molding techniques and the use of optimum molding conditions have significant influence. With regard to quality of injection molded parts, the most important variables are injection speed, injection pressure, clamping pressure, melt temperature, mold temperature, and cycle time. These variables are discussed in detail below.

Injection Speed. Normally, high injection speeds are used when molding polypropylene and copolymers because fast filling speed results in a relatively uniform temperature of the material as it fills the cavity. If the filling rate is slow, particularly in molding thin sections, the first material entering the cavity may cool much more rapidly than the subsequent material, resulting in an incomplete fill, lamination, and possible warpage of the part. This filling problem is present with any thermoplastic, but it is accentuated in molding polypropylene, which has a relatively high crystalline

melting temperature and solidifies quickly in the cavity.

It may be necessary to reduce injection speed to control the uniformity of the flow and maintain a good surface finish when parts with thick cross sections are molded with small gates.

Injection Pressure. The injection pressure should be maintained at the minimum level required to fill the mold. Molding shrinkage may be reduced and sink marks minimized by increasing the injection pressure, but this results in packing the material into the mold cavity. Such packing may cause difficulty in ejecting the piece from the mold and warpage of thin sections.

Normally, the stiffness of a part molded from polypropylene or a copolymer increases slightly as the injection pressures increase, particularly when low melt temperatures are used. Changes in injection pressure do not significantly affect the impact strength of a molded part. The effects of increasing injec-

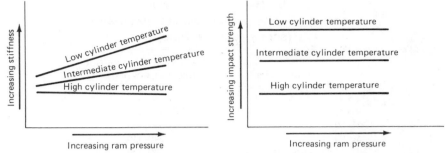

Fig. 23-17   Effects of injection pressure.

tion pressure are shown qualitatively in Fig. 23-17. This figure also shows the effects of cylinder temperature, which will be discussed later.

Clamping Pressure. Clamping pressure is the pressure needed to hold the mold closed against the opposing pressure exerted by the molten plastic under force of the injection and holding pressure.

The pressure transmitted to the mold cavity depends primarily on the type of injection unit used. For example, injection molding machines with any type of preplasticator in which the shooting ram works against molten polymer are very efficient in transmitting applied pressure to the cavity. Screw-ram machines are also very efficient, transmitting up to 90 percent of the applied ram pressure to the molten polymer in the mold cavity. A plunger-type injection unit, in which the plunger acts against unmelted pellets, is less efficient in transmitting applied pressure to the mold cavity than are screw-ram machines or machines with preplasticators.

Restrictions in the nozzle, runner system, or gates retard the flow of molten polymer and limit the transmission of injection pressure to the cavity. A web gate can be used to good advantage when molding polypropylene and copolymers because it gives a more effective area for transmitting pressure than other gates and will still freeze off when the flow stops.

Determining the exact clamping pressure required in a particular molding situation requires knowledge of the variables discussed above. In deciding how much projected area can be tolerated on a particular molding machine, Table 23-9 will serve as a guide.

It is recognized that thick molded sections remain molten longer than thin sections, and a greater proportion of the applied pressure can be transmitted to the mold cavity. However, thin section moldings require higher injection pressures and speeds due to resistance to filling the mold. Consequently, the clamp pressure requirements are higher for thin sections.

Melt Temperature. Processing temperatures for polypropylene vary more with the characteristics of the processing equipment and its accessories (mold, die, etc.) than they do with the formulation, but in any given processing situation, the optimum temperature may vary somewhat with the flow rate of the material. The best melt temperatures for processing various formulas for Tenite polypropylene will usually fall within the ranges shown in Table 23-10.

As the melt temperatures increase, there is a decrease in stiffness and impact strength of molded polypropylene and copolymers. The decrease in stiffness caused by the increase in the melt temperature is greatest when a high injection pressure is used. Below certain

Table 23-9   Clamping pressure requirements.

WALL THICKNESS, IN.	CLAMPING PRESSURE REQUIRED, TONS/IN.^{2}
Less than $\frac{1}{32}$	5–7
$\frac{1}{32} - \frac{1}{16}$	3–4
Greater than $\frac{1}{16}$	2–3

Table 23-10 Melt temperatures for processing Tenite polypropylene materials (screw-ram machine).

FLOW RATE	MELT TEMPERATURE FOR INJECTION MOLDING, °F
4.5	380–450
9.0	370–440
18–30	360–430
50–70	340–410
Talc-filled formulas	380–450
Calcium carbonate–filled formulas	380–450

minimum melt temperatures, severe stresses in the molded part can occur with a resultant loss of impact strength.

At normal injection molding temperatures, around 450–470°F, there is no significant difference in deflection temperature caused by changes in melt temperature. At extremely high temperatures, about 500–550°F, an increase is noted. High melt temperatures along with long residence time at melt temperature may result in increased flow rate (material breakdown) and reduced toughness. It is desirable that the shot size utilize one-half or more of the cylinder capacity to limit melt residence time.

Generally, an increase in the cylinder temperature makes it possible to use lower injection pressure and produce a better surface finish; but it also tends to increase drooling from the nozzle, may make flashing a problem, and increases the time for cooling.

Mold Temperature. Close control of mold temperature is important in molding any thermoplastic, but it has increased significance in the molding of polypropylene and copolymers because of the highly crystalline nature of these plastics.

Mold temperature affects the properties of copolymers less than it does those of polypropylene, and a relatively tough part can be molded from Tenite polyallomer copolymer with only limited mold cooling. Mold temperatures up to 90°F are often satisfactory.

Tenite polyallomer copolymer crystallizes more slowly than prolypropylene, and a portion of the crystallinity in the molded part forms after the part is removed from the mold.

It is for this reason that mold temperature has less effect on the properties of polyallomer than it does on the properties of general-purpose polypropylene.

In molding polypropylene and copolymer parts, it is usually desirable to obtain maximum impact strength rather than maximum stiffness. This indicates that low mold temperatures should be used, normally in the range of 30–60°F. A cold mold cools the material rapidly and causes the formation of a fine crystalline structure.

Some molded parts may require maximum stiffness with impact strength of secondary importance. If this is the case, mold temperatures in the range of 110–130°F are desirable. The general trends in stiffness and impact strength as related to mold temperature are shown qualitatively in Fig. 23-18.

High mold temperatures result in high shrinkage in the molded part. For example, on a $\frac{1}{8} \times \frac{1}{2} \times 5''$ test bar produced at a mold temperature of 60°F, the shrinkage varied from 1.9 to 2.3 percent. On an identical test piece produced at a mold temperature of 120°F, the shrinkage was 2.7 to 3.1 percent.

In molding articles of heavy cross sections, where high mold temperatures may be necessary, it may be advantageous to cool the articles in an ice water bath immediately after ejecting them from the mold. This allows the article to be ejected while still hot and thus shortens the cooling portion of the molding cycle. Cooling the parts in ice water also achieves the quick quenching necessary for good impact strength, and it hardens the surface sufficiently to prevent sink marks from forming.

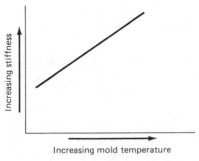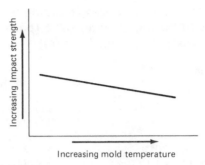

Fig. 23-18   Effects of mold temperature.

Cycle Time. Cycle time is largely dependent on section thickness, machine conditions, machine heating capacity, and injection capacity. The overall cycle time can vary from approximately 5 seconds for thin articles to 60 seconds or more for thick articles.

The ram is in the forward position usually one-fourth to one-third of the total cycle time. If the booster pump is used, it is normally set to cut out when the ram stops moving on the forward stroke of the cycle.

COPOLYESTERS

Copolyesters include a wide range of materials with widely differing processing parameters and properties. Probably the most frequently encountered injection-moldable copolyester is Eastman Chemical's Kodar PETG copolyester 6763. This material is a glycol-modified poly(ethylene terephthalate). The modification is made by adding a second glycol, cyclohexanedimethanol (CHDM), during polymerization. The second glycol is added in the proper proportion to produce an amorphous polymer. Kodar PETG copolyester 6763 will not crystallize, and thus offers wider processing latitude than conventional crystallizable polyesters. Plasticizers or stabilizers are not required in this polymer.

Kodar PETG copolyester 6763 offers an excellent combination of clarity, stiffness, and toughness. It is lawful for use as an article or component of articles intended for use in contact with food, subject to the provisions of Food Additive Regulations 21 C.F.R. 177.1315 and 21 C.F.R. Part 174, published by the U.S. Food and Drug Administration.

Kodar PETG copolyester 6763 is not lawful for use as an article or a component of articles intended for use in contact with carbonated beverages, beer, or containers for food that will be subject to thermal treatment, or under conditions of fill or storage exceeding 120°F.

Typical injection molding applications for PETG include toys, chairs, protective covers, medical device parts, face shields, brush backs, display racks, ice scrapers, containers, and appliance parts.

Injection Molding Variables

The processing variables that affect the quality of articles molded of PETG 6763 include injection speed, screw speed, back pressure on the screw, injection pressure, clamping pressure, melt temperature, nozzle temperature, mold temperature, and cycle time. The typical conditions for molding PETG 6763 are shown in Table 23-11.

The shot size of items to be molded should utilize at least 50 percent of the machine's plasticating capacity and, preferably, 75 to 80 percent.

Injection Speed. Low injection speeds are desirable from the standpoint of part appearance. If splay (visible flow lines generally radiating from the gate) is encountered, it can generally be minimized by reducing the injection speed. If the molding machine is equipped with programmed injection, an initial slow injection rate can be used until some material has entered the cavity; then a more rapid fill rate can be used without causing the splay effect.

Table 23-11 Typical molding conditions for Kodar PETG copolyester 6763 [part thickness = 0.125 in. (3.2 mm)].

Cylinder temperatures, °F (°C)	
Rear	420 (216)
Center	450 (323)
Front	470 (243)
Nozzle	470 (243)
Melt temperature, °F (°C)	480 (249)
Mold temperature, °F (°C)	80 (27)
Injection pressure, psi (MPa)	12,000 (82.7)
Cycle time, sec.	
Inject	20
Cooling	20
Recycle	2
Overall	42
Screw speed, rpm	60
Injection speed	Slow

Screw Speed. A screw speed of 30 to 60 rpm is suggested for processing PETG 6763. High screw speeds result in frictional heating, and this has the same effect as increasing the cylinder temperature. If frictional heating is excessive, polymer degradation may occur.

Back Pressure on the Screw. Back pressure is normally not required for molding PETG. However, if large reground particles are present in the feed or color concentrates or dry coloring agents are being used, it may be necessary to use a small amount of back pressure to keep the screw full to plasticate the material properly and to aid in color dispersion.

Injection Pressure. Injection pressure is not critical when molding PETG 6763. Relatively high injection pressures are usually required because the material is somewhat viscous at normal molding temperatures.

Melt Temperature. PETG 6763 can be molded with stock temperatures ranging from 380°F to 525°F. At the low end of the tem-

Table 23-12 Thermal properties of Kodar PETG Copolyester 6763.

PROPERTY, UNITS		ASTM METHOD	VALUE
Deflection Temperature			
at 264 psi (1.82 MPa) fiber stress,	°F	D 648	145
	°C		63
at 66 psi (0.46 MPa) fiber stress,	°F		158
	°C		70
Vicat Softening Point,	°F	D 1525	180
	°C		82
Thermal Conductivity, cal/cm·s·°C (W/m·K)		C 177	7.8×10^{-3} (0.32)
Glass Transition Temperature,	°F	—	178
	°C		81
Specific Heat, cal/g·°C (J/kg·K)			
at 10°C			0.26 (1090)
30°C			0.27 (1130)
50°C			0.28 (1170)
70°C			0.30 (1260)
100°C			0.37 (1550)
120°C		D 2766	0.38 (1590)
140°C			0.40 (1670)
180°C			0.41 (1710)
200°C			0.42 (1760)
280°C			0.47 (1970)
300°C			0.49 (2050)
Coefficient of Linear Thermal Expansion, Units/°C (−30 to +40°C)		D 696	5.1×10^{-5}
Melt Density at 250°C, g/ml		—	0.98

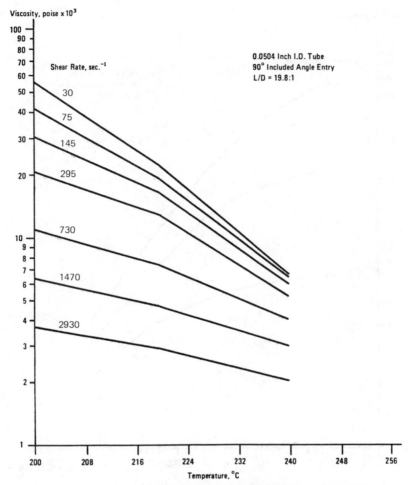

Fig. 23-19   Viscosity vs. temperature, Kodar PETG Copolyester 6763.

perature range, the material is extremely difficult to push into the mold. At the high end
of the range, material breakdown can easily
occur if long cycles or an oversized machine
results in long dwell time of the material in
the heated cylinder. For the most part, temperatures in the range of 420°F to 500°F are
desirable.

Mold Temperature.   Mold temperature is
not critical to the production of acceptable
parts from PETG. Where fast cycles are desirable, a cold mold can be used. Where cavity
filling is a problem, a warmer mold can be
used. Mold temperatures higher than 130°F
should be avoided because parts may tend to
stick in the cavities.

Purging

PETG 6763 is easily purged from molding
machiens. Polypropylene, high-density polyethylene, or an acrylic purging compound will
readily purge a machine filled with PETG
6763. In general, PETG 6763 can be removed
with any material that follows it in the machine.

Shutdown and Start-up

After the molding machine is purged, normal
shutdown procedures should be employed.
The screw should be moved to its forward
position and the machine operated as an extruder until no more plastic is being extruded.

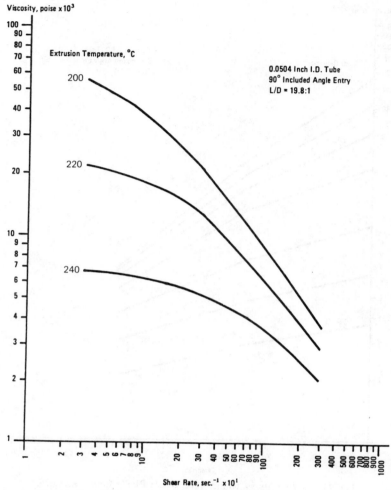

Fig. 23-20 Viscosity vs. shear rate, Kodar PETG Copolyester 6763.

The heat can then be turned off and the screw stopped. When processing is to be started again, the heater should be turned on and allowed to heat the machine to operating temperature. Then, the injection switch should be activated, even though the screw is already in its forward position. The small motion that results will free the non-return valve slip ring, so that it will operate properly. Rotation of the screw can be begun.

Thermal and Rheological Properties

As PETG 6763 is amorphous, its thermal properties are essentially determined by its glass-transition temperature of 178°F. This copolyester has deflection temperatures of 145°F at fiber stress of 264 psi and 158°F at 66 psi. Deflection temperatures and other thermal properties are given in Table 23-12.

At temperatures encountered in injection molding 450–525°F), the viscosity of PETG 6763 decreases rapidly with increasing temperature and at high shear rates. The effects that temperature, shear rate, and shear stress have on the viscosity of PETG 6763 are shown in Figs. 23-19, 23-20, and 23-21. Shear stress vs. shear rate is given in Fig. 23-22.

Drying

Successful injection molding of Kodar PETG copolyester 6763 requires that the pellets be

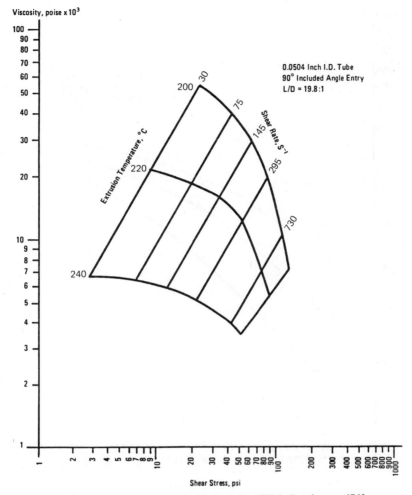

Fig. 23-21   Viscosity vs. shear stress, Kodar PETG Copolyester 6763.

dried before processing, as is the case with all thermoplastic polyesters. PETG 6763 is subject to hydrolysis when it is in the molten state during processing. This hydrolysis results in a decrease in molecular weight that is reflected by a lowering of physical properties, especially toughness. To prevent hydrolysis during the injection molding process, PETG 6763 must be thoroughly dried. Drying the material in a dehumidifying dryer at a temperature of 150°F for 4 hours is normally sufficient to reduce the moisture content to a level (approximately 0.04 percent) that will prevent significant hydrolysis in processing equipment operating at 380°F to 525°F. The dryer temperature should not exceed 150°F to prevent the pellets from softening and sticking together in the hopper.

Type of Dryer

The most practical production system for drying pellets is a hopper-dryer system. Modern dryers use a molecular sieve desiccant material in small bead form. The desiccant is placed in the return air stream of the dryer in two or more beds or canisters. One bed is "on-stream" removing moisture from the return processed air while the second bed is being regenerated (dried) at high temperature by secondary heaters. Several companies supply dryers to the plastics industry.

It is important that the dryer purchased have the correct design requirements for the plastic resin to be processed. Kodar PETG copolyester 6763 should be dried at a maximum temperature of 150°F. Care should be

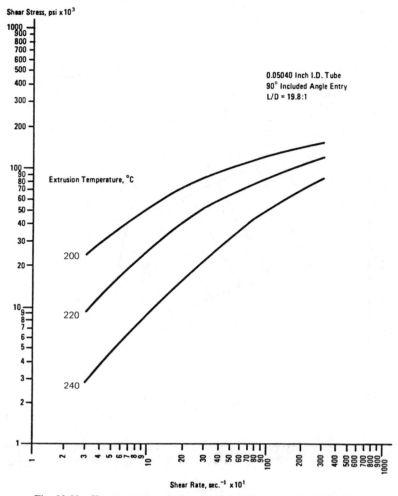

Fig. 23-22   Shear stress vs. shear rate, Kodar PETG Copolyester 6763.

taken to obtain a dryer that can be accurately controlled to deliver an air temperature of 150°F. Some dryers tend to cycle uncontrollably hotter. An inlet temperature gauge is a useful addition to the drying hopper.

The use of dry air provides significantly more efficient drying than does ambient air, particularly during humid summer months. Consequently, a desiccant-type dryer is required. This type of dryer provides drying air having a dewpoint of −20°F or lower. Dewpoint is an absolute measure of air moisture and is independent of air temperature. Dewpoint, not relative humidity, should always be specified regarding air dryness. It is worth the small additional cost to incorporate a dew-

point monitor into the dryer used; this is an option available on modern dryers.

The drying hopper should be of sufficient size to allow 4 hours of drying time for the pellets. For example, if the typical throughput of the molding machine is 100 pounds per hour, a 400-pound capacity hopper is used. A good drying hopper has the screen and cone system on the bottom to ensure uniform air flow through the pellet bed. It is also designed to provide an even flow of pellets from the top to the bottom (plug flow) as well as provide an air-lock loading system. Insulated hoppers should be used to maintain the pellet beds at the inlet air temperature.

The dryer and drying hopper can be conve-

Table 23-13   Mechanical properties of Kodar PETG Copolyester 6763 (injection-molded specimens).

PROPERTY, CONDITIONS, UNITS	TEST METHOD	VALUE OF PROPERTY
Specific Gravity, at 23/23°C	ASTM D 792	1.27
Tensile Strength, at Fracture, psi		4000
MPa		28
at Yield, psi	ASTM D 638	7100
MPa		49
Elongation at Fracture, %		180
Flexural Modulus of Elasticity, 10^5 psi		2.9
MPa		2000
Flexural Strength, psi	ASTM D 790	10,000
MPa		69
Rockwell Hardness, R scale	ASTM D 785	105
Izod Impact Strength		
notched, at 73°F (23°C), ft-lb/of notch		1.7C
J/m of notch		91C
at −40°F (−40°C), ft-lb/in. of notch		0.7C
J/m of notch	ASTM D 256	37C
unnotched, at 73°F (23°C)		No break
at −40°F (−40°C)		No break
Plaque Impact Strength, ft-lb	Eastman	>42
J	ECD-1072	>60

All tests not otherwise noted were run at 73°F (23°C) and 50% relative humidity.

niently located on the floor near the molding machine. The drying hopper can be placed in the production line after the regrind blending system with vacuum loaders being used to feed the regrind and pellets from the loader/ blender to the drying hopper and then to the molding machine hopper. Another approach is to use a central dryer/blender system to serve several molding machines. The material from a central system can be transferred auto-

Table 23-14   Electrical properties of Kodar PETG Copolyester 6763.

PROPERTY, UNITS	ASTM TEST METHOD	VALUE
Dielectric Constant		
at 1 kHz		3.5
at 10 kHz		3.4
at 1 MHz		3.2
Dissipation Factor	D 150	
at 1 kHz		0.01
at 10 kHz		0.02
at 1 MHz		0.3
Arc Resistance, sec	D 495	150
Volume Resistivity, Ohm·cm	D 257	7×10^{13}
Dielectric Strength, $\frac{1}{8}$ in. (3.2 mm)		
Specimen, step-by-step, V/mil (V/mm)		400 (15,750)
At 500 V/s rate-of-rise, V/mil (V/mm)		400 (15,750)
Dielectric Strength, $\frac{1}{16}$ in. (1.6 mm)	D 149	
Specimen, step-by-step, V/mil (V/mm)		560 (22,050)
At 500 v/s rate-of-rise, V/mil (V/mm)		550 (21,650)

matically to the hoppers of several machines, or the dried material may be loaded in the cleaned drums and moved to individual machines.

Dryers are specified according to their blower capacity in cubic feet of air delivered per minute (CFM). A good dryer design criterion is to allow 1.0 CFM per pound per hour of material to be extruded. For example, if 100 pounds per hour of material is to be extruded, a minimum blower capacity of 100 CFM is used. A dryer with blower capacity of 100 CFM should be considered the smallest size for any use—smaller-capacity dryers should not be purchased. Smaller blowers may not have sufficient air flow to prevent excessive heat loss.

Mechanical Properties

The most outstanding mechanical properties of Kodar PETG copolyester 6763 are high stiffness and good impact strength, particularly at low temperatures. These properties are easily obtainable using the drying and molding conditions given above. The typical mechanical properties of PETG 6763 are given in Table 23-13.

Electrical Properties

The electrical properties of Kodar PETG copolyester 6763 are given in Table 23-14.

Chemical Resistance

Unstressed tensile bars molded of Kodar PETG copolyester 6763 exhibit good resistance to dilute aqueous solutions of mineral acids, bases, salts, and soaps, and to aliphatic hydrocarbons, alcohols, and a variety of oils. Halogenated hydrocarbons, low-molecular-weight ketones, and aromatic hydrocarbons dissolve or swell the plastic. The chemical resistance of PETG 6763 is the subject of an Eastman publication that is available upon request (Publication TR-59).

Weatherability

Kodar PETG copolyester 6763 is not suggested for use in applications requiring outdoor exposure.

Color

Kodar PETG copolyester 6763 may be colored by using color concentrates, dry colors, or liquid colorants. Compounded colors are not available. Color concentrates in a PETG 6763 base to provide material compatability when mixed are available from Eastman. Most color concentrates are custom-made, but a few standard color concentrates are available from stock. The mixing ratio for PETG 6763 and color concentrate is usually 20:1.

For injection molding PETG 6763 mixed with color concentrate, it is sometimes necessary to use mixing nozzles or Venturi plates to obtain proper color dispersion.

The color concentrates should be kept as dry as the PETG 6763 to minimize hydrolytic degradation of the materials during processing.

Chapter 24
Injection Molding Polyvinyl Chloride

D. V. Rosato

INTRODUCTION

It is important to understand the molding characteristics of polyvinyl chloride (PVC) and how they relate to the design of processing equipment. Injection molding of flexible vinyl compounds is relatively easy, but special modifications to standard sprue machines are suggested for processing rigid vinyls. Low-shear screws are necessary to avoid overheating the melt. Special surface treatment, to protect machine surfaces against the corrosion PVC can cause, is also available. Plating, high-nickel steels, or stainless steels should be specified in the mold designs. Heavy metals such as copper must be protected because they are adversely affected by PVC, even at levels as low as 2 ppm. Finally, internal mold passageways should be designed without sharp corners or other restrictions where material may stagnate and degrade.

PVC homopolymers have a melting point between 198 and 205°C (388 and 401°F) and begin to decompose rapidly at 200°C (392°F). Here lies the challenge to the processor. Co-polymers are somewhat more forgiving. They begin to melt between 140 and 175°C (284 and 347°F) and offer significant processing advantages; decomposition remains at the high level.

There are molders, and designers of molds, who have no difficulty in processing PVC. As an example, even hot-runner molding of PVC is used even though it represents a difficult job for some molders and for the moldmakers. To appreciate just how troublesome the material can be, one should first understand something about the physical and processing character of PVC.

PVC is chemically inert, and it is water-, corrosion-, and weather-resistant. It has a high strength-to-weight ratio. It is an electrical and thermal insulator, and it maintains its properties over long periods of time. Perhaps more important than these other considerations, it has demonstrated good price stability in the market.

Currently, PVC applications are concentrated in construction-related products. A large portion of those applications, and some of the most familiar, are PVC pipe and fittings. Recent developments in removal of unreacted vinyl chloride monomer promise to reopen the rigid-food-packaging markets. This could vastly expand the use of and the demand for efficient processing equipment to be used with PVC.

It is little wonder that the industry has restricted PVC to cold-runner gating systems and three-plate molds. However, it is recognized that the use of cold or surface runners is relatively inefficient because of the wasteful generation of runner scrap and because additional clamp tonnage is often needed to contain the extreme pressure generated by the large cross-sectional areas. This technology, as noted, has been the standard for distributing melt within a multicavity PVC mold, however.

Three-plate molds, which convey melt between the primary plate and an auxiliary plate, offer more flexibility of gate location than cold-runner systems. These molds, however, produce as much runner scrap as cold-runner designs—often weighing as much as the part being molded. Further, the time required to open the second plate for runner extraction can add considerably to the overall molding cycle.

Hot-runner systems, which keep the polymer heated throughout its passage from the machine nozzle to the mold cavity, have been used very successfully with most polymers. Temperature-control problems and the tendency in such systems toward complex and sometimes restricted flow paths have made it difficult to apply hot-runner technology to PVC, however.

Significant developments in resin stability and major strides in the technology of hot-runner heating have been made, and in many cases the problems mentioned above have been minimized or have been avoided entirely. It is important to remember that there is little difference between PVC's melting temperature and the temperature at which it degrades and burns. Even heat distribution and tight temperature control are required for both flexible and rigid materials.

In any hot-runner system, the melt will be contained within the system under precise temperature control until it reaches the gate area. More conventional hot-edge gating systems isolate the hot sprue from the cold mold using some of the molten polymer as a thermal barrier. Since insulating material would not be flushed with every shot, it would eventually degrade and could break away in small pieces to block the gate or mar the appearance of the product.

The use of a titanium insert at the gate makes for a hotter gate and eliminates the need for polymer insulation and the possibility of degradation. There is no place for material to hang up, and the system produces a cosmetically acceptable gate.

Despite the success of hot-runner systems for flexible PVC, comparatively few systems for rigid materials have been produced. As noted, rigid PVC is far more heat-sensitive than the flexible product, and therefore heat control and flow-path design are more critical for rigid PVC. Hydraulic valve-gating systems have been successfully applied.

Valve gates are particularly appropriate because they offer a larger gate than other techniques, while they also produce a cosmetically clean gate. The larger gate subjects the material to lower shear levels, and unlike conventional sprue gates (which could be just as large) it does not leave objectionable marks or drool between shots.

RIGID PVC DRY BLENDS

R. G. Hale and K. G. Mieure

Conoco Chemicals
Ponca City, Oklahoma

Rigid PVC compounds in dry blend form have been used for a variety of extrusion applications for quite some time because of favorable economics and processing. The use of dry blends for injection molding rigid PVC products has become increasingly popular. Typical items that are commercially molded from rigid PVC dry blend include fittings for pipe, pipe furniture, and conduit (1).

Pelletized rigid PVC compounds were used for the initial commercial production of most current rigid PVC applications. PVC pipe is an example. As the pipe manufacturing process became more sophisticated and competition intensified, the PVC compound form evolved from pellets to dry blend, primarily to take advantage of the improved economics that result from the elimination of the pelletizing step. Thus, such large PVC markets as pipe and vinyl siding currently use dry blend almost exclusively. We believe that other markets for rigid PVC will evolve from pellets to dry blend (although pellets will likely retain major portions of some markets because of differences in handling and processing characteristics between pellets and powder). Our information indicates that the injection molded PVC fittings market has recently converted to a mostly dry blend market.

We estimate the total PVC injection molded fittings market at about 130 mm pounds per year currently. At least 70 percent of this market is supplied by the ten largest PVC fittings producers. Of the total market volume, 55 to 60 percent is molded from dry blend, while about 65 percent of the volume of the ten largest producers is molded from dry blend. Thus, dry blend accounts for the majority of the PVC injection molded fittings market. This represents a tremendous increase in dry blend usage since the mid-1970s, when only about 10 percent of the fittings volume was molded from dry blend. The driving force for this conversion has been the lower raw material costs for dry blend compared to pellets. Improved dry blend formulations have helped to make this conversion possible.

The major portion of the PVC fittings market produces fittings for PVC potable water, sewer, and DWV (drain, waste, and vent) pipe. PVC gutter system accessories are also injection molded from rigid PVC. A smaller portion of the market includes electrical and communications conduit fittings. In addition to the more common couplings, elbows, etc., conduit fittings include more diverse items such as outlet boxes and other replacements for metal items for electrical wiring systems. A small but rapidly growing application for PVC injection molded fittings is to connect round or square extruded PVC profiles to fabricate PVC casual or patio furniture.

Various sizes of fittings for these applica-

tions have been successfully molded from dry blend, covering a range from $\frac{3}{8}$-in. diameter to at least 8 in. The demands on the PVC dry blend with respect to processing and physical properties vary considerably depending on the size of the molded part and the intended application. For example, potable water fittings require high burst strength for resistance to water pressure, while electrical conduit fittings require excellent low temperature impact to meet the standards of such organizations as UL (Underwriters Laboratories Inc.) and CSA (the Canadian Standards Association). Gutter and patio furniture fittings require excellent resistance to the effects of long-term outdoor exposure. The fittings producers are able to meet these various criteria largely because of adjustments in the PVC formulation.

Formulations

As with any rigid PVC formulation, the primary considerations for an injection molding dry blend formulation are cost, ease of processing (molding), and providing the necessary stiffness, impact strength, weatherability, appearance, and other physical properties in the finished part. However, molding from dry blend also requires that other factors be considered. Formulations must be designed to ensure rapid fusion; attempting to inject cold or poorly fused material can result in degradation problems due to high "melt" viscosity at the high shear rates encountered, as well as poor physical properties in the finished part. Volatile components in the dry blend must be minimized, since there is no pelletizing step to help remove volatiles. Additives must be selected so as to avoid any potential for segregation of dry blend components during dry blend handling and transfer. For example, ingredients should be selected so that their particle size distributions and bulk densities match fairly closely. Fine particles can be undesirable because they can contribute to housekeeping problems with poorly sealed material transfer systems.

A potential advantage in formulating dry blends as compared to pelletized compounds can be in the use of stabilizer. The required minimum level of this expensive ingredient may be lower in the case of dry blends, since a pelletized compound has already experienced one "heat history" before the compound reaches the hopper of the injection molder.

A typical starting-point formulation for a PVC pipe fitting dry blend is shown in Table 24-1. For high impact applications, the impact modifier concentration should be increased to a level such as 8 to 12 phr. For applications where optimum weathering resistance is required, the titanium dioxide level should be increased to perhaps 8 to 12 phr, and the impact modifier used should be an acrylic or EVA type.

The preparation of PVC dry blends for injection molding is similar to other dry blend applications. High-intensity mixers should be used. Care should be taken to avoid excessive fluffing or generation of a static electric charge in the cooling blender or dry blend conveying system, since air entrapment in the melt can result during processing. Properly formulated and blended injection molding dry blend can be shipped and handled in either bulk railcar, bulk truck, or 1000-pound boxes without problems.

Equipment Design Considerations

Rigid PVC in one form or another (i.e., dry blend, pellets, or regrind) has been processed on nearly every conceivable type of reciprocating screw molding machine. These machines have been equipped with numerous types of screws and mold configurations. Successful field evaluations have been conducted using ABS screws and molds. Shot sizes have ranged from 20 percent to nearly 100 percent of rated capacity. Various types of gates, such as submarine gates, have been successfully used. However, making a good part and maximizing productivity and efficiency are not necessarily synonymous. Let's take a look at some of the well-known and not so well-known aspects of the molding equipment used for maximizing productivity and efficiency with PVC dry blend.

Molding Machine

The molding machine shown in Fig. 24-1 highlights items that are necessary as well as some that are not necessary but definitely advantageous when molding PVC dry blend.

Shot Size. Unfortunately it is not always possible to size the mold with the machine. For maximum productivity and ease of processing the shot size should be approximately 75 percent of rated machine capacity. The lower the shot size as a percent of machine capacity, the longer the residence time. This means that degradation is more difficult to control, and quality problems due to greater

Table 24-1 PVC dry blend formulations.

INGREDIENT	COMMON TYPE	CONCENTRATION
PVC suspension resin	Homopolymer, 0.68–0.74 IV	100.0 phr (parts/hundred)
Tin stabilizer	Mercaptide, 13–20% tin	1.2–2.0
Processing aid	Methacrylate copolymer	1.5–3.0
Co-stabilizer/lubricant	Calcium stearate	0.5–2.0
Filler	Calcium carbonate, 1–3 micron	0–5
Pigment/UV stabilizer	Titanium dioxide	1–2
Impact modifier	ABS or MBS polymer	0–5
Lubricant	Paraffin wax or fatty acid amide or fatty acid esters	0.5–1.5

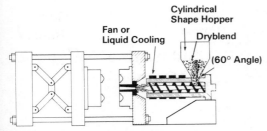

Fig. 24-1   Injection molding machine for dry blend PVC.

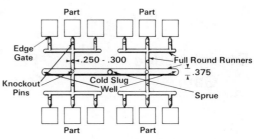

Fig. 24-2   Mold design.

temperature gradient of the melt off the screw are more apt to occur.

Clamp Requirements. It is generally recognized that a clamping force of 3 to 4 tons per square inch of projected part area is needed. With less force you run the risk of the mold's parting during the injection cycle. It should be noted that rigid PVC requires a relatively high injection pressure because it is one of the most viscous of all the thermoplastics.

Barrel Cooling. Few molding machines in the field today that are processing rigid PVC are equipped with barrel cooling. Temperature override, particularly on the front barrel zone, is not uncommon. Therefore, cooling of this zone offers more control and reduces the risk of degradation, thus giving more processing latitude. A liquid (water or oil) cooling system is more efficient; however, a fan is generally adequate.

Hopper Design. A properly designed hopper can eliminate powder flow (bridging) problems. The optimum angle is 60 degrees; anything less than 45 degrees will lead to problems. The best design is a round cone shape, but a square design is adequate. The primary consideration is the angle of the bottom portion of the hopper.

Mold and Nozzle

The primary considerations for the mold and nozzle design are outlined in Figs. 24-2 through 24-6.

Cold Slugs. The importance of properly placed cold slug wells is sometimes overlooked. The small amount of cold material in the nozzle can create problems if cold slug wells are not used or are incorrectly placed. The cold material entering the cavity can result in poor physical properties and/or blush on the surface of the part.

Runners. The main consideration in a multicavity mold is to balance the flow so that the melt reaches the gates to all the cavities at the same time. Round runners with a diameter of 0.375 in. are recommended. The secondary runner should be slightly smaller: 0.250 to 0.300 in. and perpendicular to the main runner.

Gates. Many types of gates have been used (Fig. 24-3). Selection of the best design depends on whether emphasis is placed on automation or productivity and ease of processing. Generally, the more open edge gate facilitates ease of processing and fast mold filling. The pin gate is obviously better for automated part and runner separation. However, the restricted pin gate can slow down the injection rate and/or require higher injection pressure which provides more opportunity for shear degradation. Restricted gates may also impair part appearance owing to more turbulent melt flow.

Nozzle. The nozzle should be reverse-tapered with a diameter of 0.300 to 0.400 in. A nozzle with a smaller diameter increases mold filling time and/or requires higher injection pressures, which can cause shear degradation. Short nozzles (2 to 4 in.) are recommended. Accurate nozzle temperature control is essential in order to avoid thermal degradation. It is best to use a temperature controller as opposed to a rheostat.

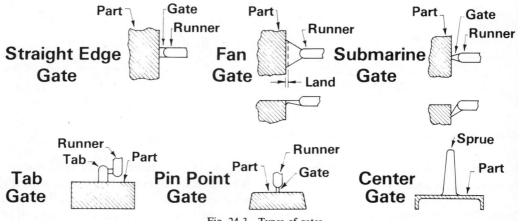

Fig. 24-3   Types of gates.

Screw Design

The common square pitch feed screw (see Fig. 24-4) with a compression ratio of 2:1 and L/D ratio of 24:1 is used extensively. The shorter L/D of 16:1 is also successfully used, but for dry blend an L/D of 24:1 is preferred. The screw should be chrome-plated and rebuilt or replaced when worn beyond acceptable limits. Often, loss of productivity and quality can be traced to an excessively worn screw. Figure 24-5 shows a suggested screw design for PVC dry blend, which consists of a barrier screw with a provision for cooling of the screw tip. The barrier screw should provide better melt homogeneity than the square pitch feed screw. The hottest portion of the melt is in the center of the melt stream that flows off the screw tip; so it makes good sense to control the temperature of the tip of the screw.

There has been considerable debate regarding the best compression ratio (CR) for PVC dry blend. In theory a CR of 2.3–2.4:1 is well suited because the bulk density of the powder is about 37 lb/cu ft, and the PVC melt is about 85 lb/cu ft. The higher CR can cause frictional heat override, but a low CR can be detrimental to the physical properties of the product because of poor fusion. Some fit-tings producers believe that a low CR (1.5:1) is needed for larger machines (750 tons) and a higher CR (2.0–2.4:1) works well for smaller machines (375 tons).

Screw Tip.   The screw tip described in Fig. 24-6 is recommended. Of primary concern is the clearance between the screw tip and the nozzle. When the screw is in its full forward position, a clearance of 0.025 to 0.035 in. ensures that a minimum amount of PVC is left in the nozzle during each injection cycle. Both the nozzle and the screw tip should be chrome-plated.

Material Transfer Equipment

Plantwide Bulk System.   Figure 24-7 shows a bulk handling system designed specifically for dry blend. Dry blend is obviously more dusty than pellets. However, a closed system such as the one described minimizes housekeeping problems.

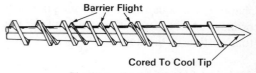

Fig. 24-5   Barrier screw.

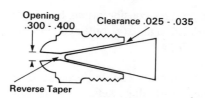

Fig. 24-6   Nozzle and screw tip.

Fig. 24-4   Standard square pitch feed screw.

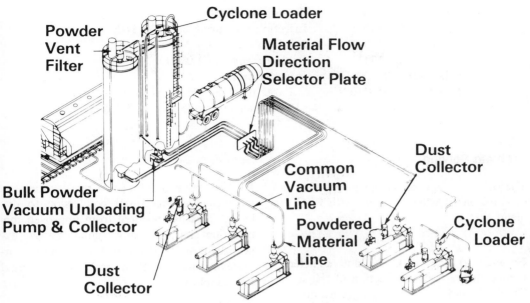

Fig. 24-7   Bulk dry blend transfer system.

Beside the Press Loader. Inexpensive vacuum powder loaders such as the one seen in Fig. 24-8 can be used successfully if the following conditions are met:

- The dry blend should not contain more than 5 percent of an additive with small particles such as calcium carbonate.
- A PVC resin that has an abundance of fine particles will blind the filter, so particle size must be regulated.

Processing Parameters

The conditions shown below can be used as a guide for molding conditions for dry blend.

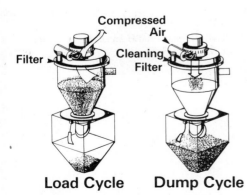

Fig. 24-8   Beside the press loader.

The optimum conditions depend on many factors:

- Dry blend formulation
- Machine size
- Screw design and condition
- Shot size relative to machine capacity
- Mold design

The following suggestions are recommended as a guide for maximizing dry blend performance:

- Maintain the back pressure at a minimum (0–100 psi).
- Run the screw as long as possible for a better melt consistency.
- Keep the injection pressure under 2000 psi (1200–1600 psi).
- Operate with a low injection speed: large machine (750 tons)—3–4 sec/in.; small machine (375 tons)—1.5–2.0 sec/in.

Note: The cooling time is the overall factor that governs productivity. The aforementioned molding conditions can be optimized to stay within the normal cycle time so that no loss in productivity occurs.

The recommended molding conditions are as follows:

Temp. profile (°F):   Screw rpm 20–50
    Zone 1 300–330   Injection pressure (psi) 1200–1600
    Zone 2 320–340   Hold pressure (psi) 800–1000
    Nozzle 320–350   Back pressure (psi) 0–100
Mold temp. 50–100°F   Injection rate:
Stock temp. 395–410°F   Large machine (750 tons) 3–4 sec/in.
                                    Small machine (375 tons) 1.5–2.0 sec/in.

Problem Solving

The three most prominent problems that can be experienced when molding with PVC dry blend are:

Splay. (a) The PVC is not sufficiently melted in the metering section of the screw. Thus, air and volatiles are injected into the mold cavities.

(b) The PVC is fused too quickly in the feed section of the screw and can trap unfused particles. Also, the premature melting of the dry blend can act as a barrier (seal) to air and volatiles that need to be forced back through the hopper.

(c) Inadequate venting of the mold exaggerates the problem.

Degradation. A discoloration on the inside of the sprue indicates that the melt off the screw is too hot. A discoloration on the outside of the sprue, runner, and part is most likely due to shear degradation.

Blush. Blush is generally caused by too large a variation in the melt temperature. In other words, cold and possibly poorly fused material is injected in the mold along with the hotter, more homogenous melt.

Summary

PVC injection molding is a market that has undergone a substantial conversion to dry blend. This has resulted from improved formulation and processing technology. We expect the trend toward use of dry blend to continue in this market, as well as other rigid PVC applications.

REFERENCES

1. Hale, R. G. and Mieure, K. G., "Injection Molding Using Rigid PVC Dry Blend," *SPE-ANTEC* (May 1983).
2. Bourland, L. G., Wambach, A. D., and Francis, P. S., "Alloys of PVC with Dylark Styrenic Copolymers Having Heat and Toughness for Use in Electrical/Electronic Enclosures," *SPE-RETEC* (October 20–23, 1983).
3. Buchdahl, R., and Nielsen, L. E., *J. Polym. Sci., 15,* 1 (1955).
4. Prud'Homme, R. E., *Polym. Eng. Sci., 22,* 90 (1982).
5. Coleman, M. M., and Zarian, J., *J. Polym. Sci.—Polym. Phys. Ed., 17,* 837 (1979).
6. Cruz, C. A., Paul, D. R., and Barlow, J. W., *J. Appl. Polym. Sci., 23,* 589 (1979).
7. Olabisi, O., *Macromolecules, 8,* 316 (1975).
8. Varnell, D. F., and Coleman, M. M., *Polymer, 22,* 1324 (1981).
9. Varnell, D. F., Moskala, E. J., Painter, P. C., and Coleman, M. M., *SPE-NATEC Proceedings,* 115 (October 25–27, 1982).
10. Fava, R. A., and Chaney, C. E., *J. Appl. Poly. Sci., 21,* 781 (1977).
11. Alland, D., and Prud'Homme, R. E., *J. Appl. Poly. Sci., 27,* 559 (1982).

Chapter 25
Polymer Structure and Injection Moldability

Rudolph D. Deanin, Piaras V. DeCleir, and Ashok C. Khokhani
Plastics Engineering Department
University of Lowell

SYNTHESIS AND STRUCTURE OF POLYMERS

The chemical composition of plastics is basically organic polymers—very large molecules composed of chains of thousands of carbon atoms:

$$-\overset{|}{\underset{|}{C}}-\overset{|}{\underset{|}{C}}-\overset{|}{\underset{|}{C}}-\overset{|}{\underset{|}{C}}-\overset{|}{\underset{|}{C}}-\overset{|}{\underset{|}{C}}-\overset{|}{\underset{|}{C}}-\overset{|}{\underset{|}{C}}-\overset{|}{\underset{|}{C}}-\overset{|}{\underset{|}{C}}-$$

generally connected to hydrogen atoms (H), and often also to oxygen (O), nitrogen (N), chlorine (Cl), fluorine (F), and sulfur (S). The first polymers were synthesized by nature— natural rubber, cellulose and starch in trees and plants, proteins in plant and animal life— and during the past century and a half were utilized by man, who modified them chemically to meet emerging industrial needs. More recently, modern industrial organic chemists have learned to synthesize a much greater variety of new polymers, controlling and varying their structures to balance processability and properties in tens of thousands of different end-products. During the first half of this century, their raw materials were most often coal tar chemicals, salt, water, and air. For the

past 40 years, petroleum chemistry has offered the easiest and most economical starting point for the manufacture of most polymers; but if petroleum supplies should become too scarce or too expensive, industrial organic chemists could make all of our commercial polymers from coal and from completely renewable raw materials such as wood and plant life.

The key step in the manufacture of polymers is the polymerization reaction, a chemical process in which many hundreds or thousands of small monomer molecules are linked together permanently to form each large polymer molecule. When all of the monomer molecule (M) is incorporated into the polymer molecule:

$$n\text{M} \longrightarrow -\text{M}-\text{M}-\text{M}-\text{M}-\text{M}-\text{M}-\text{M}-\text{M}-\text{M}-$$

we speak of addition polymerization; when only part (M) of the monomer molecule (XMY) is incorporated into the polymer molecule, and the remainder (XY) is formed as a by-product:

$$n\text{XMY} \longrightarrow -\text{M}-\text{M}-\text{M}-\text{M}-\text{M}-\text{M}-\text{M}-\text{M}-\text{M}- + n\text{XY}$$

we speak of condensation polymerization. While monomers are generally quite reactive (polymerizable), we usually use catalysts, initiators, pH control, heat, and vacuum to speed and control the polymerization reaction and thus optimize the manufacturing process and

This chapter is abstracted from Deanin, R. D., *Polymer Structure, Properties, and Applications*, Cahners Books, Boston, 1972, and updated from M. S. Plastics Engineering theses by P.V. DeCleir and A. C. Khokhani at the University of Lowell.

595

final product. When pure monomers can be converted directly to pure polymers, we call the process bulk polymerization; but often it is more convenient to run the polymerization reaction in an organic solvent (solution polymerization), in a water emulsion (emulsion polymerization), or as organic droplets dispersed in water (suspension polymerization). Since the pioneer work of Ziegler and Natta 30 years ago, we now often choose catalyst systems that exert precise control over the structure of the polymer they form, and we refer to these as stereospecific systems.

The greatest tonnage of polymers used today is in the form of large, linear, stable molecules:

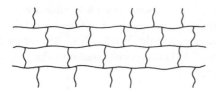

which soften when heated and solidify when cooled, and are therefore called thermoplastics. Offering the greatest variety and often the highest performance, however, are thermosetting materials, which are polymerized in two or more steps—first to small soluble fusible molecules of high reactivity:

which can still be melt-processed into the desired shape; and finally to highly cross-linked structures of almost infinite dimensions:

which cannot be dissolved or fused again, and are therefore called thermoset. Efficient economical processing of thermosets remains a major challenge to the polymer chemist and the plastics processor.

While polymers thus form the structural backbone of plastics, they are rarely used in pure form. In almost all plastics, other chemical ingredients are added to modify and optimize the properties for each desired process and application. These additives most often include stabilizers, fillers and reinforcements, and colors; and often also processing aids, plasticizers, flame retardants, blowing agents, cross-linking agents, and more specialized types of additives. All of these additives affect both processability and end-use properties in many ways.

Structures of the more common polymers are presented in Table 25-1. Beyond the structure of these individual polymer molecules, many polymer systems form larger-scale structures that have important effects on processing and properties. Thus a mulecule of regular structure may fold back and forth upon itself to form a submicroscopic crystallite, and these tiny crystals may further group radially into microscopic circular structures called spherulites; this crystallization process can greatly speed the molding cycle, and the resulting structures can greatly harden and strengthen the final product. Also unidirectional flow through a gate or thin wall section, or along a mold surface, can orient polymer molecules and crystallites axially in the flow direction (parallel vs. perpendicular to the flow direction), and thus produce anisotropic properties. On a still larger scale, fillers and reinforcements, polymer blends, and foamed plastics separate into phases on a microscopic scale; these multiphase structures are influenced by molding conditions, and in turn greatly modify composite properties.

For a detailed study of the effects of polymer structure on injection moldability, this chapter will consider polymer structure in order of increasing size and complexity, based on the following rationale:

1. *Individual atoms and functional groups* in the polymer molecule determine thermal stability of thermoplastics during melt processing, and reactivity and cure processes of thermosets.

2. *Molecular weight* determines resistance to melt flow during melt processing.

3. *Molecular flexibility* determines melt fluidity and crystallizability (both rate and extent of crystallization).

4. Intermolecular order (*crystallinity and orientation*) is affected by the conditions of

the molding process, and rate of crystallization in turn determines the length of the molding cycle.

5. *Intermolecular bonding* determines molding temperature and restriction to melt flow; and cure cycles during thermoset molding are controlled by process temperature, pressure, and time.

6. *Additives* may improve or hinder moldability, and may also modify the molding process in a variety of more specialized ways. Accidental impurities, particularly atmospheric moisture, are also frequently involved.

INDIVIDUAL ATOMS AND FUNCTIONAL GROUPS

Injection moldability is affected by structural features much smaller than the individual polymer molecule, which may conveniently be called submolecular structural features. These are principally the primary bonds between pairs of atoms and functional groups within the polymer molecule. The strength of these bonds is important to the injection molder in two ways: (1) strong bonds provide stability of thermoplastics during hot melt processing; (2) weak bonds provide reactivity of thermosetting plastics, leading to cross-linking and cure during the molding cycle.

The chemist approaches this fundamental relationship by way of the periodic table (see Table 25-2), in which atoms from left to right range from positive to negative and from large to small, while atoms from top to bottom become larger and more positive. These combined effects explain the relative stability of the bonds in polymer molecules (Table 25-3), and these stabilities in turn are very helpful in understanding the thermal stabilities of thermplastics and the cure cycles of thermosets in injection molding.

Thermal Stability of Thermoplastics

C—C Single Bond. The single bond between two neutral carbon atoms is fundamentally balanced and small, making it stable enough to provide the basis for organic chemistry, living systems, and organic polymers in general. It usually provides the thermal stability needed for injection molding of thermoplastics. When thermal degradation does occur, it is usually due to destabilizing factors such as the following:

1. Adjacent groups such as C=C, branching, or polar atoms all reduce the stability of the C—C bond. These groups may be present in the original polymer, or they may form during the early stages of thermal degradation.
2. Quaternary carbon atoms (with four bulky groups attached to them) are so sterically hindered that they push the C—C bond apart by shear repulsion.
3. Mechanical shear of a very viscous system can actually provide enough energy to break the C—C bond.
4. Free radicals in a degrading system are often reactive enough to participate in exchanges within the electron cloud that forms the C—C bond.

Any of these factors can reduce thermal stability during injection molding, producing changes in processability and usually also degradation of end-use properties.

C=C Double Bond. The four electrons in the C=C double bond repel each other enough to make the bond unstable, and also tend to destabilize the adjacent —CH$_2$— group. When butadiene elastomers are incorporated into high-impact thermoplastics such as impact styrene, ABS (2), and rigid PVC, this instability can cause thermal degradation during injection molding, and require addition of strong stabilizers. In thermoplastic elastomers based on styrene/butadiene/styrene sandwich block copolymers, the instability is so serious that it may be necessary to replace the polybutadiene mid-block by a saturated ethylene/butene copolymer (3). Even small traces of C=C bonds, occurring as incidental impurities in PVC and other polymers, can act as loci of instability from which degradation can spread and auto-accelerate. Particularly in PVC, the resulting growth of —(CH=CH)$_n$— extended conjugated unsaturation can rapidly cause severe discoloration during processing, making the product worth-

Table 25-1 Structures of common polymers.

High-density polyethylene	$-(CH_2CH_2)-$
Low-density polyethylene	$-(CH_2CH_2)-(CH_2CH)-(CH_2CH)-(CH_2CH)-$
	$\qquad\qquad\qquad\qquad C_2H_5 \qquad C_4H_9 \qquad C_nH_{2n+1}$

Ionomer

$$-(CH_2CH_2)-(CH_2\overset{\displaystyle CH_3}{\underset{\displaystyle O=C-O^-\ Na^+}{C}})-$$

Ethylene/vinyl acetate copolymer

$$-(CH_2CH_2)-(CH_2CH)-$$
$$\qquad\qquad\qquad\qquad \underset{\displaystyle O}{\overset{\displaystyle OCCH_3}{|}}$$

Polypropylene

$$-(CH_2CH)-$$
$$\qquad\quad CH_3$$

Polystyrene

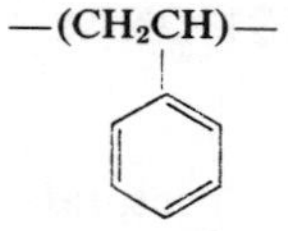

$$-(CH_2CH)-$$

Styrene/acrylonitrile copolymer

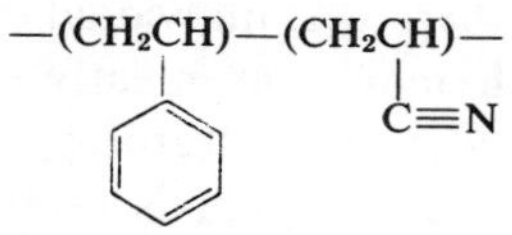

$$-(CH_2CH)-(CH_2CH)-$$
$$\qquad\qquad\qquad\qquad C\equiv N$$

Impact styrene

$$-(CH_2CH=CHCH)-$$
$$\qquad\quad (CH_2CH)-$$

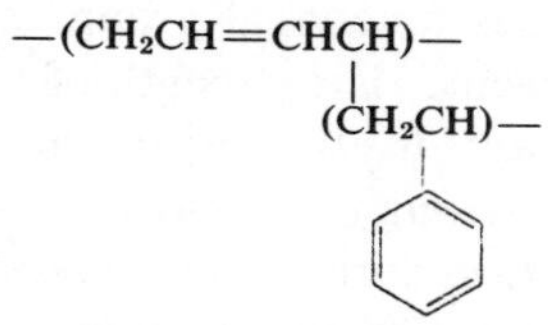

Acrylonitrile/butadiene/styrene terpolymer (ABS)

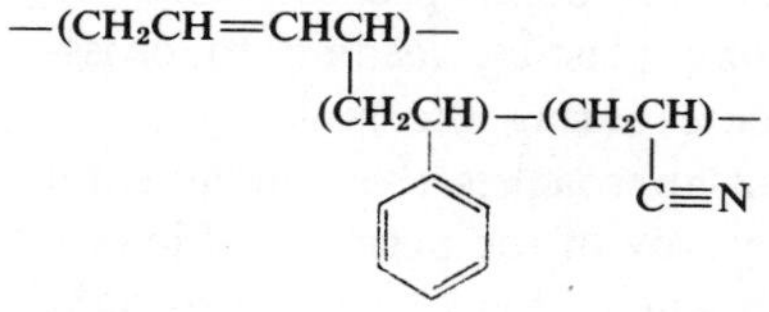

$$-(CH_2CH=CHCH)-$$
$$\qquad\quad (CH_2CH)-(CH_2CH)-$$
$$\qquad\qquad\qquad\qquad\qquad C\equiv N$$

Polyvinyl chloride

$$-(CH_2CH)-$$
$$\qquad\quad Cl$$

Vinyl chloride/vinyl acetate copolymer

$$-CH_2CH)-(CH_2CH)-$$
$$\qquad Cl \qquad\quad \underset{\displaystyle O}{\overset{\displaystyle OCCH_3}{|}}$$

Polyvinylidene chloride

$$\qquad\quad Cl$$
$$-(CH_2C)-$$
$$\qquad\quad Cl$$

Polytetrafluoroethylene	$-(CF_2CF_2)-$
Fluorinated ethylene/propylene copolymer	$-(CF_2CF_2)-(CF_2CF)-$
	$\qquad\qquad\qquad\qquad CF_3$

Polychlorotrifluoroethylene

$$-(CF_2CF)-$$
$$\qquad\quad Cl$$

Polyvinylidene fluoride

$$-(CH_2CF_2)-$$

Table 25-1   (continued)

Polymethyl methacrylate	$-(CH_2C)-$ with CH_3 above and $O=COCH_3$ below
Polyacrylonitrile	$-(CH_2CH)-$ with $C\equiv N$ below
Phenoxy resin	$-(CH_2CHCH_2O-\langle\bigcirc\rangle-C(CH_3)_2-\langle\bigcirc\rangle-O)-$ with OH below
Epoxy resin	$CH_2\!\!-\!\!CHCH_2O-\langle\bigcirc\rangle-C(CH_3)_2-\langle\bigcirc\rangle-OCH_2CH\!\!-\!\!CH_2$ (terminal epoxide rings)
Poly(2,6-dimethylphenylene oxide) (PPO)	$-(\langle\bigcirc\rangle-O)-$ with two CH_3 groups
Polysulfone	$-(\langle\bigcirc\rangle-C(CH_3)_2-\langle\bigcirc\rangle-O-\langle\bigcirc\rangle-SO_2-\langle\bigcirc\rangle-O)-$
Polyoxymethylene (acetal)	$-(CH_2O)-$
Cellulose triacetate	glucose ring with CH_2OCCH_3 (O), and acetate groups CH_3C-O, $O-CCH_3$
Ethyl cellulose	glucose ring with $CH_2OC_2H_5$, OH, OC_2H_5
Polycarbonate	$-(\langle\bigcirc\rangle-C(CH_3)_2-\langle\bigcirc\rangle-OCO)-$
Poly(ethylene terephthalate)	$-(CH_2CH_2OC-\langle\bigcirc\rangle-CO)-$ with two O
Unsaturated polyester (cured)	$-(CHCH_2OC\;\;\langle\bigcirc\rangle\;\;COCHCH_2OCCHCHCO)-$ with CH_3, O, O, CH_3, $(CH_2CH\langle\bigcirc\rangle)$

Table 25-1   (continued)

Diallyl phthalate

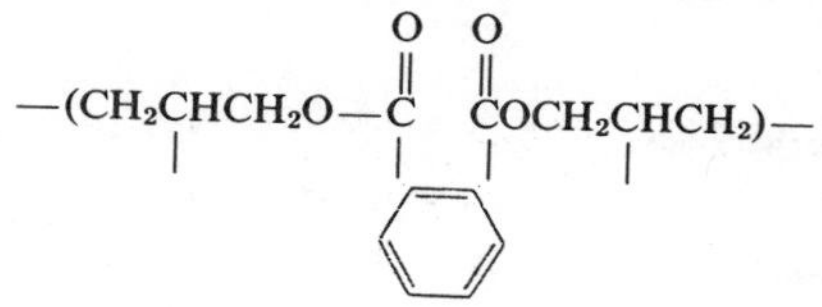

$$-(CH_2CHCH_2O-\overset{O}{\overset{\|}{C}}\quad\overset{O}{\overset{\|}{C}}OCH_2CHCH_2)-$$
(with phenyl ring bridging the two carbonyls)

Alkyd

$$-(\overset{O}{\overset{\|}{C}}\quad\overset{O}{\overset{\|}{C}}OCH_2CHCH_2O)-$$
(with phenyl ring bridging the two carbonyls)
$$OC(CH_2)_7CH=CHCH_2CH=CH(CH_2)_4CH_3$$
(carbonyl O on side)

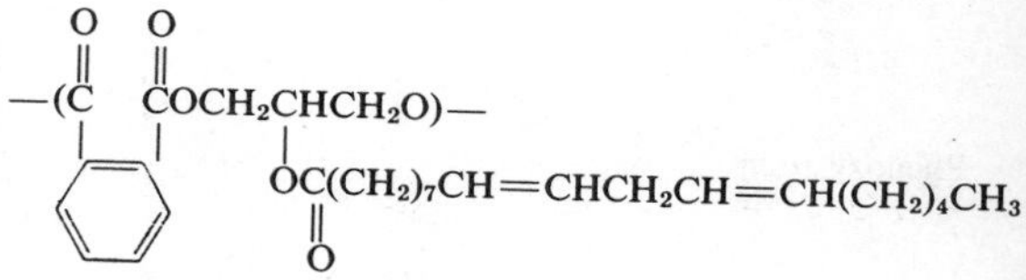

Polyether polyurethane

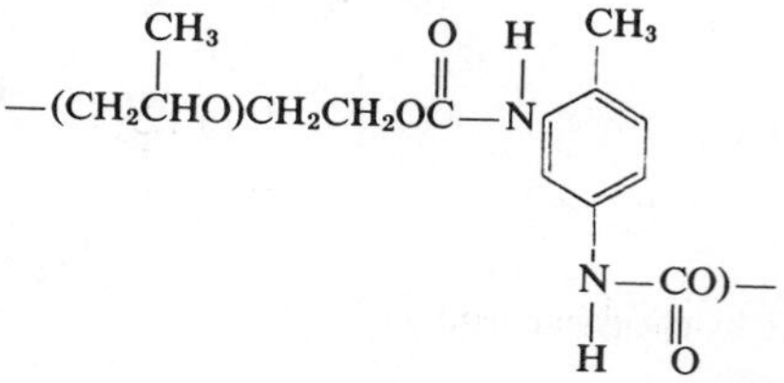

$$-(CH_2\overset{CH_3}{\overset{|}{C}}HO)CH_2CH_2O\overset{O}{\overset{\|}{C}}-\overset{H}{\overset{|}{N}}\;\;(CH_3 \text{ on ring})$$
$$\overset{|}{N}-CO)-$$
$$\overset{|}{H}\;\;O$$

Polyester polyurethane

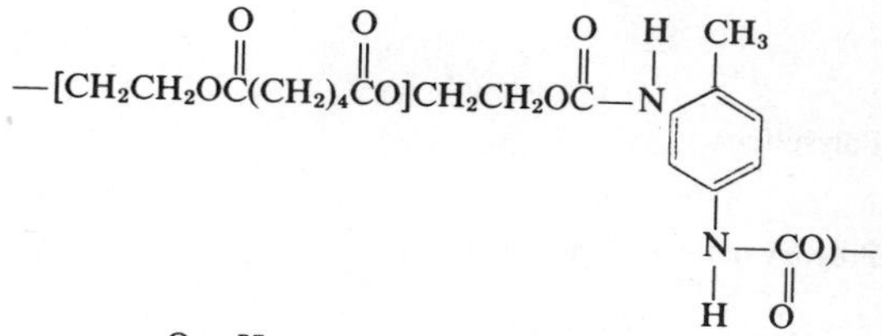

$$-[CH_2CH_2O\overset{O}{\overset{\|}{C}}(CH_2)_4\overset{O}{\overset{\|}{C}}O]CH_2CH_2O\overset{O}{\overset{\|}{C}}-\overset{H}{\overset{|}{N}}\;\;CH_3$$
$$\overset{|}{N}-CO)-$$
$$\overset{|}{H}\;\;O$$

Nylon 6

$$-[(CH_2)_5\overset{O}{\overset{\|}{C}}-\overset{H}{\overset{|}{N}}]-$$

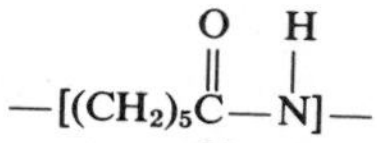

Nylon 66

$$-[(CH_2)_6\overset{H}{\overset{|}{N}}-\overset{O}{\overset{\|}{C}}(CH_2)_4\overset{O}{\overset{\|}{C}}-\overset{H}{\overset{|}{N}}]-$$

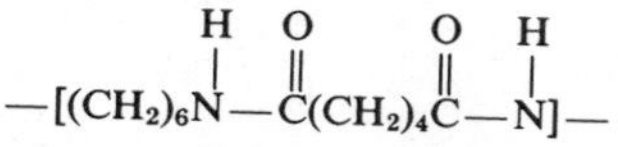

Nylon 610

$$-[(CH_2)_6\overset{H}{\overset{|}{N}}-\overset{O}{\overset{\|}{C}}(CH_2)_8\overset{O}{\overset{\|}{C}}-\overset{H}{\overset{|}{N}}]-$$

Nylon 11

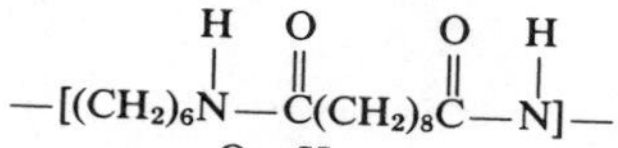

$$-[(CH_2)_{10}\overset{O}{\overset{\|}{C}}-\overset{H}{\overset{|}{N}}]-$$

Nylon 12

$$-[(CH_2)_{11}\overset{O}{\overset{\|}{C}}-\overset{H}{\overset{|}{N}}]-$$

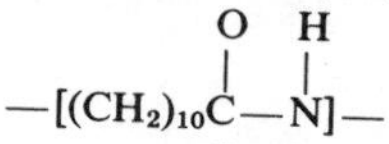

Polyimide

(fused-ring imide structure with carbonyl groups, benzene ring, N, and an N-bridged aromatic ring)

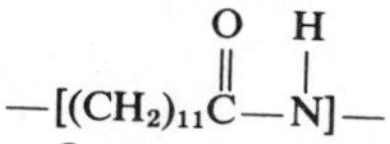

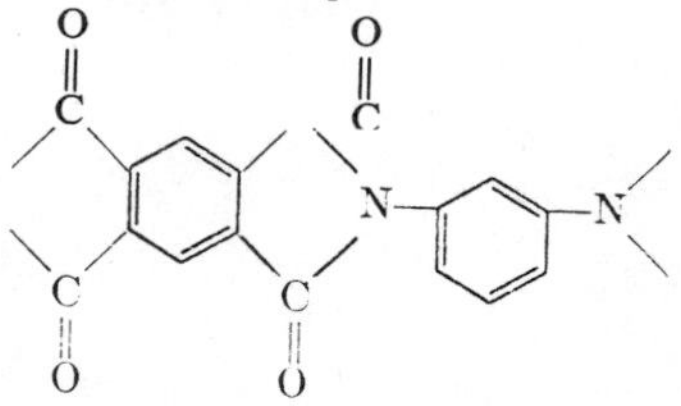

Urea-formaldehyde

$$-(N-\overset{O}{\overset{\|}{C}}-N-CH_2)-$$
$$\overset{|}{C}H_2-$$

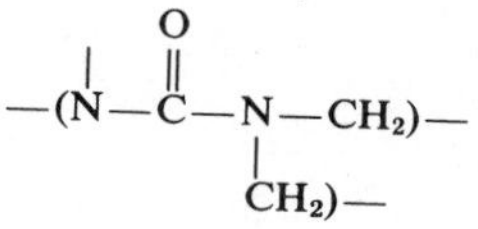

Table 25-1 (concluded)

Melamine-formaldehyde	(structure: triazine ring with $-(N-C$, $C-N-CH_2)-$, N, $CH_2)-$, and $-(CH_2-N-CH_2)-$ linkages)
Phenol-formaldehyde	(structure: benzene ring with OH, $CH_2)-$, and CH_2 groups)
Silicone	$-(Si-O)-$ with CH_3 groups above and below the Si

Table 25-2 Polymer chemist's periodic table.

			GROUP				
PERIOD	I	II	III	IV	V	VI	VII
2				C	N	O	F
3	Na	Mg		Si	P	S	Cl
4		Ca					Br

Table 25-3 Stabilities of bonds in polymers, expressed as bond energies in kcal/mole (1).

BOND	ENERGY	BOND	ENERGY
C—F	116	C_{ar}—H	100
C—H	97	C_{ar}—N	110
Si—O	106	C_{ar}—O	107
C—C	83		
C_{ar}—C_{ar}*	100		

Subscript ar denotes aromatic.

less even before there is any significant change in engineering properties.

Paradoxically, the C=C unsaturation in the aromatic benzene ring

produces resonance stabilization which is the source of enhanced thermal stability in most of our new ultra-high-temperature plastics. Thus it is important to remember the distinction between the instability of the aliphatic C=C double bond and the resonance stabilization of the aromatic bonds in the benzene ring.

C—H Bond. While the electron pair between carbon and hydrogen is usually considered a stable covalent bond between two neutral atoms, hydrogen is actually more positive than carbon, and this makes the bond less balanced and less stable than the C=C bond. While the C—H bond in polymers is still stable enough to survive most thermoplastic processing, the combination of heat and atmospheric oxygen is often enough to break the bond, creating free-radical chain reactions that produce oxidation to unstable peroxide intermediates and thence on to alcohols and carboxylic acids, as well as cleavage or cross-linking of whole molecules. The C—H bond is further destabilized by adjacent groups such as C=C, tertiary branches, oxygen, and other polar atoms, all of which activate the thermal oxidation reaction. Thus when a hydrocarbon polymer suffers thermal degradation during injection molding (4), the initial attack is generally at the least stable C—H bond, and then produces groups that can destabilize the adjacent C=C bonds as well. Similar mechanisms often initiate thermal degradation of nonhydrocarbon polymers as well (5, 6).

C—O—C Bonds. Oxygen is much more negative than carbon, so the C—O bond is quite polar. Its thermal stability during injection molding depends upon the specific type of C—O group and the interaction of atmospheric contaminants. Consider the following four classes of C—O—C bonds:

1. Aromatic ethers contain the —O— between two resonating benzene rings. The resonance extends beyond the individual benzene rings, involving the oxygen as well in a larger extended resonance structure and thus producing increased thermal stability, a prominent advantage in many new engineering polymers such as PPO (7), polyetheretherketone (8), and polysulfones (9, 10).

2. Aliphatic —CH$_2$—O— groups have a tendency toward atmospheric oxygen attack on the —CH$_2$— group, particularly in aliphatic ethers, generally resulting in peroxidation leading to cleavage and acidity. This may be an intermediate mechanism in thermal degradation of acetals and polyesters during injection molding.

3. The acetal linkage —C—O—C—O—C— in cellulosics and in polyformaldehyde acetal resin is quite sensitive to acid hydrolysis. Thus thermal oxidation produces acidity, atmospheric moisture produces hydrolysis, cellulosics cleave to lower molecular weight, and acetal resin suffers unzipping depolymerization directly to gaseous formaldehyde monomer (5). Protection is generally obtained by adding various stabilizers, copolymerizing formaldehyde with ethylene oxide, and careful predrying before injection molding.

4. The —CO$_2$— (ester) group is particularly likely to hydrolyze, producing acidity and lower molecular weight. Absorption of atmospheric moisture, thermal oxidation to carboxylic acid, and the heat of injection molding thus all act together to make further hydrolysis autocatalytic (6, 11). Predrying is helpful, but some thermal oxidation generally occurs, and is particularly evident during reprocessing (6).

Peroxides. Kinetic studies on thermal degradation during melt processing generally point to the intermediate formation of peroxides (R—O—O—R) and hydroperoxides (R—O—O—H) due to free-radical chain reactions with atmospheric oxygen. Since the

—O—O— bond is thermally unstable, it in turn breaks down readily during melt processing, initiating further free-radical chain reactions that do the real practical damage to both processability and end-use properties. Thus most techniques for protection against thermal oxidative degradation involve polymer structures and/or additives that can prevent or at least delay these reactions (12).

C—N Bonds. Nitrogen is more negative than carbon, so the C—N bond is fairly polar. Its thermal stability during injection molding depends upon the specific functional group and atmospheric conditions. In aromatic heterocycles it is extremely stable, so these are used in many of the new ultra-high-temperature plastics. In urethanes and nylons, strong hydrogen-bonding leads to absorption of moisture from the atmosphere, whereupon the heat of processing leads to hydrolytic degradation. In urethanes this is irreversible, but in nylons, careful reprocessing can remove the moisture again and rebuild the original molecular weight to produce quality equal to that of the virgin resin. Careful predrying is obviously essential. The worst stability is observed in the quaternary ammonium (R_4N^+ X^-) structures of some antistatic additives, which cannot survive high processing temperatures and must be withheld for postmolding application to the finished product.

Nitriles. The nitrile side-group in polyacrylonitrile barrier resins has a strong tendency toward side-chain polymerization during the heat of injection molding:

and the resulting extended conjugated unsaturation produces a reddish discoloration that is undesirable in consumer products. Copolymerization and possibly additive inhibitors may be helpful in retarding this discoloration.

Fluorine. Fluorine is a very small atom, and when it shares electrons with carbon to form a covalent C—F bond, they both hold these electrons very tenaciously, so the C—F bond is inherently short and very strong. This produces very high thermal stability, particularly in perfluorinated structures where all the hydrogen has been replaced by fluorine. In partially fluorinated structures, thermal stability is generally related directly to the F/H ratio. Fortunately for injection molders, molding temperatures are also related directly to the F/H ratio, so polyvinylidene fluoride and ETFE provide relatively easy moldability and good process stability.

Chlorine. Chlorine is larger than fluorine, giving a longer C—Cl bond, and chlorine is also negative enough to make the bond polar. These factors make the C—Cl bond somewhat unstable to heat, and less than marginally reliable for injection molding. The C—Cl bond is further destabilized by adjacent activating groups such as —C=C—, tertiary branching, and peroxides resulting from polymerization initiators and from thermal atmospheric oxidation during drying and melt processing. Any of these factors can cause PVC to lose HCl:

$$-CH_2-CH-CH_2-CH-CH_2-CH-$$
$$\qquad\quad |\qquad\qquad |\qquad\qquad |$$
$$\qquad\quad Cl\qquad\quad Cl\qquad\quad Cl$$

$$\downarrow$$

$$-CH=CH-CH_2-CH-CH_2-CH- \; + \; HCl$$
$$\qquad\qquad\qquad\qquad |\qquad\qquad |$$
$$\qquad\qquad\qquad\quad Cl\qquad\quad Cl$$

$$\downarrow$$

$$-CH=CH-CH=CH-CH_2-CH- \; + \; HCl$$
$$\qquad\qquad\qquad\qquad\qquad\qquad |$$
$$\qquad\qquad\qquad\qquad\qquad\quad Cl$$

$$\downarrow$$

$$-CH=CH-CH=CH-CH=CH- \; + \; HCl$$

producing the —C=C— group, which then autocatalyzes further loss of HCl. As the —(CH=CH)$_n$— conjugated unsaturated chain grows longer, autocatalysis accelerates. When n exceeds 6, the conjugated resonance begins to absorb visible light, and the polymer discolors from white to yellow to red to brown to black, making it worthless for consumer products. Even before this, the HCl liberated corrodes process equipment, causing severe damage and rapid wear. Experimentally, it is possible to eliminate the activating groups that initiate loss of HCl and thus improve the stability of the polymer somewhat. For all practical purposes, it is essential to add strong selected stabilizer systems to retard the degradation process, and to keep melt processing temperatures, times, and shear to a minimum, including avoidance of dead spots, down time, and any other source of incipient degradation. With such careful compounding and processing, injection molding of PVC is a satisfactory commercial process; but even so, it requires constant vigilance.

Bromine. Continuing with the halogen family in the periodic table, bromine is larger than chlorine and still less firmly bonded to carbon. The C—Br bond can be built into polymers and additives, and is very useful for flame retardance, but thermal stability is borderline at best, requiring great care in both molecular design and melt processing. The best stability is generally obtained by attaching the bromine directly to the benzene ring, particularly in polycarbonate and polyesters.

Sulfur. From its position directly below oxygen in the periodic table, chemists would expect the behavior of sulfur to be fairly similar to that of oxygen. Comparing C—O—C and C—S—C bonds, this is generally true, with behavior more pronounced for the sulfur bonds. In aromatic polymers such as polyphenylene sulfide, resonance stabilization of the benzene rings extends to include the sulfur as well, producing very high thermal stability. In aliphatic sulfides, the sulfur is very liable to thermal atmospheric oxidation or, worse, to splitting out as H_2S gas.

In polysulfones, extended resonance between the benzene rings and the sulfone groups produces broad resonance stabilization which is responsible for the high-temperature stability of aromatic polysulfone engineering thermoplastics in general.

Silicones. The Si—C bond is remarkably strong, making silicones among the most thermally stable polymers in commercial use. Most silicone processing is carried out at relatively low temperature, which does not challenge the bond strength at all. When silicone resins such as CH_3SiX_3 are baked at high temperature to "burn out" the remaining organic content and produce "glass resins," it is probably the C—H bond that oxidizes first, and the Si—C bond is the last to go.

Cross-Linking and Cure of Thermosets

Ideally thermosetting plastics should combine (1) low molecular weight during processing, to provide easy melt fluidity; and (2) infinite molecular weight in the end-product, to provide maximum end-use properties. The organic polymer chemist has myriad functional groups and reactions to produce this paradoxical combination of properties, and a number of them are actually in commercial use. Before considering them individually, however, it is best to start by noting that the injection molding of thermosetting plastics has encountered a number of practical difficulties that have limited the rate of growth of this technology. These difficulties are based in the following conflicting requirements:

The process engineer, first of all, would like materials that have unlimited shelf life (warehouse storage before use), pot life (working time after the reactive components are mixed), and process working time in general (resistance to premature cross-linking between cycles, in dead spots, and during down time in general); in fact, ideally, he would like a one-part system, which means that a mixture of all the reactants would be stable indefinitely. All these requirements spell *low* reactivity.

On the other hand, once he has flowed the melt into the mold, he would like the fastest possible reaction to produce final cure and a short process cycle for maximum process economy. This clearly means *high* reactivity.

Considering the total irreconcilability of these two conflicting demands, it is remarkable how far the ingenuity of organic polymer chemists has gone toward producing some reasonable compromises, and the range of balance in these compromises has increasingly diversified with general progress in the field. A variety of techniques are used:

1. Mixing of reactants as they are injected into the mold has been most highly developed in RIM technology.
2. Thermal activation can combine stability at low temperature with high reactivity at high temperature.
3. Mixtures of solids are stable at room temperature, but melt and cure rapidly at molding temperature.
4. Microencapsulation of the catalyst or curing agent can produce a stable one-part system that is activated by crushing or melting of the encapsulant during molding.
5. Latent catalysts are stable at room temperature, but are liberated or otherwise activated at molding temperature.
6. "Blocked" reactants are stable at room temperature; at molding temperature they "unblock," liberating the reactant to permit cure. This is most commonly practiced in urethanes.

A third difficulty in many thermosetting systems is due to the fact that they are condensation reactions which liberate gases or volatile liquids that must be vented to permit production of solid flaw-free parts. Venting is an established practice in much compression molding of thermosets, but may be much clumsier when adapted to normal injection molding techniques.

Despite these difficulties, nearly all conventional thermosetting plastics are potentially adaptable to injection molding, and some of them have met moderate acceptance in commercial practice (13–16).

The following paragraphs discuss the individual functional groups and cure reactions most commonly used.

C=C Double Bond.

The reactivity of the C=C double bond is most commonly used in making the commodity thermoplastics—polyolefins, styrenics, and vinyls. But it is also used for thermoset cure of two commodity polymers—rubber and polyester (17)—and also for smaller specialties such as EPDM (18), 1,2-polybutadiene resins (19), Hercules H-Resin (20, 21), DAP diallyl phthalate, bis-maleimide, and some silicones. Some of these require catalysis, others occur spontaneously at molding temperature, and all are addition reactions that do not liberate any by-products. They are all conventionally practiced by compression molding. In theory all can be converted to injection molding, but commercial development is still quite small.

Aromatic C—H.

In phenols (and anilines), the ortho and para C—H bonds are extremely activated. This reactivity is commonly used in the polymerization and cure of phenolic resins. Even though this cure produces water vapor, and often also formaldehyde and ammonia, phenolic molding powders have been successfully adapted to injection molding and represent one of the major penetrations of this type (22). This could be a significant factor in making phenolics economically competitive with commodity thermoplastics while still delivering their superior thermoset end-properties—rigidity, heat resistance, and chemical resistance.

Epoxy.

The $-\overset{\displaystyle O}{\underset{}{C\diagup\diagdown C}}-$ group in epoxy "resins" contains acute angles that are much smaller than the normal positions of the electron pairs. This structure is therefore "strained" and unstable—reactive. And this reactivity permits easy cure by a variety of "hardeners," particularly active-hydrogen sources such as amines, acids, and alcohols, and even homopolymerization by acidic or basic catalysts. This high reactivity and easy cure is widely used in casting, impregnation, coat-

ings, and adhesives. With the growth of interest in thermoset injection molding, clever formulators have devised a number of one-part systems using all the techniques described earlier, permitting small-scale specialty transfer and injection molding to become commercial realities.

Peroxides.

Organic peroxides (R—O—O—R) and hydroperoxides (R—O—O—H) contain the O:O bond which is unstable and tends to separate into free radicals:

$$RO:OR \rightarrow RO^{\cdot} + {^{\cdot}OR}$$
$$RO:OH \rightarrow RO^{\cdot} + {^{\cdot}OH}$$

which in turn are very useful for initiating thermosetting cure reactions. These cure reactions fall into two distinct classes:

1. Addition polymerization of vinyl groups is used to cure unsaturated polyesters and DAP diallyl phthalate resins.
2. Hydrogen abstraction from saturated polymers:

$$RO^{\cdot} + {-CH_2-} \rightarrow ROH + {-\overset{\cdot}{C}H-}$$

produces unstable radicals on the polymer chains. As soon as such radicals appear on adjacent chains, they couple to produce cross-links:

This technique is frequently used to cross-link polyethylene and to cure ethylene/propylene EPR rubber and silicone rubber.

To adapt the cure reaction to different polymers and different molding temperatures, the formulator can manipulate two primary variables:

1. The R group of the peroxide can be chosen from a wide variety of organic structures, giving a wide range of stability/instability and thus a wide range of processing temperatures.
2. The peroxide decomposition is activated not only by heat but also by addition of transi-

tion metal ions and reducing agents, giving an even wider range of processing temperatures and rates.

Given all these choices, peroxide cure has found a number of specialty uses in injection molding of thermosets.

O—H Group. The hydroxyl group is important in two types of thermosetting polymerization/cure reactions:

1. Reaction with isocyanate:

$$R-O-H + O=C=N-R' \rightarrow R-O-\overset{\overset{\textstyle O}{\|}}{C}-\overset{\overset{\textstyle H}{|}}{N}-R'$$

produces polyurethanes. This is an addition polymerization reaction widely used in RIM (reaction injection molding).

2. Polymerization and cure of urea, melamine, and phenolic resins is based on addition of formaldehyde to form the —CH₂OH (methylol) group:

$$H-\overset{\overset{\textstyle H}{|}}{N}-\overset{\overset{\textstyle O}{\|}}{C}-\overset{\overset{\textstyle H}{|}}{N}-H + CH_2O \rightarrow H-\overset{\overset{\textstyle H}{|}}{N}-\overset{\overset{\textstyle O}{\|}}{C}-\overset{\overset{\textstyle H}{|}}{N}-CH_2OH$$

$$H-\overset{\overset{\textstyle}{|}}{N}-H \qquad\qquad H-\overset{\overset{\textstyle}{|}}{N}-CH_2OH$$
$$\overset{\overset{\textstyle}{|}}{C} \qquad\qquad\qquad \overset{\overset{\textstyle}{|}}{C}$$

$$H-\overset{\overset{\textstyle H}{|}}{N}-\overset{\overset{\textstyle N}{\|}}{C} \quad \overset{\overset{\textstyle N}{|}}{C}-\overset{\overset{\textstyle H}{|}}{N}-H + CH_2O \rightarrow \quad H-\overset{\overset{\textstyle H}{|}}{N}-\overset{\overset{\textstyle N}{\|}}{C} \quad \overset{\overset{\textstyle N}{|}}{C}-\overset{\overset{\textstyle H}{|}}{N}-H$$
$$N \qquad\qquad\qquad\qquad\qquad N$$

followed by further condensation of this methylol group with urea, melamine, and phenol and with other methylol groups. These reactions liberate water, and sometimes also formaldehyde, which must be vented carefully to prevent bubbles and cracks in the cured product; such venting techniques are conventional in compression molding, but are more difficult to adapt to conventional injection molding technique. Nevertheless, these reactions are fast, controllable, and capable of producing optimum thermoset properties with

improved process economics. Thus injection molding of phenolics has been quite widely adopted (22), and injection molding of urea resins has also been described (23).

N—H Group. Organic amines:

$$R-\overset{\overset{\textstyle H}{|}}{\underset{\underset{\textstyle H}{|}}{N}}:$$

contain both the basic unshared electron pair on the nitrogen, which has strong catalytic effects; and also the reactive acidic hydrogens, which add readily to electron pairs in a wide variety of co-reactants. This provides the most common basis for cure of epoxy resins, the fastest and firmest cure of polyurethanes, and

the basic mechanism for polymerization and cure of urea and melamine resins. For commercial development of thermoset injection molding, the cure of urethanes in RIM has been the most widely used, while the others have been used primarily in smaller specialty applications.

Isocyanate. The high reactivity of the isocyanate group with active hydrogen:

$$\underset{\text{Alcohol}}{R-N=C=O + HOR'} \rightarrow R-\overset{\overset{\textstyle H}{|}}{N}-\overset{\overset{\textstyle O}{\|}}{C}-\underset{\text{Urethane}}{O-R'}$$

$$R-N=C=O + H_2NR' \rightarrow R-N-C-N-R'$$
$$\quad\quad\quad\quad\quad\quad \text{Amine} \quad\quad\quad\quad\quad \text{Urea}$$

$$R-N=C=O + HOH \rightarrow R-NH_2 + CO_2$$
$$\quad\quad\quad\quad\quad \text{Water} \quad\quad \text{Amine} \quad\quad \text{Gas}$$

is the basis for the success of polyurethanes in general and for the ease of conversion to RIM specifically. While most injection molding uses two-part systems which must be mixed immediately before injection into the mold, formulators have been applying their maximum ingenuity to the development of one-part systems based on all of the principles discussed earlier (24). Even among so-called thermoplastic urethanes, some of the best depend upon side-reactions that form allophanate and biuret cross-links in the hot mold to produce somewhat thermoset products with maximum end-use properties.

S—H Group. The active hydrogen of the mercaptan end-group in polysulfide rubber oligomers is generally used in two types of thermosetting systems:

1. Polysulfide elastomers are cured by oxidative coupling of the mercaptan groups:

$$2 R-S-H + (O) \rightarrow R-S-S-R + H_2O$$

Evolution of water could thus be a problem to which the injection molder would have to adapt as he had to with phenolics.

2. Epoxy resins are sometimes cured with polysulfides in order to reduce their brittleness. Since this is a simple addition reaction, it should be readily adaptable to injection molding.

Commercial use of either of these systems has not yet been reported.

In summary, these are the ways in which submolecular chemical composition is important to the injection molder, in determining either thermal stability of thermoplastics or the chemical reactivity of thermosetting systems.

MOLECULAR WEIGHT

Plastics require very large molecules to provide the cohesion necessary for most end-use properties; but these large molecules disentangle and flow only with difficulty during melt processing. Thus thermoplastic injection molding requires a compromise molecular weight, low enough for reasonably easy processing but high enough for reasonably good end-use properties.

For simplicity we speak of a polymer as if it had a certain molecular weight; but in reality any polymer is a mixture of large molecules of different sizes, which we call the molecular weight distribution (Fig. 25-1). The polymerization mechanism and conditions, and to some extent the compounding and molding conditions, determine whether the distribution is narrow or broad, normal, or skewed, or occasionally even bimodal or multimodal. Both (1) average molecular weight and (2) molecular weight distribution will have specific effects on both injection moldability and end-use properties. Two related considerations are (3) branching of the polymer molecule and (4) plasticizers added in compounding. Each of these four effects can be considered separately.

Average Molecular Weight

Melt Viscosity. Ask any polymer scientist or engineer about the relationship between polymer structure and injection moldability, and his first answer will most likely be: "Increasing molecular weight produces increasing melt viscosity and more difficult injection molding." Countless studies have documented this relationship, many of them quantitatively. A typical scan of the literature reveals such data for polyethylene (25), polypropylene (26),

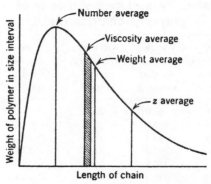

Fig. 25-1 Typical molecular weight distribution and averages.

impact styrene (27), ABS (28), and PVC (Table 25-4) (29, 30). Similarly, whenever melt processing causes thermal or hydrolytic degradation to lower molecular weight, the reprocessed material has lower melt viscosity; typical studies have observed such effects in polypropylene (4), polycarbonate (31, 32), and PBT polybutylene terephthalate (11).

Quantitatively the relationship takes the form:

$$\eta = KM_w^a$$

meaning that melt viscosity η is proportional to some exponential function of weight-average molecular weight M_w. The proportionality constant K depends on the flexibility and intermolecular attraction of the polymer molecules, and on processing conditions—temperature, pressure, and shear rate. The exponent a, the slope of the log–log plot (Fig. 25-2), is equal to one at low molecular weights, meaning that the melt viscosity is simply proportional to the size of the molecules (33, 34). At high molecular weights, however, the exponent rises to 3.4–3.5, meaning that increasing molecular weight will have a much more severe effect on melt viscosity and injection moldability because of the difficulty of disentangling large polymer molecules to permit melt flow. The transition from $a = 1$ to $a = 3.4$–3.5 occurs at some critical molecular weight M_c which is generally between 5000 and 15,000, depending on molecular flexibility (33, 35), and which may even be predictable from basic theoretical considerations (36).

Table 25-4 Molecular weight, thermal stability, and melt viscosity of polyvinyl chloride (29).

MOLECULAR WEIGHT IN 10³		THERMAL STABILITY, min./ 200°C	MELT VISCOSITY, 10⁴ POISES/SEC AT 200°C
NUMBER AVERAGE	WEIGHT AVERAGE		
26.8	60.0	20	0.95
51.6	105.0	29	5.98
69.9	149.3	31	13.00
103.0	207.3	31.5	17.79
109.5	274.9	36	25.20
163.0	340.0	32	28.50

Since melt viscosity is a linear function of both molecular weight and temperature, graphical and mathematical analysis can apply the superposition principle to produce simple master curves and shift factors, permitting the use of limited experimental data to make broad predictions about processing conditions and processability (26).

Increasing molecular weight not only increases melt viscosity, but also rubbery melt flow—the inability of molecules to disentangle completely within the limited temperature, pressure, shear rate, and time span of the process, producing an elastic melt that can result in a variety of injection molding problems such as die swell, melt fracture (37), and post molding shrinkage, warping, and cracking. The injection molder can compensate for these problems by increasing temperature and time, decreasing shear rate, or changing to a lower-molecular-weight grade of resin.

While the effect of molecular weight on melt viscosity is the most important relationship in injection molding, molecular weight may also affect other properties that are important in injection molding, such as thermal stability, thermal conductivity, coefficient of thermal expansion, and melting/crystallization phenomena.

Thermal Stability. It is occasionally observed that increasing molecular weight produces increased thermal stability, which in turn gives the injection molder increased latitude in processing. There are two reasons why stability increases with molecular weight: (1) most polymerization mechanisms leave unstable structures on the ends of the polymer molecules, so the concentration of these unstable structures is inversely related to molecular weight; (2) most chemical reactions, including degradation reactions, require molecular mobility, which is inversely related to molecular weight. Thus increasing molecular weight reduces both (1) the concentration of unstable groups that would initiate degradation, and (2) the molecular mobility which controls the kinetics of the degradation reaction. Practically, the effect has been noted most often in rigid PVC (Table 25-4) (29), where the in-

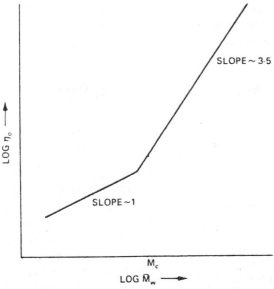

Fig. 25-2 Log–log plot of melt viscosity vs. molecular weight (34).

jection molder must compromise between the low molecular weight needed for melt flow and the high molecular weight needed for thermal stability.

Thermal Conductivity. Once the polymer melt has filled the mold, the injection molder wants to cool it as rapidly as possible in order to shorten the molding cycle. Here thermal conductivity of the polymer is his prime concern. In a molten polymer, conduction is primarily due to convection, which depends on molecular mobility, and is therefore inversely related to molecular weight (38). On the other hand, as the cold wall of the mold solidifies the outer layer of polymer, further conduction through this solid polymer is required to complete the cooling of the hot interior. Conduction through this layer of solid polymer is no longer by convection, but by atomic vibration. These vibrations are transmitted much more efficiently down the length of a polymer molecule than they are through the spaces between polymer molecules. Thus, in the solid polymer, conduction is directly related to molecular weight. Basic research on these effects remains to be done, but it should illustrate clearly how the two successive phenomena contribute to overall cooling in the injection mold.

Coefficient of Thermal Expansion. Once the mold has been filled, and the polymer proceeds to cool, decreasing thermal vibration produces decreasing free volume, and the practical result is mold shrinkage, which must be compensated by foresight in mold design. Since the ends of the polymer molecules have the greatest mobility, they play the major role in free volume and therefore in shrinkage. Increasing molecular weight decreases the concentration of end groups, and therefore the shrinkage during cooling.

Melting and Crystallization. When the injection molder heats a crystalline resin to melt it, he finds that higher molecular weight requires higher melting temperatures and longer times, which may increase the molding cycle somewhat. Then, once the melt has filled the mold, he must cool it until it crystallizes before he can open the mold and begin another cycle. Here he finds that lower molecular weight provides the molecular mobility needed for polymer molecules to fit into the growing crystal lattice structure and thus to hasten crystallization and shorten the molding cycle (Table 25-5) (39). Of these two conflicting factors, fast crystallization during mold cooling is usually the more critical, so low molecular weight favors faster molding cycles.

Table 25-5 Molecular weight and crystallization: polyethylene terephthalate (39).

MOLECULAR WEIGHT[b] (NUMBER AVERAGE)	HALFTIME OF CRYSTAL-LIZATION MINUTES
11,200	3.5
13,600	9.0
14,000	15.0
15,200	17.5
15,800	18.5

[a] At 118°C, starting with an amorphous sample.
[b] Obtained from osmotic pressure data.

Molecular Weight Distribution

Up to this point we have considered the effects of *average* molecular weight. Any real polymer is composed of a range of molecules from low to high molecular weight; and many polymer scientists believe that the shape of this molecular weight distribution curve—narrow or wide, normal or skewed, or even multimodal—may have critical effects on injection moldability. A typical sampling from the research literature indicates that the subject is too complex and too obscure for complete understanding at the present time. In some cases there is fairly general agreement; in others there are mysterious conflicts between theories in their present state. A few examples will illustrate the present state of our knowledge:

Melt Viscosity. Broadening the molecular weight distribution (MWD) decreases the melt viscosity of polyethylene (40) and impact styrene (27), but increases the melt viscosity of ABS (28) and PVC (41). Confusion may result from different ways of expressing molecular weight averages and distributions, and/or the complicating effects of branching.

Shear Sensitivity. Most polymer melts are pseudoplastic—at increasing shear rates they become less viscous, probably because of parallel orientation of polymer molecules during unidirectional flow (42). However, the relative degree of shear sensitivity varies greatly from one polymer to another. Generally broadening

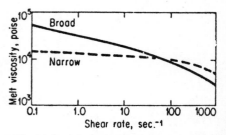

Fig. 25-3 Molecular weight distribution and shear sensitivity in melt processing: pseudoplasticity of polyethylene.

MWD produces increasing shear sensitivity (Fig. 25-3) (43), most typically and frequently studied in polyethylenes (44). An important consequence has been recognized but needs to be more widely understood: measurement of melt viscosity at low shear rate should not be used to predict injection molding at high shear rate because viscosity vs. shear rate curves can diverge, converge, or even cross, giving very poor predictability and wasting much engineering time and money.

Pressure Sensitivity. Broadening MWD increases the sensitivity of melt viscosity to changes in pressure, as illustrated in studies on polyethylenes (Fig. 25-4) (45). Since increasing pressure produces increasing shear rate, this may simply corroborate what has already been said about pseudoplasticity.

Temperature Sensitivity. It is generally recognized that increasing temperature increases atomic vibration and molecular mobility and thus reduces melt viscosity (42). Thus whenever an injection molding engineer finds a polymer too viscous, his first reaction is to increase the temperature of the melt. The effect of MWD on this relationship is not yet clear. In polyethylenes, broadening MWD decreases the sensitivity of melt viscosity to temperature (45), whereas in polystyrene broadening the MWD increases temperature sensitivity (46). Here again, methods of expressing molecular weight averages and distributions, and the complicating effects of branching, may be responsible for the discrepancy.

Melt Elasticity. Rubbery melt flow is due to persistent residual entanglement of large

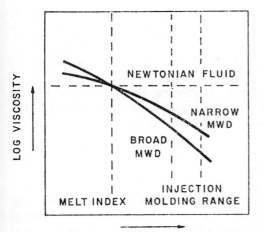

LOG SHEAR STRESS (PRESSURE)

Fig. 25-4   Molecular weight distribution and pressure sensitivity in melt processing (44).

Branching

Most people think of the thermoplastic polymer molecule as a long linear chain, and this is reasonably precise for most thermoplastics. In the high-pressure polymerization of LDPE, however, and to some extent in PVC as well, there is considerable branching:

leading to what Tobolsky early dubbed the "pin-cushion molecule." In general, viscosity seems to depend on the overall diameter of the molecule, whereas end-use properties often depend on the total molecular weight of the molecule. Thus at equal molecular weights (and presumably equal end-use properties), the branched molecule would have a smaller overall diameter and therefore a lower viscosity and easier processability. Or conversely, at equal viscosity and processability, the branched molecule could incorporate much more molecular weight and therefore deliver higher end-use properties. Phillips actually capitalized on this theory by purposely synthesizing multibranched styrene/butadiene block copolymers which proved to be thermoplastic elastomers of easier melt processability and superior end-use properties (48).

Branching has other significant effects on injection moldability of polyethylene and impact styrene (49). In addition to reducing melt viscosity as described above, it also increases sensitivity to temperature and pressure. Studies on polyethylenes generally find that increased branching produces increased broadening of the MWD, so it is often difficult to distinguish which of the two structural features is actually basic for explaining the practical effect on properties.

polymer molecules, which refuse to disentangle and flow at the temperature and shear rate of the process. Thus it would be expected that broad MWD, containing some very large molecules, would be most sensitive and most likely to show such melt elasticity. This has been observed experimentally in both polyethylenes (Fig. 25-5) (40) and polystyrene (47).

At this point it is evident that the effects of MWD on injection moldability require a general sophistication in measurement and expression of the shape of the MWD curve, and in the correlation and explanation of its effect on properties in different polymers.

Plasticizers

Addition of low-molecular-weight liquid or low-melting solid plasticizers is very effective in reducing melt viscosity and increasing melt

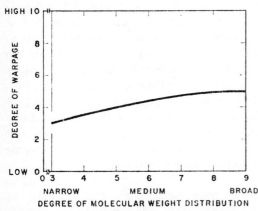

Fig. 25-5   Molecular weight distribution and melt elasticity: warpage in injection molding of high-density polyethylenes (44).

flow of many polymers, and is widely practiced commercially in polystyrene, PVC, and cellulosics to improve injection moldability. This may be understood theoretically as an extreme broadening of the MWD at the low-molecular-weight end, lowering the average molecular weight of the total material. For those who prefer the effect of end-groups on molecular mobility and free volume, every molecule of monomeric plasticizer adds two more end-groups, and thus produces a tremendous increase in total end-group concentration and therefore free volume, space in which the polymer molecules can then move more freely and thus exhibit lower melt viscosity. Presumably mathematical theory could easily be extended to encompass plasticization in the general relationships between molecular weight and injection moldability.

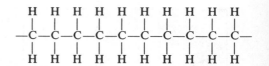

is quite flexible and can be taken as a good starting point for this discussion. Introducing oxygen atoms into the main-chain:

$$\sim\!\!\sim\!\!\sim \ddot{O} \sim\!\!\sim\!\!\sim$$

may make it ever more flexible; but most other groups tend to stiffen the polymer backbone. Putting rings in the main-chain has a distinct stiffening effect; especially when they are unsaturated resonating benzene rings, they form very flat rigid units in the polymer molecule. When the benzene ring is adjacent to atoms or groups that have extra electrons:

MOLECULAR FLEXIBILITY

The structure of the polymer molecule determines its inherent flexibility/rigidity, and this molecular flexibility in turn affects its injection moldability. The basic theory is generally accepted and used in this way, even though direct quantitative studies are rarely reported. It is best described by first explaining the effects of molecular structure on molecular flexibility, and then discussing the effects of molecular flexibility on injection moldability.

Effects of Molecular Structure on Molecular Flexibility

Molecular structural features are best divided into (1) main-chain structure of the polymer backbone and (2) side-chain structure of the groups attached to the main-chain.

Main-Chain Structure.
The polyethylene chain:

the resonance tends to extend beyond the benzene ring and involve these adjacent atoms or groups as well, thus forming larger flat rigid structures and having an even greater stiffening effect on the polymer molecule. Many of our newer engineering thermoplastics have such extended resonance structures in the polymer main-chain; and while this stiffening effect is beneficial for end-use properties, it makes melt processability more difficult, as will be discussed below. The most extreme examples of extended resonance structures and molecular stiffening are seen in the ultra-high-temperature plastics, where series of fused-ring structures keep the polymer molecule rigid up to very high temperatures, but simultaneously make it difficult to impossible to use the molecule in melt processing.

Side-Chain Structure.
The flexibility of the polymer main-chain may be severely restricted by the side-groups attached to it. Even in polyethylene the hydrogen atoms have a

slight restrictive effect. All other side-groups are larger and/or more polar and therefore have greater restricting effects on the flexibility of the main-chain.

Size and Shape. The larger and bulkier the side-group, the greater stiffening effect it will have; this is called steric hindrance. The methyl groups in polypropylene:

$$-CH_2-CH-CH_2-CH- \atop \qquad\; CH_3 \qquad\quad CH_3$$

and the benzene rings in polystyrene:

$$-CH_2-CH-CH_2-CH-$$

are typical of such effects. In particular, when there are four bulky groups attached to the same carbon atom in the main-chain, the stiffening effect on such a quaternary carbon atom can be very marked; typical examples would be acrylic plastics (polymethyl methacrylate):

$$\begin{array}{cc} CH_3 & CH_3 \\ | & | \\ -CH_2-C-CH_2-C- \\ | & | \\ CO_2CH_3 & CO_2CH_3 \end{array}$$

and the bisphenol group in polysulfones and aromatic polyesters:

$$-\langle\!\!\bigcirc\!\!\rangle-\overset{\displaystyle CH_3}{\underset{\displaystyle CH_3}{C}}-\langle\!\!\bigcirc\!\!\rangle-$$

Such stiffening effects can be very beneficial for high modulus, creep resistance, and heat deflection temperature, but they can also produce very viscous melts that are difficult to injection-mold.

Polarity. A polymer molecule naturally tends to form a flexible random coil. When the side-groups attached to it are very polar, they repel each other and force the random coil to open up almost to a rodlike structure, in order to place the polar groups as far from each other as possible. Such a structure is very stiff, and makes the polymer melt much more viscous and difficult to injection-mold. Some typical examples are rigid polyvinyl chloride:

$$-CH_2-CH-CH_2-CH- \atop \qquad\; Cl \qquad\qquad Cl$$

polyacrylonitrile barrier plastics:

$$-CH_2-CH-CH_2-CH- \atop \qquad\; CN \qquad\qquad CN$$

and polytetrafluoroethylene (PTFE):

$$\begin{array}{c} F\;\; F\;\; F\;\; F\;\; F\;\; F\;\; F\;\; F\;\; F\;\; F \\ |\;\; |\;\; |\;\; |\;\; |\;\; |\;\; |\;\; |\;\; |\;\; | \\ -C-C-C-C-C-C-C-C-C-C- \\ |\;\; |\;\; |\;\; |\;\; |\;\; |\;\; |\;\; |\;\; |\;\; | \\ F\;\; F\;\; F\;\; F\;\; F\;\; F\;\; F\;\; F\;\; F\;\; F \end{array}$$

Effects of Molecular Flexibility/Rigidity on Injection Moldability

Flexible polymer molecules can uncoil readily and give a fluid melt that flows quickly and uniformly in injection molding. Increasing molecular stiffness, by rings in the main-chain, or by bulky or polar side-groups, will generally increase melt viscosity and slow the injection molding process, requiring higher temperature and/or pressure for successful molding (Table 25-6). Furthermore, increasing molecular stiffness makes the polymer melt more sensitive to changes in temperature (Table 25-7) and pressure. As molecular stiffness increases, the rheology of melt flow becomes more non-Newtonian, with increasing pseudoplastic and thixotropic effects: at low shear rates and low shear times, the entangled polymer molecules flow only with great difficulty, whereas at higher shear rates and shear times, they disentangle and align parallel to the direction of flow, and flow relatively easily. Thus, in general, molecular flexibility favors fast uniform

Table 25-6   Molecular stiffening by rings in the main-chain: effect on injection molding temperature (50).

POLYMER	AVERAGE INJECTION MOLDING MELT TEMPERATURE
Polyether sulfone*	700°F
Polysulfone*	688
Polycarbonate*	548
Styrene/acrylonitrile	500
Acrylonitrile/butadiene/styrene	500
Ionomer	475
Polystyrene	468
Polymethyl methacrylate	453

* These polymers have benzene rings in the main-chain.

Table 25-7   Temperature-sensitivity of melt viscosity: activation energies of melt flow.

POLYMER	ACTIVATION ENERGY, KCAL/GM MODE
Polyethylene	11.0–12.8
Polyisobutylene	15.7–16.4
GR-S rubber	20.8
Polystyrene	22.0–23.0
Polyvinyl butyral	25.9
Polyvinyl chloride acetate	35.0 and 60.0
Cellulose acetate	70.0

melt flow in injection molding, whereas molecular stiffness tends to produce more difficult injection molding, requiring higher temperatures and pressures and very often wider flow channels as well.

Another possible relationship between molecular flexibility and injection moldability may be the effect on mold shrinkage. Generally a flexible polymer molecule has more mobility, empty space, and free volume in the molecular structure than a rigid polymer molecule, and this free volume is more sensitive to changes in temperature, producing a higher coefficient of thermal expansion. Thus it would be expected that molecular flexibility would produce greater mold shrinkage, and that molecular stiffness would produce less mold shrinkage.

Another relationship of considerable impor-

tance involves mold cooling time and ejection temperature. Generally molecular stiffness correlates with glass transition temperature, which in turn correlates with heat deflection temperature in amorphous thermoplastics (Table 25-8). When the injection molder tries to shorten mold cooling time, until the product is rigid enough to be ejected from the mold without distortion, he finds that the stiffer molecular structure, with higher glass transition and higher heat deflection temperature, can be ejected at a higher temperature and thus at a shorter cooling time. Thus molecular stiffness would provide definite benefits here to the injection molder.

In practice these effects are best seen in amorphous thermoplastics. Where cyrstallization, polar effects, and multiphase systems are involved, as discussed later, they can mask the basic effects of molecular flexibility as discussed above. In such complex systems, molecular flexibility is still very important, but in more complex ways than the straightforward relationships considered here.

CRYSTALLINITY AND ORIENTATION

Intermolecular order refers to the geometric arrangement of adjacent polymer molecules in the solid mass. There are three distinct types of intermolecular order:

1. *Amorphous random coils* are isotropic and entangled, and their properties depend primarily on molecular flexibility, Polystyrene, acrylic plastics, polyphenylene oxide, polysulfone, and polycarbonate are typical examples.

Table 25-8   Molecular stiffness and heat deflection temperature (50).

POLYMER	AVERAGE HDT
Polysulfone	210°C
Polycarbonate	146
Styrene/acrylonitrile	96
Acrylonitrile/butadiene/styrene	95
Polymethyl methacrylate	91
Polystyrene	89
Polyvinyl chloride	68
Ionomer	47

Talbe 25-9   Heat capacity of polymers (51).

POLYMER	ORDER	HEAT ENERGY 150–500°F
Low-density polyethylene	Crystalline	265 Btu/lb
High-density polyethylene	Crystalline	280
Nylon 6	Crystalline	270
Polystyrene	Amorphous	170
High-impact polystyrene	Amorphous	170

2. *Crystalline polymers* have their molecules arranged in a very regular repeating lattice structure, so precise that every atom of the polymer molecule must recur at very specific points in the repeat structure. While no polymer is completely crystalline, very regular polymers can be mostly crystalline with only small amorphous areas remaining between the crystallites. Typical highly crystalline polymers are high-density polyethylene, polypropylene, polyformaldehyde (acetal resin), thermoplastic polyesters, and polyamides (nylons). Less regular polymers may be partly crystalline (low-density polyethylene) or only slightly crystalline (polyvinyl chloride).

3. *Oriented polymers* have some of their molecules lying fairly parallel to each other. When crystalline polymers are oriented, their crystallites lie fairly parallel to each other. Such orientation produces anisotropic properties, very different when tested in the direction of orientation from those properties observed perpendicular to it.

Intermolecular order has important effects on the injection molding process, and the injection molding process in turn has important effects on intermolecular order. Amorphous order has already been described in the discussion of molecular flexibility. It remains for us to discuss crystallinity and orientation.

Crystallinity: Effects on Injection Molding

Effects of crystallinity on injection moldability can be divided into effects due to (1) the heating phase of the molding cycle, (2) the cooling phase of the molding cycle, and (3) mold shrinkage.

Heating Phase.   Crystallinity has a number of effects on the time and energy required to heat and melt the solid polymer and bring it to the proper viscosity for injection into the mold.

Heat Transfer.   The transfer of heat from the molding machine into the polymer is a slow inefficient process that lengthens the molding cycle. However, packing the polymer molecules tightly into a crystalline structure does improve the transfer of thermal vibrations between them, and thus increases thermal conductivity (51), so that it is often as much as twice as high as for amorphous polymers.

Heat Capacity.   Heating of amorphous thermoplastics simply requires a gradual input of energy to increase the thermal vibration of atoms and molecules. By contrast, melting of crystalline thermoplastics requires a much greater input of energy at the melting point, in order to overcome the attractive forces holding the crystals together (Table 25-9; Fig. 25-6) (52). All this additional energy must be supplied by the molding machine to produce a melt that is ready for injection.

Amorphous Softening vs. Crystalline Melting. When an amorphous polymer is heated to process temperature, it softens gradually from rigid to leathery to rubbery to a reasonably liquid state. By contrast, when a crystalline polymer is heated, it remains solid until it reaches the precise temperature at which thermal vibrations destroy the crystal lattice structure; at that point it changes suddenly from a crystalline solid to a molten liquid.

Melting Point.   The temperature at which this equilibrium melting process occurs de-

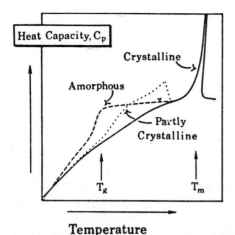

Temperature

Fig. 25-6 Heat capacity of amorphous and cyrstalline polymers.

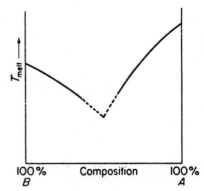

Fig. 25-7. Copolymerization and melting point.

pends upon (1) the regularity of the polymer molecule and the polar attractions between polymer molecules, which act to hold the crystal structure together, vs. (2) the molecular flexibility, which determines how readily the polymer molecule can unzip from the crystal lattice and transform into a random coil of higher entropy. Thus molecular stiffness and high polarity favor a high melting point, whereas molecular flexibility and low polarity favor a low melting point (Table 25-10). Gen-

erally, random copolymerization breaks up a regular polymer structure, decreasing its ability to crystallize and lowering its melting point (Fig. 25-7) (5, 53).

Melt Viscosity. Above the melting point, the crystalline polymer changes to a random amorphous coil, just like any normal amorphous polymer, and melt viscosity depends on molecular weight and molecular flexibility. Thus crystalline polymers of low molecular weight and high molecular flexibility can be molded a little above their melting temperature; but crystalline polymers that have very high molecular weight (UHMWPE) or very stiff molecules (PTFE, high-temperature engineering thermoplastics) may require much higher molding temperatures to produce good melt flow.

Table 25-10   Polymer melting point.

POLYMER	CRYSTALLINE MELTING POINT
Polytetrafluoroethylene	327°C
Polyacrylonitrile	317
Cellulose triacetate	306
Polyethylene terephthalate	267
Nylon 66	265
Isotactic polystyrene	240
Polybutylene terephthalate	232
Nylon 610	227
Nylon 6	225
Polycarbonate	220
Polychlorotrifluoroethylene	220
Polyvinyl chloride	212
Polyvinyl fluoride	200
Polyvinylidene chloride	198
Nylon 11	194
Polyoxymethylene (acetal)	181
Polypropylene	176
Isotactic polymethyl methacrylate	160
Polyethylene	137

Cooling Phase. Cooling is a major portion of the molding cycle because it requires removal of all the enthalpy of melting, and the growth of crystalline rigidity until the product can be removed from the mold without distortion (Fig. 25-8). The cooling phase also determines the type of crystalline structure that forms, and its effect on end-use properties.

Heat Capacity. All the thermal energy that was put into the crystalline polymer to overcome the attractions in the crystal lattice and produce melting must now be removed to permit the polymer molecules to reenter the crystal lattice and re-form the crystalline structure in the final product. This is a major part of the total cooling requirement (52).

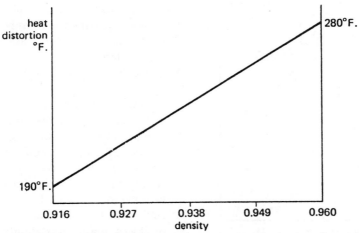

Fig. 25-8   Crystallinity and Distortion Temperature. (*Note:* Polyethylene density is a direct measure of percent crystallinity.)

Nucleation.   As the polymer melt cools, and the regular polymer molecules develop a tendency to pack into a regular crystal structure, they must find nuclei that can initiate such crystal structures. These nuclei are probably small ordered areas in the molten polymer, small areas of regularity that act as incipient nascent crystallites. The cooler the melt, and the less the thermal vibrations, the more of these tiny ordered nuclei that will form (Fig. 25-9). Thus low temperature favors the nucleation that is a prerequisite for crystallization.

Crystal Growth.   The second stage of crystallization is the movement of polymer molecules, from the random coils in the melt onto the surface of the growing crystal, fitting each atom into the precise position required in the crystal lattice. Such movement requires molecular mobility, which increases with temperature. Thus it is often noted that increasing mold temperature produces an increasing rate of crystal growth (39, 54–56).

Overall Rate of Crystallization.   Since low temperature favors faster nucleation but high temperature favors faster crystal growth, it is not surprising that overall rate of crystallization is the result of these opposing functions, and peaks at some intermediate temperature. Also, since high temperature favors crystal growth, but high temperature also produces thermal vibrations that cause crystal melting, it is not surprising that overall rate of crystallization is the result of these opposing functions, and peaks at some intermediate temperature. While basic research has not often tried to distinguish between these two pairs of opposing mechanisms, the phenomenon of maximum crystallization rate of optimum intermediate temperature has often been observed in practice (Fig. 25-10) (57). It is sometimes estimated at $\frac{9}{10}$ of the melting point in ^{0}K, other times at the midpoint between the glass transition temperature and the melting temperature. A molder can generally use either estimate

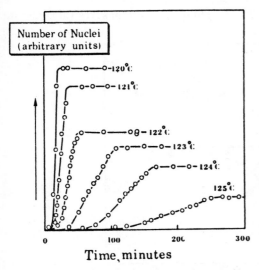

Fig. 25-9   Rate of nucleation vs. temperature (decamethylene terephthalate).

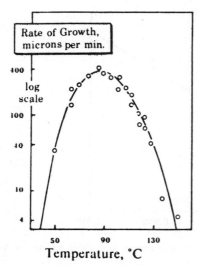

Fig. 25-10   Crystallization rate vs. temperature (polyoxymethylene [acetal]).

as a starting point in seeking the optimum mold temperature for the fastest crystallization and thus the shortest molding cycle.

Nucleants.   Since low temperature favors fastest nucleation but high temperature favors fastest crystal growth, the molder is forced to compromise in choosing the optimum mold temperature for the shortest molding cycle. One way to avoid this conflict is to add a nucleant to the molding powder. A nucleant is either a trace of high-melting polymer which will crystallize first on cooling of the melt, or a finely powdered inorganic whose sharp edges and corners can act as nuclei for crystal growth. Such techniques have been very successful in polymers such as polypropylene and nylon, permitting nucleation even at high mold temperatures, which then provide the molecular mobility for fast crystal growth.

Transverse Structure.   Since polymers are poor conductors of heat, injecting a hot melt into a cold mold produces sharp thermal gradients and time–temperature profiles. Since temperature influences crystallization, the resulting moldings show different crystallinity in different parts of the product. Thus fast cooling at the skin and far end of the mold produces lower crystallinity, whereas slow cooling in the interior and near the gate produces higher crystallinity (58). Furthermore, there are more complex changes in the structures of the crystallites produced, as one examines them from the surface down into the interior (59, 60).

Postannealing.   To shorten the molding cycle as much as possible, the molder will often eject the product as soon as crystallization has produced sufficient rigidity to permit ejection without distortion. This may be long before crystallization is complete or optimum. To maximize final end-use properties, it is then often desirable to postanneal the product in a warm oven, to facilitate molecular mobility and thus completion of crystallization (Fig. 25-11). This of course still saves valuable time

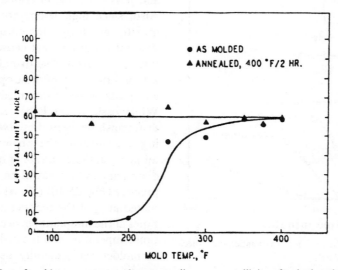

Fig. 25-11   Effect of mold temperature and postannealing on crystallinity of polyphenylene sulfide (54).

in the molding cycle and thus improves molding economics without sacrificing end-use properties.

Mold Shrinkage.   Precise design of product and mold dimensions requires a knowledge of the mold shrinkage during the cooling process. Since organic polymers always have a higher coefficient of thermal expansion than metal molds, shrinkage is always a problem that must be taken into account. In amorphous plastics this is simply a gradual process (Fig. 25-12; Table 25-11). In crystallizable plastics, on the other hand, the change from random amorphous coils and high free volume in the melt down to neatly packed chains in the crystal lattice produces a much more marked decrease in volume, resulting in much higher mold shrinkage (51, 65). Thus crystalline engineering plastics require very careful consideration of this factor in precise design of dimensions.

Orientation during Melt Flow in Injection Molding

When the polymer melt is injected into the mold, the randomly coiled molecules must disentangle and orient themselves parallel to the flow axis in order to slide past each other easily. This is the mechanism of pseudoplastic and thixotropic rheology typical of most non-

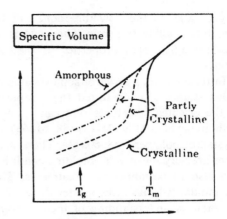

Fig. 25-12   Thermal expansion of amorphous and crystalline polymers.

Newtonian polymer melts. It is most pronounced when high-speed flow through narrow gates and thin wall sections produces the maximum rate of shear and therefore maximum orientation. After entry into the mold, high temperature and thicker volumes permit turbulence and Brownian randomization to reduce or eliminate the orientation with a return toward the random-coil isotropic condition. On the other hand, oriented molecules flowing along a cold wall film will solidify so that the orientation is permanently frozen into the final solid product. Likewise, if crystallization occurs simultaneously with orientational flow, the crystallites themselves will be oriented axi-

Table 25-11   Mold shrinkage of crystalline and amorphous polymers (51).

MATERIAL	POLYMER TYPE	APPROXIMATE MOLD SHRINKAGE IN./IN.	RELATIVE VOLUME AT 400°F
Nylon 66*	Crystalline	0.020	—
High-density polyethylene	Crystalline 80–93%	0.020–0.050	1.27
Low-density polyethylene	Crystalline 40–55%	0.015–0.030	1.22
Polystyrene	Amorphous	0.002–0.006	1.11
High-impact polystyrene	Amorphous	0.002–0.008	1.11
Styrene acrylonitrile Copolymer	Amorphous	0.002–0.005	1.16

* No relative volume data.

ally parallel to the flow direction, and this orientation will be frozen-in. These effects will produce anisotropic properties in the end-product.

In recent years a series of sophisticated detailed studies has illustrated the complexity of the oriented structures that can result in both amorphous (Fig. 25-13) (58, 61–63) and crystalline polymers (58–60, 64). These structures are very sensitive to molding conditions and also to postannealing treatments. Thus the molder will have to consider these effects in detail as understanding of them becomes more complete.

Generally orientation improves many properties in the direction of flow, but correspondingly sacrifices them in the transverse direction (51). In a limited number of products, molders have succeeded in orienting these effects to improve their products, notably in blow-molding of bottles; but in most cases, orientation has produced serious weakening of properties in important transverse directions, harming end-use properties and requiring modification of design or process to minimize such harmful effects. One of the most frequent problems is dimensional instability, particularly thermal shrinkage as the low-entropy oriented state tends to revert to a high-entropy random amorphous state. Another

frequent problem is splitting along lines parallel to the flow axis, due to weak attractive forces between molecules lying mainly parallel to the flow axis.

Wherever flow orientation during injection molding may prove to be harmful to end-use properties, the injection molder can try to eliminate or at least minimize the problem in several ways:

Higher Mold Temperature. Generally raising the wall temperature of the mold will increase Brownian randomization and coiling of the molecules in the polymer melt, and reduce freezing-in of orientation.

Lower Flow Rate. Often reducing the rate of mold filling will reduce parallel orientation of molecules during flow, and permit them to randomize again as they fill the mold at a more leisurely rate. This will of course slow the molding cycle and make it less economical.

Wider Flow Channels. Increasing the gate diameter and wall thicknesses reduces shear rate and orientation during liquid flow and permits molecules to retain their random-coil conformation. Such thicker cross-sections may or may not be acceptable in the end-product.

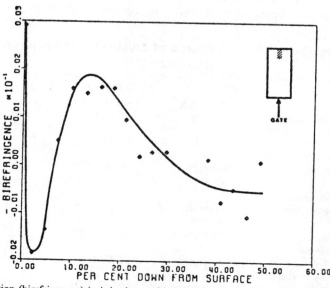

Fig. 25-13   Orientation (birefringence) in injection molded polystyrene: variation from surface into interior (63).

Redesign of the Flow Pattern. Ingenuity in mold design can often change flow patterns enough to minimize high-shear flow orientation and to permit Brownian rerandomization into amorphous coils and isotropic properties. More rarely it is even possible to use orientation to positive advantage, designing it to strengthen the critical properties in the critical directions and actually produce improved products. Detection and solution of such orientation problems is a prime need of mold designers and injection molders.

INTERMOLECULAR BONDING

Entanglement of amorphous random coils, parallel bundles of oriented molecules, and even neat tight packing in crystallites are not enough to explain the remarkable mechanical and thermal properties of plastics. It is the attractive forces between polymer molecules, which hold them firmly together and make them able to resist mechanical and thermal stress, that give them their useful properties. On the other hand, when the injection molder wants to melt these polymers and make the liquids flow and fill the mold, these same attractive forces can make the injection molding process much more difficult.

There are a variety of such intermolecular attractive forces. They may be arranged from weakest to strongest in the following order: London dispersion forces, polarity, hydrogen-bonding, orientation and crystallinity, ionic bonding, and permanent primary covalent cross-linking. It is useful to consider the effect of each of them individually.

London Dispersion Forces

Linear hydrocarbon polymers have no polarity, and there is no obvious reason why their molecules should be attracted to each other. Nevertheless their mechanical strength and resistance to melt flow show that even here some attractive forces must exist. They are vaguely ascribed to dynamic fluidity and mobility of the valence electron cloud in the polymer molecule, producing transient unsymmetrical states of electrical imbalance and thus momen-

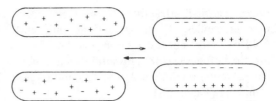

Fig. 25-14 Schematic model for London dispersion forces.

tary polarity (Fig. 25-14). These weak fugitive polar attractions draw polymer molecules to within 3 to 5 Å of each other with a force of 1 to 2 kcal/mole. While one such attraction would be negligible, the large number of such attractions between large polymer molecules accumulate to produce a large total effect.

For amorphous flexible molecules such as atactic polypropylene, polyisobutylene, and unvulcanized rubber, such attractions produce viscous liquids to tacky gums to soft rubbery solids; when heated, they liquefy and flow easily. For amorphous stiff molecules such as polystyrene, these attractions produce rigidity and strength up to 80 to 90°C; but they melt out easily to permit melt processing. In crystalline polymers such as polyethylene and isotactic polypropylene, the same London dispersion forces recur so frequently in the tightly packed crystal lattice that their total is very significant, producing rigidity and strength up to heat deflection temperatures as high as 120°C, and requiring the injection molder to go to melting points as high as 176°C in order to melt out the attractions and produce liquid flow.

These are polymers in which there are no intermolecular attractions other than London dispersion forces. In all other polymers, where there are also polar and hydrogen-bonding attractions, there are still these same London dispersion attractions, and they still form a considerable portion of the total intermolecular attraction in all of them.

Polarity

When polymers contain other atoms in addition to carbon and hydrogen, they are generally more negative than carbon and hydrogen (Table 25-12), and the resulting polarity produces electrical attractions between the polymer molecules. These attractions pull and hold polymer molecules close together. They are most often measured by solubility in different solvents, quantified as the "solubility parameter" (Table 25-13).

These polar attractions improve many end-use properties. They generally require more heat to overcome them and permit melt processing; so melting points and processing temperatures are generally higher than for less polar polymers. The overenthusiastic polymer chemist, keeping his eye only on end-use properties, can easily synthesize polymers with so much polar attraction that they will not melt and flow even when they finally reach the decomposition temperature. Thus processability vs. properties generally requires some compromise level of polarity. A typical method of balancing these requirements is to increase the

Table 25-12 Electro-negativities of atoms in polymers.

Sodium	0.9
Silicon	1.8
Hydrogen	2.1
Phosphorus	2.1
Carbon	2.5
Sulfur	2.5
Bromine	2.8
Nitrogen	3.0
Chlorine	3.0
Oxygen	3.5
Fluorine	4.0

Table 25-13 Solubility parameters of polymers.

Polytetrafluoroethylene (PTFE)	6.2
Polydimethyl siloxane (silicone)	7.3
Polypropylene	7.9
Polyethylene	8.0
Polystyrene	9.1
Polymethyl methacrylate (acrylic)	9.2
Polyvinyl chloride	9.5
Polycarbonate	9.5
Polyethylene terephthalate	10.7
Epoxy resin	11.0
Polyoxymethylene (acetal)	11.1
Cellulose diacetate	11.35
Nylon 66	13.6
Polyacrylonitrile	15.4

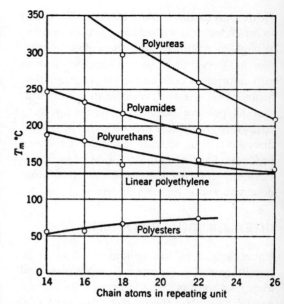

Fig. 25-15 Effect of polarity on melting point: —(CH_2)— spacing between polar groups in condensation polymers.

length of the —$(CH_2)_n$— chains between polar groups (Fig. 25-15).

Here again, one polar attraction would be quite negligible, but the many such attractions between large polymer molecules easily accumulate to produce a very large total effect.

Hydrogen-Bonding

Negative atoms in the polymer molecule can draw electrons away from adjacent hydrogens, leaving them electron-deficient:

$$H^+ \qquad\qquad H^+$$
$$|\qquad\qquad\qquad$$
$$A^- \qquad\qquad\quad A^-$$

Nitrogen, oxygen, fluorine, and chlorine are the negative atoms that most often do this in commercial polymers. Meanwhile, negative atoms in polymers do not use all of their outer shell of electrons to form covalent bonds, and the remaining electrons are available for sharing:

$$\ddot{B}$$

In commercial polymers the negative atoms that most often do this are oxygen and particularly nitrogen:

$$-\overset{\cdot\cdot}{\underset{\cdot\cdot}{O}}- \qquad -\overset{\displaystyle |}{\underset{\displaystyle \parallel}{C}}- \qquad -\overset{\displaystyle |}{\underset{\cdot\cdot}{N}}- \qquad \overset{\displaystyle |}{\underset{\underset{\cdot\cdot}{N}}{\overset{\parallel\parallel\parallel}{C}}}$$

Whenever both of these conditions exist in a polymer system, there is a strong tendency for negative atom B to share an electron pair with the positive hydrogen atom:

This shared pair of electrons is called a hydrogen-bond. It is a somewhat stronger attraction than ordinary polarity, generally drawing the two molecules to within about 3 Å of each other with a strength of 3 to 7 kcal/mole. This contributes handsomely to many end-use properties. It also requires much more heat to separate such bonds and produce free melt flow for injection molding. Fortunately the hydrogen-bonds in most commercial polymers, such as nylons and polyurethanes, do melt out at reasonable processing temperatures (Fig. 25-15), permitting a gratifying compromise between processability and end-use properties.

In ethylene copolymers with acrylic acids (66), the carboxylic acid groups hydrogen-bond readily with each other (Fig. 25-16), and the hydrogen-bonds melt out gradually with

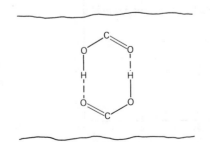

Fig. 25-16   Hydrogen-bonding in ethylene/acrylic acid copolymers (66).

increasing processing temperatures (Fig. 25-17); but increasing their concentration clearly increases the melt viscosity, pseudoplasticity, and activation energy for melt flow (Fig. 25-18).

Here again, as with London dispersion and polarity, an individual hydrogen-bond is quite negligible, but the many such attractions between large polymer molecules easily accumulate to produce a very large total effect.

Orientation and Crystallinity

Orientation and crystallinity were discussed earlier in terms of intermolecular order. What remains is to point out that the main effect of such intermolecular order is to pack polymer molecules parallel and close together, and thus greatly increase the frequency with which intermolecular attractions can come into play (Fig. 25-19). This is why weak individual attractions, such as London dispersion, polarity, and hydrogen-bonding, can accumulate into such large total forces that they greatly affect the properties or oriented and especially crystalline polymers. Fortunately, orientation melts out above the glass transition temperature, and crystallinity melts out above the melting point, so such polymers can then undergo easy melt processing. Thus crystallinity combines many of the end-use properties of thermoset plastics along with most of the easy processability of thermoplastics, and can be considered as a form of "thermoplastic cross-linking."

Ionic Bonding

In recent years the introduction of ionic bonding in plastics has offered a new balance of properties that have proved to be a pleasant surprise. On the one hand, clustering of ionic groups into dispersed domains has provided a moderate form of cross-linking that contributes strength and adhesion to end-use properties. On the other hand, these ionic domains have proved remarkably fusible and thermoplastic, as compared with the extremely high melting points of inorganic ionic compounds, thus still permitting or even facilitating nor-

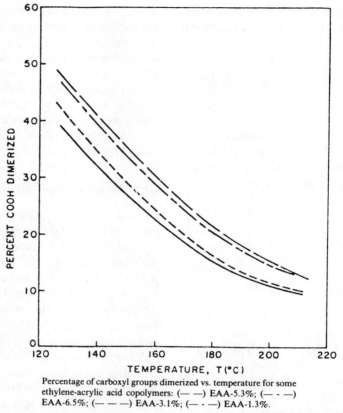

Percentage of carboxyl groups dimerized vs. temperature for some
ethylene-acrylic acid copolymers: (— —) EAA-5.3%; (— · —)
EAA-6.5%; (— — —) EAA-3.1%; (— · —) EAA-1.3%.

Fig. 25-17   Melting of hydrogen-bonds in ethylene/acrylic acid copolymers (66).

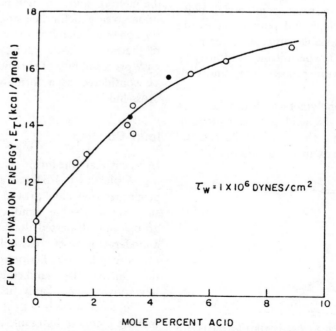

Fig. 25-18   Activation energy for melt flow: ethylene/acrylic acid copolymers (66).

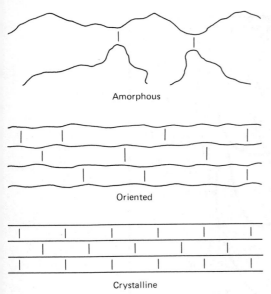

Fig. 25-19   Intermolecular order and intermolecular attractions.

mal thermoplastic processing. This combination of good thermoplastic processing with strong intermolecular attraction, typically 2 to 3 Å in length and 10 to 20 kcal/mole in strength, offers new dimensions in balance of properties that the plastics industry will undoubtedly use with increasing skill in the future.

Cross-Linking

Cross-linking in vulcanized rubber and thermoset plastics is the ultimate intermolecular bond, drawing molecules together with bonds 1 to 2 Å in length and typically 60 to 100 kcal/mole in strength, the same as the bonds within the original molecules themselves. It provides the ultimate improvements in end-use properties such as creep resistance, high temperatures, and chemical resistance. At the same time, it presents great problems to the injection molder accustomed to stable thermoplastics. Thermosetting materials all involve problems of balancing long shelf life and stable working conditions with fast cure time once they have been injected into the mold.

The first few cross-links that form do not immediately prevent melt flow or injection molding, but they do increase the melt viscos-

ity and complicate the melt rheology in more complex ways. Effects of such very light cross-linking have been detailed in recent studies on polyethylenes (4), rubber particles in high-impact polystyrene (27), and polyphenylene sulfide (Fig. 25-20) (54). They simply require the injection molder to adapt to such complex rheology.

High cross-linking in thermosetting plastics presents the injection molder with more serious problems. Thermosetting resins were designed originally for compression or at most transfer molding. Faced by competition from the more economical injection molding of thermoplastics, makers of thermosetting resins have redesigned and reformulated these resins sufficiently to permit development of injection molding techniques for almost all of them. Thus successful injection molding has been reported for EPDM (18, 67), 1,2-polybutadiene (19), epoxy resins (14, 67), unsaturated polyesters (17), alkyds (14), diallyl phthalate resins (13, 14), urea-formaldehyde (13, 23) resins, melamine-formaldehyde (13, 14) resins, phenol-formaldehyde resins (13, 14, 22, 67), H-resin (Fig. 25-21) (20), and silicone rubber (16). In-depth studies have detailed the relationships between cross-linking kinetics, time-dependent rheology, and balance of molding conditions required for optimum performance (Fig. 25-22) (22, 67). It is hoped that such thorough studies make injection molding of thermosetting resins more successful and useful in the future.

Reaction injection molding (RIM) has developed rapidly in the past several years because it combines the storage stability of two-part systems, the fluidity of low-molecular-weight oligomers, and the fast cure of highly reactive systems once they have been mixed during the injection process. Since the materials, equipment, and process conditions are so different from conventional injection molding, the two fields have remained quite separate from each other, losing the potential benefits of cross-fertilization. One suggestion for bridging the gap is the development of one-part systems for reaction injection molding, based on the use of solid or blocked reactants and/or catalysts (Table 25-14) to provide long pot-

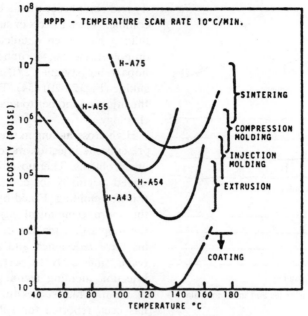

Fig. 25-20   Very light cross-linking in polyphenylene sulfide (54).

life but still activate rapidly when heated in the injection molding process (24). Hopefully more such developments will accelerate progress in this intermediate technology in the next few years.

ADDITIVES

Aside from the structure of the polymer itself, most injection molding compounds contain many additives that have important effects on injection moldability. These additives may be grouped according to whether they improve or hinder melt flow or have other effects on processability.

Additives That Increase Melt Flow

Plasticizers. Monomeric liquids or low-melting solids, of low volatility, are commonly

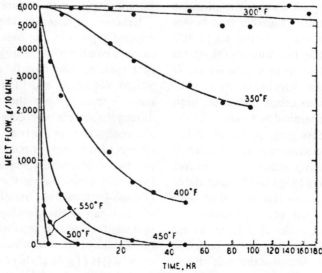

Fig. 25-21   Melt viscosities of thermosetting H-resins (20).

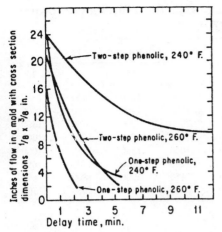

Fig. 25-22   Melt flow of phenolic molding compounds (22).

added to polymers to improve melt flow. Two common theories to explain their action are: (1) they intercept polymer–polymer attractions, setting the individual polymer molecules free to flow ("Lubricity Theory"); and (2) they lower the average molecular weight of the composition and thus increase melt flow. Practically, they lower the melt viscosity and/or permit lower molding temperatures. They are most commonly used in vinyls and cellulosics (and in synthetic rubber). Related effects have been reported from "melt flow additives" in 1,2-polybutadiene (19); and from calcium stea-

rate and fatty amide lubricants and liquid chlorinated organic flame-retardant in ABS (2).

Polymers.   Polymers that have high molecular flexibility and/or low molecular weight, and that are miscible with the major polymer, can also act as polymeric plasticizers to improve melt flow of the major polymer. Some typical examples are low-MW polyethylene in high-MW polyethylene; reclaim rubber in SBR; SBS thermoplastic elastomer in polystyrene (68); SAN in ABS (Table 25-15) (69); chlorinated polyethylene, ethylene/vinyl acetate (EVA) (Table 25-16) (70), nitrile rubber, and polycaprolactone (71) in polyvinyl chloride; and ionomers in nylon. Actually this technique should be more widely recognized and extended to a much greater variety of polymers.

Table 25-15   SAN as polymeric plasticizer in ABS (69).

SAN % IN ABS	MELT FLOW, G/10 MIN, 220°C, 10 KG
60	0.42
70	1.92
75	2.60
80	5.20

Table 25-14   Typical one-part polyurethane RIM systems (24).

SYSTEM	COMPONENT	DESCRIPTION	SUPPLIER
Solid isocyanate	Hylene TU	Isocyanate	Du Pont Company
	Ethylene glycol	Chain extender	Eastman Chemical Products, Inc.
	NIAX 11-34	Polyether triol, SAN graft	Union Carbide Corp.
	T-9	Stannous octoate catalyst	M&T Chemicals, Inc.
Complexed amine	E-1000	TDI-terminated prepolymer	Wilco Chemical Corp.
	E-0038	Complexed amine	Wilco Chemical Corp.
Solid catalyst	Isonate 181	Modified MDI	Polymer Chemicals Div., The Upjohn Co.
	HQEE	Chain extender	Eastman Chemical Products, Inc.
	P-581	Polyether triol, SAN graft	BASF Wyandotte Corp.
	Zinc stearate	Catalyst	Allied Chemical Corp., or Diamond Shamrock Chemical Co.

Table 25-16 EVA as polymeric plasticizer in rigid PVC: Brabender torque in dyne-cm (70).

VINYL ACETATE IN EVA	MELT INDEX OF EVA	%EVA IN RIGID PVC					
		0	5	10	15	20	25
28%	3.0	2500	2350	1900	1600	1400	1300
33	48	2500	2100	1700	1400	1250	1000
40	7.5	2500	2300	2000	1700	1450	1300
45	0.5–4.0	2500	2300	2100	1700	1500	1400
45	7.5	2500	2000	1650	1350	1250	1000
52	3.5	2500	2500	2250	2100	1900	1650
55	7.5	2500	2250	2150	2000	1800	1450
60	3.5	2500	2400	2250	2100	1950	1700

Additives That Produce Other Improvements

Additives can produce a variety of other improvements in injection moldability. Several may be mentioned as typical examples.

Polymers. Selected polymers have functioned as low-shrink modifiers in poly-1,2-butadiene (19). Special acrylic modifiers improve melt elasticity for blow molding of rigid vinyls (72).

Fillers. Fillers often contribute to frictional heating, and to thermal conduction during heating and cooling portions of the injection molding process, thus speeding the molding cycle—an effect particularly noted with iron and copper powders (73). Fillers often reduce rubbery melt flow (74–76). They also generally reduce mold shrinkage (77, 78) and often postwarpage.

Lubricants and Antistats. Whenever it is difficult to release the molding from the mold, the molding cycle is lengthened. This problem is commonly relieved by use of external lubricants, occasionally by antistats.

Additives That Hinder Melt Flow

Particulate fillers and especially fibrous reinforcements generally increase viscosity and impede melt flow. The effect increases with fiber length and L/D ratio, and requires higher temperatures and/or pressures to permit satisfactory injection molding. Recent observations have included high-density polyethylene (74, 79), polypropylene (76, 79, 80), polystyrene (74, 81), SAN (75), acetal (77), polycarbonate (32, 82), and nylon (76, 83).

Another additive that of course interferes with melt flow is the type used to produce cross-linking. This may be a catalyst, initiator, or actual co-reactant in the thermosetting process, as described earlier. For example, when peroxide is used to cross-link polyethylene during injection molding, various ingenious tricks are required to permit complete mold filling before cross-linking (84).

Other Problems Due to Additives

Use of fillers, and particularly reinforcing fibers, also introduces or accentuates a number of other processing problems. Several of these may be noted as follows:

Non-Newtonian Flow. Polymer melt flow is generally non-Newtonian, often pseudoplastic. Addition of glass fibers generally accentuates this behavior (79, 83). The mechanism here is reasonably straightforward, the practical effects being qualitatively similar to those in pure polymers.

Separation of Fillers during Flow. When a suspension of solid particles in a liquid flows through a channel, the solid particles tend to concentrate at the front of the flow. This was verified experimentally using glass spheres in polyethylene and in nylon, in a spiral flow mold. Glass concentrations at the end of the spiral were as high as double the average concentration at the beginning of melt flow (Fig. 25-23). With finer powders or short fibers, the effects were relatively mild (85).

Orientation of Fibers During Flow. Short fibers in a polymer melt tend to orient in the direction of flow during injection molding, but this depends on the specific flow patterns (86, 87). In convergent flow (e.g., in a capillary), they align parallel to the axial flow. In divergent flow (e.g., the entrance from a gate into a mold cavity), they align perpendicular to the major flow direction, as the melt moves transversely to fill the sides of the mold. In shear flow, particularly at low flow rates, they tend to lose alignment and distribute more randomly. Effects of fiber orientation on end-use properties generally resemble those of molecular orientation.

Fiber Degradation. While molten polymer can be injection-molded without serious

change in structure and properties, short-glass-fiber reinforcement suffers serious degradation during the injection molding process, and its reinforcing ability suffers accordingly. Recent studies have quantified the effect in polycarbonate (32), polybutylene terephthalate (11), and nylon (88). In the last study, the effects of melt temperature (Table 25-17), screw speed and back pressure (Fig. 25-24), and number of recycles (Table 25-18) were each detailed individually. These are effects the injection molder must be much more aware of and careful to minimize as much

Table 25-17 Effect of melt temperature on length of glass fibers in nylon 66 after injection molding (88).

CYLINDER REAR TEMPERATURE, °F	AVERAGE FIBER LENGTH, MILS
550	21.9
525	21.4
500	20.2
475	19.7
450	19.5

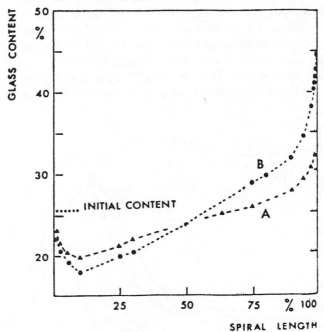

Fig. 25-23 Concentration gradients for glass spheres in polyethylene after injection into a spiral flow mold (85). The variation of the concentration of glass spheres is shown along the length of the spiral for two spiral lengths. LDPE, MI 70, 25.7 percent large glass spheres (50–100 μm in diameter). A: 0.7 m, B: 1.6 m spiral length.

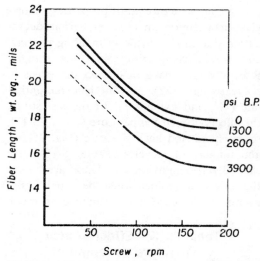

Fig. 25-24 Effect of screw speed and back pressure on length of glass fibers in nylon 66 after injection molding (88).

as possible, in order to retain maximum reinforced properties in the final product.

Machine Wear. Since inorganic fillers and reinforcing fibers are almost as hard as steel, they cause severe abrasive wear as they flow through screws, nozzles, gates, and molds in general, particularly around pins and projecting edges. Some materials suppliers claim that certain coupling agents can reduce wear. Generally soft fillers and glass spheres cause less wear, while the sharp ends of glass fibers are particularly harmful (89). While short fibers

Table 25-18 Effect of recycling on length of glass fibers in nylon 66 after injection molding (88).

Virgin resin	31.4 mils
First molding	21.7
Second molding	19.5
Third molding	17.0
Fourth molding	14.7
Reground after fourth molding	13.6

have more ends, long fibers cause higher viscosity and therefore even more severe wear.

Impurities

Aside from additives used purposely to improve properties, accidental impurities also often affect injection moldability. Most often absorption of water from the atmosphere causes hydrolytic degradation of polyesters, polyurethanes, and polyamides, lowering molecular weight, increasing melt flow, and degrading end-use properties. Recent studies have detailed this effect in polycarbonate (Fig. 25-25) (31, 78), polyethylene terephthalate (39), and polybutylene terephthalate (Table 25-19) (11, 90). All these studies emphasize the importance of predrying molding resins before use and particularly during recycle.

Aside from moisture, one careful study on ABS found that volatile impurities in general produced plasticization and thus increased

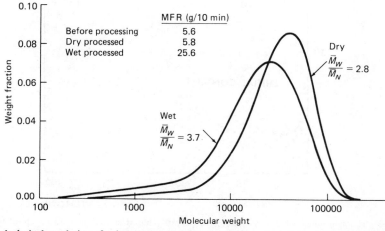

Fig. 25-25 Hydrolytic degradation of polycarbonate in injection molding: importance of drying to remove atmospheric moisture (78).

Table 25-19   Hydrolytic degradation of polybutylene terephthalate in injection molding: importance of drying to remove atmospheric moisture (11).

DRYING	MOLDING TEMPERATURE	MELT FLOW RATE g/10 min
	Unmolded	68
None	252°C	132
4 hr/121°C	252	107
None	271	182
4 hr/121°C	271	176

melt flow; but because they were volatile, the effect was undependable and disappeared during recycle (2).

In conclusion, this chapter has shown that the injection molder can benefit in many ways from a detailed understanding of the relationships between polymer structure and injection moldability, both in solving current problems and in development of new improved process technology.

REFERENCES

1. Stivala, S. S., and L. Reich, "Structure vs Stability in Polymer Degradation," Polym. Eng. Sci., 20 (10), 654 (mid-July 1980).
2. Blyler, L. L., Jr., "The Influence of Additives on the Flow Behavior of ABS," Polym. Eng. Sci., 14 (11), 806 (November 1974).
3. Shell Chemical Company, technical bulletins on Kraton G.
4. Mitterhofer, F., "Processing Stability of Polyolefins," Polym. Eng. Sci., 20 (10), 692 (mid-July 1980).
5. Grinblat, V. N., and Lapshin, V. V., "Moulding Machines for Processing Polyoxymethylene and Oxymethylene Copolymers," Soviet Plastics, 9, 25 (August 1970), RAPRA 242C371–831, UDC 678.644'141+678.644'141–139).057.
6. Kotrelev, V. N., Kovarskaya, B. M., Tsvetkova, Yu. I., and Levantovskaya, I. I., "Stabilisation of Polycarbonate during Processing," Soviet Plastics, 6, 26 (1969).
7. General Electric, technical bulletins on Noryl PPO.
8. ICI, technical bulletins on PEEK.
9. Union Carbide, technical bulletins on polysulfones.
10. ICI, technical bulletins on polysulfones.
11. Kelleher, P. G., Bebbington, G. H., Falcone, D. R., Ryan, J. T., and Wentz, R. P., "Thermal and Hydrolytic Stability of Poly(Butylene Terephthalate)," SPE ANTEC, 25, 527 (1979).
12. Hawkins, W. L., Polymer Stabilization, Wiley-Interscience, New York, 1972, Chapters 1 and 2.
13. Gardner, H. M., and Burges-Short, M. G., "The Injection Molding of Thermosetting Plastics," SPE ANTEC, 12, VIII-4 (1966).
14. Hoffman, K. D., "Thermoset Resins for Injection Molding," Plastics Des. Proc., 9 (10), 25 (October 1969).
15. Zecher, R. F., "Where is Thermoset Liquid-Resin Molding Going?" SPE J., 26 (5), 45 (May 1970).
16. Dudinova, N. P., Kongarov, G. S., Baturova, E. V., and Chertkova, V. F., "Injection Moulding of Siloxane-Based Rubber Compounds," Kauchuk i Rezina, No. 1, p. 14 (1977), Int. Polym. Sci. Tech., 4 (5), T/70 (1977).
17. Stankoi, G. G., Trostyanskaya, E. B., Kazanskii, Yu. N., Mikhasenok, O. Ya., and Okorokov, V. V., "Injection Moulding of Polyester Moulding Compositions," Soviet Plastics, p. 42 (March 1968), UDC 678.644–13.027.74, RAPRA 43D1–831.
18. Knox, R. E., "Rapid Molding of Large, Paintable EPDM Rubber Parts," Mod. Plastics, 49 (2), 56 (February 1972).
19. Leland, J. E., "Peroxide Cured, Injection Moldable Thermoset," SPE ANTEC, 26, 155 (1980).
20. Cessna, L. C., Jr., and Jabloner, H., "A New Class of Easily Moldable Highly Stable Thermosetting Resins," SPE ANTEC, 19, 57 (1973).
21. Elias, H.-G., New Commercial Polymers 1969–1975, Gordon and Breach, New York, 1977, p. 26.
22. Fishberg, L. D., and Longstreet, D. C., "New Data on Injection Molding Phenolics," Mod. Plastics, 46 (6), 92 (June 1969).
23. Nakamura, T., "Urea Injection Molding Material—Properties and Product Designing," Jap. Plastics Age, 17 (168), 33 (July–August 1979).
24. Cox, H. W., and Iobst, S. A., "Molding Polyurethane Parts on Standard Injection Machines," Plastics Eng., 34 (5), 49 (May 1978), SPE ANTEC, 24, 161 (1978).
25. Brydson, J. A., Plastics Materials, 4th ed., Butterworth, London, 1982, p. 203.
26. Thomas, D. P., "The Influence of Molecular Weight on the Rheology of Polypropylene," Polym. Eng. Sci., 11 (4), 305 (July 1971).
27. Nikitin, Yu. V., Aleksandrova, L. M., and Moskovskii, S. L., "Influence of Molecular and Structural Characteristics on Viscous Flow of HIPS," Plasticheskie Massy, No. 7, p. 58 (1982), Int. Polym. Sci. Tech., 9 (7) (1982), Abstract PM 82/07/58, Transl. Serial No. 8675.
28. Kubota, H., "Flow Properties of ABS (Acrylonitrile–Butadiene–Styrene) Terpolymer," J. Appl. Polym. Sci., 19, 2299 (1975).
29. Collins, E. A., "Relationship of Poly(Vinyl Chloride) Stability to Flow," Polym. Eng. Sci., 18 (16), 1240 (December 1978).
30. Murrey, J. L., and Dito, A. J., "Injection Molding PVC: New Resins Make It Easier than Ever," Plastics Tech., 27 (12), 79 (November 1981).
31. Long, T. S., and Sokol, R. J., "Molding Polycarbo-

nate: Moisture Degradation Effect on Physical and Chemical Properties," *Polym. Eng. Sci., 14* (12), 817 (December 1974).

32. Abbas, K. B., "Reprocessing of Thermoplastics. II. Polycarbonate," *Polym. Eng. Sci., 20* (5), 376 (March 1980).

33. Pearson, G. H., and Garfield, L. J., "The Effect of Molecular Weight and Weight Distribution upon Polymer Melt Rheology," *Polym. Eng. Sci., 18* (7), 583 (May 1978).

34. Brydson, J. A., op. cit., p. 155.

35. Shaw, M. T., and Miller, J. C., "The Rheology of Polysulfone," *Polym. Eng. Sci., 18* (5), 372 (April 1978).

36. Bueche, F., *J. Chem. Phys., 20,* 1959 (1952).

37. Brydson, J. A., op. cit., p. 157.

38. Ramsey, J. C., III, Fricke, A. L., and Caskey, J. A., "Thermal Conductivity of Polymer Melts," *J. Appl. Polym. Sci., 17,* 1597 (1973).

39. Burke, L. R., and Newcome, J. M., "Tips on Molding Modified PET Parts," *Plastics Eng., 38* (4), 35 (October 1982).

40. Brydson, J. A., op. cit., p. 204.

41. Park, I. K., and Riley, D. W., "The Effect of Molecular Weight Blending on PVC Melt Rheology," *SPE ANTEC, 26,* 393 (1980).

42. Bernhardt, E. C., *Processing of Thermoplastic Materials,* Reinhold, New York, 1959, pp. 554–679.

43. Kamal, M. R., Tan, R., and Ryan, M. E., "Injection Molding: A Critical Profile," in Suh, N. P., and Sung, N. H., *Science and Technology of Polymer Processing,* MIT Press, Cambridge, Mass., 1979, p. 34.

44. Leegwater, M. G., "Effects of Molecular Weight Distribution on HDPE Container Molding," *SPE J., 25* (11), 47 (November 1969).

45. Mills, D. R., Moore, G. E., and Pugh, D. W., "The Effect of Molecular Weight Distribution on the Flow Properties of Polyethylene," *SPE ANTEC, 6,* 4 (1960).

46. Hagan, R. S., Thomas, D. P., and Schlich, W. R., "Application of the Rheology of Monodisperse and Polydisperse Polystyrenes to the Analysis of Injection Molding Behavior," *Polym. Eng. Sci., 6* (10), 373 (October 1966), *SPE ANTEC, 12,* XI-3 (1966).

47. Thomas, D. P., and Hagan, R. S., "The Influence of Molecular Weight Distribution on Melt Elasticity, Processing Behavior and Properties of Polystyrene," *SPE ANTEC, 14,* 49 (1968), *Polym. Eng. Sci., 9* (3), 164 (May 1969).

48. Phillips Petroleum, technical bulletins on Solprene.

49. Spaak, A., and Weir, C. L., "Molding Linear Polyethylene," *Mod. Plastics, 35* (8), 122 (April 1958).

50. Howard, M. J., *International Plastics Selector,* Cordura Publications, La Jolla, 1977, pp. 553–567.

51. Staub, R. B., "Effects of Basic Polymer Properties on Injection Molding Behavior," *SPE ANTEC, 7,* 11–1 (1961), *SPE J., 17* (4), 345 (April 1961).

52. Warfield, R. W., Petree, M. C., and Donovan, P., "The Specific Heat of High Polymers," *SPE J., 15* (12), 1055 (December 1959).

53. Ajroldi, G., Stea, G., Mattiussi, A., and Fumagalli, M., "Thermal Behavior of Nylon 6 Copolyamides Containing Aromatic Rings," *J. Appl. Polym. Sci., 17,* 3187 (1973).

54. Hill, H. W., Jr., and Brady, D. G., "Properties, Environmental Stability, and Molding Characteristics of Polyphenylene Sulfide," *Polym. Eng. Sci., 16* (12), 831 (December 1976).

55. Hill, H. W., Jr., "Polyphenylene Sulfide—An Unusual Engineering Plastic," *SPE ANTEC, 26, 501* (1980).

56. Osborn, C. W., and Walker, J. H., "The Effects of Molding Conditions on Physical Properties of Parts Molded from Mineral Filled Polyphenylene Sulfide," *SPE ANTEC, 23,* 117 (1977).

57. Brydson, J. A., op. cit., p. 46.

58. Bakerdjian, J., and Kamal, M. R., "Distribution of Some Physical Properties in Injection-Molded Thermoplastic Parts," *Polym. Eng. Sci., 17* (2), 96 (February 1977).

59. Djurner, K., and Rigdahl, M., "The Structure of High Molecular Weight Linear Polyethylene Injection Molded at High Pressures," *Int. J. Polymeric Mater., 6,* 125 (1978).

60. Katti, S. S., and Schultz, J. M., "The Microstructure of Injection-Molded Semicrystalline Polymers: A Review," *Polym. Eng. Sci., 22* (16), 1001 (November 1982).

61. Han, C. D., and Villamizar, C. A., "Development of Stress Birefringence and Flow Patterns During Mold Filling and Cooling," *Polym. Eng. Sci., 18* (3), 173 (February 1978).

62. Kamal, J. R., and Tan, V., "Orientation in Injection Molded Polystyrene," *SPE ANTEC, 24,* 121 (1978).

63. Kamal, M. R., Tan, V., and Ryan, M. E., "Injection Molding: A Critical Profile," in Suh, N. P., and Sung, N. H., Int. Conf. on Sci. & Tech. of Polym. Proc., Cambridge, Mass., Mass. Inst. Tech., August 1977, p. 34.

64. Kazaryan, L. G., Zezina, L. A., Abramova, I. M., Gumen, R. G., Kuznetsov, V. V., and Kuznetsova, I. G., "Structural Processes in the Injection Moulding of Crystalline Polymers," *Sov. Plastics, 7,* 24 (1973), UDC 678.644′141:678.027.74)01:53, RAPRA 9112.

65. Nitschke, C. C., "Thermoplastic Polyesters," *SPE RETEC* on "Injection Molding," Chicago, May 19–20, 1974, p. 39.

66. Blyler, L. L., and Haas, T. W., "The Influence of Intermolecular Hydrogen Bonding on the Flow Behavior of Polymer Melts," *J. Appl. Polym. Sci., 13,* 2721 (1969).

67. Mussatti, F. G., and Macosko, C. W., "Rheology of Network Forming Systems," *Polym. Eng. Sci., 13* (3), 236 (May 1973).

68. Kassa, A., Dreval, V. E., Kerber, M. L., and Borisenkova, E. K., "Study of the Rheological Properties of Blends of Polystyrene with Thermoplastic Elastomers," *Izvestiya VUZ, Khimiya i Khimicheskaya Tekhnologiya,* No. 12, p. 1500 (1979), *Int. Polym. Sci. Tech., 7* (4), 105 (1980).

69. Bley, J. W. F., and Mohammed, S. A. H., "Performance Characteristics of ABS Resins Prepared by Melt Blending," *SPE ANTEC, 28,* 151 (1982).

70. Deanin, R. D., and Shah, N. A., "Polyblends of Polyvinyl Chloride with Ethylene/Vinyl Acetate Copolymers," *ACS Org. Coatings & Plastics Chem., 45,* 290 (August 1981).

71. Deanin, R. D., and Zhang, Z. B., "Polycaprolactone as Permanent Plasticizer for Polyvinyl Chloride," *ACS Org. Coatings & Appl. Polym. Sci. Proc., 48,* 799 (March 1983).

72. Rohm & Haas, technical bulletins on Acryloid K-120-N.

73. Kubat, J., and Rigdahl, M., "Reduction of Internal Stresses in Injection Molded Parts by Metallic Fillers," *Polym. Eng. Sci., 16* (12), 792 (December 1976).

74. Chan, Y., White, J. L., and Oyanagi, Y., "Influence of Glass Fibers on the Extrusion and Injection Molding Characteristics of Polyethylene and Polystyrene Melts," *Polym. Eng. Sci., 18* (4), 268 (March 1978).

75. Agarwal, P. K., Bagley, E. B., and Hill, C. T., "Viscosity, Modulus, and Die Swell of Glass Bead Filled Polystyrene–Acrylonitrile Copolymer," *Polym. Eng. Sci., 18* (4), 282 (March 1978).

76. Crowson, R. J., and Folkes, M. J., "Rheology of Short Glass Fiber-Reinforced Thermoplastics and Its Application to Injection Molding. II. The Effect of Material Parameters," *Polym. Eng. Sci., 20* (14), 934 (September 1980).

77. Okada, T., "Processing and Applications of Polyacetals Filled with Glass Beads or Carbon Fibers," *Jap. Plastics Age, 15* (154), 22 (March–April 1974).

78. Newcome, J. M., "Effect of Processing Conditions on Mechanical properties of Injection Molded Polycarbonate," *SPE RETEC* on "Injection Molding of Engineering Thermoplastics," June 8–9, 1977, p. 45.

79. Moskal, E. A., "Gate Geometry and Rheology of Glass-Filled Polymer Melts," *SPE ANTEC, 25,* 25 (1979).

80. Schmidt, L. R., "Glass Bead–Filled Polypropylene. Part II: Mold-Filling Studies During Injection Molding," *Polym. Eng. Sci., 17* (9), 666 (September 1977).

81. Tee, T. T., and Yap, C. Y., "Rheological Properties of Thermoplastic Melt Composite," *SPE ANTEC, 24,* 262 (1978).

82. Hadden, C. W., and Talbot, H. M., "Top-Performance Polycarbonate Molding," *Plastics Tech., 20* (5), 33 (May 1974).

83. Pisipati, R., and Baird, D. G., "Correlation of Rheological Properties of Filled Nylon Melts with Processing Performance," *SPE ANTEC, 27,* 32 (1981).

84. Menges, G., and Rheinfeld, D., "Processing of Crosslinkable Molding Compounds," *SPE ANTEC, 19,* 83 (1973).

85. Kubat, J., and Szalanczi, A., "Polymer–Glass Separation in the Spiral Mold Test," *Polym. Eng. Sci., 14* (12), 873 (December 1974).

86. Folkes, M. J., and Russell, D. A. M., "Orientation Effects During the Flow of Short-Fibre Reinforced Thermoplastics," *Polymer, 21,* 1252 (November 1980).

87. Crowson, R. J., Folkes, M. J., and Bright, P. F., "Rheology of Short Glass Fiber-Reinforced Thermoplastics and Its Application to Injection Molding. I. Fiber Motion and Viscosity Measurement," *Polym. Eng. Sci., 20* (14), 925 (September 1980).

88. Filbert, W. C., Jr., "Glass Reinforced 66 Nylon— The Effect of Molding Variables on Fiber Length and the Relation of Fiber Length to Physical Properties," *SPE ANTEC, 14,* 394 (1968).

89. Mahler, W. D., *Kunststoffe, 67* (4), 224 (1977).

90. Avery, J. A., and Kramer, M., "Effect of Molding Conditions on the Practical Performance of Reinforced Thermoplastic Polyester Parts," *SPE ANTEC, 20,* 356 (1974).

Chapter 26
The Role of Rheology in Injection Molding

J. M. Dealy

McGill University
Dept. of Chemical Engineering
Montreal, Canada

INTRODUCTION

The rheological properties of a melt govern the way it deforms and flows in response to applied forces as well as the decay of stresses when the flow is halted. These properties therefore play a central role in the injection molding process. In mold filling, it is the viscosity, along with the thermal properties, that governs the ability of the melt to fill the mold, that is, the pressure required to force the melt through the runner and gate and into the cavity. After filling, it is the relaxation of stresses in the melt that determines residual orientation in the finished part, and this can have an important effect on its mechanical properties. For these reasons, it is important to the molder as well as to the manufacturer of molding resins to know something about melt rheology and to be able to perform rheological tests on melts.

This chapter begins with a discussion of melt viscosity and how it is affected by shear rate, temperature, pressure, and molecular weight. This is followed by an explanation of the role of viscosity in the mold filling process and a description of a method for measuring viscosity. The chapter closes with a brief discussion of the role of melt elasticity in the relaxation of residual orientation stresses.

STEADY SHEAR FLOW OF MOLTEN PLASTICS

Simple Shear Flow

It is convenient to discuss shear flows by referring to the simplest type of shear flow. Simple shear is defined as the flow between two parallel plates, one of which is stationary while the other moves in a straight line with a velocity V. Referring to Fig. 26-1, the velocity distribution is given by:

$$v_1 = \frac{V}{h} \cdot x_2 \qquad (26\text{--}1)$$

The shear rate at each point in the fluid is:

$$\dot{\gamma} = \frac{dv_1}{dx_2} = \frac{V}{h} \qquad (26\text{--}2)$$

Thus, the shear rate is uniform throughout the fluid. If V does not change with time, we have a steady simple shear flow.

If F is the total force required to move the upper plate (equal to the force required to hold the lower plate stationary), and A is the surface area of the plate that is in contact with the liquid, then the shear stress, σ, is given by:

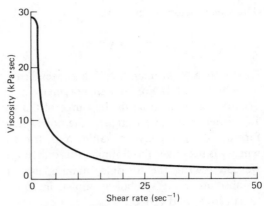

Fig. 26-1 Simple shear flow between parallel plates.

$$\sigma = F/A \qquad (26\text{--}3)$$

This is the tangential force per unit area required to produce the shear rate, $\dot{\gamma}$.

The Viscosity

Of course for a given shear rate, the force or stress required varies from one material to another. To describe the tendency of a fluid to resist steady simple shearing, we define the viscosity, η, as the ratio of the shear stress to the shear rate:

$$\eta \equiv \sigma/\dot{\gamma} = Fh/AV \qquad (26\text{--}4)$$

For low-molecular-weight, single-phase liquids such as water, glycerine, and syrup, the viscosity depends on the temperature and the pressure but not on the shear rate. Such liquids are said to be Newtonian. The viscosity of a Newtonian liquid decreases sharply as the temperature rises and increases (less sharply) as the pressure rises.

A thorough treatment of the rheological properties of molten polymers and their role in plastics processing is given the book by Han (1). We will summarize here only those aspects of melt rheology that are important in the injection molding process.

Effect of Shear Rate. Molten plastics have a rheological behavior that is much more complicated than that of a Newtonian liquid. For example, if we perform a simple shear experiment, as described above, we find that the viscosity depends not only on the temperature and pressure but also on the shear rate, $\dot{\gamma}$. In fact, the viscosity of a melt decreases as the shear rate increases, as shown in Fig. 26-2. Fluids that behave in this way are said to be shear thinning or pseudoplastic.

Fig. 26-2 Typical viscosity–shear rate curve for a molten polymer showing the zero shear rate viscosity, η_0.

At sufficiently low shear rates, the viscosity normally becomes independent of shear rate. The constant viscosity that prevails at low shear rates is called the zero shear viscosity or the low shear rate limiting viscosity, and is given the symbol η_0.

If viscosity vs. shear rate data are plotted on a double logarithmic scale, that is, if we plot $\log \eta$ vs. $\log \dot{\gamma}$, the points for high shear rates often fall very close to a straight line, as shown in Fig. 26-3. This suggests the use of an empirical "power law" formula to describe the dependence of viscosity on shear rate:

$$\eta = K\dot{\gamma}^{n-1}$$

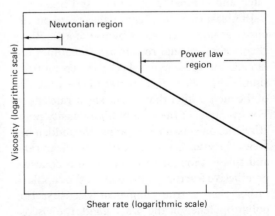

Fig. 26-3 Plot of the logarithm of the viscosity vs. the logarithm of the shear rate showing power law behavior at high shear rates and Newtonian behavior at low shear rates.

The shear stress is then given by:

$$\sigma = K\dot{\gamma}^n$$

Obviously a Newtonian liquid is a special case for which $n = 1$. For molten polymers, n is usually observed to be in the range of 0.3 to 1.0. When n is not an integer, the power law formula does not give a value of σ (or η) when $\dot{\gamma}$ is negative. To eliminate this difficulty, we can rewrite the formula, making use of an absolute value symbol as shown in equations (26–5) and (26–6):

$$\eta = K|\dot{\gamma}|^{n-1} \qquad (26\text{–}5)$$

$$\sigma = K|\dot{\gamma}|^{n-1}\dot{\gamma} \qquad (26\text{–}6)$$

It must be kept in mind that the simple power law is only an empirical expression that fits high-shear-rate data for a number of polymers. A number of other empirical viscosity–shear rate expressions have been proposed that have one or more advantages over equation (26–5). The Carreau equation, for example, tends to Newtonian behavior in the limit of very low shear rates (2):

$$\eta = \frac{\eta_0}{[1 + (\lambda\dot{\gamma})^2]^p} \qquad (26\text{–}7)$$

Other viscosity expressions have been tabulated and compared by Elbirli and Shaw (3).

In measuring the effect of shear rate on viscosity at high shear rates by use of a capillary rheometer, unusual results are sometimes obtained, and these may reflect a deviation from simple shear flow or a change in the structure of the melt, rather than a true shear rate effect. For example, in the case of high-density polyethylene, one observes an apparent sudden decrease in viscosity at a particular shear rate, and this is thought to result from a cohesive or adhesive fracture of the melt in the capillary rather than a true change in viscosity. For polypropylene, on the other hand, the viscosity suddenly increases at some shear rate, and this is thought to reflect a change in the structure of the melt.

Effect of Molecular Weight. Both the magnitude of the viscosity and the shape of the viscosity curve vary significantly from one resin to another. For a given polymer, the zero shear viscosity, η_0, increases with the 3.5 power of the molecular weight for many linear, monodisperse polymers as shown by equation (26–8):

$$\eta_0 \, \alpha \, M^{3.5} \qquad (26\text{–}8)$$

At higher shear rates, however, the effect of molecular weight is less pronounced, as illustrated in Fig. 26–4.

Broadening the molecular weight distribution results in a viscosity curve for which the transition from the low-shear-rate Newtonian behavior to power law behavior begins at a lower shear rate and extends over a wider range of shear rates. Branching has a similar effect.

Some polymers are much more non-Newtonian than others. Polyethylenes, for example, have viscosities that are strongly dependent on shear rate, with power law indexes in the range of 0.3 to 0.4. Polycarbonates, nylons, and polyesters, on the other hand, are often Newtonian over a wider range of shear rates.

Effects of Fillers. The presence of a filler or reinforcing fiber in the molding compound can have a strong effect on its rheological properties (4). At typical loadings, 20 to 40 percent by volume, the viscosity is increased significantly, and a sharp rise in the curve that is often observed as the shear rate is decreased suggests the presence of a yield stress. However, in a study of the mold filling behavior of a glass-bead-filled polypropylene (5) it was observed that although the beads increased the viscosity quite markedly, they did not have much effect on the flow pattern in the cavity.

Coupling agents that are used with certain fillers can alter the behavior of the filled resin. For example, in a study of the effect of a titanate coupling agent on the behavior of a polypropylene filled with calcium carbonate (50 wt %), it was found that the presence of the coupling agent reduced the viscosity dramatically while increasing the first normal stress difference (6).

Effects of Temperature and Pressure.
The viscosity of a liquid decreases with temperature, often rather sharply. A simple expression often used to describe this effect is given by equation (26–9):

$$\eta(T) = Ae^{E/RT} \qquad (26\text{--}9)$$

In this equation, R is the gas constant, and E is an "activation energy for viscosity." Clearly a material with a higher activation energy will have a viscosity that is more sensitive to temperature. This equation has been found useful to describe the low shear rate viscosity, η_0, of molten plastics, and E is found to vary considerably from one polymer to another. For example, for polyethylene it is about 12 kcal/mole, while for polystyrene it is in the range of 23 kcal/mole, and for cellulosic polymers it can be much higher still.

Another equation often used to describe the temperature dependence of viscosity is shown below:

$$\eta(T) = \eta(T_0)e^{-b\left(\frac{T-T_0}{T_0}\right)} \qquad (26\text{--}10)$$

This suggests that if the logarithm of viscosity is plotted as a function of temperature, a straight line will result whose slope is $(-b)$. This is found to be a good approximation over temperature ranges of 50°F. As we have seen, viscosity also depends strongly on shear rate, and a plot of the logarithm of the viscosity at 1000 sec^{-1} vs. the temperature is often used as a guide to the moldability of a resin.

Several software packages are now available that can be used as aids in the designing of molds. Clearly, any such model must incorporate an equation that describes the dependence of viscosity on both shear rate and temperature. Since most of the stages of the flow in a mold involve shear rates in the power law region, a combination of one of the above expressions for $\eta(T)$ with the power law is usually used in mold filling programs. One example of such an expression is given by equation (26–11):

$$\eta = Ae^{E/RT}|\dot{\gamma}|^{n-1} \qquad (26\text{--}11)$$

The viscosity of a liquid also depends on pressure, increasing as the pressure increases. It is thought that the variations of viscosity with both temperature and pressure are manifestations of its dependence on a more fundamental quantity, the free volume. This is the volume of the space in a melt that is not actually occupied by molecules and is thus available to permit the mobility of the molecules. Obviously, the greater the free volume, the easier it will be for the molecules to adjust to deformations, and this will be reflected in a lower viscosity. We know that increasing the temperature results in thermal expansion and thus an increase in free volume. This explains the decrease of viscosity as the temperature increases. Increasing the pressure, on the other hand, results in compression and thus a decrease in free volume and an increase in viscosity.

Since polymers are not very compressible, the dependence of free volume, and thus viscosity, on pressure is not nearly so important as its dependence on temperature. Cogswell (7) noted that those polymers most sensitive to temperature are often also most sensitive to pressure. This observation led him to define a temperature–pressure equivalence coefficient as shown in equation (26–12):

$$N_c = -\left(\frac{\Delta T}{\Delta P}\right)_\eta \qquad (26\text{--}12)$$

ΔT is the temperature decrease required to cause the same decrease in viscosity as a pressure increase of ΔP. In other words, if the temperature of the melt is changed by $-\Delta T$ while the pressure is simultaneously increased by an amount ΔP, the viscosity will remain unchanged. Cogswell reports that N_c does not vary a great deal from one polymer to another, being about 3×10^{-7} °C/Pa for PVC, nylon 66, and PMMA; about 4.5×10^{-7} °C/Pa for polyethylene; and about 8×10^{-7} °C/Pa for polypropylene.

The Melt Index

As is explained in a later section of this chapter, the accurate measurement of the viscosity over a wide range of shear rates requires the use of a fairly sophisticated capillary rheometer, and analyzing the raw data to yield the $\eta(\dot{\gamma})$ curve is not a trivial task. For purposes of quality control the much simpler "melt indexer" is often used. This instrument is described in detail later; here we wish only to note that some caution must be exercised in using the melt index to compare molding resins. ASTM standard test method D1238 prescribes the exact procedure for measuring the melt index, which is the mass, in grams, of polymer extruded from a die of specified design during a period of 10 minutes, as a result of a prescribed driving force.

Obviously, a higher melt index implies a lower resistance to flow and thus a lower viscosity. Injection molding resins are generally "high flow" materials having a melt index much larger than those typical of film and blow molding resins. This is so because a low resistance to flow is desirable to ensure the rapid, uniform filling of the mold cavity. For example, polyethylene molding resins often have melt index values in the range of 30 to 60 g/10 min, while polyethylenes for film and bottle applications have values in the range of 1 to 5 g/10 min.

However, considerable caution must be exercised in using the melt index test to compare resins because of the strong variation of melt viscosity with shear rate. The melt index is a measure of the resistance to flow of a melt at rather low shear rates, often in the range of 10 to 20 sec^{-1}, while the shear rates that occur in the injection molding process are in the range of 10^3 to 10^5 sec^{-1}. In comparing resins of the same family (i.e., resins having nonintersecting viscosity curves of the same general shape), the ranking of a series of resins according to melt index will correspond to the ranking according to viscosity. Even in this case, however, differences between resins

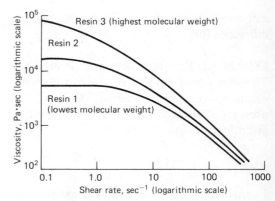

Fig. 26-4 Viscosity vs. shear rate curves for several linear polymers having different molecular weights. The higher the molecular weight, the higher the viscosity and the broader the shear-rate range for non-Newtonian behavior.

may be exaggerated by considering only low shear rate behavior, as shown by Fig. 26-4. Moreover, if there is a significant difference in molecular weight distribution between two resins, their viscosity curves can actually cross. In this case, a ranking based on melt index would be totally inappropriate as a measure of ease of flow in a mold.

The important point here is that the melt index is a convenient tool for ranking resins of the same family, but it does not provide a complete rheological characterization of a resin.

Isothermal Flow in Tubes

It will be convenient in discussing the flow of melts in molds as well as in capillary rheometers to make reference to the equations that describe isothermal flow in a tube. These tube-flow equations can be used to analyze capillary rheometer data and to describe the flow in runners.

For any liquid, the magnitude of the shear stress at the wall of a tube, σ_w, is related to the pressure gradient, $-\Delta P/L$, as follows:

$$\sigma_w = (-\Delta P/L)R/2 \qquad (26\text{--}13)$$

The minus sign on the pressure drop is inserted for convenience, to yield a positive value of σ_w, since ΔP itself ($P_{\text{out}} - P_{\text{in}}$) is a negative quantity.

The velocity profile and the wall shear rate for fully developed flow (i.e., away from the ends of the tube) depend on the relationship between the velocity and the shear rate. For example, for a Newtonian fluid the magnitude of the shear rate at the wall is given by:

$$\dot{\gamma}_w = \frac{4Q}{\pi R^3} \qquad \text{(Newtonian fluid)}$$

where Q is the volumetric flow rate and R is the radius of the tube. This equation is no longer valid in the case of fluids with a shear-rate-dependent viscosity. However, the quantity $(4Q/\pi R^3)$ is still found to be useful and is sometimes referred to as the apparent shear rate at the wall:

$$\dot{\gamma}_A \equiv \frac{4Q}{\pi R^3} \qquad (26\text{--}14)$$

For the particular case of power law behavior, it can be shown that:

$$\begin{aligned}\dot{\gamma}_w &= \left(\frac{3n+1}{4n}\right)\dot{\gamma}_A \\ &= \left(\frac{3n+1}{4n}\right)\left(\frac{4Q}{\pi R^3}\right)\end{aligned} \qquad (26\text{--}15)$$

The quantity $[(3n+1)/4n]$ tells us the error involved in using the apparent wall shear rate in place of the true value for a power law fluid. For example, when the power law index n is 0.5, the true wall shear rate is 25 percent greater than the apparent value, $\dot{\gamma}_A$; whereas when n is equal to 0.25, the true value is 75 percent greater than the apparent one.

From the definition of the viscosity:

$$\sigma_w = \eta \dot{\gamma}_w$$

Therefore, from equations (26–13) and (26–15), we have, for the isothermal flow of a power law fluid:

$$\left(\frac{-\Delta P}{L}\right)\frac{R}{2} = K\left(\frac{3n+1}{4n}\right)^n \left(\frac{4Q}{\pi R^3}\right) \qquad (26\text{--}16)$$

Viscous Heating

Whenever a viscous fluid is deformed, at least some of the work that goes into causing the deformation is dissipated, that is, converted irreversibly into heat. For an inelastic liquid and for a viscoelastic liquid undergoing steady state shearing, all the work goes into viscous dissipation. If the heat generated is not permitted to flow away (adiabatic case), the rate of increase of temperature is:

$$\frac{dT}{dt} = \frac{\sigma \dot{\gamma}}{\rho C_p} = \frac{\eta \dot{\gamma}^2}{\rho C_p} \qquad (26\text{--}17)$$

It is clear that viscous dissipation can cause significant increases in temperature when the viscosity and shear rate are large.

If the flow channel is not insulated from the environment, heat will flow out of the fluid as the temperature builds up at a rate that depends on the thermal conductivity, k, of the melt as well as that of the channel walls.

Time-Dependent Phenomena

The viscosity is defined as the ratio of the steady-state shear stress to a shear rate that is held constant in an experiment. However, since polymers are viscoelastic, this stress is not attained instantaneously when shearing is started, but is reached only over some finite length of time. This time increases with increasing molecular weight and decreases with increasing shear rate. Likewise, if shearing is suddenly halted, the shear stress does not suddenly fall to zero, as it would for an inelastic liquid, but relaxes over a period of time. As is discussed in more detail later in this chapter, this relaxation process is the mechanism by which residual stresses in the cavity are relieved between the end of the filling stage and the solidification of the polymer. However, time-dependent effects can also play a role in mold filling.

Once the viscosity of a melt has been reduced by shearing at a high strain rate, if the shear rate is reduced to a lower value, the viscosity does not move immediately to the higher value appropriate to the new low shear rate, but attains that value only over a period of time. Thus, when a melt is sheared at a high rate, in an extruder or at the gate of a mold cavity, the melt continues to be affected by this process for some time into the succeeding lower-shear-rate stage of the process. As explained above, however, the viscosity reduction at high shear rate is itself time-dependent; and if the duration of the high-shear-rate process is very short, the full reduction of the viscosity to its high-shear-rate value will not be achieved.

First Normal Stress Difference

In a fluid at rest there is no shear stress; the normal components of stress are all of equal magnitude, and this magnitude is simply the hydrostatic pressure. If we take tensile stresses to be positive and compressive stresses negative, this situation can be described as shown below:

$$\left.\begin{array}{c} \sigma = 0 \\ \sigma_{11} = \sigma_{22} = \sigma_{33} = -P \end{array}\right\} \quad \text{(fluid at rest)}$$

When a Newtonian fluid is subjected to simple shearing, a shear stress appears, but the normal stresses continue to be equal as shown above.

However, for molten polymers in steady shear, differences between the normal components of stress appear. In describing the behavior of such viscoelastic liquids it is convenient to define the first and second normal stress differences, N_1 and N_2, as follows:

$$N_1 = \sigma_{11} - \sigma_{22} \qquad \text{(26–18a)}$$
$$N_2 = \sigma_{22} - \sigma_{33} \qquad \text{(26–18b)}$$

where the σ_{11} component of the normal stress is measured in the direction of flow, and the σ_{22} component is measured in the direction of the velocity gradient, as indicated in Fig. 26-1. The first normal stress difference is significantly larger than N_2 and is more easily measured.

The appearance of normal stress differences in a sheared melt is a direct effect of the orientation of polymer molecules that occurs as a result of the shearing deformation. This orientation tends to occur in the direction of flow, and it can lead to residual tensile stresses in injection molded parts. Therefore, measurement of the buildup and relaxation of normal stresses in melts would provide valuable information about the role of residual orientation stresses in injection molded parts. However, no reliable technique is currently available for the measurement of normal stress differences during the start-up and relaxation stages of deformations involving homogeneous shearing at high strain rates.

ROLE OF RHEOLOGY IN MOLD FILLING

Melt Flow in the Injection Molding Process

The objective of the injection molding process is to produce a product that is free of voids and sink marks, is not subject to warpage, and has sufficient strength and stiffness for its end use. This requires that the melt flow freely into the mold cavity and that the final part be reasonably free of residual stresses. At the same time, the product must be produced at minimum cost, and this implies the shortest possible cycle time and minimal waste of resin and energy. The challenge facing the molder, then, is to produce a good-quality product at a minimum cost, and an understanding of melt rheology is a valuable tool in meeting this challenge.

To see more clearly the role of rheology in injection molding, it will be useful to examine the various stages of the process with special attention to shear rates and stresses. First the resin is melted and plasticated, often in a reciprocating screw extruder. This stage of the process is neither unique to injection molding nor critical in terms of product quality and cost. The injection stage, on the other hand, is crucial, and we will look at that in some detail.

When the screw or ram moves forward, it forces the molten resin through the nozzle and sprue to the runner system. Runners are normally designed to allow the melt to reach the cavity while contributing as little as possible to the overall pressure drop between the cylinder and the end of the cavity. A round runner gives the lowest pressure drop for a given flow rate, but other cross sections are often used because they facilitate waste removal and are easier to fabricate than the round design.

The injection pressure required to fill the runners is generally rather low. However, the pressure drop in the runner can be important in the case of a multicavity mold because it is highly desirable to have equal flow rates to all cavities (if the cavities are identical). One approach to this problem is simply to make all the runners of equal length and diameter. However, the use of such a naturally balanced runner system often involves long runners, with a high pressure drop and a large amount of regrind.

In designing an artificially balanced runner system, one wants to adjust the flow to each cavity so that all cavities fill at about the same time. We can get some idea of what is involved here by rearranging equation (26–16) so that it gives the flow rate as a function of the pressure drop:

$$Q = \frac{\pi R^3}{.4} \left(\frac{4n}{3n + 1} \right) \left(\frac{-\Delta P \cdot R}{2LK} \right)^{1/n} \quad (26\text{--}19)$$

For this simple case of isothermal flow of a power law fluid, the flow rate for a given pressure decreases with L and increases rather sharply with R. Of course, for the quantitative design of a runner system the temperature dependence of the viscosity and the heat transfer must also be taken into account, and computer programs are available commercially for solving this type of complex flow problem.

While the shear rate experienced by the melt in flowing from the injector system to the gate is usually around 1000 sec^{-1}, that in the gate is much higher. The gate cross section is very small to facilitate waste removal and to minimize outflow when the pressure is released. The shear rate in the gate often reaches 10^5 sec^{-1}, and this will reduce the viscosity. Furthermore, once the viscosity is reduced by shearing at high rates, the melt will continue to flow with this reduced viscosity for a time as it enters the cavity, even though the shear rate in the cavity is much lower. However, it must be kept in mind that the melt experiences this high shear rate in the gate for only a brief period of time, as gates are generally short in order to minimize pressure drop. At the present time, therefore, it is not clear how much reduction in viscosity actually occurs in the gate, or how much this influences the cavity flow near the gate.

Cavity Filling

The objective in filling the cavity is to achieve complete filling without short shots while avoiding sink marks, warpage, sticking in the mold, flash, and poor mechanical properties. This is accomplished by delivering the correct amount of resin to the cavity while avoiding overpressurization, high thermal stresses, and high residual orientation. Some of the factors that favor complete filling, however, also promote overpressurization and residual stresses, so care must be taken in selecting operating conditions for a given mold and resin.

As melt flows into the cavity, the situation cannot be described in terms of pressure flow between parallel plates with a gap equal to the mold clearance, because a frozen layer forms immediately at the cavity wall (8). Moreover, the melt in the center has a lower viscosity due to its higher temperature, and as a result, the maximum shear rate occurs not at the surface of the frozen layer but closer to the center (9). The shear rate in the cavity is generally in the range of 8,000 to 15,000 sec^{-1}.

Another important phenomenon that causes the flow to deviate from two-dimensional flow between parallel plates is termed the "fountain effect." Thus, the melt does not reach the wall or the surface of the frozen wall layer by simple forward advance but rather tends to flow down the center of the cavity to the melt front and then flow out toward the wall. This can have an important effect on the direction of the flow-induced orientation of the polymer molecules.

If the melt must flow around an obstacle of any kind in the cavity, a weld line will result. Once the melt is separated into two streams by such an obstacle as a slot or an insert, it loses its structural integrity along the line of separation, and this can only be regained by a rather slow diffusional process.

A phenomenon that can lead to a complex pattern of weld lines is "jetting." This term refers to the tendency of the melt to spurt into the cavity without wetting the walls near the gate, and the result is that the cavity fills by a piling up of the jet at the end of the cavity rather than by the smooth advancement of a melt front starting at the gate. Oda et al. (10) suggest that jetting occurs when the melt swell at the exit from the gate is insufficient to cause immediate contact between the melt and the cavity walls. The condition for jetting can thus be expressed as shown by equation (26–20):

$$\frac{BD}{h} \leqslant 1 \qquad (26\text{--}20)$$

where B is the swell ratio, D is the gate diameter, and h is the cavity thickness.

It is generally accepted that viscosity is the key rheological property in mold filling. It should be noted that elasticity does make some contribution to the pressure drop, particularly at points in the flow path where the cross section decreases or where streams divide, and the net effect of elasticity is always to increase the pressure drop above what it would be for an inelastic fluid of the same viscosity. However, viscous resistance to flow is the dominant effect. Furthermore, as can be seen from a consideration of the nature of the flow in the runners, gate, and cavity, it is, in particular, the high shear rate viscosity that is of direct importance. The viscosity at low shear rates and the melt index, therefore, are not of direct relevance.

Finally, in the packing stage, flow in the delivery system falls almost to zero, and the pressure at the gate rises to approach the injection pressure. This maintains a small flow into the cavity to compensate for thermal contraction resulting from cooling and freezing. Once the gate freezes, no more flow can occur. If the injection pressure is released before this

happens, there will be some discharge of melt back out of the cavity.

Residual stresses and orientation are present in molded parts as a result of the rapid cooling that takes place in the mold. These effects can cause warpage, delamination, and poor mechanical properties, particularly low impact strength. It is therefore desirable to keep them at a low level. Orientation effects are a direct result of molecular orientation that occurs in response to melt flow in the cavity, and are thus governed by the rheological properties of the melt. The shear stresses that occur in the cavity during filling provide a rough guide to the level of orientation that has been generated. Thus, a high viscosity or a high injection pressure will usually mean high orientation. These decay as long as the resin is still molten, but the rate of decay falls rapidly as the temperature falls. Stress relaxation is discussed further in a later section of this chapter.

Filling and Cooling Times

The principal components of the cycle time are the filling time and the cooling time. The relative importance of the two, and thus the role played by rheology as a factor in the cycle time, depends on the shape of the cavity. For a thin cavity, the filling time will be long compared to the cooling time. This is due to slow filling as a result of the large resistance to flow in the cavity and the rapid cooling associated with the thin section. In fact, as a result of the slow filling and enhanced heat transfer, much of the cooling will take place during the filling stage, and this makes it particularly difficult to avoid a short shot. In addition, the rapid cooling will result in high residual stresses, and thus in a low heat distortion temperature. This will require that the part be cooled to a lower temperature than usual before the mold is opened.

In a thick mold, on the other hand, the filling process makes a minor contribution to the cycle time. Little cooling occurs during filling; and even, complete filling and avoidance of overpressurization can be accomplished entirely by means of viscosity control.

The key factors governing whether a cavity is to be considered thin or thick are the wall thickness of the molded part and the distance from the gate to the end of the cavity (the flow length). Clearly, when the wall is thin and the flow length is long, it is particularly desirable to use a resin with a low viscosity at high shear rates.

Effects of Temperature and Pressure

Once a mold and a resin have been selected, the molder still has some flexibility in the selection of operating conditions that can help him to optimize his process, the key variables being temperature and pressure. The mold temperature is usually between 125°F and 300°F. It must be lower than the softening point of the resin, but if it is too low, high thermal stresses can cause poor part appearance and performance. The situation with regard to melt temperature is more complex, as this temperature has a strong effect on both the rheological properties of the melt and the thermal phenomena. For example, one can increase the melt temperature to eliminate a problem with short shots. This is effective because it decreases the viscosity, the extent of the decrease depending on the value of E in equation (26–9), and increases the flow time available before the gate freezes. Thus, for a given injection pressure, the flow rate will be higher. The higher temperature also leads to faster relaxation of orientation and a longer time available for relaxation. On the other hand, an increase in melt temperature lengthens the cycle time, increases energy costs somewhat, and can lead to sticking in the mold.

Increasing the pressure is another way to achieve faster flow into the mold. Moreover, as is illustrated by equation (26–19), because the melt is shear thinning the flow goes up with the $1/n$ power of the pressure. Since n is less than one for most melts (never greater than one), the flow rate goes up at a higher than proportional rate with pressure. Some of this gain, of course, is lost because of the increase of viscosity with pressure, but the net effect will usually be at least a proportional increase of flow rate, especially in the initial stages of the filling process. As in the case of temperature increases, there are limitations on injection pressures. Increasing this pressure means a higher clamp force and higher energy consumption. It can also cause sticking, flash, and high residual stresses, especially near the gate.

The molding of thin parts is a special challenge because the viscous resistance to flow is large, and solidification occurs quickly. Obviously an especially low viscosity is essential in this case.

Resin Selection

It is possible to decrease the viscosity and increase the flow rate to some extent by increasing the temperature and/or the pressure. However, we have seen that there are limits on how far one can go in compensating for a resin with a relatively high viscosity. This may mean that the cavity has to be designed to produce a part that is thicker than required by strength or stiffness considerations in order to avoid molding problems. Particularly in the case of parts with thin sections, there is thus a strong incentive to use low-viscosity resins, but this may involve a sacrifice in terms of mechanical properties.

For a given polymer, a wide variety of resins are available that differ in molecular weight, molecular weight distribution, copolymer content and distribution, extent of long-chain branching, and additives. All of these factors can influence flow properties and part performance. Polyethylene, for example, is available in grades having melt indexes (M.I.) varying from 0.5 to 100 g/10 min. Even if we restrict our attention to injection molding grades, the M.I. range is from 30 to 100. Of course, melt index doesn't tell the whole story, as it is possible to have two resins with M.I. values as different as 60 and 100 that still have similar viscosities at 1000 sec^{-1}. Still, the range of M.I. values does indicate that a great variety of resins are available.

Decreasing the molecular weight reduces the viscosity but may also have an adverse effect on the mechanical properties of the molded part. Broadening the molecular weight distribution causes the melt viscosity to be more shear-rate-sensitive, and this will usually mean a lower viscosity at high shear rates. However, a narrow molecular weight distribution usually yields greater toughness in the molded part.

It is obviously necessary to select a resin that provides acceptable strength, stiffness, and appearance in the part and that can be molded easily and reliably. It is equally obvious that this selection may be a difficult one because of the need to balance moldability with good part strength.

Likewise, after a resin has been selected that provides both adequate ease of molding and acceptable part quality, it is important to avoid large day-to-day variations in resin properties. Some batch-to-batch variations in resin properties are inevitable because of the complexity of polymerization reactions, but if these variations are sufficiently large, they can cause molders a lot of headaches.

It is therefore desirable both in the selection of a resin for a given application and in resin quality control to make use of a rheological measurement. A later section describes measurement techniques that have proved useful in the injection molding industry.

Relative Importance of Rheological and Thermal Properties

Both rheological and thermal properties of melts are important in the injection molding process, and the specific role played by each depends on the details of the mold design. However, it is more useful in resin selection and quality control to measure rheological rather than thermal properties. The reason for this is that rheological properties are much more sensitive to molecular weight and molecular weight distribution than are the thermal properties. Moreover, the thermal properties are not so strongly dependent on temperature and shear rate as rheological properties. For these reasons, one can rely on standard hand-book values for the thermal properties for many purposes, while this is a highly unreliable approach in the case of rheological properties.

MEASUREMENT OF RHEOLOGICAL PROPERTIES RELEVANT TO MOLD FILLING

A complete treatment of experimental methods for the rheological characterization of molten plastics has been published in an SPE book (11). Here, attention will be focused on those techniques that have been found useful in the evaluation of injection molding resins.

Moldability Tests vs. Viscosity Measurement

As was mentioned in the previous section, the resistance to flow of a melt is of central importance in mold filling, particularly flow at high shear rates. There are two general approaches to the measurement of the moldability of a resin. One approach involves simultaneous flow and cooling in a simple mold geometry. In this case, one simulates, in some sense, the actual mold filling process.

The advantage of such empirical tests is that they are simple to perform and provide direct evidence of the ability of the resin to fill a mold. The disadvantage is that the shear rate, temperature, and pressure all vary during the test in an uncontrolled way and the resulting moldability index is a complex function of several rheological and thermal properties. As a result, there is no straightforward way to scale up the results so that they are quantitatively relevant to the filling of a specific mold. In going from the test mold to the production mold, for example, the relative roles played by the rheological and thermal properties can be altered. Or, the role played by various regions of the viscosity–shear rate curve may be changed. As a result, one may obtain a ranking of several resins that is not a correct indication of the relative ease with which they can fill a given production mold. The empirical moldability tests are therefore most useful when used to compare several resins of the same family, for example, several linear polyethylenes having similar molecular weight distributions but different average molecular weights.

The second approach to the problem of characterizing molding resins is to measure the viscosity, or at least to measure the isothermal resistance to flow in a simple standard flow geometry. In this case, the results depend only on rheological properties at a given temperature and are independent of thermal properties. As we have seen, it is the high-shear-rate viscosity that is of most interest in connection with mold filling, and capillary rheometers are the instruments almost universally used to measure this property. The following sections, therefore, describe, in turn, moldability tests and capillary viscometers.

Moldability Tests

One of the earliest devices for evaluating molding resins was the Rossi-Peakes flow tester described in ASTM standard test method D569–59, "Measuring the Flow Properties of Thermoplastic Molding Compounds." The Rossi-Peakes test is carried out as follows: A premolded, cylindrical specimen is placed at room temperature in a "charge chamber" (barrel) at the bottom of a steam-heated block that also contains a capillary with a diameter of 0.125 in. and a length of 1.5 in. A piston is immediately raised into position and presses against the polymer, causing it to flow upward into the capillary. A specified upward driving pressure is supplied by means of weights suspended from pulleys. The rate of rise of polymer in the capillary is monitored by means of a follower rod that rests on the surface of the melt. The quantity reported is the temperature at which the flow in the first 2 minutes is 1 in.

A commercial version of the Rossi-Peakes tester, the Bakelite Flow Tester, is available from Tinius Olsen. There are two models, both equipped with recorder and bench. One of these uses 150-psi steam for heating, and the other is heated electrically. Similar to the Rossi-Peakes device is the Koka Flow Tester made by Shimadzu in Japan. This instrument is somewhat more sophisticated than the Rossi-Peakes, but is designed for the same sort of moldability study in which the melting occurs under load.

A moldability test widely used by molders is to inject the resin into a standard mold having a simple geometry involving a long flow path. The moldability index in this case is simply the length of mold filled before freeze-up under standard filling conditions. Discs, spirals, snakes, and bar molds have all beeen used in this way. In a disc mold the flow passage is long and narrow, and the moldability index is the length, measured along the flow path, of the filled section of the mold.

The spiral mold is perhaps the most popular of these test molds. While ASTM standard test method D3123–72 describes a spiral flow mold for use with thermosetting molding compounds, there appears to be no universal standard for themoplastics. One commercially available spiral flow mold is shown in Fig. 26-5, but a variety of other designs are also to be found in molding shops. Since there is no universally accepted mold design or set

Fig. 26-5 A commercially available spiral test mold manufactured by MUD (see chapter appendix for address). The mold is constructed with a 65-in. flow distance and is engraved in 1-in. increments. The mold can be constructed for center or parting line injection.

of molding conditions, the moldability index that is determined only has meaning within a given company. Furthermore, as pointed out above, this method of rating resins is most useful in comparing resins of the same family or as a quality control tool. In comparing different batches of the same resin or of similar resins from different manufacturers, spiral flow tests have been found to correlate with the viscosity at high shear rates (12).

Another approach to evaluating the processability of a molding resin is to measure the pressure required to fill a standard mold. To facilitate this type of test and to carry out a wide range of related tests, a small, highly instrumented injection molding machine is useful. A commercial apparatus of this type is shown in Fig. 26-6. When mold filling tests are used to evaluate resins, it is sometimes desirable to mount pressure transducers directly in the wall of the cavity. A transducer especially designed for this application has been developed by Dynisco.

Torque Rheometers

A popular type of testing equipment in the plastics industry is the torque rheometer, which consists of a horizontally mounted, heavy duty motor drive together with a torque sensor. A mixing head is coupled to the drive shaft; this consists of a steel-walled cavity in which the melt is sheared by a set of rotors or blades. A variety of rotor shapes are available, including sigma, roller, cam, and Banbury types. The flow in the cavity is neither uniform nor controllable, the details of the flow depending on the rheological properties of the melt being sheared. Also, because there is alternate shear and relaxation as the rotor turns, elasticity is likely to play some role in the material response; and temperature nonuniformities can be significant, since the cavity is not thin. However, the torque rheometer does measure the resistance to flow of the melt, usually at shear rates in the range of 30 to 300 sec^{-1}.

A small extruder can also be coupled to the drive shaft, and by using an appropriate die and measuring the pressure drop and flow rate, a torque rheometer can be used as a capillary rheometer.

Torque rheometers are available from Brabender, Haake-Buchler, Hampden, and Toyoseiki. The addresses of these companies are given in the appendix to this chapter.

Melt Indexers

The extrusion plastometer, or melt indexer, is the most widely used rheological test device

Fig. 26-6 A highly instrumented laboratory injection molding machine, the "Injectometer," made by Göttfert.

in the plastics industry. It is not a true viscometer, in the sense that a reliable value of the viscosity cannot be calculated from the flow index that is normally measured. On the other hand, it does measure isothermal resistance to flow using an apparatus and test method that are standard throughout the Western world. Standard test methods based on the use of the extrusion plastometer include the following:

United States	ASTM D1238
United Kingdom	BS 2782–105C
West Germany	DIN 53735
France	AFNOR T51–016
Netherlands	NFT 51–006, 51–061
Italy	UNI 5640–74
Japan	JIS K7210
International	ISO R1133, R292

The essential features of the melt indexer are shown in Fig. 26-7. The polymer (6) is contained in a barrel equipped with a thermometer (3) and surrounded by an electrical heater (4) and an insulating jacket (5). A weight (1) drives a plunger (2), which forces the melt through the die (7). The standard die has a diameter of 2.095 millimeters and a length of 8 millimeters.

The standard procedure involves the determination of the amount of polymer extruded in 10 minutes. The flow rate, expressed as g/10 min (equivalent to dg/min), is the result reported. Obviously, this flow rate increases as the viscosity decreases. In the ASTM test method (D1238), and in some of the others listed above, two methods are described for measuring the amount of polymer extruded. In Procedure A, the extrudate is allowed to accumulate at the exit of the die and is simply cut off and weighed at the end of 10 minutes. For materials with large flow rates (i.e., those with low viscosities) or when automation is desired, Procedure B provides for the use of a switch and timer that automatically determine the time required for the piston to fall through a specified distance. The standard flow rate is then calculated as follows:

$$\text{Flow rate (g/10 min or dg/min)} = KL\rho/t$$
$$(26\text{–}21)$$

where K = a constant, depending on the effective barrel diameter; L = distance through which piston falls in time t; and ρ = density of the melt. For the standard instrument dimensions and if L is expressed in centimeters and t in seconds, K is equal to 427.

Although there is only one standard geometry, no fewer than 20 combinations of temperature and load are specified to accommodate a wide variety of materials. Condition E, often used for polyethylene, involves a temperature of 190°C and a load of 2.160 kg, from which the nominal driving pressure can be calculated to be 298.2 kPa. Of course, the actual driving pressure will differ somewhat from this value because of piston friction and the weight of the polymer in the barrel. ASTM Designation D2473, "Standard Specification for Polycarbonate Plastic Molding. Extrusion and Casting Materials," includes a flow test that is to be carried out according to D1238, this time using condition O. When condition

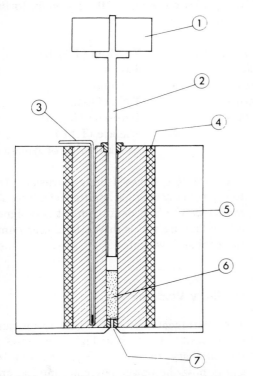

Fig. 26-7   Essential elements of the extrusion plastometer described in ASTM D1238. Reproduced from J. L. Leblanc (21) with permission of Editions CEBEDOC.

E is used to characterize polyethylene, the flow rate, in g/10 min, is commonly called the melt index, and devices designed according to ASTM D1238 are often called melt indexers.

Because of their simplicity and relatively low cost, testers designed in accordance with ASTM D1238 are widely used for quality control and for distinguishing between members of a single family of polymers. However, without extensive modification, these testers cannot give reliable values of the viscosity, and the ASTM standard contains this statement: "The flow rate obtained with the extrusion plastometer is not a fundamental polymer property. It is an empirically defined parameter critically influenced by the physical properties and molecular structure of the polymer and the conditions of measurement." The reference to conditions of measurement is reinforced by a warning: "Relatively minor changes in the design and arrangements of the component parts have been shown to cause differences in results between laboratories."

The wall shear rate in an extrusion plastometer varies from one material to another, since it is the driving force rather than the flow

flow at a significantly higher shear rate. He then defines a stress exponent, S.E., as follows:

$$S.E. \equiv \frac{1}{0.477} \log \left[\frac{\text{wt. extruded with 6.480 kg}}{\text{wt. extruded with 2.160 kg}} \right]$$

$$(26\text{--}22)$$

It is not difficult to demonstrate that if the flow in the capillary of a melt indexer obeys equation (26–16) and if the load is equal to the pressure drop, $-\Delta P$, then the stress exponent is the reciprocal of the power law index. However, these criteria are generally not satisfied for melt index flow; so the stress exponent must be looked upon as an empirical measure of the variation in the resistance to flow with flow rate.

A list of companies manufacturing extrusion plastometers is shown below. In the case of overseas manufacturers, the name of the U.S. sales agent or the country of origin is noted in parentheses. The addresses of these companies are listed in the appendix to this chapter.

CEAST (CEAST of America)
Custom Scientific
Davenport (T.M.I.)
Karl Frank (Germany)
Göttfert (Automatik)
Kayeness
Monsanto

Ray-Ran (T.M.I.)
Seiscor
Slocumb
Tinius Olsen
Toyoseiki (Japan)
Wallace (T.M.I.)
Zwick (Zwick of America)

rate that is fixed. However, for many common molding resins, the shear rate is in the range of 10 to 20 sec^{-1} when condition E is used. Clearly, the melt index is a single-point test that provides information on resistance to flow only at a single shear rate. Thus, since variations in branching or molecular weight distribution can alter the shape of the viscosity curve, the melt index may give a false ranking of resins in terms of their high-shear-rate resistance to flow. To overcome this problem, extrusion rates are sometimes measured for two loads. For example, Elston (13) suggests the use of two loads, 2.160 kg, the standard condition E load, and 6.480 kg, which produces

Figure 26-8 shows a basic extrusion plastometer. Many optional additional features are available. The Seiscor instrument is designed for on-line use and is driven by a gear pump rather than a weight.

Capillary Viscometers

The instrument used universally to measure the viscosity of a melt at high shear rates is the capillary rheometer. Equation (26–16) shows how the power law constants, K and n, can be determined by measuring the pressure drop $(-\Delta P)$ for various flow rates (Q):

$$\frac{(-\Delta P)R}{2L} = K \left(\frac{3n+1}{4n}\right)^n \left(\frac{4Q}{\pi R^3}\right)^n \quad (26\text{--}16)$$

Thus, if we make a plot of:

$$\log\left[\frac{(-\Delta P)R}{2L}\right] \text{ vs. } \log\left(\frac{4Q}{\pi R^3}\right)$$

we will obtain a straight line, the slope and intercept of which can be used to determine values for K and n.

If an equation relating the viscosity to the shear rate, such as the power law, is not available, then we cannot derive expressions for the velocity profile and the shear rate at the wall in terms of material constants. However, there is a way to calculate values for the viscosity, provided that a large number of data are available over a range of experimental variables. In particular, it can be shown that:

$$\dot{\gamma}_w = \frac{4Q}{\pi R^3}\left(\frac{3+b}{4}\right) \quad (26\text{--}23a)$$

where:

$$b \equiv \frac{d\ln\left(\dfrac{4Q}{\pi R^3}\right)}{d\ln\left(\dfrac{-\Delta P \cdot R}{2L}\right)} = \frac{d\log\left(\dfrac{4Q}{\pi R^3}\right)}{d\log\left(\dfrac{-\Delta P \cdot R}{2L}\right)}$$

$$(26\text{--}23b)$$

This equation has been variously attributed to Weissenberg, Rabinowitsch, and Mooney, while the bracketed term involving b is usually called the Rabinowitsch correction. By comparison with equation (26–14) we see that this term represents the deviation from Newtonian behavior. For a power law fluid, it is obvious from equation (26–15) that $b = 1/n$.

Once the shear rate is known, the viscosity can be calculated using equation (26–13) and the definition of the viscosity:

$$\eta = \sigma_w / \dot{\gamma}_w$$

End Corrections. The pressure drop $(-\Delta P)$ can be determined by measuring the wall pressure at two axial positions, both in

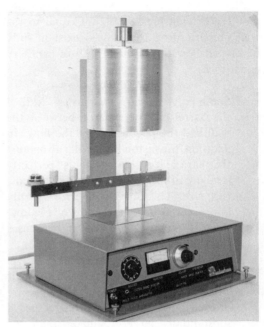

Fig. 26-8   Basic melt index apparatus made by Custom Scientific. Courtesy of manufacturer.

the fully developed flow region. However, a more common procedure is to measure the driving pressure, P_d, in the barrel and to assume that the pressure at the outlet of the capillary is equal to the ambient pressure, P_a. If the pressure is monitored by means of a load cell that measures the load on a driving piston (plunger), then the measured quantity is the driving force, F_d, and this is related to the driving pressure as follows, if friction between the plunger and the barrel is neglected:

$$P_d = \frac{F_d}{\pi R_b^2} \quad (26\text{--}24)$$

Thus, the pressure drop $(-\Delta P)$ in equation (26–16) is replaced by the pressure difference $(P_d - P_a)$ or, since for melts P_d is nearly always much larger than P_a, the pressure drop is simply replaced by P_d. However, this is clearly not the wall pressure drop that one would observe for fully developed flow in a length of capillary equal to L. It is, therefore, necessary to make some kind of end correction when using P_d to calculate σ_w.

First, let us examine the various reasons why the driving pressure is not equal to the wall pressure drop that would exist in fully developed flow through a length of capillary equal to L:

1. Some polymer may stick to the wall of the barrel and be sheared between the wall and the piston driving the flow. In addition, the piston itself will rub against the barrel wall unless it is perfectly straight, properly aligned, and of the correct size. This source of error is only important at low shear rates or for low-viscosity materials, because in these cases the wall shear stress is small.

2. The flow of polymer through the barrel will have associated with it a wall shear stress and a corresponding pressure drop. However, since the barrel has a substantially larger diameter than the capillary, this pressure drop is usually small compared to that resulting from the wall shear stress in the capillary.

3. In the neighborhood of the entrance, the wall shear stress is larger than it would be for fully developed flow, and this gives rise to a wall pressure gradient larger than that corresponding to fully developed flow. The excess pressure drop resulting from velocity rearrangements at the entrance is called the entrance pressure drop, while the length of capillary required for the velocity profile to approach its fully developed form is called the entrance length. The entrance pressure drop can make a substantial contribution to the driving pressure, especially in the case of short capillaries (i.e., when L/D is less than 20). A similar end effect can be present near the exit, where the velocity profile begins to change in anticipation of the end of the capillary. The type of wall pressure distribution actually observed for capillary flow of molten polymers is sketched in Fig. 26-9.

The phenomena described under 1 and 2 above are usually neglected in the treatment of capillary flow data. Item 3 involves the ef-

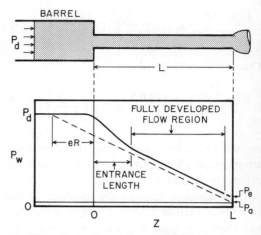

Fig. 26-9   Wall pressure distribution for flow in a capillary rheometer.

fects of velocity redistribution at the ends of the capillary, and the usual way of accounting for them is to use the Bagley end correction, e, defined by equation (26–25):

$$\sigma_w = \frac{P_d}{2(L/R + e)} \qquad (26\text{–}25)$$

The product of e and R is the length of fully developed capillary flow having a pressure drop equal to the excess pressure drop resulting from end effects. Its significance is illustrated in Fig. 26-9. (The product, eR, is not to be confused with the entrance length.) A variety of capillaries having different L/R values are used, and P_d is plotted as a function of L/R with the apparent wall shear rate as a parameter, as shown in Fig. 26-10. If straight lines are obtained, the end correction is independent of L/R, and this implies that all the

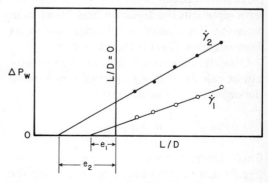

Fig. 26-10   Bagley plots for two apparent shear rates.

capillaries used were sufficiently long that fully developed flow obtained over some length of the capillary. The end correction appropriate for each apparent shear rate is then determined by extrapolating these lines to zero driving pressure and noting the intercept on the horizontal axis.

For sufficiently short capillaries, some curvature will be present, and the extrapolation should be made from the straight portion of the curve. On the other hand, curvature of the lines on a Bagley plot at large values of L/R, especially at high shear rates, may indicate a viscous heating problem or can result from the pressure dependence of viscosity.

Once the end correction has been found for each apparent wall shear rate, the actual shear rate at the wall can be found by use of equation (26–23).

For quality control applications, where high absolute precision is not required, the end corrections are often neglected. When L/D is greater than 20, the correction is usually quite small in any event.

To further simplify data analysis, the raw Q vs. ΔP_d data are sometimes used to calculate an apparent viscosity by using the apparent rather than the actual wall shear rate. From equations (26–14) and 26–25), we can show that this quantity is given by:

$$\eta_A = \frac{\pi P_d R^4}{8LQ} \qquad (26\text{–}26)$$

This quantity is not a true rheological property, but it can be useful in comparing resins that have the same power law index when end effects are not large.

Flow Instabilities. The equations given above for calculating shear rate and viscosity from capillary flow data are based on the assumptions that the flow is steady and that the velocity is a smooth function of radius, going to zero at the wall (the "no slip" condition). However, under certain conditions, the flow of melt in a capillary can become unsteady, leading to a severe distortion in the shape of the extrudate. This tends to occur at high values of shear stress, so that increases

in shear rate or viscosity or a decrease in temperature will increase the likelihood of its occurrence. This phenomenon, sometimes called melt fracture, has been studied extensively because it poses a severe restriction not only on the use of capillary rheometers but on certain industrial extrusion processes as well (1).

Another phenomenon that is sometimes observed in capillary flow of molten polymers, especially high-density polyethylene, is a sudden increase of flow rate when the driving pressure reaches a certain level. This seems to be related to a slip of polymer at or near the wall, resulting from a loss of adhesion between the polymer and the wall or a loss of cohesion along a cylindrical surface within the polymer but very near the wall. An unstable situation can develop in which the polymer flow rate alternates between higher and lower values; this has been referred to as the stick–slip phenomenon. It results in a marked irregularity in the shape of the extrudate. It has been sometimes observed that at flow rates somewhat above those at which the stick–slip phenomenon occurs, a smooth extrudate is again produced. However, it is likely that this results from the permanent detachment of the melt from the wall.

It is unsteady flow at the inlet to the capillary that is thought to be the ultimate origin of most, if not all, extrudate distortion. Its occurrence can usually be deferred by the use of a very smoothly shaped transition from the barrel to the capillary in place of the simpler "flat" entry usually used.

When extrudate distortion or slip flow is observed, capillary flow data cannot be used to calculate reliable values for rheological properties. However, the conditions under which the distortion occurs should be recorded, as this information may be of practical importance.

Temperature and Pressure Effects; Degradation. Viscous heating will increase the temperature of the melt. As was explained earlier in this chapter, the extent of the resulting temperature rise increases with viscosity and shear rate and decreases with thermal conductivity. This temperature rise results in a deviation between the set-point temperature of the thermal control system and the actual melt temperature, and thus an error in the calculated viscosity. Very long capillaries will allow more time for the temperature to build up and should be avoided. If the Bagley plot is linear, viscous heating is not a problem.

The pressure also affects viscosity, and for high flow rates in long capillaries, the driving pressure can be quite high. However, under normal circumstances pressure effects are not a serious source of error.

If the polymer being studied is subject to rapid degradation in the presence of oxygen or moisture, special precautions may be necessary to minimize the exposure of the melt to these gases. For polycarbonate resin that is to be used in a melt flow test, ASTM D2473 specifies that "the material shall be dried 16 h at 125°C in a circulating air oven immediately prior to testing." In order to facilitate their studies of PET, another moisture-sensitive polymer, Wissbrun and Zahorchak (14) developed a technique for blanketing the melt with an inert gas.

Commercial Capillary Rheometers.

Commercial capillary rheometers include the following types, classified according to the method in which the driving force is applied:

1. Gas pressure
2. Electromechanical
3. Hydraulic
4. Gear pump
5. Extruder

The essential elements of a capillary rheometer designed to be used in conjunction with a standard mechanical testing machine are shown in Fig. 26-11. The assembly is mounted on a support (8) that fits between the columns (10) of the testing machine. It consists of a barrel (4), electrical heaters (6), and thermocouples (5). The melt (2) is forced through the die (1) by a piston (3) connected to the load cell (9). Capillaries (dies) can be easily changed so that a wide range of L/D values can be employed, facilitating the determination of the end correction and allowing varia-

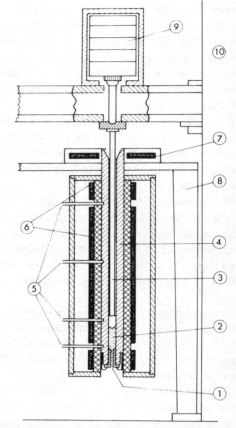

Fig. 26-11 Essential elements of a capillary rheometer designed for use with a standard mechanical testing machine. Reproduced from J. L. Leblanc (21) with permission of Editions CEBEDOC.

tions in wall shear rate between 1 and 12,000 sec^{-1}.

Table 26-1 lists manufacturers of capillary rheometers. The address of each company is given in the appendix to this chapter. In the case of overseas manufacturers, the U.S. sales agent is also noted. When there is no U.S. sales agent, the country of origin is noted. In general, the gas pressure instruments are the least expensive, while the servohydraulic units are most costly, reflecting their greater versatility. In recent years the trend has been toward automation, with preprogramming of a sequence of experiments and immediate data analysis and plotting of results (15).

The gear pump units are designed to monitor the viscosity of a process stream on a continuous basis. Melt under pressure is taken as a side stream from the flow line, usually

Table 26-1 Manufacturers of Capillary Rheometers for Molten Plastics.

MANUFACTURER* (COUNTRY OF ORIGIN) (U.S. AGENT)	DRIVE METHOD
Brabender	Extruder
CEAST (Italy)	Electromechanical
(CEAST of America)	
Davenport (U.K.)	Electromechanical
(T.M.I.)	Gas Pressure
Karl Frank (Germany)	Servohydraulic
Göttfert (Germany	Electromechanical
(Automatik)	Servohydraulic
	Gear Pump
	Extruder
Haake-Buchler	Extruder
Instron	Electromechanical
Killion	Extruder
MTS	Servohydraulic
Monsanto	Gas pressure
Porpoise (U.K.)	Gear pump
Seiscor	Gear pump
Tinius Olsen	Gas pressure
(Imass, Inc.)	
Toyoseiki (Japan)	Electromechanical

* See appendix to this chapter for addresses.

near the exit of an extruder, conditioned to the desired temperature and fed to a metering pump. The metered flow then passes to a capillary, where the pressure drop is measured by means of pressure transducers.

ROLE OF MELT ELASTICITY IN INJECTION MOLDING

It has already been mentioned that elasticity makes a modest contribution to mold filling pressure and causes the melt to have time-dependent rheological properties. However, the most important role played by melt elasticity in injection molding is in the relaxation of orientation-related stresses prior to solidification.

Residual Stresses

Residual stresses are of two types: thermal stresses and orientation stresses. Thermal stresses result from the fact that solidification takes place in the presence of a steep temperature gradient and from the temperature dependence of density. Orientation stresses, on the other hand, are a direct result of melt deformation that occurs during the mold filling process. This deformation has both shear and extensional components, although the shear stresses are thought to be the major contributor to residual orientation stresses (16). The relative importance of thermal stresses and orientation depends on the shape of the cavity, but this subject is not well understood at the present time, and thermal stresses are often ignored in modeling the injection molding process. In any event, we will be concerned here only with orientation.

As the melt flows into the cavity, it is undergoing shear at high rates, and, to a lesser extent, extension. Since the velocity gradient is predominantly in the direction normal to the cavity wall, the molecules tend to be oriented in planes parallel to the wall. The direction of orientation in this plane is governed by the direction of flow in the cavity, which can be rather complex. The level of orientation is indicated by the magnitude of the shear stresses (17), and it has been suggested (18) that the power law viscosity model is adequate for the prediction of the shear stresses developed during filling. The relaxation of these stresses is associated with the return of the molecules to random coils and thus with the disappearance of orientation. The rate of relaxation of stresses decreases with temperature, and is nearly zero below the melting point.

It is the stress relaxation process, then, that governs the extent to which orientation remains in the molded object. Near the surface, where cooling occurs most rapidly, we would expect to find the highest levels of orientation, and this has been demonstrated by measuring birefringence in thin sections cut from molded plates (16).

Clearly, a quantitative description of residual orientation requires an understanding of stress relaxation in molten polymers.

Viscoelastic Properties of Molten Plastics

When the deformation is very small or very slow, the mechanical behavior of polymeric liquids can be described in terms of the theory of linear viscoelasticity. According to this theory, the response of the material to any type of deformation can be predicted from knowledge of a single material function, for example, the relaxation modulus, $G(t)$. Methods of determining this function and of using it to calculate the response of the material to many types of deformation have been described by Ferry (19).

The method usually used to determine the linear viscoelastic properties of a melt is to subject it to oscillatory shear between a disc and a cone, or between two discs, in a rotational rheometer (11). The major manufacturers of rotational rheometers suitable for use with molten plastics are Instron, Rheometrics, and Sangamo (see chapter appendix for addresses).

However, the deformations that occur in mold filling are neither small nor slow. Therefore, the theory of linear viscoelasticity is not very helpful in modeling mold flow. There is at present no unified theory of nonlinear viscoelasticity. Many measurable material functions have been defined (20), and each is independent of the other. The nonlinear material function most relevant to the relaxation of orientation stresses in mold cavities is the normal stress decay function, $N^-(t, \dot{\gamma})$, defined as the time-dependent first normal stress difference in an experiment in which the melt is sheared at a steady shear rate until the stresses are all constant and the motion is suddenly stopped at time $t = 0$. For the special case of very low shear rates, this function becomes independent of shear rate and can be calculated from the relaxation modulus, as shown in equation (26–27):

$$\eta^-(t) = \int_t^\infty G(s)\,ds \qquad (26–27)$$

However, when the shear rate is high, as in mold filling, the linear theory is no longer valid, and the decay function must be measured independently at the shear rate of interest. It is known that the shear stress decays more rapidly than N_1, and that both functions decay more rapidly when the temperature or the shear rate is increased. However, our ability to measure the viscoelastic response of melts to deformations involving high strain rates is at present quite limited (11). Efforts are currently being made to overcome these limitations by developing techniques for generating uniform deformations involving high strain rates (15).

REFERENCES

1. Han, C. D., *Rheology in Polymer Processing*, Academic Press, New York, 1976.
2. Carreau, P. J., Ph.D. thesis, Chem. Eng., Univ. of Wisconsin, 1969.
3. Elbirli B., and Shaw, M. T., *J. Rheol.*, *22*, 561 (1978).
4. Han, C. D., *Multiphase Flow in Polymer Processing*, Academic Press, New York, 1981.
5. Schmidt, L. R., *Polym. Eng. Sci.*, *17*, 666 (1977).
6. Han, C. D., Van den Weghe, T., Shete, P., and Haw, J. R., *Polym. Eng. Sci.*, *21*, 196 (1981).
7. Cogswell, F. N., *Plastics and Polymers*, 39 (Feb. 1973).
8. Janeschitz-Kriegl, H., *Rheol. Acta*, *16*, 327 (1977).
9. Van Vijngaarden, H., Dijksman, J. F., and Wesseling, P., *J. Non-Newt. Fl. Mech.*, *11*, 175 (1982).
10. Oda, K., White, J. L., and Clark, E. S., *Polym. Eng. Sci.*, *16*, 585 (1976).
11. Dealy, J. M., *Rheometers for Molten Plastics, A Practical Guide to Testing and Property Measurements*, Sponsored by SPE, Van Nostrand Reinhold Co., New York, 1982.
12. Rubin, I. I., *Injection Molding, Theory and Practice*, John Wiley & Sons, New York, 1972.
13. Elston, C. T., U.S. Patent 3,645,992, assigned to DuPont Canada Ltd. (1972).
14. Wissbrun, K. F., and Zahorchak, A. C., *J. Polym. Sci., A-1*, *9*, 2093 (1971).
15. Dealy, J. M., *Plastics Eng.*, 57 (March 1983).
16. Kamal, M. R., and Tan, V., *Polym. Eng. Sci.*, *19*, 558 (1979).
17. Wales, J. L. S., *Rheol. Acta*, *8*, 38 (1969).
18. Isayev, A. I., and Hieber, C. A., *Rheol. Acta*, *19*, 168 (1980).
19. Ferry, J. D., *Viscoelastic Properties of Polymers*, 3rd ed., John Wiley & Sons, New York, 1980.
20. Dealy, J. M., *J. Rheol.*, *28*, 181 (1984).
21. Leblanc, J. L., *Rhéologie Expérimentale des Polymères à l'Etat Fondu*, Editions CEBEDOC, 2 rue A. Stévart, Liège, Belgium, 1974.

NOMENCLATURE

A = area of shearing plate in equation (26–3) or constant in equation (26–9).

b = constant in equation (26–10) or slope defined by equation (26–23b)

B = swell ratio; diameter of swollen extrudate divided by diameter of gate.

D = diameter of gate

e = 2.303 or end correction

E = activation energy for viscosity

F = force on plate

F_d = driving force for capillary flow

h = spacing between plates in simple shear or between cavity walls

k = thermal conductivity of melt

K = constant in power law equation

L = length of capillary

n = power law index

N_c = isoviscosity coefficient (equation 26–12)

N_1 = first normal stress difference

p = constant in equation (26–7)

P = pressure

P_d = driving pressure in capillary rheometer

Q = volumetric flow rate

R = gas constant (equations 26–9 and 26–11) or radius of tube or capillary

R_b = radius of barrel feeding capillary rheometer

T = temperature

T_0 = reference temperature in equation (26–10)

v_i = velocity in x_i direction

V = velocity of plate in equations (26–1) and (26–2)

x_i = spatial coordinate (i = 1, 2, or 3)

$\dot{\gamma}$ = shear rate

$\dot{\gamma}_A$ = apparent shear rate in capillary

$\dot{\gamma}_w$ = shear rate at the wall of a capillary

ΔP = final pressure minus initial pressure

η = viscosity

η_0 = zero shear viscosity

λ = characteristic time in equation (26–7)

σ = shear stress

σ_w = shear stress at the wall of a capillary

σ_{ii} = normal stress in the x_i direction

APPENDIX: ADDRESSES OF MANUFACTURERS AND THEIR AGENTS

Automatik (U.S. agent for Göttfert)
Automatik Machinery Corporation
9724-A Southern Pine Blvd.
Charlotte, NC 28210

Brabender
C. W. Brabender Instruments, Inc.
P.O. Box 2127
So. Hackensack, NJ 07607

CEAST
CEAST S.p.A.
Via Asinari di Bernezzo, 70

10146 Torino
Italy

CEAST of America
Division of Testing Equipment of America
48 Saybrook Road
P.O. Box 466
Essex, CT 06426

CUSTOM SCIENTIFIC
Custom Scientific Instruments, Inc.
P.O. Box A
Whippany, NJ 07981

DAVENPORT (see also T.M.I.)
Davenport (London), Limited
Tewin Road
Welwyn Garden City
Herts AL7 1AQ
U.K.

DYNISCO
Dynisco
Ten Oceana Way
Norwood, MA 02062

KARL FRANK
Karl Frank GmbH
Postfach 1320
D-6940 Weinheim
Fed. Republic of Germany

Göttfert (see also Automatik)
Göttfert Werkstoff-prüfmaschinen GmbH
Postfach 1220
6967 Buchen
Fed. Republic of Germany

Haake-Buchler
Haake Buchler Instruments Inc.
P.O. Box 549
Saddle Brook, NJ 07662

Hampden
Hampden Test Equipment, Ltd.
Rothersthorpe Avenue
Northampton, NN4 9JH
U.K.

Imass (Sieglaff-McKelvey capillary rheometer)
Imass, Inc.
P.O. Box 134
Accord (Hingham), MA 02018

Instron
Instron Corporation
100 Royall Street
Canton, MA 02021

Iwamoto
Iwamoto Seisakusho Co., Ltd.
354 Hazukashi Furukawacho
Kyoto, Japan

Kayeness
Kayeness, Inc.
RD 3, Box 30
Honeybrook, PA 19344

Killion
Killion Extruders Inc.
56 Depot Street
Verona, NJ 07044

MTS
MTS Systems Corporation
P.O. Box 24012
Minneapolis, MN 55424

Monsanto
Monsanto Industrial Chemicals
947 West Waterloo Road
Akron, OH 44314

MUD
Master Unit Die Products, Inc.
866 Fairplains Street
P.O. Box 194
Greenville, MI 48838

Porpoise
GEC Electrical Projects, Marine and Offshore
Broughton Road
Rugby, Warwickshire
U.K.

Ray-Ran (see also T.M.I.)
Ray-Ran Engineering
Dawson House
Bennetts Road
Keresley CV7 8HY
U.K.

Rheometrics
Rheometrics, Inc.
One Possumtown Road
Piscataway, NJ 08854

Sangamo
Sangamo Schlumberger
Rheology Division
North Bersted
Bognor Regis
Sussex, PO229BS
U.K.

1875 Grand Island Blvd.
Grand Island, NY 14072

Seiscor
Seiscor Division
P.O. Box 1590
Tulsa, OK 74102

Shimadzu
Shimadzu Seisakusho, Ltd.
14–5, Uchikanda 1-Chomo
Chiyoda-Ku
Tokyo 101, Japan

Slocumb
F. F. Slocumb Corporation
P.O. Box 1591
Wilmington, DE 19899

T.M.I.
Testing Machines, Inc.
400 Bayview Avenue
Amityville, NY 11701

Tinius-Olsen
Tinius-Olsen Testing Machine Co., Inc.
Easton Road
Willow Grove, PA 19090

Toyoseiki
Toyo Seiki Seisaku-sho, Ltd.
15–4, 5-Chome
Takinogawa, Kita-ku
Tokyo 114, Japan

WALLACE (see also T.M.I.)
H.W. Wallace & Co., Ltd.
St. James Road
Croydon, U.K.

ZWICK
Zwick GmbH & Co.
Postfach 4350
D-7900 Ulm
Fed. Republic of Germany

Zwick of America
P.O. Box 20
Old Saybrook, CT 06475

Section VII
PROBLEM SOLVING

Chapter 27
Molding Problems and Solutions

D. V. Rosato

INTRODUCTION

Throughout this book there are reviews on "why problems develop" during injection molding and how they can be either eliminated or kept to a minimum. To do the best job of eliminating or reducing problems, one must understand the complete molding operation. For example, all that may be required is some degree of incoming inspection of molding material or replacement of a worn-out screw. This book reviews the different parameters so that all the factors that influence molding performance may be understood.

When we discuss troubleshooting, it is important that the terms used to identify a problem be understandable and clear. As an example, the word "flaw" could have any of the following meanings:

Blush	Discoloration caused by plastic flow during molding.
Burn	Discoloration caused by thermal decomposition.
Discoloration	Any change from original color or unintended, inconsistent part color.
Fill-in	An excess of ink that alters the form of a screened feature, affecting clarity and legibility.
Flow marks (plastic)	Wavy or streaked appearance of a surface.
Flow marks (silk screen)	Waviness of edge or excessive linear surface texture of screened areas.
Glossiness	An area of excessive or deficient gloss.
Gouge	Indentation that can be felt (dents).
Haze	Cloudiness of an otherwise transparent part.
Inconsistency	Variation of gloss, thickness of line, or surface texture not called for by master artwork.
Marks	Pits, sanding, machining, or other marks on part surface that are unacceptable.
Misalignment	The failure of the screened graphics to align with the part or its features.
Nonadhesion	Lack of proper sticking of the coating to the surface (chipping, orange peel).
Nonuniform (coverage)	Areas that have an insufficient or excessive coating.

Pit	Small crater on a surface.
Porosity	Holes or voids (blow holes, pits, or underfills).
Protrusion	A raised area on a surface (blister, bump, ridge).
Runs	Excessive coating that causes drips.
Scratches	Shallow grooves.
Sink	A depression on a surface (shrink mark).
Smearing	The presence of ink on areas not called for by master artwork.
Speck	An included substance that is foreign to its intended composition (bubble, inclusion).
Void	Failure of a plastic to completely fill a cavity.
Weld line	A visible line or mark on a surface, caused by plastic flow molding.

The usual problem can be resolved with one or just a few changes in the complete molding operation. Simplified guides to troubleshooting are given in Tables 27-1, 27-2, and 27-3. Table 27-4 lists "errors" in molding and product design that can lead to problems for the molding process and/or the molded part. (See Chapters 6 and 7 on designing parts and molds.) A guide to processing temperature ranges for injection molding general-purpose grades of thermoplastics is given in Table 27-5. A more detailed guide to troubleshooting is included in the following section.

TROUBLESHOOTING

It is important to use the proper approach in eliminating molding problems. The following review can help the molder find the cause and probable remedy for problems that result in unsatisfactory molded parts. In the detailed troubleshooting guide presented below, practical possible remedies have been classified according to: (1) materials, (2) mold, (3) molding cycle, and/or (4) machine performance.

Basically, faulty or unacceptable molded parts usually result from problems in one or more of three areas of operation:

1. Premolding: material handling and storage
2. Molding: conditions in the molding cycle
3. Postmolding: parts handling and finishing operations

Problems occurring in (1) and (3) include those involving contamination, color, static dust collection, painting, and vacuum metalizing. The solutions to these problems are usually quite obvious, or they are very specialized. This review discusses primarily the solution of problems encountered in the molding cycle. These faults can be attributed to the following:

1. Machine
2. Molds
3. Operating conditions (time, temperature, pressure)
4. Material
5. Part design
6. Management

The analysis of most molding problems focuses on the molding cycle. The molding cycle can best be described by what happens to the polymer in terms of:

1. Fill time
2. Packing time/rate
3. Cooling time
4. Ejection time
5. Open time
6. Mold temperature
7. Sprue and runner design
8. Gate size and location
9. Section thickness
10. Length of flow path

This differs somewhat from the molding machine operating cycle, which is commonly divided into (1) plunger forward time, (2) mold closed time, and (3) mold open time, a division that is convenient for setting machine controls.

Molding cycle problem analysis is con-

Table 27-1   Simplified guide to troubleshooting.

SUGGESTED REMEDIES	Drooling at Nozzle	Short Shot	Screw Does Not Return	Sink Marks	Burning	Surface Blemishes	Flashing	Dull Surface	Laminations	Part Sticks in Mold	Runner Breaks	Parts Distort	Discoloration of Sprue	Flow Lines	Brittle Parts	Wavy Surfaces	Worm Tracks On Part	Melt Temperature Too High	Streaks On Part	Voids in Part
Increase injection pressure		●												●		●				
Decrease injection pressure							●						●							
Increase stock temperature		●			●		●	●					●	●	●	●	●		●	
Decrease stock temperature				●			●			●			●			●		●	●	●
Increase holding pressure and time				●												●				●
Decrease holding pressure and time							●			●	●	●				●			●	
Increase nozzle temperature		●	●		●			●									●			●
Clear nozzle		●																		
Clear shutoff valve	●	●																		
Increase screw r.p.m.			●			●		●	●							●	●			
Decrease screw r.p.m.																		●	●	
Tighten nozzle or shutoff valve													●						●	
Inject with rotating screw		●				●		●	●				●		●	●				
Increase clamping pressure							●													
Start injection later	●																			
Decrease injection speed				●	●	●	●	●	●	●		●		●	●		●	●		●
Increase injection speed		●														●	●			
Increase back pressure		●				●		●	●				●		●	●		●		
Decrease back pressure			●	●														●		
Enlarge nozzle orifice		●	●	●				●	●					●	●		●	●		
Increase mold temperature		●				●		●	●	●				●	●	●				●
Decrease mold temperature				●		●	●	●		●	●	●				●	●			●
Polish mold and break corners						●		●		●										
Rework mold							●			●										
Polish sprue, runners and gates						●							●							
Increase size of gates		●		●	●	●		●	●					●	●	●	●			●
Provide vents in mold		●			●											●				
Enlarge cold slug well						●								●	●	●	●			
Use dry material			●			●		●	●						●	●				
Use uncontaminated material								●	●											
Fill hopper or remove obstruction			●																	
Increase feed		●		●										●		●				●
Use mold release										●										
Adjust nozzle pressure	●																			
Check radius of nozzle, & of sprue bushing	●									●	●									
Reduce nozzle temp., break sprue later	●																			
Reduce temperature—rear zone*			●																	
Balance mold filling; rework runners					●															
Provide air for ejection										●										
Lengthen cooling and mold-open time												●	●				●		●	
Shorten cooling and mold-open time																		●		

*exception: increase temp. for nylon

cerned with the three major elements in the molding operation, as follows:

1. Injection molding machine: Is it adequate in clamping capacity in pounds per hour capacity, in shot capacity, etc.?

2. Mold: Does it function properly; is there an engineering design deficiency, etc.?

3. Material: Is the polymer formulation correct for the part specification, for the molding cycle adjustment limitations, etc.?

Table 27-2 Troubleshooting guide for clear plastics molding.

Legend:
- **+** Increase
- **−** Decrease
- **∗** Check

Problem	Eliminate contamination	Remove moisture—increase drying temperature, time, or air flow	Adjust melt temperature	Adjust barrel temperature profile	Injection pressure	Injection rate	Injection hold time	Booster time	Backpressure	Improve flow pattern—relocate gates	Check nozzle, gate, runner and sprue obstructions, dimensions, layout	Mold temperature	Improve mold venting	Mold cooling time	Other Possible Remedies
Splay and splay marks, silver streaking	∗	∗	−	∗	+				+/−		∗		∗	+	Reduce cycle time
Low gloss, dull or rough surface			+		+	+					∗	+			Clean and polish cavity surfaces
Surface lamination, peeling	∗		−			+				∗	∗	+			Possibly caused by contamination by other resins
Sinks		∗			+	+	+	+				−	∗	+	Reduce cooling time in mold by using water bath. Even out part cross-sections, if possible
Blisters		∗	−		+								∗	+	Decrease screw speed
Bubbles, shrinkage voids		∗	+		+	−	+	+	+		∗	+		−	Cool more slowly—use hot water bath
Cloudiness, haze	∗		+	∗	+				+			+			
Weld lines, flow marks					+	+	+	+		∗	∗	+	∗		Equalize filling rate between cavities. Reduce clamp pressure. Vent at parting line or weld point. Check core positioning
Jetting			+/−							∗	∗	+			Check nozzle opening
Black spots or streaks	∗		+/−	∗	+/−	−					∗		∗		Look for hot spots. Check screw clearance

Table 27-3   Troubleshooting: simplified approach for "cause due to plastic material."

PROBLEMS	TOO HIGH A MOISTURE CONTENT	TOO LITTLE LUBRICANT	TOO MUCH MONOMER	CONTAMINATED GRANULES	TOO HIGH A PROPORTION OF MATERIAL TO BE GROUND	TOO LONG PREHEATING	TOO HIGH A DRYING TEMPERATURE	UNEVEN ADDITION OF COLORANT	UNEVEN GRANULE SIZE	TOO MUCH FINES	UNEVEN GRANULE FEED	VARIATIONS IN GRANULE PREHEATING TEMPERATURE	VARIATIONS IN MOISTURE CONTENT
sink marks	X	X	X								X		
flow marks	X			X									
brittleness	X			X	X								
discoloration	X					X	X	X					
surface blemishes				X					X	X	X	X	
varying shrinkage	X								X		X		
varying dimensional stability	X								X		X		
sticking to the mold	X	X											
varying strength												X	X

The performance of these three operating elements is influenced by three major variables: (1) time, (2) pressure, and (3) temperature. Most of the difficulties occurring during the molding cycle are corrected by adjustment of these three variables, each of which may be adjusted to a varying degree in each of the operating elements. As they are all interrelated, attention must be directed to each during the analysis of molding problems.

Finding the Fault

Before correcting a fault, one must find it. To find a fault, good quality control is necessary. Quality control should not start when a customer returns rejects. It should be a continuing process that starts when the raw material is ordered and follows each operation until the product is shipped. Also, unless the equipment is adequate and subjected to a continual, effective maintenance program, consistent-quality injection molding is not possible. Molds must be kept in good operating condition. Auxiliary equipment, such as mold heating and refrigeration units, grinders, finishing tools, and gauges, must be readily available.

The cause of a problem may be obvious, and the problem corrected by an adjustment in the three major variables. If the area of difficulty is not apparent, however, then each set of adjustment variables must be examined and corrections made where necessary. When a molder is starting up a new mold using a material on which he has certain data, he uses past experience on similar molds and materials to set up an approximate cycle. If the moldings are not perfect on this cycle, he will vary the pressure, temperature, and time sequences by adjusting the machine conditions until he obtains good pieces. Adjustments are always made in the machine variables first (use the "mold-area-diagram" approach, see page 331).

Adjustments are always made in the machine variables first (use the "mold-area-diagram" approach, see page 331).

If acceptable pieces are not produced after machine conditions have been changed, then the design of the mold should be examined. Any changes in the mold design can affect the temperature, pressure, and time sequences, but these interrelations are difficult to calculate and predict.

Table 27-4   Errors in mold and product design with possible consequences for process and/or molded part.

FAULTS	POSSIBLE PROBLEMS
Wrong location of gate	Cold weld lines, flow lines, jetting, air entrapment, venting problems, warping, stress concentrations, voids and/or sink marks
Gates and/or runners too narrow	Short shots, plastics overheated, premature freezing of runners, sink marks and/or voids and other marks
Runners too large	Longer molding cycle, waste of plastics, and pressure losses
Unbalanced cavity layout in multiple-cavity molds	Unbalanced pressure buildup in mold, mold distortion, dimensional variation between products (shrinkage control poor), poor mold release, flash, and stresses
Nonuniform mold cooling	Longer molding cycle, high after-shrinkage, stresses (warping), poor mold release, irregular surface finish, and distortion of part during ejection
Poor or no venting	Need for higher injection pressure, burned plastic (brown streaks), poor mold release, short shots, and flow lines
Poor or no air injection	Poor mold release for large parts, part distortion, and higher ejection force
Poor ejector system or bad location of ejectors	Poor mold release, distortion or damage in molding, and upsets in molding cycle
Sprue insufficiently tapered	Poor mold release, higher injection pressure, and mold wear
Sprue too long	Poor mold release, pressure losses, longer molding cycle, and premature freezing of sprue
No round edge at end of sprue	Notch sensitivity (cracks, bubbles, etc.) and stress concentrations
Bad alignment and locking of cores and other mold components	Distortion of components, air entrapment, dimensional variations, uneven stresses, and poor mold release
Mold movement due to insufficient mold support	Part flashes, dimensional variations, poor mold release, and pressure losses
Radius of sprue bushing too small	Plastic leakage, poor mold release, and pressure losses
Mold and injection cylinder out of alignment	Poor mold release, plastic leakage, cylinder pushed back, and pressure losses
Draft of molded part too small	Poor mold release, distortion of molded part, and dimensional variations
Sharp transitions in part wall thickness and sharp corners	Parts unevenly stressed, dimensional variations, air entrapment, notch sensitivity, and mold wear

For details on this subject see Chapters 6 and 7.

Most molding problems are solved by varying machine conditions, and a few more are solved by additional changes of mold conditions; but if problems remain on which both of these approaches are unsuccessful, their cause and possible solution may be found by examining polymer variables such as:

1. Flow characteristics: melt viscosity at molding temperature and change in viscosity at different flow rates (shear dependence of viscosity).

2. Thermal properties: heat distortion (setup temperature), specific heat, heat of fusion, thermal conductivity, and

Table 27-5   Processing temperature ranges for general-purpose grades of thermoplastics.

MATERIAL	PROCESSING TEMPERATURE RANGE (°C)	(°F)
ABS	180–240	356–464
Acetal	185–225	365–437
Acrylic	180–250	356–482
Nylon	260–290	500–554
Polycarbonate	280–310	536–590
Polyethylene		
low density	160–240	320–464
high density	200–280	392–536
Polypropylene	200–300	392–572
Polystyrene	180–260	356–500
Polyvinyl chloride, rigid	160–180	320–356

crystallization induction time (the delay before crystallites start to form).

3. Granulations: granulation size and shape and granulation lubrication.

Checklist

Basic rules for problem solving are:

1. Have a plan.
2. Watch molding conditions.
3. Change one condition at a time.
4. Allow sufficient time at each change.
5. Keep an accurate log of each change.
6. Check housekeeping.
7. Narrow down the problem to a particular area—that is, machine, mold, operating conditions, material, part design, or management. Some tips are:
   - Change the material. If the problem remains the same, it probably is not the material.
   - Changing the type of material may pinpoint the problem.
   - If the trouble occurs at random, it is probably a function of the machine, the temperature control system, or the heating bands. Changing the mold from one press to the other permits a determination of whether it is in the machine, mold, and/or powder.
   - If the problem appears, disappears, or changes with the operator, look for differences in the action of the operators.
   - If the problem appears in about the same position of a single-cavity mold, it is probably a function of the flow pattern and system from the front of the plunger through the nozzle, sprue, runner, and gate. It might also indicate a scored cylinder or some hang-up therein.
   - If the problem appears in the same cavity or cavities of a multicavity mold, it is in the cavity or gate and runner system.
   - If machine operation malfunctions, check hydraulic or electric circuits. As an example, a pump makes oil flow, but there must be resistance to flow to generate pressure. Determine where fluid is going. If actuators fail to move or move slowly, the fluid must be bypassing them or going somewhere else. Trace it by disconnecting lines if necessary. No flow (or less than normal flow) in a system will indicate that the pump or pump drive is at fault. Machine instruction manuals will provide details concerning correcting malfunctions.
8. Set up a procedure to "break in" a new mold:
   - Obtain samples and molding cycle information if the mold is new to the shop but has been run before.
   - Clean the mold.
   - Visually inspect the mold. Obvious corrections, such as improving the polish or removing undercuts, should be done before the mold is put in service.
   - Check out actions of the mold: try cams, slides, locks, unscrewing devices, and other devices on the bench.
   - Install safety devices.
   - Operate the mold in the press, and move it very slowly under low pressure.
   - Open the mold and inspect it.
   - Dry cycle the mold without injecting

material. Check knockout stroke, speeds, cushions, and low pressure closing.

- After the mold is at operating temperature, dry cycle it again. Expansion or contraction of the mold parts may affect the fits.
- Take a shot using maximum mold lubrication and under conditions least likely to cause mold damage. These are usually low material feed and pressure.
- Build up slowly to operating conditions. Run until stabilized, at least 1 to 2 hours.
- Record operating information.
- Take the part to quality control for approval.
- Make required changes.
- Repeat the process until it is approved by quality control and/or the customer.

Troubleshooting Guide

What follows is a detailed listing of the problems that are encountered during injection molding. The probable cause and/or possible remedy for each problem is also given. Note that there may be several causes for each difficulty, as well as several possible remedies for each cause. Any one remedy may solve the problem, but it may be necessary to try several remedies. The target is to determine specifically what action should be taken to correct problems. If the time (and initial expense) was spent to properly evaluate how to operate the machine with the mold and material, you can set up a specific plan on how to eliminate a problem, since it was already developed during the setup time. At that time corrective action was taken so that a simplified troubleshooting guide for the operator could be prepared such as the guide in Table 27-3.

The first section of the following guide concerns the basic molding machine operation, listing causes and remedies of problems related to the machine, mold, material, and/or molding cycle. After this listing, Table 27-6 provides troubleshooting information as it relates to auxiliary equipment, namely, dryers, granulators, conveying equipment, metering/proportioning/feeding equipment, and chillers/mold temperature controllers; and Table 27-7 covers dehumidifying dryer performance.

CAUSE	POSSIBLE REMEDY
Black Specks (also see Black Streaks)	
• Flaking off of burned plastic on cylinder walls (especially polyethylene). • Airborne dirt.	• Contamination from degradation of other resins previously in cylinder: clean cylinder. • Thermal degradation of material on cylinder wall: clean cylinder wall. • Purge heating cylinder. • Purge through a stiffer molding compound to scour cylinder walls. • Avoid holding plastic for long periods at high temperatures. • Cover hopper. • Keep cover on virgin material.
Black Spots	
• Air trapped in mold causing burning.	• Vent mold properly. • Redesign part. • Relocate gate. • Reduce injection pressure or speed. • Alter flow pattern in mold by raising or reducing cylinder and mold temperature.
Black Streaks	
• Frictional burning of cold granules against one another and/or the cylinder walls.	• Relocate plunger and allow sufficient tolerance to permit air to escape back around the plunger.

CAUSE	POSSIBLE REMEDY
• Plunger off center; frictional burning of material between plunger and cylinder wall. • Burning in nozzle that is too hot. • Wide cycling of nozzle temperature.	• Avoid finely ground material that can get between plunger and wall. • Use externally lubricated plastic. • Lubricate reground. • Raise rear cylinder temperature. • Reduce nozzle temperature. • Avoid "on–off" controller. Use variable transformer.

Brittleness

• Degradation of the material during molding.
 • Accentuated by a part designed at the low limits of mechanical strength.
	• Materials
	   • Contaminated material: clean.
	   • Wet material: dry.
	   • Volatiles in material: use material with lower volatile content.
	   • Too much reground: reduce the amount of reground.
	   • Low-strength materials: increase strength of material (e.g., add more rubber to high impact polystyrene).
	 • Mold
	   • Part design too thin: redesign.
	   • Gate too small: change.
	   • Rubber too small: change.
	   • Add reinforcement (ribs, fillets).
	 • Molding
	   • Low cylinder temperature: increase cylinder temperature.
	   • Low nozzle temperature: increase nozzle temperature.
	   • If material is thermally degrading, lower cylinder and nozzle temperature.
	   • Increase injection speed.
	   • Increase injection pressure.
	   • Increase injection forward time.
	   • Increase injection boost.
	   • Low mold temperature: increase mold temperature.
	   • Part stressed: mold so that part has minimum stress.
	   • Weld lines: mold to minimize weld line.
	   • Screw speed too high, degrading the material: adjust speed.
	 • Machine
	   • Machine plasticizing capacity too low for the machine: change as needed.
	   • Cylinder obstruction degrading the material: change as needed.

Brown Streaks (also see Black Streaks)

• Hang-up in cylinder or nozzle causing burning. • Either general or local overheated cylinder.	• Purge cylinder; remove nozzle and clean; or if necessary, remove cylinder and clean. • Nozzle temperature too high: adjust as needed. • Cylinder temperature too high: adjust as needed.

Bubbles (also see Sinks)

• Nonuniform mold temperature. • Moisture on granules. • Short shot (insufficient) plastic in the mold to prevent excessive shrinkage caused by:	• Dry granules before molding. Avoid drastic temperature changes before molding. • Rearrange water lines to obtain good mold temperature uniformity.

CAUSE	POSSIBLE REMEDY
• Heavy sections, bosses, and ribs. • Injection pressure too low. • Plunger forward time too short. • Insufficient feed.	• Short shot: change as needed. • Increase feed. • Insufficient injection pressure: adjust as needed. • Insufficient injection time: adjust as needed. • Excessive feed buildup in cylinder (cushion): change as needed. • Stock temperature too high: adjust. • Excessive restriction in plastic flow due to under-sized gates, sprues, runners, or part design: correct design. • Improper gate location: change as needed. • Gate land length too long: change as needed. • Machine undersized for shot size: make changes needed.

Charred Area

• Insufficient mold venting.

Cracking/Crazing

• Mold temperature too low.
• Improper mold draft or undercuts.
• Ejector pins or ring poorly located.
• Packing excess plastic into mold.

Delamination

• Resin contaminated.

Dimensional Variation

• Inconsistent machine control.
• Incorrect molding conditions.
• Poor part design.
• Variations in materials.

• Increase venting.

• Raise mold temperature.
• Rework mold.
• Locate for balanced removal force. It's better to push off than pull off.
• Reduce feed.
• Reduce injection pressure.

• Clean resin.

• Machine—make corrections for:
  • Malfunctioning feed system in a plunger machine.
  • Inconsistent screw stop action.
  • Inconsistent screw speed.
  • Malfunctioning non-return valve.
  • Worn non-return valve.
  • Uneven back pressure adjustment.
  • Malfunctioning thermocouple.
  • Malfunctioning temperature control system.
  • Malfunctioning heater band.
  • Insufficient plasticizing capacity.
  • Inconsistent cycle, machine-caused.
• Molding
  • Uneven mold temperature: make changes required.
  • Low injection pressure: increase injection pressure.
  • Insufficient fill or hold time:
    • Increase injection forward time.
    • Increase injection boost time.
  • Too high barrel temperature: lower barrel temperature.
  • Too high nozzle temperature: lower nozzle temperature.
  • Inconsistent cycle: eliminate.
• Mold—make corrections for:
  • Incorrect mold dimensions causing parts to appear out of tolerance.
  • Distortion during ejection.

CAUSE	POSSIBLE REMEDY
	• Uneven mold filling.
	• Interrupted mold filling.
	• Incorrect gate dimensions.
	• Incorrect runner dimensions.
	• Inconsistent cycle, mold-caused.
	• Materials—make corrections for:
	• Batch-to-batch variation.
	• Irregular particle size.
	• Wet material.
Discoloration	
• Burning of plastic.	• Temperature
• Degradation of plastic.	• Cylinder temperature too high: decrease temperature.
• Material contamination.	• Nozzle temperature too high: decrease temperature.
	• Machine
	• Clean nozzle.
	• Inspect nozzle and sprue bushing for burrs.
	• Purge cylinder.
	• Reseat nozzle.
	• Clean cylinder and check for burrs.
	• Check for cracked cylinder.
	• Dirty machine: clean.
	• Dirty hopper dryer: clean.
	• Dirty atmosphere; colorants can float in the air and settle in the hopper and grinder. Take necessary precautions.
	• Injection end of machine too large: adjust as needed.
	• Thermocouple not functioning: adjust as needed.
	• Temperature control system not functioning: adjust as needed.
	• Heater band not functioning: adjust as needed.
	• Cylinder obstruction degrading the material: change as needed.
	• Molding
	• Decrease screw speed.
	• Decrease back pressure.
	• Reduce clamp pressure.
	• Decrease injection pressure.
	• Decrease injection forward time.
	• Decrease injection boost time.
	• Slow down injection rate.
	• Decrease cycle.
	• Mold
	• Vent mold.
	• Increase gate size.
	• Increase runner–sprue–nozzle system.
	• Change gating pattern.
	• Remove lubricant and oil from mold.
	• Investigate mold lubricant.
	• Materials—make corrections for:
	• Contamination.
	• Material that is not dry.
	• Too many volatiles in the material.
	• Material degrading.
	• Colorant degrading.
	• Additives degrading.

CAUSE	POSSIBLE REMEDY

Drooling

- Overheated material. (The objection to nozzle drooling is that it introduces solidified material into the part which causes surface defects. It may also interfere with the flow and mechanical properties.)

- Nozzle
    - Use positive-seal type nozzle.
    - Use reverse taper nozzle.
    - Reduce nozzle bore diameter.
- Molding
    - Reduce nozzle temperature.
    - Increase suckback.
    - Use sprue break.
    - Decrease material temperature.
    - Reduce injection pressure.
    - Reduce injection forward time.
    - Reduce injection boost time.
- Mold
    - Increase cold slug well.
    - Increase run-off.
- Materials
    - Check for contamination.
    - Dry the material.

Ejection Poor

- Mold part remains in mold.

- Make adjustments for roughness or undercuts in mold.
- Eliminate excessive mold packing.
- Change inadequate knockout system.
- Correct insufficient taper or draft.

Erratic Cycle

- Holding mold open various lengths of time.
- Erratic pressures.
- Erratic feed.
- Nonuniform mold temperature.
- Nonuniform cylinder temperature (cycling).

- Maintain an overall constant cycle time by the use of mold open timers.
- Ensure sufficient pressure to fill the mold consistently.
- Check pressure system for leaks, etc.
- Check feeding mechanism.
- Mold
- Use mold temperature control.
    - Provide proper waterlines in the mold.
    - Allow proper venting of the mold.
    - Provide proper hookup for water through the mold.
- Cylinder
    - Check temperature controls to ensure proper operation.
    - Use best temperature controls available.
    - Check line voltage to the machines for consistency.
    - Ensure that heater bands are working properly.
    - Have material temperature constant from one drum to another before placing material in the hopper.

Flashing

- Material too hot.
- Pressure too high.
- Excessive feed.
- Erratic feed.
- Poor parting line or mating surfaces.

- Adjust material flow that is too soft for parts.
- Temperature—make correction for:
    - Cylinder temperature too high.
    - Nozzle temperature too high.
    - Mold temperature too high.

CAUSE	POSSIBLE REMEDY

• Mold deficiency.
• Erratic cycle time.
• Insufficient clamp.
	• Molding—adjust for:
	  • Clamp pressure too low.
	  • Injection pressure too high.
	  • Injection time too long.
	  • Boost time too long.
	  • Injection feed too fast.
	  • Unequalized filling rate in cavities.
	  • Interrupted flow into cavities.
	  • Feed setting too high.
	  • Inconsistent cycle, operator-caused.
	• Machine—adjust for:
	  • Projected area of the molding parts too large for clamping capacity of machine.
	  • Machine set incorrectly.
	  • Mold put in incorrectly.
	  • Clamp pressure not maintained.
	  • Machine platens not parallel.
	  • Tie bars unequally strained.
	  • Inconsistent cycle, machine-caused.
	• Mold—make correction for:
	  • Cavities and cores not sealing.
	  • Cavities and cores out of line.
	  • Mold plates not parallel.
	  • Insufficient support for cavities and cores.
	  • Mold not sealing off because of foreign material (flash) between surfaces.
	  • Something other than flash keeping the mold open (e.g., foreign material in leader pin bushing so that leader pin is obstructed when entering the bushing, keeping the mold open).
	  • Insufficient venting.
	  • Vents too large.
	  • Land area around the cavities too large, reducing the sealing pressure.
	  • Inconsistent cycle, mold-caused.

Flow Lines and Folds • Material temperature too low. • Mold temperature too low. • Gates too small, causing jetting. • Nonuniform section thickness.	• Increase plastic temperature. • Increase mold temperature. • Enlarge gates and reduce injection speed. • Redesign part to obtain greater uniformity of section thickness. • Eliminate heavy bosses and ribs.

Gate (splay, blush, lamination, dull spots)
• Melt fracture as material expands entering mold.
• Material too cold.
• Mold too cold.
• Slow injection speed.
• Insufficient pressure.
• Plunger dwell too long.
• Contamination of material.
• Excessive mold lubricant.
• Runners and gates too large or too small.
• Excessive mold heat, particularly at sprue or center gates.
	• Molding
	  • Mold temperature too low: adjust.
	  • Nozzle temperature too low: adjust.
	  • Injection speed too fast: adjust.
	  • Increase injection pressure.
	  • Change injection forward time.
	  • Use minimum lubricant.
	  • Change lubricant.
	• Mold
	  • Raise mold temperature.
	  • Increase gate size.
	  • Change gate shape (tab or flare gate).
	  • Increase cold slug well.

CAUSE	POSSIBLE REMEDY

	• Increase runner size.
	• Change gate location.
	• Increase venting.
	• Radius gate at cavity.
	• Material
	  • Dry material.
	  • Remove contaminants from the material.

Granules Unmelted

• Too low plastic temperature.
• Too fast a cycle for cylinder capacity.
• Insufficient restriction to flow.

    • Increase plastic temperature.
    • Lengthen cycle.
    • Use restricted nozzle.

Insert Cracking

• Insufficient material around insert.

    • Poor part design: change as needed.
    • Contamination: remove.

Jetting

• Resin too cold.

    • Injection too fast: adjust.
    • Gate too small: change.
    • Gate land too long: change.

Long Cycles

• High material temperatures.
• Mold temperature excessive.
• Erratic cycle time.
• Insufficient heating capacity.
• Inadequate cooling of local heavy section.
• Excessive flash requiring operator trimming (see Flashing).
• Excessive delay in machine operation.
• Slow setup in mold.

    • Lower temperatures.
    • Reduce mold temperature.
    • Maintain a constant overall cycle time by the use of mold open timer.
    • Change mold to larger press and/or preheat the material.
    • Locate bubbles to cool area. Use quenching bath.
    • Reduce machine dead time as much as possible.
    • Reduce mold temperature.
    • Try more heat-resistant grade of material.

Low Heat Distortion Temperature

• Variations in section thickness.
• Too low mold temperature.
• Incorrect cylinder temperature relative to mold temperature.
• Excessive feed.
• Excessive pressure.
• Excessive plunger dwell.
• Excessive mold temperature variation between front and back.
• Gate slow freezing.

    • Maintain as uniform a section thickness as possible.
    • Increase the mold temperature.
    • Select proper cylinder temperature and mold temperature.
    • Reduce feed and weigh starve feed if possible.
    • Reduce pressure.
    • Use a minimum plunger forward time and dwell.
    • Have both front and rear mold temperatures as nearly the same as possible.
    • Reduce size of gate.

Short Shot

• Cold material.
• Cold mold.
• Insufficient pressure.
• Nonuniform mold temperature.
• Insufficient feed.
• Entrapped air.
• Insufficient external lubricant.
• Insufficient plunger forward time.
• Improper balance of plastic flow in multiple-cavity molds.
• Insufficient injection speed.
• Small gates.
• Shot size larger than machine capacity.

    • Molding condition causes—correct for:
      • Injection pressure too low.
      • Loss of injection pressure during cycle.
      • Injection forward time too short.
      • Injection boost time too short.
      • Injection speed too low.
      • Unequalized filling rate in cavities.
      • Interrupted flow in cavities.
      • Inconsistent cycle, operator-caused.
    • Temperature related causes
      • Raise cylinder temperature.
      • Raise nozzle temperature.

CAUSE	POSSIBLE REMEDY

	• Check pyrometer, thermocouple, heating bands system.
	• Raise mold temperature.
	• Check mold temperature equipment.
	• Mold-related causes—correct for:
	  • Runners too small.
	  • Gate too small.
	  • Nozzle opening too small.
	  • Improper gate location.
	  • Insufficient number of gates.
	  • Cold slug well too small.
	  • Insufficient venting.
	  • Inconsistent cycles, mold-caused.
	• Machine causes—correct for:
	  • No material in the hopper.
	  • Hopper throat partially or completely blocked.
	  • Feed control set too low.
	  • Feed control set too high which can cause lowering of injection pressure in a plunger machine.
	  • Feed system operating incorrectly.
	  • Plasticizing capacity of machine too small for the shot.
	  • Inconsistent cycles: machine-caused, operator-caused, or mold-caused.
	  • Malfunctioning of return valve on tip of screw. This is usually indicated by screw turning during injection.

Shrinkage (also see Warpage)

• Excessive shrinkage and warpage are usually caused by design of the part, the gate location and molding conditions. Orientation and high stress levels are also factors.

Silver Streaks

CAUSE	POSSIBLE REMEDY

• Excessive nozzle, torpedo, or cylinder temperatures.
• Exceeding plasticizing capacity of machine in pounds per hour.
• Variation in temperature of material being placed in hopper.
• Plastic temperature too high.
• Injection pressure too high.
• Air trapped between granules in cold end of machine.
• Mold temperature too low.
• Injection speed too fast.
• Intermittent flow in the cavity.
• Moisture on granules.
• Lack of external lubrication.
• Excessive external lubrication.
• Nonuniform external lubrication.
• Mixture of coarse and fine granules (as with reground).
	• Machine
	  • Reduce nozzle temperature first, then cylinder temperature.
	  • Lengthen cycle or operate mold in machine with larger heating capacity.
	  • Preheat material or install hopper dryers to maintain material temperature.
	  • Reduce injection pressure.
	  • Reduce rear cylinder temperature, and avoid use of reground.
	  • Operate with no cushion of material ahead of plunger.
	• Mold
	  • Raise mold temperature.
	  • Vent mold.
	  • Balance gates.
	  • Relocate gates.
	  • Maintain uniform mold temperature.
	  • Obtain as uniform a section thickness as possible.
	• Material
	  • Dry material prior to use, or use hopper dryer. Avoid exposing material to drastic temperature changes prior to molding.

CAUSE	POSSIBLE REMEDY
	• Add zinc stearate, often necessary with reground. • Avoid use of nonuniform material, or screen to give uniform granule size. • Blend with nonlubricated or reground material. • Allow longer blending time, or add a little more lubricant and blend.

Sink Marks (also see Bubbles)

- Insufficient plastic in mold to allow for shrinkage due to:
    - Thick sections, bosses, ribs, etc.
    - Not enough feed.
    - Injection pressure too low.
    - Plunger forward time too short.
    - Unbalanced gates.
    - Injection speed too slow.
- Plastic too hot.
- Piece ejected too hot.
- Variation in mold open time.
- No cushion in front of injection ram with volumetric feed.
- Too much cushion in front of ram.

- Material
    - Dry material.
    - Add lubricant.
    - Reduce volatiles in material.
- Changes in cooling conditions
    - Piece cooled too long in mold, preventing shrinking from the outside in: make changes needed.
    - Shorten mold cooling time.
    - Cool part in hot water.
- Molding
    - Insufficient feed: adjust.
    - Increase the injection pressure.
    - Increase the injection forward time.
    - Increase the boost time.
    - Increase the injection speed.
    - Increase the overall cycle.
    - Change method of molding (intrusion).
    - Inconsistent cycles, operator-caused: correct.
- Machine changes
    - Increase plasticizing capacity of the machine.
    - Make cycle consistent.
- Temperature—adjust for:
    - Material too hot causing excessive shrinkage.
    - Material too cold causing incomplete filling and packing.
    - Mold temperature too high so the material on the wall does not set up quickly enough.
    - Mold temperature too low preventing incomplete filling.
    - Local hot spots on the mold.
    - Mold temperature control system malfunctioning.
- Mold
    - Increase the gate size.
    - Increase the runner size.
    - Increase the sprue size.
    - Increase the nozzle size.
    - Vent mold.
    - Equalize filling rate of cavity.
    - Prevent interrupted flow into the cavities.
    - Put gate in thick sections.
    - Reduce uneven wall thickness where possible. Use cores, ribs, and fillets.
    - Inconsistent cycle, mold-caused: correct.

Sprue Sticking

- Undercuts in mold.
- Mold rough surface.
- Excessive pressure.
- Hot material.
- Excessive size of sprue.

- Molding
    - Use sprue break (machine moves back slightly, breaking contact between nozzle and sprue).
    - Increase suckback.
    - Reduce feed.

CAUSE	POSSIBLE REMEDY

• Insufficient draft.
• Improper fit between sprue bushing and nozzle.
• Too much feed.
• Long plunger dwell.
• Vacuum under deep draw part.
• Variation of mold open time.
• Core shifting.
• Unbalanced gates in multicavity molds or single-cavity molds with two or more gates.
	• Reduce injection pressure.
	• Reduce ram forward time.
	• Reduce injection boost time.
	• Reduce material temperature.
	• Reduce cylinder temperature.
	• Reduce nozzle temperature.
	• Use more mold release agent.
	• Use proper mold release agent.
	• Reduce material feed.
	• Reduce injection pressure.
	• Reduce injection forward time.
	• Reduce injection boost time.
	• Reduce mold temperature.
	• Increase overall cycles. This lowers temperature making the part more rigid and increases the amount of shrinkage.
	• Make inconsistent cycles (operator-caused) consistent.
	• Materials
	  • Remove contamination in material
	  • Add lubricant to the material.
	  • Dry the material.
	• Machine
	  • Repair any malfunctioning of the knockout system.
	  • Lengthen insufficient knockout travel distance.
	  • Make inconsistent cycles (machine-caused) consistent.
	  • Check to see if platens are parallel.
	  • Check the tie-rod bushings.
	• Mold
	  • Increase the pull-out force of the sprue–puller system.
	  • Reduce mold temperature.
	  • See if sticking is caused by short shot not engaging knockout system.
	  • Remove undercuts.
	  • Remove burrs, nicks, and similar irregularities.
	  • Remove scratches and pits.
	  • Improve the mold surface.
	  • Restone and polish using movement only in the direction of ejection.
	  • Increase the taper.
	  • Increase the effective knockout area.
	  • Decrease the gate size.
	  • Add additional gates.
	  • Relocate the gates.
	  • Equalize the mold filling rate.
	  • Prevent interrupted filling.
	  • Determine whether the part is strong enough for ejection.
	  • Radius and reinforce parts, giving greater rigidity.

Surface Defects
• Slow injection.
• Unbalanced flow in gates and runners.
• Poor flow within mold cavity.
	• Molding
	  • Reduce screw speed.
	  • Reduce back pressure.

CAUSE	POSSIBLE REMEDY

• Cold material.
• Mold too cold.
• Injection pressure too low.
• Water on mold face.
• Excess mold lubricant on mold.
• Not enough plastic into mold to contact mold metal at all points.
• Excessive internal or external lubricants.
• Poor surface on mold.
	• Alter injection speed.
	• Increase injection pressure.
	• Increase injection forward time.
	• Increase booster time.
	• Increase cycle.
	• Temperature
	  • Too low or too high cylinder temperature, depending on problem: change temperature profile of cylinder.
	  • Too low mold temperature: raise mold temperature.
	  • Nonuniform mold temperature: check.
	• Material
	  • Use uniform-size particles.
	  • Reduce the amount of fines.
	  • Use minimum amount of lubricant.
	  • Change type of lubricant.
	• Mold
	  • Increase runner extension.
	  • Increase runner.
	  • Polish sprue runner and gate.
	  • Open gate or change gate to tab.
	  • Change gate location. If jetting, flare gate or use tab or flared gate.
	  • Increase venting.
	  • Improve mold surface.
	  • Clean mold surface.
	  • Water caused by leaks and condensation: remove.
	  • Flow over depressions and raised section: change the part design.
	  • Try localized gate heating.
	• Machine
	  • Check nozzle for partial obstruction.
	  • Check sprue–nozzle–cylinder system for restrictions and burrs.

Tearing • Mold part tears.	• Inadequate core cooling: correct. • Hot core pins: make needed changes.
Voids (also see Bubbles) • Trapped gases.	• Vent cavities. • Provide for thin to thick transition. • Mold surface too cold: correct. • Resin too hot: correct. • Lack of pressure: open gate, runner and decrease gate land length.

Warpage (also see Shrinkage)
• Part ejected too hot.
• Plastic too cold.
• Variation of section thickness or contour of part.
• Too much feed.
• Unbalanced gates on parts having more than one gate.
• Poorly designed or operated ejection system.
• Mold temperature nonuniform.
• Excessive material discharged from or packed into the area around the gate.
	• Molding
	  • Increase cycle time.
	  • Increase injection pressure without excessive packing.
	  • Increase injection forward time without excessive packing.
	  • Increase injection boost time without excessive packing.
	  • Increase the feed without excessive packing.

CAUSE	POSSIBLE REMEDY

- Lower the material temperature.
- Keep packing at a minimum.
- Increase injection speed.
- Slow down ejection mechanism.
- Annealing parts after molding may reduce warping.
- Cool in shrink fixture.
- Cool in water.
- Make cycle consistent.
- Material
  - Use quicker-curing material.
- Mold
  - Change gate size.
  - Change gate location.
  - Add additional gates.
  - Increase knockout area.
  - Keep knockouts even.
  - Have sufficient venting, especially for deep parts.
  - Strengthen part by increasing wall thickness.
  - Strengthen part by adding ribs and fillets.
  - If differential shrinking and warping caused by irregular wall section, core, if possible, or change the part design.
  - To reduce warpage, reduce mold temperature to stiffen the outer surface.
  - To decrease shrinkage, raise mold temperature to increase packing.
  - Check mold dimensions. Wrong mold dimensions may cause parts to appear to have shrunk excessively.

Weak Parts

- Part "breaks."

  - Excessive moisture in resin: remove.
  - Stock temperature too high: reduce.
  - Contamination: remove.
  - Poor welds: correct.

Weld Lines/Flow Marks

- Plastic too cold.
- Excess mold lubricant on mold.
- Weld line too far from gate.
- Air unable to escape from mold fast enough.
- Section thickness variation within part.
- Mold too cold.
- Insufficient pressure.
- Slow injection speed.
- Gas trap.

  - Molding—correct for:
    - Injection pressure too low.
    - Injection feed too slow.
  - Temperature—correct for:
    - Too low cylinder temperature.
    - Too low nozzle temperature.
    - Too low mold temperature.
    - Too low mold temperature at spot of weld.
    - Uneven melt temperature.
  - Mold
    - Insufficient venting of the piece: add run-off at weld.
    - Runner system too small: correct.
    - Gate system too small: correct.
    - Sprue opening too small: correct.
    - Mold shifting, causing one wall to be too thin: make necessary adjustments.
    - Part too thin at weld: thicken it.
    - Unequal filling rate: equalize it.
    - Interrupted filling: correct.

CAUSE	POSSIBLE REMEDY
	• Nozzle opening too small: correct.
	• Gate too far from the weld: add additional gates (these might add additional welds but put them in a less objectionable location).
	• Wall section too thin, causing premature freezing: make changes needed.
	• Core shifting, causing one wall to be too thin: make necessary adjustments.
	• Machine
	• Plasticizing capacity too small for the shot: make changes needed.
	• Excessive loss of pressure in the cylinder (plunger machine): correct.
	• Materials
	• Contaminated material, which can prevent knitting properly: purify.
	• Poor material flow: lubricate material for better flow.

Table 27-6   Troubleshooting guide to auxiliary equipment. (Courtesy of *Plastics Technology.*)

Dryers

Problem	Possible Causes	Suggested Solutions
Process air temperature too low	Incorrect temperature selected on control panel	Dial in correct temperature
	Controller malfunction	Check electrical connections; replace controller if necessary
	Process heating elements	Check electrical connections; replace elements if necessary
	Hose connections at wrong location	Check to make sure delivery hose is entering bottom of hopper
	Supply voltage different from dryer voltage	Check supply voltage against name-plate voltage
Process air temperature too high	Thermocouple not located properly at inlet of hopper	Secure thermocouple probe into coupling at inlet of dryer
Material not drying	Process and/or auxiliary filter(s) clogged	Clean filter(s)
	Incorrect blower rotation	Check rotation
	Regeneration heating elements inoperative	Check electrical connections; replace elements if necessary
	Desiccant assembly not rotating	Check motor electrical connections. Replace motor if necessary. Check drive assembly for slippage; adjust
	Material residence time in hopper too short	Drying hopper too small for material being processed; replace with larger model
	Moist room air leaking into dry process air	Check all hose connections and tighten if required. Check hoses for cracks; replace as necessary. Check filter covers for tightness; secure
	Desiccant contaminated	Replace desiccant cartridge

Table 27-6 Continued

Granulators

Problem	Possible Causes	Suggested Solutions
Stalled machine	1. Overloading	Feed material slower
	2. Worn, damaged or improperly set knives.	Readjust or replace as required
	3. Screen and/or blower chute blockage	Check to see if line is clogged and check rotation of blower
	4. Drive-belt slippage	Check tensioning
	5. Loss of power	Check power supply, electrical hook-up and safety switches
	6. Motor running in reverse	Check direction of rotation and rewire per diagram if necessary
Material overheating	See items 1, 2, 3, & 6	Same as above
	Screen too small	Change to screen with larger diameter holes
Too many fines in material	Worn, damaged or improperly set knives	Readjust or replace as required
Bearing overheating	Failure to lubricate properly	Check frequency of lubrication and type of grease used
	Too much tension on drive belts	Adjust tensioning
Excessive knife wear	Highly abrasive material	Change to higher alloy knives
Knife breakage	Tramp metal in scrap	Check scrap material for foreign matter
	Loose or stretched bolts	Check bolts and retorque per specification sheet
	Uneven knife seats	Inspect and clean seat surfaces
Screen breakage	Improperly seated	Check that screen is fitted correctly
Motor won't start	Power supply failure	Check main power supply and fuses
Motor won't start	Overheated motor	Allow motor to cool, reset starter overloads
	Starter failure	Check for burned-out contacts, replace if necessary
	Inoperative safety switches	Check that hand guard is secure and all contact points closed. Replace if necessary

Conveying Equipment

Problem	Possible Causes	Suggested Solutions
Spiral Conveyors		
Excessive wear	System may have kinks, sags, sharp or compound bends, or contact with sharp surfaces	Re-position drive-end tube supports or feed hopper
	System may exhibit excessive vibration	Inspect for proper feeding into inlet (Consult supplier if material is bridging)
	Are abrasive materials being conveyed?	Consult supplier
Excessive spiral wear or breakage	There may not be sufficient clearance at inlet end of conveyor for spiral to expand, taking into consideration length and inclination of conveyor, and bulk density of material	Shorten spiral
Low delivery rate	Material may not be flowing properly into inlet	Rotate outer tube so that inlet opening is aligned with hopper feed
		Install bin vibrator and/or agitator
		Adjust spiral length as per manual instructions
		Reverse spiral direction if incorrect
		Seal openings in system if material is hygroscopic and system is installed in high-humidity environment
Belt Conveyors		
Slipping clutch	Oil contamination or excessive lubrication	Clean system, adjust clutch properly

Table 27-6 Continued

Improper belt tracking	Belt not properly tightened when changed or installed	Check alignment and tighten
	System damaged in delivery	
Electrical malfunctions	Motor may be exposed to excessive heat under molding machine	Install overload-protection temperature limit switches
Parts jam up or fall off	Transition points not long enough	Contact supplier
	Parts too wide for machine	Identify to supplier what's being conveyed before purchase
Belt speed too slow or too fast	Improper adjustment of variable-speed drive motor	Change sprockets
		Install variable-speed options or speed-adjusting kit
	Damaged in delivery	Specify desired speed range to supplier before purchase
Pneumatic Parts Conveyors Parts won't move	Air orifices blinded	Inspect regularly for contamination and clean. Replace fan belt
Pneumatic Materials Conveyors Material won't move	Filters clogged Filters improperly sized Blower blinded System improperly sized to suit plant layout	Inspect filters regularly and frequently. Specify to supplier exactly what types of materials are to be conveyed. Check vacuum pressure gauges. Replace filters. Indicate to supplier anticipated future growth plans if possible. Check vacuum seals.
Blower overheating	Blower blinded Excessive ambient-temperature exposure	Check filters. Make sure they are properly sized. Install temperature limit switches
Blower too noisy		Install muffler. Enclose in a well ventilated sound enclosure. Place blower in a sound-proofed room

Metering/Proportioning/Feeding Equipment

Problem	Possible Causes	Suggested Solutions
Dry-Solids Metering Weight distortion	Dust or adhesive dry solids accumulation	Install a dust-exhaust system or hardware to "tare-off" dust accumulation in critical areas. Clean belts, trays, augers periodically
Mechanical and electrical component failure—improper weight signals	Dust, environmental conditions, materials adhering to underside of belt or other system components	Evaluate several systems in production trials before purchase. Clean system components periodically
Belts stretching and mistracking	Material buildup or proximity of system to moisture	Inspect regularly. Consult supplier. Install corrective recalibration devices
Liquids Metering Materials won't flow	Improperly sized system—can't handle highly viscous materials	Install drum pump. Specify to supplier what type of material to be conveyed. Make sure tubing is properly sized
Proportioning Loaders Sluggish loading, excessive loading time needed to fill hopper	Clogged filter in either dust collector or pump	Clean filter (replace if necessary)
	Material line clogged	Clean line
	Vacuum leak in either material line or vacuum line	Seal lines
	Hopper lid or hopper receiver not sealed	Clamp lid or replace hopper seals if necessary
	Valves not sealed	Clear obstruction. Check for proper air pressure.
	Air-to-material ratio not correct for feed tubes (too much air)	Adjust feed tubes

Table 27-6 Continued

Excessive dust carryover into the dust collector	Valves not sealing	Check for obstruction and proper air pressure
	Improper feed tube setting—too much air —not enough material creates high velocities	Adjust feed tube to give highest obtainable vacuum and smoothest flow
	Excessive fines and dust in material or improper blending	Consult material supplier.

Continued on next page

Chillers/Mold Temperature Controllers

Problem	Possible Causes	Suggested Solutions
Cooling water lines frozen	Thermostat set too low (i.e., below 40 F without antifreeze)	Check thermostat and reset if necessary to 40 F or higher
	Insufficient antifreeze in process cooling water	Add antifreeze
System shuts down or cools slowly or poorly while refrigerant pressure is low or drooping; bubbles in refrigerant-level sight glass; oily-looking moisture on coolant-circulating tubes or floor nearby	Refrigerant leak	Replace refrigerant, plug leak source, clean condenser
System shuts down while refrigerant pressure is high	Condenser not getting enough cooling water because constricted or blocked by dirt	Clean condenser
Low cooling-water pressure	Leak in process-water circulating lines	Plug leak
	Empty water-storage tank	Fill tank
	Broken pump motor	Repair or replace (hermetically sealed motor/pump assemblies must usually be replaced)
	Broken pump seal	Replace
Slow or inadequate cooling of molds, high water pressure	Water flow constricted by dirt or mineral scale	Clean water circulating system
		Treat water
		Install intermediate heat exchanger for mold-cooling water (optional)
Slow or inadequate mold cooling, low pressure	Water-circulating line leak	Repair leak
	Pump seal failure	Replace seal or tighten packing gland
Process water heats slowly or insufficiently	Heating element encrusted with dirt, mineral scale or, in oil-circulating systems, carbonized oil	Clean or replace heating element, treat water, replace oil

Table 27-7 Dehumidifying dryer performance. (Courtesy Upjohn Co.)

SYMPTOM	POSSIBLE CAUSE(S)	CURE
1. Cannot attain desired air inlet temperature.	Heater failure	Check process air or after heaters—regeneration heaters play no part in this aspect of operation.
	Hose leakages and excessive length air inlet side.	Locate and repair—if the hose is old and brittle, replace. Shorten all hose to minimum lengths.
	Line, hopper, filter blockage.	Check for collapsed or pinched lines, valves that are closed (some makes have air flow valves located on the air inlet side of the hopper). Filters should be changed or cleaned frequently—a good trial period is every four weeks until experience dictates a shorter or longer period.
2. Dew point as measured at air inlet to the hopper is unacceptable.	Loss of regeneration heaters in one or both beds or line fuses.	These can be checked with a volt meter at the control panel.
	Loss of timer or clock motor switching ability from one head to the other, i.e., continuous operation on only one desiccant bed.	Check clock motor for movement by observing either function indicators or valve-shifting mechanisms. Note that loss of regeneration heaters may occur if the clock motor or shifting mechanism malfunctions.
	Desiccant has deteriorated or been contaminated.	Most manufacturers suggest checking the desiccant annually and replacing when it does not meet test criteria. Typically two to three years is a reasonable interval, depending upon the severity of service.
	Loss of power to one or both desiccant beds.	During regeneration cycle exterior of the desiccant bed should be hot to touch. Check contacts on relays or printed circuit board for flaws, check line fuses if so equipped.
3. Low or nonexistent air flow.	Fan motor burnout.	Replace.
	Loose fan on motor shaft.	Tighten.
	Clogged filter(s).	Change.
	Restricted or collapsed air lines.	Correct and relieve restrictions.
	Blower motor is reversed.	Use of a pressure gauge or flow meter is suggested. Proper rotation is that in which the highest flow is indicated.

SUMMARY

The key to understanding troubleshooting is to gain as complete as possible a knowledge of what the machine is doing to the plastic, what the plastic is doing to the mold, etc. This book describes the complete process so that you can obtain an in-depth understanding of all the parameters involved.

Chapter 28
Testing and Quality Control

D. V. Rosato

INTRODUCTION

Properties of plastics are directly dependent on temperature, time, and environmental conditions (see Fig. 28-1). These conditions can be related to raw material performance, processing performance, and part performance. (In our discussions plastics are also called resins or polymers.) This interrelationship provides unique characteristics to the plastics processor and the designer of the part. In fact it provides the means to set up logical testing and quality control procedures to meet zero defects, particularly when compared to the use of other materials. We can have complete control of "the complete process."

Unfortunately there is no single set of rules that designates which tests are to be conducted in order to manufacture a part repeatedly with zero defects. Tests are dependent on the performance required. For example, if a part is to operate where any type of failure could be catastrophic to life, then extensive and very expensive testing is required. This situation is similar to setting up a machine to perform at maximum efficiency (and at the lowest cost) or designing a part to meet performance requirements (and at the lowest cost). What is required is a knowledge of what is available and how to apply the knowledge.

Since testing or quality control is important to part production, this chapter has been prepared to make you aware of some of the differ-

ent tests that are available. How deeply you get involved in testing depends on your performance requirements. If all you need to do is weigh the part, that is all you do. However, there usually is an opportunity to set up a test/control that permits meeting the same performance requirements but producing parts at a lower cost (the value analysis approach). Perhaps you can produce a thinner wall or mold the part to tighter tolerances, thus reducing plastic consumption and cost of the part. A cost advantage can still exist even though you now require more expensive testing/controls.

Testing and quality control are two of the most talked about, yet often least understood, facets of business and manufacturing. Many companies spend a high percentage of each sales dollar on quality control. Usually, this involves inspection of components and parts as they complete different phases of manufacturing. Parts that are in specification proceed, while those that are out of specification are either repaired or scrapped. The workers who made the out-of-spec parts are notified that they produced defective parts and that they should correct their mistakes.

This is an after-the-fact type of quality control. All defects caught in this manner are already present in the piece being manufactured. While this type of QC will usually catch defects resulting from special causes, it does little to correct basic problems in production.

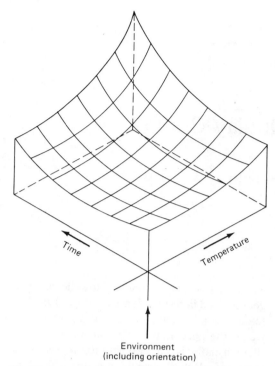

Fig. 28-1   Properties interrelate to temperature, time, and environment.

One of the problems with add-on quality control systems of this type is that they constitute one of the least cost-effective ways of obtaining a high-quality part. Quality must be built into a product from the beginning; it cannot be inspected in. The closest any add-on, after-the-fact quality control system can come to improving the quality built into the part is to point out manufacturing defects to the departments and persons responsible for a particular phase of the manufacturing operation.

The add-on system is the basic approach of many companies. The result, however, is often less than satisfactory. Often the desired quality level can only be achieved with heavy inspection costs. Workers who are continually badgered about their quality even when they are doing their best get discouraged and become antagonistic toward the quality control department. In the extreme, this can result in defects being deliberately hidden and a further degradation of quality.

After-the-fact quality control is necessary. Yet, by itself it does little to improve the qual-

ity level of an operation. The object must be to control quality before a part becomes defective. If this can be done, then after-the-fact QC can be minimized. Rework, scrap, and production costs will also be minimized. In addition, overall QC costs will be reduced. With plastics there are many different types of tests/controls that can be readily used when required. The goal here is to use controls when they are required.

The widespread use and rapid growth of plastics result largely from their versatility and desirable mechanical properties, as they range from soft elastomers to rigid or high-strength polymers. Because of this widespread use of plastics, people with widely differing backgrounds and interests must have knowledge of their mechanical properties. For this reason, the mechanical properties (see Fig. 28-2) can be considered the most important properties when compared to others (physical, chemical, permeability, electrical, etc.). There are a great many structural factors that determine the nature of the mechanical behavior of these polymers (see chapter 25). Factors that influence properties are polymer composition (fillers, molecular weight distribution, cross-linking and branching, crystallinity and crystal morphology, etc.; see Figs. 28-3, 28-4, and 28-5), the method used in molding parts, and part performance environments.

MECHANICAL PROPERTIES

The most important mechanical property of a plastic material is its stress–strain curve (see Fig. 28-2). This curve is obtained by stretching a sample in a tensile testing machine and measuring the sample's extension and the load required to reach this extension.

Since polymeric materials show viscoelastic behavior (a combination of elastic and viscous behavior) that is highly sensitive to temperature and, in some materials, to relative humidity variations, it is important to use samples of standard shapes, preconditioned at a constant and standard temperature and relative humidity before testing. Also they are stretched at a constant speed if the results are to be comparable to other tests.

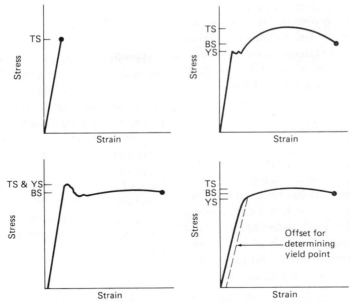

Fig. 28-2   Tensile stress–strain curves. TS = tensile strength; YS = yield strength; BS = breaking strength.

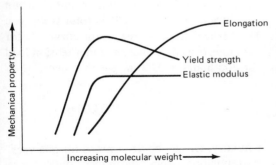

Fig. 28-3   Example of mechanical properties vs. molecular weight of plastics.

The stress–strain curve provides information about the modulus of elasticity (Young's modulus), which is related to the material's stiffness or rigidity. This curve also provides information about the yield point, tensile strength, and elongation at break. The curve defines toughness (the area under the curve), which is the energy per unit volume required to cause the sample to fail. Thus the stress–strain curve reveals much about a material's mechanical behavior.

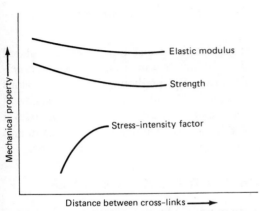

Fig. 28-4   Effect of distance between cross-link sites on compression properties.

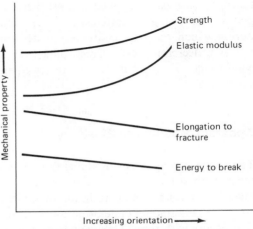

Fig. 28-5   Effect of uniaxial orientation on tensile properties tested in the direction of orientation.

For plastic foams used for cushioning, the stress–strain curve is obtained in compression rather than in tension, because this is the usual mode of loading the material in use. The information obtained from compression measurements is almost the same as in tension.

Creep Data

A very important mechanical property that must be considered for a plastic, as well as other materials, is its behavior when it is subjected to a continuous load, or creep. It takes into account the total deformation under stress after a specific time in a given environment (such as temperature). This property is very useful in the design of plastics (see Chapter 6).

A plastic material subjected to continuous load experiences a continued deformation with time (creep). With most plastics, deformation can be significant even at room temperature. However, with certain plastics, and particularly reinforced thermoset plastics (composites), elevated-temperature creep resistance is excellent.

Creep is the total deformation under stress after a specified time in a given environment beyond the instantaneous strain that occurs immediately upon loading. Independent variables that affect creep are time under load, temperature, and load or stress level.

Initial strain or deformation occurs instantaneously as a load is applied to most thermoplastics. Following this initial strain is a period during which the part continues to deform but at a decreasing rate. Creep data over a wide range of temperatures are reviewed in Chapter 6.

Apparent Modulus of Elasticity

The concept of apparent modulus is a convenient method for expressing creep because it takes into account initial strain for an applied stress plus the amount of deformation or strain that occurs with time. Thus, apparent modulus E_A is:

$$E_A = \frac{\text{Stress (psi)}}{\text{Initial strain} + \text{Creep}} \quad (28\text{--}1)$$

Because parts tend to deform in time at a decreasing rate, the acceptable strain based on service life of the part must be determined—the shorter the duration of load, the higher the apparent modulus and the higher the allowable stress. Apparent modulus is most easily explained with an example.

As long as the stress level is below the elastic limit of the material, modulus of elasticity E is obtained from equation (28–1). For example, a compressive stress of 10,000 psi gives a strain of 0.015 in. per in. for FEP resin at 73°F; then:

$$E = \frac{10,000}{0.015} = 667,000 \text{ psi}$$

If the same stress level prevails for 200 hours, total strain will be the sum of initial strain plus strain due to time. This total strain can be obtained from a creep data curve. If, for example, total deformation under tension load for 200 hours is 0.02 in. per in., then:

$$E_A = \frac{10,000}{0.02} = 500,000 \text{ psi}$$

Similarly, E_A can be determined for one year. Extrapolation from the creep data curve (which is a straight line) gives a deformation of 0.025 in. per in., and:

$$E_A = \frac{10,000}{0.025} = 400,000 \text{ psi}$$

When plotted against time, these calculated values for apparent modulus provide an excellent means for predicting creep at various stress levels. For all practical purposes, curves of deformation vs. time eventually tend to level off. Beyond a certain point, creep is small and may be neglected for many applications.

CHARACTERIZING PROPERTIES

A number of physical parameters, including density, morphology, molecular structure, and mechanical and thermal properties, influence a plastic's performance.

Density

Density is defined as weight per unit volume. Each material has a specific density range that is frequently used as an auxiliary identification method. For semicrystalline polymers, density depends on the degree of crystallinity estimation. In addition, other properties (at low strains) such as the modulus of elasticity, yield stress, and hardness of semicrystalline polymers also depend on the degree of crystallinity and are thus related to density.

The permeability of semicrystalline polymers to gases and vapors also depends on the degree of crystallinity; as percent crystallinity increases, permeability decreases. For a specific material such as polyethylene, its hardness, modulus of elasticity, and yield stress will usually be higher for a high-density grade, whereas the permeability will be lower.

Morphology/Amorphous and Crystalline

Morphology is concerned with the molecular structure of polymers, ranging from amorphous to crystalline. The degree of crystallinity has direct effects on the mechanical properties of plastics. Since crystallinity can be varied widely, the structure can range from a flexible to a very high strength polymer.

Crystalline plastics have uniform and compact molecules, a structure attributed to the formation of crystals (which can be of different sizes) having definite geometric and orderly form. The amorphous plastics are just the opposite: they lack an orderly form. Practically all plastics are normally in the amorphous stage during heat-melting. This characteristic morphology of plastics can be identified by tests (to be reviewed later). It provides an excellent control as soon as material is received in the plant, during processing, and after it is molded.

As an example, highly crystalline polymers such as polypropylene have a complex morphological structure. The polymer chains generally appear to fold into a laminar structure.

Between the layers are amorphous-like chain folds and some chains that go from one layer to the next to tie the whole structure together (1).

Molecular Structure

The weight of a molecule of a substance is referred to that of an atom of oxygen as 16.000, and is the sum of the atomic weights of the atoms in the molecule. The molecular weight of a monomer (basic material to produce the polymer) is a definite figure calculated from its composition. In polymers the number of units making up the molecule varies considerably, and molecular weights are generally stated as averages. The determination of the molecular weights of polymers can be performed by different techniques (to be reviewed later in this chapter).

As opposed to most simple organic compounds, polymers do not have a uniform molecular weight (MW), but rather a molecular weight distribution (MWD). In a polymeric material, molecules of varying molecular weights are found. A common way of dealing with such a molecular weight distribution is to use averages. Although each distribution has an infinite number of averages, only a limited number are important because they affect properties significantly. These averages include the number average molecular weight ($\bar{M}_n$), weight average molecular weight ($\bar{M}_w$) and the z-average molecular weight ($\bar{M}_z$).

$\bar{M}_n$ can be determined from measuring properties of dilute polymeric solutions. The methods include: (1) cryoscopic measurements (freezing point depression); (2) ebuliometric measurements (boiling point elevation); (3) vapor pressure measurements; and (4) osmotic pressure measurements.

$\bar{M}_w$ is usually determined by light scattering, while the z-average is determined from ultracentrifuge measurements or calculated from the MWD. The MWD can be determined by fractionation method. Presently, MWD is usually determined by gel permeation chromatography (GPC). Since the entire molecular weight distribution is determined from GPC measurements, any average can be calculated by this method (information on GPC will be reviewed later).

Molecular weight averages (primarily $\bar{M}_n$ and $\bar{M}_w$) and the MWD influence mechanical properties of polymeric materials, especially high strain properties such as tensile strength and ultimate elongation. However, additional properties such as environmental stress crack resistance also depend on molecular weight averages and the MWD. Further, flow properties of polymeric melts (during processing) are highly sensitive to $\bar{M}_w$ and, to a lesser extent, to the MWD. Viscoelastic properties, on the other hand, are very much dependent on $\bar{M}_z$ and higher averages. The higher the molecular weight average and the narrower the MWD, the better the polymer's mechanical properties. However, the flow of polymeric melts becomes more difficult with an increase in $\bar{M}_w$ and a decrease in MWD.

Two additional quantities often measured and reported for MW estimation are intrinsic viscosity and melt flow index (MFI). From the intrinsic viscosity, the viscosity average molecular weight ($\bar{M}_v$) can be calculated to obtain an estimate of $\bar{M}_w$. The MFI, a measure of the polymer's fluidity in the molten state, can also be related to molecular weight. The weight average molecular weight of a polymer sample is related inversely to its fluidity; therefore the lower the MFI, the higher the $\bar{M}_w$.

However, these two quantities can only be used to estimate $\bar{M}_w$ for linear polymers and for comparison between different grades of the same polymer, such as high-density polyethylene (HDPE). For branched polymers, such as low-density polyethylene (LDPE), these quantities do not give a good estimate for $\bar{M}_w$. They can, however, be used for comparison between different grades.

Mechanical Properties

The testing of plastics is generally carried out for the same reasons as the testing of other materials, for example, to determine their suitability for a particular application, for quality control purposes, or to obtain a better understanding of their behavior under various conditions. It is also necessary for the manufacturer of a new plastic to be able to measure

performance compared to that of other materials, including other plastics (2).

Because of the diversity of polymers, copolymers, and modifiers, the range of properties is extensive. An understanding of some of the basic attributes of tests is helpful in determining whether or not to employ a plastic in a given application. One test may measure a single property or several properties at once. In every case the test has been devised to be as accurate as possible. After many years of work by thousands of technical specialists in the plastics industry, the tests presently used are generally regarded as suitable. Nevertheless, further improvement is constantly sought.

Tests are not ends in themselves, but rather means of extracting knowledge about materials. Most production plants have laboratories for routine quality control tests, and similar tests are conducted separately at research and development facilities. Both types of test facilities characterize materials for sales descriptions and as reference points in quality-improvement programs.

Standard tests such as those described in this chapter are frequently used in government and industry specifications to spell out properties required in a material. Plastics processors are generally interested in all test values, but they particularly watch those that affect the handling qualities of materials in production equipment.

The real test of a material comes with actual service. Once a plastic product is taken home and used by the consumer, it no longer matters whether tensile strength is 5000 to 50,000 psi. The product either succeeds entirely or it fails. To assure success of toys, housewares, industrial products, and automotive components, the properties of likely materials are studied by design engineers who, through experience and judgment, balance material characteristics and service requirements against the amounts of material needed in parts to give adequate safety margins. Hence, the tests are tested.

In a certain case a service requirement may be so complex that suitable material can be determined only in actual service. For example, plastic for pipe (injection molded fittings and extruded pipe) is tested by making parts of it, attaching it to a pressurized waterline, and seeing what happens.

Tests are meaningful provided they are used properly. There are several considerations here: (1) Data-sheet properties are used only for initial screening of materials for an application, with the knowledge that the elastic modulus is usually less affected by materials and process variables than is strength. (2) It is necessary to obtain structural properties on materials made by the process that will be used in the final manufacture of the component; all details of the process should be considered. (3) For thermoplastics, some of the parameters considered are process temperatures and cooling methods, pressures, flow patterns of the molten plastic, knit lines, processing aids and additives, proportions of regrind, moisture content of raw materials, and any stress concentrations envisioned for the product. (4) For reinforced plastics many of the variables discussed above are appropriate for consideration; other important process variables include the effects of coupling agents, additives for flexibility, fire-retardant additives, modification of viscosity, proportions of catalysts, method of consolidation and compaction of the laminate and means for removing entrapped air, placement of oriented layers (see Fig. 28-6), spray-up patterns and procedures, laps at joints in preformed reinforcements, and so on. (5) Short-term or basic stress, strain, and strength behavior, which are frequently the only structural properties available from data sheets, are perhaps least affected by materials and process variables. Unless these points are recognized, it is highly likely that the limits set for the plastic structural component may be exceeded. Impact tests and tests performed under sustained stress, either at elevated temperatures or in aggressive environments, prove to be meaningful and economical methods for evaluating such effects.

Mechanical Test Equipment

The most important mechanical characterization of a polymer is its stress–strain relationship. These data can show the presence or absence of a yield point and the slope of the initial curve, which is the elastic modulus in tension and is related to stiffness and resistance to deformation. The mechanical properties observed depend on the rate of testing. Polymers are tested in tension, flexure, and compression and at impact speeds.

Low-shear-rate testing of polymers is considered to be in the range of 0.02 to 50 in./min, and a variety of instruments are available to measure the forces over a wide range through the use of interchangeable load cells, amplifier circuits, and recorders. Suitable test fixtures and couplings provide the required mode of loading. A polymer may be tough and ductile when tested at slow rates of loading, while it may show a more brittle type of failure when tested at high rates of loading (see Fig. 28-7).

Electrical Tests

Plastics are used in electrical applications mainly because they are excellent electrical insulators. The most significant dielectric properties of a plastic are dielectric strength, dissipation factor, dielectric constant, and resistivity. These properties are affected by many factors including time, temperature, moisture content, electrode size, and test frequency. Furthermore, these factors can interact in a complex manner. New test instruments for automatic dielectric measurement provide the capability for using changes in dielectric properties as a meter to monitor plastics processing. It is important to recognize that electrical tests can be conducted that relate to mechanical properties.

Thermal Properties

Thermal properties of plastics include thermal conductivity, heat capacity, and thermal expansion. In addition, plastics are tested for heat resistance, heat deflection point, melting point, and flammability to establish high-temperature service conditions. Thermal conductivity can be measured under static or dynamic conditions.

Dilatometers are instruments used for measuring the thermal expansion or contraction

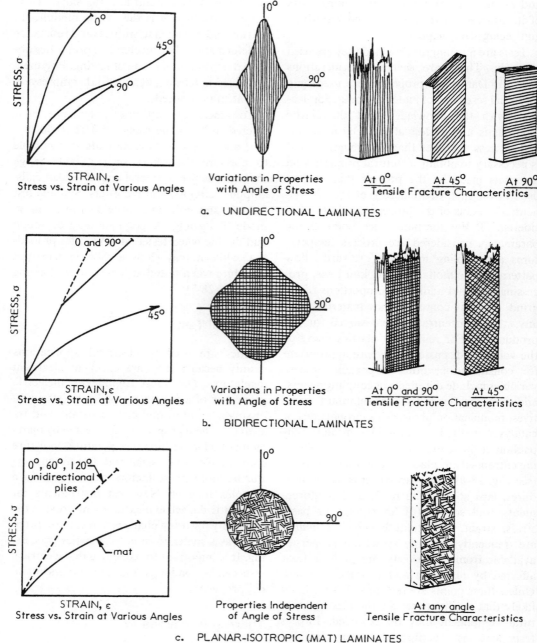

a.   UNIDIRECTIONAL LAMINATES

b.   BIDIRECTIONAL LAMINATES

c.   PLANAR-ISOTROPIC (MAT) LAMINATES

Fig. 28-6   Tensile properties based on orienting composite plastics (3).

of liquids or solids. Several types exist, with the accuracy of observations depending on the size and uniformity of the measuring capillary, the precision of the temperature control, and the efficiency of heat transfer.

Thermogravimetric analysis (TGA) is a process that continuously measures sample weight as the reaction temperature is programmed at a linear rate of heating. Instruments are available that simultaneously record temperature and sample weight loss along with the derivative thermogravimetric (DTG) and the differential thermal analysis (DTA) curves. TGA is used to study pyrolysis, reac-

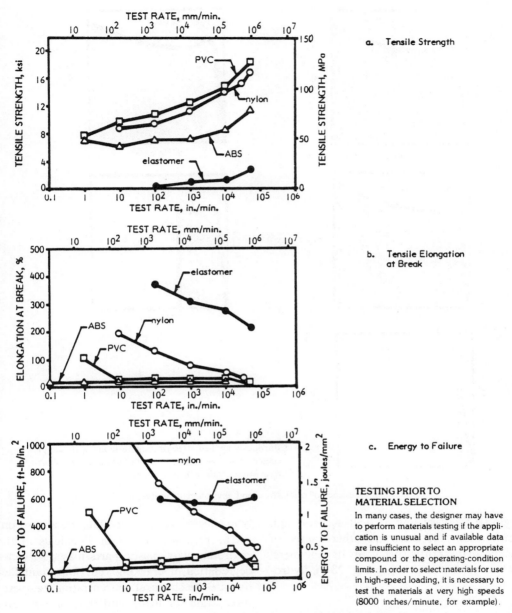

a. Tensile Strength

b. Tensile Elongation at Break

c. Energy to Failure

TESTING PRIOR TO MATERIAL SELECTION

In many cases, the designer may have to perform materials testing if the application is unusual and if available data are insufficient to select an appropriate compound or the operating-condition limits. In order to select materials for use in high-speed loading, it is necessary to test the materials at very high speeds (8000 inches/minute, for example).

Fig. 28-7 Behavior of several plastics during high-speed tests (3).

tion kinetics, thermal stability, and thermal degradation behavior.

Chemical Properties

Test equipment is available to measure the resistance of a plastic to moisture, acid, alkali, and other chemicals. The tests are chiefly immersion tests with measurement of swelling and accompanying loss in mechanical proper-

ties. Also environmental tests are conducted on specimens that have experienced prior stress (see Fig. 28-8). Some plastics can self-destruct when under stress and immersed in certain solutions.

Cracks can grow very slowly under small sustained loads in ductile materials. Amorphous polymers are brittle in impact if the part is too thick in comparison to a notch or corner radius. Polymers exhibit low strain

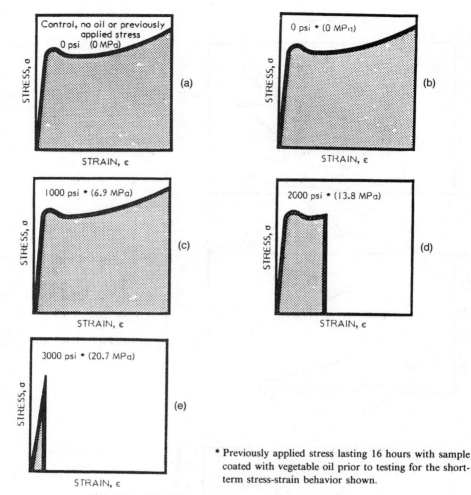

* Previously applied stress lasting 16 hours with sample coated with vegetable oil prior to testing for the short-term stress-strain behavior shown.

Fig. 28-8 Influence of prior stress and environment on ductility (3).

at break for very low as well as very high strain rates. But there are other causes of unexpectedly brittle behavior of polymers, such as solvent attack (see Fig. 28-9), hydro-lytic effects, aging effects, fatigue, and so on.

The presence of certain chemicals in the environment of plastic parts may also cause degradation or cracking. The important factors are the chemical concentration, temperature, strain (or load), and time of exposure. The chemical may have no effect on the polymer unless it is applied while the sample is under load. In small concentrations, the chemical may evaporate off the polymer before the cracking can occur and thus not appear to be a stress-cracking agent. Moreover, the state of stress as well as its magnitude is frequently quite important.

The biaxial data in Fig. 28-10 illustrate the time dependence inherent in all environmental stress-cracking phenomena. The biaxial-failure stress for polyethylene is substantially reduced over the corresponding value in air at

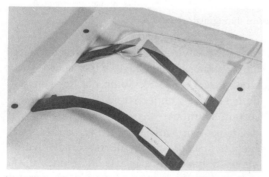

Fig. 28-9 Two test bars under the same stress were sprayed with acetone. One cracked quickly; the other did not fail. Different plastics were used.

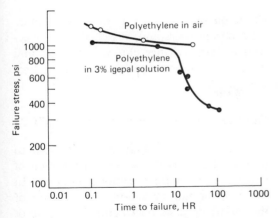

Fig. 28-10   Time to failure of polyethylene at constant biaxial stress at 73°F (4).

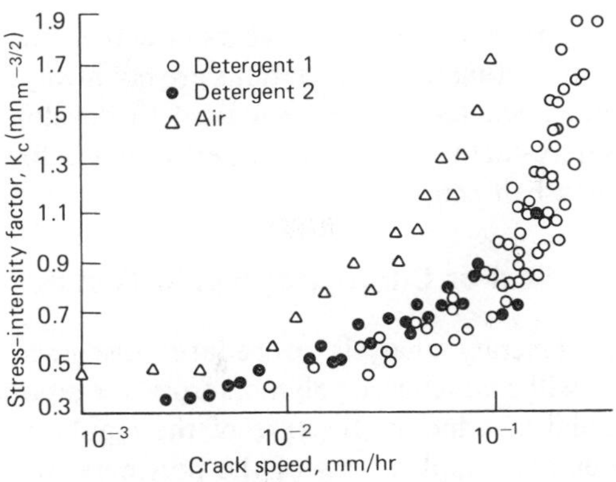

Fig. 28-11   Effects of different environments on crack growth in polyethylene (4).

the longer times by the detergent (Igepal) solution. As for the long-term brittle failure of plastics in benign environments, these phenomena are best understood through the concepts of fracture mechanics (4).

Crack growth in polymers is often described by:

$$(1 - v^2) D \left(\frac{\beta K_1^2}{\dot{a}}\right) = \frac{8\gamma}{K_1^2} \quad (28\text{--}2)$$

where v is Poisson's ratio, D is the creep compliance, $\dot{a} = da/dt$ is the crack speed, K_1 is the applied stress-intensity factor, γ is the fracture energy for propagation, and β is a failure-zone property having to do with the stress distribution in the failing zone of material at the tip of the crack. If the creep compliance can be approximated by the power law:

$$D(t) = D_0 t^n \quad (28\text{--}3)$$

then:

$$K_1 = \left[\left(\frac{8\gamma}{(1 - v^2)\beta^n D_0}\right)^{\frac{1}{2(1 + n)}}\right]\left[\dot{a}^{\frac{n}{2(1 + n)}}\right] \quad (28\text{--}4)$$

We see from equation (28–2) that when the crack speed is fast, D is evaluated in the glassy region of the material, and the slope n in equation (28–3) is small. When the crack speed is smaller, the response of the material ahead of the crack tip is evaluated at later times

where n is larger. Thus, in general, the slope $n/(2(1 + n))$ on the K_1–$\dot{a}$ curve will increase with increasing crack speed, as shown in Fig. 28-11.

The shapes of these curves are all about the same, indicating from equation (28–2) that the detergent is affecting crack growth by its influence on the polyethylene failure-zone properties γ and β and not by altering its viscoelastic properties.

So, the environment can influence crack growth either by changing the material's fracture energy γ and the stress measure β of the failure zone (plasticizing the crack tip), or by affecting the viscoelastic properties of the plastic. The former is primarily responsible for so-called static fatigue (which is really creep crack growth) in inorganic glasses, whereas one or both of these phenomena play a role in the environmental behavior of polymers.

Climatic chambers are used to simulate the effects of sunlight and weathering on plastics. Test chambers are designed to simulate the influence of: solar radiation; moisture; atmospheric pollutants including oxides of nitrogen, hydrocarbons, and ozone; and a variety of solid particles on the service life of plastics. Such a test should provide accelerated aging by a factor of 1000 so that a one-week test would represent about 20 years of actual life; but, unfortunately, correlations are not always good. Such tests have been used to develop a safety factor for expected performance for over a half century.

Analytical or Chromatographic Tests

The generaly analysis procedures discussed here will characterize polymers from a compositional standpoint. Because of the emphasis on end-use applications of the polymers, the results of analysis should be correlated with traditional physical property measurements required in most material specifications.

The characterization of polymers and their additives can be quickly and accurately accomplished by the use of chemical analyses and the appropriate instrumentation. In many instances, the chemical composition of the polymer and additive package can be used to predict physical property data that could not be obtained for months, even with the use of accelerated life testing. The timeliness and relevance of the chemical characterization data offer a strong incentive for the increased use of this type of polymer evaluation.

Solution viscosity is the most widely used analytical test for characterizing the molecular structure of the polymer. There are a number of capillary viscometers available which are used to measure the flow time of a polymer solution against that of pure solvent at specified conditions of temperature and concentration. The resulting viscosity ratio, which is a dimensionless number, can be correlated with the average molecular weight. The test is used to determine changes in polymer brought about by aging, processing, or exposure.

Liquid Chromatography (LC)

This technique separates components of a mixture by differences in their rates of elution arising from interactions between the sample and the column-packing material. There are four principal mechanisms by which components are separated in liquid chromatography. These are differences in partition coefficients (liquid–liquid chromatography), absorption

effects on surfaces such as silica gel (liquid–solid chromatography), dissociation of electrolytes (ion-exchange chromatography), and molecular size or shape (size exclusion chromatography).

Gel Permeation Chromatography (GPC)

Gel permeation chromatography is used to provide the molecular weight distribution of a polymer by a fractionation technique. In the final forming operation, the behavior of the plastic depends on whether the range of species is wide or narrow and whether or not the distribution is skewed. GPC (size exclusion) separates molecules in solution by size. The effective size of a molecule in solution is related closely to the molecular weight.

The separation is accomplished by injecting the sample solution into a continuously flowing stream of solvent that passes through highly porous, rigid gel particles, closely packed together. The pore sizes of the gel particles cover a wide range. As the solution passes through the gel particles, molecules with small effective sizes will penetrate more pores than will molecules with larger effective sizes, and therefore will take longer to emerge and to be detected.

Gas Chromatography (GC)

Gas chromatography separates, characterizes, and quantifies the vaporized components of samples using both conventional and pyrolysis techniques. This procedure is used for identification of plastics and elastomers by GC fingerprinting, compositional analysis of copolymers and blends, and determination of residual monomers and highly evaporative agents. GC can be used to identify a polymer or the products of a degradative process, to monitor purity of monomers, to follow reaction rates and polymerizations, and to determine residual monomers.

This method separates volatile components of a mixture by differences in the rates of elution arising from absorption or partition interactions between the sample and the column-packing material. The term "gas chromatography" indicates that the moving phase is a gas. Gas–solid chromatography refers to the use of an active solid absorbent as the column packing. Gas–liquid partition chromatography refers to the use of a liquid distributed over the surface of a solid support as the column packing.

Ion Chromatography (IC)

This procedure is a chromatographic technique that utilizes the principles of ion exchange to separate mixtures of ionizable materials and, in most instances, a conductivity detector to sense the components resolved (5).

In practice, a liquid sample is introduced at the head of an appropriate separator column into a stream of ionic eluant which then carries the mixture through the column and toward a detector. The rate of travel of each sample component through the system depends upon its particular affinity for the column packing under the conditions of analysis. If migration rates are sufficiently different, each elutes from the column as a discrete band, ready for measurement.

The time at which a component exits the column is a clue to its identity, whereas the size of the peak is related to concentration. The heart of an IC system is the separator column. Usually, this is a tube packed with an ion exchange resin designed to separate either anions or cations. The resin is generally a pellicular material with a known charge and exchange capacity, tailored to meet specific application requirements.

Presently, two conductometric techniques are used for sensing analytes emerging from a column (see Fig. 28-12). In the first, the "single column" approach, the column effluent enters the detector directly, the electrical conductivity due to the eluant alone is electronically suppressed, and then sample peaks are measured above this baseline. Theoretically, this technique is relatively simple and inexpensive. It minimizes dead volume within the instrument, thus maximizing sample resolution, and makes it easier to analyze the anions of weak acids.

Using the second procedure, the column ef-

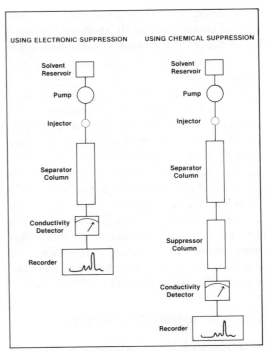

Fig. 28-12   Two techniques used in ion chromatography (5).

fluent first enters a high-capacity, ion exchange column of opposite charge, called a "suppressor," where the eluant is chemically modified to a less conductive form. At the same time, analyte ions are converted into highly conductive acids or hydroxides. The chief benefit of this method is an enhanced sensitivity for most species. With either system the result is a plot of conductivity vs. time, a chromatogram.

Samples for analysis by ion chromatography must be in solution form. Water-soluble liquids and solids are simply dissolved in deionized water, then diluted as needed. Water-insoluble samples often can be leached or extracted with water to obtain the impurities of interest. Gases that produce ionic species in water can be analyzed after absorption in an appropriate medium. Generally, the only other sample treatment needed is micro-filtration to remove any insoluble material that might damage pumps or plug columns.

Ion chromatography is a powerful analytical tool. Because it is a separation technique, IC allows the analyst to determine a number of components with a single sample injection. This, of course, permits the qualitative screening of samples, minimizes interferences when one is analyzing for specific impurities, and enhances analyte sensitivity.

IC is also fast and simple, and can often replace several tricky, time-consuming, and costly procedures with one method of analysis. Because of its inherent sensitivity, IC not only permits the determination of trace impurities but also makes the analysis of small samples practical. This feature is particularly useful when one is sample-limited. Ion chromatography can also be automated, freeing the analyst for other duties and allowing unattended round-the-clock operation. However, IC equipment is expensive to buy and maintain, and requires special training for operation.

Thermoanalytical (TA) Methods

Thermoanalytical methods characterize a system, either single or multicomponent, in terms of the temperature dependencies of its thermodynamic properties and its physiochemical reaction kinetics. Techniques involved are thermogravimetric analysis (TGA), differential scanning calorimetry (DSC), and thermomechanical analysis (TMA).

Thermogravimetric Analysis (TGA)

This method measures the weight of a substance heated at a controlled rate as a function of time or temperature. To perform the test, a sample is hung from a balance and heated in the small furnace on the TGA unit according to the predetermined temperature program. Because all materials ultimately decompose on heating, and the decomposition temperature is a characteristic property of each material, TGA is an excellent technique for the characterization and quality control of materials (see Figs. 28-13 and 28-14).

Properties measured include thermal-decomposition temperatures, relative thermal stability, chemical composition, and effectiveness of flame retardants. TGA is commonly used to determine the filler content of many thermoplastics.

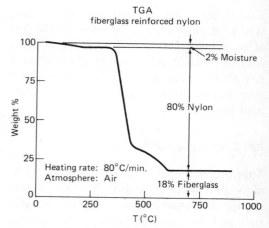

Fig. 28-13 Using TGA to determine amount of fiberglass reinforcement in nylon (6).

Thermal analysis is also useful in the quality control of thermosets. It characterizes curing profiles, which can be used to optimize curing conditions to achieve the desired degree of cure with the optimal combination of time and temperature. One can also check the curing profiles of samples from incoming lots of materials to make sure that materials from various lots are acting in the same manner.

One typical application of TGA is compositional analysis. For example, a particular polyethylene part contained carbon black and a mineral filler. Electrical properties were important in the use of this product and could be affected by the carbon black content. TGA was used to determine the carbon black content and mineral-filler content for various lots that were considered acceptable and unacceptable. The samples were heated in nitrogen to volatilize the PE, leaving carbon black and mineral-filler residue. Carbon content was then determined by switching to an air environment to burn off the carbon black. Weight loss was a direct measure of the carbon black content.

Differential Scanning Calorimetry (DSC)

Differential scanning calorimetry directly measures the heat flow to a sample as a function of temperature. A sample of the material weighing 5 to 10 grams is placed on a sample pan and heated in a time- and temperature-

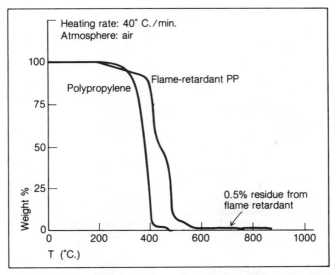

Fig. 28-14   Characterizing flame retardant in polypropylene using TGA.

controlled manner. The temperature is usually increased linearly at a predetermined rate. DSC is used to determine specific heats (see Fig. 28-15), glass-transition temperatures (see Figs. 28-16, 28-17, and 28-18), melting points (see Fig. 28-19) and melting profiles, percent crystallinity, degree of cure, purity, thermal properties of heat-seal packaging and hot-melt adhesives, effectiveness of plasticizers, effects of additives and fillers (see Fig. 28-20), and thermal history (see Fig. 28-21).

DSC is also used to determine percentage of crystallization (see Fig. 28-19). A significant consideration for polyolefins is their susceptibility to crystallization. The molder needs to know how rapidly material crystallizes as it is cooled. A comparison of materials from different lots will indicate whether they will crystallize in the same manner under the same molding conditions. (Polyolefins are provided in both nucleated and nonnucleated grades. A nucleating agent is added to a material to increase the material's rate of crystallization, a factor bearing on the performance of parts molded from that material.)

DSC is also a very useful technique for monitoring the level of antioxidant in, for example, polyolefins such as polypropylene. Polypropylene is among the materials most susceptible to oxidation, which causes brittleness and cracking to a degree that depends partly on the end-use of the molded part. Antioxidants

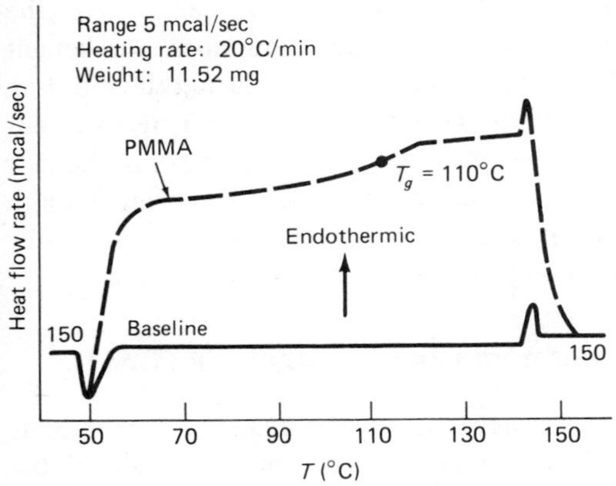

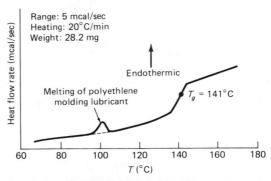

Fig. 28-15   DSC used to determine heat capacity of PMMA near the glass-transition temperature (6).

Fig. 28-16   DSC related to glass-transition temperature and detection.

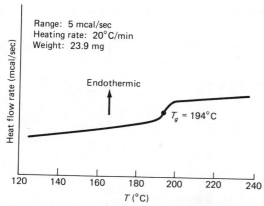

Fig. 28-17 DSC used to determine glass transition temperature of polysulfone (6).

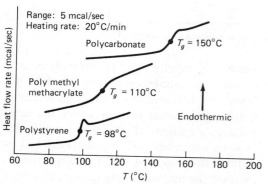

Fig. 28-18 DSC identifies glass transition temperature for amorphous plastics PC, PMMA and PS; indicated minimum temperature for processing the polymers (6).

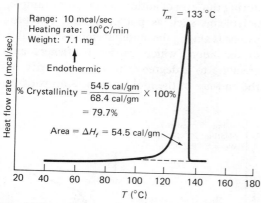

$$\% \text{ Crystallinity} = \frac{54.5 \text{ cal/gm}}{68.4 \text{ cal/gm}} \times 100\%$$

$$= 79.7\%$$

Area $= \Delta H_f = 54.5$ cal/gm

Fig. 28-19 DSC determines melting point and percent crystallinity of HDPE (6).

are added to extend service life and to protect material during the molding operation. However, the antioxidants are sacrificially oxidized to protect the polymer during the molding operation; and once the antioxidants are de-

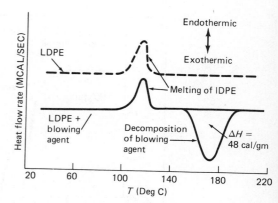

Differential scanning calorimetry is used for determining the effects of additives and fillers from a process and quality-control point of view. The above graph characterizes LDPE foam.

Fig. 28-20 DSC relates to the effects of additives and fillers that can be used in quality control for LDPE foam (6).

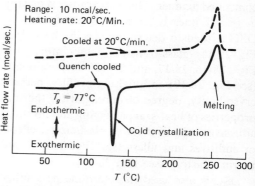

Fig. 28-21 DSC is used for determining effect of thermal history for thermoplastic polyester (6).

pleted, the material is vulnerable to oxidation. The client (end-user of the part) needs the antioxidant protection and does not benefit from antioxidants used up during the molding operation. Therefore, the molder needs to ensure that sufficient antioxidants are in the raw material before processing and that enough antioxidants remain in the material after molding to meet the customer's needs.

Thermomechancial Analysis (TMA)

This system measures dimensional changes as a function of temperature. Dimensional behavior of the material can be determined precisely and rapidly on small samples in any form—powder, pellet, film, fiber, or molded part. Parameters measured by thermomechanical analysis are coefficient of linear thermal expansion, glass-transition temperature (see Fig. 28-22), softening characteristics, and degree of cure. Also among the applications of TMA are the taking of compliance and modulus measurements and the determination of deflection temperature under load.

Tensile-elongation properties and melt index can be determined by using small samples such as those cut directly from a part. Part uniformity can be determined by using samples taken from several areas of a molded part. Samples can also be taken from an area where failure has occurred or continues to occur. This permits comparisons of material properties in a failed area with properties measured either at an unfailed section or from a sample of new material. Samples may also be taken from within a material blend to ensure that a uniform blend is being supplied. The results of such testing can be used either to evaluate part failure or in the acceptance testing of incoming materials or parts.

In basic mechanical testing, mechanical characteristics that can be tested include expansion, penetration, extension, flexure, and compressive compliance. Photoelastic-stress analysis allows stress distribution to be visually displayed, and strain gauging allows stress distribution to be approximated. Residual stress, also known as molded-in stress, can be measured by a variety of techniques.

Dynamic Mechanical Analysis (DMA)

The DMA procedure measures the viscoelastic properties (modulus and damping) of a material as functions of time and temperature. The material is deformed under a periodic resonant stress at a low rate of strain. Microprocessor data-reduction techniques provide graphical and tabular outputs of these properties as functions of time or temperature. The values determined, the modulus and damping data, aid in establishing realistic structural design criteria; the speed of analysis provides high throughput and low labor cost; precise temperature control can be used to simulate processing conditions; the breadth of material types ranges from rubbery to very high stiffness; and the data obtained correlate both structure–property and property–processing characteristics.

The DMA instrument can be calibrated to provide quantitative accuracy and precision in the range of ±5 percent coefficient of variation. To achieve this level of accuracy, the analyst considers several factors in the mathematical treatment of data: instrument compliance (i.e., the measurement system is not infinitely stiff), length compensation (to counteract end-effects at the clamps), Poisson's ratio (the ratio of lateral to axial strains for mixed shear/flexure deformation or interconversion, G' to E'), and shear distortion (for shear deformation in a flexural mode) (7, 8).

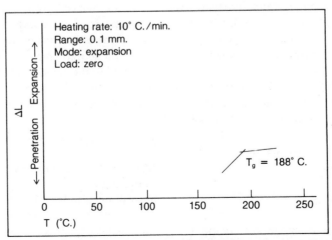

Fig. 28-22   TMA determines coefficient of expansion and glass-transition temperature of epoxy-graphite composite plastic.

Infrared (IR) Spectroscopy

Infrared spectroscopy records spectral absorptions in the infrared region using pyrolysis, transmission, and surface-reflectance techniques. Exposing the sample to light in the infrared range and recording the absorption pattern yield a "fingerprint" of the material. Infrared spectroscopy is used for identification of plastics and elastomers, polymer blends, additives, surface coatings, and chemical alteration of surfaces.

This is one of the most common analytical techniques used with plastics. The easy operation and the availability of this type of equipment have contributed to its popularity. Although the infrared spectrum characterizes the entire molecule, certain groups of atoms give rise to absorption bands at or near the same frequency, regardless of the rest of the molecule's structure. The persistence of these characteristic absorption bands permits identification of specific atomic groupings within the molecular structure of a sample.

For accurate interpretation of an infrared spectrum, the following criteria must be met:

1. The spectrum must be adequately resolved, and absorption bands must be of adequate intensity.
2. The spectrum should be of a reasonably pure compound. For example, the infrared spectrum of a polymer blend is often quite similar to the corresponding copolymer of comparable monomers ratio. Further, the presence of high levels of additives (i.e., plasticizers, stabilizers, slip agents, etc.) can also provide easily misconstrued information.
3. The spectrophotometer should be calibrated so that absorption bands are observed at their proper frequencies or wavelengths. Proper calibration can be made with an appropriate standard, such as polystyrene film.

X-ray Spectroscopy

This method identifies crystalline compounds by the characteristic X-ray spectra produced

when a sample is irradiated with a beam of sufficiently short-wavelength X radiation. Diffraction techniques produce a "fingerprint" of the atomic and molecular structure of a compound, and are used for identification. Fluorescence techniques are used for quantitative elemental analysis.

Nuclear Magnetic Resonance (NMR) Spectroscopy

Proton magnetic resonance spectroscopy characterizes compounds by the number, nature, and environment of the hydrogen atoms present in the molecule. Identification is possible because of the characteristic absorptions of radio-frequency radiation in a magnetic field as a result of the magnetic properties of nuclei. NMR techniques are used to solve problems of crystallinity, polymer configuration, and chain structure. Test instruments can provide fields of 50,000 gauss at frequencies of 60 MHz, so that the nuclei of polymer molecules can be made to resonate to provide NMR spectra.

Atomic Absorption (AA) Spectroscopy

Atomic absorption spectroscopy is one of the most sensitive analytical methods available for the determination of metallic elements in solution. The element of interest in the sample is not excited, but is merely dissociated from its chemical bonds and placed into an unexcited, un-ionized "ground" state. In this state, it is capable of absorbing characteristic radiation of the proper wavelength which is generated in a source lamp containing the sample element as the anode. The usual method of dissociation is by burning the sample in a flame of the appropriate gas or gases.

Raman Spectroscopy

Most molecular motions that cause Raman scattering of ultraviolet light also produce IR absorption bands. Macromolecular motions that are uniquely accessible to Raman analysis include accordionlike stretchings of chains in lamellar regions. Lamellae are sheetlike regions of crystalline ordering that coexist in many polymers with amorphous regions. Raman spectroscopy thus is important in determining maximum theoretical extents to which polymers may be drawn when high-tensile-modulus fibers are made.

Transmission Electron Microscopy (TEM)

Transmission electron microscopy is a technique to greatly magnify images of objects by means of electrons. Electron microscopes serve two purposes; (1) they permit the visual examination of structures too fine to be resolved with light microscopes, and (2) they permit the study of surfaces that omit electrons. In its simplest form, a transmission electron microscope consists of a source supplying a beam of electrons of uniform velocity, a condenser lens for concentrating the electrons on the specimen, a specimen stage for displacing the specimen that transmits the electron beam, an objective lens, a projector lens, and a fluorescent screen on which the final image is observed.

Optical Emission Spectroscopy (OES)

This technique characterizes most of the metallic ions, in addition to certain nonmetals, in terms of the emission spectra produced when electrons are excited by an arc or by other means.

Summary of Characterizing Properties

Listed above are many of the techniques available to the processor. They can be used from the time that the plastic raw materials (additives, fillers, color, reinforcements, etc.) arrive in the plant, during the time that the materials are processed, to control regrind performance, and for quality control of the finished part.

Most of the testing performed continues to be predominantly mechanical rather than these analytical systems. Based on the most pervasive trend in analytical instrumentation with increased computerization, more analytical testing will be conducted.

In the past, such analytical techniques have moved out of the rarefied atmosphere of university and corporate chemical-research laboratories and into the workaday world of formulating and quality control in manufacturing shops. This transition has been aided by advances in microelectronics that have tended to bring prices down, make testing much less time-consuming and labor-intensive, and render instruments much easier to operate by non-specialists. Other advances in instrument technology have made possible new types of determinations of polymer composition or performance that were more difficult or impossible previously.

Use of analytical techniques has two important effects: One is microcomputer control of the instrument itself, providing automatic running of preprogrammed test routines, allowing non-experts to run tests by pushing a button, and allowing the operator to walk away while the system sequentially tests virtually any number of samples. The other result is data management: automatically converting test data into usable form, performing calculations, drawing graphs, and storing data for retrieval.

Gone is the need to search manually through voluminous paper files. With laboratory information management systems (LIMS), test data on a lot of outgoing products can be called up on a CRT, with accompanying information about who performed the analysis and when, who the customer was, what the lot numbers were of raw materials from which the product was made, and who supplied them. Even more impressive are the data-manipulation capabilities being offered for such techniques as infrared spectroscopy.

In Table 28-1, a condensation is presented only on the performance characteristics of the more widely used analytical systems. Table 28-2 summarizes the relative merits of the analytical instruments used to characterize plastics. Included in these tables are melt flow tests (MFT) and rheological mechanical instruments (RMI), which will be discussed later in this chapter. Of particular importance will be real-time process control during the complete injection molding process. The on-

702 VII / PROBLEM SOLVING

Table 28-1   Typical instruments used to characterize plastics.

DMA (Dynamic Mechanical Analysis)

Applications:	Principally applicable to processed end product.
	Dynamic mechanical analysis usually measures the stress response of the material subjected to a strain that is a periodic function of time. It involves the determination of the dynamic mechanical properties of polymers and their assemblies. Dynamic modulus, loss modulus, mechanical damping or internal friction, and others are determined from this analysis.
Limitations	Dissimilar physically composite systems (multilayer constructions) are not readily analyzable.
	Mechanical frequency of operation is the least sensitive of rheological systems.
Data output:	Quantitative.
Sample: type/sample size/ time for measurement:	Solid/5 to 10 g/15 mins.
Method of analysis:	Measures resonant frequency (such as 0 to 10 Hz) and energy dissipation characteristics over a wide temperature range (such as 0 to 300°C).

DSC (Differential Scanning Calorimeter)

Applications:	Total QC capability for thermoplastics and thermosets; from raw material, through processing, to end product. Basically provides continuous measurement of the heat absorbed or given off by a sample while it is being heated at a controlled rate. Measures heat flow, melting profile or T_g, processing energy, percent crystallinity, curing profile (thermoset), additive analysis (mold release, antistat, etc.), etc.
Limitations:	Thermally analyzing a liquid solvent system may be misleading in terms of heat of cure.
Data output:	Quantitative.
Sample: type/sample size/ time for measurement:	Solid and Liquid/0.01 to 0.5 g/15 to 30 min.
Method of analysis:	Controlled enthalpy (heat analysis); measured chemical and thermal reactivity in the area of the polymer's glass transition (T_g) through a wide temperature range (0 to 300°C).

IR (Infrared Spectroscopy)

Applications:	Development tool; can be used as secondary interruptive tool to LC.
	Provides surface analysis; such as coatings, adhesives, films, etc.; also plasticizer. Thorough chemical structure identification; polymer composition migration, silicone release migration, etc.
Limitations:	Highly qualitative tool that requires extensive interpretative capability by user.
Data output:	Qualitative.
Sample: type/sample size/ time for measurement:	Solid, liquid, and gas/1 to 2 g/5 min.
Method of analysis:	IR absorption analyzing organic chemical structure.

LC (Liquid Chromatography) and GPC (Gel Permeation Chromatography)

Applications:	Total QC capability; from raw material, through processing, to end product. Analyze amount of antioxidants, plasticizers, lubricants, polymer molecular weight distribution, etc.
	When operated in gel permeation mode (GPC) separates polymers and other compounds in order of decreasing molecular weight. Useful for separating additives in the low-molecular-weight samples from prior GPC separation.
	Note: LC is separation of solutes by chemical affinity or polarity using various combinations of solvents and column packings.
	Note: GPC is a special size separation technique employing a three-dimensional

Table 28-1 *(Continued)*

gel network as the LC packing; the "molecular sieve" effect separates molecules by molecular weight. It is also called "exclusion chromatography," since larger molecules are excluded from the pores of the gel structure, and having a shorter path, elute first.

Limitations:	Sample must be dissolvable in common laboratory solvents.
	Not directly applicable to cured thermosets.
	To evaluate inorganic additives, the additive must be chemically bonded to the organic substance; otherwise evaluation is null. To evaluate nonbonded inorganics, atomic absorption spectra photometer is applicable. There are exceptions; e.g., silicone (inorganic) when in an elastomer can be identified with LC.
	Fillers must be filtered or centrifuged out of sample solution.
Data output:	Quantitative.
Sample: type/sample size/ time for measurement:	Solid and liquid/1 to 2 g/$\frac{1}{2}$ hr or less; includes time to dissolve sample.
Method of analysis:	Molecular weight distribution via refractive index (RI) detection or:
	Absorbance ratioing via UV absorption (used for non-IR materials where passage of natural light is not possible)—after a sample is separated into its components by the above column technique.

MI (Melt Index or Melt Flow Tests per ASTM)

Applications:	ASTM D569: Thermoplastic molding material ($\frac{3}{8}''$ dia. × $\frac{3}{8}''$ long specimen subjected to a pressure and time in a specific mold).
	ASTM D621: Compression deformation ($\frac{1}{2}''$ cube specimen).
	ASTM D648: Flexural deformation ($\frac{1}{8}''$ to $\frac{1}{2}'' \times \frac{1}{2}'' \times 5''$ specimen subjected to a pressure and temperature).
	ASTM D1238: Thermoplastic extrusion plastometer (time to move melt through a die of specific length and diameter at prescribed temperature, load, and piston pressure.
	ASTM D1703: Thermoplastic capillary flow.
	ASTM D3123: Spiral flow of thermosets.
	ASTM D3364: Flow rate of rheologically unstable thermoplastics and others.
Limitations:	Very specific applications.
	Characteristics of plastic melts depend on a number of variables; since the values of the variables occurring in these tests may differ substantially from those in large-scale processes, test results may not correlate directly with processing behavior. Use tests for intended purpose per ASTM review.
Data output:	Quantitative.
Sample: type/sample size/ time for measurement:	Solid/2 to 50 g/5 to 75 min.
Method of analysis:	Methods vary (moving melt through die, mechanical deflection, etc.); see ASTM standards for details.

RMS (Rheological Mechanical Spectrometer)

Applications:	Total QC capability; from raw material, through processing, to end product. Most sensitive rheological instruments available. Rheological methods directly relate chemical structure to physical properties—whereas others measure only key variables such as molecular weight, glass transition, etc., which then have to be interpreted as physical properties.
Limitations:	Familiarity with rheology required.
Data output:	Quantitative.
Sample: type/sample size/ time for measurement:	Solid and Liquid/5 to 10 g/5 min.
Method of analysis:	Measures viscous and elastic response in terms of dynamic viscosity over a wide temperature (0 to 300°C) and mechanical frequency range. Nearly all viscometric systems fall into two basic classes; (1) those in which flow is caused by a difference in pressure from one part of the liquid to another, such as

Table 28-1 (*Continued*)

capillary types; and (2) those in which flow is caused by controlled relative motion of the confining solid boundaries of the liquid, such as rotational, sliding plate, falling ball, and vibrating reed types. The capillary is the oldest and most widely used.

TGA (Thermal Gravimetric Analysis)

Applications:	Principal use on processed end items, but also used on processed plastics.
	Applicable in specialized weight loss analysis using solvents, moisture, and other liquids.
	Accelerates lifetime testing—one-day test could relate to one or two years of oven-aging tests.
	Measures percent volatiles, percent plasticizers, percent carbon black, percent inert material, degradation profiles, percent glass content, etc.
Limitations:	Specialty test to measure weight loss.
Data output:	Quantitative.
Sample: type/sample size/ time for measurement:	Solid and Liquid/0.01 to 0.5 g/30 min.
Method of analysis:	Weight loss measurement as a function of time and over a wide temperature range (such as 0 to 300°C).

TMA (Thermal Mechanical Analysis)

Applications:	Principal use on processed end items.
	Highly sensitive deformation measurement.
	Measures dimensional changes, thermal expansion, softening point, heat distortion temperature, thermal orientation, shrink "from mold," flexural strength and modulus, tensile strength and modulus, creep data, etc.
Limitations:	Only good where highly sensitive deformation measurement is necessary to product quality.
	Deformation measurement is only directly related to type of probe design.
Data output:	Quantitative.
Sample: type/sample size/ time for measurement:	Solid/0.01 to 0.5 g/30 min.
Method of analysis:	Millimeter displacement measurement against sample over a wide temperature range (such as 0 to 300°C).
	ASTM tests basically duplicate deformation type measurement of TMA; ASTM penetration, impact, flexural, etc., tests.

TR (Torque Rheometer)

Applications:	Standardization in extrusion and particularly high-intensity compounded material.
Limitations:	More sensitive rheological instruments available, if required.
Data output:	Quantitative.
Sample: type/sample size/ time for measurement:	Solid and liquid/5 to 30 g/10 min.
Method of analysis:	Measure temperature of fusion, time to fusion, and torque (work) required.
	Auxiliary capability allows gas evolution measurements (cc/g) useful in chemical blowing agent studies, pollution control emission measurements, etc.

Note: Typical materials that can be analyzed by various instruments: *LC*—antioxidants (phenols, thioesters, phosphites, etc.), plasticizers, lubricants, polymer MW distribution, etc.; GPC—residual monomers, nonpolymeric compounds/oils/plasticizers, etc.; IR—polymer composition, additives (qualitative, quantitative), phosphates, etc.; TGA—fillers, lubricants (molybdenum disulfide, etc.), polymer MW (degradation), PE cross-linking, etc.; X-ray—fillers (talc, mica, etc.), flame retardants (alumina trihydrate, antimony trioxide, etc.), stabilizers (organotin, etc.), etc.; NMR—polyesters, silicones, phenols, mineral oil, etc.; microscopy—contaminants, surface films (continuity, etc.), crystallinity, etc.; wet chemistry—lubricants, flame retardants, catalyst residues, etc; GC (gas chromatography)—residual monomers, nonpolymeric compounds, oils, plasticizers, etc.; and others.

A flow diagram for polymer identity could be as follows:

Table 28-1 (*Continued*)

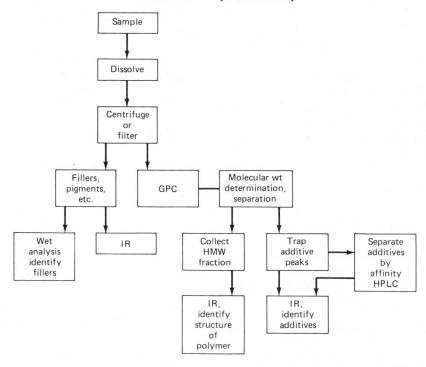

line rheometer gives the processor rapid, real-time data that can be used to improve product quality, increase throughout, and reduce down time and scrap.

TYPES OF TESTS

By far the most important tests conducted continue to be the mechanical tests. These tests, conducted under procedures established principally by the American Society for Testing and Materials (ASTM), are a means of extracting basic knowledge about materials, but never were thought of as yielding precise property values of fabricated products in service. Thus, values generated from ASTM tests give a great deal of extremely useful data (both absolute and comparative), but by no means

Table 28-2   Relative merits of typical instruments used to characterize plastics (No. 1 indicates lowest cost or best capability).

CHARACTERIZATION OF INSTRUMENT*		CAPABILITY			
	COST	MATERIAL INCOMING QC	IN-PROCESS QC	FINISHED PRODUCT QC	SAMPLE TIME AND INTERPRE- TATIVE TIME
DMA	4	3	3	3	5
DSC	5	6	6	6	2
IR	7	7	7	7	4
LC	6	5	5	5	1
MI	1	8	8	8	8
RMS	8	1	1	1	7
TGA	3	6	6	6	2
TMA	3	6	6	6	2
TR	4	4	4	4	3

* *See Table 28-1 for explanation of abbreviations.*

should be taken as guaranteed property values that will at all times and under all conditions be generated by a given material.

Several ASTM tests referred to in "The International Plastics Selector" are summarized here in condensed versions, arranged numerically (9). They are presented as general information. The full test procedures may be obtained from the American Society for Testing and Materials, 1916 Race Street, Philadelphia, PA 19103 U.S.A.

Selected ASTM Tests

C177 Thermal Conductivity

Specimen: Two identical specimens whose ratio of thickness to area are such that they give a true average representation of the material. The specimens should be smooth so they achieve good thermal contact with the testing apparatus.

Procedure: The two specimens are put next to a central heater; cooling elements are on the other side of the specimens, and thermocouples are inserted at appropriate places to measure temperatures and temperature differentials. There are two kinds of hot plates, one for a hot surface up to 550°K, the other from 550° K to 1350°K. Both are insulated around the edges to prevent heat loss and to achieve uniform heat distribution.

Significance: The thermal conductivity is the time rate of heat flow, under steady-state conditions, through unit area, per unit temperature gradient in the direction perpendicular to an isothermal surface. Thermal conductivity may be affected by moisture or other conditions, and it may change with time or high temperature.

C351 Specific Heat

Specimen: At least three randomly chosen specimens are taken and pressed in a hollow cylinder with a close fitting plunger. Prior to the test, all specimens shall be dried to constant weight at 100°C (or in a desiccator if this temperature would damage the specimens).

Procedure: The approach is to add a known mass of the material under test at a known high temperature to a known mass of water at a known lower temperature. The equilibrium temperature is determined, the heat absorbed by the water and the containing vessel is calculated, and from this the heat given up by the test material (and therefore its specific heat) may be calculated.

Significance: The mean specific heat (or the quantity of heat required to change the temperature of a unit

mass of substance by 1 degree centigrade) is an essential property of all insulating materials when used under conditions of unsteady or transient heat flow. It is part of the parameter generally known as thermal diffusivity which governs the rate of temperature diffusion through insulation. It is a basic thermodynamic property of all substances, the value of which depends upon chemical composition and temperature. It cannot be calculated theoretically for most solid substances.

D149 Dielectric Strength

Specimen: Specimens are thin sheets or plates having parallel plane surfaces and of a size sufficient to prevent flashing over. Dielectric strength varies with thickness and therefore specimen thickness must be reported.

Since temperature and humidity affect results, it is necessary to condition each type material as directed in the specification for that material. The test for dielectric strength must be run in the conditioning chamber or immediately after removal of the specimen from the chamber.

Procedure: The specimen is placed between heavy cylindrical brass electrodes which carry electrical current during the test. There are two ways of running this test for dielectric strength:

1. **Short-Time Test:** The voltage is increased from zero to breakdown at a uniform rate—0.5 to 1.0 kv/sec. The precise rate of voltage rise is specified in governing material specifications.
2. **Step-By-Step Test:** The initial voltage applied is 50% of breakdown voltage shown by the short-time test. It is increased at rates specified for each type of material and the breakdown level noted.

Breakdown by these tests means passage of sudden excessive current through the specimen and can be verified by instruments and visible damage to the specimen.

Significance: This test is an indication of the electrical strength of a material as an insulator. The dielectric strength of an insulating material is the voltage gradient at which electric failure or breakdown occurs at a continuous arc (the electrical property analogous to tensile strength in mechanical properties). The dielectric strength of materials varies greatly with several conditions, such as humidity and geometry, and it is not possible to directly apply the standard test values to field use unless all conditions, including specimen dimension, are the same. Because of this, the dielectric strength test results are of relative rather than absolute value as a specification guide.

The dielectric strength of polyethylenes is usually around 500 volts/0.001 in. The value will drop sharply if holes, bubbles, or contaminants are present in the specimen being tested.

The dielectric strength varies inversely with the thickness of the specimen.

D150 Dielectric Constant and Dissipation Factor

Specimen: The specimen may be a sheet of any size convenient to test, but should have uniform thickness. The test may be run at standard room temperatures and humidity, or in special sets of conditions as desired. In any case, the specimens should be preconditioned to the set of conditions used.

Procedure: Electrodes are applied to opposite faces of the test specimen. The capacitance and dielectric loss are then measured by comparison or substitution methods in an electric bridge circuit. From these measurements and the dimensions of the specimen, *dielectric constant* and *loss factor* are computed.

Significance: *Dissipation factor* is a ratio of the real power (in phase power) to the reactive power (power 90° out of phase). It is defined also in other ways:

Dissipation factor is the ratio of conductance of a capacitor in which the material is the dielectric to its susceptance.

Dissipation factor is the ratio of its parallel reactance to its parallel resistance. It is the tangent of the loss angle and the cotangent of the phase angle.

The dissipation factor is a measure of the conversion of the reactive power to real power, showing as heat.

Dielectric Constant is the ratio of the capacity of a condenser made with a particular dielectric to the capacity of the same condenser with air as the dielectric. For a material used to suppport and insulate components of an electrical network from each other and ground, it is generally desirable to have a *low* level of dielectric constant. For a material to function as the dielectric of a capacitor, on the other hand, it is desirable to have a high value of dielectric constant, so the capacitor may be physically as small as possible.

Loss Factor is the product of the dielectric constant and the power factor, and is a measure of total losses in the dielectric material.

D256 Izod Impact

Specimen: Usually $\frac{1}{8}$ x $\frac{1}{2}$ x 2 inches.

Specimens of other thicknesses can be used (up to $\frac{1}{2}$ inch) but $\frac{1}{8}$ inch is frequently used for molding materials because it is representative of average part thickness.

A notch is cut on the narrow face of the specimen.

Procedure: A sample is clamped in the base of a pendulum testing machine so that it is cantilevered upward with the notch facing the direction of impact. The pendulum is released, and the force consumed in breaking the sample is calculated from the height the pendulum reaches on the follow-through.

Significance: The Izod Impact test indicates the energy required to break notched specimens under standard conditions. It is calculated as ft-lb per inch of notch and is usually calculated on the basis of a one-inch specimen (although the specimen used may be thinner in the lateral direction).

The Izod value is useful in comparing various types or grades of a plastic. In comparing one plastic with another, however, the Izod Impact test should not be considered a reliable indicator of overall toughness or impact strength. Some materials are notch-sensitive and derive greater concentrations of stress from the notching operation. The Izod Impact test may indicate the need for avoiding sharp corners in parts made of such materials. For example, nylon and acetal-type plastics, which in molded parts are among the toughest materials, are notch sensitive and register relatively low values on the notched Izod Impact test.

D257 Direct Current Resistance or Conductance

Specimen: The measurement is of greatest value when the test specimen has the shape, the electrodes and the mountings it will have in actual use. The specimen forms most commonly used are flat plates, tapes, rods, and tubes.

Procedure: The resistance or conductance of a material or of a capacitor is determined from a measurement of the current or of the voltage drop under specified conditions. By using appropriate electrode systems, surface and volume resistance or conductance may be measured separately. The resistivity or conductivity can then be calculated when the required specimen and electrode dimensions are known.

In the test, electrical current is passed through a specimen at fixed voltage, and the transmitted current is measured.

Significance: Insulating materials are used to isolate components of an electrical system from each other and from ground, as well as to provide mechanical support for the components. Thus, for the intended purpose, it is generally desirable to have the insulation resistance as high as possible consistent with the acceptable mechanical and chemical and heat-resisting properties.

Insulation resistance or conductance combines both volume and surface affects. Surface resistance or conductance changes rapidly with humidity, while volume resistance or conductance changes slowly with humidity (although the final change may eventually be greater).

Resistivity or conductivity may be used to predict indirectly the low-frequency dielectric breakdown and dissipation factor properties of some materials.

Specific Definitions:

a) The *insulation resistance* between two electrodes that are in contact with or embedded in a specimen is the ratio of the direct voltage applied to the electrodes to the total current between them. It is dependent upon both the volume and surface resistances of the specimen.

b) The *volume resistance* between two electrodes that are in contact with or embedded in the specimen is the ratio of the direct voltage applied to the electrodes to that portion of the current between them is distributed through the volume of the specimen.

c) The *surface resistance* between two electrodes that are on the surface of the specimen is the ratio of the direct voltage applied to the electrodes to that portion of the current between them which is primarily in a thin layer of moisture or other semiconducting material that may be deposited on the surface.

d) The *volume resistivity* of a material is the ratio of the potential gradient parallel to the current in the material to the current density.

e) The *surface resistivity* of a material is the ratio of the potential gradient parallel to the current along its surface to the current per unit width of the surface.

D395 Compression Set

Specimens: These are to be cylindrical discs cut from a laboratory prepared slab of between 0.49 and 0.51 inch (12.5 and 13.0 millimeters).

Procedure: The test is designed to measure the residual deformation of a test specimen after it has been stressed under either a constant load or a constant deflection. A dial micrometer measures the deformation remaining thirty (30) minutes after the removal of the loads. The constant-load method species a force of 1.8 kN (400 pounds); the constant-deflection procedure calls for a compression of approximately 25%.

Significance: The compression set (i.e., residual deformation) measures the ability of compounds to retain elastic properties after prolonged action of compressive stresses. Compression-set tests should be limited to those involving static loading—i.e. hysteresis effects confuse the results in dynamic-stress testing.

D412 Tension Testing of Vulcanized Rubber

Specimen: Test specimens may be made in three different forms. Dumbell and ring specimens are prepared from standard dies while straight specimens are of sufficient length to permit their installation in the grips of the test apparatus. Bench marks are placed on the dumbell and straight forms for use as measuring points.

For ring forms, measurement is made using the apparatus grips holding the specimen.

Procedure: Tension tests are made on a power-driven machine equipped with a suitable dynamometer and recording device for measuring the applied force within ±2%, and the response of the specimen to the force. Specimens are symmetrically placed within the grips of the machine. Stress is measured at the elongation specified for the material and at rupture. Slightly different procedures are used to measure tension responses of ring specimens.

Significance: This method covers testing for the following:

1. **Tensile Stress**—The applied force per unit of original cross sectional area.
2. **Tensile Strength**—The maximum tensile stress applied while stretching a specimen to rupture.
3. **Elongation or Strain**—Extension of a uniform section of a specimen, produced by a tensile force applied to the specimen, expressed as a percentage of original length of section.
4. **Ultimate Elongation**—Maximum elongation prior to rupture.
5. **Tensile Stress at Given Elongation**—Tensile stress required to stretch a uniform section of a specimen at a given elongation.
6. **Tension Set**—The extension remaining after a specimen has been stretched and allowed to retract, expressed as a percentage of original length.
7. **Set After Break**—Tension set of a specimen stretched to rupture.

This method is not applicable to the testing of material classified as ebonite or hard rubber.

D471 Changes in Properties Resulting from Immersion in Liquids (Solvent Swell)

Specimens: Rectangular specimens 1 x 2 x 0.08 inches (25 x 50 x 2 millimeters) are to be used; results of specimens from different thicknesses can not be compared.

Procedure: The test describes the method for exposing specimens to the influence of liquids under standard conditions and then measuring the resulting deterioration by noting changes in physical properties before and after immersion. Three grades of liquids are described—ASTM oils, ASTM reference fuels and certain service fluids. Descriptions are given on how to check for changes in weight, changes in volume, changes in dimensions, and changes in various mechanical properties—e.g., tensile strength, elongation and hardness.

Significance: The method is not to be used in testing cellular materials, porous compositions or compressed asbestos fibers. And, because of the wide variation in service conditions, the test is not intended to give any direct correlation with eventual end use.

D495 High-Voltage, Low-Current, Dry Arc Resistance of Solid Electrical Insulation

Specimens: Test specimens shall be 0.125 ± 0.01 in. (3.17 ± 0.25mm) in thickness and during the test no part of the arc is closer than $\frac{1}{4}$ in. (6.6 mm) to the edge or closer than $\frac{1}{2}$ in. (12.7 mm) to a previously tested area. Surfaces should be clean.

Procedure: Electrodes are applied and internal current steps are applied until failure occurs. The failure is defined as the point at which a conducting path is formed across the sample and the arc completely disappears into the material.

Significance: The test is a high voltage-law current test which simulates those existing in AC current circuits at low current. Types of failure for plastics and elastomers include ignition, tracking and carbonization.

D542 Index of Refraction

Specimen: The (clear) test specimens shall fit conveniently on the face of the fixed half of a standard refractometer prism; a size of $\frac{1}{2}$ inch by $\frac{1}{4}$ inch (12.7 millimeters by 6.3 millimeters) on one face is usually satisfactory. The surfaces in contact with the prisms will be flat and have a good polish.

Procedures: Two procedures, the refractomatic and the microcopical, are described with the former being preferred wherever applicable. In it the specimen is placed in firm contact with the surface of the refractometer prism in the Abbé refractometer. The instrument is used in a standard fashion to determine the index of the fraction for the sodium D line.

In the microscopical method, the travel of the microscope lens from the top to the bottom of the surface of the specimen is used to give a measure of the index of refraction.

Significance: This test measures a fundamental property of matter useful for the control of purity and composition, for simple purposes of identification, and for the design of optical parts. It can be measured extremely precisely in fact, with much greater precision than is ordinarily required.

D543 Resistance to Chemical Reagents

Specimen: A wide variety of shapes and sizes are possible, the main criteria being that the specimens have smooth and accurately known dimensions so that any changes in size, appearance, etc. can be recorded.

Procedure: The full test lists 50 reagents together with a variety of balances, micrometers, containers and testing devices to measure the changes in weight and dimension, and in mechanical properties.

Significance: As can be inferred, there is an almost infinite variety of combinations of material, chemical reagents, and affects. The full ASTM test specifies the conditions as a basis for standardization, and serves as a guide to investigators wishing to compare the relative resistance of various plastics to chemical reagents.

D570 Water Absorption

Specimen: For molding materials the specimens are discs 2 inches in diameter and $\frac{1}{8}$ inch thick. For sheet materials the specimens are bars 3 inches x 1 inch x thickness of the material.

The specimens are dried 24 hours in an oven at 50°C, cooled in a dessicator, and immediately weighed.

Procedure: Water absorption data may be obtained by immersion for 24 hours or longer in water at 73.4°F. Upon removal, the specimens are wiped dry with a cloth and immediately weighed. The increase in weight is reported as percentage gained.

For materials which lose some soluble matter during immersion—such as cellulosics—the sample must be re-dried, re-weighed, and reported as "percent soluble matter last." The % gain in weight + % soluble matter lost = % water absorption.

Significance: The various plastics absorb varying amounts of water, and the presence of absorbed water may affect plastics in different ways.

Electrical properties change most noticeably with water absorption, and this is one of the reasons that polyethylene, since it absorbs almost no water, is highly favored as a dielectric.

Materials which absorb relatively larger amounts of water tend to change dimension in the process. When dimensional stability is required in products made of such materials, grades with less tendency to absorb water are chosen.

The water absorption rate of acetal type plastics is so low as to have a negligible effect of properties.

D618 Conditioning Procedure

Procedure: Procedure A for conditioning test specimens calls for the following periods in standard laboratory atmospher (50 ± 2% R.H., 73.4 ± 1.8°F):

Specimen Thickness, inch	Time, hr.
0.25 or under	40
Over 0.25	88

Adequate air circulation around all specimens must be provided.

Significance: The temperature and moisture content of plastics affects physical and electrical properties. This standard has been established to get comparable test results at different times and in different laboratories.

In addition to Procedure A, there are other conditions set forth to provide for testing at higher or lower levels of temperature and humidity.

D624 Tear Resistance

Specimen: The test describes the sizes and shapes of three specimens, each of them with curve and contour. Two of them have a slit cut in the edge.

Procedure: The specimen is clamped in the jaws of a testing machine and the jaws then separated at a speed of 500 millimeters (20 inches) per minute. After rupture of the specimen, the breaking force in newtons (pounds force) is noted from the scale in the test machine. The resistance to tear is calculated from the force and the median thickness of the specimen. Values are given in newtons per meter, or in pounds force per inch for tearing the specimen of one meter (or one inch) in thickness.

Significance: This method determines the tear resistance of the usual grades of vulcanized rubber, but not of hard rubber. Since tear resistance may be affected to a large degree by a mechanical fibering or the rubber under stress as well as by stretch distribution, by strain rate, and by the size of the specimen, the results obtained in the test can be regarded only as a measure of the resistance under the conditions of the test rather than necessarily as having any direct relation to service value.

D638 Tensile Properties

Specimen: Specimens can be injection molded or machined from compression molded plaques. Typically $\frac{1}{8}$ inch thick, their size can vary; the center portion is less thick than the ends, which are held by the testing equipment.

Procedure: Both ends of the specimen are firmly clamped in the jaws of an Instron testing machine. The jaws may move apart at rates of 0.2, 0.5, 2, or 20 inches a minute, pulling the sample from both ends. The stress is automatically plotted against strain (elongation) on graph paper.

Significance: Tensile properties are the most important single indication of strength in a material. The force necessary to pull the specimen apart is determined, along with how much the material stretches before breaking.

The elastic modulus ("modulus of elasticity" or "tensile modulus") is the ratio of stress to strain below the proportional limit of the material. It is the most useful tensile data because parts should be designed to accommodate stresses to a degree well below this.

For some applications where almost rubbery elasticity is desirable, a high ultimate elongation may be an asset.

For rigid parts, on the other hand, there is little benefit in the fact that they can be stretched extremely long.

There is great benefit in moderate elongation, however, since this quality permits absorbing rapid impact and shock. Thus the total area under a stress-strain curve is indicative of overall toughness. A material of very high tensile strength and little elongation would tend to be brittle in service.

D648 Deflection Temperature

Specimen: Specimens measure 5 x $\frac{1}{2}$ inch x any thickness from $\frac{1}{8}$ to $\frac{1}{2}$ inch.

Procedure: The specimen is placed on supports 4 inches apart and a load of 66 or 264 psi is placed on the center. The temperature in the chamber is raised at the rate of $2° \pm 0.2°C$ per minute. The temperature at which the bar has deflected 0.010 inch is reported as "deflection temperature at 66 (or 264) psi fiber stress."

Significance: This test shows the temperature at which an arbitrary amount of deflection occurs under established loads. It is not intended to be a direct guide to high-temperature limits for specific applications. It may be useful in comparing the relative behavior of various materials in these test conditions, but it is primarily useful for control and development.

D695 Compressive Properties

Specimen: Prisms $\frac{1}{2}$ x $\frac{1}{2}$ x 1 inch or cylinders $\frac{1}{2}$ inch diameter x 1 inch.

Procedure: The specimen is mounted in a compression tool between testing machine heads which exert a constant rate of compressive movement. An indicator registers loading.

The compressive strength of a material is calculated as the psi required to rupture the specimen or deform the specimen a given percentage of its height. It can be expressed as psi either at rupture or a given percentage of deformation.

Significance: The compressive strength of plastics is of limited design value, since plastic products (except foams) seldom fail from compressive loading alone. The compressive strength figures, however, may be useful in specifications for distinguishing between different grades of a material, and also for assessing, along with other property data, the over-all strength of different kinds of materials.

D696 Coefficient of Linear Thermal Expansion

Specimen: The specimen is between 2 inches and 5 inches long (50 millimeters to 125 millimeters). Its cross section is round, square or rectangular, and should fit easily into the outer tube of the dilatometer equipment without excessive play or friction. The specimens shall be prepared so that they give a minimum of strain anisotropy.

Procedure: The specimen is placed at the bottom of the outer dilatometer tube with the inner tube resting on it. The measuring device, which if firmly attached to the outer tube, is in contact with the top of the inner tube; it indicates variations in the length of the specimen with changes in temperature. Temperature changes are brought about by immersing the outer tube in a liquid bath at the desired temperature. A vitreous silica dilatometer is commonly used.

Significance: The thermal expansion of a plastic is composed of a reversible component on which are superimposed changes of length due to changes in moisture content, curing, loss of plastisizer or solvents, release of stresses, phase changes, etc. This particular test method attempts to eliminate all other forces except linear thermal expansion. The measure is obtained by dividing the linear expansion per the unit length by the change in temperature. Frequently a phase change in the plastic is accompanied by a change in the coefficient of the linear thermal expansion, so preliminary investigations should be conducted to determine any such possible phase changes.

D732 Shear Strength

Specimens: These shall be either a 2 inch (50 millimeter) square or a 2 inch (50 millimeter) diameter disc cut from sheet material that is 0.005 to 0.500 inch (0.125 to 12.5 millimeters). A hole approximately $\frac{5}{16}$ths of an inch (11 millimeters) in diameter shall be drilled through the specimen at it's center.

Procedure: A testing machine allows the precise measurement of load, and the means to move ahead at a constant rate until the specimen is sheared such that the moving portion has completely been separated from the stationary portion. The hole in the specimen is placed over a punch and the apparatus moved until shearing has taken place.

Significance: The test gives the maximum load measured in either mego newtons per square meter or pounds per square inch to shear the specimen. It is calculated by dividing the total load by the area of the sheared edge; this is taken as the product of the thickness of the specimen and the circumference of the punch.

D746 Brittleness Temperature

Specimen: Pieces $\frac{1}{4}$ inch wide, 0.075 inch thick, and $1\frac{1}{4}$ inch long. The apparatus chills the specimen and then strikes it to establish the temperature at which it fractures.

Procedure: The conditioned specimens are cantilevered from the sample holder in the test apparatus which has been brought to low temperature (that at which specimens would be expected to fail). When the specimens have been in the test medium for 3 minutes, a single impact is administered and the samples are examined for failure. Failures are total breaks, partial breaks, or any visible cracks. The test is conducted at a range of temperatures producing varying percentages of breaks. From these data, the temperature at which 50% failure would occur is calculated or plotted and reported as the brittleness temperature of the material according to this test.

Significance: This test is of some use in judging the relative merits of various materials for low temperature flexing or impact. However, it is specifically relevant only for materials and conditions specified in the test, and the values cannot be directly applied to other shapes and conditions.

The brittleness temperature does not put any lower limit on service temperature for end-use products. The brittleness temperature is sometimes used in specifications.

D747 Stiffness in Flexure

Specimen: The specimens must have rectangular cross section, but dimensions may vary with the kind of material.

Procedure: The specimen is clamped into an apparatus that holds it at both ends, and measures both the load used to attempt to bend it and the specimen's response; a 1% load is first applied manually and the deflection scale set at zero. The motor is engaged and the loading increased, with deflection and loading figures recorded at intervals. A curve is drawn of deflectoin versus load, and from this is calculated stiffness in flexure in pounds per square inch.

Significance: This test does not distinguish the plastic and elastic elements involved in the measurement and therefore a true elastic modulus is not calculable. Instead, an apparent value is obtained and called "stiffness in flexure." It is a measure of the relative stiffness of various plastics and taken with other pertinent property data is useful in material selection.

D759 Determining the Physical Properties of Plastics At Subnormal and Supernormal Temperatures

This method presents recommended practice for determining the various physical properties of plastics at temperatures from −452 to 1022°F (−269 to 550°C).

Specimens: Test specimens shall conform to the applicable ASTM method.

Procedure: All parts of the test equipment which are exposed to the test temperature shall be adjusted to function normally. An insulated test chamber shall be used to enclose the specimen and adequate circulation shall be provided to ensure uniform temperature. Temperature measuring equipment capable of the required equipment accuracy of ±3°C (±5°F) from −70 to 300°C (−94 to 572°F) and ±2% over 300°C (572°F) and ±4% below −70°C (−94°F) shall be used.

Specimens shall be preconditioned either in a preconditioning chamber or in the test chamber. Transfer time from a preconditioning chamber to the test fixture and chamber should not exceed 30 seconds. Time to establish thermal equilibrium should be 1.3 times the period required for the control specimen.

D785 Rockwell Hardness

Specimen: Sheets or plaques at least $\frac{1}{4}$ in thick. This thickness may be built up of thinner pieces, if necessary.

Procedure: A steel ball under a minor load is applied to the surface of the specimen. This indents slightly and assures good contact. The gauge is ten set at zero. The major load is applied for 15 seconds and removed, leaving the minor load still applied. The indentation remaining after 15 seconds is read directly off the dial. This value is preceded by a letter representing the Rockwell hardness scale used.

The size of the balls used and loadings vary (giving rise to several ranges of Rockwell hardness); values obtained with one set cannot be correlated with values from another set.

Significance: Rockwell hardness can differentiate relative hardness of different types of a given plastic. But since elastic recovery is involved as well as hardness, it is not valid to compare hardness of various kinds of plastic entirely on the basis of this test.

Rockwell hardness is not an index of wear qualities or abrasion resistance. For instance, polystyrenes have high Rockwell hardness values but poor scratch resistance.

D790 Flexural Properties

Specimen: Usually $\frac{1}{8}$ x $\frac{1}{2}$ x 5 inches. Sheet or plaques as thin as $\frac{1}{16}$ inch may be used. The span and width depend upon thickness.

Procedure: The specimen is placed on two supports spaced 4 inches apart. A load is applied in the center of a specified rate and the loading at failure (psi) is the flexural strength. For materials which do not break, the flexural property usually given is Flexural Stress at 5% strain.

Significance: In bending, a beam is subject to both tensile and compressive stresses. Compressive at the

concave surface, zero in the center, tensile at the convex surface of the bend.

Since most thermoplastics do not break in this test even after being greatly deflected, the flexural strength cannot be calculated. Instead, stress at 5% strain is calculated—that is, the loading in psi necessary to stretch the outer surface 5%.

D792 Specific Gravity and Density

Specimen: The volume of the specimen must be not less than one cubic centimeter (0.06 cubic inches), and its surface and edges are to be smooth.

Procedure: The specimen is first weighed in air, then immersed in a fluid (either water or another substance—both are described in the full test) and then weighed in this other medium. The value is determined by calculating the ratio of the apparent weight of the specimen in air and the apparent weight when completely immersed in fluid.

The full test describes methods for testing plastics that are heavier than water, lighter than water, and are of large and irregular shapes.

Significance: Density and specific gravity have almost exactly the same numerical values; specific gravity is however a dimensionless unit because it is the ratio of the weight in air of a unit volume of the material compared to the weight in air of an equal volume of distilled water at the same temperature. Density is the weight in air in grams per cubic centimeter. There is a very slight difference in the two values because water at the specified temperature (23°C) weighs $0.99756g/cm^3$; thus $D = SG \times 0.99756$.

Either value gives a means of identifying a material, of following any physical changes in it, and of indicating the degree of uniformity in a product. Changes in the property can be brought about by changes in crystallinity, loss of plasticizer or absorption of solvent. Specific gravity is a strong element in the price factor and thus has great importance. Beyond the price/volume relationship, however, specific gravity is used in production control, both in raw-material production, and in molding and extrusion. Polyethylenes, for instance, may have density variation, depending upon the degree of "packing" during molding, or the rate of quench during extrusion.

D945 Mechanical Properties of Elastomeric Vulcanizates Under Compressive or Shear Strains by the Mechanical Oscillograph

Specimen: At least two specimens are tested. Test specimens for compression measurements are right circular cylinders, chosen from standardized dimensions. Each specimen is conditioned by exposure to the test temperature for sufficient time to ensure temperature equilibrium.

Test specimens for shear are rectangular sandwiches consisting of two blocks of the composition to be tested adhered between parallel metal plates having standardized dimensions. Each specimen is allowed to reach the test temperature equilibrium.

Procedure: The Yerzley mechanical oscillograph is used for measuring mechanical properties of elastomeric Vulcanizates. These properties include compression and shear testing. Specimens are loaded by an unbalanced lever and the resultant deflections are recorded on a chronograph.

Significance: Elastomeric properties measured by this procedure are important for the isolation and absorption of shock and vibration. These properties are identifiable with the physics of polymeric materials as a basis of quality control, development and research. In applying this data though, a shape factor must be incorporated into the mathematical transferral to the application.

D955 Mold Shrinkage

Specimen: The full test describes detailed methods of preparing specimens of various bar and disc shapes in a series of compression molds, injection molds, transfer molds, etc.

Procedure: The materials are molded under carefully controlled conditions (sizes, rates of heat, etc.) discharged from the mold, cooled for a short period of time, and then measured. The difference in dimension size and mold size is recorded as the mold shrinkage.

Significance: The test is record initial shrinkage—i.e., not for any shrinkage after the first 48 hours. Under any of the standard methods of molding, the mold shrinkage will vary according to design and operation of the mold. Some further comments:

a) Compression molding. Shrinkage will be at a minimum where there is a maximum of material being forced solidly into the mold cavity, and vice versa. The plasticity of the material may affect shrinkage insofar as it effects the retention and compression of the charge given during the molding.
b) Injection molding. In addition to type, size and thickness of the piece, mold shrinkage here will vary on the nozzle size of the mold, the operating cycle, the temperature, and the length of time that follow-up pressure is maintained. As with compression molding, shrinkages will be much higher where the charge must flow into the mold-cavity but does not receive enough pressure to be forced firmly into all of the recesses.
c) Transfer molding. The comments for compression and injection molding also apply; it should be noted that the direction of flow is not as an important a factor as would be expected.

D1044 Resistance of Transparent Plastics to Surface Abrasion

Specimens: Test specimens are clean transparent disks 102 mm (4 in.) in diameter, or plates 102 mm (4 in.) square, having both surfaces plane and parallel. Thicknesses shall not exceed 12.7 mm (0.50 in.) A 6.3 mm (0.25 in.) hole is centrally drilled in each specimen.

Procedure: The apparatus consists of a Taber abraser, constructed so that wheels of several degrees of abrasiveness may be used. The grade of "Calibrase" wheel designed C5-10F is used. Loads on the wheels may be selected from 250 g, 500 g and 1000 g. Conditioning and testing shall be carried out at 23 ± 2°C (73.4 ± 3.6°F). Degree of abrasion is measured on transparent materials by a photometric method.

Significance: Resistance to abrasion is an important factor in many plastics including transparent thermoplastics. The principal limitation of this test is the poor reproducibility. Lab to lab variation is significant although intralab data has been fairly good.

D1054 Impact Resilience and Penetration of Rubber by the Rebound Pendulum

Specimen: Test specimens are rectangular blocks, 25 ± 0.5 mm by 50 ± 1 mm (1 ± 0.02 in. by 1 ± 0.02 in. by 2 ± 0.04 in.) prepared from sheets of uncured compounded rubber (mixed and cured per ASTM D15). Identification marks are placed on either the top or bottom of the block as it lies in the mold.

Procedure: The test specimen maintained at 23 ± 1° C (73.4 ± 2°F) for at least 60 minutes before testing is placed in an apparatus consisting of a free-swinging rebound pendulum supported by ball bearings and carrying a striking hammer. An angular scale enables measurement of the angle of rebound after the pendulum strikes the test sample. The penetration of the pointer is determined from the observed deflection.

Significance: This method covers the determination of impact resilience and penetration of rubber by means of the Goodyear-Healey rebound pendulum. Dynamic stiffness is a factor that influences impact resilience. Penetration measurements present a convenient index of stiffness.

D1238 Flow Rate (Melt Index)

Specimen: Any form which can be introduced into the cylinder bore may be used, e.g., powder, granules, strips of film, etc.

The conditioning required varies, being listed in each material specification.

Procedure: The apparatus, an Extrusion Plastometer, is a cylinder in which the material is melted at a known temperature, and then extruded through a standard orifice; it is preheated to 190°C for polyethylene. Material is put into the cylinder and the loaded piston (approx. 43.25 psi) is put into place. After 5 minutes, the extrudate issuing from the orifice is cut off flush, and again one minute later. These cuts are discarded. Cuts for the test are taken at 1, 2, 3, or 6 minutes, depending on the material or its flow rate. The melt index is calculated and given as grams/10 minutes.

Significance: The melt index test is primarily useful to raw material manufacturers as a method of controlling material uniformity. While the data from this test is not directly translatable into relative end-use processing characteristics, the melt index value is nonetheless strongly indicative of relative "flowability" of various kinds and grades of polyethylene.

The "property" measured by this test is basically melt viscosity or "rate of shear." In general, the materials which are more resistant to flow are those with higher molecular weight.

D1418 Rubber and Rubber Latices Nomenclature

The ASTM has recommended a standardized terminology system classifying all forms of elastomeric materials which is based upon the chemical composition of the polymer's backbone chain.

The "M" Class

These elastomers have saturated main polymer chains and are usually prepared from ethylenic or vinyl type monomers containing one double band.

ACM-	Copolymers of an acrylate and a small amount of other monomer which provides vulcanizability.
ANM-	Copolymers of an acrylate and acrylonitrile.
CM-	Chloro-polyethylene.
CFM-	Polychloro-trifluoro-ethylene.
CSM-	Chloro-sulfonyl-polyethylene.
EPDM-	Terpolymers of ethylene, propylene and a nonconjugated diene which results in pendant unsaturation (not in the main chain.)
EPM-	Copolymers of ethylene and propylene.
FKM-	A polymer with a saturated main chain with substituents of fluorine, perfluoroalkyl, or perfluoroalkoxy.

The "O" Class

These elastomers have oxygen in the main chain.

CO-	Polyepichlorohydrin
ECO-	Copolymer of ethylene oxide and epichlorohydrin.
GPO-	Copolymer of propylene oxide and allyl glycidyl ether.

The "R" Class

These elastomers contain unsaturation in the main chain. The letter immediately before the "R" designates the conjugated diene which is used in its synthesis (except natural rubber.)

ABR-	Copolymer of acrylate and butadiene.
BIIR-	Copolymer of bromoisobutene and isoprene.
BR-	Polybutadiene.
CIIR-	Copolymer of chloroisobutene and isoprene.
CR-	Polychloroprene.
IIR-	Copolymer of isobutene and isoprene.
IR-	Polyisoprene (synthetic only).
NBR-	Copolymer of acrylonitrile and butadiene.
NCR-	Copolymer of acrylonitrile and chloroprene.
NIR-	Copolymer of acrylonitrile and isoprene.
NR-	Natural Rubber (poly-cis-isoprene).
PBR-	Copolymer of vinyl pyridine and butadiene.
PSBR-	Terpolymer of vinyl pyridine, styrene and butadiene.
SBR-	Copolymer of styrene and butadiene.
SCR-	Copolymer of styrene and chloroprene.
SIR-	Copolymer of styrene and isoprene.
X-	Prefix indicated carboxyl substitution.

The "Q" Class

These elastomers have silicone in the main chain. Prefixes indicate the following types of substitution:

M—methyl
V—vinyl
P—phenyl
F—fluorine

The "U" Class

Elastomers with carbon, nitrogen and oxygen in the main chain—typically polyurethanes:

AU—Polyester based polyurethanes.

EU—Polyether based polyurethanes.

"Y" Designation

The "Y" prefix indicates a thermoplastic rubber which requires no vulcanization.

D1525 Vicat Softening Point

Specimen: Flat specimens must be at least $\frac{3}{4}$ inch wide and $\frac{1}{8}$ inch thick. Two specimens may be stacked, if necessary, to get the thickness, and the specimens may be compression or injection molded.

Procedure: The apparatus for testing Vicat Softening Point consists of a temperature-regulated oil bath with a flat ended needle penetrator so mounted as to register degree of penetration on a gauge.

A specimen is placed with the needle resting on it. The temperature of the bath (preheated to about 50°C lower than anticipated Vicat Softening Point) is raised at the rate of 50°C/hr or 120°C/hr. The temperature at which the needle penetrates 1 mm. is the Vicat Softening Point.

Significance: The Vicat softening temperature is a good way of comparing the heat-softening characteristics of polyethylenes; it also may be used with other thermoplastics.

D1646 Viscosity and Curing Characteristics of Rubber by the Shearing Disk Viscometer —Mooney Viscosity

Specimen: The sample consists of two pieces of the elastomer specimen having a mass of 27 ± 3 g and shall be cut to fit the die cavities of the viscometer. The die cavity has the following dimensions . . . 50.93 ± 0.13 mm in diameter and 10.59 ± 0.13 mm in depth.

Procedure: A rotating disk is used to determine the viscosity of elastomeric materials. Vulcanization can be detected by a change in observed viscosity. Using a specified rotor speed of 2 rpm with a load of 11500 N, the torque required is measured (usually at 100°C).

Significance: Viscosity values depend on the size and configuration of the polymer molecule. With proper interpretation, the viscosity and the molecular weight or molecular size can be correlated.

D1709 Impact Resistance of Polyethylene by the Dart Impact Method

Specimen: They are to be large enough to extend outside the clamp gasket of the specimen at all points.

Procedure: The method describes the determination of the energy that causes polyethylene film to fail under specified conditions of impact of a free-falling dart. This energy is expressed in terms of the mass of the missile falling from a specified height which will result in 50% failure of the specimens tested. There are two kinds of darts: method A has a dart of 1.5 inches (38.1 millimeters) in diameter; method B has a dart of 2 inches (50.8 millimeters) in diameter. Weights are added to the darts until 50% failure rates of the polyethylene have been attained.

Significance: There is no correlation between the results obtained by methods A and B nor by other tests employing different conditions of missile velocity, dart diameter, etc. The impact resistance of polyethylene, while partly dependent on thickness, has no simple correlation with it. Hence, impact values cannot be normalized over a range of thickness without producing misleading data.

D1895 Apparent Density, Bulk Factor and Pourability

Specimen: The plastic powder or granules received from the manufacturers are dried prior to test (Method D618).

Procedure: A given amount of the powder is poured through a funnel and its volume and time to flow through the orifice are measured.

Significance: The *apparent density* is the weight per unit volume of a material including the voids inherent in the material's manufacture.

The *bulk factor* is the ratio of the volume of any given quantity of loose plastic material to the volume of the same quantity of the material after molding or forming. The bulk factor of the material is also equal to the ratio of the density after forming to the apparent density before forming.

Pourability is the measure of the time required for a standard quantity of material to flow through a funnel of specified dimension.

Apparent density is thus a measure of the fluffiness of the material; bulk factor is the measure of the volume change that may be expected in fabrication; and pourability characterizes the handling of properties of a finely divided plastic material.

D1921 Particle Size (Sieve Analysis)

Specimen: The material as received from the manufacturer is conditioned according to tests D618.

Procedure: The test describes four methods for shaking particles through a series of nested sieves, using a different range of sieves, or pulling the material through the sieves with a vacuum. In each instance, the material is shaken through the sieves, and a determination made of the percentage of the material caught on each layer. The test can also be used with just one sieve, and the values given of the amounts retained or passed through.

Significance: The test describes only dry sieving methods, and so the lower limit of measurement is considered to be about 38 micrometers (number 400 sieve). For plastics of smaller particle sizes, sedimentation methods are recommended.

D2117 Melting Point

Specimens: The test is for semi-crystalline polymers and powdered samples must first be heated and melted to generate a semi-crystalline condition. Molded, pelletized, film or sheet samples shall be prepared so that the specimens have an approximate diameter of $\frac{1}{18}$th of an inch (1.6 millimeters) and $\frac{1}{64}$th of an inch (0.04 millimeters) thick.

Procedure: The samples are heated by a hot stage unit mounted under a microscope. The specimen is viewed through crossed-polar prisms, and the melting point indicated by the disappearance of the prisms characteristic double refraction.

Significance: This is an extremely accurate test useful for specimen acceptance manufacturing control, etc. Note that only materials capable of forming at least a two dimensional intromolecular order are suitable for this procedure. A spread in particle size will have a noticeable effect on the melting point, as will the presence of heat, air or anisotropic crystals.

D2240 Indentation Hardness of Rubber and Plastic by Means of a Durometer

Specimen: The flat specimen must be at least 6mm ($\frac{1}{4}$ inch) thick and wide enough to enable measurement of at least 12mm ($\frac{1}{2}$ inch) in any direction from the indentor point to the edge of the specimen.

Procedure: Five measurements are to be made at least 6mm ($\frac{1}{2}$ inch) apart. Place the specimen on a hard, flat surface. As rapidly as possible, apply the pressure foot of the Durometer without shock. The scale is read within 1 second after contact is made to record the penetration of the indicator into the material.

Significance: Two types of Durometers are available, A and D, which allow measurement of soft and hard rubbers. This test is primarily used for control purposes since no relationship exists between indention hardness determined by this method and any fundamental property of the material.

D2471 Gel Time and Peak Exothermic Temperature of Reacting Thermosetting Resins

Specimen: All components of the test—i.e., specimens, container, etc. are conditioned for at least four hours.

Procedures: The method covers the determination of the time from the initial mixing of the reactants of a thermosetting plastic to the time when solidification commences under conditioned approximating those in use. The method also provides a means for measuring tne maximum temperature reached by a reacting thermosetting composition, as well as the time from initial mixing to the time when this peak exothermic temperature is reached. This method is limited to reacting mixtures exhibiting gel times greater than five minutes.

In the test, the reactants are slowly mixed together, a sample taken, poured in a container and then its temperature recorded. The end of the reaction is recorded when material no longer adheres to the end of a clean probe—i.e., the "gel time." The time and temperature are recorded until the temperature starts to drop—i.e., until the peak exothermic temperature is reached.

Significance: Since both gel time and peak exothermic temperature vary with the volume of material, it is essential that the volume be specified in any determination. Test results can be extrapolated for application to reaction conditions, quality control, and final material characteristics. For the most useful result, the dimensions of the test apparatus should be in the same proportion as the production equipment.

D2583 Indention Hardness by Means of a Barcol Impressor

Specimen: They are to be smooth, free from mechanical damage, and large enough to ensure minimum distance of three millimeters ($\frac{1}{8}$ inch) in any direction from the indentor point to the edge of the specimen.

Procedures: The samples are struck by an indentor of hardened steel; this is the shape of a truncated cone having an angle of 36 degrees with a flat tip of 0.157 millimeter (0.0062 inches). There is an indicating dial with 100 divisions on it, each division representing a depth of .0076 millimeters (0.0003 inches) penetration. The higher the reading, the harder the material.

The Barcol impressor is equipped with hard and soft standard aluminum alloy discs for calibration.

Significance: The Barcol impressor is portable, and therefore suitable for testing the hardness of fabricated parts and individual test specimens for production control. Statistical procedures, including the number of readings for each material, and the variance in readings, are given in the full test description to generate the final Barcol hardness value.

D2632 Impact Resilience of Rubber by Vertical Rebound

Specimen: The standard test specimen is 12.5 ± 0.5 mm (0.50 ± 0.02 in.) in thickness and cut so that the point of a plunger falls a minimum distance of 14 mm (0.55 in.) from the edge of the specimen. This may be a molded specimen or cut from a slab.

Procedure: A plunger, 28 ± 0.5 g (1 ± 0.01 oz.), is suspended 400 ± 1 mm (16 ± 0.04 in.) above the specimen by an apparatus designed to release the plunger and measure its rebound height. The plunger is dropped, guided by a vertical rod, and rebounds are measured against a scale of 100 equally spaced divisions. Recordings are taken of the fourth through the sixth rebound. The resilience is equal to the average rebound height of the 4th, 5th, and 6th impacts.

Significance: Resilience is sensitive to temperature changes and the depth of penetration of the plunger. It is also dependent upon the dynamic modulus and internal friction of the rubber. Resilience values from one type of apparatus may not be predicted from results on another type of apparatus. This test is not applicable to cellular rubbers or coated fabrics.

D2863 Flammability Using the Oxygen Index Method

Specimen: The area and thickness of specimens will depend upon whether the plastic is self-supporting, cellular, a film, etc. If moisture content is suspected (it will affect the flammability rating), specimens should be conditioned prior to test.

Procedure: The test measures the minimum concentration of oxygen in a mixture of oxygen and nitrogen flowing upwards in the test column that will just support combustion measured under equilibrium conditions of candle-like burning. The equilibrium is established by the relation of the heat generated from the combustion of the specimen and the heat lost to the surroundings; it is measured by one or the other of two arbitrary criteria—a time of burning or the length of specimen burned. This point is approached from both sides of the critical oxygen concentration in order to establish the oxygen index.

The apparatus consists of a columnar glass tube in which the specimen can be suspended and through which a stream of gas of variable oxygen content may be passed. The test description gives criteria for deciding when the material does or does not support combustion.

Significance: This standard should be used solely to measure and describe the properties of a material in response to heat and flame under controlled laboratory conditions. It should not be considered or used for the description, appraisal or regulation of the fire hazard of materials, products, or systems under actual fire conditions.

Viscoelastic Properties

It is important to recognize that the properties and performance of plastics depend strongly on temperature, time, and environmental conditions. This situation is also true in work with other materials such as metals. The major difference is that plastics processing is easier to control than metals processing. Temperature, time, and environmental conditions are important during the manufacture of the resin, when the resin is being injection molded into a molded part, and when the molded part is put into service. During the manufacture of the resin and when it is being injection molded, the environmental conditions refer to factors such as pressure, rate of movement of material, etc.; with the molded part, service conditions (environmental conditions) can include abrasion resistance, static or dynamic loads, electrical requirements, etc.

This strong dependence of properties relates to an important behavior of plastics called viscoelasticity. The viscoelastic behavior of plastics under various loading and environmental conditions can be characterized so that conventional well-understood engineering methods can be used. Viscoelasticity implies behavior similar both to viscous liquids, in which the rate of deformation is proportional to the applied force, and to purely elastic solids, in which the deformation is proportional to the applied force.

In viscous systems all the work done on the system is dissipated as heat, whereas in elastic systems all the work is stored as potential energy, as in a stretched spring. It is this dual nature of plastics that makes their behavior interesting and useful. The great variety of mechanical tests available and the numerous factors that make plastics useful, such as exposure to all kinds of environments, would make the study of their mechanical properties very complex if it were not for some general phenomena and rules-of-thumb that greatly simplify the subject. The range of viscoelastic behavior permits the use of plastics ranging from those that are very flexible to those that are extremely strong and rigid as well as those that can operate at extremely low temperatures to those used at extremely high temperatures; and so on regarding other properties.

The dynamic mechanical tests (fatigue, etc.) generally provide a large amount of useful information about a plastic. However, the basically static mechanical tests theoretically can give the same information.

Rheology, Viscosity and Flow

Rheology is the science that deals with the deformation and flow of matter under "environmental conditions." The rheology of plastics is complex because these materials exhibit properties that combine those of an ideal viscous liquid (pure shear deformations) with those of an ideal elastic solid (pure elastic deformations). The mechanical behavior of plastics is dominated by viscoelastic phenomena that are often the controlling factors in tensile strength, melt viscosity, elongation at break, and rupture energy. The viscous attributes of polymer melt flow are important considerations in plastics manufacturing and fabrication. Thus, we try to consider separately the viscous and elastic effects of plastic resins that undergo flow in the molten state.

Rheometers are the instruments used to obtain characteristic flow curves of shear stress as a function of shear rate for viscous materials.

Absolute viscosity measurement in centipoise can be obtained in rotational viscometers, which are generally of two types: (1) coaxial cylinder systems and (2) cone and plate systems.

Viscometers that operate with only a simple rotor with no breaker to provide a fixed gap can give only relative viscosity measurements for plastics. Polymer rheology can also be studied by capillary extrusion techniques. The

polymer melt is forced through a fine bore tube under isothermal conditions, and the volumetric flow rate is measured as a function of the extrusion pressure. (Details on melt rheometers are given in Chapter 26.)

On-Line Viscoelastic Measurements for Polymer Melt Processes

This section concerns a very important quality control instrument that applies dynamic mechanical measurements to thermoplastics (10, 11). To date most of the work has been with extruders; so information presented that concerns extruders can be applied to injection molding. It is widely understood that most thermoplastics processing techniques involve high shear rates; thus it is very common for engineers to study their thermoplastics at comparable shear rates. By simulating the process shear rates, it is often thought that one can best explain problems and understand the key properties of the material being used. Also, in working with a diversity of materials processed with a variety of techniques, additional information can be very helpful.

Although very high rates of shear are experienced during some plastics processing steps, many steps occur over a much longer time scale. For example, mold or die swell, distortion, foaming, and surface roughness may occur over time scales of seconds. Viscoelastic measurements corresponding to low shear rates are very sensitive to the polymer behavior in such cases. Rough time scales for a variety of plastics processing steps are shown in Fig. 28-23.

Several examples have been selected to illustrate these observations. In each of these cases measurements were made with an oscillatory shear technique, which is much easier than steady shear. The oscillatory results match the steady shear results very accurately when the frequency (in radians/sec) is the same as the shear rate (in sec^{-1}). It is important to note that low frequencies of oscillatory shear correspond to processes that occur slowly (seconds or minutes) and high frequencies correspond to fast processes (fractional seconds).

Two examples involving blow molding will

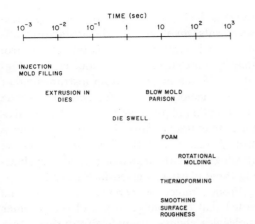

TIME (sec)

INJECTION
MOLD FILLING

EXTRUSION IN
DIES

BLOW MOLD
PARISON

DIE SWELL

FOAM

ROTATIONAL
MOLDING

THERMOFORMING

SMOOTHING
SURFACE
ROUGHNESS

Fig. 28-23 Approximate time scales of typical thermo-plastics processes.

be considered first. A processor observed that one batch (A) of polymer processed better than another batch (B); batch A gave higher throughput during parison extrusion, and it had better sag resistance during the blowing step. The polymer manufacturer said both batches were the same. Measurement of the viscosities (η^*) of these two batches as a function of frequency helped the processor to verify and understand the differences between them (see Fig. 28-24). Sagging is a slow process that should relate to low frequencies. At low frequencies batch A actually has a higher viscosity. The throughput at constant pressure

during parison molding is a high shear rate (i.e., high frequency) process. It is observed that at high frequencies the A batch has the lower viscosity. The rheological measurements are quite consistent with the processor's observations. Indeed A and B are different, both at low frequencies and at high frequencies. This behavior is typical of two polymers with different molecular weight distributions.

In the blow molding of milk bottles it was observed that some HDPE produced bottles with defective handles. Viscosity measurements as a function of frequency are shown in Fig. 28-25. The viscosities of the two materials are virtually identical throughout the frequency range. The elastic modulus (G') is the key here. At low frequencies the good material has a higher elastic modulus, which results in increased die swell during extrusion. This produces a tube with a larger diameter, which allows the handle part of the mold to catch the tube and make the proper bottle shape. Here the die swell (slow process) differences match nicely with low frequency measurements of the elastic modulus.

During pipe extrusion an uneven surface was observed. Measurement of the viscoelastic properties of the polymer at low frequencies clarifies the problem. The polymer that produces the smooth surface has a lower elastic modulus, which causes more stable extrusion and faster relaxation of any surface roughness that might develop.

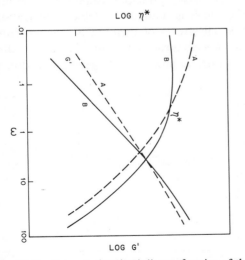

Fig. 28-24 Polymer viscosity (η^*) as a function of the frequency (ω) as measured in oscillatory shear. Curve A: good blow molding performance. Curve B: poor blow molding performance.

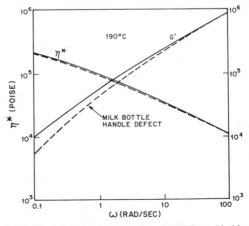

Fig. 28-25 Viscoelastic properties of HDPE used in blow molding of milk bottles.

The injection molding of record discs is another interesting example of the use of low-frequency information. The moduli (G) of two PVC samples were measured at 6 radians/sec over a range of temperatures (Fig. 28-26). One sample produced good discs, whereas the other gave voids in the record grooves. The PVC with the higher modulus did not flow as rapidly as the other. As a result stresses were frozen into the polymer before it could relax completely and fill all the voids.

Another example involves the compounding of carbon black into rubber. Two rubbers that had identical Mooney values behaved very differently in a Banbury mixer. Figure 28-27 shows the power load as a function of time for the two rubbers. Careful study of the viscosities of the two rubbers at the mixing temperature indicates subtle differences as a function of frequency. The key is that at very low frequencies (i.e., long process times) the good rubber will have a lower viscosity. This very-low-frequency difference is what accounts for the compounding differences.

It is clear from these examples that the processing of thermoplastics involves a wide variety of time frames. It has been shown that rheological measurements are very sensitive to this wide variety of processes. At times high-frequency measurements are appropriate; however, more often low-frequency measurements or a combination of the two types is necessary for proper characterization of the thermoplastic. Thus the ability to vary the frequency of the measurement over a broad range is the key to process control. Such measurements can often be made satisfactorily off-line in a laboratory, whereas process control may be more effective when done on-line. Figure 28-28 is a diagram of an instrument that allows rheological measurements to be made as a function of frequency on-line. This instrument gives unique capabilities for process control. A diagram of the Rheometrics on-line rheological testing instrument used in evaluating these plastics is given in Fig. 28-29.

From a molecular perspective it should be noted that the configurations of the polymer molecules change dramatically during many processing steps. The rapid changes in configuration induced by the high shear fields correlate very well with high-frequency viscoelastic measurements. The subsequent relaxation of the molecules after the removal of these fields is considerably slower; molecular motions during relaxation correlate very well with low-frequency viscoelastic measurements. Most thermoplastics processes involve both high shear and relaxation. Thus, a variety of frequencies are useful for an understanding of different materials. Interestingly, it has been determined that many processing problems and concerns involve the relaxation steps; thus, the low-frequency measurements are generally the most useful.

The basic principle behind the on-line rheometer is simple. In a typical application, a small stream of material is diverted from the process through a small tube directly after the melt is produced. A gear pump directs

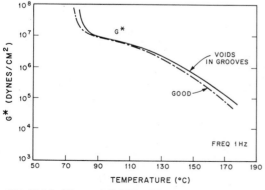

Fig. 28-26   The moduli (G*) of PVC resins as measured in oscillatory shear.

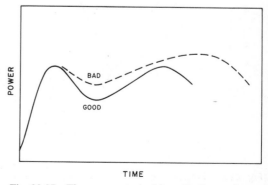

Fig. 28-27   The power required by a Banbury mixer as carbon black is compounded into two rubbers with equal Mooney values.

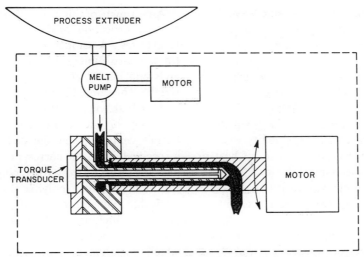

Fig. 28-28 An on-line viscometer that allows measurement of viscoelastic properties as a function of frequency.

the flow of the polymer into the sample chamber. In a short time (one to ten minutes, depending on flow rate), the material fills the small space between two concentric cylinders. At this point, the gear pump either stops, cutting the flow of material to the sample chamber, or continues to pump material through the sample chamber. A DC torque motor then

drives the outer cylinder in an oscillatory manner, with the oscillating cylinder imposing a sinusoidal strain on the material. The corresponding oscillatory stress is measured by monitoring the torque on the inner cylinder.

Because the polymer is not completely elastic, it dissipates some of the input energy. A phase shift develops between the input (strain)

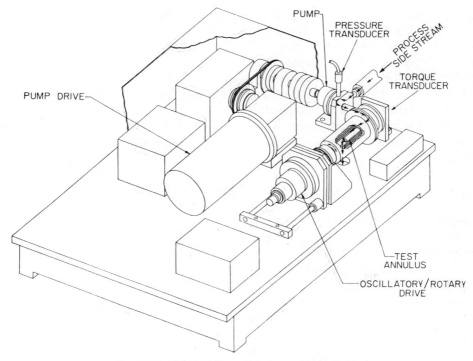

Fig. 28-29 Rheometrics on-line rheometric diagram.

sine wave and output (stress) sine wave. This phase shift along with the amplitudes of stress and strain is used to determine:

1. The elastic modulus (G'), which represents how much energy a material stores.
2. The viscous modulus (G''), which indicates a material's ability to dissipate energy in the form of heat or other mechanical means.
3. The dynamic viscosity (η), which is related to steady shear viscosity.

To minimize operator attention, a high-speed microprocessor is used. The microprocessor accepts operator commands, controls test conditions (velocity, position, temperature), and simultaneously analyzes the strain and stress output signals. These raw data are then reduced into standard engineering units that can be printed, plotted, or sent via a link to an external computer or remote control center.

THERMAL PROPERTIES

The thermal properties of plastics that can readily be examined by different test procedures are: (1) useful temperature range, (2) transition temperatures (glass transition, T_g, and melt temperature, T_m), (3) thermal conductivity, (4) heat capacity, (5) coefficient of linear thermal expansion, and (6) temperature dependence of mechanical properties. These important properties are defined in the following paragraphs.

Useful Temperature Range

The upper temperature limit at which a polymeric material can be used for a prolonged period of time depends on the polymer's structure and internal forces holding the chains together. When temperature increases, these forces become weaker in comparison to the thermal energy of the molecules, allowing relatively large structural deformations. Temperatures above which large deformations start to form are usually not recommended for prolonged use.

An estimate of this temperature comes from the heat deflection, or distortion, temperature (HDT) test. A sample in the form of a beam (of standard dimensions) is supported at the ends and loaded at the center by a constant weight. The sample is immersed in an oil bath, and the bath's temperature is raised gradually, resulting in increasing deflections at the beam's center. When the deflection reaches a specified value, the corresponding bath temperature is recorded as the polymer's HDT. For semicrystalline polymers, the maximum allowable temperature will also depend on the polymer's melting range. The HDT can be used as a guide to the temperature limit at which the polymer can be used, which is based on using 50 percent of the HDT and room temperature.

A polymer's lower temperature limit is dictated by the temperature at which the polymer becomes brittle, which depends on the glass transition temperature.

Transition Temperatures (Glass Transition and Melt Temperature)

The most important of the transition temperatures are the glass transition temperature, T_g, and the melting temperature (or, better, melting temperature range), T_m (see Tables 28-3 and 28-4).

T_g is the temperature below which the polymer behaves similarly to glass, being

Table 28-3   Glass transition (T_g) values for various polymers.

	°F	°C
Polyethylene	−184	−120
Polypropylene	−6	−22
Polybutylene	−13	−25
Polybutadiene	−112	−80
Polyvinyl fluoride	−4	−20
Polyvinyl chloride	185	85
Polyvinylidene chloride	−4	−20
Polystyrene	203	95
Polyacetal	−112	−80
6-Nylon	158	70
66-Nylon	122	50
Polyester	230	110
Polycarbonate	302	150
Polytetrafluoroethylene	−175	−115
Silicone	−193	−125

Table 28-4 Melting temperatures (T_m) for various crystalline polymers*

	°F	°C
Low density polyethylene	230	110
High density polyethylene	266	130
Polypropylene (isotactic)	347	175
6-Nylon	419	215
66-Nylon	500	260
Polyester	500	260
Polytetrafluoroethylene	626	330
Polyarylamides	716	380

* Amorphous polymers exhibit a softening range of temperatures.

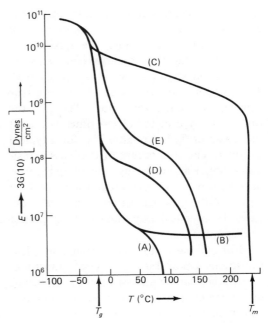

Fig. 28-30   Curves relate storage modulus to temperature (T_g): (A) linear amorphous; (B) cross-linked; (C) semi-crystalline; (D, E) polyester urethanes.

strong, very rigid, but brittle. Above this temperature, the polymer is not so strong and not so rigid as glass; however, it is also not so brittle. A polymer's glass transition temperature can be determined by measuring the change in its density or specific volume with temperature or by such methods as differential scanning calorimetry (DSC) and differential thermal analysis (DTA).

Most polymers are either completely amorphous or have an amorphous-like component even if they are crystalline (1). Such materials are hard, rigid glasses below a fairly sharply defined temperature, the glass transition temperature, T_g. At temperatures above the glass transition temperature, at least at slow to moderate rates of deformation, the amorphous polymer is soft and flexible and is either an elastomer or a very viscous liquid. Mechanical properties show changes in the region of the glass transition. For instance, the elastic modulus may decrease by a factor of over 1000 times as the temperature is raised through the glass transition region.

Figure 28-30 shows typical storage modulus data for several representative polymer systems. Below T_g, the glassy state prevails with modulus values of the order of 10^{10} dynes/cm^2 for all materials. A rapid decrease of modulus is seen as the temperature is increased through the glass transition region (above $-50°$C for these polymers). A linear amorphous polymer that has not been cross-linked (curve A) shows a rubbery plateau region followed by a continued rapid drop in modulus. Cross-linking (curve B) causes the modulus to stabilize with increasing temperature at about three decades below the tempera-

ture of the glassy state. In block copolymers (curves D and E) an enhanced rubbery plateau region appears where the modulus changes little with increasing temperature. Another rapid drop in modulus occurs when the temperature is increased to the hard-segment transition point (12).

Thus, T_g can be considered the most important material characteristic of a polymer as far as mechanical properties are concerned. Many other physical properties change rapidly with temperature in the glass transition region. These properties include coefficients of thermal expansion, heat capacity, refractive index, mechanical damping, nuclear magnetic resonance behavior, and electrical properties. Elastomeric or rubbery materials have a T_g, or a softening temperature, below room temperature. Brittle, rigid polymers have a T_g above room temperature. Glass transitions vary from $-123°$C for polydimethyl siloxane rubber to 100°C for polystyrene and on up to above 300°C or above the decomposition temperature for highly cross-linked phenol formaldehyde resins and polyelectrolytes.

The glass transition temperature is generally measured by experiments that correspond to a time scale of seconds or minutes. If the experiments are done more rapidly so that the time scale is shortened, the apparent T_g is raised. If the time scale is lengthened to hours or days, the apparent T_g is lowered. Thus, as generally measured, T_g is not a true constant but shifts with time. Changing the time scale by a factor of ten times will shift the apparent T_g by roughly 7°C for a typical polymer. The true nature of the glass transition is not clear, and many conflicting theories have been proposed. Although the theoretical nature of the glass transition is subject to debate, the practical importance of T_g cannot be disputed.

Most polymers show small secondary glass transitions below the main glass transition. These secondary transitions can be important in determining such properties as toughness and impact strength.

Cross-linking increases the glass transition of a polymer by introducing restrictions on the molecular motions of a chain. Low degrees of cross-linking, such as found in normal vulcanized rubbers, increase T_g only slightly above that of the uncross-linked polymer. However, in highly cross-linked materials such as phenolformaldehyde resins and epoxy resins, T_g is markedly increased by cross-linking.

It is important to recognize that crystalline polymers do have sharp melting points; however, part of the crystallites, which are small or imperfect, melt before the final melting point is reached. This melting point action must be considered with respect to the melting of the plastic in the plasticator as well as the rate of cooling of the hot melt in the mold.

Thermal Conductivity

This is the rate at which heat can be transferred through a material. For example, in packaging, this property may become important in food-freezing applications or in thermal processing such as pasteurization and sterilization. It is important in evaluating the rate of heating plastic melts during screw plasticating and the rate of cooling plastic melts in the mold.

Heat Capacity

The heat capacity is an indicator of how much heat has to be added to a material in order to raise its temperature by 1°C. This property is important, in principle, in the same areas as thermal conductivity and can be measured by DSC, DTA, or other calorimetric methods.

Coefficient of Linear Thermal Expansion

This property measures expansion upon heating and contraction upon cooling. The coefficient of linear thermal expansion is expressed as the relative change in length per degree of temperature change of a material. If the sample has a rodlike configuration with one dimension larger than the other two, surface expansion will take place; the coefficient of surface expansion is approximately twice the coefficient of linear expansion. For volume expansion (when no dimension is small relative to the others), the coefficient of volume expansion is usually taken as three times the coefficient of linear expansion.

Experience is still a basic requirement for mold design with regard to the determination of cavity dimensions. The costs for changing mold cavities are high—even when similar moldings are to be produced.

Until now, theoretical efforts to forecast linear shrinkage have been unsuccessful because of the number of existing variables. One way to solve this problem is by simplification of the mathematical relationship, leading to an estimated but acceptable assessment. This means, however, that the number of processing necessary changes will also be reduced (13).

As a first approximation, a superposition method can be used to predict mold shrinkage (see Fig. 28-31). However, problems arise in measuring the influencing variables because they are often interrelated—such as variations in the pressure course in the mold with varying wall thickness.

The parameters of the injection process must be provided. They can either be estimated or, to be more exact, taken from the thermal and rheological layout. The position

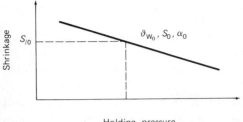

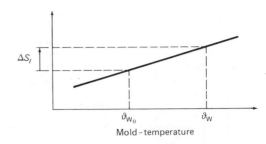

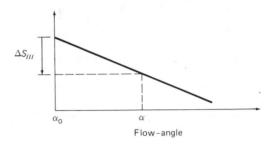

$$S_1 = S_{l0} + \Delta S_l + \Delta S_{ll} + \Delta S_{lll}$$

Fig. 28-31   Superposition of shrinkage (13).

of a length with respect to the flow direction is in practice a very important influence. This is so primarily for glass-filled material but also for unfilled thermoplastics, as is shown in Fig. 28-32. The difference between a length parallel (0°) and perpendicular (90°) to the flow direction depends on the processing parameters. Measurements with unfilled PP and ABS have shown that a linear relationship exists between these points.

Regarding this relationship, it is necessary to know the flow direction when designing the mold. To obtain this information, a simple flow pattern construction can be used (see Fig. 28-33). The flow direction, however, is not constant. In some cases the flow direction in the filling phase differs from that in the holding phase. Here the question arises of whether this must be considered using superposition.

In order to get the flow direction at the end of the filling phase and the beginning of the holding phase (representing the onset of shrinkage), an analogous model was developed that leads to the flow direction at the end of the filling phase.

For a flow with a Reynolds number less than 10, which is valid regarding the processing of thermoplastics, equation (28-5) can be used:

$$\Delta \phi = 0 \qquad (28\text{-}5)$$

For a two-dimensional geometry with quasi-stationary conditions, equation (28-6) is valid:

$$\frac{\partial^2 \phi}{\partial x^2} + \frac{\partial^2 \phi}{\partial y^2} = 0 \qquad (28\text{-}6)$$

Instead of the potential ϕ, it is possible to introduce the flow-stream function ψ for two-dimensional flow. The streamlines ($\psi = $ const.) and the equipotential lines are perpendicular to each other. To express this, the Cauchy-Rieman differential equations can be used (equation 28-7):

$$\frac{\partial \phi}{\partial x} = \frac{\partial \psi}{\partial y} \qquad \frac{\partial \phi}{\partial y} = -\frac{\partial \psi}{\partial x} \qquad (28\text{-}7)$$

Differential equation (28-6) has the same form as is used for a stationary electrical potential field (equation 28-8):

$$\frac{\partial^2 U}{\partial x^2} + \frac{\partial^2 U}{\partial y^2} = 0 \qquad (28\text{-}8)$$

as it can be realized with an unmantled molding out of resistance paper and a suitable voltage.

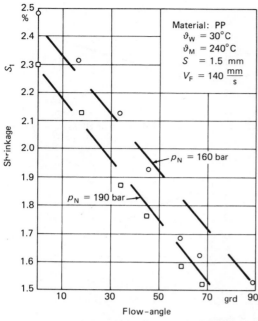

Fig. 28-32   Influence of flow-angle on processing-shrinkage (13).

To control the theoretically determined flow with respect to orientation direction, a color study was made. The comparison between flow pattern, color study, and analogous model is shown in Figs. 28-34 and 28-35.

For a simple geometry the flow pattern method describes the flow direction in the fill-ing phase as well as in the holding phase (see Fig. 28-34).

This description changes when a core is added and the flow is disturbed (see Fig. 28-35). In this case, the flow at the beginning of the holding phase differs from the flow pattern, as it is shown in the color study as well as in the analogous model. Even welding lines are broken in the holding phase so that at this place another flow direction than in the filling phase is found. With further measurements this influence has to be tested by using more complex moldings.

Temperature Dependence of Mechanical Properties

The key to understanding the mechanical properties of plastic materials at different temperatures is a knowledge of their behavior in

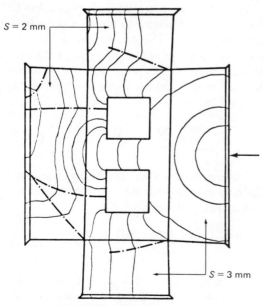

Fig. 28-33   Flow pattern (13).

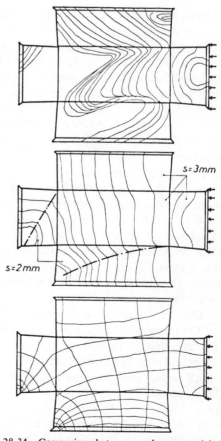

Fig. 28-34   Comparison between analogous model, flow pattern, and color studies (13).

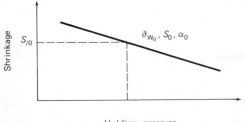

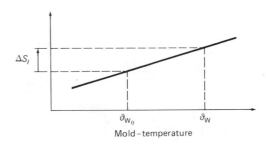

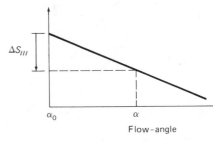

$$S_1 = S_{/0} + \Delta S_I + \Delta S_{II} + \Delta S_{III}$$

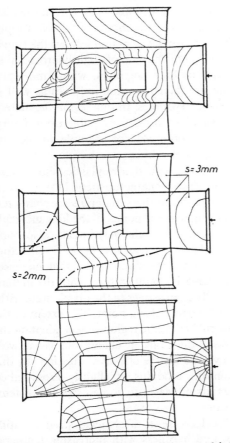

Fig. 28-35   Comparison between analogous model, flow pattern, and color studies with core added (12).

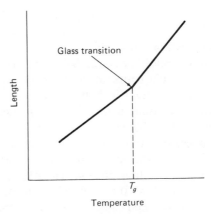

Fig. 28-36   Glass transition temperature.

the transition region between the distinct phase of glass temperature and the crystalline melting point (14). Below the second order transition temperature, characterized by the glass temperature, lies the hard elastic condition usually typified by high brittleness. Micro movement (see below), still possible down to the glass temperature, has ceased (see Fig. 28-36). The position of the second order transition temperature range is influenced by the strength of the secondary bonding; that is, the more effective these forces are, the higher this temperature is.

From the nature of their secondary bond forces the polyolefins are known as dispersion type plastics. The dispersion forces are small by comparison with polar bonds (polyvinyl chloride, polyoxymethylene) and hydrogen bridges (polyamide). Accordingly the glass temperature in the case of polyethylene is −70

to −100°C, depending on the degree of crystallinity; in polyisobutylene, −70°C, and in polypropylene, −32°C. Polyvinyl chloride has a second order transition temperature at +65°C, polyamide 6 at +40°C, and polyamide 6,6 at +50°C.

By blending materials of low second order transition temperatures with other materials of higher second order transition temperatures, the brittle temperature region can be raised. This also holds for polymerization with suitable comonomers.

On further heating, after the brittle-elastic phase and the glass temperature, there follows the workable, tough-elastic phase, which in the case of the polyolefins is the temperature range most commonly used in practical applications. This range and the combination of properties associated with it are characteristic for high polymers. The micro movement reveals itself more and more.

The macromolecule chains have mobility around the bond axes, and chain segments can change place and diffuse in the micro regions. The temperature-independent dispersion forces remain fully active and prevent transposition over great distances.

In Fig. 28-37, the phase regions of an amorphous and a partially crystalline thermoplastic are shown in relation to the temperature-dependent elastic modulus. Curves of this nature are obtained by evaluation of torsional vibration tests, converting the shear modulus G into the dynamic modulus of elasticity E' according to the equation;

$$E' = 2G(1 + \mu) \approx 3G$$

where μ is Poisson's ratio.

Elastic moduli measured below the second order transition temperature T_g, also termed glass moduli, attain in both amorphous and partially crystalline material classes values of around $4 \cdot 10^3$ N/mm^2.

Temperature dependency is not significant; minor steps in the curves indicate secondary molecular relaxation phenomena. Above the glass temperature the molecular chains of thermoplastics and also of slightly cross-linked high polymers are mobile in the amorphous regions. Apart from the dispersion forces, kinks in the chains, entanglements, or in some cases (elastomers and duromers) areas of cross-linking prevent movement over large distances. Thus, above the glass temperature amorphous high polymers do not suddenly melt like low molecular weight materials, but gradually soften over a wide temperature range, without losing the character of a solid material in the process. This phase for high polymers extends over a range of up to 50°C. The elastic modulus of amorphous substances falls in this transition region by about 10^3.

With partially crystalline materials such as polyethylene and polypropylene, the potential large loss of modulus is limited by the stiffening effect of the reinforcing crystallites. Depending on the strength of their active secondary bonds these materials retain a horny character almost up to the crystalline melting point, T_s. Only well above T_g (i.e. in the plastic region) does the modulus fall off steeply.

The rubber elastic region follows the transition region in the case of slightly cross-linked amorphous high polymers, whose molecular chains are linked by atomic forces as well as by secondary forces. In this region the materials display a strongly reversible extensibility, as may be observed, for example, in the case of vulcanized rubber.

The amorphous thermoplastics show quasi-rubber elastic characteristics. In the cross-linking positions these materials are characterized by inter-nodal points of the long molecule chains and by relatively weak dispersion forces. The quasi-elastic behavior is therefore overlain by a measure of flow that increases with increasing temperature, until the materials finally change over to the plastic state without a definite melting point. The extent of the quasi-rubber elastic region is dependent on the length of the molecule chains, increasing with the average degree of polymerization. In this region the modulus of slightly cross-linked materials increases to some extent with temperature.

The decrease in mechanical strength, stiffness, and hardness with increasing temperature is in no way confined to plastic materials. Metals, in spite of their quite different structure (when superficially regarded) behave similarly. Whereas with plastic materials mobility of the macromolecule chains increases with an increase in temperature and changes over

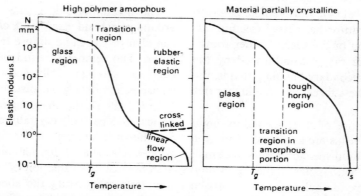

Fig. 28-37 Temperature dependency of the dynamic modulus of elasticity of amorphous and partially crystalline thermoplastics.

from micro movement into the macro-Brownian state (the secondary bonds being gradually overcome in the process), with metals, the crystallite mobility in the sliding planes increases with temperature.

The essential difference between plastic materials known at present and the metals (with the exception of some nonferrous metals) is, however, that fall-off in mechanical strength with metals does not occur until considerably higher temperatures have been reached than those it occurs with for plastics. Briefly, as far as plastic materials are concerned, it may be said that high polymers under mechanical stress normally show a particularly strongly marked viscoelastic character in comparison with the majority of other construction materials; that is, the deformation that occurs is partly elastic (reversible) and partly yield, or plastic (irreversible). In consequence, when plastics are used as construction materials, relevant data such as elastic modulus, shear modulus, and, other important mechanical properties of high polymers depend not only on temperature, but, among other factors, also on the rate and duration of stress loading.

Diffusion/Transport Properties

The ability of a plastic to protect and preserve products in storage and distribution depends, in part, upon the diffusion (i.e., transport) of gases, vapors, and other low-molecular-weight species through the materials. A substance's tendency to diffuse through the polymer bulk phase is the diffusivity or diffusion coefficient (D); the rate of diffusion is related to the resistance, within the polymer wall, to the movement of gases and vapors.

Two important aspects of the transport process are permeability and migration of additives. Possible migrants from plastics can include residual monomer, low-molecular-weight polymer, catalyst residues, plasticizers, antioxidants, antistatic agents, chain transfer agents, light stabilizers, FR agents, polymerization inhibitors, reaction products, decomposition products, lubricants and slip agents, colorants, blowing agents, residual solvents, etc.

Permeability

The driving force for gases and vapors penetrating or diffusing through, as an example, permeable packages is the concentration difference between environments inside and outside the package. A diffusing substance's transmission rate is expressed by mathematical equations commonly called Fick's first and second laws of diffusion:

$$F = -D\frac{dC}{dX} \qquad (28\text{-}9)$$

$$\frac{dC}{dt} = D\frac{d^2C}{dX^2} \qquad (28\text{-}10)$$

where F = flux (the rate of transfer of a diffusing substance per unit area), D = diffusion coefficient, C = concentration of diffusing substance, t = time, and X = space coordinate measured normal to the section.

To measure gas and water vapor permeability, a film sample is mounted between two chambers of a permeability cell. One chamber holds the gas or vapor to be used as the permeant. The permeant then diffuses through the film into a second chamber, where a detection method such as: infrared spectroscopy; a manometric, gravimetric, or coulometric method; isotopic counting; or gas–liquid chromatography provides a quantitative measurement. The measurement depends on the specific permeant and sensitivity required.

Three general test procedures used to measure the permeability of plastics films are:

1. The absolute pressure method
2. The isostatic method
3. The quasi-isostatic method

The absolute pressure method (ASTM D1434-66, "Gas Transmission Rate of Plastic Film and Sheeting"), is used when no gas other than the permeant in question is present. Between the two chambers, a pressure differential provides the driving force for permeation. Here, the change in pressure on the volume of the low pressure chamber measures the permeation rate.

With the isostatic method, the pressure in each chamber is held constant by keeping both chambers at atmospheric pressure. In the case of gas permeability measurement, there must again be a difference in permeant partial pressure or a concentration gradient between the two cell chambers. The gas that has permeated through the film into the lower-concentration chamber is then conveyed to a gas-specific sensor or detector by a carrier gas for quantitation. Commercially available isostatic testing equipment has been used extensively for measuring oxygen and carbon dioxide permeability of both plastic films and complete packages.

The quasi-isostatic method is a variation of the isostatic method. In this case, at least one chamber is completely closed, and there is no connection with atmospheric pressure. However, there must be a difference in penetrant partial pressure or a concentration gradient between the two cell chambers. The concentration of permeant gas or vapor that has permeated through into the lower-concentration chamber can be quantified by a technique such as gas chromatography.

Three related methods are used to measure the permeability based on the quasi-isostatic method. The most commonly used technique allows the permeant gas or vapor to flow continuously through one chamber of the permeability cell. The gas or vapor permeates through the sample and is accumulated in the lower-concentration chamber. At predetermined time intervals, aliquots are withdrawn from the lower cell chamber for analysis; the total quantity of accumulated permeant is determined and plotted as a function of time. The slope of the linear portion of the transmission rate profile is related to the sample's permeability.

Migration

Migration is a complex process depending in part (if no chemical reaction takes place) on the migrating species' diffusivity. Diffusivity, or the diffusion coefficient (D), is the tendency of a substance to diffuse through the polymer bulk phase. Migration, therefore, also can be considered a mass transport process under defined test conditions (i.e., time, temperature, and the nature and volume of the contacting phase).

The driving force for migration is the concentration gradient, where dissolved species diffuse from a region of higher concentration (i.e., polymer) to a region of lower initial concentration (i.e., contact phase). The diffusion rate is related to the resistance against the movement of migrant within the polymer bulk phase.

Thus, if migration from a package to a contact phase is to occur, the migrant has to undergo two processes in succession: diffusion through the polymer bulk phase to the polymer surface, and dissolution or evaporation to the contact phase.

Under current federal regulation, extractability of packaging material components is one of the most important characterization parameters for plastics used in package foods and pharmaceuticals. Both migration of base material and migration of trace constituents such as residual solvents can affect the packaged product's quality.

MICROTOMING/OPTICAL ANALYSIS

Quality control of plastic molded parts can use optical techniques. In the procedure thin slices of the material are cut from the part and microscopically examined under polarized light transmitted through the sample. Study of the microstructure by this technique enables rapid examination of quality-affecting properties. This kind of approach can provide the molder with information for failure analysis, part and mold design, and processing optimization (15, 16).

Thin sectioning and microscopy are old techniques, having been applied to biological samples for many years. Furthermore, metallurgists have used similar techniques in the microstructural analysis of metals to determine their physical and mechanical properties and to aid in failure analysis.

Microtoming enables slices of plastic to be cut from opaque parts. These slices are so thin (under 30 μm) that light may be transmitted through them. The sample can then be analyzed under a microscope. Another useful technique is to use the microtome to slice down through a specimen until the specific level to be examined is reached. This method reveals a series of sequential levels, each smooth enough for viewing without need for polishing. The usual method is to cut, mount, and polish. When a series of cuts is needed, it becomes necessary to regrind and repolish. The microtome technique eliminates these tedious steps.

The two pieces of equipment are required for microstructural examination—a microscope and a microtome. Both of these facilities should be of good quality. The microscope must be equipped as a light transmission microscope fitted with a polarizer analyzer and variable intensity light source. The microtome must be a substantial, rugged machine capable of slicing ultrathin sections from a wide variety of materials without flexing of the frame. It must have a well-made slide-bearing surface in order to have accurate smooth action. The specimen-holding vise must be substantial and securely attached to the sled.

An attractive aspect of the microtome analysis procedure is the speed with which results can be obtained. Generally the sample can be rough-cut from the product with a hacksaw and secured in the microtome vise, although in some cases it is necessary first to embed the sample in a block of epoxy. The slicing is a simple procedure. Usually slices 8 to 15 μm thick will be produced. These are mounted on a microscope slide using mounting cement and a cover plate.

Polymers are often categorized as either amorphous or crystalline. Some can exist in either or both forms, and thus it is common to discuss degree of crystallinity when referring to the microstructure of a part. Often the effects of molding are clearly exhibited by observing the transition from the amorphous skin of a part to the crystalline core.

Much of the analysis of plastics microstructures is fairly straightforward. It is easy to tell whether you are dealing with a crystalline or an amorphous material by observing the sample using polarized light. Amorphous areas appear black, while crystalline areas can be clearly examined. The explanation for this effect is that in the case of crystalline polymers the molecules crystallize and fold together in a uniformly ordered manner, whereas the amorphous polymers do not produce crystallites and occur randomly positioned. Thus, under polarized lighting crystalline materials exhibit multicolored patterns, whereas amorphous materials appear black. In this way the crystalline microstructure can be examined. Features of the crystalline polymers are readily discerned, whereas those of the amorphous polymers are not.

It is interesting to notice differences in the cyrstalline structure found with different materials. A comparison was made between a nylon 6/6 micrograph and that produced from one of acetal homopolymer. The acetal has a characteristic structure that is very different from the square crystallites seen in the structure of the nylon. This difference is related to the propensity of nylon to supercool to a greater extent than most crystalline materials, whereas acetals crystallize much more rapidly.

Optical techniques can be used for both quality control and failure analysis. Stress concentration can for a variety of reasons be a principal failure mode. One of these reasons relates to the use of contaminated or mixed materials, which may be caused by the presence of foreign materials or improper machine cleaning. Incorrect regrinding procedures, improper dry coloring methods, and use of the wrong pigment are additional causes of this condition. Stress concentrations that result from material contamination can be detected by observing the break area by reflected light. Particle size and dispersion can be found by examination under transmitted polarized light. Using polarized light it is possible with crystalline materials to identify residual stresses caused by incorrect gating and sharp corners emanating from poor part design. Impact, bending, and other physical stresses imparted to the part during service can also be identified.

Generally it is necessary to know whether or not you are dealing with a stressed-in-service part. Then it is possible to determine whether residual stresses resulted from service, or whether they occurred in molding. Stresses imposed in the molding process usually appear as regular patterns in the flow line direction, whereas those that result from imposed stresses created in service tend to exhibit semicircular arc-shaped configurations. Another source of stress involves the use of the microtome itself, since with some materials induced stresses are not difficult to create. These are usually found along the edges of the sample, and frequently the microstructure becomes smeared in these areas. Fortunately, stress due to the microtome is not difficult to detect when viewing the specimen.

It is particularly important to ensure that the sample being microtomed is securely supported during the cutting operation so that the imposition of cutting stresses is held to a minimum. In some cases it is necessary to fill holes or slots with epoxy to avoid tearing of edges and the development of vibration at the surface of the specimen during cutting.

Optical examination of the microstructure will determine whether or not correct mold temperatures were used in the production of parts from crystalline or partially crystalline polymers. With these thermoplastics the degree of crystallinity achieved depends on the temperature of the mold, temperature of the melt, and time that the pressure on the melt is maintained.

In the case of acetal, the use of a cold mold results in fast dissipation of heat from the melt into the mold wall. Consequently, the threshold limit for the formation of crystallization nuclei is quickly reached, and a skin is formed on the parts that has an amorphous appearance but is actually crystalline, although to a much lower extent that the spherulitic region that is formed below the skin. The thickness of the "amorphous" zone is dependent on the mold and melt temperatures and screw forward time.

From the micrographs it is possible to judge the extent of what might be loosely described as the amorphous skin, the transcrystalline zone, and the spherulitic core. From their relative proportions it is not only possible to esti-

mate the processing conditions that were employed to produce the parts, but also to predict part performance. Particles may remain unmelted within the molten mass of plastic. These particles inhibit the formation of crystallization nuclei. Since the thermal conductivity of plastics is poor, the length of time for the material to cool controls to a large extent the length of the molding cycle. Reducing melt temperature to shorten cycles and increase production reduces the quality of the product. Mold temperature, melt temperature, and screw forward time all interact to influence part quality.

Screw forward time refers to the injection time plus the time that injection pressure is exerted on the material. It is a very important factor affecting the structural quality of molded parts. Upon entry into the mold the material will rapidly freeze where it contacts the mold wall surfaces. Material in the interior, however, remains molten much longer. Thus more material can be forced into the molten interior of the part although the skin has frozen. This process may be carried out until the gate has frozen off; when pressure on the melt is reduced prematurely (i.e., before gate freeze-off), changes occur in the part structure. The freezing point of the material is a function of pressure as well as temperature. Sudden removal of screw pressure on the material will lower the freezing point, so that a change occurs in the rate of crystallization of the molten material in the interior. In the case of acetals this results in a discontinuity in the structure. Removal of this pressure may also permit a backflow of melt through the gate, which in turn creates another discontinuity zone. These discontinuities can act as stress concentrators to lower elongation properties. The effects of loss of pressure at this critical moment are a reduction in part weight and material density and increased shrinkage.

From a microtomed specimen it is possible to obtain processing history information that can be immediately transmitted to the manufacturing area. It is possible to do most of the following:

- Identify the polymer, fillers, reinforcements, and pigments.
- Examine distribution and orientation of fillers, reinforcements, and pigments.
- Determine the presence of molded-in and subsequently imposed stress concentrations.
- Determine whether contamination is present.
- Reveal excessive use of reground material.
- Study weld lines and material flow characteristics.
- Determine variations in melt temperature and mold temperature.
- Show the effects of gate size and position.
- Study improvements to part and tool design.

As with any technique it is necessary to acquire the skills to recognize what is seen under the microscope. Much of the time, comparison of good and bad parts is a considerable aid to understanding the situation. The provision of a file of micrographs is an effective way to aid problem diagnosis.

NONDESTRUCTIVE TESTING (NDT)

In the familiar form of testing known as destructive testing, the original configuration of a specimen is changed, distorted, or even destroyed for the sake of obtaining such information as the amount of force the specimen can withstand before it exceeds its elastic limit and permanently distorts (usually called yield strength) or the amount of force needed to break it (tensile strength). The data collected in this instance are quantitative and could be used to design an airplane wing to withstand a certain oscillating load or a highway bridge subject to wind storms or heavy traffic usage. However, one could not use this specimen in the wing or bridge. One would have to use another specimen and hope that it would behave exactly like the one that was tested (17).

Nondestructive testing (NDT), on the other hand, examines a specimen without impairing its ultimate usefulness. It does not distort the test specimen's configuration, but provides a different type of data. NDT allows suppositions about the shape, severity, extent, configuration, distribution, and location of such internal and subsurface defects as voids and pores, shrinkage, cracks, and the like.

Most materials contain some flaws. This may or may not be cause for concern. Flaws that grow under operating stresses can lead to structural or component failure. Other flaws present no safety or operating hazards. Nondestructive evaluation provides a means for detecting, locating, and characterizing flaws in all types of materials, while the component or structure is in service, if necessary, and often before the flaw is large enough to be detected by more conventional means. The following is a brief guide to nondestructive evaluation methods.

Radiography

Radiography is the most frequently used nondestructive test method. X rays and gamma rays passing through a structure are absorbed distinctively by flaws or inconsistencies in the material. Cracks, voids, porosity, dimensional changes, and inclusions can be viewed on the resulting radiograph.

Ultrasonics

In ultrasonic testing the sound waves from a high-frequency ultrasonic transducer are beamed into a material. Discontinuities in the material interrupt the sound beam and reflect energy back to the transducer, providing data that can be used to detect and characterize the flaws.

When an electromagnetic field is introduced into an electrical conductor, eddy currents flow in the material. Variations in material conductivity caused by cracks, voids, or thickness changes can alter the path of the eddy current. Probes are used to detect the current movement and thus describe the flaws.

When flaws or cracks grow, minute amounts of elastic energy are released and propagate in the material as an acoustic wave.

Sensors placed on the surface of the material can detect these acoustic waves, providing information about location and rate of flaw growth. These principles form the basis for the acoustic emission test method.

Although commercially available for the past 20 years or so, ultrasonic detectors never really caught on as a diagnostic or maintenance tool. The biggest problem with ultrasonic detectors was their inability to produce measurements as accurately or as consistently as could many competing devices for nondestructive testing. The advent of microprocessing is dramatically improving the ability of ultrasonics to detect the wall thickness of metal and plastic pipes and process vessels; to determine particle dispersion in suspensions; and to detect potential leakage and faulty parts in pumps, steam traps, and valves (18).

Liquid Penetrants

The liquid penetrant method is used to identify surface flaws and cracks. Special low-viscosity fluids containing dye, when placed on the surface of a part, penetrate into the flaw or crack. When the surface is washed, the residual penetrants contained in the part reveal the presence of flaws.

Acoustics

In acoustical holography, computer reconstruction provides the means for storing and integrating several holographic images. A reconstructed stored image is a three-dimensional picture that can be electronically rotated and viewed in any image plane. The image provides full characterization and detail of buried flaws.

Photoelastic Stress Analysis

Photoelastic stress analysis is a way to determine why a part broke and how to prevent similar failures in the future. Parts ranging from structural glass fiber–reinforced boat hulls to tiny thermoplastic heart valves can all be tested easily. The test method is also

a valuable tool for predicting where prototype parts may fail (19).

Manufactures of plastic products want to be sure that their parts will withstand service stresses, especially since they are now faced with increasingly rigorous safety requirements, strict liabilities, and extended product warranties. Mechanical failure due to thermal or mechanical stress is a strong possibility if any of three manufacturing functions—design, processing conditions, or assembly techniques—is mishandled. Poorly designed features such as corners, ribs, or holes are common causes of failure. So are improper processing conditions, including excessive injection pressure, poor mold design, or inconsistent mold temperature. Careless assembly techniques such as overtightening of a bolt can also cause part failure.

Photoelastic analysis, one of several related testing techniques, is easy to use and usually more economical and positive than computer analysis. From the information it provides, the test can lead to better-designed, lower-cost products. Traditionally used to test the integrity of metal parts, photoelastic analysis is now being used to physically test thermoplastics as well as thermosets. For transparent plastics, the analysis can be made directly on the plastic. For nontransparent plastics, a transparent coating is used. Actual parts and representative models can be tested by a simple procedure. The former may be stressed under actual use conditions, whereas models are tested under simulated conditions.

Although theoretical analytical methods such as finite element analysis offer a chance to solve complex stress problems, there are many causes of strain in parts that cannot be reliably tested by these expensive computer-oriented techniques. For instance, strains due to assembly of components and those caused during processing are extremely difficult problems to analyze without physically testing the part.

Photoelastic analysis is more than just another pretty experimental stress test. When examined under a polariscope, the colorful interference pattern can be used to survey stress distribution and the degree of strain. This analysis ultimately leads to pinpointing which manufacturing function—design, processing conditions, or assembly techniques—led to part failure or might do so in the future. Interference patterns for coatings and models are analyzed in the same way. The photoelastic color sequence shows stress distribution in the part (see Fig. 28-38). In order of increasing stress, the sequence is black, gray, yellow, red, blue-green, yellow, red and green. Black and gray areas show low strains, while a continued repetition of red and green color bands indicates extremely high concentrations of stress. An area with uniform color is under a uniform stress.

The degree of strain is indicated by a fringe order, which is simply a collection of black bands appearing in close proximity to each other between colors in the stress pattern. As the stress concentration increases, the number of black bands in a fringe order does, too.

DRYING HYGROSCOPIC PLASTICS

Thermoplastics such as polyurethanes, nylons, polycarbonates, acrylics, ABS, etc., are categorized as hygroscopic (see "Dryers" section in Chapter 15). Polymers of this type absorb moisture, which has to be removed before they can be converted into acceptable finished products. This is true of thermoplastic polyurethanes, especially those processed in excess of 160°C. Very low moisture concentrations can be achieved through the utilization of an efficient drying system and proper handling of the dried material prior to and during the molding or extrusion operation. Drying hygroscopic resins should not be taken casually. Simple tray dryers (so-called pizza ovens) or mechanical convection hot air dryers, while adequate for some materials, simply are not capable of removing water to the degree necessary for proper processing of hygroscopic polymers, particularly during periods of high ambient humidity (20).

The effect of excess moisture content upon thermoplastic molding- and extrusion-grade resins manifests itself in various ways, depending upon the process being employed. Splays, nozzle drool between shots, foamy melt, bubbles in the part, poor shot size control, sinks, and/or lower physical properties are the results of high water content during processing operations. Effects seen during extrusion can also include gels, trails of gas bubbles in the extrudate, arrowheads, waveforms, surging, lack of size control, and poor appearance.

The most effective and efficient drying system for hygroscopic polymers is one that incorporates an air-dehumidifying system in the material storage/handling network, which can consistently and adequately provide moisture-free air in order to dry the "wet" polymer (21). Although this type of equipment is expensive initially, it results in improved production rates and lower reject levels in the long

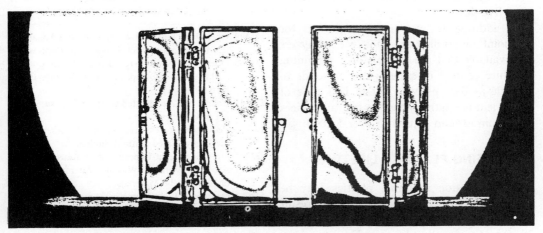

Fig. 28-38   Photoelastic stress pattern for these two parts molded during the same production run shows that the processing conditions changed.

run. There are a variety of manufacturers and systems from which to choose. While all systems are designed to accomplish the same end (i.e., dry polymer), the approaches to regeneration of the desiccant beds vary widely. Years of field experience with these systems have shown that breakdowns in performance are not usually the fault of the equipment, but are due to the user's lack of attention to preventive maintenance details as outlined by the manufacturer.

Determining Moisture Content

In order to determine the effectiveness of the system, some method of determining the moisture content of the air in the drying system is recommended. The installation and monitoring of a dewpoint meter in the drying arrangement is a worthwhile investment. Equipment performance can be easily monitored by both visual signal (telltale indicators) and recordings. Dewpoint monitors can be purchased from most of the dryer manufacturers and be installed at the time of purchase or retrofitted at a later time. Also available are portable types that can be used to spot-check various sections of the material-handling network. Although the investment is somewhat high ($500 to over $3,000), the payback, when there are problems during production, is incalculable in time and material savings. As to the type of installation, the processor must decide what is best for his particular needs as well as his pocketbook.

In addition to instruments designed for dewpoint determination, moisture analyzers are available that are capable of determining moisture content of either gases or solids to as little as 0.01 percent water. This type of equipment is relatively easy to use, and prices vary from around $2,000 to over $8,000.

ESTIMATING PLASTICS LIFETIME

During injection molding, plastics may be subjected to an overload of heat. The result can be immediate decomposition (see Table 28–5) and a very short lifetime. For a practical determination of their lifetime, plastic molded parts generally must go through a time period

in actual service so reliable data can be obtained. However, tests (usually per ASTM) are used that have a degree of reliability based on experience or as presented in an ASTM standard (see Table 28–6). If proper material controls and proper process controls are used, the parts might outlast predictions.

Plastic molded parts (and plastics processed by other techniques) have been used for long time periods—some beyond their expected lifetime—for the past century. Military, industrial, and commercial parts have done their jobs; examples are many, such as parts for aircraft, automobiles, electronics, agriculture,

Table 28-5 Decomposition temperatures (T_d) for various plastics.

	°F	°C
Polyethylene	645–825	340–440
Polypropylene	610–750	320–400
Polyvinyl acetate	420–600	215–315
Polyvinyl chloride	390–570	200–300
Polyvinyl fluoride	700–880	370–470
Polytetrafluoroethylene	930–1020	500–550
Polystyrene	570–750	300–400
Polymethyl methacrylate	355–535	180–280
Polyacrylonitrile	480–570	250–300
Cellulose acetate	480–590	250–310
Cellulose	535–715	280–380
6-Nylon	570–660	300–350
66-Nylon	610–750	320–400
Polyester	535–610	280–320

Table 28-6 ASTM standards for testing weatherability of plastics.

D1499	Operating Light- and Water-Exposure Apparatus (Carbon-Arc Type) for Exposure of Plastics
D2565	Operating Xenon-Arc Type (Water-Cooled) Light- and Water-Exposure Apparatus for Exposure of Plastics
D4141	Conducting Accelerated Outdoor Exposure Testing of Coatings
E838	Performing Accelerated Outdoor Weathering Using Concentrated Natural Sunlight
G23	Operating Light-Exposure Apparatus (Carbon-Arc Type) With and Without Water for Exposure of Nonmetallic Materials
G26	Operating Light-Exposure Apparatus (Xenon-Arc Type) With and Without Water for Exposure of Nonmetallic Materials
G53	Operating Light- and Water-Exposure Apparatus (Fluorescent UV-Condensation Type) for Exposure of Nonmetallic Materials

tanks/containers, telephones, etc. Unfortunately the information generally perceived about these parts (particularly in news accounts) is examples of "what went wrong."

For a more objective appraisal, there are procedures used to estimate plastic lifetime rather quickly and realistically. An example is the use of TGA decomposition kinetics (22).

QC BEGINS WHEN PLASTICS RECEIVED

Although care is taken by materials manufacturers to assure consistency, subtle variations exist in their products. In most general applications, these variations have little effect on finished part properties, but in more stringent cases, these irregularities can present problems. To simplify the task of assuring that the physical properties of a system are in specification, simple techniques can be used in incoming, in-process, and outgoing quality control. Use of these procedures by companies concerned with maintaining critical properties can keep a tight rein on product quality and provide documented qualification.

In the continuous pursuit of improvement, much work has been done with equipment such as temperature controls. Automation has been developed to control speed of press closing, clamping pressure, and breathing cycles. Tool designers and moldmakers have become more effective in designing and building high-quality tools. Special tool steels have been developed to meet such needs (23).

Operators have been trained in the operation of this complex group of machines. Quality control technicians have been equipped with sophisticated checking fixtures and gauges. Maintenance people have been sent to training courses to learn how to cope with repair problems. Even management personnel have been given courses in the skills of management.

All of this activity has served to narrow the gap between pounds of material purchased and pounds of product shipped, which is, after all, one of the key factors in determining the profitability of a plastics processor. However, one serious problem often exists in molding plants, namely, control of the quality of the incoming raw material.

Some will say that the material suppliers have done such a good job that incoming or receiving inspection of material is only a waste of time. It is true that material suppliers have generally improved the quality of their materials, as well as the consistency of that quality. Yet problems continue to develop somewhere in the molding process. Why? If all the equipment is operating satisfactorily, if the setup has been made correctly, if the mold is in good condition, what is responsible for sudden (or gradual) changes in the finished quality of a molding?

Suppliers need a performance standard they cannot misunderstand. It is up to management, at all levels, to provide that standard. Unfortunately, that usually doesn't happen. Some companies use the word "excellence" when they talk about quality. It has a nice ring, and it looks good in ads. But what does it really mean? "Bring this back when it is excellent." Could you be certain that an employee or supplier understood that command? Everyone has different ideas about it.

No More A-B-Cs. Some years ago, it was fashionable to establish a "classification of characteristics." It began with hardware and migrated to paperwork and software. Every requirement was classified as to its importance: A, B, or C. All A requirements had to be met; they were not negotiable (unless they were downgraded). All B requirements ought to be met, but they could have some variation as long as it didn't affect form, fit, or function. All C requirements were easily disposed of, since they were primarily cosmetic.

As a result of this plan, the whole world was negotiable! People ran around all day long asking, "Is this good enough?" Management and quality control engineers redesigned products on a daily basis. Quality, clearly, was a distant third after schedule and cost.

Need for Dependability

The biggest problem of a wavering performance standard is that we cannot depend on one another. If what we receive from another department or supplier doesn't have to be the way we said it should, then we can't do what we were going to. Everyone has to be resourceful, but individuals do not know enough about the complete system to be able to make performance decisions, and they never did. The whole system of a company depends on being able to know what someone else, including a supplier, is going to do.

People sometimes have a problem with the words "zero defects." But the words merely symbolize the idea of "doing it right the first time." Some companies use "defect-free," a perfectly acceptable substitute. But nonspecific words such as "good" or "pride" or "excellence" mean you are not being specific about quality.

Quality Auditing

Some organizations have a documented quality assurance program that includes an audit program. A quality assurance program usually contains three tiers of documentation: the quality assurance manual, system-level procedures, and instructions. The purpose of an audit program is to evaluate the existence and adequacy of the QA program and to ensure that the manufacturer's operations are in compliance with it (24).

Putting a program in writing does not ensure that it will be followed, nor does it, in and of itself, provide the feedback necessary to correct and update programs and processes. The audit fills both of these gaps. By monitoring product, process, and system, and by rating performance against a predetermined scale, the auditor determines the need for corrective measures. By investigating, in turn, the underlying reasons for nonconformance, he isolates the causes and provides sufficient feedback to ensure that the causes, not just the symptoms, are corrected. Finally, through partial and follow-up audits, the auditor ensures that both symptoms and causes have been eliminated. In this way the quality auditing system provides a foundation for satisfactory development and a means of ensuring the existence of a sound program of managerial control.

Many people have difficulty distinguishing between audits and inspections, believing that an audit is designed to verify compliance only. The distinction between the two is related to their objectives. The primary objective of an inspection is to accept or reject a particular product or process. The primary objective of an audit, on the other hand, is to evaluate the existence of, compliance with, and adequacy of a documented QA program. An audit that verifies compliance with an inadequate GMP (general manufacturing procedures) quality program is worse than useless; it is misleading. A competent auditor has the training, experience, and skill to develop an adequate quality program and can, therefore, assess the effectiveness of the program under review. Other members of the audit team provide the expertise necessary to assess the adequacy of the program's technical aspects.

According to the requirements for auditing, an audit must: (1) be planned and periodic, (2) verify compliance with and effectiveness of the quality program, (3) be performed in accordance with written procedures or checklists, (4) be performed by qualified individuals who are independent of the area being audited, (5) be followed by appropriate measures and corrective action, and (6) be reviewed by management. These six elements are in fact stipulated by FDA in the medical device GMP's (21 CFR 820).

ECONOMIC SIGNIFICANCE OF QUALITY

There tends to be a positive correlation between the quality of the goods offered by a company (i.e., fulfillment of explicit requirements or implicit customer expectations) and its profit margin. The well-known PIMA study (25) indicates that the return on investment (ROI) as a yardstick for a company's profit depends not only on market share but above all on product quality. The notion "quality first . . . profit is its logical consequence," constantly expressed by Japanese entrepreneurs, has to be interpreted in this sense.

The customer is only in a position to assess a few of the quality features at the instant of purchase; so Purchase is and remains a matter of trust. An endeavor to improve the market share calls for strengthening this trust. Above all, customer loyalty, as defined by the proportion of customers who will buy the same make of product again, largely depends on the customer's experiences with products of that make. Sales promotion can be used to good effect, but in the long run it cannot overcome the impressions made by inferior products on an ever more critical market.

Clearly there is a close connection between quality and cost-effective production. Inspection of products can identify faults and can serve as a basis for their correction, although it does not prevent the occurrence of faults in the first place. These faults must be prevented, however, usually with an investment in methods and personnel. This investment must be profitable, like any other investment. The return on the investment in this case is the non-occurrence of faults. Success can be measured by the reduction of failure costs.

The notion that the production planning department is solely responsible for costs, the production department for delivery date, and the inspection department for quality is clearly outmoded. Quality (i.e., the fulfillment of explicitly specified requirements and/or implicit customer expectations within the framework of delivery and cost schedules) can only be ensured by collaboration among all the departments in a company. Quality assurance is an interdepartmental responsibility with the objective of preventing faults. It serves to improve a product's chances of success on the market and reduces the risk of warranty and other claims. Thus it is part of the company strategy. Management must initiate, implement, and continually adapt the quality assurance system in the light of changing conditions.

It is sound practice to create a "quality control department" responsible to top management. This function should suggest, coordinate, and analyze quality-related measures and inform all concerned about them without relieving the line managers of responsibility for the quality of the work performed. A quality assurance system is not an end in itself. It serves to ensure and improve quality in the light of steadily more exacting market require-

ments and the necessity to reconsider both hitherto taken-for-granted design margins and every production process in the light of cost considerations. Promotion of quality must have high priority within the system.

INDUSTRY TESTING STANDARDS

Test procedures and plastic materials are subject to change, and it is essential that you keep up to date on the changes. You may obtain the latest issue on any test (such as a simple tensile test or molecular weight test) by contacting the organization that previously issued the test/standard/qualifying list, etc. As an example, ASTM issues new annual standards that include all changes. The 1984 *Annual Book of ASTM Standards* contained more than 7000 standards published in 66 volumes. They include different materials, products, etc. There are four volumes on plastics standards: *Volume 08.01/Plastics I,* which lists C177 to D1600 standards; *Volume 08.02/ Plastics II,* which lists D1601 to D3099 standards; *Volume 08.03/Plastics III,* which lists D3100 to the latest standards; and *Volume 08.04/Plastic Pipe and Building Products,* with 165 standards.

IDENTIFICATION OF PLASTICS

To identify a specific plastic, the characterization techniques described in this chapter can be used, as well as the more conventional chemical analysis/synthesis methods that are routinely performed in various laboratories. To provide a quick way of identifying plastics refer to Table 28–7. This table is only a guide, not foolproof. The detailed chart covers a wide range of plastics (26).

Although the chart may appear to be somewhat formidable at first glance, only three simple tests are necessary to identify all of the plastics shown. No special equipment is needed—just water, matches, and a hot surface—and the only sensors required are one's eyes and nose.

The first step is to try to melt the material to determine whether it is a thermoset or a thermoplastic. This is usually done with a

Table 28-7 Plastics identification Chart.

Flowchart:

PLASTICS MATERIALS → Press a soldering iron or a hot rod (500°F) against the sample

- **Softens** → **Thermoplastics** → Drop a small sample in water
  - **Sinks** → All others → Burn a small corner of the sample
    - **Continues to burn** → Drips?
      - Yes → ABS, Acetals, Cellulose acetate, Cellulose acetate butyrate, Cellulose propionate, Cellulose nitrate, Polystyrene, Polyurethane, Polyester
      - No → (see group below)
    - **No flames** → PTFE, CTFE, PVF, FEP → Drips?
      - Yes
      - No
  - **Floats** → PP, PE
- **Thermosets** → Burn a small corner of the sample
  - **Self-extinguishing** → Dap, Melamine formaldehyde, Phenol formaldehyde, Urea formaldehyde
  - **Continues to burn** → Polyester, Silicone, Epoxy
  - **Self-extinguishing** → Nylon, Polycarbonate, PPO, Polysulfone, PVC → Drips?
    - Yes → Nylon, Poly-sulfone
    - No → Poly-carbonate, PPO, PVC

Thermosets table

Material \ Observations	Dap	Melamine formaldehyde	Phenol formaldehyde	Urea formaldehyde	Polyester	Silicone	Epoxy
Color of flame	Yellow	Yellow with blue tip	Yellow	Yellow with greenish blue edge	Yellow with blue edges	Bright yellow	Yellow
Odor	Faint odor of phenol	Fish like	Phenol	Formaldehyde	Sour cinnamon	None	Pungent amine
Other characteristics	Black smoke	Swells and cracks	May or may not be self-exiting	Swells and cracks	Black smoke with soot	Continues to burn	Black smoke

Self-extinguishing thermoplastics/thermosets table

Material \ Observations	Nylon	Poly-sulfone	Poly-carbonate	PPO	PVC
Color of flame	Blue with yellow tip	Orange	Orange or yellow	Yellowish orange	Yellow with green edges
Odor	Burnt wool or hair	Odor of sulphur	Phenol	Phenol	Hydrochloric acid
Speed of burning slow <3 inches fast >3 per min.	Slow	Fast	Slow	Slow	Slow
Other characteristics	Froths	Black smoke with soot	Black smoke with soot	Difficult to ignite smoke	White smoke

Continues to burn table

Material \ Observations	ABS	Acetal	Cellulose acetate	Cellulose acetate butyrate	Cellulose propionate	Poly-styrene	Polyester	Cellulose nitrate	Poly-urethane
Color of flame	Blue with yellow edges	Blue	Yellow with sparks	Yellow with blue tip	Yellow	Yellow	Yellow with blue edges	Pale yellow	Yellow
Odor	Acrid	Formaldehyde	Vinegar	Rancid butter	Burnt sugar	Illuminating gas or marigold	Burning rubber	Camphor	Faint apple
Speed of burning	Slow	Slow	Slow	Slow	Fast	Fast	Fast	Fast	Fast
Other characteristics	Black smoke with soot	No smoke	Black smoke with soot	Some smoke with soot	Some black smoke	Dense smoke with soot	Black smoke with soot	Sample burns completely	Slight black smoke

Fluoropolymer table

Material \ Observations	FEP	CTFE	PTFE	PVF
Color of flame	…	…	…	Acidic
Odor	Burnt hair	Acetic acid	Burnt hair	
Speed of burning	…	…	…	…
Other characteristics	…	…	…	…

PP, PE table

Material \ Observations	PE	PP
Color of flame	Blue with yellow tip	Blue with yellow tip
Odor	Paraffin	Acrid or diesel fumes
Speed of burning slow <3 inches fast >3 per min.	Fast	Slow
Other characteristics	Melts & drips	…

Table 28-8   Some characteristics of common plastics (courtesy of U.S.I. Chemicals).

	Specific Gravity	Flame Color *(Copper Wire)		Color	Smoke Density	Odor	Solvents	Comments
		As Is	Melts/Soft					
Polypropylene	0.85-0.9	Blue-Yellow	Yes (trans)	White	Very little	Heavy	†Toluene (slowly-slight)	Drips, Swells
LDPE	0.91-0.93	Blue-Yellow	Yes (trans)	White	Very little	Candle Wax	†Dipropylene Glycol	Drips, Swells
HDPE	0.93-0.96	Blue-Yellow	Yes (trans)	White	Very little	Candle Wax	†Toluene	Drips, Swells
Epoxy	1-1.25	Orange Yellow (Green)	No	Black		Phenolic		Some Soot
Chlorinated PE	1-2.4	Green	Yes				†Toluene	
Polystyrene	1.05-1.08	Orange Yellow	Yes	Black	Dense	Sweet Marigolds	†Diethyl Benzene	Soot, No Drip
Polyvinyl Butyral	1.07-1.08	Blue Mantle Yellow	Yes (trans)			Rancid Butter		Drips, Swells
Nylon	1.09-1.14	Blue Mantle Yellow	Yes			Burnt Hair		Swells, Froths
Ethyl Cellulose	1.1-1.16	Blue White	Yes			Sweet	Sec-amyl Alcohol	Drips
Polyester	1.12-1.46	Yellow	No	Black	Dense	Sweet (Resinous)		Softens
Vinyl Chloride	1.15-1.65	(Green) Yellow-Orange	Yes Softening	White to Green	Little	Acrid Chlorine	†Toluene	No Drip
Arcylic	1.18-1.19	Blue Mantle Yellow-Orange	Yes (trans)	Some Black		Floral burnt fat	†Toluene	Clear Bead
Vinyl Acetate	1.19	Dark Yellow	Yes	Black		Acetic	Sec-Hexyl Alcohol Cyclo-hexanol Acetio-nitrile	Some Swell
Polycarbonate	1.20	Orange-Yellow	No	Black		Phenolic Sweet	†Toluene	Chars
Cellulose Acetate	1.27-1.34	Dark Yellow Mauve Blue	Yes	Black		Acetic Vinegar	Furfuryl Alcohol & Acetio-nitrile	Burns, Charred Bead
Casein	1.35	Yellow	No	Gray		Burnt Milk		Swells, Chars
Cellulose Nitrate	1.35-1.40	Intense White	Yes			No Odor	Dipro-pylene Glycol & Acetio-nitrile	
Acetal	1.41-1.42	Blue Mantle Yellow	Yes			Formal-dehyde		Drips
Urea Formaldehyde	1.47-1.52		No			Urinous		
Melamine Formaldehyde	1.50-2.20		No			Fish		
Phenol Formaldehyde	1.55-1.90		No			Phenolic		
Saran	1.58-1.75		Yes					
Vinylidene Chloride	1.62-1.72	(Green) Yellow	Yes			Sweet		Heavy Black
Chlorinated Rubber	1.64		Softens	Black	Dense	Rubber		
Alkyd	1.80-2.24		No					
Tetrafluoroethylene	2.1-2.3		No			Burnt Hair		Chars
Neoprene		(Green) Orange	Softens	Black		Rubber		

*Test for Halogen (Chlorine)
†Hot

(courtesy of U.S.I. Chemicals)

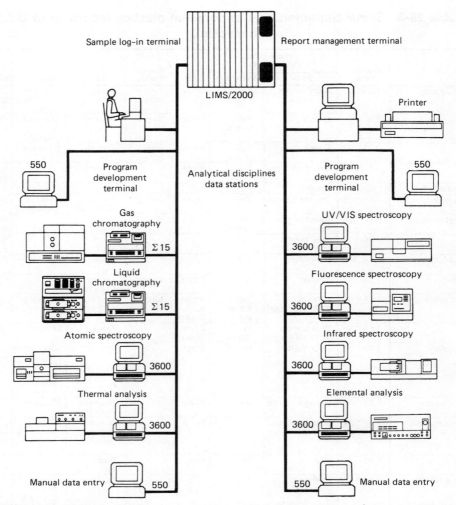

Fig. 28-39   Examples of plastics identification in a computer-aided chemistry laboratory. Courtesy of Perkin-Elmer.

soldering iron, but any implement with a temperature of approximately 500°F could be used. If the material softens, it is a thermoplastic; if it does not, it is a thermoset.

If the material is found to be a thermoplastic, the next step is to find out whether its specific gravity is greater than or less than 1. This is done simply by dropping a sample in water. If the material floats, its specific gravity is less than 1; if it sinks, its specific gravity is greater than 1. The thermoplastics that have specific gravity of less than 1 are the polyolefins—polypropylene and polyethylene.

The final step for both thermosets and thermoplastics is a burn test, which should, of course, be performed in a well-ventilated area. The material should be held with pliers or clamps, and ignited with long wooden matches or a Bunsen burner. If there is only a small piece of material to test, it is best to break it into several parts, as it might take several tries to identify the odor and to observe the other effects noted on the chart.

The major difficulty in interpreting the burn test is that the burn rate and color of the flame of many plastics are affected by fillers, fire retardants, and other additives. However, in most cases the odor is not affected by these additives. It is recommended that you first perform the tests on a styrene drinking glass, a polyethylene milk bottle, or some other known plastic. This practice will prove invaluable when it is time to identify an unknown material.

Another summary of the characteristics of common plastics that can help in their identification is given in Table 28–8.

The identification tests reviewed in this section are only a quick way to possibly get information about the type of plastic. They should not replace laboratory analysis and testing of the material for definitive identification (Fig. 28-39).

REFERENCES

1. Nielsen, L. E., *Mechanical Properties of Polymers and Composites,* Vol. 1, Marcel Dekker, Inc., New York, 1974.
2. Rosato, D. V., Testing and Quality Control Seminar Workbook, University of Lowell, Lowell, MA, 1984.
3. Heger, F. J., Chambers, R., and Dietz, A. G. H., *Structural Plastics Design Manual,* U.S. Government Printing No. 023–000–00495–0, Washington, DC, 1978.
4. Brockway, G. S., "Preventing Fatigue and Environmental Stress Cracking," *Plastics Design Forum,* 73–83 (May/June 1983).
5. Marmion, D. M., "An Overview of Ion Chromatography," *ASTM Standardization News,* 22–25 (February 1983).
6. Brennan, W. P., "Characterization & Quality Control of Engineering Plastics by Thermal Analysis," Perkin-Elmer Corp., CT, 1983.
7. Gill, P. S., and Lear, J. D., *Proceedings* of the 11th NATS Conference (1981).
8. Gill, P. S., "Dynamic Mechanical Analyses Assures Composite Quality," *Industrial Research & Development,* 104–107 (March 1983).
9. "Plastics, The International Plastics Selector Guide," A. Cordura Co., San Diego, CA, 1983.
10. Starita, J. M., Orwoll, R. D., and Macosko, C. W., "Applications of Dynamic Mechanical Measurements to Thermoplastics Processing," *SPE-ANTEC* (May 4, 1983).
11. Zeichner, G. R., and Macosko, C. W., "On-Line Viscoelastic Measurements for Polymer Melt Processes," *SPE-ANTEC* (May 1982).
12. West, J. C., and Cooper, S. L.
13. Menges, G., and Hoven-Nievelstein, W. B., "Studies on the Shrinkage of Thermoplastics," SPE-IMD Tech. Conference, October 24–26, 1983.
14. American Hoechst Corp., Plastics Div., Leominster, MA 1983.
15. Miller, H. L., "Optical Methods for Improving Quality Control of Plastic Moldings," ASME Conference, Chicago, IL, April 27–30, 1981.
16. Bell, R. G., "Microtoming: An Emerging Tool for Analzing Polymer Structures," *Plastics Engineering* (August 1979).
17. Moyer, R. B., "Committee E-7 on Nondestructive Testing: An Overview," *ASTM Standardization News* (November 1982).
18. "Microcomputers Smarten Up Ultrasonic Testing," *Chemical Week* (May 18, 1983).
19. Vishay Intertechnology Inc., Measurements Group, Malvern, PA, 1983.
20. "Drying Thermoplastic Polyurethanes and Hygroscopic Polymers," The Upjohn Co., D. S. G. Report No. 20, 1983.
21. Waldman, F. A., and Von Turkovich, R., "On-Line Measurement of Moisture in Polymers," SPE-IMD Tech. Conference, October 24–26, 1983.
22. Blaine, R. L., "Estimation of Polymer Lifetime by TGA Decomposition Kinetics," Du Pont Co., Thermal Analysis Application Brief No. TA-84.
23. Formo, J. L., "Quality Control Begins With Material," *SPE Thermoset Newsletter* (Spring 1982).
24. Marash, S. A., "Quality Auditing," *M. D. & D. I.* (October 1982).
25. Shoeffler, S., Buzzell, R. W., and Heany, W. F., "Impact of Strategic Planning on Profit Performance," *Harvard Business Review* (March/April 1974).
26. Burrow, S. W., "A Three-Step Method for Fast Identification of Plastics," *M. D. & D. I.* (October 1983).

Section VIII
DIFFERENT MOLDING TECHNIQUES

Chapter 29
Specialized Injection Molding Machines

D. V. Rosato

In 1872, the first U.S. patent was issued for an injection molding machine. Changes in machine design have occurred constantly since then, utilizing the basic principle of melting a plastic and forcing it into the cavity of a mold. So specialized injection molding machines have existed for a long time. In fact, many machines that could be called "specialized" are actually well-documented machines that produce large quantities of molded products. An example is injection blow molding machines, which will be reviewed later in this chapter. (See references 1–40.)

In this chapter a few specialized machines will be reviewed. Obviously what is developed as a specialized machine is based on market requirements. In the extremely competitive lid-and-container field, for example, specialized thinwall presses exist that reduce cycle times by just seconds—which, in turn, result in large cost savings. These machines incorporate very advanced techniques to increase speed of injection into the cavity, temperature and pressure sensors placed directly on the cavity wall, microprocessors to operate functions of the machine more accurately, etc. (1).

These special machines permit us to save money, produce quality parts with zero defects, meet very tight tolerances and reduce plastic use, reduce energy consumption required for their operation, etc. Examples of some special machines are reviewed in Table 29-1 (2).

STRETCHING TECHNIQUES

A very important aspect of specialized molding is that special machines can be designed and built in order to mold certain parts such as stretched injection blow molded containers. For example, the market for the popular two-liter carbonated beverage bottle has developed since the late 1970s. These approximately 65-gram thermoplastic polyester (PET) bottles are now consuming over one-half billion pounds of PET annually in the United States.

Basically, by stretching the plastic near its plastic melt temperature (very accurately with respect to temperature and time) significant improvements in properties can be achieved (Table 29-2). More information will be presented on this subject in the section on blow molding. An especially important technique for stretching has been developed by Dow Chemical U.S.A. (also very important in extrusion molding), it is called Molding with Rotation (MWR) (3). A brief description follows.

The Dow Molding with Rotation "MWR" process is patented, licensable technology designed to help injection and blow molders of thermoplastic resins achieve the best balance of available properties in their fabricated end products, and do so economically. MWR is the result of many years' investment of time, dollars, and experience by Dow Plastics Research laboratories.

Table 29-1 Some specialized injection molding machines (2).

Supplier and machine size	How modified[b]	Productivity gain[c]	Materials	Applications/markets
Battenfeld: 22-ton through 170-ton	1.2:1 low-compression screw; 30,000 p.s.i. high-pressure injection cylinder; accumulator	Thinner wall sections and more complex parts possible	Polyamide-imide	Electrical connectors
750-ton	Two parallel injection units: 180-deg. rotary table on movable platen	Eliminates secondary operations	Two materials, e.g., engineering/commodity resins	Medical parts
50-ton[d] through 170-ton	Two 1-oz., 45-deg. injection units; special mold has hydraulically actuated stays to admit sequential shots from each cylinder while clamp is shut	Eliminates secondary operations; faster clamp cycle than rotary table	Two materials, e.g., rigid/flexible PVC	Pipe fittings, rain gutters
1300-ton (2 x 650)	Rectangular platen; accumulator; two 650-ton clamp units per machine	Fewer rejects resulting from platen deflection; fast fill and higher part yield	Rubber-modified PP[g]	Automotive bumpers dash strips, trim, rocker panels
Epco: 275-ton	Supplied with three interchangeable injection units (5 oz., 14 oz., 28 oz.); electroless nickel-plated screw	Eliminates materials degradation with small molds and permits rapid cycling of larger molds	Teflon TFE	Chemical process pipe fittings
HPM: 700-ton	Higher watt-density (1000°F.) heater bands; 150 cu. in./sec. injection accumulator; 75 oz., 26,000 p.s.i. injection units	Longer heater service life; fewer complex part rejects; increased thin-walling ability	Ultem polyetherimide	R&D, foamed thinwall parts, thinwall TV cabinets
1000-ton	Blower-cooled injection cylinder; auger starve-feeder; double wave screw	Lower materials costs vs. pellet	PVC dry-blend	Pipe fittings
1500-ton	Stacked 28-oz. injection units; rotary mold indexing table	Eliminates secondary operations	Acrylic, PC	Two-material/color automotive tail lights
Husky: 225-ton	Robotized "cooling conveyor" receives hot preforms without deformation	Reduces cycle times to 20 from 30 sec.	PET	Bottle preforms
225 through 600 ton[e]	More plasticating capacity via 25D extrusion screw and higher-wattage heater bands; can be equipped with extra accumulator for rapid fill of 5-gal. pails and other large containers	Allows higher mold cavitation to increase part yield per shot; permits thinwalling large container size	High-flow PE and PP	Containers
200-ton[e]	Overhead label conveyor and photoelectric sensors on mold face	Eliminates secondary labeling	PE, PP, PS	In-mold labeled containers
Impco: 400-ton[e]	24D screw for faster shot recovery;	18-22 mils part thinwall-	HDPE	PET bottle base cups

Table 29-1 (Continued)

Supplier and machine size	How modified[b]	Productivity gain[c]	Materials	Applications/markets
	30,000 p.s.i. injection pressure; accumulator injection rate 150 cu. in./sec.	ing with 8-cavity molds (single face)		
400-ton[e]	Identical to above with clamp-synchronized support for center stack molds	16-cavity stack mold	HDPE	PET bottle base cups
500-ton	Three pumps include variable displacement unit for optimum energy utilization/part yield	Tailors energy use to piece-part yield	Impact-modified PS	Cutlery
Klockner Windsor: 440-ton	Parting-line injection; vertical clamp; 2-station shuttle	Eliminates secondary operations	Flexible PVC	Automotive window frame bezels
375-ton	Vertical tie-bar spacing extended to accommodate 16-cavity stack mold with a shuttle plate between each pair of parting lines	Eliminates secondary operations	HDPE	Paperboard/plastics 1-qt. oil cans
Ludwig Engel: 275-ton 385-ton	Reinforced platens; "coining" clamp;[f] high-speed, high-pressure injection control closed-loop CC80 process control	Reduces part stress, warpage and rejects	Acrylic	Home entertainment and computer video disks
70-ton through 165-ton	"Coining" clamp,[f] accumulator	Remelt gate area, eliminates turbulence lines and improves yield	Acrylic, SAN, engineering thermoplastics	Optical lenses, precision parts
Natco: 750-ton	Injection speed increased to 110 cu.in./sec.; hp. of pump motors increased to 100	Yield increased 50% via ability to fill two-cavity mold	PE, PP	5-gal. containers
Newbury: 30-ton	1000°F. cartridge heaters; thicker injection cylinder walls; injection pressure increased to 40,000 p.s.i.; accumulator injection	Longer machine and heater service life; ability to thinwall part sections to 60 mils	Polyarl sulfone	Aerospace electronics
Stokes: 375-ton	Separate hydraulics for unscrewing rack; larger (38 oz.) injection units and screw (3½, 25D); 12 unscrewing sequences	Improved plastication at lower melt temperatures shortens cooling cycles	PE, PP	Threaded closures
Van Dorn: 300-1000-ton	Press-integrated quick mold changer	Cuts setup time	NA[h]	Short and medium volume production runs
75 through 1000-tons	Sealed hydraulics; nylon tie-rod bushings and platen support shoe faces	Prevents rejects from part contamination	NA	"Clean room" medical, electronic, and aerospace components

a. Representative listing of injection presses that have been designed or modified to perform better within a more restricted range of parts/materials (than a general-purpose unit) or to increase productivity with value-added functions   b. Or applicable design features if new machine   c. Resulting from modifications   d. New machine   e. Two-stage machine   f. Clamp modified for injection/compression sequencing   g. Current application, other materials feasible   h. NA = not applicable

Table 29-2   Example of increasing tensile strength and modulus for polypropylene thin construction.

a.   Effect on Stretching on Tensile Strength of Polypropylene Film

		STRETCH (%)			
PROPERTIES	NONE	200	400	600	900
Tensile Strength, psi	5,600	8,400	14,0000	22,000	23,000
Elongation, at break, %	500	250	115	40	40

b.   Directional Orientation vs. Balanced Orientation of Polypropylene Films

PROPERTIES	AS CAST	UNIAXIAL ORIENTATION	BALANCED ORIENTATION
Tensile Strength, psi			
MD*	5,700	8,000	26,000
TD**	3,200	40,000	22,000
Modulus of Elasticity, psi			
MD	96,000	150,000	340,000
TD	98,000	400,000	330,000
Elongation at break, %			
MD	425	300	80
TD	300	40	65

* *MD* = Machine Direction
** TD = Transverse Direction, and Direction of Uniaxial Orientation

MWR technology uses existing fabrication equipment and commercial thermoplastic resins. Special mold design modifications are required. Supportive details on equipment and mold engineering, resin rheology, and end product design are part of the license package.

Use of the technology is most effective when employed with: (a) articles having a polar axis of symmetry; (b) articles having reasonably uniform wall thickness; and (c) articles whose dimensional specifications and part-to-part trueness are important to market acceptance. *Note:* Within these requirements are many parts having variable surface and wall geometries.

Initial "target" applications have been in the bottle and jar market areas. However, use of this technology is not restricted to those particular shapes or markets. Practically any article, container, or blown or injection molded part having one surface reasonably rotationally symmetrical can be fabricated by MWR.

The MWR process asks no sacrifice of either cycle time or surface finish. Both laboratory and early commercial runs identify good potentials for reducing cycle time; for either reducing the amount of resin required or improving properties with the same amount of resin, or both; and for substituting less expensive resin while achieving adequate properties in the fabricated part.

The MWR process is a fabrication method using a rotating mold element in the injection molding machine. The end product can come directly from the injection molding machine mold, or can be a result of two-stage fabrication: making a parison, blow molding the parison. The two-step process can be "integrated": in-line injection blow, or separate operations for injecting molding of parisons with reheating and blowing at separate stations.

Orientation of the molecules within a thermoplastic mass has a direct effect on molded end properties, and is the subject of many articles and reports. Injection molders commonly try to minimize the unidirectional orientation resulting from essentially linear mold fill. Sheet producers may induce lateral orientation by different stretching-tentering techniques. Blow molders anticipate and plan for certain structural improvements that result from biaxial orientation occurring during the blowing process.

Plastic fabricators know that minimizing unidirectional flow orientation usually results in better-performing end products. The know-how of polymer rheology and processing temperature, implemented with varying mold fill techniques and end product design geometries—all are used to minimize problems associated with uniaxial orientation.

The MWR process developed by Dow took a radically different approach. Instead of seeking to minimize uniaxial orientation, or to minimize its adverse affects, Dow research sought a practical technique by which controlled multiaxial orientation could assure maximum properties (for the resin used) in the fabricated end product.

The MWR process permits fabricators to control molecular orientation and thus produce top-performance end products

The MWR process permits a balancing of resin temperature, resin rheology, pressures, time, and either mold core or mold surface rotation, to achieve a carefully controlled degree of multiaxial orientation within the thermoplastic resin mass.

During fabrication using the MWR process, two forces act on the polymer: injection (longitudinal) and rotation (hoop). The targeted "balanced orientation" is a result of those forces. As the part wall cools, additional high-magnitude, "cross-laminated" orientation is developed—frozen-in—throughout the wall thickness. *Note:* Orientation on molecular planes occurs as each "layer" cools after injection.

This orientation can change direction and magnitude as a function of the wall thickness. The result is analogous to plywood—and the strength improvements are as dramatic. *Note:* In the MWR process, there are an *infinite* number of "layers," each of which has its own controlled direction of orientation. By appropriate processing conditions, both the magnitude and direction of the orientation can be varied and controlled throughout the wall thickness.

MWR technology produces parts having greatly increased tensile strength compared to the same parts conventionally molded. The relative degree of improvement is plotted in Fig. 29-1 for SAN (styrene acrylonitrile), GPPS (general purpose polystyrene), and HIPS (high impact polystyrene) resins. Because MWR-type improvements are based on balanced multiaxial orientation, the gain in tensile strength also directly correlates with gain in practical toughness.

In a gross sense, stress crack agents cause failure of molded plastic parts by attacking the chemical bonds of the molecules. Failure normally occurs as a crack perpendicular to the direction of greatest weakness. With MWR technology, the internal structural bonding of the plastic part is greatly improved through multiaxial "laminar" orientation of the molecules. This results, often, in a measurable improvement in stress crack resistance of the molded part.

Note: Stress crack behavior is dependent on so many variables—resin used, part thickness, part shape, stress crack agent, environment of use, etc.—that each part must be analyzed carefully in its own right. In any case, the stress crack resistance of a part molded with MWR technology can be improved to a commercially significant degree.

Parts molded of polymers that normally exhibit crazes as a predecessor to catastrophic failure can be improved significantly by fabrication with MWR. For common styrenics, yield strengths of parts having MWR balanced orientation are significantly higher than are

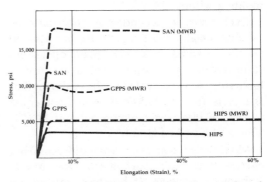

Fig. 29-1   Stress–strain response, oriented vs. unoriented (MWR vs. non-MWR) (3).

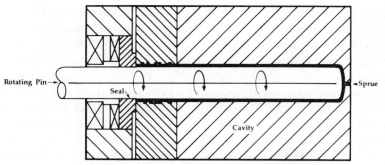

Fig. 29-2 Typical mold schematic for MWR.

yield strength of conventionally molded parts.

Additionally, the mode of failure may become shear yielding because of the high (balanced) orientation provided by MWR. When this is accomplished, crazes, as a form of failure, will not occur.

Mold with Rotation Procedure

Standard on-the-floor injection equipment can be used. The equipment "key" to MWR success is in the mold area. A simplified schematic is shown in Fig. 29-2.

The effect of cycle time on injection molding economics is great. With MWR, one potential for reducing cycle time relates to the ability to get satisfactory end-product performance with less polymer. This can result in a shorter cycle because less polymer = less heat = less cooling time = shorter cycle.

Another potential for reduced cycle time occurs because injection molding with MWR is most effectively done when the plastic melt is at a much lower temperature ($\approx 100°$F lower for styrenics) than would be used for injection molding without MWR.

Cycle times are dependent not only on plastic shot weight and temperature but also on all the variables in a given plant operation. It is reassuring therefore that a number of laboratory tests on cycle time have shown that the cycle time with MWR is at least equal to that of injection molding without MWR. This is shown in Fig. 29-3.

Mold design, while somewhat different from current practice, is part of the Dow MWR technology package. It has been readily ac-

quired by several commercial injection mold builders working with Dow and/or licensees.

As is common with conventional injection molding, MWR also results in parts having excellent dimensional properties. In addition, MWR permits parts with high length-to-diameter ratios to be molded without problems of core deflection and consequent thinner–thicker sections in the part wall.

With core rotation during MWR, the pin "self centers," and part wall uniformity is excellent. The final molded part therefore is uniformly strong about its circumference. This fact has particular value if the molded part is a parison. Parisons fabricated with MWR can be reheated and blown without problems caused by wall eccentricity. In injection molding with MWR, part designers and engineers should keep in mind that significant part wall thickness variation and surface geometry variation are possible, if desired, with MWR.

A basic profile of injection molding conditions to be used with MWR is given below:

- Any orientable injection-moldable plastic resin
- Temperature—100°F lower

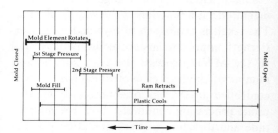

Fig. 29-3 Standard plastics injection molding cycle.

- Injection pressure—high
- Hold pressure—high
- Rotation—before, during, after mold fill
- Rotation and injection controls

CONTINUOUS MOLDING OF VELCRO STRIPS

The Velcro fastener consists of two mating strips. One strip is covered with nearly microscopic hooked or barbed spines, the other with tiny loops. When the two strips are pressed together, their projections become entangled to produce the gripping action.

Peeling the strips apart deflects the spines, disengaging them from the loops. Because of the resilience of the materials, the projections snap back to their original geometry so that the strips can be used repeatedly.

For years Velcro had been made by a slow, complex textile process in which the loops are woven through the back of a flexible base strip; for the male strip, the loops are cut to create the hooks. Seeking a more economical alternative, Velcro USA (Manchester, NH) engineers wondered if the fastener could be produced by injection-molding the projections integral with the base strip in a single, continuous operation (5).

Foster-Miller Associates, Waltham, MA, which specializes in designing and building one-of-a-kind machines, took on the project of developing equipment to mold the male half of the Velcro system. The engineering firm subsequently received a patent (assigned to Velcro USA, Inc.) on the resulting molding machine.

A few details about the fastener will underscore the formidable molding problems that the designers had to solve. The spines are almost too small to see: they project about $\frac{1}{16}$ in. from the 0.010-in.-thick base strip, are 0.020 in. wide at their base, and taper to about 0.012 in. at the tip (Fig. 29-4). They are very closely spaced, on approximately 0.050-in. centers; a single square inch of the strip contains more than 250 of these projections. Moreover, they are not simple, needle-like shapes, but are triangular in cross section and

have a microscopic hook or other type of barb at the tip. Dimensions must be held to 0.0015 in., and no flash is permissible anywhere on the strip.

Molding Technique

The molding line developed by Foster-Miller produces the Velcro in a continuous process. The equipment molds the strip from the resin, trims it, conditions it for flatness, applies an adhesive backing, and winds it on a reel.

The key elements are an extruder and a rotating "ferris wheel" mold. The extruder runs continuously, feeding the melt into the continuously rotating mold through a special adapter mounted on the extruder-barrel outlet. The 2-foot-diameter mold turns at about 10 rpm, delivering the Velcro at 60–70 ft/min. The extruder is basically standard; the most innovative features of the installation are the rotating mold and the adapter.

Feeding a melt onto a rotating mold is not a brand new idea; the same concept has been applied to making items such as shoe soles and similar, relatively simple, flat shapes. What makes this installation so unusual is the precision of the product and the ingenuity of the mold. The mold contains more than 15,000 cavities, each about $\frac{1}{16}$ in. deep and arrayed less than $\frac{1}{16}$ in. apart in parallel rows around its circumference (Fig. 29-5a).

Besides the task of designing a tool to mold the Velcro, the design firm had to figure out how to strip it from the mold. Remember that each spine has a hook or other projection that

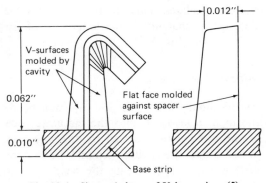

Fig. 29-4   Size and shape of Velcro spines (5).

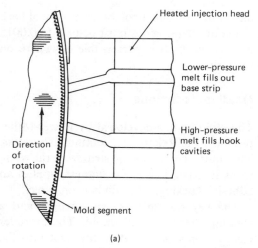

(a)

Fig. 29-5a   Two orifices feed melt onto rotating mold.

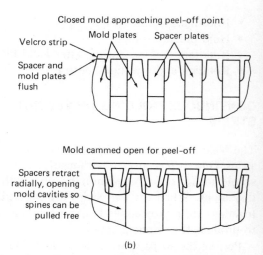

(b)

Fig. 29-5b   How Velcro strip is peeled from mold.

must be disengaged without damage from the undercut at the base of its cavity as the strip is being peeled from the mold (Fig. 29-5b).

They devised a wheel-shaped mold consisting of several dozen, thin (0.060 in.), round plates bolted together. The plates are of two types, which alternate across the thickness of the "wheel." One, the cavity plate, contains a ring of molding cavities for the projections on both sides of the plate at its outer edge. Between each cavity plate is a blank spacer plate. Being in intimate contact with the cavity plate, the spacer plate acts to seal off the open side of the cavities.

The set of alternating spacer plates is designed to slide in and out (radially) as a group; the cavity plates have no radial motion.

During most of the cycle the spacer plates are extended to the full diameter of the mold so that their edges line up with the edges of the cavity plates. This alignment creates the flat surface that molds the inner face of the Velcro strip (Fig. 29-6).

Injection Process

As the mold rotates past the injection head, the melt is injected onto the circumference of the mold and is forced by the injection pressure into the cavities. Enough additional melt is supplied to create the base strip at the same time.

By the time the mold has completed about one-half a revolution, the plastic has solidified and cooled. At that point the spacer plates are pulled down by a circular cam.

The effect is like a piano keyboard with every other key pushed in. The retracted plates open up the sides of the cavities to give the spines room to deflect outward and release from the cavity undercuts as the Velcro strip is pulled off the mold by the winder. After passing the peel-off point, the spacer plates move back out to their original position (Fig. 29-7).

Multiple separately mounted segments, each individually free to move radially, make up each spacer plate. For manufacturing convenience the cavity plates likewise are constructed from individual segments. To make electrodes of such small size and high precision, Foster-Miller had to develop special techniques. Company engineers decline to describe the process beyond saying that it consisted of several steps and that the working electrodes were made by electro-forming.

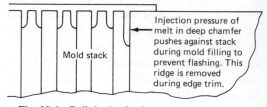

Fig. 29-6   Built-in "melt piston" compresses mold.

Because of its complicated construction, the mold is cooled externally. Mold temperature is controlled by an air plenum located about one-quarter of a revolution before the injection point; the molded Velcro itself is cooled by another plenum ahead of the strip-off point. The air flow is adjusted to keep the nylon sufficiently warm and flexible so that the projections can pull free from the undercuts in the cavities without damage during strip-off.

The injection head that delivers the melt to the mold has a concave face with the same curvature as the periphery of the mold. The head is mounted on the extruder and is separated from the mold face by a gap equal to the thickness of the base strip (about 10 mils).

The injection head contains two orifices, each fed by a separate gear pump. The lower orifice, which fills the projections, operates at a relatively high pressure (equivalent to a typical injection pressure in a closed mold) to force the melt into the blind cavities. The second orifice, just above, supplies additional melt at lower pressure to fill out the base strip.

The clearance between the injection head and the mold, which determines the strip's thickness, is set by a fine gear-reduction drive and is measured by four air-gauge sensors, one at each of the four corners of the head to ensure perfect parallelism. A 1-mil error in the 10-mil gap can result in as much as a 50 percent variation in injection pressure.

Cartridge heaters in the head maintain the melt at the proper viscosity for injection; temperature control is within ±5°F.

The sides of the gap between the injection head and the mold are not enclosed. Seepage is prevented by a careful balance of melt temperature, mold temperature, injection pressure, and mold velocity. The as-molded edges are, of course, uneven, but the edges are squared off by trimming the strip in a downstream operation after conditioning.

One of the most critical requirements in the mold design is to prevent flash. Foster-Miller used two approaches to avoid this problem. One was to control the geometry of the mold plates to extremely close tolerances. Every plate in the mold stack is surface-ground to be flat within 0.002 in. across its 2-foot

Fig. 29-7 Rotating mold to produce Velcro strips "continuously" in a specially designed injection molding machine.

diameter. Also, plate thickness, which determines the spacing between adjacent rows of spines as well as the quality of the seal between adjacent plates, is controlled to within 0.0001 in.

The second strategy against flashing is to prevent the edges of the mold plates from flexing outward as the melt is injected into the cavities. To supplement the tie-rods through the molds, Foster-Miller devised a simple way to resist potential flexing. The melt itself is used to supply a hydraulic squeezing action on the mold stack. (Fig. 29-7).

At the outermost mold plate in the stack, and beyond the nominal width of the strip, Foster-Miller cut a deep chamfer aroun the edge of the plate. As the mold is being filled, the pressurized melt also flows into this chamfer. The resulting sidewise force against the side of the plate tends to compress the mold stack and prevent the plates from spreading.

Another key to reliable production is ensuring trouble-free radial movement of the spacer segments during mold opening, and particularly during closing. Any significant galling or binding between the spacers and mold plates could prevent the spacers from returning to their "home" position, flush with the edges of the mold plates. The resulting offset would produce thickness steps across the base of the Velcro strip, and probably flashing as well. The engineering firm avoided this problem by applying a low-friction coating, in the form of an internally lubricated polymer, to the sliding faces.

CONTINUOUS ROTARY INJECTION MOLDING MACHINE

Introduction

Conventional "reciprocating" injection molding machine systems are being improved upon continually. New mold designs, machine control devices, part handling systems, and "Quick-Mold" change are being developed to reduce labor costs and increase productivity. But, all these improvements are on the same basic process.

Molders can improve productivity by:

- Reducing runners—mold hot runner systems and CAD/CAM design.
- Reducing cycle time—molding process.
- Automating parts handling—reducing manual labor and controlling parts.

Definition

Conventional injection molding material melt is accomplished by the machine screw. The machine ram injects material to mold a part. Conventional molding machines are a cyclic injection molding process. Parts handling equipment must be added to enhance part ejection control. This molding system is analogous to a reciprocating engine.

CRIMM (Continuous Rotary Injection Molding Machine) is a rotary system using a continuous-extrusion concept. A turret carrying from 10 to 48 radially mounted, wedged-shaped molds, continuously rotating at a constant speed, and continuously fed by a nonreciprocating extruder describes the system (6).

The molds rotate on the turret. Each of the lower stationary molds contains a receiving pot to accept material melt from an extruder. A plunger injects material to mold a part by simple transfer molding. Rotary molding produces parts on a continuous cycle. The movable molds open and then pivot out from the machine. Access to parts during ejection complements parts handling. This continuous molding system is analogous to a rotary turbine engine.

Injection Molding Process

Injection Process	Molding System
MATERIAL	CONVENTIONAL–ROTARY
Machine ⟶ MELT	⟶ Machine screw —Extruder
Machine & Mold ⟶ MOLD	⟶ Machine ram —Transfer
Manufacturing System ⟶ PART	⟶ Cyclic —Continuous

Rotary Machine Status

A production rotary molding machine is being developed for continuous injection molding. Its description is in Patent #4,370,124 issued on January 25, 1983 for a "Modular Rotary Molding Machine" (Fig. 29-8).

Modular Construction

The modular construction pertains to common elements that are used to manufacture machines of different sizes. The size is dictated by the number of stations contained in the turret assembly. A machine of "any" fixed number of stations will use common: turret side panels, mold sets, clamping units, extruder material delivery.

Machine Description

A rotary machine (Fig. 29-9) has a fixed number of stations on a turret to retain common side panel inserts. The lower stationary molds will have slots to match retaining keys in the side panels. A receiving pot in each of the lower molds receives material from an extruder. Material from the extruder is fed from a rotating drum with slots to match the mold pots. Injection of material into the mold is by plungers on hydraulic cylinders. Injection

of material, into cavities next to the mold's receiving pot, is by simple transfer molding.

The upper molds are attached to clamping units contained in a tubular frame. Mold opening–closing is actuated by a four-bar linkage system, driven by a hydraulic cylinder. The actuating linkage is harmonic and a self-locking system. Each clamping unit has an adjustable mold clamp tonnage system. Initial mold opening is in a straight-line draw; then the mold pivots out from the machine for access to the product during part ejection.

Machine Operation and Molding Features

Extruder delivers material to the lower molds as the rotating turret molds pass over a continuous material feed drum system. The lower mold receiving pot is filled with material. The material is delivered from an extruder. The rotating feed drum has slots that match the mold pot, so that material is fed from the extruder with a uniform melt and a low fill pressure. Material is now in the mold pot ready for injection.

Material injection by a plunger, driven by a hydraulic cylinder, enters the mold pot. A portion of the material in the pot is injected

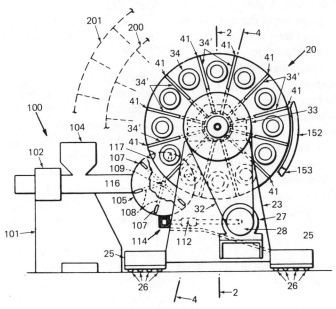

Fig. 29-8 Patent figure.

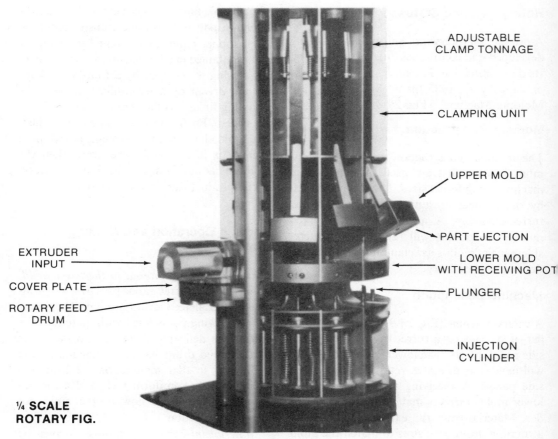

ADJUSTABLE
CLAMP TONNAGE

CLAMPING UNIT

UPPER MOLD

PART EJECTION

LOWER MOLD
WITH RECEIVING POT

PLUNGER

INJECTION
CYLINDER

EXTRUDER
INPUT

COVER PLATE

ROTARY FEED
DRUM

¼ SCALE
ROTARY FIG.

Fig. 29-9   Rotary machine.

into the cavity or cavities, by simple transfer molding. Material pressure is determined by the hydraulic pressure of the injection cylinder times the ratio of injection cylinder area to receiving pot area. A nominal 20,000 psi material pressure is designed for a 2000 psi hydraulic pressure. Higher or lower material pressures can be set by changing the pot diameter or injection cylinder hydraulic pressure. Each station's mold set can have a different material pressure for the same injection cylinder hydraulic pressure. A consistent material melt and very small pressure loss between pot and cavity improves the process, to mold a part.

Mat'l Press.

$$= \text{Plunger Force/Mold Pot Area}$$
where
Plunger Force
$$=\text{Hyd. Press.} \times \text{Cyl. Piston Area}$$
therefore

Mat'l Press.
$$= \text{Hyd. Press.} \times \text{Cyl. Pistol/Mold Pot}$$
(Ratio of Cylinder Pistol/Mold Pot
$$- \text{Areas or Diameters})$$

Material shot size: volume is determined by the mold thickness and receiving pot diameter. Molding shot size is determined by the part shot size (which is fixed) plus runner left in the mold pot. The mold pot runner material serves two distinct functions. First, it acts as a puller during part ejection. On initial start-up there will be "short-shots" in the mold. The pot runner acts as a self-purging system for flow of material through the machine. Second, it minimizes the amount of excess pot runner material. By adjusting the extruder rate and/or the turret's revolutions per minute rate the amount can be varied. A standard cold runner mold's product and runner system is fixed and cannot be compensated.

Clamping units are modular in construction. A four-bar linkage assembly is contained inside a rectangular tubing frame. Each unit has an adjustable mold clamp tonnage system. Loading is applied to the center of the mold. This means that only clamp tonnage needed to mold "flash free" is applied to the mold.

Part ejection: molded parts move into a fixed location at the machine for ejection. The upper mold starts its opening in a straight-line draw and then moves outside the turret diameter. The motion is determined by a cam path in the clamping unit assembly. An ejector system will eject the part and a controlled slug of material from the mold receiving pot.

Molds are wedge-shaped and fit in the turret opening. The taper of the molds depends upon the number of stations contained on the machine. Because of the similarity in dimensions molds can be fabricated in high volume. The molds are designed for changeable inserts and material receiving pots. This results in a significant reduction in tooling costs for new product. Changing of inserts for new product or to replace defective inserts is quick because of accessibility.

A 12-station rotary machine mold base tapers from 10 in. to 6 in. and is 8 in. deep. The thickness is from 3 to 4 in. There can be one or more pots to increase the amount of material delivered to mold product.

Turret assembly is formed by common side panels nesting into circular retaining rings. The side panels support the lower mold. The lower molds are inserted into these nesting frame stations. The turret assembly contains rotary joints for the hydraulic, water, and air services for the machine/mold cylinders and molds.

Number of stations and productivity: the rotary machine's production capability is shown for different numbers of stations (Table 29-3). The number of stations contained on a rotary machine is determined by the volume of produce needed. The ability to mix the product being produced does not limit the rotary machine only to high-volume users.

Rotary Machine Characteristics

Material melt consistency: because the material is delivered to receiving pots from an extruder, instead of a reciprocating screw, an improved melt is fed to each mold. Better molded part consistency can be expected.

Process consistency: the injected material is in closer proximity to the cavities than a conventional mold/machine system. There is a smaller material pressure loss. This is another enhancement of molding product not readily incorporated in standard injection molding machines and molds.

Molding inserts: inserts will fit into standard rotary machine mold bases. This will substantially reduce tooling costs. The ability to quickly change inserts on the machine will improve mold change time and reduce the need to continue running with defective cavities.

Installation: a rotary machine is unique in construction and molding capabilities. It integrates a machine and mold into each station of the turret. A substantial reduction in floor space and energy requirements is achieved. Smaller molds and inserts simplify handling during fabrication and production. This all leads to increased productivity.

Molding System Comparison

A 12-station rotary machine that is 40 in. in diameter holds molds that are 64 sq in. of projected area each. This represent 12 conventional molds and machines. If conventional equipment is already in place, the justification for a rotary machine has to be for other capabilities.

MACHINE	CONVENTIONAL	ROTARY
Molding System	12 50-Ton Machines	12-Stations
Floor Space	12 Machines	1 Machine
Energy Requirement	12 Barrels	1 Extruder
Clamp Application	Through Platens	Central on Mold
Mold Handling	Mold Base	Inserts

Table 29-3   Rotary machine product capability.

Table A — PART PER STATION

NO. OF STATIONS / APPROX. O.D.	°/STATION	RPM No.	SECS. PER	1	2	3	4	5	6
10 — 3 FT. / P.D. 26.67, O.D. 34.67 INCHES	36°/STATION	1	60	10	20	30	40	50	60
		2	30	20	40	60	80	100	120
		3	20	30	60	90	120	150	180
		4	15	40	80	120	160	200	240
		5	12	50	100	150	200	250	300
		6	10	60	120	180	240	300	360
12 — 3½ FT. / P.D. 32.00, O.D. 40.00 INCHES	30°/STATION	1	60	12	24	36	48	60	72
		2	30	24	48	72	96	120	144
		3	20	36	72	108	144	180	216
		4	15	48	96	144	192	240	288
		5	12	60	120	180	240	300	360
		6	10	72	144	216	288	360	432
15 — 4 FT. / P.D. 40.00, O.D. 48.00 INCHES	24°/STATION	1	60	15	30	45	60	75	90
		2	30	30	60	90	120	150	180
		3	20	45	90	135	180	225	270
		4	15	60	120	180	240	300	360
		5	12	75	150	225	300	375	450
		6	10	90	180	270	360	450	540
18 — 4⅔ ft. / P.D. 48.00, O.D. 56.00 INCHES	20°/STATION	1	60	18	36	54	72	90	108
		2	30	36	72	108	144	180	216
		3	20	54	108	162	216	270	324
		4	15	72	144	216	288	360	432
		5	12	90	180	270	360	450	540
		6	10	108	216	324	432	540	648

PARTS PER MINUTE

Table B — PARTS PER STATION

NO. OF STATIONS / APPROX. O.D.	°/STATION	RPM No.	SECS. PER	1	2	3	4	5	6
24 — 6 FT. / P.D. 64.00, O.D. 72.00 INCHES	15°/STATION	1	60	24	48	72	96	120	144
		2	30	48	96	144	192	240	288
		3	20	72	144	216	288	360	432
		4	15	96	192	288	384	480	576
		5	12	120	240	360	480	600	720
		6	10	144	288	432	576	720	864
30 — 7¼ FT. / P.D. 80.00, O.D. 88.00 INCHES	12°/STATION	1	60	30	60	90	120	150	180
		2	30	60	120	180	240	300	360
		3	20	90	180	270	360	450	540
		3½	17+	105	210	315	420	525	630
		4	15	120	240	360	480	600	720
		4½	13+	135	270	405	540	675	810
36 — 8⅔ FT. / P.D. 96.00, O.D. 104.00 INCHES	10°/STATION	1	60	36	72	108	144	180	216
		2	30	72	144	216	288	360	432
		3	20	108	216	324	432	540	648
		3½	17+	126	252	378	504	630	756
		4	15	144	288	432	576	720	864
		4½	13+	162	324	486	648	810	972
48 — 11⅓ FT. / P.D. 128.00, O.D. 136.00 INCHES	7½°/STATION	1	60	48	96	144	192	240	288
		2	30	96	192	288	384	480	576
		3	20	144	288	432	576	720	864
		3½	17+	168	336	504	672	840	1008
		4	15	192	384	576	768	960	1152
		4½	13+	216	432	648	864	1080	1296

PARTS PER MINUTE

Patent Application
Frederick J. Buja, P.E.

INJECTION MOLDING METALS

Over a century ago the plastics injection molding machine advanced the state of the art by taking advantage of machines that had been operating in molding diecast/metal parts. The diecast machines "injection mold" their "melt" at above 1000°F.

Now another process for the injection molding of metal powder debuts (with plastic binder). The proprietary process, Injectalloy, combines the best of plastics processing and powder-metal technologies of Remington Arms Co., Inc. This emerging process holds great promise for the production of complex precision parts (7).

Specifically, upper size restriction is about 2-in. diameter (50 mm) and 0.250-in. wall thickness (6.35 mm) on cross sections. Parts are "near net shape," at 94 to 98 percent of theoretical density. Tolerances are generally from 0.003 to 0.005 in./in. (0.076 to 0.127 mm/mm).

The process, still in development, uses spherical powder less than 10 microns in diameter (compared with 30 to 200 microns for conventional PM processing). Fine powders are said to promote rapid diffusion during sintering, producing near-complete homogenization. In addition, smaller size allows more intricate part geometry, thinner walls, sharper edges, and low porosity.

The particles are mixed with a plastic binder and injection molded in commercial thermoplastics molding equipment. The binder improves flow and assures uniform fill of mold cavities. Binder is removed through solvent extraction or heating, and the parts are then sintered.

The alloys offered include iron with 2 to 50 percent nickel, and a proprietary Fe–Ni–Co alloy with 90,000 psi tensile strength (620 MPa) and 25 percent elongation.

BLOW MOLDING

A summary is presented on the technology/performance/economics of blow molding. Different processes are used to produce blow molded plastics products. This review will per-

mit you to appreciate the differences that exist in the basic extrusion and injection blow molding techniques as well as stretch-blow molding (8).

Hollow plastics products such as squeeze bottles, milk bottles, fuel tanks, toys, oil containers, chemical tanks, furniture, electrical housings, etc., are blow molded. Different processes are used, but the majority are basically similar.

The basic process involves producing a plastics parison or preform (tube, pipe, or test tube plastics shape), placing this parison or preform into a closed two-plate mold (cavity in the mold represents the outside shape of the part to be produced), injection of air into the heated parison to blow it out against the mold cavity, cooling of the expanded parison, opening the mold, and removing the rigid blow molded part.

Blow molding techniques can be divided into three major categories, namely, the extrusion blow molding process which principally uses an unsupported parison, and the injection blow molding process which uses a preform supported on a metal core pin. In turn these processes can be called conventional or unstretch-blow molding processes. The third major category is called the stretch-blow molding process. Stretch-blow molding can start with either the extrusion or injection blow molding processes. By stretching at prescribed temperatures the properties of many plastics can be significantly improved providing cost/performance advantages.

These processes provide different advantages to produce all types of products; so it is necessary to examine the process to be used based on product/performance requirements, materials (plastics) performance, and production quantity. As an example, the plastic bottle does more than hold the product. It combines safety, light weight, design freedom, appealing colors, convenience, ease of dispensing, multidecorating, and low energy usage. Other factors to be examined in a blown container can be the desired shelf life, moisture barrier, oxygen barrier, drop strength, heat distortion, compatibility of plastic and product, top load, environmental stress cracking, clarity requirements, coloring of the plastic material, and cost.

In extrusion blow molding the advantages include high rate of production, low tooling cost, blown handle ware, wide selection of machine manufacturers, etc. Disadvantages are usually high scrap rate, recycling of scrap, limited wall thickness control or material distribution, the fact that trimming can be accomplished in the mold for certain type molds or secondary trimming operations have to be included in the production lines, etc.

With injection blow molding, the major advantages include the fact that no scrap or flash is molded, best wall thickness and material distribution control, best surface finish of parts, low-volume production quantities which are economically feasible, etc. Disadvantages are high tooling cost, no handle ware, the fact that based on the cost to produce large extruded blow molded parts the injection blow molded is limited to smaller sizes, etc. Advantages and disadvantages are similar for stretched injection blow molding. The major advantage is that cost/performance can be significant for certain sizes (and quantities) of products such as the carbonated beverage bottle.

Important factors to consider when examining the blow molding process to be used usually start with part size, number to be manufactured, design/shape, and cost limitations.

Information is presented here to highlight past, present, and future developments in blow molding. The blow molding techniques to be reviewed have to take into consideration the melt process that occurs in the extruder or injection molding machine to produce the "ideal" parison or preform. So, in addition to analyzing the actual manufacturing process to be used in blow molding, it is extremely important to provide the properly designed machine to properly melt (and control the rheological/melt characteristic of) the parison or preform.

Blow molding of thermoplastics is now in its second century. The first U.S. patent was filed May 22, 1880, published February 1, 1881, and issued to the Celluloid Manufacturing Company of New York. A fabricated cellulose nitrate tube was produced prior to heat/ blowing. Later this tube was identified as a parison. A ram extruder was used to fabricate a tubular-shaped sheet that was cemented along a scarf joint. Other methods to form the "parison" included the preparation of a high-viscosity lacquer solution used to overcoat a mandrel. Vials and other containers were produced by these original concepts.

Major new developments occurred in blow molding during the 1920s and 1930s with the advent of extrudable cellulose acetate, ethyl cellulose, polystyrene, acrylic, and, most important, polyvinyl chloride. With the castable plastics, dip-coating of parisons became popular. Composite materials were used (similar to obtaining property/performance of our present co-extrusion techniques) such as PVC adhesively bonded to pearlescent cellulose nitrate.

By the late 1930s, major new developments involved the use of a controlled parison softening rate and relating the temperature profile to improving blow molding efficiencies. Prior technology involved basically the extrusion of a "pipe" that was positioned in 10 to 20 blow molds operating in-line. The target was to provide sufficient heat to the pipe so that in-line bottles could be blown within certain time periods. With controlled parison temperature the present era of single and multiple blow molding started.

Multiwalled blow-molded containers were produced during the late 1930s using double-walled tubes, stacked sheets, or combinations of tube/sheet. Component adhesively assembled parts were produced; now the technique has been extended with the use of more thermoplastic solvent systems and sonic bonding. Rubber forming bags were used to produce blow-molded double- and single-walled containers. This rubber bag technique started during the 1920s to produce single-wall blown parts. By the late 1930s, plants were producing different-shaped pipes (different-shaped water traps such as the "S") by blowing PVC against a mold using a rubber-coated spring; the spring would be positioned to provide the shape prior to blowing.

During the early 1940s, polystyrene extruded blow molded parts were very popular. Major commercial, large-production parts were started during the 1950s using low-density polyethylene to make squeeze bottles. By the early 1960s, work started in using high-density polyethylene to produce extruded blow molded milk bottles. By the late 1970s, HDPE milk bottles reached a yearly U.S. production of over three billion bottles.

Since the 1940s, developments have occurred in producing injection blow molded bottles and other containers. Development of marketable, large-production containers did not occur until the 1970s with stretch-blow molding of two-liter carbonated beverage bottles. Most of the original work started by Monsanto used acrylonitrile (AN); however, production by others was with polyethylene terephthalate (PET). By 1980 over two billion PET carbonated bottles were produced annually by companies such as Amoco Chemical, Continental Group, Hoover Universal, Imco, National Can, Owens-Illinois, Sewell Plastics, etc.

Extrusion Blow Molding

In extrusion blow molding, a parison is formed by an extruder. The plastic pellets are melted by heat which is transferred from the barrel and by the shearing action of the extruder screw as they pass through the extruder. The helical flights of the screw change configuration along its length from input to output ends to assure a uniformly homogeneous melt (Fig. 29-10).

Turning continuously, the screw feeds the melt through the die-head as an endless parison or into an accumulator. The size of the part and the amount of material necessary to produce the part (shot size) dictate whether or not an accumulator is required. The non-accumulator machine offers an uninterrupted flow of plastic melt.

With the accumulator, flow of parison through the die is cyclic. The connecting channels between the extruder and the accumulator, and within the accumulator itself, are designed rheologically to prevent restrictions that might impede the flow or cause the melt to hang up. Flow paths should have low resistance to melt flow to avoid placing an unnecessary load on the extruder.

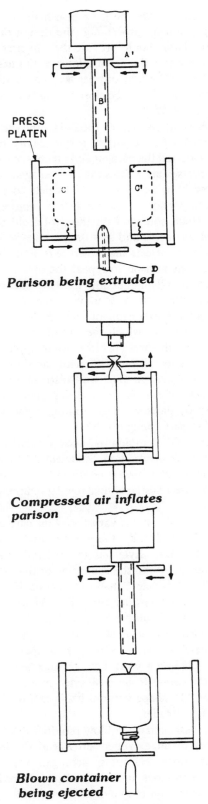

PRESS PLATEN

Parison being extruded

Compressed air inflates parison

Blown container being ejected

Fig. 29-10 Extrusion blow molding.

To ensure that the least heat history is developed during processing, the design of the accumulator should provide that the first material to enter the accumulator is the first to leave when the ram empties the chamber; and the chamber should be close to totally emptied on each stroke.

When the parison or tube exits the die and develops a preset length, a split cavity mold closes around the parison and pinches one end. Compressed air inflates the parison against the hollow blow mold surfaces, which cool the inflated parison to the blow mold configuration. Upon contact with the cool mold wall, the plastic cools and sets the part shape. The mold opens, ejects the blown part, and closes around the parison to repeat the cycle.

Various techniques are used to introduce air into the parison. It may be accomplished through the extrusion die mandrel, through a blow pin over which the end of the parison has dropped, through blow heads applied to the mold, or through blowing needles that pierce the parison. The wall distribution and thickness of the blown part are usually controlled by parison programming, blow ratio, and the part configuration.

The mold clamping methods are hydraulic and/or toggle actuation. Sufficient daylight in the mold platen area is required to accommodate parison systems, unscrewing equipment, etc.

Clamping systems vary based on part configuration. Basically there exist three types. The "L-shape" style has the parting line at an angle of 90° to the centerline of the extruder. The "T-shape" has the parting line inline with the extruder centerline. Mold opening is perpendicular to the machine centerline. The third method is the "gantry" type. The extruder/die unit is arranged independently of the clamping unit. This arrangement permits the clamp to be positioned in either the "L" or "T" shape without being tied directly into the extruder.

The basic extrusion blow molding machine consists of an extruder, crosshead die (and accumulator), clamping arrangement, and mold. Variations include multiple extruders for co-extrusion of two or more materials, parison programmer to shape the parison to match complex blown part shapes, and multiple station clamp systems to improve output through the use of multiple molds.

Injection Blow Molding

In injection blow molding, a preform (or parison), is formed by a conventional injection molding machine plasticator. The injection molding machine injector provides an optimum plastics melt, with a uniformly homogeneous melt that is repeatable. This plastics melt is injected into the preform cavity forming the preform around the core rod. A completely finished injection molded neck is formed at this station (Fig. 29-11).

In tool design the core rod and parison are very important. Each container to be blow molded has its own unique parison and core rod design.

The second stage consists of transferring the injection molded preform, via either the core rod or the neck ring, into the blow mold. At this station compressed air enters through the core rod or seal ring, and the preform is blown to the blow mold configuration. It is held in the cold blow mold until the material is set, and then the air is exhausted and the blown bottle ejected (Fig. 29-12).

Machines are used with from two to six stations. In the two-station machine, the finished container is ejected after the blow mold opens at the blow station by air pressure or by mechanical means. In the more conventional three-station machine, the finished container stays with the core pin as it is indexed to the third station where ejection takes place. Four-, five-, and six-station machines are available. These additional stations are used for further processing of the containers, such as decorating, positioning of blown parts, filling, etc.

There have been several other types of machines available with different methods of transporting the core rods from one station to another. These include the shuttle, two-position rotary, axial movement, and rotary with three or more stations used in conventional injection molding clamping units.

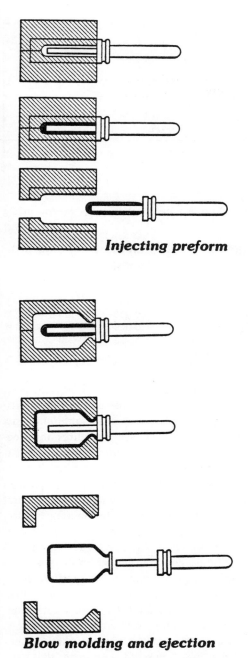

Injecting preform

Blow molding and ejection

Fig. 29-11 Injection blow molding.

The injection blow molding process is used for containers that have very close tolerance threaded necks, wide mouth openings, and highly styled shapes. It features good materials distribution and the production of finished containers without the need for trimming and reaming.

Stretch-Blow Molding

Injection-Stretch. Since the late 1970s stretch-blow (injection) molding has become an important high-speed, high-production technique, largely because of the two-liter carbonated beverage bottle that uses PET resin. A variety of other bottling applications exist such as foods, cosmetics, etc. Stretched polypropylene (extrusion) blow molded bottles have been used since the early 1970s principally for packaging detergents (Fig. 29-13 and Table 29-4).

The stretch-blow process can give many resins improved physical and barrier properties. In biaxial orientation, bottles are stretched lengthwise by an external gripper, or by an internal stretch rod, and then stretched radially by blow air to form the finished container against the mold walls. This process aligns the molecules along two planes, providing additional strength and, even more important, better barrier properties than are possible without biaxial orientation. Other advantages include better clarity, increased impact strength, or toughness, and reduced creep. The actual increase is dependent on the ratio of blow-up in each direction.

The carbonated beverage/soft drink bottles start with a tube or an injection molded preform. This preform is heated, stretched, and blown to its final shape. Because of the stretch and flow at a precise temperature, the molecules in the side wall become biaxially oriented; thus the term oriented. Generally the top and bottom of these containers are not biaxially oriented; so these parts do not have the same physical properties as the side walls.

Ratios that are used to achieve the best properties in a PET bottle are 3.8 hoop by 2.8 axial. This will yield a bottle with hoop tensiles approximately 29,000 psi and axials of 14,000 to 16,000 psi. Moisture barrier and CO_2 barrier are improved as a result of the process, as is drop impact.

Stretch blow molding is possible for thermoplastic materials such as PET, PVC, polystyrene, acrylonitrile, polypropylene, and acetals. The amorphorous materials with a wide range of thermoplasticity are easier to stretch blow

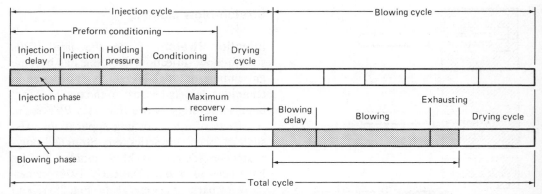

Fig. 29-12 Time sequence in injection blow molding. The colored bar indicates the status of the material during the various stages.

than the partially crystalline types. With the partially crystalline type (such as polypropylene), if the crystallizing is too rapid, the bottle is virtually destroyed.

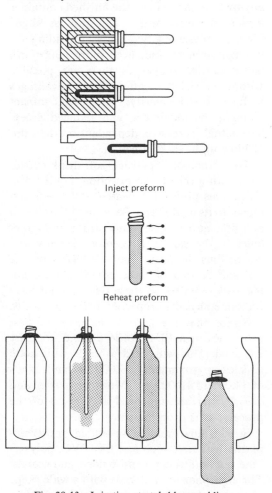

Fig. 29-13 Injection stretch blow molding.

Table 29-4 Stretch blow molding processing characteristics.

	PVC	PET	AN
Melting, °F	400–450	475–510	475–525
Glass-transition, °F	170–180	150–180	220–230
Orientation, °F	175–225	180–210	260–290
Specific gravity	1.4	1.4	1.1

Stretch-blow processing can be separated into two categories: in-line and two-stage. In-line processing is done on a single machine, while two-stage processing requires an injection line to produce preforms, and a reheat-blow machine to make the finished bottles.

In the in-line, an injection molded parison passes through conditioning stations that bring it to the proper orientation temperature. A rather tight temperature profile is held in the axial direction of the preform. Advantages of in-line systems are that heat history is minimized (crucial for temperature-sensitive materials), and the preform can be programmed for optimum material distribution if it is maintained under continuous control.

With the two-stage, the process uses extruded or injection molded preforms that have been cooled, and indexes them through an oven that reheats them to the proper orientation-blow temperature. (The temperature profile is held accurately in the axial direction of the preform.) Advantages of these processes can be the fact that scrap production is minimized, improved thread finish, higher output rates, and the capability to stockpile preforms.

In-line injection stretch-blow systems, while

offering more flexibility from a material stand-point, do not give the degree of parison pro-gramming available in two-stage systems. An extruded parison can be heat-stabilized, and then the parison is grabbed externally and pulled to give axial orientation while the bottle is blown radially. The process orients the en-tire bottle, including the thread area. Threads are then post-mold finished.

The original-design two-stage injection ma-chines have higher output rates than single-stage, while keeping scrap at a minimum. Pre-forms are injection molded on multicavity molds, including the thread finish, then cooled to ambient temperatures before going to the reheat-blow molding machine. Preforms are heated in a reflective radiant heat oven with provisions being made to shield preform thread areas.

After the oven heat exposure, preforms pass through an equilibration zone to allow the heat to disperse evenly through the preform wall. At the blow station, a center rod is in-serted into the preform, stretching it axially, and to properly center the preform in the bot-tle mold. High-pressure air is then introduced to blow the preform to shape against the mold walls and give the bottle its radial orientation.

Based on the work conducted in stretch-blow molding two-liter containers of PVC, PET, and AN, average wall thicknesses of 0.012 to 0.015 in. have been used. Normal blow molding would use an average wall thick-ness of 0.018 to 0.022 in., with no less than 0.009 to 0.010 in. in the thinnest point. The latter is used in either stretch-blow or standard blow molding. The material is so thin that, even though it will not fracture, it is just too thin to prevent dimpling.

The relationship of wall thickness to degree of orientation is important. Control of the bot-tle's wall thickness can be as important as, if not more important than, the amount of biaxial orientation. Soft drink bottles are pres-surized to about 60 psi, so the side walls of 0.009 to 0.015 in. cannot collapse because of internal pressure. However, a product such as mouthwash does not have internal pressure, and thin walls of 0.009 to 0.015 in. would experience paneling or wall collapse due to

permeation losses. Figure 29-14 shows how to include a handle on an injection blow molded container.

Extrusion-Stretch. Extrusion processes also uses programmed parisons and two sets of molds. The first mold is used to blow and temperature-condition the preform (actually cool it to orientation temperature), and the second mold can use an internal rod to stretch and blow the bottle.

When the two-stage, extrusion stretch-blow molding process is used, the preforms or pari-sons used are open-ended. They are reheated, then stretched by pulling one end while the other end is clamped in the blow mold. Mini-mum scrap is produced, and the compression molded threads provide a good neck finish. This system utilizes a conventional extrusion line and a reheat-blow machine whose oven contains quartz lamps to provide the proper temperature profile in order to obtain the proper container wall distribution (Fig. 29-15).

This review principally has included injec-tion stretched blow molding, since major new developments in stretched blow molding were accelerated by using the basic injection mold-ing process. Also important and available are different types of equipment, with rather spec-tacular developments occurring in extrusion stretched blow molding.

Materials

Practically any thermoplastics can be blow molded, particularly the commodity type plas-tics. Different properties can be obtained with the different available and useful plastics that are now used to blow mold different-size and -shaped products. Properties can include strength, toughness, clarity, gas barriers, chemical resistance, heat resistance, color, and many others. Different combinations of prop-erties can be achieved by combining two or more plastics. Parisons can be co-extruded, coated, dipped, co-injected, etc., in order to provide the combinations of different proper-ties available in different plastics. Stretch-blow molding also provides an important processing

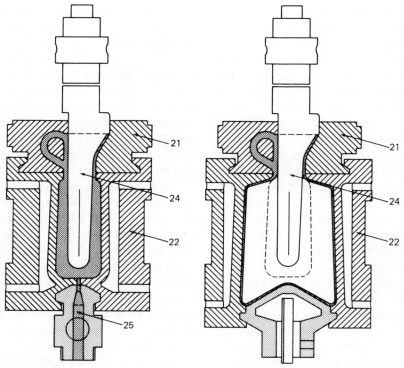

Fig. 29-14   One proposal for incorporating a carrying handle is anticipated in the French patent, number 1,192,475, granted to the Italian company Manifattura Ceramica Pozzi SpA. The handle is molded as part of the preform and is undisturbed when the container is blown. A direct extrapolation to stretch-blow molding technology would incorporate the jug handle immediately below the neck finish of an injection molded preform. It would be necessary to mold such a preform in a split mold which is suited to production on certain rotary type injection stretch-blow equipment.

technique to provide improved properties and reduce the quantity of plastics to improve cost/performance characteristics.

Examples of properties that can be obtained with the different plastics are as follows: *clarity*—crystal polystyrene, acrylonitrile, styrene acrylonitrile, polyvinyl chloride, stretched PET, polysulfone; *moisture barrier*—polypropylene, low-density polyethylene, high-density polyethylene, transparent nylon, polymethylpentene; *oxygen barrier*—PET, polyvinyl chloride, polyacrylonitrile, polysulfone, transparent nylon; etc.

Extrusion blow molding represents the biggest outlet for blow molded containers with high-density polyethylene being the major plastic used, about 80 percent of all plastic consumed. There are different grades of HDPE. With increased density, the stiffness, strength, hardness, creep resistance, softening temperature, and gloss improve. However, toughness, stress crack resistance, and permeability decrease. It resists most acids and salts but is attacked by strong oxidizing agents.

Polyvinyl chloride (PVC) is the second largest resin used in poundage in the United States today. Because of its low oxygen transmission, water-clear clarity, and relative ease of processing, it has captured the hair care, toiletries, cosmetics, household chemical, and edible oil markets. It is now being challenged by stretch-blown polypropylene and stretch-blown PET.

This brief review of the different plastics is used to emphasize the importance of using the correct plastics to meet specific requirements. The real test of any material is that it is blown, either unoriented or stretch-oriented. Unfortunately, in the past, many simulated "fast" tests to evaluate plastic bottle performance have had "bad" records. But there are "fast" tests that have been developed to simulate actual performance.

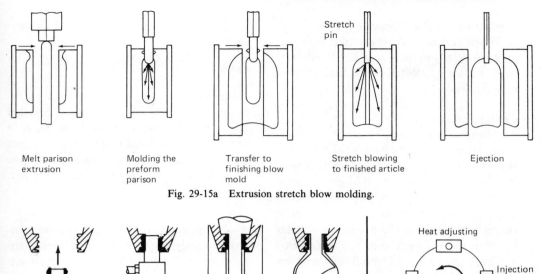

| Melt parison extrusion | Molding the preform parison | Transfer to finishing blow mold | Stretch blowing to finished article | Ejection |

Fig. 29-15a Extrusion stretch blow molding.

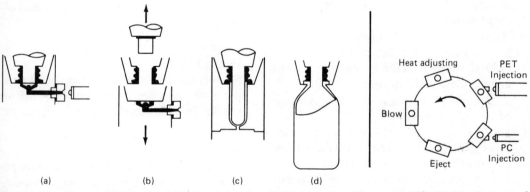

(a) (b) (c) (d)

Fig. 29-15b Heat-set PET bottle with PC neck insert. Conventional PET bottle filling is at ambient or lower temperatures, but with this insert bottles can withstand filling at 185–195°F. Courtesy of Nissei.

(a) (b) (c) (d)

Fig. 29-15c Heat-set PET bottle with PC co-injected neck. These meet requirements of Fig. 29-15b. Courtesy of Nissei.

Guide to Troubleshooting Injection Blow Molding

The information is divided alphabetically by problem and is divided by parison or part defects. Causes and solutions are denoted with "C" and "S," respectively. It is further divided by three processes; extrusion-blow, injection-blow, and stretch-blow molding. The next section reviews troubleshooting stretch-blow molding (9). Information on troubleshooting conventional injection molding is given in Chapter 27.

Cocked Necks
S—Movable bottom plug is stuck in. • Reset stripper plate. • Reduce inject portion of cycle. • Increase blow air time/pressure.

Color Streaks—Flow Lines
S—Raise back pressure. • Increase injection pressure. • Increase melt temperature. • Open nozzle orifice. • Change color mix. • Change color batch. • Add mixing pin to screw. Reduce injection speed. • Increase parison mold temperature. • Dry material.

Contamination (Oil/Grease on Part)
S—Wash parison and blow molds with solvent (especially in neck rings). • Wash core rods. • Clean air filter. • Replace O-rings between molds.

Cracked Necks
S—Increase melt temperature. • Increase neck temperature in parison mold. • Reduce core rod cooling. • Increase neck temperature in blow mold. • Reduce retainer grooves in core rods. • Movable bottom plugs are stuck in (rigid materials only). • Increase injection speed. • Balance nozzles for even fill. • Check core rod alignment. • Check operation of mold temperature controller. • Check stripper location and speed (especially in styrene). • Open nozzle orifice. • Replace O-rings in face blocks.

Dimensional Problems*
H Dimension = Height
S—Increase by moving parison and/or blow mold out (add shim). • Reduce by moving parison and/or blow mold in (remove shim).
S Dimension = Neck finish
S—Increase by moving parison mold out (add shim). • Reduce by moving parison and/or blow mold in (remove shim).
T Dimension (*To Raise T*) (*T* = Average of two dimensions)
S—Lower parison mold neck temperature. • Increase injection time. • Increase stabilize time. • Lower blow mold neck temperature. • Increase core rod cooling (internal).
T Dimension (*To Lower T*)
S—Increase parison mold neck temperature. • Increase blow mold neck temperature. • Reduce core rod cooling. • Hot line blow mold neck. (*Note:* In some cases the opposite will happen when above is done when the T is being blown out in the blow mold.)

---

* Dimensional problems:
  H = Height
  I = Minimum dimension of bottle
  S = Neck finishes
  T = Average of two measurements
  E = Taken across major and minor axes

E Dimension = Across major and minor axes
S—(Refer to T dimension)

Distorted Shoulder
S—Increase blow air pressure. • Shorten cycle. • Wash out vents in blow mold. • Increase parison mold temperature. • Clean or replace (plugged) core rods. • Increase blow air time. • Reset or replace stripper.

Engraving
S—Increase body mold temperature controller temperature. • Increase melt temperature. • Sandblast molds. • Adjust engraving depth and width. • Adjust vents. • Increase blow air pressure. • Clean out engraving. • Increase blow delay. • Balance nozzles for even fill. • Clean, check, and set all gaps evenly in core rod valve. • Check mold temperature controller operation. • Increase venting on blow mold. • Adjust parison mold temperature. • Adjust blow mold temperature. • Increase core pin cooling.

Heavy in Center
S—Increase temperature of gate mold temperature controller (parison mold). • Move nozzles in. • Decrease core rod cooling air.

Hot Spots
S—Increase core rod cooling. • Reduce injection pressure. • Reduce melt temperature. • Lower parison mold temperature in affected area. • Reset stripper for localized external cooling of core rod. • Check mold temperature controller operation.

Inconsistent Shot Size
S—Increase back pressure. • Check hydraulic injection pressures for variations. • Increase screw recovery time. • Balance nozzles. • Check for loose or bad thermocouples. • Check for broken element in mixing nozzle. • Check mold temperature controller operation. • Adjust screw rpm.

Nicks

S—Replace damaged blow mold. • Clean plastic or dirt out of blow mold cavity. • Repair or replace damaged parison mold. • Replace or repair damaged core rod. • Remove strings from nozzles. • Remove burrs from parison and blow molds. • Repair or replace damaged parison and blow molds. • Wash out vents in blow mold. • Sandblast blow mold. • Reset mold in die set.

Nozzle Freeze-off

S—Remove contaminated material from nozzle. • Raise gate mold temperature controller temperature. • Increase manifold temperature. • Increase melt temperature. • Reduce cycle. • Open nozzle orifice. • Check manifold heaters, fuses, and wiring.

Ovality of T&E Dimensions

C—Uneven shrinking.
S—Increase temperature in shoulder/body area parison mold. • Reduce temperature in neck of parison mold. • Lower blow mold neck temperature. • Increase melt temperature. • Check operation of temperature control unit.

Push-up Depth

S—Increase blow time/pressure. • Adjust mold/bottom plug cooling. • Lower melt temperature. • Clean and check core rod valve; set all gaps evenly. • Check movable plug if used. • Increase core rod cooling. • Decrease air pressure. • Adjust nozzle. • Increase tip cooling.

Saddle Finish (Usually apparent with oval T&E—try to correct oval T&E)

S—Parison not packed up tight enough; add inject or screw time and/or screw time and/or pressure. • Increase retainer grooves in core rods. • Reduce parison mold neck temperature. • Increase or decrease cushion. • Increase holding pressure. • Increase cooling time.

Short Shots

S—Clean nozzle. • Open nozzle orifice. • Increase high injection pressure. • Increase packing pressure. • Increase injection time. • Increase melt temperature. • Raise all parison mold temperatures. • Increase back pressure. • Increase screw recovery stroke. • Lengthen cycle time. • Check and clean non-return valve on end of extruder screw. • Adjust screw rpm. • Increase screw speed. • Increase cushion. • Clean out hopper and throat.

Sticking of Parison to Core Rods (Core Rod Too Hot)

S—Decrease cushion. • Reduce melt temperature. • Pack parison harder (increase injection pressure/time). • Reduce screw speed. • Add stearate (release agent). • Increase back pressure. • Adjust core rod temperature. • Check mold temperature controller operation. • Check for folds in bottom I. • Increase internal and external air cooling.

Stripping Difficulties

S—Increase stripping pressure. • Check core pins for burrs. • Add lubricant to polymer. • Adjust stripper bar alignment. • V groove is too deep.

Sunken Panels

S—Increase blow time. • Reduce blow mold temperature. • Shorten inject/transfer portion of cycle. • Reduce core rod cooling. • Add sink correction to blow mold. • Check mold temperature controller operation.

Tom Parts

S—Lower gate mold temperature controller temperature in parison mold. • Check core rod lock-off in parison mold. • Check nozzle seats. • Reset nozzles. • Check parison mold part line. • Lower melt temperature. • Move nozzles out. • Add injection and/or screw time. • Replace nozzles. • Check to see that mold temperature controller is functioning properly.

White or Black Marks on Neck Finish Caused by Gas Burning

S—Lower melt temperature. • Reduce injection speed. • Lower temperature in neck of parison mold. • Vent (relief) neck ring of parison mold. • Open nozzle orifice. • Check mold temperature controller operation. • Reduce ram speed.

Guide to Troubleshooting Stretched Injection Blow Molding

Air Bubbles in the Preform
C—Air entrapment due to too much decompression in plastifier.
S—Check dryer settings. • Increase back pressure slightly.

Bands of Thick and Thin Sections in Part Wall
C—Improper settings on heat zones. (Zones colder than others.)
S—Ensure uniform temperature from capping ring to tip.
C—Not enough time for equilibration of preform before blowing.
S—Equilibration time. • Change heat zones.

Bands or Vertical Stripes on the Preform
C—Too much heat at specific area on preform.
S—Reduce heat.
C—Improper rotation for vertical bands.
S—Check rotation speed.

Blemishes on Part
C—Dirt in molds.
S—Blow molds should be cleaned.
C—Water droplets forming in molds or sweating causing condensation.
S—Increase mold temperature slightly to alleviate condensation.

Cloudiness
C—Melt temperature too low.
S—Increase melt temperature slightly.
C—Moisture in injection molds.
S—Check for condensation on cores or in cavities. • Increase water temperature in injection molds.

Drag Marks on Preform
C—Injection mold damaged or scratched.
S—Polish and possibly rechrome cores and cavities.

Fish Eyes or Zippers
C—Scratch marks on preform surface. (Normally caused by preforms contacting each other after being ejected from injection mold while still hot.)
S—Minimize contact of preforms after injection and prior to cooling.

Folds in Neck Area
C—Center rod stretching preform too early.
S—Synchronize center rod stretch with air delay timers to get blow air to enter at correct time.

Heavy Material in Bottom of Part
C—Improper preform design.
S—Redesign preform.
C—Improper cooling in molds causing heavy amount of material to shrink back.
S—Improve mold cooling.

Knit Lines Appearing in Preform
C—Melt not being injected fast enough or hot enough.
S—Increase temperature. • Raise injection pressure.

Long Gates
C—Valve gates in mold operating improperly.
S—Clean mold.
C—Incorrect temperature in hot runner system.
S—check thermocouples.

Mismatch Lines on the Preform
C—Mold misalignment.
S—Check cores and cavities for alignment.

Off-Centered Gates
C—Could be caused by not using center rods.
S—Use center rods.
C—Poor concentricity in the preform. (Preform concentricity should be held to 0.005 in. max.)
S—Check injection mold for concentricity of core rod and cavity. • Reduce injection speed and injection pressure.

Pearlescense (Haze in Container)
C—Preform stretching too fast for heat in preform.

S—Increase heat in area showing pearlescence going back to preform.

Preform Drooling

C—Valve gates are too warm.
S—Decrease temperature to valve gates.
C—Not enough packing pressure or time.
S—Increase hold time.

Radial Rings on Preform

C—Moisture condensation on core rods.
S—Increase water temperature in molds.

Scratches on Part

C—Possible drag marks on preforms from cavity or core of preform.
S—Polish core and cavity of preform mold.

Soft Necks or Deformed Capping Rings on Finished Container

C—Too much heat in top area of preform.
S—Reduce heating in affected zone. Heat shield may be added to shield capping rings.

Undersized Parts

C—Not enough high pressure air blow time. (Container not being blown to side wall and held under high pressure to freeze material and set outline of mold.)
S—Check mold cooling. • Check blow pressures and time. • Check vents on mold.

Yellowing of Preform (Indicating oxidization through excessive heating during drying)

C—Check drying temperature and time.
S—Adjust drying time as required.

Summary

Hollow plastics products are produced, ranging in volume from parts of an ounce to thousands of gallons. They are used to contain liquids, pastes, powders, and other solid particles. Use also extends into many other applications such as housings for tools, electronic hardware, wheels and other parts of toys, medical devices, platforms, building panels, etc. Just in the market for bottles, over 15 billion thermoplastics bottles were produced during 1980.

Even though blow molding is considered a major production process in the United States, new markets, as well as expanded present markets, are rather unlimited. Examples of these markets include medical devices, automobile gasoline tanks, small carbonated beverage bottles, food containers, etc. These markets are also expanding because of more action in developing co-extruded or co-injection materials to improve cost/performance of containers.

Past/Present/Future of Blow Molding

Blow molding has kept pace with the times for about 4000 years. Up until the past century the only material used for blow molding was "thermoplastic" glass. About a hundred years ago the glass blow molding technique was extended to thermoplastic plastics. Blow molding of plastics basically has been in production only since the 1930s.

Important developments in this century can be highlighted during the following periods of research and production development:

1900–1930

Celluloid formulations are molded into various products by heat.

1930–1950

Basic technological development work is carried out in the United States (injection blow molding preferred as alternative to glass processing) (Plax Corporation/Monsanto).

1950–1960

Polyethylene is introduced, thus permitting the broad application of the blow molding process. The technology of extrusion blow molding is developed, opening up entirely new fields of application for plastics.

1960–1970

Crystal-clear, inert PVC becomes a favored material in the packaging industry. The adaptation of the blow molding process to this material makes it necessary to redesign the machines. A new generation of blow molding machinery is developed. Wall thickness control of the parison makes its mark, with the result that large markets develop in the use of blow molded products, with about 90 percent of blown products being extrusion blow molded.

1970–1980

High-molecular-weight polyethylene captures the market for large-size hollow articles and compels industry to carry out thorough and extensive technological spadework. Two "raw materials crises" result in plastics (PVC and AN) being reassessed.

Over-capacity in machinery construction and in the processing sector intensifies competition, reduces earnings, and slows down the rate of development to an equal extent. As growth rates drop, specialization commences. (The 1975 "depression" occurs).

Electronic controls, servohydraulics, and microprocessors rapidly begin to influence development work on the control of blow molding machines.

Major new development goes into "big production by the late 1970s, namely, PET injection stretched blow molding.

1980–1990

Major new development and new market penetration will occur in the use of the more conventional plastics (PVC, PS, PP, PC, etc.) and new plastics and alloys, with significant advances in machinery to provide major gains in blow molding better-quality containers (of all sizes, from ounces to thousands of gallons) at faster rates and lower costs. The present major development is in stretched extrusion blow molding. Simultaneously, more action is occurring in stretched injection blow molding of different-size containers.

REACTION INJECTION MOLDING (RIM)

Reaction injection molding (RIM) is a process that involves the high pressure impingement mixing of two or more reactive liquid components and injecting into a closed mold at low pressure. With RIM technology, cycle times of 2 minutes and less have been achieved in production for molding large and thick (4 in.) parts (Fig. 29-16). Principal plastic used is

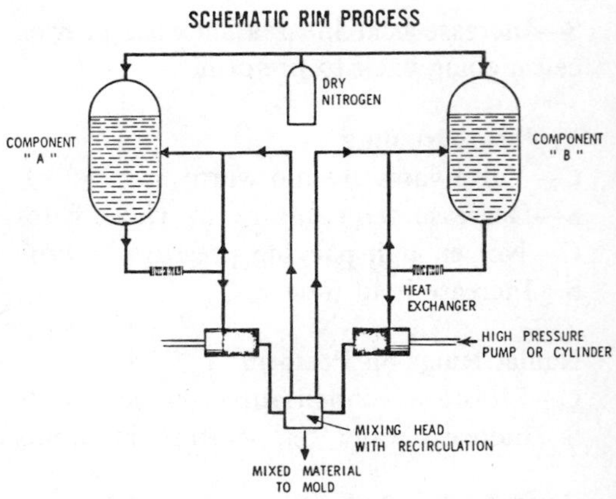

SCHEMATIC RIM PROCESS

Fig. 29-16   Typical polyurethane RIM process involves precise metering of two liquid components under high pressure from holding vessels into the static impingement mixhead. The co-reactants are homogenized in the mixing chamber and injected into a closed mold, to which the mixhead is attached. The heat of reaction of the liquid components vaporizes the blowing agent, beginning the foaming action which completes the filling of the mold cavity.

thermoset polyurethane (PUR). Other materials used are thermoplastic nylon; thermoset polyester and epoxy; etc. This RIM review only includes PUR.

The advantages of RIM over injection molding include the molding of parts larger than 10 pounds; they can be made on a production basis using thinner walls because of lower processing viscosities, or using very thick walls because curing is uniform throughout the part. There are problems associated with RIM, however. The lack of a suitable internal release has made the RIM process labor-intensive, but changes are now occurring to significantly reduce or eliminate this problem.

The molded polyurethane faithfully reproduces the surfaces of the mold and tends to stick to them. Originally the application of mold-release agents was necessary with each cycle of the RIM technology. After polymerization, if the mold is not covered with a mold-release agent, the part will adhere to the mold, making it difficult to remove from the mold. In addition, a film will remain on the mold surface, which will impair the appearance of the product. In view of these occurrences, the mold material should be highly polishable and platable with nickel, since this coating has proved to be most effective in product removal.

RIM is experiencing considerable growth due to savings in producing large parts. Most of the RIM processes in operation use flexible or semirigid thermoset polyurethanes (PUR). Other materials being used or being developed include thermoplastic nylons, thermoset polyesters, thermoplastic polystyrenes, thermoset epoxies, thermoplastic acrylics, etc. (11–22).

In the processes of injection molding of thermoplastic, injection molding thermosets, structural foam molding, and expandable polystyrene molding, we are dealing with materials that are chemically complete compounds, ready for conversion into a finished product. The materials are received from suppliers with certain properties based on test bar information and recorded in material processing data sheets. The processors are expected to convert these materials into products with similar mechanical, electrical, and environmental characteristics, as indicated on the data sheets. The processors are also furnished with a range of molding parameters that should be optimized to attain the desired product properties. In brief, they are given a material along with guidelines for its conversion; but they can do little to change the processing behavior of the material, since they are dealing with a finished raw material that is fully prepared for conversion into a finished product by application of time, temperature, and pressure.

In reaction injection molding (RIM), the starting point for the conversion process are liquid chemical components (monomers, not polymers). These components are metered out in proper ratio, mixed, and injected into a mold where the finished product is formed. In reality, it is a chemical and molding operation combined into one system of molding in which the raw material is not a prepared compound but chemical ingredients that will form a compound when molded into a finished part. The chemicals are highly catalyzed to induce extremely fast reaction rates. The materials that lend themselves to the process are urethane, epoxy, polyester, and others that can be formulated to meet the process requirement.

The system is composed of the following elements:

1. Chemical components that can be combined to produce a material of desired physical and environmental properties. Normally, this formulation consists of two liquid chemical components that have suitable additives and are supplied to the processor by chemical companies (three or more are also used).
2. A chemical processing setup, which stores, meters, and mixes the components ready for introduction into the mold.
3. To facilitate smooth continuous operation, a molding arrangement consisting of a mold, mold-release application system, and stripping accessories.

The success of the overall operation will depend on the processor's knowledge of: (1) the chemistry of the two components and how to keep them in good working order; (2) how to keep the chemical adjunct in proper functioning condition so that the mixture entering the mold will produce the expected result; and (3) mold design, as well as the application of auxiliary facilities that will bring about ease of product removal and mold functioning within a reasonable cycle (e.g., 2 minutes).

RIM molding is energy-conserving as compared to conventional injection molding. The two liquid urethane components are injected generally at room temperature, and a typical mold temperature is 150°F. Also, since the material is expanded after injection, very low clamp pressures (100 psi) are required.

Since internal mold pressures would not normally exceed 100 psi, the clamping requirements for RIM are substantially lower than for thermoplastic processing. Calculations have been done on a part and show that a clamp requirement of 2500 to 5000 tons necessary to produce a part from conventional injection molded thermoplastic polyurethane can be reduced to less than 100 tons for RIM.

The production of polyurethane elastomers

involves the controlled polymerization of an isocyanate, a long-chain-backbone polyol and a shorter-chain extender or cross-linker. The reaction rates can be controlled through the use of specific catalyst compounds, well-known in the industry, to provide sufficient time to pour or otherwise transfer the mix, and to cure the polymer sufficiently to allow handling of the freshly demolded part. The use of blowing agents allows the formation of a definite cellular core (thus the term "microcellular elastomer") as well as a nonporous skin, producing an integral sandwich-type cross section.

In RIM, all necessary reactive ingredients are contained in two (or more) liquid components: an isocyanate component, A, and a resin component, B.

The choice of isocyanate, as well as variations within isocyanate families, exerts a profound effect on the processing and final properties of the elastomer. The chemical structures of two of the major diisocyanate types, 4,4' diphenyl methane diisocyanate (MDI) and toluene diisocyanate (TDI), is commonly supplied in an 80/20 mixture of the 2,4 and 2,6 isomers. Early in the development of RIM systems, the MDI family was chosen over TDI, based on the following considerations:

1. Reactivity: Given the same set of co-reactants, MDI and MDI types are more reactive than TDI. This can be used to advantage when short cycles are required.
2. Available co-reactants: The high reactivity of the MDI types also makes available a larger number of co-reactants. For example, where hindered aromatic amines yield a given level of reactivity, a variety of glycols can give equivalent reactivity—thus allowing more formulation versatility.
3. Handling: The MDI materials offer excellent handling characteristics due to comparatively low vapor pressure.
4. Green strength: The ortho–isocyanate groups of TDI are less reactive than the para-groups. Thus, at the end of the reac-

tion to form a polymer, the rate of reaction slows, resulting in green strength problems upon demolding. MDI does not suffer this deficiency.

Reaction injection molding involves very accurate mixing and metering of two highly catalyzed liquid urethane components, polyol and isocyanate. The polyol component contains the polyether backbone, a chain extender or cross-linking agent, and a catalyst. A blowing agent is generally included in either the polyol or isocyanate component.

In order to achieve the optimum in physical properties and part appearance, instantaneous and homogeneous mixing is necessary. Insufficient mixing and/or lead/lag results either in surface defects on the part or, at the time of postcure, delamination or blistering.

The urethane liquid components are stored at a constant temperature in a dry air or nitrogen environment. These components are delivered to high-pressure metering pumps or cylinders that dispense the respective materials at high pressure and accurate ratios to a mixing head. The materials are mixed by stream impingement. Additional mixing is generally encouraged via a static mixer (tortuous material path) incorporated into the runner system of the mold. Following the injection of the chemicals, the blowing agent expands the material to fill the mold.

The preferred route for high-volume RIM manufacturing is multiple clamps fed from a single metering pumping unit, the logic being that this is the most efficient way to utilize the capacity of the mold-filling equipment.

The Mold

Since one of the ultimate objectives of the RIM process, for its major market of automotive exterior part production, was a cycle time of 2 minutes or less, a great deal of effort was applied to mold construction and design. Continuous automatic operation of a molding station without interruption required improvements in mold release and mold surface technology. Originally, mold preparation following a shot was required due to the buildup of external release agents, which were necessary to enable easy removal of the part from the mold. This problem was approached from the material side, through a search for suitable internal releases, and through the development of improved external mold release compounds. From the equipment side, the development of automatic molds was required if the RIM process was to compete with classical injection molding with respect to mold cycle times and efficient production.

General Motors Corporation constructed such a mold for a production trial of the 1974 Corvette fascia (which actually started the development of RIM). This mold was tool steel with a highly polished nickel-plated surface. Most of the mold seals were elastomeric, to prevent excessive flash (up to 10 percent, by weight, of flash can occur; and PUR can not be reused, since it is a thermoset) due to leakage of the low-viscosity thermoset polyurethane reacting material. This was possible because of the low internal mold pressures encountered in the RIM process, less than 100 psi. This evaluation was highly successful in demonstrating the capability of total automation of the RIM process.

In the construction of molds for RIM processing, it must be kept in mind that part quality and finish are roughly equivalent to the quality and finish of the mold surface itself. A common misconception is that because the clamp tonnage for a RIM setup is relatively low, low-quality tools can be used. This, however, is true only insofar as the pressure requirements for the mold are concerned. Experience has shown that the finish on the part surface is a direct function of the mold finish, and that the mold finish is a direct function of the quality of the mold material. Excellent results have been obtained using high-quality, nickel-plated, tool steel molds and electro-formed nickel shells.

For production runs of 50,000 parts per year, a P-20, P-21, or H-13 steel would be most appropriate, not only because of these steels' homogeneous nature, but also because of their excellent polishability and adaptability for a good plating job. The prehardened grades of 30 to 44 RC are preferable because of the degree of permanency that they impart to a tool. After machining, a stress-relieving operation is very important in order to avoid possible distortions or even cracking.

Nickel shells that are electro-formed or vapor-formed when suitably backed up and mounted in a frame are also excellent materials for large-volume runs. For activities of less than 50,000 parts per year, aluminum forgings of Alcoa grade No. 7075-T73 machines to the needed configuration will perform satisfactorily. They have the advantage of good heat conductivity, an important feature in RIM.

Cast materials are used for RIM molds with reasonable success. One such material is Kirksite, a zinc alloy casting material. Kirksite molds are easy castable, are free from porosity, will polish and plate well, and have been used with favorable results.

For consistent quality and molding cycles with PUR, the mold temperature should be maintained within $\pm 4°F$. The mold temperatures will range from $101°F$ to $150°F$, depending on the composition being used.

The cooling lines should be so placed with respect to the cavity that there is a $\frac{3}{4}$-in. wall from the edge of the hole to the cavity face. The spacing between passages should be 2.5 to 3 diameters of the cooling-passage opening. These dimensions apply to steel; for materials with better heat conductivity, the spacing can be increased by one hole size.

As with the chemical components, it is necessary to maintain constant surface temperatures in the mold for a reproducible surface finish and constant chemical reactivity. This temperature varies according to the chemical system being used and has been determined empirically.

The mold orientation should be such as to allow filling from the bottom of the mold cavity, allowing escape of air through a top flange at a hidden surface. This allows controlled venting, and positioning of vent pockets that can be trimmed from the part at a later time.

Process Controls

The chemical systems for RIM all have one characteristic in common: they require a RIM machine to convert liquid raw materials into quality plastic products. Assuming a properly formulated chemical system, the quality of the end product results from the ability to measure, control, and adjust temperature, ratio, pressure, and other essential process parameters of the RIM dispensing machine. Such exacting control leads to a reduction in start-up time, minimal rejects and touch-up work, reproducible product quality, and the ability to pinpoint changes in product properties (22).

In the high-temperature RIM processing of nylon, temperatures are monitored and controlled within ±2°F using both electrical heat tracing and hot oil jacketing. The controllers contain high–low set-points; all temperature zones must be at the required settings to permit machine operation. A graphic diagnostic panel, with light-emitting diodes (LED), associated with all key switches, valves, and pressures, aids in troubleshooting; if a malfunction occurs, the cause is pinpointed by a blinking light. Low- and high-pressure circulation is monitored by transducers and displayed digitally; high/low pressure limits, if exceeded, will abort the RIM cycle for safety reasons.

LIQUID INJECTION MOLDING (LIM)

The process of liquid injection molding (LIM) has been used longer than RIM. From a practical view these two processes, LIM and RIM, are similar. Their concept of automated low-pressure processing of liquid thermosets in converted injection machines has conclusively demonstrated advantages of faster cycles, lower labor costs, lower capital investment, energy savings, and space savings relative to conventional potting, encapsulation, compression transfer processes, and conventional injection molding.

The term used is usually determined by the processor, machinery manufacturer, and/or material supplier. LIM usually refers to the processing of liquid thermoset silicones, whereas RIM refers to thermoset polyurethanes (PUR). However, prior to the late 1970s many processors of PUR called their equipment and processing LIM. With thermoplastic nylon 6 becoming popular, its original

identification with LIM has changed to an identification with RIM, the more widely used term.

A major application for the LIM-silicones continues to be encapsulating electrical/electronic devices. The usual LIM system basically uses two or more pumps that move the components of the liquid system (such as catalyst and plastic) to a mixing head, before they are forced into the heated mold cavity. There are systems in which screw mixing is used, similar to conventional injection molding. Also, one-part materials are used (all components combined) and in turn forced into the heated mold cavity to develop an exothermic chemical reaction when the material is heated. A fast-operating LIM system that can mix up to four separate components is shown in Fig. 29-17. The equipment in this figure processes nylon material.

STRUCTURAL FOAM

A great variety of foam products are available from plastic, but basically there are just two types, flexible and rigid. The flexible type generally identifies the very large market of principally extruded polyurethane foam for cushioning (for chairs, mattresses, etc.); about 5 percent of all plastic goes into these flexible foams. Another important flexible type can include the expandable polystyrene (EPS) that is used in special injection molding machines (steam curing is used). The EPS market is about another 7 percent of all plastic used.

Within the rigid types are the important structural foam (0.2 percent of plastic) and reaction injection molding (0.3 percent of plastic) types, both of which are involved in basically injection molding.

The generally accepted definition for structural foam is a plastic product having integral skins, a cellular core, and a high-enough strength-to-weight ratio to be designated "structural" (Fig. 29-18). What minimum value this strength-to-weight ratio must have to be classified as "structural" depends entirely upon the application (23).

The first rigid foams used polyurethane, and were developed in 1937 by Dr. Otto Bayer

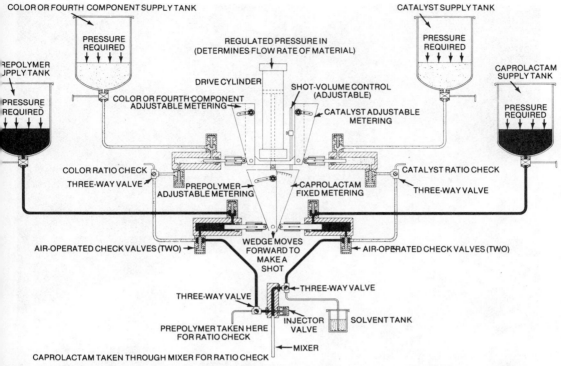

Fig. 29-17 Liquid injection molding (or reaction injection molding) machine that can process different liquid plastics (e.g., nylon) with an accuracy of 0.1 percent. Mixing action is achieved with the "flying wedge" technique. Courtesy of Amplan Inc., Middlesex, NJ.

(Germany). This type of structure, in sandwich construction, was used in many military applications by the Germans during World War II (aircraft, tanks, submarines, etc.). In the United States during 1943 and 1944, the first all-plastic airplane included structural foam components. This all-plastic U.S. Air Force airplane was successfully flight-tested in 1943 at Wright-Patterson Air Force Base. Different sandwich design constructions were used that included glass fiber/polyester reinforced plastic facings, glass cloth/polyester reinforced honeycomb core, and cellular cellulose acetate structural foam (24–31). By 1945, the laboratories at this Air Force Base also designed and fabricated radomes (radar antenna structural windows) of structural foam using polyurethane (6).

Performance

The use of structural foam (SF) molding is interesting principally because it provides a three- to fourfold increase in rigidity over a solid plastic part of the same weight. (This three- to fourfold advantage, or a greater advantage, can be designed into many applications with solid plastic by using the basic engineering rib design for molds.) SF also permits molding large parts with the cost advantages that injection molding (solid parts) offers to smaller parts. Thus, large parts with a high degree of rigidity can be molded. The self-expanding nature of SF results in low-stress parts with dimensional stability and less tendency to warpage and sink marks. It also offers thermal and acoustic insulation.

Other advantages exist for the use of SF, but there are also disadvantages. Molding cycle time will at least increase by the square of the thickness increase. Moreover, most SF parts are made by a low-pressure technique that causes a surface finish that visually resembles the splay marks found in injection molding. This surface condition, called swirl, is the result of broken bubbles in the surface; and

Fig. 29-18   Cross section of structural foam.

techniques such as counter-pressure must be used to significantly remove the swirl finish. Thus, producing conventional low pressure SF parts can result in higher finishing costs and longer cycle times (23, 31–33).

Plastic Materials

Polystyrene, polyethylene, ABS, and polypropylene represent about 90 percent of the resin used for SF. The remaining engineering resins provide the usual advantages of performance such as increased mechanical creep resistance, and heat resistance properties. Of these engineering resins (polycarbonates, nylon, ABS, PBT, PPO, and acetal), the principal choice has usually been polycarbonate.

Applications for SF are found in computer and business machine housings, appliances, building products, and so on.

Characteristics of Foam

A density reduction of up to 40 percent can be obtained in SF parts. The actual density reduction obtained will depend on part thickness, design, and flow distance. Low-pressure structural foam parts will have the characteristic surface splay patterns; however, the utilization of increased mold temperatures, increased injection rates, or grained mold surfaces will serve to minimize or hide this surface streaking. Finishing systems (e.g., sanding, filling, painting) for structural foam are readily available and have proved to be capable of completely eliminating surface splay. It should be noted, however, that utilization of techniques to minimize splay can very often result in reduced finishing costs.

High-pressure structural foam parts have generally been found to require little or no postfinishing. Although high-pressure foam parts may exhibit visual splay, surface smoothness is maintained, and no sanding or filling is required.

Structural foam parts expanded with chemical blowing agents will exhibit increased stiffness because they are normally thicker than solid moldings. Their lower density also provides a higher strength-to-weight ratio when compared to solid moldings. Because of the foamed core within the part as well as its greater thickness, acoustical and insulating properties are enhanced.

Foaming a polymer does not change its chemical structure or its resistance to chemical attack, provided the proper chemical blowing agent and processing conditions are used. Mechanical properties such as tensile and impact strength (Fig. 29-19) and flexural modulus will be lower in foam parts because of their low densities.

The cell structure of structural foams varies quite widely for the various molding processes. In the expansion cast molding process (similar to cold compression molding; not SF molding), the products to be foamed are placed in a cold mold. Then the mold is heated, and expansion takes place relatively slowly, making for slow growth of the cell structure; this results in quite a uniform cell structure. This

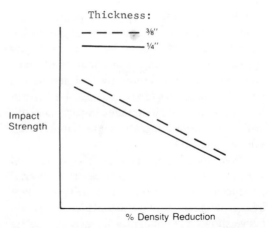

Fig. 29-19 Impact strength of structural foam based on part thickness and weight.

holds not only for thermoplastic foams produced by rotomolding and the foundry process but also for polyurethanes produced by expansion cast molding.

In injection molding, the cell structure of the molded foam varies markedly for the various processes. Where the mold is filled with a short shot accompanied by *low* mold pressure, the cell structure shows a wide distribution in cell size across the part cross section. In fact, small random voids may occur in the structure. When the mold is filled under low pressure, the foam density shows a gradient along the flow path, with the highest density at the end of the flow path and the lowest density near the runner. Where the mold is filled under *high* injection pressure, no foaming occurs in the mold until a solid skin has been formed. Then the mold pressure is intentionally reduced by either melt egression (Allied) or mold expansion (Dow, ICI, and USM), permitting the still molten core to foam. These techniques make for uniform cell structure not only across the part thickness but over the entire part.

Design Analysis

For structural foam, mold pressures of approximately 600 psi (4.1 MPa) are required, compared to typical pressures of 5000 psi and greater (34.5 MPa) in injection molding. As a result, large, complicated parts, 50 pounds and up (22.7 kg), can be produced using multinozzle equipment, or up to 35 pounds (15.9 kg) with single-nozzle equipment and hot-runner systems. Part size, in fact, is limited only by the size of existing equipment, while part complexity is only limited by tool design and material properties. Part cost can be kept in line through such advantages as parts consolidation, function integration, and assembly labor savings (32).

When an engineering plastic resin is used with the structural foam process, the material produced exhibits predictable behavior over a large range of temperatures. Its stress/strain curve shows a significantly linearly elastic region like other Hookean materials, up to the proportional limit.

However, since thermoplastics are viscoelastic in nature, their properties are dependent on time, temperature, and strain-rate. The ratio of stress and strain is linear at low strain levels of 1 to 2 percent, and, therefore, standard elastic design principles can be applied up to the elastic transition point.

Large and complicated parts will usually require more critical structural evaluations to allow better predictions of load-bearing capabilities under both static and dynamic conditions. Thus, predictions require a careful analysis of the structural foam cross section.

The composite cross section of a structural foam part contains an ideal distribution of material with a solid skin outer region and a foamed core. The manufacturing process distributes a thick, almost impervious solid skin, which is in the range of 25 percent of the overall wall thickness at the extreme locations from the neutral axis (29-20a). These are the regions where the maximum compressive and tensile stresses occur in bending.

The simply supported beam has a load applied centrally. The upper skin goes into compression while the lower skin goes into tension, and a uniform bending curve will develop (Fig. 29-20b). However, this only happens if the shear rigidity or shear modulus of the cellular core is sufficiently high. If this is not the case, then both skins will deflect as independent members, thus eliminating the load-bearing capability of the composite structure (Fig. 29-20c).

The fact that the cellular core provides resistance against shear and buckling stresses implies an ideal density for a given foam wall thickness. This optimum thickness is critically important in the design of complex, stressed parts.

At a $\frac{1}{4}$-in wall (6.4 mm), for example, both modified polyphenylene oxide and polycarbonate resin exhibit the best processing, properties, and cost—in the range of 25 percent weight reduction. Laboratory tests show that with thinner walls, about 0.157 in. (4.0 mm), this ideal weight reduction decreases to 15 percent. When wall thickness reaches approximately 0.350 in. (8.9 mm), weight can be reduced 30 percent.

However the structural foam cross section is analyzed, its composite nature still results in a two-fold increase in rigidity, compared to an equivalent amount of solid plastic, since rigidity is a cubic function of wall thickness. This increased rigidity allows large structural parts to be designed with minimal distortion and deflection when stressed within recommended values for a particular foamable resin. Depending on the required analysis, the moment of inertia can be evaluated three ways.

In the first approach, the cross section is considered to be solid material (Fig. 29-21). The moment of inertia, I_x, is then equal to:

$$I_x = bh^3/12$$

where b = width and h = height. This commonly used approach provides acceptable accuracy when load-bearing requirements are minimal—for example, in the case of simple stresses—and when time/cost constraints prevent more exact analysis.

The second approach ignores the strength contribution of the core, and assumes that the two outer skins provide all the rigidity (Fig. 29-22). The equivalent moment of inertia is then equal to:

$$I_x = b(h^3 - h_1^3)/12$$

This formula results in conservative accuracy, since the core does contribute to the stress-absorbing function. It also adds a built-in

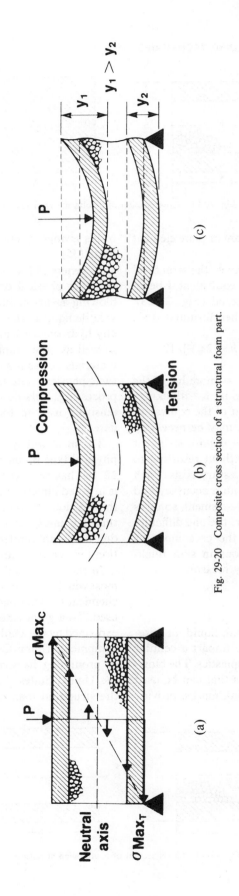

Fig. 29-20 Composite cross section of a structural foam part.

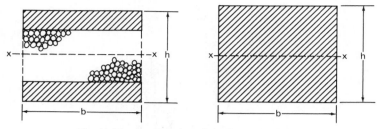

Fig. 29-21   Cross section of a solid material (32).

safety factor to a loaded beam or plate element when safety is a concern.

A third method is to convert the structural foam cross section to an equivalent I-beam section of solid resin material (Fig. 29-23). The moment of inertia is then formulated as:

$$I_x = [bh^3 - (b - b_1)(h - 2t_x)^3]/12$$

where $b_1 = b(E_c)/(E_s)$; E_c = modulus of the core; E_s = modulus of the skin; t_s = thickness of the skin; and h_1 = height of the equivalent web (core). This approach may be necessary where operating conditions require stringent load-bearing capabilities without resorting to overdesign, and thus unnecessary costs. Such an analysis produces maximum accuracy and would be suitable for finite-element analysis on complex parts. However, the one difficulty with this method is that the core modulus and the as-molded variations in skin thicknesses cannot be accurately measured.

Blowing Agents

Blowing agents, be they solid, liquid, or gaseous substances, are used to impart a cellular structure to molded thermoplastics. The blowing agent is a source of gas that can be used by the molder to control sink marks, provide resins savings, or manufacture structural foam parts (35).

In general, blowing agents can be classified as either physical or chemical. The physical blowing agents include compressed gases and volatile liquids. The volatile liquids are generally hydrocarbons such as hexane or pentane as well as other aliphatic hydrocarbons. The materials act as a source of gas by changing their physical state from liquid to gas during processing. Volatile liquids have not been extensively used in foaming thermoplastics to date.

The most widely used blowing agent of the physical type is compressed nitrogen. (Freon-22 gas has seen very limited use.) Nitrogen is injected directly into the polymer melt prior to injection. Advantages of nitrogen gas are that it is inert, leaves no decomposition residue, and is not limited to a specific decomposition temperature range.

In the United States nitrogen is by far the most widely used blowing agent, but in Europe chemical blowing agents are the most widely used. To a great extent this difference can be explained by the availability of structural foam equipment. Union Carbide has been a leader in promoting its structural foam process in the United States. In recent years, however, the structural foam equipment designed and

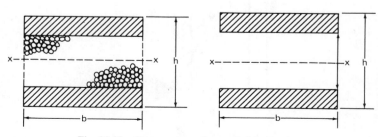

Fig. 29-22   Cross section of a sandwich structure.

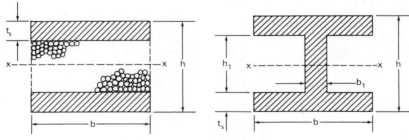

Fig. 29-23   Cross section of an I-beam.

used with chemical blowing agents in Europe has been introduced in the United States.

A comparison of the effectiveness of nitrogen vs. chemical blowing agents is a complicated and controversial subject. Therefore, we will not go into the details of such a comparison. However, it is worth noting that the cost involved in finishing a structural foam part is a primary factor to consider when you choose a structural foam process. It has been widely reported that chemical blowing agents can provide superior cell structure and surface appearance relative to nitrogen. The resulting effects on finishing costs are obvious.

Chemical blowing agents (CBA's) are generally solid materials that decompose when heated to a specific temperature, yielding one or more gases and a solid residue. Chemical blowing agents also can be divided into the organic and the inorganic types. The most common inorganic chemical blowing agent is sodium bicarbonate, which is being used to some extent in the production of foam parts. The major advantage of sodium bicarbonate is its low cost. The major disadvantage is that sodium bicarbonate decomposes over a very broad temperature range as compared to organic chemical blowing agents, so that its decomposition cannot be controlled as readily as that of the organic chemical blowing agents.

Organic chemical blowing agents are solid materials designed to decompose over specific temperature ranges. Therefore, the primary criterion used to select a chemical blowing agent is the processing temperature of the plastic to be foamed. The Tables 29-5, 29-6, and 29-7 give typical data for CBA's.

Direct injection of nitrogen requires special equipment to allow for the nitrogen to be in-

jected into the extruder cylinder, whereas incorporation of chemical blowing agents can be accomplished by a variety of techniques, depending of available equipment and the volume of foam to be processed.

The most common means of incorporating chemical blowing agents is by drum tumbling with the resin. A wetting agent such as white mineral oil is commonly used to allow for good adhesion of the blowing agent to the resin pellet. This is particularly important if the material is to be air-conveyed to the hopper. The second method of incorporating chemical blowing agents is to use one of many hopper metering and blending units available. This method, while it requires special equipment, eliminates the labor involved in drum tumbling. The third method of incorporation involves a liquid dispersion of the chemical blowing agent. The powdered blowing agent is dispersed in a medium compatible with the resin being processed, and is pumped directly into the throat of the machine.

The fourth, and probably the fastest-growing method of incorporating chemical blowing agents, is through the use of blowing agent–resin concentrates. These concentrates are becoming increasingly popular and are available in a variety of resins. The concentrates can then be tumbled or meter-blended with virgin resin prior to entering the machine. This method can be used to completely eliminate the need for handling the powdered chemical blowing agents. Concentrates give excellent dispersion of the blowing agent prior to injection.

Any blowing agent selected must fill the usual requirements for polymer additives; that is, the blowing agent as well as its decomposi-

Table 29-5   Celogen® selection guide for injection molding. (Courtesy of Uniroyal Chemical, CT.)

BLOWING AGENTS	FEATURES	STOCK TEMPERATURE RANGE, °F (°C)	PROCESS		TOOLING		USAGE LEVEL, % (1)		POLYMERS
			LOW PRESSURE	HIGH PRESSURE	All Except Be-Cu	Be-Cu	FOAM	SINKS	
Celogen OT	Low Temperature No Ammonia	300-350 (149-177)	Yes	Yes	Yes	Yes*	.8	.15	LDPE*, EVA
Celogen AZ Grade 130	Low Cost Most Efficient	390-450 (200-232) (350 with activation)	Yes	No	Yes	No	.5	.1	Acetal, HIPS, ABS, HDPE PP, Thermoplastic Rubber (LDPE, EVA*, Vinyl* with Activation) Noryl
Grade 199	Slower decomposition rate Low cost Most Efficient	Same as Above	Yes	Yes	Yes	No	.5	.1	Same as Above
Celogen AZNP Grade 130	Non-plate out type	Same as Above	Yes*	No	Yes	No	.6	.1	Acetal*, HIPS*, ABS*, HDPE* PP*, Thermoplastic Rub.* Noryl* (LDPE, EVA Vinyl with activation)
Grade 199	Non-plate out type slower decomposition rate	Same as Above	Yes	Yes*	Yes	No	.6	.1	Same as Above
Celogen RA	Intermediate to high temperature	420-500 (216-260) 375 with activation	Yes*	Yes	Yes		.8	.15	ABS, HIPS, HDPE PP, Thermoplastic Rubber, Noryl*, Nylon
Celogen HT-500	High Temperature	480-520 (249-271)	Yes	Yes*	Yes	Yes	.5	.1	Polyester, Nylon, ABS*
Celogen HT 550	High Temperature	550-650 (288-343) 500 with activation	Yes*		Yes	Yes	.4	.1	Polyester, Polycarbonate*, Polysulfone*, Noryl, Nylon*

USE OF GUIDE: 1. Using the polymer column first, select the polymer being molded each place where it is listed in the column.
2. The possible blowing agents that can be used to foam the particular polymer are listed in the blowing agent column.
3. The most widely used blowing agent for the polymer is indicated by the asterisked polymer.
4. Using the most common blowing agent as a first choice, move from left to right to determine if temperature, process, and tooling conditions warrant use of one of the alternatives. Preferred alternatives are asterisked.
Example: HIPS is to be foam molded at 475°F using the low pressure process.

(Courtesy of Uniroyal Chemical, CT.)

Table 29-6 Typical formulations and operating conditions for expanded injection molded thermoplastics. (Courtesy of Naugatuck Chemical, CT.)

	Poly-propylene	Poly-propylene	High Density Poly-ethylene	Low Density Poly-ethylene	High Impact Poly-styrene	High Impact Poly-styrene	Poly-ester	ABS	Poly-phenylene oxide	Poly-acetal	6/6 Nylon (filled)	Poly-phenylene sulfide	Poly-carbonate	Poly-sulfone
Unexpanded S.G.	0.91	0.91	0.95	0.92	1.05	1.05	1.5	1.05	1.05	1.42	1.5	1.6	1.35	1.24
Formulation (PHR)														
CELOGEN® OT	0.5	—	—	0.8	—	—	—	—		—	—	—	—	—
CELOGEN® AZ	—	0.8	0.5	—	0.5	—	—	0.5	0.5	0.5	—	—	—	—
CELOGEN® RA	—	—	—	—	—	0.8	—	—	—	—	—	—	—	—
CELOGEN® HT-500	—	—	—	—	—	—	0.5	—	—	—	—	—	—	—
CELOGEN® HT-550	—	—	—	—	—	—	—	—	—	—	0.5	2.0	0.25	1.0
Expanded S.G.	0.70	0.70	0.75	0.75	0.75	0.75	0.79	0.75	0.80	1.0	1.1	0.95	0.94	0.744
Compounding														
Tumble	15 min.	15 min.	15 min.	15 min.	15 min.	15 min.	15 min.	15 min.	15 min.	15 min.	15 min.	15 min.	15 min.	15 min.
Temperatures (°F.)														
Barrel, Back	390	390	380	320	375	380	460	380	520	380	510	600	545	550
Barrel, Front	420	430	400	340	400	410	470	410	550	400	520	625	555	575
Nozzle	430	440	410	350	410	430	480	420	560	410	530	650	560	600
Mold	65-100	65-100	65	65	65	65	175	65	120	—	180	100	180	160

Cycles — For structural foam parts at 1/4 inch thickness, cycle times typically range from 60 to 120 seconds depending on the polymer being molded.

Pressures — Injection pressure in structural foam molding should be high enough to enable maximum injection speed on the machine being used.

(Courtesy of Naugatuck Chemical, CT.)

Table 29-7   Chemical blowing agents. (Courtesy of Naugatuck Chemical.)

APPLICATIONS		CHEMICAL BLOWING AGENTS (CBA's) (operating temperature range, °F)											
		CELOGEN XP-100	CELOGEN TSH (220-270)	CELOGEN OT (290-350)	CELOGEN AZ130 (330-450)	CELOGEN AZ3990 (330-450)	CELOGEN AZNP130 (330-450)	CELOGEN AZ760 (330-350)	CELOGEN AZRV (320-390)	CELOGEN RA (375-500)	CELOGEN HT500 (486-520)	CELOGEN HT550 (500-650)	CELOGEN AZ754
INJECTION MOLDING—STRUCTURAL FOAM	EVA				◆					◇			●
	LDPE			●	◇								◇
	HDPE				○		●	●					
	PP				○		●	●					
	HIPS				○		●	●					
	ABS				○		●	●			●		
	Thermoplastic Rubber				○		●	●					
	Acetal				●								
	Flexible PVC		●			●							
	Rigid PVC				◇				●				
	PPO Based				○		●	●		○			
	Nylon									○	●	●	
	Polyester										●	○	
	Polycarbonate											●	
	Polysulfone											●	

● Primary Recommendation   ○ Secondary Recommendation   ◆ Primary Recommendation with suitable Activation   ◇ Secondary Recommendation with suitable Activation   ASP-4384 R1

(Courtesy of Naugatuck Chemical)

tion products should not negatively influence the physical-mechanical and toxicological properties of the plastic material.

The agents must homogenize with the polymer—it cannot have corrosive or other side effects. The temperature range of decomposition has to coincide with the temperature range of the plastic melt. Blowing agents with an Azodicarbonamid (ADC) base are widely used. Compared to other agents, they display a finer cell structure and therefore improve mechanical properties. Unfavorable properties of the Azodicarbonamid (i.e., the strong smell of ammonia separation, the yellowing of the plastic material, corrosive effects on some metals) are not present with the multiple component injection process, since the foamed plastic is enclosed by a compact external skin.

In regard to the use of Freon-22 gas instead of traditional nitrogen, this is principally being done in various laboratories. The target is to significantly reduce the finishing cost on low pressure "swirl" molded parts. Even though Freon is at a nominal disadvantage compared to nitrogen (about 50 percent more gas required), the results suggest that the extra material cost will be largely offset by savings elsewhere in the manufacturing process. Those savings will come from lower finishing costs, reduced cycle time, and possibly greater density reductions than are typical with nitrogen (37).

Methods of Processing SF with Chemical Blowing Agents

Injection molded structural foam parts may be produced by both low and high pressure processes. In this context, low or high pressure refers to the mold cavity pressure. Nitrogen gas and chemical blowing agents are widely used in both processes (11).

Some of the specialized structural foam processes and equipment are patented and may require licensing. The processor is advised to ascertain the patent situation before employing any of these specialized techniques.

Low Pressure Foam.   Injection molded (structural) foam is produced by incorporating the selected chemical blowing agent with a resin and injecting a short shot (less than the volume of the mold cavity) into the tool. Gases released by decomposition of the blowing agent expand the polymer to fill the cavity. Since the mold cavity is not completely filled with resin, the pressure in the tool is only that generated by the blowing agent.

Low pressure foam is produced on a variety of equipment with internal cavity pressures ranging from 200 to 600 psi.

Foam molding on conventional machines (Fig. 29–24) requires some modifications to produce good-quality parts. The most important of these is the use of a positive shut-off

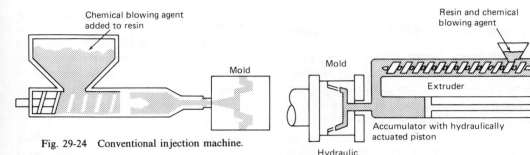

Fig. 29-24 Conventional injection machine.

Fig. 29-25 Two-stage injection molding structural foam.

nozzle to prevent drooling of the expandable melt, which causes variation in part weight as well as nozzle freeze-up. The shut-off nozzle may be mechanically, spring-, or hydraulically activated.

While a shut-off nozzle is essential, other modifications can be made to improve part quality and increase capacity of the machine. These include an intensifier to increase injection speed and an accumulator to increase shot size. Conversion kits are commercially available for all of the above modifications.

This approach allows the molder to convert a standard injection machine from solid to foam (or the reverse) without difficulty and requires a relatively small capital investment.

Special machines, similar to high-speed two-stage injection molding machines have been specifically designed and built for the production of low pressure foam moldings. Typically, these machines offer the advantage of high-speed injection rates, large shot capacity, and large platens. Because of the lower clamp tonnages used, less expensive tooling is required. Both in-line reciprocating screw and two-stage screw-plasticating/ram-injection (Figs. 29–25 and 29–26) units are available.

Chemical blowing agent is also used in low pressure foam systems where compressed nitrogen is the primary blowing agent. The addi-

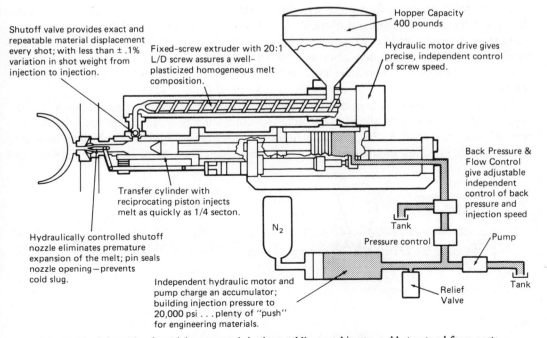

Fig. 29-26 Schematic of special two-stage injection molding machine to mold structural foam parts.

tion of the CBA in this process facilitates cell formation and uniformity in the molded parts.

High Pressure Foam. Chemical-blowing-agent-expanded products can also be made on specialized foam machines using the high pressure technique. A full shot of expandable plastic is injected at pressures normal for the resin involved. A skin of solid plastic is formed by cooling at the mold surface, and expansion of the core occurs by moving one or more plates to enlarge the mold cavity (Fig. 29–27).

This process provides a more distinct skin than the low pressure systems, better reproduction of cavity detail, and a surface that may be essentially free of splay if the correct combination of chemical blowing agent and processing conditions is used.

With this process it is possible to vary density by controlling the mold expansion motion so that essentially solid sections are obtained where high strength is required and weight reduction is limited to noncritical areas.

Co-injection (sandwich) machines are also available that are capable of injecting both solid and foam polymers. Simultaneous injection occurs, resulting in a solid outer layer surrounding a foam core (see next section). Since the solid polymer forms the exterior skin, parts have an excellent out-of-mold appearance and require little or no postfinishing operations. Different resins may also be combined in the same part to maximize cost/performance.

The multiple-component injection molding process with blowing agent allows the production of parts that are 5 to 30 percent lighter than compact injection molded parts.

Processing SF with Gas Blowing Agent

Nitrogen gas blowing agent, when introduced into a molten polymer, requires specialized equipment. Such equipment was extensively developed by Union Carbide Corp. and consists of a continuously running extruder, a gas inlet into the cylinder, one or more accumulators to hold the foam mixture, and a mold. All of these are connected by suitable pipes and one or more injection nozzles that feed the mold. The multiple-nozzle arrangement is necessary because of the limited flow length of the polymer and blowing agent mixture, and facilitates the use of multicavity molds and the making of large objects.

The extruder thoroughly mixes the gas and material, and feeds a prescribed volume of material and foam mixture into one or more accumulators, where it is kept under pressure to prevent premature expansion. When the proper volume of the mixture is reached, a valve opens, and a piston in the accumulator quickly forces the material into the mold. The stroke of the piston determines the volume of material delivered to the mold. The mold is only partially filled. At this point, the valve closes, and the expanding gas fills the mold and exerts pressure on the forming skin to prevent sink marks. With a high melt temperature of the polymer, rapid delivery of the material to the mold, and 25 percent of the circumference of the parting line devoted to equally spaced vents, a smooth surface finish can be attained (Fig. 29–28).

There are other processes either in the development stage or in use for specialized applications (Table 29–8). Most of them, including those described herein, involve patents, and

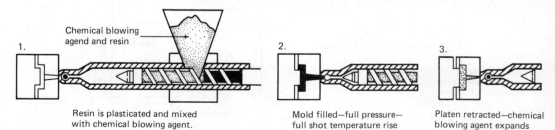

1.
Chemical blowing
agend and resin

Resin is plasticated and mixed
with chemical blowing agent.

2.
Mold filled—full pressure—
full shot temperature rise
to decompose chemical
blowing agent as melt
passes through nozzle.

3.
Platen retracted—chemical
blowing agent expands
resin to fill enlarged cavity.

Fig. 29-27   High pressure foam process.

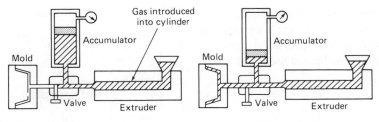

Fig. 29-28   Injection molding SF with nitrogen gas blowing agent.

the owners of such patents look for licensing arrangements. The patent question is another aspect of structural foam molding that requires attention and analysis before one makes a move toward application of the system of structural foam molding.

Table 29-8   Some of the different patented techniques for molding structural foam.

Union Carbide: Injection molding using extruder with blowing agent (usually inert nitrogen) and an accumulator. The mold cavity is "underfilled," which identifies this system as "low pressure" (most popular that was previously patented; latter patent cancelled).

USM: System using basically conventional type injection molding machines with expansion mold (or special mold).

ICI: Injection molding with two or more screw plasticizers to obtain integral skin; used where skin and core material can be of different materials.

Mobay (Bayer): Durometer process, in which a two-liquid-component urethane is injected into a closed mold; referred to as chemical reaction molding.

Allied Chemical: Similar system to conventional reciprocating high pressure screw machines except that after full shot load enters cavity, excess material escapes from the cavity, going back into a special manifold. This excess material is reinjected during the next shot.

Phillips Petroleum: Engelit low pressure process, which takes melted resin pellets from a revolving turntable with the blowing agent metered into an extruder/injection unit.

Cincinnati Milacron: Urethane foam that provides self-skinning, in fire retardant.

Hoover Universal: Special screw injection machine with specially designed mold that includes venting system.

Upjohn: Isoderm process which provides for a mix of two-part isocyanite materials.

Rubicon Chemicals (jointly owned by ICI and Uniroyal): Rubicast process which uses special integral skinning urethane foam.

Marbon: Use of ABS for expansion casting.

Hercules: Use of polypropylene bead with blowing agent for application in processes other than injection.

The patented Cashiers Structural Foam with counterpressure, which can practically eliminate the usual swirl finish associated with low pressure molding, is described in Fig. 29–29 (38).

Another important patented process, by Hoover Universal and Union Carbide, is called the structural-web molding technique (Figs. 29–30 through 29–34). The structural-web process is so named because of the part's interior configuration. The idea behind the process is to inject gas into a molten polymer in the mold such that the gas–polymer interface is deformed into a wavelike corrugation—using the principle of the hydrodynamic instability of viscous fingering. The structural-web process has molded such parts as painted tote boxes. It clearly has potential for applications where a high strength-to-weight ratio is desired. Economy of material recommends it for other applications (39).

The process consists of these steps:

- Passing molten plastic material into a mold cavity until it is partially filled.
- Injecting pressurizing gas (usually nitrogen) into the melt.
- Coordinating the gas-injection rate, pressure, and other variables so that the gas/polymer interface is deformed into a wavelike corrugation, and the movement of the gas/polymer interfacial flow front is divergent.
- Maintaining a positive pressure inside the part until it is self-supporting.
- Releasing the gas pressure so pressure inside the part is reduced to the atmospheric pressure.
- Removing the molded structural-web part from the mold.

Allied Chemical has a high pressure patented injection molding process for producing structural foam. In this process a standard injection molding machine is used with a spe-

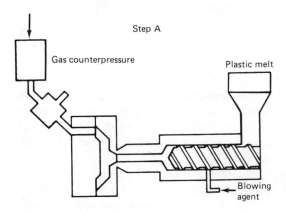

Step A

Gas counterpressure

Plastic melt

Blowing agent

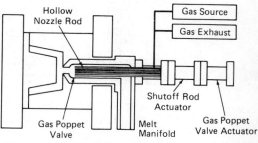

Structural Web Injection Nozzle

Hollow Nozzle Rod

Gas Source

Gas Exhaust

Shutoff Rod Actuator

Gas Poppet Valve

Melt Manifold

Gas Poppet Valve Actuator

Fig. 29-30 Hoover/Carbide web process features injection nozzle modified to handle both melt and gas (left).

cially designed mold. Plastic melt is permitted to egress (Fig. 29–35) from the fully packed mold, and thus the pressure within the mold is reduced, allowing foaming to occur.

Tooling

For low pressure foam applications, molds can be less expensive because of the lower clamp forces used. Molds for low pressure foam systems may be constructed from forged aluminum, supported cast aluminum, Kirksite, or steel. For foam molding on high pressure systems or modified conventional machines, steel molds are used because of the high clamp tonnages utilized. It is not recommended that Azodicarbonamide (Azobisformamide) be used with beryllium–copper molds because of

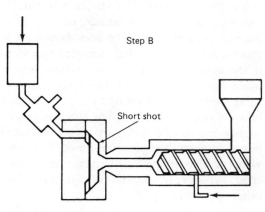

Step B

Short shot

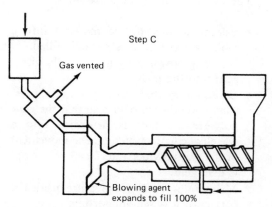

Step C

Gas vented

Blowing agent expands to fill 100%

Fig. 29-29 In counterpressure, the cycle begins with gas pressurization of the mold cavity (A), followed by injection of a short shot (B). Back pressure prevents the blowing agent from expanding until part skin forms, at which time it is vented (C), and expansion fills the mold 100%. Photo courtesy Cashiers Structural Foam.

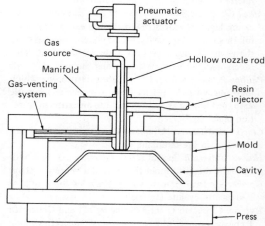

Pneumatic actuator

Gas source

Manifold

Gas-venting system

Hollow nozzle rod

Resin injector

Mold

Cavity

Press

Fig. 29-31 Resin is injected by an extruder, extruder-accumulator, or injection-molding equipment. Process is controlled by vertical movement of the hollow nozzle rod.

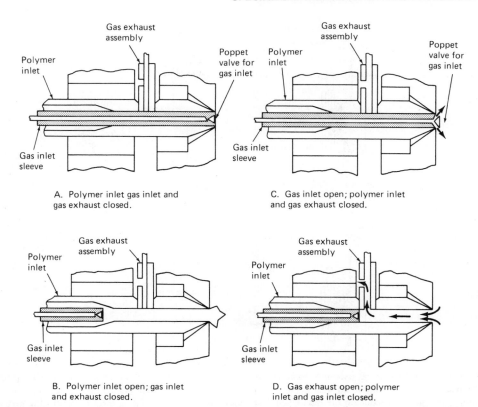

Fig. 29-32 Gas-delivery and exhaust system is shown horizontally (vertically in Fig. 29-31). The nozzle rod, or gas-inlet sleeve, has an integral poppet valve. In initial position, *A*, all passages are closed. Retracting the nozzle rod, *B*, allows resin to enter the mold. With the nozzle rod reinserted, *C*, gas pressure opens the poppet valve. Partially retracting the rod keeps resin inlet closed, *D*, but allows gas to exit through gas-exhaust assembly.

corrosion caused by prolonged runs with this blowing agent. Chrome plating of beryllium–copper molds has been used to reduce the degree of corrosion but has not proved to be the ultimate solution.

Molds should be designed for efficient cooling when molding foams to minimize cycle times. This is especially important where the part has thick sections.

Adequate venting of mold cavities is essen-

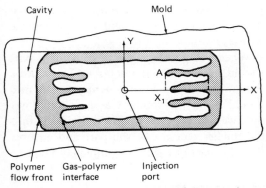

Fig. 29-33 Distances are measured from central gate (injection port) on coordinate axes. Length of finger, *A*-*B*, equals $x_2 - x_1$. Flow fronts progress to left and right along *x* axis.

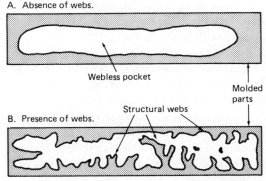

Fig. 29-34 Sketches based on photographs show absence of webs, *A*, when incipient interfacial flow instability is absent. Presence of incipient interfacial flow instability produces webs as at *B*.

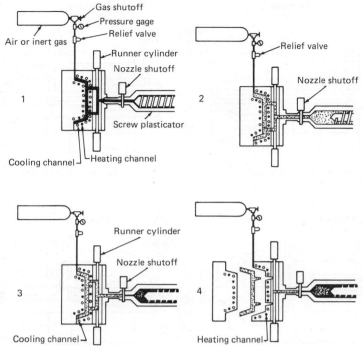

Fig. 29-35—The Allied Chemical patented structural foam process operates as follows: (1) the reciprocating screw has just advanced and filled the mold with polymer under full pressure; (2) after the skin of desired thickness has formed, the screw retracts, reducing the internal mold pressure as the excess melt egresses back into the manifold and plasticator cylinder; (3) as soon as the desired degree of foaming has occurred in the core of the molding, polymer egression is stopped by runner cylinders advancing to close off the egression ports; and (4) in the last step of the process, the molded part is removed from the mold.

tial to allow excess gas to escape and to enable complete filling of the mold as the plastic expands. Inadequate venting will result in unfilled parts and can also cause "burning" of the part in the vent area.

Usually vents from 0.005 in. to 0.010 in. are suitable, but actual experimentation with the mold using metal shims should be done to determine where vents should be placed and what their depth should be.

Sprues, runners, and gates are usually made as generous as possible, but should not be so large as to cause an increase in cycle time or in the amount of regrind generated.

Sprues are usually tapered (going from the machine nozzle to the tool) to help keep expansion of the melt to a minimum. Length should also be kept to a minimum so as not to interfere with cycles.

Runners should be generous to allow for fast injection rates. Care should be taken, however, that runners are designed so that

pressure will be maintained on the melt to prevent expansion. Runner systems for multi-cavity molds should be designed so that fill rates to each cavity are balanced.

Gates should be sized so that fast and complete fill of the part is facilitated. Usually width and thickness of gates are smaller than part thickness. This provides easy removal from the molded part, and no interference with cycle times occurs.

Whenever possible, gating of a foam part should be in the thinnest area. This allows the low-pressure melt to flow more easily into should be in the thinnest area. This allows the low-pressure melt to flow more easily into the thicker sections of the part and ensures that thin sections will be completely filled.

Start-up for Molding

Over 98 percent of all structural foam molding to date has been with the low pressure techniques, and it is likely that the major technique will continue to be low pressure.

Factors to consider when molding low pressure foamed parts may be inferred from the guidelines reviewed in Chapter 27. The following suggestions especially apply (34):

1. Injection pressure should be set high enough to enable the maximum injection speed obtainable. High-speed injection provides improved surface quality. Back pressure should be used (100–200 psi) for consistent, even filling during plastication. Screw speeds of 20–50 rpm are normally used.
2. Shot size should be adjusted to approximately 25 percent less than cavity volume. *Note:* Shot size setting should be such that the screw completely bottoms out during injection; i.e., no cushion is used.
3. Processing temperatures should be chosen that are consistent with the polymer and blowing agent being used. An increasing profile is preferred, with the rear zone temperature set lower than the decomposition point of the blowing agent. This ensures that blowing agent efficiency will not be lost by degassing through the hopper (for recommended blowing agent and processing temperatures, refer to Tables 29–5, 29–6, and 29–7).
4. Mold temperatures affect surface finish, skin thickness, and cycle time. Hot molds will yield a more glossy surface, thin skins, and longer cycle time. Cool molds, on the other hand, yield a duller finish and thicker skins with shorter cycles. Mold temperatures will normally range from 60°F to 140°F, but higher or lower temperatures are not uncommon. It is sometimes advantageous to include both heating and cooling channels in the mold to obtain an improved surface (heating) and short cycle times (cooling). Quenching the part in water immediately upon demolding may also be helpful in reducing post-expansion and cycle times. This is particularly true

for molded parts containing thick sections which would require a long cooling cycle.
5. Cycle times typically range from 60 to 120 seconds, but are dependent upon the polymer being foamed, part thickness, and mold temperature.
6. Venting should be determined by experimentation with the mold, using metal shims before cutting the mold.

Table 29–9 is a troubleshooting chart for foamed parts.

CO-INJECTION

Co-injection basically means that two or more different plastics are "laminated" together. These plastics could be the same except for color. When different plastics are used, they must be compatible in that they provide proper adhesion (if required), melt at approximately the same temperature, etc. Table 29-10). Two or more injection units are required, with each material having its own injection unit. The materials can be injected into specially designed molds—rotary, shuttle (Fig. 29-36), etc.

The term co-injection can denote different products, such as sandwich construction, double-shot injection, multiple-shot injection, structural foam construction, two-color molding, inmolding, etc. Whatever its designation, a sandwich configuration has been made in which two or more plastics are "laminated" together to take advantage of the different properties each plastic contributes to the structure.

This form of injection has been in use since the early 1940s, and in the past decade has become more commercial. Many different advantages exist; for example: (1) it combines performance of materials; (2) it permits use of a low-cost plastic such as a regrind; (3) it provides a decorative "thin" surface of an expensive plastic; (4) it includes reinforcements, etc. Co-injection molding is being redefined today in light of the approaches now available for molding multicomponent parts (automotive taillights, containers, business machine housings, etc.).

There are three techniques offered for multiple-component injection, called the one-channel, two-channel, and three-channel techniques. In the one-channel system the plastic melts for the compact skins and for the foam core are injected into the mold one after another by shifting a valve (Fig. 29-37). Because of the flow behavior of the plastic melt in the

Table 29-9   Troubleshooting structural foam parts (34).

CORRECTION OF FOAM DEFECTS

Mold not completely filled

Cause:
  a) Shot size too small.
  b) Insufficient blowing agent or inefficient use of blowing agent.

  c) Insufficient venting of mold.

Corrective Action:
  a) Increase shot size.
  b) Use additional 0.5% blowing agent or increase stock temperature.
  c) Increase size and number of vents if a and b do not correct fault.

Rough Surface

Cause:
  a) Mold temperature too low.
  b) Injection rate too low.
  c) Injection pressure too low.
  d) Poor resin flow.

Corrective Action:
  a) Increase mold temperature.
  b) Increase injection rate.
  c) Increase injection pressure.
  d) Increase stock temperature or use higher melt flow resin.

Post Mold Swelling of Parts

Cause:
  a) Cooling cycle too short.
  b) Mold temperature too high.
  c) Shot size too large.

Corrective Action:
  a) Increase cooling cycle time.
  b) Reduce mold temperature.
  c) Reduce shot size.
  d) If all of above fail, use post-mold quenching of part in water.

Density Too High

Cause:
  a) Shot size too large.
  b) Insufficient blowing agent or inefficient use of blowing agent.

  c) Blowing agent decomposing too early.

  d) Intrusion molding during plastication.

Corrective Action:
  a) Reduce shot size.
  b) Use additional 0.5% blowing agent or increase stock temperature.
  c) Lower temperature in rear zones or use higher temperature blowing agent.
  d) Install shut-off nozzle; increase screw forward time.

Cycle Too Long

Cause:
  a) Mold temperature too high.
  b) Stock temperature too high.

  c) Insufficient cooling of mold.

  d) Blowing agent level too high.

Corrective Action:
  a) Reduce mold temperature.
  b) Reduce stock temperature. (Do not go below decomposition temp. of blowing agent).
  c) Increase flow of cooling medium through mold. Use post-mold quenching.
  d) Reduce level of blowing agent.

Cell-Size too large—Non-uniform

Cause:
  a) Injection speed too slow.
  b) Melt viscosity too low.
  c) Density of part too low.
  d) Blowing agent decomposed too early in cylinder.
  e) Expansion taking place in cylinder or nozzle.

Corrective Action:
  a) Increase injection speed.
  b) Lower temperature profile.
  c) Increase shot size.
  d) Use nucleating agent.

  e) Use lower melt index resin.

  f) Install shut-off nozzle on machine.

Table 29-10 Compatibility of materials for co-injection.[a]

Materials	ABS	Acrylic ester acrylonitrile	Cellulose acetate	Ethyl vinyl acetate	Nylon 6	Nylon 6/6	Polycarbonate	HDPE	LDPE	Polymethylmethacrylate	Polyoxymethylene	PP	PPO	General-purpose PS	High-impact PS	Polytetramethylene terephthalate	Rigid PVC	Soft PVC	Styrene acrylonitrile
ABS	+	+	+				+	–	–	+		–		–	–	+	+	0	+
Acrylic ester acrylonitrile	+	+		+						–					0				+
Cellulose acetate	+		+	–															
Ethyl vinyl acetate		+	–	+				+	+			+		+			+	0	
Nylon 6					+	+	–	–				–				–			
Nylon 6/6					+	+	–	–	–			–				–			
Polycarbonate	+				–		+					–	0						+
HDPE	–		+	–	–		+	+	+	–	0								
LDPE	–		+	–	–		+	+	+	–		+						–	
Polymethylmethacrylate	+							–	–	+				–	0		+	+	+
Polyoxymethylene								–	–	+	–								
PP	–	–	+	–	–		0	+			–	+		–			–	–	
PPO													+	+	+				
General-purpose PS	–		+				–	–	–			–	+	+	+				
High-impact PS	–	0			–	–	0			0		–	+	+	+			–	
Polytetramethylene terephthalate	+															+			+
Rigid PVC	+		+							+		–					+	+	
Soft PVC	0		0				–			+		–		–	–		+	+	+
Styrene acrylonitrile	+	+					+			+				–	–	+	+	+	+

[a] + = good adhesion, – = poor adhesion, 0 = no adhesion, blank indicates no recommendation (combination not yet tested). The addition of fillers or reinforcements leads to a deterioration of adhesion between raw materials for skin and core.

Source: Battenfeld

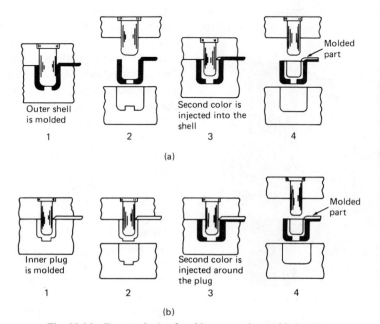

(a)

1 — Outer shell is molded
2
3 — Second color is injected into the shell
4 — Molded part

(b)

1 — Inner plug is molded
2
3 — Second color is injected around the plug
4 — Molded part

Fig. 29-36 Two methods of making two-color molded parts.

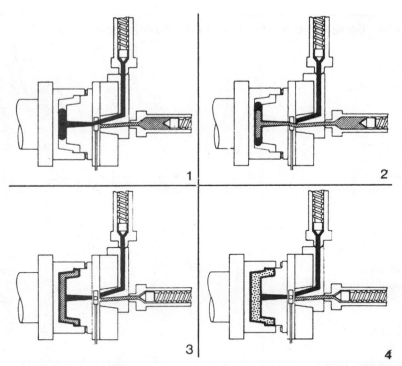

Fig. 29-37   In the one-channel co-injection system the sequence of mold filling starts with the skin being injected, then the core. In the third stage the skin polymer is injected again to clear the sprue and seal the skin on the injection side of the part. In this application a foam core is used. Up to stage three all melts have been injected at the conventional high pressure of injection molding. After the skin solidifies, the mold opens to a preset amount and permits the core to foam as shown in stage 4. Courtesy of ICI.

tool and because the first injected plastic for the compact skin cools off under the cooler mold surface, a closed compact skin and a core are formed under proper parameter set-

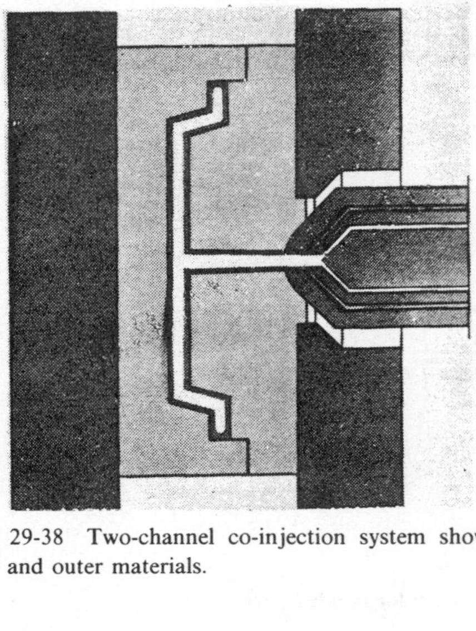

Fig. 29-38   Two-channel co-injection system showing core and outer materials.

tings. The thickness of the compact skin may be changed by varying the process parameters. This single-channel technique can incorporate either a solid or a foamed core. The technique's major advantage is to include a foamed-in-place core as described by Fig. 29-37. Thus this type of construction produces a structural foam (see earlier section in this chapter on structural foam).

The two-channel system (Fig. 29-38) allows the formation of the compact skin and the core material simultaneously. With this technique the thickness of the compact skin in the gate area can be easily controlled (a difference from the one-channel system).

The three-channel system (Fig. 29-39) allows simultaneous injection, using a direct sprue gating, of the compact skin and core (foamable or solid). The wall thickness of the compact skin may be influenced on both sides of the part. With this system the foamed core progresses farther toward the end of the flow

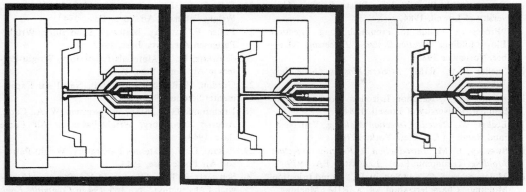

Fig. 29-39   The three-channel co-injection system simultaneously injects two different plastic melts. Courtesy of Battenfeld.

path, compared to the one-channel or two-channel technique. Also parts can be designed to be lighter in weight for the structural foam product.

In Fig. 29-40 a three-channel system is used to process three different plastics.

At present the markets are not ready for the co-injection structures because most designers are not familiar with their advantages. Machines and different parts are being molded, principally with the three-channel system (Fig. 29-39). This system can provide the designer with different approaches to meeting performance requirements at low cost.

REFERENCES

1. Rosato, D. V., Plastics Industry/Injection Molding Plastics Seminar, University of Lowell, 1984.
2. Sneller, J. A., "Need to Boast in Productivity? Try Specialized Injection," *Modern Plastics*, 59–59 (November 1983).
3. Mold With Rotation Licensing, Dow Chemical U.S.A., 2040 Dow Center, Midland, MI 48640.
4. Coor's Beer Bottle Development.
5. "No Off-The-Shelf Solutions Here: Continuous Molding," *Plastics World*, 64–70 (December 1980).
6. Buja, F. J., "Continuous" Rotary Injection Molding Machine, CRIMM B-Des., Inc., 104 Shale Dr., Rochester, NY 14615 (SPE-ANTEC, May 1984).
7. "Injection Molding Metals: Metals News," *Materials Engineering*, 9 (June 1983).

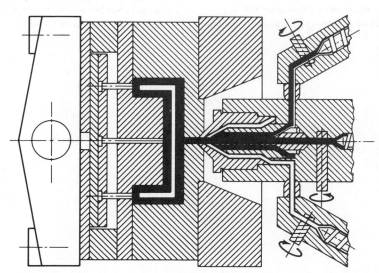

Fig. 29-40   A three-channel co-injection nozzle assembly developed by Billion that is injecting three different plastic melts.

8. Rosato, D. V., Blow Molding Plastic Seminar, University of Lowell, 1984.

9. "Processors' Guide to Troubleshooting Injection Blow Molding," *Plastics Design & Processing* (October/November 1983).

10. Sneller, J. A., "RIM," *Modern Plastics,* 66–68, July 1984.

11. Rosato, D. V., Reaction Injection Molding Plastic Seminar, University of Lowell, 1984.

12. Becker, W., *Reaction Injection Molding,* Van Nostrand Reinhold Co., New York, 1978.

13. Sweeney, F. M., *Introduction to Reaction Injection Molding,* Technomic Publ., Lancaster, PA, 1979.

14. Malguarnera, S. C., and Suh, N. P., "Liquid Injection Molding: Characterization of a RIM Machine," *Polymer Engineering & Science* (February 1977).

15. Manzione, L. T., and Osinski, J. S. (Bell Labs, Murray Hill, NJ), "Predicting Reactive Fluid Flow in RIM and RTM Systems," *Modern Plastics* (February 1983).

16. Lehnert, A. B., "New Technology in Reinforced Reaction Injection Molding (RRIM) Bulk Blending Tank Farms," Admiral Equipment Co./Upjohn Co., Akron, OH, 1983.

17. "RIM Materials Take Quantum Leap Forward," *Plastics Technology,* 37–44 (March 1983).

18. "Nylon-RIM Takes Off," *European Plastics News,* 4 (December 1983).

19. "Internal Mold Releases, Key to RIM Productivity," *Plastics Technology* (January 1983).

20. Farris, R. D., La Mare, H. E., Overcashier, R. H., and Gottenberg, W. G., "Structural Parts from Epoxy RIM Using Preplaced Reinforcements," *International Journal of Vehicle Design* (1984).

21. "Liquid Injection Molding Is Coming of Age," *Plastics Technology* (May 1983).

22. Peters, G. M. (Battenfeld, RI), "Exacting RIM Process Control Help Yield Higher-Quality Plastic Parts," *Plastics Engineering* (November 1983).

23. Society of the Plastics Industry, Structural Foam Division, 1984.

24. Fell, C. L., Lt., Air Corps, Aircraft Laboratory, Wright-Patterson Air Force Base, 1943.

25. Fuller, F. B., Maj., Materials Laboratory, Wright-Patterson Air Force Base, 1943.

26. Schwartz, R. T., Materials Laboratory, Wright-Patterson Air Force Base, 1943.

27. Gordon, H. R., Structures Unit, Naval Air Experimental Station, 1943.

28. Rheinfrank, G. B., Maj., and Norman, W. A., Capt., Aircraft Laboratory, Wright-Patterson Air Force Base, 1944.

29. Rosato, D. V., Materials Laboratory, Wright-Patterson Air Force Base, 1944.

30. Schwartz, R. T., and Rosato, D. V., "Structural Sandwich Construction," pp. 165–194 in *Composite Engineering Laminate Book,* edited by A. G. H. Dietz, MIT Press, Cambridge, MA, 1969.

31. Rosato, D. V., "Designing with Structural Foam," SPI Structural Foam Annual Meeting (April 1982).

32. Deslorieux, A. M., "Structural Foam," *Materials Engineering,* 41–45 (February 1983).

33. Wendle, B. C., *Engineering Guide to Structural Foam,* Technomic Publ., Lancaster, PA, 1976.

34. "Injection Molded Structural Foam & Compression Molding Using Celogen Blowing Agents," Uniroyal, Inc., ASP-5550.

35. Lacallade, R. G., "Blowing Agents for Foam Molding," Plastics Seminar, Univ. of Lowell, 1975.

36. Rosato, D. V., Reaction Injection Molding and Structural Foam Molding Plastics Seminars, University of Lowell, 1984.

37. "An End to Finishing Problems? Structural Foam," *Plastics World* (August 1981).

38. Colangelo, M., "Structural Foam: Where It's Headed," *Plastics Technology* (June 1983).

39. Olagoke, O., "Structural-Web Molding," *Plastics Engineering,* 25–28 (October 1983).

40. Hettinga, S., "A Review of a Few Basic Theories; Change Is Not Necessarily Progress," *European Plastics News,* 52–53 (October 1983).

Chapter 30
Competitive Processes

D. V. Rosato

Many fabricating processes are employed to produce plastic products. Many of them can compete directly with injection molding, particularly if only a relatively small quantity must be fabricated. For small quantities, competing processes include compression molding, rotational molding, casting, stampable reinforced plastics, etc. For large production runs, competitive processes include a combination of extrusion with thermoforming (e.g., for drinking cups), extrusion blow molding, die casting, reinforced plastics (thermosets), etc.

The ways in which plastics can be processed into useful products are as varied as the plastics themselves (see Chapter 18). While the processes differ, however, there are elements common to many of them. Which process to use depends upon the nature and requirements of the plastic to be processed, the properties required in the final product, the cost of the processing, speed, and the volume to be produced. Figures 30-1 and 30-2, with Tables 30-1 through 30-5, provide examples of the types and performance of a few processes. A simplified selection procedure for evaluating processes is given in Tables 30-6 through 30-8. More details on these as well as other processes follow (1–12).

BLOW MOLDING

See Chapter 29 (section on blow molding) for a review of the extrusion and injection mold-

ing processes used for blow molding. Blow molded parts cannot meet the tight tolerances achieved with conventional injection molding; however, blow molding permits the production of complicated hollow shapes. Sections of a complicated part could individually be injection molded, and a secondary operation could bond them together (adhesives, ultrasonics, spin welding, etc.); but the cost of such secondary operations must be carefully studied. Expect more action in blow molding with an extruder, which will make it competitive with conventional injection molding and injection blow molding, since it provides a basically lower cost of operation than the latter two processes.

CASTING

Resins for casting emerged a little over a half century ago, but formulations suitable for increasingly widespread use date back only about 30 years.

Casting may be used with both thermoplastics and thermosets to make products, shapes, rods, and tubes, by pouring a liquid monomer–polymer solution into an open or closed mold where it finishes polymerizing into a solid. Film and sheeting can also be made in this way by casting directly into a flat open mold, casting onto a wheel or belt, or precipitation in a chemical bath.

One essential difference between casting and molding is that pressure need not be used in

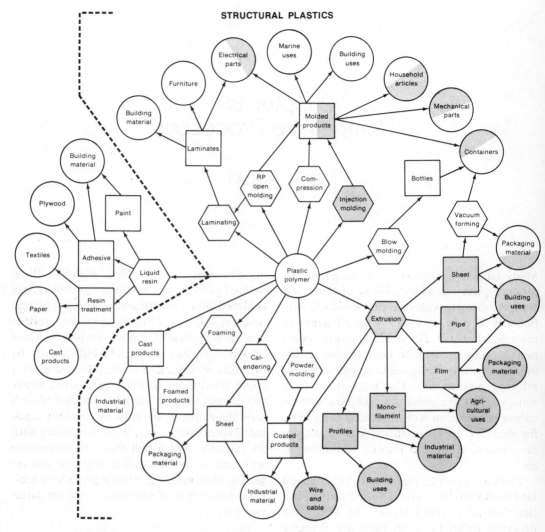

Source: Adapted from the United Nations, *Studies in the Development of Plastics Industries.*

Fig. 30-1   Interrelationship among methods of plastics processing, end products, and applications. Adapted from the United Nations, *Studies in the Development of Plastics Industries.*

casting (although large-volume, complex parts can be made by pressure-casting methods). Another difference is that the starting material is usually in liquid form rather than solid (such as pellets, granules, flakes, powder, etc.). A third is that the liquid is often a monomer rather than the polymers used in most molding compounds.

A variation on casting is known as liquid injection molding (LIM) and involves the proportioning, mixing, and dispensing of liquid components and directly injecting the resul-

tant mix into a mold which is clamped under pressure (see Chapter 29).

COLD FORMING

In contrast to the conventional processing methods for thermoplastics which take place with the material in the melt condition or with semi-finished product in the plastic state, new process methods starting from cold preforms or material heated below melt temperature but still in the solid state have recently been re-

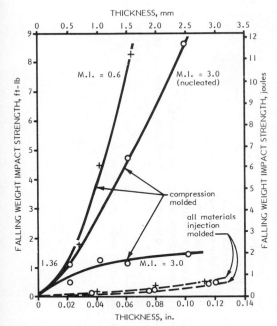

Fig. 30-2   Effect of processing conditions on the impact strength of several polypropylene homopolymers. (M.I. = melt index.)

ported on a number of occasions. These cold forming processes are chiefly suitable for thick-walled parts, since, as is well known, the cooling time increases as the square of the wall thickness. In addition the following advantages are achieved:

- Reduction in machine and mold costs by 65 to 75 percent as compared with injection molding,
- Improvement in impact strength by a factor of 10,
- Improvement in transparency,
- Elimination of finishing operations,
- Elimination of gate marks and weld-lines.

Table 30-1   Plastics consumption by processes.

Extrusion	36%
Injection	32%
Calendering	6%
Coating	5%
Compression	3%
Powder	2%
Others (includes extrusion & inj. blow mold).	16%

Cold forming can be performed at room temperature; with preheating the shaping forces required are considerably less. Plastics that have hitherto been successfully solid-phase-formed include HDPE, PP, ABS, PVC, PTFE, CAB, and polysulfone.

A further great advantage is that the processability is no longer adversely affected by high molecular weight. It is particularly important to take into account the springback or recovery forces. The temperature of the preform and of the mold must be controlled at optimum level. In this type of wrought processing the thinnest section of the finished shaping determines the pressing force required. If, for example, for a minimum wall thickness of 2 mm, 30 tons is required, then for 2.5 mm, 10 tons suffices.

The die in the conventional punch and die method can be replaced by a rubber pad. This method is mainly used for large area moldings, where pressures of about 300 bar are employed. As a rule cycle times amount to 20 to 40 seconds for each molding. Unlike drape formed moldings, deep-drawn articles are free from thin-walled corner areas.

COMPRESSION MOLDING

Compression molding is one of the oldest processing techniques for plastics. It is employed in both the production of finished molding and semi-finished products from thermosetting plastics and from thermoplastic materials, in the form of powder, chips, or granules.

In laminated plastics manufacture the prepared layers are stacked between bright polished platens and bonded together by means of heat and pressure; packs of thin thermoplastic film that are to be pressed to thicker sheet are similarly processed (Figs. 30-3 and 30-4).

Plastic molding powder, mixed with such materials or fillers as woodflour, cellulose, etc., to strengthen or give other added qualities to the finished product, is put directly into the open mold cavity. (It is often essential to preheat first by radio frequency dielectric techniques.)

Compression molding is the most common

Table 30-2 U.S. plastics processing plants. Type/number of plants (est.).

BLOW MOLDING	BONDING/ FASTENING	CALENDERING	CASTING/ ENCAPSULATING/ POTTING	COATING (EXCLUDING EXTRUSION COATING)	COMPOUNDING	COMPRESSION/ TRANSFER MOLDING	DECORATING AND FINISHING	EXTRUSION (INCLUDING COATING, FILM/SHEET, PROFILES AND OTHER)
2,396	8,833	1,790	5,577	6,271	4,659	3,592	6,796	8,035

FOAM PROCESSING (INCLUDING POLYSTYRENE, URETHANE AND OTHER)	INJECTION MOLDING (INCLUDING THERMOPLASTICS AND THERMOSETS)	LAMINATING	MACHINING	MOLD/ DIE MAKING	REINFORCED PLASTICS PROCESSING (INCLUDING FILAMENT WINDING, HAND LAY-UP MOLDING, MATCHED-DIE MOLDING, PULTRUSION, SPRAY-UP MOLDING AND OTHER)	ROTATIONAL MOLDING	THERMOFORMING
5,888	7,904	4,386	8,608	6,996	4,037	962	3,963

Note: Many plants perform more than one type of processing. Total number of plants in the United States that process plastics is estimated at about 20,000.

Table 30-3 Cost comparison of plastic parts based on different processes (Cost factor × material cost = Purchase price of part).

PROCESS	COST FACTOR	
	OVERALL	AVERAGE
Blow Molding	$1\frac{3}{8}$ to 5	$1\frac{1}{2}$ to 3
Calendering	$1\frac{1}{2}$ to 5	$2\frac{1}{2}$ to $3\frac{1}{2}$
Casting	$1\frac{1}{2}$ to 3	2 to 3
Centrifugal Casting	$1\frac{1}{2}$ to 4	2 to 4
Coating	$1\frac{1}{2}$ to 5	2 to 4
Cold Pressure Molding	$1\frac{1}{2}$ to 5	2 to 4
Compression Molding	$1\frac{3}{8}$ to 10	$1\frac{1}{2}$ to 4
Encapsulation	2 to 8	3 to 4
Extrusion Forming	$1\frac{1}{16}$ to 5	$1\frac{1}{8}$ to 2
Filament Winding	5 to 10	6 to 8
Injection Molding	$1\frac{1}{8}$ to 3	$1\frac{3}{16}$ to 2
Laminating	2 to 5	3 to 4
Match-Die Molding	2 to 5	3 to 4
Pultrusion	2 to 4	2 to $3\frac{1}{2}$
Rotational Molding	$1\frac{1}{4}$ to 5	$1\frac{1}{2}$ to 3
Slush Molding	$1\frac{1}{2}$ to 4	2 to 3
Thermoforming	2 to 10	3 to 5
Transfer Molding	$1\frac{1}{2}$ to 5	$1\frac{3}{4}$ to 3
Wet Lay-Up	$1\frac{1}{2}$ to 6	2 to 4

Table 30-4 Examples of manufacturing methods and products.

Compression molding	wiring devices, closures, sheets
Expansion bead molding	ice chests, packaging
Extrusion blow molding	hollow objects, bottles
Extrusion	sheets, rods, tubes and profiles
Fluidized bed	plastic-coated metal parts
Forging	thermoplastic uniform thick sections
Hand layup	boats, auto bodies, structural sections
Injection molding	thermoset and thermoplastic products
Injection blow molding	bottles and simple shapes
Liquid resin casting	tanks, novelties, encapsulations
Reaction impingement molding	auto bodies and high-volume large parts
Rotational molding	tanks, balls, housings, dolls
Spray-up molding	furniture, boats, automobile components
Structural foam molding	business machines, beams, sheets, furniture
Slush molding	novelties, balls, dolls
Sheet thermoforming	
Vacuum forming	blister packages, domes, trays
Pressure forming	furniture, signs, domes
Trapped sheet forming	boxes, machine covers, furniture
Steam pressure forming	ping pong balls, novelties, dolls
Transfer molding	complex thermoset pieces, delicate inserts

Table 30-5   Process considerations.

MATERIAL	EXTRUSION	BLOW MOLDING	FIBER	FILM	FOAM	INJECTION MOLDING	COMPRESSION MOLDING	THERMOFORMING
ABS	X	X	—	—	X	X	X	X
SBR	X	X	—	—	X	X	X	X
EVA	X	X	—	X	X	X	X	—
Melamine	—	—	—	—	—	—	X	—
Nylon	X	X	X	—	X	X	X	X
Phenolic	—	—	—	—	—	—	X	—
PC	X	X	—	—	X	X	X	X
LDPE/HDPE	X	X	—	X	X	X	X	X
PET	—	—	X	X	—	—	—	—
Acetal	X	X	—	—	X	X	X	—
Methyl pentene	X	X	—	—	—	X	X	X
PP	X	X	—	X	X	X	X	X
GPS	X	X	—	—	X	X	X	X
TFE	—	—	—	—	—	—	X	—
Urethanes	—	—	X	X	—	—	—	—
PVC	X	X	—	X	X	X	X	X
Ureas	—	—	—	—	—	—	X	—
PVC/PVA	X	X	—	X	—	X	X	X
Acrylic	X	—	X	—	—	X	X	X

method of forming thermosetting materials. Until the advent of injection molding, it was the most important of plastics processes.

EXTRUSION

Extrusion basically provides a process for producing large quantities of products where capital equipment and operating costs are low when compared to other "large quantity" processes such as injection molding (or any process). Obviously for the shapes that are feasible in injection molding, cost with IM is lower (as reviewed throughout this book).

Extrusion molding is the method employed to form thermoplastic materials into continuous sheeting, film, tubes, rods, profile shapes, and filaments, and to coat wire, cable, and cord.

In extrusion, dry plastic material is first loaded into a hopper, then fed into a long heating chamber through which it is moved by the action of a continuously revolving screw. At the end of the heating chamber the molten plastic is forced out through a small opening or die with the shape desired in the finished product. As the plastic extrusion

comes from the die, it is fed onto a conveyor belt where it is cooled, most frequently by blowers or by immersion in water (Fig. 30-5).

In the case of wire and cable coating, the thermoplastic is extruded around a continuing length of wire or cable that, like the plastic, passes through the extruder die. The coated wire is wound on drums after cooling.

In producing wide film or sheeting, the plastic is extruded in the form of a tube. This tube may be split as it comes from the die and then stretched and thinned to the dimensions desired in the finished film.

In a different process, the extruded tubing is inflated as it comes from the die, the degree of inflation of the tubing regulating the thickness of the final film. In this process, known as blown film manufacturing, the extruded tubing of film is inflated with air as it comes from the die to form a bubble of the volume necessary to produce film of the desired width and thickness. The bubble is then slit and stretched out.

More recently, there has been interest in still another variation on extrusion that involves the simultaneous extrusion, or co-ex-

PROCESS	DESIGN FLEXIBILITY	STRUCTURAL INTEGRITY	SECONDARY OPERATIONS	RELATIVE TOOLING COST	ASSEMBLY FLEXIBILITY
Structural foam molding	• Due to low pressures, significant design flexibility possible (parts consolidation, etc.) • High rigidity allows for high load-bearing structural members • No sink marks with integral function	• Good structural integrity • Low-molded-in stress provides low warp, dimensionally stable parts	• Sprue removal • Painting req. for appearance surfaces	• Lower tooling cost; aluminum tools possible	• Vibration welding, ultrasonic bonding, self-tapping screws, ultrasonic inserts, adhesive bonding possible • Many parts can be integrally molded
Injection molding	• some flexibility possible, but due to high pressures, large complex parts not cost effective • Ribs required for high load-bearing parts • Sink marks in thick sections	• Good structural integrity	• Sprue removal • Class A finish	• Higher pressures req. expensive steel tools: high-strength, pre-hardened	• In thermoplastics, vibration welding, ultrasonic bonding, self-tapping screws, ultrasonic inserts and adhesive bonding possible • Parts can be consolidated
Sheet molding compound	• Fiber orientation and resin-rich areas may occur in complex, load-bearing areas • Lower fatigue strength limits complex, dynamic parts • Limited deep draws on complex/large surfaces	• Possible nonuniform physical properties • Lower impact strength	• Deflashing • Large or small openings must be trimmed or cut out	• Steel tools required	• Thermoset materials: requires molded-in inserts
Sheet metal	• Only simple shapes and contours possible • Requires multiple dies for part complexity • Inferior dimensional control	• Minimal integral component strength due to multiple component assembly	• Multiple assembly operations: drilling, tapping, welding	• Low cost tooling • Complex deep-draw dies etc. • High piece part cost	• Screws, nuts, bolts, rivets, welding • Parts consolidation nearly impossible
Die casting	• Limited complex-part capability • Large parts are heavier	• Good structural integrity • Lower impact strength	• Trim dies required • Machining of critical surfaces	• High tooling cost tool maintenance required due to potential wear damage	• Hardware assembly
Reaction injectin molding	• Good design flexibility due to low pressures but large complex structural parts not feasible due to lower material properties • Thicker sections required • Batch process	• Lower properties can prohibit complex high-stressed features • Lower impact and creep resistance	• Flashing must be trimmed	• Low tooling cost	• Thermoset process such that thermal fastening techniques not possible

Table 30-7 How structural foam compares with five other processes.

	FOAM VS. SHEET METAL	FOAM VS. DIE CASTING	FOAM VS. SHEET MOLDING COMPOUND	FOAM VS. HAND LAYUP FIBERGLASS	FOAM VS. INJECTION MOLDING
ADVANTAGES	1. Fabrication economy: less assembly time; tighter dim. tolerances; increased product integrity; less final product-inspection time. 2. Fewer parts required for assembly. 3. Dent resistance. 4. Elimination of oil canning. 5. Greater design freedom. 6. Better sound damping. 7. Reduced damage from shipping. 8. Reduced tooling costs for complex configurations.	1. Much lower tooling costs. 2. Longer tool life, lower maintenance. 3. No trim dies required. 4. Lighter weight. 5. Higher impact resistance. 6. Better sound damping. 7. Better strength to weight capability. 8. Better impact resistance.	1. Uniform physical properties throughout the part. 2. Warping & sink marks reduced or eliminated. 3. No resin-rich areas to cause configuration problems. 4. Higher impact resistance. 5. Greater inherent structural capabilities. 6. Lower shipping costs. 7. Large parts more economical. 8. Lower tooling costs. 9. Better sound damping.	1. More consistent part reproduction. 2. Lower labor. 3. Simplified assembly. 4. Better dimensional stability. 5. More design freedom. 6. More uniform physical properties. 7. Better sound damping.	(Many process similarities exist) 1. Flexibility for functional engineering. 2. Better low to medium volume economics. 3. Lower tooling costs. 4. Better large-part capability. 5. Better sound damping. 6. Lower internal stresses. 7. Sink marks reduced or eliminated. 8. Inherent structural strength.
LIMITATIONS	1. Smaller variety of finishes available, such as chrome or baked enamel. 2. No R.F.I. and grounding capabilities. 3. Harder to retrofit to frame or skins. 4. Thicker wall. 5. Higher tool costs than with brake-forming.	1. No heat sink capabilities. 2. No R.F.I. and grounding capabilities. 3. Fewer available finishes for cosmetic appearance. 4. Higher finishing costs. 5. Thicker walls. 6. Possible internal voids.	1. Increased finishing costs (surface swirl). 2. Heat distortion. 3. Thicker wall. 4. Lower physical properties. 5. Possible internal voids.	1. More prone to heat distortion. 2. Poorer economies of part size vs. quantity. 3. Thicker walls. 4. Higher tooling costs.	1. Poorer surface finish. 2. Application of cosmetic detail for appearance parts. 3. Longer cycle time. 4. Thicker walls. 5. Poorer high-volume economics. 6. Less equipment available for various shot sizes.
	Structural foam offers potential savings of 50% or more	Structural foam offers potential savings of 50% to 30%.	Structural foam offers potential savings up to 30%.	Structural foam offers potential savings of 50% or more.	Structural foam offers potential savings of 15% to 20% (depending on unit

Table 30-8 Economic comparisons of plastics processes.

	STRUCTURAL FOAM	INJECTION MOLDING	SHEET MOLDING COMPOUND
Typical minimum number of parts vendor is likely to quote on for a single setup	250 (using multiple nozzle equip. with tools from other sources designed for the same polymer and ganged on the platen.)	1000 to 1500	500
Relative tooling cost, single cavity	Lowest. Machined aluminum may be viable depending on quantity required.	20% more. Hardened steel tooling.	20% to 25% more. Compression-molding steel tools.
Average cycle times for consistent part reproduction	2 to 3 minutes ($\frac{1}{4}$ in. nominal wall thickness)	40 to 50 seconds	$1\frac{1}{2}$ to 3 minutes
Is multiple-cavity tooling approach possible to reduce piece costs?	Yes	Yes. Depends on size and configuration, although rapid cycle time may eliminate the need.	Not necessarily. Secondary operations may be too costly and material flow to difficult.
Are secondary operations required except to remove sprue?	No	No	Yes, e.g., removing material where a "window" is required. (Often done within the molding cycle.)
Range of materials that can be molded.	Similar to thermoplastic injection molding.	Unlimited; cost depends on performance requirements.	Limited; higher cost
Finishing costs for good cosmetic appearance	40 to 60 cents per sq ft of surface (depending on surface-swirl conditions)	None, if integrally colored; 10–20 cents per sq ft if painted.	None, if secondary operations such as trimming are not required. Otherwise 20 to 30 cents per sq ft of surface.

trusion, of multiple molten layers of plastic from a single extrusion system. As used in the marketplace, co-extrusion has been adapted to the production of products such as packaging films that incorporate in a single film structure several layers of different plastics, each offering varying degrees of moisture resistance, gas barrier properties, adhesive qualities, economics, etc.

FOAM MOLDING

The manufacture of foam plastics parts cuts across most of the processing techniques cov-

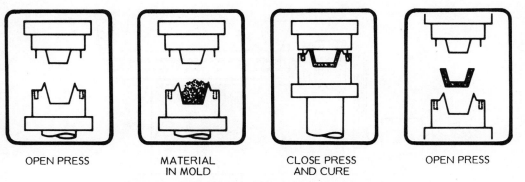

OPEN PRESS   MATERIAL IN MOLD   CLOSE PRESS AND CURE   OPEN PRESS

Fig. 30-3 Compression molding cycle, single-cavity mold.

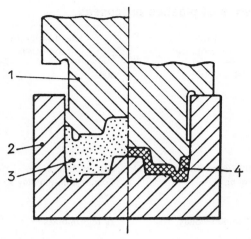

Fig. 30-4 Compression molding with a positive mold: (1) upper half of mold (top force), (2) lower half of mold (die), (3) loading cavity, (4) finished molding volume.

ered in this section. Foams can be used in casting, calendering, coating, rotational molding, blow molding, even injection molding (as reviewed in the structural foam and co-injection sections of Chapter 29), and extrusion. Typical requirements in such instances are for the incorporation of blowing agents in the resin that decompose under heat to generate the gasses needed to create the cellular structure and for various controls to accommodate the foaming action.

There are, however, some techniques unique to foamed plastics. In working with expanded polystyrene beads, for example, to produce cups, picnic dishes, etc., various "steam-chest" molding methods are used. The

application of steam causes the beads to expand and fuse together (Fig. 30-6).

In working with urethene foams, it is possible to use spray guns or mixing tanks.

MECHANICAL PROCESSING

"Mechanical processing" is an overall designation for processing used to machine, cut, seal, and form plastics. Mechanical processing can be used for prototyping, in which the machined part generally does not provide the properties achieved in molding, as a means to conduct field tests, etc. (For small quantity requirements it may be the most economical method. For details on prototyping injection molded parts, see Chapter 9.) Forming involves different techniques, such as simple bending and joining of plastic parts, hot stamping, and the major method called thermoforming (discussed later in this chapter).

Plastics, whether thermoplastic or thermosetting, can be machined by the same techniques used for processing wood or light alloys. They can be turned, milled, drilled, nailed, screwed, cut with shears, sawn with a saw or, for expanded materials, a hot wire. They can be finished superficially with rasps, planes, buffing machines, etc. They can be glued with particular adhesives selected according to the type of resin and purposes of the operation. There are pressure, aqueous emulsion, solution, two-component, and thermowelding adhesives.

As regards welding, certain polymers can

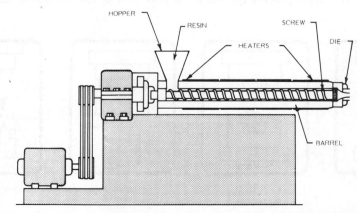

Fig. 30-5 Extrusion. Resin feeds into heated barrel and is forced by screw through die.

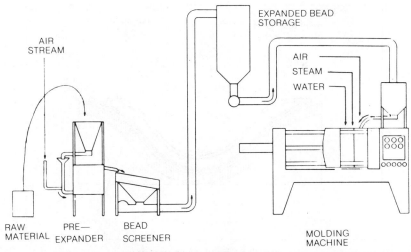

AIR
STREAM

EXPANDED BEAD
STORAGE

AIR

STEAM

WATER

RAW
MATERIAL

PRE—
EXPANDER

BEAD
SCREENER

MOLDING
MACHINE

Fig. 30-6 Styrene foam molding. In this operation, the expandable beads, containing a blowing agent, are pre-expanded with steam, then screened to remove large clumps. The expanded beads are next blown into a storage hopper and allowed to dry and stabilize. From here, they feed into the final mold where steam is again used to complete expansion of the beads so that they fill the mold and fuse together. Water is used for cooling, prior to opening the mold and removing the finished foamed styrene part.

be treated with normal welding torches, heated tools, a hot bar (for film), and ultrasonically. The starting semi-finished products may be slabs, sheets, laminates, films, solid or perforated bars, cylinders, or tubes. However, some processing suggestions should be borne in mind: for turning, in view of the low thermal conductivity of these materials, an efficient cooling system is recommended (with emulsified oils, pure water, or compressed air) in order to avoid harmful local heating; for drilling, it is advisable to advance slowly without trying to force the pace; for milling, use soft metal high-speed millers (400 m/min), with slow feed and abundant cooling; for saw-

ing, band saws are preferable. In some resins, thermosetting types in particular, the surface film has a precise physical function, and its destruction may release internal stresses that will damage the material.

REACTION INJECTION MOLDING

Details on this molding process are reviewed in Chapter 29.

REINFORCED PLASTICS PROCESSING

Like foam processing, RP (reinforced plastics) cuts across almost all processing techniques (an example is shown in Fig. 30-7). Reinforced plastics are composites in which resins (acting as a binder material) are combined with reinforcing materials (usually in a fibrous form) to produce products that have exceptional strength-to-weight ratios and outstanding physical properties. The resins may be either thermosets or thermoplastics. However, the thermosetting resins, which were the first plastics to be adapted to this concept (polyester resins, in particular), dominate the field.

Reinforced thermoplastics can be injection molded, rotationally molded, or extruded on conventional equipment. There are even rein-

Ⓐ MALE MOLD (ALUMINUM)

Ⓑ PLASTIC MOLD

Ⓒ TOGGLE CLAMPS

Ⓓ INJECTION PROBE

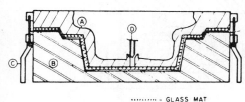

·········· - GLASS MAT
━━━━ - POLYESTER RESIN

Fig. 30-7 Resin injection: Cold Molding.

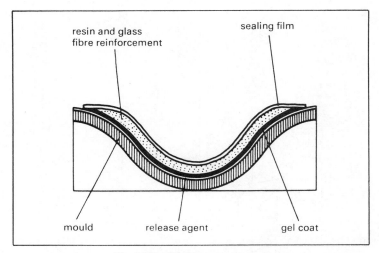

Fig. 30-8   Hand lay-up process.

forced thermoplastics sheets that can be "cold" stamped into shape using matching metal molds that form the parts. It is called cold stamping because the molds are kept at or slightly above room temperature. The sheets, however, must be preheated. (More details on cold stamping are given later in this chapter.)

It is also possible to use modified injection molding techniques with reinforced thermoset resins, although, by and large, the processing of reinforced thermoset resins has developed a technology all its own. For example, it is possible to process these reinforced plastics into extremely large parts (e.g., boat hulls,

storage tanks, etc.), using little or no pressure. One such system is known as hand lay-up (Fig. 30-8) and simply involves the manual placement of the reinforcing fibers (in the form of fabrics, woven roving, or mat) into a mold, impregnating the lay-up with liquid resin, squeezing out entrapped air with a squeegee, and then allowing the part to cure, with or without heat, into a finished product.

Most high-volume reinforced plastics processes in use today, however, do involve the application of pressure. The most popular of these is known as matched metal die molding and is basically a compression molding system (Fig. 30-9) in which the reinforced thermoset

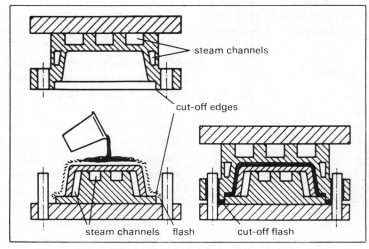

Fig. 30-9   Compression molding.

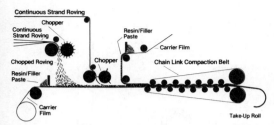

Fig. 30-10 Method of manufacturing sheet molding compound (SMC) with directional properties.

plastics are placed between matching heated molds and held under pressure until the fiber-filled resin polymerizes (i.e., hardens into a solid). Where differences do exist, it is in the way in which the reinforced plastic material is fed into the mold.

The three available forms include: (1) pre-forms shaped from the fibers to the approximate shape of the parts to be made and placed into the mold (more uniformity in product, less waste); (2) bulk molding compounds made up by mixing resin fibers, catalyst, etc., together into a puttylike mass that can be placed in the mold as is or extruded into a ropelike form for easier handling; and (3) sheet molding compounds (SMC), in which fiber, resin, and other ingredients are precombined into a sheet form that can easily be loaded into the mold, either manually or with automatic techniques (Fig. 30-10).

Other unique reinforced plastics processing methods include: pultrusion, a method for making continuous shapes (e.g., stair railings) by pulling resin-impregnated fibers through shaping dies and curing operations (it's the reinforced plastics industry's counterpart to thermoplastic extrusion), and filament winding (Fig. 30-11), a method for making cylindrical shapes (e.g., rocket motor cases) by winding resin-impregnated fibers around a mandrel, curing the part, and removing the mandrel (11).

Reinforced plastic products now consume approximately 6 percent of all plastics. No product yet marketed is produced in quantities on the order of the typical mass-produced injection molded products; however, changes will occur. It is possible to include in an injection mold reinforcements that have all types of directional properties, and then to inject

the plastic. This idea is not new; it has been used to a limited extent since the 1940s. The different reinforcements discussed in this section (and others) are applicable. The industry is already involved in using reinforced injection molding compounds made from thermoplastics or thermosets. What lies ahead is to take advantage of significantly increasing mechanical properties and developing specific directional properties. The review in this section can give you ideas about "marrying" reinforced plastics fabrication with injection molding. Figures 30-12 through 30-18 will provide additional ideas, including processes that started in 1944.

STAMPABLE REINFORCED THERMOPLASTICS

This technique can be considered a part of the overall reinforced plastics (RP) processing

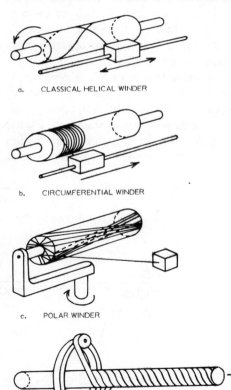

a. CLASSICAL HELICAL WINDER

b. CIRCUMFERENTIAL WINDER

c. POLAR WINDER

d. CONTINUOUS HELICAL WINDER

Fig. 30-11 Schematic representation of basic methods and types of filament placement.

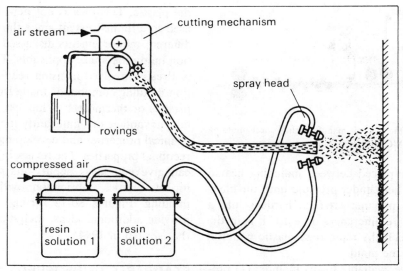

Fig. 30-12   Spray molding process.

industry. As mentioned, the RP industry is using practically all thermoset resins, whereas the stampable reinforced plastics are thermoplastic resins. Most of the work to date has been with nylon 6 and polypropylene; however, many others can be used.

These stampable plastics can achieve weight and/or cost reduction in parts that conform to a stampable shape. This technique principally competes with metal stamping. Plastic stamping differs from most metal fabrication in that it is a flow process; a one-step process

produces a completely finished part such as an automotive engine oil pan, etc.

ROTATIONAL MOLDING

Essentially, rotational molding consists of charging a measured amount of resin into a warm mold which is rotated in an oven about two axes. Centrifugal force distributes the plastic evenly throughout the mold, and the heat melts and fuses the charge to the shape of the cavity. After the mold is removed and

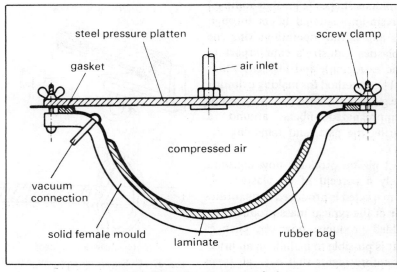

Fig. 30-13   Pressure bag method.

MODIFIED VACUUM BAG MOLDING

- (A) FORMED PVA BAG
- (B) FRAME
- (C) FIXTURE
- (D) SEAL

VAC

TRAP

⌒⌒⌒ GLASS
====== RESIN
xxxxxxx BLEEDER

Fig. 30-14 Modified vacuum bag molding.

cooled, the finished part is extracted. Rotational molding is particularly suited to the production of very large parts, size being limited only by the capacity of the oven (Figs. 30-19 and 30-20).

The advantages of this type of molding are: low mold cost, strain-free parts, uniform wall thickness, and use of cross-linked PE to mold large parts (20,000 gallons and larger).

Closed and open top hollow articles are made by this process from polymeric material in paste or powder form. The processing of

pastes, principally vinyl plastisols, for the manufacture of beach balls, floating animals, and other toys, as well as for industrial items, has long been known. The processing of plas-

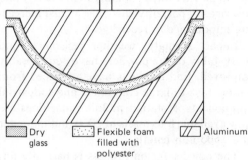

| Dry glass | Flexible foam filled with polyester | Aluminum |

Fig. 30-17   Foam reservoir molding.

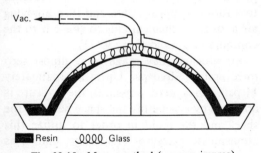

Vac.

■ Resin   ԼԼԼԼ Glass

Fig. 30-15   Marco method (vacuum impreg.).

(A) Male mold (aluminum)
(B) Plastic mold
(C) Jacks
(D) Clamps

Oroglas Dr

▨ Fiberglass reinforced plastic

Fig. 30-16   Squeeze molding.

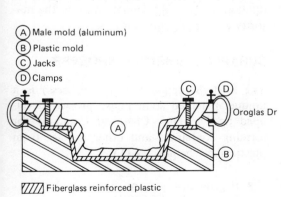

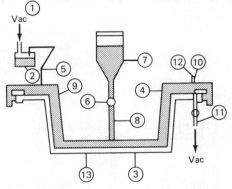

Vac

Vac

1 Vacuum 27° Hg.
2 Resin trap
3 Lower mold
4 Upper mold
5 Vacuum line to resin suction
6 Resin cut off valve
7 Resin supply reservoir
8 Injection line
9 Resin suction channel
10 Rubber gasked
11 Vacuum 5° Hg.
12 Vacuum mold closing channel
13 FRP molding

Fig. 30-18   Vacuum injection molding or resin transfer molding.

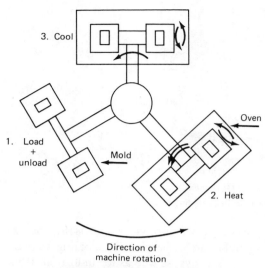

Fig. 30-19   Rotational molding.

tics in powder form—mainly high-density and low-density polyethylene—has gained increasing importance in recent years.

Impact strength, weather resistance, and workability of the powders have considerably improved. End-products have advanced from simple toys to large containers and industrial components. Rotation molding machines are now available for vessels of 2000 mm diameter and 2500 mm length.

The powder fusion process is basically unchanged over recent years. Thermoplastic materials in powder form—principally LDPE,

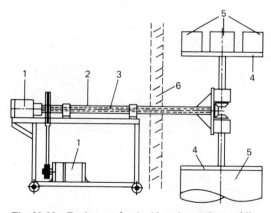

Fig. 30-20   Equipment for double-axis rotation molding: (1) driving motors and variable speed transmission; (2) hollow (outer) shaft; (3) inner shaft; (4) mold platforms; (5) molds; (6) wall of heated chamber. These machines can have single or multiple platforms to hold molds; here two platforms are used.

HDPE, and plasticized PVC—are charged in preweighed amounts into one half of the mold; the two mold halves are then clamped together and rotated in two planes at right angles to one another in a heated chamber.

After a predetermined melting time, the rotor is cooled. The parts can then be taken from the molds, and a new cycle begun. Figure 30-20 shows an installation with two mold platforms.

THERMOFORMING

This process can produce single products in small quantities or multiple-cavity formed parts in large amounts equivalent to output quantities in injection molding, with equipment that has higher output rates. This technique is very competitive with injection molding for certain size and shape products. Also this technique is competitive with blow molding hollow parts; two halves can be thermoformed, followed by secondary operations of bonding.

Thermoforming of plastic sheet has developed rapidly in recent years. Basically, this process consists of heating thermoplastic sheet to a formable plastic state and then applying air and/or mechanical aids to shape it to the contours of a mold (Fig. 30-21).

Air pressure may range from almost zero to several hundred psi. Up to approximately 14 psi (atmospheric pressure), the pressure is obtained by evacuating the space between the sheet and the mold in order to utilize this atmospheric pressure. This range, known as *vacuum forming,* will give satisfactory reproduction of the mold configuration in the majority of forming applications.

SCRAPLESS FORMING PROCESS

The SFP (Scrapless Forming Process) is a unique, patented solid phase process developed by The Dow Chemical Company for forming containers and other articles from plastic sheet or powder (13).

- It generates no trim scrap.
- It is used with most thermoplastics.

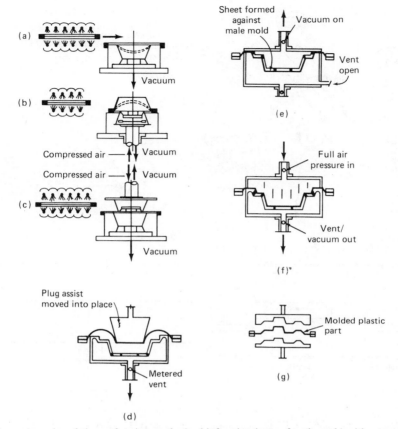

Fig. 30-21   Some examples of thermoforming methods: (a) forming into a female mold without prestretching; (b) forming over a male mold with prestretching; (c) forming into the female mold with prestretching; (d) billow/plug assist forming; (e) vacuum snap back forming; (f) pressure forming; (g) mechanical forming with matched mold.

• It is used with both single and multilayer sheet structures.

• It provides a high degree of molecular orientation, resulting in improved part toughness and stress crack resistance.

• It is a "solid phase" process, forming the plastics at temperatures below their melt temperature.

• It is a high-speed process (less heat in; less heat to be removed).

• The required equipment is available, worldwide, from a major plastics fabricating machinery supplier.

• It combines the advantages of thinwall injection molding and thermoforming— with none of their disadvantages—and with excellent material, fabrication, and end-product economy.

The process for forming these parts is shown in Fig. 30-22. Effects of forming techniques on physical properties are shown in Fig. 30-23. (Campbell Soup using process.)

TRANSFER MOLDING

Transfer molding is most generally used for thermosetting plastics. This method is like compression molding in that the plastic is cured into an infusible state in a mold under heat and pressure. It differs from compression molding in that the plastic is heated to a point of plasticity before it reaches the mold and is forced into a closed mold by means of a hydraulically operated plunger (Figs. 30-24 and 30-25) resulting in lower cavity pressure.

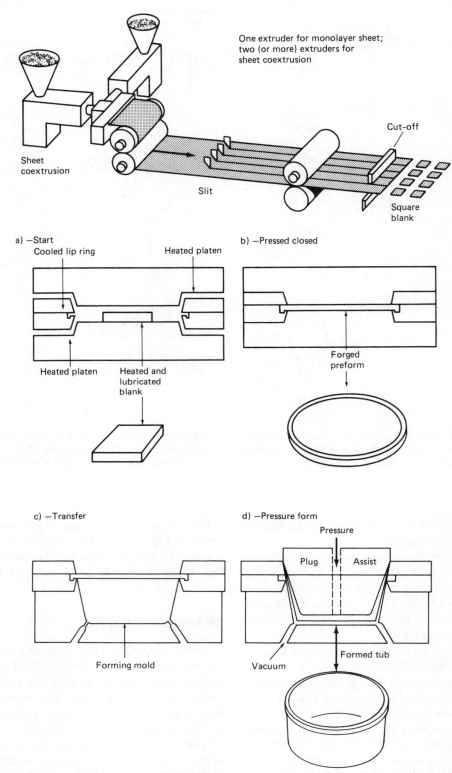

One extruder for monolayer sheet; two (or more) extruders for sheet coextrusion

Sheet coextrusion

Slit

Cut-off

Square blank

a) —Start
Cooled lip ring
Heated platen
Heated platen
Heated and lubricated blank

b) —Pressed closed
Forged preform

c) —Transfer
Forming mold

d) —Pressure form
Pressure
Plug
Assist
Vacuum
Formed tub

Fig. 30-22   Scrapless Forming Process (13) technique starts with extruding a sheet followed by basically four steps. Controlled biaxial orientation of properties is obtained.

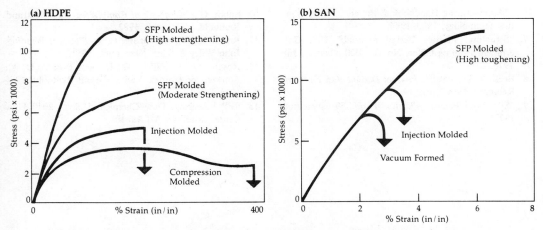

Fig. 30-23 Effects on physical properties of molded SFP, compression, and injection techniques.

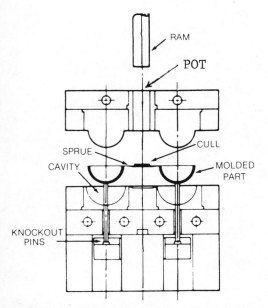

Fig. 30-24 In transfer molding the plastic is transferred from a pot through runners and gates into cavities retained in a closed heated mold.

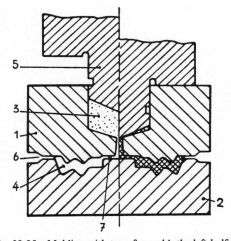

Fig. 30-25 Molding with transfer mold; the left half has plastic in the fill space or pot (3), and in the right side melt is transferred into the cavity. Parts of mold are: (1) center part of mold, (2) lower part of mold (die), (3) fill space or pot, (4) finished molding volume, (5) upper part of mold (transfer plunger), (6) venting channel, and (7) transfer channel.

Transfer molding was developed to facilitate the molding of intricate products with small deep holes or numerous metal inserts. The dry molding compound used in compression molding sometimes disturbs the position of the metal inserts and the pins that form the holes. The liquefied plastic material in transfer molding flows around these metal parts without causing them to shift position.

REFERENCES

1. Rosato, D. V., Plastic Industry—Fundamentals Plastic Seminar, University of Lowell, 1984.
2. "Plastics," International Plastics Selector, Inc., 1983.
3. "The Story of the Plastics Industry," Society of Plastics Industry, May 1977.
4. Frados, J., Plastics Engineering Handbook of SPI, Van Nostrand Reinhold, New York, 1976.
5. "Structural Foam," Materials Engineering, 44 (February 1983).

6. "Manufacturing Handbook & Buyers' Guide," Plastics Technology, 1983–1984.

7. "Structural Plastics Design Manual," U.S. Government Printing Office, No. 023 000 00495–0, July 1978.

8. Beck, R. D., *Plastics Product Design*, Van Nostrand Reinhold, New York, 1980.

9. "Structural Foam," *Plastics World*, 39 (September 1977).

10. Lubin, G., *Handbook of Composites*, Van Nostrand Reinhold, New York, 1982.

11. Rosato, D. V., and Grove, C. S., *Filament Winding*, John Wiley & Sons, New York, 1964.

12. Rosato, D. V., Fallon, W. K., and Rosato, D. V., *Markets for Plastics*, Van Nostrand Reinhold, New York, 1969.

13. SFP Licensing, Dow Chemical U.S.A., 2040 Dow Center, Midland, MI 48640.

Section IX
SUMMARY

Chapter 31
Financial Management

Terrence R. Ozan
Partner, Ernst & Whinney

This chapter summarizes the basic financial management concepts and techniques and raises issues for attention by management.

The material discussed is excerpted or summarized from the *Financial Management Manual for Plastics Processors* and related seminar and workshop materials. The *Manual* is published by the Society of the Plastics Industry, Inc. Both the *Manual* and seminars were developed by Ernst & Whinney under the sponsorship of SPI's Financial Management Committee. The *Manual,* consisting of four volumes, describes in great detail the concepts and techniques needed for effective financial management of an injection molding plant. The reader should obtain a copy of the *Manual* from SPI for complete conceptual and procedural discussions.

THE IMPORTANCE OF FINANCIAL MANAGEMENT

Economic forecasters are predicting that the injection molding industry will be among the top growth industries in the future. The expected growth can and should mean more business and profit for all molders. But, industry growth will also cause competition to increase as more companies attempt to take advantage of the market's opportunities. What then are the prospects for molders?

Some molders will be very successful, while others will just muddle along. What will make

the difference? Three things will determine success:

1. The market served by the company.
2. The company's technical capabilities.
3. Attention to the fundamentals of financial management.

Without all three, a molder's profits are likely to be mediocre, even in the best economic conditions.

Most companies continually challenge their marketing thrust by asking such questions as:

- Are we serving growth markets?

- Are our customers the leaders of their industries?
- Do we have as much business from each customer as we can expect?
- Who is trying to get our piece of the business?

Most companies are also prepared to challenge their technical skills:

- Should we use microprocessor controls?
- Are our Q.C. methods consistent with the markets we want to serve?
- Should we get into CAD/CAM?

On the other hand, few companies continually challenge the management techniques used to control the operation of the business on a day-to-day and year-to-year basis. Why? Marketing, engineering, and quality are the ways to get new orders, but managing the business operation is the way to make profit on those orders. However, management is not as glamorous or exciting as thinking of ways to expand the business. Since orders are being produced and shipped, managers, although aware of some operational deficiencies, often put attention to them on the "back burner." Unless there is a crisis, unfortunately, little attention is paid.

To improve financial management of a plant, attention should be given to three fundamental areas:

- Cost management
- Profit planning and budgeting
- Materials management

The issues related to each are the subject of this chapter.

COST MANAGEMENT

The issue of obtaining timely, accurate cost information has challenged molders in the past and will continue to do so in the future. As long as material prices, utility rates, equipment, labor costs and the like continue to change, molders need the capability to respond quickly. The company must be able to recalculate product costs as the elements that make up that calculation change. It must also be able to pose "what if . . ." costing questions and get reliable answers to estimate product costs effectively.

In addition to the traditional treatment of material, labor, and overhead, the molder's system must account for the cost impact of material mixes, regrind, family mold usage, and movable auxiliary equipment.

The costing system should be able to identify excess cost as it occurs by part number, by job or order, and by work center. The reasons for excess cost should be isolated so the molder knows whether he is dealing with excess scrap, slow machines, or breakdowns.

If a molder can routinely generate reliable cost information, isolate cost overruns, and

quickly recalculate product costs, he is well on the road to profitability.

Companies that understand profitability can direct their own performance in forceful and creative ways. They can turn around unprofitable trends by directing marketing emphasis to products produced in underutilized work centers. If all else fails, they know when to cut their losses and withdraw or at least de-emphasize a particular product. They also gain new insights into their customers and can distinguish those who are truly profitable from those who merely exhibit the appearance of profitability because of high order volumes.

The companies that make very high profits as a percent of sales are those that continually monitor their products, increasing those that are profitable and eliminating those that prove to be unprofitable despite all efforts. These companies also are not reluctant to de-emphasize a customer who is not providing sufficient overall profits to the company. These are all hard decisions to make, however, and they generally do not get made without reliable cost information.

What then is an effective cost management system? Fundamentally, it is one that can assign dollar values to both expected and actual engineering and production information in ways that support a variety of management analyses and decisions.

Figure 31-1 illustrates a cost system and its uses. The information can be divided into four major categories:

- General accounting
- Control of day-to-day operations
- Planning and decision processes
- Special managerial analyses

General accounting functions primarily include inventory valuation and cost of goods sold determination. Almost any cost system can provide this information reasonably well.

Control of day-to-day operations includes performance reporting of the manufacturing function as well as actual-to-budget expense reporting of support departments. Reporting includes the cost impact of labor and machine efficiency and productivity, machine utiliza-

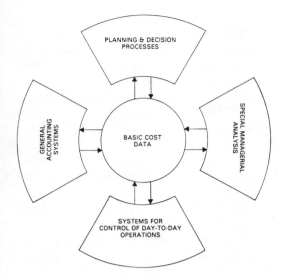

Fig. 31-1 Cost system and uses.

tion, scrap and rework, as well as other analyses.

Planning and decision functions include cost estimating for pricing, resource requirements planning (inventory, manpower, capacity), profitability analyses of various types, and support of profit planning, budgeting, and forecasting.

Special analyses are unique by their nature, but may include analysis of the best production location for a product, capital expenditure justification support, and determining the impact of volume on cost estimating and pricing.

The basic cost data include engineering and production information as well as costs and expenses. Engineering and production data consist of:

- Material types, quantities, and processing losses.
- Mold specifications, including cavitation and family groupings.
- Operation times including cycle times, operator-hour-to-machine-hour ratios, and expected time to complete secondary operations.

Cost and expense data consist of:

- Material prices, labor rates, machine-hour rates, and special overhead rates such as for material preparation.

The first step in establishing an effective cost management system is to align the organization in such a way that responsibilities for cost control are clearly defined.

Once the organization has been structured to specify responsibilities clearly and permit effective monitoring of those responsibilities, the next step demanding top-management attention is to select those cost management concepts that will permit routine identification of problem areas. In selecting the concepts, management must develop answers to three basic questions:

- *Which manufacturing costs should be associated with products?* Two options exist. Under the first option only the variable production costs such as material direct labor and the variable elements of manufacturing overhead are associated with the individual product. Alternatively, all production costs including fixed items such as depreciation are associated with products. The decision regarding these two options should be based on whether profitability analyses are more meanful with only the variable costs or with all production costs included. In practice, the cost management system can be designed to permit both types of analyses.
- *In what manner will costs be monitored?* Two options exist again. First cost can be monitored for the plant or subsections of the plant (e.g., work centers) without regard to specific production orders. Alternatively, the monitoring of cost can be done on an order-by-order basis. The first alternative, process costing, generally requires a simpler reporting system, but provides little information about each production or customer order. Although job costing is slightly more complex than process costing, it should be strongly considered when profitability information on an order-by-order basis would be beneficial for control of the company.
- *How will the identified cost be monitored?* Two options exist here also. First, production costs actually incurred can be associated with production. Alternatively,

predetermined expected costs can be associated with production and routinely compared with actual costs to determine where excess manufacturing cost or savings have occurred. The ability to routinely assess performance makes the second alternative (standard costing) the more effective method of controlling cost and identifying problem areas. It also permits product profitability analyses that are undistorted by manufacturing efficiencies, and enables a simpler compilation of inventory values and cost of goods sold.

Information Necessary for Product Costing and Cost Control

After the most appropriate cost concepts for a company have been determined, the detailed information, reporting procedures, and control reports must be developed. There are many options regarding these system elements that need to be considered as a company refines its manufacturing cost control system. As management considers each of the options, it should attempt to design a system that will meet its control objectives in the simplest manner possible.

As discussed, the information necessary for product costing and cost control includes the material makeup of each product, the operations that will be performed to produce it, and the tooling requirements. These requirements may then be expressed as product costs by:

1. Valuing materials at expected or actual purchase prices.
2. Valuing the machine and labor hours required in the manufacturing process at expected or actual hourly rates.
3. Allocating the tooling costs to production when appropriate.

The development of machine-hour rates involves many detailed consideration. Fundamentally, it can be described as a seven-step process:

1. Identify the expense categories to be included in the rate.
2. Determine the anticipated amount of expense in each category as a part of the profit planning/budgeting process.
3. Identify the appropriate production centers (e.g., machines, machine groups, secondary work centers) for which individual rates are to be developed.
4. Distribute the expenses among the production centers (via direct charges or allocations).
5. Split distributed costs into their fixed and variable elements.
6. Determine the practical capacity and expected production hours (per business plan) for each production center.
7. Calculate the hourly rates.

Figure 31-2 illustrates the completion of steps 1–4. To keep the illustration simple, only three production centers are considered. It is often appropriate to break the costs down into more centers.

A machine-hour rate (excluding direct labor) can then be calculated by estimating the expected production hours and dividing those hours into the production center costs. For example, if 20,000 production hours were expected for the large presses, the overhead rate for the center would be $32.58 per hour. This rate could be used in combination with material and labor costs to determine the expected cost of a product run in that center.

One consideration of particular importance in estimating costs and setting prices is the "capacity overhead rate." The use of this rate permits two types of control information to be routinely developed:

• Product costs that do not include the cost of lack of orders or unscheduled equipment downtime (idle capacity costs).
• Variable costing profitability analyses while maintaining a full costing system.

Use of capacity overhead rates permits a company to determine the unit product cost that would be achieved if the company were able to generate volumes approaching its prac-

	COST DISTRIBUTION BASIS	BREAKDOWN OF OVERHEAD BY PRODUCTION CENTER			
		LARGE PRESSES	SMALL PRESSES	ASSEMBLY/ FINISHING	TOTAL OVERHEAD
Large presses indirect labor	Direct	$ 48,151			$ 48,151
Small presses indirect labor	Direct		$ 48,965		48,965
Assembly/finishing indirect labor	Direct			$ 31,444	31,444
Total Indirect Labor— Production Departments		48,151	48,965	31,444	128,560
General manager's staff	1'3 each	16,977	16,977	16,978	50,932
Personnel	Total empl.	13,680	8,550	11,970	34,200
Cast accounting	Total empl.	15,748	9,843	13,779	39,370
Material control	No. of items	26,004	17,335	43,340	86,679
Engineering	No. of items	13,607	9,071	22,678	45,356
Quality assurance	No. of items	21,985	14,657	36,643	73,285
Purchasing	No. of items	8,183	5,456	13,639	27,278
Maintenance	Analysis	71,970	35,985	11,995	119,950
Receiving and shipping	No. of items	18,467	12,311	30,779	61,557
Total Indirect Labor— Service Departments		206,621	130,185	201,801	538,607
Labor connected expenses (except O/T)	Total payroll	108,576	77,952	91,871	278,399
Overtime premium	Analysis	23,460	10,330	9,100	42,890
Total Labor Connected Expenses		132,036	88,282	100,971	321,289
Electricity	Analysis	23,450	14,070	9,380	46,900
Telephone	Purch & GM	2,736	1,824	4,560	9,120
All other utilities	Analysis	3,769	2,931	1,675	8,375
Total Utilities		29,955	18,825	15,615	64,395
Depreciation	Analysis	110,100	78,200	6,800	195,100
All other facilities costs	Floor space	13,489	8,710	30,908	53,107
Total Facilities Cost		123,589	86,910	37,708	248,207
Maintenance materials	Analysis	39,420	19,710	6,570	65,700
Mold maintenance and amortization	Analysis	46,918	11,730		58,648
All other supplies	D. L. empl.	14,665	12,570	14,665	41,900
Total Supplies and Mold Costs		101,003	44,010	21,235	166,248
Computer services	Analysis	1,050	1,050	1,400	3,500
Rental of (computer) equipment	Analysis	3,180	3,180	4,240	10,600
Travel expenses	Analysis	4,190	4,190	4,320	12,700
All other office expenses	⅓ each	1,917	1,917	1,916	5,750
Total Office Expenses		10,337	10,337	11,876	32,550
Total Overhead		$651,692	$427,514	$420,650	$1,499,856

Fig. 31–2 Breakdown of overhead by production cost center.

tical capacity. Routine availability of this type of information is invaluable, since it permits management to determine the price levels below which increased volume will not have a significant positive impact on profitability.

Computation of this rate involes steps 5 and 6. (Figure 31-3 illustrates the differences between theoretical capacity, practical capacity, and expected production volume.)

Applying this concept to the example yields the different capacities and the resulting capacity overhead rate. (See Tables 31-1 and 31-2).

Compare the two rates yields a difference or capacity differential of $3.18 per hour (Table 31-3).

The capacity differential represents the value (on an hourly basis) of increasing vol-

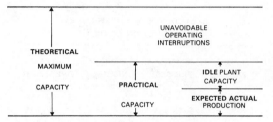

Fig. 31-3   Capacity Terminology.

good benchmark for pricing decisions in a competitive environment.

Reporting from the Production Floor and Management Control Reports

ume. Unless the manufacturing process is improved, the practical capacity rate represents the lowest possible hourly cost. Using this rate to calculate estimated costs will provide a

There are many production/warehouse reporting forms and procedures that can be implemented with any of the system concepts selected. The choice of detailed information to be gathered and procedures and formats for gathering and summarizing the information should be made to minimize the clerical

Table 31-1   Calculation of hours for two volume levels.

	THEORETICAL CAPACITY HOURS	MAXIMUM UTILIZATION	PRACTICAL CAPACITY HOURS	BUSINESS PLAN	
				UTILIZATION	EXPECTED PRODUCTION HOURS
Large press machine hours	30,770	76%	23,380	65%	20,000
Small press machine hours	62,000	80	49,900	72	44,900
Assembly labor hours	93,000	70	65,100	60	56,000
	185,770		138,380		120,900

Table 31-2   Practical capacity, calculation of machine-hour costing rates (large presses).

	PRACTICAL CAPACITY		
	TOTAL	FIXED	VARIABLE
Manufacturing overhead	$651,692	$440,000	$211,692
Machine hours		23,380	20,000
Machine-hour based overhead rate	$ 29.40	$ 18.82	$ 10.58

Table 31-3   Differential between practical capacity rate and the business plan rate.

	AT PRACTICAL CAPACITY	AT BUSINESS PLAN VOLUME	CAPACITY DIFFERENTIAL
Machine-hour based overhead rate	$29.40	$32.58	$3.18

demands on production personnel, reduce the chances of error, and provide for multiple uses of the data (e.g., production scheduling, inventory control, and cost management). The design should also consider the personnel available, personal management preferences, and, of course, the control objectives to be met.

Whatever reporting procedures are developed, the fundamental requirement for effective manufacturing cost control is that reported production quantities be reliable. There are many reasons for unreliable production quantity reporting, ranging from simple employee error to underpacking or overpacking of resin in mold cavities. Great care must be taken to assure reliability in production counts. Generally, the foreman must take responsibility for reliability of the production counts within his department in addition to his other duties. This responsibility consists of:

- Overseeing the production counting procedures to assure that the methods and paperwork requirements are conducive to reliable recording.
- Establishing proper counting procedures.
- Verifying the counts.

The control reports included in the system should routinely provide management with the information necessary to monitor results and identify problems. In accomplishing this, however, proliferation of reports should be avoided. Such proliferation of paperwork is usually counterproductive.

PROFIT PLANNING AND BUDGETING

Profit planning and budgeting are closely related but are not precisely synonymous. Profit plans include the strategies, tactics, and specific actions the company is planning to adopt to achieve specified profit and return-on-investment goals. A budget is the financial representation of a plan, including estimates of revenues, costs, and expenses. When developed in sufficient detail, the budget serves as the document against which the various management functions are measured during a year.

Development of a profit plan and budget is a cycling process in which top management establishes goals and the rest of the organization determines whether the goals are feasible, and, if so, what resources (i.e., capital, people, equipment, etc.) will be required to achieve those goals. The cycle can be viewed in three phases:

- Phase I begins with the establishment of goals for the planning period (e.g., one year). The feasibility of the goals is then determined on a preliminary basis by using overall sales, cost, and expense projections. Little detail is developed at this time, so that analysis of alternatives can be done in a relatively short time.
- Phase II includes the preparation of detailed budgets by all functions consistent with the goals. Since it is possible that the detailed sales, cost, and expense projections may be significantly different from the preliminary estimates, the feasibility of the goals is once again reviewed.
- Phase III includes the finalization of the detailed budgets that will support the overall plan.

This phased process is an effective method of planning and budgeting in that it limits the time invested in developing detailed budgets until there is general belief that the overall goals can in fact be achieved.

Gathering the Data for Profit Planning and Budgeting

To provide historical perspective and realism for the profit planning and budgeting process, a company should gather and summarize all relevant information regarding sales, production, and costs/expenses during the previous and current years.

Establishing Profit, Goals and Sales Forecasts

As stated, Phase I of the profit planning and budgeting cycle includes the establishment of goals and the determination of the feasibility of those goals. This is the most crucial phase of the profit planning cycle and is also the most difficult. The process of establishing the goals and determining their feasibility may consist of the following steps:

- Analysis of the financial and operating ratios of the company to determine its financial and operating strengths and weaknesses as compared with the industry as a whole and to establish profitability and sales goals based on improvements in certain other ratios such as return-on-investment and return-on-sales.
- Development of sales forecasts based on the judgments of marketing personnel and/or extrapolations of current sales trends.
- Comparison of the sales goals established via the financial and operating ratio analyses with forecasted sales based on knowledge of the marketplace and development of the initial sales plan.
- Determination of the anticipated gross profit that would result from the sales plan.
- Comparison of the gross profit that is likely to result from the sales plan with the goals established.
- Identification of actions that would bring the anticipated profitability in line with the initially established goals.

In order to form a solid foundation, the forecast must simultaneously be realistic and aggressive. There are fundamentally two approaches to forecasting sales: the top-down and the bottom-up techniques. The top-down technique first forecasts sales in total for a company and then divides the aggregate forecast among the various product lines and individual products. The bottom-up technique forecasts each product or product line individually and then summarizes the individual forecasts into an aggregate or total forecast. As a general rule, neither technique is always better than the other. In fact, effective profit planning is best accomplished when both types of forecasts are made and are then judgmentally reconciled to a final forecast both in aggregate and individually for each product or product line.

Developing the Detailed Plans and Budgets

Once it is believed that the sales projections, marketing plans, and anticipated profitability results are satisfactory and attainable, the company should began Phase II of the planning process. This includes the development of the detailed requirements for production, inventory, purchasing, personnel, and the cost and expenses associated with each. The overall procedure for developing the detail includes the following steps:

- Break the annual forecast for each product line into anticipated monthly sales. This is to take account of any seasonality of the business.
- Establish finished goods inventory levels based on customer service requirements and the availability of cash to finance any inventory buildups.
- Develop the total annual production requirements and the monthly requirements by product line using the sales and inventory requirements.
- Develop the requirements for purchased materials and personnel (manpower) required.
- Analyze the expected utilization of machinery and personnel and determine whether additional capacity would be required to produce the sales forecast.
- Using the sales, production, inventory, purchasing, and personnel plans, develop the initial budget for costs and expenses related to manufacturing.
- Develop the initial budgets for selling, with general and administrative expenses consistent with the marketing plans assumed in the sales forecast and with the supervisory and administrative requirements necessary to manage the business at the anticipated sales level.
- Develop pro forma financial statements reflecting the budget detail. Review them

against the initial profitability goals, and either adjust the plan detail to conform with the goals or revise the goals where it is not feasible to achieve them.

- After the final plan and budget are adopted, establish monthly departmental budgets to be used for monitoring the costs and expenses during the year (Phase III).

Flexible Budgeting

The budgets and plans developed according to the above procedures remain constant unless revisions are made in response to some unusual circumstance. These budgets can be used to monitor monthly performance in total and by department. There are, however, some limiting aspects of a fixed or constant budgeting system.

Fixed budgets are based on the specific definitive conditions and results assumed in the planning process. In actual practice, such conditions are rarely precisely predictable, thus making the comparison of actual expenditures to the fixed budgets somewhat distorted. As a result, for month-to-month measurement of departmental performance against budget, the technique of flexible budgeting is often employed.

With flexible budgeting, the cost and expense levels allowed (budgeted) are adjusted to reflect the actual volume during the period. Thus, the comparison of actual expenditures with the budgeted expenditures is based on the same set of circumstances.

Flexible budgets are not always substituted for the fixed budget. Many companies use the fixed budget for monitoring the income statement and balance sheet results and the flexible budget for monitoring departmental performance.

The procedure for developing a flexible budget includes the identification of those costs that are considered fixed (unchanging over reasonable fluctuations in volume) and those costs that are considered variable (changing in direct proportion to volume fluctuations). Using the identified fixed and variable costs and expenses, a flexible budget formula is developed whereby it is possible to compute an allowed expenditure level given any actual volume within a company's normal operating range. In practice, the allowed amount is computed after a month is completed when the actual volume is known. The actual expenditure level is then compared to the flexible budget amount to determine whether there is a favorable or an unfavorable difference.

The development of flexible budget information has some very important side benefits:

- Realistic measurement of management's ability to perform under various conditions.
- Measurement of expenses incurred due to idle capacity (underutilization) within each department and for the plant.
- Determination of a production break-even point by department and for the plant.
- Calculation of standard or expected product cost at various volume levels.

MATERIALS MANAGEMENT

The concept of materials management centralizes responsibility for the four distinct but interdependent functions of inventory control, production control, purchasing, and shipping/receiving. Provision is also made for close coordination with the related functions of sales order entry, credit, billing, and collection. Because of the wide variety of products and processing methods, it is not possible to recommend a single uniform system for all companies.

The overall objective of materials management is to meet customer requirements at the lowest product cost and the lowest inventory investment. Stated in another fashion, the objective is to establish a high customer service level at an appropriate cost. To achieve a good customer service level, a company must:

- Usually deliver quality products on a timely basis.
- Establish honest and complete communications with customers.

The first point is really a definition of customer service. The functions involved in delivering quality products on a timely basis are those of purchasing, production scheduling, and inventory control. In addition to providing good service, a company achieves a high service reputation by providing reliable information to its customers regarding expected shipping dates and the extent and reasons for any difficulties in meeting the date originally promised. The order processing and customer contact functions fulfill the latter objective.

The types of products manufactured and sold by a molder dictate certain procedural variations in the control of material. Custom products and some proprietary products are generally made to order. Producing to customer order needs keeps the finished goods inventory at a low level but causes a company to respond more slowly to customer demands. Alternatively, proprietary products may be produced to stock in anticipation of customer orders. This allows a company to respond more quickly to customer demands but requires more capital investment in inventory and may result in slow-moving products that must be disposed of at a low profit or loss.

The elements of the product environment—proprietary vs. custom products, make-to-order vs. make-to-stock—influence how operations should be managed to achieve the lowest possible cost for a good customer service level.

Order Processing

Order entry is the link between the customer and the company. The order entry function records customer requirements, communicates this information to the functions involved in filling the order, and gathers from these same functions the information necessary to report back to the customer on order status. Although order entry may be narrowly defined as a clerical function and, strictly speaking, is not a materials management function, its role as a communication center for information on open orders makes it a key factor in achieving both good customer service and operating efficiency.

The key elements in developing an effective order entry and complete order processing system include:

- Developing an order entry form that permits quick and accurate communication of the customer's requirements to all operating departments within the company.
- Effective and quick analysis of the creditworthiness of the potential customer.
- Frequent communications from production and inventory control personnel as to the ability of the molder to meet the customer's due date.

To fulfill these functions, a procedure for collecting appropriate order status information should be developed and summarized in a manner that permits customer contact personnel to communicate effectively with customers. Also, the order processing system should make provisions for the treatment of various types of orders including blanket, standing, sample, and export. Each of these requires slightly different treatment by the company, and the order processing procedures should be developed to handle them in a routine manner.

Inventory Control

As part of the profit planning and budgeting process, management makes basic decisions about the level of inventory needed to satisfy production and sales goals without straining the cash resources of the company. Once guidelines have been established, management must set controls for maintaining the appropriate level of inventory on a day-to-day basis. Management's task is twofold:

- To establish a system of policies and procedures that will assist personnel to determine how much and when to order or produce (managerial control).
- To protect the inventory asset from loss or misstatement through a system of accurate records and physical safeguards (physical control).

The criterion for achieving both managerial and physical control of inventories is that the cost and effort required to maintain the system does not exceed the benefits. The basic issue in inventory control is how much to order to produce and when. There are two approaches to answering this question: (1) order point/order quantity and (2) requirements planning.

The principle behind the order point/order quantity approach is "order when stock drops below a certain level." The fixed level is called the order point. The order quantity may also be fixed or may vary according to the amount of stock on hand at the time that the order point is reached.

The principle behind the material requirements planning approach is "order in time to meet production." The materials required to meet production during a given time period are analyzed, and the ordering process is phased so that the company carries only enough to meet production requirements.

The order point/order quantity systems are relatively flexible and simple to operate. For example, the management that decides to refill the silo whenever the resin drops below a fixed level is implementing a simple order point/ order quantity system.

The requirements planning approach is more complex. Because it is so closely tied to production, there is little margin for error. Precise record keeping, and computer support, is required to implement a requirements planning system effectively.

The choice of approach depends on the value of the inventory, the types of products produced by the company, and whether these products are stocked or made to order. The key element in selecting the system is not to become more sophisticated than is necessary.

Production Scheduling and Control

The process of developing one production schedule can be quite simple. All one need do is: (1) determine what needs to be made, (2) assign the orders to the proper work stations, and (3) sequence the orders according to need.

However, the influx of new orders, difficulties with existing orders, and constantly changing priorities complicate the job. There

may be insufficient capacity to produce all orders at the right time. Priorities will change to meet new or revised customer requests. Production on an order may not begin at the specified time because orders in process are delayed because of mold or machine breakdown, excessive employee absenteeism, or quality problems. Delays in receiving the materials or new molds may also make it impossible to start production on an order at the scheduled time.

Because of all the possible changes, it is necessary to revise the production schedule continuously. The continuing need to update, change, and expand the schedule makes the simple scheduling task difficult. Thus there is the need to formalize the scheduling process to some degree.

There are two objectives for production scheduling and control regardless of the size of the company or the complexity of the manufacturing operation:

- Produce the product on time.
- Keep production as level as possible at each machine and work center.

Producing to meet customer due dates or established lead times for stocked items keeps customers happy and helps to bring in additional orders. Leveling production keeps the work force stable, and aids in keeping productivity high and training costs low.

Schedules are established to control the production process. Overall schedules extending several weeks or months into the future indicate when materials, molds, etc., must be made available to production. The longer-range production loads also give management the information to determine such things as the best hourly personnel levels and the amount of overtime. They also provide the information about existing orders necessary to quote reliable promise dates on new orders.

Detailed schedules covering shorter time periods (e.g., a week or day) provide the foremen with information needed to decide what order to run next to meet customer due dates.

Without some type of schedule it would not be possible to coordinate the control of inventory, the efficiency of production, and the meeting of customer due dates.

Scheduling Approaches

The approach to scheduling and controlling production adopted by a company must be one that can accommodate the many changes in production order status and customer priorities that occur each day.

Two basic approaches can be taken to developing schedules:

- Because of all the likely changes, do not attempt to plan the specific timing and sequence of orders at each operation. Schedule production orders only to the extent of indicating the week that production will start. Then sequence and expedite orders on the floor in a manner that keeps each operation backlogged and gets the orders out on time.
- Prepare detailed schedules based on meeting the due dates and balancing the workloads. Then closely control all production-related activities against the schedule.

The objectives of each approach are the same: to produce the product on time and keep production as level as possible at each machine or work center. However, the manner of achieving the objectives differs.

With the first approach, relatively large backlogs are maintained at each work center. The backlogs serve as a constant supply of work. They reduce the possibility that no work will be available at a work station because the scheduled order is delayed. Production orders are released to the machines as they become available, and are placed in the backlog of the appropriate machines. As each order is completed on a machine, it is transferred to the backlog of the appropriate secondary operation. The sequence of orders produced at each operation is largely determined by the foreman. When an order waits too long in a backlog or the due date is changed, production control personnel expedite individual orders to meet the due date.

The usual results of this approach are:

- Production levels at each work center remain relatively constant (assuming there is sufficient overall volume).
- Customer due dates are generally met by expediting or "pulling" orders through the plant.
- The amount of expediting necessary to meet due dates is relatively high.
- The average time interval on the production floor for an order is relatively long because of the wait time between operations.
- The lead time for a specific order is difficult to predict because the wait time is not easily predicted.

With the second approach, production scheduling personnel exert strong guidance on the time an order is released to the floor and the sequence of orders at each machine and secondary operation. Great care is taken to schedule orders in a fashion that allows orders to move quickly from one operation to the next. Overall, this approach attempts to "push" orders through the plant in a sequence that keeps each work center busy and meets the customer due dates.

The usual results of this approach are:

- Production levels at each work center remain relatively constant.
- Customer due dates are generally met by "pushing" the orders through in the appropriate sequence.
- Expediting is required, but the amount is generally less than with the first approach.
- The average time interval on the production floor is relatively short because small backlogs are maintained at each operation.
- The lead time for an order is more predictable than in the first approach because the wait time is shorter and less variable.

In determining which of the two approaches is most appropriate for a company, two issues should be considered:

- How much control of the schedule should rest with the foreman vs. production scheduling personnel?
- Is it cheaper to do a great deal of expediting or to prepare more detailed schedules?

Giving foremen the authority to select and sequence orders for the backlog can result in efficient production. Effectively sequencing orders at each operation can minimize setup costs and thereby reduce overall production costs. However, the sequence selected by the foreman for a molding machine may not sufficiently take account of the need to keep a constant flow of work to secondary operations or meet the customer due dates.

Regarding the second issue, expediting is expensive. But, it may not cost as much as attempting to develop and continuously update detailed schedules for each operation that take account of the need to keep a constant flow of work to secondary operations and meet the due dates.

The difficulty of meeting the scheduling objectives increases with: (1) the number of work centers, (2) the average number of operations per order, and (3) the total number of outstanding production orders. As the difficulty increases, there is a greater need for an overview perspective to achieve sufficient control. Production scheduling personnel are generally in a position to have such an overview. Consequently, smaller companies with few or no secondary operations can usually operate quite well with the first approach. As companies grow or add secondary operations, the more disciplined second approach becomes more appropriate.

Purchasing

The amount of the sales dollar devoted to purchased goods and services ranges from 40 to 60 percent in a majority of companies. Too often, little management attention is focused on the potential *profit-making* functions of a purchasing department.

Profitable buying is not a simple process. It requires an understanding of market conditions and continuous contact with reputable suppliers. It must be preceded by internal research into quality specifications, supported by analysis of past purchases and vendor performance, and followed up by good ordering procedures. Profitable buying is actually a three-stage process:

- *Requirements determination.* The purchasing process begins, for production-related purchases, with the generation of order quantities by the inventory control system.
- *Procurement decision.* This is the heart of the buying process. In this second stage, the purchasing agent (buyer) applies quantitative and analytical skills to evaluate the alternative sources of goods. Identifying and analyzing vendors, negotiating price, delivery, and terms, and, finally, selecting the best vendor are the activities that contribute to profit.
- *Procurement process.* This is the mechanical stage of issuing the purchase order, following up on delivery of the goods, and payment of the vendor's invoice. Procedures should be established to reduce the effort involved in this stage of the process so that purchasing personnel can devote the bulk of their time to analytical profit-making activities.

Too often, purchasing personnel spend most of their time on the clerical or mechanical functions of the third stage. While this is needed to actually acquire the material, no real profit ensues. More time and effort should be spent on the procurement decision. By selecting and negotiating with vendors for improved quality at a fair price, purchasing can add directly to a company's profitability.

CONCLUSION

This chapter has set forth many areas and issues for review and improvement. Each area is complex. Improving any one of the areas requires a deep commitment of resources by management to analyze their current effectiveness, select the best concepts and techniques, and implement them in the organization. Doing this is not really fun or glamorous; but, if done right, it can be a great source of profits.

Chapter 32
Summary—Mold with Profit

D. V. Rosato

INTERFACING MACHINE PERFORMANCE

In order to injection mold all sizes, shapes, and weights of parts to meet all types of performance requirements, the plastics industry has made steady progress in advancing the state of the art and science of injection molding over the past century. This book has reviewed many new developments that have improved the complete injection molding process. These advances have been based on knowledge gained in understanding the parameters involved in meeting part performance requirements. These parameters include:

1. Setting up specific performance requirements.
2. Evaluating material requirements and molding characteristics.
3. Designing parts based on the material molding characteristics.
4. Designing and manufacturing molds based on part design.
5. Setting up and operating the complete injection molding machine line so as to meet mold and material processing requirements.
6. Testing and providing quality control of incoming materials, materials during processing, and molded parts.
7. Interfacing all these parameters by using the simplified computerized program(s) available.

Injection molding machines and all types of auxiliary equipment used in the complete molding line can be installed with computerized controls to meet manufacturing requirements. Terms such as open and closed loop, analogue, proportional, digital, servo hydraulics, and process control, as well as the product names used by the machinery producers, tend to confuse the molder rather than provide clear technical definition. This book provides information to eliminate this confusion.

The molder should clearly define the requirements that injection molding machines must fulfill according to production requirements. On this basis, selection of equipment is made with the appropriate control system (1). In most cases, these requirements have changed over the last few years. In the past, the technical solution of a production problem was often the main consideration, and the production costs were of lesser importance. Today, the molder, as usual, is faced with continuously rising costs that can rarely be transferred completely to his customers. Hence, rationalization and cost reduction in the production area have become very topical.

In pursuit of such a policy, total automation in the manufacturing of bulk products has become essential. In this respect it is necessary to consider not only the single production machine, but the complete production area. The injection molding machine should be regarded as one element within an interconnected production system, which must be operated with a minimum of people; this is of even greater importance when three-shift manning is used.

Fully automated production from raw material to the finished product requires not only a very high degree of reliability of the machine and its component parts but also the use of monitoring functions, which are used not only to monitor machine performance but also component quality. Ideally, auxiliary equipment should also be included in the monitoring systems, as it forms as essential a part of the complete production unit as the machine.

In addition, optimum utilization of the existing production equipment becomes even more important. This can mean that the mold clamping force and other machine parameters are utilized to their maximum values, and that cycle times are shortened as much as possible without allowing any reduction in the quality of the articles. However, the closer one gets to these limits, the more important is the high consistency of the moldings produced.

Only a few years ago it was acknowledged as "state of the art technology" to set hydraulic parameters manually at the hydraulic station of the machines by means of individual valves. Today, many machines used for production still have such a requirement. The set pressures and speeds have to be measured with manometers and stopwatches, which is not possible with sufficient precision during fast phases of the cycle. Consequently, mistakes occur during these measurements and during the resetting of the machine for a new production run with the same mold. Normally it is necessary to optimize the production settings each time mold is loaded into the machine, even when a known mold is used.

Today's technical standard is the digital setting of speeds, pressures, temperatures, and times centrally at the control cabinet. In this way the setting data are clearly defined and recorded on setting sheets. The risk of making mistakes when a machine is reset for a mold used earlier is greatly reduced, and renewed optimization of the setting data is hardly ever necessary. This leads to a reduction of both setting time and start-up time, which means

that highly skilled setters have more time for other tasks, and that unproductive times are reduced. Thus the use of digital setting techniques dramatically reduces reliance on a "specialist" to set each job.

The digital setting of hydraulic parameters at the control cabinet requires hydraulic valves that are controlled electrically. The kind of valve to be used is determined both by the required reproducibility from shot to shot over long production periods, including accurate resetting, and by the cost of components. Depending on the technical requirements of the components to be produced, a choice among proportional, digital, and servo-hydraulic systems can be made.

For smaller injection molding machines especially, proportional technology offers a solution for the digital setting of pressures and speeds at the control cabinet that is less expensive than pure digital valves or servo-hydraulics (see Chapters 2 and 11).

The electrical input current causes an alteration of the position of the armature that is proportional to the height of the current. The armature is fixed to the cone of the pressure setting valve or to the spool of the flow control valve. Furthermore, the armature is fixed to a displacement pickup that measures the position of the armature and feeds this information back to an electronic controller. If the valve cone or the spool deviates from its set position, a correction by closed-loop control takes place.

This is not closed-loop control in the generally accepted sense, and the use of such valves must not be confused with servo valves that offer a complete closed-loop solution. Practical experience with these proportional valves has established that they are a good technical solution for a wide range of applications. However, regarding precise repeatability from cycle to cycle, reproducibility when a machine is reset to the same values as before, and long-term consistency, proportional hydraulics does not reach the precision of pure digital hydraulics.

For medium-sized and larger machines there is not such a big difference between the costs for proportional and digital hydraulics, and, furthermore, the additional costs are relatively small in comparison with the machine price.

The principle of digital hydraulics is the addition of single fixed values to give a total combined value. As the single steps are clearly defined as fixed values, and as they are only switched fully on or fully off, optimum conditions exist for the precise repeatability and long-term consistency of such systems. The advantages of digital hydraulics are, however, not limited to applications where the highest precision is required. The simple and robust construction of the cartridge valves provides, in addition, a high degree of reliability and production safety.

The precision of a machine for the production of articles with normal quality requirements becomes more important, the closer the machine is operated to its limits. It is, for instance, possible to mold an acrylic lighting dome with dimensions of 1 m $\times$ 1 m by using an injection molding machine of only 1600 tons clamping force generating only 160 kN/cm². The condition for this, however, is that the mold filling phase, the switchover from injection to follow-up pressure phase, and the height of the follow-up pressure be exactly reproduced by the digital hydraulics from cycle to cycle in order to avoid flash.

In comparison with proportional hydraulics, closed-loop control of hydraulic parameters by means of servo valves also improves precision and repeatability. Extensive tests with a specially prepared injection molding machine that could be operated alternatively with both pure digital and servo-hydraulics showed that with both systems the same high quality of moldings could be achieved. Also, during start-up of production and after interruptions it was possible to produce good articles after only a few shots with both systems.

It must be borne in mind, however, that servo valves are prone to hysteresis and long-term drift, and this tendency, coupled with their complexity and increased oil filtration requirements, places fully closed-loop systems at a distinct technical disadvantage. Furthermore, the energy consumption with a closed-loop servo system is approximately 30 percent higher than that of a machine utilizing the pure digital system.

Very often there is no choice between these two methods of control, as all injection molding machines incorporate both systems, depending on the function that must be controlled. The speed of the moving platen and the nozzle sealing force are, for instance, normally controlled by an open loop, while the cylinder heating zones, of course, have closed-loop control.

Closed-loop controls with servo valves are used in relatively small numbers. In most cases, these systems are used only to control hydraulic parameters, such as pressures and flow rates, at a constant value. When judging these systems, one must remember that they are expensive and require high energy consumption. In comparison, digital hydraulic systems offer a less expensive and very reliable solution to ensure right from the start that the hydraulic parameters have a high repeatability and long-term consistency.

In addition, closed-loop servo systems sometimes offer the possibility of correcting deviations of the actual mold cavity pressure curve from a preset pressure curve. These deviations can have many different causes, but the only means of compensating for them is adjustment of the hydraulic pressure of the machine. It is very doubtful that such systems should be applied, as the solution is totally illogical.

It is more logical, instead, to monitor actual values or to install individual closed-loop control circuits and to eliminate disturbances right at the root of the problem. Examples can be used to illustrate the point.

A deviation of the cavity pressure caused, for instance, by an alteration of the mold temperature can of course be compensated by the hydraulic pressure; but the cooling conditions for the article change at the same time, which means for semi-crystalline materials that the crystallinity and consequently the after-shrinkage will change as well. This in turn means that articles that seem to be of good quality will be sent back by the customer after some weeks because of changes of dimensions. Also, other quality criteria, such as the surface finish, are influenced by the mold temperature.

This example shows that it is far more logical to operate with individual control circuits than with process control circuits. It is better in this case to control the mold temperature directly and monitor a possible deviation of the mold temperature than to try to compensate for such deviations by inadequate and totally illogical means.

Also, deviations of the cavity pressure due to a change of the material viscosity should preferentially be monitored by measuring an appropriate signal, such as the maximum cavity pressure. Automatic illogical compensation can have the disadvantageous effect of altering conditions to the detriment of other quality criteria. In effect, it is easy to end up "chasing the error," with no warning that an unsalable product is on its way to the dispatch bay.

The mere fact of using bit-processors and microprocessors in the control systems of injection molding machines will neither improve article quality nor reduce cycle time. In the first instance they are just different modules that control the same functions as their predecessors. The difference is that we now need only very few modules in comparison to the conventional discrete electronic control systems. The enormous reduction of the number of components and wire connections is the reason for the high degree of reliability of such systems. Furthermore, microprocessors, because of their high storage capacity, offer the possibility of supplying machine control systems of high sophistication at an economical price (1).

Rules to Remember

1. Injection molding is the marriage of machine, mold, material, process, and operator, all working together in a constant cycle.
2. The plastics business is a profit-making business, not a charitable organization.
3. Heat always goes from hot to cold through any substance at a fixed rate.
4. Hydraulic fluids are pushed, not pulled, at a rate that depends on pressure and flow.

5. The fastest cycle that produces the most parts uses:
   - Minimum melt temperature for fast cooling.
   - Minimum pressure for lowest stress.
   - Minimum time—fast fill, fast cool, fast and early ejection, minimum delay between cycles.
6. All problems have a logical cause. Understand the problem, solve it, and then allow the machine to equalize its cycle to adjust to the change.
7. *If it doesn't fit, don't force it.*

"Rules" to Forget

1. If a little bit does a little good, a whole lot does a whole lot of good.
2. If you twist enough knobs, the problem will go away.
3. The machine has a mind of its own.
4. It takes a genius to operate a molding machine.
5. All problems are caused by bad part design, bad tooling, or bad setup.
6. My job is secure.

Automatic Pressure and Temperature Monitoring

Quantitative information, although important, is not enough to achieve the overall objective. Knowing that a part is made in most cases does not ensure that it is a good part. Dimensional variation, flash, and short shots, which are the main causes of rejects in an injection molding plant, should be detectable and accounted for automatically.

In order to achieve these objectives, a straightforward conceptional approach can be implemented. This approach involves putting a mold pressure sensor at the last point to fill in each cavity of the injection mold. This cavity pressure sensor is connected to a mold monitor device that, on each cycle, can detect the presence or absence of cavity pressure and whether or not that cavity pressure falls within a preset range. Using cavity pressure as the variable to be sensed has many advantages.

First, the presence or absence of cavity pressure at the last point to fill is a totally reliable method of determining whether or not a part has been made in that cavity. Cavity pressure can normally be reliably detected by sensing the force on existing ejector pins in the cavity without changing the characteristics of the molded part. In cases where ejector pins cannot be conveniently used, flush-mount cavity pressure sensors can be installed to accurately measure the pressure under even the most adverse conditions.

In addition, cavity pressure sensing at the end of fill can detect all of the changes in the molding process that reflect on the quality of the molded part. Cavity pressure at the end of fill will vary because of temperature variations, fill rate variations, variations in the hydraulic systems pressures, variations in mold temperature, and variations in the raw materials used in the process. These are virtually all of the variations from the plastics point of view. Even variations in such things as back pressure during plasticizing will be detected in cavity pressure during the next cycle. This makes the sensing of cavity pressure the most comprehensive approach to intelligent molding with a minimal amount of complexity.

In large parts, where long flow distances and the need for high dimensional accuracies exist, two or more sensors may be put in each mold cavity. Having a sensor near the gate end of the part and one near the end of fill allows the cavity pressure profile across the part to be monitored. This cavity pressure profile monitoring provides the ability to detect the qualitative aspect of dimensions and weight in a plastics part. Cavity pressure profile across a mold cavity totally detects the molecular distribution of the material across the plastics part. This profile will not only predict the overall dimensional integrity of the part, but the integrity of all of the areas of the part as well. In other words, if both cavity pressures, the one at the end of fill and the one near the gate, are duplicated for each shot, the plastics parts made in that cavity must be identical. This ultimate concept of two sensors in each cavity is only necessary on large parts of long flow length where extreme dimensional accuracies are important. Normally, a single sensor in each cavity is sufficient.

On hot runner molding of large parts where multiple drops are used in each cavity, the ultimate approach to the intelligent mold concept is to have a sensor to monitor each zone. This is not necessary in all cases. However, the ultimate in predictability can be achieved by utilizing such a scheme.

Pinpoint temperature accuracy is also essential to successful molding, including runnerless molding, particularly of engineering materials. In order to achieve it, microprocessor-based temperature controllers use a proportional-integral-derivative (PID) control algorithm acknowledged to be the most accurate in tuning gate heaters to process variations (see PID section in Chapter 11).

The three terms of this software equation provide a series of repetitive corrective actions over a given time period (usually 2 seconds): the "P" term adjusts the percentage of total power available that is fed to heaters proportionally to the margin of difference between the temperature readout and previously established setpoint value; the "I" is a corrective factor applied to "P" that prevents over- or undershoot by tuning heater power to equalize readouts and setpoints; and "D" is a final check factor that adjusts both these values to process swings in order to achieve the most accurate application of heater boost.

But while all microprocessor temperature controllers now use PID software programs, the reason some controllers are more accurate than others is that PID control programs must be written according to each controller's level of computational ability. Small microprocessors take too long (in terms of correcting actual process fluctuations) to handle all the factors in the necessarily complex PID equation. Simplified PID programs are therefore written for them to reduce the number of mathematical terms they handle and increase their speed of reaction to mold temperature changes. But this program simplification results in an inability to check temperature swings within the narrower temperature windows being called for by molders of many heat-sensitive resins.

Compromises Must Frequently Be Made

Since modifying resin properties or machine conditions occasionally affects some end-product properties (and also certain processing factors) favorably and others unfavorably, frequent compromises are inevitable in injection molding.

One such case is the influence that a number of resin properties and machine conditions exert on flow, warpage, and shrinkage. A decision may have to be made as to which of these three consequences is the most disturbing and should be decreased (or, occasionally, increased).

Mold cycle time is often considered the most important factor in determining both resin type and operating conditions. Obviously, the faster the molding cycle time, the more economical the molding process, other factors remaining equal. However, desired properties of the molded item must be considered; frequently, a compromise must be found.

Since gloss and piece detail on the one hand and economy on the other are caused to move in opposite directions by varying certain factors such as melt temperature and mold time, it is often essential to make a compromise between gloss and maximum economy (minimum mold cycle time).

Another compromise concerns the temperatures involved—both melt and mold temperatures. Whereas generally higher melt temperatures improve appearance, they also increase mold cycle time. This gain makes a compromise necessary between product appearance and economy.

Generally, gloss and resin processability move in one direction when strength properties and environmental stress crack resistance move in the other. This often requires a compromise.

One such compromise has to do with the resin melt index. As an example, a low melt index could mean high resin viscosity and thus reduced processability; but it also improves environmental stress crack resistance and impact strength (toughness). However, low flow resins, though having high inherent resistance to environmental stress cracking, are more likely to acquire residual stresses in the molding process. This might make higher melt index resins preferable. Thus, here too a compromise must be reached before the molder decides which resin melt index is most suitable for mass-producing a molded item. Frequently, only a real test with the molded product in use can answer the question of whether the melt index chosen will yield the desired properties.

In this book, numerous problems have been discussed and solutions offered. However, each case must be handled individually. With the innumerable variations in equipment and resins that exist today, even a seemingly straightforward problem could easily be complicated by application of an improper solution (1–28).

DECORATING

Decorating of molded plastics can be functional as well as decorative. Functional improvements include resistance to wear, scratching, ultraviolet light, chemicals, etc. Decoration can provide part identification and product information. All decorating depends on applying a surface marking or paint coating permanently to the plastic surface, without any damage to the original shape, properties, and appearance. Virtually every type of molded plastic can be decorated, using many different techniques. (See Tables 32-1a and 32-1b for most of the methods used.)

Problems that can occur during decorating result in damage to the decoration, the molded part, or both the decoration and the part. Some of the common causes of failure are mold lubricants, additives that have migrated to the molded plastic surface, surface moisture, frozen-in stresses that remained during molding, improperly dried molding material, material contamination, regrind, etc. (Many possible problems have been reviewed throughout this book.) To eliminate problems it is important to set up quality control procedures for the decorating material as well as the process used to apply the decoration. A guide to setting up procedures is reviewed in Chapter 28.

Table 32-1a Printing and decorating systems. Courtesy of *Plastics Technology.*

THE PROCESS	WHAT IT'S ABOUT	EQUIPMENT	APPLICATIONS	EFFECT
Painting				
1. Conventional Spray	Paint's sprayed by air or airless gun(s) for functional or decorative coatings. Especially good for large areas, uneven surfaces or relief designs. Masking used to achieve special effects.	Spray guns, spray booths, mask washers often required; conveying and drying apparatus needed for high production.	Can be used on all materials (some require surface treatment).	Solids, multi-color, overall or partial decoration, special effects such as woodgraining possible.
2. Electrostatic Spray	Charged particles are sprayed on electronically conductive parts; process gives high paint utilization; more expensive than conventional spray.	Spray gun, high-voltage power supply; pumps; dryers. Pretreating station for parts (coated or preheated to make conductive).	All plastics can be decorated. Some work, not much, being done on powder coating of plastics.	Generally for one-color, overall coating.
3. Wiping	Paint is applied conventionally, then paint is wiped off. Paint is either totally removed, remaining only in recessed areas, or is partially removed for special effects such as woodgraining.	Standard spray-paint setup with a wipe station following. For low production, wipe can be manual. Very high-speed, automated equipment available.	Can be used for most materials. Products range from medical containers to furniture.	One color per pass; multicolor achieved in multistation units.
4. Roller Coating	Raised surfaces can be painted without masking. Special effects like stripes.	Roller applicator, either manual or automatic. Special paint feed system required for automatic work. Dryers.	Can be used for most materials.	Generally one-color painting, though multicolor possible with side-by-side rollers.
Screen Printing	Ink is applied to part through a finely woven screen. Screen is masked in those areas which won't be painted. Economical means for decorating flat or curved surfaces, especially in relatively short runs.	Screens, fixture, squeegee, conveyorized press setup (for any kind of volume). Dryers. Manual screen printing possible, for very low-volume items.	Most materials. Widely used for bottles; also finds big applications in areas like tv and computer dials.	Single or multiple colors (one station per color).
Hot Stamping	Involves transferring coating from a flexible foil to the part by pressure and heat. Impression is made by metal or silicone die. Process is dry.	Rotary or reciprocating hot stamp press. Dies. Highspeed equipment handles up to 6000 parts/hr.	Most thermoplastics can be printed; some thermosets. Handles flat, concave or convex surfaces, including round or tubular shapes.	Metallics, wood grains or multicolor, depending on foil. Foil can be specially formulated (e.g., chemical resistance).
Heat Transfers	Similar to hot stamp but preprinted coating (with a release paper backing) is applied to part by heat and pressure.	Ranges from relatively simple to highly automated with multiple stations for, say, front and back decoration.	Can handle most thermoplastics. A big application area is bottles. Flat, concave or cylindrical surfaces.	Multi-color or single color; metallics (not as good as hot stamp).

Process	Description	Equipment	Materials/Applications	Characteristics
Electroplating	Gives a functional metallic finish (matte or shiny) via electrodeposition process.	Preplate etch and rinse tanks; Koroseal-lined tanks for plating steps; preplating and plating chemicals; automated systems available.	Can handle special plating grades of ABS, PP, polysulfone, filled Noryl, filled polyesters, some nylons.	Very durable metallic finishes.
Metallizing 1. Vacuum	Depositing, in a vacuum, a thin layer of vaporized metal (generally aluminum) on a surface prepared by a base coat.	Metallizer, base- and top-coating equipment (spray, dip or flow), metallizing racks.	Most plastics, especially PS, acrylic, phenolics, PC, unplasticized PVC. Decorative finishes (e.g., on toys), or functional (e.g., as a conductive coating).	Metallic finish, generally silver but can be others (e.g., gold, copper).
2. Cathode Sputtering	Uniform metallic coatings by using electrodes.	Discharge systems—to provide close control of metal buildup.	High-temperature materials. Uniform and precise coatings for applications like microminiature circuits.	Metallic finish. Silver and copper generally used. Also gold, platinum, palladium.
3. Spray	Deposition of a metallic finish by chemical reaction of water-based solutions.	Activator, water-clean and applicator guns; spray booths, top- and base-coating equipment if required.	Most plastics. For decorative items.	Metallic (silver and bronze).
Tamp Printing	Special process using a soft transfer pad to pick up image from etched plate and tamping it onto a part.	Metal plate, squeegee to remove excess ink, conical-shaped transfer pad, indexing device to move parts into printing area, dryers, depending on type of operation.	All plastics. Specially recommended for odd-shaped or delicate parts (e.g., drinking cups, dolls' eyes).	Single- or multi-color—one printing station per color.
In-the-Mold Decorating	Film or foil inserted in mold is transferred to molten plastics as it enters the mold. Decoration becomes integral part of product.	Automatic or manual feed system for the transfers. Static charge may be required to hold foil in mold.	Most plastics, especially polyolefins and melamines. For parts where decoration must withstand extremely high wear.	Single- or multi-color decoration.
Flexography	Printing of a surface directly from a rubber or other synthetic plate.	Manual, semi- or automatic press, dryers.	Most plastics. Used on such areas as coding pipe and extruded profiles.	Single- or multi-color.
Offset Printing	Roll-transfer method of decorating. In most cases less expensive than other multicolor printing methods.	Ranges from low-cost hand presses to very expensive automated units. Drying, destaticizers, feeding devices.	Most plastics. Used in applications like coding pipe.	Multi-color print or decoration.
Valley Printing	Uses embossing rollers to print in depressed areas of a product.	Embosser with inking attachment or special package system.	Used largely with PVC, PE for such areas as floor tiles, upholstery.	Generally two-color maximum.
Labeling	From simple paper labels to multicolor decals and new preprinted plastic sleeve labels.	Equipment runs the gamut from hand dispensers to relatively high-speed machines.	Can be used on all plastics. Used mostly for containers and for price marking.	All sorts of colors and types.

Table 32-1b   Guide to plastic-decorating methods. Courtesy of *Plastics Technology*.

Done in the Mold

	ECONOMICS	AESTHETICS	PRODUCT DESIGN	CHEMISTRY	MANU-FACTURING	COMMENTS
1. Engraved mold	*Unit cost:* low *Labor cost:* low *Investment:* moderate	Limited	Unrestricted	Not critical	No extra operations	Best for simple lettering and texture.
2. In-mold label	*Unit cost:* high *Labor cost:* high *Investment:* none to moderate	Unlimited	Somewhat restricted	Good durability Critical Good durability	Longer molding cycles	Good for thermoplastics and thermosets. Automatic loading equipment becoming available.
3. Inserted nameplates	*Unit cost:* high *Labor cost:* high *Investment:* moderate	Partially limited	Restricted	Not critical Good durability	Longer molding cycles	Allows three-dimensional as well as special effects.
4. Two-shot molding	*Unit cost:* high *Labor cost:* high *Investment:* moderate to high	Limited	Somewhat restricted	Not critical Good durability	Two molding operations	Good where maximum abrasion resistance necessary.

Done after Molding

	ECONOMICS	AESTHETICS	PRODUCT DESIGN	CHEMISTRY	MANU-FACTURING	COMMENTS
1. Applique	*Unit cost:* high *Labor cost:* high *Investment:* moderate to high	Somewhat limited	Unrestricted	Not critical Good durability	Hand operation	Allows unusual effects.
2. Electrostatic	*Unit cost:* low to moderate *Labor cost:* low *Investment:* moderate to high	Limited	Somewhat restricted	Critical Moderate to good durability		Dry process, no tool contact with product.
3. Flexographic	*Unit cost:* low *Labor cost:* low *Investment:* moderate to high	Somewhat limited	Restricted	Critical Moderate durability	Automates well	Wet process, tool contacts product. Sometimes requires top coat.
4. Hand painting	*Unit cost:* high *Labor cost:* high *Investment:* low	Somewhat limited	Unrestricted	Critical Good durability	Hand operation	Wet process, tool contacts product.
5. Heat transfer	*Unit cost:* low to moderate *Labor cost:* low to moderate	Unlimited	Somewhat restricted	Critical	Requires little floor space	Dry process, tool contacts product.

Table 32-1b   (Cont.)

	ECONOMICS	AESTHETICS	PRODUCT DESIGN	CHEMISTRY	MANU-FACTURING	COMMENTS
6. Hot stamping	*Investment:* low to moderate *Unit cost:* low *Labor cost:* low to moderate	Limited	Somewhat restricted	Good durability Critical	Requires little floor space	Multicolor graphics. Dry process, tool contacts product.
7. Labeling	*Investment:* low to moderate *Unit cost:* low to moderate *Labor cost:* low to moderate *Investment:* low to high	Unlimited	Somewhat restricted	Good durability Less critical Moderate to good durability	Adaptable to many situations	Produces bright metallics. Dry process, no tool contact with product at times.
8. Metallizing	*Unit cost:* moderate to high *Labor cost:* moderate to high *Investment:* high	Limited	Somewhat restricted	Critical	Requires special technological know-how	Multicolor graphics. Wet and dry process, no tool contact with product.
9. Nameplates	*Unit cost:* high *Labor cost:* moderate to high *Investment:* low to moderate	Unlimited	Somewhat restricted	Good durability Less critical Good durability	Adaptable to many situations	Produces bright metallics. Dry process, tool contacts product. Multicolor graphics.

10. Offset	*Unit cost:* low *Labor cost:* moderate *Investment:* high	Unlimited	Restricted	Critical Moderate to good durability	Automates well	Wet process, tool contacts product.
11. Offset intaglio	*Unit cost:* low *Labor cost:* moderate *Investment:* moderate	Limited	Unrestricted	Critical Moderate to good durability	Requires little floor space	Multicolor graphics. Wet process, tool contacts product.
12. Silk screen	*Unit cost:* moderate *Labor cost:* moderate *Investment:* moderate	Somewhat limited	Somewhat restricted	Critical Good durability	Flexible operation	New process. Wet process, tool contacts product.
13. Spray	*Unit cost:* moderate *Labor cost:* moderate *Investment:* moderate to high	Limited	Unrestricted	Critical Good durability	Requires much floor space	Wet process, no tool contact with product.
14. Woodgraining	*Unit cost:* high *Labor cost:* high *Investment:* moderate to high	Specialized	Specialized	Critical Good durability	Mostly hand operated	Wet process, tool contacts products.

Co-injection can be considered a method for decorating. Chapter 29 includes a section on co-injection that provides information on its operation.

An important, relatively new decorating procedure is diffusion printing. Diffusion printing binds an image to plastic, down to a depth of 6 mils. The process produces a scratch-resistant, multicolored image, and is competitive with the more established techniques such as painting and co-injection. This process is giving good results in penetrating co-injection molded key caps for typewriters and other equipment that uses key caps. The image durability is about seven times that of the two-color (co-injected) mold caps that have been popular for over 25 years. Production advantages exist such as being able to produce multiples of entire keyboard arrays in a single operation.

Based on work that General Electric Co., Pittsfield, MA, has conducted, the process works with plastics such as PC, nylon, and PPO, but works best with PBT (polybutylene teraphthalate). The quality of image derived from the diffusion printing process depends on a resin's molecular structure and heat deflection temperature.

Two basic methods of diffusion printing—dry and wet—produce similar results via different application techniques. The wet method can use conventional pad transfer equipment; the dry method requires special machinery.

Diffusion dyes have a strong affinity for the substrate, based on the molecular weight and the polarity of the dye. The transfer of the image, from either a pad in the wet method or a carrier film in the dry process, occurs under varying combinations of time, heat, and pressure.

The dry process dyes sublimate upon heated contact with the plastic substrate—that is, they change directly from a solid form to a gas and back into a solid again without going through a liquid phase. Attractive forces between the dye and the resin molecules "pull" the image into the part, to a depth of 4–6 mils.

In the wet process, the dye is transferred from an inked plate to the part via conventional stamping pad equipment. In this process, colors are applied one at a time.

For successful wet transfer, the part surface must be precleaned to ensure that no contaminants interfere with the printing process. In addition, the parts must be postcured to set the images.

Dry diffusion decorating requires unique equipment and materials. Processors purchase carrier film onto which the images have been preset. Customized transfer equipment heats the plastic part and the carrier film to 300–310°F. The image is transferred as the film is pressed against the part with a 1–2 psi force for 30 seconds.

Dry diffusion requires no prior part cleaning—the dye transfer can take place through most surface contaminants. It also requires no postcuring. In addition, since dry dye crystals are transferred from a pattern *preset* on the carrier film, multicolored images can be transferred in a single pass.

One drawback to both methods is that dyes used for diffusion decorating are translucent. Thus, images can only be printed onto light substrates. About eight basic colors can be transferred via the process, yellow being the lightest.

Plastics processors interested in diffusion printing should consider the relative merits of dry vs. wet. The dry process requires a capital investment for equipment dedicated to diffusion decorating. But, dry decorating is simpler; it does not require part precleaning and postcuring. Finally, the dry method transfers multicolored images in a single pass.

COMPUTERS AND STATISTICS

Computers make statistics a more flexible tool and help prevent the "cookbook" approach (the blind application of the same standard techniques no matter what problem exists). Basically, a statistical perspective can be a simple route to substantially increasing productivity, quality, and profit. Statistics is concerned with design of efficient experiments and with transformation of data into information—in other words, with asking good questions and getting good answers. For most people, the word "statistics" conjures up endless tables of uninteresting numbers. But modern statistics has very practical applications and, thanks to computers, is no dreary science of number-crunching drudgery (11).

Statistical methods should be applied to decision making at all stages of production, from incoming materials to outgoing products. For example, statistics can help with forecasting, a problem managers face every day: should raw materials be reordered; should marketing and advertising techniques be changed? The data used to make these decisions represent random variation—white noise—as well as real changes, such as drops in sales or increases in production.

Quality control is an area where that management strategy can be applied easily. In the past, quality control simply meant throwing out bad products, and management regarded it as a trade-off with productivity. That meant quality control was being exercised too late. Quality control should mean learning about the variability of all aspects of production, including maintenance, purchasing, marketing, and design. Traditionally quality control has been the exclusive concern of engineers. It should be the concern of all employees, and quality control data should be displayed prominently for workers, engineers, and managers to examine and discuss.

Statistics is also concerned with designing experiments. Poorly designed experiments give no useful information no matter how sophisticated the statistical techniques used to analyze their results. Most companies need to run experiments to develop new processes, but experimentation is expensive. Factorial experimental design is the way to get the most information for the least expenditure.

Many experimenters still believe that one variable must be examined at a time. That variable is varied while other conditions are held constant. Besides requiring an enormous number of runs, this method of experimentation does not reflect nature.

If you vary one factor at a time, you assume that nature behaves as if variables operate independently. They don't usually. Raising the temperature may have one result at low mix-

Table 32-2 Statistical analysis; interactions of several variables.

RUN	T	C	K		T	C	K
1	−	−	−	1	0	0	0
2	+	−	−	t	1	0	0
3	−	+	−	c	0	1	0
4	+	+	−	tc	1	1	0
5	−	−	+	k	0	0	1
6	+	−	+	tk	1	0	1
7	−	+	+	ck	0	1	1
8	+	+	+	tck	1	1	1

ing speeds, an opposite effect at high mixing speeds. Interactions of several variables as well as the effects of changing a single variable can be examined in factorial experiments. A simple example is a 2^3 factorial experiment: three variables—temperature, concentration, and catalyst—are examined at two levels (+, −). All combinations can be examined in only eight runs, as shown in Table 32-2.

Production results for each set of conditions can be plotted as corners of a cube, as shown in Fig. 32-1. Factorial experiments produce an impressive quantity of information. In the figure, three main effects—from increases in concentration, temperature, and change in catalyst—can be found, with each main effect being discovered by comparing the means of two sets of four trials. Production with catalyst A, for example, is examined at high temperature, high concentration (180°C, 40 per-

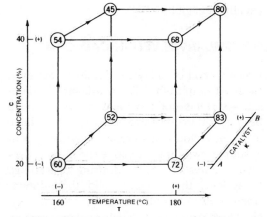

Fig. 32-1 A 2^3 factorial experiment can produce an impressive quantity of data. Here temperature, concentration, and catalyst are examined for their effects on production.

cent); at high temperature, low concentration (180°C, 20 percent); at low temperature, high concentration (160°C, 40 percent); and at low temperature, low concentration (160°C, 20 percent). The mean of these production levels (the numbers circled in the front corners of the cube) is compared with the mean of production levels when catalyst B is used (the numbers circled in the back corners of the cube) under the same conditions of temperature and concentration.

Interactions of variables can also be detected. An increase in temperature affects production differently when catalyst B replaces catalyst A. A third-order interaction—change in the two-factor interaction when the third factor is varied—can also be obtained. It is suggested that sometimes it is better to study four or five variables. Four variables require 2^4 or 16 runs; five require 2^5 or 32.

Frequently in practice, and especially in early stages of process development, more than five factors must be examined. But a full factorial experimental design with 2^n runs is usually unnecessary. Fractional factorial designs, in which only a carefully selected portion of the possible combinations of experimental conditions is run, can still provide an enormous amount of information.

Fractional factorial designs are especially useful for finding the main factors that will affect production. They miss interactions of several variables, but these interactions are usually negligible.

BECOME INVOLVED IN CAD/CAM

It is a marvel of today's technology that even the most simple of CAD—computer-aided design—systems not too long ago would have been considered nothing short of miraculous. Yet today, you can purchase anything from a simple two-dimensional system that uses "wire figures" to assist drafting operations, to systems that present solid three-dimensional views of components, in color if need be. These components, such as shown in Fig. 32-2, can be combined with others and made to move on the screen to check for clearances and stress (see Chapters 11, 12, and 13). They can be rotated so you can look at the front, side, and back of the object being designed, as well as the inside, should you choose to.

In addition to CAD, there is CAM—computer-aided manufacturing. The idea behind this concept is that once you design a product on a CAD system, you can have it generate a tape that is then fed into a numerical-controlled (NC) machine. The machine then follows the computer's directions and produces the mold. CAM can also be used to schedule the machines in a shop so that optimal use is made of them.

And there is CAT—computer-aided testing. Here, the computer tests the product. Also available is CAE—computer-aided engineering (see Fig. 32-3).

You can have CAD without CAM and CAT; in fact, most installations do not integrate these elements. However, they see limited use in some form. Their integration should come in the next 10 to 20 years.

Together, CAD/CAM and CAT provide a totally integrated factory system. The day has arrived when these computerized systems will keep track of workers' time, determine how much material is used, and reorder automatically when necessary. If you really want to get fancy, at the front end you can have computer-aided marketing.

While the marriage of CAD and CAM generally is still rare in injection modling, it is not uncommon in the electronics industry where complex semiconductor chips can have as many as 50,000 circuit elements. In this field, even CAT is currently being used. "In fact," says one CAD/CAM expert, "you could not design a modern computer without other computers—it's just too complicated" (12).

PRODUCTIVITY: SET PLANT IN ORDER

A manufacturing company is a system composed of the complete molding operation. Its maximum productivity can be attained only if the whole system works effectively and efficiently. Also, the whole system must be responsive to change, and it must evolve and improve with time. But today's approach to

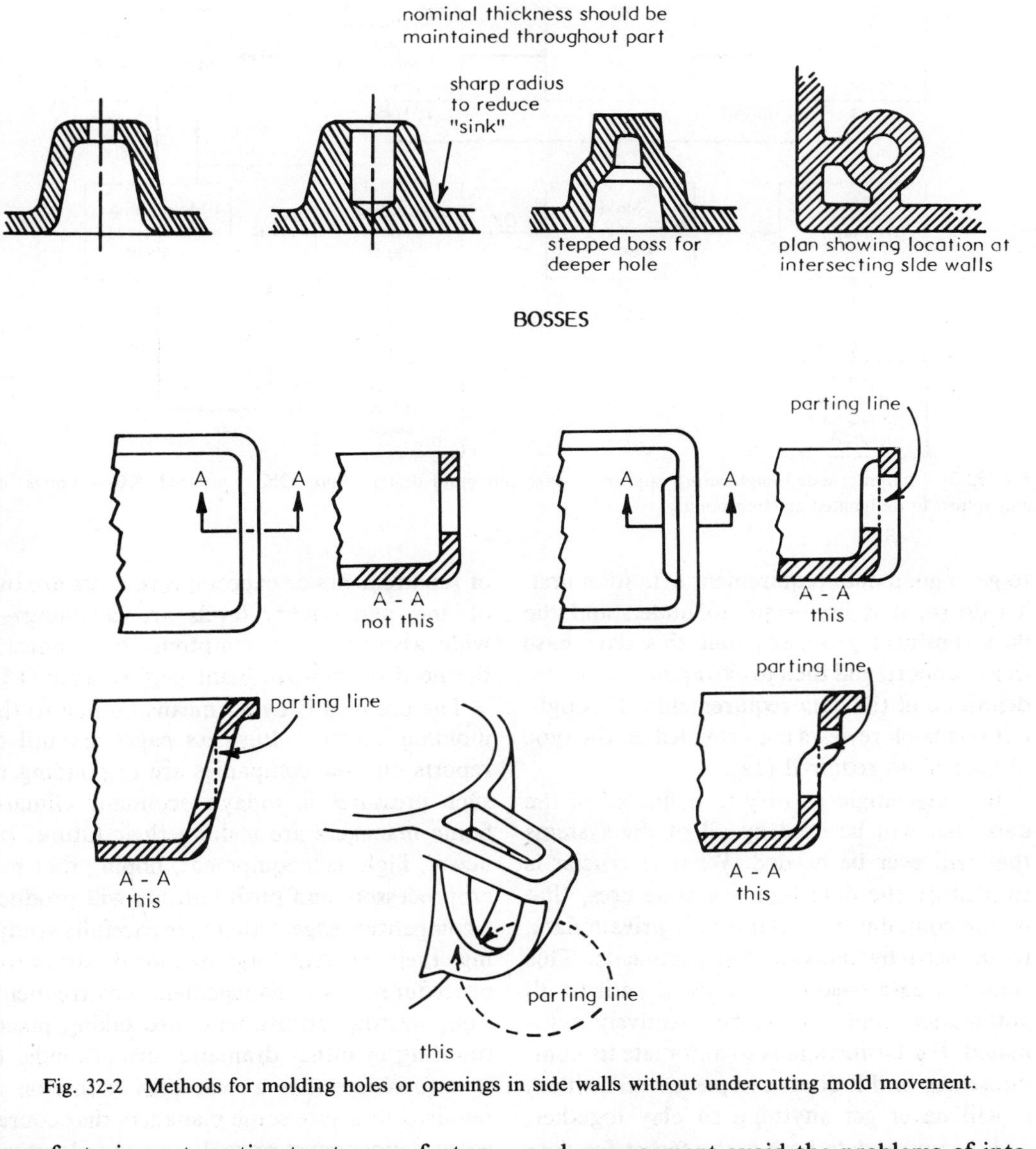

nominal thickness should be
maintained throughout part

sharp radius
to reduce
"sink"

stepped boss for
deeper hole

plan showing location at
intersecting side walls

BOSSES

parting line

A   A

A - A
not this

A   A

A - A
this

parting line

parting line

A - A
this

A - A
this

parting line

this

Fig. 32-2   Methods for molding holes or openings in side walls without undercutting mold movement.

manufacturing automation treats manufacturing as a conglomeration of individual systems such as inventory, purchasing, shop control, and accounts payable. In fact, there are over 50 individual areas of manufacturing that can be profitably automated. The problem is to get everything to play together—to integrate the pieces into a whole that is larger than the sum of the pieces (13).

Across all manufacturing systems, there is only one basic common denominator—data. Planning for manufacturing automation must focus on data as the key to systems integration. Only in this way can manufacturing engineers

and management avoid the problems of integration as an after-the-fact phenomenon of almost impossible magnitude. This is especially true if management intends to buy standard software to perform individual system functions. Data requirements definition is indispensable to successful automation planning.

The natural outcome of data requirements planning is a data base definition. This definition—if it is properly developed—can first be implemented on a data-base management system (DBMS) and then interfaced with the multitude of current and future application systems needed by the users. The problem is

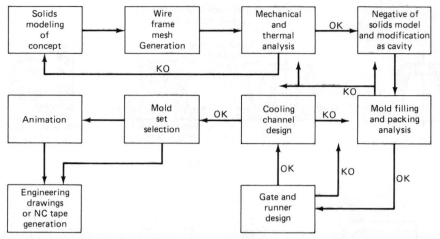

Fig. 32-3 Computer-aided engineering applied to part and mold design. *Note:* OK = proceed; KO = knock out and return to designated starting point.

to get a good data requirement definition first. To do so, it is important to understand the basic business processes that this data base must support, and then to extrapolate a precise definition of the data requirements. Throughout this book reviews are provided on the type of data bases required (14).

It is meaningless to try to define all of the data that will be used by all of the systems that will ever be needed. What is crucial is to identify the data that everyone uses, that is, the common data. All else is private data, to be used by individual departments. This common data base is the critical path to all automation, and it must be effectively automated. If a business fails to automate its common data, or if it does so in piecemeal fashion, it will never get anything to play together.

A common data base is the target for data requirements definition. But what do those data look like? To understand that, we must delve into the inner workings and hidden mechanisms of manufacturing planning and control systems. We must understand how data evolve in manufacturing from an elementary manufacturing control system to a fully automated factory. To accomplish that, we will look at both manufacturing planning and the complete molding operation.

The need for improved productivity sometimes reflects a need for better management. Backlogs are building up, output falls short

of requirements or expectations, costs are out of line, and quality levels are declining—a wide assortment of symptoms that indicate the need to improve plant performance (15).

The problem is by no means limited to the molding industry. Business pages are full of reports on how companies are responding to such pressures in today's economic climate. Some managers are staking their futures on newer, high-tech equipment, hoping that microprocessors and push buttons will produce a competitive edge. Others are carefully studying their current organizational structure, procedures, and management environment. Long-overdue adjustments are taking place, resulting in rather dramatic turn-arounds. It is unfortunate that a near crisis situation is required to assure some managers that courageous actions must prevail over complacency.

Be aware that bringing a new high-tech machine into a poorly managed environment will only guarantee that it too will suffer the same delays and poor handling as the ones already in place.

For example, many molding operations are scheduled on a three-shift basis with the potential for optimum utilization of equipment capabilities. Yet some can be readily classified as true round-the-clock operations, while others are merely running for three consecutive single shifts.

What makes the difference? It all boils down

to management controls. If the machines can achieve an uninterrupted transition from shift to shift and continue to run through rest and lunch breaks, they are indeed in a position to attain their optimum potential. On the other hand, if the machines shut down 15, 20, or even 30 minutes before the end of the shift for reasons of report writing, clean-up, or lack of incentive, with a 15-minute or so delay in getting started on the next shift, lost momentum and output can never be regained. The same is true for idle break periods.

Poor time management is often associated with lack of recognition of the "one best way" to do a job and inadequate training in consistent working procedures. Between-shift shutdowns do not provide an opportunity for the two operators to exchange information about machine conditions or problems in running an order. In fact, the oncoming operator may delay the start of his shift further by adjusting or modifying the previous setup. To make matters worse, operating procedures may fail to require a new approval of output when production is resumed on the new shift.

Managers who seek optimum utilization of their resources must be constantly aware of plant work habits, which sometimes drift away from their objectives without constant monitoring. Some operations lend themselves to such simple procedures as a worker watching another's machine when he is away from his station. Others train operators of related tasks to step in at such times. Some plants employ floating relief operators. Still other plants schedule an extra quarter- or half-hour overlap period for their machine or line people so there can be an orderly transition and exchange of information. The relieving operator comes in early and takes over to keep the machine running without interruption. Remember, transforming a lost hour into a productive one will gain about 300 hours per shift annually with a value of $9,000 per shift or $27,000 for a three-shift operation on a machine at a $30/hr rate.

Also, there exists the often mentioned need for production standards, based upon realistic utilization and output expectations, to signal unacceptable performance. Reliable, timely reporting and hands-on observations would monitor nonproductive delays. Unfortunately, this time management problem of missing existing productivity improvement opportunities is quite common. Potential advantages to be gained by asking questions, keeping informed, taking nothing for granted, and willingness to make changes in current operations may at times approach those to be gained by upgrading equipment.

In any event, buying new equipment will not change a poor working environment. In the order of priorities, setting one's house in order first will usually require less capital than new equipment and guarantee faster, more lasting returns. You cannot solve the problem of a poorly managed shop with a new machine.

Despite the growth and prosperity of the plastics industry, which includes injection molding, many "wrong turns" have been made to produce parts, which have resulted in added expenses and usually limited use of the product. There is an unfortunate tendency to jump from theory to theory while supposedly solving each molding problem as it arises, rather than to evaluate the entire system to see why the problem existed in the first place. There is a practical solution: a logical, back-to-basics approach (as reviewed in this book) can be used.

VALUE ANALYSIS

Immediately after the molded part goes into production, the next step (a very important one) is to use the "value engineering analysis" approach: produce parts that will meet the same performance requirements but are molded at a lower cost. There has to be room to reduce costs. If you do not take this approach, your competitors are sure to do so!

Reevaluate all the parameters used in part design. Use less plastic; or use a lower-cost plastic with similar processing costs; or, very important, use a plastic with a higher cost that processes much faster resulting in total lower cost. Check hardware performance—and all the other parameters described in this book.

The trouble with value analysis is that it

sounds too good to be true. Thus too many people give it little more than lip service. But it is good, and it is true. Value analysis (VA) is like money in the bank—and very often that *is* the problem. Many VA programs are set up to provide guaranteed savings rather than to earn a maximum return on investment. Or to phrase it another way, VA is an organized study of function—but with some programs a little more organized than others.

Value analysis is the most effective, all-purpose technique in your professional tool kit. It is not exclusively a cost-cutting discipline. With VA, you literally can do it all—reduce costs, enhance quality, and boost productivity.

Value analysis sounds too easy. Like sports, singing, and writing, we all think VA is something we are naturally good at. Not so. For real results, VA (like the other three) demands hard, disciplined work. It must be a systematic, formal effort, endorsed and strongly supported by top management.

Here's a fast self-test, which is published by *Purchasing Magazine* (16) annually to help you determine whether your department really has a working VA program. The questions are:

1. Is your top management committed to VA? Is there a written statement spelling out that commitment?
2. Does the person who heads up your program have any formal VA training?
3. Do your VA teams include people from a variety of departments and disciplines?
4. Have key members of your VA teams received any formal value analysis training?
5. How are VA projects or targets chosen?
6. Are progress reports made on team meetings?
7. Is the emphasis consistently focused on function?
8. Is there a VA manual? If not, is there a VA section in the purchasing manual?
9. What VA targets or goals have been set over the past several years? What were actual results?
10. Do you look for VA-oriented suppliers?
11. Is supplier VA help encouraged? Are suppliers included on VA teams? Are they rewarded for their contributions?

12. Is your program a continuing effort, or is it a crash cost-cutting response to bad times?

TROUBLESHOOTING BY REMOTE CONTROL: SERVICE AND MAINTAIN EQUIPMENT

Throughout this book troubleshooting information and guides have been included. It includes analyzing troubles through microprocessors. To aid the manufacturing plants "remote troubleshooting" is available from different equipment manufacturers. Users of certain microprocessed equipment need not be concerned about their plant personnel's ability to service and maintain the equipment. Via telephone link from your controller to a central service computer, a specialist at the controls supplier's office can immediately check out conditions in your controller and in the entire processing line or machine.

Different techniques are used but basically the equipment has installed, or one can add an interfacing link that goes from the machines to the telephone. The adapter (or modem) provides a remote diagnostic link so that the machine manufacturer's service personnel could directly provide corrective action to the equipment with no interference in production; no down time. When required instructors will be provided.

This diagnostic link can be used to set-up preventative maintenance programs (more information on "remote control" is included in the next section).

PROFIT THROUGH TECHNOLOGICAL ADVANCES

Charles E. Waters
AEC, Inc.
Elk Grove Village, IL

The idea of profit through technological advances covers a wide variety of activities. In order to gain some perspective on the overall plastics industry, one should compare an overview of activities that took place in 1960 to activities that occurred in 1980.

Refer to Fig. 32-4, which shows that in

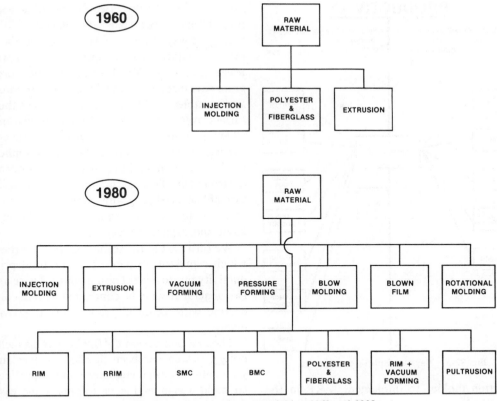

Fig. 32-4 Plastics industry activities, 1960 and 1980.

1960, raw materials were being processed by injection molding, extrusion, and polyester and fiberglass lay-up schemes. There was some activity in vacuum forming, pressure forming, and blow molding. Looking next at the 1980 schematic, it can be seen that raw materials were being processed into finished goods in a wide variety of methods. This book covers injection molding; however, it is important to have some perspective on the overall processing capabilities of the plastics industry.

Realizing that there is great complication in the proliferation of processing methods and the development of raw materials, it becomes evermore important to orient around a common objective. Profit is the ultimate objective but in order to generate profit, it is necessary to have this common objective and it is suggested that perhaps the term *Productivity Improvement* is an all-encompassing term. The productivity improvement opportunities concerning labor and capital are about 41% of the total opportunity. Productivity improvement opportunity or potential on the technical

side is about 59% of the total. There are orderly, well-organized and well-defined programs for maximizing productivity in the two areas.

Refer to Fig. 32-5 and notice that the productivity opportunities relating to labor and capital are pretty much covered by four activities:

The first box is *Cost of Quality*. In any operation, zero defect management and minimum cost of quality will usually result in a superior performance. There are specific programs available for identifying and managing cost of quality. These programs include hardware/software packages that can be applied in the injection molding industry in order to minimize defective performance throughout the injection molding organization.

The second box covers *Material Resource Planning*. This is another well-defined activity that has been in the American management scheme of things for about ten years. There are many programs available, including hardware/software packages that would allow the

PRODUCTIVITY

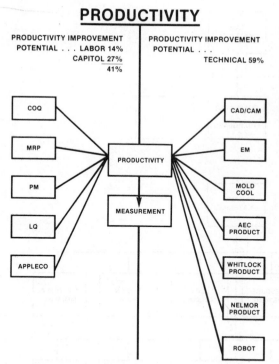

PRODUCTIVITY IMPROVEMENT
POTENTIAL . . . LABOR 14%
CAPITOL 27%
41%

PRODUCTIVITY IMPROVEMENT
POTENTIAL . . .
TECHNICAL 59%

COQ

MRP

PM

LQ

APPLECO

PRODUCTIVITY

MEASUREMENT

CAD/CAM

EM

MOLD
COOL

AEC
PRODUCT

WHITLOCK
PRODUCT

NELMOR
PRODUCT

ROBOT

Fig. 32-5 Ways to improve productivity.

injection molder to implement MRP in the overall plant activity. One of the first results of MRP is to greatly reduce finished goods and raw material inventories. This, in itself, represents a large-enough savings to more than offset the costs of MRP implementation.

The third box is *Preventive Maintenance* (PM). There are two ways of carrying on maintenance. One, unfortunately the more common one, is to put out fires as they occur. The better way is to have a regular preventive maintenance program that involves maintenance and exchange of parts and materials on a programmed basis. Every mechnical device has a known life, and by identifying that life and making maintenance changes before the life expires, one can minimize downtime costs related to firefighting maintenance tactics. The preventive maintenance programs are another well-defined area of management. Hardware/software packages are available that can be directly applied to the injection molding operation.

The fourth box under labor and capital productivity improvement opportunities is *Lead Qualification* (LQ). This relates to the

sales activities of the injection molding venture. Many, many leads are generated in a variety of ways through advertising, word-of-mouth, feature articles, publicity, news releases, and so on. The lead qualification program sifts through all this kind of information and identifies "Class A" leads. These are then turned over to the sales force for speedy and effective follow-up. The hardware/software package is available and can be readily applied in the injection molding machine operating environment. This discussion of four readily available overall programs completes the coverage of productivity improvement through labor and capital management.

We turn next to the technical opportunities, which represent 59 percent of the total productivity improvement opportunity. There are two general areas to be concerned with in connection with technical productivity improvement.

The first box covers *CAD/CAM*. This technology, which has been developing over the years, allows design engineers, and in particular mold design engineers, to greatly increase their effectiveness by using electronic design device as opposed to the old drafting method with pencil and paper. Systems are available for application in-house on very large multiple machine installations or, alternatively, on a time-sharing basis. A number of vendors are setting up regional productivity centers that will be available to the injection molding machine plant operator on a time-sharing basis.

The second box is for the *Interface with AEC Division MOLDCOOL Programs*. This is a very important part of technical productivity improvement. The MOLDCOOL program was developed by the AEC Companies over a period of eighteen years. It deals with the injection molding machine cycle and effectively minimizes cycle time and optimizes machine productivity. About 75 to 80 percent of the typical injection molding machine cycle is devoted to stationary, motionless situations waiting for the molten plastic to turn into a solid. The MOLDCOOL program deals with this 75 to 80 percent of the cycle time and reduces it to a minimum. Productivity improvement is usually on the order of 10 to

20 percent. The program can be used for any size of injection molding machine operations by utilization of one of four paths to the same productivity improvement number:

- Path Number 1 involves consulting work by the AEC MOLDCOOL Division. The work is done by AEC, working from prints or drawings and other input from the user.
- Path Number 2 involves a time-sharing program where the user does all of the work but has access to a central computer over telephone lines. He uses his own terminal on site.
- Path Number 3 allows the user to take the AEC MOLDCOOL program and install it on his own hardware computing system.
- Path Number 4 is a combination hardware/software package that is placed into the user's facilities with complete instructions on application. All materials are then totally proprietary to the individual user.

The seven boxes under technical productivity improvement opportunities concerns hardware. There are three broad areas of auxiliary equipment hardware that can contribute immensely to productivity improvement in the injection molding machine environment. They are:

1. Water cooling, reclaim, recycling, and handling products, including temperature control, and environmental control, manufactured by the AEC Division in Wood Dale, Illinois.
2. Size reduction and reclaim products manufactured by the Nelmor Division in Uxbridge, Massachusetts.
3. Material handling, blending, drying, storage, and coloring utilizing products manufactured by the Whitlock Division located in Farmington, Michigan.

Each of these technical programs and hardware packages provides an opportunity for incremental productivity improvement. Every-

thing discussed in Fig. 32-5 can be measured so that the progress and profitability attached to each program related to the investment in that program can be accurately measured.

Another good opportunity for improvement on the technical side, giving higher levels of productivity, involves plant automation. (See Fig. 32-6.) It can be seen that the upstream side, starting with raw material arriving in a railcar, has been well-automated. It is possible to transport, store, blend, dry, and transport to the molding machine on a completely automatic basis. On the downstream side there are many difficulties because of the varied nature of finished goods. Nevertheless, by a judicious assembly of products on the market, downstream automation can, in fact, be achieved; and it is now being accomplished in many injection molding machine shop operations. The totally automated shop is coming very rapidly and provides the technical innovator with an enormous opportunity to increase productivity and reduce costs that are generally associated with the inaccuracies and unreliabilities of human labor.

Another productivity improvement opportunity available to injection molders involves total energy management. Injection molding machine operations consume raw material and energy. Water is usually reclaimed and recycled. The large energy costs can be greatly reduced by proper design and application of products that have been available to the industry for many years. Refer to Fig. 32-7, a scheme for energy management that is generally divided into two parts—the left hand part involving hardware, the right-hand part involving services—all directed toward the reduction of energy costs.

On the left-hand side of Fig. 32-7 are: insulation of existing roofs; utilization of AEC S-Units and AEC W-Units for reclaiming water and transporting energy back into the space for supplemental or total plant heating; energy management systems that provide a readout, giving the user a picture of the success of the various energy management programs; heat exchangers utilizing waste heat to heat makeup air or plant heating air; and various mechanical devices including polymer

AUTOMATION

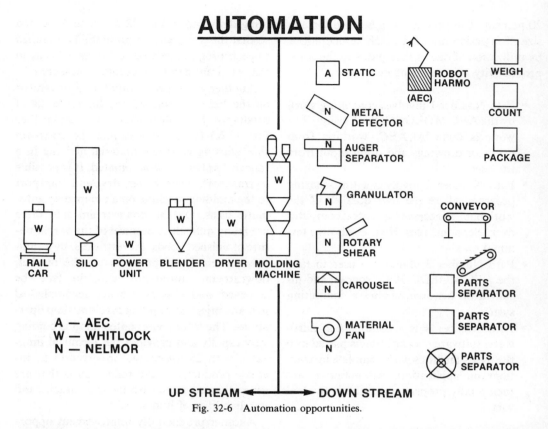

Fig. 32-6 Automation opportunities.

pumps that reduce the energy used to perform polymer handling activities.

On the right-hand side of Fig. 32-7 are listed a variety of services that impact on energy management if they are implemented and properly planned. They include: service on mechanical equipment; proper design, which includes utilization of the CAD/CAM tool and mold analysis tools; contracting, which covers efficient installation of various systems; finance, a very important tool relating to energy management because quite frequently the cost of an improvement can be financed at a rate that produces a total carrying charge far lower than the energy costs that are being incurred; and, finally, preventive maintenance, mentioned earlier in this section (the concept of a planned replacement maintenance program as opposed to the fire-hose program where maintenance is performed after a failure occurs).

Thus, in general, because of the very high energy requirements of an injection molding plant, it is most important that a proper analy-

sis of the plant be developed and that all opportunities for technical innovation be employed.

What is perhaps the latest development in the industry, and really the cutting edge of technology, involves automatic diagnostics (a) at the machine site, (b) remote on-site, which would be the plant manager's office or some other manager's office, and, finally, (c) remote off-site. Figures 32-8 and 32-9 describe a typical network system that ties together mechanical equipment for water chilling, material handling, blending, coloring, drying, and size reduction; robotics for downstream automation; the molding machine itself, the extrusion machine, and any process control devices that may be employed on the system. By using solid state microprocessor technology, it is possible now to provide a visual and audible signal at the machine site. In other words, the particular piece of equipment will speak in plain language as well as indicating visually when a problem exists. This same information can be passed on to the remote on-site location

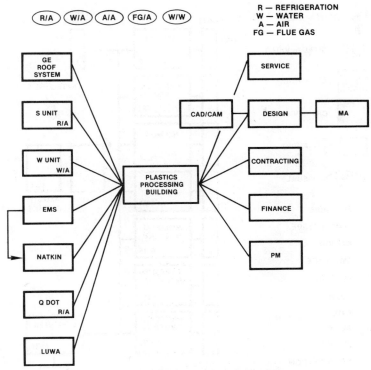

R — REFRIGERATION
W — WATER
A — AIR
FG — FLUE GAS

Fig. 32-7   AEC Division—energy management.

where a manager can index to obtain information from various pieces of mechanical equipment, and the same information can be passed through the phone lines to a central diagnostic site. That central diagnostic site would be the AEC service department, the Whitlock service department, or the Nelmor service department. Thus by using up-to-date technology and transferring the information through phone lines, many expensive service calls can be avoided and very expensive downtime minimized. All of the data being collected can be sent in to a local on-site computer for storage and computation and then printed out on a conventional printer, thereby providing management with a total picture of all machine operations. Transfer of control is passed into the management area and is taken away from the floor. This is an important transfer of responsibility because frequently information developed on-site is not accurate and results in low levels of productivity on particular machine operations.

There is a certain degree of complexity attached to the many programs discussed in this section. However, every single program has been fully developed and commercialized and is available to the sophisticated manager. Utilization of this information will maximize productivity, maximize profitability, and tend to eliminate the less advanced operators. The industry is developing so rapidly that only the sophisticated, well-informed manager will be able to exist in the competitive environment.

ANALYSIS OF POLYMERS: Instrumental analysis affecting business strategies in real-world situations

Jonathan L. Rolfe

Sooner or later, someone is going to show up with a better product than yours, or one that sells for less, or one that is suspiciously similar to yours. Then the questions invariably start: "Why are *they* having a low failure rate with this product?"; "What are *they* making this device out of that allows them to fabricate it so cheaply?"; or, worst of all, "How did *they* find out what we use for raw materials,

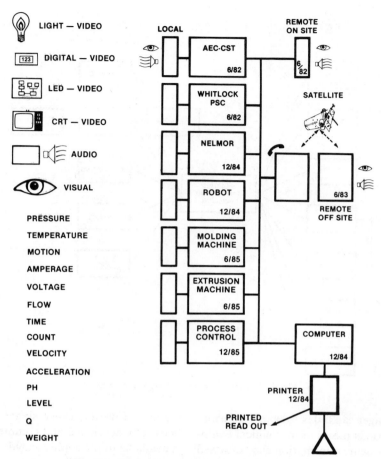

LIGHT — VIDEO

DIGITAL — VIDEO

LED — VIDEO

CRT — VIDEO

AUDIO

VISUAL

PRESSURE

TEMPERATURE

MOTION

AMPERAGE

VOLTAGE

FLOW

TIME

COUNT

VELOCITY

ACCELERATION

PH

LEVEL

Q

WEIGHT

Fig. 32-8   Network development, mech. diagnostics.

where we get them, and how we blend them?"

Polymer analysis is a definitive science that produces answers to these and other disturbing questions—rapidly, accurately, and routinely (see Chapter 28).

The following are some examples of puzzles that were easily dispatched by application of some common analytical techniques; some of them are not without humor.

Example One

After Company A had introduced and begun marketing a high-performance specialty polymer for a few years at a premium price, a new company no one had ever heard of showed up with a supposedly similar product for the same market at less than half the price. Terror and gloom rippled through Company A.

Efforts to discover the new company's secret were frustrated by the new company's touting a "Secret Formula," refusing even to patent the new material. Company A prided itself on being a progressive and innovative firm, and a few years before had invested in some rudimentary instruments for a small analysis laboratory. A tiny sample of the new material was obtained with no small effort, and the little laboratory was asked one of "The Questions."

A solvent was readily found that dissolved the sample. The solution was run through the gel permeation liquid chromatograph and was found to be a blend of two polymers differing greatly in molecular weight. The fractions were trapped as they came out of the chromatograph and qualitatively analyzed by evaporating the solvent and placing the resulting films in the infrared absorption spectropho-

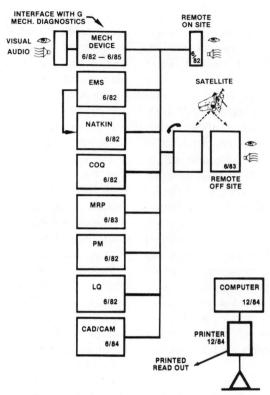

Fig. 32-9 Network development, software/systems.

tometer. The two major components were identified in minutes beyond any doubt, though no one at first wanted to believe the answer. With a great deal of hilarity, Company A realized they had nothing to fear from the competitor; the new company had reintroduced some 30-year-old technology and was certainly doomed in the marketplace as the shortcomings of its product become apparent. It was humorously easy to see why they guarded their "secret" so well.

The determination of this composition took a little over an hour. How much marketing retrenching and sudden product redevelopment time was saved?

Example Two

Octopus Chemicals, a colossal corporation, had a division that had manufactured PVC plastisols and a variety of fabric-supported vinyls for many years. The PVC resin was purchased from another equally huge corporation for as long as anyone could remember. All

was going very well until Octopus decided to go into the resin business so it could sell itself raw materials, for a number of excellent reasons.

Unfortunately, more than half the first year's production had to be discarded because it seemed impossible to obtain consistent results from day to day. Quality control would report good physical test results, the huge machines would start up and be committed, and, as the day wore on, the physical properties, and even the cure temperatures, of the plastisols would become nonsense. It became impossible to prepare a formulation whose properties could even be guessed at, and there began to be corporate casualties as Octopus became more and more exasperated with this bungling division, which, in the past, had been profitable and productive.

Finally, someone asked the Right Question: "What about molecular weight?" Octopus did not have its own GPC facilities, but there was a consulting lab in the area, so a number of tests were initiated. The division insisted the peak and average molecular weights were very similar to those of the material that was formerly purchased from the other corporation. The consulting lab, after a dismayingly short time, agreed.

But they had more to say. In short, a resin with an average molecular weight of 290,000 is not really the same as a blend of two resins, one with a peak molecular weight of 1,400,000 and the other with a peak molecular weight of 2,000! Octopus said, "Well they have the same average molecular weight, don't they?" The consulting lab explained that, if a person were to write a unreasonably large overdraft on his or her checkbook, then tried to explain to the bank that the year-long average was enough to cover the check, the results would be similar to what happened to the plastisol.

The Right Question asked a year earlier, and involving polymer analysis, would have saved enough money to buy many polymer analysis laboratories, and would have saved a few careers as well.

Example Three

A company was in the business of producing medical catheters. In use, these thin flexible tubes were positioned by X-raying the patient; a radio-opaque filler was incorporated into the formulation. To provide the right combination of stiffness for insertion with flexibility for bending easily around corners, the resin was modified by blending an elastomer with another resin. The radio-opaque filler chosen was bismuth trioxide, which was rather expensive but provided excellent radio-opacity.

It was learned that a competitor was charging considerably less for a product that appeared identical, and of course the Inevitable Question was asked.

A small sample of the competing product was dissolved in chloroform, and the mineral fillers were centrifuged out. The dissolved polymer was separated by GPC, and the fractions were collected and analyzed by infrared absorption spectrophotometry. The identities of the two polymers in the competitor's blend were readily determined. By comparisons of the infrared spectra, running commercial resin

samples, even the identity of the suppliers of the resin were found with reasonable certainty, because so few manufacturers were supplying resin to this market. Even so, this didn't explain why the product cost so little; their competitor had to be using the expensive bismuth compound because its distinctive yellow color was evident in both products.

An ultraviolet absorption detector was used in the GPC separations, and it was noticed that the detector response was much more sensitive to the competitor's sample, even though great care was used to prepare the dissolved samples at the same dilution as the company's. A suspicion formed, and the centrifuged filler was examined; it was white.

Simple wet analysis showed the filler in the competitive product to be the much less costly barium sulfate. The competitor had added an organic yellow tint to the product so that it mimicked the color of the more expensive bismuth filler! The yellow tint accounted for the strong ultraviolet absorption.

Obviously, the company could have used the cheaper filler in the first place because their competitor's product worked fine.

Even the most abbreviated polymer laboratory, with little more than a liquid chromatograph and an infrared absorption spectrophotometer, can save a company a lot of money and anquish by answering questions rapidly and with certainty. Just as quality control applications of these instrumental techniques can prevent problems before they occur, surveillance of the marketplace can make tremendous contributions to a competitive edge.

In commercial plastics and resins, as well as many other fields, the days of "secret formulas" are essentially over. There are not many "secret formulas" that will withstand even two hours of scrutiny in a reasonably well-equipped instrumental analysis laboratory.

New instruments employ dedicated microprocessors, enormous memory, and rapid data access. It has never been easier to keep track of the marketplace, and, in many instances, it borders on being fun. Product formulation decisions, marketing strategies, and failure analyses can be approached with more awareness and confidence than ever before.

Conclusion: Knowledge is power.

WAREHOUSING

There is wide variety of procedures for warehousing. Raw materials, additives, auxiliaries, spare parts, molds, tools, and often the finished articles as well have to be stored economically. Various systems have been developed, such as the unit warehouse which makes use of pallets, cages, and similar equipment. It employs a certain organizational scheme for integrating order-picking and transportation. The system is perfected by integration of the inward and outward flow of goods, the factory administration, and process control.

The unit warehouse acts as a buffer between the goods-in and the goods-out. In the direction of materials flow, it embraces the identification point, the transportation system, racks with handling equipment, and the control point. In conventional installations, forklift trucks are used for transportation and for inserting goods into and removing them from the racks. In automatic installations, the goods loaded into units are transported from the identification point to the rows of racks by a handling system, and they are handled within the aisles by rack stackers. The most common units for loading are cassettes, containers, and pallets. Their common feature is that given dimensions are retained.

Unit warehouses are classified into flow warehouses and high-bay warehouses. The former are served with forklifts; and the latter with either forklifts or continuous conveyor-feeders together with rack stackers. The characteristic disadvantage of flow warehouses and high-bay warehouses served by forklift trucks is that they cannot be automated. This is not the case with continuous conveyors and rack stackers. The automatic rack stackers run on rails within the aisles of the high-bay warehouse. The pallets to be stored are carried on continuous conveyors past the identification point, where their route is further programmed. A continuous conveyor system then takes them to the station where they have to be transferred onto the rack stackers, which bring them to their allotted racks. Delivery from the warehouse proceeds in the reverse sequence up to the control point. The system is controlled from the identification point. If a punched card system is adopted, the available storage space is given by a card index system. If the control is by computer, the data are fed accordingly.

If the amounts to be taken from stores are less than a unit, order picking must be resorted to. In this case, the goods must be taken from the rack either by a storeman with a ladder or by an order picker in high-bay warehouses. Alternatively, the pallets concerned may be brought to a collecting point with the aid of a rack stacker and then returned after the goods required have been removed.

Many of the substances to be stored are stable at room temperature and atmospheric humidity. Apart from the normal heating system, air conditioning is unnecessary. If flammable substances, such as paints, oils, adhesives, and solvents, are stored separately from others, there is no need for sprinkler installations. A smoke warning device will suffice (e.g., an ionization installation), provided that the local firefighting force can be brought on the spot immediately.

Fluctuations in the amounts to be stored demand a flexible system, referred to as chaotic warehousing. It ensures a high degree of space utilization. By means of the card index system, each pallet delivered is allotted its place in the warehouse shortly before it arrives. This place is noted in the card index system. Thus maximum and minimum storage times and delivery from the warehouse by the first in, first out principle can be controlled.

The following warehouse layout has proved successful in plastics processing factories:

- Racks for storing pallets on which molding compounds, packaging materials, and large amounts of accessories are stacked.
- Flow warehouse for small parts.
- Separate store for inflammable products.
- Goods-in with check.
- Goods-out with order-picking section.

For many factories of small and medium size, the most favorable solution is the conventional warehouse with forklift trucks. In large factories, automated high-bay warehouses save space and offer advantages in rationalization.

Parts Handling

A feature of most injection molding factories is the versatility of the production program. It must allow for many different types of compounds, moldings of various sizes, the location of the machining station, the degree of precision required, and the size of the production run. For this reason, ideal solutions that are valid for all or many factories simply cannot be found. In fact, each stage in the process must be studied individually and afterwards integrated into the operations as a whole.

A number of important steps take place after injection molding. They include deflashing with subsequent regrinding and recycling the sprue, cooling under pressure in the mold if necessary, machining, conditioning articles produced from certain molding compounds, bonding with adhesives, finishing, and preparing for shipment from the molding factory. In addition, there are transportation and storage between the different stations, and intermediate and final quality checks. Tooling must also be included in the overall plan for the materials flow after injection molding. It embraces the location and layout of the storeroom for the molds, maintenance and repair of the molds, means of conveyance during retooling, hoisting and tensioning devices on the injection molding machine, and quick mold change equipment.

RECYCLING PLASTICS

While many still view plastics waste as an economic and environmental nuisance, others, including plastics processors, recyclers, and equipment manufacturers, are cashing in on the profits of recycling it. All types of plastics are being recycled, and being turned into synfuels or chemicals, or reused in plastics manufacturing (17).

A market exists for many reclaimed plastics because they are as much as 50 percent less expensive than virgin plastic, and often contain physical properties comparable to those of newly produced materials.

A big recycling area is in polyethylene terephthalate (PET) bottles, those plastic soft drink bottles on store shelves. Today, 500 million pounds of PET containers are produced yearly, making it an attractive material for recycling.

Some firms are developing systems that convert incinerated PET containers into steam, hot water, or hot air for use in bottling and other plants.

Many chemical companies are developing methods for reusing the valuable chemicals in PET bottles such as terephthalic acid (TPA) and ethylene glycol.

Cryogenic separation is a successful method for recycling support vinyl fabric, and plated ABS (acrylonitrile butadiene styrene rigid plastic)—to name a few plastic-based products. The technique employs liquid nitrogen, a compound widely used for ultrafine plastic grinding operations.

In one application, Ford Motor Company's Saline, Michigan, plant reclaims more than 1.2 million pounds of chrome-plated ABS scrap yearly from automotive grilles. Ford operates a cryogenic separation and reclaiming system, developed by Air Products and Chemicals, Inc., that granulates the ABS substrate and vacuum-forces it into an enclosed nitrogen-cooled auger. As the chrome and ABS expand differently in the supercold, most of the chrome separates from the ABS. Magnets separate the chrome, ultimately yielding a 99 percent metal-free ABS. The system then pelletizes the ABS into minute particles that are reused in injection molding.

Air Products notes that the cryogenic process can economically recycle other composite scrap materials that cannot be recycled at ambient temperatures. These include supported vinyls, reinforced PVC garden and industrial hose, polyester/copper printed circuit boards, non-cross-linked polyethylene covered cable, glass-fiber-reinforced engineered thermoplastics, and other plated thermoplastics.

The largest market for cryogenic separation is recycling supported vinyl fabrics. Cryogenic recycling works because thermoplastics such as PVC become more brittle than other materials at ultracold temperatures.

When supported vinyl fabrics are induced into a cryogenic grinding and conveying system containing −300°F liquid nitrogen, the embrittled PVC component is ground into a powder and separated from its fabric backing. The backing remains relatively flexible, resisting grinding into fine particles. The system then screens and segregates the PVC and backing.

Old cars never die—they become reborn as mixed plastic scrap converted into pellets for injection molding. Dr. Martin J. Cooper, a principal of a small West Coast scientific firm, developed a junkyard scrap plastic reclaim method using existing scrap-pelletizing lines with capacities averaging 1000 pounds per hour. Cooper's process pelletizes about 75 of the 160 pounds of plastic in the typical junk car, which consists mainly of polyproplyene, sheet molding compound, and ABS in addition to other thermoplastics. Uses for the recycled material include shipping endcaps for pipe, fence posts, flower pots, and battery cases. (The material is not recommended for engineered-type applications.)

Another firm, Upjohn Company, La Porte, Texas, designed a process that turns automotive plastic scrap into particle board for construction. The method granulates scrap into fine particulate, and presses it into test boards. Upjohn reports these synthetic particle boards absorb less water than conventional wood-based samples. One three-layer piece, consisting of 50 percent mixed auto plastic scrap between two aspen outer layers, was recommended for building because of its high stiffness. Because the wood content is reduced 50 percent, the polymer particle boards offer cost savings to processors.

A logistics problem with this technology is stripping the plastic off cars for rapid processing. Upjohn projects that this problem could be resolved by setting up local reprocessing centers for regrinding and pressing the components into boards.

In addition, General Motors developed a process for recycling polyurethane foam seats. The process involves thermal and hydrolytic degradation to reclaim the foam's largest component, polyol, without complex purification. GM's method grinds the polyurethane foam, and reclaims the polyol by reactor-heating the foam to temperatures between 540 and 550°F. Five percent reclaimed polyols can then be blended with 95 percent virgin polyols to produce new foam car seats and cushions. To date, however, the technology has not been applied.

Admittedly most of this recycling technology remains under- or unutilized because of the current low price of oil and other energy forms the recycled products intend to replace. But this situation is expected to change as oil prices stabilize and begin to rise.

As a futuristic thought for recycling plastic, consider the paper presented by J. Karthigesan and B. S. Brown in the *Journal of Chemical Technology and Biotechnology*, which describes converting waste plastics, such as polyethylene, polystyrene, and polypropylene, into food protein via pyrolysis and fermentation. Most underdeveloped countries are more concerned with converting plastic waste to energy and/or fuels than to food. However, in areas where food is scarce this technique may prove a viable source of protein. (Plasticburger, anyone?)

Disposing efficiently of the vast amount of solid waste generated in the United States has become a massive social and economic problem. The plastics industry is, of course, part— a relatively minor but highly visible part— of that problem. We nevertheless recognize that we have a responsibility to do our share to help achieve effective solid waste management practices and to work for a better environment (18).

Because of their utility in many applications, plastic products have made extraordinary contributions to the contemporary world. But as concern about environmental pollution has mounted, our industry has become a target of uninformed criticism. Plastics have been called a litter menace. Both polyvinyl chloride (PVC) and polystyrene have been attacked erroneously as serious air pollutants and have come under threat of restrictive legislation. Most of this criticism has been inaccurate or grossly exaggerated.

To respond effectively to these developments, which could eventually affect not only markets for packaging items but for all plastic products, The Society of the Plastics Industry, Inc. (SPI) has undertaken two tasks: an effort to help find solutions to solid waste problems and a program of information and education on the role of plastics in solid waste.

In the area of research, SPI has commissioned a major study of plastics in incineration, believing that proper incineration is the most feasible method of solid waste disposal now, and that it will be for the foreseeable future.

In the long term, if we are to conserve the country's resources and preserve our environment, we must learn how to reclaim the solid wastes we help to generate. This is not something that the plastics industry, or any industry, can do alone. The importance and complexity of the task require a national effort and heavy expenditure of both private and public funds.

COSTING

Costing is an indispensable economic aid for management (Figs. 32-10, 32-11, and 32-12). Efforts must be made to ensure correct booking. Costs include raw material and production costs, production overheads, and administration and running costs. Raw material costs include the weight of the moldings, the sprue, rejects, losses on starting up, and recycled material. They include materials chargeable to

overheads for storage, transportation, and depreciation. Production costs are split into wages and machinery costs. Costs for molds, production aids, sampling, and retooling must be included.

Cost variation may be due to one or more of the following factors:

1. Improper performance requirements.
2. Improper design of part.
3. Improper selection of plastic.
4. Improper hardware selection.
5. Improper operation of the complete line.
6. Improper setup for testing/quality control/ troubleshooting (assuming one is being used).

The sum of the raw material costs, running costs, and production overheads represents the production costs. Administration and selling costs are broken down into wages, machinery charges, and charges incurred by the lot sizes. Packaging and freight costs must also be added. The sum of all these is the total costs except for the amounts allowed for commissions, risks, and profits.

A distinction must be made between single-cavity and multicavity molds. In the latter case, consideration must be given to the data for the machine, the demands imposed on quality, delivery times, lot sizes, etc. The most economical number of mold cavities is attained when the production and tooling costs represent a minimum. In principle, the optimum number of mold cavities increases with the total amount produced. For instance, if a single-cavity mold were to pay off under certain conditions for a production run of 100,000 or less, the optimum number of cavities for one million moldings would be eight; and for ten million, twenty-four. An important factor in costing is the cycle time. Useful figures are obtained from a careful study of the process in the light of data on the machinery and the raw materials.

Estimating Part Cost

What one aspect of the entire custom injection molding operation is absolutely critical to suc-

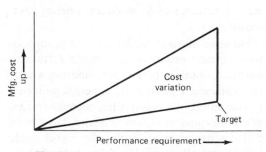

Fig. 32-10   Costing manufactured products.

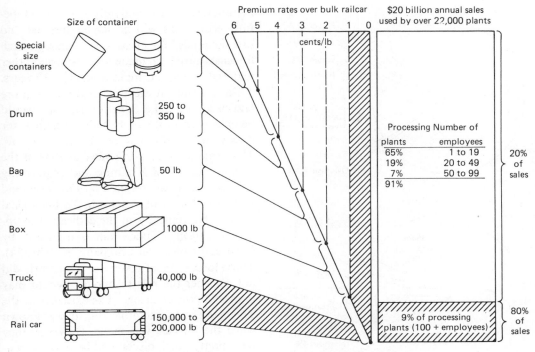

Fig. 32-11   Plastic $ purchases by plant size and size of container.

cess or failure, and yet is practiced many times with considerable lack of logic—is shrouded in mystery and rarely discussed among molders—is thought of as the dullest of topics (19, 20)?

It is an extraordinary thing—if one estimate in ten produces a successful bid, that's a good percentage. That is, 90 percent failure is terrific. No wonder estimating seems like some bizarre sacrificial rite. That does not include all those estimates you just go through the motions of preparing because you know the requests from companies going through the motions of getting three bids, and you have no chance at all of landing the jobs.

But what more directly represents the heart and soul of your business than estimating? You're pulling together every facet of your operation, distilling it, putting numbers on it, and then putting yourself and your company on the line and saying: This is what we can do, and this is what we must charge to make a profit.

There are probably as many estimating techniques as there are estimators. (See Table 32-3.) Just who does the estimating varies

widely. It could be the company president, the sales manager, the production manager, the treasurer; or it could be a person or a department devoted to the task.

Much of the estimating done today follows very vaguely defined procedures. The number of factors assembled to reach the appropriate numbers is sometimes alarmingly minimal; many companies do not consider such matters as scrap, colorant, and setup time, to mention only a few of the more obvious factors. Some estimates are created by determining part weight, cost of resin, and machine time; scribbling down some numbers; and adding a fudge factor. Some companies do not even use a standard (to their organization) estimating form.

MARKETS

The plastics industry still has many opportunities for growth, in both the displacement of other materials and the development of new products and applications. Predicted growth patterns are summarized in Figs. 1-2 and 1-3. More detailed information is given in

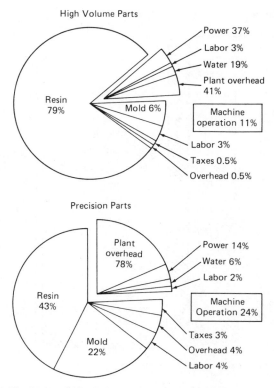

High Volume Parts

Resin 79%

Mold 6%

Power 37%
Labor 3%
Water 19%
Plant overhead 41%

Machine operation 11%

Labor 3%
Taxes 0.5%
Overhead 0.5%

Precision Parts

Resin 43%

Plant overhead 78%

Power 14%
Water 6%
Labor 2%

Machine Operation 24%

Mold 22%

Taxes 3%
Overhead 4%
Labor 4%

Fig. 32-12   Share of cost to mold high volume parts and precision parts.

Figs. 32-13 through 32-15. (See also references 21–23.)

Spectacular or revolutionary technical developments cannot be expected. It will be the usual numerous refinements and improved details that will attract future markets. One of the major tasks of marketing experts should be to convince the consumer that plastics are not conducive to environmental pollution, but are essential for preserving the quality of life. In fact, they basically require much less energy to produce than many other products, from the obtaining of raw materials through all operations to the molding of parts.

Plastics will continue to be indispensable constituents of many innovations and technological developments in our industrialized society. Oil crises and other problems affecting raw material prices will lead to restructuring rather than brake the onward march of plastic products. Other raw materials besides oil, gas, and coal can be used to produce plastics. There is no problem in using a wide variety of raw materials that are abundantly available. Hopefully economists will no longer forecast "no raw materials" for plastics, even though such forecasts have been made for plastic and other materials at various times in the last 50 years.

For example, it was predicted that iron ore bodies in the Lake Superior District would be exhausted in the early 1970s. Chances for discoveries of big new ore bodies were nil. That grim forecast made news in 1938 at the annual meeting of a mining and engineering association. It was recalled by Dr. Raymond L. Smith, President, Michigan Technological University, ironically, at the dedication of upper Michigan's Tilden mine, managed by Cleveland-Cliffs Iron Company in 1970 (24). Those forecasters of yesteryear apparently did not consider the role technology would play in meeting the challenge. The existence of the Tilden ore body, and others like it, was known when the forecast was made. However, they were pretty much discounted because no economically feasible process was available to tap low-iron-content ore bodies.

The mining industry, however, was not inclined to believe statements that it was digging its own grave. Pelletizing was perfected, and that development made it feasible to mine low-iron-bearing ores. Concentrating techniques were researched. Cleveland-Cliffs, working with the U.S. Bureau of Mines and steel companies, invested millions of dollars and 25 years in perfecting a chemical technique to concentrate the Tilden's nonmagnetic iron deposits. These pioneering efforts paid off. As Dr. Smith said, "Here we are, not closing shop but celebrating a grand opening."

ANALYZE FAILURES

You do not have to wear rose-colored glasses to view reverses as a route to eventual success. Putting failures under the microscope of an objective critique, in fact, is far better than playing pollyanna. You may not want or need to schedule a full-scale inquest every time. But even a quick postmortem on a project that has foundered may keep you from fumbling another one.

Table 32-3 Guide to preparing an estimate.

Customer __ABC CO.__ Date __12/15/81__ Quote # __225__
Part name __CONNECTOR__ RFQ¹ __326__ Estimator __PMR__
Part number __12345__ Print #² __A__ Approval __MRS__

Tooling costs

Date	Initial	Vendor	Delivery date	Description	External cost	Internal cost	Total cost	Markup	Customer price
1/5/82	J. W.	Walker	16 wks.	4 cav, 3 plate	$16,000	$1500	$17,500	10%	$19,250

Resin and additive costs

Compound	Color	Specific gravity	Volume	Gm/part	Purchase quantity	$/lb	Source³	(Resin cost in $/lb is included in formula determining part cost)
Ryton R-4	black	1.6	.114 (coin) 4		10,000	$3.33	list	

Part cost

Excess material⁴	Cavities	Part weight	Waste factor⁵	Conversion from gm to lb	Lb/1000 pieces	$/lb	Factory cost/ 1000 pieces
6 gm	4	3 gm	1.05	2.2 / 2.2	10.4	$3.33	$34.62⁶

Molding costs

Operation	Press	Automatic/ Operator	Cavities	Cycle time	Pieces/ hr	Rate, $/hr	
Insert	Engel	Op	4	40 sec	306	$25	$81.70⁸

Secondary-operations costs

Operation	Machine	Cycle time	Waste	Pieces/ hr	Rate, $/hr	
grind gate	wheel	7 sec	10%	468	$10/hr	$21.37⁹

Purchased-items costs

Date	Initial	Supplier	Delivery	Description	Purchase quantity	Cost	Markup	
1/5/82	J.C.	Hex Nut	3 wks	insert	100,000	$14.76/M	10%	$16.24⁷

Packaging/ shipping costs

Date	Initial	Supplier	Delivery	Materials	Purchase quantity	Cost	Pieces/item	
1/5/82	C.G.	Polybag	Stock	10×12 bag	1000	$30/M	250	$0.12/M¹⁰
1/5/82	C.G.	Box Co.	stock	10×10×10 box	500	$0.27ea	250	$1.08/M¹¹

¹Request for quote
²Print revision number
³Source of cost figure (for example, inhouse list, vendor's quote, vendor's price list)
⁴Excess material is that which will not be reused.
⁵Waste refers to material used in rejects or machine startup and that for other reasons will not wind up as finished parts. Assuming 5 percent waste, the waste factor used is 1.05; for 10 percent waste, the factor would be 1.10.
⁶To determine part cost, divide the excess material by the number of cavities, and add the part weight. Multiply the total by the percentage of waste. Multiply that total by 2.2 to get lb/ 1000 parts, which will give you the resin cost/1000 parts when multiplied by the cost/lb. Using the example entered on this form:

$$[(6 gm - 4 cavities) + 3 gm] \times 1.05 = 4.725$$
$$4.725 \times 2.2 = 10.4 lb/1000 pieces$$
$$10.4 \times \$3.33/lb = \$34.62/1000 pieces$$

⁷To determine purchased-item costs, add the amount of markup per 1000 to the cost per 1000.
$$\$14.76 + \$1.48 = \$16.24$$
⁸The following example can be used to determine molding cost (assuming 85 percent efficiency):

Factory cost/1000 pieces	$ 155.13
General and administrative costs	$ 27.38
Profit	$ 20.28
Sales commissions	$ 10.67
Selling price	$213.46/M¹²

$$(3600 seconds/hour \times 0.85) \div 40 seconds/cycle = 76.5 cycles/hour$$
$$76.5 \times 4 cavities = 306 parts/hour$$
$25/hour ÷ 306 parts/hour = $0.08170/part or $81.70/1000
⁹To determine secondary cost, assuming 7 seconds/part and 10 percent inefficiency, use 7.7 seconds/part.
3600 seconds/hour ÷ 7.7 seconds/part = 468 parts/hour
$10/hour ÷ 468 parts/hour = $0.02137/part or $21.37/1000
¹⁰To determine bag cost, first determine that one polybag will hold 250 parts.
$30/1000 bags ÷ 250 parts/bag = $0.12/1000 parts
¹¹To determine box cost, determine that one box will hold one bag of 250 parts.
$0.27/box ÷ 250 parts/box = $0.00108/part = $1.08/1000
¹²Add administrative costs, profit and sales commissions. Some molders use a percentage of the factory cost/1000.

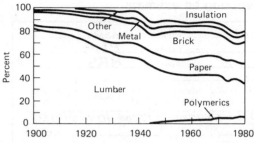

Fig. 32-13 Solids usage, cumulative volume percent, U.S.

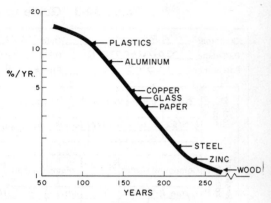

Fig. 32-15 Growth rate vs. age.

Here are some areas to cover, and some questions to ask:

Scope. Were you overly ambitious in establishing your original goals for the project? Should you perhaps have lowered your sights in terms of financial or other targets? Maybe the way you expressed your goals was inappropriate. Would it have been better to set a percentage-of-purchase-dollars target for example, rather than a flat dollar figure? Or vice versa?

Money. Was the project sufficiently funded, with budgeted money actually available when needed? It is often necessary to spend money to make money. That truism is not limited to hardware or tooling; it also applies to promotional efforts such as vendor days where you invite suppliers in en masse. You cannot run such programs on the cheap. Vendors may

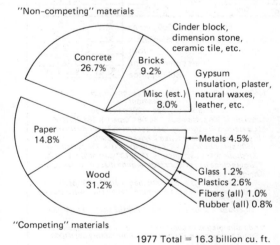

Fig. 32-14 U.S. consumption of materials, volume percent.

not expect red-carpet treatment at such affairs, but you can't treat them like carpetbaggers, either.

People. Were the right in-house staffers assigned to your project? How much say did you have in selecting them? What changes would you make if you had it all to do over again? *Tip:* Remember that those who have shared the experience of failing with you may have learned something from it, too. Perhaps they should be exactly the ones to try again with you. For starters, find out if they are critiquing the failed project just as you are.

Vendors. Was the right supplier assigned to the effort? Did they know exactly what was expected of them—whether taking on a new commodity or providing technical troubleshooting to users? Sure, parceling out business and assignments to suppliers is what the buying job is all about. But it often takes on a special meaning when a particular exercise fizzles. *Key question:* Does the supplier know that it fizzled?

Structure. Was the original project overly complicated? Did its success hinge on too many intangibles and imponderables all meshing like precision gears? Remember the value analysis principle of "simplification" (the KISS approach). Unwieldy programs often collapse of their own weight.

Timing. Was the timing right? This means the whole "climate" surrounding the project: the environment of people, systems, business conditions, etc. Even an excellent idea can fail if it falls on barren or rocky ground. For example, it might not be a good plan to introduce a new method for handling rush or small orders, during a period of many new-hires at the shop level.

Information. Were the data with which you worked accurate, timely, and valid? How recently were they gathered? Were they gathered especially for your project? Do you now consider this a plus or a minus? Be aware that data pulled together just for one project may be recent, but they could be biased toward the hoped-for result of the project, which makes it less accurate.

Salvaging. Should an attempt be made to salvage something from a wrecked plan? Are there successful portions of an overall failure that can be lifted out and applied somewhere? Or should you go back to square one and start all over? Will the eventual benefit of success be worth doing this?

Graphics. Would it be helpful to chart the course of the project? This can be a big help when a failed system is under analysis. It helps you identify problems along the line. This kind of graphic treatment can also be used on less methods-oriented programs. If nothing else, you can indicate dates along a charted axis, and identify time periods in which things started to go wrong.

Comparisons. What makes this particular failure different from a previous success? For best results, this kind of comparison has to pair projects in the same general area, of course.

Motivation. Who else could help sell—or resell—this idea or plan? Again, would it be worth it? Take a figurative look in the mirror and ask yourself one more question: "Would I buy a used idea from him or her?"

Using Failure To Be Successful. Believe it or not—it can be beneficial to plan failure. Thus you can gradually improve the product

so that the minimum cost is applicable to molding parts.

INNOVATE

Just as the computer is a superior control and filing device, it is also a management tool. Even today, the more sophisticated captive and custom molders rely on their multicolor CRT screens and daily computer-generated management reports to check the pulse of their business. The leaders of the industry are so busy today that they could not possibly work any harder. They have to work smarter. Thus they will increasingly rely on the computer to provide them with the information they need to intelligently settle the disposition of manpower, machines, and materials. They will let the product designers worry about part design, the plant engineers and machinery builders be concerned about science, the production people see about molding the part, etc.—while they run the business.

Today, these people are seen as innovators. They are busy further automating their plants, in some cases integrating molding with the entire manufacturing operation. They are looking for ways to improve molding quality by controlling temperature and humidity in their buildings and saving energy by using the heat that escapes from the molding process.

An example of productive innovation is the following review on injection molded printed circuits developed at Bell Laboratories, Murray Hill, New Jersey (28). Figure 32-16 displays the conventional circuit boards. Copper-clad materials are shown in the bottom two layers, for printed circuit boards in computers, communications equipment, consumer electronics devices, business machines, instrumentation, and other electronics equipment. Fabricators machine the laminate, etch on the circuitry, and apply the electrical and other electronics components for devices such as this radio.

Injection Molded Circuit Boards, by P. Hubbauer et al.

Injection molded substrates for printed wiring boards have the potential of providing signifi-

Fig. 32-16 Conventional printed circuit boards that are copper-clad, laminated, and etched.

cant cost savings over standard epoxy/glass substrates. Three-dimensional features such as spacers, stand-offs, soldering sites, and holes for through-hole connections can be molded in rather than added in costly handling operations. These features can all be incorporated with the high precision characteristic of the injection molding process (28).

The recent introduction of high-temperature thermoplastic materials has made this emerging technology feasible, since conventional injection molding materials cannot withstand the high temperatures encountered during soldering operations and actual use. Molded printed circuit boards (MPCB) must be manufactured of high-strength, high-impact materials, since they are expected to withstand dropping and other abuse without breakage. Circuit board materials must be chemical-resistant to tolerate the various cleaning and processing steps as well as fire-retardant to meet all relevant codes and requirements. Molded boards have the added advantage of superior properties with regard to conductive anodic filament growth (CAF).

Recent work by material suppliers, plating shops, and board manufacturers has made important advances in the metallization and processing of MPCB. This review provides a brief overview of MPCB technology, including materials, molding, and subsequent processing. It highlights the opportunity this emerging technology holds for the electronics industry and for injection molders.

Materials. Materials for MPCB applications fall into the general category of engineering thermoplastics. There are currently four classes of materials that were evaluated for this demanding application: polyphenylene sulfide (PPS), polysulfone (PS), polyether sulfone (PES), and polyetherimide (PEI). Various filled, reinforced, and blended formulations of these base resins have also been tested and often show important improvements in properties and metallization. All these materials show high heat distortion temperatures with attendant high process melt temperature requirements. All require heated mold cavities with temperatures near or in excess of 300°F.

The high melt viscosities require unusually high clamping pressures. A listing of material and process parameters for the several thermoplastics evaluated is presented in Table 32-4.

Glass fibers and inert mineral fillers can be used to provide additional rigidity, and it has been shown that reinforcements may be needed to prevent warpage during high-temperature wave soldering.

Processing. Printed circuit board substrates are produced in a conventional injection molding operation. In injection molding, high-temperature polymer melts of the various engineering thermoplastics discussed above are injected into precision steel molds in a complex, dynamic process. High molding pressures (often approaching 20,000 psi) are required for these materials, and the flow lengths are relatively short. The molding of larger boards (>100 in²) and the use of multi-cavity tools, though difficult, is not impossible with proper mold design and multiple gating.

The injection of the high-viscosity melt into a narrow cavity can cause significant molded-in stresses due to orientation effects of the polymer flow. Although these residual stresses can be minimized with careful gating and mold design, the MPCB often requires annealing to prevent warpage. A schematic diagram of accepted gating geometries for MPCB is presented in Fig. 32-17. Even slight imperfections in surface planarity are debilitating for PCB applications, since the plating will be defective. Metallization is also influenced by orientation in a manner unrelated to planarity of the board. The vicinity of the gate is more

Table 32-4 Plastic materials information.

MATERIAL	COST ¢/IN²	GLASS TRANSITION TEMPERATURE (°F)	BARREL TEMP. (°F)	MOLD TEMP. (°F)	INJECTION PRESSURE (PSI)
Polysulfone	19.1	375	710/740	325	13,500
Polyetherimide	19.2	425	740/770	308	17,500
Polyether sulfone	20.4	446	680/700	305	14,000
Polyphenylene sulfide	13.2	545	670/670	200	11,000

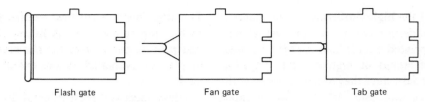

Flash gate   Fan gate   Tab gate

32-17   Recommended gating for injection molding printed circuit board substrates.

highly stressed, and it is found to be more easily etched than the static end.

As with all injection molding, detailed features can be incorporated into the tool. This is the principal advantage of MPCB technology, since drilling, punching, and handling of the boards to attach various connectors and other 3-D features add significantly to the cost of conventional epoxy/glass boards. These more intricate molds, however, can be expensive and add to already high capital costs required to produce MPCB. Manufacturing lines to produce and process epoxy/glass boards are already in place in many of the major PCB manufacturers, so the significant investment in molding machinery and tooling often appears prohibitive. As such, this early stage of the technology appears most suited for high-production runs to minimize the capital costs per unit. The situation, though, presents a near windfall opportunity for injection molders who are prepared to accommodate this new important process.

Metallization. The most important question for the technical feasibility of MPCB is the ability to successfully and economically metallize. As described earlier, the injection molding process often creates residual stresses that can cause warpage or nonuniform etch rates. Nonmolded circuit boards produced from extruded sheets of some of the same materials listed in Table 32-4 are commercially available and for the most part can be metallized using conventional processes, but they do not provide the important benefits of injection molding. Fortunately, continued progress in plating technology should soon make MPCB metallization routine (Fig. 32-18).

The molded substrate can be metallized by a semi-additive or fully additive plating pro-

cess. The fully additive process for MPCB is a patented process whereby a photographic image is formed by a UV-induced reduction of a copper complex to metallic copper. The circuit, including the sides of holes, is built up to the required thickness by extended electroless plating. The semi-additive process for MPCB employs an electroless plate of the entire board followed by pattern plating of the circuitry and flash etching to remove unwanted material. These additive processes are generally more expensive than subtractive processes.

Swell and etch treatments common to both additive processes are required to produce a chemically and mechanically altered surface to which copper can adhere. There are rigid criteria for the adhesion of copper to printed circuit board substrates, and most MPCB materials can meet or exceed these specifications. Improved adhesion is not simply a matter of surface roughness, though, since overetching can decrease peel strengths, while other abrasive methods such as sandblasting are prohibitively expensive. The addition of fillers and reinforcing agents has been found to improve metallization in some instances where all else had failed.

Summary. Injection molded circuit boards are a promising new technology, since they offer a number of important advantages. Three-dimensional features such as connectors, spacers, standoffs, board stiffeners, and large numbers of holes can be molded in, thereby eliminating some parts and assembly costs. Significant savings in comparison to epoxy/glass boards can be realized, but more important is the opportunity for PCB designers. Injection molding enables the widespread implementation of 3-D and contoured boards

Fully additive process

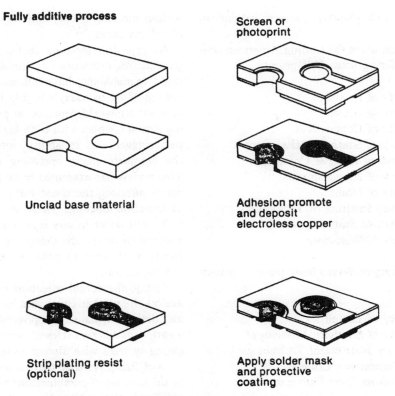

Unclad base material

Strip plating resist
(optional)

Screen or
photoprint

Adhesion promote
and deposit
electroless copper

Apply solder mask
and protective
coating

Fig. 32-18   New metallizing process creates conductor paths.

that were impractical or expensive with conventional technology. In addition, these thermoplastic molding materials exhibit excellent resistance against CAF growth as well as low-loss dielectric properties. The principal applications for MPCB are therefore likely to be: (1) high-production runs, (2) hostile environments, (3) high frequency, low-loss boards, and (4) three-dimensional boards.

The principal drawback is that the technology for injection molded circuit boards is not yet in place. A fairly large capital investment is required for machinery and tooling, and the start-up time needed to design and produce boards is long compared to extant epoxy/glass technology. Also, plating costs are high. These problems, however, are related to the early stage of the technology and do not reflect any serious disadvantages with the process itself. Continued development and the acceptance of production MPCB by the burgeoning electronics industry should provide an important growth market for injection molders.

EDUCATION

Courses in polymer science and plastics engineering are offered at more than 100 American colleges and universities. The University of Lowell, formerly the Lowell Technological Institute, was the first American university offering a formal program in plastics engineering. Founded in 1954 and graduating its first class in 1958, the university followed with a Master of Science in plastics and a doctoral degree in polymer science and plastics engineering. Lowell's program was accredited in 1977 by the Accrediting Board for Engineering and Technology (formerly Engineers Council for Professional Development). The University of Lowell, through the Continuing Education Department, provides the industry with intensive plastics seminars in the United States and other parts of the world, as well as in manufacturing plants. Seminars include the subjects of injection molding, extrusion, blow molding, reaction injection molding, reinforced plastics, designing with plastics, process validation, and so on.

Listed below are the leading American universities offering formal programs:

University of Lowell
University of Akron
University of Connecticut
University of Southern Mississippi
Case Western Reserve University
University of Massachusetts
University of Utah
New Jersey Institute of Technology
North Dakota State University
University of Wisconsin

The leading undergraduate plastics centers are:

University of Lowell
University of Southern Mississippi
New Jersey Institute of Technology
Bronx Community College
North Dakota State University

Principal graduate-level programs in plastics include those at:

University of Lowell
University of Akron
University of Massachusetts
Case Western Reserve University
University of Wisconsin

CONCLUSION

Aside from the fact that a company will do everything in its power to make products that are serviceable, correctly shaped, and attractive, company management still must make a profit and make sure there are enough employees to get the necessary jobs done. Although economic success essentially depends upon external influences such as the market situation and competition and ecological considerations, management's attention must also be directed toward the possibilities existing inside its own factory. This review summarizes the many different questions—both technical and organizational—that arise in plastics in-

jection molding shops, and ways and means of solving them.

As injection molding technology has expanded, requirements for precision injection molding machinery have followed suit. Profitable molding in today's highly technical and competitive world depends on precise clamping action coupled with both high plasticizing performance and reliability, integrated with the complete plant operating performance. This review has examined basic physical processes affecting the speed for productivity of an injection machine. These are the processes that must occur in any injection machine regardless of its specific design or of the refinements it contains in control systems or hydraulic circuitry.

Today, the actual injection machines that are on the market come in a myriad of sizes and designs. Because it is possible to combine a variety of different sizes of injection or plasticating systems with almost as many different sizes of clamp systems, there is almost no limit to the number of combinations available, considering both the domestic and foreign sources of equipment. However, in every case, each machine must provide the necessary facilities to melt the plastic, inject it, and cool it.

REFERENCES

1. Wilhelm Elbe I., "Are Today's Injection Molding Machines Overcontrolled?," *European Plastics News* (September 1983).
2. Cole, N. V., "Optimization of Injection Molding Using a Data Acquisition System," *SPE-ANTEC* (May 1984).
3. AI13 A/D Converter Manual, Interactive Structures, Inc., 1981.
4. Ajay, S., "Dynamic Modeling and Control of Injection Molding Machines," Ph.D thesis, Carnegie-Mellon University, 1978.
5. " 'PID' Injection Pressure Control: Why Today's Machines Need It," *Plastics Technology* (January 1983).
6. Coates, P. D., Sivakumar, A. I., and Johnson, A. F., "RRIM-Process Measurements On A Computerized Small Scale Machine," University of Bradford, England, 1983.
7. "Data Acquisition Moves into the Realm of PCs," *Electronic Design* (September 29, 1983).
8. Osinki, J. S., "Low Cost Data Acquisition System for Instrumented Thermoset Molding," *Proceedings SPE ANTEC* (1984).

9. "The Significance and Detection of Cavity Pressure for Control in Injection Molding Machines," *Kunststoffe—Plastics* (February 1979).

10. Hunkar, D. B., "On Line Process Diagnostics: A Tool for Technical Management of an Injection Molding Plant," *SPE-ANTEC* (May 1983).

11. Hunter, W. G., Box, G. E. P., and Joiner, B. L., "Statistics in Industry," University of Wisconsin–Madison College of Engineering Seminar, 1982.

12. Rosato, D. V., Plastic Industry—Fundamentals Workbook, University of Lowell, 1984.

13. Appleton, D. S., "Four Steps to Productivity: Setting Up A Manufacturing Data Base," *M. D. & D. I.* (October 1983).

14. Lardner, J. F., "Design—A Key to Productivity," ASME Design Engineering Conference, March 28–31, 1983.

15. Abrams, H. H., "Better Management Key to Improved Productivity," *Paper, Film & Foil Converter* (August 1983).

16. O'Conner, J. F., "Value Analysis," *Purchasing,* 37 (March 31, 1983).

17. "Plastic Waste: The Second Time Around," *Compressed Air,* 12–17 (December 1983).

18. "The Plastics Industry & Solid Waste Management," SPI, 1980.

19. Rosato, D. V., "Industry and Fundamentals of Plastics," University of Lowell Plastic Seminars, 1984.

20. Richards, P. M., "Estimating: Putting Method in the Madness," SPE-IMD Technical Conference, October 24–26, 1983.

21. Rosato, D. V., Fallon, W. K., and Rosato, D. V., *Markets for Plastics,* Van Nostrand Reinhold, New York, 1969.

22. "Have Plastics Peaked Out?," Shell Chemical, *Trend,* 8 (4) (1982).

23. "Fracture Costs U.S. $119 Billion a Year," National Bureau of Standards, *U.S. Dept. of Commerce News,* Washington, DC (March 1, 1983).

24. Modic, S. J., " 'Less' Isn't the Answer," *Industry Week* (August 25, 1975).

25. Byrne, J. R., "Expanding the Bubble Theory of Mold Filling," SPE-IMD Technical Conference, October 24–26, 1983.

26. Groleau, R. J., "Intelligent Molds," *SPE-ANTEC* (May 1984).

27. Manzione, L. T., "Cycle Time in Injection Molding," *SPE-ANTEC* (May 1984).

28. Hubbauer, P., Poelzing, G. W., and Manzione, L. T., "Injection Mold Circuit Boards," Bell Laboratories, *SPE-ANTEC* (May 1983).

Index

Index

879

Metal (*Continued*)
  molding, 761
  screen guard, 117
  shim venting, 795
Metalizing, 843, 874
Metering length-screw, 60
Metering pump, 776
Metering zone-screw, 66. *See also* Screw melting zone
Microprocessor control, 288, 747. *See also* Process control
Microprocessor reliability, 282
Microtoming, 730
  basics in, 730
Migration, 730
Milk bottles, 761
Missile impact. *See* Falling dart impact
Mixing, 74. *See also* Compounding
  head, 774, 778
  in screw, 74. *See also* Screw
  pins, 74
Mobius strip, 325
Mode of operation. *See* Machine operation
Model to drawing relationship, 332
Modeling methods, 320
Modeling tools, 320
Modified trapezodial runner, 175. *See also* Runner
Modular construction, 757
Modulus of elasticity, 134, 135, 685, 782
Modulus vs. temperature, 723
Moisture determination, 736
Moisture level, 77, 95, 391, 443, 523, 536, 761
Moisture vs. degradation, 551. *See also* Hygroscopic
Moisture vs. granulating, 532
Mold, 47, 95, 160, 358
  accuracy, 318
  air, 186
  automatic operation, 87
  "break in," 665
  breakaway, 40
  buying, 263
  cavities, custom molded, 238
  changes, 47, 400
  clamp tonnage, 225, 777
  clamping, 95, 757
  coating, 223
  collapsible core, 247
  components, 235, 241
  compression, 809
  configuration, 161
  connection, 196
  contact area, 162, 172, 530
  cool channel positions, 190
  cooling, 88, 189, 237, 241, 318, 355, 358, 456, 541, 795, 856. *See also* Residual stress
   basics in, 358, 456, 466, 642
   clearance, 192
   heat pins, 241
   time, 196, 358, 518, 554
  corrosion, 461
  cost, 268. *See also* Cost

Mold (*Continued*)
  data record, 19, 264
  deformation, 228
  delivery schedule revised, 272
  delivery time, 235
  design, 160, 236, 316, 344, 515, 531, 664, 851
   basics in, 127, 160, 316, 335, 344
   CAD/CAM benefits, 317, 344. *See also* Computer-aided design
   check list, 268
   guide for, 164, 266
   undercut, 530
  dimensioning, 331, 518
  drawing generation, 316
  evaluation, 265
  expansion, 781
  eyebolt, 109, 232
  file, 354
  filling, 58, 162, 229, 294, 309, 311, 336, 352, 374, 644, 752
  finish. *See* Mold polishing
  flashing, 100, 187, 287, 535, 670
  foam, 792
  functions, 162
  gate freezing, 58, 100
  geometry, 505, 555
  growth, 308
   basics in, 307
  heat balance, 196
  heat pipe, 196
  heat treating, 220, 224
  in-mold decorating, 844
  inspection, 96, 262
  installation, 109
  leakage, 180
  life, 88
  machining, 316
  maintenance, 270
  manufacturers, 236
  manufacturing flow chart, 334
  material, 186, 189, 207, 274
   selection, 207, 365
   guide for, 210
  materials vs. type plastics, 210
  melt flow, 332. *See also* Melt flow
  mounting, 47, 107
  moving sections, 108
  opening, 87
  operation, 87, 313
  packing, 310, 503
  parting line. *See* Parting line
  parting sensors, 755
  pillar supports, 226
  placement, 105, 108
  plate tolerance, 755
  plating, 220
  polishing, 216, 269
   polishing SPI-SPE kit, 217
  polyvinyl chloride, 587
  pot, 757

Mold (*Continued*)
  pre-engineered, 208, 235, 253
  pressure, 90, 162
    vs. time, 310, 312
  problems, 23. *See also* Troubleshooting
  progress report, 267
  protection, 106, 232
  prototyping, 250
  purchase, 263
  quotation, 263
    guide for, 209, 263
  redimensioning, 354
  release, 529, 540, 774
  RIM. *See* Reaction injection molding
  rotates, 754, 757, 795
  set-up, 89, 96, 117
  shop selection, 259
  shrinkage, 168, 199, 203, 524, 526, 538, 619. *See also*
    Shrinkage
    test, 713
  side action, 199
  side wall deflection, 230
  size, 225, 227
  slots, 95
  spacing, 44
  stack, 754
  standardization, 235
  start-up time, 318
  steel change vs. heat treatment, 213
  strength, 225
  stress level, 226
  surface, 107, 502
    temperature, 531
  temperature, 85, 94, 502, 505, 518, 560, 578, 620. *See
    also* Cooling mold
    controller troubleshooting, 681, 839
    vs. cool time, 197
  thermal fatigue, 193
  tolerance analysis, 331
  troubleshooting, 660, 664. *See also* Troubleshooting
  types, 164, 388
  vent diagram, 187
  venting, 186, 793. *See also* Venting
  wear, 220
  weight, 107
  with profit, 836
  with rotation, 752, 754
  with undercut, 198
Mold-Masters Ltd., 235
Moldability tests, 645
Moldability vs. viscosity, 644
Moldcool analysis, 362
Molded circuit boards, 871
Molded-in inserts, 149, 155. *See also* Ultrasonic
Molded-in stresses. *See* Orientation
Molded parts, 200, 252
Moldflow Australia, 349
Molding, 3, 84, 660, 794
  area diagram, 313, 663
  clamping pressure determination, 103

Molding (*Continued*)
  coinjection, 795
  conditions guide, 501, 537, 538, 575, 803
  continuously, 753
  cycle time, 22, 25, 50, 70, 73, 85, 196, 203, 205, 278,
    528, 643, 660, 752, 756, 841
  data record, 19
  flash free, 754
  for electroplating, 512
  force-remove part, 197
  guide for, 96, 285, 313, 501, 517
  machine functions, 10
  machine operation, 84
  metal melt, 761
  operation basics, 4
  operation principal, 89. *See also* Injection molding op-
    eration
  optimization, 111
  orientation, 619. *See also* Melt flow and Orientation
  parameters, 96
  powder, 803
  set up improved, 98
  specialities, 747
  stretched blow, 765
  terms, 6
  variables vs. machine settings, 502, 513, 579
  volume diagram, 313
  vs. crystallinity, 615. *See also* Crystalline plastics
  vs. molecular structure, 613. *See also* Molecular weight
  vs. rheology, 640. *See also* Rheology
  with rotation, 747, 754
Moldmaker tolerances, 270
Molds interchangeability, 47
Molecular flexibility, 612
Molecular orientation, 750. *See also* Orientation
Molecular stiffening, 614
Molecular structure, 595, 612, 687
Molecular weight, 543, 607, 636, 685
  compound, 486
  distribution, 610, 687
Molecule, 92, 486, 595. *See also* Plastics
Moment of inertia, 134
Monitoring costs, 825
Monitoring mold parameters, 287, 839. *See also* Process
  control
Monomer, 93, 595, 775. *See also* Plastics
Moog, 289
Mooney viscosity, 715
Morphology, 687. *See also* Amorphous and Crystalline
Motivation, 871
Motors, 56, 85
Mottled product, 77
Mounting mold, 107
Moving platen. *See* Clamping
MUD. *See* Master unit die
Multi-cavity mold, 160, 236
Multiple shot injection, 795. *See also* Coinjection
Multiwalled blow molding, 762
MVD. *See* Molding volume diagram
MWD. *See* Molecular weight distribution